T0200503

Statistical Methods
for Business and Economics

Statistical Methods for Business and Economics

Gert Nieuwenhuis
Tilburg University
The Netherlands

Mc
Graw
Hill
Education

London Boston Burr Ridge, IL Dubuque, IA Madison, WI New York San Francisco
St. Louis Bangkok Bogotá Caracas Kuala Lumpur Lisbon Madrid Mexico City
Milan Montreal New Delhi Santiago Seoul Singapore Sydney Taipei Toronto

Statistical Methods for Business and Economics
Gert Nieuwenhuis
ISBN-13 9780077109875
ISBN-10 0077109872

Published by McGraw-Hill Education
Shoppenhangers Road
Maidenhead
Berkshire
SL6 2QL
Telephone: 44 (0) 1628 502 500
Fax: 44 (0) 1628 770 224
Website: **www.mheducation.co.uk**

British Library Cataloguing in Publication Data
A catalogue record for this book is available from the British Library

Library of Congress Cataloging in Publication Data
The Library of Congress data for this book has been applied for from the Library of Congress

Acquisitions Editor: Rachel Gear
Head of Development: Caroline Prodger
Marketing Manager: Mark Barratt
Production Editor: James Bishop

Text design by Hard Lines
Cover design by Paul Fielding
Printed and bound in Great Britain by Ashford Colour

ISBN-13 9780077109875
ISBN-10 0077109872

Brief Table of Contents

Detailed Table of Contents

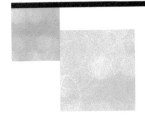

Preface

Statistics has to do with variation, variability. The gross national product changes from year to year; people differ in opinion; sales on the market vary daily. Therefore the main theme of this book is **variation**. Statistics tries to describe and analyse variation, and above all, to explain it. Variation is the reason for statistics.

Why I wrote this book

During the past two decades, new directions in (international) economics came into existence. The growing importance of the European market and the accompanying internationalization of many organizations caused a serious need for research and knowledge about internationally oriented economics and business. The increased competition gave rise to quantification: to measure the quality of products, to explore the risks of new investments, to learn about the market and the competitors, to learn about other countries and their possibilities for investments.

At the economic faculties of universities, the process sketched above, the disappearance of the boundaries between the EU member states and the introduction of the euro stimulated the creation of new study opportunities: international business, international economics, international finance, business studies, etc. Many universities in Europe opened their doors to students from abroad, while domestic students are encouraged to do a part of their study at other universities in Europe. These developments have several consequences for the courses offered to students in economics and business. New courses on international competition prepare the students for the new situation in the European market. Other courses are adapted to include new ideas and results. Students are challenged and encouraged to widen their horizons.

Apart from the use of the computer in textbooks, introductory statistics courses for students in business and economics have hardly changed during the past thirty years. Although the growing international character should stimulate students to learn as much as possible about new ideas and methods, the courses in elementary statistics remained more or less the same. The introduction of the computer even had a serious negative side effect: statistics partly degenerated into a push-the-button science. Students learn to do the trick, but they are not encouraged to learn why this trick is a good one. It would appear that computers are so impressive that calculation is more important than understanding. Furthermore, the (often American) textbooks do not counterbalance this development. Although the need for critical and creative quantitatively oriented economists is great, students are hardly encouraged to understand the things they are doing. Books on introductory statistics do not offer a step-by-step path that students can follow to learn what statistical procedures are and how they can be used to solve problems in business and economics. Practice is that most students just use the formulae and often apply them without any understanding.

In this book I have tried to stop, and partly reverse, this process. Of course, the computer is very important for an economist and it really is indispensable for this book too. But a computer is only a powerful calculator, and a statistical computer package is no statistician. It is primarily the **understanding** of the statistical procedures that statistics in economics and business has to be about. The technical knowledge about how to perform the statistical methods with a computer is also important, but very much secondary. Students have to be challenged to understand these methods, to stimulate their creativity. It is not enough that they know the buttons to be pushed; they also have to know why. They have to be challenged to reach as high as they can. The present competitive situation in Europe demands creative and motivated economists and managers.

What distinguishes this book from others

In this book, students are challenged to understand the **statistical thinking** behind the **methods**. To accomplish this, the following guidelines are used:

- There is no reluctance to express methods as formulae.
- However, only the formulae that really increase understanding are presented.
- New methods are analysed thoroughly, until complete understanding is achieved.
- To increase understanding, emphasis is on the common elements of many seemingly different methods.
- Basic statistical methods, such as hypothesis tests, are presented as step procedures.
- Many examples are used to increase understanding of the statistical methods.

Indeed, formulae are slightly more important than in many other introductory books on statistics. But on the other hand, much more effort than usual is made to teach the ability to read the formulae and to emphasize that a formula is shorthand notation for an idea that can be expressed in words as well. The underlying aim is to explain why a formula looks as it does, to avoid the 'learning it by heart' and 'treating it as a black box'.

Much understanding can also be gained by emphasizing the common form and common ingredients of many statistical methods. To start with, many formulae about population variables in descriptive statistics and random variables in probability are basically identical; it is a waste and a shame not to point out and make use of these similarities. As a second example, the test statistics of many hypothesis tests have a common basic form. By emphasizing this underlying common structure, many formulae turn out to be similar. To stress the common features of many basic statistical methods, some of them are presented as multiple-step procedures. For instance, a hypothesis test is presented as a five-step procedure.

Many examples and exercises are about European circumstances, about EU countries or enterprises in the EU. Many of the datasets originally come from institutions such as Eurostat, OECD, World Bank and the European Central Bank. However, examples about non-economic topics, for example games and sports, can also be very stimulating. The book also contains examples using data from Statistics Netherlands, from other international statistical agencies and from my private archives. Such examples are usually European in nature: similar data might have been obtained in other countries as well.

Traditionally, introductory books on statistics offer introductions to the four sub-fields of descriptive statistics, probability theory, sampling theory and inferential statistics, treated in this order. This book also has this useful subdivision. Part 1 'Descriptive Statistics' discusses how to summarize a dataset by way of tables, graphs and statistics. If the dataset consists only of measurements on a part (sample) of the population (i.e. all objects of interest), the descriptive findings of this sample dataset are used in inferential statistics (the subject of Part 4) to draw conclusions about the whole population. It is important to note that these general conclusions are valid only if the sample is obtained in a very precise way. The sub-field of sampling theory (Part 3) discusses sampling procedures that allow such general conclusions. As usual in introductory texts on statistics, only random sampling is treated here in detail. The sub-field of probability theory (Part 2) is partly independent, but it also has to build a bridge between descriptive statistics and inferential statistics: based on the sample information and the sampling procedure it shows how to draw valid conclusions and to ascertain the precision of these conclusions.

When compared with other introductory books, this book pays more attention to the sub-fields of descriptive statistics and probability theory. Furthermore, the links between the four sub-fields and their main similarities – such as their joint purpose to describe variation of variables – are emphasized.

Introductory descriptive statistics is traditionally the least challenging part of statistics. It is heavily based on computer work and hence the underlying intentions easily get lost in viewing so many data. To overcome this, its preparatory role with respect to inferential statistics is emphasized.

For instance, in Chapter 5 the basic idea behind regression analysis – the wish to understand why a variable shows variation – is considered (and partly worked out).

Indeed, probability theory is an independent science and offers elegant, stimulating examples. But its role as intermediary between descriptive statistics and inferential statistics must also be emphasized. In many introductory books on statistics, this role does not become clear; the emerging difficulties are avoided. Discussion of probability theory often constitutes an island in isolation. In the present book, I have tried to demystify the role of probability. On the one hand, this is done by looking back to descriptive statistics and putting emphasis on the experiment 'random observation'. On the other hand, the gap with inferential statistics is bridged by looking forward and by considering probability results that are basic for inferential statistical methods. Any emerging theoretical difficulties are tackled by carefully explaining all steps and by giving examples. Some of the basic probabilistic results that underlie the theory of confidence intervals and hypothesis testing are treated in the parts of the book that deal with the sub-fields of probability and sampling. This is done to make the intermediate roles of these sub-fields more transparent and to facilitate the introduction of the statistical procedures in inferential statistics.

As mentioned at the beginning, the book concentrates on variation. This concept is crucial for economists and managers since it is often the variation of datasets and variables that is of interest. In studies regarding incomes or GDPs, measures of variation give information about income inequality. In research on product satisfaction (as in marketing) or on political opinions, little variation refers to consensus. In studies regarding investment, variation is often related to risk. The underlying purpose of many papers in economics and business is to detect the factors that, at least partially, cause the variation of the variable of interest. That is why it is extremely important to have a good understanding of the concept 'variation' and its complicated measures (such as variance, standard deviation, standard error), and of their importance for inferential statistics. In my opinion, it is not possible to inform students about similarities and differences between the many related concepts on variation without occasionally being a bit formal.

In brief, the objectives of this book are:

- to stimulate the students to reach as high as they can;
- to challenge, to increase the understanding, to make the learning by heart unnecessary;
- to demonstrate the coherence of the four sub-fields of statistics;
- to demonstrate the importance of the concept 'variation';
- to illustrate the methods with European examples.

Special notes for students and instructors

Computer packages

Most of the graphs and printouts in the book are created with Excel or SPSS. However, within the text, examples and exercises, references to these computer packages are omitted. This is done to make it possible to use the book with other computer packages as well.

For students and instructors who do prefer to use Excel and/or SPSS, the explanations of techniques are placed in Appendix A1 and put on the internet. In this appendix, the subdivision into sections is such that, for instance, A1.8 is about Excel and SPSS techniques for Chapter 8 of the book. Among Sections A1.1–A1.25, the package Excel is most important in the first sections and SPSS in the last. The reasons for putting emphasis on Excel in the first half of Appendix A1 are:

- Excel is more accessible than SPSS;
- many students have already used Excel at school or college;
- Excel is less a 'black box' than SPSS and hence fits better with the objectives of this book;
- Excel has nice options that allow data manipulations (such as the Fill Handle, which enables data to be filled into adjacent cells).

The reasons for increasing the role of SPSS throughout Appendix A1 are:

- SPSS has standard (built-in) statistical procedures;
- SPSS is especially suitable for inferential statistics.

But again, it is possible to use these packages otherwise and even to use other packages.

Traditionally, probabilities for distributions are determined with tables. I believe that tables are incomplete and outdated, and that their use has to be discouraged. However, in tutorials not all students have access to a computer, while graphical calculators can usually only deal with the normal distribution. That is why I have decided to include some tables in the book and to put other tables on the internet. However, in the text of the book, probabilities are calculated with a computer.

Sometimes a probability can be calculated just by using common sense. But in other cases the computer is needed to calculate probabilities that come from special families of distributions. In this book I have used the icon (*) to indicate that a computer is used in the calculation of a probability.

Exercises

Each of the 25 chapters ends with an exercise section: some simple exercises to practise the mechanics and to better understand the theory, some exercises to apply the theory, some more advanced exercises to challenge the reader.

Some exercises are based on datasets, others are not. For some exercises a computer is necessary to summarize the data; these exercises are marked '(computer)'. In other exercises the underlying dataset is added but not really needed to answer the questions since the data are already summarized in the text of the exercise. If wanted, such exercises can also be used on a computer practical by inviting the students first to check the summarized results.

Internet

For students, written solutions of the odd-numbered exercises and of most case studies are available on the internet. For the instructors, all solutions are available. All datasets are placed on the internet. In the datasets the decimal point is used; not the decimal comma.

Also PowerPoint files are available on the internet, one file for each of the 25 chapters. These ppt files summarize the chapters and can be used by instructors.

Although I did my utmost to avoid them, the book will probably contain errors and mistakes. I invite students and instructors to mail all errors as soon as they are detected. A file will be posted on the internet that contains the list of errors found so far. If necessary, it will be regularly updated. Of course I am also interested in general opinions about the book. Please contact me for discussion.

Cases

The book contains many cases, one at the start of each chapter (except Chapter 1) and usually one or even more at the end. They are meant to motivate and illustrate the contents of the chapters and can be used by instructors during their lectures. In each chapter, the solution of the initial case is given in the course of the chapter; the solutions of other cases are available on the internet.

Special notes for students

From the many years of my experience I know that a considerable number of students try to learn statistics by doing only the exercises. This approach will not work! The text (theory) is an essential part of the book since it explains the methods. If only the exercises are done, students will get lost in the seemingly enormous number of formulae and tricks; they will have a horrible time. But if the text is read before the exercises are attempted, the methods of the exercises are revealed and become easy to remember.

The book makes use of many symbols and letters, including Greek letters. A list of those used in the book is given in Appendix A3.

Special notes for instructors

I have tried to follow international notations as much as possible. However, I noticed that common notations are not always consistent. Since I believe that students have to learn right from the start to distinguish between the methods and the realizations that are the results of applications of the methods, I have decided to be slightly more consistent than the authors of many other books. In this book, random variables and test statistics are usually denoted by capitals (X, Y, T, G) and their realizations by small letters (x, y, t, g). Furthermore, population statistics (parameters) are usually denoted by Greek letters; sample statistics by suitable Latin letters. However, I have decided not to be too provocative and to write p for a population proportion (although π would have been more consistent). For the random sample proportion and its realization, I use the respective notations $\hat{P}$ and $\hat{p}$.

There is one concept for which I have introduced a private naming: the number that in a sense lies between the null hypothesis and the alternative hypothesis, the number that SPSS calls the test value. Since I do not know of another common name for it and since 'test value' is not suitable since it is often confused with value of the test statistic or critical value, I have called it 'hinge'.

The level of mathematics needed to read this book is the ordinary level of those who finished secondary school with the intention to do a further university education in business or economics. In Chapter 8 (on probability distributions, expectations, variances), the mathematical topic differentiation is cautiously used. Integration is also used, but only for those who are familiar with it. In my experience, students learned about the summation operator at secondary school but many of them forgot about it. That is why this topic is intensively (but separately) considered in Appendix A2.

The book has 25 chapters, slightly more than most other books. Some of the chapters are small but others are rather large. If wanted, some chapters can be combined and treated in one lecture, for instance Chapters 6–7 and Chapters 12–14. I have decided to place the definitions of probability and the probability rules in different chapters (6 and 7). The main reason is that Chapter 6 is rather philosophical and, being not too large, offers the opportunity to recover from being confronted with so many descriptive statistics in Chapters 1–5.

Some sections and subsections are optional, for instance Sections 9.3 (Poisson distributions) and 10.2 (exponential distributions). If wanted, Sections 22.5 (instrumental variables) and 22.6 (logit model) can be omitted too. Even the whole of Chapter 22 (model violations for regression) can, if wanted, be omitted, since elementary residual analysis is also part of Chapters 19 (simple linear regression) and 21 (multiple linear regression: extension).

The order of the chapters is not always strict. For instance, it is possible to treat Chapters 24 and 25 immediately after Chapter 18.

Guided Tour

Introduction

Each chapter opens with an outline of the main techniques and methods covered in the chapter, summarizing what knowledge, skills or understanding readers should acquire once they have read it.

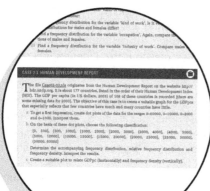

Real-life case studies to apply statistics to business

The book includes chapter case studies designed to test how well you can apply the main techniques learned. The initial case study is revisited within the chapter so that you can see how to arrive at solving the problems. There is also a selection of longer cases at the end of most chapters for extra examples.

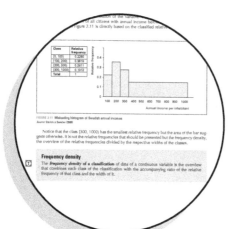

Key terms and key equations – highlighting what you need to know

Key terms are highlighted throughout the chapter in bold italic, with page number references at the end of each chapter so they can be found quickly and easily. Key equations and formulae are also highlighted in the book, and symbols listed at the end of each chapter too. An ideal tool for last minute revision or to check key formulae as you read.

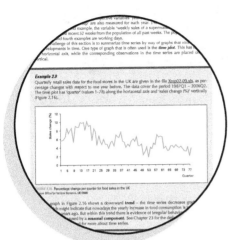

Packed with examples

Each chapter includes lots of short examples. They aim to show how a particular concept or statistical technique is used in practice, by providing data and examples showing how statistics can be applied in a business or economics context.

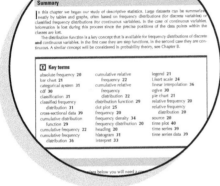

A useful chapter summary

This briefly reviews and reinforces the main topics you will have covered in each chapter to ensure you have acquired a solid understanding of the key topics. Use it as a quick reference to check you've understood the chapter. Each summary also includes a list of key terms in statistics.

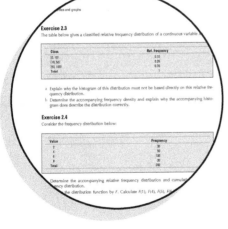

Plenty of exercises

These questions encourage you to review and apply the knowledge you have acquired from each chapter. They are a useful revision tool to check that you have mastered statistical techniques; they can also be used by your lecturer as assignments or practice exam questions.

Technology to enhance learning and teaching

Visit **www.mheducation.co.uk/textbooks/nieuwenhuis** *today*

Online Learning Centre (OLC)

After completing each chapter, log on to the supporting Online Learning Centre website. Take advantage of the study tools offered to reinforce the material you have read in the text, and to develop your knowledge in a fun and effective way.

Resources for students include:
- Solutions to the odd-numbered exercises, to allow students to check their progress as they work through the exercises
- Solutions to selected case study problems
- Datasets from the text

Also available for lecturers:
- Chapter by chapter PowerPoint for use in presentations or as handouts
- All solutions to the exercises
- Other additional material and updates

Custom Publishing Solutions: Let us help make our **content** your **solution**

At McGraw-Hill Education our aim is to help lecturers to find the most suitable content for their needs delivered to their students in the most appropriate way. Our **custom publishing solutions** offer the ideal combination of content delivered in the way which best suits lecturer and students.

Our custom publishing programme offers lecturers the opportunity to select just the chapters or sections of material they wish to deliver to their students from a database called Create at http://create.mheducation.com/uk

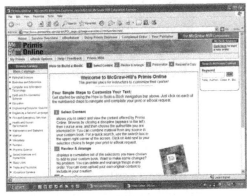

Primis contains over two million pages of content from:

- textbooks
- professional books
- case books – Harvard Articles, Insead, Ivey, Darden, Thunderbird and BusinessWeek
- Taking Sides – debate materials

Across the following imprints:

- McGraw-Hill Education
- Open University Press
- Harvard Business School Press
- US and European material

There is also the option to include additional material authored by lecturers in the custom product – this does not necessarily have to be in English.

We will take care of everything from start to finish in the process of developing and delivering a custom product to ensure that lecturers and students receive exactly the material needed in the most suitable way.

With a Custom Publishing Solution, students enjoy the best selection of material deemed to be the most suitable for learning everything they need for their courses – something of real value to support their learning. Teachers are able to use exactly the material they want, in the way they want, to support their teaching on the course.

Please contact *your local McGraw-Hill representative* with any questions or alternatively contact e: custom.publishing@mheducation.com

About the author

About the Author

Gert Nieuwenhuis is associate professor of probability and statistics at Tilburg University. He works at the Faculty of Economics and Business Administration, at the department of Econometrics and Operations Research. He has more than 30 years experience of teaching basic probability and statistics, regression analysis, time series forecasting, actuarial sciences, risk theory and basic econometrics to both undergraduate and graduate business and economics students. Together with Hans Moors and Maarten Janssens he has also written a series of four books, *Statistics for Economics* (in Dutch). In his spare time Professor Nieuwenhuis enjoys reading and listening to rock music, and likes to run and cycle through the holms of the river Maas and the hills of Nijmegen.

About Tilburg University

Tilburg University is a compact institution for higher education, specialized in human and social sciences and located in the southern part of the Netherlands. It has an outstanding international track record for teaching and research excellence. Its business and economics institute GentER is a world-class research institute.

Acknowledgements

Many people have contributed to the realization of this book. I want to thank all colleague lecturers and all students of Tilburg University who gave me their fruitful comments. Also many referees have given me useful comments; I want to thank them all. Many of their suggestions are incorporated in the final text: In particular, I want to thank Noud van Giersbergen for his detailed comments.

I also want to thank McGraw-Hill for giving me the opportunity to publish my educational ideas about statistics and its relation to business and economics.

I want especially to thank my colleagues and friends Hans Moors and Gé Groenewegen. Cooperation with Hans in a former book project, where we were co-authors, was very stimulating for me and helped me to develop my ideas. I also want to thank Hans for reading a part of the manuscript and for permitting me the use of previously published material and ideas. Gé was my anchor during the often exhausting process of writing the book. Apart from his critical reading of parts of the manuscript, he handed me many datasets and ideas for examples and exercises. I really want to thank him for that.

I also want to thank our children Gijs, Nienke, Martijn, Bas and Lonneke. I want to thank them for accepting that I was not always accessible, not even in the few cases when I was physically present.

But most of all I want to thank Ineke. She really was great, although it sometimes must have been a hard job to find the real Gert in the abstract world of statistics. She always remained sweet, careful, enthusiastic and supporting, even when confronted with so much physical and spiritual absence. I love you and I promise to do better.

Gert Nieuwenhuis
g.nieuwenhuis@uvt.nl
Malden
September 2008

A note from the Publisher
Every effort has been made to trace and acknowledge ownership of copyright and to clear permission for material reproduced in this book. The publishers will be pleased to make suitable arrangements to clear permission with any copyright holders whom it has not been possible to contact.

CHAPTER 01

Introduction and basic concepts

Chapter contents

This chapter begins with a brief explanation of what statistics is about. Students are then introduced to the basic notions of statistics, the concepts that have to be named right at the start in order to enable theory to be explained. In this book, the first use of a statistical concept is usually accompanied by its definition, while the name of the concept is highlighted by way of ***bold italic*** type and the indicator 🔑 is placed in the margin. When a word is printed **bold**, the underlying intention is to emphasize it.

1.1 What is statistics?

A study in economics often uses measurements collected from a previously specified set of objects. The set of objects under investigation is called the ***population***; the objects themselves are the (***population***) ***elements***. The measurements (called ***data***) are made on the elements and reflect some individual characteristic of the elements. But often the population is so large that it is too expensive and/or time consuming to obtain measurements for all population elements. That is why many studies in economics consider only a part of the population of interest, a so-called ***sample*** from the population. In that case, the data are measurements of the elements in the sample. So, statistics is about data. These data contain hidden **information** that has to be detected by statistics in order to become **knowledge**.

To illustrate, we start with three examples:

1 An economist studying welfare in the European Union (EU) might be interested in:

  a a comparison of the overall levels of prosperity **between** the EU member states;

  b the individual levels of prosperity for households **within** each of the EU member states.

Regarding (a), the population is the set of all 27 member states in the EU and the elements are the member states. Interest is in the 'level of prosperity' of each member state, which can be measured by the gross national income (GNI) per capita. The data will be numbers that give the 27 GNIs per capita. Regarding (b), there are 27 populations, one for each EU member state. For instance, population 1 (or 27) might be the set of all households in Austria (or the UK). So,

1

the elements of a population (and hence the objects of the research) are the households of the corresponding member state. Interest might be in the 'level of prosperity' of each household, which can be measured by annual household income; the data are numbers that reflect these annual household incomes. With respect to (a), the economist wants information on each element of the population, that is, the economist wants the GNIs for all member states of the EU. With respect to (b), it will be impossible to obtain the level of prosperity for each individual household in all EU member states. Instead, the economist has to collect sample data.

2 A researcher studying female labour force participation will be interested in the population of all adult women. Within this population, the elements are the adult women and interest is in 'labour force participation'. The measurements might be the numbers of hours that the women work outdoors. However, the measurements might also be the answers (yes or no) to the question: "do you participate in the labour market"? In the last case, the measurements are not ordinary numbers but the categories yes / no.

3 A financial analyst would like to predict the state of the economy in the coming year based on the state of the economy during the past 30 years. In his analysis, he distinguishes the following categories: regressive (r), normal (n) and expansive (e). His dataset (of the past 30 years) consists of a series of 30 measurements r, n or e. In this study, the population is the set of all years; the elements are the years. The analyst is interested in the 'state of the economy' and measures it for a sample of 30 years, in terms of the categories r, n, e.

So, measurements (data) often are ordinary numbers, but they can also be categories (as in the last two examples above). In general, data have to be **collected** first, so one needs (clever) ways to do that. Having collected the data, one needs methods to extract the hidden information from the (often enormous amount of) data: the data have to be **summarized** in terms of informative numbers and pictures. Often the measurements come from a **sample**, usually a small part of the population. What do these sample measurements tell us about the whole population? What **conclusions** can be drawn for the whole population?

Statistics includes the following actions: collecting the data, summarizing them, and (in case the data come from a sample) drawing conclusions for the whole population. Of course, there is some theory involved that tells us how to perform those actions in a clever way. Such theoretical considerations also belong to the field of statistics. Since many of them are essentially probabilistic reflections, **probability** is very important for statistics.

Example 1.1

A Spanish economist wants to study the present prosperity and poverty in her country. She is especially interested in income dissimilarities in Spain. To find answers to her questions, she needs data on the annual income per household, preferably for the whole population of interest: all Spanish households.

Maybe the National Statistics Institute of Spain can provide the data she needs, especially the annual incomes of all households in Spain. If this is the case, she can start up a *census* study: a study based on the data of all population elements. Table 1.1 contains data on household incomes.

Class	Number of households (× 1000)
Up to €9000	2628.7
€9000 – €14 000	2267.6
€14 000 – 19 000	2271.2
€19 000 – 25 000	2199.4
€25 000 – 35 000	2538.2
More than €35 000	2050.5
No data recorded	732.3
Total	14 687.8

TABLE 1.1 Annual household income in Spain, 2003
Source: National Statistics Institute, Spain (2004), www.ine.es

But assume that she does not have recent data and that she wants to collect the data herself. Since it is too time consuming to interview all Spanish households, she decides to choose a sample of households and record their annual incomes. The information that this sample gives her about prosperity and poverty will then be used to draw conclusions about the whole population. Below, several problems are considered that this economist may encounter.

Apparently, she wants to measure the annual incomes of households in Spain. But what exactly does she mean by 'annual income'? Gross or net incomes? Are the measurements in euros or are they in terms of (previously defined) categories such as 'very low', 'low', 'medium', 'high', 'very high'? Let us assume that she measures annual gross household incomes in euro. But several new questions come up:

- How large should this sample of Spanish households be and how should it be chosen (drawn)?

- She wants to use the sample measurements of gross annual incomes to approximate (estimate) accurately the unknown average of the gross annual incomes of **all** Spanish households. Furthermore, she wants to use the sample measurements to obtain information about the degree of income dissimilarity in the whole population of Spanish households. But how to draw the sample in such a way that it contains as much information as possible? Does there exist a clever sampling method that returns a sample with a maximal amount of information? The size of the sample will somehow be related to the level of precision that is reached for the approximation of the overall average annual household income. How about this relationship?

- There will be huge differences between various regions in Spain since the country has poorer and richer parts. The sample should reflect this. How can it be organized?

- And when the data are observed, how can the relevant information be released from the enormous amount of data? Can she construct nice pictures that offer visual impressions of the data?

- It seems reasonable to use the average of the annual household incomes in the sample to approximate the overall average of the household incomes in the population. But how can information about income dissimilarities be extracted from the data? The degree of variation (spread) within the sample of annual incomes seems to be an indicator. But how to measure the degree of variation in a dataset? Of course, within the dataset the difference between the largest measurement and the smallest measurement tells something about income dissimilarity, but it does not tell the whole story.

- The economist's ultimate goals are the average annual income and the degree of income dissimilarity in the population of all households in Spain. The sample average will differ from the population average. But will it be far off? How accurate is this sample average? How large is the probability that the distance between the sample average and the population average is more than €1000? Furthermore, she wants to draw conclusions about the degree of income dissimilarity in Spain. But this has to be done on the basis of only a sample of household incomes, so without measuring the larger part of the households in Spain. This automatically implies that there is a chance that a conclusion is incorrect. How large is the probability that a conclusion is incorrect?

All the above questions and problems are part of the statistical problems encountered by the economist. Obviously, she cannot blithely measure family incomes and feed them into a computer. She needs to have knowledge about collecting data, how to summarize them, and how to use the summarized results to draw conclusions about the whole population. To be able to do so, she needs good knowledge of probability.

1.2 Subdivision of statistics

Statistics can be divided into four sub-fields: probability theory, sampling theory, descriptive statistics and inferential statistics. The sub-fields are introduced below in the order in which they will be treated in this book.

Descriptive statistics includes collecting the data, and summarizing and presenting them by means of tables, graphs, and distinctive numbers (such as average and variance). Collecting the data involves making observations (measurements, counts, interviews, enquiries, polls, etc.) that follow from experiments. Often the number of data is very large, which makes it difficult to get a good overview. The data have to be ordered and maybe reduced. In any case, they have to be summarized in such a way that their relevant information is maintained as much as possible, and presented clearly by means of tables and graphs. At this stage of the research, it is also important to use good distinctive numbers to summarize the data. Of course, summarizing often leads to loss of information. However, the important challenge here is to reduce this loss of information as much as possible.

The sub-field of *probability* (or *probability theory*) studies the behaviour and the laws of chance and probability in experiments that allow more than one outcome. One simple example of such an experiment is the throw with a fair die, where the outcomes are 1, 2, ..., 6. Random experiments and probability are very important in life, not only in discussions that analyse the chances of Ajax Amsterdam in its match with Manchester United or in the analysis of the chances of a gambler, but in particular since probability is applied in so many fields. Insurance business is an example: for instance, the premium that has to be paid for car insurance is the result of a thorough study in probability. Statistics also leans heavily on probability theory. This is because the precision of statistical procedures (that use samples to draw conclusions about populations) is always expressed in terms of probabilities.

Sampling theory studies methods of sampling and their properties. One important sampling method is random sampling, where the elements of the population have the same chance of being chosen in the sample. Sampling is closely related to probability, since the qualities of a sampling method are determined by its probabilistic properties.

Inferential statistics studies and applies methods to draw conclusions about distinctive numbers (such as mean and variance) of the whole population of interest by considering only a sample. Again, probability plays an important role since the qualities of such methods depend heavily on the way the sample was obtained.

The economist in Example 1.1 will run into all four sub-fields. Of course, before starting she needs to analyse the planned research thoroughly. She has already decided to take the Spanish households as objects and to measure their degree of prosperity by way of gross annual household income. Before starting her research, she also needs to know the current value of the household poverty level in Spain, that is, the amount of money that households really need to be able to buy their minimal necessaries during a period of one year. To find this out, she may need information from other sources or the knowledge of a specialist. Then she can get started.

She first wants a good sampling method to obtain the sample. This brings her into the field of **sampling theory**. She has to understand the merits of different sampling methods and to choose the one that seems to have the best probabilistic properties in view of the things she wants. So, the field of **probability** is implicitly visited. She will probably choose a sampling method that enables proper coverage of the various regions and of the rural and urban areas. But how large should the sample be? It seems intuitively clear that larger sample sizes will yield higher levels of precision when approximating (estimating) the overall average annual household income. Again, **probability** enters the study.

Suppose that she manages to draw the (large) sample in a clever way and that she records the gross annual income of each sampled household. Then she is saddled with an enormous amount of data. The dataset is so large that it is impossible to get a good overview. She needs to enter the field of **descriptive statistics**; she needs to summarize the data with graphs and tables. Furthermore, particular numbers will be calculated that give information about central locations of the ordered family incomes in the sample, about the income dissimilarity in the sample, about the percentage of the sampled households with annual income below the household poverty level in Spain, and so on. But how can the central location of a dataset be measured? Obviously, the data show variation around such a central location. How can the degree of variation be measured? What does it tell about income dissimilarity?

Having finished the descriptive part of her research she wants to find out what her data tell about the population of **all** Spanish households. What can be concluded about the level of prosperity of all households in Spain? And what can be concluded about the percentage of the Spanish households living below the poverty line? Obviously, she is in the field of **inferential statistics**: drawing conclusions about populations on the basis of samples.

1.3 Variables

Each study or research aims at elements, characteristics and variables. The elements are the objects of the study. They can be persons, material objects, time epochs, events or experiments. Elements can be compared by way of a ***characteristic***: a feature of interest that can be used to compare the elements. Examples are: the age of a person, the size of a company, the GDP of a year, the education level of an adult. There usually is more than one way to observe (measure) a characteristic. A (***population***) ***variable*** is a well-defined prescript for observing (measuring) a characteristic. The set of ***values*** of the variable are the **different** outcomes when measuring is done at the elements. Often the values are ordinary numbers, but sometimes they are categories. When a variable is measured at an element, the measurement (outcome) is called an ***observation*** or an ***observed value***. It is possible that two observations are the same value. Observations are also referred to as ***data*** or ***data points***.

Variables are denoted by Latin capitals (X, Y, ...); for their values lower-case letters (x, y, ...) are used.

If a researcher is interested in overweight adult men, the elements are the adult men and the characteristic of interest is 'weight'. The variable X = 'weight in kg (two decimals)' attaches weights – amounts in kilograms to two decimal places – to the adult men. It is possible that two adult men have the same attached weight, although this is rather unusual when measurements contain two decimals. Another variable measuring the same characteristic is Y = 'categorized weight (with categories underweight / normal / overweight)'. This variable also attaches values to the adult men, but the possible values are now the three categories. Of course, the categories of the last variable have to be defined carefully in advance. It is obvious that for this variable a lot of the sampled men will have the same value.

In Example 1.1 the elements are the households. The economist is interested in the characteristic 'annual household income' (or even 'level of prosperity') and wants to compare the elements on the basis of this concept. But then she needs a prescript about how to measure it. One possibility is to measure this characteristic through a ***qualitative*** (or ***categorical***) ***variable***, a variable with categorized values. In this case, the categories might be: very low, low, medium, high, very high (that have to be defined carefully in advance). However, a ***quantitative*** or ***numerical variable*** with values that are ordinary numbers is more common. One quantitative variable that measures the characteristic 'annual household income' is X = 'gross annual household income in rounded euro'; the variable Y = 'gross annual household income in thousands of dollars' is another one. Both variables attach numbers to the elements. The first yields amounts of euro, the second amounts of thousands of dollars. The value 12 842 of the first variable corresponds to €12 842; the value 12 842 of the second variable corresponds to $12 842 000.

Quantitative variables can be discrete or continuous. A variable is called ***discrete*** if its set of possible values can be counted. If the set of possible values consists of all real numbers in an interval, the variable is called ***continuous***. For this distinction discrete / continuous, the underlying population is of no importance. Even if the variable X = 'weight (kg)' is considered on a population of only ten people, it is still considered to be continuous.

The variable 'number of children (< 18) per household' is a discrete variable: the set {0, 1, 2, ..., 25} counts 26 elements and will contain all possible values of the variable. The number line below describes the discrete character of the variable by marking the values.

However, the set of values of the variable 'gross annual household income in rounded euro' is so large that its values cannot be marked effectively in a similar way. Instead, the set of values might be described as follows:

(Although very uncommon, household incomes can be negative.) That is why this variable is continuous: there are so many different values which possibly lie so close together that we stylize its presentation and use an interval of real numbers to capture its values, in this case the interval (−50 000, 200 000).

Strictly speaking, all variables are discrete, since – because of measure precision – measurements have only a finite number of decimals and hence the set of values can be counted. Still, continuous variables are important. This is because, for conceptual reasons, a variable is often **assumed to be** continuous: if its set of possible values is very large and/or these values can lie close together, the variable is stylized in the sense that one assumes that **all** numbers in an interval – even those with many decimals – may occur as values. Continuous variables are used to stylize reality, just to make theory easier. For instance, the variable 'weight in kg (two decimals) of an adult man' is in essence discrete since only several thousands of values are possible. However, the possible values lie very close together and are numerous. That is why one usually considers the variable as continuous. In other words, it is assumed that values with more than two decimals are also possible and allowed. Describing the values of weight by way of the interval (0, 150) is easy:

Also the variable 'gross annual household income in rounded euro' is, strictly speaking, discrete since it can take only a countable number of values. Furthermore, the values are integers, so they seem to be isolated. Still, this variable will usually be assumed to be continuous, because of the enormous number of values that it can take.

In Example 1.2 below, the notation EU25 is used to denote the collection of the 25 EU member states when excluding Bulgaria and Romania (which entered the EU in 2007). The ten new member states that entered the EU in 2004 are denoted as NMS10 and the fifteen 'old' member states as EU15.

Example 1.2

On 1 May 2004, the EU grew from 15 to 25 member states. The entry of the ten new member states has affected many characteristics of the former EU. For Eurostat, the statistical department of the European Commission, the extension had substantial consequences since the data of the new countries had to be incorporated into the datasets. Below, the characteristic 'degree of economic development' is considered which is especially important since it enables economists to compare the economic activities of the 25 EU states.

One can think of many variables to measure this characteristic. The variable 'gross domestic product (in billions of euros)' – GDP for short – will yield important information about the economic development of the individual EU countries. The same holds for the variable GDPpc = 'GDP per capita in thousands of euros'. Both quantitative variables are considered to be continuous since they can take so many different values.

However, neither variable takes into account the different price levels between the countries. To overcome this problem, economic organizations such as the OECD (Organisation for Economic Co-operation and Development) often use the purchasing power standard (PPS) to correct for different price levels. Table 1.2 contains (among others) the values of the variable X = 'GDPpc in PPS (index)'. Notice that the observation of X for the EU25 (as a whole) is put equal to 100. That is, the GDPs of the 25 countries are added up and the result is divided by the total number of EU25 inhabitants; this amount (in euro) is expressed in PPS and the result is put equal to 100 (per cent). The observation (percentage) 122 of Austria then means that the GDPpc (in PPS) of Austria is 22% larger than the GDPpc (in PPS) for the EU25 as a whole. In this way individual member states are compared with the EU25 average.

Country in EU15	GDPpc (in PPS)	Country in NMS10	GDPpc (in PPS)
Austria	122	Cyprus	83
Belgium	118	Czech Republic	69
Denmark	123	Estonia	49
Finland	113	Hungary	61
France	111	Latvia	41
Germany	108	Lithuania	46
Greece	81	Malta	75
Ireland	133	Poland	46
Italy	107	Slovakia	52
Luxembourg	215	Slovenia	77
Netherlands	121		
Portugal	74		
Spain	98		
Sweden	115		
UK	118		

TABLE 1.2 GDP per capita in PPS (index, EU25 = 100)
Source: Eurostat news release STAT/04/145, 3 December 2004

X is considered to be a continuous variable but the reported values are in fact whole numbers in the interval [41, 215]. One may decide to divide the EU25 into only four categories, making things less accurate but, of course, useful for global overview. This can be done as follows:

- The high-income group (category 4): the countries with observation of X in [120, ∞);
- The high-middle-income group (category 3): the countries with observation of X in [100, 120);
- The low-middle-income group (category 2): the countries with observation of X in [50,100);
- The low-income group (category 1): the countries with observation of X below 50.

Table 1.3 shows the resulting categorization.

Income ranking	Country
High	Austria, Denmark, Ireland, Luxembourg, Netherlands
High-middle	Belgium, Finland, France, Germany, Italy, Sweden, UK
Low-middle	Cyprus, Czech Republic, Greece, Hungary, Malta, Portugal, Slovakia, Slovenia, Spain
Low	Estonia, Latvia, Lithuania, Poland

TABLE 1.3 Global ranking of EU25 by income

The variable Y = 'GDPpc category' that assigns the category to each of the 25 member states, is a qualitative variable.

Notice that four of the ten new EU members fall into category 1; the others fall into category 2. But remember that this is an ordering within the EU25. Compared with countries in Sub-Saharan Africa, even the poorest EU country, Latvia, will be considered as rich.

The values of a quantitative variable are ordinary numbers; the values of a qualitative variable are categories or codes to describe the categories. The values of the variable Y in Example 1.2 are categories, so Y is indeed a qualitative variable. Even if these categories are coded 1, 2, 3 and 4, the variable remains qualitative since these coded values are no ordinary values (we might have used the codes A, B, C and D as well). Qualitative variables that can take only two values (yes or no, man or woman, success or failure, correct or false) are called **alternative** or **dichotomous variables**. If one of the two values of an alternative variable is coded as 1 and the other as 0, the variable is called a **dummy variable**.

The difference between qualitative and quantitative variables can also be described by subjecting the values of variables to the arithmetic operation 'take the difference'. For a quantitative variable, differences between its values make sense; for a qualitative variable such differences are meaningless.

A dummy variable takes an intermediate position: in a sense, it is both qualitative and quantitative. The dummy variable X that takes the value 1 if the answer of a respondent to a certain yes/no question is yes, and 0 if the answer is no, can also – rather cryptically – be formulated as: number of times that the respondent answers yes to that question. The difference between the values 1 and 0 is 1, which means one more yes-answers.

The set of qualitative variables can be divided into two classes by subjecting their values to the arithmetic operation 'ordering'. If the values of a qualitative variable cannot be ordered in a natural way, the variable is called **nominal**. The variable 'gender' is nominal: its two possible values (man and woman) cannot be placed in a natural order. If the values of a qualitative variable **can** be ordered naturally, the variable is called **ordinal**. A simple example is the size of a t-shirt, with values XS, S, M, L, XL, XXL, 3XL, 4XL. The variable Y in Example 1.2 is also an ordinal variable: the coded values 1 to 4 represent categories that can be ordered naturally.

Also the set of quantitative variables is often divided into two classes, by subjecting their values to the arithmetic operation 'take the ratio'. With respect to the variable X = 'temperature (in °C)', it cannot be stated that the temperature in Athens (30°C) is two times as large as in Copenhagen (15°C); the ratio 30/15 is meaningless. However, with respect to the variable Y = 'number of visitors per hour to a certain internet site' it does make sense to state that the number of visitors (2000) between 10p.m. and 11p.m. was twice as high as the number of visitors (1000) between 11p.m. and midnight. The reason is that for X the value 0 (that is, 0°C) does not mean that there is no temperature, but for Y the value 0 does mean that there were no visitors. If the ratio of two values of a quantitative variable is meaningless, the variable is called an **interval** variable; if this ratio **does** make sense, the variable is called a **ratio** variable. Although the distinction interval / ratio is a standard one, it will rarely be used in this book. Usually it is sufficient to know that the variable of interest is quantitative.

Example 1.3

Table 1.4 refers to ten simple studies. In each of the studies, elements, characteristics and variables are considered. Furthermore, the variables are classified as discrete / continuous (d or c), as qualitative / quantitative (qual or quant), and as nominal / ordinal (n, o).

Study	Element	Characteristic	Variable	qual/quant	d/c	n/o
1	company	size	number of employees at 1 January 2005	quant	d	–
2	company	size	gross sales (thousands of euro) in 2005	quant	c	–
3	company	age	year of first registration	quant	d	–
4	adult	gender	man / woman	qual	–	n
5	adult	height	number of cm	quant	c	–
6	adult	height	small / middle / large	qual	–	o
7	adult	colour of eyes	green / blue / brown	qual	–	n
8	traffic accident	seriousness	number of deaths	quant	d	–
9	throw of a die	number of eyes	odd / even	qual	–	n
10	sample of polluted water	degree of pollution	boiling temperature (°C)	quant	c	–

d = discrete, c = continuous, n = nominal, o = ordinal

TABLE 1.4 Variables for nine studies

A characteristic can be measured by several variables. It is clear that the variable in study 8 is discrete: it (hopefully) will not take many values. However, regarding the variable in study 3 there might be some doubt or discussion about the classification discrete / continuous. Whether the variable in study 1 is discrete or continuous may depend on the underlying population. If the population consists of all small enterprises, the variable obviously is discrete. But if the population is the set of **all** companies, then the set of possible values is very large and the variable will usually be considered continuous. Later, such questionable choices will turn out to be a consequence of the 'model' that one takes.

Apart from the variables in studies 10 and 3 (which are interval variables), all quantitative variables are ratio variables.

1.4 Populations versus samples

As discussed before, a standard economic research aims at a population – the set of objects (elements) under study – and a variable (or maybe two or more). Sometimes a variable of interest is observed at **all** elements of the population; the research is then called a **census**. In that case, the resulting dataset is called a **population dataset**; it contains all possible information that can be gathered as far as that variable and that population are concerned. Often, however, it is too expensive and/or too time consuming to observe the whole population, since the population is too large. Instead of observing all elements, only a part of the population – a subset, usually chosen in an intelligent way – is considered and the accompanying observations of the variable are collected. The subset of the population is called a **sample** and the resulting dataset is a **sample dataset**.

Being interested (as in Example 1.2) in the economic development of the EU countries, the elements are the 25 EU member states of EU25 and the variable of interest might be X = 'GDPpc in PPS (index)'. Since the population is small and well documented, it is not too difficult to measure X for each individual country of the EU; see Table 1.2 above. The dataset is a population dataset, consisting of the 25 measurements of X in columns 2 and 4 of the table.

If a researcher is interested in income dissimilarity among EU25 inhabitants, he would prefer to have the gross annual incomes of **all** households in the EU25 but will instead have to be satisfied with the gross annual incomes of a carefully chosen **sample** of households. That is, he has to use a sample dataset, and the population dataset remains unobserved. This sample dataset of annual household incomes will be used to obtain information about **all** gross annual household incomes in the EU25.

Usually, interest is not primarily in the dataset of observations of the variable but in special, distinctive numbers or qualifications describing important features of the dataset as a whole. Examples are its mean value (the average) and variance (see Chapters 3 and 4), and the percentage of the observations that falls above some fixed level. Such distinctive numbers are called **statistics**. If, for a certain variable, the dataset is a sample dataset, the statistics are called *sample statistics* or just *statistics*; if the dataset is a population dataset the statistics are called *population statistics*. Often the variable is observed on only a part of the population, so the population dataset itself is unobserved. Still, the statistics of this population dataset (and hence of the accompanying variable) are called population statistics. Under these circumstances, population statistics are often unknown numbers called *parameters*.

In Example 1.2, the percentage of the low-income (category 1) countries in the EU25 is a population statistic. It describes a feature of the population dataset of the variable Y = 'GDPpc category', namely its percentage number of 1-values. Obviously, this statistic is equal to 16 since 4 out of the 25 countries are category 1 countries.

The concepts 'statistic' and 'variable' are often confused since they both **measure** features regarding a set of objects. Still it is important to be clear about their differences. A statistic measures some **overall** feature of a set of objects; this overall feature is a fixed number. A variable measures some **individual** feature that can take different values with different individual objects. For instance, when studying the weight of adults, a variable and a statistic of interest might respectively be 'weight (kg) of an adult' and 'mean weight of the adults'. The latter is a fixed – possibly unknown – number that concerns the whole set of adults; the former can take several values, which can be different for different adults.

In cases where the population dataset is unobserved, a sample dataset is often used to draw conclusions about the whole population. That is why it is important to use different notations for statistics of a population dataset and of a sample dataset. Often, Greek letters are used to denote statistics of a population dataset; they remain unknown if the population dataset is not observed. Latin letters are usually used to denote statistics of a sample dataset; they can be calculated as soon as the dataset is obtained. Since so many Greek letters will be used in this book, they are listed in Appendix A3.

Example 1.4

Interest is in a large population of households and in the variable X which measures the annual household expenditure on housing (in thousands of euros, to two decimal places). Special interest is in the average value (usually called **mean**) of the observations of X for the whole population. Since it is too expensive and time consuming to collect all these observations, 300 households are selected completely arbitrarily and their annual expenditures on housing are recorded. So, the resulting sample dataset contains 300 observations of X; see the file Xmp01-04.xls. For instance, the value 10.03 in this dataset refers to a sampled household whose annual expenditure on housing is €10030.

Being interested in the mean of the annual household expenditures on housing, it is important to notice that this study actually involves two means. The first mean is the average of the annual housing expenditures of **all** households. This statistic refers to the whole population and is denoted by μ (the Greek letter mu). Although μ is unknown, it is perfectly clear how it would be calculated if the population dataset were observed: add up all observations of X for the whole population and divide this total by the size of the population, that is, divide it by the total number of households. The second mean refers to the sample. This statistic is just the average of the annual expenditures on housing for the 300 households **in the sample**. Since the sample data are known, this average can be calculated: add up all 300 observations and divide the result by 300. Since this mean comes from the sampled 300 observations of the variable X, it is denoted by $\bar{x}$; the bar represents the averaging symbol.

Obviously, μ is unknown but $\bar{x}$ can be calculated since the data are given. Using a computer, it is an easy job to do the calculation (see Appendix A1 for Excel or SPSS). It turns out that $\bar{x} = 11.3652 \approx 11.37$, which indicates that the mean annual income of the **sampled** 300 households is equal to €11 370. But what are the inferences for the (unknown) mean μ? To answer this question, the field of inferential statistics is entered. Obviously, the true value of μ will not be exactly equal to 11.37, but it seems reasonable to conclude that it lies close to 11.37. But how close? Can it be inferred that it is pretty likely that μ lies between 10.5 and 12? This would indicate that the mean annual housing expenditure of the whole population is likely to be between €10 500 and €12 000. But how likely? With the theory of Chapters 15 and 16 more can be said about this likeliness.

Let me stress again that it is very important to distinguish between the population and the sample. Above, it was noted that often Greek letters are used for population statistics whereas Latin letters indicate sample statistics. If a population is finite, that is, if it contains a finite number of elements, its *size* (that is, its number of elements) is denoted by capital N. By contrast, the size of a sample is denoted by lower-case n.

All populations that one comes across in practice are finite. Still, infinite populations are also important. This is because populations are sometimes **assumed** to be infinite, to stylize reality in (often) theoretical situations. For instance, when considering the population of all rounding errors when rounding is done to the nearest integer, it seems natural to take the interval $(-\frac{1}{2}, +\frac{1}{2})$ for it. (The number 18.50001 is rounded to 19 and the rounding error is -0.49999; the number 18.49999 is rounded to 18 and the rounding error is $+0.49999$.) But this interval contains an infinite number of elements. Unless stated otherwise, the populations that are considered in this book are assumed to be finite. Note that, by definition, samples are always finite.

When interest is in a population that is too large to observe a variable X at all elements, a sample is taken from the population to obtain information needed to draw conclusions about the whole population. This is the principle of inferential statistics. But then it is obvious that this sample has to be taken in some intelligent way. To obtain information about the mean of all annual household expenditures, the collection of households in a rich neighbourhood does not form a wisely chosen sample. It will be intuitively clear that here the drawing of the sample can best be done in such a way that the result in a sense is a cross-section of the population. Such a sample arises for example if the n population elements are drawn completely arbitrarily; it is then called a *random sample*. This sampling procedure is very important for inferential statistics, and its merits and properties are considered in Chapters 12 and 13. Example 1.5 shows what might go wrong.

Example 1.5

A chain of toy shops is interested in opening a new outlet in a shopping centre in a particular district of a city. It is obvious that knowledge about the statistic μ = 'mean number of schoolchildren per household in that district' is important to enable an initial decision to open a new outlet. For that, 150 children are chosen at random from the only primary school in the district and they are asked about the number of schoolchildren of the household they belong to; the resulting average 2.8 of these 150 numbers is used as an approximation (estimate) of μ.

Taking a closer look at this problem yields that 2.8 probably overestimates μ, that is, it is likely that μ is smaller than 2.8. There are two reasons for this assertion. First, the district will also contain households without children. They do not get a chance to be represented in the sample. Secondly, households with more than one schoolchild have a larger chance to belong to the sample since such households often send all children to the same school. Essentially, it is not the sampling procedure that is wrong here but it is the choice of the population. The sample should have been drawn from the population of all households in the neighbourhood instead of from the schoolchildren.

Summary

A part from a first attempt to instil some primary feeling for statistics, this first chapter introduced some basic concepts. Some of these concepts are considered only superficially; a more precise description will follow later.

The practical importance of statistics reveals itself especially in inferential statistics: drawing conclusions about the whole population on the basis of only a sample. Although the other three sub-fields – descriptive statistics, probability and sampling theory – are also of interest in their own right, they are of fundamental importance for inferential statistics.

The most important concept of this introductory chapter is the concept '(population) variable'. Variables are divided into several subgroups, as illustrated in Figure 1.1.

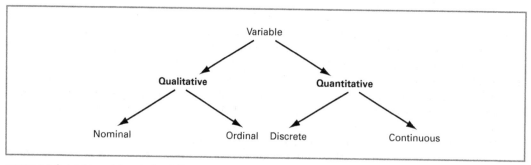

FIGURE 1.1 Subdivision of the set of variables

❓ Exercises

Exercise 1.1

In your own words, describe the difference between descriptive statistics and inferential statistics.

Exercise 1.2

In your own words, give definitions of the following concepts: population, sample, variable, population dataset, sample dataset, population statistic and sample statistic.

Exercise 1.3

According to the definition, a variable assigns values to each of the population elements. Describe how this works for the variable X below.

A chain of supermarkets has 100 outlets denoted by 1, 2, ..., 100. The outlets 1, ..., 30 have 25 employees each, the outlets 31, ..., 80 have 30 employees and the outlets 81, ..., 100 have 40 employees. Interest is in the variable X = 'number of employees of the outlet'.

Exercise 1.4

a Qualitative and quantitative variables can be distinguished by subjecting their values to the arithmetic operation 'take the difference'. How?

b Qualitative variables can be divided into two classes on the basis of an arithmetic operation. Which classes? Which operation? How does it work?

c Quantitative variables can also be divided into two classes on the basis of an arithmetic operation. Which classes? Which operation? How does it work?

Exercise 1.5

A quantitative variable can be discrete or continuous. Explain the difference between these concepts in your own words.

Exercise 1.6

The website www.europa.eu/abc/european_countries/index_en.htm presents many facts about the 27 EU member states. By clicking the countries on the map, the values of the variables below can be determined for each EU country:

- Year of EU entry
- Political system (values: republic, federal republic, constitutional monarchy)
- Capital city
- Total area (in km²)
- Population (in millions)
- Currency

a What is the population of interest? What are the elements?

b Say, for each of the variables, whether the variable is qualitative (nominal / ordinal) or quantitative (discrete / continuous).

c (computer, internet) Determine the values of these variables for Lithuania.

Exercise 1.7

The website http://uk.shopping.com offers consumers the opportunity to compare several groups of products that can be bought in the UK. For instance, MP3 and digital media players are compared on the basis of:

- Price range (below £40, £40–£60, £60–£80, £80–£110, £110–£170, above £170)
- Brand (Apple, Creative Labs, Sony, Samsung, Archos, …)
- Storage capacity (number of GBs)
- Family line (iPod nano, iPod video, Zen, Sansa, …)
- Audio format (MP3, WMA, WAV, …)
- Number of songs
- Adjustable playback speed (yes / no)
- Weight (in ounces)
- Voice recorder (1 = yes, 0 = no)

The products are also rated by consumers for special qualities, scale 1, 2, 3, 4, 5:

- Sound (1, 2, 3, 4, 5)
- Ease of use (1, 2, 3, 4, 5)
- Durability (1, 2, 3, 4, 5)
- Portability (1, 2, 3, 4, 5)
- Battery life (1, 2, 3, 4, 5)
- Overall rating (number of stars: 0.5, 1, 1.5, …, 4.5, 5)

These items are well defined and can be considered as variables.

a Describe a population that underlies all variables. Explain why the above items really are variables.

b Say, for each of the variables, whether the variable is qualitative (nominal / ordinal) or quantitative (discrete / continuous).

c Which variables are alternative variables? Which are dummy variables?

Exercise 1.8

A consumer in the UK is interested in an iPod of a certain type (60 GB). According to the site of Example 1.7, the price of this iPod varies among stores from £215 to £310. To find the mean price, the consumer plans to sample eight stores randomly and to record the respective prices of the iPod.

a Describe the population that underlies the research of this consumer. Is the variable of interest discrete or continuous?

b Before actually drawing the sample, the consumer consults a textbook on statistics and ascertains that there is a high probability that the mean value of the eight prices will fall between £220 and £275. To which of the four sub-fields of statistics does this statement belong?

The mean of the eight prices turns out to be £250.

c To which of the four sub-fields does this result belong?

d On the basis of the mean price of the eight sampled stores, the consumer concludes that the mean price of the iPod for the whole population of stores probably falls in the interval (240, 260). To which of the four sub-fields does this conclusion belong?

e Suppose that – using more information and more advanced statistical procedures – it can be concluded that there is 95% confidence that the mean price of the whole population falls in the interval (240, 260). Use your common sense to give your opinion on the following statements:

i The mean price of the whole population is above £230.

ii The mean price of the whole population lies below £265.

iii The mean price of the whole population is above £245.

Exercise 1.9

Consider the following three characteristics:

- The temperature of the water in a swimming pool.
- The amount of savings.
- The direction of the wind.

a Information is rather poor. For each of the cases, think of a suitable population.

b The three characteristics do not offer precise definitions. In each of the cases, think of two variables of different types (nominal, ordinal, quantitative) that measure the characteristic.

Exercise 1.10 (computer)

This exercise is about European football. Especially, it is about the Champions League cup, formerly called European Cup I. Real Madrid has won this cup nine times. But does this mean that the cup was exactly nine times won by a Spanish team? Or were there more Spanish winners?

The file Xrc01-10.xls records all winners of the cup; take a look at its contents. It contains data for the period 1956–2007. It is your task to determine an overview of the number of times the cup was won by all individual European countries.

a Formulate a population that is suitable for this study.

b Is the dataset a sample dataset or a population dataset? Why?

c Consider the variable X = 'country where the winner of that year comes from'. Is it qualitative (nominal / ordinal) or quantitative?

d Recall that a variable attaches a data point (observation) at each individual population element. Describe how this works for the variable X.

e Consider the dummy variable (called Y) that takes the value 1 if the winner is Dutch. Determine the proportion of ones in the dataset of all observations of Y.

f Let Z be the dummy variable that takes the value 1 if the winner is Spanish. Determine the proportion of ones in the dataset of all observations of Z.

Exercise 1.11

Eurostat's *Yearbook 2004* (p. 79) states the following about EU25 and Japan:

> Throughout almost the entire European Union, there are more women than men among tertiary students. Exceptions are Germany, where male students are slightly more numerous than female students, and The Netherlands and Czech Republic with a balanced proportion. In Japan, the number of male tertiary students significantly exceeds that of female students.

Actually, Eurostat compared the 26 proportions of male / female students among all tertiary students in the 25 individual EU countries and Japan. It is likely that Eurostat received the proportions directly from the 26 statistical agencies and simply compared them.

a Carefully describe the population and the variable that was used by the Federal Statistical Office of Germany to obtain the data necessary for calculating the proportions of male / female tertiary students in Germany.

b Carefully describe the population and the variable that was used by Eurostat once it had received the proportions from all 26 statistical agencies.

CASE 1.1 TRADING PARTNERS OF THE EU25

Every year, the WTO (World Trade Organization) publishes a voluminous report about the merchandise flows of the previous year. The report contains information on the trading activities of many countries and economies around the world. For the EU25, the WTO report *International Trade Statistics 2005* contains – among much else – information about destinations of exports and origins of imports; see the table in the file <u>Case01-01.xls</u>. As so often when tables are considered, it takes some time to deduce the whole story behind the numbers. The solution can be found on the McGraw-Hill website.

a Determine the total amounts of exports and imports of the EU25 in 2004.

b For both exports and imports two populations are considered called 'World' and 'Economies'. List their population elements, both for exports and for imports.

c Do these populations cover the whole world? Why (not)?

d Carefully describe all the variables that are measured in the table.

e For both the exports and the imports another population can be defined that has only the two elements 'top 5 economies' and 'the rest'. Describe (for both the destination case and the origin case) these elements and determine – if possible – the values that the variables of part (d) take at these elements.

PART 1
Descriptive Statistics

As pointed out in Chapter 1, descriptive statistics is one of the basic fields of statistics. The starting point is a dataset, a set of observations of one variable or more. Although the way the data are collected certainly forms part of descriptive statistics, this will not be considered right now; some of the ideas behind methods of sampling are treated in Chapter 12. Below, in Chapters 2 to 5, emphasis will be on **summarizing** and **presenting** the data. Many statistical tools will be considered to extract relevant information from the mass of data.

Chapter 2 gives an overview of useful tables and graphs that summarize data of one variable. These tools offer visual presentations of the data.

In Chapters 3 and 4, data of one variable are summarized in terms of statistics. Although summarizing a dataset usually leads to loss of information, it will turn out that from a suitably chosen list of statistics a surprisingly large amount of information can be regained.

Chapter 5 is about datasets originating from more than one variable. Methods are presented to deduce relevant information about the relationships **between** two variables. Again, graphs as well as statistics will be used to visualize and quantify the relationships.

CHAPTER

Tables and graphs

Chapter contents

Datasets are often large, containing a lot of information on many population elements. This information is often hard to survey because of the large amount of data. Hence, there is a need for suitable tables and graphs that present the relevant information pictorially. A manager who wants to present the yearly figures of the company in, say, a historical perspective, wants tables and graphs that show nicely the important features hidden in the data.

In this chapter, data of nominal, ordinal and quantitative variables will be considered. Most tables present overviews of frequencies. Important concepts are frequency distribution and distribution function. Bar charts, histograms and scatter plots are used to present data graphically.

Appendix A1 explains how Excel and SPSS can be used to create a graph or to obtain the contents of a table.

CASE 2.1 COMMITMENT TO DEVELOPMENT INDEX 2006

Each year, the Center for Global Development (CGD) crunches thousands of numbers to compute the Commitment to Development Index. This index rates 21 rich countries on how much they help poor countries to build prosperity, good government and security. Each rich country receives a score in seven policy areas, and these are then averaged for an overall score. The areas are: aid, trade, investment, migration, environment, security and technology. See the website of CGD (www.cgdev.org) for information about these areas and the way the scores are determined.

The file Case02–01.xls contains a table that gives an overview of the scores of the 21 rich countries for each of the seven areas, jointly with the overall scores. The objective is to present the data in one or more attractive charts whose strong visual impact will encourage competition between the countries. See the end of Section 2.3 for a solution.

2.1 Nominal variables

Emphasis will be on a dataset of observations of a nominal variable. So, the data are categories or coded categories and there is no natural way to put them in ascending or descending order; adding or subtracting the data points makes no sense.

In general, the observed values of the variable will not all be different. The number of times that a certain value occurs in the dataset is called the (**absolute**) **frequency** of that value. Dividing the frequency of a value by the total number of observations yields the **relative frequency** of that value. The relative frequency of a value is just the proportion of all observations in the dataset with that value; multiplication with 100 yields the accompanying percentage. An overview of all different values in the dataset jointly with the accompanying frequencies is called **frequency distribution**. If relative frequencies accompany the different values, the overview is called **relative frequency distribution**.

Example 2.1

Table 2.1 divides employed persons of the EU25 into eight industries.

Industry	Frequency (× 100)	Relative frequency
Manufacturing	340 058	0.302
Wholesale and retail trade; repair of motor vehicles, motorcycles and household goods	264 675	0.235
Real estate, renting and business activities	202 143	0.179
Construction	117 810	0.104
Transport, storage and communication	105 908	0.094
Hotels and restaurants	75 151	0.067
Electricity, gas and water supply	15 196	0.013
Mining and quarrying	6 710	0.006
Total	1 127 651	1

TABLE 2.1 Employed persons in the EU25, by industry, 2001
Source: Eurostat Yearbook 2005, p. 235

The table's accompanying text in the *Yearbook* reveals that the population of interest is 'all persons working in the various industries: employees and non-employees (e.g. family workers, delivery personnel) with the exception of agency workers'. The variable of interest is X = 'industry of the employed person'. This variable is indeed qualitative, and even nominal since the values cannot be ordered in a natural way. It has eight values: the eight industries of Table 2.1. The dataset (that is, the list of all employed EU25 citizens and the industries they are working in) that underlies Table 2.1, is not given. (Probably, Eurostat drew information from each of the 25 national statistical agencies and amalgamated the figures into one table.)

To increase the readability of the table, the order of the industries is determined by the frequencies and the totals are also included. Note that the table has a **heading** (that briefly describes its contents) and a **source** (that explains where the (original) contents come from).

The overview offered by the first two columns yields the frequency distribution of X and its underlying dataset. However, do not forget to multiply all numbers in column 2 by 100 to obtain the frequencies. The total number of persons in the population is $N = 112\,765\,100$, a bit less than 112.8 million people. Columns 1 and 3 jointly present the relative frequency distribution of X; the relative frequency of each industry arises by dividing the corresponding number in the frequency column by the total 1 127 651.

Interpretation of the table yields that most people work in the sector 'manufacturing', whereas the sector 'mining and quarrying' has the smallest frequency. The accompanying relative frequencies 0.302 and 0.006 of these two industries are **proportions**; they correspond to the **percentages** 30.2 and 0.6.

The (relative) frequency distribution of a dataset gives a **quantitative** overview. For a more **visual** overview, a **_bar chart_** can be used. The values (categories) are placed horizontally and equally spaced; vertical bars represent the (relative) frequencies that belong to the values. These bars are separated from each other, and they all have the same width. So, both the height and the area of a bar represent the (relative) frequency. Since the variable is nominal, the order of the values on the horizontal axis can be chosen arbitrarily. However, the order is often taken by decreasing or increasing frequency, to make the picture most expressive. In Figure 2.1 both the frequency distribution and the relative frequency distribution of the variable X = 'industry of the employed person' are pictured by means of a bar chart.

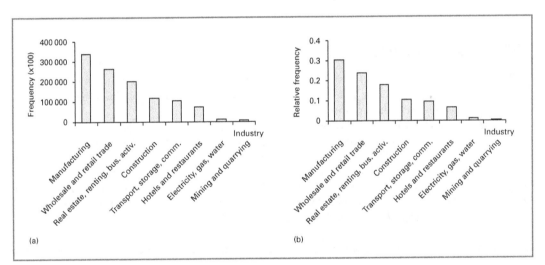

FIGURE 2.1 Bar charts of the frequency distribution and relative frequency distribution

The two charts differ only because of the scaling on the vertical axis. The right-hand chart immediately reveals that about 30% of the employed EU25 citizens work in the sector 'manufacturing'.

Pie charts are also used to visualize a relative frequency distribution. A pie is divided into segments that refer to the different values. The magnitudes of the segments represent the corresponding relative frequencies. The respective percentages can be included in the pie, as in Figure 2.2 which shows the pie chart for the variable 'industry of the employed person'.

The bar chart invites the reader to compare the sizes of the industries; the pie chart puts more emphasis on the different industries in relation to the total.

There are several ways to present a bar or a pie chart. For instance, the bars of a bar chart can be graphed horizontally, whereas the pie can be given a three-dimensional effect. Furthermore, percentages can be included or left out; the **_legend_** (which contains information about the colours or shading) can be omitted or adapted. The reader is invited to use the file Xmp02-01.xls to reproduce the above charts and to try some alternatives.

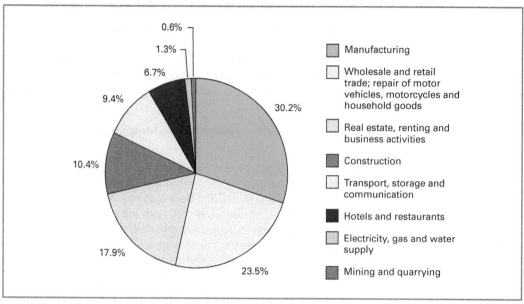

FIGURE 2.2 Pie chart of the relative frequency distribution

2.2 Ordinal variables

Consider a dataset of observations of an ordinal variable. Hence, the observations in the dataset are categories and they can, in a natural way, be placed in ascending order. However, taking differences between two values of the variable still makes no sense.

Again, it is possible to create an overview of the different values in the dataset jointly with their (relative) frequencies. This (relative) frequency distribution can again be presented in a table, bar chart and pie chart. Since the different values can now be put in increasing order, the frequencies up to and including certain values can also be considered. Such frequencies are called **cumulative frequencies**. Relative frequencies up to and including certain values are called **cumulative relative frequencies**. The overview of all different values combined with the respective cumulative (relative) frequencies is called **cumulative (relative) frequency distribution**.

Example 2.2

In many surveys, a five-point opinion scale is used to measure a characteristic of a population.

The Education Office of the Faculty of Economics and Business Administration of a university asked the BBA students of two statistics classes to evaluate the performance of the instructors A and B. To measure the characteristic 'quality of instructor's performance', the variable X = 'answer to the question: instructor explains things very clearly' was considered. Its possible values were coded:

1 = strongly disagree;   2 = disagree;   3 = neutral;   4 = agree;   5 = strongly agree

The dataset Xmp02-02.xls consists of two columns. The first column contains the ordered opinions of the students in the class of instructor A; the second column contains the ordered opinions of the students in the class of instructor B. Notice that the two columns do not have the same length. The file Xmp02-02.sav has organized the data in a different way. All (ordered) opinions (for both instructors) are in the first column while the second column records the corresponding observations of the alternative variable Y = 'instructor' with values A and B. Parts of both files are presented in Table 2.2.

Xmp02-02.xls		Xmp02-02.sav	
1	1	1	A
1	1	1	A
3	1	3	A
3	1	3	A
4	1	4	A
4	1	4	A
4	2	.	.
4	2	.	.
.	.	5	A
.	.	5	A
5	4	1	B
5	4	1	B
	4	1	B
	5	.	.
	5	.	.
	5	4	B
	5	5	B
	5	5	B
	5	5	B
		5	B
		5	B

TABLE 2.2 Parts of the files Xmp02-02.xls **and** Xmp02-02.sav
Source: Education Office (2005)

In the first file, two numbers in the same row refer to different students; the observations in a row of the second file refer to one student.

The data are summarized in Tables 2.3 and 2.4; they can easily be checked from both files. Columns 1 and 2 of Table 2.3 (respectively 2.4) present the frequency distribution of the opinions regarding instructor A (respectively B). The combination of the columns 1 and 4 in each of the two tables gives the corresponding relative frequency distributions. Since the five different values of the variable can now be ordered in a natural way, it is possible to consider, for each value, the cumulative (relative) frequency up to and including that value. For instance, for the class of instructor B, the cumulative relative frequency at 4 is just the sum of the relative frequencies up to and including 4 (that is, at 1, 2, 3 and 4). The two overviews offered by the columns 1 and 5 are just the cumulative relative frequency distributions, one for the class of A and one for the class of B.

'Instructor explains things clearly'	Frequency	Cumulative frequency	Relative frequency	Cumulative relative frequency
1	2	2	0.03	0.03
2	0	2	0.00	0.03
3	2	4	0.03	0.06
4	27	31	0.44	0.50
5	31	62	0.50	1.00
Total	62	–	1.00	–

TABLE 2.3 Opinion of BBA students of the class of instructor A

'Instructor explains things clearly'	Frequency	Cumulative frequency	Relative frequency	Cumulative relative frequency
1	6	6	0.09	0.09
2	13	19	0.19	0.28
3	19	38	0.28	0.55
4	25	63	0.36	0.91
5	6	69	0.09	1.00
Total	69	–	1.00	–

TABLE 2.4 Opinion of BBA students of the class of instructor B

The charts in Figure 2.3 show the two relative frequency distributions in a combined and a stacked bar chart.

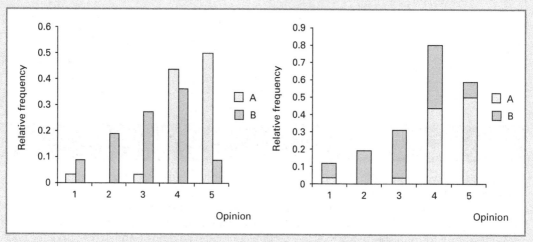

FIGURE 2.3 Combined and stacked bar chart of relative frequency distributions of A and B

Both charts demonstrate that the distribution of instructor A is more concentrated on the higher values than is the distribution of instructor B. From the cumulative relative frequency distribution of A it follows that only 6% of the students were neutral or negative regarding the teaching skills of the instructor. For instructor B the corresponding percentage was 55%, but only half of the students were really negative.

 The type of response scaling used in the above example is commonly found in questionnaires; the scale is then called a **_Likert scale_**, after Rensis Likert who, in 1932, published a report on its use. Respondents specify their level of agreement to each of the statements in the questionnaire by choosing one of the five options of the five-point scale.

2.3 Quantitative variables

In this section, the datasets contain observations of quantitative variables. That is, the observations are now ordinary, uncoded numbers that measure how much or how many. As a consequence, most graphs and figures discussed in the present section have the usual line of real numbers as their horizontal axis. Some figures (dot plots, bar charts) immediately show the original observations, whereas others present a division of the data in classes accompanied by the corresponding (relative) frequencies. In cases where the variable is **discrete**, each different value forms a class if there are not too many. But if the variable is considered **continuous**, the classes are usually adjoining intervals.

The first subsection below is about graphs that give visual presentations of the original data. The graphs in the second and third subsections present frequency distributions for, respectively, data from a discrete variable and classified data from a continuous variable.

2.3.1 Original observations

Some charts will be discussed that directly visualize the original observations of a dataset of a quantitative variable.

Since the observations are now ordinary numbers, they can be situated along the horizontal line of real numbers by marking a dot or a small vertical stroke. The resulting chart is called a ***dot plot***.

Example 2.3

To study welfare within the EU25, the gross national income (GNI) of the member states is of interest. In particular, the GNI per capita (GNIpc, the GNI divided by the total number of citizens of the country) is an instrument to compare the EU members. For 2002, these GNIpcs (measured in units €1000) are shown in Table 2.5; see also Xmp02-03.xls. Note that the rows in the table are ordered by GNIpc.

Country	GDP(€bn)	GNI total (€bn)	GNIpc (× €1000)	Rounded
Luxembourg	22.4	20.2	44.9	45
Denmark	183.7	180.3	33.6	34
Sweden	255.4	253.8	28.3	28
UK	1660.1	1690.6	28.2	28
Ireland	129.3	104.7	27.0	27
Netherlands	444.6	435.5	27.0	27
Finland	139.7	139.4	26.8	27
Austria	218.3	216.3	26.6	27
Belgium	260.0	264.5	25.7	26
Germany	2110.4	2108.8	25.6	26
France	1520.8	1527.8	25.5	26
Italy	1258.3	1246.3	21.5	22
Spain	696.2	687.6	17.1	17
Cyprus	10.8	10.8	14.1	14
Greece	141.4	141.6	13.4	13
Portugal	129.3	126.6	12.6	13
Slovenia	23.3	23.3	12.0	12
Malta	4.1	4.0	10.0	10
Czech Republic	73.9	74.2	7.2	7
Hungary	68.9	65.2	6.5	7
Poland	199.9	197.9	5.1	5
Estonia	6.9	6.6	4.6	5
Slovakia	25.1	24.6	4.6	5
Lithuania	14.7	14.5	4.0	4
Latvia	8.9	8.9	3.8	4

TABLE 2.5 GDP and GNI for the EU25, 2002
Source: Statistics Netherlands (2004)

The dot plot in Figure 2.4 shows the positions of the observations along the line. Although this picture is very simple, it presents a good view of the way the observations are situated relative to each other.

FIGURE 2.4 Dot plot of GNIpcs in the EU25

Three clusters of observations can be distinguished: one around 5, one between 10 and 15, and one between 25 and 30. The smallest observation is about 4; the largest lies close to 45, which means that the GNIpc of the country involved is about €45 000. Notice that this picture does not tell that this maximal GNIpc belongs to Luxembourg. Apparently, some information is lost; the connection of the observations with the individual countries within the EU25 cannot be regained from the dot plot.

In Figure 2.5, the connection between a country and its GNIpc **is** maintained. The countries are placed vertically, while the GNIpcs can be read horizontally as the lengths of bars.

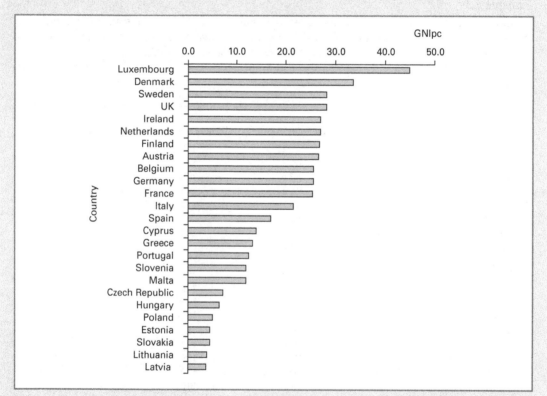

FIGURE 2.5 Bar chart of GNIpcs in the EU25

This picture preserves the connection between GNIpc and country. For instance, it tells us that Latvia has the lowest GNIpc (about €4000). The detection of the three clusters is with this chart a bit harder but it does have the advantage that the countries involved can be read from it: the countries grouped around Slovakia, Portugal and Finland respectively have similar levels of GNIpc.

2.3.2 Data of discrete variables

In this subsection the data are measurements of a discrete variable. A set of such data usually consists of a limited number of different values, some of them occurring more than once in the dataset. Think of the variable X = 'number of children per household' being measured for a sample of 500 European households. The smallest value that X can take is 0, whereas it is hard to say in advance what the largest value will be. The *Guinness Book of Records* (1996) mentions that a Russian woman – the wife of a peasant – gave birth to 69 children in 27 confinements, in the period 1725 to 1765. But, nowadays, it seems acceptable to state that the different values of X will be integers between 0 and 30. Within the dataset of the 500 observations, some of the values in this range will have frequencies larger than 1.

It is natural to summarize the dataset in terms of an overview that lists the different values jointly with their frequencies. As before, this overview is called the **frequency distribution** of the dataset. If the frequencies in the overview are replaced by the relative frequencies, the overview is (again) called the **relative frequency distribution** of the dataset. Instead of combining each different value in the overview with its relative frequency, it might as well be combined with its **cumulative relative frequency**: the summation of the relative frequencies of all different values up to and including that value. For instance, in the above example regarding X = 'number of children per household', the cumulative relative frequency of the value 2 is just the proportion of all households (in the sample of size 500) with at most 2 children and is calculated by summing the relative frequencies of the values 0, 1 and 2. An overview that combines the different values with their cumulative relative frequencies is again called the **cumulative relative frequency distribution** of the dataset. Bar charts can be used to graph these types of distributions.

Some statistical packages cannot list frequency distributions that have too many different values. For instance, for the above variable X, it might be problematic to present the 31 categories and their (cumulative, relative) frequencies in a nice table or picture. The solution is to make a summarizing category 'more than 10' and to present the accompanying frequency instead of all individual frequencies that belong to the values 11, …, 30.

Example 2.4

The variable Y = 'size of household' is important in many areas of economic research. Table 2.6 gives the frequency distributions for three countries for 2005; the values 5, 6, etc. are comprised in one class.

	Unit	Denmark	Germany	Netherlands
Total number of households	1000	2499	39 178	7091
1-person households	1000	949	14 695	2449
2-person households	1000	829	13 266	2318
3-person households	1000	296	5 477	906
4-person households	1000	289	4 213	973
Households with at least 5 persons	1000	136	1 527	445

TABLE 2.6 Frequency distributions of 'size of household' in three countries
Sources: Statistical agencies of Denmark, Germany and the Netherlands

For instance, for Germany the frequency distribution of Y is the overview of columns 1 and 4.

However, the three **relative** frequency distributions are more suitable for comparison. Table 2.7 lists these distributions jointly with the corresponding cumulative relative frequency distributions.

	Denmark		Germany		Netherlands	
	Rel. freq.	Cum. rel. freq.	Rel. freq.	Cum. rel. freq.	Rel. freq.	Cum. rel. freq.
1-person households	0.379	0.379	0.375	0.375	0.345	0.345
2-person households	0.332	0.711	0.339	0.714	0.327	0.672
3-person households	0.118	0.829	0.140	0.854	0.128	0.800
4-person households	0.116	0.945	0.108	0.962	0.137	0.937
Households with at least 5 persons	0.054	1.000	0.039	1.000	0.063	1.000
Total	1	–	1	–	1	–

TABLE 2.7 **(Cumulative) relative frequency distributions for three countries**

The three relative frequency distributions follow easily by dividing the frequencies by the respective totals. The three cumulative relative frequency distributions follow by cumulatively adding up the respective relative frequency distributions. For instance, for Denmark the cumulative relative frequency at the value 2 is just the summation of the relative frequencies at 1 and 2; the cumulative relative frequency at 3 arises from this sum by adding the relative frequency at 3 to it, etc. It follows from the three relative frequency distributions that the Netherlands has the smallest percentage of 1-person households (34.5%) but the largest percentage of households with at least 5 persons. When compared with Germany, the proportion of the Dutch households with at least 5 persons is 61.5% larger than the corresponding proportion for Germany, since:

$$\frac{0.063 - 0.039}{0.039} = 0.615$$

When a household with at most 3 persons is characterized as 'small' and a household with at least 4 persons as 'large', then Table 2.7 can be summarized as shown in Table 2.8.

	Households		
	Denmark	Germany	Netherlands
	Rel. frequency	Rel. frequency	Rel. frequency
Small households	0.829	0.854	0.800
Large households	0.171	0.146	0.200
Total	1	1	1

TABLE 2.8 **Relative frequencies for small and large households in three countries**

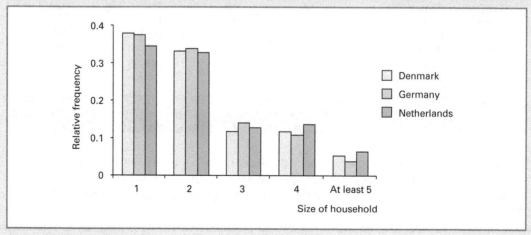

FIGURE 2.6 Bar charts of 'size of household' for three countries

Notice that, for each of the countries in Table 2.8, the relative frequency for 'small' households is just the cumulative relative frequency of Table 2.7 at the value 3. The proportion of large households in the Netherlands is 37% greater than the corresponding proportion for Germany.

The graphs in Figure 2.6 offer a visual comparison of the three distributions. This combined bar chart also shows that, for the Netherlands, the variable Y is more concentrated on the classes 4 and 'at least 5' than it is for Denmark and Germany.

The cumulative relative frequency distribution can be used to determine the percentage of the population elements (or sample elements) for which the variable of interest takes a value at or below a certain fixed value. For instance, from Example 2.4 the percentage of Dutch households that have at most 4 members is 93.7 since 0.937 is just the cumulative relative frequency of Y at the value 4. It is also useful for solving reversed problems such as: determine a number x_0 such that for at least 75% of the population the variable of interest takes a value $\leq x_0$ while for less than 75% of the population it takes a value $< x_0$. To solve such problems, the concept 'cumulative relative frequency distribution' has to be adapted.

Cumulative distribution function when the variable is discrete

Consider a discrete variable. The cumulative distribution function of a dataset of observations of the variable gives, for each real number b, the relative frequency of the observations that are at most b. To say it more formally:

(Cumulative) distribution function for a discrete variable

The *cumulative distribution function* (or *distribution function* for short) of a dataset of observations of a discrete variable is the function F such that, for all real numbers b:

$$F(b) = \text{relative frequency of the observations} \leq b \tag{2.1}$$

For the values b that are present in the dataset, $F(b)$ is just the cumulative relative frequency at b. For values b not in the dataset, $F(b)$ is equal to the cumulative relative frequency at the largest value smaller than b that **is** present in the dataset. The following example illustrates.

Example 2.5

The manager of a computer shop wants to make an inventory of the daily numbers of Inspiration 5100 notebooks sold during the last year. The dataset <u>Xmp02-05.xls</u> contains the sales (#) for 300 days. Special interest is in the distribution function of the dataset.

The bar chart in Figure 2.7 shows the relative frequency distribution of the dataset of the variable 'daily number of Inspiration 5100 sold'; the table lists the cumulative relative frequency distribution.

On 10.7% of the days not a single Inspiration 5100 was sold. Furthermore, it is remarkable that during the past 300 days it never happened that precisely 1 notebook was sold. For the distribution function F of the dataset it holds that – at the values $b = 0, 2, 3, 4, 5$ – $F(b)$ is just the cumulative relative frequency at b; see the second column of the table in Figure 2.7. For other values b the accompanying $F(b)$ follows by looking at the largest value of the dataset that is smaller than b. For instance, $F(1)$ is equal to $F(0) = 0.107$ and $F(4.3)$ equals $F(4) = 0.897$. To state things more formally: for b **between** two successive values of the dataset, $F(b)$ is just the cumulative relative frequency at the smallest of the two values. Apparently F does not increase **between** two successive values of the dataset. Of course, for $b < 0$ it holds that $F(b) = 0$ while $F(b) = F(5) = 1$ for $b > 5$.

The table in Figure 2.8 gives an overview of the function F; the graph of F is plotted in the figure.

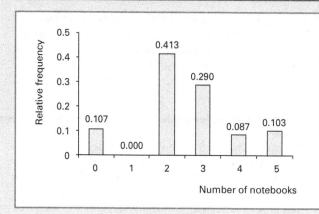

Sales (#) per day	Cum. rel. frequency
0	0.107
2	0.520
3	0.810
4	0.897
5	1.000

FIGURE 2.7 (Cumulative) relative frequency distribution of the dataset

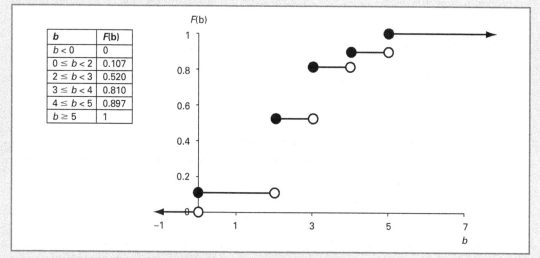

b	*F*(b)
$b < 0$	0
$0 \leq b < 2$	0.107
$2 \leq b < 3$	0.520
$3 \leq b < 4$	0.810
$4 \leq b < 5$	0.897
$b \geq 5$	1

FIGURE 2.8 Distribution function of the dataset of 'number of notebooks sold per day'

The distribution function F is a step function. It takes only six (vertical) levels; see the values in the second column of the table. The solid dots of the graph at the values $b = 0, 2, 3, 4, 5$ indicate that at these values F jumps to a higher level; the jump sizes are just the relative frequencies at these values. The open dots indicate that the corresponding vertical levels are **not** the levels that belong to the values 0, 2, 3, 4 and 5.

It follows immediately that F does not take the (vertical) level 0.5; it **passes** this level at $b = 2$. Hence, on more than 50% of the 300 days (that is, on more than 150 days) at most two Inspiration 5100 notebooks were sold daily while on less than 50% of the 300 days fewer than two such notebooks were sold. Similarly, F passes the level 0.75 at $b = 3$. On more than 75% of the 300 days of the dataset, three or fewer notebooks were sold daily; fewer than three notebooks were sold on less than 75% of the days.

A (cumulative) distribution function will shortly be denoted by **cdf**; its graph is occasionally called **ogive**. If the variable is discrete, the cdf has the following properties:

Properties of the distribution function for a discrete variable

- It is a non-decreasing step function.
- It jumps to higher vertical levels at the different values of the dataset.
- The jump sizes are just the relative frequencies of the different values in the dataset.

So far, frequency distributions have been connected to **datasets** of discrete variables. However, frequency distributions are also connected to the discrete **variable** itself. The (relative / cumulative / absolute) frequency distribution of a discrete variable is defined as the (relative / cumulative / absolute) frequency distribution of the accompanying population dataset. Moreover, the distribution function of a variable is defined as the cdf of the accompanying population dataset of observations. Even if this population dataset is not observed, it makes sense to infer the frequency distribution and the distribution function of the variable albeit that it is not known.

2.3.3 Data of continuous variables

When the values in the dataset come from a continuous variable, it usually makes no sense to create an overview that combines the different values with their frequencies, as in the previous subsection. This is because the observations of a continuous variable only rarely coincide. Nevertheless, one still wants to obtain a good understanding of the distribution of the data over the real line. This can be achieved by dividing the range of the values (that is, the interval between the smallest and the largest value) into classes and to combine these classes with the absolute, relative or cumulative frequencies of the data per class.

Let X be the grade of an International Business student for the Final Exam of the course Statistics 2, measured on the scale 0–10 with a precision of one decimal. We consider X as continuous. As a subdivision of the range 0–10, one might choose the classes as the intervals between two successive integers:

$[0, 1], \quad (1, 2], \quad (2, 3], \quad \ldots, \quad (8, 9], \quad (9, 10]$ **(Classification I)**

Here the class $(2, 3]$ includes the number 3 but **not** the number 2, while the class $[0, 1]$ encloses both numbers 0 and 1. Next, for each interval the frequency of the enclosed data points is determined. The resulting frequency distribution combines classes and accompanying frequencies. It summarizes the dataset and presents an overview of it. Note, however, that some of the information is lost: within a class we do know the number of the observations that falls in it, but we lose the precise positions of these observations. Also note that the choice of the classes (and hence the frequency distribution) is **not unique**: other reasonable divisions in classes could have been chosen as well; see the example below.

A division of the range of the dataset into classes has to form a **categorical system**, that is, the classes do not have common values and they cover the whole range. As a consequence, each of the different values in the dataset falls in exactly one of the classes. The term **classification** will be used to denote that the data are classified in a categorical system. A frequency distribution that gives an overview of the chosen classification and the respective frequencies is called a **classified frequency distribution**.

A classified frequency distribution can be pictured by way of a **histogram**: a bar chart with **connected** bars. The classes of the chosen classification are located on the horizontal axis, while the heights of vertical bars are such that their **areas** reflect the (relative) frequencies of the classes. In a histogram the bars are **not** separated. The bars cover the classes completely, to indicate that the data come from a continuous variable. The area of a bar has to reflect the importance of the class within the classified frequency distribution. If all classes in the classification have the same width, this indeed is accomplished by giving the bars the heights of the corresponding frequencies or relative frequencies. But if the classes do not all have the same width, the heights of the bars must be taken as the (relative) frequencies **divided by the corresponding width of the classes**.

Example 2.6

The Final Exam of the course Statistics 2 was taken by 258 students in International Business and International Economics and Finance. The variable Y measures their grades on the scale 0–10, to one decimal place; see the file Xmp02-06.xls. To get a good overview of the data, the above classification I was chosen and the accompanying frequencies (absolute, relative and cumulative) were determined: see Table 2.9 – note especially that all classes have the same width. The histogram in Figure 2.9 gives a graphical presentation of the distribution of the data.

Class	Freq.	Rel. freq .	Cum. rel. freq.
[0, 1]	21	0.081	0.081
(1, 2]	9	0.035	0.116
(2, 3]	21	0.081	0.197
(3, 4]	24	0.093	0.291
(4, 5]	37	0.143	0.434
(5, 6]	30	0.116	0.550
(6, 7]	42	0.163	0.713
(7, 8]	37	0.143	0.856
(8, 9]	21	0.081	0.937
(9, 10]	16	0.062	1.000
Total	258	1	–

TABLE 2.9 Frequency distribution of Y with classification I
Source: Private archives (2005)

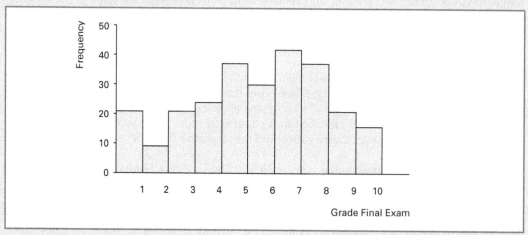

FIGURE 2.9 Histogram for classification I

Notice that 55% of the participants scored at most 6, as can be read immediately from the cumulative relative frequency distribution. This corresponds to $0.55 \times 258 = 142$ students.

But what about the percentage of the participants with 'insufficient' scores below 5.5? From the **original observations** it easily follows that 117 students (45.35%) had scores in the interval [0, 5.5). However, from the above frequency distribution this percentage cannot be calculated exactly, since this overview 'forgets' the positions of the observations within the classes. Since 5.5 lies precisely in the middle of the class (5, 6], and since 43.4% scored 5 or less and 55.0% scored 6 or less, we guess that about $(43.4 + 55.0)/2 = 49.2\%$ of the participants scored less than 5.5. Hence, by using the (cumulative relative) frequency distribution we cannot calculate exactly the percentage of the participants with scores below 5.5, but we can approximate it. For this approximation, we essentially assumed that the 'forgotten' grades in the class (5, 6] are homogeneously (that is, evenly) spread out over the interval.

It is important to **interpret** the histogram carefully, that is, to study its form and the consequences for the population of the participants. The top of the histogram belongs to the class (6, 7], which indicates that a relatively large number of students scored just above the level 6. However, the classes (4, 5] and (7, 8] also have large frequencies. Quite a lot of the students (56.6%) scored in the central part (4, 8] of the distribution. However, the histogram has another local top at the interval [0, 1]: it seems that a relatively large number of students (8.1%) hardly prepared themselves.

The classified frequency distribution above is not unique for the dataset but depends on the classification that was taken. The classes of the following classification are also equally wide; they correspond to the qualifications very poor, poor, median, sufficient, very good:

[0, 2],   (2, 4],   (4, 6],   (6, 8],   (8, 10]   **(Classification II)**

The accompanying frequency distribution (Table 2.10) and histogram (Figure 2.10) can be constructed by combining pairs of two classes from Table 2.9.

Class	Freq.	Rel. freq.	Cum. rel. freq.
[0, 2]	30	0.116	0.116
(2, 4]	45	0.174	0.290
(4, 6]	67	0.260	0.550
(6, 8]	79	0.306	0.856
(8, 10]	37	0.143	1.000
Total	**258**	1	–

TABLE 2.10 Frequency distribution of Y with classification II

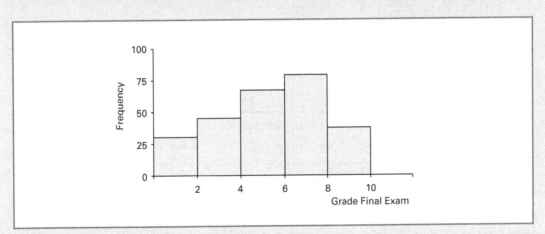

FIGURE 2.10 Histogram for classification II

With the frequency distribution of classification II even more information is lost than with the frequency distribution of classification I. For instance, the percentage of participants with scores at most 5 cannot be calculated exactly any more; we have to approximate it by the average 42.0 of the numbers 29.0 and 55.0 that correspond to the percentages with scores at most 4 and at most 6, respectively.

The second classified frequency distribution in essence summarizes the first. The top of the histogram falls at the interval (6, 8]. Furthermore, it again follows that the interval (4, 8] is an important central part of the frequency distribution.

Which classification should be chosen? There is no unique answer to that question. If possible, one chooses classes that are in some sense natural and that all have the same width. The bounds of the classes are usually round numbers, while, as a rule of thumb, the number of classes is often close to the square root of the number of data points. In the previous example, the number of data points is 258 and its square root is about 16. The choice of 10 classes in the first frequency distribution is the most in accordance with this rule of thumb. The reader is invited to construct for Example 2.6 the frequency distribution with accompanying histogram for a classification of 20 classes. Comparison with the above frequency distribution with ten classes is then interesting; see also Exercise 2.20 at the end of the chapter.

Sometimes there is a good reason to take a classification with classes that do not all have the same width. For instance, many national statistical agencies use classes with unequal class widths to present national income distributions.

Statistics Sweden uses the classification [0, 100), [100, 200), [200, 300), [300, 1000) to present statistics about the frequency distribution of the variable 'annual income of an adult (in 1000 kronor)' on the population of all citizens with annual income below 1 million kronor. The misleading histogram in Figure 2.11 is directly based on the classified relative frequency distribution of the variable.

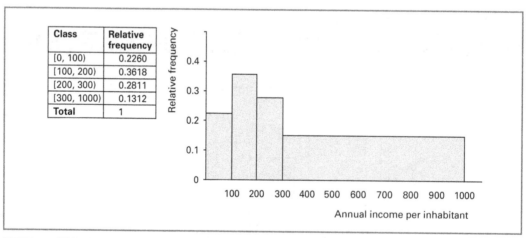

Class	Relative frequency
[0, 100)	0.2260
[100, 200)	0.3618
[200, 300)	0.2811
[300, 1000)	0.1312
Total	1

FIGURE 2.11 Misleading histogram of Swedish annual incomes
Source: Statistics Sweden (2006)

Notice that the class [300, 1000) has the smallest relative frequency but the area of the bar suggests otherwise. It is not the relative frequencies that should be presented but the frequency density, the overview of the relative frequencies divided by the respective widths of the classes.

Frequency density

The *frequency density* **of a classification** of data of a continuous variable is the overview that combines each class of the classification with the accompanying ratio of the relative frequency of that class and the width of it.

Example 2.7

The exam regulations for a certain course in statistics prescribe that only students who score more than 5.5 for the Midterm exam halfway through the course, are directly qualified to sit the Final Exam when the course is finished. Students with Midterm scores in the interval (4.5, 5.5] have to redo the Midterm within one week, and are then qualified to sit the Final Exam if the score for this retake is more than 5.5. Students with Midterm scores in the interval [0, 4.5] are not allowed to take the Final Exam.

Let X be the variable that measures the grade for the (regular) Midterm, to one decimal place, on the scale 0–10. See the file Xmp02-07.xls for the scores of the 227 participants last year. To present a summary of the data of X that reflects only the consequences of the above exam rules, it seems natural to base a frequency distribution on the following classification with classes of unequal width:

[0, 4.5],   (4.5, 5.5],   (5.5, 10]   **(Classification III)**

Table 2.11 summarizes the dataset.

Class	Freq.	Rel. freq.	Cum. rel. freq.	Freq. density
[0, 4.5]	51	0.225	0.225	0.050
(4.5, 5.5]	17	0.075	0.300	0.075
(5.5, 10]	159	0.700	1.000	0.156
Total	227	1	–	–

TABLE 2.11 Frequency distribution of *X* with classification III
Source: Private archives (2006)

The frequency density arises by dividing the relative frequencies by the corresponding class width. For instance, the relative frequency 0.700 of the class (5.5, 10] has to be divided by the width 4.5; the resulting density is 0.156. The histogram in Figure 2.12 gives a correct picture of the frequency distribution of the variable X.

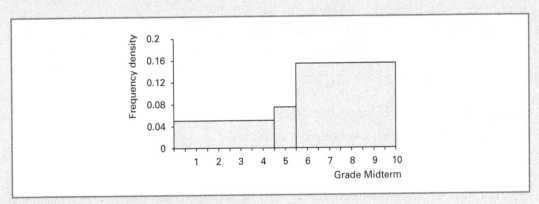

FIGURE 2.12 Histogram of the frequency density for classification III

Since area is obtained as density times width, the areas of the three vertical bars are indeed equal to the corresponding relative frequencies.

Cumulative distribution function when the variable is continuous

The concept 'distribution function' can also be defined for the relative frequency distribution of a classification of data from a **continuous** variable. In this case, $F(b)$ is the proportion of the observations that **according to the classified frequency distribution** is smaller than or equal to b. This is explained further below.

 If b is the upper bound of a class in the classification, then $F(b)$ is just the cumulative relative frequency up to and including that class. If b is smaller than the lower bound of the first class in the classification or larger than the upper bound of the last class, then it is clear that $F(b)$ respectively equals 0 or 1. But how is $F(b)$ defined for b in a class $(l, u]$ of the classification with lower bound l and upper bound u? Note that $F(u)$ is just the cumulative relative frequency up to and including that class while $F(l)$ is the cumulative relative frequency up to and including the preceding class in the classification. Hence, $F(u) - F(l)$ is the relative frequency for the class $(l, u]$. For b in a class $(l, u]$, the value $F(b)$ of F at b follows by putting the pairs $(l, F(l))$ and $(u, F(u))$ as dots into a two-dimensional system of axes, connecting them with a straight line and using this line to define $F(b)$. This method is called **_linear interpolation_** and is illustrated in Figure 2.13.

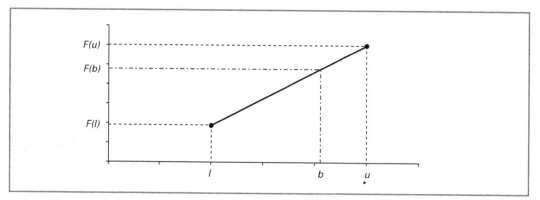

FIGURE 2.13 Linear interpolation

By doing this for all classes $(l, u]$ of the classification, the distribution function F arises.

(Cumulative) distribution function of a classified frequency distribution

The **(_cumulative_) _distribution function_** F of a classified frequency distribution is the function that arises from the cumulative relative frequencies of the lower and upper bounds of the classes in the classification by using the method of linear interpolation.

Property

F is not a step function but a **continuous** and non-decreasing function that on each class $(l, u]$ of the classification goes from $F(l)$ to $F(u)$ by way of a straight line.

Recall that the classified frequency distribution 'forgets' the original observations within the classes. The straight-line behaviour of the cdf in each of the classes reflects the assumption that **within** all classes the observations are evenly distributed. $F(b)$ is derived from the frequency distribution of the **classification** and hence is not necessarily precisely equal to the proportion of the original observations smaller than or equal to b. If b is an upper or a lower bound of a class, then $F(b)$ **is** precisely equal to the cumulative relative frequency up to and including b; in other cases $F(b)$ is only an approximation.

Example 2.8

Recall the classifications I and II of the Final Exam data of 258 students of the course Statistics 2. (See also Example 2.6 and the original data in the file Xmp02-06.xls.) Their relative and cumulative relative frequency distributions are repeated in Table 2.12:

Classification I	Rel. freq.	Cum. rel. freq.	Classification II	Rel. freq.	Cum. rel. freq.
[0, 1]	0.081	0.081	[0, 2]	0.116	0.116
(1, 2]	0.035	0.116	(2, 4]	0.174	0.290
(2, 3]	0.081	0.197	(4, 6]	0.260	0.550
(3, 4]	0.093	0.290	(6, 8]	0.306	0.856
(4, 5]	0.143	0.434	(8, 10]	0.143	1.000
(5, 6]	0.116	0.550	Total	1	–
(6, 7]	0.163	0.713			
(7, 8]	0.143	0.856			
(8, 9]	0.081	0.937			
(9, 10]	0.062	1.000			
Total	1	–			

TABLE 2.12 Two classified frequency distributions for Y = 'grade for the Final Exam'

Denote the cumulative distribution functions of the classifications I and II respectively by F and G. The values of these functions at the bounds of the classes follow immediately from Table 2.12; see Table 2.13.

Classification I	$F(b)$	Classification II	$G(b)$
$b = 0$	0	$b = 0$	0
1	0.081	2	0.116
2	0.116	4	0.290
3	0.197	6	0.550
4	0.290	8	0.856
5	0.434	10	1.000
6	0.550		
7	0.713		
8	0.856		
9	0.937		
10	1.000		

TABLE 2.13 Values of the cumulative distribution functions F and G at the bounds of the classes

For each class $(l, u]$ of a classification, the value of the cdf at u is the cumulative relative frequency up to and including the upper bound u. Below, it is demonstrated how general values of a cdf can be calculated from the linear interpolation technique. The cdf G is used to illustrate this.

Consider classification II. For the lower bound 2 and the upper bound 4 of the class $(2, 4]$, it holds that $G(2) = 0.116$ and $G(4) = 0.290$; see the table. The aim is to determine $G(2.8)$ with the linear interpolation technique; see Figure 2.14.

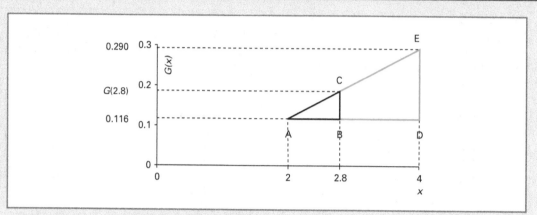

FIGURE 2.14 Illustration of the linear interpolation technique

Notice that the triangle ABC fits in angle A of triangle ADE, and that BC and DE are parallel. By a well-known mathematical rule, the ratios AB/AD and BC/DE are equal. Hence:

$$\frac{2.8 - 2}{4 - 2} = \frac{AB}{AD} = \frac{BC}{DE} = \frac{G(2.8) - 0.116}{0.290 - 0.116}$$

It follows that $G(2.8) = 0.1856$.

The graph of F can be created by connecting, for each class $(l, u]$, the pairs $(l, F(l))$ and $(u, F(u))$ by straight lines. Construction of the graph of G goes similarly (Figure 2.15).

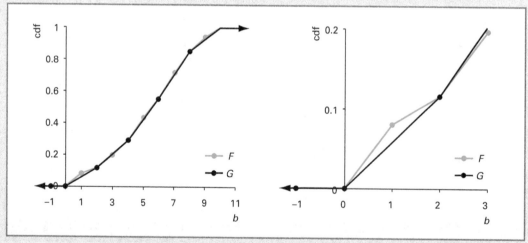

FIGURE 2.15 Graphs of the cdfs F and G

In the left-hand graph it is hard to distinguish F and G. By zooming in to the interval $[-1, 3]$, the right-hand graph arises. This graph shows that F is indeed a refinement of G.

CASE 2.1 COMMITMENT TO DEVELOPMENT INDEX 2006 – SOLUTION

The data can be presented graphically in many ways. However, the challenge is to find a presentation that gives a clear and stimulating image of the data.

Of course, the CGD wants to put emphasis on the ranking within the group of 21 rich countries, to promote the competition between the countries. That is why the overall ranking of the countries in column 1 of the dataset can be used to put the countries and the dataset in order. In the figure below, the seven policy areas are all included in one chart. The choice is made to place the countries vertically. By doing so, it is easier to compare the countries.

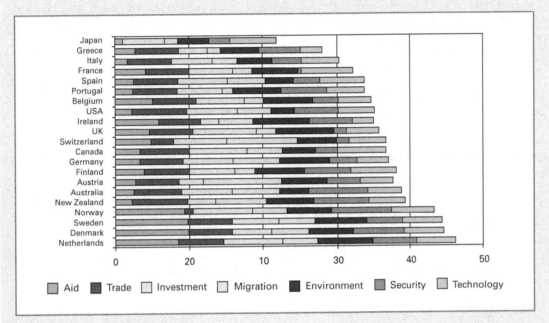

Stacked bar chart of the seven policy areas

Although, strictly speaking, the column with the overall averages is not included in this bar chart, the stacking of the bars has the same effect. The 'heights' of the (horizontal) bars are just the totals of the seven scores for the countries; the comparison of these heights is equivalent to the comparison of the overall averages. Thus, the figure clearly shows the high overall scores of Norway, Sweden, Denmark and the Netherlands. By picking out one colour, or pattern of shading, the contributions of the countries to an individual area can be compared. For instance, for the area Aid it turns out that the above four countries contribute much more than other countries.

2.4 Time series data

Most datasets considered so far have been measurements made at (more or less) one moment in time. Such data are called ***cross-sectional data***.

Grades for a final exam of a class of students are scores that are determined within a short period of time, say one week. The GNIs per capita of the 25 EU member states for 2002 are figures dealing with the same year. Obviously, the respective datasets contain cross-sectional data.

Time series data are measurements of a single variable at successive periods or moments in time. The sequence of successive data is called a ***time series***. Studying such sequences offers possibilities to observe developments over time. Examples are:

- The yearly GDPs of a country for the period 1960–2007.
- The weekly sales of a supermarket during the most recent year.
- The daily sales of Inspiration 5100 notebooks by a computer shop during the previous year.
- The daily closing prices of a stock at a stock exchange during the previous month.
- The yearly numbers of employees of a firm on 1 January during the period 1990–2007.
- The yearly winners of the Champions League cup for the period 1956–2007.

Notice that the population of a time series consists of time epochs. Measurements of a variable at such time epochs constitute the time series. For instance, in the first example above, the population is a set of years; the variable 'yearly GDP' is measured for each of the years 1960 to 2007. In the second-last and last examples, the respective variables 'yearly number of employees on 1 January' and 'yearly winner of the cup' are also measured for each year. (Notice that the last variable is nominal.) In the second example, the variable 'weekly sales of a supermarket' is measured for the sample of the most recent 52 weeks from the population of all past weeks. The population elements of the third and fourth examples are working days.

 The challenge of this section is to summarize time series by way of graphs that nicely present the developments in time. One type of graph that is often used is the *time plot*. This has time on the horizontal axis, while the corresponding observations in the time series are placed on the vertical.

Example 2.9

Quarterly retail sales data for the food stores in the UK are given in the file Xmp02-09.xls, as percentage changes with respect to one year before. The data cover the period 1987Q1 – 2006Q2. The time plot has 'quarter' (values 1–78) along the horizontal axis and 'sales change (%)' vertically (Figure 2.16).

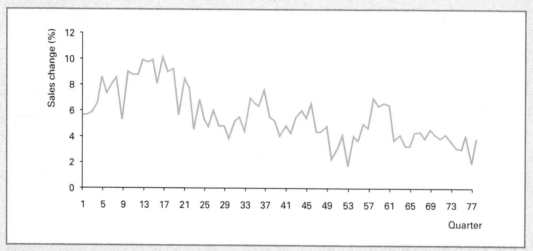

FIGURE 2.16 Percentage change per quarter for food sales in the UK
Source: Office for National Statistics, UK (2006)

The graph in Figure 2.16 shows a downward **trend** – the time series decreases gradually in time – which might indicate that nowadays the yearly increase in food consumption is less sizeable than it was 15 years ago. But within this trend there is evidence of irregular behaviour that, at least partially, may be caused by a **seasonal component**. See Chapter 23 for the definitions of trend and seasonal component, and for more about time series.

Summary

In this chapter we began our study of descriptive statistics. Large datasets can be summarized neatly by tables and graphs, often based on frequency distributions (for discrete variables) or classified frequency distributions (for continuous variables). In the case of continuous variables, information is lost during this process since the precise positions of the data points within the classes are lost.

The distribution function is a key concept that is available for frequency distributions of discrete **and** continuous variables. In the first case they are step functions, in the second case they are continuous. A similar concept will be considered in probability theory, see Chapter 8.

🔑 Key terms

absolute frequency **20**
bar chart **21**
categorical system **31**
cdf **30**
classification **31**
classified frequency
  distribution **31**
cross-sectional data **39**
cumulative distribution
  function **29**
cumulative frequency **22**
cumulative frequency
  distribution **36**

cumulative relative
  frequency **22**
cumulative relative
  frequency
  distribution **22**
distribution function **29**
dot plot **25**
frequency **20**
frequency density **34**
frequency distribution **20**
heading **20**
histogram **31**
interpret **33**

legend **21**
Likert scale **24**
linear interpolation **36**
ogive **30**
pie chart **21**
relative frequency **20**
relative frequency
  distribution **20**
source **20**
time plot **40**
time series **39**
time series data **39**

❓ Exercises

For (parts of) some of the exercises below you will need a computer. If necessary, you can use the guidelines in Appendix A1.

Exercise 2.1

Bar charts and histograms are both used as graphical presentations of (relative) frequency distributions.

a When do we use a bar chart and when do we use a histogram?
b What are the differences between a bar chart and a histogram?

Exercise 2.2

In the following questions, F is a distribution function.

a Suppose that F is the distribution function of a dataset of measurements of a discrete variable. Formulate important properties of F.
b Suppose that F is the distribution function of a classified frequency distribution of a dataset of measurements of a continuous variable. Formulate important properties of F.

Exercise 2.3

The table below gives a classified relative frequency distribution of a continuous variable X:

Class	Rel. frequency
[0, 10)	0.10
[10, 50)	0.20
[50, 100)	0.70
Total	1

a Explain why the histogram of this distribution must not be based directly on this relative frequency distribution.

b Determine the accompanying frequency density and explain why the accompanying histogram does describe the distribution correctly.

Exercise 2.4

Consider the frequency distribution below:

Value	Frequency
2	30
4	50
6	100
8	20
Total	200

a Determine the accompanying relative frequency distribution and cumulative relative frequency distribution.

b Denote the distribution function by F. Calculate $F(1)$, $F(4)$, $F(5)$, $F(6.5)$ and $F(9)$. Create the graph of F.

Exercise 2.5

Consider the classified frequency distribution below:

Class	Frequency
[0, 1)	30
[1, 3)	50
[3, 6)	100
[6, 10)	20
Total	200

a Determine the accompanying relative frequency distribution, the frequency density and the cumulative relative frequency distribution.

b Denote the distribution function by F. Calculate $F(1)$, $F(4)$, $F(5)$, $F(6.5)$ and $F(9)$. Create the graph of F.

Exercise 2.6

A dataset of 540 observations of a certain variable has the following distribution function F:

$$
\begin{aligned}
F(b) &= 0 && \text{for all } b < 2 \\
&= 1/12 && \text{for all } b \text{ in } [2, 5) \\
&= 3/5 && \text{for all } b \text{ in } [5, 8.3) \\
&= 7/10 && \text{for all } b \text{ in } [8.3, 12) \\
&= 8/9 && \text{for all } b \text{ in } [12, 13) \\
&= 1 && \text{for all } b \geq 13
\end{aligned}
$$

The aim is to calculate the (absolute) frequency distribution of this dataset.

a Create the graph of F.

b Determine the different values of the dataset.

c Determine the jump sizes and express them as relative frequencies.

d Determine the (absolute) frequency distribution of the dataset.

e Calculate the average value of the dataset.

f Determine the percentage of the data points that are smaller than or equal to that average value.

Exercise 2.7

Market researchers often try to measure consumer opinion by asking consumers to complete a questionnaire which tests several aspects of a product or service.

Travel agency XA Travels organizes holidays to sunny places in Mediterranean countries. To check the experienced quality, all customers are asked to fill in a questionnaire as soon as they finish their vacations. This questionnaire contains several statements that have to be qualified with a Likert-type evaluation (see Example 2.2 for Likert scaling). Among others, it contains the following questions, jointly with the rating scheme:

1 The flight was very pleasant   1 2 3 4 5

2 The hotel was great   1 2 3 4 5

3 The services offered by XA Travels were very good   1 2 3 4 5

Here are the summarized answers of some clients.

	Strongly disagree	Disagree	Neutral	Agree	Strongly agree	Total
Rel. freq. for 1	9	21	45	18	7	100
Rel. freq. for 2	3	7	20	53	17	100
Rel. freq. for 3	5	15	20	45	15	100

Relative frequency distributions (%)
Source: Archives of XA Travels (2005)

a Define the population and the three variables of interest. Are the variables nominal, ordinal or quantitative?

b Interpret the three relative frequency distributions and compare them.

c Do cumulative relative frequency distributions make sense? If so, determine them.

d Regarding question 1 on the questionnaire, determine the percentage of the clients who are neutral, disagree or strongly disagree.

e Suppose that the size of sample dataset 3 is 500. Determine the (absolute) frequency distribution of dataset 3.

Exercise 2.8

Frequency distributions of the variable 'age at death' are often used in researches that study developments over time in a country or to compare developing countries with developed countries.

In recent research in the UK, the ages at death were considered for all people who died in the periods 1900–02, 1950–52 and 1999. They are summarized in the table below. The numbers in columns 2 and 3 are averaged frequencies (per year) for the respective periods of three years.

Age	1900–02	1950–52	1999
Under 1	156 012	24 398	4 045
1–4	73 998	4 683	794
5–19	40 781	6 354	2 569
20–44	101 512	34 949	19 864
45–74	210 519	298 933	206 638
75 and over	79 889	229 591	398 152
Total	662 711	598 908	632 062

Classified frequency distributions of X = 'age at death'
Source: Carel Press (2002)

It is good exercise to answer the following questions **without** the help of a computer.

a Describe the three populations that underlie these distributions.

b Create the frequency densities of the three classified frequency distributions and present them in a table as above; take 95 as upper bound for the last class. *Warning*: describe the classes and their widths; for instance, note that the class '1–4' corresponds to the interval [1, 5).

c Create accompanying histograms; keep in mind that they have the same scales along the axes. Compare them and give your comments.

Exercise 2.9

The following results come from research by the UK Department of Health. An impression was wanted of the time spent (in hours) participating in sports, by young men and women. Several people were interviewed and asked about their gender, their age and the number of hours per week they spend playing sport. The results are summarized in the table below. For instance, 38% of the interviewees in the category 'males aged 16–24' spend less than 1 hour on sports.

Time spent	Males aged 16–24 %	Females aged 16–24 %	Males aged 25–34 %	Females aged 25–34 %
Less than 1 hour	38	59	58	69
Between 1 hour and less than 3	22	22	20	19
Between 3 hours and less than 5	13	10	10	8
Between 5 hours and less than 7	9	4	5	2
7 or more hours	18	5	7	2
Total	100	100	100	100

Time spent on sports
Source: Department of Health, *Health Survey for England* (2001)

a Define the population and the variables of interest. Are the variables qualitative / quantitative?

b What are the overviews in columns 2, 3, 4, 5 called?

c Is the underlying dataset (which is not given here) a sample dataset or a population dataset?

d Suppose that the interviewed males were equally distributed over the two categories 16–24 and 25–34. However, for the interviewed females, 60% belonged to the category 16–24 and 40% to 25–34. Determine the classified relative frequency distribution for 'time spent on sports' for young males aged 16–34. Do the same for young females aged 16–34.

e Determine frequency densities for the two distributions in part (d); take 12 as upper bound for the last – open – class. Present the two distributions in histograms; create them by hand. Compare the histograms.

Exercise 2.10

You want to compare the sizes of the 717 135 companies that on 1 January 2005 were registered at the Chamber of Commerce. Although there are several ways to measure the size of a company, you decide to use the integer-valued variable 'number of employees on 1 January 2005'. In *Statistical Yearbook 2005* of Statistics Netherlands you find the figures shown in the table below.

Number of employees	Frequency
1	365 435
2	115 020
3, 4	111 000
5–9	59 225
10–19	33 365
20–49	19 710
50–99	6 685
100–149	2 150
150–199	1 110
200–249	675
250–499	1 360
500–999	695
1000–1999	375
≥ 2000	330
Total	717 135

Number of companies per class
Source: Statistics Netherlands (2005)

a Describe the population you are interested in. Which characteristic is of interest? Which variable do you use to measure it?

b In the course of your research, will you consider the variable to be discrete or continuous? Why?

c Determine the accompanying classified relative frequency distribution; take (0.5, 1.5], (1.5, 2.5], (2.5, 4.5], etc as classes. Take 5000.5 as upper bound for the last class. Also determine the accompanying frequency density.

d Denote the distribution function of this classified frequency distribution by F. Calculate $F(8)$ and $F(2200)$. Interpret these numbers carefully.

Exercise 2.11

Although poverty is less visible in west European countries when compared with, for instance, African countries, it does exist in Europe too. Each European country has its own definition of the poverty line, an abstract line that separates really poor people from others. This poverty line is hard to define and is adjusted every year. It usually depends on the disposable income level of a low-income group.

In France, the poverty line of the year 2002 for (for instance) 'ordinary' households was set at €625. The table below is based on a survey of the individuals in 16 000 households, where (among others) the variables 'disposable household income per month (in euro)', 'gender' and 'age at 31 December of income year' were measured. A person is qualified as poor if the corresponding monthly disposable household income lies below the poverty line. Incidentally, only households whose reference person (say, head of household) is **not** a student are considered in the research.

	Women		Men	
	Number of poor (× 1000)	Rate (%)	Number of poor (× 1000)	Rate (%)
Under 18 years	556	8.6	553	8.1
18–29 years	340	8.1	310	7.3
30–49 years	562	6.5	488	5.9
50–59 years	250	6.4	259	6.9
60–74 years	119	2.9	86	2.4
75 years and older	137	5.1	34	2.1
Total	1964	6.6	1730	6.1

Number and percentage of people living below poverty line, by age
Source: National Institute of Statistics and Economic Studies, (INSEE), France (2003)

a This table is about individuals. Describe this population of individuals more precisely. Is the underlying dataset a sample dataset or a population dataset?

Notice that the results in the table are statements about **all** population elements.

b Which age category suffers most under poverty? Which suffers least? Answer these questions for both the females and the males.

c Column 2 gives a classified frequency distribution of the variable 'age' for a subpopulation of individuals. Which sub-population?

Exercise 2.12

Reconsider Exercise 2.11.

a According to column 3 of the table, 8.6% of females under 18 are poor. But how many millions (poor or not poor) of people does this age category contain? Give your answer to three decimal places.

b Use column 3 to determine the classified frequency distribution of the variable 'age' for all women in the population.

c Same question as in part (b) but now for column 5 and for the men.

d Suppose that the population consists of 50% men and 50% women. Determine the relative frequency distribution of the qualitative variable X = 'the individual is poor (yes or no)' for the whole population of individuals.

Exercise 2.13 (computer)

The file Xrc02-13.xls gives information about the number of visitors (in millions) to the most visited cultural and recreational sites in France in 2003. Present the results in an attractive bar chart with horizontal bars.

Exercise 2.14

Two identical American football balls, one air-filled and one helium-filled, were used outdoors on a windless day at the Ohio State University's athletic complex. Each football was kicked 39 times and the two footballs were alternated with each kick. The experimenter recorded the distance (in yards) travelled by each ball; the table below contains the original observations; see the data in the columns 2 and 3 of the file Xrc02-14.xls. Our goal is to obtain good overviews for each of the balls separately and to compare the results.

Trial	Air	Helium	Trial	Air	Helium	Trial	Air	Helium	Trial	Air	Helium
1	25	25	11	25	12	21	31	31	31	27	26
2	23	16	12	19	28	22	27	34	32	26	32
3	18	25	13	27	28	23	22	39	33	28	30
4	16	14	14	25	31	24	29	32	34	32	29
5	35	23	15	34	22	25	28	14	35	28	30
6	15	29	16	26	29	26	29	28	36	25	29
7	26	25	17	20	23	27	22	30	37	31	29
8	24	26	18	22	26	28	31	27	38	28	30
9	24	22	19	33	35	29	25	33	39	28	26
10	28	26	20	29	24	30	20	11			

Distances (in yards) travelled by two types of footballs
Source: M.B. Lafferty, 'OSU scientists get a kick out of sports controversy', *The Columbus Dispatch*, 21 November 1993, p. B7

a Give a complete definition of the two variables of interest (called X and Y). Are they continuous or discrete?

b The dataset contains – apart from the first column – two other columns with data, one with observations of X and one with observations of Y. Are they sample data or population data? Why?

c Create (**by hand**, on two horizontal axes with the same scale 0, 1, …, 40) dot plots of the two sets of observations. In the case of multiple occurrences, put the dots above each other. Compare and interpret the two dot plots, for instance by considering their central position and their variation.

Exercise 2.15 (computer; can be done without)

Reconsider Exercise 2.14.

a Consider the classification (10, 15], (15, 20], …, (35, 40]. Notice that it is indeed a categorical system and that the classes have the common width 5. Use it to determine classified frequency distributions for each of the two variables. Put your results in one table that also contains – for both balls – the overviews of the relative frequency distributions, the cumulative relative frequency distributions, and the frequency densities.

b Present the two classified frequency distributions in attractive histograms. Compare and interpret them.

c Denote the distribution functions of the two classified frequency distributions by F and G, respectively. Create attractive graphs in one figure. Give your comments.

d Calculate $F(26)$ and $G(26)$. Interpret your answers and compare them.

Exercise 2.16 (computer)

The exam of the course Statistics 2 for International Business students consists of three parts: Midterm, Team Assignment (TA) and Final Exam. For the TA, the students have to write, in teams of five persons, a statistical report on a topic from the field of business. The grades that the students get for the TA are integer-valued scores within the range 0, 1, ..., 10. Last year's grades are recorded in the file Xrc02-16.xls.

Let X denote the grade for the TA measured as an integer-valued evaluation over the population of the 180 participants. Obviously, the dataset is a population dataset and the different values are the integers 0, 1, 2, ..., 10 (although they may not all occur).

a Determine the frequency distribution, the relative frequency distribution and the cumulative relative frequency distribution of X.

b Create the bar chart of the relative frequency distribution of X.

c What are the relationships between the relative frequency distribution and the distribution function F of X?

d Determine $F(5.5)$.

Exercise 2.17 (computer)

The file Xrc02-17.xls contains the numbers of children (aged <18) in two samples of 300 arbitrarily chosen households. Both samples are drawn from a common large population of households. To compare the two sample datasets, they will be summarized in terms of frequency distributions; their distribution functions will also be compared.

a The file contains the original observations of 600 arbitrarily chosen households: two sample datasets, each of size 300. Describe the variable where the data come from. Is the variable discrete or continuous?

b Determine the frequency distributions of each of the two datasets. Create one table for both frequency distributions, the accompanying relative frequency distributions and the accompanying cumulative relative frequency distributions.

c Create a clustered bar chart that pictures both relative frequency distributions.

d Let F and G be the cumulative distribution functions (cdfs) of the datasets in columns 1 and 2, respectively. It can easily be calculated that the average numbers of children for the two datasets are respectively 0.8733 and 0.8667. Determine $F(0.8733)$ and $G(0.8667)$ and interpret the answers.

e Determine b such that $F(b)$ is at least 0.90 while $F(c)$ is smaller than 0.90 for all c smaller than b.

Exercise 2.18 (computer)

Reconsider the file Xrc01-10.xls of Exercise 1.10.

a Determine the frequency distribution of the variable X = 'country of the winner in that year'. (Place the countries in alphabetic order.)

b Determine the relative frequency distribution of X. Also determine the percentage of the years in the period 1956–2007 that the cup was won by a team from Spain.

c Determine the cumulative relative frequency distribution.

d Present the frequency distribution in an attractive bar chart.

e Also present it in a pie chart.

Exercise 2.19 (computer)

a How much of the current rise in global temperatures is due to man and how much is part of normal variation in global temperatures, is very much in dispute. The dataset Xrc02-19a.xls (time series) contains the global mean temperatures from 1866 to 1996.[†] There are intriguing non-random patterns in the data. The variable names are:

[†] *Source*: Compiled by Worldwatch Institute from James Hansen and Reto Ruedy, Goddard Institute for Space Studies, 14 January 1997.

Year

Global Mean Temperature (degrees Celsius)

Create a suitable graph to present the developments of the variable 'global mean temperature (degrees Celsius)' over time. Are there any intriguing non-random patterns in the data?

b The dataset <u>Xrc02-19b.xls</u> (time series) contains data on global temperature and atmospheric carbon dioxide (CO_2) concentrations for the last 159 000 years.[‡] Due to there being no year 0, the number of years prior to the present was used instead of an ordinary variable 'year'. The temperature given is relative to the present-day temperatures. The variable names are:

Year = 'thousands of years before present'

Temp = 'temperature change (degrees Celsius)'

CO2 = 'carbon dioxide concentration (parts per million)'

Use suitable graphs to present the developments of the variables 'global mean temperature (degrees Celsius)' and 'carbon dioxide concentration' over time. Are there any intriguing patterns in the data?

c Give your comments on the findings in parts (a) and (b).

Exercise 2.20 (computer)

In Section 2.3.3, two classified frequency distributions were considered for the Final Exam grades of the course Statistics 2; see <u>Xmp02-06.xls</u> for the 258 original data points. Classification I contained the classes [0, 1], (1, 2], ..., (9, 10], a refinement of classification II with the five classes [0, 2], (2, 4], ..., (8, 10]. In the present exercise, a refinement of the frequency distribution of classification I will be considered.

a Determine a suitable classification with 20 classes that all have width 0.5.

b Determine the accompanying relative frequency distribution; put it in a table that also contains the relative frequency distribution of classification I.

c Explain why this new relative frequency distribution indeed refines the relative frequency distribution of classification I. Compare the two distributions.

d Create histograms of the two relative frequency distributions. Compare them.

e Also determine the distribution function H of the new frequency distribution.

f Calculate $H(7.3)$.

CASE 2.2 THE ECONOMY OF TOKELAU

Most countries have statistical agencies whose websites are often useful sources of information about the countries, their population and their economy.

Take Tokelau, a group of three atolls (Atafu, Fakaofo and Nukumonu) 500 kilometres north of Samoa in the central South Pacific. The Statistics Unit of Tokelau has a website: www.spc.int/prism/country/tk/stats/. We will use the website to learn more about the country.

Here are the questions you are required to answer for the most recent year for which information can be found.

---

‡ *Sources*: Compiled by Worldwatch Institute from J. M. Barnola et al. 'Historical CO_2 record from the Vostok ice core', in Thomas A. Boden et al., eds, *Trends '93: A Compendium of Data on Global Change* (Oak Ridge, TN.: Oak Ridge National Laboratory, 1994); J. Jouzel et al., 'Vostok isotopic temperature record', in Boden et al., op. cit.; Timothy Whorf, Scripps Institution of Oceanography, La Jolla, CA, private communication, 2 February 1995.

a About the population:

  i What is the size of the population? How many males and how many females?

  ii Find a classified frequency distribution of the variable 'age'.

  iii Find the frequency distribution of the variable 'highest qualification gained at school'. Compare the separate distributions for males and females.

b About the economy, trade:

  i Which currency is used?

  ii About the variable 'value of imports': report the time series of the most recent observations. How is the observation of the year 2002 distributed over main categories?

  iii What can be said about the variable 'value of exports'?

c About the economy, labour:

  i Find a frequency distribution for the variable 'kind of work'. Is it very informative? Do the distributions for males and females differ?

  ii Find a frequency distribution for the variable 'occupation'. Again, compare the distributions of males and females.

  iii Find a frequency distribution for the variable 'industry of work'. Compare males and females.

CASE 2.3 HUMAN DEVELOPMENT REPORT

The file Case02-03.xls originates from the Human Development Report on the website http://hdr.undp.org. It is about 177 countries, listed in the order of their Human Development Index (HDI). The GDP per capita (in US dollars, 2003) of 168 of these countries is recorded (there are some missing data for 2003). The objective of this case is to create a suitable graph for the GDPpcs that especially reflects that few countries have much and many countries have little.

a To get a first impression, create dot plots of the data for the ranges 0–60000, 0–10000, 0–2000 and 0–1000. Interpret them.

b On the basis of these dot plots, choose the following classification:

   (0, 500], (500, 1000], (1000, 2000], (2000, 3000], (3000, 4000], (4000, 5000], (5000, 10000], (10000, 15000], (15000, 20000], (20000, 25000], (25000, 30000], (30000, 60000]

Determine the accompanying frequency distribution, relative frequency distribution and frequency density. Interpret the results.

c Create a suitable plot to relate GDPpc (horizontally) and frequency density (vertically).

Measures of location

In the previous chapter, tables and graphical methods were used to summarize a set of measurements of a variable. By doing so, we tried to detect information that is present but hidden in the (often large) dataset.

There is another way to release hidden information from data. Recall that (population or sample) statistics describe special features of the dataset of a variable. The calculation of suitable statistics offers another way to extract information from a dataset. For instance, when studying overweight adult European men, a researcher may want to summarize a sample dataset of the variable $X =$ 'weight (kg)'. The statistics 'mean value' and 'percentage overweight' are of interest; they help summarize the dataset.

In the present chapter, we are interested in a group of statistics describing the ***location*** of the variable, of the data and of an accompanying frequency distribution. 'Location' refers to some central position of the dataset and its distribution, but is not yet defined precisely. Several statistics (such as mean, median, mode) are considered in order to measure central positions in a precise and well-defined way. They are called ***measures of location***.

The concept 'mean' presupposes that the data in the dataset can be summed (and divided by the size of the dataset), so it makes sense only for data that come from quantitative variables. As we shall see, the 'median' is (roughly speaking) the middlemost of the ordered observations, so it presupposes that the data can be put in order. Hence, the median exists only for datasets of ordinal or quantitative variables. That is why the present chapter will again (as Chapter 2) be divided into three parts: location measures for nominal variables, for ordinal variables and for quantitative variables.

It is important to notice that in this chapter we use the convention for notation that was described in Chapter 1. Population statistics are usually indicated by Greek letters, even when the population data are not all observed. If the dataset is a sample dataset, then Latin letters will be used to denote the statistics. We will talk about **population statistics** and **sample statistics**. However, they will often be called simply **statistics** if it is clear from the context which of the two types is meant.

To understand the theory in the sections below, you really have to know about the **summation operator** Σ (Greek capital sigma). Read Appendix A2 first if you need more practice with this operator. Appendix A2 also explains the **absolute value operator** $|\cdot|$.

CASE 3.1 THE GENDER GAP IN EMPLOYMENT RATES

One of the EU's Lisbon targets, set in 2000, is to achieve a reduction in the gender gap for employment rates by 2010. More specifically, the aim is to achieve an average 60% employment rate for women by that date. In 2002, the employment rate for women in the EU15 was 55.5%. See also www.SEO.nl, especially the 2007 publication *Mind the Gap*.

What progress has been made, now that we are more than halfway through to 2010? The situation of 2005 will be studied for a special group of countries. We will base our investigations on the data in the table below; see also Case03-01.xls.

Country	Women	Men	Country	Women	Men
Austria	62	75	Sweden	71	75
Belgium	54	68	UK	66	77
Denmark	71	80			
France	58	69	Canada	68	77
Germany	59	71	Czech Republic	56	73
Ireland	58	76	Hungary	51	63
Italy	45	70	Japan	58	80
Luxembourg	54	73	Poland	46	58
Netherlands	66	80	Switzerland	70	84
Portugal	62	73	Turkey	24	68
Spain	51	75	USA	66	78

Employment rates for women and men, 2005
Sources: Eurostat Labour Force Survey (lfsq_ergan, 2006); OECD Labour Force Survey (2006)

Emphasis will be on ways to measure the **mean** employment rates of subgroups of countries. Should we simply take normal averages of rates from the table or should the relative contributions of the individual countries to the overall labour force also be taken into account? See Section 3.3.1 for a solution.

3.1 Nominal variables

Data from a nominal variable cannot be ordered; differences between two of the data points make no sense. However, (relative) frequencies of the different values in the dataset **can** be considered. If we want to characterize a central, crucial position within the dataset and its frequency distribution, a natural thing to do is to take the value(s) with the highest frequency.

Consider some nominal variable (say X) and a dataset of observations of X. The value (or, possibly, values) within the dataset that has the highest frequency is called a **mode** or **modal value** of the dataset and the frequency distribution. If the dataset is a population dataset, the mode is denoted by μ_{mode}; if the dataset is a sample dataset of measurements of X, the mode is denoted by x_{mode}. The **mode of the variable** X is the mode of the accompanying population dataset and hence is denoted by μ_{mode} too. Often, the variable X is not observed over the whole population; μ_{mode} is then unknown. In that case, the mode χ_{mode} of an observed sample dataset of X can be used as an approximation (estimate) of μ_{mode}.

If a dataset has only one mode, it is called **_unimodal_**. If it has two modal values, it is called **_bimodal_**. If there are more than two modal values, we have a **_multimodal_** situation. In such a case the list of all modes will be of little use.

Figure 3.1 below shows that the value 'manufacturing' of the variable X = 'industry of the employed person in EU25' (refer to Example 2.1) has the largest frequency.

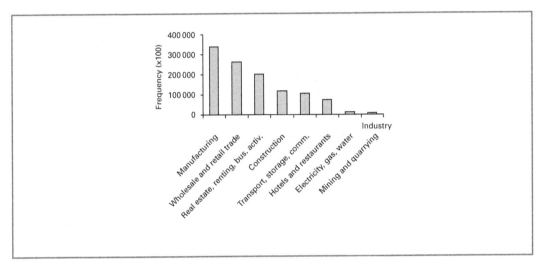

FIGURE 3.1 Bar chart of the frequency distribution of X

The frequency distribution graphed here is about the whole population of employed people in the EU25 (see Example 2.1 for the precise definition of the population). Hence, the modal value is the manufacturing industry and it has to be denoted by μ_{mode}.

Example 3.1

Figure 3.2 comes from research by Statistics Netherlands into labour force participation within husband–wife couples who have children aged under 13 years. The summarized results derive from large **samples** of such husband–wife couples.

Figure 3.2 not only contains three relative frequency distributions but also shows developments in time. It does not tell us about the sizes of the three underlying samples, but the samples are drawn from three populations: the 1990, 1995 and 2001 population of husband–wife couples with children under 13. The variable of interest (say X) is 'labour participation'; it is nominal with the following possible values: both working, only husband works, only wife works, neither works. Notice that similarly shaded bars reflect, from left to right, the relative frequency distribution of the sample data for the years 1990, 1995 and 2001.

Figure 3.3 reorganizes the data to illustrate these distributions more clearly. With this graph the three relative frequency distributions can be compared easily. For 2001, the mode is the value 'both working'; however, the modes of the 1995 and the 1990 samples are both 'only husband works'. In the course of time the mode has changed.

The graphs can be used to loosen more information from the sample(s) and the population(s) under study. For instance, 57% of the wives within the 2001 sample of husband–wife couples work outdoors. If we transform the variable into the variable Y that measures whether the wife (within

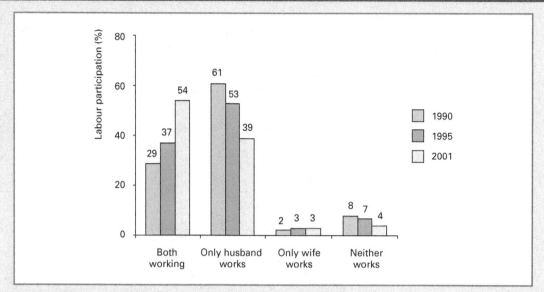

FIGURE 3.2 Labour participation (%) within couples with children aged under 13
Source: Statistics Netherlands (2006)

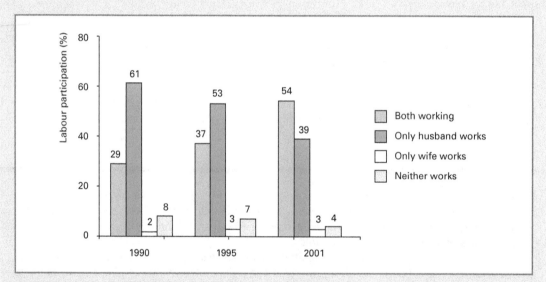

FIGURE 3.3 Labour participation (%) for the years 1990, 1995 and 2001

	1990 %	1995 %	2001 %
Wife does not work	69	60	43
Wife does work	31	40	57
Total	100	100	100

TABLE 3.1 Relative frequency distribution of 'wife does not / does work'

the couple) does not / does work, the three relative frequency distributions for *Y* are as shown in Table 3.1.

For 2001, the mode of this frequency distribution is 'wife does work' and the accompanying sample proportion is 0.57. We see that the mode of the sample data from the variable Y has changed in the course of time while the sample proportion of 'wife does work' increased considerably from 0.31 in 1990 to 0.57 in 2001.

3.2 Ordinal variables

For datasets of ordinal variables, the mode and median are measures of location; they are considered below.

First consider some ordinal or quantitative variable X and a dataset of observations of X. Place these data in their natural order. If the number of data points is odd, then the **median** of the dataset is defined as the value of the middlemost observation of the ordered data. If the number of data points is even, the middlemost **pair** of the ordered data can be determined. When these two observations turn out to be equal, the common value is defined to be the median. But when the two observations differ, there will be no median in the case of ordinal data. For data of a quantitative variable, we shall find a way out in Section 3.3.

In a sense, the median takes a strategic, central position within the range of the ordered data points. It splits the ordered dataset into two equal-sized groups of observations in such a way that all data in one group are not larger than the data in the other group.

If the dataset is a population dataset, its median (if it exists) is denoted by μ_{median}; if the dataset is a sample dataset of observations of the variable X, its median is denoted by x_{median}. The median of the variable X itself is defined as the median of the corresponding (possibly not observed) population dataset and hence is denoted as μ_{median} too.

From now on, assume that X is ordinal.

Example 3.2
Reconsider the Likert-type variable $X =$ 'instructor explains things clearly' of Example 2.2. This variable was observed on two classes of students, leading to 62 observations for class 1 (with instructor A) and 69 observations for class 2 (with instructor B). To calculate the medians of the two datasets, the datasets both have to be put in order.

For the dataset of class 2 with size 69, the middlemost observation is the 35th. According to the cumulative frequencies in Table 2.4, the first 19 observations were equal to the coded values 1 or 2 while the first 38 observations were 1, 2 or 3. Consequently, the 35th observation has to be 3 = neutral. Hence, the median of the second dataset is 'neutral'. From the table it follows immediately that the mode is 4 = agree.

The size of the dataset of class 1 is the even number 62. The middlemost pair of observations is formed by the 31st and the 32nd of the ordered data. From the cumulative frequencies in Table 2.3 it follows that the first 31 observations are equal to the coded values 1, 2, 3 or 4; the last 31 observations are all equal to 5. Hence, observation 31 is 'agree' and observation 32 is 'strongly agree'. The median falls between these two values and is **not** defined. Note, however, that the mode **does** exist: it is equal to 5 = strongly agree.

3.3 Quantitative variables

For quantitative variables, several measures of location are important: mode, median, arithmetic mean, geometric mean, weighted mean. They are considered below. The section is divided into three subsections that are based, respectively, on the dataset itself (the set of the original observations), the frequency distribution of (the dataset of) a discrete variable, and a classified frequency distribution of (the dataset of) a continuous variable.

3.3.1 Original observations

In this subsection, the original observations of a quantitative variable are considered, that is, the observations are not yet summarized in a frequency distribution.

Median

Recall that, in the case of an ordinal variable and an even number of observations, we **can** determine the middle pair of the ordered data, but there is not always a well-defined median. This may happen when these middlemost observations are different; see Section 3.2.

Now, however, the variable is quantitative. If the size of the dataset is odd, the median is well defined. If the size of the dataset is even, the middle **pair** of observations (x_{m1}, x_{m2}) splits the dataset as follows:

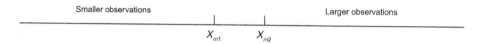

Recall that the intention behind the median is that:

50% of the data $\leq$ median   and   50% of the data $\geq$ median

Any value in the open interval (x_{m1}, x_{m2}) satisfies these criteria. Since we nevertheless want to identify a unique median, we interpolate between x_{m1} and x_{m2}. So,

$$\text{median} = \frac{x_{m1} + x_{m2}}{2} \quad \text{if the size of the dataset is even}$$

Two things have to be noted:

- this definition does not work for ordinal data since then $x_{m1} + x_{m2}$ is not well defined;
- for quantitative data it is possible that the median is not one of the different values of the variable.

Other important measures of location are described next.

(Arithmetic) Mean

Suppose that the dataset is the **population** dataset of all measurements of a quantitative variable X on the population of interest. Possibly, the dataset is not observed. The population elements can usually be numbered as 1, 2, ..., N while the observations of X in the dataset are formally denoted as $x_1, x_2, ..., x_N$. Here x_1 is the observation of X at population element 1, x_2 is the observation of X at population element 2, and so on. In short, x_i is the observation of X at population element i ($i = 1, 2, ..., N$).

To obtain the *(arithmetic) mean* μ of the population dataset, the observations $x_1, x_2, ..., x_N$ have to be summed and the result has to be divided by N:

Population mean

The **population mean** of a quantitative variable X and its population dataset is:

$$\mu = \frac{1}{N} \sum_{i=1}^{N} x_i \qquad\qquad (3.1)$$

Note that this definition is very intuitive and has already been used several times.

Example 3.3

Let X be the GNI per capita (GNIpc, measured in units of €1000), considered on the population of the $N = 25$ countries in the EU25; see also Example 2.3 and the accompanying (population) dataset in column 4 of the file Xmp02-03.xls. If the countries are ordered alphabetically, then x_1 is the GNIpc of Austria, x_2 the GNIpc of Belgium, and x_{25} is the GNIpc of the UK. It follows that:

$$\mu = \frac{1}{25} \times (26.6 + 25.7 + \ldots + 28.2) = 18.228$$

Consequently, the mean of the GNIpcs of the 25 EU countries is €18228.

To calculate the median, we place the 25 observations in ascending order (Table 3.2).

Position	1	2	...	12	13	14	...	25
Ordered observation	3.8	4.0	...	14.1	17.1	21.5	...	44.9

TABLE 3.2 Ordered GNIpcs of EU25, with positions

It follows immediately that the middlemost of the ordered observations (the 13th) is 17.1. This is the GNIpc of Spain. Consequently, the median is $\mu_{median} = 17.1$. Since only two observations have frequency larger than 1 (both 4.6 and 27 occur twice), the mode of the dataset is not very useful. From the dot plot in Figure 3.4 it follows that the values 17.1 and 18.228 (median and mean) indeed take a position in the centre of the dataset.

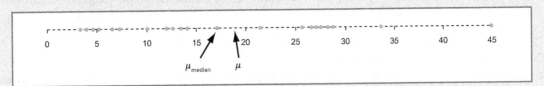

FIGURE 3.4 Dot plot of the GNIpcs of EU25 countries with mean and median

The large observation (44.9) of Luxembourg seems to draw the mean away from the median.

Now let the dataset be a **sample** dataset of observations of a quantitative variable X. The sampled population elements are numbered as $1, 2, \ldots, n$ while the observations of X in the dataset are formally denoted by $x_1, x_2, \ldots, x_n$. Under these circumstances, x_1 is the value of X observed at the first element in the sample, x_2 is the value of X observed at the second element in the sample, etc. In short, x_i is the value of X measured at the ith sampled element; $i = 1, 2, \ldots, n$.

To obtain the mean $\bar{x}$, the observations $x_1, x_2, \ldots, x_n$ have to be summed and the result divided by n.

Sample mean

The *sample mean* of a sample dataset of a quantitative variable X is:

$$\bar{x} = \frac{1}{n}\sum_{i=1}^{n}x_i \tag{3.2}$$

Take care to notice the difference between equations (3.1) and (3.2), and the different meaning of the terms x_i in these two formulae.

Example 3.4

The manager of the computer shop in Example 2.5 not only wants tables and graphs to summarize last year's daily sales (#) of notebooks Inspiration 5100, but also wants some measures of location. Let X be the variable 'daily number of Inspiration 5100 sold'. See Xmp02-05.xls for the dataset of the sales for the most recent 300 working days. Table 3.3 contains (a part of) the ordered observations, jointly with the accompanying positions. For instance, the 157th ordered observation is equal to 3.

Position	1	...	32	33	...	156	157	...	243	244	...	269	270	...	300
Ordered sales (#)	0	...	0	2	...	2	3	...	3	4	...	4	5	...	5

TABLE 3.3 Ordered daily sales (#) of Inspiration 5100, with positions

The following sample statistics will be calculated: mean $\bar{x}$, median x_{median} and mode x_{mode}.

To determine the sample mean, the original observations x_i of the dataset have to be added up and the result has to be divided by 300. The answer can be obtained with the help of a computer: $\bar{x} = 2.56$. (It can also easily be calculated by hand, by using the frequencies of the different values; see also Section 3.3.2 below.)

To determine the median, notice that the sample size is even and that the middlemost pair of observations is at positions 150 and 151. It follows from Table 3.3 that these observations are both equal to 2, so $x_{median} = (2 + 2)/2 = 2$.

Figure 2.7 shows that the value 2 has the highest frequency, so $x_{mode} = 2$. It would appear that the sample mean is the largest of the three measures of location that are considered here.

Occasionally, formulae (3.1) and (3.2) are used in a slightly different (but equivalent) form:

$$\sum_{i=1}^{N}x_i = N\mu \quad \text{and} \quad \sum_{i=1}^{n}x_i = n\bar{x} \tag{3.3}$$

Notice that these equations say that the sum of all observations in the dataset is equal to the product of the size of the dataset and the mean of the dataset.

Proportion of successes

The variable of interest now is an alternative variable. Recall that such a variable is nominal and that it takes only two different values; these values are frequently called success and failure. Often, 'success' is coded as 1 and 'failure' as 0. Such 0-1 variables are in a sense intermediate between the qualitative variables and the quantitative variables. For qualitative variables the mean of a dataset is **not** defined, but for a quantitative variable it **is** defined. Below, it will be shown that the mean of observations of a 0-1 variable **does** make sense and is equal to the proportion of ones in the dataset.

Consider a 0-1 variable X that measures whether population elements do or do not have a certain quality called 'success'. In the dataset, 'success' is coded 1 and 'failure' 0. We are interested in the proportion of the observations that are equal to 1.

In the case of a population dataset, this proportion is denoted by p, a number between 0 and 1. (Indeed, this is not a Greek letter but we follow the international notation.) Hence, p is the proportion of **successes** in the population. Since the population dataset $x_1, \ldots, x_N$ of the variable X consists of ones and zeros, the sum $\sum_{i=1}^{N} x_i$ of all observations is just the total number of ones in the dataset. This is because the zeros in the summation do not contribute. Hence, the ratio $\left(\sum_{i=1}^{N} x_i\right) / N$ is nothing but the **proportion** of population elements that have a 1. Hence, p is just the mean of the 0-1 observations in the dataset.

In case of a sample dataset $x_1, \ldots, x_n$, the summation $\sum_{i=1}^{n} x_i$ is just the total number of ones for the sample elements, that is, the total number of sampled elements with the quality. But the sample proportion of successes in the dataset is equal to the total number of ones divided by n, so it equals $\left(\sum_{i=1}^{n} x_i\right) / n$.

Property of a 0-1 variable X with 1 = success

population proportion successes $= \dfrac{1}{N} \sum_{i=1}^{N} x_i$

sample proportion successes $= \dfrac{1}{n} \sum_{i=1}^{n} x_i$

(3.4)

Example 3.5

Recall that in Example 3.1, for the year 2001, the 0-1 variable Y measures whether the wife in husband–wife couples works outdoors or not. Here, the value 1 (success) of Y refers to a couple with a wife working outdoors and the value 0 (failure) to a couple where the wife does not work outdoors. The file Xmp03-05.xls has 10 000 observations, only zeros and ones. It contains the data that underlie the 2001 frequency distribution for Y in Example 3.1.

With a computer, it can easily be calculated that the mean of the 0-1 observations in the dataset is 0.57, which indeed is the proportion of successes mentioned in Example 3.1.

Weighted mean

In practice, there are many situations where the mean that is asked for is **not** the arithmetic mean of the observations. One has to be particularly careful when the mean of grouped data is of interest. For instance, the stacked bar chart in Figure 3.5 is taken from an article that claims that 'men consume about 76% of all alcoholic beverages'.

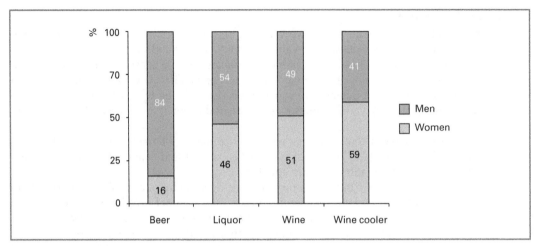

FIGURE 3.5 Percentages of alcoholic beverages consumed by gender
Source: www.darmouth.edu

However, when we take the arithmetic mean of the four percentages for the men, we get 57% and not 76%. What goes wrong?

To determine the overall percentage drunk by men, the above four types of drink are not equally important; for instance, the total amount of beer (in litres) will certainly be more than the total amounts (in litres) of the other types of drinks. To obtain the claimed 76% of the alcoholic beverages drunk by men, the following calculations have to be performed:

- The percentages 84, 54, 49 and 41 have to be multiplied by the respective amounts (in litres, men and women) consumed in the categories beer, liquor, wine and wine cooler.
- The results have to be summed.
- The resulting sum has to be divided by the total amount (in litres) of the four categories.

Unfortunately, the amounts for the four categories are not given, so we cannot check the 76% claim.

Such means are called weighted means; weights give the relative importance of the observations. In the beverages example, the weights are just the four amounts (in litres) that are consumed in the four categories. Here is the precise definition:

Weighted mean

For fixed non-negative constants w_1, w_2, ..., w_n, the **weighted mean** of observations x_1, x_2, ..., x_n is defined as:

$$\frac{\sum_{i=1}^{n} w_i x_i}{\sum_{j=1}^{n} w_j} \tag{3.5}$$

The constants w_i are called **weights**.

Of course, things are similarly for observations x_1, x_2, ..., x_N of the whole population.

In the beverage example, $n = 4$ and the observations x_1, ..., x_4 are the percentages 84, 54, 49 and 41; the weights w_1, ..., w_4 are the amounts drunk in the respective categories.

Example 3.6

Product reviews offering guidance to consumers are widely available in magazines and on the internet. In Figure 3.6 mean ratings are recorded for the printers Epson Stylus Photo 935 (€300) and Canon S830D (€390). They are based on reports of several reviewers, who all gave ratings to six 'qualities': quality of pictures, equipment, quality of text, speed of printing, costs of printing, and ergonomics. The figure contains, for both products and for each quality, the mean score of the ratings of the reviewers (scale 1–100). It also gives **overall** mean ratings for both printers.

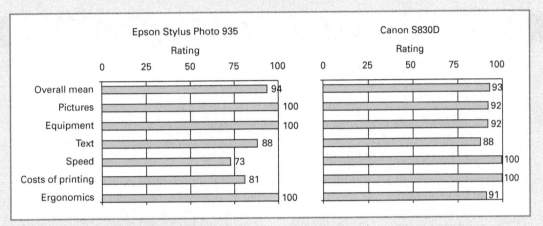

FIGURE 3.6 Mean ratings for two types of printer
Source: CHIP magazine, October 2003

An interesting question is how, for each of the two printers, the overall mean can be deduced from the six individual mean ratings. Notice that the ordinary, arithmetic means of the two groups of six scores respectively are 90.3 and 93.8, so the Canon printer would get the highest score. However, Epson's overall score is 94, slightly greater than Canon's score of 93. Apparently, the six qualities are not equally important for the determination of the overall score.

The article in *CHIP* also contains the pie chart shown in Figure 3.7. This chart shows the relative importance of each of the six qualities for the determination of the overall rating. For instance, the rating for 'quality of pictures' counts 45%, 'quality of text' counts 10%, and so on. These percentages are the weights $w_1, \ldots, w_6$.

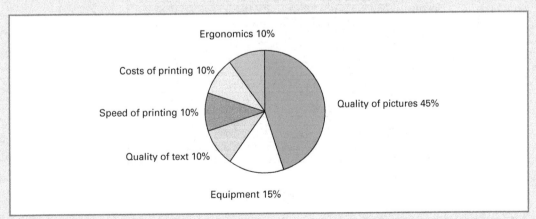

FIGURE 3.7 Pie chart for the weights (%) of six qualities

For the Canon printer, it follows (since $\sum_{j=1}^{6} w_j = 100$) that

$$\text{overall mean rating} = \frac{45 \times 92 + 15 \times 92 + \ldots + 10 \times 91}{100} = 93.1$$

Similarly, the mean rating for the Epson printer turns out to be 94.2. These scores are in accordance with the scores 93 and 94 of the bar charts above.

CASE 3.1 THE GENDER GAP IN EMPLOYMENT RATES – SOLUTION

As so often in official reports, the Lisbon report is not very clear regarding the precise definitions of the concepts it uses. Several questions come up, including:

■ What is the definition of **employment rate** that is used in the report?

■ What is meant by **mean** employment rate? Is it the average of the employment rates of the countries involved (so, just the arithmetic mean)? Or is it the overall percentage of the employed for all the involved countries together?

We will follow the report *Mind the Gap* of SEO Economic Research (2007) and take the following definition: the employment rate of a population is the percentage of the employed persons within the potential labour force (the subpopulation of the persons with age 15–64).

The employment rates in the file Case03–01.xls can be used to create the following – very transparent – figure for 21 countries. Notice that the countries are sorted by the employment rates of the men. The bar chart of the women overlays the bar chart of the men, while the numbers are, respectively, the percentages for the women and for the gender gap.

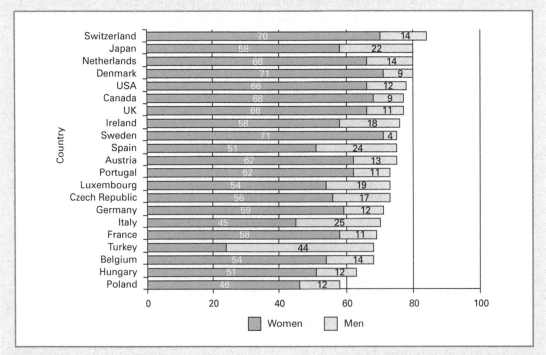

Bar charts of the 2005 employment rates of women and men
Source: SEO Economic Research (2007)

The gender gap is especially large for Turkey, Italy, Spain and Japan; the gap is relatively small in Sweden, Denmark and Canada. Also note that in some countries the female employment rate is already above the 60% target.

To compare the 2005 situation of the 'mean' female employment rate with the 60% target for 2010, we calculate means for thirteen EU15 countries; Finland and Greece are excluded.

The calculation of the arithmetic mean of the thirteen countries (ordered alphabetically) is straightforward:

$$\frac{62 + 54 + \ldots + 66}{13} = 59.8$$

If this mean is taken as the benchmark, the 60% target rate is – already in 2005 – almost achieved. However, it seems more realistic to take – for each of the thirteen countries – the relative size of the potential labour force into account. That is, we ought to calculate the **weighted** average of the thirteen female employment rates using the sizes of the populations aged 15–64 as weights. (The weights can be found in the file.) Since the sum of the sizes of these subpopulations is 121.849 million, it follows that this weighted mean equals:

$$\frac{2770 \times 62 + 3417 \times 54 + \ldots + 19546 \times 66}{121\,849} = \frac{7\,011\,550}{121\,849} \approx 57.5\%$$

If this weighted mean is taken as the benchmark, the 60% target is still far away.

As a comparison, the corresponding means for the men are respectively 74.0% and 72.7%.

Geometric mean

A friend was bragging to a mathematician how two-day volatility in the stock market had treated his holdings rather kindly. He chortled: 'Yeah … yesterday I gained 60% but today I lost 40% for a net gain of 20%.' The mathematician sat in horrified silence. He finally mustered all his courage and said: 'My good friend, I'm sorry to inform you but …'

What is the sad news the mathematician has to tell his friend?

Suppose that you invest €500 in some three-year project. Let the variable R denote the yearly (rate of) return (as a proportion, not a percentage). That is, at year t, the variable R measures the relative change of the value (price) of the invested amount when compared with the year before. Suppose that in the first year the rate of return is 100%, but in the second it is −20% and in the third −37.5%. That is, $r_1 = 1$, $r_2 = -0.2$ and $r_3 = -0.375$. You might call this a time series of length 3 with interest rate observations that correspond to 100%, −20% and −37.5%. Consequently, the invested amount of €500 increases to €1000 after the first year, but decreases to €800 after the second and €500 after the third. However, the mean rate of return is equal to:

$$\bar{r} = \frac{1 + (-0.2) + (-0.375)}{3} = 0.1417$$

This number is not showing us what the result after three years is. The invested amount did not change at all, so the 'mean rate of return' should be 0.

After one year, the €500 doubles into $500 \times 2 = $ €1000, which after two years reduces to $500 \times 2 \times 0.8 = $ €800. After three years the initial amount, €500, is back again:

$$500 \times 2 \times 0.8 \times 0.625 = 500$$

Slightly more generally: an amount of A euro turns into $A(1 + r_1)$ after one year, into $A(1 + r_1)(1 + r_2)$ after two years, and into $A(1 + r_1)(1 + r_2)(1 + r_3)$ after three years. Apparently, a suitable average interest rate (rate of return) r_g must satisfy:

$$A(1 + r_g)^3 = A(1 + r_1)(1 + r_2)(1 + r_3)$$
$$(1 + r_g)^3 = (1 + r_1)(1 + r_2)(1 + r_3)$$
$$1 + r_g = \sqrt[3]{(1 + r_1)(1 + r_2)(1 + r_3)}$$
$$r_g = \sqrt[3]{(1 + r_1)(1 + r_2)(1 + r_3)} - 1$$

Geometric mean

The **geometric mean** of observations $r_1, r_2, \ldots, r_n$ of a variable R is defined as

$$r_g = \sqrt[n]{(1 + r_1)(1 + r_2) \ldots (1 + r_n)} - 1 \tag{3.6}$$

It has the following **property**:

$$A(1 + r_g)^n = A(1 + r_1)(1 + r_2) \ldots (1 + r_n) \tag{3.7}$$

The property expressed by formula (3.7) states that an (invested) amount A that is subject to the successive periodic returns $r_1, r_2, \ldots, r_n$ has, after n periods, turned into $A(1 + r_g)^n$. Notice that in the above €500 example:

$$r_g = \sqrt[3]{(1 + 1)(1 + (-0.2))(1 + (-0.375))} - 1 = 0$$

which nicely reflects the fact that the value of the investment after three years has not changed. Try to answer for yourself the question raised at the start of this subsection on geometric means.

Example 3.7
The file Xmp03-07.xls contains the levels (prices) of the AEX share price index on business days (at closing times) during the period 25 April 2004 – 24 April 2005. In total there were 256 business days in this period yielding the observations $x_1, x_2, \ldots, x_{256}$ of the variable $X = $ 'level of AEX index at closing time'; see column 2 of the dataset. The first three and the last two are also given in Table 3.4.

Day t	Value AEX at day t	Rate of return r_t
1	354.57	–
2	353.99	–0.0016
3	347.92	–0.0172
⋮	⋮	⋮
255	353.89	–0.0077
256	353.26	–0.0018

TABLE 3.4 Daily prices and returns of the AEX, 25 April 2004 – 24 April 2005
Source: Euronext (2005)

Notice that the first observation (354.57) in the one-year period is slightly larger than the last observation (353.26), so the 'mean' of the daily percentages of the changes should be close to 0 but slightly negative.

But what kind of mean is meant here? Let R denote the relative daily change of the AEX with respect to the day before, measured as a proportion (not percentage). That is, the daily return r_t on business day t is equal to the change $x_t - x_{t-1}$ divided by x_{t-1}. So:

$$r_t = \frac{x_t - x_{t-1}}{x_{t-1}}$$

Since $x_1 = 354.57$ and $x_2 = 353.99$, it follows that:

$$r_2 = \frac{353.99 - 354.57}{354.57} = -0.0016$$

That is, on the second business day of the period, the AEX index closed 0.16% lower than on the first business day. Since $x_3 = 347.92$, it follows analogously that $r_3 = -0.0172$ which represents another decrease, of 1.72%, on the third business day. Apparently, transformation of the observations x_t into the returns r_t means a 'loss' of one observation: r_1 cannot be calculated since the level of the AEX index on the last business day before 25 April 2004 is not given. The dataset with the observations of R contains only 255 returns. Of course, a computer was used to calculate the third column of the file.

Having got the 255 returns, it is easy to obtain the arithmetic mean $\bar{r}$ and the geometric mean r_g, at least with the help of a computer. It follows that:

$$\bar{r} = 1.6852 \times 10^{-5} = 0.000016852; \quad r_g = -1.45154 \times 10^{-5} = -0.0000145154$$

The fourth column of the above file is used to calculate r_g using formula (3.6).

Notice that many decimals are used to express these means: the two means respectively correspond to 0.0016852% and −0.00145154%, a positive outcome for the arithmetic mean and a negative one for the geometric mean. As we saw above, the geometric mean is the best way to describe the 'average daily change'. Since the end value is smaller than the starting value, we need the average to be negative. The geometric mean satisfies this condition, the arithmetic mean does not.

Using the geometric mean we get:

$$354.57 \times (1 + r_g)^{255} = 354.57 \times 0.9963054 = 353.26$$

So, this geometric mean indeed transforms 354.57 on business day 1 into 353.26 on business day 256, which is 255 business days later.

3.3.2　Data of discrete variables

In this subsection we demonstrate how to calculate measures of location from a frequency distribution of a discrete variable.

Start with a discrete variable (say X) and suppose that a dataset of observations of this variable is presented in a frequency distribution. That is, an overview of all different values jointly with their frequencies is given. To obtain the mean value of the dataset in the way formulae (3.1) and (3.2) prescribe, one first has to **add** all original observations x_i. Consequently, each different value has to be summed as often as its frequency prescribes. That is:

Step 1:　Multiply all different values with their corresponding frequencies and sum the results.

Step 2:　The mean follows by dividing this sum by the number of observations.

Example 3.8

Table 3.5 gives information about the sizes of households in Denmark.

Size of household	Frequency	Rel. frequency	Cum. rel. frequency
1	965 653	0.384	0.384
2	832 087	0.331	0.714
3	293 208	0.117	0.831
4	290 103	0.115	0.946
5	100 049	0.040	0.986
6	23 969	0.010	0.995
7	6 523	0.003	0.998
8	5 090	0.002	1
Total	2 516 682	1	–

TABLE 3.5 Distribution of 'size of household' in Denmark, 1 January 2006
Source: Statistics Denmark (2006)

The frequency distribution is used to calculate the mean household size in Denmark with the steps above:

Step 1: $965\,653 \times 1 + 832\,087 \times 2 + \ldots + 5090 \times 8 = 5\,400\,303$

Step 2:   The mean household size is $\frac{5\,400\,303}{2\,516\,682} \approx 2.15$

Since the value 1 has the largest frequency, the mode of the frequency distribution is 1. To calculate the median with the original observations, the middlemost pair of the 2 516 682 ordered observations is needed; that is: the 1 258 341st and 1 258 342nd observations. From the frequency distribution it follows that these observations are both 2, so the median is 2. Consequently, mean, mode and median are ordered as follows:

mode < median < mean

The mean is drawn to the right because of a few relatively large observations.

The median of a frequency distribution of a dataset can also be determined with its cumulative distribution function F. In essence, the median is a solution of the equation $F(x) = \frac{1}{2}$. This is shown below for the discrete case.

Suppose that the dataset comes from a discrete variable. Recall that the median divides the ordered dataset into two equal-sized groups in such a way that the observations in one group are smaller than or equal to all observations in the other group. But this means that at the median the function F reaches the level $\frac{1}{2}$ or passes it. To understand this last sentence completely, Figure 3.8 shows the two possible situations: F takes the value $\frac{1}{2}$ (Figure 3.8b) or passes it (Figure 3.8a).

For Figure 3.8(a), the following holds:

less than 50% is < a **and** less than 50% is > a

To create a situation such that precisely 50% is ≤ a **and** precisely 50% is ≥ a, a certain percentage of the a observations (the observations equal to a) has to be added to the group of observations < a

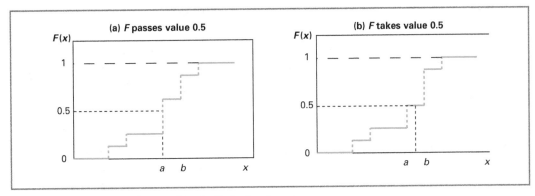

Figure 3.8 Finding the median by way of the distribution function

and the complementary percentage has to be added to the group of observations $> a$. (Notice that possibly a non-integer number of observations equal to a has to be joined with the $< a$ group.) In any case, the median is equal to a.

In Figure 3.8(b), the two successive values a and b are such that $F(a) = \frac{1}{2}$ while $F(b) > \frac{1}{2}$. Consequently, exactly 50% of the observations are $\leq a$. That is, the group with the smallest 50% of the ordered observations has a as its largest value. Since there are no observations between a and b, the group with the largest 50% of the ordered observations has b as its smallest value. Hence, the median falls between a and b and is – in accordance with the arrangement made in Section 3.3.1 – equal to $(a + b)/2$.

Here are some examples. Suppose that the discrete variable X takes only the values 1, 2 and 4 on a population of ten elements. The table below contains the (cumulative relative) frequencies.

Value of X	Frequency	Cum. rel. frequency
1	3	0.3
2	2	0.5
4	5	1
Total	10	–

Note that $F(2) = \frac{1}{2}$, so F **takes** the value $\frac{1}{2}$. Since 4 is the smallest value larger than 2, it follows that the median $\mu_{median} = (2 + 4) / 2 = 3$. Indeed, the population dataset is {1, 1, 1, 2, 2, 4, 4, 4, 4, 4} and the middlemost observations are 2 and 4.

For the households in Denmark in Example 3.8, the cumulative relative frequency distribution in the table reveals that $F(1) = 0.384$ and $F(2) = 0.714$. Hence, the distribution function F passes the level 0.5 at 2 and Figure 3.8(a) describes the situation. Consequently, the median is equal to 2, which is the same conclusion as in Example 3.8.

3.3.3 Data of continuous variables

Now suppose that the underlying dataset comes from a continuous variable. Often – for instance in articles in newspapers and other publications – the original dataset is not given. Instead, a frequency distribution is given with some classification of the data. Since the precise values of the data are not known any more, it is not possible to find the exact mean of the data if only a classified frequency distribution is given. However, the classified frequency distribution can be used to **approximate** the mean (and also the median) of the underlying dataset. This will be demonstrated below.

Example 3.9

With the help of a computer the mean of the grades for the Final Exam in Example 2.6 can easily be calculated; see the file Xmp02-06.xls. It turns out to be 5.3709. Now suppose that the original observations are not given anymore, but that instead only the frequency distribution of classification I is given; see Table 3.6.

Class	Frequency	Rel. frequency	Cum. rel. frequency
[0, 1]	21	0.081	0.081
(1, 2]	9	0.035	0.116
(2, 3]	21	0.081	0.197
(3, 4]	24	0.093	0.291
(4, 5]	37	0.143	0.434
(5, 6]	30	0.116	0.550
(6, 7]	42	0.163	0.713
(7, 8]	37	0.143	0.856
(8, 9]	21	0.081	0.937
(9, 10]	16	0.062	1.000
Total	258	1	–

TABLE 3.6 Frequency distribution of Y for classification I

Of course, it will not be possible to calculate the mean of the grades precisely from this frequency distribution, but the exact mean can be approximated by assuming that within each class the observations are evenly distributed and consequently the means of the observations per class are all located in the respective centres of the classes. To be precise: for the class [0, 1] it is assumed that its 21 observations are located in such a way that the mean value of these observations is just the centre 0.5 of the class, so their sum is equal to 21 × 0.5. Similarly, the 9 observations of the class (1, 2] are assumed to be distributed such that their mean value falls in the centre 1.5 and their sum is equal to 9 × 1.5. And so on.

The total of **all** observations (which is needed to calculate the mean) is equal to the total of all summed observations per class, which – as explained above – can be approximated by:

$$21 \times 0.5 + 9 \times 1.5 + \ldots + 16 \times 9.5 = 1373 \tag{3.8}$$

Hence, the mean of all original observations can be approximated by 1373 / 258 = 5.3217. The value 5.3217 is called the **mean value of the classified frequency distribution**.

Consequently, based on this frequency distribution alone, it is concluded that the mean of the underlying dataset is about 5.32. (Comparison with the precise value 5.3709 calculated before indicates that this approximation 5.32 is quite good.)

Notice that the calculation method in formula (3.8) is similar to the calculation method in Section 3.3.2 for the discrete case: the homogeneity assumption comes down to assuming that the different values of the variable are just the centres 0.5, 1.5, ..., 9.5.

For a classified frequency distribution, a **modal class** is a class with the largest frequency **density**. When all classes have an equal width, a modal class is just a class with the highest frequency. In the former example the class (6, 7] is the modal class.

The median of a classified frequency distribution can be determined with the accompanying distribution function F. Since F is now an increasing function, the procedure is simple: the median is the unique solution of the equation $F(x) = \frac{1}{2}$; see Figure 3.9.

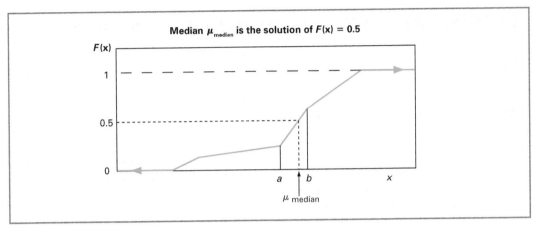

FIGURE 3.9 Determination of the median by way of *F*

To find the median of the classified frequency distribution in the above example, note that

$$F(5) = 0.4341 \quad \text{and} \quad F(6) = 0.5505$$

So, the median falls in the class (5, 6]; see Figure 3.10.

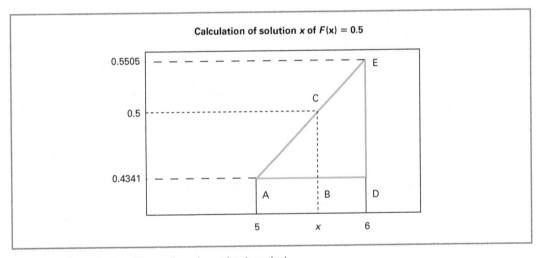

FIGURE 3.10 Determination of the median using a triangle method

Notice that triangle ABC fits in angle A of triangle ADE, and that BC and DE are parallel. By a well-known mathematical rule, the ratios AB/AD and BC/DE are equal. Hence

$$\frac{x - 5}{6 - 5} = \frac{0.5 - 0.4341}{0.5505 - 0.4341}$$

Consequently, x = 5.5662. The median of the classified frequency distribution is equal to 5.5662; the median of the underlying dataset is approximately equal to 5.5662. For the exact median, the dataset Xmp02-06.xls is needed; it turns out to be 5.6. The approximation 5.5662 is quite precise.

3.4 Relationship between mean/median/mode and skewness

Recall that frequency distributions are meant to summarize the datasets. Consequently, the rough form of a frequency distribution as pictured in a bar chart or a histogram has to be interpreted in terms of the observations in the dataset. This is the topic of the present section.

The form of a frequency distribution can give important information about the underlying dataset. A frequency distribution is called **symmetric** if its bar chart or histogram is symmetrical about a vertical line, as in the histogram of Figure 3.11(b): the images at the right side and the left side of that vertical line are reflections of each other. In essence, the line of symmetry of a symmetric frequency distribution divides the underlying ordered dataset into two equal-sized groups of observations, so the median of the frequency distribution is just the symmetry point on the horizontal axis. But because of this symmetry, the mean of such a symmetric frequency distribution is also equal to this symmetry point, so median and mean coincide.

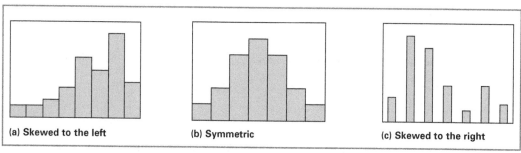

(a) Skewed to the left   (b) Symmetric   (c) Skewed to the right

FIGURE 3.11 Symmetric and skewed distributions with one major peak.

If a frequency distribution is not symmetric, it is called **skewed**.

The bar chart in Figure 3.11(c) represents a skewed distribution of a discrete variable. Because of its 'tail' at the right side, the distribution is said to be **skewed to the right**. The few relatively large values that cause that tail force the mean to be large in the sense that relatively few observations fall above the mean and relatively many fall below it. As a consequence, the mean tends to be larger than the median.

A frequency distribution is said to be **skewed to the left** if its bar chart or histogram looks like the histogram in Figure 3.11(a). Some relatively small observations force the distribution to stretch to the left. As a consequence, relatively few observations are smaller than the mean while relatively many are larger than the mean: the mean tends to be smaller than the median.

The mode of a symmetric unimodal frequency distribution as in Figure 3.11(b) is also equal to the symmetry point, so mean = median = mode. On the other hand, if mean, median and mode of a dataset are close to equal, this is often considered an indication that a frequency distribution (that summarizes the dataset) will be unimodal and more or less symmetric.

Although it is not a strict law, the measures of location of a right-skewed unimodal frequency distribution are often arranged as shown in figure 3.12.

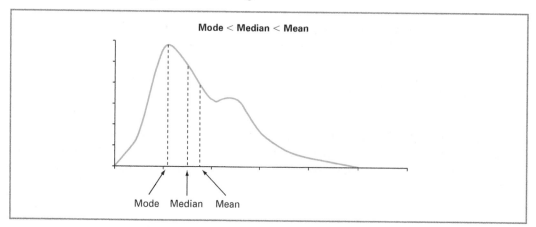

Mode < Median < Mean

Mode   Median   Mean

FIGURE 3.12 Frequent ordering of mean, median and mode for right-skewed distributions

Conversely, if the measures of location of a unimodal dataset are ordered in this way, a summarizing frequency distribution is often right-skewed.

Analogously, the measures of location of a left-skewed distribution are often ordered as shown in Figure 3.13.

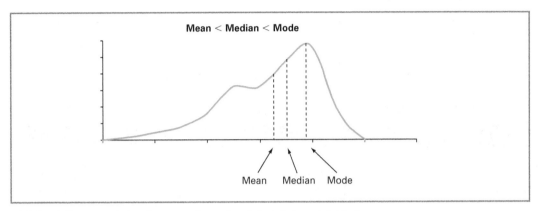

FIGURE 3.13 Frequent ordering of mean, median and mode for left-skewed distributions

On the other hand, if the three measures of location are ordered in this way then this is a strong indication that a summarizing frequency distribution is skewed to the left.

Figure 3.14 shows two distributions (of a discrete variable) that do **not** follow the above orderings.

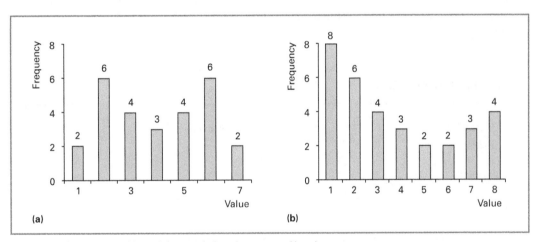

FIGURE 3.14 Distributions with recalcitrant ordering of measures of location

The distribution in Figure 3.14(a) is symmetric; mean and median are both equal to 4 (check it yourself). However, the distribution has two modes: 2 and 6. In this case, not only are the measures of location of the whole dataset important, but the measures of location of the smallest half and the largest half seem to be interesting too.

The distribution in Figure 3.14(b) has a major peak at 1 and a minor peak at 8. Its mean 3.7 is smaller than its median 4; the mode is 1. Still this distribution stretches out to the right; it might be called right-skewed. Again, apart from mean, median and mode, more statistics are needed to describe the special form of this distribution.

Example 3.10

Table 3.7 offers an overview of the measures of location and the type of skewness of frequency distributions of three examples considered above.

Example	Skewness of freq. distribution	Mean	Median	Mode	Arrangement
Notebooks	right	2.56	2	2	mode = median < mean
Size of households in Denmark	right	2.15	2	1	mode < median < mean
Final Exam grades (classification I)	left	5.32	5.57	–	mean < median

TABLE 3.7 Overview of skewness versus arrangement of measures of location

For each of the three examples, the ordering of the measures of location is in accordance with the skewness of the distribution.

Summary

The location of a (dataset of a) variable and a frequency distribution is a very relevant characteristic that can be measured with several statistics. The most important measures of location are mean, median and mode. The table below indicates the variable types and their corresponding statistic(s).

Measures of location

Variable	Measure of location
Nominal	Mode
Ordinal	Mode, median
Quantitative	Mode, median, mean

The median of a frequency distribution for a discrete variable is most easily determined from the accompanying distribution function (cdf), since the cdf reaches or passes the level 0.5 at the median. If the frequency distribution comes from a continuous variable, linear interpolation is used to find its median.

Some variants of the (arithmetic) mean are also important. A weighted mean is used when an average is wanted that assigns different weights to the observations since they do not all have the same importance. A geometric mean is sometimes used when the observations come from a time series, especially when the observations are interest rates. In a population or sample of successes and failures, the proportion of successes can be considered as the arithmetic mean of the dataset that arises by coding success as 1 and failure as 0.

Measures of location describe central positions of the dataset and its distribution. To describe the strength of the **variation around** these measures of location, **measures of variation** are important. Such measures will be considered in the next chapter.

List of symbols

Symbol	Statistic
μ_{mode}, x_{mode}	Mode
μ_{median}, x_{median}	Median
μ, $\bar{x}$	(Arithmetic) Mean
r_g	Geometric Mean
p	Population proportion

🔑 Key terms

absolute value operator **52**
arithmetic mean **56**
bimodal **53**
geometric mean **64**
location **51**
measures of location **51**
median **55**
modal class **68**

modal value **52**
mode **52**
mode of a variable **52**
multimodal **53**
population mean **57**
sample mean **58**
skewed **70**
skewed to the left **70**

skewed to the right **70**
summation operator **52**
symmetric **70**
unimodal **53**
weight **60**
weighted mean **60**

❓ Exercises

Exercise 3.1

Consider the following five sample observations of a variable X:

$$x_1 = 3, \quad x_2 = 5, \quad x_3 = 3, \quad x_4 = 7, \quad x_5 = 4$$

a Calculate $\sum_{i=1}^{5} x_i$ and the sample mean $\bar{x}$. Also calculate the median x_{median}.

b Calculate $\sum_{i=1}^{5}(3x_i - 2)$.

c Calculate $3 \times \left(\sum_{i=1}^{5} x_i\right) - 10$. Compare with part (b).

d The equality of the answers in parts (b) and (c) is a consequence of the following important property of the summation operator Σ that is considered and proved in formula (A2.3) of Appendix A2.

$$\sum_{i=1}^{n}(ax_i + by_i + c) = a\sum_{i=1}^{n}x_i + b\sum_{i=1}^{n}y_i + nc \tag{A2.3}$$

What are – for the present situation – the values of the constants a, b and c?

Exercise 3.2

Reconsider Exercise 3.1.

a Calculate $\sum_{j=1}^{5} x_j^2$ and $\left(\sum_{j=1}^{5} x_j\right)^2$. Notice that the answers are different!

b The numbers $x_1 - \bar{x}, x_2 - \bar{x}, ..., x_5 - \bar{x}$ are called the **deviations** with respect to the sample mean $\bar{x}$. Calculate them. The summations below respectively are the sum of the deviations and the sum of the squared deviations. Calculate these summations:

$$\sum_{k=1}^{5}(x_k - \bar{x}) \quad \text{and} \quad \sum_{k=1}^{5}(x_k - \bar{x})^2$$

c The numbers $|x_1 - \bar{x}|, |x_2 - \bar{x}|, ..., |x_5 - \bar{x}|$ are called the **absolute deviations** with respect to the sample mean $\bar{x}$. For instance, $|x_2 - \bar{x}|$ measures the **distance** between x_2 and the mean $\bar{x}$. Calculate these absolute deviations. Also calculate the sum of the absolute deviations:

$$\sum_{i=1}^{5}|x_i - \bar{x}|$$

Exercise 3.3

For each member of a group of six people, the number of kilometres cycled yesterday was recorded. The mean of the six numbers turned out to be 12. When taking a closer look at the data, it became clear that two of the six people did not cycle at all.

Determine the mean number of kilometres cycled yesterday by the four people in the group who **did** cycle.

Exercise 3.4 (computer)

Consider the following ten sample observations of a variable X:

4.2   7.1   −0.6   −2.4   4.3   −1.8   2.1   7.2   9.6   0.7

Use a computer (not in part (f)) to answer the questions below:

a Calculate $\sum_{i=1}^{10} x_i$, $\bar{x}$ and x_{median}.

b Calculate $\sum_{i=1}^{10} 5x_i$ and also $5 \times \sum_{i=1}^{10} x_i$. Compare your answers.

c Calculate $\sum_{j=1}^{10} x_j^2$ and $\left(\sum_{j=1}^{10} x_j\right)^2$. Compare your answers.

d Calculate $\sum_{i=1}^{10} 1/x_i$ and $1/\left(\sum_{i=1}^{10} x_i\right)$. Compare your answers.

e Calculate $\sum_{i=1}^{10}(5x_i - 4)$ and $5 \times \left(\sum_{i=1}^{10} x_i\right) - 40$. Compare your answers.

f Compare your answers in parts (b) and (e) with property (A2.3) of the summation operator in Appendix A2. Determine for both cases the values of the constants a, b and c.

g Calculate:

$$\sum_{k=1}^{10}(x_k - \bar{x})^2 \quad \text{and} \quad \sum_{i=1}^{10}|x_i - \bar{x}|$$

Interpret the answers in terms of **deviations**.

Exercise 3.5

Consider the following small sample dataset of a discrete variable X:

1   1   2   2   2   3   3   3   3   4

a Calculate the mean and the mode. Also calculate the median by ordering the data.

b Determine the frequency distribution and the cumulative relative frequency distribution of the dataset.

c Use the accompanying distribution function F to calculate the median again.

d Comment on the skewness of the distribution and the ordering of the three measures of location.

Exercise 3.6

Consider the following classified relative frequency distribution of a population dataset of a continuous variable X:

Class	Relative frequency
(0, 1]	0.2
(1, 5]	0.2
(5, 8]	0.5
(8, 10]	0.1
Total	1

a Calculate the mean of this classified frequency distribution.
b Determine the frequency density and the cumulative relative frequency distribution.
c Calculate the median of this classified frequency distribution; also determine the modal class.

Exercise 3.7

An electronics firm has two outlets, I and II. The mean weekly sales per employee for the outlets are respectively €2000 and €2500. Calculate the mean weekly sales per employee for the firm as a whole if you know additionally that outlet I has 20% more employees than outlet II. What is such an overall mean called?

Exercise 3.8

Suppose that you inherit €10 000. You decide to invest this money in a risky three-year project. During the first year, the value of your investment increases to €12 000, but in the second year it decreases to €10 500. After three years you get €9500 back.

a Determine the three yearly returns (proportions).
b Calculate the geometric mean r_g of these returns.
c Check that with the constant return r_g the value of the invested amount after three years indeed decreases to €9500.
d Compare the values of r_g and the arithmetic mean.

Exercise 3.9

This exercise enables you to get some practice with proofs similar to the one used for property (A2.3) in Appendix A2.

Consider a population with two variables X and Y. For a sample of n population elements, both variables are observed. Denote the n observations of X by $x_1, x_2, \ldots, x_n$ and the n observations of Y by $y_1, y_2, \ldots, y_n$. Below, b, c and d are constants, that is, they do not depend on the summation index. Prove – in the same way as was done in the proof of (A2.3) in Appendix A2 – that:

a $\sum_{i=1}^{n}(x_i + y_i) = \sum_{i=1}^{n}x_i + \sum_{i=1}^{n}y_i$. Interpret this result.

b $\sum_{i=1}^{n} bx_i = b \cdot \sum_{i=1}^{n} x_i$. Interpret this result.

c $\sum_{i=1}^{n} (cx_i + dy_i) = c \cdot \sum_{i=1}^{n} x_i + d \cdot \sum_{i=1}^{n} y_i$. Interpret this result.

Exercise 3.10

The necessity of 'lifelong learning' in a knowledge-based economy is often remembered by employers. But how many employees actually attend training courses sponsored by their employer? The table below lists the percentages of employed males and females from the private sector attending a course sponsored by their employer in 1995 and 2004.

	1995	2004
Female	24	35
Male	27	44
Total	26	41

Percentage of employees attending employer-sponsored training

The population is the set of all employees in the private sector, respectively in 1995 and 2004. The variable of interest is X = 'did (yes/no) attend at least one course sponsored by my employer', an alternative variable with values yes and no. In both years, samples of 10000 men and 5000 women were asked to respond, resulting in 15000 answers yes/no for each of the two years.

a Create a table that contains, for 1995 as well as for 2004, the relative frequency distributions for the subpopulation of females, the subpopulation of males and the whole population. Create a similar table for the accompanying (absolute) frequency distributions.

b Determine the modes of all frequency distributions.

c Check that for both subpopulations – from 1995 and 2004 – the relative frequencies of the modes have decreased. Interpret this result.

Exercise 3.11

Recent research shows that most young people (aged 18–24) believe that it is important to decrease extreme differences between incomes. To be precise, the following results were obtained regarding the question: Do you think that decreasing extreme income differentials is important?

Opinion	Rel. frequency %	Cum. rel. frequency %
Extremely important	5	5
Very important	9	14
Important	47	61
Am uncertain about it	24	85
Unimportant	15	100
Total	100	–

Opinion about decreasing income differentials
Source: Statistics Netherlands (2002)

Suppose that the sample size was 1000.

a Determine the variable of interest. What are its values? Is the variable nominal, ordinal or quantitative?

b Determine the mode, and also the median if it exists.

Exercise 3.12 (computer)

Reconsider the dataset Xrc02-16.xls that reports the (integer-valued) grades (x) for a Team Assignment of the course Statistics 2; see Exercise 2.16. The dataset is a population dataset of 180 observations of the variable X. Note that $x_1 = 9$, $x_2 = 8$, ..., and that $x_{180} = 7$.

a Calculate the population mean μ.

b Calculate the median by ordering the data.

Exercise 3.13

The table below is about the frequency distribution of the integer-valued grades for the Team Assignment in Exercises 2.16 and 3.12.

Grade	2	5	6	7	8	9	10	Total
Frequency	5	14	5	20	78	54	4	180
Cum. freq.	5	19	24	44	122	176	180	–
Cum. rel. freq.	0.028	0.106	0.133	0.244	0.678	0.978	1	–

a Use this distribution to calculate the mean of the TA grades.

b Use the accompanying distribution function F to determine the median.

c Calculate the mode of X.

d Compare the ordering of these three measures of location with the skewness of the frequency distribution of X.

Exercise 3.14

According to recent research by Statistics Netherlands, 'car ownership in The Netherlands is still rising: between 1995 and 2003 the number of cars rose by 22 percent to nearly 7 million'. To study the development over time of the number of cars per household, consider the table below.

Year	1995	2000	2002	2003
Households with more than two cars	1.0	1.7	2.1	2.1
Households with two cars	13.3	17.1	18.9	19.0
Households with one car	59.5	56.0	55.6	55.9
Households with no cars	26.2	25.2	23.4	22.9
Total	100	100	100	100

Car ownership per household (% of total number of households)
Source: Statistics Netherlands, *Statistical Yearbook 2005*

a Determine the population and variable of interest. What are the values of the variable? Explain why the variable is ordinal and not quantitative.

b For each of the four years considered, determine the relative frequency distribution of the variable.

With respect to the category 'more than two cars', fix the value of the variable at **exactly** 3 thus making the variable quantitative. All further questions are about the year 2003.

c Determine the distribution function F of the relative frequency distribution. Use it to calculate the median number of cars per household.

d Also calculate the mode.

e Calculate the mean number of cars per household. Do you need the size of the 2003 population for that?

Exercise 3.15

Reconsider Exercise 3.14. The table below gives an overview of medians, modes and means for the years 1995, 2000, 2002 and 2003.

	1995	2000	2002	2003
Median	1	1	1	1
Mode	1	1	1	1
Mean	0.8910	0.9530	0.9970	1.0020

a Check the results for 2002.

b Is the development of these measures of location over time in accordance with the statement: 'the location of the number of cars per household has increased'? Give your comments. Which measure of location is the most suitable?

c For each of the four years, give the ordering of the three measures of location.

d (computer) Create the bar charts of the four relative frequency distributions in one figure. Give your comments on the developments over time of the frequency distribution of the variable 'number of cars per household'.

Exercise 3.16

'Afternoon rush hours and weekends claim most traffic victims.' This was the headline of an article in Statistics Netherlands' *Web Magazine* (30 August 2005). The table below is derived from a graph in that article.

	Monday	Tuesday	Wednesday	Thursday	Friday	Saturday	Sunday
Average number of deaths per day	2.20	2.25	2.20	2.40	2.55	2.75	2.60

Road deaths by day of accident, 2004
Source: Statistics Netherlands, *Web Magazine*, 30 August 2005

For convenience, assume that the year 2004 had exactly 52 Mondays, Tuesdays, …, Sundays. Consider the population of all deaths due to road accidents in the Netherlands in 2004 (recorded in the order the accidents occurred) and the variable Y = 'day of the week on which the accident (that caused the death) took place'.

a Calculate the size of the population. (Round your answer.)

b Is Y nominal, ordinal or quantitative? Determine its frequency distribution.

c The week starts on Monday. Determine the mode and median of the frequency distribution of Y. Do you have an idea why the mode is in one of the weekend days, although the non-weekend days usually have the most traffic?

d Consider the dummy variable D that assigns the value 1 to all road deaths for which the accident occurred during the 'weekend' days Friday, Saturday and Sunday. Determine the relative frequency distribution of D and its mean.

Exercise 3.17

Reconsider the grades of the Final Exam for the course Statistics 2. Frequency distributions were determined for the classifications I and II; see Example 2.8. Mean and median were calculated for the frequency distribution of classification I; it was found that the mean is 5.3217 and the median is 5.5662. The modal class is (6, 7].

Below, measures of locations will be determined for the frequency distribution of classification II.

a Calculate the mean, median and modal class of the frequency distribution of classification II. Is the arrangement of these measures of location in accordance with the skewness of the classified frequency distribution as it follows from the histogram in Example 2.6?

b Let G be the distribution function of the frequency distribution of classification II. Determine the value of G at the mean of this classified frequency distribution. Is this value larger or smaller than the level $\frac{1}{2}$? Is this in accordance with the arrangement of mean and median considered in part (a)?

Exercise 3.18

Quarterly figures on GDP growth in the UK for the period 1994Q1–2003Q3 are given in the table below. (See also the file Xrc03-18.xls.) This exercise examines whether these quarterly growth data (%, with respect to the quarter before) justify the conclusion that, in the long run, one of the four quarters has a higher mean growth than the other three.

Year	Quarter			
	Q1	Q2	Q3	Q4
1994	1.37	1.38	1.00	2.05
1995	1.01	1.41	1.18	1.63
1996	1.62	1.76	1.33	0.90
1997	2.32	1.43	0.91	1.92
1998	1.23	1.45	2.13	0.86
1999	0.86	1.36	1.64	1.21
2000	1.92	0.55	1.20	0.63
2001	1.89	1.09	0.27	1.51
2002	1.40	1.08	1.49	1.32
2003	1.39	1.25	1.26	–

Quarterly growth (%) of GDP in the UK, 1994Q1–2003Q3.
Source: Biz/ed (2005)

a Regarding the Q1 data, what is the underlying population? What is the variable of interest?

b Use a calculator to determine the mean and the median of the growth data for Q1.

c The means of the data for Q2, Q3 and Q4 are, respectively, 1.276%, 1.241% and 1.337%. The respective medians are 1.37%, 1.23% and 1.32%. Give your opinion on the statement: 'the UK economy tends to grow fastest in the first quarter, Q1'.

d Denote the respective means per quarter by $\bar{x}_1$, $\bar{x}_2$, $\bar{x}_3$ and $\bar{x}_4$. Is the overall mean quarterly growth during the period 1994Q1–2003Q3 just the average of the four means per quarter? Determine this overall mean and compare it with the (arithmetic) mean of $\bar{x}_1$, $\bar{x}_2$, $\bar{x}_3$ and $\bar{x}_4$.

Exercise 3.19

Enterprises can be classified on the basis of their numbers of employees. A classified frequency distribution of the variable 'number of employees per enterprise' for the German manufacturing sector is given below.

Total	1–9	10–49	50–249	250 and over
281 187	215 021	47 468	14 821	3877

Manufacturing enterprises in Germany by size classes of employees
Source: Federal Statistical Office, Germany (2006)

In the questions below, treat the variable 'number of employees per enterprise' as continuous. Take 999.5 as upper bound for the last class. Take the classes as (0.5, 9.5], (9.5, 49.5], (49.5, 249.5] and (249.5, 999.5] with centres 5, 29.5, 149.5 and 624.5.

a Create a table that contains the classified relative frequency distribution, the cumulative relative frequency distribution and the frequency density.
b Determine the mean and the modal class of the classified frequency distribution.
c Determine the median of the classified frequency distribution.
d Calculate the value of the distribution function at the mean. The answer is much larger than 0.5. What does this tell you about the distribution?

Exercise 3.20 (computer)

Interest is in the employment rates in Belgium for the period 1956–2004. In the file Xrc03-20.xls, employment and population data are given for this period.[†]

a Create a single figure that contains time plots of both the employment data and the population data; put 'year' along the horizontal axis. Are both time series strictly increasing?
b Determine the mean employment size and the mean population size in Belgium during the period 1956–2004.
c Also determine the corresponding medians. Which years belong to these medians? Is the answer for the population data surprising?

To find out whether the development of the employment data over time behaves irregularly, it seems more interesting to study them as percentages of the population data.

d Determine, for each of the years 1956–2004, the percentage of the Belgian people that was employed.
e Determine the mean and the median employment percentage over the period 1956–2004. Is the mean equal to the ratio of the two means of part (b)?

[†] *Source*: Groningen Growth and Development Centre and The Conference Board, Total Economy Database, August 2005.

f Determine the years that the employment percentages were minimal and maximal. Also determine the difference between the maximum and minimum values of these employment rates.

g To obtain a better understanding of the development of the time series of part (d), present it in a time plot; put 'year' along the horizontal axis. Interpret this plot.

Exercise 3.21 (computer; can be done without)

The possession of a personal computer nowadays seems to be taken for granted, at least in rich countries. For poor countries the situation is very different. In this exercise and the next we will compare the numbers of computers per 1000 inhabitants for the countries in the EU10, EU15 and the world as a whole. Such investigations are often used by organizations such as the Worldbank to analyse the distribution of wealth.

To learn the techniques with a small dataset, the present exercise considers only the EU10 countries. The variables of interest are $X = $ 'number of PCs per 1000 people' and $Y = $ 'number of inhabitants per country'.

Country	x	y
Cyprus	269.89	764 967
Czech Republic	177.44	10 200 000
Estonia	210.33	1 358 000
Hungary	108.35	10 200 000
Latvia	171.75	2 338 000
Lithuania	109.75	3 469 000
Malta	255.05	397 000
Poland	105.65	38 200 000
Slovakia	180.36	5 379 000
Slovenia	300.60	1 994 000

Observations of X and Y for the EU10, 2002
Source: Worldbank (2005)

a Calculate – with your pocket calculator – the mean and the median of the dataset of the x observations.

b What is the meaning of this mean? Is it the number of PCs per 1000 people for the whole of EU10? Which country has its value of X closest to the mean and the median? Which countries have the maximal and the minimal values in the x dataset?

c To get a first impression of the variation in the x dataset for EU10, calculate the difference between the largest and the smallest observation of X.

d The number of PCs per 1000 people for the whole of EU10 can be considered as a weighted mean of the x dataset. What are the weights? Which country has its value of X closest to this average?

Exercise 3.22 (computer)

Emphasis is again on the variables X and Y defined in Exercise 3.21. In the present exercise the numbers of PCs within the EU15 and the whole world will be summarized. See Xrc03-22a.xls (for EU15) and Xrc03-22b.xls (world) for the data.

a For EU15:

i. Calculate the mean and the median of the x observations. To which country does this median belong? Which country is closest to the mean?

ii. To get a first impression of the variation in the x dataset for EU15, calculate the difference between the largest and the smallest observations.

iii. Similarly to Exercise 3.21, the number of PCs per 1000 people for the whole of EU15 can be considered as a weighted average of the x observations. First determine the weights and then calculate the number of PCs per 1000 people for the whole of EU15.

Unfortunately, for some countries in the world the 2002 figures for X = 'number of computers per 1000 people' and Y = 'number of inhabitants' are not yet known; the original dataset has **missing values**. After having omitted the countries for which an x and/or y observation is missing, 164 countries are left; see Xrc03-22b.xls.

b Answer the same questions as in part a., but now for the whole world.

c Compare the results for the EU10, EU15 and world. Give your comments on differences in measures of location and variation.

d The Worldbank states that – in 2002 – there were 6.2×10^9 people and 377.828×10^6 PCs on earth. Use this information to calculate again the mean number of PCs per 1000 people. Is the answer in accordance with the corresponding answer in part (b) where a small part of the countries was omitted because of missing values? Explain why the answer in (b) is larger / smaller.

Exercise 3.23

According to an article in the newspaper *De Volkskrant* (4 June 2005), some important economic indicators of the 12 countries that adopted the euro fell far behind those of the EU countries that kept their own currency. The arguments were illustrated by showing data for budget surpluses, inflation rates, unemployment rates and GDP growth – see the table below.

Country	Budget surplus (% of GDP)	Inflation (%)	Unemployment (%)	Growth GDP (%)
Finland	2.5	1.2	10.0	3.6
Germany	−3.9	1.7	11.8	1.1
Austria	−1.2	2.6	4.6	2.2
Greece	−3.4	3.4	10.4	4.0
Italy	−2.9	1.9	8.0	−0.2
France	−3.4	1.8	10.2	1.7
Spain	−0.2	3.5	10.2	3.3
Portugal	−4.1	2.1	7.5	0.6
Luxembourg	−2.1	2.5	4.3	3.6
Belgium	−0.2	2.5	8.0	1.8
Netherlands	−3.0	1.5	6.6	0.1
Ireland	unknown	2.2	4.2	2.8
Eurozone	−2.6	2.1	8.9	1.3
Sweden	0.3	0.3	5.9	2.0
Denmark	1.2	1.8	5.9	0.8
UK	−3.0	1.9	4.7	2.7

Economic indicators for several EU countries, 2003
Source: Bloomberg (2005)

Do these data support the claim that the triple Denmark–Sweden–UK does better than the eurozone? This question is considered below.

We first use 'growth GDP (%)' to compare. According to the table, two of the three countries of the triple have better growth figures than the eurozone as a whole. To determine an average that represents the GDP growth of the triple as a whole, we need for each of the three countries the 2003 percentages of the GDP when compared with the total GDP of the triple. These percentages are:

Country	Sweden	Denmark	UK
GDP weight (%)	12.17	8.75	79.08

a Calculate a suitable average for the GDP growth of the triple as a whole. Compare it with the corresponding eurozone growth percentage and give your comments regarding the claim in the newspaper.

We next use budget surplus, inflation and unemployment to compare.

b Budget surplus: does the triple do better than the eurozone? (The budget surplus for Ireland is not known; take it as 0.)

c Regarding inflation, the eurozone inflation rate 2.1% is the arithmetic mean of the 12 inflation rates involved. Does the triple do better than the eurozone?

d With respect to unemployment, you need more information to calculate a suitable average for the unemployment data of the triple. Although you do not have that information, it is still possible to answer the question whether the triple did better than the eurozone. Do so!

Consider the variable that measures whether a country's GDP growth is larger than its inflation rate. Suppose that 1 means yes and 0 means no.

e Determine the three relative frequency distributions of this variable on the populations eurozone, the triple Denmark–Sweden–UK, and the EU15. Also determine the modes, medians and means of the three frequency distributions. Give your comments.

Exercise 3.24 (computer)

In Example 3.3, the arithmetic mean (€18 228) of the GNIpcs of the 25 EU countries was calculated. Note, however, that this number is **not** the GNIpc for the entire EU25. The reason is that not all EU countries have the same number of inhabitants. For each country, the relative number of inhabitants determines the importance of its GNIpc in finding the GNIpc for the EU25 as a whole.

According to the file Xmp02-03.xls (bottom) the GNIpc for the whole of the EU25 is equal to €21 100. Can we find this value ourselves by starting from the GNIpcs of the member states and using only the data of the file?

Let the variables X, Y and W respectively denote the GNIpc (in euro) of a country, the GNI (in euro) of that country, and the number of inhabitants of that country. See the file Xmp02-03.xls for the x and y data. Note that:

$$Y = X W = (GNIpc) \times (number\ of\ inhabitants)$$

So,

$$W = \frac{Y}{X}$$

Consequently, the numbers of inhabitants of the 25 EU countries can be obtained by dividing the observations of Y (in column 3) by the respective observations of X (in column 4). In this way the observations $w_1, w_2, \ldots, w_{25}$ of W can be calculated with a computer:

$$w_1 = 8\,131\,579, \quad w_2 = 10\,291\,829, \quad \ldots, \quad w_{25} = 59\,950\,355$$

See the file Xrc03-24.xls. For instance, Belgium had (at the end of 2002) about 10 291 800 inhabitants.

   a Create yourself the column with the observations of W.

   b Determine the GNIpc for the entire EU25.

Exercise 3.25 (computer)

The file Xrc03-25.xls contains the percentages of the yearly growth of the GDP in China for the period 1979–2005.[†]

   a Calculate the geometric mean and the arithmetic mean.

   b In 1978, the GDP of China was 3645.2×10^8 yuan. Calculate the GDP level of 2005. (On 7 December 2007, the value of 1 yuan was \$0.135329.)

CASE 3.2   THE PARADOX OF MEANS (SIMPSON'S PARADOX)

Even the simple concept 'mean' can lead to mysterious results, as is illustrated in the following example (which is adapted from the paper 'Simpson's paradox' by Gary Malinas and John Bigelow (2004) in *Stanford Encyclopedia of Philosophy*).

The European football players van Nistelrooij and Makaay are having a dispute about their effectiveness as forwards. Roy Makaay informs his colleague that during his 5 most recent matches for the Dutch national team he scored 1 goal, while in the 8 most recent matches for Feyenoord he scored 6. Ruud van Nistelrooij smiles and claims that he is the best: he scored 2 out of 8 for the Dutch team and 4 out of 5 for Real Madrid, so he beats Roy on both fields. But Roy Makaay is a clever guy; he notices nicely that they both played 13 matches in total, and that he scored 7 times while van Nistelrooij only scored 6.

But is that possible? Can it be that van Nistelrooij has the highest mean number of goals on both counts (in the Dutch national team and in the team of the private employer) while Makaay has the highest overall success rate? It seems counterintuitive, even more since the two samples do have the same size (both 13).

The above example is an instance of **Simpson's paradox**, which states that there exist integers (whole numbers) $a_1, a_2, b_1, b_2, c_1, c_2, d_1, d_2$ such that:

$$\frac{a_1}{b_1} < \frac{a_2}{b_2} \quad \text{and} \quad \frac{c_1}{d_1} < \frac{c_2}{d_2} \quad \text{while} \quad \frac{a_1 + c_1}{b_1 + d_1} > \frac{a_2 + c_2}{b_2 + d_2}$$

Answer the following questions.

   a Determine the values of the integers in the above example and check the paradox.

   b Of course, samples of size 13 are only small. Does the result change if all numbers in the above example are multiplied by 10?

[†] *Source*: National Bureau of Statistics of China (2007).

c Check that the above example is indeed about means. What is (are) the population(s)? What is the variable from which the underlying (but not mentioned) datasets are derived and what are the means?

Some people say that Simpson's paradox is not a paradox at all; it is just a matter of incomprehension and having a wrong intuition. Still, it can give rise to results that sabotage a certain measure or policy, as in the following example.

In accordance with the policy to increase the percentage of female staff members, the Department of Social Economics of a university tries to discriminate in favour of women when hiring personnel for some new positions. Suppose that, traditionally, this university hires (very aggressively) its economics staff from the economics department of Harvard, while its sociology staff members come from the sociology department of Oxford. From Harvard, 70 of the 100 male economists (70%) and 15 of the 20 female economists (75%) are invited for an interview. From Oxford, 5 of the 20 male sociologists (25%) and 35 of the 100 female sociologists (35%) are invited. That is, the university tries to initiate its policy by inviting higher percentages of females than of males from both Harvard and Oxford. But even so, the university has to conclude that 75 of the 120 male candidates (62.5%) and only 50 of the 120 female candidates (41.7%) are invited for an interview.[†]

d Determine the values of the integers in the last example and check the paradox again.

e Check that this example is also about means. What is (are) the populations? What is the variable from which the underlying (but not mentioned) datasets are derived and what are the means?

CASE 3.3 DID THE EURO CAUSE PRICE INCREASES?

On 1 January 2002 the euro was introduced in some of the EU countries. However, many people complain that, from that point onwards, life became much more expensive. To investigate this complaint, price development in Germany will be studied by way of the consumer price index (CPI) and the inflation rate of four economic sectors.

Parts (a) and (b) below will be used to learn the mechanics of transforming CPIs into inflation rates. The table below contains the **overall** CPI for Germany for 2000 to 2005.

Year	2000	2001	2002	2003	2004	2005
CPI	100	102.0	103.4	104.5	106.2	108.3

CPI of Germany, 2000–2005 (2000 = 100)
Source: Federal Statistical Office, Germany

a Transform these six observations into five inflation rates, that is, determine for each of the years 2001 to 2005 the relative change (%) of CPI with respect to the year before.

---

† See also 'The paradox of averages' by J. E. Paulos on ABCnews, January–June 2001 (http://abcnews.go.com).

b Treat these inflation rates as proportions, not percentages. Use a pocket calculator to determine their arithmetic mean and their geometric mean. (Since all rates are close to zero, the two means will be almost equal.)

The file Case03-03.xls contains the CPIs of Germany for the years 1991 to 2005. Again, the base year is 2000. The following four economic sectors are considered:

1 = Food and non-alcoholic beverages
2 = Alcoholic beverages and tobacco
3 = Clothing and footwear
4 = Housing, water, electricity, gas and other fuels

c Transform the data into inflation rates (proportions, no percentages), for the overall category and for the four sectors; see also Appendix A1.3.

d For these inflation data, calculate all accompanying arithmetic means and geometric means.

e Which sectors are the main perpetrators of inflation during the period 1992–2005?

f When compared with 2000, the overall CPI of 1991 is 91.9. Derive the (overall) CPI of 2005 (which is 108.3) from the CPI of 1991 and the geometric mean.

g Create one figure that contains all five scatter plots of the inflation rates, with 'year' on the horizontal axis.

h Can you see any effects of the introduction of the euro?

Measures of variation

Variation refers to the spread (fluctuation) within a dataset of observations of a variable. In essence, variation is the reason for doing statistics. We want to draw knowledge out of some data. If the data points were all equal, it would be a trivial exercise. But as they generally vary, it is necessary to quantify the variation as well as to overcome it in order to learn anything from the data.

One wants to measure and understand the variation in a dataset. When studying the hourly wage Y (in euro), one inevitably notices that the observed hourly wages **vary**; that they are not all the same, that the dataset of hourly wages exhibits *variation*. An immediate question then is: how much variation does the dataset have? Followed by: which factors cause this variation? In the present chapter the first question will be considered. The second question is considered briefly in Chapter 5 and extensively in Chapters 19–22.

In this chapter we will work only with quantitative variables. A dataset of observations of a variable is said to **show variation** if it contains more than one different value, that is, if the data are not all the same. Other words to describe the phenomenon variation are *variability*, *spread* and *fluctuation*. In Figure 4.1, dot plots are given of two datasets. It is immediately clear that not only do the locations of the two datasets differ, but also the variations in the two diagrams are different, although it is not yet clear how to measure the degree of the variation.

It is the **degree** of variation economists are really interested in. This is because many interesting concepts essentially concern degrees of variation. Here are some examples.

- Income dissimilarity: refers to the degree of variation of a dataset of incomes. A large variation refers to large income dissimilarity.

- Risk of an asset: can be studied by considering the degree of variation within a dataset of historical returns of that asset. If this dataset shows much variation, one often concludes that it is risky to invest in that asset.

■ Consensus regarding the support for a certain proposal: can be studied by measuring the degree of variation within a dataset of responses (for instance, on the scale 0–100) to the statement 'I really do support the proposal'.

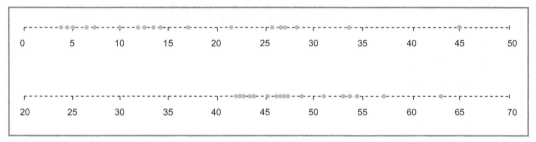

FIGURE 4.1 Two dot plots with different locations and variations

In the field of business and economics, variation is equal to uncertainty, inequality or instability. That is why there is a need to express the degree of variation as a number: one wants to **measure** the degree of variation. A good measure should be such that its measurements can be compared when applied to datasets of a common variable. If two datasets of one variable are given and their degrees of variation are measured with the same measuring method, the dataset with the largest measurement should have more variation than the other dataset.

 The aim of the present chapter is to find and study measuring methods, so-called *measures of variation*. These are statistics that express the degree of variation as a non-negative number in such a way that a large number corresponds to much variation and a small number corresponds to less variation.

Two types of measures of variation are studied, both for datasets of quantitative variables. The first type is based on relative positions within the **ordered** dataset and measures the variation around the median. The interquartile range, which measures the range of the middlemost 50% of the ordered observations, is the most frequently used representative. This measure is studied in Section 4.1. The second type is based on the deviations of the observations with respect to their (arithmetic) mean and measures the variation around the mean. Well-known representatives are the variance and the standard deviation. This type will be studied and interpreted in Sections 4.2–4.4. In Sections 4.5 and 4.6 special cases are considered.

In general, measures of variation are based on differences between observations. That is why the present chapter considers only quantitative variables.

Statistics that measure variation in a population dataset will be denoted by Greek letters. Suitable Latin symbols are used to denote statistics of sample datasets.

CASE 4.1 ERICSSON SHARES VERSUS CARLSBERG SHARES

An experienced investor is thinking of investing money in either Ericsson shares or Carlsberg shares, both traded on the Copenhagen Stock Exchange. Of course, he wants high returns. However, he is also aware of running risk when investing in shares. But how large is the risk of the shares Ericsson and Carlsberg?

To compare the two shares, he visits the site www.cse.dk and gathers the daily returns (%) for the period 15 February 2001 – 30 June 2007; see Case04–01.xls. He realizes that the spread of the data in the two columns gives information about the risk of the two shares, since a large spread indicates that the share is very volatile. But how can he measure the spread? See Section 4.2 for the solution.

4.1 Measures based on quartiles and percentiles

Recall that the intention behind the definition of the median was to divide the ordered data into two equal-sized groups. This idea will now be generalized: we want to divide the dataset into **four** equal-sized groups, into **ten** equal-sized groups, etc.

In Section 4.1.1, the starting point is the original dataset. Sections 4.1.3 describes what to do if only a (classified) frequency distribution is given. Box plots are studied in Section 4.1.2.

4.1.1 Using the original observations

For the division into four 'equal-sized' groups, we need three real values called **quartile 1**, **quartile 2** and **quartile 3**, ordered from small to large, such that:

25% of the data ≤ quartile 1

and   25% of the data ≥ quartile 1 but ≤ quartile 2

and   25% of the data ≥ quartile 2 but ≤ quartile 3

and   25% of the data ≥ quartile 3

This is illustrated below:

Defining quartile 2 is easy: just take the median. But things are more complicated for the other quartiles. Even computer packages do not agree about unique definitions. Table 4.1 gives some examples of the disagreement of Excel and SPSS.

| Dataset | Excel | | SPSS | |
	Quartile 1	Quartile 3	Quartile 1	Quartile 3
2, 3, 5, 7, 9	3	7	2.5	8
3, 5, 7, 9	4.5	7.5	3.5	8.5
1, 2, 2, 2, 3	2	2	1.5	2.5
1, 2, 2, 2, 2	2	2	1.5	2

TABLE 4.1 Disagreement of Excel and SPSS about quartiles 1 and 3

Apparently, Excel and SPSS work with their own definitions of the quartiles 1 and 3.

In the present book, we will not specify the definitions of these quartiles any further. Instead, we will follow the definition of the package we are using to determine quartiles 1 and 3. For quartile 2 there is no disagreement. All packages use the definition of the median that was given in Chapter 3.

The three quartiles of a **population** dataset are denoted as κ_1, κ_2 and κ_3, also if this dataset is not observed. As a consequence, there are two notations for the median of a population dataset: μ_{median} and κ_2. The (first, second, third) **quartile of a variable** is defined as the corresponding quartile of its (possibly unobserved) population dataset and denoted by the corresponding notation. If the dataset is a **sample** dataset, the three quartiles are denoted as k_1, k_2 and k_3.

The dataset of Example 2.5 contains for 300 days the daily numbers of Inspiration 5100 notebooks sold by a computer shop; the median was found to be 2. That is, $k_2 = 2$. With Excel it follows that $k_1 = 2$ and $k_3 = 3$, but other packages may give other answers.

The distance between the third and the first quartile measures the range of the middlemost 50% of the observations. It is called **_interquartile range_** (IQR) and denoted as δ (population dataset) or d (sample dataset):

$$\delta = \kappa_3 - \kappa_1 \quad \text{and} \quad d = k_3 - k_1$$

(Notice that the value of δ or d may also depend on the statistical package that is used.) Indeed, the IQR is a measure of variation since more spread in the middle half of the ordered observations leads to a larger measurement for δ or d. The quartiles and the IQR have the same dimensions as the underlying variable. If the dataset contains measurements of the variable X = 'weight (kg) of an adult', the dimension of the variable is kg. The quartiles and IQR are also measured in kg.

The (population) dataset of Example 2.3 contains the 25 GNIpcs of the EU25 countries. In Example 3.3, it was found that $\mu_{\text{median}} = \kappa_2$ equals 17.1; all computer packages will agree. Excel yields $\kappa_1 = 7.2$ and $\kappa_3 = 26.8$, and obtains 19.6 for the IQR. The central 50% of the observations covers a range of 19.6 × 1000 euro.

Since the dimension of the IQR is just the dimension of the variable, we cannot use it to compare two datasets with different dimensions. To overcome this problem, one often makes this measure of variation dimensionless by dividing it by the absolute value of the median. The measures of location

$$\frac{\delta}{|\kappa_2|} \quad \text{and} \quad \frac{d}{|k_2|}$$

are called the **_relative interquartile range_**, respectively for population data and sample data. These measures are **dimensionless** since the dimensions of the numerator and the denominator are the same, so they cancel out.

Using Excel, the relative IQR of the GNIpc data is $1.1462 \approx 1.15$, which is dimensionless. The Inspiration data have relative IQR: $(3 - 2)/2 = 0.5$. So, the variation of the GNIpc data is larger than the variation of the Inspiration data, at least as far as their middlemost part is concerned.

Quartiles and interquartile range can be used to detect **_outliers_**, observations that are extremely large or extremely small. The point here is what **extremely** means in this context. We will follow Tukey's definition and call an observation of a population dataset an outlier if it is smaller than $\kappa_1 - 1.5\delta$ or larger than $\kappa_3 + 1.5\delta$. This definition of outlier will be called the **_1.5δ-defintion_**.

Such outliers have to be reconsidered or even omitted if they are 'incorrectly observed'. But if an outlier is a correct observation of some relatively eccentric element, it is less obvious what should be done. It is important to realize that the three quartiles are **not** influenced by the sizes of outliers.

Example 4.1

The file Xmp04-01.xls contains the total GNIs (variable Y, in billions of euro) and the numbers of inhabitants (variable S, in millions) of the EU25 countries. The table below summarizes (using Excel) the data in terms of quartiles.

Variable	κ_1	κ_2	κ_3	δ = IQR	Rel. IQR	$\kappa_1 - 1.5\delta$	$\kappa_3 + 1.5\delta$
Y	23.3	139.4	264.5	241.2	1.7	−338.5	626.3
S	3.6250	8.9682	16.1296	12.5046	1.3943	−15.1319	34.8865

When using the relative IQR as measure of variation, it follows that the variability of the total GNIs is larger than the variability of the population sizes. From the dataset it follows that the data of Y have five outliers: the total GNIs of France, Germany, Italy, Spain and the UK. For the data of S there are even six outliers: the numbers of inhabitants of the countries mentioned before Poland.

Of course, there is no reason to have any doubts about the reliability of these outlier observations. They should not be excluded from the dataset; they are just observations that fall far from the central part of the data.

The outliers do not influence the quartiles. The only fact that matters is that they are in the upper part of the ordered dataset.

As mentioned before, the IQR measures the degree of variation of the middlemost 50% of the data. One may want to use another measure, for instance for the degree of variation of the central 80% of the data. But then the dataset has to be divided into ten equal-sized groups, and the total range of the central eight groups can be used as a measure of variation.

The intention is to define, for each integer m between 0 and 100, a real number called the **mth percentile** or **m% percentile** such that:

m% of the data is ≤ that number   and $(100-m)$% is ≥ that number

The illustration below demonstrates the ideas behind the 65% percentile:

65%		35%

65% percentile

Again, statistical packages do not agree on the precise definition. That is why we will not adopt a precise definition of the concept 'percentile'. Instead, we will follow the definition of the package that we use (without specifying this definition any further).

Fortunately, for large datasets the quartiles and percentiles as calculated with different statistical packages are usually not much different. However, you should always be aware that answers may be package-dependent.

The above ideas suggest that the 25%, 50% and 75% percentiles are respectively equal to the first, second and third quartiles. This indeed is the case.

Example 4.2
The dataset Xmp02-06.xls contains the grades of the 258 participants of the Final Exam for the course Statistics 2. Suppose that, to get direct permission to do the MPhil in Business, the Final Exam grade for Statistics 2 has to be at or above the 90% percentile. Which students got this permission?

Excel yields that the 90% percentile is 8.40; SPSS yields 8.41. Apparently, the number 8.4 divides the dataset into two groups such that (about) 90% of the grades are ≤ 8.4 and 10% are ≥ 8.4. Student who scored at least 8.4 for the final exam got direct permission to do the MPhil in Business.

4.1.2 Box plot

Some statistical packages offer a special graphical technique called **box plot**. The technique is based on the quartiles and can be used to get a rough impression of the distribution of the variable and especially of its variability and skewness. It is particularly useful when datasets of one variable (for different populations) are compared.

A box plot graphs five statistics: the minimum and the maximum of the observations, and three quartiles. Furthermore, it shows the outliers (if present). The box plot in Figure 4.2 comes from the GNIpc data; it can be created with SPSS.

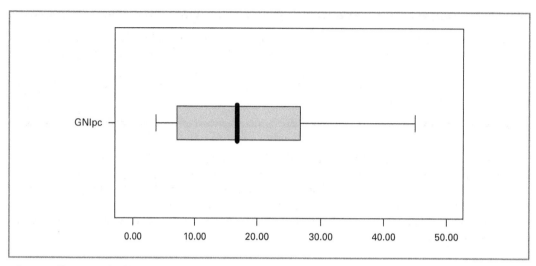

FIGURE 4.2 Box plot of the GNIpc data

The three vertical lines of the box – including the thick line at the median – are called **Tukey's hinges**; they are located at positions called 'quartiles' although the positions of the outer hinges may deviate (slightly) from the first and third quartiles. In the GNIpc dataset the first and third quartiles (with SPSS) differ slightly from the positions 6.85 and 26.9 of the hinges. Still, the length of the box is called the interquartile range and it is said that the 'middle' 50% of the data falls in the box; the presence of outliers is checked with the length of the box. Although they may be different, the locations of the hinges are often called the quartiles. The two lines that extend to the right and left of the box are called the **whiskers**. The length of a whisker is at most 1.5 times the interquartile range (that is, 1.5 times the length of the box); it ends at the most extreme observation that is not an outlier.

The length of the box represents the interquartile range; it measures the variability of the central half of the ordered dataset. Notice that for the GNIpc data the second quartile falls more or less in the centre of the box, which suggests that the central half of the data falls equally around the median. Apparently, there are no outliers; otherwise they would have been indicated. The whiskers end at the most extreme observations. Because of the form of this box plot it is expected that the frequency distribution of the dataset will be skewed to the right, since the right-hand whisker is much longer than the left-hand whisker.

Example 4.3

The box plots in Figure 4.3 belong, respectively, to the GNIs and the numbers of inhabitants of the EU25 countries; see Xmp04-01.xls and Xmp04-01.sav.

The locations of Tukey's hinges for GNI and number of inhabitants are respectively 23.3, 139.40, 264.5 and 3.625, 8.9682, 16.1296; see Appendix A1.4 for their calculation with SPSS.

For the GNI data, notice that there are five outliers. The right-hand (left-hand) whisker shows the position of the maximal (minimal) observation that is not an outlier. The box is slightly asymmetric. According to this box plot we expect the frequency distribution to be skewed to the right.

For the dataset of number of inhabitants, there are actually six outliers, although the diagram suggests that there are only five. Apparently, the maximal observation that is not an outlier falls in the box since there is no right-hand whisker. Again, the box plot suggests that the frequency distribution is skewed to the right.

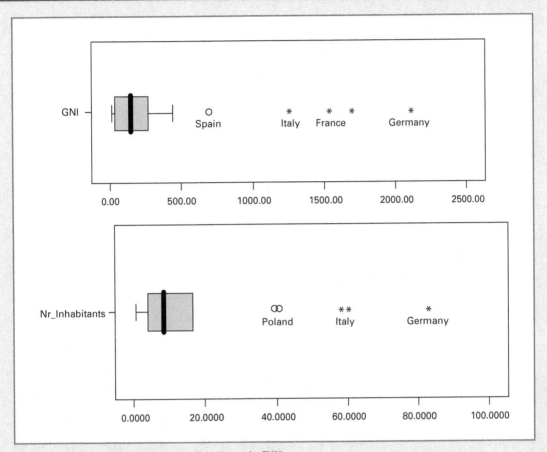

FIGURE 4.3 Box plots of GNI and number of inhabitants for EU25

Box plots are often used to compare two sets of observations of one variable.

Example 4.4

The file <u>Xmp02-02.sav</u> contains the opinions of students in the classes of instructor A, respectively B about the question 'instructor explains things very clearly', with possible answers:

1 = strongly disagree;   2 = disagree;   3 = neutral;   4 = agree;   5 = strongly agree

Although the variable of interest is ordinal and hence not quantitative, the data can be considered as observations of a quantitative variable that assigns a score from the integer scale 1–5 to the instructor. In total, 131 students gave their opinion, either about instructor A or about instructor B. The two sets of observations – one for instructor A and one for instructor B – can be summarized separately into two box plots; see Figure 4.4.

The upper box plot suggests that the scores for instructor B lie equally around the median at 3, since the median lies in the middle of the box and the two whiskers have more or less equal length. The lower box plot seems to summarize a frequency distribution that is skewed to the left. Notice that there are two outliers, both equal to 1. Comparison of the two box plots shows that the data for B have more variability than the data for A. Furthermore, the location of the B dataset is much more to the left when compared with the A dataset.

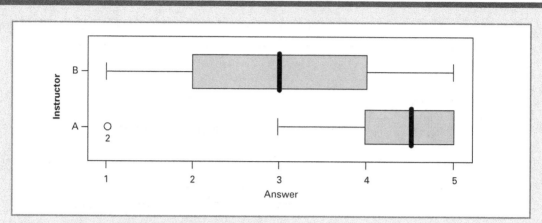

FIGURE 4.4 Box plots of the opinions about instructors A and B

To check whether the above ideas about the frequency distributions are valid, the accompanying bar charts are graphed in Figure 4.5.

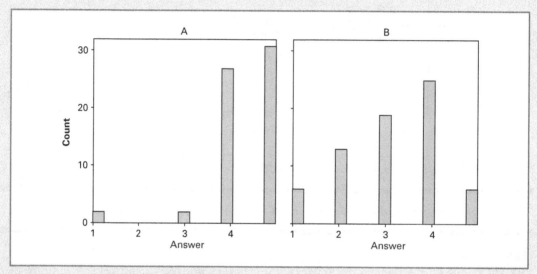

FIGURE 4.5 Bar charts of the two frequency distributions

The frequency distribution for B shows details that were not clear from the upper box plot. For instance, when compared with the frequency of 2, considerably more students answered 4; the distribution is **not** symmetric. The frequency distribution for A is indeed skewed to the left.

4.1.3 Using the cumulative distribution function

Suppose, first, that the underlying data come from a discrete variable. However, these original data are not given any more but they are summarized in a frequency distribution. How to determine the quartiles and the (relative) IQR?

Computer packages do not have a standard option to find quartiles from a frequency distribution; they want the original data. In principle it is possible to trace the original observations back from the frequency distribution of a discrete variable. However, this gives extra work. Instead, we will use the distribution function (see Section 2.3.2) to find the quartiles and the (relative) IQR.

Recall the (cumulative) distribution function; we denote this cdf by F. In Section 3.3.2 it was noted that the median can be found by solving $F(x) = \frac{1}{2}$. In a similar way, the first and third quartiles can be obtained by respectively solving:

$F(x) = \frac{1}{4}$ and $F(x) = \frac{3}{4}$

That is, if F **passes** the level $\frac{1}{4}$ at x_0, then x_0 is taken as the first quartile; but if F exactly **reaches** the level $\frac{1}{4}$ at the value x_0, then it is the average of x_0 and the subsequent value that is taken as the first quartile. Solving $F(x) = \frac{3}{4}$ goes analogously.

The first and third quartiles that result from this method may differ from the answers that are obtained using a computer package for the underlying **original** dataset. Usually, though, the differences are small, especially when the dataset is large.

Example 4.5

The sizes of the households in Denmark were considered in Example 3.8. Suppose now that only the frequency distribution is given (Table 4.2).

Size of household	Frequency	Rel. frequency	Cum. rel. frequency
1	965 653	0.384	0.384
2	832 087	0.331	0.714
3	293 208	0.117	0.831
4	290 103	0.115	0.946
5	100 049	0.040	0.986
6	23 969	0.010	0.995
7	6 523	0.003	0.998
8	5 090	0.002	1
Total	2 516 682	1	–

TABLE 4.2 Frequency distribution of 'size of household' in Denmark, 1 January 2006

It follows immediately from the last column of Table 4.2 that the accompanying distribution function F passes the level 0.25 at 1, the level 0.50 at 2, and the level 0.75 at 3. Hence, $\kappa_1 = 1$, $\kappa_2 = 2$ and $\kappa_3 = 3$; so the interquartile range δ equals 2 and the relative interquartile range equals 1. As a consequence:

$\kappa_1 - 1.5\delta = -2$ and $\kappa_3 + 1.5\delta = 6$

So, there are $6523 + 5090 = 11613$ outliers, households whose sizes are extremely large when compared with the sizes of the majority of households in Denmark.

Suppose now that the underlying data come from a continuous variable but only a classified frequency distribution is given. Then, you cannot expect to find the exact values of the median and the quartiles. The reason is that only the frequencies of the observations in the classes are known but not their precise positions. However, by assuming that they are evenly (homogeneously) spread over the respective classes, the median and other quartiles can be approximated.

Recall that the cdf now grows linearly on the classes. In Example 2.6, the grades of the Final Exam of the course Statistics 2 were summarized by way of the frequency distribution on classification I. The corresponding cumulative relative frequency distribution and the cdf F are repeated in Figure 4.6 below.

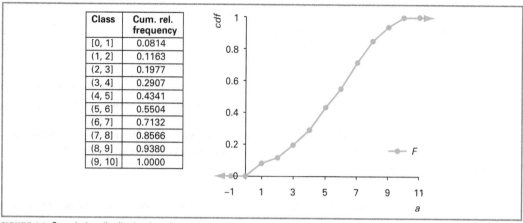

FIGURE 4.6 Cumulative distribution function of the variable 'grade for Final Exam'

To solve $F(x) = 0.25$, note that $F(3) = 0.1977$ and $F(4) = 0.2907$, so κ_1 lies between 3 and 4. Since

$$\frac{x - 3}{4 - 3} = \frac{0.25 - 0.1977}{0.2907 - 0.1977} \quad \text{and hence} \quad \frac{x - 3}{1} = \frac{0.0523}{0.0930} = 0.5624$$

it follows that $\kappa_1 = 3.5624$. Since $F(7) = 0.7132$ and $F(8) = 0.8566$, the third quartile κ_3 falls between 7 and 8. Solving $F(x) = 0.75$ yields $x = 7.2566$, which is κ_3. At the end of Section 3.3.3 it was derived that $\kappa_2 = 5.5662$. The IQR and the relative IQR follow immediately: 3.6942 and 0.6637. Check that there are no outliers yourself.

To summarize: the quartiles and the IQR that were obtained above, belong to the classified frequency distribution. They can be considered as approximations of the quartiles and IQR of the underlying dataset.

4.2 Measures based on deviations from the mean

Since the arithmetic mean is the most frequently used measure of location, statistics that measure variation around this mean are important. Below, the statistics variance, standard deviation and coefficient of variation are studied.

4.2.1 Variance and standard deviation

Again, data are assumed to be observations of a quantitative variable X. In the case of a population dataset the observations are formally denoted by $x_1, x_2, \ldots, x_N$, in the case of a sample dataset by $x_1, x_2, \ldots, x_n$. Recall that x_i is the value of X measured at the ith population element, respectively the ith sample element.

If the dataset is a **population** dataset, its mean is denoted by μ. The observations $x_1, x_2, \ldots, x_N$ are located around this mean. The *deviations*

$$x_1 - \mu, \quad x_2 - \mu, \quad \ldots, \quad x_N - \mu$$

describe the locations of the observations with respect to their average location. A negative (respectively positive) deviation corresponds to a location at the left-hand (respectively right-hand) side of μ.

To illustrate things, consider the population of the five households in Baker Street and their annual expenditures $x_1, \ldots, x_5$ (in euro) on food and clothing:

5620   21370   22180   9770   8810

Since $\mu = 13\,550$, the corresponding deviations are:

-7930 7820 8630 -3780 -4740

Three of these deviations are negative and two are positive. That is, three observations lie at the left-hand side of the central position μ and two at the right-hand side (see Figure 4.7). For instance, the observation 9770 of household 4 is €3780 **smaller** than the mean.

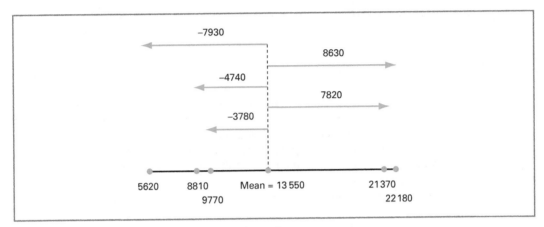

FIGURE 4.7 The deviations relative to the mean of the observations

At first sight, the average of the deviations might seem to be a good measure for variation. However, closer inspection reveals that this average is equal to zero:

$$\frac{1}{N}\sum_{i=1}^{N}(x_i - \mu) = \frac{1}{N}\left(\sum_{i=1}^{N}x_i - \sum_{i=1}^{N}\mu\right) = \frac{1}{N}\sum_{i=1}^{N}x_i - \frac{1}{N}\sum_{i=1}^{N}\mu = \mu - \mu = 0$$

The negative deviations neutralize the positive deviations. To overcome this problem, one might take the average of the **absolute** deviations. (Recall that taking the absolute value makes negative values positive and leaves positive values unchanged; see also Example A2.4.) If done so, the following formula arises:

$$\frac{1}{N}\sum_{i=1}^{N}|x_i - \mu|$$

This measure of variation is called **_mean absolute deviation_** (MAD). Although MAD is used occasionally, the absolute value operator makes this measure less appealing.

Another way is to use squared deviations. This is what is usually done.

Population variance

The (**_population_**) **_variance_** of the population data $x_1, x_2, \ldots, x_N$ and the variable X is defined as:

$$\sigma^2 = \frac{1}{N}\sum_{i=1}^{N}(x_i - \mu)^2 \tag{4.1}$$

Notice that first the squares of all deviations are calculated and that next the results are summed, not the other way around.

Since σ^2 is a mean of squares, it is non-negative itself. In practice it will be strictly positive, since this mean of squared deviations can only be 0 if all squared deviations are 0 and hence all deviations themselves are 0 too. However, if all deviations are 0, then all observations x_i are equal to their mean μ and all data points are the same. But this would mean that there is no variation – an uncommon and statistically uninteresting case.

The square root of σ^2 yields a related measure of variation:

Population standard deviation

The (*population*) *standard deviation* of the population data $x_1, x_2, \ldots, x_N$ and the variable X is defined as:

$$\sigma = \sqrt{\sigma^2} \tag{4.2}$$

The advantage of using σ^2 is its simple interpretation as mean of the squared deviations. However, the dimension of σ^2 is the square of the dimension of the data – if the data are measured in kg then σ^2 is measured in kg^2 – which is a disadvantage. The dimension of σ is the same as the dimension of the data, which explains its popularity as a measure of variation.

In the Baker Street example above, σ^2 is the mean of the squared deviations:

$$\sigma^2 = \frac{1}{5}\left[(-7930)^2 + (7820)^2 + (8630)^2 + (-3780)^2 + (-4740)^2\right]$$

So, $\sigma^2 = 47\,054\,040$ euro^2 and $\sigma = 6859.59$ euro.

For a **sample** dataset of observations of a quantitative variable X, the mean is denoted by $\bar{x}$. The observations $x_1, x_2, \ldots, x_n$ are located around this mean. The deviations

$$x_1 - \bar{x}, \quad x_2 - \bar{x}, \quad \ldots, \quad x_n - \bar{x}$$

express the relative positions of the observations around their mean.

Sample variance and sample standard deviation

The **sample variance** and the **sample standard deviation** of the observations $x_1, x_2, \ldots, x_n$ are respectively defined by:

$$s^2 = \frac{1}{n-1}\sum_{i=1}^{n}(x_i - \bar{x})^2 \quad \text{and} \quad s = \sqrt{s^2} \tag{4.3}$$

Notice the factor $1/(n-1)$ instead of $1/n$. Strictly speaking s^2 is **not** a mean of squared deviations, although for large sample sizes it hardly makes any difference that $1/(n-1)$ is used instead of $1/n$. The choice of the factor $1/(n-1)$ is based on an estimation argument: if the sample dataset and s^2 are used to obtain information about the unknown population variance σ^2, then it turns out that the formula with the factor $1/(n-1)$ yields – at least on average – better approximations of σ^2 than the formula with $1/n$. See also Chapter 14.

Example 4.6

Consider the variable X = 'weekly sales (in units of $100\,000$)' for a supermarket in the USA; see also Example A2.1. For the most recent eight weeks, the observations $x_1, x_2, \ldots, x_8$ of X are given in the second column of Table 4.3. A manager of the supermarket decides to use the sales figures of these eight weeks as a sample to gain an impression of the overall mean weekly sales μ and of the accompanying variance σ^2.

Since the dataset consists of sample observations, the formulae (3.2) and (4.3) have to be used for the calculation of the sample mean $\bar{x}$, the sample variance s^2 and the sample standard deviation s. The results can be used as approximations of the unknown population statistics μ, σ^2 and σ. To illustrate the formulae in (4.3), the calculations are done 'by hand'. That is, only a pocket calculator is used to fill in the columns 3 and 4 of the table.

Week i	x_i	$x_i - \bar{x}$	$(x_i - \bar{x})^2$
1	12.0	−0.975	0.9506
2	13.2	0.225	0.0506
3	11.1	−1.875	3.5156
4	14.6	1.625	2.6406
5	13.9	0.925	0.8556
6	12.3	−0.675	0.4556
7	14.8	1.825	3.3306
8	11.9	−1.075	1.1556
Total	103.8	0	12.9548

TABLE 4.3 Calculation of s^2 by hand

The formula for s^2 involves the sample observations x_i, the deviations $x_i - \bar{x}$ from their mean, and the squares of these deviations. That is why these figures have their own columns in Table 4.3. It follows easily that $\sum_{i=1}^{8} x_i$ is equal to 103.8 (the total of the second column), so $\bar{x} = 12.9750$. All individual deviations from this mean can then be calculated; see column 3. The squared deviations are put in column 4; by summing them the sum of the squared deviations is obtained (12.9548). Hence:

$$s^2 = \frac{1}{8-1}\sum_{i=1}^{8}(x_i - \bar{x})^2 = \frac{12.9548}{7} = 1.8507 \approx 1.85$$

and

$$s = \sqrt{s^2} = \sqrt{1.8507} = 1.3604 \approx 1.36$$

To interpret these numbers, recall that the sales data are measurements in units of 10^5 dollars. Hence, the mean of the eight sales is 1 297 500 dollars with a variance 18 507 000 000 dollars^2 and standard deviation 136 040 dollars.

Table 4.4 lists the observations of a second sample dataset, for the eight-week period preceding the period mentioned above.

Week j	x_j	$x_j - \bar{x}$	$(x_j - \bar{x})^2$
1	11.5	−0.8125	0.6602
2	12.4	0.0875	0.0077
3	11.9	−0.4125	0.1702
4	11.6	−0.7125	0.5077
5	13.2	0.8875	0.7877
6	13.8	1.4875	2.2127
7	12.2	−0.1125	0.0127
8	11.9	−0.4125	0.1702
Total	98.5	0	4.5291

TABLE 4.4 Calculation of s^2 for another period of eight weeks

Check that sample mean, sample variance and sample standard deviation of the second dataset of observations of X are respectively equal to:

$$\bar{x} = 98.5/8 = 12.3125; \quad s^2 = 4.5291/7 = 0.6470; \quad s = \sqrt{0.6470} = 0.8044$$

Since the two datasets come from the same variable, it is possible to compare their locations and their degrees of variation. Firstly, take sample mean and standard deviation to measure location and variation. Obviously, the mean of the first dataset is larger than the mean of the second. The same holds for the standard deviations. It follows that, when compared with the eight-week period that preceded it, the most recent eight-week period had not only a larger mean sales but also a larger standard deviation.

To find out whether a similar conclusion can be drawn for the median and interquartile range, we next calculate these measures of location and variation for both datasets. With Excel it follows that the triples of the quartiles of the two datasets are respectively 11.975, 12.750, 14.075 and 11.825, 12.050, 12.600. Hence, the most recent eight-week period has median 12.75 and IQR 2.10; the preceding eight-week period indeed has smaller median and IQR: 12.05 and 0.775.

Notice that both the mean–standard deviation approach and the median–IQR approach yield the same conclusion: location and variation of the sales in the most recent eight-week period are both larger than location and variation in the eight-week period before it.

The standard deviation is a measure of variation that has the same dimension (unit) as the mean and the variable itself. Just as with the relative IQR in Section 4.1, we can create a measure of relative variation that is dimensionless:

Coefficient of variation

The *coefficient of variation* of a population and a sample dataset are respectively defined as:

$$\frac{\sigma}{|\mu|} \quad \text{and} \quad \frac{s}{|\bar{x}|}$$

The coefficient of variation is the ratio of the standard deviation and the absolute value of the mean. Since it is dimensionless, it offers the opportunity to compare the variations of different datasets, even if these datasets use different dimensions. For the population dataset in the Baker Street example, the coefficient of variation is $6859.5947 / 13\,550 = 0.5062$. The above-mentioned sample dataset of the weekly sales of the supermarket for the most recent eight weeks, has coefficient of variation equal to 0.1048, which is considerably smaller.

All computer packages provide options to calculate the mean, variance and standard deviation of a dataset. Since the formulae for population variance and sample variance are different, it is important always to check whether the package returns a population or a sample statistic. See also Appendix A1.4.

4.2.2 Short-cut formula for variance

Of course, calculations 'by hand' (that is, with a simple pocket calculator only) as in the above example are very time consuming or even impossible if the sample size is large; computers can do the work much more easy. However, in practice it is also important to know how values of s^2 can be calculated by hand. Being a manager, you may want to get a quick impression of the mean, variance and standard deviation of a large sample dataset by calculating them for only a small part of the data; because you are curious or do not want to wait until the statistical department presents the values in a report. This is one reason why the ***short-cut formula*** (calculation formula) is important: it makes the calculation of the summation in the formula of s^2 easier.

Derivation of the short-cut formula

The summation rules of Appendix A2 are used to derive the formula, as follows:

$$\sum_{i=1}^{n}(x_i - \bar{x})^2 = \sum_{i=1}^{n}(x_i^2 - 2\bar{x}x_i + \bar{x}^2) = \sum_{i=1}^{n}x_i^2 + \sum_{i=1}^{n}(-2\bar{x}x_i) + \sum_{i=1}^{n}(\bar{x}^2)$$

$$= \sum_{i=1}^{n}x_i^2 - 2\bar{x}\sum_{i=1}^{n}x_i + \sum_{i=1}^{n}(\bar{x}^2)$$

$$= \sum_{i=1}^{n}x_i^2 - 2n\bar{x}^2 + n\bar{x}^2 = \sum_{i=1}^{n}x_i^2 - n\bar{x}^2$$

It follows that the sum of the squared deviations is equal to the sum of all squared **observations** minus n times the squared mean of the observations; s^2 then follows by dividing the result by $n - 1$. In the case of a population dataset, the deviations are calculated with respect to the **population** mean μ, and a similar short-cut formula for the sum of squared deviations can be derived:

Short-cut formulae

$$\sum_{i=1}^{n}(x_i - \bar{x})^2 = \sum_{i=1}^{n}x_i^2 - n\bar{x}^2 \quad \text{and} \quad s^2 = \frac{1}{n-1}\left(\sum_{i=1}^{n}x_i^2 - n\bar{x}^2\right) \tag{4.4}$$

$$\sum_{i=1}^{N}(x_i - \mu)^2 = \sum_{i=1}^{N}x_i^2 - N\mu^2 \quad \text{and} \quad \sigma^2 = \left(\frac{1}{N}\sum_{i=1}^{N}x_i^2\right) - \mu^2 \tag{4.5}$$

Notice that the **population** variance is equal to the mean of the squares minus the square of the mean.

Example 4.7

On 1 May 2004, ten countries joined the EU; they are referred to as the EU10. The question arises, how the mean, variance and standard deviation of the GNIpcs of the EU10 compare with the mean, variance and standard deviation of the 15 countries (EU15) that constituted the EU before 1 May 2004. The GNIpc data for the EU10 and EU15 are recorded in Table 4.5; see also the files Xmp04-07A.xls and Xmp04-07B.xls. Below, the mean, variance and standard deviation will be calculated for both sets of countries. To illustrate the above short-cut formulae, a table and a pocket calculator will be used for the EU10 data. For the EU15 data, a computer package can be used to obtain the values of the formulae.

The first question that comes up is: do we have samples or populations? Since the purpose is to compare statistics of two sets of countries, it is natural to consider them as populations.

As before, the variable X measures GNIpc in units of 1000 euro. The EU10 dataset is used to illustrate the formula in (4.5). That is why the x_i and their squares are put into columns in Table 4.5 (a).

First, Table 4.5(a). The second column contains the GNIpcs of the ten new EU countries (in units of €1000). Notice that $\sum_{i=1}^{10}x_i = 71.9$, so $\mu = 7.19$. That is, the mean of the GNIpcs is €7190. The third column contains the squares of the observations of each country; their sum is $\sum_{i=1}^{10}x_i^2 = 635.67$. The variance (say σ_{10}^2 to denote that it is about EU10) can easily be calculated using formula (4.5):

(a)

Country EU10	x_i	x_i^2
Cyprus	14.1	198.81
Czech Republic	7.2	51.84
Estonia	4.6	21.16
Hungary	6.5	42.25
Latvia	3.8	14.44
Lithuania	4.0	16.00
Malta	10.0	100.00
Poland	5.1	26.01
Slovakia	4.6	21.16
Slovenia	12.0	144.00
Total	71.9	635.67

(b)

Country EU15	GNIpc ($\times$ €1000)
Austria	26.6
Belgium	25.7
Denmark	33.6
Finland	26.8
France	25.5
Germany	25.6
Greece	13.4
Ireland	27.0
Italy	21.5
Luxembourg	44.9
Netherlands	27.0
Portugal	12.6
Spain	17.1
Sweden	28.3
UK	28.2

TABLE 4.5 GNIpc ($\times$ €1000) for the EU10 and EU15

$$\sigma_{10}^2 = \frac{1}{10}(635.67 - 10 \times (7.19)^2) = \frac{1}{10} \times 118.7090 = 11.8709$$

$$\sigma_{10} = \sqrt{11.8709} = 3.4454$$

That is, the GNIpcs of the ten new EU countries have mean 7190 euro, with variance 11 870 900 euro^2 and standard deviation 3445.40 euro.

A computer is used to get similar results for the EU15 data of Table 4.5(b). Take care that the **population** variance and **population** standard deviations are returned, that is, use the correct commands. Table 4.6 gives an overview.

Statistic	EU10	EU15
Mean μ	7.19	25.59
Variance σ^2	11.87	57.59
Standard deviation σ	3.45	7.59
Coefficient of variation $\sigma/|\mu|$	0.48	0.30

TABLE 4.6 Overview of statistics for the EU10 and EU15

The mean GNIpc in the EU15 is more than three times the mean GNIpc in the EU10. On the other hand, the standard deviation of the GNIpcs in the EU10 is less than half the standard deviation of the GNIpcs in the EU15. If prosperity of a country is measured by way of its GNIpc, these comparisons seem to indicate that, in the EU15, prosperity varies more than in the EU10. Put another way: in the EU10 the degree of prosperity seems to be more stable but on a much lower level. However, the coefficients of variation indicate that for the EU10 countries the standard deviation is a larger part of the mean than for the EU15 countries.

Recall that the formulae for the variance are based on the squares of the deviations. Since large positive and small negative deviations $x_i - \mu$ (or $x_i - \bar{x}$) grow rapidly when squares are taken, such deviations have a large influence on the value of the variance. This makes the variance a measure of variation that is very unstable by comparison with the interquartile range. The same holds, although less strongly, for standard deviation and coefficient of variation.

CASE 4.1 ERICSSON SHARES VERSUS CARLSBERG SHARES – SOLUTION

There are many ways to measure the risk of a share; most often the variance or the standard deviation of its historical returns is used. The table below gives an overview of some statistics regarding the daily returns of the Ericsson and Carlsberg shares.

	Ericsson	Carlsberg
Mean	−0.0171	0.0375
Maximum	19.0476	14.9068
Minimum	−25.9804	−14.1434
Variance	12.2130	3.3545
Standard deviation	3.4947	1.8315

Some statistics for the shares of Ericsson and Carlsberg, 15 February 2001 – 30 June 2007
Source: Copenhagen Stock Exchange (2007)

When comparing the two shares on the basis of their mean daily returns, Carlsberg scores much better: with (about) 300 business days in a year, the results of this share correspond to an annual return of $300 \times 0.0375 = 11.25\%$. If the investor is risk averse, it is obvious that he will choose the Carlsberg share to invest his money in. The reason is that this share has not only the larger historical mean daily return but also the smaller accompanying standard deviation. As the investor is risk averse, this relatively low standard deviation will convince him that he will run less risk by investing his money in the Carlsberg share.

However, if the investor likes to take risks, he might choose to invest his money in the Ericsson share. This share has the higher maximum daily return (more than 19%) and offers special opportunities because of its large standard deviation. If the investor knows the market of Ericsson very well, the high maximum and large standard deviation might be very attractive to him.

Most investors are risk averse and will choose the share Carlsberg. This will probably also be the case for this investor, who wants to choose between the two shares.

4.3 Interpretation of the standard deviation

Starting with the observations in the dataset, the mean and standard deviation can be calculated. Below, things will be considered the other way round: what do the mean and the standard deviation tell us about the observations in the dataset?

There is a famous result that is named after its discoverer: Chebyshev's theorem. One special instance of this theorem states that, in the case of a population dataset, the interval bounded by $\mu - 2\sigma$ and $\mu + 2\sigma$ contains **at least** three-quarters of the observations. This result is always valid, irrespective of the dataset one is considering. For instance, for the GNIpc dataset of the EU15 countries it holds that:

$$\mu - 2\sigma = 25.5867 - 2 \times 7.5887 = 10.4094$$

$$\mu + 2\sigma = 25.5867 + 2 \times 7.5887 = 40.7640$$

So, according to Chebyshev's rule, at least 75% of the data points (that is, at least 12 out of 15) have to fall in the interval (10.4094, 40.7640). Studying the data reveals that 14 observations fall in it.

   Chebyshev's theorem applies to the general case as follows. Below, k is some real number that is at least 1.

Chebyshev's theorem (population version)

The interval $(\mu - k\sigma, \mu + k\sigma)$ contains at least the proportion $1 - 1/k^2$ of the observations in the population dataset.

Chebyshev's theorem (sample version)

The interval $(\bar{x} - ks, \bar{x} + ks)$ contains at least the proportion $1 - 1/k^2$ of the observations in the sample dataset.

For $k = 2.5$, it follows that at least the proportion $1 - 1/(2.5)^2 = 0.84$ of the data falls in the interval $(\mu - 2.5\sigma, \mu + 2.5\sigma)$, respectively $(\bar{x} - 2.5s, \bar{x} + 2.5s)$.

   If the frequency distribution of the data is symmetric and bell-shaped, the percentage of the data captured by the interval $(\mu - 2\sigma, \mu + 2\sigma)$ will be larger than the minimum value of Chebyshev's theorem. It turns out that then approximately 95% of the data fall in it, instead of the 'at least 75%' that is guaranteed by Chebyshev's rule. The rules in such a case of a symmetrical distribution are called *empirical rules*. For the population case, both types of rules are summarized in Table 4.7.

k	Interval	Chebyshev	Empirical
1	$(\mu - \sigma, \mu + \sigma)$	At least 0%	$\approx 68\%$
1.5	$(\mu - 1.5\sigma, \mu + 1.5\sigma)$	At least 55.56%	$\approx 87\%$
2	$(\mu - 2\sigma, \mu + 2\sigma)$	At least 75%	$\approx 95\%$
2.5	$(\mu - 2.5\sigma, \mu + 2.5\sigma)$	At least 84%	$\approx 99\%$
3	$(\mu - 3\sigma, \mu + 3\sigma)$	At least 88.89%	$\approx 99.5\%$

TABLE 4.7 Chebyshev's rules and empirical rules

Example 4.8

Reconsider the Final Exam grades of the course Statistics 2; see the file Xmp02-06.xls for the data. With a computer it follows easily that:

$$\mu = 5.3841 \quad \text{and} \quad \sigma = 2.5026$$

With these results, Table 4.7 can be filled in; see Table 4.8. The last column contains the percentages of the grades that actually fall in the intervals. They follow by sorting the data and counting the observations that fall in the intervals.

   If $k = 1.5$, the interval (1.6302, 9.1380) contains at least the proportion $1 - 1/(1.5)^2 = 0.5556$ of the observations. If the frequency distribution of the data were bell-shaped, the empirical rule would ensure that about 87% of the observations fall in this interval. However, from (for instance) the histogram in Example 2.6 it follows that the frequency distribution that belongs to classification I is not completely symmetric. Sorting the grades and counting the observations in the interval reveals that 218 out of 258 observations fall in the interval, that is, 84.5%.

k	Interval	Chebyshev	Empirical	Reality
1	(2.8815, 7.8867)	At least 0%	≈ 68%	65.5%
1.5	(1.6302, 9.1380)	At least 55.56%	≈ 87%	84.5%
2	(0.3789, 10]	At least 75%	≈ 95%	98.06%
2.5	[0, 10]	At least 84%	≈ 99%	100%
3	[0, 10]	At least 88.89%	≈ 99.5%	100%

TABLE 4.8 Chebyshev's rules and empirical rules

As pointed out in Section 4.1, outliers are observations that are extremely large or extremely small when compared with most other observations in the dataset. In Tukey's 1.5δ-definition an observation is designated an outlier if it is smaller than $\kappa_1 - 1.5\delta$ or larger than $\kappa_3 + 1.5\delta$.

Other studies use the mean and standard deviation to find outliers. They qualify an observation as an outlier if it is smaller than $\mu - 3\sigma$ or larger than $\mu + 3\sigma$ (smaller than $\bar{x} - 3s$ or larger than $\bar{x} + 3s$ in case of a sample dataset). This definition of an outlier is called the **3σ-definition**. For the 25 GNIs of EU25 in the second column of Xmp04-01.xls, it follows easily that:

$$\mu - 3\sigma = -1404.4628 \quad \text{and} \quad \mu + 3\sigma = 2170.3828$$

Since 2108.8 (Germany) is the largest observation in the dataset, there are no outliers in the sense of the 3σ-definition. Recall that the 1.5δ-definition yielded five outliers; see Example 4.1.

Note: In the case of a skewed distribution (as with the GNI data of the EU25), extreme observations in the long tail push the mean in the direction of that tail, making it harder for an observation to become an outlier in the sense of the 3σ-definition. This effect is strengthened since the standard deviation is also influenced heavily by very large or very small observations. However, the first and third quartiles are **not** influenced by extreme observations; the same holds for the IQR. That is why the 1.5δ-definition is more suitable for uncovering extreme observations than is the 3σ-definition.

4.4 z-Scores

The standard deviation of a dataset is also used to describe the relative position of values when compared with that dataset.

Consider a dataset of observations of a variable X, with mean $\bar{x}$ (or μ) and standard deviation s (or σ). For some value x_0 (not necessarily belonging to the dataset) of X, its **z-score** is calculated by subtracting the dataset's mean and dividing the result by the standard deviation. Since we have to distinguish between sample data and population data, there are two analogous formulae:

Sample z-score and population z-score

The **sample z-score** of x_0 with respect to a sample dataset with mean $\bar{x}$ and standard deviation s is defined as:

$$z_0 = \frac{x_0 - \bar{x}}{s}$$

The **population z-score** of x_0 with respect to a population dataset with mean μ and standard deviation σ is defined as:

$$z_0 = \frac{x_0 - \mu}{\sigma}$$

For the sample dataset {2, 3, 8, 4, 7, 5, 6} the mean is 5 and the standard deviation is 2.1602. The (new) observation $x_0 = 5.5$ has z-score $z_0 = (5.5 - 5)/2.1602 = 0.2315$. That is, the relative position of 5.5 is $0.2315 \times s$ units to the **right** of the mean of the dataset. The observation $x_0 = 4.3$ has z-score $z_0 = (4.3 - 5)/2.1602 = -0.3240$; with respect to the dataset it lies $0.3240 \times s$ units to the **left** of the mean.

For an arbitrary observation, the z-score measures the number of standard deviations that the observation is separated from the mean of the dataset. If this number is positive, the observation lies to the right of the mean. If it is negative, it lies to the left of the mean.

Example 4.9

In the steel industry, the mean monthly salary for workers is €2000 with a standard deviation of €200. Steelworker Clinton is always complaining about his salary. Although the z-score of his salary is 1.7, he is not satisfied at all. He claims that more than 40% of the steelworkers have a salary at least as high as his, although in his opinion this is unfair because of the many years of experience he has. Is it possible that he is right?

The positive z-score 1.7 corresponds to a salary of €2340. By Chebyshev's rule with $k = 1.7$ it follows that at least 65.4% of the steelworkers have salaries in the interval (1660, 2340). That is, **at most** 34.6% of the salaries fall outside this interval. In the most 'unfavourable case', 34.6% of the salaries are $\geq$ €2340. This contradicts Clinton's claim.

4.5 The variance of 0-1 data

Recall that a 0-1 variable, with 1 referring to 'success', is intermediate between qualitative and quantitative. In Section 3.3.1 it was shown that it makes sense to consider the mean of observations of such a variable and that this mean is just the proportion of successes among the observations. The question arises whether the variance of a set of observations of a 0-1 variable also makes sense and how it is related to the proportion of successes. This is the topic of the present section.

Let X be a 0-1 variable, where the value 1 corresponds to success and 0 to failure. Interest is in the proportion of ones in a dataset. For instance, if the dataset {0, 1, 1, 0, 0, 0} is a population dataset, we write $p = 1/3$; if it is a sample dataset, the proportion is denoted by $\hat{p}$ and we write $\hat{p} = 1/3$.

First suppose that the dataset of observations of X is a population dataset; let p denote the population proportion of successes and $1 - p$ the population proportion of failures. Even if this dataset is not observed, it can be denoted by $x_1, \ldots, x_N$ (a sequence of zeros and ones) and its mean by μ. Recall from Section 3.3.1 that $\mu = p$. As in formula (4.1) it **is** possible to consider the average of the squared deviations with respect to the mean μ; call it σ^2, the **variance** of the 0-1 variable. Furthermore, equality (4.5) remains valid. Also notice that the operation 'take square' does not affect the values 0 and 1, since $0^2 = 0$ and $1^2 = 1$. That is why all observations x_i in the dataset of X satisfy $x_i^2 = x_i$. This will be used in the second equality below, to express σ^2 in terms of p. By (4.5) it follows that:

$$\sum_{i=1}^{N}(x_i - \mu)^2 = \sum_{i=1}^{N}x_i^2 - N\mu^2 = \sum_{i=1}^{N}x_i - N\mu^2 = N\mu - N\mu^2$$
$$= Np - Np^2 = Np(1 - p)$$

To obtain σ^2, the above results have to be divided by N. It follows that the variance of a 0-1 variable and the population dataset is equal to the product of the proportion of successes and the proportion of failures.

Next suppose that the dataset is a sample dataset of observations of the 0-1 variable X. As before, denote this dataset by $x_1, \ldots, x_n$, a sequence of zeros and ones. We follow the international notation and denote the sample proportion of successes (ones) by $\hat{p}$, so the sample proportion of failures is $1 - \hat{p}$. In Section 3.3.1 it was shown that the mean $\bar{x}$ of the sample is equal to the sample proportion of successes, so $\bar{x} = \hat{p}$. It makes sense to consider the sum of the squares of the deviations with respect to the sample mean $\bar{x}$ (see 4.3). Since $x_i^2 = x_i$ and since the short-cut formula (4.4) remains valid, it follows that:

$$\sum_{i=1}^{n}(x_i - \bar{x})^2 = \sum_{i=1}^{n}x_i^2 - n\bar{x}^2 = \sum_{i=1}^{n}x_i - n\bar{x}^2 = n\bar{x} - n\bar{x}^2$$
$$= n\hat{p} - n\hat{p}^2 = n\hat{p}(1 - \hat{p})$$

To obtain the sample variance s^2, the above result has to be divided by $n - 1$.

Properties of a 0-1 variable and its datasets

The population variance σ^2 and the population standard deviation σ of a 0-1 variable can be expressed in the population proportion p of successes:

$$\sigma^2 = p(1 - p) \quad \text{and} \quad \sigma = \sqrt{p(1 - p)}$$

(4.6)

The sample variance s^2 and the sample standard deviation s of a sample dataset of a 0-1 variable can be expressed in the sample proportion $\hat{p}$ of successes:

$$s^2 = \frac{n}{n - 1}\hat{p}(1 - \hat{p}) \quad \text{and} \quad s = \sqrt{\frac{n}{n - 1}\hat{p}(1 - \hat{p})}$$

(4.7)

For large n, the factor $n/(n - 1)$ approaches 1 and its influence on the formulae in (4.7) is negligible.

Example 4.10

Reconsider the 0-1 variable Y that measures for the year 2001 whether the wife in a husband–wife couple works outdoors (1) or not (0). The file Xmp03-05.xls records 10000 observations of this variable. In Example 3.5 it was shown that the mean of this sample dataset equals 0.57, so the sample proportion of the 10000 husband-wife couples where the woman works outdoors is 0.57. From the above theory it follows that the sample variance and standard deviation are equal to:

$$s^2 = \frac{10000}{9999} \times 0.57 \times 0.43 = 0.2451 \quad \text{and} \quad s = \sqrt{0.2451} = 0.4951$$

Using a computer, the sample variance can also be calculated directly. Notice that the factor $n/(n - 1)$ equals 1.0001, so it hardly influences the product 0.57×0.43.

4.6 The variance of a frequency distribution

Suppose that – as often occurs in newspapers and articles – the dataset of observations of a discrete variable is not given, but only a summary in terms of a frequency distribution. That is, only an overview is given of the **different values** $x_{(j)}$ in the dataset jointly with their frequencies f_j. The question then is how the mean and the variance can be calculated. Regarding the mean, the answer has already been given in Sections 3.3.2 and 3.3.3. Formally, the products $fx_{(j)}$ have to be calculated and added up, and the result has to be divided by N or n.

It will now be demonstrated how the **variance** can be calculated from a frequency distribution of a population dataset of a discrete variable. Formula (4.1) prescribes how to find σ^2 when the data are given: for the observations x_i, the squared deviations $(x_i - \mu)^2$ have to be summed and the sum has to be divided by N. But some of these observations x_i have the same value, so the same holds for the corresponding squared deviations $(x_i - \mu)^2$. Hence, for each different value $x_{(j)}$, the squared deviation has to be calculated and the result has to be multiplied by the frequency f_j. All products have to be summed and divided by N.

Table 4.9 below shows all the constituent parts. The products in the third column are needed to find μ, the products in the last column are needed for σ^2.

Value $x_{(j)}$	Freq. f_j	Freq. × Value $fx_{(j)}$	Squared dev. $(x_{(j)} - \mu)^2$	Freq. × Squared dev. $f_j(x_{(j)} - \mu)^2$

TABLE 4.9 Notations and meanings for a population dataset

Formulae for μ and σ^2 for a discrete frequency distribution

$$\mu = \frac{1}{N}\sum f_j x_{(j)} \quad \text{and} \quad \sigma^2 = \frac{1}{N}\sum f_j(x_{(j)} - \mu)^2 \tag{4.8}$$

Notice that $\sum f_j = N$, so μ and σ^2 are both weighted means (with weights f_j) of respectively $x_{(j)}$ and the squared deviations $(x_{(j)} - \mu)^2$; see also Section 3.3.1.

For a sample dataset, similar formulae can be given for $\bar{x}$ and s^2: replace μ by $\bar{x}$, and N by n in the left-hand formula and by $n - 1$ in the right-hand formula.

Example 4.11

Recall the frequency distribution of the sizes of the households in Denmark; see Example 3.8.

Size of household $x_{(j)}$	Frequency f_j	Freq. × size. $f_j x_{(j)}$	Squared dev. $(x_{(j)} - \mu)^2$	Freq. × Squared dev. $f_j(x_{(j)} - \mu)^2$
1	965 653	965 653	1.3129	1 267 771.56
2	832 087	1 664 174	0.0213	17 688.93
3	293 208	879 624	0.7297	213 939.95
4	290 103	1 160 412	3.4380	997 387.61
5	100 049	500 245	8.1464	815 043.23
6	23 969	143 814	14.8548	356 055.53
7	6 523	45 661	23.5632	153 702.94
8	5 090	40 720	34.2716	174 442.56
Total	2 516 682	5 400 303	–	3 996 032.31

$$\mu = \frac{5\,400\,303}{2\,516\,682} = 2.1458 \approx 2.15$$

$$\sigma^2 = \frac{3\,996\,032.31}{2\,516\,682} = 1.5878 \quad \text{and} \quad \sigma = 1.2601$$

TABLE 4.10 Frequency distribution of X = 'size of household in Denmark'

The calculation of the mean size of the households is repeated from the earlier example in the second last row of Table 4.10: the total of column 3 is divided by the total of column 2. To obtain the sum of all squared deviations, the numbers in the last column have to be added up. To obtain σ^2, the result has to be divided by N (the total of column 2); see the last row of the table.

Now suppose that only a **classified** frequency distribution of a dataset is given; this is often the case for a continuous variable. Since under these circumstances the values of the data points are lost, the precise determination of mean and variance is impossible. However, by assuming that all observations per class are evenly distributed over that class, the mean and variance of the underlying dataset can be approximated. This comes down to assuming that the average of all observations in a class is located at the centre of the class. The approximations of μ and σ^2 can be obtained with the formulae of (4.8), by taking the $x_{(j)}$ as the centres of the classes.

Example 4.12

Suppose that classification I, jointly with the corresponding frequencies, is all there is about the 258 grades for the Final Exam of the Statistics 2 course. In Example 3.9 it was found that 5.3217 is the mean of the classified frequency distribution; it can be considered as an approximation of the mean grade of the underlying dataset. To obtain an approximation of the variance of the underlying dataset, it is again assumed that for each class the observations are evenly distributed over that class and consequently that, on average, they fall in its centre. The resulting frequency distribution is set out in Table 4.11.

j	1	2	3	4	5	6	7	8	9	10
Centre $x_{(j)}$ of class j	0.5	1.5	2.5	3.5	4.5	5.5	6.5	7.5	8.5	9.5
Frequency f_j	21	9	21	24	37	30	42	37	21	16

TABLE 4.11 Frequency distribution of classification I with centres

The sum of the squared deviations of the **original** observations in the dataset, can be approximated by:

$$(0.5-5.3217)^2 \times 21 + (1.5-5.3217)^2 \times 9 + \ldots + (9.5-5.3217)^2 \times 16 = 1617.79845$$

As a consequence, the variance and standard deviation of this classified frequency distribution are, respectively:

$$\frac{1617.79845}{258} = 6.2705 \quad \text{and} \quad 2.5041$$

They can be considered as approximations of the variance σ^2 and the standard deviation σ of the underlying dataset. (Incidentally, the underlying dataset is Xmp02-06.xls, and $\sigma^2 = 6.2632$ and $\sigma = 2.5026$.)

Summary

Measures of variation, such as interquartile range (IQR) and standard deviation, quantify the 'degree of variability' of a dataset. They express the degree of variability of the data as numbers.

The IQR is the difference between the 'upper' and the 'lower' quartiles and measures the range of the 'middle' 50% of the ordered data. Computer packages can be used to calculate these quartiles, but often the packages do not agree about their precise values. When the data are summarized in a frequency distribution, the distribution function can be used to calculate quartiles and interquartile range.

The variance and the standard deviation are the most important measures of variation. The box below sets out the formulae for mean and variance.

Extremely large and extremely small observations in a dataset are called outliers. Two precise definitions of 'outlier' were considered, the 1.5δ-definition and the 3σ-definition. According to Chebyshev's inequality, only relatively few observations can lie very far from the mean.

		Mean	**Variance**
Original observations	Population	$\mu = \frac{1}{N}\sum_{i=1}^{N} x_i$	$\sigma^2 = \frac{1}{N}\sum_{i=1}^{N}(x_i - \mu)^2$
	Sample	$\bar{x} = \frac{1}{n}\sum_{i=1}^{n} x_i$	$s^2 = \frac{1}{n-1}\sum_{i=1}^{n}(x_i - \bar{x})^2$
Frequency distribution	Population	$\mu = \frac{1}{N}\sum f x_{(i)}$	$\sigma^2 = \frac{1}{N}\sum f_i(x_{(i)} - \mu)^2$
	Sample	$\bar{x} = \frac{1}{n}\sum f x_{(i)}$	$s^2 = \frac{1}{n-1}\sum f_i(x_{(i)} - \bar{x})^2$

Formulae for μ, $\bar{x}$ and σ^2, s^2

List of symbols

Symbol (population, sample)	Statistic
κ_i, k_i	ith quartile ($i = 1, 2, 3$)
δ, d	Interquartile range $\kappa_3 - \kappa_1$, $k_3 - k_1$
σ^2, s^2	Variance
σ, s	Standard deviation

🔑 Key terms

box plot **91**
Chebyshev's theorem **104**
coefficient of variation **100**
deviation **96**
empirical rules **104**
fluctuation **87**
interquartile range **90**
mean absolute deviation (MAD) **97**
measures of variation **88**
mth percentile **91**
m% percentile **91**

outlier **90**
(population) standard deviation **98**
(population) variance **97**
population z-score **105**
quartile **89**
quartile of a variable **89**
relative interquartile range **90**
sample standard deviation **98**
sample variance **98**

sample z-score **105**
short-cut formula **101**
show variation **87**
spread **87**
Tukey's hinges **92**
variability **87**
variation **87**
whiskers **92**
z-score **105**
1.5σ-definition **90**
3σ-definition **105**

Exercises

Exercise 4.1

Use formula (4.3) to calculate the variance and the standard deviation of the following sample dataset:

   3 8 6 4 7 6

Exercise 4.2

Use formulae (4.1) and (4.2) to calculate the variance and the standard deviation of the following population dataset:

   12 18 15 19 21 15 18 25

Exercise 4.3

Use a short-cut formula (see (4.4) and (4.5)) to calculate the variance of the following sample dataset:

   8 15 22 4 −6 13 12 16 −9

Exercise 4.4

A friend contacts you and states that the standard deviation of the daily return of a certain share is −0.07%. What will be your response?

Exercise 4.5

Create a sample of five numbers with the following statistics:

  a Mean 4 and standard deviation 0.
  b Mean 4 and standard deviation 1.

Exercise 4.6

A large dataset has mean 20 and standard deviation 2.

  a What can be said about the percentage of data that deviate from the mean by less than 5?
  b What can be said about the percentage of the data that are at least 25?
  c Answer the questions in parts (a) and (b) again if it is now given that the frequency distribution of the dataset is bell shaped.

Exercise 4.7

Consider the following frequency distribution of a sample dataset of observations of a discrete variable X:

Value	Frequency
2	12
4	18
6	13
8	7
Total	50

a Calculate the mean, standard deviation and coefficient of variation.

b Use the accompanying distribution function F to calculate the interquartile range and the relative interquartile range.

Exercise 4.8

Consider the following classified frequency distribution of a population dataset of observations of a continuous variable X:

Class	Frequency
(0, 2]	12
(2, 4]	18
(4, 6]	13
(6, 8]	7
Total	50

a Calculate the mean, standard deviation and coefficient of variation.

b Use the accompanying distribution function F to calculate the interquartile range and the relative interquartile range.

Exercise 4.9

a The value 20 has z-score 2 with respect to a dataset with mean 12. Interpret this z-score and calculate the standard deviation of the dataset.

b A sample dataset of observations of a 0-1 variable contains 50 ones and 150 zeros. Calculate the mean and the standard deviation of that dataset.

Exercise 4.10

In the health insurance market, there are five main insurance companies. Their sizes can be measured by way of the variables X = 'number of insured ($\times$ 10^6)' and Y = 'number of employees ($\times$ 1000)'. The results are shown in the table.

Company	Number of insured ($\times$ 10^6)	Number of employees ($\times$ 1000)
1	4.2	3.7
2	3.5	3.2
3	2.6	2.3
4	1.9	2.6
5	1.2	1.5

a Calculate the variances and the coefficients of variation of X and Y.

b Companies 5 and 2 decide to merge. Calculate the variances and the coefficients of variation for this new situation.

c Compare the spread of the sizes of the main insurance companies before the merger with the spread after the merger.

Exercise 4.11 (computer)

In the WO-monitor 2004/2005 of VSNU (Association of Universities in the Netherlands), a total of 12 855 recently graduated masters students were asked the following question about their study:

Did you (yes/no) try to get highest grades possible?

The summarized results (percentages) are set out in the table.

Country	Percentage yes
Austria	54
Finland	44
France	62
Germany	68
Italy	63
Netherlands	34
Norway	65
Spain	71
Switzerland	58
UK	64

a Which country has the highest percentage? Which has lowest?

b Use a computer package to calculate the median and the IQR of the observations of X = 'percentage yes per country'.

c Use a computer package to calculate the mean and the standard deviation of X.

d Are there any outliers? Use the 1.5δ-definition as well as the 3σ-definition to answer this question.

Exercise 4.12

WO-monitor 2004/2005 of VSNU (2007) also gives the results of an enquiry carried out in ten European countries about the length of time it took newly graduated masters students to find their first job. Note that a row of the table gives a classified relative frequency distribution (percentages) of the variable X = 'search duration for job (in months)' for the country involved. Suppose that the classes 1–3 and 4–6 essentially refer to the intervals [1, 4) and [4, 7).

Country	Less than 1 month	1–3 months	4–6 months	More than 6 months	Total
Austria	50	35	8	7	100%
Finland	56	30	8	6	100%
France	52	28	10	10	100%
Germany	65	22	7	6	100%
Italy	38	35	14	13	100%
Netherlands	55	34	8	3	100%
Norway	58	30	7	5	100%
Spain	34	31	14	21	100%
Switzerland	46	41	8	5	100%
UK	46	38	8	8	100%

a Calculate the mean, variance and standard deviation of the frequency distribution of Norway. (For the last – open – class, take 9 as centre.)

b Use the accompanying distribution function to determine – again for Norway – the median, the first and third quartiles and the IQR.

c The table below gives an overview of the means, variances and standard deviations of the other nine countries. Compare the ten countries on the basis of their mean search duration and their spread of the search duration.

Country	Mean	Variance	Standard deviation
Austria	2.1950	5.5845	2.3631
Finland	2.0100	5.2549	2.2924
France	2.4100	7.1969	2.6827
Germany	1.8000	5.2750	2.2967
Italy	3.0050	8.0175	2.8315
Netherlands	1.8350	3.7453	1.9353
Norway			
Spain	3.6050	10.2715	3.2049
Switzerland	2.1450	4.5465	2.1322
UK	2.3400	5.9144	2.4320

Exercise 4.13 (computer)

The instructor of the course Statistics 2 for Economics was in June 2006 very dissatisfied about the grades and the study attitude of his students. To improve things, he took the following actions for the 2007 version of the course:

■ Making use of another – more challenging – textbook.

■ Maintaining the Team Assignment and the Final Exam, but offering two midterms instead of one.

■ A threshold of 5 for the **final** grade, in the following sense: a sufficient final grade (≥ 6) is decreased to 5 if the grade for the Final Exam is less than 5.

In June 2007, as soon as the final grades were available, the instructor compared the final grades of 2007 with those of 2006 to see whether the alterations were successful. See Xrc04-13.xls and Xrc04-13.sav.

a Calculate, for both years, the percentage of the sufficient grades.

b Calculate for both years: the mean, the standard deviation, the three quartiles and the IQR.

c Create box plots in a single figure; see Appendix A1.4 if necessary.

d Create, for each of the two years, a suitable histogram of the relative frequency distribution. (Note: The examination rules demand that the final grade 5.5 is not assigned; not yet rounded final grades in the intervals (5, 5.5) and [5.5, 6) are respectively rounded to 5 and 6.)

e Compare the above summaries of the 2006 scores and the 2007 scores, and give your comments.

Exercise 4.14

An enthusiastic Canadian Smart-rider (Smart-cdi cabrio, 2005) logged the diesel consumption (in litres per 100 km) of her car each time just before she refilled the tank. Moreover, she recorded whether she (yes/no) filled up with 'VP-power diesel' and on what types of road she had driven since the last fill-up:

1 = city / country roads
2 = motorway / city / country roads
3 = motorway / city

4 = country roads
5 = city

The data for the period 14 January 2005 – 24 June 2007 are in the files Xrc04-14.xls and Xrc04-14.sav. (See also www.spritmonitor.de.) The sample statistics in the table below summarize the data.

| | Total | Yes/no VP-power | | Types of road | | | | |
		1	0	1	2	3	4	5
Nr of observations	159	144	15	144	2	5	5	3
Mean	3.870	3.987	3.858	3.851	3.910	4.146	3.854	4.363
Variance	0.098	0.087	0.099	0.095	0.016	0.017	0.166	0.015
Standard dev.	0.314	0.295	0.314	0.308	0.127	0.129	0.408	0.123
Median	3.850	4.000	3.845	3.845	3.910	4.090	3.830	4.330
IQR	0.440	0.590	0.460	0.440	–	0.240	0.610	–

a Give your comments on the consequences of using VP-power.

b Give your comments on the influences of the types of road.

c (advanced) Use the above results to calculate the mean and the variance after having combined classes 3 and 5 of types of road.

Exercise 4.15

Datasets of continuous variables *A*, *B*, *C* and *D* are summarized by way of the histograms shown below.

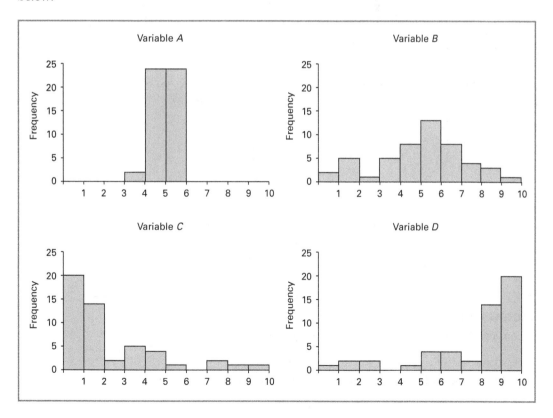

Also some statistics are calculated for the datasets. Each of the rows in the table below belongs to one of the variables *A, B, C* and *D*. Which histogram belongs to which row of statistics? Why?

Dataset	Mean	Standard deviation	Range (max − min)	Nr of observations
1	2.17	2.29	9.72	50
2	4.91	0.69	2.75	50
3	7.74	2.46	9.31	50
4	5.14	2.19	9.66	50

Exercise 4.16

Reconsider the histograms in Exercise 4.15. Each of the box plots below belongs to one of the histograms? Which? Explain why.

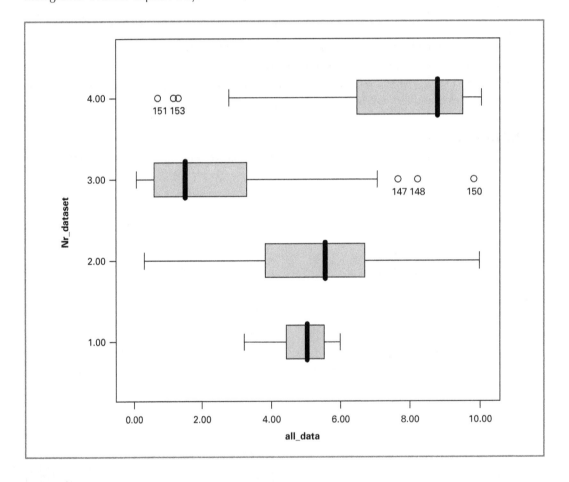

Exercise 4.17

Reconsider the five sample observations of Exercise 3.1:

$$x_1 = 3, \quad x_2 = 5, \quad x_3 = 3, \quad x_4 = 7, \quad x_5 = 4$$

a Use a table, as in Example 4.6, to calculate the variance and the standard deviation of this dataset.

b Calculate the data $y_i = 3x_i - 4$ for $i = 1, 2, 3, 4, 5$. Calculate the variance and the standard deviation of the y dataset. Compare your answers with the answers of part (a).

c Calculate the data $w_i = -2x_i + 5$ for $i = 1, 2, 3, 4, 5$. Calculate the variance and the standard deviation of the w dataset. Compare your answers with those of part (a).

Exercise 4.18 (computer)

In Exercise 3.21 the variables X = 'number of PCs per 1000 people' and Y = 'number of people per country' were considered for the EU10; see the table below.

Country	x	y
Cyprus	269.89	764 967
Czech Republic	177.44	10 200 000
Estonia	210.33	1 358 000
Hungary	108.35	10 200 000
Latvia	171.75	2 338 000
Lithuania	109.75	3 469 000
Malta	255.05	397 000
Poland	105.65	38 200 000
Slovakia	180.36	5 379 000
Slovenia	300.60	1 994 000

Observations of X and Y for the EU10, 2002
Source: World Bank (2005)

It was calculated that $\mu_X = 188.917$. See also Xrc04-18.xls.

a Consider the observations of X in column B. Create all deviations $x_i - \mu_X$ in column D and their squares in column E. Use the squares in column E to calculate the variance σ_X^2 and the standard deviation σ_X with formulae (4.1) and (4.2).

b Calculate σ_X^2 and σ_X directly with a computer package; compare with part (a).

c Also determine the three quartiles, the IQR and the relative IQR.

d Determine the number of the x observations that according to Chebyshev's rule and the empirical rule fall in the interval $(\mu_X - 1.5\sigma_X, \mu_X + 1.5\sigma_X)$. Also calculate the actual number that fall in it.

e For the y observations, use both the 1.5δ- and the 3σ-definition to search for outliers.

Exercise 4.19 (computer)

In Exercise 3.18 a study was started up to investigate whether GDP growth in the UK is on average largest in the first quarter of each year. The study is based on data from the period 1994Q1–2003Q3, for which it was found that $\bar{x}_1 = 1.501\%$, $\bar{x}_2 = 1.276\%$, $\bar{x}_3 = 1.241\%$ and $\bar{x}_4 = 1.337\%$. See the file Xrc04-19.xls; the data are repeated below. Consider these observations as a **sample**.

	Quarter			
Year	**1**	**2**	**3**	**4**
1994	1.3700	1.3800	1.0000	2.0500
1995	1.0100	1.4100	1.1800	1.6300
1996	1.6200	1.7600	1.3300	0.9000
1997	2.3200	1.4300	0.9100	1.9200
1998	1.2300	1.4500	2.1300	0.8600
1999	0.8600	1.3600	1.6400	1.2100
2000	1.9200	0.5500	1.2000	0.6300
2001	1.8900	1.0900	0.2700	1.5100
2002	1.4000	1.0800	1.4900	1.3200
2003	1.3900	1.2500	1.2600	–

GDP growth (%) in the UK, 1994Q1–2003Q3

a To learn the mechanics of the short-cut formula, we will use it in combination with a computer to calculate the variance and the standard deviation of the observations for the first quarter of the years.

In column F, create the squares of the observations of column B. Use these squares to determine the variance and the standard deviation of the Q1 observations.

b For each of the four quarters, determine the interquartile range with a computer package.

c The variances for the second, third and fourth quarters are, respectively, 0.1028, 0.2373 and 0.2374; the respective standard deviations are 0.3206, 0.4871 and 0.4872. Which quarter has the lowest variation?

d Do you believe that the above considerations justify the statement: 'on average, GDP growth in the UK is largest in the first quarter'?

Exercise 4.20

Reconsider the four relative frequency distributions of the variable X = 'number of cars per household' (values 0, 1, 2, 3); see also Exercise 3.14.

	1995	2000	2002	2003
Households with three cars	1.0	1.7	2.1	2.1
Households with two cars	13.3	17.1	18.9	19.0
Households with one car	59.5	56.0	55.6	55.9
Households with no car	26.2	25.2	23.4	22.9
Total	100	100	100	100

Car ownership per household (as % of total number of households)

a For each of the four years, use the distribution function to calculate the interquartile range of the corresponding relative frequency distribution.

b For each of the four years, calculate the 80% percentile. That is, find out where the cdfs reach or pass the level 0.80. Interpret and compare the results.

c Recall that the respective means of the four relative frequency distributions are 0.8910, 0.9530, 0.9970 and 1.0020. Calculate the accompanying variances and standard deviations. Interpret and compare the results.

d In 2003, the household Doubledutch had two cars. Determine the z-score and interpret it.

e For 1995, determine the percentage of the households with at least $\mu + 1.5\sigma$ cars. Compare your result with Chebyshev's rule.

Exercise 4.21 (computer)

In Exercise 3.22, the variables X = 'number of PCs per 1000 people' and Y = 'number of people per country' were considered on the populations of the countries in the EU15 and in the whole world. Variances and standard deviations of variables (such as X) that measure the computer density per country are often used to show the dissimilarities in development between countries.

a Recall the dataset Xrc03-22a.xls that contains – for the EU15 countries – the observations of X and Y. Calculate the mean, variance and standard deviation for each of these two variables.

b Determine, for the EU15, the percentage of the countries with X at least 1.5 standard deviations away from the mean value of X. Compare this actual result with the results of the empirical rule and Chebyshev's rule.

c Consider the dummy variable V that takes the value 1 if the country has an above-average number of PCs per 1000 people. Determine the mean, the variance and the standard deviation of the resulting set of observations. Also interpret these statistics in terms of the proportion of EU15 countries with an above-average number of PCs.

Next, consider the dataset Xrc03-22b.xls which contains – for 164 countries in the world – the observations of X and Y.

d Answer the questions in parts (a)–(c) again, but now for these 164 countries.

e The mean, variance and standard deviation for the x observations of the EU10 (see Exercise 4.18) are $\mu_{10} = 188.917$, $\sigma_{10}^2 = 4405.331$ and $\sigma_{10} = 66.37267$.
   Compare the means, standard deviations and coefficients of variation of the x datasets of the EU15, EU10 and world. Give your interpretations in terms of levels of development and dissimilarities with respect to PC density.

Exercise 4.22 (computer)

The file Xrc04-22.xls contains the weekly prices (in dollars) of a barrel of crude oil (the 'all countries spot price FOB weighted by estimated export volume') for the period 3 January 1997–9 September 2005.[†]

a Construct a time plot of these data. Give your comments.

   Transform these data into percentage changes from one week to the next.

b Construct a time plot of these change data. Compare it with the plot of part (a).

c Determine the mean, variance and standard deviation of these change data; treat the dataset as a population dataset. What are the respective dimensions?

d Determine the percentage of these change data that are at least two standard deviations from the mean value. Is this in accordance with the empirical rule?

e Create a histogram of the change data; take $(-20, -18]$, $(-18, -16]$, ..., $(16, 18]$, $(18, 20]$ as classification. (Notice that the underlying variable is continuous so the bars have to be adjoining.) Is the histogram bell shaped?

f Calculate the z-scores of the original crude oil prices and determine a suitable classified frequency distribution of them (take about 20 classes). Is the histogram bell shaped?

[†] *Source*: US Department of Energy, Energy Information Administration, 14 September 2005.

Exercise 4.23

The table below contains yearly returns (percentages with respect to the year before) of the Standard and Poor's index; see also Xrc04-23.xls.

Year	R	Year	R	Year	R	Year	R	Year	R	Year	R	Year	R	Year	R
1928	38	1938	25	1948	−1	1958	38	1968	8	1978	1	1988	12	1998	27
1929	−12	1939	−5	1949	10	1959	8	1969	−11	1979	12	1989	27	1999	20
1930	−28	1940	−15	1950	22	1960	−3	1970	0	1980	26	1990	−7	2000	−10
1931	−47	1941	−18	1951	16	1961	23	1971	11	1981	−10	1991	26	2001	−13
1932	−15	1942	12	1952	12	1962	−12	1972	16	1982	15	1992	4	2002	−23
1933	44	1943	19	1953	−7	1963	14	1973	−17	1983	17	1993	7	2003	26
1934	−5	1944	14	1954	45	1964	18	1974	−30	1984	1	1994	−2	2004	−2
1935	41	1945	31	1955	26	1965	9	1975	32	1985	26	1995	34		
1936	28	1946	−12	1956	3	1966	−13	1976	19	1986	15	1996	20		
1937	−39	1947	0	1957	−14	1967	20	1977	−11	1987	2	1997	31		

Percentage returns, Standard and Poor's index, 1928–2004
Source: www.delen.be, nieuwsbulletin, 7 September 2005

Straightforward calculations give the means and volatilities (that is, sample standard deviations) of the returns for several periods.

	1928–2004	1950–2004	1990–2004	1990–1999	1995–1999	1995–2004	2000–2004
Mean	7.4%	9.2%	9.2%	16.0%	26.4%	11.0%	−4.4%
Volatility	19.62%	16.65%	18.22%	14.45%	6.35%	20.85%	

We give an analysis of these means. First, notice the 'long-run' mean 7.4%. The poor performance of the stock market (−4.4%) in 2000–04 can be considered as a correction of the market itself, a reaction on the huge returns during the periods 1990–99 (16%) and 1995–99 (26.4%). The means of the returns for 1990–2004 and 1995–2004 are 'corrections' into, respectively, 11.0% and 9.2%.

The above analysis is about the mean of the returns and illustrates the **mean-reversion** of the stock market. But what about the **volatility** of the stock market? Calculate the volatility of the stock market for the period 2000–04 and give an analysis of the volatility of the returns similar to the analysis given above for the mean of the returns.

Exercise 4.24

In recent research, Statistics Netherlands studied (for the period 1996–2004) the frequency distribution of the times (during a day) at which traffic accidents with one or more deaths occur. The twelve intervals (0, 2], (2, 4], ..., (20, 22], (22, 24] were taken as classification. In this study, the population of interest is the collection of all road deaths in the period 1996–2004 and the variable of interest is X = 'time of the day at which the accident took place'.

a Determine the relative frequencies (percentages) and the mean if the times of the accidents had been evenly (homogeneously) distributed over the interval (0, 24]. Use the distribution function to determine the median and the IQR for such a homogeneous frequency distribution.

The table below contains the actual classified relative frequency distribution.

Period	Relative frequency (%)
0–2	5.1
2–4	5.0
4–6	3.9
6–8	7.4
8–10	6.7
10–12	8.0
12–14	12.1
14–16	13.3
16–18	13.6
18–20	10.0
20–22	7.7
22–24	7.2

Road deaths by time of accident, 1996–2004
Source: Statistics Netherlands, *Nieuwsflits*, 1 September 2005

b Calculate the mean, variance and standard deviation of this relative frequency distribution (treat the underlying dataset as a population dataset).

c Also determine the modal class, the median and the IQR.

d Compare the above results for the actual frequency distribution with the corresponding results for the homogeneous frequency distribution.

Exercise 4.25

The table below is about all unemployed people, aged 15–64, for the years 1995, 2000, 2003 and 2004.

	1995	2000	2003	2004
Unemployed labour force	533	269	396	480
Men	253	113	206	247
Women	280	156	190	233
15–24 yr	118	59	97	119
25–34 yr	171	69	102	112
35–44 yr	142	70	103	123
45–54 yr	86	55	65	88
55–64 yr	16	16	29	38

Unemployed labour force (× 1000)
Source: Statistics Netherlands, *Statistical Yearbook 2005*

Consider the variables X = 'dummy female' (with 1 = female and 0 = male) and Y = 'age' on the population of all unemployed people between 15 and 65.

a Determine, for each of the four years, the relative frequency distribution of X. Calculate the respective means and standard deviations. Comment on the development over time of this relative frequency distribution and its statistics.

b Determine, for each of the four years, the relative frequency distribution of Y. Calculate the respective means and standard deviations. Comment on the development over time of this relative frequency distribution and its statistics.

c Determine, for each of the years 1995 and 2004, the 95% percentile of Y. Compare and interpret the two answers.

Exercise 4.26

Nowadays, households in the Netherlands are quite small. The table below concerns households with fewer than six persons.

	1995	2000	2003	2004
1 person	2109	2272	2384	2424
2 persons	2058	2242	2293	2305
3 persons	903	897	904	906
4 persons	957	944	965	969
5 persons	441	446	451	449
Total	6468	6801	6997	7053

Households by size (× 1000), 1 January
Source: Statistics Netherlands, *Statistical Yearbook 2005*

a For each of the four years, determine the mean and the standard deviation of the variable 'household size'. Give your comments.

b Also determine the median and the IQR for 2004.

c (advanced) The mean of all households turns out to be 2.28 in 2004. By assuming that the households in the category 'more than 5 persons' all have exactly 6 persons, calculate the number of households in this category.

Exercise 4.27

The dataset Xrc02-16.xls reports the (integer-valued) grades of a Team Assignment for the course Statistics 2; see Exercises 2.16 and 3.12. Recall that these data come from a discrete variable X.

a Construct (manually) the graph of the distribution function F.

b Use it to calculate the three quartiles, the IQR and the relative IQR.

c Use the frequency distribution of X to calculate the mean, variance and the standard deviation.

Exercise 4.28

In a recent US report it was stated that, in 2002, the mean of the GDPpcs of 'all' countries in the world was equal to $6193, with a standard deviation of $9564.05. (In fact, the research was about 177 countries, but for three of them (including Cuba) no observation was known.) Is this enough information to get some idea about the percentage of the 174 countries with GDPpc at least $25 000, a bound that is often used to classify a country as very wealthy? We will try to answer that question.

a Determine k such that $\mu + k\sigma = 25\,000$.

b Why can we not use the empirical rule to find an answer to the above question?

c Use Chebyshev's rule to answer the question.

CASE 4.2 THE REIGNS OF BRITISH KINGS AND QUEENS

The data (in rounded years) in the file Case04-02.xls go back to the year 1066, the year that William the Conqueror became the king of England. Note that Elizabeth II, the present queen, is not included since her reign has not ended yet: in 2007 her reign had lasted 55 years. But is that a record?

a Determine some basic population statistics.

b Are you sure that the statistics in part (a) are population statistics?

c Which king / queen holds the maximum record? Who holds the minimum record? How should this last value be interpreted?

d How many years should Elizabeth II stay on to become an outlier? Answer this question with the 3δ-definition as well as with the 1.5δ-definition of the concept 'outlier'.

CASE 4.3 FOOD INSECURITY IN THE WORLD

The report *State of Food Insecurity in the World 2005* by the Food and Agriculture Organization (FAO) of the United Nations, sketches the state of affairs with respect to the Millennium Development Goals (MDG). One of these MDGs is to halve, between 1990 and 2015, the proportion of people who suffer hunger. The report considers the situation at the halfway stage.

Among others, the report compares the 1990/92 population of the developing countries with more than 2.5% undernourished, with the corresponding 2000/02 population. The table presents some summarized results for the variable X = 'percentage undernourished per country'.

		1990–92				2000–02		
Class	Freq.	Rel. freq. (%)	Cum. rel. freq. (%)	Freq. density	Freq.	Rel. freq. (%)	Cum. rel. freq. (%)	Freq. density
2.5–4	8	9.3	9.3	0.0465	12	13.6	13.6	0.0680
5–9	9	10.5	19.8	0.0210	11	12.5	26.1	0.0250
10–19	18	20.9	40.7	0.0209	19	21.6	47.7	0.0216
20–34	29	33.7	74.4	0.0225	28	31.8	79.5	0.0212
35–74	22	25.6	100	0.0064	18	20.5	100	0.0051
Total	86	100	–	–	88	100	–	–

Classified frequency distribution for percentage undernourished per country
Source: FAO (2006)

For the determination of the frequency density, the respective classes are taken as (2.5, 4.5], (4.5, 9.5], (9.5, 19.5], (19.5, 34.5], (34.5, 74.5]. Check the frequency density yourself.

a Determine and graph the respective cdfs F and G for the 1990/92 and 2000/02 population of developing countries with more than 2.5% undernourished. Compare the cdfs.

b For each of the two populations, find approximations of the mean and the median of X. What are the modal classes?

▶
    c For each of the two populations, find approximations of the standard deviation of X.

    d Give your comments on the results. For your information: between 1990/92 and 2000/02 the overall proportion of undernourished in the world decreased from 0.20 to 0.17.

CHAPTER

Pairs of variables

05

So far, we have studied methods to summarize data of **one** variable. Tables and graphs were used to obtain a good impression of the distribution of the data of that variable; statistics such as mean and variance were used to describe special features of the data. However, in many (economic) studies, emphasis is not primarily on the characteristics of individual variables but more on the relationships between variables. This chapter is about the relationships between **two** variables (often) called X and Y.

In principle, three cases have to be distinguished:

1 Both variables are quantitative.
2 Both variables are qualitative.
3 One is quantitative and the other is qualitative.

With respect to case 3, you will have to wait until Chapter 21 (for instance, ANOVA). In the present chapter, we shall start to look at cases 1 and 2 (especially case 1). Case 2 will be reconsidered in Section 18.4 (comparison of two proportions), and case 1 in Chapters 18–22.

In this chapter, one of the following three primary objectives is aimed at (depending on the problem of interest):

1 Positioning the population elements with respect to each other by comparing the datasets of the two variables
2 Studying the combined relationship of X and Y with time (if the two datasets are time series)
3 Studying the dependence of the data of one variable (say Y) on the data of the other variable.

Situations where only objective 1 or 2 is aimed at will be considered only in a few examples or exercises. For instance, it is essentially 2 that is the objective in Case 5.1.

Most attention will be paid to situations where the objective is 3. The question then is **whether**, **how** and **to what extent** the data of Y depend on the data of X. In particular, linear dependence will be considered. A graph (the so-called scatter plot) depicts the pairs (x, y) of the dataset in a system of axes, thus visualizing the relationship between the x and y data; see Section 5.1. The relationship is summarized numerically in Section 5.2 by way of the equation of the straight line that fits best through that plot. The degree of concentration of the pairs (x, y) around this line is expressed in two statistics: the covariance and the correlation coefficient; see Section 5.1. In Section 5.3 the consequences of changing the dimension (unit) of Y or X are studied.

Sections 5.1–5.3 are about quantitative variables; Section 5.4 is about graphical methods to depict the relationship if the variables are both qualitative.

CASE 5.1 WOMEN'S WORLD RECORDS APPROACH MEN'S WORLD RECORDS

Sports offer interesting problems about relationships between variables. For instance, given the fact that improvements in women and men's world records in athletics are becoming smaller and smaller, several questions arise. Does a limit exist for the performances of men and women? Will, in the long run, the records of the women close up to meet the records set by the men?

The present case is about the 100 metres and the marathon in athletics, two extremes in human sports. For the 100 metres it is explosiveness that is important; for the marathon, endurance and perseverance. Do women and men possess these qualities equally, and will this in the long run lead to comparable records?

Based on the historical data of the period 1930–2007 in the file Case05–01.xls, it will be shown that adopting the straight line as best-fitting curve in the scatter plots leads to the conclusion that women and men will have the same world records in 2015 (marathon) and 2054 (100 metres). To be continued in Section 5.3.

5.1 Scatter plot, covariance and correlation

Economists often are searching for factors that cause the variation in the data of a certain quantitative variable. For instance, when studying the variation of the variable 'annual expenditures on recreation for a household', an important factor seems to be the variable 'annual income of that household'. Several questions then arise. Does the first variable really depend on the second? And if so, can this dependence somehow be expressed by a mathematical equation (linear, quadratic, …)?

Obviously, more than one variable is involved; in this chapter we consider the case of two variables. The variation of one of the two variables (the ***dependent variable***, often denoted as Y) is the topic of investigation. The other variable (the ***independent variable***, often denoted by X) might, at least partially, cause this variation. Research is aimed at finding out whether (and how) Y is related to X.

Here are some examples:

■ Is GDP related to inflation rate? And if so, what does this relation look like?

■ How are the weekly sales of a company related to the amounts of money that is spent on advertisements? If the weekly amount on advertisements is increased by €1000, how will this affect the level of the weekly sales?

- If the price of all brands of packs of coffee in a supermarket increases by €0.25, will this affect the sales of tea?
- It is generally expected that the price of fuel will become at least €2 per litre in the near future. How will this affect the sales of new cars?
- If the Amsterdam Stock Exchange closes 1% below yesterday's level, what consequences can be expected for an individual stock?

To study questions like these, one needs data of two variables. For the first problem above, you might use the time series of GDPs and inflation rates of one country during the past 30 years. But a cross-sectional research is also possible: use the 25 pairs of GDP and inflation rate in a recent year for the EU25 countries. To study the third problem, you might use the quarterly prices of packs of coffee and those of tea for the past 10 years.

5.1.1. Scatter plot

Interest is in a dependent variable Y and an independent variable X, both quantitative. Suppose that a dataset of pairs (x, y) contains measurements of X and Y of a group of population elements. To obtain a first impression of the way the y data depend on the corresponding x data, one usually constructs a **scatter plot of y on x**. That is, one puts the n (or N, in case of a population dataset) pairs of data into a two-dimensional system of axes, where the (horizontal) x-axis refers to the independent variable and the (vertical) y-axis refers to the dependent variable. In the (x, y)-plane, a cloud of dots results. This cloud is called a **population cloud** or a **sample cloud**, depending on whether the dataset is a population dataset or a sample dataset.

Scatter plots are used to get a visual idea of the relationship between two variables. It is important to note that scatter plots do not tell which variable – in a further study – should be taken as dependent variable and which as independent variable. This has to be determined from the economic context or theory.

Example 5.1

An employee of the Ministry of Tourism observes that household expenditures on recreation tend to vary rather strongly. She wants to find reasons that explain this variation. An obvious reason seems to be that households usually have different incomes: households with relatively high incomes often spend more on recreation than households with relatively low incomes. The employee believes that the variation in household incomes partially explains the variation in recreation expenditures.

To test her ideas in a quick study, she randomly samples 20 households and measures the variables X = HINC = 'annual income per household ($\times$ €1000)' and Y = RECREXP = 'annual expenditure on recreation per household ($\times$ €1000)'. That is, she obtains a dataset with two columns of data: column 1 contains the 20 observations of X and column 2 the accompanying 20 observations of Y; see file Xmp05-01.xls and Table 5.1. Notice that row 1 of the dataset is about household 1. It contains the pair (x_1, y_1), the annual income of household 1 and the accompanying annual expenditure on recreation. Row 2 contains the pair (x_2, y_2) of annual income and annual expenditure on recreation of household 2; etc.

Household i	Annual income x_i ($\times$ €1000)	Recreation expenditures y_i ($\times$ €1000)
1	26.90	2.13
2	43.30	3.59
3	44.68	3.71
4	15.78	1.32
5	30.81	2.79
6	34.78	3.03
7	22.39	1.93
8	14.98	1.34
9	38.29	3.12
10	30.34	2.22
11	37.61	3.00
12	14.00	1.15
13	20.58	1.64
14	34.83	3.08
15	38.44	3.19
16	27.14	2.28
17	31.00	2.58
18	30.92	2.70
19	29.83	2.37
20	13.82	1.12

TABLE 5.1 Twenty observations of (X, Y)
Source: Private research (2006)

Some statistics of, respectively, the x data and the y data can easily be calculated:

$$\bar{x} = 29.0210; \quad s_x = 9.54203; \quad \bar{y} = 2.4145; \quad s_y = 0.80047$$

Figure 5.1 gives a first visual impression of the relationship between the y data and the corresponding x data. See Appendix A1.5 for its construction. Notice that y is put vertically and x horizontally since we are interested in the way y depends on x (and not the other way around).

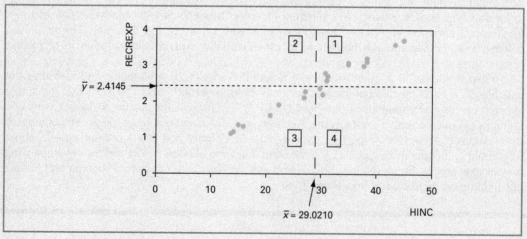

FIGURE 5.1 Scatter plot of RECREXP on HINC

The scatter plot also contains the **grid-lines** with equations $x = \bar{x}$ and $y = \bar{y}$, that is, the vertical line $x = 29.0210$ and the horizontal line $y = 2.4145$. They divide the (x, y)-plane into four **quadrants**. The north-east quadrant is called quadrant 1; the north-west, south-west and south-east quadrants are called quadrants 2, 3 and 4, respectively. Notice that most dots of the scatter plot fall in the quadrants 1 and 3, where the deviations $x_i - \bar{x}$ and $y_i - \bar{y}$ are either both positive or both negative. For a large majority of the sampled households, those with above-average income also show above-average expenditure on recreation, while households with below-average income often also show below-average expenditure on recreation. The pairs (x_i, y_i) tend to concentrate around an imaginary, increasing straight line.

The relationship between y data and x data can be of several types.

Figure 5.2(a) is the scatter plot of Example 5.1. In that plot, the relationship between the y data and the x data is best described by an **in**creasing straight line; one says that the y data and the corresponding x data are **positively linearly related**, where 'positively' refers to the fact that the best-fitting straight line is increasing. Relatively large (small) observations of X often combine with relatively large (small) observations of Y, in such a way that the pairs seem to concentrate strongly around an increasing straight line.

In Figure 5.2(b), the y and x data are measurements of the variables Y = 'budget deficit (% of GDP)' and X = 'GDP growth (%)' for the twelve eurozone countries. The relationship between the y data and the corresponding x data seems to follow a **de**creasing straight line, although this relationship does not look very strong. Large (small) observations of X are relatively often paired with relatively small (large) observations of Y, while the dots seem to concentrate somewhat around a decreasing straight line. One says that the y data and the corresponding x data tend to be **negatively linearly related**.

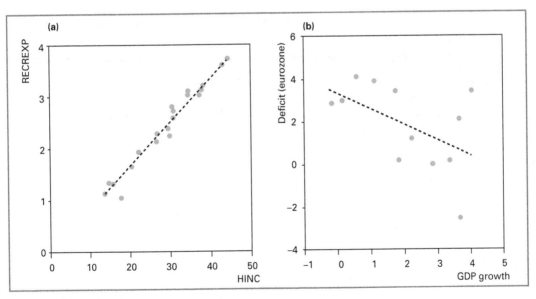

FIGURE 5.2 Scatter plots of y on x
Sources: (a) Private research (2006); (b) Bloomberg (2005)

In some cases, a non-linear mathematical equation can better describe the dependence between the y data and the corresponding x data. Figure 5.3(a) is the scatter plot of 40 pairs of observations of the variables X = 'age of an adult' and Y = 'hourly wage of the adult' on a certain unspecified population. Notice that this graph seems to have a downward curvature for larger observations of X, which indicates that it might be better to use a quadratic (a mountain-shaped parabolic) relation to describe this picture. Although relatively large observations of X are often combined with relatively large observations of Y, this effect seems to die out above $x = 35$ years. The y data and the corresponding x data have, at least to some extent, a *quadratic relationship*.

Figure 5.3(b) presents observations of X = 'GNI per capita' and Y = 'index of happiness (scale 0–100)' for 16 countries, some poor and some rich. The mathematical relationship is not obvious. The plot seems to suggest that only very low values of GNIpc influence feelings of (un)happiness. The mathematical relationship does not seem to be linear or quadratic. In the plot, a logarithmic curve is fitted.

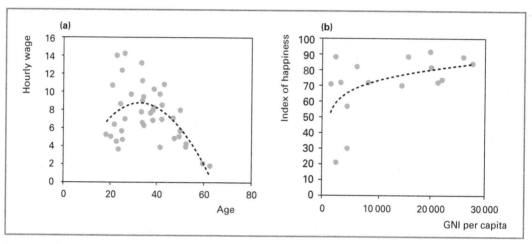

FIGURE 5.3 Scatter plots of y on x
Sources: (a) Private research (2005); (b) World Values Survey (2005)

The question arises how a scatter plot will look like if the underlying variables are **un**related. Such an example is considered below.

Example 5.2

Consider the variables Y = HINC = 'annual income (× €1000) of a household' and X = HEIGHT = 'height (in cm) of the reference person of the household'. Since it is hard to think of reasons why X and Y would be related, we expect that a sample cloud in a scatter plot will reflect this.

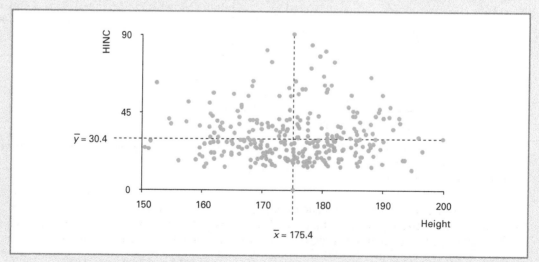

FIGURE 5.4 Scatter plot of annual income on height (cm)
Source: Private research (2005)

The scatter plot in Figure 5.4 shows pairs of observations for a sample of 300 households; see the file Xmp05-02.xls for the data. Included are the grid-lines at $\bar{x} = 175.4$ and $\bar{y} = 30.4$; they represent, respectively, the mean height of the reference persons and the mean annual incomes. Indeed, there does not seem to be an obvious relation between the y data and the corresponding x data. Above-average observations of X (that is, observations larger than 175.4) are sometimes combined with above-average observations of Y and sometimes with below-average observations of Y. The same holds for observations of X that are below average. That is: the y data and the corresponding x data do not jointly increase or decrease; the (x, y) data prefer neither the NE–SW quadrants nor the NW–SE quadrants.

Although there are many types of ways in which two variables might be related, this chapter is devoted mainly to the **linear** relationship between two variables. By creating a scatter plot one gets a visual impression of the linear relationship between the y data and the accompanying x data. The concentration of the dots around some imaginary, 'best-fitting' straight line gives an idea of the **degree** of this linear relationship (but be aware of the scale on the y axis). However, it is not wise to rely on visual effects only since the observation of visual effects may be subjective. Therefore we want to develop a statistic that expresses the degree (strength) of linear relationship as a number in such a way that a positive number means a positive linear relationship, zero means no linear relationship and a negative number means a negative linear relationship. Furthermore, a larger positive (or smaller negative) number should mean a stronger positive (or negative) linear relationship. Such statistics are developed below.

5.1.2 Covariance

Interest is in the linear relationship between a dependent variable Y and an independent variable X; both variables are quantitative. A dataset is available, containing pairs of observations of these two variables. In the case of a sample dataset, these pairs are denoted by

$$(x_1, y_1), (x_2, y_2), \ldots, (x_n, y_n)$$

Here, (x_i, y_i) refers to the observations of the variables X and Y measured at the ith sample element. There are two types of deviations: the deviations $x_i - \bar{x}$ of the x observations with respect to the mean $\bar{x}$, and the deviations $y_i - \bar{y}$ of the y observations with respect to their mean $\bar{y}$. When the dataset is a population dataset, the pairs are denoted by

$$(x_1, y_1), (x_2, y_2), \ldots, (x_N, y_N)$$

where the (x_i, y_i) are the observations of X and Y measured at the ith population element. Now the two types of deviations $x_i - \mu_X$ and $y_i - \mu_Y$ are considered with respect to the respective population means μ_X and μ_Y.

A statistic that measures the degree of linear relationship between y data and the accompanying x data will, in the case of a sample dataset, be based on the deviations $x_i - \bar{x}$ and $y_i - \bar{y}$, and in the case of a population dataset on $x_i - \mu_X$ and $y_i - \mu_Y$.

Example 5.3

Reconsider Example 5.1 and the scatter plot of y on x in Figure 5.1. The grid-lines divide the (x, y)-plane into four quadrants. The dots in quadrant 1 represent pairs (x_i, y_i) for which the deviations $x_i - \bar{x}$ and $y_i - \bar{y}$ are both positive and hence the products $(x_i - \bar{x})(y_i - \bar{y})$ are positive too. The dots in quadrant 3 represent pairs (x_i, y_i) for which the deviations $x_i - \bar{x}$ and $y_i - \bar{y}$ are both negative and hence the products $(x_i - \bar{x})(y_i - \bar{y})$ are positive. For the dots in quadrants 2 and 4, the corresponding pairs (x_i, y_i) have deviations $x_i - \bar{x}$ and $y_i - \bar{y}$ for which one is positive and the other is negative, so the products $(x_i - \bar{x})(y_i - \bar{y})$ are negative. Consequently, quadrants 1 and 3 give positive products while quadrants 2 and 4 give negative products. Notice that most dots of the scatter plot lie in quadrants 1 and 3. On the one hand, this indicates that above-average (respectively below-average) observations of X are often combined with above-average (respectively below-average) observations of Y. On the other hand, it also ensures that the summation

$$\sum_{i=1}^{n}(x_i - \bar{x})(y_i - \bar{y}) \tag{5.1}$$

of **all** products of the deviations will be positive. Apparently, the sign of the summation in (5.1) gives information about the linear relationship between the y data and the accompanying x data. A positive result informs us that the positive products $(x_i - \bar{x})(y_i - \bar{y})$ of the dots in quadrants 1 and 3 have the upper hand. So, the pairs (x_i, y_i) tend to concentrate around an **increasing** straight line.

If the summation in (5.1) had been negative, the negative products $(x_i - \bar{x})(y_i - \bar{y})$ deriving from data pairs (x_i, y_i) in quadrants 2 and 4 would have been dominant; the pairs (x_i, y_i) would have tended to concentrate around a **de**creasing straight line.

The summation in (5.1) is an important indicator to decide about a positive or negative linear relationship between the y data and the corresponding x data. So, let us calculate this summation for the present example.

i	x_i	y_i	$x_i - \bar{x}$	$y_i - \bar{y}$	$(x_i - \bar{x})(y_i - \bar{y})$
1	26.90	2.13	−2.12	−0.28	0.6034
2	43.30	3.59	14.28	1.18	16.7850
3	44.68	3.71	15.66	1.30	20.2862
4	15.78	1.32	−13.24	−1.09	14.4923
5	30.81	2.79	1.79	0.38	0.6718
6	34.78	3.03	5.76	0.62	3.5447
7	22.39	1.93	−6.63	−0.48	3.2127
8	14.98	1.34	−14.04	−1.07	15.0871
9	38.29	3.12	9.27	0.71	6.5393
10	30.34	2.22	1.32	−0.19	−0.2565
11	37.61	3.00	8.59	0.59	5.0289
12	14.00	1.15	−15.02	−1.26	18.9941
13	20.58	1.64	−8.44	−0.77	6.5376
14	34.83	3.08	5.81	0.67	3.8659
15	38.44	3.19	9.42	0.78	7.3044
16	27.14	2.28	−1.88	−0.13	0.2530
17	31.00	2.58	1.98	0.17	0.3275
18	30.92	2.70	1.90	0.29	0.5422
19	29.83	2.37	0.81	−0.04	−0.0360
20	13.82	1.12	−15.20	−1.29	19.6777
Total	580.42	48.29	0	0	143.4610

TABLE 5.2 Calculation of the summation in (5.1)

The data in the second and third columns yield $\bar{x} = 29.0210$ and $\bar{y} = 2.4145$. Columns 4 and 5 contain the deviations $x_i - \bar{x}$ and $y_i - \bar{y}$. Column 6 is obtained by multiplying columns 4 and 5. The total of column 6 is the result of formula (5.1):

$$\sum_{i=1}^{20}(x_i - \bar{x})(y_i - \bar{y}) = 143.4610$$

This summation is positive, which is consistent with our visual observation that the dots tend to fall in quadrants 1 and 3. The y data and the accompanying x data tend to be positively linearly related and to concentrate around an **increasing straight line**.

In practice, one does not use the sum of all products of the two types of deviation but instead the **mean** of these products. In the case of a population dataset of pairs, this special mean is denoted by $\sigma_{X,Y}$ and is called the **population covariance** of the x data and the accompanying y data (or also the **covariance of Y on X**). In the case of a sample dataset of pairs, the **sample covariance** is denoted by $s_{X,Y}$. It is obtained by summing all n products of the deviations $x_i - \bar{x}$ and $y_i - \bar{y}$ and then dividing the result by $n - 1$.

Population covariance $\sigma_{X,Y}$ and sample covariance $s_{X,Y}$

$$\sigma_{X,Y} = \frac{1}{N}\sum_{i=1}^{N}(x_i - \mu_X)(y_i - \mu_Y) \quad \text{and} \quad s_{X,Y} = \frac{1}{n-1}\sum_{i=1}^{n}(x_i - \bar{x})(y_i - \bar{y}) \tag{5.2}$$

The reason that one takes the $(n-1)$ version to define the sample covariance will become clear later. In essence, it is because this $(n-1)$ version $s_{X,Y}$ does better (at least on average) in approximating the usually unknown $\sigma_{X,Y}$. (We have seen this before in formula (4.3) for the sample variance and standard deviation.)

Note that replacing Y and y by X and x in the formulae in (5.2) yields the population covariance $\sigma_{X,X}$ and the sample covariance $s_{X,X}$. However, the formulae are just the population variance and the sample variance, respectively. That is:

$$\sigma_{X,X} = \sigma_X^2 \quad \text{and} \quad s_{X,X} = s_X^2$$

In Example 5.3, the sample covariance follows immediately by dividing the result 143.4610 of the summation by $20 - 1 = 19$. That is $s_{X,Y} = 7.5506$. Since X and Y are both measured in units of 1000 euros, $s_{X,Y}$ is measured in units of 10^6 euros². This positive sample covariance informs us that the y data and the corresponding x data are positively linearly related.

To calculate the covariance with only a simple calculator, there is some need for a **short-cut formula** that facilitates the calculation of (5.1) and (5.2). To derive it, the summation rules that follow from (A2.3) are crucial.

Derivation of short-cut formula for covariance

$$\sum_{i=1}^{n}(x_i - \bar{x})(y_i - \bar{y}) = \sum_{i=1}^{n}(x_iy_i - \bar{x}y_i - \bar{y}x_i + \bar{x}\bar{y})$$

$$= \sum_{i=1}^{n}x_iy_i + \sum_{i=1}^{n}(-\bar{x}y_i) + \sum_{i=1}^{n}(-\bar{y}x_i) + \sum_{i=1}^{n}\bar{x}\bar{y}$$

$$= \sum_{i=1}^{n}x_iy_i - \bar{x}\sum_{i=1}^{n}y_i - \bar{y}\sum_{i=1}^{n}x_i + \sum_{i=1}^{n}\bar{x}\bar{y}$$

$$= \left(\sum_{i=1}^{n}x_iy_i\right) - n\bar{x}\bar{y} - n\bar{y}\bar{x} + n\bar{x}\bar{y}$$

$$= \left(\sum_{i=1}^{n}x_iy_i\right) - n\bar{x}\bar{y}$$

It follows that the sum of all products of the deviations $x_i - \bar{x}$ and $y_i - \bar{y}$ is equal to the sum of all products of the **observations** x_i and y_i **minus** n times the product of the two means. A similar result follows for the population case. To obtain $s_{X,Y}$ and $\sigma_{X,Y}$ the formulae have to be divided by $n - 1$ and N, respectively:

Short-cut formulae for $\sigma_{X,Y}$ and $s_{X,Y}$

Basic results: $\quad \sum_{i=1}^{N}(x_i - \mu_X)(y_i - \mu_Y) = \sum_{i=1}^{N} x_i y_i - N\mu_X\mu_Y$ (5.3)

$$\sum_{i=1}^{n}(x_i - \bar{x})(y_i - \bar{y}) = \sum_{i=1}^{n} x_i y_i - n\bar{x}\bar{y}$$

Short-cut: $\quad \sigma_{X,Y} = \dfrac{1}{N}\sum_{i=1}^{N} x_i y_i - \mu_X\mu_Y \quad$ and $\quad s_{X,Y} = \dfrac{1}{n-1}\left(\sum_{i=1}^{n} x_i y_i - n\bar{x}\bar{y}\right)$ (5.4)

Notice that the formula for $\sigma_{X,Y}$ has an easy interpretation: the **population** covariance is equal to the mean of the products minus the product of the means.

Example 5.4

To illustrate how the short-cut formula works in the case of a sample dataset of pairs, the 20 households of Example 5.1 are reconsidered, with the pairs of observations of the variables X = 'annual income' and Y = 'annual expenditure on recreation'.

i	x_i	y_i	$x_i y_i$
1	26.90	2.13	57.30
2	43.30	3.59	155.45
3	44.68	3.71	165.76
4	15.78	1.32	20.83
5	30.81	2.79	85.96
6	34.78	3.03	105.38
7	22.39	1.93	43.21
8	14.98	1.34	20.07
9	38.29	3.12	119.46
10	30.34	2.22	67.35
11	37.61	3.00	112.83
12	14.00	1.15	16.10
13	20.58	1.64	33.75
14	34.83	3.08	107.28
15	38.44	3.19	122.62
16	27.14	2.28	61.88
17	31.00	2.58	79.98
18	30.92	2.70	83.48
19	29.83	2.37	70.70
20	13.82	1.12	15.48
Total	580.42	48.29	1544.89

TABLE 5.3 Calculation of the covariance with the short-cut formula

Notice that the xy column in the table is just the product of the x and y columns. Summation of all numbers in the xy column yields 1544.89, which is the sum of the products. Since $\bar{x} = 29.0210$ and $\bar{y} = 2.4145$ were calculated before, it follows that

$$\sum_{i=1}^{20} x_i y_i - 20\overline{xy} = 1544.89 - 20 \times 29.0210 \times 2.4145$$

which – apart from rounding errors – is indeed equal to 143.4610. The sample covariance $s_{X,Y} = 7.55$ is obtained by dividing by 19.

The concept of covariance is not immediately suitable to measure the strength of a linear relationship. In the above example, the sample covariance 7.5506 is a positive number from which it can be deduced that the y data and the corresponding x data show a positive linear relationship. However, it is not clear whether this number reflects a strong or a weak linear relationship. This is because a reference point is missing.

Another disadvantage of using the covariance as a measure of the strength of a linear relationship is that it depends on the dimensions (the units in which the variables are measured). In the example, the dimensions of both variables are thousands of euro, so the covariance has unit 10^6 euro^2. A dimensionless measure would be better because it would enable us to compare the measured results of different examples. Such a statistic is considered below.

5.1.3 Correlation coefficient

The statistic correlation coefficient is a measure of linear relationship that does not have the disadvantages described above. Again, we distinguish between sample data and population data.

In the case of a sample dataset of observations of X and Y, the sample correlation coefficient r (or $r_{X,Y}$) is defined as the ratio of the sample covariance and the product of the two sample standard deviations. For a population dataset, the **population** correlation coefficient is the ratio of the population covariance and the product of the two population standard deviations.

Sample correlation coefficient and population correlation coefficient

The **sample correlation coefficient** of pairs of observations of the variables X and Y is defined as:

$$r = r_{X,Y} = \frac{s_{X,Y}}{s_X s_Y}$$

The **population correlation coefficient** of X and Y and the accompanying population dataset is defined as:

$$\rho = \rho_{X,Y} = \frac{\sigma_{X,Y}}{\sigma_X \sigma_Y}$$

Recall that the two standard deviations in both denominators above have the same dimensions as X and Y, respectively. So, the dimension of the denominators is just the product of these two dimensions. Since the dimension of the numerators is also the product of the dimensions of X and Y, this dimension drops out in each of the two ratios. Hence, r and ρ are dimensionless.

Often the population dataset of **all** pairs of observations of X and Y is not observed. In such a case, the population correlation coefficient ρ cannot be calculated. However, if we **are** in possession of a sample dataset, we **can** calculate r.

Notice that dividing the covariance by the (positive) product of the standard deviations does not change its sign. Hence, just as with the covariance, the sign of r (and ρ) gives information about the sign of the linear relationship. However, it turns out that a correlation coefficient can never be larger than $+1$ or smaller than -1. That is, the absolute value of both types of correlation coefficient is at most 1.

Property of sample and population correlation coefficients

$$|r| \le 1 \quad \text{and} \quad |\rho| \le 1$$

The values $+1$ and -1 only appear in very extreme situations: the value $+1$ in the unusual case that all pairs (x_i, y_i) fall precisely on one increasing straight line, the value -1 in the case that they all lie on one decreasing straight line. So, in these unusual cases the linear relationship is perfect. Apparently, the correlation coefficient has reference points $+1$ and -1, where $+1$ is the largest possible strength of positive linear relationship while -1 describes the largest possible strength of negative linear relationship. If a sample dataset has its correlation coefficient r close to 1 (or -1), the y data and the corresponding x data of the sample are strongly positively (or negatively) linearly related, so the pairs (x_i, y_i) tend to concentrate strongly around an increasing (or decreasing) straight line. If the value of r is close to 0, the y data and the corresponding x data are barely linearly related. If $r = 0$ (or $\rho = 0$), one says that the x data and the y data are ***uncorrelated***.

Example 5.5

In the examples about the 20 pairs of observations regarding household income X and household expenditure on recreation Y it was found that:

$$s_{X,Y} = 7.5506; \quad s_X = 9.54203; \quad s_Y = 0.80047$$

So, it follows that:

$$r = \frac{7.5506}{9.54203 \times 0.80047} = 0.9885$$

It would appear that the 20 pairs of y data and corresponding x data tend to concentrate strongly around an increasing straight line. Between annual household recreation expenditure Y and annual household income X there seems to be a strong positive linear relationship.

Of course, statistical packages have options to calculate the covariance and the correlation coefficient. For Excel and SPSS, see Appendix A1.5.

In practice the correlation coefficient r never reaches the level 0 precisely. However, there are many examples where the correlation coefficient is close to 0, indicating that a linear relationship is hardly an issue.

Figure 5.5 shows some graphs of types of linear relationships, all for samples of size 300. The first scatter plot is the graph of Example 5.2 that visualizes variables that are unrelated. Indeed, the correlation coefficient turns out to be -0.0028, which is close to 0 and hence reflects that the x data and the accompanying y data are not **linearly** related. The middle pair of graphs shows moderately positive ($r = 0.558$) and negative ($r = -0.558$) linear relationships. The last two graphs illustrate strongly positive ($r = 0.958$) and negative ($r = -0.958$) linear relationships.

The interpretation of the correlation coefficient as degree of linear relationship is not always useful in the case of time series. The reason is that for many time series the relationship between Y and X is partially indirect because of the variable $T = $ 'time'. Example 5.7 shows problems that may arise when interpreting the correlation coefficient of two time series. See also objective 3 at the start of this chapter. But first objective 1 is demonstrated in Example 5.6.

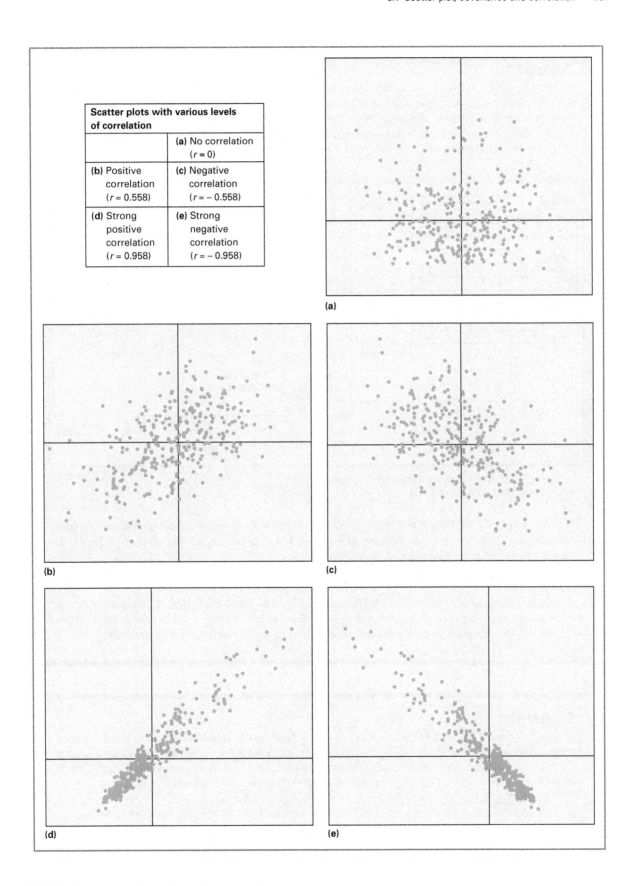

FIGURE 5.5 Some graphical illustrations of the concept of correlation

Example 5.6

Scatter plots are also used to demonstrate the positions of population elements with respect to other population elements; see objective 1 at the start of this chapter. The table in Figure 5.6 lists for 2003 the exports and imports (in billions of euro) for the EU10; the scatter plot – with dots accompanied by the names of the countries – shows the relative position of each individual country with respect to the other countries.

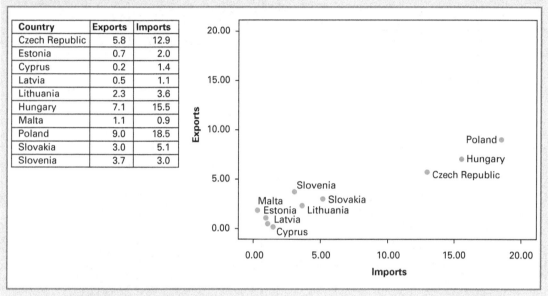

Country	Exports	Imports
Czech Republic	5.8	12.9
Estonia	0.7	2.0
Cyprus	0.2	1.4
Latvia	0.5	1.1
Lithuania	2.3	3.6
Hungary	7.1	15.5
Malta	1.1	0.9
Poland	9.0	18.5
Slovakia	3.0	5.1
Slovenia	3.7	3.0

FIGURE 5.6 Exports and imports in the EU10 (billions of euro)
Source: Eurostat, *Statistical Yearbook 2005*, p. 190

The covariance and the correlation coefficient of the variables X = 'imports (billions of euro) of a country' and Y = 'exports (billions of euro) of that country' on the population of the EU10 countries can easily be calculated with a computer:

$$\sigma_{X,Y} = 15.796 \quad \text{and} \quad \rho = 0.966$$

Apparently the governments of most EU10 countries did not – as so many other countries – succeed in creating a trade balance. It is not the line $y = x$ that fits best to the cloud of dots. Instead, there is a strong concentration around some other straight line; see Section 5.2 for its equation.

Example 5.7

In financial theory it is well known that the prices of many stocks and bonds tend to behave oppositely: if the price of a stock increases, then the price of a bond often decreases, and vice versa. But what about funds of stocks and bonds? Do they behave similarly? In this example we consider a stock investment fund and a bond investment fund, both issued by one bank.

The file Xmp05-07.xls contains the daily prices and returns (%) of the SNS Euro Stocks Fund and the SNS Euro Bonds Fund for the period 22 November 2004 to 17 November 2006. Figure 5.7 shows the time plots and the scatter plot of the prices of the two funds.

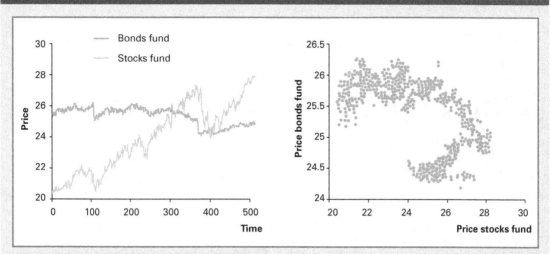

FIGURE 5.7 Time plots and scatter plot of bonds fund and stocks fund
Source: Euronext (2006)

Of course, it is possible to calculate the correlation coefficient of the prices of the two funds (−0.5960), but interpreting it as the degree of linear relationship between the prices of the bonds fund and the prices of the stocks fund is useless. The reason is that the two datasets are time series, so things change over time and the relationship between the prices of the two funds is much more complicated than just a linear relationship. This also follows from the right-hand scatter plot, where the role of time is unclear but causes strange 'wings'.

Transition to returns often largely reduces influences of time. That is why we will consider the variables X = 'daily return (%) of the SNS Euro Stocks Fund' and Y = 'daily return (%) of the SNS Euro Bonds Fund', and use the dataset and the computer to calculate the correlation coefficient of the x and the y data. From the scatter plot in Figure 5.8 it becomes clear that, at best, a weak positive linear relationship can be expected.

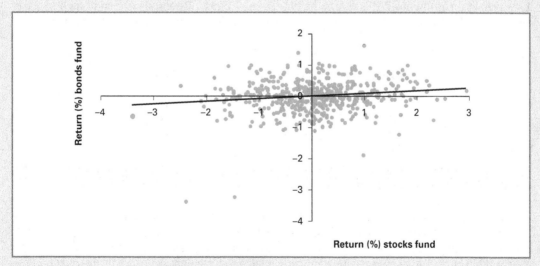

FIGURE 5.8 Scatter plot of returns bonds fund on returns stocks fund

Using a computer package the statistics set out in Table 5.4 can be calculated; see also Appendix A1.5.

	Mean	Standard deviation	Covariance	Correlation
Bonds fund	−0.0036	0.5059	0.0702	0.1547
Stocks fund	0.0647	0.8968		

TABLE 5.4 Some statistics of the returns of the two funds, 23 November 2004–17 November 2006

The mean daily return of the stocks fund is much larger than the mean daily return of the bonds fund. However, the risk – as measured by the standard deviation – of the stocks fund is also larger. The returns of the two funds are weakly positively correlated.

Regarding the strength of linear relationship, two more things have to be noted.

Firstly, a scatter plot with an obvious curvature usually does not have a correlation coefficient (for **linear** relationship) equal to 0. For instance, in Figure 5.3(a) the 'best-fitting' smooth graph through the plot seems to be a part of a parabolic curve. Still, the best-fitting **straight** line also seems to fit quite well; to some extent, there is a linear relationship too.

Secondly, the correlation coefficient measures the degree of **linear** relationship between two variables. If r is equal to 0, the y data and the corresponding x data are not linearly related but this does not mean that they are completely unrelated. Another kind of relationship can exist, such as a quadratic one, as in the following example.

Example 5.8

In Baker Street, the five households of the population all consist of husband–wife couples with small children. A woman (who lives here) believes that the men in Baker Street are too fat. She has read in the newspaper that the mean height of the adult men in the whole of the city is 176 cm and that the mean weight of these men is 73 kg. As a part of her study, she measures the (centred) variables Y = 'weight (in kg) minus 73' and X = 'height (in cm) minus 176' for all five adult men in Baker Street. Her table of data and scatter plot are shown in Table 5.5 and Figure 5.9, respectively.

i	Height	x_i	Weight	y_i
1	174	−2	82	9
2	176	0	74	1
3	177	1	73	0
4	178	2	74	1
5	180	4	82	9

TABLE 5.5 Heights and weights of the men in Baker Street

She easily calculates that $\sigma_{X,Y} = 0$ and hence $\rho_{X,Y} = 0$. In Baker Street, the variables X and Y are uncorrelated: there is no linear relationship. Still, X and Y are not unrelated. Essentially, Y is heavily dependent on X since $Y = (X - 1)^2$, so Y is quadratically related to X. Put another way, the weight W (in kg) of a man is strictly quadratically related to his height H (in cm) by the relation $W = (H - 176)^2 + 73$.

(Whether the men in Baker Street are indeed too fat is a matter of opinion. Their weights obviously are above average, but not by very much.)

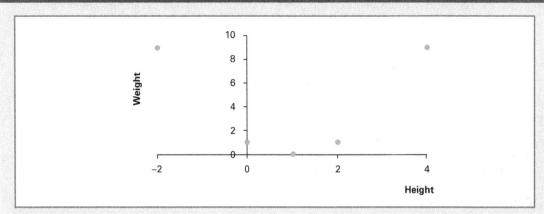

Figure 5.9 Scatter plot of *y* on *x*

5.2 Regression line

In the previous section, the 'best-fitting' straight line in a scatter plot was mentioned several times but no explanation of it was given. We shall do that now.

Again, interest is in a quantitative variable called *Y*, especially in its variation. Suppose that it is expected that another variable (called *X*) partially causes this variation. But **how** does *Y* depend on *X*? The study of this problem is called ***regression of Y on X***. For example, it is expected that the variable *X* = 'annual income of a household' will be an important factor when studying the variation of the variable *Y* = 'annual recreation expenditure of that household'.

Below, only linear dependence of *Y* on *X* will be considered. The correlation coefficient measures the strength of the concentration of the (x, y) dots in the scatter plot around an imaginary straight line. But which straight line is meant here? Although there are several approaches to find such a line, only the ***least-squares method*** will be considered. This method is based on the following ideas. Start with a general line with equation $y = a + bx$ in a **sample** scatter plot; the intercept a and the slope b are fixed. At $x = x_i$, the *y* level of that straight line is $a + bx_i$; see Figure 5.10. Notice that the pair (x_i, y_i) has vertical deviation $y_i - (a + bx_i)$ with respect to that line, as indicated in the figure.

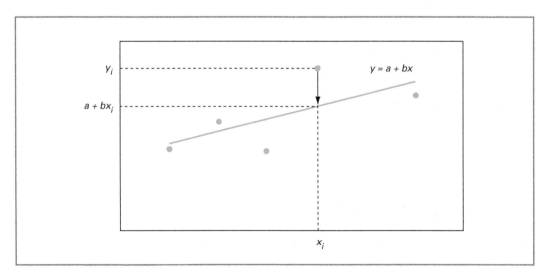

FIGURE 5.10 A general straight line in a sample scatter plot

So, the vertical deviations between the dots in a sample cloud and the straight line $y = a + bx$ can formally be written as:

$$y_1 - (a + bx_1), \quad y_2 - (a + bx_2), \quad \ldots, \quad y_n - (a + bx_n)$$

The least-squares (LS) method determines the constants a and b such that the sum of the **squared** deviations is as small as possible. That is, in the **LS method** the constants a and b are chosen such that

$$\sum_{i=1}^{n}(y_i - (a + bx_i))^2 \quad \text{is minimal} \tag{5.5}$$

(In the case of a population cloud, n has to be replaced by N.) By choosing a and b as in (5.5), the straight line with equation $y = a + bx$ 'fits best' through the cloud of dots. This line is the 'imaginary' line or 'best fitting' line we were talking about.

Derivation of a and b

To find a and b as indicated in (5.5), we denote the above sum of squared vertical deviations by $L(a,b)$ and set its partial derivatives equal to 0. That is:

$$\frac{\partial}{\partial a} L(a,b) = 0 \quad \text{and} \quad \frac{\partial}{\partial b} L(a,b) = 0 \tag{5.6}$$

From mathematical theory it follows that:

$$\frac{\partial}{\partial a}(y_i - a - bx_i)^2 = -2(y_i - a - bx_i)$$

$$\frac{\partial}{\partial b}(y_i - a - bx_i)^2 = -2x_i(y_i - a - bx_i)$$

Hence, the equations of (5.6) turn into:

$$\sum_{i=1}^{n} -2(y_i - a - bx_i) = 0 \quad \text{and} \quad \sum_{i=1}^{n} -2x_i(y_i - a - bx_i) = 0$$

Multiplying both equations by $-\frac{1}{2}$ and getting rid of the parentheses in the second summation yields:

$$\sum_{i=1}^{n}(y_i - a - bx_i) = 0 \quad \text{and} \quad \sum_{i=1}^{n}(x_i y_i - ax_i - bx_i^2) = 0$$

By the summation rules of (A2.3) it follows that:

$$\left(\sum_{i=1}^{n} y_i\right) - na - b\sum_{i=1}^{n} x_i = 0 \quad \text{and} \quad \sum_{i=1}^{n} x_i y_i - a\sum_{i=1}^{n} x_i - b\sum_{i=1}^{n} x_i^2 = 0$$

Hence:

$$a = \bar{y} - b\bar{x} \quad \text{and} \quad an\bar{x} + b\sum_{i=1}^{n} x_i^2 = \sum_{i=1}^{n} x_i y_i$$

Substitution of the left-hand expression into the right-hand equation gives:

$$n\bar{x}\bar{y} - bn\bar{x}^2 + b\sum_{i=1}^{n} x_i^2 = \sum_{i=1}^{n} x_i y_i$$

so

$$b\left(\sum_{i=1}^{n} x_i^2 - n\bar{x}^2\right) = \sum_{i=1}^{n} x_i y_i - n\bar{x}\bar{y}$$

But recall that

$$\sum_i x_i^2 - n\bar{x}^2 = \sum_i (x_i - \bar{x})^2 \quad \text{and} \quad \sum_i x_i y_i - n\bar{x}\bar{y} = \sum_i (x_i - \bar{x})(y_i - \bar{y})$$

(see (4.4) and the second basic result in (5.3)). As a consequence,

$$b\sum_{i=1}^{n} (x_i - \bar{x})^2 = \sum_{i=1}^{n} (x_i - \bar{x})(y_i - \bar{y})$$

By dividing both sides of the last equation by $n - 1$, it follows that $b = s_{X,Y}/s_X^2$; so the solution b is determined. Since $a = \bar{y} - b\bar{x}$, the solution a is determined too.

The LS method yields exactly one pair (a, b) of constants. This pair minimizes the sum of the squared vertical deviations.

For the sample case, these optimal constants are respectively denoted by b_0 and b_1 and are called **sample regression coefficients**. A similar derivation can be conducted for a population dataset. The resulting constants (a, b) are then denoted by β_0 and β_1 and are called **population regression coefficients**.

Sample and population regression coefficients

The **sample regression coefficients** b_1 and b_0, and the **population regression coefficients** β_1 and β_0 are defined as:

$$b_1 = \frac{s_{X,Y}}{s_X^2} \quad \text{and} \quad b_0 = \bar{y} - b_1\bar{x}, \qquad \beta_1 = \frac{\sigma_{X,Y}}{\sigma_X^2} \quad \text{and} \quad \beta_0 = \mu_Y - \beta_1\mu_X \qquad \textbf{(5.7)}$$

The straight lines with equations

$$y = b_0 + b_1x \quad \text{and} \quad y = \beta_0 + \beta_1x$$

are respectively called the **sample regression line** and **population regression line**. They respectively 'fit best' through the sample cloud and the population cloud. Instead of regression line, one can also use the term **least-squares line**. The population regression line is also called the **line of means** (you will have to wait until Chapter 19 to find out why).

The constants b_0 (or β_0) and b_1 (or β_1) are respectively called the **intercept** and the **slope** of the sample (or population) regression line. The slope of the straight line that fits best through the cloud of dots is the ratio of the covariance and the variance with respect to x.

Notice that in many practical situations the variables X and Y are measured only for a sample of elements. Then, the statistics μ_X, μ_Y, σ_X^2, $\sigma_{X,Y}$ are usually unknown, and hence the population regression coefficients β_0 and β_1 and the population regression line are also unknown. However, if a sample of x data and corresponding y data is available, it is possible to calculate the sample regression coefficients b_0 and b_1. The sample regression line is then considered to be an approximation (estimate) of the unknown population regression line. In this context, the equation $y = b_0 + b_1x$ of the sample regression line is often written as

$$\hat{y} = b_0 + b_1x$$

to distinguish it better from the population regression line.

Recall that the sample regression equation is based on the pairs of observations $(x_1, y_1), \ldots, (x_n, y_n)$. If a new population element has x value denoted by x_p but its y value is not measured, then $\hat{y}_p = b_0 + b_1x_p$ gives an approximation (called **prediction**) of the y value. That is why the sample regression line is also called **prediction line**. If x_p is in the range of the x data (that is, minimum $x_i \le x_p \le$ maximum x_i), then the prediction can usually be trusted; this is called **interpolation**. However, be careful with making a prediction if x_p is not in the range of the x data (**extrapolation**, $x_p <$ minimum x_i or $x_p >$ maximum x_i).

The slope β_1 of the population regression line has the following interpretation: if x increases by 1, then the y level of the population regression line goes from $\beta_0 + \beta_1x$ to $\beta_0 + \beta_1x + \beta_1$ and hence changes by β_1. Similarly, b_1 is the change of the vertical level of the **sample** regression line if x increases by 1.

It seems natural to interpret the intercept β_0 (or b_0) of the population (or sample) regression line as the y level that belongs to $x = 0$. However, this interpretation is allowed only if 0 is in the range of the x data. Consequently:

Interpretation of the regression coefficients

- β_1 is the change in the y level of the population regression line if x increases by 1.
- If 0 belongs to the range of the population x data, β_0 is just the y-level of the population regression line that belongs to $x = 0$.
- b_1 is the change in the y level of the sample regression line if x increases by 1.
- If 0 falls in the range of the sampled x data, b_0 is just the y level of the sample regression line that belongs to $x = 0$.

A sample regression line has an important property that should be emphasized: the line passes through the pair $(\bar{x}, \bar{y})$ formed by the two means of the x data and the y data. This result follows immediately from (5.7):

$$b_0 = \bar{y} - b_1\bar{x}, \quad \text{so} \quad \bar{y} = b_0 + b_1\bar{x}$$

A similar result holds for the population regression line.

A property of regression lines

- A sample regression line passes through $(\bar{x}, \bar{y})$.
- A population regression line passes through (μ_x, μ_y).

Example 5.9

Recall the example about the variables X = 'annual household income' and Y = 'annual expenditure on recreation'. The 20 observed pairs (x_i, y_i) were summarized before:

$$\bar{x} = 29.0210; \quad \bar{y} = 2.4145; \quad s_x^2 = 91.0503; \quad s_{X,Y} = 7.5506$$

Using these results, it immediately follows from (5.7) that:

$$b_1 = \frac{7.5506}{91.0503} = 0.0829 \quad \text{and} \quad b_0 = 2.4145 - 0.0829 \times 29.0210 = 0.0079$$

So, the equation of the straight line that fits best to the left-hand sample cloud of Figure 5.2 is:

$$\hat{y} = 0.0079 + 0.0829x$$

To interpret the coefficients, recall that X and Y are both measured in units of €1000. So, 1 unit corresponds to €1000. The slope $b_1 = 0.0829$ is just the estimated increase of y if x increases by 1. That is, if the household income increases by €1000, the household expenditure on recreation is estimated to increase by $0.0829 \times 1000 = €82.90$.

For the interpretation of $b_0 = 0.0079$, notice that the range of the x data used to create the regression line goes from 13.82 to 44.68, so $x = 0$ is not included. Hence, the intercept of the regression line is not interpreted.

The regression equation can be used to predict the recreation expenditure of a household for which the income is known. A household with an annual income of €18 500 is predicted to spend

$$\hat{y}_p = 0.0079 + 0.0829 \times 18.5 = 1.5416 \ (\times 1000)$$

that is, about €1542, on recreation.

But how precise are such predictions? It will intuitively be clear that the predicting performance of the regression line has something to do with how well the line fits the data in the sample cloud of Figure 5.2(a). Hence, the vertical deviations between the actual observed y levels in the sample and the corresponding y levels of the regression line are of interest. Household 1 in the sample has income $x_1 = 26.90$ ($\times$ 1000) and spends $y_1 = 2.13$ ($\times$ 1000) on recreation. The accompanying y level of the regression line is denoted by $\hat{y}_1$ and equals:

$$\hat{y}_1 = 0.0079 + 0.0829 \times 26.90 = 2.2379$$

The corresponding vertical deviation is called residual e_1; it is equal to:

$$e_1 = y_1 - \hat{y}_1 = 2.13 - 2.2379 = -0.1079$$

Notice that $\hat{y}_1$ can be considered as the **prediction** of the recreation expenditure of a household with the same annual income as household 1. Similarly, the predicted annual recreation expenditures $\hat{y}_2$ of a household with the same income as household 2 can be calculated; the accompanying vertical deviation (residual) $e_2 = y_2 - \hat{y}_2$ then follows immediately. And so on. Since, for the 20 sampled households, the annual expenditures y_i on recreation are known, they can be compared with the predictions $\hat{y}_i$ delivered by the regression line, to establish the predicting performance of the line; see Table 5.6.

i	x_i	y_i	$\hat{y}_i$	e_i	e_i^2
1	26.90	2.13	2.2379	−0.1079	0.0116
2	43.30	3.59	3.5975	−0.0075	0.0001
3	44.68	3.71	3.7119	−0.0019	0.0000
4	15.78	1.32	1.3161	0.0039	0.0000
5	30.81	2.79	2.5620	0.2280	0.0520
6	34.78	3.03	2.8912	0.1388	0.0193
7	22.39	1.93	1.8640	0.0660	0.0044
8	14.98	1.34	1.2497	0.0903	0.0081
9	38.29	3.12	3.1821	−0.0621	0.0039
10	30.34	2.22	2.5231	−0.3031	0.0919
11	37.61	3.00	3.1258	−0.1258	0.0158
12	14.00	1.15	1.1685	−0.0185	0.0003
13	20.58	1.64	1.7140	−0.0740	0.0055
14	34.83	3.08	2.8953	0.1847	0.0341
15	38.44	3.19	3.1946	−0.0046	0.0000
16	27.14	2.28	2.2578	0.0222	0.0005
17	31.00	2.58	2.5778	0.0022	0.0000
18	30.92	2.70	2.5712	0.1288	0.0166
19	29.83	2.37	2.4808	−0.1108	0.0123
20	13.82	1.12	1.1536	−0.0336	0.0011
Total	–	–	–	0.0152	0.2774

TABLE 5.6 Predictions $\hat{y}_i = 0.0079 + 0.0829x_i$ and residuals $e_i = y_i - \hat{y}_i$

The residuals are the vertical deviations between the truly observed values of Y and the corresponding predicted values of Y according to the sample regression line. By summing the sixth column, the **sum of the squared residuals** is obtained:

$$\sum_{i=1}^{20}(y_i - \hat{y}_1)^2 = \sum_{i=1}^{20} e_i^2 = 0.2774$$

This sum of squares plays an important role in modern regression analysis. In a sense, it measures the quality of the regression line as a predictor of y. If this sum of squares is relatively small, then the predictions $\hat{y}_i$ are close to the truly observed values y_i, so the predicting performance of the regression line seems to be good.

The analysis of residuals, started up in the former example, will now be generalized. The sample dataset (x_1, y_1), (x_2, y_2), ..., (x_n, y_n) yields the sample regression line $\hat{y} = b_0 + b_1 x$. By substituting x_i for x we obtain $\hat{y}_i = b_0 + b_1 x_i$ that will be considered as a prediction of y_i. The vertical deviations $y_i - \hat{y}_i$ are denoted by e_i and are called **residuals** or **errors**. See also Figure 5.11.

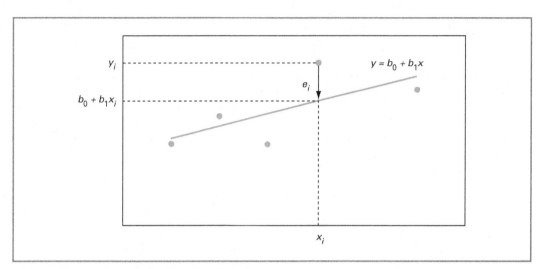

FIGURE 5.11 Scatter plot, regression line and residuals

These residuals e_1, ..., e_n give an idea of the concentration of the dots around the regression line and give information about the predicting performance of the regression line. Although it will not be proved here (but see Exercise 5.12 at the end of this chapter), it is not too difficult to show that the sum of the residuals is always 0:

$$\sum_{i=1}^{n} e_i = 0 \qquad \text{(5.8)}$$

(In the former example this sum was 0.0152, which differs from 0 because of rounding errors.) As a consequence, this sum of the vertical deviations is not a good measure for the predicting performance of the regression line. Squaring turns out to be a good solution to this problem.

Sum of squared errors

The **sum of squared errors (SSE)** is defined as:

$$\text{SSE} = \sum_{i=1}^{n} (y_i - \hat{y}_i)^2 = \sum_{i=1}^{n} e_i^2$$

The SSE is used to measure variation around the regression line and hence the predicting performance of the regression line. Since $(y_i - \hat{y}_i)^2 = (y_i - (b_0 + b_1 x_i))^2$, it follows that SSE is just the minimum value of the summation in (5.5); recall that $a = b_0$ and $b = b_1$ give the minimal value of that summation.

The SSE is very important in regression analysis. This is because it measures the amount of variation around the regression line. The smaller SSE is, the better the regression line fits to the data and the better the predicting performance of the line is.

Example 5.10

An 'ice day' is a day where the maximum temperature stays below 0. The Royal Netherlands Meteorological Institute believes that the number of ice days in the second half of a year influences the number of ice days in the first half of the year after it. To find out to what extent this belief is valid, measurements for the period 1945–88 are used; see <u>Xmp05-10.xls</u>.

Interest is in the variables V = 'number of ice days in the **second** half of a year' and W = 'number of ice days in the first half of the year after it'. The sample dataset contains the pairs of observations (v_i, w_i), for $i = 1, ..., 43$. Here v_1 and w_1 give the numbers of ice days in the second half of 1945 and in the first half of 1946, respectively. Below, they will be termed the numbers of ice days in the first and the second half of the winter of 1945/46. Similarly, the pair (v_{43}, w_{43}) gives the numbers of ice days in the first and the second half of the winter of 1987/88. The intention is to regress the w data on the v data. The scatter plot in Figure 5.12 gives a visual impression.

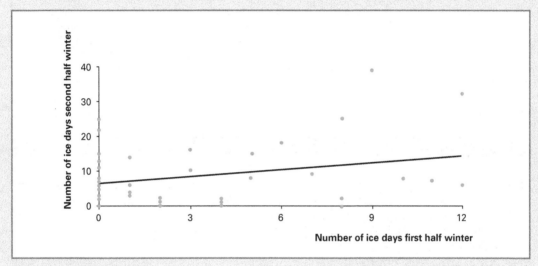

FIGURE 5.12 Scatter plot of *w* on *v*
Source: Royal Netherlands Meteorological Institute

With the help of a computer it immediately follows that:

$$\bar{v} = 3.3256; \quad \bar{w} = 8.2326; \quad s_V^2 = 15.6058; \quad s_{V,W} = 10.1606; \quad r = 0.2853$$

Notice that the v data and the w data are weakly positively correlated.

With these results, the equation of the sample regression line can easily be calculated:

$$b_1 = \frac{s_{V,W}}{s_V^2} = \frac{10.1606}{15.6058} = 0.6511$$

$$b_0 = \bar{w} - b_1\bar{v} = 8.2326 - 0.6511 \times 3.3256 = 6.0674$$

So, the equation of the regression line is $\hat{w} = 6.0674 + 0.6511v$. It is this line that is included in the scatter plot.

We next give interpretations of the slope and the intercept. For each additional ice day during the second half of a year, the number of ice days in the first half of the next year is estimated to increase by 0.6511. Since 0 falls in the range of the data of the independent variable, it can be used to interpret the intercept of the regression line: the intercept 6.0674 is just the estimated number of ice days in the first half of a year for which there were no ice days in the second half of the year before it.

Computer packages can be used to determine the precise regression coefficients, which are set out in Table 5.7. (See Appendix A1.5 for the creation of regression printouts with Excel.)

Regression Statistics	
Multiple R	0.285291923
R Square	0.081391482
Adjusted R Square	0.058986396
Standard Error	8.745494171
Observations	43

ANOVA

	df	SS
Regression	1	277.8440184
Residual	41	3135.8304
Total	42	3413.674419

	Coefficients	Standard Error
Intercept	6.067343173	1.751920288
X Variable 1	0.651078626	0.341599532

TABLE 5.7 Part of Excel printout for the regression of w on v

At the moment, only the highlighted numbers in Table 5.7 are important. The correlation coefficient is reported under Multiple R. In the column Coefficients we see that $b_0 = 6.0673$ and $b_1 = 0.6511$, in accordance with the calculations before. Although the printout does not mention the predictions $\hat{w}_i$ and the residuals $e_i = w_i - \hat{w}_i$, it **does** mention SSE, the sum of the squared **residuals** in the column SS:

SSE = 3135.8304

This statistic measures the variation of the dots in the scatter plot around the regression line.

In the **second** half of 1988 there were no ice days, that is, $v_{44} = 0$. However, the number of ice days w_{44} in the first half of 1989 is not known. By using the regression line we can predict w_{44}:

$\hat{w}_{44} = 6.0673 + 0.6511 \times v_{44} = 6.0673 + 0.6511 \times 0 = 6.0673 \approx 6$

Enquiry at the Royal Netherlands Meteorological Institute yields that the first half of 1989 had 3 ice days, which is different from the prediction 6.

CASE 5.1 WOMEN'S WORLD RECORDS APPROACH MEN'S WORLD RECORDS – SOLUTION

Interest is in the relationships between the variables W = 'world record of women at a certain moment in time' and M = 'world record of men at that moment in time'. Will, in the long run, the observations of these variables have more or less the same level? We will consider the 100 metres and the marathon.

It is intuitively clear that it is not M that drives W, at least not directly. It makes more sense to introduce the variable T = 'time', since time represents the improving training conditions and scientific developments with respect to knowledge about the human body and training improvements. The relationship between W and M evolves indirectly by way of the external developments that in a sense are included in time.

The Figures below depict the world records over time.

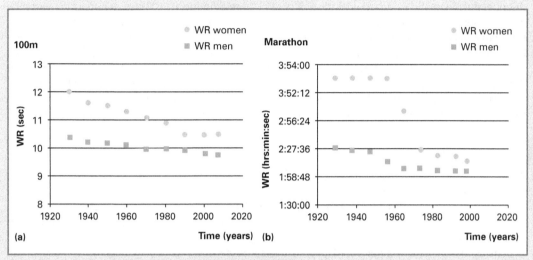

World records in (a) the 100 metres and (b) the marathon, 1920–2007

To forecast the future with respect to these world records, we have to find a 'best-fitting' curve and to extrapolate the data.

When fitting a linear curve as described in this chapter, the equations of the linear regression lines of w and m on time t are:

	Women	**Men**
100 m	$w = -0.0201618t + 50.80427184$	$m = -0.00759651t + 24.9937026$
Marathon	$w = -0.000958858t + 2.014684394$	$m = -0.000241896t + 0.56991462448$

(*Note*: For the marathon, the unit of w and m in the equation is **day**, so 1 unit corresponds to 24:00:00 and, for instance, 0.1 corresponds to 8640 seconds and hence to 2:24:00.) For both the 100 metres and the marathon, the regression line of the women decreases faster than the regression line of the men, so they will finally intersect. The intersection times can be calculated by equating the two expressions. Here are the results:

| 100m: | Intersection in the year 2054 | World record: 9.39 |
| Marathon: | Interssection in year 2015 | World record: 1:58:45 |

Intersection will occur once. For the marathon, it will happen relatively soon.

However, didn't we stylize things too much by adopting straight lines as best-fitting curves? Of course, the results will be different if other functions are chosen. For instance, the marathon scatter plots seem to suggest that third-order functions ($f(t) = a + bt + ct^2 + dt^3$) fit the plots well for extrapolation in the near future. Moreover, the 100 metre scatter plots indicate that the women's record did not change during 1990–2007. Does that mean that the women more or less reached their limit? Shouldn't we adopt a curve that takes that into account?

Of course, we cannot answer these questions only on the basis of the few data in the dataset.

5.3 Linear transformations

The unit (dimension) that is used to measure a certain characteristic is of great importance. For instance, the characteristic GDP is, in the EU, usually measured by the variable X = 'GDP in billions of euro' whereas the USA will use the variable Y = 'GDP in billions of dollars'. As a consequence, the mean and standard deviation of data of X are amounts in 10^9 euro whereas the variance is an amount in 10^{18} euro^2. However, the mean, standard deviation and variance of data of Y are, respectively, amounts in 10^9 dollars, 10^9 dollars and 10^{18} dollars^2. The question now arises how statistics (such as mean, standard deviation and variance) regarding one currency can be transformed into the corresponding statistics of another currency. Since 1 euro is worth 1.3 dollars (November 2006), the linear transformation $Y = 1.3X$ will play an important role in the above GDP example.

Here is another example. Temperatures can be measured in the units 'degrees Celsius (°C)' or 'degrees Fahrenheit (°F)'; the first dimension is usually used in Europe and the second in the USA. The relationship is linear, as follows:

$$F = 32 + \frac{9}{5}C$$

Again, it is important to know how sample or population statistics that are measured in °C are transformed into statistics in °F.

 In this section, emphasis is on **_linear transformations_** $Y = a + bX$, where X and Y are variables and a and b are constants. We investigate the properties of such transformations with respect to statistics of datasets of X.

The upper dot plot in Figure 5.13 depicts the 25 observations of X = 'GNIpc ($\times$ €1000)' of the EU25 countries. To show how transformations work, these x data are transformed by way of $y = 1.3x$ into a dataset of the variable $Y = 1.3X$. Note that Y measures GNIpc in thousands of dollars. For instance, the observation 26.6 of Austria transforms into $1.3 \times 26.6 = 34.58$. The resulting y data are shown in the lower dot plot.

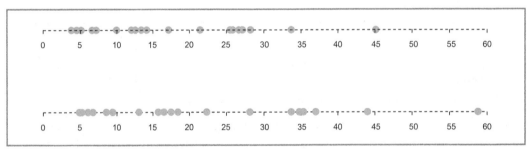

FIGURE 5.13 Dot plots of x and y data with $y = 1.3x$

The centres of location in the upper dot plot are somewhere around 18, whereas the centres of the lower dot plot are larger. Furthermore, the lower dataset is more spread out and seems to have more variation. Mathematical arguments will be used to find the relationships for the general transformation problem.

Suppose that a population dataset of a variable X is transformed with the linear transformation $y = a + bx$. We first consider what happens to the median of X.

The observations in the dataset are formally denoted as $x_1, x_2, \ldots, x_N$. By the linear transformation $y = a + bx$ they transform into

$$a + bx_1, \quad a + bx_2, \quad \ldots, \quad a + bx_N$$

the observations of the variable $Y = a + bX$. Notice that:

- For positive b, the ordering of the x data does not change with multiplication by b or by adding the constant a to the results.

- For negative b, the ordering of the x data is completely **reversed** by multiplication with b and this reversed ordering is left unchanged by adding the constant a to the results.

Hence, both for positive and negative b, the middlemost observation (or pair of observations) of the ordered y data is just the transformed observation (or pair of observations) of the ordered x data. Consequently, the median of the y data arises by applying the linear transformation $y = a + bx$ to the median of the x data. This also holds for $b = 0$ since in this case all x data transform into the same number a, and the median of a dataset with all observations equal to a, is a. A similar relation holds for a sample dataset.

Derivation of transformation rules for mean and variance

Again, start with a population dataset of X. Note that μ_Y is the mean of $a + bx_1$, $a + bx_2$, ..., $a + bx_N$. So:

$$\mu_Y = \frac{1}{N}\sum_{i=1}^{N}(a + bx_i) = \frac{1}{N}\left(\sum_{i=1}^{N}a + \sum_{i=1}^{N}bx_i\right) = \frac{1}{N}\sum_{i=1}^{N}a + \frac{1}{N}\sum_{i=1}^{N}bx_i$$

$$= \frac{1}{N}\sum_{i=1}^{N}a + b\frac{1}{N}\sum_{i=1}^{N}x_i = a + b\,\mu_X$$

Consequently, the mean of the y data arises by applying the linear transformation $y = a + bx$ to the mean of the x data.

Note that σ_Y^2 is the mean of the squared deviations of the y data with respect to μ_Y, that is, the mean of the squares of the deviations:

$$a + bx_1 - \mu_Y, \quad a + bx_2 - \mu_Y, \quad ..., \quad a + bx_N - \mu_Y$$

From the properties of the summation operator Σ it follows that:

$$\sigma_Y^2 = \frac{1}{N}\sum_{i=1}^{N}(a + bx_i - \mu_Y)^2 = \frac{1}{N}\sum_{i=1}^{N}(a + bx_i - (a + b\mu_X))^2$$

$$= \frac{1}{N}\sum_{i=1}^{N}(a + bx_i - a - b\mu_X)^2 = \frac{1}{N}\sum_{i=1}^{N}(bx_i - b\mu_X)^2$$

$$= \frac{1}{N}\sum_{i=1}^{N}b^2(x_i - \mu_X)^2 = b^2\frac{1}{N}\sum_{i=1}^{N}(x_i - \mu_X)^2 = b^2\,\sigma_X^2$$

In the case of sample datasets, similar relations can be derived.

It follows that the mean of the y data arises by applying the linear transformation $y = a + bx$ to the mean of the x data. The variance of the y data arises from the variance of the x data by multiplication with b^2. In particular, the variance of the y data depends only on the constant b and **not** on the constant a. This intuitively makes sense: adding a constant to all observations in a dataset shifts the complete dataset to another location, but the variation within the dataset remains unchanged.

To obtain the standard deviation σ_Y, the square root of $b^2\,\sigma_X^2$ has to be taken. For negative b, the answer is **not** $b\sigma_X$ (since this is negative) but $|b|\sigma_X$. These relationships are summarized in Table 5.8.

	Population dataset	**Sample dataset**
Location	$\mu_Y = a + b\mu_X$	$\bar{y} = a + b\bar{x}$
	$\mu_{Y\text{median}} = a + b\mu_{X\text{median}}$	$y_{\text{median}} = a + bx_{\text{median}}$
Variation	$\sigma_Y^2 = b^2\sigma_X^2$	$s_Y^2 = b^2 s_X^2$
	$\sigma_Y = |b|\sigma_X$	$s_Y = |b|s_X$

TABLE 5.8 Transformation of statistics under $y = a + bx$

For the GNIpc example with $Y = 1.3X$, recall that median, mean, standard deviation and variance of the x data are respectively:

$$\mu_{\text{Xmedian}} = 17.1; \quad \mu_x = 18.228; \quad \sigma_x = 11.2048; \quad \sigma_x^2 = 125.5479$$

The centres of location μ_{Ymedian} and μ_y can easily be calculated; they are respectively equal to 1.3 × 17.1 = 22.23 and 1.3 × 18.228 = 23.6964. For the variance and the standard deviation of the y data, we obtain:

$$\sigma_Y^2 = (1.3)^2 \times 125.5479 = 212.1760$$

$$\sigma_Y = |\ 1.3\ | \times 11.2048 = 1.3 \times 11.2048 = 14.5662$$

Next, suppose that a population dataset is given with **pairs** of observations, of variables X and Y. Then, the covariance $\sigma_{X,Y}$ and the correlation coefficient $\rho_{X,Y}$ of the x data and the y data can be considered. Suppose, further, that both the x data and the y data are linearly transformed: the x data by the linear transformation $V = a + bX$ and the y data by the linear transformation $W = c + dY$, where a, b, c and d are constants. The relationship between $\sigma_{V,W}$, $\rho_{V,W}$ and $\sigma_{X,Y}$, $\rho_{X,Y}$ will be studied. Again, the properties of the summation operator Σ are important.

Derivation of transformation rules for covariance and correlation

Recall that the population dataset can be denoted as:

$$(x_1, y_1), (x_2, y_2), \ldots, (x_N, y_N)$$

Since the x data and the y data are transformed by the linear transformations $v = a + bx$ and $w = c + dy$, the observations of V and W can be denoted, respectively, by:

$$a + bx_1, a + bx_2, \ldots, a + bx_N \quad \text{and} \quad c + dy_1, c + dy_2, \ldots, c + dy_N$$

Since $\mu_V = a + b\mu_x$ and $\mu_W = c + d\mu_y$, the deviations of the v data (with respect to μ_V) and the w data (with respect to μ_W) can be written as:

$$a + bx_i - \mu_V = a + bx_i - (a + b\mu_x) = b(x_i - \mu_x)$$

and

$$c + dy_i - \mu_W = c + dy_i - (c + d\mu_y) = d(y_i - \mu_y)$$

Recall that the covariance is the mean of the products of the two types of deviations:

$$\sigma_{V,W} = \frac{1}{N} \sum_{i=1}^{N} b(x_i - \mu_x)d(y_i - \mu_y) = bd \frac{1}{N} \sum_{i=1}^{N} (x_i - \mu_x)(y_i - \mu_y) = bd\sigma_{X,Y}$$

To obtain $\rho_{V,W}$ the last result has to be divided by $\sigma_V\sigma_W$, that is, by the product of $|b|\sigma_x$ and $|b|\sigma_y$:

$$\rho_{V,W} = \frac{\sigma_{V,W}}{\sigma_V\sigma_W} = \frac{bd\sigma_{X,Y}}{|b|\sigma_x|d|\sigma_y} = \frac{bd}{|bd|}\rho_{X,Y}$$

The constants a and c are not included in the results that transform $\sigma_{X,Y}$ and $\rho_{X,Y}$ into $\sigma_{V,W}$ and $\rho_{V,W}$. The correlation coefficient of the v and w data is the same as the correlation coefficient of the original x and y data if b and d are both positive or both negative. If b is positive and d is negative (or the other way around), then $\rho_{V,W} = -\rho_{X,Y}$. These relationships are summarized in Table 5.9.

	Population datasets	**Sample datasets**
Covariance	$\sigma_{V,W} = bd\,\sigma_{X,Y}$	$s_{V,W} = bd\,s_{X,Y}$
Correlation coefficient	If $bd > 0$: $\rho_{V,W} = \rho_{X,Y}$	$r_{V,W} = r_{X,Y}$
	If $bd < 0$: $\rho_{V,W} = -\rho_{X,Y}$	$r_{V,W} = -r_{X,Y}$

TABLE 5.9 Transformation of covariance and correlation by $v = a + bx$ and $w = c + dy$

Example 5.11

Consider the population of all husband–wife couples where both people have annual incomes of at least €3000. Let X denote the annual income in euros of the husband within the couple and let Y denote the annual income in euros of his wife. Suppose that the government has decided to stimulate the financial positions of the wives within the couples by letting them pay less tax: the husband pays 30% of his annual income above €3000 while his wife only pays 25% of her annual income above €3000. That is, if V and W denote the amounts of taxes paid by the husband and his wife, respectively, then it holds that:

$$V = 0.3 \times (X - 3000) = 0.3X - 900$$

$$W = 0.25 \times (Y - 3000) = 0.25Y - 750$$

Suppose that X and Y are rather strongly linearly related. What are the consequences for the correlation coefficient of V and W?

From table 5.8 it follows that $\sigma_V = 0.3\sigma_X$ and $\sigma_W = 0.25\sigma_Y$. By applying Table 5.9 with $b = 0.3$ and $d = 0.25$, it follows that:

$$\sigma_{V,W} = 0.3 \times 0.25 \times \sigma_{X,Y} = 0.075\,\sigma_{X,Y} \quad \text{and} \quad \rho_{V,W} = \rho_{X,Y}$$

Example 5.12

GDPs of neighbouring countries are often strongly correlated. The file Xmp05-12.xls contains the GDPs of Belgium and the Netherlands (in US dollars) for 1999–2003. Means, standard deviations, variances, covariance and correlation coefficient can easily be calculated (Table 5.10).

Country	Mean	St. dev.	Variance	Covariance	Correlation coeff.
Belgium	250.63×10^9	30.478×10^9	928.9×10^{18}		
Netherlands	416.63×10^9	55.926×10^9	3127.7×10^{18}	1659.55×10^{18}	0.974

TABLE 5.10 Summarized results for GDPs of Belgium and the Netherlands, 1999–2003
Source: World Bank (2005), devdata

But what about the amounts spent on education during the same period in the two countries?

Suppose that the government of the Netherlands spends 4.99% of GDP on education each year and that Belgium spends 6.12%. Then, the variables

V = 'education expenditure in the Netherlands (in dollars)'
W = 'education expenditure in Belgium (in dollars)'
X = 'GDP (in dollars) in the Netherlands'
Y = 'GDP (in dollars) in Belgium'

are related by $V = 0.0499X$ and $W = 0.0612Y$. The statistics for the v and the w data can easily be calculated with the rules for linear transformations in Tables 5.8 and 5.9. Here are the results:

Belgium: $\quad \bar{w} = 15.34 \times 10^9 \qquad s_w = 1.865 \times 10^9 \qquad s_w^2 = 3.479 \times 10^{18}$

Netherlands: $\bar{v} = 20.79 \times 10^9 \qquad s_v = 2.791 \times 10^9 \qquad s_v^2 = 7.788 \times 10^{18}$

Furthermore, $s_{v,w} = 5.068 \times 10^{18}$ and $r_{v,w} = 0.974$.

The correlation coefficient of the expenditures on education for Belgium and the Netherlands is exactly the same as the correlation of the GDPs of the two countries.

For regression of the w data on the v data, the regression coefficients follow easily:

$$b_1 = \frac{s_{v,w}}{s_v^2} = 0.6507 \quad \text{and} \quad b_0 = \bar{w} - b_1 \times \bar{v} = 1.8110 \times 10^9$$

Hence, the sample regression line for the (v, w) data has equation $\hat{w} = 1.8110 \times 10^9 + 0.6507v$.

5.4 Relationship between two qualitative variables

Now suppose that the two variables are both qualitative. It is obvious that the concepts 'covariance' and 'correlation coefficient' make no sense. Furthermore, in computer packages the option 'scatter plot' often cannot be applied. However, it **is** possible to create a **contingency table** (or **cross-classification table**) that gives the joint frequencies of the (data of the) two variables. These concepts will be illustrated by way of examples.

Example 5.13

The dataset Xmp05-13.xls contains information about the winners of the Champions League cup and their home countries, for the period 1956–2007. It turns out that the 52 cups were won by 21 teams from 10 countries (see Table 5.11). See also Appendix A1.5.

Count of Year	Country										
Winner	England	France	Germany	Italy	Netherlands	Portugal	Romania	Scotland	Serbia	Spain	Total
AC Milan				7							7
Ajax					4						4
Aston Villa	1										1
Barcelona										2	2
Bayern Munich			4								4
Benfica						2					2
Borussia Dortmund			1								1
Celtic								1			1
FC Porto						2					2
Feyenoord					1						1
Hamburg			1								1
Inter Milan				2							2
Juventus				2							2

continued

continued

Count of Year	Country										
Winner	England	France	Germany	Italy	Netherlands	Portugal	Romania	Scotland	Serbia	Spain	Total
Liverpool	5										5
Manchester United	2										2
Nottingham Forest	2										2
Olympique Marseille		1									1
PSV Eindhoven					1						1
Real Madrid										9	9
Red Star Belgrade									1		1
Steaua Bucharest							1				1
Grand Total	10	1	6	11	6	4	1	1	1	11	52

TABLE 5.11 Contingency table (Excel) for frequencies of country by winner
Source: http://www.xs4all.nl

The variables of interest are 'winning team of the year' and 'winning country of that year'; they are considered on the population of the years 1956 to 2007. The values of the first variable are in the rows of the table and the values of the second in the columns. The table gives an immediate overview of the countries and the teams that won the cup. For instance, it follows immediately that the cup was won six times by a German team: four times by Bayern Munich and once by Borussia Dortmund and Hamburg.

Example 5.14

A firm that markets goods targeting youth in several European countries asked the marketing company MarketView to investigate whether young consumers from different countries have a common view with respect to ethical questions. (Prior to the EU the firm had different policies for each country, as it had found that in some countries its consumers were less ethical in their dealing with the firm than in others.)[†]

Among others, MarketView interviewed randomly sampled undergraduate business students at universities in several countries about their opinion on the ethical question: 'drinking a can of cola in a supermarket without paying for it' (which was called quest2). The possible values were 1, 2, 3, 4 and 5, where 1 corresponds to 'strongly believe it is wrong' and 5 to 'strongly believe it is **not** wrong'. The country of the respondents was also recorded. The dataset Xmp05-14.xls contains the results of the research. Here, 'country' takes the (coded) values 1 to 8:

Country:	1 = Portugal	2 = Spain	3 = Denmark	4 = Scotland
	5 = Germany	6 = Italy	7 = Greece	8 = Netherlands

Table 5.12 gives the joint frequencies for the 40 country–quest2 combinations (also called cells). The omission of a number in a cell means that the frequency of that cell is 0. For instance, in Denmark (country 3), 151 business students were interviewed; the accompanying frequency distribution of quest2 can be found in the row of 3.

[†] See also 'Consumer ethics in the EU: a comparison of northern and southern views', *Journal of Business Ethics* (2001), 117–130.

country	quest2					
	1	**2**	**3**	**4**	**5**	**grand total**
1	82	27	9	4	1	123
2	52	17	12	2	1	84
3	114	29	6	1	1	151
4	69	61	10			140
5	54	19	10	4	4	91
6	73	24	7	7	8	119
7	53	76	18			147
8	78	11	2			91
grand total	575	264	74	18	15	946

TABLE 5.12 Contingency table for frequencies of quest2 by country
Source: http://u2.newcastle.edu.au

Of course, the firm is especially interested in a comparison of the distributions per country. That is, the firm wants to compare the **relative** frequency distributions that belong to the rows of the table. Figure 5.14 presents the graphs.

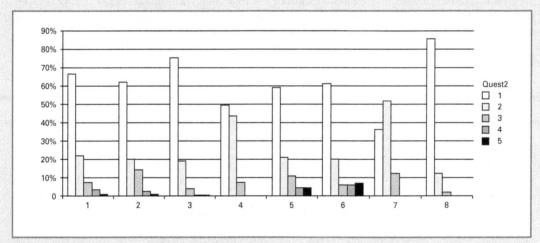

FIGURE 5.14 (Excel) Relative frequency distributions (%) of quest2 by country

The eight relative frequency distributions are obviously different. In particular, the results of the Netherlands and Denmark are much more concentrated on the value 1 than are the results of the other countries.

Classified frequency distributions of continuous – and hence quantitative – variables can be considered as frequency distributions of the accompanying **qualitative** variables with the classes of the classification as values. Sometimes – especially in articles and newspaper publications – interest is in the relationships between two quantitative variables but a dataset of observations is available only in terms of frequencies in a cross-classification table. See Exercise 5.22. Such cases are in essence examples of the present section: the row and column variables are just the classified versions of the original quantitative variables.

Summary

In this chapter, emphasis was mainly on a combined dataset of a **pair** of quantitative variables, often called X (independent variable) and Y (dependent variable). Scatter plots are very important for getting a first impression of the relationship between the x data and the y data, and hence of the relationship between X and Y. To measure the degree of **linear** relationship, the measures 'covariance' and 'correlation' were introduced. In particular, the coefficient of correlation is useful because it has a reference point: a strongly positive (or negative) linear relationship corresponds to a correlation coefficient close to 1 (or -1). On the other hand, a correlation coefficient close to 0 suggests that the x data and the y data are barely linearly related. Notice that covariance and correlation measure only the degree of **linear** relationship.

The equation of the straight line that 'fits best' to the scatter plot can easily be determined with the help of a computer. The formulae of intercept and slope of this **regression line** are given in the box below. Notice, especially, the difference between the formulae for ρ and β_1, and r and b_1.

Formulae for covariance, correlation and regression coefficients

	Covariance	Correlation coeff.	Slope	Intercept
Population	$\sigma_{X,Y} = \dfrac{1}{N}\sum_{i=1}^{N}(x_i - \mu_X)(y_i - \mu_Y)$	$\rho = \dfrac{\sigma_{X,Y}}{\sigma_X \sigma_Y}$	$\beta_1 = \dfrac{\sigma_{X,Y}}{\sigma_X^2}$	$\mu_Y - \beta_1\mu_X$
Sample	$s_{X,Y} = \dfrac{1}{n-1}\sum_{i=1}^{n}(x_i - \bar{x})(y_i - \bar{y})$	$r = \dfrac{s_{X,Y}}{s_X s_Y}$	$b_1 = \dfrac{s_{X,Y}}{s_X^2}$	$\bar{y} - b_1\bar{x}$

If the variables X and Y are both linearly transformed into variables V and W, then statistics of V and W follow easily from the corresponding statistics of X and Y; see Tables 5.8 and 5.9. The importance of the results in these tables is that it is not necessary to calculate the datasets of V and W to get their statistics.

List of symbols

Symbol (population, sample)	Statistic
$\sigma_{X,Y},\ s_{X,Y}$	Covariance
$\rho_{X,Y},\ r_{X,Y}$	Correlation coefficient
$\beta_0,\ b_0$	Intercept of regression line
$\beta_1,\ b_1$	Slope of regression line

🔑 Key terms

contingency table **154**
correlation coefficient **135**
covariance of Y on X **133**
cross-classification
    table **154**
dependent variable **126**
errors **146**
extrapolation **143**
grid-lines **129**
independent variable **126**
intercept **143**
interpolation **143**
least-squares line **143**
least-squares method **141**
line of means **143**
linear transformation **150**

negatively linearly
    related **129**
population cloud **127**
population correlation
    coefficient **135**
population covariance **133**
population regression
    coefficient **143**
population regression
    line **143**
positively linearly
    related **129**
prediction **143**
prediction line **143**
quadrant **129**
quadratic relationship **130**

regression of Y on X **141**
residuals **146**
sample cloud **127**
sample correlation
    coefficient **135**
sample covariance **133**
sample regression
    coefficient **143**
sample regression line **143**
scatter plot of y on x **127**
short-cut formula **133**
slope **143**
sum of squared errors
    (SSE) **146**
uncorrelated **136**

Exercises

Exercise 5.1

In each of the following statements something is wrong. **What** is wrong? Why?

a The covariance and the correlation coefficient of the variables X and Y are respectively equal to 2.420 and -0.267.

b The correlation coefficient of the x data and the y data is 1.851.

c For the variables X and Y it is given that $\sigma_{X,Y} = -45.1276$, $\sigma_X^2 = 25.0111$ and $\sigma_Y^2 = 64.2288$.

d The covariance of the uncorrelated variables X and Y is 12.465.

Exercise 5.2

Two variables X and Y are considered on a common population. The observations for a sample of size 5 are:

x	14	18	13	21	18
y	90	88	91	64	80

a Calculate the covariance and the correlation coefficient by using the definitions of these concepts in Section 5.1. (Use a table as in Example 5.3, but also include columns for the two types of squared deviations.)

b Interpret these values.

c Determine the equation of the sample regression line of y on x.

d Interpret the intercept and the slope.

Exercise 5.3

You want to explain the variation in the variable Y = 'daily number of hours studied' from the variation in the variable X = 'number of hours worked for your outdoor job' by doing a simple linear regression. For a random sample of 7 days you measure both X and Y. Here are the results.

x	8	5	4	6	2	5	3
y	1	3	6	3	7	2	5

a Create the scatter plot of y on x. What kind of relationship do you observe in the plot?

b Use the short-cut formulae to calculate the sample variances s_X^2 and s_Y^2, and the sample covariance $s_{X,Y}$.

c Calculate the sample correlation coefficient. Interpret the answer.

Exercise 5.4

A sample dataset of observations of the variables X and Y gives the following summarized results:

$\bar{x} = 12.2; \quad \bar{y} = 58.4; \quad s_X^2 = 3.2; \quad s_Y^2 = 10.8; \quad r_{X,Y} = 0.722$

The linear transformations $V = -2X + 4$ and $W = 3Y - 5$ are used to transform the x data and the y data; it leads to a sample dataset of v data and w data.

a Calculate $\bar{v}$, $\bar{w}$, s_V^2 and s_W^2.

b Also calculate $r_{V,W}$ and $s_{V,W}$.

Exercise 5.5

In a certain country, all adult inhabitants earn at least €3000 per year. Consider the population of all husband–wife couples in that country. Let X denote the annual income (in euro) of the husband within a couple and Y the annual income (in euro) of his wife. It is given that the mean and standard deviation of X are €20 000 and €6000, respectively; for Y the mean and the standard deviation are €15 000 and €5000. Furthermore it is given that the correlation coefficient of X and Y is 0.75.

To stimulate the financial position of the wives, the government has decided to let the wife within a couple pay a lower percentage for taxes than her husband. To be precise, the husband pays 30% of his annual income above €3000 whereas his wife only pays 25% of her annual income above €3000. Below, the variables V and W respectively denote the amounts of taxes (in euro) paid by the husband and his wife.

a Calculate the means and the standard deviations of V and W.

b Calculate the covariance and the correlation coefficient of V and W.

c The median of V is €5100. Calculate the median annual income of the men in husband–wife couples.

d Calculate, for the men in the couples as well as for the women, the mean and the variance of the annual **net** income (after taxes).

Exercise 5.6

You want to explain how the variable Y = 'daily number of hours that you study' varies with the variable X = 'daily number of hours that you watch television' by doing a simple regression of Y on X on the basis of a random sample of 7 days. For these days, you have the following observations of X and Y:

x	3	5	6	4	3	7	6
y	4	2	0	2	2	0	1

a Calculate the sample means $\bar{x}$ and $\bar{y}$, the sample variances s_x^2 and s_y^2, and the sample covariance $s_{x,y}$ and correlation coefficient r.

b Determine the equation of the sample regression line of y on x and interpret its slope in the context of the problem of this exercise.

c For tomorrow, you plan to watch TV for only 1.5 hours. Give a prediction of the number of hours that you will study tomorrow.

d Calculate the residual e_3 (that belongs to the third pair of observations). Interpret the answer.

e The other six residuals are 1, 0.5385, −0.2308, −1, 0.0769 and 0.3077. Calculate SSE and interpret the answer.

Exercise 5.7

The manager of a supermarket wants to regress Y = 'sales per week of the company (in units of 10^5 dollars)' on X = 'costs of advertisements in that week (in units of 10^4 dollars)'. A quick study, for six arbitrary weeks, gives the following statistics:

$$\bar{x} = 3.5; \quad s_x^2 = 3.5; \quad \bar{y} = 8.3333; \quad s_y^2 = 31.8667; \quad r_{X,Y} = 0.7007$$

a Calculate the covariance of the underlying x data and y data.

b Calculate the slope of the sample regression line of y on x and interpret it in the context of this exercise.

Suppose that €1 is worth $1.30. Let W and V respectively measure the weekly sales (in 10^5 euro) and the weekly costs of advertisements (in 10^4 euro), so W and V are linear transformations of respectively Y and X.

c Calculate the means, variances, covariance and correlation coefficient of the v and the w data that follow from the x and y data.

d Calculate the slope of the sample regression line of w on v.

Exercise 5.8

To investigate the general relationship between inflation rate and GDP growth, the variables Y = 'inflation (%)' and X = 'growth GDP (%)' are measured for the six countries Belgium, France, Germany, Luxembourg, the Netherlands and UK. The results for this sample are in the table.

Country	Inflation (%)	Growth GDP (%)
Germany	1.7	1.1
France	1.8	1.7
Luxembourg	2.5	3.6
Belgium	2.5	1.8
Netherlands	1.5	0.1
UK	1.9	2.7

Economic indicators for six countries
Source: Bloomberg (2005)

Below, use only a pocket calculator.

a Construct the scatter plot of y on x. Give your comments.

b Calculate the mean and the variance of both the x data and the y data.

c Calculate the covariance and the correlation coefficient of the x and y data. Interpret them.

d Determine the regression line of y on x. Check that the pair of means indeed falls on the regression line.

Exercise 5.9

Consider two variables X and Y on a (common) large population. Suppose that the means of X and Y are respectively 2 and 7, their variances are 9 and 16, and their covariance is 10. Interest is in the linear transformations $V = 4 + 3X$ and $W = 5 - 2Y$.

a Calculate the means and standard deviations of V and W.

b Calculate the correlation coefficient of X and Y. Also calculate the covariance and the correlation coefficient of V and W.

c Determine the population regression coefficients of Y on X. Also determine the population regression line of W on V.

Exercise 5.10

Reconsider Exercise 5.9. Suppose that 50 population elements are arbitrarily chosen and their values of X and Y are observed. The resulting sample dataset has the following statistics:

$$x = 1.8; \quad \bar{y} = 7.2; \quad s_X^2 = 8.4; \quad s_Y^2 = 16.8; \quad s_{X,Y} = 10.8$$

a Calculate the sample correlation coefficient of the x data and the y data.

b Determine the equation of the sample regression line of y on x.

c Since V and W are linear transformations of respectively X and Y, the dataset of (x, y) observations could be transformed into a dataset of (v, w) observations if the first dataset were given. However, this dataset is **not** given. Still, you can calculate some statistics of the (v,w) dataset. Calculate

$$\bar{v} \quad \bar{w} \quad s_V^2 \quad s_W^2 \quad s_{V,W} \quad r_{V,W}$$

and the equation of the sample regression line of w on v.

Exercise 5.11 (computer, to learn mechanics of the package and the formula)

The variable Y = 'number of personal computers per 1000 people' is often used to measure the degree of prosperity in a country. The question arises whether this computer density of a country is linearly related to the variable X = 'number of inhabitants per country'. To get some impression of the problem, the EU10 data below are reconsidered.

Country	Y	X
Cyprus	269.89	764 967
Czech Republic	177.44	10 200 000
Estonia	210.33	1 358 000
Hungary	108.35	10 200 000
Latvia	171.75	2 338 000
Lithuania	109.75	3 469 000
Malta	255.05	397 000
Poland	105.65	38 200 000
Slovakia	180.36	5 379 000
Slovenia	300.60	1 994 000

Observations of Y and X for the EU10, 2002
Source: World Bank (2005)

a Use the short-cut formulae and a computer package to calculate, **with a table as in Example 5.4** (but include columns for x^2 and y^2), the sample covariance and the two sample variances. Also calculate the sample correlation coefficient of the x data and the y data.

b Determine the equation of the (sample) regression line; interpret the regression coefficients; use enough decimals.

c For the Czech Republic, calculate the predicted number of PCs per 1000 people. Also calculate the accompanying residual.

Exercise 5.12

Show that $\bar{e}$, the mean of the residuals $e_1, e_2, \ldots, e_n$ of the sample regression line, is always equal to 0. (*Hint:* $e_i = y_i - \hat{y}_i$ and $\hat{y}_i = b_0 + b_1 x_i$; show that $\bar{e} = \bar{y} - b_0 - b_1\bar{x}$ and use the formula of b_0 in (5.7).)

Exercise 5.13

a Explain why it is impossible for two variables X and Y to have covariance 7 and variances 4 and 9.

b Explain why it is impossible for two variables X and Y to have covariance -7 and variances 4 and 9.

c Consider a variable X and a linear transformation $Y = a + bX$. Show that $\sigma_{X,Y} = b\sigma_X^2$ and that $\rho_{X,Y} = 1$ or -1. What are the consequences for the scatter plot of the population dataset (if observed)?

d Consider a variable X for which $\sigma_X^2 = 4$. Determine a linear transformation Y of X such that $\sigma_Y^2 = 9$ and $\rho_{X,Y} = 1$. Same question for $\rho_{X,Y} = -1$.

Exercise 5.14

When the economy is growing, finance ministers often fear that inflation will be growing too. In this exercise we will study whether this fear is justified.

The file Xrc05-14.xls contains data from the Federal Statistical Office of Germany about X = 'GDP growth (%) of a country in 2005' and Y = 'inflation rate (%) of that country in 2005'. The dataset contains measurements of both X and Y for 165 countries. We will use the data of 164 of these countries to study the relationship between Y and X; the observations of Zimbabwe are excluded since the inflation rate of 302.2 of this country would influence the results too much. Here is a part of the Excel regression printout.

SUMMARY OUTPUT

Regression Statistics

Multiple R	0.218451494
R Square	0.047721055
Adjusted R Square	0.04184279
Standard Error	5.103699849
Observations	164

ANOVA

	df	*SS*
Regression	1	211.4614088
Residual	162	4219.735847
Total	163	4431.197256

	Coefficients	*Standard Error*
Intercept	4.52417856	0.717391373
X Variable 1	0.348854463	0.12243733

a Why do we take inflation rate as dependent and GDP growth as independent variable (and not the other way around)?

b Use the printout to determine the equation of the regression line. Interpret the regression coefficient.

c In 2005, Zimbabwe had a GDP growth of -7.1%. Calculate the predicted inflation rate for Zimbabwe. Is this extrapolation? (The smallest x observation is -3.6).

d Calculate the predicted inflation for Denmark, with a GDP growth of 3.2% and inflation 1.7%. Also calculate the accompanying residual.

e Determine the correlation coefficient of inflation rate and GDP growth from the printout. Interpret the result.

f Determine SSE from the printout and interpret your answer. Explain how you could have derived this number yourself (if you had had enough time).

Exercise 5.15

Temperatures in Europe are usually measured in degrees Celsius (°C), in the USA they are measured in degrees Fahrenheit (°F). The relationship is as follows:

$$F = \frac{9}{5}C + 32$$

The file Xrc02-19a.xls contains the yearly global mean world temperatures for the period 1866–1996, in degrees Celsius. With a computer it follows easily that mean, median, variance and standard deviation of these data (call them x data) are respectively equal to:

$$\bar{x} = 14.8915; \quad x_{\text{median}} = 14.9100; \quad s^2 = 00559; \quad s = 0.2364$$

a Determine the dimensions of these sample statistics.

b Calculate the mean, median, variance and standard deviation after the data are transformed into degrees Fahrenheit. (Denote the transformed data by y.)

c Determine the coefficient of variation for the original data. Notice that this statistic is dimensionless. Does this mean that the coefficient of variation of the y data is the same as the coefficient of variation of the original data?

Exercise 5.16

Consider the population of all married couples (man, woman) in Germany. Let X and Y denote the gross annual incomes (2004) of the man and his wife, respectively. In this population, n couples are selected, and X and Y are measured for each. The resulting n pairs (x_1, y_1), (x_2, y_2), ..., (x_n, y_n) are studied.

Denote the sample variance of the x data by s_x^2; denote the sample covariance and the sample correlation coefficient of the n pairs respectively by $s_{X,Y}$ and $r_{X,Y}$. Below, you may take the 2004 inflation level in Germany as 1.7% (compared with 2003). Denote the real annual income in 2004 (that is, the annual income of 2004 measured in 2003 euro) for man and woman by U and V, respectively.

(*Hint*: Because of inflation, one 2003-euro has the same value as 1.017 2004-euro. Hence, x 2004-euros are worth $x/1.017$ 2003-euros.)

a Express U in X; express V in Y.

b Express s_U^2 in s_X^2; express s_U in s_X.

c Express $s_{U,V}$ in $s_{X,Y}$; express $r_{U,V}$ in $r_{X,Y}$.

Exercise 5.17 (computer)

In Figure 5.2(b) the relationship between the variables Y = 'budget deficit (% of GDP)' and X = 'GDP growth (%)' for the 12 eurozone countries has been depicted. The plot seems to suggest that there exists a (rather weak) negative linear relationship between the y data and the corresponding x data. To obtain more information about the strength of this linear relationship, the data that underlie the figure will be used; see Xrc05-17.xls. (The percentage budget deficit for Ireland was unknown; the value 0 is just as an approximation.) All questions below concern the eurozone.

   a Check that – in the eurozone – only Finland has a negative budget deficit, so a positive budget remainder.
   b Calculate the mean and the standard deviation of the x data. Do the same for the y data.
   c Calculate the sample covariance and the sample correlation coefficient of the x and y data.
   d Determine the sample regression line of y on x. Interpret the regression coefficients.
   e Now consider U = 'budget **remainder** (% of GDP)' instead of Y. Answer parts (b)–(d) again when Y is replaced by U; do not use a computer.

Exercise 5.18 (computer)

The file Xrc05-18.xls contains data about mileage (Y, in miles) and age (X, in years) of 22 cars.[†]

   a Create the scatter plot of the mileage data on the age data. Include the regression line in the plot. Also determine the equation of the line. Interpret the slope of the line.
   b Calculate the covariance and the correlation coefficient. Interpret the results.

   It is additionally given that the first eleven cars in the dataset are driven by petrol and the last eleven by diesel.

   c Create, in one figure, the separate scatter plots for petrol and diesel cars. Determine the regression equations for both types of cars separately. Add the lines to the plots.
   d Interpret and compare the slopes of the two lines. Give your comments.
   e Determine, for each of the two types of cars, the correlation coefficients of the age data and the mileage data. Give your comments.

Exercise 5.19

The share of e-commerce in European commercial activity is gradually increasing. To learn more about the relationship between e-commerce activity and broadband connection, we consider the following two variables on the population EU25:

   ■ percentage of individuals that have ordered/bought goods or services for private use over the internet in the last three months;
   ■ percentage of households with a broadband connection.

The data in the file Xrc05-19.sav come from Eurostat and are measurements for 2006.

   a In a regression study, which of the two variables will you choose as dependent variable Y?
   b The correlation coefficient of the two variables is equal to 0.749367. Interpret this number.
   c The (population) variances of the y and the x data are equal to 176.82560 and 333.68640, respectively. Calculate the covariance and the slope of the regression line of Y on X.
   d Furthermore, μ_Y = 18.12 and μ_X = 34.44. Determine the equation of the regression line. Interpret the regression coefficients.

[†] *Source*: The Mathematical Association (November 2003).

Exercise 5.20

Reconsider Exercise 5.19. The printout below is obtained with SPSS, the scatter plot with Excel.

Model Summary

Model	R	R Square	Adjusted R Square	Std. Error of the Estimate
1	.749(a)	.562	.542	9.17991

a Predictors: (Constant), PercHBroadband

ANOVA(b)

Model		Sum of Squares	df	Mean Square	F	Sig.
1	Regression	2482.413	1	2482.413	29.458	.000(a)
	Residual	1938.227	23	84.271		
	Total	4420.640	24			

a Predictors: (Constant), PercHBroadband
b Dependent Variable: PercEComm

Coefficients(a)

Model		Unstandardized Coefficients		Standardized Coefficients	t	Sig
		B	Std. Error	Beta		
1	(Constant)	−.667	3.918		−.170	.866
	PercHBroadband	.546	.101	.749	5.427	.000

a Dependent Variable: PercEComm

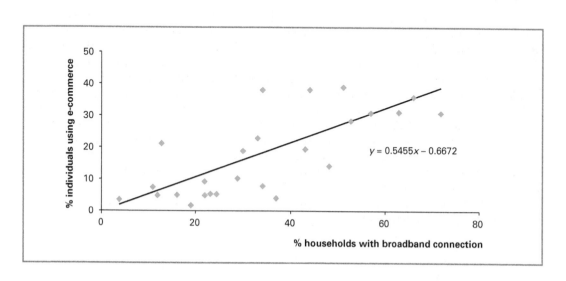

$y = 0.5455x − 0.6672$

a Use the **printout** to check the equation of the regression line in the scatter plot and your own answers to part (d) of Exercise 5.19.

b Determine the SSE and interpret it.

c The dot with the largest positive residual has $x = 34$ (which is close to the mean of the x data) and belongs to Germany. What does this tell about the relative position of Germany within the EU25 (as far as these variables are concerned)?

Exercise 5.21 (computer)

The US economy is the world's leading economy, so economic growth in the USA will also be good for economies in Europe. To study the relationship between the US economy and the Dutch economy, we will compare the GDP per capita figures – and also the GDPpc **growth** figures – of the USA and the Netherlands for the period 1950–2004. Let Y be the GDPpc (in dollars) in the Netherlands and X the GDPpc (in dollars) in the USA. The file Xrc05-21.xls contains the data.[†]

a Construct, in a single figure, time plots of the time series of X and Y. Give your comments.

b Construct the scatter plot of y on x. Give your comments.

The dependence of Y on X will be partly indirect since both time series develop over time. It can be expected that the growth figures (with respect to the year before) of the GDPpc data of the two countries behave much more independently of time. To find out whether this indeed is the case, the data of W = 'percentage GDPpc growth in the Netherlands' will be regressed on the data of V = 'percentage GDPpc growth in the USA'.

c Determine the columns with the w data and the v data. Notice that both columns have only 54 observations.

d Construct time plots of V and W in one figure. Also construct the scatter plot of w on v; add the regression line to this plot. Give your comments.

e Calculate the covariance and the correlation coefficient of the v and the w data. Interpret these results.

f Regress the w data on the v data. Determine the equation of the regression line. Also determine SSE. Interpret these results.

Exercise 5.22

In newspaper articles and other publications the relationship between two quantitative variables X and Y is sometimes described by way of frequencies in a cross-classification table, as in the table below. The following questions arise: How should such tables be interpreted? Although the precise data are not given, can we use the frequencies in the table to find approximations of the covariance and the correlation coefficient of the underlying dataset? An example is considered.

Rich countries with many inhabitants often have large GDPs. On the other hand, poor countries with many inhabitants have relatively small GDPs. To study the linear relationship between Y = 'GDP (billions of 2002 US dollars)' and X = 'midyear population ($\times 10^6$)' for comparable countries, the 24 countries with pre-1994 OECD membership are considered as a sample. Unfortunately, the exact observations are no longer given; the observations of both variables are classified and 'cross-tabled'. For instance, for four of the 24 countries the x observations fall in the class 50–100 and the y observations in the class 1000–3000.

By assuming that the observations in each of the classes are located in the centre, we still can learn something about important statistics of the (x, y) dataset.

[†] *Source*: Groningen Growth and Development Centre (August 2005).

		0–100	100–200	200–500	500–1000	1000–3000	3000–12000	Total
					y			
x	0-5	3	2					5
	5-10		2	3				5
	10-20		1	3	1			5
	20-50				2			2
	50-100				1	4		5
	100-500						2	2
	Total	3	5	6	4	4	2	24

Cross-classification table for 24 OECD members
Source: Groningen Growth and Development Centre and The Conference Board, Total Economy Database (August 2005)

a Use the frequencies in the table to approximate the means and standard deviations of the underlying x and y data.

b Find approximations of the covariance and the correlation coefficient of the underlying dataset of (x, y) observations.

c Also approximate the equation of the regression line of y on x. Interpret the coefficients.

d What do the statistics calculated in parts (a)–(c) tell about the underlying dataset of the original x and y data?

e (advanced) Suppose that the variable U measures midyear population in units of 1000 people; the variable V measures GDP in units of trillions of dollars. Express U and V in X and Y, and find approximations for the means, standard deviations, covariance, correlation coefficient, regression coefficients of the (u, v) dataset.

Exercise 5.23 (computer)

File Xmp03-07.xls contains the levels of the AEX index (on business days at closing time) for the period 25 April 2004 – 24 April 2005; see also Example 3.7. In total there were 256 business days in this period yielding the observations $x_1, x_2, \ldots, x_{256}$ of the variable $X =$ 'level AEX index at closing time'; see column 2 of the dataset.

The objective is to predict the AEX level X_t at business day t from the AEX level X_{t-1} at business day $t-1$. For that, the column of the dependent variable contains the AEX values **except the first one** (the y data) and the column for the independent variable contains the AEX values **except the last one** (the x data).

a Create the columns that you need. Next, create the scatter plot of y on x. Regress the y data on the x data, that is, create the printout.

b Determine the regression line and SSE. Interpret the result.

c Also calculate the correlation coefficient. Interpret it.

d Use the regression equation to predict the AEX level on 25 April 2005, the first business day after the last observation (on 21 April 2005).

Exercise 5.24 (computer)

Reconsider Exercise 5.23. Next, we want to predict the return r_t (a proportion, with respect to the day before) at business day t from the return r_{t-1} at business day $t-1$; see Example 3.7 for details about these returns.

a Create the columns that you need, construct a suitable scatter plot, and determine the printout of the regression of r_t on r_{t-1}.

b Determine the regression line and SSE.

c Also calculate the correlation coefficient. Give your comments.

d Interpret the results of parts (b) and (c). Compare the two regressions of this exercise and the former.

e Use the regression equation to predict the AEX level on 25 April 2005, the first business day after 21 April 2005 (when the AEX level was 353.26).

Exercise 5.25 (computer)

Reconsider the firm in Example 5.14. This firm is interested not only in differences between the views on the ethical question quest2 of young people from different countries but also in differences between males and females. Since the dataset Xmp05-14.xls also contains the gender variable 'sex', (0 = male and 1 = female) of the respondents in the eight countries, this problem can be solved in a way similar to Example 5.14.

a Create the contingency table for frequencies of quest2 by sex.

b Construct, in one graph, charts of the two relative frequency distributions for males and females.

c Give your comments.

CASE 5.2 MERCER QUALITY OF LIVING SURVEY

Mercer Human Resource Consulting makes an annual Worldwide Quality of Living Survey to help governments and multinational companies place employees on international assignments. For major cities all over the world they compose an index that is based on an evaluation of 39 criteria, including political, social, economic and environmental factors, personal safety and health, education, transport, and other public services.

Cities are ranked against New York as the base city, which has an index score of 100. For 2006, Zurich ranks as the world's top city for quality of living, according to the survey.[+] The city scores 108.2 and is only marginally ahead of Geneva, which scores 108.1, while Vancouver follows in third place with a score of 107.7. In contrast, Baghdad is the lowest-ranking city in the survey, scoring just 14.5.

Some of the results of the survey for 2005 and 2006 have been put in Case05-02.xls. Amsterdam is deleted from the original list, because we want to use it in one of the questions. That is why there are only 214 cities in the file, while the ranking ends with number 215.

The dataset will be used as an opportunity to repeat concepts of Chapters 2–5. In Chapter 5 we looked at correlation and regression.

a Create a scatter plot of the 2006 index on the 2005 index. Why is the 2006 index the dependent variable?

b Calculate the correlation coefficient for index 2005 with index 2006. Write down what this number indicates.

c Three cities are notable because their quality of living index for 2006 is clearly higher than it was in 2005. Find the names of these three cities.

d We now tell you that the quality of living index for 2005 for Amsterdam is equal to 105.7. Use the regression analysis technique introduced in Chapter 5 to estimate the 2006 index for Amsterdam.

[+] *Source*: Mercer Human Resource Consulting, *2006 Worldwide Quality of Living Survey*; see www.mercerhr.com.

We also want a frequency distribution for the 2006 data in order to repeat some of the techniques of Chapter 2.

e Create (for the 214 data from 2006) the classified frequency distribution of the classification with standard class width as shown in the table:

Class	Frequency	Percentage	Cumulative Percentage
(105, 110]			
(100, 105]			
⋮			
Total	214	100.0 %	−

Complete the table.

f Create the accompanying histogram.

We will also summarize the 2006 dataset by considering suitable statistics of Chapters 3 and 4.

g Give a five-number summary for the variable 'quality of living' by calculating: the minimum, the three quartiles and the maximum. Are there any outliers?

h Give a two-number summary by calculating the mean and the standard deviation. Are there any outliers?

CASE 5.3 ANSCOMBE'S QUARTET

When two variables X and Y are observed on n population elements, the equation of the regression line can be determined and several regression statistics can be calculated, such as the covariance, correlation coefficient, SSE, etc. The other way around, the question arises whether (and to what extent) essentials of a dataset of X and Y can be recovered from their statistics. Anscombe's Quartet shows four datasets that demonstrate that you have to be very careful when drawing conclusions about the dataset on the basis of statistics.

The table below shows four sample datasets of x and y data; see also Case05-03.xls.

x	y	x	y	x	y	x	y
10	8.04	10	9.14	10	7.46	8	6.58
8	6.95	8	8.14	8	6.77	8	5.76
13	7.58	13	8.74	13	12.74	8	7.71
9	8.81	9	8.77	9	7.11	8	8.84
11	8.33	11	9.26	11	7.81	8	8.47
14	9.96	14	8.1	14	8.84	8	7.04
6	7.24	6	6.13	6	6.08	8	5.25
4	4.26	4	3.1	4	5.39	19	12.5
12	10.84	12	9.13	12	8.15	8	5.56
7	4.82	7	7.26	7	6.42	8	7.91
5	5.68	5	4.74	5	5.73	8	6.89

Anscombe's datasets

Source: F. J. Anscombe, 'Graphs in statistical analysis', American Statistician, 27 (February 1973), 17–21

▶

a Use a computer to show that the statistics below are valid for **each** of the four datasets.

regression line of y on x:	$\hat{y} = 3.00 + 0.50x$
covariance:	5.50
correlation coefficient:	0.82
SSE:	13.8
variance of x data:	11
standard deviation of x data:	3.17
variance of y data:	4.1
standard deviation of y data:	2.03
mean of x data:	9
mean of y data:	7.50

b However, scatter plots of the datasets reveal distinct relations. Check it and give your comments.

PART 2
Probability Theory

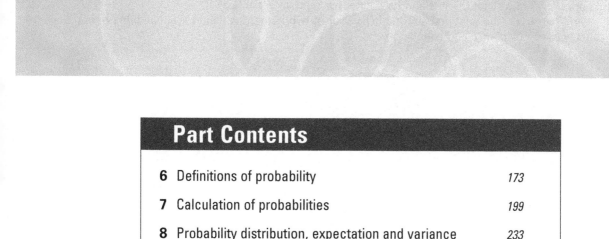

Part Contents	

The field of probability theory is very important for statistics, especially for inferential statistics. One of the main reasons for its importance is that probability arguments determine the precision of the conclusions about the whole population even though only a sample is observed. That is why some people consider probability theory to be a sub-field of statistics. However, probability is a science on its own; its results are also applied in many other fields, such as queuing theory and actuarial sciences.

The concept 'probability' makes sense only if **chance** and **uncertainty** are involved. That is why there has to be some underlying experiment for which the precise outcome is uncertain in advance. If a meteorological institute informs us that the probability of rain tomorrow is 80%, the underlying experiment is observing tomorrow's rainfall, with possible outcomes 'there will be rain' and 'there will be no rain'; apparently the probabilities of these outcomes are respectively 0.80 and 0.20.

There are two experiments in probability theory that are frequently considered: throwing a die and flipping a coin. In the first experiment the possible outcomes are 1, 2, 3, 4, 5 and 6, in the second experiment T (tail) and H (head). Indeed, in advance the outcomes of both experiments are uncertain. Although these experiments at first sight are rather dull, they turn out to serve as an example for many more complicated experiments.

There will be special emphasis on an experiment that bridges the gap between the fields of descriptive statistics and probability theory, between relative frequencies and probabilities. It is called **random drawing**; it chooses completely arbitrarily one element from the population of interest. Again, the outcome (one of the population elements) of this experiment is uncertain in advance. It will turn out that probabilities regarding this experiment correspond to relative frequencies in descriptive statistics.

Chapter 6 is about random experiments and the definition(s) of the concept 'probability', while Chapter 7 considers rules for probabilities and conditional probabilities.

In Chapter 8, **random** variables and overviews (called probability distributions) of their probabilities are studied from a general perspective. Concepts such as expectation and variance of a random variable enter the scene. Chapters 9 and 10 are about special families of probability distributions, respectively for discrete and for continuous random variables.

In Chapter 11, the joint probabilistic behaviour of **two** random variables is studied. Important concepts are covariance and correlation.

CHAPTER

Definitions of probability

06

Chapter contents

For probabilities, chance and uncertainty are necessary. That is why the concepts 'random experiment' and 'set of possible outcomes' have to be defined first. Since interest is often in special subsets of outcomes, some principles of set theory are also needed. Next, the concept 'probability' is considered from a historical perspective, leading to Kolmogorov's definition of probability.

CASE 6.1 CHANCES OF POSITIVE OR NEGATIVE RETURNS ON A PORTFOLIO

Investors tend to look with Argus' eyes at the daily movements of the important stock market price indexes. The reason is that recent daily prices and returns of these indexes give information about the current financial climate with respect to investments.

Consider an investor holding a balanced portfolio of Euronext 100 stocks. To analyse the chances of tomorrow's return of her portfolio, she decides to base the analysis completely on the daily returns of the Euronext 100 index during the most recent two years. During this period of 509 (business) days, the return was 0 or less on 211 days and on 179 of these days it was larger than −1%. On the positive side there were 53 days with a return of more than 1%.

From these considerations, the investor concludes that there is a 58.5% chance that tomorrow's return will be positive and 10.4% chance that it will even be more than 1%. However, there is 6.3% chance that tomorrow's return will be at most −1%.

How does the investor come to these conslusions? What are her precise arguments? See the end of Section 6.4 for the answers.

6.1 Random experiments

A **(*random*) *experiment*** is an experimentation or an uncontrollable phenomenon for which more than one outcome is possible. The result of a random experiment is not known with certainty in advance; it is – at least partially – determined by **chance**. The possible results of a random experiment are called **(*possible*) *outcomes***. The set of all possible outcomes is the **sample space** and will be denoted by Ω (capital omega).

The random experiment 'throw of a die' is an investigation that will lead to one of the six possible outcomes 1, 2, 3, 4, 5, 6; so $\Omega = \{1, 2, 3, 4, 5, 6\}$. In this notation, the left and right braces indicate that a set is involved; the outcomes 1 up to and including 6 are the **elements** of the set. Chance determines which of the six outcomes will occur.

If the six sides of the die are equivalent in the sense that their physical properties are exactly the same (so the die is perfectly symmetric and hence fair), then chance is completely free in determining the actual outcome and all six sides have the same likelihood of becoming the outcome.

But suppose a swindler manages to create a fair-looking die for which the outcome 6 is three times as likely as the outcome 1 while the outcomes 2, 3, 4 and 5 have the same likelihood as for a fair die. For this die the chances of the sides 1, 2, 3, 4, 5, 6 have the ratios 2 : 4 : 4 : 4 : 4 : 6.

'Flipping a coin' is an experimentation that has two possible outcomes: T (tail) and H (head). Chance determines which of the two will occur. When the referee flips a coin at the start of a football match, the two team captains assume that the two sides of the coin are alike (equivalent) in the sense that they are equally likely to become the actual outcome of the experiment.

The random experiment that observes whether there will be rain tomorrow is an observation of an uncontrollable phenomenon, not an experimentation. The sample space has two elements: the outcomes 'there will be rain' and 'there will be no rain'. Which of the two will occur is determined by many meteorological laws and partially also by chance.

> The weather man is never wrong. Suppose he says that there's an 80% chance of rain. If it rains, the 80% chance came up; if it doesn't, the 20% chance came up.
>
> *Dr Saul Barron*

In statistics, interest usually is in some population, say with size N. The random experiment that chooses an element from that population has that population as sample space. The role of chance in this experiment is determined by the way the population element is chosen: making the choice completely arbitrarily means that a method is applied that guarantees that all N outcomes (that is, all population elements) are equally likely to become the outcome that actually occurs. It is like throwing a fair die with N sides (if such a die exists).

Some philosophers think that chance does not exist. Some believe that the paths that people follow are completely determined in advance, that there is no chance or coincidence involved. All things that seem to happen by chance are actually predestined, and people can influence their own future and predestination by positive affirmations. Also see the cartoon below.

Others have more scientific reasons for their points of view. They say that it is essentially a lack of knowledge that forces people to talk about chance and probability. If the experimenter who throws the die on the table were able to measure all relevant physical quantities like velocity of the throw, the side that is up when throwing, and the angle of the die with the table at the moment it is released, then it would be possible to predict the actual outcome precisely.

Often, interest is in a special **subset** of the sample space Ω of the random experiment. For instance, if the outcomes 5 and 6 of a die make you the winner, you will especially be interested in the subset $\{5, 6\}$. That is why the subsets of the sample space have their own name: they are called **events**. A certain event (so, a subset of Ω) is said to **occur** if the actual outcome of the experiment belongs to that event. Events with exactly one outcome are called **single events**, while events with more than one outcome are called **multiple events**. Notice that Ω itself is an event too, and that this event will always occur since the actual outcome of the experiment will always fall in Ω.

Coincidence, by Scott Adams
Source: http://www.stat.psu.edu/~resources/Topics/prob.htm

There is one event that needs special attention: the subset of Ω that does not contain any outcomes, the **empty event**, denoted by $\varnothing$. That the empty set is a subset of Ω (and hence can be called event) is a matter of logic. To be a subset of Ω, all elements of the subset have to be an outcome (in Ω). Since $\varnothing$ has no elements, this logically is also the case for the subset $\varnothing$.

Example 6.1

Two gamblers (a and b) play a game with the two unusual ('obscure') dice below.

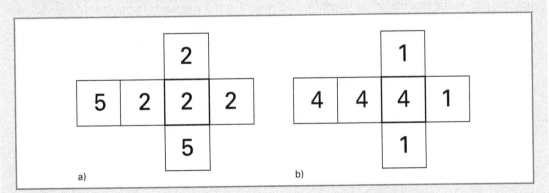

The sides of both dice are all equivalent. Player a throws the left die, player b the right one. The player with the highest score wins. Which player has the best chances?

Since both dice have six sides, they can be combined to give 36 outcomes, jointly constituting the sample space Ω of this random experiment. The sample space is represented by the 36 squares in the grid below. All squares also contain the corresponding winners.

b							
4	b	b	b	b	a	a	
4	b	b	b	b	a	a	
4	b	b	b	b	a	a	
1	a	a	a	a	a	a	
1	a	a	a	a	a	a	
1	a	a	a	a	a	a	
	2	2	2	2	5	5	a

Since all sides of the dice are equivalent (the two dice are fair), all squares of the grid are equally likely to occur. The random experiment is like throwing one fair die with 36 sides.

Notice that a wins for 24 outcomes. The event 'player a wins' is just the subset of the outcomes for which the score with the left die is larger than the score with the right die. This event has 24 elements, represented in the above grid by the squares with a in them. Similarly, the event 'player b wins' is just the set of outcomes represented by the squares with b. But there are other interesting events. For instance, the event 'player a throws 5' contains the twelve outcomes in the last two columns of the grid; the event 'both players throw the smallest possible score' is just the 3×4 south-west sub-diagram with twelve outcomes. Notice that the event 'the outcome is a draw' is empty. Finally, it is intuitively clear that the chances of a are twice as great as the chances of b. Before the experiment is done, it can already be said that the chances of player a are better.

A random experiment does not determine the sample space uniquely. In the above example, the set $\widetilde{\Omega} = \{(2,1), (2, 4), (5, 1), (5, 4)\}$ is also a possible choice for the sample space. For instance, the outcome (2, 1) corresponds to the event that a throws 2 and b throws 1. In contrast to the square-outcomes of Ω, the outcomes of $\widetilde{\Omega}$ are **not** equivalent: (2, 1) corresponds to 12 square-outcomes while (5, 4) corresponds to 6, so (2, 1) is twice as likely as (5, 4). Notice that the event 'player a wins' is just the subset $\{(2, 1), (5, 1), (5, 4)\}$ of $\widetilde{\Omega}$. The event 'player b wins' contains only the outcome (2, 4) and can be written as $\{(2, 4)\}$.

The preferred choice of the sample space depends on the actual situation. In Example 6.1, the sample space with the equivalent (square) outcomes certainly has advantages: the determination of the probability that a certain event occurs is reduced to counting the number of outcomes of the event (as will become clear in Section 6.3).

Example 6.2

Interest is in the population of all 39.178 million German households and the frequency distribution of the variable X = 'size of household'.

Specification	Unit	2005
Households	1000	39 178
1-person households	1000	14 695
2-person households	1000	13 266
3-person households	1000	5 477
4-person households	1000	4 213
Households with 5 or more persons	1000	1 527

TABLE 6.1 Frequencies of 'size of household' in Germany (× 1000)
Source: Federal Statistical Office, Germany; www.desstats.de/e_home.htm

Consider the random experiment that arbitrarily (randomly) chooses a household. Its sample space is the set of all households in Germany, which is just the population mentioned above. Interesting events are:

- A: this arbitrarily chosen household has size 1
- B: this household has size 2
- C: this household has size 3
- D: this household has size 4
- E: this household has 5 or more persons
- F: this household has at least 3 persons
- G: this household has at most 4 persons

(Note that events are denoted by capital letters.) The event A, B, C, D and E contain respectively 14.695, 13.266, 5.477, 4.213 and 1.527 million outcomes. To get the outcomes (households) of F, the outcomes of C, D and E have to be joined; to get the outcomes of G, the outcomes of A, B, C and D have to be joined. So, F has 11.217 million outcomes while G has 37.651 million.

Notice that all households in the sample space are equivalent in the sense that they are equally likely to occur. This is a consequence of the way the experiment is performed. In a sense, this random experiment is like throwing a die with 39.178 million sides. Also notice that event A has a higher chance of occuring than event B since A contains more of these equivalent outcomes than B. Event G has the highest chance of occuring since it contains the largest number of outcomes.

In this example as well, another sample space is possible. Since the above events of interest are all formulated in terms of size, the set $\tilde{\Omega} = \{1, 2, 3, 4, \geq 5\}$ is another possible choice. Here, the outcomes 1, 2, 3, 4 refer to the events that the randomly selected household respectively has 1, 2, 3 or 4 members. The outcome '≥ 5' refers to the event that this household has at least 5 members. In terms of these outcomes, the events A, B, ..., G are equal to the following subsets of $\tilde{\Omega}$:

$$A = \{1\} \qquad B = \{2\} \qquad C = \{3\} \qquad D = \{4\}$$
$$E = \{\geq 5\} \qquad F = \{3, 4, \geq 5\} \qquad G = \{1, 2, 3, 4\}$$

Notice that the outcomes of this alternative sample space are **not** equivalent. For instance, the outcome 1 has a higher chance of occurring than the outcome ≥ 5.

6.2 Rules for sets

In probability theory, interest is in random experiments and in the relationship between chance and the events. Since events are nothing but subsets of a sample space of the random experiment, it is important to have a good understanding of the rules regarding sets. Below, emphasis will be on the creation of new events from basic ones.

Although all concepts and results below can be formulated in terms of general subsets, it will always be assumed that the subsets are events of a random experiment with sample space Ω. Most definitions below are illustrated in Figure 6.1.

Start with subsets A and B of Ω. If all elements of A also belong to B, we write $A \subset B$ and say that A is a **subset** of B. That is, if A occurs (because the actual outcome of the random experiment happens to fall in A), then B automatically also occurs.

The set of elements of Ω that do **not** belong to A is called the **complement** of A and is denoted by A^c. Indeed, A^c is another subset (event) of Ω.

The elements of A and the elements of B can be joined together into a new set. This new subset of Ω is called the **union** of A and B and is denoted by $A \cup B$. This event is also characterized as 'A and/or B' since its elements belong to A or B or both.

Also, the common part of A and B is important: the set of the elements of A that also belong to B. This subset of Ω is denoted by $A \cap B$, the **intersection** of A and B. Sometimes this event is characterized as 'A and B' since its elements belong to both A and B. It is possible that A and B do not have common elements. Then their intersection is empty and hence equal to the empty event, that is, $A \cap B = \varnothing$. If A and B have an empty intersection, they are called **disjoint** or **mutually exclusive**: it is not possible that both A and B will occur.

In the case of one throw with an ordinary die, the sample space is $\Omega = \{1, 2, 3, 4, 5, 6\}$. The event $A = \{1, 3\}$ is a subset of the event B = 'odd number of eyes' = $\{1, 3, 5\}$, so $A \subset B$. Note that $A^c = \{2, 4, 5, 6\}$. Furthermore, $B^c = \{2, 4, 6\}$, the event that the number of eyes is even. The union of A and B is just $\{1, 3, 5\}$, which is B; the union of A and B^c is $\{1, 2, 3, 4, 6\}$. The intersection of A and B is $\{1, 3\}$, which is A; the intersection of A and B^c is empty. That is:

$$A \cup B = \{1, 3, 5\} \quad \text{and} \quad A \cup B^c = \{1, 2, 3, 4, 6\}$$

$$A \cap B = \{1, 3\} \quad \text{and} \quad A \cap B^c = \varnothing$$

Of course, the operations 'union' and 'intersection' are not restricted to two subsets. The union $A \cup B \cup C$ of the subsets A, B and C arises by joining together their elements into one set. The order in which this is done is of no importance: for instance, first joining together A and B, and next joining the resulting set together with C gives the same result as joining together A and the union of B and C. That is:

$$A \cup B \cup C = (A \cup B) \cup C = A \cup (B \cup C)$$

Similarly, the intersection $A \cap B \cap C$ is just the subset of Ω that contains the common elements of A, B and C; the order of the intersections is irrelevant:

$$A \cap B \cap C = (A \cap B) \cap C = A \cap (B \cap C)$$

A collection $D_1, \ldots, D_s$ of subsets of Ω is called a **partition** of Ω if each element of Ω belongs to exactly one of the subsets. That is, to be a partition, the union $D_1 \cup \ldots \cup D_s$ of all the subsets has to be Ω while the intersection of each pair of subsets has to be empty. In other words:

$$D_1 \cup \ldots \cup D_s = \Omega \quad \text{and} \quad D_i \cap D_j = \varnothing \quad \text{for all } i \neq j$$

In the above throw with an ordinary die, also consider the events $C = \{2\}$, $D = \{4, 6\}$ and $E = \{1, 2, 5\}$. Joining together the events A, C and D yields $\{1, 2, 3, 4, 6\}$; joining together the events A, B, C, D and E yields Ω. The intersection of A, B and E is $\{1\}$, while the intersection of A, B, C, D and E is empty. That is:

$A \cup C \cup D = \{1, 2, 3, 4, 6\}$ and $A \cup B \cup C \cup D \cup E = \Omega$

$A \cap B \cap E = \{1\}$ and $A \cap B \cap C \cap D \cap E = \varnothing$

Notice that the collection A, B, C, D, E is **not** a partition, but the collection B, C, D **is**.

To clarify the above concepts, ***Venn diagrams*** (after J. Venn, 1834–1923) can be used. In such diagrams, a rectangle is used to represent Ω while the inner parts of closed curves (such as ovals or rectangles) represent subsets. This is illustrated in Figure 6.1 and interpreted for the case that Ω is the sample space of a random experiment and the subsets are events.

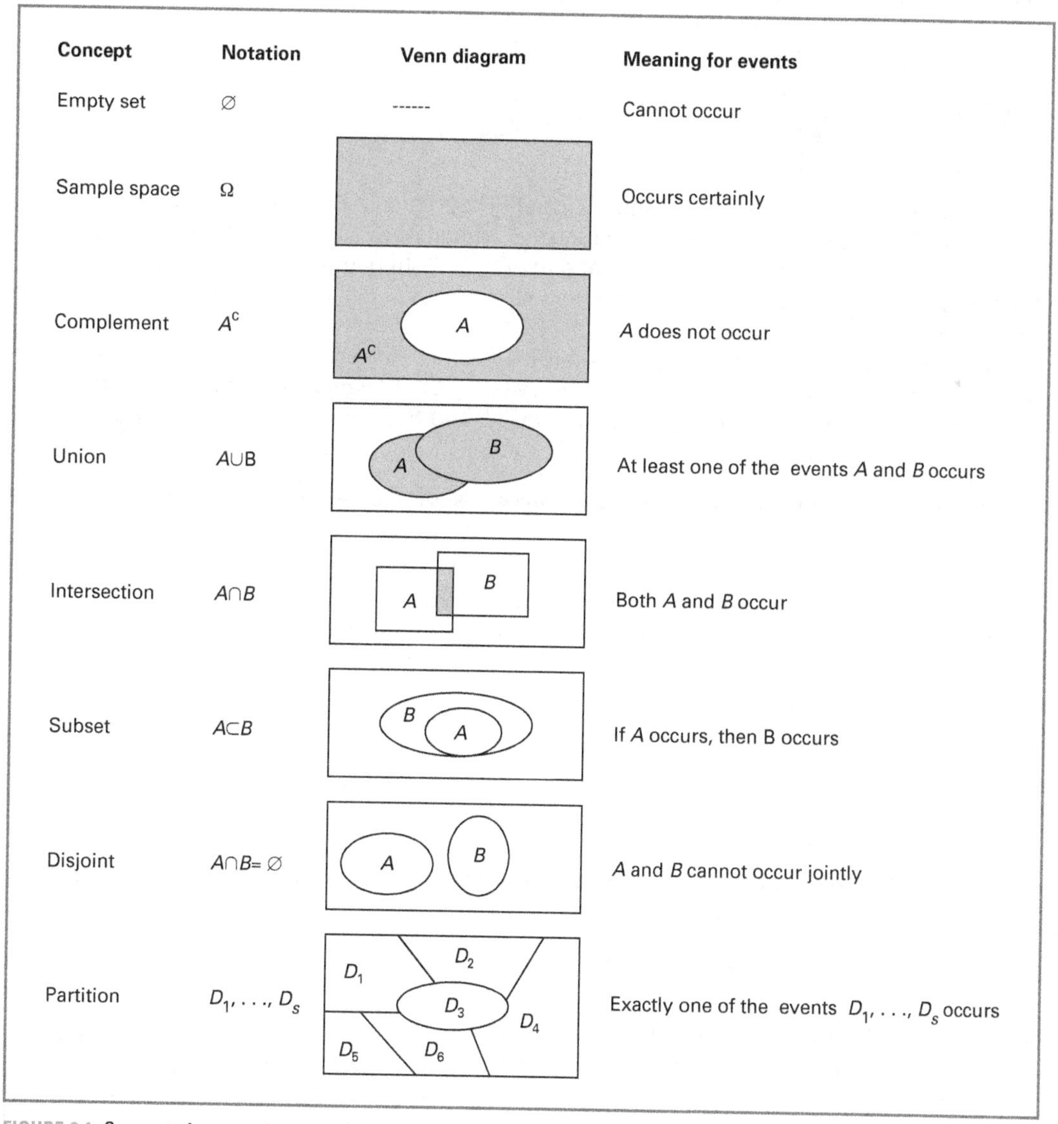

Concept	Notation	Venn diagram	Meaning for events
Empty set	$\varnothing$	------	Cannot occur
Sample space	Ω		Occurs certainly
Complement	A^c		A does not occur
Union	$A \cup B$		At least one of the events A and B occurs
Intersection	$A \cap B$		Both A and B occur
Subset	$A \subset B$		If A occurs, then B occurs
Disjoint	$A \cap B = \varnothing$		A and B cannot occur jointly
Partition	$D_1, \ldots, D_s$		Exactly one of the events $D_1, \ldots, D_s$ occurs

FIGURE 6.1 Concepts from set theory and their meaning in terms of events

Example 6.3

Reconsider the throw with the two obscure dice in Example 6.1. The sample space Ω has 36 elements; the outcomes are the 36 squares in the grid of Example 6.1. The events A = 'player a wins' and B = 'player b wins' are disjoint since the corresponding subsets of Ω have no common elements. As a matter of fact, in this game exactly one of the two players will win because the obscure dice do not allow the same number of eyes. Apparently, the collection A, B is a partition of Ω since:

$$A \cup B = \Omega \quad \text{and} \quad A \cap B = \varnothing$$

The event A^c is just the event that player a does **not** win. It consists of the 12 squares that do **not** contain the a; so A^c is equal to B.

Also consider the following events:

D_1: player a throws 5
D_2: player a throws 2
D_3: player b throws 4
D_4: player b throws 1

Notice that these events are indeed subsets of Ω. They contain 12, 24, 18 and 18 outcomes, respectively. For instance, D_1 consists of the squares in the last two columns of the grid in Example 6.1. Also notice that the collections D_1, D_2 and D_3, D_4 are both partitions of Ω: in both cases, each element of Ω belongs to exactly one of the two subsets. Also notice that D_1, D_2, D_3, D_4 is **not** a partition.

The event 'at least one of the two players throws maximal' means: a throws 5, **or** b throws 4, **or** a throws 5 and b throws 4. So, it is equal to the event $D_1 \cup D_3$, which contains 24 outcomes. The event 'both players throw maximal' means: a throws 5 **and** b throws 4. So, it is the intersection $D_1 \cap D_3$ with 12 outcomes.

Between the concepts 'complement', 'subset', 'union', 'intersection' and 'empty set', all kinds of relations are valid. To detect and illustrate them, Venn diagrams can be helpful.

Since the elements of $A \cap B$ belong both to A and to B, this intersection is a subset of A and also a subset of B. In terms of events: if A and B both occur, then A will certainly occur (and the same holds for B):

$$(A \cap B) \subset A \quad (A \cap B) \subset B$$

Since $A \cup B$ arises by joining together A and B, both A and B are enclosed in this union:

$$A \subset (A \cup B) \quad B \subset (A \cup B)$$

If A is a subset of B, then the common elements of A and B are just the elements of A, while joining B to A has no effect. That is:

If $A \subset B$, then $A \cap B = A$ and $A \cup B = B$

The shaded region in the diagram on the right represents $A \cup B$. The white area represents the set of outcomes outside this union, which is $(A \cup B)^c$. On the other hand, this white area can also be considered as the set of the outcomes that do not belong to A **and** do not belong to B. That is, the white area is just the set of the outcomes in $A^c \cap B^c$. These considerations lead to the following rule for subsets A and B:

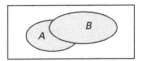

$$(A \cup B)^c = A^c \cap B^c \tag{6.1}$$

Apparently, taking the complement of the union of A and B gives the same result as taking the intersection of their complements.

A similar result can be found for the complement of an intersection. On one hand, the white area in the diagram on the right represents $(A \cap B)^c$, the set of outcomes outside $A \cap B$. On the other hand, the white area is just the union of A^c and B^c. So, the following rule follows:

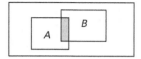

$$(A \cap B)^c = A^c \cup B^c \qquad \text{(6.2)}$$

Taking the complement of the intersection of A and B gives the same result as taking the union of their complements.

Now consider **three** events A, B and C. Since the union $B \cup C$ is an event in Ω, the intersection of A and this union makes sense; the resulting event is just $A \cap (B \cup C)$. In the Venn diagram, A, B and C are rectangles. Note that $A \cap (B \cup C)$ is the total shaded region. But this region can also be considered as the union of $A \cap B$ (the two leftmost shaded rectangles) and $A \cap C$ (the two upper shaded rectangles). That is:

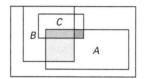

$$A \cap (B \cup C) = (A \cap B) \cup (A \cap C) \qquad \text{(6.3)}$$

The rule that arises from (6.3) by interchanging $\cap$ and $\cup$ is also valid, that is:

$$A \cup (B \cap C) = (A \cup B) \cap (A \cup C) \qquad \text{(6.4)}$$

Again, it can be illustrated by a Venn diagram. Again A, B and C are rectangles. Notice that $B \cap C$ is the region of the two very small shaded rectangles; so the union with A just gives the whole shaded region. But this shaded region is also the common part (the intersection) of $A \cup B$ and $A \cup C$.

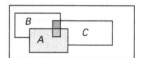

The union and the intersection of events A_1, A_2, ..., A_m can be denoted in shorthand:

$$\bigcup_{i=1}^{m} A_i = A_1 \cup A_2 \cup \ldots \cup A_m \quad \text{and} \quad \bigcap_{i=1}^{m} A_i = A_1 \cap A_2 \cap \ldots \cap A_m$$

With a partition D_1, D_2, ..., D_s, the Venn diagram of the sample space Ω turns into a tessellation as in the last rectangle of Figure 6.1. When studying all outcomes in another event A (indicated on the right by the interior region of the triangle) within Ω, it is possible instead to consider the intersections $A \cap D_1$, $A \cap D_2$, ..., $A \cap D_s$ and take their union. That is, the union of the intersections $A \cap D_i$ returns A:

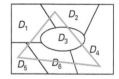

$$\bigcup_{i=1}^{s} (A \cap D_i) = A \qquad \text{(6.5)}$$

The following box repeats the most important rules.

Rules for sets and events

$(A \cup B)^c = A^c \cap B^c$ and $(A \cap B)^c = A^c \cup B^c$

$A \cap (B \cup C) = (A \cap B) \cup (A \cap C)$ and $A \cup (B \cap C) = (A \cup B) \cap (A \cup C)$

$\bigcup_{i=1}^{s} (A \cap D_i) = A$ for each partition D_1, D_2, ..., D_s

6.3 Historical definitions of probability

People usually have some intuitive understanding of the concept 'probability'. When sports commentators analyse the football match between Manchester United and Bayern Munich and conclude that the probabilities that Manchester, respectively Bayern, wins are both 30% and the probability of a draw is 50%, everyone knows that there is something wrong. Even the old Indo-Germanics had some knowledge of probability, which follows from the fact that the males had a predilection for playing dice (and gambling away their wives, if necessary). But it was not until the end of the eighteenth century that the concept was formalized by Laplace, and then only for the case that all outcomes of the random experiment are equally likely.

Below, the definitions of the concept 'probability' are considered from a historical perspective. The classical definition of Laplace is followed by the empirical definition and the subjective definition; the general definition of Kolmogorov is postponed until the next section. All definitions have in common that there necessarily has to be some underlying random experiment. Furthermore, probability is denoted as P and the probability that event A will occur as $P(A)$. But what exactly is 'probability'?

6.3.1 Classical definition of probability

The first definition is applicable only in the case of random experiments and finite sample spaces (with size N) for which all outcomes are equally likely. For instance, the throw with a fair die with sample space $\{1, 2, \ldots, 6\}$ and the throw with the two obscure dice in Example 6.1 with the 36 squares as sample space are random experiments that both belong to this category. The **classical definition of probability** defines the probability of event A as the proportion of the outcomes that fall in A. If $N(A)$ denotes the number of outcomes in A, then this proportion is the ratio $N(A)/N$. That is:

Classical definition of probability

$$P(A) = \frac{N(A)}{N} \quad \text{for all events } A \tag{6.6}$$

Requirement: **equally likely outcomes**.

It is important to emphasize that the definition in (6.6) presupposes that all outcomes are equally likely. In the case of dice, this means that the dice are supposed to have perfect physical symmetry. Hence, even in this simple example, reality is stylized since such perfect physical symmetry can only be reached in theory.

In a throw with a fair die, the $N = 6$ outcomes of the sample space $\Omega = \{1, 2, \ldots, 6\}$ are indeed equally likely. So, all single events have probability 1/6. Since the event $A =$ 'even' is just the subset $\{2, 4, 6\}$ of Ω, the probability $P(A)$ that A will occur is equal to $N(A)/N = 3/6 = 0.5$. The event 'odd' is equal to A^c, which is $\{1, 3, 5\}$. So, $P(A^c) = N(A^c)/N = 3/6 = 0.5$. The event $B = \{5, 6\}$ has $N(B) = 2$ outcomes, so $P(B) = 2/6 = 1/3$.

When flipping a fair coin, the sample space $\Omega = \{H, T\}$ has $N = 2$ outcomes. This sample space has only four subsets: the single events $\{H\}$ and $\{T\}$, the empty event $\varnothing$ and the Ω. Their probabilities are respectively: 0.5, 0.5, 0 and 1.

Example 6.4

The throw with the two obscure dice in Example 6.1 has equally likely outcomes if the set of the 36 squares is taken as sample space. The event D_1 = 'player a throws 5' contains $N(D_1)$ = 12 squares, so $P(D_1)$ = 12/36. The following table gives an overview of the probabilities of the events D_1, D_2, D_3 and D_4; see Example 6.3 for their definitions.

Event	D_1	D_2	D_3	D_4
$N(D)$	12	24	18	18
$P(D)$	1/3	2/3	1/2	1/2

The event A = 'player a wins' contains 24 squares. So $P(A)$ = 24/36 = 2/3.

There is one very important application of the classical probability definition. In the random experiment that arbitrarily chooses one of the elements in a population of interest, the outcomes are just the elements of the population and they are equally likely. But then the probability that this arbitrarily chosen element will fall in a certain subset of the population is just the relative frequency of this subset.

Example 6.5

In Example 6.2 the population of interest was the set of all 39.178 million households in Germany. The experiment that arbitrarily chooses a household has this population as sample space. All households in the population are equally likely since they all have the same chance of being chosen. But this means that the requirement for application of the classical probability definition is satisfied. As a consequence, all single events of the random experiment (such as the event that the household of the Chancellor is chosen) have probability 1/39178000. In Example 6.2 several events were considered that deal with the variable 'size of household'. For instance, the event A = 'this arbitrarily chosen household has size 1' contains 14.695 million households, so:

$$P(A) = N(A)/N = 14.695/39.178 = 0.375$$

But this is just the relative frequency of the 1-person households in Germany. Likewise for the other events in Example 6.2, the probabilities are equal to the corresponding relative frequencies; check it yourself.

Event	A	B	C	D	E	F	G
Number of outcomes ($\times 10^3$)	14695	13266	5477	4213	1527	11217	37651
Probability	0.375	0.339	0.1400	0.108	0.039	0.286	0.961

From the classical definition of probability several properties can be derived. Since the number of outcomes $N(A)$ of A cannot be negative, division by the positive number N does not change that and $P(A)$ is non-negative. Furthermore, the number of outcomes in A is at most equal to N, the number of outcomes in Ω. That is, $N(A) \leq N$ and division by N yields that $P(A)$ is at most 1. Hence:

$$0 \leq P(A) \leq 1 \qquad \text{for all events } A \tag{6.7}$$

Since $N(\Omega)$, the number of outcomes that fall in Ω, was previously defined as N, it also follows that:

$$P(\Omega) = 1 \tag{6.8}$$

If events A and B are disjoint, they do not have common outcomes and the number of outcomes in $A \cup B$ is just the number of outcomes in A plus the number of outcomes in B, so $N(A \cup B) = N(A) + N(B)$. By dividing by N, the third property follows:

$$\text{If } A \text{ and } B \text{ are disjoint, then } P(A \cup B) = P(A) + P(B) \tag{6.9}$$

These three properties of the classical probability definition motivated Kolmogorov to formulate his **general** definition of probability; see Section 6.4.

When throwing a fair die, the events $A = \{1, 2, 3\}$ and $B = \{5, 6\}$ have probabilities 1/2 and 1/3, respectively. Since $A \cap B = \emptyset$, it follows from (6.9) that $P(A \cup B) = P(A) + P(B) = 1/2 + 1/3 = 5/6$. Indeed, $A \cup B$ contains 5 out of 6 equally likely outcomes.

6.3.2 Empirical definition of probability

The foregoing definition is applicable only for special experiments. So, another definition is needed for other situations. We now consider experiments that are identically repeatable as often as one wants. That is, it is assumed that the experiment of interest can be repeated infinitely often, under completely identical circumstances. Obviously, this is no problem for throwing a false die: you can create yourself a laboratory situation such that for every throw the circumstances are identical. Also the random experiment that arbitrarily chooses an element in the population of interest falls into this category of experiments: the choice can be made as often as one wants as long as the formerly chosen element is put back into the population. But not all experiments are identically replicable. For instance, the amount of rain on 1 January 2015 can only be observed once; the state of the German economy in 2020 can only be measured one time.

Suppose that the random experiment is identically repeatable. Then the **law of large numbers** holds: if the experiment is repeated on and on, then the relative frequency of the occurrence of a fixed event approaches a constant. Let n denote the number of repetitions (**trials**) of the experiment and let A be an event. In each of the n experiments, A occurs or does not occur; let $n(A)$ denote the number among the n repetitions for which A does occur. According to the law of large numbers, the ratio $n(A)/n$ approaches a constant as n becomes larger and larger. It is this constant that is defined as the probability $P(A)$ that A will occur. So, the **empirical definition of probability** states that:

Empirical definition of probability

$$\frac{n(A)}{n} \to P(A) \quad \text{as } n \to \infty \tag{6.10}$$

Requirement: the experiment is **independently and identically repeatable**.

One says that $P(A)$ is the **limit** (as $n \to \infty$) of the ratios $n(A)/n$.

Of course, it is not possible to go on and on repeating the experiment to determine the probability $P(A)$. One's life is only finite, so you will have to stop after a finite number of (say) n repetitions. In practice, findings about the ratio $n(A)/n$ measured after n repetitions are used to model the probability $P(A)$, as in the following example.

Example 6.6

The swindler – who entered the scene at the start of this chapter – has thrown his manipulated die $n = 10\,000$ times to test its properties. All trials were done under identical circumstances. Here are the results:

Number of eyes	1	2	3	4	5	6
Observed frequency	830	1603	1598	1670	1620	2679
Observed relative frequency	0.0830	0.1603	0.1598	0.1670	0.1620	0.2679

Let A_j be the event that one throw with the false die returns j eyes ($j = 1, 2, \ldots, 6$). The second row of the table gives $n(A_j)$, the number of the n trials for which event A_j occurs. The third row shows the relative frequencies of the event A_j, the ratios $n(A_j)/n$. For instance, 2679 of the 10 000 trials returned 6 eyes. That is, the relative frequency of A_6 is $2679/10\,000 = 0.2679$. Similarly, the relative frequency of A_1 is 0.0830, which is about one-third of the relative frequency of A_6. Also notice that the relative frequencies of A_2, A_3, A_4 and A_5 are all close to 1/6.

The above considerations about relative frequencies after 10 000 trials motivated the swindler to use the following **model** for the probabilistic behaviour of his die:

j	1	2	3	4	5	6
$P(A_j)$	1/12	1/6	1/6	1/6	1/6	3/12

But what about other probabilities, for instance the probability that a throw with this die returns an even number? Let A be the event 'even' $= \{2, 4, 6\}$. Since $A = A_2 \cup A_4 \cup A_6$ is a union of three pairwise disjoint events, the conclusion might be that $P(A) = P(A_2) + P(A_4) + P(A_6) = 7/12$ because of Equation (6.9). But whether (6.9) is also valid in the case of identically repeatable experiments is not proved yet; this will be done below.

Equations (6.7)–(6.9) are also valid in the case that the probabilities are empirically determined by (6.10). Since $n(A)$ is at least 0 and at most n, the ratio $n(A)/n$ falls in the interval [0, 1] and hence the limit $P(A)$ also falls in it. This proves (6.7). Since Ω occurs in every trial, $n(\Omega) = n$ and the ratio $n(\Omega)/n$ is always equal to 1 (whatever the number of trials is). But then the limit $P(\Omega)$ is also equal to 1, which proves (6.8). For disjoint events A and B, the number of times that the outcome falls in A or B (among the n trials) is just the number of times that the outcome falls in A plus the number of times that the outcome falls in B. That is, $n(A \cup B) = n(A) + n(B)$. Consequently, the ratio $n(A \cup B)/n$ is equal to the sum of the ratios $n(A)/n$ and $n(B)/n$. Hence, the limit $P(A \cup B)$ is equal to the sum of the limits $P(A)$ and $P(B)$.

Example 6.7

A wallet contains twenty coins: three of 2 cents, five of 5 cents, nine of 10 cents and three of 20 cents. A thief grabs, completely arbitrarily, one of the coins from the wallet. What is the probability that he will have at most ten cents?

The sample space of this experiment consists of the twenty coins; all coins have the equal chance 1/20 to be grabbed. So, the experiment is like throwing a fair die with 20 sides: three sides have 2 eyes, five have 5, nine have 10 and three have 20.

Let A, B, C and D respectively be the event that this randomly taken coin is a 2, 5, 10 and 20 cent coin. We can use the classical definition to calculate their probabilities. Since there are three 2 cent coins in the wallet, $P(A) = 3 \times 1/20 = 0.15$ because of (6.9). Similarly:

$P(B) = 0.25$; $P(C) = 0.45$; $P(D) = 0.15$

But then the probability that the thief will have stolen at most ten cents is just $P(A \cup B \cup C) = P(A) + P(B) + P(C) = 0.85$, which again is a consequence of (6.9).

This experiment is also identically repeatable, at least with a computer. To let the thief grab (say) 20000 coins from the wallet that each time has the same content, will in practice be a massive task. However, this experiment can easily be simulated by instructing the computer to choose randomly (arbitrarily) 20000 numbers from a 20-number range containing three values 2, five values 5, nine values 10 and three values 20. When the 20000 simulated values are obtained, the relative frequencies of A, B, C and D after $n = 20000$ trials can easily be calculated. See Appendix A1.6 for details and Xmp06-07.xls for the simulated data. Table 6.2 below gives an overview of such relative frequencies after $n = 100, 200, 500, 1000, 2000, 5000, 10000$ and 20000 trials. The column at the right-hand side contains the actual probabilities.

n	100	200	500	1000	2000	5000	10000	20000	P
Rel. freq. for A	0.17	0.190	0.148	0.140	0.1550	0.1430	0.1487	0.1507	**0.15**
Rel. freq. for B	0.17	0.205	0.252	0.258	0.2475	0.2412	0.2453	0.2483	**0.25**
Rel. freq. for C	0.48	0.440	0.456	0.455	0.4525	0.4620	0.4528	0.4507	**0.45**
Rel. freq. for D	0.18	0.165	0.144	0.147	0.1450	0.1538	0.1532	0.1504	**0.15**

TABLE 6.2 Simulation study of randomly grabbing a coin from the wallet
Source: Private research

Notice that, after 100 trials, the relative frequencies of A and B are still the same, but after 200 trials they are different. Also note that already after 2000 trials rounding to two decimals gives results that are close to the actual probabilities. After 10000 trials rounding to two decimals gives exactly the results of the last column, while after 20000 trials the approximation is even more precise.

It is very instructive to do the above simulation experiment yourself. Of course, your relative frequencies will be different from the above, but your table will also show the convergence to the right-hand column.

There exist random experiments that neither have equally likely outcomes nor are identically repeatable. For instance, the random experiment 'attempt by Mr X to commit suicide by jumping from the tenth floor of a building' can hardly be repeated and has two outcomes (survive, not survive) that are not equally likely. In cases like this the two preceding definitions of probability are insufficient.

6.3.3 Subjective definition of probability

The subjective definition of probability can be applied for all random experiments, but it has the big disadvantage in that it is subjective in the sense that different people may obtain different probabilities for a common event.

According to the *subjective definition of probability*, the probability $P(A)$ reflects how strongly an individual believes in the (future) occurrence of event A. To avoid nonsensical values, these probabilities have to satisfy the rules (6.7)–(6.9). The following example nicely illustrates the subjective character of this definition.

Example 6.8

Suppose that the football match Manchester United v Bayern Munich will take place next Wednesday. Three persons (1, 2 and 3) are analysing the chances of the two teams and assign their private subjective probabilities to events such as M = 'Manchester wins', B = 'Bayern wins' and D = 'draw'. Person 1 is a Manchester supporter; he believes that $P(M) = 0.7$, $P(B) = 0.1$ and $P(D) = 0.2$. Person 2 is a Bayern supporter; he thinks that $P(M) = 0.1$, $P(B) = 0.7$ and $P(D) = 0.2$. Person 3 is a neutral journalist. His opinion is that $P(M) = P(B) = 0.3$ and $P(D) = 0.4$. Each individual has created a private model to describe the chances of the three outcomes (Manchester wins, Bayern wins, draw) of Ω.

Notice that each of the three persons divided a total of 1 among the three single events M, B and D. It may be expected that they also keep to the rules (6.7)–(6.9) when assigning their probabilities to other events. The union of the disjoint events M and D (that is, the event $M \cup D$ = 'Manchester does not lose') will get probability $P(M) + P(D)$, but the answer will differ for each individual (0.9, 0.3 and 0.7, respectively). In a similar way, the subjective probabilities of the events $B \cup D$ = 'Bayern does not lose', $M \cup B$, and Ω follow. The models used by the three persons are summarized in the table below. Here prob$_i$ refers to the subjective probabilities of person i.

Event	M	B	D	$M \cup B$	$M \cup D$	$B \cup D$	Ω	$\varnothing$
Prob$_1$	0.7	0.1	0.2	0.8	0.9	0.3	1	0
Prob$_2$	0.1	0.7	0.2	0.8	0.3	0.9	1	0
Prob$_3$	0.3	0.3	0.4	0.6	0.7	0.7	1	0

For each of the three persons, the table offers an overview of the probabilities of all possible events. (Since Ω has three elements, there are exactly eight events.) The numbers in the columns 5, 6 and 7 are just consequences of rule (6.9). Column 8 is indeed in accordance with rule (6.8). The empty set can be written as $M \cap B$ and all three persons will agree that this event has probability 0.

There is an additional – important – fact that we can learn from the last example: a probability (or model or probability model) is nothing but an overview of events and corresponding numbers between 0 and 1, in such a way that rules (6.7)–(6.9) are satisfied. In such a probability overview, a number $P(A)$ is assigned to event A, and $P(A)$ is considered to be the probability that A occurs. From a mathematical point of view, a probability model is just a function that assigns values from the interval [0, 1] to all events A.

6.4 General definition of Kolmogorov

It was A. N. Kolmogorov (1903–1987) who eventually stated the general definition of the concept 'probability' and who derived several important corollaries. It was his idea to consider a probability P as a special mathematical function that satisfies certain *axioms*, some minimal properties that have to hold. For finite sample spaces, these axioms are just the equations (6.7) – (6.9).

Consider a random experiment with finite sample space Ω.

Probability measure P (Kolmogorov's definition)

A *probability measure* P is a prescription that assigns real numbers $P(A)$ to all subsets A of Ω, such that the following axioms hold for all subsets A and B of Ω:

$P(A) \geq 0$	**(6.11)**
$P(\Omega) = 1$	**(6.12)**
If A and B are disjoint, then $P(A \cup B) = P(A) + P(B)$	**(6.13)**

(The term 'prescription' is used as an alternative for function.) Instead of probability measure, one also talks about *probability model*, just *model* or even just *probability*.

In the above definition, $P(A)$ is read as the *probability of A* or the *probability that A occurs* or the **probability that A will occur**. Notice that the three probability definitions of the last section indeed yield probability measures since (6.7)–(6.9) are identical to (6.11)–(6.13). The only difference seems to be that (6.7) also claims that $P(A) \leq 1$, but Section 7.1 will show that this automatically follows from (6.12).

Example 6.9

Below, three random experiments are considered: flipping a coin, throwing a fair die and throwing the die of the swindler.

A referee flips a coin, which leads to the sample space $\Omega = \{H, T\}$. The four subsets can easily be enumerated: Ω, $\emptyset$, $\{H\}$ and $\{T\}$. Now suppose that the referee is not sure whether the coin is fair, so he assumes for the time being that the probability that a head occurs is p (instead of 0.5), some number in the interval [0, 1]. But this assumption completely determines the model P for this experiment:

Event	Ω	$\emptyset$	$\{H\}$	$\{T\}$
Probability	1	0	p	$1 - p$

Note that $\{H\} \cup \{T\} = \Omega$. By first applying axiom (6.12) and then applying axiom (6.13) to the disjoint sets $A = \{H\}$ and $B = \{T\}$, it follows that:

$$1 = P(\Omega) = P(\{H\} \cup \{T\}) = P(\{H\}) + P(\{T\}) = p + P(\{T\})$$

Hence, the probability $P(\{T\})$ that tail occurs is $1 - p$. Similarly, again by (6.13):

$$1 = P(\Omega) = P(\Omega \cup \emptyset) = P(\Omega) + P(\emptyset) = 1 + P(\emptyset)$$

This yields $P(\emptyset) = 0$. Apparently, fixing $P(\{H\})$ as p determines the complete overview of subsets and probabilities. Hence, the **model** P is determined.

Next, consider a throw with a fair die with sample space $\Omega = \{1, 2, 3, 4, 5, 6\}$. Below, the events will be counted on the basis of their numbers of elements. Of course, there is only one subset with zero elements (the empty subset) and only one with six elements (Ω itself). The table below enumerates all 15 subsets with two elements. Notice that the subsets $\{3, 6\}$ and $\{6, 3\}$, for instance, are equal.

$\{1, 2\}$	$\{1, 3\}$	$\{1, 4\}$	$\{1, 5\}$	$\{1, 6\}$
	$\{2, 3\}$	$\{2, 4\}$	$\{2, 5\}$	$\{2, 6\}$
		$\{3, 4\}$	$\{3, 5\}$	$\{3, 6\}$
			$\{4, 5\}$	$\{4, 6\}$
				$\{5, 6\}$

Similarly, it can be determined that there are 20 subsets with three elements. When counting the number of subsets with four elements, time can be saved by counting their complements (which we have already done). A similar argument holds for the number of subsets with five elements. See the first two rows in the table below.

Number of elements per subset	0	1	2	3	4	5	6
Number of subsets	1	6	15	20	15	6	1
Probability of each subset	0	1/6	2/6	3/6	4/6	5/6	1

To find the probability measure P, it seems reasonable to start with

$$P(1) = P(2) = P(3) = P(4) = P(5) = \frac{1}{6}$$

(since the die is fair). Since $P(\Omega)$ has to be 1, it automatically follows that $P(6) = 1/6$ too. By applying axiom (6.13), it then follows that subsets with two elements get probability $2 \times 1/6 = 2/6$ and subsets with three (four, five) elements get probability 3/6 (4/6, 5/6).

When throwing the false die of the swindler, the sample space and its subsets are the same as in the fair case. However, P is completely different. Recall that:

$$P(2) = P(3) = P(4) = P(5) = \frac{1}{6}; \quad P(1) = \frac{1}{12}, \quad P(6) = \frac{3}{12}$$

Not all subsets with two elements get the same probability. For instance, the subset {2, 3} gets probability 2/6, but {2, 6} gets 1/6 + 3/12 = 5/12. The event A = 'at least three' is {3, 4, 5, 6}, which gets probability 1/6 + 1/6 + 1/6 + 3/12 = 9/12 = 3/4.

CASE 6.1 CHANCES OF POSITIVE OR NEGATIVE RETURNS ON A PORTFOLIO – SOLUTION

The investor is confronted with the random experiment of forecasting tomorrow's return of the portfolio. She uses the returns of the msot recent 509 business days to build a model P for the experiment. For the events

  A: tomorrow's return is positive
  B: tomorrow's return is more than 1%
  C: tomorrow's return is at most −1%

she concludes that $P(A) = 0.585$, $P(B) = 0.104$ and $P(C) = 0.063$. We now explain her arguments for those conclusions.

Essentially, the investor considers the returns of the most recent 509 business days as the results of independent and identical repetitions, and she uses the empirical definition to formulate her model. Since the return was negative or 0 on 211 days, it was positive on 509 − 211 = 298 days. Hence, $P(A) = 298/509 = 0.585$. There were 53 days with a return of more than 1%, so $P(B) = 53/509 = 0.104$. For 179 of the 211 days with non-positive returns, the return was larger than −1%. So, on 211 − 179 = 32 days the return was at most −1%, which yields: $P(C) = 32/509 = 0.063$.

Summary

A (probability) model is a function P on the set of all subsets of Ω such that axioms (6.11)–(6.13) are satisfied. The set Ω is the sample space of a random experiment and, for events A, the numbers $P(A)$ have to be read as 'probability that A will occur'. In practical situations, it has to be decided which of the following definitions must be used:

- classical definition of probability
- empirical definition of probability
- subjective definition of probability

The three axioms (6.11)–(6.13) of Kolmogorov constitute the foundation for the field of probability.

List of symbols

Symbol	Description
P	Probability (measure), model
Ω	Sample space
$\varnothing$	Empty set, empty event
$\subset$	Subset
$\cup$	Union
$\cap$	Intersection
$\cdot^c$	Complement

🔑 Key terms

axioms **187**
chance **174**
classical definition of
    probability **182**
complement **178**
disjoint **178**
elements **174**
empirical definition of
    probability **184**
empty event **175**
events **174**
experiment **174**
intersection **178**

Kolmogorov's
    definition **188**
law of large numbers **184**
limit **184**
model **188**
multiple event **174**
mutually exclusive **178**
occur **174**
outcomes **174**
partition **178**
possible outcomes **174**
probability **188**
probability measure **188**

probability model **188**
probability of A **188**
probability that A
    occurs **188**
random experiment **174**
sample space **174**
single event **174**
subjective definition of
    probability **186**
subset **174, 178**
trials **184**
union **178**
Venn diagram **179**

❓ Exercises

Exercise 6.1

For a certain random experiment, the sample space Ω is {2, 3, 4, 5, 6, 7, 8}. Consider the events A = {3, 5, 7}, B = {4, 5, 6} and C = {3, 4, 5, 6, 7}. Determine:

a $A \cup B$ and $A \cap B$

b $A \cap C$ and $A \cap C^c$

c C^c and $A \cup B \cup C$

d $A \cap B \cap C$

e Are A and B disjoint? Why (not)?

Exercise 6.2

Suppose that A, B and C are events of a certain random experiment. Express the following statements in terms of A, B and C.

a A does not occur.

b A and B both occur.

c A and/or B occur.

d A and B do not occur jointly.

e A does occur but B does not.

f None of the three events A, B and C occurs.

Exercise 6.3

In a certain random experiment the possible outcomes 1, 2, ..., 99 are all equally likely. Let P be the obvious model for this situation.

a Is this a classical probability model? Empirical? Subjective? Explain why.

b Calculate the probability that the outcome will be odd. Explain carefully how your answer is obtained.

c Calculate the probability that the outcome will be a multiple of 3.

Exercise 6.4

To check the precision of a new machine, each of 10 000 products that are produced by that machine are divided into one of the three (mutually exclusive) categories: irreparably defective, reparably defective and acceptable. It turns out that 94.5% of these products are acceptable, 3.5% are defective but reparable, and 2% are irreparably defective.

a Use this information to construct a model for the quality of a product that is produced by that machine.

b What is the underlying random experiment?

c Is this model a classical probability model? Empirical? Subjective? Explain why.

Exercise 6.5

You use the historical facts (period 1956–2007) in Table 5.11 to predict the country where next year's winner of the Champions League comes from.

a Describe the random experiment, the possible outcomes and the model that is used.

b Calculate the probability that next year's winner will come from a Mediterranean country.

c Is this model a classical probability model? Empirical? Subjective? Explain why.

d Which of the axioms (6.11)–(6.13) did you use in part (b)?

Exercise 6.6

A certain random experiment has five possible outcomes coded as 1, 2, 3, 4 and 5. For the model P it is given that $P(j) = j \times P(1)$, for all $j = 1, 2, \ldots, 5$. In all questions below, mention the axiom(s) that you use.

a Calculate $P(1)$.

b Calculate $P(\{1, 2, 3\})$.

c Calculate $P(\{1, 2, 3\} \cup \{3, 4\})$.

Exercise 6.7

Suppose that the events $D_1, D_2, \ldots, D_{50}$ form a partition of the sample space Ω of a certain random experiment. For a model P it is given that:

$P(D_i) = 0.01$ for $i = 1, 2, \ldots, 25$

$P(D_i) = 0.03$ for $i = 26, 27, \ldots, 50$

a Are these numbers in accordance with P being a model? Why (not)?

b Calculate $P(D_{17} \cup D_{27} \cup D_{48})$. Which axiom did you use?

Exercise 6.8

a The outcomes of the random experiment 'throwing a red and a blue die' can be written as pairs, with the number of eyes of the red (blue) die on the first (second) position. For instance, the outcome (2, 5) means that the red and blue dice showed 2 and 5 respectively. Determine all possible outcomes and present the sample space in a diagram as in Example 6.1.

b Suppose that in the experiment of part (a) interest is only in whether the separate dice show even (e) or odd (o). Determine an alternative sample space for the experiment.

c A coin and a die are thrown simultaneously. Determine a sample space.

d You decide that, tomorrow, you will sell your shares of stock A if tomorrow's price of the stock decreases more than 2%. If it increases more than 2%, you will buy more shares of A. If the change is between −2% and +2%, you will do nothing. Describe the underlying random experiment and determine a sample space.

Exercise 6.9

Suppose that A, B and C are events of a certain random experiment. Express the assertions below in terms of A, B and C.

a A and B cannot occur jointly.

b If A occurs, then B occurs.

c If A occurs, then both B and C occur.

d If A occurs, then B and/or C occur.

e B and C can occur jointly.

Exercise 6.10

Use Venn diagrams to answer the following questions about events A, B, C and D of a random experiment with sample space Ω. If the answer is no, use one of the experiments 'throwing a die' and 'flipping a coin' to find a counterexample.

a Is the intersection of A and B a subset of A?

b Is B a subset of the union of A and B?

c Is A a subset of the union of the events $A \cap B$ and C?
d Is the complement of the union of B and C equal to the intersection of the complements of B and C?
e Suppose that A, B, C is a partition of Ω. Is it then true that D is equal to the union of the three events $A \cap D$, $B \cap D$ and $C \cap D$?

Exercise 6.11

Three fair dice are thrown.

a Use triples to describe the sample space Ω of this experiment. Determine the sample size N.

Assign names to the events below, write them as sets of outcomes, and determine their probabilities. Use the notation you have learned in the present chapter.

b All dice have the same number of eyes.
c The numbers of eyes of the three dice are all different.
d The total number of eyes is at most 5.
e Which probability definition did you use in parts (b)–(d)? Why is this allowed?

Exercise 6.12

In a computer store at the campus of a university the daily demand for computers is 2.62 on average. The table below gives the relative frequency distribution of the daily demand.

Daily demand for computers	Relative frequency
0	0.05
1	0.16
2	0.31
3	0.20
4	0.18
5	0.08
6	0.02

This relative frequency distribution is used to model the demand forecast for the next business day.

a Describe the random experiment, the sample space and the model that is used.
b Consider the events A = 'the demand will be more than three computers' and B = 'the demand will be at most four computers'. Write them with set notations and determine their probabilities.
c At the moment, the store has only four computers in stock. Determine the probability that the demand for computers cannot be met immediately.

Exercise 6.13

A company will soon choose a new chief executive officer (CEO). The candidates are: Abramovic, Baker, Chung, Duffer and Emanuelson. According to an industry analyst, the candidates Abramovic, Baker, Chung and Duffer have the same chance of becoming the winner. However, the chances of Emanuelson are three times as great as the chances of Baker.

a Describe the random experiment that the analyst is modelling. Also determine a suitable sample space.

b Determine a model P that is in accordance with the beliefs of the analyst.

c Is this a classical probability model? Empirical? Subjective? Explain why.

Exercise 6.14

In a special Eurobarometer survey of March 2007, the European Commission published the results of an investigation about the attitudes towards alcohol among European citizens. In total, 16 450 people from the EU25 (at least 15 years old) were interviewed. One of the questions asked was:

On a day when you drink beer, wine or spirit, how much do you usually drink?

The table contains the results in percentages.

	Class							
	<1 drink	1–2 drinks	3–4 drinks	5–6 drinks	7–9 drinks	≥10 drinks	DK, refusal	Total
Rel. freq. (%)	11%	59%	18%	6%	2%	2%	2%	100%

(DK means don't know.) To answer the questions below, assume that the 16 450 people were chosen arbitrarily from the population of all EU25 citizens (at least 15 years old).

a Use the results of the table to create a model P for the random experiment that arbitrarily chooses a 15+ citizen of the EU25 and records the answer to the above question. What are the outcomes of the experiment?

b The events M, H and V are defined as follows:

M: Moderate drinker   <1 or 1–2 drinks or DK/refusal

H: Heavy drinker   3–4 or 5–6 drinks

V: Very heavy drinker   At least 7 drinks

Calculate the probabilities of these events.

c Explain why the model in part (a) comes from an empirical probability definition.

Exercise 6.15

In its Sustainability Report 2006, the European multinational Royal Philips Electronics gives the following information about the numbers of employees per sector and the sales per sector (FTE means full-time equivalent):

Sector	Sales (10^6 euro)	Employees (in FTE)
Medical Systems	6742	32 843
Domestic Appliances and Personal Care (DAP)	2645	10 953
Consumer Electronics	10576	14 486
Lighting	5466	47 739
Other activities	1547	13 347
Unallocated	–	2 364
Total	26976	121 732

a Determine the accompanying relative frequency distributions of the variables X = 'number of FTEs per sector' (where 'unallocated' is considered to be one of the sectors) and Y = 'sales (in millions of euros per sector)' for 2006.

Philips has announced that next year, to modernise affairs, €1 billion will be invested in one of its sectors, but it is not announced which sector.

b A financial market analyst believes that the chances of the five sectors to undergo the modernization are in accordance with their sales. Describe the underlying random experiment and determine the probabilities of each of the sectors.

c Another analyst thinks that the chances of the five sectors are determined by the numbers of FTEs. Determine the probabilities of the sectors if the 2364 unallocated FTEs are assigned proportionally to the five (real) sectors.

Exercise 6.16

Below, (probability) models have to be used to find the respective probabilities. On which probability definition (classical, empirical, subjective) will each model be based?

a The probability that a newborn child is a girl.

b The probability that during the winter of 2020/21 the 'Elfstedentocht' (Eleven Cities Tour, a speed skating tour in the Netherlands) will take place.

c The probability that in 2015 the developing countries will be less poor than they are now.

d The probability that the lifespan of a television set will be greater than five years.

Exercise 6.17

In a study regarding next year's state of the economy, only three possible outcomes are considered: expansive (e), recessive (r) and neutral (n).

a Determine all events of this random experiment.

b Consider the probability model P for which:

$P(\{e\}) = 0.45,\quad P(\{r\}) = 0.25,\quad P(\{n\}) = 0.30$

Use the axioms (6.11)–(6.13) to determine the probabilities of all events in part (a).

Exercise 6.18

An entrepreneur believes that next year's rounded net profit of her company (measured in millions) has only the following possible outcomes: $-3, -2, -1, 0, 1, 2, 3$. Here, -1 refers to a loss of 1 million, etc. Since the economic policy of the company is rather aggressive and risky, the entrepreneur believes that the probabilities of having positive or negative profits are more or less equal, and that the probabilities of profit x is quadratic in x. She chooses the model below, where x stands for one of the outcomes above and the constant a has to be determined later.

$P(\{x\}) = \frac{a}{20} x^2 \quad \text{for } x = 0, \pm 1, \pm 2, \pm 3$

a Use axioms (6.11)–(6.13) to determine the constant a such that this formula indeed defines a probability model.

b Is this model a classical, empirical or subjective model?

c Calculate the probability that the profit will be strictly positive. Which of axioms (6.11)–(6.13) do you use here?

Exercise 6.19

The managing director of a chain of supermarkets wants to get insight into next year's net profit of the whole chain. A team of three experienced managers (called 1, 2 and 3) of individual supermarkets is asked to propose models that describe the probabilities regarding the profit of the **whole** chain. As a starting point, the team decides to measure the net profit in units of millions of euro. They agree that the prospects are such that only the outcomes −1, 0, 1, 2, 3 and 4 have to be considered. That is, their sample space is $\Omega = \{-1, 0, 1, 2, 3, 4\}$.

Manager 1 believes that the distribution of the profit will be as follows. (Here x stands for an individual outcome and $P_1(\{x\})$ is the probability that, according to manager 1, the profit will be x.)

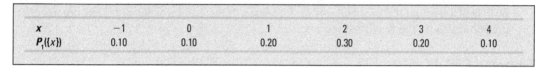

x	−1	0	1	2	3	4
$P_1(\{x\})$	0.10	0.10	0.20	0.30	0.20	0.10

The other two managers broadly agree with this model. However, they both believe that a negative profit (a loss) can be excluded, so −1 should have probability 0. Manager 2 is of the opinion that the released amount 0.10 has to be distributed equally over the other five outcomes. However, manager 3 thinks that this amount of 0.10 should be proportionally added to the probabilities of the other outcomes.

a Determine tables similar to the above to describe the probability measures P_2 and P_3 of the managers 2 and 3.

b Which of the three models assigns the highest probability to the event A = 'superprofit' = $\{3, 4\}$?

c Let B_i denote the event that next year's net profit will be **at most** i million euro, for $i = -2, -1, 0, \ldots, 4, 5$. Calculate, for each of the three models, the accompanying probabilities.

d Compare the three models in terms of caution and optimism.

Exercise 6.20

In the seventeenth century, the following question was asked of Galilei by a colleague scientist:

When three fair dice are thrown, then there are six outcomes that all have 10 eyes in total: 631, 622, 541, 532, 442 and 433. To obtain a total of 9 eyes, there also are six outcomes: 621, 522, 531, 441, 432 and 333. Still it turns out that the total 10 happens more frequently than the total 9. Dear colleague, how is this possible?

a Solve this so-called paradox of Galilei.

b What type of probability definition did you use? What type of probability definition did Galilei's colleague use?

Exercise 6.21

A recent publication by Statistics Netherlands about the 8.294 million employed persons on the Dutch labour market contained the following statistics:

Employees	7141
Self-employed	1153
Men	4554
Women	3740
Agriculture and fishery	275
Manufacturing and construction	1572
Commercial services	3708
Non-commercial services	2739

Labour market (× 1000)
Source: Statistics Netherlands (2005)

From the population of all 8 294 000 employed persons, one person is chosen at random.

a Describe the sample space Ω of this random experiment. What are the outcomes? How large is N?

b Are the outcomes equally likely?

Interest is in the events that correspond to the categories in the first column of the table. Denote them by their first letter. For instance, E is the event that this randomly chosen employed person is an employee.

c Define the events S, M_1, W, A, M_2, C, N in a similar way.

d Calculate the probabilities of the nine events (including E and Ω).

e Mention three partitions of the sample space.

f Describe the events $W \cap C$, $S \cap A$, $N \cup W$, $S \cap M_2 \cap W$ and $(S \cap M_2) \cup W$.

g Calculate the probabilities of the events $A \cup M_2$ and $C \cup N$.

Exercise 6.22 (computer)

Recall that the swindler of Example 6.6 uses the following model for the events A_j = 'the false die returns j eyes' for $j = 1, 2, ..., 6$.

j	1	2	3	4	5	6
$P(A_j)$	1/12	2/12	2/12	2/12	2/12	3/12

Below, the experiment of throwing this false die will be simulated 20 000 times; see also Example 6.7 and Appendix A1.6.

a Notice that the model is such that in a sequence of 12 throws the following ordered results are expected:

1, 2, 2, 3, 3, 4, 4, 5, 5, 6, 6, 6

Use this information to simulate a sequence of 20 000 throws with the false die.

b Create – as in Example 6.7 – a table that contains the relative numbers of times that A_j occurred (for all $j = 1, 2, ..., 6$) after $n = 100, 200, 500, 1000, 2000, 5000, 10000$ and 20000 simulations.

c Compare the relative frequencies for A_j (obtained in part (b)) with the probabilities that follow from the model.

CASE 6.2 THE MYSTERIOUS DICE

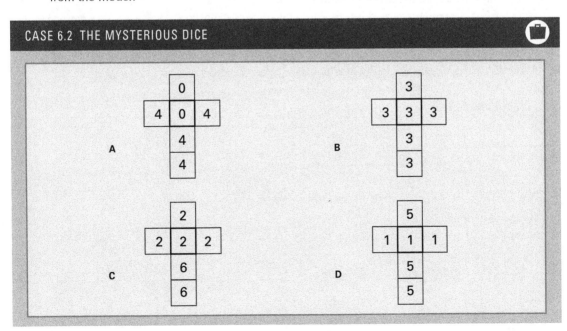

I want to play a game with you that is based on the above four dice. The rules are as follows:

- You first roll any die you like.
- Next, I roll a die I like.
- Highest roll scores one point.
- Repeat until one of us reaches ten points.

Without bragging too much, I can tell that it is very likely that I will win.

CASE 6.3 A THREE-QUESTION QUIZ ABOUT RISK (PART I)

In his column 'Everyday Economics' in *Slate*, Steven E. Landsburg asks the reader to answer a three-question quiz, to find out how 'rational' people are. (See www.slate.com.) Similar questions are considered below, to offer you a first acquaintance with **uncertainty** and **risk**; to give you some idea of your private 'risk aversion parameter'. None of the answers is better than the other; but which would you choose? Can you detect some consistency about risk aversion from your three answers? Compare your answers with those of your teammates (if any).

Q1 Choose one of the following two options:

  a Receive €10 000 cash and use it for free.

  b Receive €10 000 and put it in an investment that gives a 10% chance of returning €50 000, a 50% chance of returning €10 000 and a 40% chance of returning nothing at all.

Q2 Choose one of the following two options:

  a Receive €10 000 cash and use it for free.

  b Receive €10 000 and put it in an investment that gives a 15% chance of returning €50 000, a 45% chance of returning €10 000 and a 40% chance of returning nothing at all.

Q3 Choose one of the following two options:

  a Receive €10 000 and invest it in a share that gives a 20% chance of returning €50 000 and an 80% chance of returning nothing at all.

  b Receive €10 000 and invest it in a share that gives a 50% chance of returning €20 000 and a 50% chance of returning nothing at all.

CHAPTER

Calculation of probabilities

Chapter contents

Probability measures turn out to have several immediate properties that follow from the axioms of Chapter 6. These basic properties will eventually lead to many more advanced results. For that, it is necessary to detect some structure in the way the counting of outcomes in events is done. Furthermore, probabilities are considered when prior information is given: so-called conditional probabilities. But sometimes such prior information does not influence the probability of an event.

CASE 7.1 INTERNET CONNECTION PROBLEMS AND THE LINKNET ROUTER

The electronics company LinkNet produces a wireless broadband router (2.4 GHz, 54 Mbps). For the owners of the router only 0.1% of the serious internet connection problems are caused by the router; a large majority of the problems are caused by the modem or by the internet service provider. Furthermore, the company employs technicians who use an electronic device to test the status (defective/non-defective) of the router in the case of internet connection problems. It turns out that this device gives the correct diagnosis in 99% of the cases. That is, there is 1% chance that the device incorrectly ascribes the problems to the router and also 1% chance that it incorrectly does not ascribe the problems to the router.

But what about the opposite probability that the router really is defective if the test device comes to that conclusion? See Section 7.5 for the answer.

7.1 Basic properties

Start with a random experiment and a sample space Ω. Furthermore, suppose that P is the probability model (probability measure) that is used, so that all probabilities satisfy the axioms (6.11)–(6.13).

Each event A is mutually exclusive (disjoint) with its complement A^c. On the other hand, the union of A and A^c is Ω. But then it follows by (6.12) and (6.13) that:

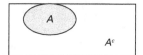

$$1 = P(\Omega) = P(A \cup A^c) = P(A) + P(A^c)$$

This leads to the first immediate consequence of the axioms:

$$P(A^c) = 1 - P(A) \tag{7.1}$$

By applying this result for $A = \Omega$, we get $P(\Omega^c) = 1 - P(\Omega) = 1 - 1 = 0$. But the complement of Ω (within Ω) is just the empty set $\varnothing$. Hence:

$$P(\varnothing) = 0 \tag{7.2}$$

Notice that (7.2) emphasizes that $\varnothing$ is justly called the empty event.

Axiom (6.13) can be generalized for the case that more than two events are involved. The Venn diagram on the right shows three (mutually) disjoint events A, B and C. That is, A and B are disjoint, A and C are disjoint, and B and C are disjoint. The probability of the union of A, B and C is equal to the summation of the individual probabilities. Of course, things can be generalized easily to the case that m disjoint events $A_1, A_2, \ldots, A_m$ are involved:

$$\text{If } A, B \text{ and } C \text{ are disjoint, then} \quad P(A \cup B \cup C) = P(A) + P(B) + P(C) \tag{7.3}$$

$$\text{If } A_1, A_2, \ldots, A_m \text{ are disjoint, then} \quad P(A_1 \cup A_2 \cup \ldots \cup A_m) = \sum_{i=1}^{m} P(A_i) \tag{7.4}$$

This last result can immediately be applied to a partition $D_1, D_2, \ldots, D_s$ since these events are (by definition) disjoint. Since their union is Ω, it follows that:

$$\text{If } D_1, D_2, \ldots, D_s \text{ is a partition of } \Omega, \text{ then} \quad \sum_{i=1}^{s} P(D_i) = 1 \tag{7.5}$$

It seems intuitively obvious that $A \subset B$ implies that $P(A) \leqslant P(B)$: if A is part of B, then $P(A)$ should not be larger than $P(B)$. But does this follow from the axioms? Since $A \subset B$, the event B is just the union of the disjoint events A and $A^c \cap B$, the two differently shaded parts in the Venn diagram on the right. By (6.13), the probability $P(B)$ is the sum of $P(A)$ and $P(A^c \cap B)$. To put it another way, the probability $P(B)$ arises by adding the non-negative number $P(A^c \cap B)$ to $P(A)$. But then $P(A)$ can never be larger than $P(B)$. That is:

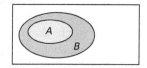

If $A \subset B$, then $P(A) \leqslant P(B)$ **(7.6)**

The probability of the union of (rectangles) A and B can be expressed in terms of $P(A)$, $P(B)$ and $P(A \cap B)$. The diagram on the right shows that the intersection $A \cap B$ belongs to both A and B. To obtain $P(A \cup B)$ we must add up $P(A)$ and the probability of B with $A \cap B$ excluded. That is, we must add up the terms $P(A)$ and $P(B) - P(A \cap B)$. To put it another way, $P(A \cup B)$ arises by adding $P(A)$ and $P(B)$ and subtracting the probability of the part $A \cap B$ that otherwise would be counted twice.

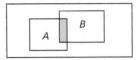

$$P(A \cup B) = P(A) + P(B) - P(A \cap B)$$ **(7.7)**

Equation (7.7) generalizes axiom (6.13): if A and B are disjoint, then $P(A \cap B) = 0$ and it indeed follows from (7.7) that $P(A \cup B)$ is then equal to $P(A) + P(B)$.

Every event falls apart into two sub-events as soon as another event enters the scene. This is illustrated by the diagram on the right: as soon as rectangle B is put on the stage, rectangle A can also be considered as the union of the two differently coloured parts. But these parts are just $A \cap B$ and $A \cap B^c$. That is: $A = (A \cap B) \cup (A \cap B^c)$. But the two intersections in this expression are disjoint, so (6.13) can be applied. It follows that:

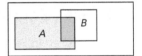

$$P(A) = P(A \cap B) + P(A \cap B^c) \quad \text{for all events } B$$ **(7.8)**

Since the events B and B^c constitute a partition, it might be expected that (7.8) can be generalized for general partitions $D_1, D_2, \ldots, D_s$. Indeed, $P(A)$ is equal to the sum of the probabilities of all intersections $A \cap D_i$, a consequence of (6.5):

If $D_1, D_2, \ldots, D_s$ is a partition of Ω then $P(A) = \sum_{i=1}^{s} P(A \cap D_i)$ **(7.9)**

Example 7.1

According to recent research by Statistics Sweden, 17.3% of Sweden's population are at least 65 years of age, 49.5% are younger than 40 and 58.8% are between 20 and 65. Consider the experiment that randomly chooses an inhabitant of Sweden. How great is the probability that this person is between 20 and 40 years old?

First, consider an informal solution in terms of percentages. Since 82.7% of the population are younger than 65 and 58.8% are between 20 and 65, it follows that 23.9% are younger than 20. Since 49.5% are younger than 40, it is obvious that $49.5 - 23.9 = 25.6\%$ are between 20 and 40. Hence, the answer to the question is 0.256.

Here is the formal solution with events and probabilities. Since the choice is made completely arbitrarily, all inhabitants have the same chance of being chosen. Hence, all outcomes (the inhabitants of Sweden) are equally likely. Consider the following events:

A = 'arbitrarily chosen inhabitant is at least 65'
B = 'arbitrarily chosen inhabitant is younger than 40'
C = 'arbitrarily chosen inhabitant is between 20 and 65'

From the facts (percentages) given above it follows that:

$P(A) = N(A)/N = 0.173;$ $P(B) = 0.495;$ $P(C) = 0.588$

Notice that $B \cap C$ = 'arbitrarily chosen inhabitant is at least 20 and younger than 40', so it is the probability $P(B \cap C)$ that is asked for. By applying (7.7) to B and C, it follows that:

$P(B \cup C) = P(B) + P(C) - P(B \cap C)$

In this equation, $P(B \cup C)$ and $P(B \cap C)$ are not known yet. But notice that:

$B \cup C$ = 'arbitrarily chosen inhabitant is younger than 65' $= A^c$

so $P(B \cup C) = P(A^c) = 0.827$ by (7.1). Hence,

$0.827 = 0.495 + 0.588 - P(B \cap C),$

so $P(B \cap C) = 0.256.$

7.2 Rules for counting

As mentioned in Section 6.1, the following random experiment is very important:

'arbitrarily choosing a sample of n elements from the population'

It is this experiment that offers the opportunity to draw conclusions about population statistics by considering similar statistics for the sample of size n. The possible outcomes of this random experiment are sequences of n population elements. They are all equally likely, so the classical probability definition can be used. Since according to this definition $P(A)$ depends on the **number of** the outcomes in A, it is important to have convenient counting rules to determine such a number of outcomes. Here is an example that illustrates this.

Example 7.2

A consumer organization called ConsumCare wants to check the contents of cartons of milk (of a certain brand) which, according to the label, should contain one litre. The manufacturer states that only 0.5% of the cartons may contain a bit less than one litre of milk. ConsumCare has doubts about that claim and wants to check it by – completely arbitrarily – choosing 100 packs from the daily production of 100 000. The manufacturer will be prosecuted if at least 1 pack contains less than one litre of milk.

ConsumCare will base the decision of yes/no prosecuting the manufacturer on a sample of only 100 packs. An interesting question now is whether the manufacturer will be prosecuted undeservedly. That is, whether the sample will contain at least one pack containing less than one litre of milk while the manufacturer is right in the sense that only 0.5% of the 100 000 packs contains too little. Or to say it in a probabilistic way: if indeed only 0.5% of the population contains too little, what then is the probability that the manufacturer will be prosecuted?

The random experiment is just the arbitrary selection of 100 packs from the population of 100 000 packs. If the population elements are, in the order of production, numbered 1, 2, ..., 100 000, then the possible outcomes of the experiment are sequences with 100 positions, each position representing a number of the pack. For instance, the following ordered sequence with 100 positions is a possible outcome:

(34 550, 92 896, 59 453, ..., 10 823)

This outcome implies that pack 34 550 was chosen first (on the first position), pack 92 896 was second and pack 10 823 was the 100th pack chosen. The sequence is **ordered** in the sense that placing the numbers in the sequence on other positions yields a different sequence. To count the possible outcomes, there are 100 000 possibilities for the first position, 99 999 for the second, 99 998 for the third, and so on. For the 100th there are 99 901 possibilities left. Hence, the total number of outcomes is:

$$N = 100\,000 \times 99\,999 \times 99\,998 \times \ldots \times 99\,901$$

It will be intuitively obvious that **arbitrarily** choosing the 100 packs guarantees that all N outcomes in the sample space Ω are equally likely. Hence, the event $A = $ 'all 100 packs contain at least one litre' has probability $N(A)/N$. Here $N(A)$ is the number of outcomes (sequences) in Ω that contain only packs with at least one litre of milk. Now suppose that the manufacturer is right, so precisely 0.5% of the packs in the population do **not** contain enough milk. Hence, 99 500 of the 100 000 packs contain at least one litre of milk. To determine $N(A)$, we have 99 500 possibilities for the first position, 99 499 for the second, 99 498 for the third, etc. For the last position there are 99 401 possibilities left. That is:

$$N(A) = 99\,500 \times 99\,499 \times 99\,498 \times \ldots \times 99\,401$$

Hence,

$$P(A) = \frac{N(A)}{N} = \frac{99\,500}{100\,000} \times \frac{99\,499}{99\,999} \times \frac{99\,498}{99\,998} \times \ldots \times \frac{99\,401}{99\,901}$$

To calculate this expression, note that each of the 100 ratios on the right-hand side is approximately equal to 0.9950. So:

$$P(A) \approx (0.9950)^{100} = 0.606$$

But this means that the probability $P(A^c)$ that the manufacturer will be prosecuted undeservedly is about $1 - 0.606 = 0.394$.

Several counting rules will be considered. The first is about situations where (say) r subsequent choices have to be made while for each individual moment of choice the number of possibilities does not depend on the actual choices made at former choice moments. Let a_k denote the number of possibilities at the kth moment of choice. Then the **multiplication rule** states that the total number of possibilities is just the product of the numbers of possibilities at all individual moments of choice. That is, the total number of possibilities is:

$$a_1 \times a_2 \times \ldots \times a_r$$

In Example 7.2 this intuitive rule was used twice: to determine N and $N(A)$. In both cases there were $r = 100$ choice moments. For N, there were 100 000 possibilities for the choice of the first position (so $a_1 = 100\,000$), 99 999 for the choice of the second position (so $a_2 = 99\,999$), etc; notice that a_r equals 99 901. At every choice moment the number of possibilities decreases by one and does depend on the number of possibilities at the former choice moment. But the number of possibilities does **not** depend on the actual choices made before, and this is what the multiplication rule requires. To determine $N(A)$, we used $a_1 = 99\,500$, $a_2 = 99\,499$, ..., $a_{100} = 99\,401$.

The following rule informs about the number of orderings (**permutations**) of k different objects. If the k objects are denoted by 1, 2, ..., k, then, to determine the orderings, k positions have to be filled with the numbers 1, 2, ..., k. For the first position, there are k possibilities. For the second position, there are $k - 1$ left; for the third $k - 2$; etc. For the last position only one possibility is left. Application of the above multiplication rule yields that the **number of permutations** of k objects is equal to $k \times (k - 1) \times \ldots \times 2 \times 1$. But this is equal to:

$$k! = 1 \times 2 \times 3 \times \ldots \times k \quad \text{and} \quad 0! = 1$$

The notation $k!$ is pronounced **k-factorial**. For example:

$$4! = 24; \quad 1! = 1; \quad 12! = 479\,001\,600; \quad 24! = 6.2045 \times 10^{23}$$

That is, there are 24 possibilities to put four different objects in order; there are more than 479 million possibilities to place twelve objects in order. In the above considerations, k has to be a positive integer. But also 0! – the number of orderings of zero objects – is occasionally needed. It is just defined that $0! = 1$.

In how many ways can k objects be chosen (sampled) from a set or population of m objects? In this experiment, the outcomes are sequences with k positions that are filled with elements from the set $\{1, 2, \ldots, m\}$. Below, such outcomes will be denoted as **k-tuples**. To answer the above question, more information is needed. Are the k-tuples ordered? That is, does it matter in which order the objects are chosen? And is it allowed that objects are chosen more than once? That is, is a chosen object put back into the population so that it can be chosen again at future moments of choice?

In Example 7.2, a sample of 100 packs of milk was chosen from 100 000 packs; the total number of possible outcomes N was calculated. Although it was not emphasized, the outcomes were ordered. Furthermore, a pack of milk chosen at one of the moments of choice was not put back into the population of 100 000 packs so it could not be chosen in future moments of choice.

It would appear that the procedure of taking k objects from a population of m objects has to be better specified. More information is needed about:

1. Are the k-tuples supposed to be ordered or unordered?

2. Is the sample of k objects taken (drawn) with replacements or without replacements?

With replacement means that after each of the k moments of choice the chosen object is put back into the population; **without replacement** means that it is not put back. Ordered k-tuples are called **variations** and are denoted as sequences $(\cdot, \cdot, \ldots, \cdot)$ of k positions with parentheses. For instance, the 4-tuple (2, 3, 1, 4) is a variation; it informs us that object 2 of the population is chosen first, object 3 is chosen second, etc. This 4-tuple is different from (4, 1, 3, 2). Unordered k-tuples are called **combinations** and denoted as sequences $\{\cdot, \cdot, \ldots, \cdot\}$ of k objects with brackets. The 4-tuple $\{1, 2, 3, 4\}$ informs us that the objects 1, 2, 3 and 4 are chosen; the order in which this was done is of no importance. That is, $\{1, 2, 3, 4\}$ is exactly the same as $\{4, 1, 3, 2\}$ and $\{2, 3, 1, 4\}$.

First, suppose that choosing the k objects from m objects is done **ordered** and **with** replacement. When filling in the k positions of $(\cdot, \cdot, \ldots, \cdot)$ there are m possibilities for the first position. Since after each moment of choice the chosen object is put back into the population of m objects, the second position also has m possibilities. The same holds for the third, etc. That is, the numbers of possibilities $a_1, a_2, \ldots, a_k$ at the k moments of choice are all equal to m. Hence, the total number of possible k-tuples is:

$$a_1 \times a_2 \times \ldots \times a_k = m^k$$

If the sampling is done **ordered** and **without** replacement, then the first position of the k-tuple offers m possibilities. Since the chosen objects are not put back, the second position has $m - 1$ possibilities, the third only $m - 2$, etc. At the last (the kth) moment of choice, only $m - (k - 1)$ possibilities are left. But this implies that:

$$a_1 \times a_2 \times \ldots \times a_k = m \times (m - 1) \times \ldots \times (m - k + 1) \tag{7.10}$$

It is obvious that, in this procedure, k cannot be larger than m. To simplify this expression, a simple trick is performed: the right-hand side of the above equation is multiplied by

$$1 = \frac{(m - k)!}{(m - k)!} = \frac{(m - k) \times (m - k - 1) \times \ldots \times 2 \times 1}{(m - k)!}$$

As a consequence, (7.10) turns into:

$$a_1 \times a_2 \times \ldots \times a_k = \frac{m!}{(m-k)!}$$

To find the number of k-tuples when the sampling is done **un**ordered and **without** replacement, we start with the total number of k-tuples in the **ordered** case without replacement. It was just found that this last number is $m!/(m-k)!$ Now recall that each k-tuple has $k!$ permutations (orderings). Transition from the ordered to the unordered case implies that groups of $k!$ different orderings become equal in the unordered case. But then the above number $m!/(m-k)!$ of k-tuples for the ordered case without replacement has to be divided by $k!$ to obtain the **un**ordered number of k-tuples without replacement. The result has its own notation (pronounced **m choose k**):

$$\binom{m}{k} = \frac{m!}{k! \times (m-k)!}$$

It is called a **binomial coefficient**; the reason will become clear in Section 9.2.

The fourth case is **unordered** sampling **with** replacement. Since this is rarely done in practice and the derivation is more complicated, this case will be omitted. The four possible ways to choose k from m objects are summarized below (note that the two cases for sampling without replacement require that $k \leq m$):

Number of possible ways to choose k objects from m objects

	With replacement	Without replacement
Ordered	m^k	$\frac{m!}{(m-k)!}$
Unordered	–	$\binom{m}{k} = \frac{m!}{k!(m-k)!}$

Example 7.3
Here are some calculation examples and immediate consequences.

$$\binom{7}{2} = \frac{7!}{2! \times 5!} = \frac{5040}{2 \times 120} = 21; \quad \binom{7}{5} \text{ gives the same answer;}$$

$$\binom{9980}{9978} = \frac{9980!}{9978! \times 2!} = \frac{9978! \times 9979 \times 9980}{9978! \times 2!} = \frac{9979 \times 9980}{2!} = 49\,795\,210$$

(Notice that 9980! and 9978! are too large for calculators to handle, so dividing out the factor 9978! is really necessary.)

$$\binom{m}{k} = \binom{m}{m-k}$$

$$\binom{m}{0} = \binom{m}{m} = 1 \quad \text{(thanks to } 0! = 1\text{)}$$

$$\binom{m}{m-1} = \binom{m}{1} = m$$

Example 7.4

Suppose that the coach of a European football team wants to use two forwards in an away match. To draw up his forward line in this way, the coach has to choose two out of his four forward players called *a*, *b*, *c* and *d*. Table 7.1 gives overviews of all possibilities for the four cases.

	With replacement	Without replacement
Ordered	aa	–
	ab	ab
	ac	ac
	ad	ad
	ba	ba
	bb	–
	bc	bc
	bd	bd
	ca	ca
	cb	cb
	cc	–
	cd	cd
	da	da
	db	db
	dc	dc
	dd	–
Unordered	aa	–
	ab	ab
	ac	ac
	ad	ad
	–	–
	bb	–
	bc	bc
	bd	bd
	–	–
	–	–
	cc	–
	cd	cd
	–	–
	–	–
	–	–
	dd	–

TABLE 7.1 All possible ways to choose 2 from 4

The NW quadrant yields 16 possible forward lines; the NE, SE and SW quadrants give respectively 12, 6 and 10 possibilities. Of course, the two with-replacement cases are included only for the sake of completion: the coach cannot make use of these approaches. By taking *k* = 2 and *m* = 4 in the table preceding Example 7.3, the answers 16, 12 and 6 can be checked.

Example 7.5

A chain of fashion houses owns 52 stores with four different names: HE, SHE, H&S and WE; each of the four types of stores has 13 outlets. The CEO of the chain arbitrarily chooses five of the 52 stores and allocates them to the young and enthusiastic manager Levi Trousers. What is the probability that all five stores are of the same type?

The outcomes of the random experiment are 5-tuples. Of course, the five stores are chosen without replacement and the order in which the choices are made is **not** important. The total number of possible 5-tuples is:

$$\binom{52}{5} = \frac{52!}{5! \times 47!} = \frac{48 \times 49 \times 50 \times 51 \times 52}{120} = 2\,598\,960$$

For how many of them are all stores of the same type? To determine the total number of 5-tuples with only HE stores, 5 have to be chosen from 13; so there are

$$\binom{13}{5} = 1287 \text{ possibilities}$$

Of course, the same number is found for SHE, H&S and WE stores. Hence, there are $4 \times 1287 = 5148$ possibilities of obtaining 5-tuples with all stores of one type. Because of the classical definition, the probability that the five stores are all of the same type is $5148/2\,598\,960 = 0.002$.

In a random experiment with a finite sample space Ω of size N, the events are just the subsets of the sample space. But how many subsets does Ω have? This question can now be answered with the above counting rules.

If the outcomes in Ω are renumbered as 1, 2, ..., N, then each subset of Ω can be characterized as an N-tuple by putting a 1 on position j if outcome j belongs to the subset and putting a 0 if j does not belong to the subset. For instance, the N-tuple of the subset {8, 12, 45} of Ω has a 1 on positions 8, 12 and 45; the other positions all contain zeros. On the other hand, the N-tuple with ones on positions 12, 38, 77, 87 and zeros on all other positions corresponds to the subset {12, 38, 77, 87}. Hence, the determination of the number of subsets of Ω leads to the same answer as the determination of the number of possibilities when zeros and ones are placed on the N positions of an N-tuple. At each of the N moments of choice, there are two possibilities. So, in total there are 2^N possibilities because of the multiplication rule. Hence:

Total number of subsets of sample space with N outcomes $= 2^N$

7.3 Random drawing and random sampling

This section is about two random experiments that have already been considered briefly in Sections 7.1 and 7.2. Although some of their properties were also mentioned before, they will be formalized now. This is done because of the extreme importance of these random experiments: they turn out to connect not only the sub-fields of probability theory and descriptive statistics, but also the sub-fields of probability theory and inferential statistics.

Start with some finite population of interest, denoted as {1, 2, ..., N}. Now suppose that one element of the population is drawn completely arbitrarily. That is, by way of a fair lottery one element of the population is chosen. This random experiment is called ***random drawing***. Its sample space Ω is just the population, so {1, 2, ..., N}. Because of the way the drawing is done, all outcomes are equally likely. But then, according to the classical probability definition, the probability $P(j)$ that population element j is drawn is equal to $1/N$. That is:

$$P(j) = \frac{1}{N} \quad \text{for each element } j \text{ in the population} \tag{7.11}$$

Now consider the set of the population elements that have a certain property a; denote their number by $M(a)$. Furthermore, let A be the event that the randomly drawn population element has property a; denote the number of outcomes in A by $N(A)$. In descriptive statistics, the relative frequency (say p_a) of the elements in the population with property a is just the ratio of $M(a)$ and N. In probability theory, the classical definition guarantees that:

$$P(A) = \frac{N(A)}{N}$$

Since the sample space is equal to the population and outcomes are population elements, $N(A)$ is just the number of **population** elements with property a (which is $M(a)$) and the probability above is equal to the relative frequency p_a. Consequently:

$$P(A) = p_a$$

This result is intuitively clear and has already been applied a few times, for instance in Example 7.1. The formal result states that a relative frequency in descriptive statistics can also be considered as a probability by way of the random experiment random drawing. In this way, the sub-field of descriptive statistics becomes part of the sub-field of probability.

How can this procedure of randomly drawing an element be carried out in practice? You might use a tombola that contains N balls with the numbers 1, 2, ..., N and chooses one of the balls completely arbitrarily. Or you might write these numbers on N pieces of paper, put them in a box, shuffle the box thoroughly and pick one of the pieces of paper with your eyes closed. However, more modern approaches make use of random generators of computer packages. For Excel, see Appendix A1.7.

Example 7.6

Suppose that the European Commission wants to give a windfall payment of €2 billion to a randomly drawn EU25 member state, to stimulate its economy. According to Example 1.2, a proportion 0.16 of the EU25 countries has the property 'low-income country'. But this relative frequency 0.16 can be considered as a probability regarding the experiment 'random drawing'. It follows that the event $B =$ 'the randomly drawn EU25 country belongs to the low income group' has probability $P(B) = 0.16$. In words: the probability that this randomly drawn EU25 country is a low-income country is 0.16.

Denote these 25 countries as 1, 2, ..., 25, in alphabetical order. Each individual country has probability 0.04 of receiving the money. The experiment 'random drawing' is simulated 25000 times; the results are in the file Xmp07-06.xls. Because of the law of large numbers (see Section 6.3) it may be expected that in this file the relative frequency of 1s, 2s, ..., 25s is in each case close to 0.04. To check that, the relative frequency distribution of the data file is given in Table 7.2.

Ctry	Rel. freq.	Ctry	Rel. freq.	Ctry	Rel. freq.	Ctry	Rel. freq.	Ctry	Rel. freq.
1	0.04152	6	0.04084	11	0.03992	16	0.04056	21	0.03960
2	0.03912	7	0.04056	12	0.04056	17	0.03828	22	0.03980
3	0.03948	8	0.04068	13	0.04084	18	0.04004	23	0.04124
4	0.03936	9	0.03976	14	0.03992	19	0.04012	24	0.03940
5	0.03920	10	0.04128	15	0.03876	20	0.03876	25	0.04040

TABLE 7.2 Relative frequency distribution of 25000 random drawings from EU25

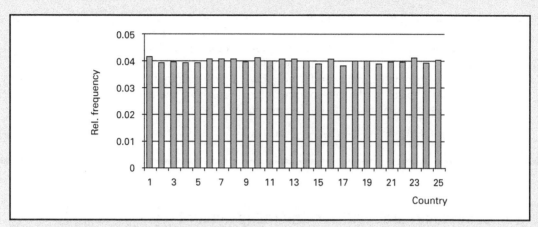

FIGURE 7.1 Bar chart of the relative frequency distribution

Indeed, the 25 relative frequencies are all close to 0.04. The random generator used seems to work well. By the way, notice that the relative frequency of the four low-income countries (6, 14, 15 and 19) is equal to 0.1596, which is indeed close to the relative frequency 0.16.

Usually it is not the drawing of only one random element that people are interested in. The combined experiment that randomly draws (say) n elements from the population of interest is called ***random sampling***; the sample that is created in this way, is called a ***random sample*** of size n. As mentioned in the former section, we have to be more specific and define the random experiment precisely. ***Random sampling with replacement*** means that after each of the n moments of choice the recently drawn element is put back into the population. On the other hand, ***random sampling without replacement*** tells us that after each partial draw the element is not put back. For practical reasons, the outcomes of random sampling (of size n) with replacements will be ordered n-tuples whereas the outcomes of random sampling without replacement will be **un**ordered n-tuples.

Example 7.7

Now suppose that the European Commission wants to use the windfall payment of €2 billion to stimulate the economy of **four** randomly drawn EU25 countries. What is the probability that only low-income countries will profit? And the probability that at least two of the low-income countries will profit?

Of course, this random sampling is done without replacement. Hence, the outcomes are unordered 4-tuples of four different countries. Let A_k be the event that exactly k of the four low-income countries are sampled, for $k = 0, 1, 2, 3$ and 4. Since all possible outcomes are equally likely, the probability $P(A_4)$ is equal to the ratio $N(A_4)/N$. Here, N is the total number of outcomes in Ω, which is 12 650 (i.e. the binomial coefficient '25 choose 4'). Furthermore, $N(A_4)$ is just the number of (unordered) 4-tuples in Ω that contain the four low-income countries. Hence, $N(A_4) = 1$ and the answer to the first question is:

$$P(A_4) = \frac{1}{12\,650} = 0.000079$$

To calculate $N(A_2)$, note that there are six possible ways (4 choose 2) to choose two countries from the four low-income countries. For each **fixed** pair of two low-income countries there are 210 outcomes that contain that pair of countries and two non low-income countries (21 choose 2, since the other two countries of the 4-tuple have to be chosen from the 21 non low-income countries). Hence, in total there are $N(A_2) = 6 \times 210 = 1260$ outcomes in A_2. There are four possible ways to choose three from the four low-income countries (4 choose 3). To create 4-tuples with a fixed triple of three low-income countries and 1 non low-income country, there are 21 possible ways (since there are 21 choices for the last country in the 4-tuple). So, there are $N(A_3) = 4 \times 21 = 84$ outcomes in A_3. Hence, the answer to the second question is:

$$P(A_2) + P(A_3) + P(A_4) = \frac{1260 + 84 + 1}{12\,650} = 0.1063$$

7.4 Conditional probabilities and independence

There are many occasions in statistics when some information about the final outcome of a random experiment is already known while the precise outcome is not yet known. For instance, it might already be known that a throw with a fair die yielded an even number of eyes but the precise outcome is still unknown. Or maybe you want to study next month's AEX price index under the condition ('pre-information') that its level will be at least as high as it is at this very moment. In such situations, several questions are of interest. Does this pre-information influence the probability of a certain event? And if so, in what way? Is the resulting probability different from the probability of the event without the pre-information?

7.4.1 Conditional probabilities

Having pre-information about the outcome of the random experiment means that a certain event is known to have occurred or is assumed to have occurred. If it is already known that a fair die returns an even number of eyes, the event 'even' has occurred while it still is uncertain whether events such as 'at most three' occurred. Interest is in the probability of an event A under the **condition** that B has occurred.

The probability of event A given the pre-information that event B has occurred can be understood by way of a Venn diagram. Again, the large rectangle represents the sample space Ω. Now suppose that it is already known that B has occurred. That is, it is already known that the outcome falls in the right-hand large rectangle in Ω that is denoted as B. So, B is the actual sample space. With respect to the event A, only the part $A \cap B$ that belongs to B can occur. That is why the probability of **A given B** is just the relative magnitude of $P(A \cap B)$ with respect to $P(B)$:

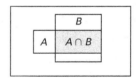

Conditional probability of A given B

The **conditional probability of event A given event B** is denoted as $P(A \mid B)$ or as $P_B(A)$ and is defined as

$$P(A \mid B) = \frac{P(A \cap B)}{P(B)}$$

(7.12)

(Of course, this ratio makes sense only if $P(B) \neq 0$.) It is the ratio of the probability of the intersection and the probability of the condition. Interchanging A and B leads (if $P(A) \neq 0$) to $P_A(B)$, the probability of B given A, which is

$$P(B \mid A) = \frac{P(A \cap B)}{P(A)} \qquad\qquad (7.13)$$

In this context, probabilities like $P(A)$ and $P(A \cap B)$ are often called **unconditional probabilities**.

The unconditional probability $P(B)$ and the conditional probability $P(B \mid A)$ are called the **prior probability** and the **posterior probability** of B. They respectively give the probability of B **before** and **after** the information about A became available.

Suppose that you throw a fair die in a poker cup. After your throw, your partner peeks under the cup and tells that the number of eyes is even. With this pre-information, it is your task to decide whether the throw is at most three. That is, you have the pre-information that the event B = 'even' has occurred. So, there are three possible outcomes left: the outcomes 2, 4 and 6 of B. Since these outcomes are equally likely, the conditional probability that A = 'at most three' will occur should be equal to 1/3. Indeed, that answer also follows from (7.12):

$$P(A \cap B) = P(\{2\}) = 1/6; \quad P(B) = 1/2; \quad P(A \mid B) = \frac{1/6}{1/2} = 1/3$$

Recall that P is a probability measure, so it satisfies the axioms (6.11)–(6.13). Below it will be shown that P_B, the conditional probability given a fixed event B, is a probability measure too. Since $P(A \cap B)$ and $P(B)$ are both $\geqslant 0$, their ratio $P_B(A)$ is also $\geqslant 0$, so axiom (6.11) with P replaced by P_B is valid. Furthermore:

$$P_B(\Omega) = \frac{P(\Omega \cap B)}{P(B)} = \frac{P(B)}{P(B)} = 1$$

So, P_B also satisfies axiom (6.12). If A and C are disjoint, then $A \cap B$ and $C \cap B$ are also disjoint and:

$$P_B(A \cup C) = \frac{P((A \cup C) \cap B)}{P(B)} = \frac{P((A \cap B) \cup (C \cap B))}{P(B)}$$

$$= \frac{P(A \cap B)}{P(B)} + \frac{P(C \cap B)}{P(B)} = P_B(A) + P_B(C)$$

Hence, P_B satisfies all three axioms and it is indeed a probability measure.

In Section 7.1 several properties of probability measures were derived. Since P_B is also a probability measure, properties (7.1)–(7.9) are also valid for P_B instead of P. For instance, (7.1), (7.7) and (7.8) respectively turn into:

$$P_B(A^c) = 1 - P_B(A)$$

$$P_B(A \cup C) = P_B(A) + P_B(C) - P_B(A \cap C)$$

$$P_B(A) = P_B(A \cap C) + P_B(A \cap C^c)$$

Example 7.8

Consider the random experiment in which a throw with a red die is followed by a throw with a blue die. Suppose that both dice are fair. The possible outcomes of this combined random experiment can be recorded in a 6 × 6 grid by placing the possible outcomes of the first (red) throw on the first position and those of the second (blue) throw on the second position; see Figure 7.2.

Blue die						
6	(1, 6,					
5	(1, 5)	(2, 5,				
4	(1, 4)	(2, 4)	(3, 4,				←
3	(1, 3)	(2, 3)	(3, 3)	(4, 3,			
2	(1, 2)	(2, 2)	(3, 2)	(4, 2)	(5, 2,		
1	(1, 1)	(2, 1)	(3, 1)	(4, 1)	(5, 1)	(6, 1,	
	1	2	3	4	5	6	Red die

FIGURE 7.2 Possible outcomes of the random experiment

Consider the events:

A = 'total number of eyes is at least seven'
B = 'maximum of the two single throws is at most five'
C = 'second single throw yields four'
D = 'maximum of the two single throws is more than five'
E = 'maximum of the two single throws is at least five'

Notice that A is just the set of outcomes while all outcomes in the grey cells belong to B. The event C is just the row (with the arrow) in the diagram. Since the outcomes of this combined experiment are equally likely, probabilities of events can easily be calculated by counting the numbers of outcomes enclosed. For instance:

$$P(A) = \frac{21}{36}; \quad P(B) = \frac{25}{36}; \quad P(A \cap B) = \frac{10}{36}; \quad P(A \cup B) = \frac{36}{36} = 1; \quad P(C) = \frac{6}{36}; \quad P(A \cap C) = \frac{4}{36};$$
$$P(A \cup C) = \frac{23}{36}$$

Now suppose that it is already given that A has occurred. How does this pre-information influence the probabilities of B, C and other events?

It is the probability measure P_A that is of interest. By using the definition of conditional probabilities it follows that:

$$P_A(B) = \frac{P(A \cap B)}{P(A)} = \frac{10 / 36}{21 / 36} = 10/21$$

$$P_A(C) = \frac{P(A \cap C)}{P(A)} = \frac{4 / 36}{21 / 36} = 4/21$$

$$P_A(B \cap C) = \frac{P(A \cap (B \cap C))}{P(A)} = \frac{3 / 36}{21 / 36} = 3/21$$

Apparently, the pre-information influences the probabilities of B and C since the unconditional probabilities $P(B)$ and $P(C)$ are **un**equal to 10/21 and 4/21, respectively. Notice that the above answers are also intuitively correct. The knowledge that A has occurred reduces the possible outcomes to the 21 outcomes of A. But only ten of these outcomes fall in B, four fall in C and three in $B \cap C$.

We next calculate other conditional probabilities by using the result that P_A is a probability measure and hence satisfies the probability rules. Since $D = B^c$, it follows that

$$P_A(D) = 1 - P_A(B) = 1 - 10/21 = 11/21$$

$$P_A(E) = P_A(E \cap B) + P_A(E \cap B^c)$$

$$= P_A(\text{maximum is five}) + P_A(\text{maximum is six})$$

$$= \frac{7 / 36}{21 / 36} + \frac{11 / 36}{21 / 36} = \frac{7}{21} + \frac{11}{21} = 18/21 = 6/7$$

Multiplying both sides of (7.12) by $P(B)$ and both sides of (7.13) by $P(A)$ yields:

$$P(A \cap B) = P(B) \times P(A \mid B) \quad \text{and} \quad P(A \cap B) = P(A) \times P(B \mid A) \tag{7.14}$$

It appears that $P(A \cap B)$ can be expressed in terms of a conditional probability; indeed it can be done in two ways. This expression can be generalized for three events A, B and C. Notice that:

$$P(A) \times P(B \mid A) \times P(C \mid A \cap B) = \frac{P(A)}{1} \times \frac{P(A \cap B)}{P(A)} \times \frac{P(A \cap B \cap C)}{P(A \cap B)}$$

Since some terms cancel out, we obtain:

$$P(A \cap B \cap C) = P(A) \times P(B \mid A) \times P(C \mid A \cap B) \tag{7.15}$$

In the case of four events A, B, C and D it follows analogously that:

$$P(A \cap B \cap C \cap D) = P(A) \, P(B \mid A) \, P(C \mid A \cap B) \, P(D \mid A \cap B \cap C) \tag{7.16}$$

Generalizations for more than four events are straightforward. Rules like (7.14)–(7.16) are called **product rules for probabilities of intersections**. They express that a probability of an intersection can be rewritten in terms of conditional probabilities.

Such product rules are frequently used in random experiments that can be considered as a combination of sub-experiments:

Example 7.9

Reconsider the random allocation of five stores to Levi Trousers, by the CEO of a chain of 52 fashion houses in Example 7.5. An alternative solution will be considered for the probability that Levi's stores will all be of one type.

The random sampling of the five stores can also be considered as a combination of five sub-experiments that all randomly draw a store. Let A_i be the event that the ith store drawn is an HE store, for $i = 1, \ldots, 5$. The product rule will be used to calculate $P(A_1 \cap A_2 \cap A_3 \cap A_4 \cap A_5)$, the probability that only HE stores are sampled. Below, step i refers to the sub-experiment of drawing the ith store:

1 $P(A_1) = 13/52$ since 13 of the 52 stores are HE stores.
2 $P(A_2 \mid A_1) = 12/51$; having obtained an HE store in the first drawing, 12 HE stores are left for the second on a total of 51.
3 $P(A_3 \mid A_1 \cap A_2) = 11/50$; having obtained HE stores in drawings 1 and 2, there are 11 HE stores left for the third on a total of 50.
4 $P(A_4 \mid A_1 \cap A_2 \cap A_3) = 10/49$.
5 $P(A_5 \mid A_1 \cap A_2 \cap A_3 \cap A_4) = 9/48$.

By the product rule we obtain the probability of getting only HE stores:

$$P(A_1 \cap A_2 \cap A_3 \cap A_4 \cap A_5) = \frac{13}{52} \times \frac{12}{51} \times \frac{11}{50} \times \frac{10}{49} \times \frac{9}{48} = 0.000495$$

Indeed, by multiplying this answer by 4, the answer of Example 7.5 follows.

7.4.2 Independence

It is easy to think of situations in which the given occurrence of a certain event does **not** influence the probability of another event. It will be intuitively clear that the results of two consecutive throws with a fair die will not influence one another. Indeed, the events $A = $ 'six eyes on the first die' and $B = $ 'six eyes on the second die' satisfy:

$$P(B) = \frac{6}{36} = \frac{1}{6} \quad \text{and} \quad P(B \mid A) = \frac{P(A \cap B)}{P(A)} = \frac{1/36}{6/36} = \frac{1}{6}$$

Consequently, the pre-information that the first die yielded six does not influence the probability that the second die will give six eyes too.

An event B is said to be independent of another event A if $P(B \mid A) = P(B)$; that is, if the probability that B will occur is not influenced by the fact that it is already given that A has occurred. But if B is independent of A, then it follows from (7.14) that $P(A \cap B) = P(A)P(B)$ and hence

$$P(A \mid B) = \frac{P(A \cap B)}{P(B)} = \frac{P(A)P(B)}{P(B)} = P(A)$$

Consequently, if B is independent of A then A is independent of B, so A and B are independent of each other. Hence, the following definition makes sense:

(Stochastically) independent events

Events A and B are called (**stochastically**) **independent** if: $P(A \mid B) = P(A)$ **(7.17)**

From the above it follows that each of the two equations in the box below is equivalent to (7.17).

Two other equivalent ways to express independence of *A* and *B*

$P(B \mid A) = P(B)$ **(7.18)**

or

$P(A \cap B) = P(A)P(B)$ **(7.19)**

To check the independence of two events A and B, it is sufficient to check only one of the three equations (7.17), (7.18) and (7.19). Notice that (7.17) requires that $P(B) \neq 0$, while (7.18) requires that $P(A) \neq 0$. Equation (7.19) has the advantage that there are no requirements.

From (7.19) it also follows that the impossible event $\varnothing$ is independent of any event A, since:

$$P(\varnothing \cap A) = P(\varnothing) = 0 \quad \text{and} \quad P(\varnothing)P(A) = 0$$

Some people claim from 'intuitive' arguments that each pair of disjoint events automatically should be independent. This is a huge misunderstanding that has to be solved once and for all. On the contrary! If two events A and B are disjoint and hence $B \subset A^c$, then the occurrence of B implies that A will surely not occur, which makes A and B 'very' **de**pendent. It is easy to find a counterexample for the misunderstanding. In a throw with a fair die, let A and B be the events that the result is even, respectively odd. Then A and B are disjoint, but **not** independent since $P(A \mid B) = 0$ while $P(A) = \frac{1}{2}$.

Independence of events is often a consequence of 'physical' independence that is present within the random experiment. For instance, the combined random experiment that first measures the number of hours of sunshine tomorrow and next records the result of throwing a die, consists of two sub-experiments that are physically independent since the result of the die will not be influenced by the number of hours that the sun shines tomorrow. But notice that this combined experiment has pairs as outcomes: the first position refers to the number of hours of sunshine tomorrow while the second position gives the number of eyes of the die. Hence, the events of this combined random experiment contain pairs of measurements. Still it seems obvious that the events A = 'less than four hours of sunshine tomorrow' and B = 'even number of eyes' are stochastically independent even though they contain outcomes of the combined experiment. This is because the descriptions of these events refer to different sub-experiments.

Independence can also be a matter of modelling. The combined random experiment that first measures the number of hours of sunshine tomorrow and next the closing value of the AEX index tomorrow is constituted by two sub-experiments for which you are not completely sure that they are physically independent. (Is the value of the AEX influenced by the number of hours of sunshine? Strictly speaking it cannot be excluded until the reverse is proved.) But things will turn out to be much more complicated if this physical independence is not valid. That is why it will just be **assumed** that the two sub-experiments are physically independent, meaning that events like A = 'less than four hours of sunshine tomorrow' and B = 'closing price of the AEX index tomorrow is at least 500', which depend on different sub-experiments, can be considered to be stochastically independent.

Example 7.10
Consider one throw with two fair dice, a red one and a blue one. The outcomes of this random experiment can be considered as pairs (i, j), where i (by convention) refers to the red die and j to the blue die. Hence, the sample space contains 36 such outcomes that are all equally likely. Furthermore, this experiment cannot be distinguished from the combined random experiment that first throws the red die (sub-experiment 1) and next the blue die (sub-experiment 2).

Consider the following events:

A = 'at least 4 with red'
B = 'at most 4 with blue'
C = 'total score is 8'

(In what follows, it may be helpful to refer back to the diagram in Example 7.8.) Since A and B contain respectively 18 and 24 outcomes, it follows that:

$$P(A) = \frac{18}{36} = 1/2 \quad \text{and} \quad P(B) = \frac{24}{36} = \frac{2}{3}$$

Event A refers only to sub-experiment 1, B to sub-experiment 2, and the two sub-experiments are physically independent. Hence, A and B are stochastically independent and:

$$P(A \cap B) = P(A)P(B) = \frac{1}{2} \times \frac{2}{3} = \frac{1}{3}$$

Notice that C gives information about both sub-experiments, so we cannot expect A and C to be independent. Since $A \cap C = \{(4, 4), (5, 3), (6, 2)\}$ and $C = \{(2, 6), (3, 5), (4, 4), (5, 3), (6, 2)\}$, we obtain:

$$P(A \mid C) = \frac{P(A \cap C)}{P(C)} = \frac{3/36}{5/36} = \frac{3}{5}$$

which is indeed unequal to $P(A) = 1/2$.

It is unwise to let intuition completely decide about the dependence or independence of two events. Intuition **can** give you the idea that events might be independent or not. But you will always have to show whether your intuition is correct.

Example 7.11
Consider a tombola with twelve balls. One ball has the number 1, one has 2, three have 3, two have 4, two have 5 and three have 6. The tombola shuffles the balls thoroughly and randomly chooses one of them. Interest is in the value of the number on the ball.

Since interest is in the number on the ball, the sample space can be taken as $\{1, 2, 3, 4, 5, 6\}$. Notice that the outcomes are not equally likely. The following table gives an overview of the probabilities of all single events.

Outcome	1	2	3	4	5	6
Probability	1/12	1/12	1/4	1/6	1/6	1/4

Consider the events $A = \{1, 4, 6\}$ and $B = \{2, 5, 6\}$. Intuition seems to fail to conclude about the (in)dependence of A and B. You might even be inclined to say that they are dependent since both events contain the relatively important outcome 6. Notice that A and B have intersection $\{6\}$ with probability $\frac{1}{4}$, and that

$$P(A) = P(1) + P(4) + P(6) = \frac{1}{2} \quad \text{and} \quad P(B) = P(2) + P(5) + P(6) = \frac{1}{2}$$

Hence, $P(A \cap B)$ and $P(A)P(B)$ are both $\frac{1}{4}$. The events A and B are **in**dependent.

If A and B are independent, then A^c and B are independent too; check it yourself. In other words, if the probability of B is not influenced by the occurrence of A, then it is also not influenced by the non-occurrence of A.

Independence can also be generalized to more than two events. Three events A, B and C are called (**stochastically**) **independent** if all four equations below are valid:

$$P(A \cap B) = P(A)P(B)$$

$$P(A \cap C) = P(A)P(C)$$

$$P(B \cap C) = P(B)P(C)$$

$$P(A \cap B \cap C) = P(A)P(B)P(C)$$

Exercise 7.33 at the end of this chapter shows that the fourth equation is not an immediate consequence of the first three. That is, independence of A and B, A and C, and B and C does **not** automatically guarantee that the three events A, B and C are independent.

7.5 Bayes' rule

The product rules in (7.14) show that there are two ways to write $P(A \cap B)$ in terms of a conditional probability:

$$P(A \cap B) = P(B) \times P(A \mid B) \quad \text{and} \quad P(A \cap B) = P(A) \times P(B \mid A)$$

By equating the two right-hand sides and dividing the result by $P(B)$, the following rule arises:

Bayes' rule

$$P(A \mid B) = \frac{P(A) \times P(B \mid A)}{P(B)} \qquad (7.20)$$

Bayes' rule expresses a conditional probability in its opposite conditional probability. The product in the numerator of the right-hand side is just the probability of the intersection of A and B (as it should be); the denominator is the probability of the condition (as it should be).

But Bayes' rule is more than a rule that reverses conditional probabilities. It enables us to use information on event B to update the prior probability of A. When rewriting Bayes' rule as

$$P(A \mid B) = P(A) \times \frac{P(B \mid A)}{P(B)}$$

it becomes clear that the posterior probability of A follows from the prior probability of A by multiplying this prior probability by a ratio that represents the value of information on event B. The more different the ratio is from 1, the more valuable it is to have information on event B.

Bayes' rule is often applied in circumstances where a partition $D_1, \ldots, D_s$ of the sample space is given. In such cases conditional probabilities $P(B \mid D_i)$ are often known while the opposite conditional probabilities are wanted. Equation (7.20) with $A = D_i$ turns into

$$P(D_i \mid B) = \frac{P(D_i) \times P(B \mid D_i)}{P(B)} \quad \text{for all } i = 1, \ldots, s \tag{7.21}$$

where, by (7.9), the denominator can be written as

$$P(B) = \sum_{j=1}^{s} P(B \cap D_j) = \sum_{j=1}^{s} P(D_j)P(B \mid D_j) \tag{7.22}$$

If the conditional probabilities $P(B \mid D_1), \ldots, P(B \mid D_s)$ and the unconditional probabilities $P(D_1), \ldots, P(D_s)$ of the partition are known, all opposite conditional probabilities can be calculated.

Example 7.12

Traditionally, only a small percentage of the women (when compared to the men) work in the sector Manufacturing and Construction. In particular, relatively few women used to work in the sub-sector Construction. The question is whether this is different nowadays. Manufacturing and Construction can be subdivided into four sub-sectors. Interest is in the percentages of female employees of the sector that work in each of the four sub-sectors.

Class	%	% Women	Event	$P(D_i)$	$P(F \mid D_i)$
Total	100	20.06			
Mineral extraction	0.57	12.55	D_1	0.0057	0.1255
Manufacturing	68.63	24.77	D_2	0.6863	0.2477
Energy and water companies	2.49	22.86	D_3	0.0249	0.2286
Construction	28.31	8.54	D_4	0.2831	0.0854

TABLE 7.3 Jobs of employees in the sector Manufacturing and Construction (2004)
Source: Statistics Netherlands, *Statistical Yearbook 2005*

The population is all employees working in the sector Manufacturing and Construction. The second column of Table 7.3 records the percentages of the four sub-sectors considered, while column 3 gives the relative percentages of the female employees. For instance, 28.31% of the employees in the population work in the sub-sector Construction; only 8.54% of them are female.

By considering the experiment 'randomly drawing an employee from the population', the figures can be converted into probabilities. Column 4 defines the four events that correspond to the four sub-sectors, while the columns 5 and 6 transform the percentages of columns 2 and 3 into probabilities. Here F denotes the event that this randomly drawn person is a woman. For instance, D_4 is the event that this arbitrarily chosen person works in the Construction sub-sector; its probability is 0.2831. If it is already given that this person works in the sub-sector Construction, then the probability that this person is a woman is 0.0854.

But what are the percentages of the female employees of the whole sector that work in each of the four sub-sectors? Or to put it another way: how large are the conditional probabilities $P(D_i \mid F)$? Notice that $D_1, \ldots, D_4$ is a partition of the sample space (which is the set of employees in the sector Manufacturing and Construction). Furthermore, the conditional probabilities $P(F \mid D_i)$ are known, the (unconditional) probabilities $P(D_i)$ are known, and $P(F) = 0.2006$ since 20.06% of the employees in the sector are female. Hence:

$$P(D_i \mid F) = \frac{P(D_i) \times P(F \mid D_i)}{P(F)} \quad \text{for all } i = 1, 2, 3, 4$$

For instance, $P(D_1 \mid F) = (0.0057 \times 0.1255)/0.2006 = 0.0036$. Similarly, it follows that $P(D_2 \mid F) = 0.8474$, $P(D_3 \mid F) = 0.0284$ and $P(D_4 \mid F) = 0.1205$. (As a check: these four conditional probabilities should – apart from rounding errors – add up to 1.) Translating things back to percentages yields the following conclusions: regarding the women in the sector Manufacturing and Construction, by far the most (84.74%) work in the sub-sector Manufacturing; only about 12% work in the sub-sector Construction.

CASE 7.1 CONNECTION PROBLEMS AND THE LINKNET ROUTER – SOLUTION

Suppose that a (randomly chosen) person with a LinkNet router has serious internet connection problems and that a technician uses the test device to determine the status (defective/non-defective) of the router. Consider the events:

  A = 'the test device indicates that the router is defective'
  D = 'the router is defective'

Notice that the question at the start of this chapter is about the conditional probability $P(D \mid A)$ and that the percentages can be rewritten as:

  $P(D) = 0.001 \quad P(A \mid D) = 0.99 \quad P(A^c \mid D^c) = 0.99$

From these, the probabilities of the complements follow immediately:

  $P(D^c) = 0.999 \quad P(A^c \mid D) = 0.01 \quad P(A \mid D^c) = 0.01$

Furthermore, by Bayes' rule:
$$P(D \mid A) = \frac{P(D) \times P(A \mid D)}{P(A)}$$
The numerator of this ratio follows immediately. The denominator is

  $P(A) = P(A \cap D) + P(A \cap D^c) = P(D)P(A \mid D) + P(D^c)P(A \mid D^c)$

Substitution yields

$$P(D \mid A) = \frac{0.001 \times 0.99}{0.001 \times 0.99 + 0.999 \times 0.01} = 0.0902$$

In only 9% of the cases where the test device indicates a defective router, is the router indeed defective. Although at first sight this result might seem surprising, this conditional probability of D **is** more than 90 times as large as the unconditional probability of D.

Summary

The topics of this chapter were heavily based on the axioms of Kolmogorov. Several rules for probabilities have been derived from them:

Important rules for probabilities

- $P(A^c) = 1 - P(A)$
- $P(\varnothing) = 0$
- If $A \subset B$, then $P(A) \leqslant P(B)$
- $P(A \cup B) = P(A) + P(B) - P(A \cap B)$
- $P(A) = P(A \cap B) + P(A \cap B^c)$ for all events B
- If $D_1, D_2, \ldots, D_s$ is a partition of Ω, then $P(A) = \sum_{i=1}^{s} P(A \cap D_i)$

The experiments 'random drawing' and 'random sampling' are random experiments that bridge the gap between probability on one hand and the fields descriptive statistics and inferential statistics on the other. Since the probabilities regarding these experiments are in essence applications of the classical model, we derived rules for counting the number of possibilities when drawing k objects from m objects:

Rules for counting when randomly drawing k objects from m

	With replacement	Without replacement
Ordered	m^k	$\dfrac{m!}{(m-k)!}$
Unordered	–	$\dbinom{m}{k} = \dfrac{m!}{k!(m-k)!}$

Furthermore, conditional probabilities were introduced. The box below summarizes some important definitions and rules:

Important formulae regarding conditional probabilities

- $P(A \mid B) = \dfrac{P(A \cap B)}{P(B)}$ **Probability of A given B**
- To show that A and B are **independent**, one of the following equations has to be checked:

  $P(A \mid B) = P(A); \quad P(B \mid A) = P(B); \quad P(A \cap B) = P(A)P(B)$
- $P(A \cap B \cap C) = P(A) \times P(B \mid A) \times P(C \mid A \cap B)$ **Product rule**
- $P(A \mid B) = \dfrac{P(A) \times P(B \mid A)}{P(B)}$ **Bayes' rule**

🔑 Key terms

Bayes' rule **216**
binomial coefficient **205**
combinations **204**
conditional probability of
   A given *B* **210**
independent **214**
k-factorial **204**
k-tuple **204**
m choose *k* **205**
multiplication rule **203**

permutations **203**
posterior probability **211**
prior probability **211**
probability tree **229**
product rules **213**
random drawing **207**
random sample **209**
random sampling **209**
random sampling with
   replacement **209**

random sampling without
   replacement **209**
stochastically
   independent **214, 216**
unconditional
   probabilities **211**
variations **204**
with replacement **204**
without replacement **204**

❓ Exercises

Exercise 7.1

Consider a random experiment with events *A* and *B*. Let *P* be some probability model. It is given that:

$$P(A) = 0.6; \quad P(B) = 0.4; \quad P(A \cap B) = 0.3$$

Calculate the probabilities below. (Remember to mention the probability rule(s) that you used; see (7.1)–(7.9).)

$$P(A^c); \quad P(A \cup B); \quad P(A^c \cap B)$$

Exercise 7.2

A chain of restaurants asks its customers to rate the quality of the food and whether they will come back. The table below gives an overview (proportions) of the results of thousands of such interviews.

		Come back?	
		Yes	No
Rating	Poor	0.02	0.08
	Fairly good	0.08	0.09
	Good	0.30	0.15
	Excellent	0.25	0.03

a Determine the proportion of the customers who give the rating excellent.
b Determine the proportion of the customers who will come back.
c Determine the proportion of the customers who say that they will come back and who give the rating excellent.
d For the customers who give the rating excellent, what is the proportion that says that they will come back?
e For the customers who say that they will come back, what is the proportion that gives the rating excellent?

Exercise 7.3

Consider an arbitrary customer of a restaurant of the chain in Exercise 7.2. Denote the event that this customer rates the food as excellent by E and the event that this customer says that he/she will return by R.

a Write the proportions of all parts of Exercise 7.2 as conditional or unconditional probabilities concerning these events.

b Calculate the complementary probabilities of the probabilities of part (a). Write them as conditional or unconditional probabilities concerning the events E, R, E^c and/or R^c.

c Are E and R independent? Why (not)?

Exercise 7.4

Consider the events A, A^c, B and B^c of a certain random experiment. The table below gives the probabilities of the four intersections:

Probability	B	B^c
A	0.1	0.2
A^c	0.4	0.3

a Calculate $P(A)$ and $P(B)$. Which of the probability rules from the Summary did you use here?

b Calculate the probabilities $P(A \cup B)$ and $P(A^c \cup B^c)$. Which of the probability rules did you use?

c Is $A \subset B$? Why (not)?

d Calculate $P(A \mid B)$ and $P(B \mid A)$.

e Are A and B independent? Why (not)?

f Use part (d) to calculate $P(A^c \mid B)$ and $P(B^c \mid A)$. Which rule did you use?

Exercise 7.5

For the events A, B and C of a certain random experiment it is given that:

   $P(B) = 0.4$; $P(A \cap B) = 0.3$; $P(C \mid B) = 0.2$; $P(C) = 0.5$; A and B are independent

a Calculate $P(A \mid B)$.

b Calculate $P(B \cap C)$.

c Calculate $P(A)$.

d Are B and C independent?

Exercise 7.6

a Calculate the number of permutations of 7 objects.

b Calculate $\binom{20}{3}$, $\binom{40}{34}$ and $\binom{40}{6}$.

c Four houses (different as far as their price and size are concerned) are arbitrarily assigned to four households from a group of six households. Calculate the total number of possibilities.

d Five men and ten women are arbitrarily sorted in a row, in such a way that only one man is standing next to a woman. Calculate the total number of possibilities.

Exercise 7.7

Suppose that 5 people are randomly drawn from a group of 100 people. Calculate the number of outcomes (5-tuples) if this is done:

a Ordered and with replacement.

b Ordered and without replacement.

c Unordered and without replacement.

Exercise 7.8

The market for DVD players is completely ruled by the three manufacturers a, b and c. Their market shares are respectively 20%, 30% and 50%. It turns out that 80% of the owners of a type a DVD player are satisfied about it; for the types b and c these figures are respectively 75% and 70%. Calculate the probability that an arbitrarily chosen DVD owner is satisfied with his/her DVD player.

Exercise 7.9

A telemarketer sells newspaper subscriptions over the telephone. The probability of getting no contact (a busy line or no answer) is 0.59. If a contact is established, the probability of selling a subscription is 0.15. Determine the probability that in any one call the marketer does not sell a subscription.

Exercise 7.10

In a certain random experiment, A is an 'ordinary' event and the events D_1, D_2, D_3 form a partition. It is given that:

$$P(A \mid D_1) = 0.7; \quad P(A \mid D_2) = 0.6; \quad P(A \mid D_3) = 0.4; \quad P(D_1) = 0.4; \quad P(D_2) = 0.5$$

a Calculate $P(A)$.

b Calculate $P(D_3 \mid A)$ and $P(D_3 \mid A^c)$.

Exercise 7.11

Consider a random experiment with events A and B. Let P be some probability model. It is given that:

$$P(A) = 0.4; \quad P(B) = 0.2; \quad P(A \cap B) = 0.2$$

When calculating the probabilities below, always mention the probability rule(s) that you used; see (7.1)–(7.9).

a Calculate $P(A \cup B)$, $P(A^c)$, $P(A^c \cap B)$ and $P(A \cap B^c)$.

b Are A and B disjoint? Why (not)?

c Are A and B independent? Why (not)?

Exercise 7.12

In some random experiment, the following is true for two events A and B:

$$P(A) = 0.5; \quad P(B) = 0.3; \quad P(A \cup B) = 0.7$$

Calculate $P(A \cap B)$, $P(A \cap B^c)$ and $P(A^c \cap B^c)$.

Exercise 7.13

For a certain random experiment with sample space Ω, the two series of events A_1, A_2, A_3 and B_1, B_2, B_3 are both partitions of Ω. The probabilities of the intersections of A and B events are given in the following two-way table:

Probability	B_1	B_2	B_3
A_1	0.06	0	0.14
A_2	0.10	0.06	0.10
A_3	0.14	0.20	0.20

a Calculate the probabilities $P(A_i)$ and $P(B_j)$, for all $i, j = 1, 2, 3$.

b Calculate the probabilities $P(A_i^c \cap B_j)$, for all $i, j = 1, 2, 3$. Place them in a two-way table as above.

c Are A_1 and B_1 independent? A_1 and B_2? A_1 and B_3? A_2 and B_1? A_1^c and B_1?

d Calculate the probabilities $P(A_i \cup B_j)$, for all $i, j = 1, 2, 3$. Place them in a two-way table.

Exercise 7.14

A European 'all stars' football team will soon have a match against an all stars team from Brazil. Coach Cruijff of the European team has selected 25 possible players, three goalkeepers and 22 field players. Suppose that Cruijff will randomly select the ten field players who take the kick off. Calculate the total number of field-teams (without keeper) that can be created if:

a the order in which the players are chosen determines their position within the team;

b the order in which the players are chosen does **not** matter since the precise positions will be determined later.

  Suppose that the players Ballack, Beckham, Ibrahimovic, van Persie **and** Zidane belong to the group of 22 field players.

c Calculate, for both selection methods, the probability that they will all be part of the field-team that takes the kick off.

d Now suppose that the complete team – including the goalkeeper – is drawn randomly; the keeper from the group of three keepers and the field players from the group of 22 field players. Answer the questions (a) and (b) again for the whole team.

e Suppose that Kahn is one of the three keepers. Calculate, for both selection methods, the probability that Ballack, Beckham, Ibrahimovic, van Persie, Zidane and Kahn are part of the team that takes the kick off.

Exercise 7.15

At the January meeting of the shareholders, the CEO of an international group of supermarkets states that she believes that there is a 90% chance that the EBIT (earnings before interest and taxes) of the current year will be more than that of last year (which was €5 billion) **if** the upward economic trend continues. However, if the economy stabilizes she thinks that there is only a 40% chance that last year's EBIT will be exceeded, and if it starts on a downward trend, there is a 10% chance of bettering last year's EBIT.

  Clearly, the future developments of the EBIT depend heavily on the state of the economy during the current year. Suppose that specialists believe that the probabilities of a continuing upward economic trend, of stabilization, and of a downward economic trend are respectively 0.80, 0.15 and 0.05.

Combine the two subjective visions to calculate the probability that this year the economic trend will continue going up if this year's EBIT of the group of supermarkets is **at most** as large as last year's.

Exercise 7.16

Consider a random experiment with events A and B. Let P be a probability model. It is given that:

$P(A) = 0.3; \quad P(B) = 0.5; \quad P(A \cap B) = 0.2$

Calculate $P(A \mid B)$, $P(B \mid A)$, $P(A^c \mid B)$ and $P(B^c \mid A)$.

Exercise 7.17

Which of the following statements about events A and B are true? If you think that the statement is true, try to show that it really is. If you think that it is not true, find a counterexample.

a If $P(A \cap B) \neq 0$ and $P(A \mid B) = P(B \mid A)$, then $P(A) = P(B)$.

b If A and B are disjoint, then A and B are independent.

c If the occurrence of A implies the occurrence of B, then $P(A \cap B) = P(A)$.

d If A and B are independent, then A^c and B^c are independent too.

Exercise 7.18

For two events A and B the following is given:

$P(A) = \frac{1}{4}; \quad P(B \mid A) = \frac{1}{2}; \quad P(A \mid B) = \frac{1}{4}$

a Calculate $P(A \cap B)$ and $P(B)$.

b Calculate $P(A^c \mid B^c)$.

c Is $A \subset B$?

Exercise 7.19

A construction company participates in a race to obtain two contracts I and II. The probability of winning contract I is 0.25. If the company wins contract I, then the probability of also winning contract II is 0.4. However, if the company does not win contract I, then the probability of winning contract II is only 0.15.

a Calculate the probability that the construction company will win both contracts.

b Calculate the probability that the company wins at least one contract.

c Are the events 'winning contract I' and 'winning contract II' independent?

Exercise 7.20

When categorizing private cars into three price ranges, it turns out that 50% of the cars belong to the cheap class, 40% to the middle class and the rest to the expensive class. It has been shown that 10% of the cars from the cheap class are regularly driven too fast; for the middle class and the expensive class the corresponding percentages are 20% and 40%.

a Calculate the percentage of the cars that are regularly driven too fast.

b Also calculate the probability that a car that is regularly driven too fast belongs to the expensive class.

Exercise 7.21

An insurance company sells policies of a certain type. Two salesmen (*a* and *b*) jointly cover the same region. They have to sell as many policies as possible. Salesman *a* visits three times as many people as salesman *b*, but *a* sells a policy in only 30% of the cases and *b* in 60% of the cases.

Calculate the probability that a policy that was recently sold in that region was sold by salesman *a*.

Exercise 7.22

Human resource management (HRM) is an important task and challenge for many companies and organizations. In the conference report "Analysis of the human resources management systems applicable to NGOs in the humanitarian sector (2002)", the HRM policies and problems of 53 European aid organizations (agencies) are studied and compared. The table below summarizes the answers to the question: does the aid organization have staff exclusively dedicated to HRM?

Size of agency	Number of agencies	Yes	No
1–10	25	12	13
11–59	18	14	4
60+	10	10	0
Total	53	36	17

Answer to the question, by size of the agency
Source: www.ec.europa.eu

a For each of the three size categories, determine the percentage answering yes.

Let *A* denote the event that an aid organization (randomly chosen from the 53 agencies) does have exclusive HRM staff. Moreover, denote the events that this aid organization belongs to the size-categories 1–10, 11–59 and 60+ respectively by D_1, D_2 and D_3.

b Write the three percentages of part (a) as conditional probabilities.
c Calculate the probabilities $P(D_i)$ for $i = 1, 2, 3$.
d Use parts (b) and (c) to calculate $P(A)$.
e With respect to the aid organizations that do have exclusive HRM staff, determine the percentage with size more than 10. Use the results of parts (b)–(d).

Exercise 7.23

The retail group Maxeda comprises clothing stores, department stores and DIY (do-it-yourself) outlets in seven European countries; 1520 stores in total. The table below gives an overview.

	Belgium	Denmark	France	Germany	Luxembourg	Netherlands	Spain	Total
Clothing	133	19	155	136	10	298	9	760
Department stores	46	–	–	3	–	378	–	427
DIY	120	–	–	–	–	213	–	333
Total	299	19	155	139	10	889	9	1520

Number of Maxeda Group stores by country
Source: www.maxeda.com, May 2007

Consider an arbitrary store of this retail group. Let *A* denote the event that this store is situated in Belgium, and let D_1, D_2, D_3 be the events that it is a store from the divisions clothing, department stores and DIY, respectively.

a Calculate the unconditional probabilities $P(D_i)$ and the conditional probabilities $P(A \mid D_i)$, for i = 1, 2, 3.

b Use the results of part (a) to calculate $P(A)$.

c Use the results of parts (a) and (b) to calculate $P(D_2 \mid A)$.

d Check your answers to parts (b) and (c) by calculating the probabilities directly from the numbers in the table.

Exercise 7.24

Suppose that 50% of the 60 participants of a certain exam will pass. Three participants of the exam are randomly drawn. Consider the events:

A = 'all of them will pass the exam'
B = 'only one of the three will pass the exam'
C = 'two of the three will pass the exam'
D = 'at least one of them will pass'
E_i = 'the ith student will pass the exam'

a Determine a suitable sample space for this random experiment in terms of the codes 1 = will pass and 0 = will not pass.

b Are the outcomes equally likely?

c Express the events A, B, C and D in terms of E_1, E_2 and E_3.

d Use the product rule to calculate $P(A)$, $P(B)$, $P(C)$ and $P(D)$.

Exercise 7.25

Use the formal definition of conditional probability to prove or disprove the following statements about events A, B and C. If necessary, think of a counterexample.

a $P(A \mid A) = 1$.

b If $A \subset B$, then $P(A \mid C) \leq P(B \mid C)$.

c If $P(A) > P(B)$ then $P(A \mid C) > P(B \mid C)$.

d If A and B are independent, then $P(A \cap B \mid C) = P(A \mid C) \times P(B \mid C)$.

Exercise 7.26

After the Second World War, it turned out that the number of newborn male children in Germany was 'significantly' larger than the number of newborn female children. Does this mean that a population has an intrinsic urge to restore the balance between the numbers of males and females if, for some reason, this balance has been disturbed? The model below incorporates this urge.

Suppose that, within families, babies are born according to the following principle: the (conditional) probability that a baby will be male if a family already has m male children and f female children is equal to $0.5 + (f - m) \times 0.01$.

For the questions below, it may be convenient to distinguish the following events: M_1 (respectively F_3) is the event that the first (respectively third) child will be male (respectively female). Similar events can be defined for other births.

a Calculate the probability that the third child will be male if the first two children were both female.

b Calculate the probability that the third child will be female if the first two children were both female.

c Prove that the (conditional) probability that a baby will be female if a family already has m male children and f female children is equal to $0.5 + (m - f) \times 0.01$.

d Calculate the probability that the first three children of a family will all be male. (*Hint*: Use the product rule (7.15).)

e Calculate the probability that at least one of the first three children of a family will be male.

f Calculate the probability that the second child will be male.

g Are the events 'first child is male' and 'second child is male' independent?

h Now consider the model that assumes that the sex of a child does **not** depend on the sex of the children that were born before. Answer parts (a), (b) and (d)–(g) also for this model.

Exercise 7.27

Consider events A and B of a certain random experiment with model P. Suppose that $P(A) = 0.5$ and $P(A \cup B) = 0.7$.

a Calculate $P(B)$ for the case that A and B are independent.

b Calculate $P(B)$ for the case that A and B are disjoint.

c Calculate $P(B)$ for the case that $P(A \mid B) = 0.5$.

d Calculate $P(B)$ for the case that $P(A \mid B) = 0.3$.

Exercise 7.28 (advanced)

A student participates in a certain multiple-choice exam with five options per question. For each question, there are two possibilities:

1 The student knows the correct answer; if so, that option is chosen.

2 The student has no idea; if so, one of the five options is chosen completely arbitrarily.

Consider an arbitrarily chosen question and the events A = 'the student answers this question correctly' and B = 'the students knows the correct answer to this question'. Let p denote the probability that the student knows the correct answer to this question.

a Express $P(B \mid A)$ in terms of p

b Show that $P(B \mid A) \geq P(B)$.

c Calculate $P(B \mid A)$ if $p = \frac{1}{2}$.

Exercise 7.29

A rabbit hutch contains twenty identical, recently born rabbits; ten males and ten females. Someone buys five of them, choosing them completely at random.

a Calculate the probability that the buyer will eventually realize that he has bought two male–female couples.

b Calculate the probability that the buyer has bought no male–female couples.

Exercise 7.30

One hundred students participated in a certain course that was offered during the autumn semester. All students took the exam in January. Those who did not pass this exam but had a grade between 4.5 and 5.5 (scale 0–10) were allowed to do the conditional resit in February. But those students who did not pass the January exam and had a grade below 4.5 had to do the resit in August. The table below gives the results of the January exam; they are categorized according to gender.

	Females	Males	Total
August resit	10	20	30
February resit	10	10	20
Passed in January	20	30	50
Total	40	60	100

Answer the questions below using the notations introduced in this chapter. For an arbitrarily chosen participant of the January exam, the following events are considered:

A = 'this student has to participate in the August resit'
C = 'this student is allowed to do the conditional resit'
S = 'this student passed (in January)'
F = 'this student is female'
M = 'this student is male'

Calculate the probability that:

a this student is female;

b this student has to do the August resit;

c this student is female and passed;

d this student did not pass the January exam;

e this student passed and/or is a female student (also calculate it with (7.7));

f this student passed if it is already given that it concerns a female student;

g this student is male given that the student passed;

h this student is allowed to do the conditional resit given that he/she did not pass.

i Check whether or not the following events are independent: F and S; F and M; M and A,

Exercise 7.31

To produce a certain product, a company uses three machines. The oldest machine produces only 10% of the total production. However, 7% of its output is defective. Besides this very old machine, a newer machine is also used; it produces 30% of the total, of which 3% is defective. The rest is produced by the third, ultra-modern machine, which produces only 1.5% defectives.

a Calculate the probability that a random product of this company is defective.

b Calculate the probability that a random, defective product is produced by the ultra-modern machine.

Exercise 7.32

The relative frequency distribution of the labour force (aged 15–64) and the unemployment rates in the Netherlands in 2004 are given in the table below.

	Labour force (%)	Unemployment (% of labour force)
Men	57.68	5.7
Women	42.32	7.3
15–24 yr	11.84	13.3
25–34 yr	25.31	5.9
35–44 yr	28.57	5.7
45–54 yr	23.70	4.9
55–64 yr	10.58	4.8

Distribution (%) labour force and unemployment
Source. Statistics Netherlands, *Statistical Yearbook 2005*

Suppose that for the present year the situation is the same as described in the table. Consider an arbitrary member of the present labour force. Let M (W) be the event that this person is a man (woman). Furthermore, let D_1 be the event that the age of this person falls in the class 15–24, ..., let D_5 be the event that the age of this person falls in the class 55–64.

   a Calculate the probability that this person is unemployed by using the partition M, W.

   b Calculate the probability that this person is unemployed by using the partition D_1, ..., D_5.

   c Calculate the probability that this person is a woman if it is already given that the person is unemployed.

   d Calculate the probability that this person's age is in the class 15–24 if it is already given that he/she is unemployed.

Exercise 7.33

A 'roulette without zero' contains 36 red (r) or black (b) numbers as indicated in the plan below:

1	2	3	4	5	6	7	8	9	10	11	12	13	14	15	16	17	18
r	r	r	r	r	b	b	b	b	r	r	r	r	b	b	b	b	b
36	35	34	33	32	31	30	29	28	27	26	25	24	23	22	21	20	19

(All 36 boxes are equally likely; for instance, the positions 8 and 29 of the roulette are both black.) Gamblers can put their money on (among others) the events:

   R = 'a red number appears'
   E = 'an even number appears'
   L = 'a low ($\leq$ 18) number appears'

   a Show that R and E, R and L, and E and L are independent.

   b Show that the triple R, E, L is not independent.

Exercise 7.34

Persons a and b play the following game. Person a throws a coin. If the outcome is tail (T), then a pays €3 to b. If the outcome is head (H), then a throws a fair **die**. If the number of eyes is different from 1, then a receives from b as many euro as the result of the die. But if the number of eyes is 1, then none of the players is paid. The **probability tree** in the diagram below summarizes the game for player a.

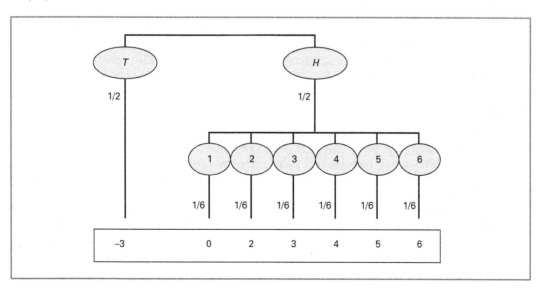

Notice that there are seven possible paths; they are the possible outcomes of the game. Along each path, the events and the accompanying probabilities are marked. For instance, the right-hand path tells us that in the first sub-experiment head appeared (with probability 1/2) and that in the second sub-experiment 6 was thrown (with probability 1/6). The left-hand path tells us that the first sub-experiment yielded tail, and that a second sub-experiment was not conducted. Additionally, the corresponding payments **to** player a are mentioned at the end of the paths.

a Determine two suitable sample spaces; the first in terms of H/T and numbers of eyes of the die, the second in terms of the payment (in euro) to player a.

b For both sample spaces, determine the probabilities of all individual outcomes.

c Determine the probabilities of the events 'player a wins', 'player b wins' and 'draw'.

d Calculate the probabilities of the events C = 'even number of eyes with the die', D = 'head' and E = 'tail'.

e Determine the events $D \cap C$, $E \cap C$, $D \cup C$ and $E \cup C$. Calculate the accompanying probabilities.

Exercise 7.35

The swindler of Example 6.6 plays the following game in a casino. Two dice are thrown by the swindler himself, first a red one and next a blue one. The swindler wins €170 (and gets his €5 stake back) if the total number of eyes is twelve. But if the total number of eyes is less than twelve, he gets nothing and even loses his stake of €5.

a Create the **probability tree** if the game is played with fair dice. Determine the probabilities of all paths. Also determine the probability that the player will win the €170.

b Now suppose that the swindler occasionally replaces the blue die with his false one. If the red die returns less than six, then it is already clear that he cannot win any more and he just throws the casino's (fair) blue die. But if the red die yields six, he secretly takes his own manipulated (blue) die and uses this die for the second throw. Create the probability tree and determine the probabilities of all paths. Also determine the probability that the swindler will win the €170.

Exercise 7.36

Reconsider Exercise 7.35. Let A_i be the event of throwing i with the first die (i = 1, 2, ..., 6). Analogously, let B_j be the event of throwing j with the second die (j = 1, 2, ..., 6).

a Write the events A_1, ..., A_6 and B_1, ..., B_6 in terms of the outcomes (i, j), where i refers to the red die and j to the blue die.

b Suppose that the game is completely fair. Explain why A_i and B_j will be independent for $i \neq j$.

c (advanced) Now suppose that the swindler occasionally replaces the blue die with his manipulated die, in the way described in Exercise 7.35. Find out which of the events A_i and B_j are still independent and which are not. Explain your answers.

Exercise 7.37

In a certain country, 0.2% of the population suffers from a very serious disease. Although there exists a simple test to spot the disease, the results of this test cannot be trusted completely: for both groups of suffering and non-suffering people, the test gives the correct diagnosis in 98% of the cases. That is, there is a 2% chance that a healthy (non-suffering) person will get a positive test result, and also a 2% chance that a person who does suffer from that disease is not detected in the case of a performed test and hence gets a negative test result. Here, a positive (negative) result of the test means that the test indicates that the person has (does not have) the disease. But how large is the opposite probability that a person with a positive test result indeed suffers from the disease?

Exercise 7.38 (computer)

a Use a computer package to simulate a throw with a fair die.

b Simulate 10 000 throws. Make a frequency distribution and a histogram.

c Simulate a delegation of ten students from the participants of the January exam in Exercise 7.30. Take care! The delegation has to consist of three **different** persons. How many of the students in your delegation passed? How many are female students?

CASE 7.2 THE MARKET FOR WOOL DETERGENT

In the market for wool detergents there are four main competitors. We will denote them by A, B, C and D. Not everybody is loyal to a certain brand. Market research has given us a model showing the transition of buyers in the course of a quarter of a year. The model is depicted below.

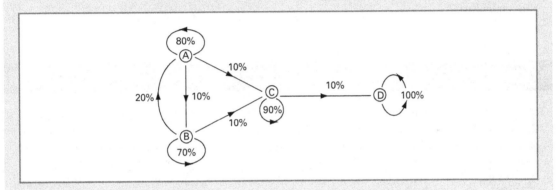

You see here, among other things, that if someone is buying brand B today there is a:

- 20% probability that he will buy brand A 3 months later,
- 10% probability that he will buy brand C 3 months later,
- 70% probability that he will still buy brand B 3 months later.

a Someone is buying brand A. Calculate the probability that he will buy brand D 6 months later.

b Someone is buying brand B. Calculate the probability that he will (still or again) buy brand B 6 months later.

c Someone is buying brand A. Calculate the probability that he will buy brand C 6 months later.

d Suppose this model remains valid for a long period of time. What will eventually happen in this market? Explain.

CASE 7.3 RUSSIAN ROULETTE

Russian Roulette is an amusement game for two risk lovers. The game is very lugubrious; it occasionally is played in films (like *The Deer Hunter* by Michael Cimino) and in comics (for instance, the comic album *Le Grand Duc* in the Lucky Luke series by Morris and Goscinny). The game is also of interest because it is often used as a metaphor to describe very risky business opportunities, especially in columns in business and financial magazines, such as the columns 'Are you playing business Russian roulette?' by Lea Strickland in the *Carolina Newswire* and 'Russian roulette' by Radhuka Chadka in *Business Line*. Below, the game will be explained and analysed.

Russian roulette is 'played' by two people, with a gun that has a cylinder with six chambers but only one of the chambers contains a bullet. The two players take turns to spin the cylinder (like a roulette wheel, so that an arbitrary chamber rests in line with the barrel), aim the gun at their head and pull the trigger. The game ends as soon as one of the players 'kicks the bucket'.

And here is the question: what is the probability that the player who starts the game will at the end be the loser?

CASE 7.4 WAS THE DRAW FOR THE UEFA EURO 2004 PLAY-OFFS FAIR?

In June 2004, the finals of the European Football Championship took place in Portugal. Being the host, Portugal was automatically admitted to this tournament, but 15 other countries had to qualify via a competition that took place between September 2002 and October 2003. For this, 50 European countries were divided into ten groups. All ten group-winners automatically qualified for the finals. But the ten runners-up had to take part in five play-off matches on a home-and-away basis. These matches took place on 15 and 19 November 2003, with the five winners progressing to the finals. Before the draw for these five matches took place, many reporters (and other people) argued that it would be a pity if any of the five stronger teams had to play each other. Financial arguments especially seemed to matter, since these strongest countries historically had the largest crowds of supporters.

In a random draw in Frankfurt in October 2003, the five play-off matches were determined. And, believe it or not, the five strongest countries on the FIFA ranking published on 19 November 2003 were coupled with the five weakest countries according to that list. Many people and reporters did not believe that an outcome like this could be the result of a fair draw. Some reporters even suggested that the five balls that represented the five stronger teams in the bowl had been warmed in a microwave oven before the draw so that the UEFA official could combine a warm with a cold ball. Some people claimed that, under fair circumstances, this outcome has only a $(1/2)^5 = 0.0313$ probability, and this would mean that the doubt about the fairness of the UEFA draw might be justified.

Is this probability 0.0313 correct? Should we really distrust the fairness of this UEFA draw?

Probability distribution, expectation and variance

In Part 1, which looked at the field of descriptive statistics, the concepts of a **variable**, its **frequency distribution**, **mean** and **variance** were considered. Similar concepts will now be introduced in the field of probability theory: **random variable**, its **probability distribution**, and its **expectation** and **variance**. It will turn out that the corresponding concepts of the two fields are linked by way of the experiment 'random drawing'.

As always in probability, emphasis is on a random experiment and accompanying sample space (called Ω). This will (often without mentioning) be assumed to be the case throughout the present chapter.

CASE 8.1 EXPECTED RETURN AND RISK OF CARLSBERG BREWERIES STOCK

Investors require information about the return they can expect on the stocks in their portfolio. However, also important to them are the levels of risk associated with these stocks. But what is meant by 'expected' return and risk? How can they be quantified?

The stock of Carlsberg Breweries is traded on the Stuttgart Stock Exchange. The dataset <u>Case08–01.xls</u> contains daily closing prices and returns (%) with respect to the business day before, for the period May 2000 – May 2007.[†] These historical data will be used to find a model for the daily percentage return R of the stock on a business day in the future. Furthermore, the expected return and the accompanying risk will be quantified and calculated. See the end of Sections 8.2.1, 8.4.1 and 10.3.5.

[†] *Sources*: www.Carlsberggroup.com and www.Euroland.com.

8.1 Random variables

A random variable is a well-defined prescript for measuring some feature of interest of the sample space; it attaches a value to each outcome of the sample space. From a mathematical point of view, a random variable is a function. From a practical point of view, a random variable is a quantity that takes real values, but its precise outcome is not known in advance and is determined by chance.

Random variable

A *random variable* is a function on the sample space of the random experiment; it assigns a value (usually a number) to each outcome of the experiment.

If the possible values of the random variable are ordinary numbers, the random variable is called *quantitative*; if the values are categories, the random variable is called *qualitative*. The possible values of a random variable (rv for short) are called the *outcomes* of the random variable, although the concept 'outcome' was used before to describe the elements of the sample space. Outcomes of the rv X are usually denoted as x, small letters. The outcome of X that actually occurs, is called the *realization* of X.

The random experiment that hands one card to a player from a carefully shuffled complete deck of 52 cards, has 52 outcomes. The rv X = 'colour of the card' has four outcomes: hearts, diamonds, spades and clubs. Notice that this random variable indeed assigns a value to each of the 52 outcomes of the sample space; 13 of the outcomes get the value hearts, 13 get diamonds, etc. The outcomes of this rv are not ordinary numbers, so this rv is qualitative. The rv in the following example is quantitative.

Example 8.1

The random experiment of throwing a red and a blue die has 36 outcomes (i, j), where i refers to the red die and j to the blue. The rv Y = 'total number of eyes' can take the values 2, 3, ..., 12. Figure 8.1 contains the outcomes of the sample space and also the outcomes of Y. Notice that Y indeed assigns a value to each outcome of the experiment by adding up the two coordinates of that outcome. (For instance, the outcomes (4, 6) and (6, 4) both get the value 10.) Hence, Y really is a random variable.

Blue die							
6	(1, 6) 7	(2, 6) 8	(3, 6) 9	(4, 6) 10	(5, 6) 11	(6, 6) 12	
5	(1, 5) 6	(2, 5) 7	(3, 5) 8	(4, 5) 9	(5, 5) 10	(6, 5) 11	
4	(1, 4) 5	(2, 4) 6	(3, 4) 7	(4, 4) 8	(5, 4) 9	(6, 4) 10	
3	(1, 3) 4	(2, 3) 5	(3, 3) 6	(4, 3) 7	(5, 3) 8	(6, 3) 9	
2	(1, 2) 3	(2, 2) 4	(3, 2) 5	(4, 2) 6	(5, 2) 7	(6, 2) 8	
1	(1, 1) 2	(2, 1) 3	(3, 1) 4	(4, 1) 5	(5, 1) 6	(6, 1) 7	
	1	2	3	4	5	6	Red die

FIGURE 8.1 Possible outcomes of the random experiment and of Y

Notice that some outcomes of the sample space get the same value. For instance, the value 5 is assigned to each of the outcomes in the set {(1, 4), (2, 3), (3, 2), (4, 1)}. This set is a subset of the sample space, so it is an event. It is denoted as {$Y = 5$}, the event that Y takes the value 5. Similarly, the event that Y takes the value 7 is denoted as {$Y = 7$} and is just the subset {(1, 6), (2, 5), (3, 4), (4, 3), (5, 2), (6, 1)} of Ω. Notice that the events {$Y = 2$} and {$Y = 12$} are single events. In Figure 8.1, all events so far are NW–SE diagonals. The event {$Y = 7$} covers the main NW–SE diagonal. The event {$3 \leq Y \leq 6$} is the event that Y takes a value in the range from 3 to 6.

The events above are of a special type since they are all dealing with the random variable Y. Not all events are as such. For instance, the event 'even with the red die' is **not** of that type.

Events that are dealing with a random variable X are called **X events**. So, X events are special subsets of Ω and the definition depends on the name of the rv. The above example is about Y events, since the rv of interest is called Y. For instance, the Y event {$Y = 5$} occurs if the outcome of Y falls in the subset {5}. Since Y only has integers as outcomes, this event can also be written as {$4 < Y < 6$}. The event {$3 \leq Y \leq 6$} occurs if the value of Y falls in the interval [3, 6].

Example 8.2

In 2005, Sudoku puzzles became a craze. Such puzzles can also be solved online and the time that it takes you to solve them can be compared with others. Suppose that – so far – solving online a certain Sudoku puzzle of the level 'heavy' took the puzzlers 25 minutes and 30 seconds (that is, 25.5 minutes) on average.

You also want to solve this puzzle and to measure the time X (in minutes, with decimals) that this will take. Obviously, X is a random variable since – at the end – one value will come out but – at the start – many outcomes are possible. Several events can be considered. The events 'it will take you less than 25.5 minutes' and 'you will need at least 30 minutes' are both X events. They can easily be written in terms of X: respectively {$X < 25.5$} and {$X \geq 30$}. The event 'the time that it will take you is less than half a minute away from the average result so far' is also an X event. The phrase 'the average result so far' refers to the 25.5 minutes; 'less than half a minute away from' 25.5 minutes means that the distance between X and 25.5 is less than 0.5 minutes. But such distances can be written in terms of absolute values (see the last part of Appendix A2). This special event can be written as {$| X - 25.5 | < 0.5$}, but also as {$25 < X < 26$}.

Recall that quantitative variables in descriptive statistics can be discrete or continuous. Similarly, a quantitative **random** variable is called *discrete* if it can take only a finite or countable number of values. It is called *continuous* if it may assume any value in some interval.

It is without doubt that the rv 'total number of eyes' in Example 8.1 is discrete; it can take only 11 values. The rv $V =$ 'tomorrow's number of visitors to the website www.youtube.com' can take many values. It is hard to find an upper bound for the number of visitors that certainly will not be reached. In cases like this, it is assumed that **each** non-negative whole number is a possible outcome of the rv. Although the resulting set of possible values is infinite, the variable is still discrete since its outcomes are clearly separated.

The rv $X =$ 'time (in minutes, with decimals) that it will take you to solve the Sudoku puzzle' is continuous. Since there are so many values possible that lie so close together, things are stylized by assuming that all real numbers in an interval are possible values.

Similar to continuous variables in descriptive statistics, a continuous random variable stylizes reality. By assuming that each real value in a certain interval can be an outcome of the rv, reality is stylized in order to facilitate theory and the calculation of probabilities. For instance, it might be convenient to consider the above-mentioned (originally discrete) rv V as continuous. The reasons will become clear in forthcoming sections.

Here are other random variables that usually are considered to be continuous: the length (in metres, two decimals) and the weight (in kg, two decimals) of a randomly chosen adult; the age of a person; the price of a product; next year's GDP of a country; the salary of an employee; next month's price of a stock; next year's sales of a company.

In the rest of the present chapter, all random variables are assumed to be **quantitative**.

8.2 Probability distributions

Recall that the X events of the rv X are subsets of the sample space Ω, so it makes sense to talk about the probability of an X event. In essence, the overview of the probabilities of **all** X events is of special interest; it is called the **probability distribution** (or **distribution** for short) of X. If X is discrete (continuous), the probability distribution is called discrete (continuous) too. Notice that such an overview may contain many probabilities. For instance, the variable $Y =$ 'total number of eyes when throwing two dice' has 11 possible outcomes, and the set of these outcomes has 2^{11} subsets that all generate one Y event, as follows from the end of Section 7.2.

Probability distribution of a random variable

The **probability distribution** of a random variable X is an overview of the probabilities for all X events.

Example 8.3

Two fair coins are tossed; the random variable Y counts the number of heads. The diagram below lists all possible outcomes of the sample space Ω and the accompanying values of Y.

(H, H)	(H, T)
2	1
(T, H)	(T, T)
1	0

Y has **three** possible outcomes (0, 1 and 2), so there basically are 2^3 (eight) Y events. Although there are more ways to express them, they can be enumerated as follows:

$\varnothing$, $\{Y = 0\}$, $\{Y = 1\}$, $\{Y = 2\}$, $\{Y = 0 \text{ or } 1\}$, $\{Y = 0 \text{ or } 2\}$, $\{Y = 1 \text{ or } 2\}$, $\{Y = 0, 1 \text{ or } 2\}$

The Y event $\{Y = 0, 1 \text{ or } 2\}$, the set of outcomes of the sample space with 0, 1 or 2 heads, is Ω itself. The empty event $\varnothing$ can also be written as $\{Y = 3\}$ so it really is a Y event. If the two coins are fair, each of the four outcomes of the sample space has probability 1/4. Hence, the probabilities of the Y events can be calculated. For instance:

$P(Y = 0 \text{ or } 2) = P(\{(T, T), (H, H)\}) = 1/2$

The overview in the table below is just the probability distribution of Y.

Y event	∅	{Y = 0}	{Y = 1}	{Y = 2}	{Y = 0 or 1}	{Y = 0 or 2}	{Y = 1 or 2}	{Y = 0, 1 or 2}
Prob.	0	1/4	1/2	1/4	3/4	1/2	3/4	1

In the above example, a simplifying notation was adopted: the probability $P(\{Y = 0 \text{ or } 2\})$ was written as $P(Y = 0 \text{ or } 2)$. Similar simplifying notations will be used in the future.

If an rv X has k possible outcomes, then the probability distribution yields 2^k probabilities. This number can be quite high, even for small k. To reduce this number, statisticians have been searching for other, and less elaborate, concepts that are equivalent to the concept of probability distribution. The solutions for the discrete case and the continuous case will be considered separately below.

8.2.1 General cdfs and discrete probability distributions

Suppose first that the rv X is discrete, so it can take only a finite or countable-infinite number of values.

(Discrete) probability density function

The **probability density function** (pdf) of a discrete random variable X is the function f defined by:

$$f(x) = P(X = x) \quad \text{for all outcomes } x \text{ of } X$$

A (discrete) probability density function is also termed (**discrete**) **density**. So, $f(x)$ is the probability that the realization of X will be x. Hence, the function f offers the overview of all outcomes x accompanied by the probabilities of the respective X events $\{X = x\}$. Since the pdf f is based on the underlying probability measure P, the axioms of Kolmogorov in (6.11)–(6.13) can be applied. The properties below follow from them.

Properties of the pdf f of a discrete X

$$f(x) \geq 0 \quad \text{and} \quad \sum_x f(x) = 1 \tag{8.1}$$

Here $\sum_x$ means that all probabilities $f(x)$ are added up for all possible outcomes x of X.

The discrete rv Y = 'number of heads in two tosses with a fair coin' can take the values 0, 1 and 2. The probability density function follows immediately from the probability distribution in the table of Example 8.3:

x	0	1	2
f(x)	1/4	1/2	1/4

Thanks to the general probability rules, other probabilities in the probability distribution can be obtained from f. For instance:

$$P(Y = 1 \text{ or } 2) = P(Y = 1) + P(Y = 2) = f(1) + f(2) = 1/2 + 1/4 = 3/4$$

The probability density function f of a discrete rv is equivalent to the probability distribution in the sense that the one can be obtained from the other and vice versa. Of course, f can be obtained from the probability distribution of X since the overview offered by f is a part of the overview offered by the probability distribution. But all probabilities in the probability distribution can also be obtained from f. For instance, the probability that X will take a value in some interval I of real numbers follows by adding up all $f(x)$ for outcomes x of X that belong to I.

As an aside, suppose that X is a **general** random variable, discrete or continuous. There exists another important function that contains the same information as the probability distribution.

(Cumulative) distribution function – general definition and properties

The (*cumulative*) *distribution function* (cdf) F of a general – discrete or continuous – random variable X is defined by:

$$F(a) = P(X \le a) \quad \text{for all real numbers } a$$

It has the following **properties**:

$$\left.\begin{array}{l} F \text{ is non-decreasing} \\ F(-\infty) = 0 \quad \text{and} \quad F(\infty) = 1 \\ F(b) - F(a) = P(a < X \le b) \quad \text{for all } a \text{ and } b \text{ with } a < b \end{array}\right\} \tag{8.2}$$

So, $F(a)$ is the probability that the realization of X will be smaller than or equal to a, that is, will fall in the interval $(-\infty, a]$. The validity of the properties will be demonstrated next.

The first property follows since $x < y$ yields that the occurrence of the event $\{X \le x\}$ automatically implies the occurrence of $\{X \le y\}$. Put another way, if $x < y$, then $\{X \le x\} \subset \{X \le y\}$ and hence

$$P(X \le x) \le P(X \le y)$$

The second property follows directly from the definition of cdf, while the third is a consequence of axiom (6.13): since $\{X \le b\} = \{X \le a\} \cup \{a < X \le b\}$, we have

$$F(b) = F(a) + P(a < X \le b)$$

The third property is especially important. This is because it illustrates that probabilities from the overview offered by the probability distribution of X can be expressed in terms of F. As a consequence, the probability distribution and its cdf are equivalent in the sense that the one is determined by the other and vice versa.

Now suppose again that X is discrete and hence takes only isolated values. Then the cdf of X is a non-decreasing **step function**: it makes jumps at the outcomes of X and is constant between two subsequent outcomes. This is because for two subsequent outcomes x and x' of X and some real number a with $x < a < x'$, it holds that the events $\{X \le a\}$ and $\{X \le x\}$ coincide and hence $F(a) = F(x)$.

The pdf f and the cdf F are also equivalent in the sense that the one follows from the other and vice versa. F follows from f since the event $\{X \le a\}$ can be written as the union of the events $\{X = x\}$ for all outcomes x that are $\le a$ and (by (6.13)):

$$F(a) = P(X \le a) = P(\bigcup_{x \le a}\{X = x\}) = \sum_{x \le a} P(X = x) = \sum_{x \le a} f(x) \tag{8.3}$$

Here $\Sigma_{x \le a}$ means that summation is done for all outcomes x of X that are not larger than a. Hence, F follows from f since $F(a)$ is just the sum of all probabilities $f(x)$ with $x \le a$. On the other hand, when x is an outcome of X and x' is the largest outcome that is smaller than x (if it exists), then the event $\{X \le x\}$ is just the union of the events $\{X = x\}$ and $\{X \le x'\}$. Hence:

$$P(X \leq x) = P(X = x) + P(X \leq x')$$

This yields immediately that

$$f(x) = F(x) - F(x')$$

(8.4)

so f follows from F and $f(x)$ is just the jump size of F at x.

Properties of the cdf F of a discrete X

- It is a non-decreasing step function.
- It is completely determined by the pdf f since:

$$F(a) = \sum_{x \leq a} f(x) \quad \text{for all real numbers } a$$

- It completely determines the pdf f since:

$$f(x) = F(x) - F(x') \quad \text{for all pairs } (x, x') \text{ of two successive outcomes of } X$$

Since the pdf of a discrete random variable completely characterizes its probability distribution, the pdf itself is often (incorrectly) called probability distribution.

Example 8.4

The cdf F of the random variable $Y =$ 'number of heads in two tosses with a fair coin' can easily be calculated from the pdf f:

$$F(0) = P(Y \leq 0) = P(Y = 0) = f(0)$$

$$F(1) = P(Y \leq 1) = P(Y = 0) + P(Y = 1) = f(0) + f(1)$$

$$F(2) = P(Y \leq 2) = f(0) + f(1) + f(2)$$

So, F is obtained by stepwise adding up the $f(x)$:

x	0	1	2
$f(x)$	1/4	1/2	1/4
$F(x)$	1/4	3/4	1

Notice that the function F also makes sense outside the set {0, 1, 2}. For instance:

$$F(1.3) = P(Y \leq 1.3) = P(Y \leq 1) = F(1) = 3/4$$

Figures 8.2 and 8.3 present the graphs of the pdf and the cdf of Y.

The cdf is a non-decreasing step function that jumps at the values 0, 1 and 2 to a higher level; the jump sizes are just the local values of the pdf. For instance, at 1 the jump size is $f(1)$.

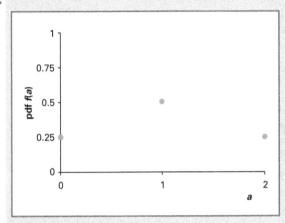

FIGURE 8.2 Graph of the pdf of *Y*

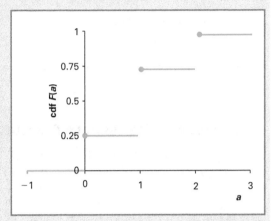

FIGURE 8.3 Graph of the cdf of *Y*

Example 8.5

Two fair dice are thrown in turn, first a red die and next a blue die. The outcomes of this random experiment can be presented as in Example 8.1. We will derive the pdf and the cdf of the variable 'total number of eyes' for two cases: the case that the dice are fair and the case that both dice are created by the swindler (Example 6.6).

Suppose that the experiment is completely fair; denote the variable by *X*. Then all 36 outcomes of the experiment are equally likely and the determination of the pdf *f* of *X* is just a matter of counting. For instance, the event $\{X = 8\}$ consists of the five outcomes (2, 6), (3, 5), (4, 4), (5, 3) and (6, 2), so $f(8) = 5/36$. The function *f* is part of Table 8.1.

a	2	3	4	5	6	7	8	9	10	11	12	Total
f(*a*)	1/36	2/36	3/36	4/36	5/36	6/36	5/36	4/36	3/36	2/36	1/36	1
F(*a*)	1/36	3/36	6/36	10/36	15/36	21/36	26/36	30/36	33/36	35/36	36/36	–
g(*a*)	1/144	4/144	8/144	12/144	16/144	22/144	24/144	20/144	16/144	12/144	9/144	1
G(*a*)	1/144	5/144	13/144	25/144	41/144	63/144	87/144	107/144	123/144	135/144	1	–

TABLE 8.1 Pdfs and cdfs of the two random experiments

The distribution function *F* can easily be obtained from the pdf *f*. For instance:

$$F(7) = P(X \leq 7) = \sum_{x=2}^{7} f(x) = \frac{21}{36} \quad \text{and} \quad F(8) = F(7) + f(8) = \frac{21}{36} + \frac{5}{36} = \frac{26}{36}$$

The function *F* is a non-decreasing step function that goes horizontal between successive outcomes of *X* and that jumps to a higher level at an outcome *x*, with jump size *f*(*x*). Also notice that the function *f* is symmetric around 7. That is, $f(6) = f(8)$, $f(5) = f(9)$, etc. See also the graphs of *f* and *F* in Figures 8.4 and 8.5.

Now suppose that the swindler created the two dice; denote the variable 'total number of eyes' in this case by *Y*. Let A_i and B_j respectively be the events that *i* eyes are thrown with the red die and *j* eyes with the blue die. Since the two throws are physically independent sub-experiments and the events A_i and B_j are dealing with different sub-experiments, the events are mutually independent. Notice that the 36 outcomes are not all equally likely any more. For instance:

$$P((1, 1)) = P(A_1 \cap B_1) = P(A_1) \times P(B_1) = \frac{1}{12} \times \frac{1}{12} = 1/144$$

$$P((1, 6)) = P(A_1 \cap B_6) = P(A_1) \times P(B_6) = \frac{1}{12} \times \frac{3}{12} = 3/144$$

$$P((2, 3)) = P(A_2 \cap B_3) = P(A_2) \times P(B_3) = \frac{1}{6} \times \frac{1}{6} = 4/144$$

$$P((6, 6)) = P(A_6 \cap B_6) = P(A_6) \times P(B_6) = \frac{1}{4} \times \frac{1}{4} = 9/144$$

(The independence property is used in all second equalities.) Let g and G denote the pdf and the cdf of Y. For g, see the fourth row in Table 8.1. For instance:

$$g(5) = P(\{(1, 4), (2, 3), (3, 2), (4, 1)\}) = 2\times2/144 + 2\times4/144 = 12/144$$

$$g(7) = P(\{(1, 6), (2, 5), (3, 4), (4, 3), (5, 2), (6, 1)\})$$

$$= 2\times3/144 + 4\times4/144 = 22/144$$

$$g(10) = P(\{(4, 6), (5, 5), (6, 4)\}) = 2\times6/144 + 1\times4/144 = 16/144$$

$$g(12) = P((6, 6)) = 9/144$$

See the table for the complete overview of the pdf g. The distribution function G of the unfair case follows easily from g.

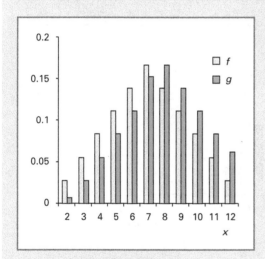

FIGURE 8.4 Graphs of the pdfs f and g

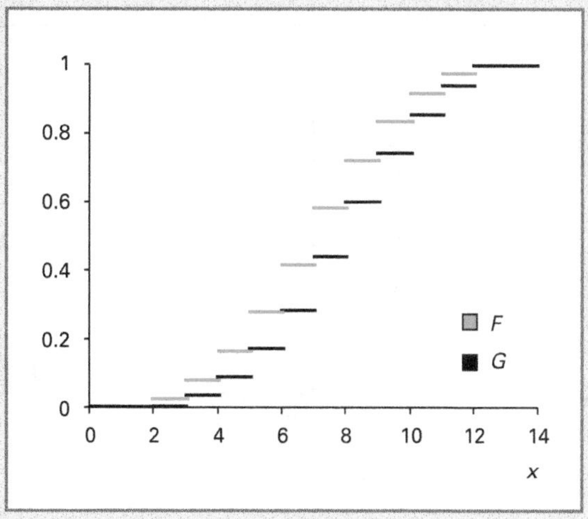

FIGURE 8.5 Graphs of the cdfs F and G

When comparing the pdfs and cdfs of X and Y, we come to the following conclusions:

- The mode of g is larger than the mode of f.
- g is more concentrated on the larger outcomes than is f.
- $F(a) > G(a)$ for all a in the interval $[2, 12]$.

Loosely speaking, a random variable is a quantity that can take real values, but the precise outcome is not known in advance and depends on chance. From this point of view a constant is not an rv; when throwing a die that has the number 6 on all six sides, the quantity $X = $ 'number of eyes of the throw' does not give an rv. Nevertheless, there are certain advantages to considering constants as rvs. In the present book, a constant is occasionally called a ***degenerated random variable***.

An rv X is called **degenerated at the constant** b if b is the only possible outcome of X. Notice that the pdf f of such an rv satisfies $f(b) = 1$ and $f(x) = 0$ for $x \neq b$. Its cdf F satisfies $F(x) = 0$ for $x < b$ and $F(x) = 1$ for $x \geq b$. This cdf makes only one jump, with jump size 1.

CASE 8.1 EXPECTED RETURN AND RISK OF CARLSBERG BREWERIES STOCK – SOLUTION, PART 1

One model for the daily return R of the Carlsberg stock is a discrete, empirical model. By rounding the 1754 returns in the dataset to integer values, the following frequency distribution results, from which the bar chart below can be created.

Value	−15	−9	−8	−7	−6	−5	−4	−3	−2	−1	0
Rel. freq.	0.0011	0.0006	0.0006	0.0006	0.0034	0.0051	0.0125	0.0348	0.0735	0.1870	0.3193

Value	1	2	3	4	5	6	7	8	9	12	–
Rel. freq.	0.2144	0.0770	0.0393	0.0148	0.0080	0.0017	0.0034	0.0011	0.0011	0.0006	–

Relative frequency distribution of the rounded returns

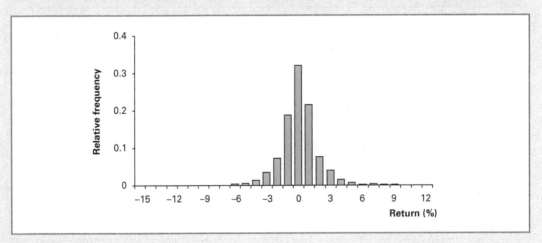

Accompanying bar chart

This relative frequency distribution can be used as a model for the daily return R of a business day in the future. The frequency distribution in the table becomes the (discrete) probability density function. Probabilities can be calculated easily. For instance, the probability that tomorrow's return will be positive is equal to:

$$P(R > 0) = 0.2144 + 0.0770 + \ldots + 0.0006 = 0.3614$$

8.2.2 Continuous probability distributions

Now suppose that the rv X is continuous, so it is assumed that **all** real numbers in some interval are possible outcomes of X. Then, the (cumulative) distribution function F with $F(x) = P(X \leq x)$ still makes sense and still completely characterizes the probability distribution of X. However, it will turn out that the concept 'probability density function' has to be reconsidered.

The cdf of a discrete rv is a step function whereas the cdf of a continuous rv is a continuous function (without jumps). Figure 8.6 illustrates this continuity and the triple of properties of (8.2).

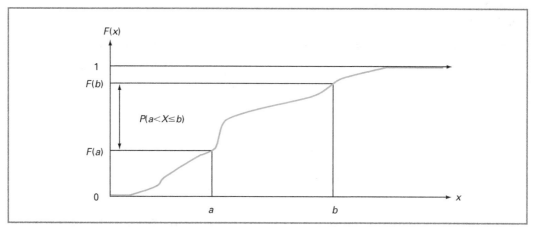

FIGURE 8.6 Graph of the cdf of a continuous random variable

Example 8.6

The high-speed train from Amsterdam to Paris departs Amsterdam central station on the hour, every hour, precisely on time. Mrs Doubledutch goes to Amsterdam central station to take this train to Paris, but she is completely ignorant of the schedule. Suppose that she measures her waiting time X (until the departure of the train) very precisely, in minutes with decimals. What can be said about the probability that her waiting time is less than 10 minutes? At most 13 minutes and 30 seconds? At most x minutes when x is some real, fixed number in the interval $[0, 60)$? Less than x minutes? Precisely x minutes?

It is obvious that the waiting time of Mrs Doubledutch is at least 0 seconds, and less than 60 minutes. Since she measures her waiting time very conscientiously, many outcomes are possible and they lie close together. That is, the variable X is continuous; **all** outcomes in the range 0–60 minutes are possible.

Since Mrs Doubledutch is completely ignorant of the schedule, all values in the interval $[0, 60)$ are equally likely to occur as her waiting time. As a consequence, the probability that X will fall between 2.5 and 12.5 minutes is equal to the probability that X will fall between 22.3 and 32.3 minutes; and so on. In particular, all six intervals $[0, 10)$, $[10, 20)$, ..., $[50, 60)$ of width 10 have equal probability that the realization of X will fall in them. Since the sum of these probabilities is one, it follows that the probability that X is less than 10 minutes, is 1/6. It would appear that the probability that X falls in the interval $[0, 10)$ is just the width of that interval divided by the width of the interval $[0, 60)$.

For the probability that X is at most 13.5, we have to divide the width 13.5 of the interval $[0, 13.5]$ by 60 and obtain $13.5/60 = 0.2250$.

The probability that X is **at most** x minutes is just the probability that X will fall in $[0, x]$, which is equal to the ratio of the width of $[0, x]$ and the width of $[0, 60)$. The probability that X is **less than** x minutes is just the probability that X falls in $[0, x)$, which is the width of $[0, x)$ divided by 60. Consequently:

$$P(X \leq x) = P(X < x) = \frac{x}{60} \quad \text{for all } x \text{ with } 0 \leq x < 60$$

Since the event $\{X \leq x\}$ is just the union of the disjoint events $\{X < x\}$ and $\{X = x\}$, it follows from axiom (6.13) that:

$$P(X = x) = P(X \leq x) - P(X < x) = 0 \tag{8.5}$$

The probability that X is exactly equal to the fixed number x, is 0. This apparently is a consequence of the fact that X is continuous.

Since X is a continuous random variable, the number of possible outcomes of X is uncountable. As a consequence, the probability that the realization of X will be equal to a fixed real number x, will be 0. That is:

Important property for continuous X

$$P(X = x) = 0 \text{ for all real numbers } x \tag{8.6}$$

So, each outcome x of X has probability 0 to become the realization. But one of the possible outcomes of the random variable has to occur, so that not all events $\{X = x\}$ are empty. Apparently, there exist non-empty events with probability 0.

At first sight, (8.6) seems to lead to a contradiction. Since the sample space Ω is equal to the union of all disjoint events $\{X = x\}$, the following statement **seems to be** valid:

$$1 = P(\Omega) = \sum_x P(X = x) = \sum_x 0 = 0$$

However, the error in this statement is that the 'summation' $\sum_x$ has no meaning since the number of x's over which the summation occurs is uncountable.

It is obvious that property (8.6) of continuous random variables makes the probability density function (as it was defined for discrete random variables) completely useless. The property also has consequences for the distribution function F of a continuous random variable. Since $\{X \leq x\} = \{X < x\} \cup \{X = x\}$, it follows that:

$$F(x) = P(X \leq x) = P(X < x) \quad \text{for all real numbers } x \tag{8.7}$$

Although the probability density function as defined in Section 8.2.1 is useless now, the concept 'pdf' can be adapted in such a way that it is suitable for a continuous rv X. The idea is to choose the pdf f in such a way that probabilities of X events become areas under the graph of f. The rough picture in Figure 8.7 illustrates the relationship between this function f and the probability distribution of X: the shaded area has to be equal to $P(a \leq X \leq b)$.

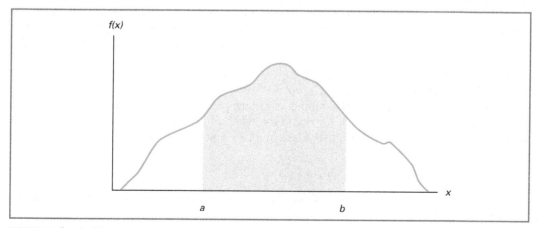

FIGURE 8.7 Graph of f

(Continuous) probability density function

The *probability density function* (pdf) of a **continuous** random variable X is a non-negative function f that is chosen such that

$$P(a \leq X \leq b) = \int_a^b f(x)dx = \text{area from } a \text{ to } b \text{ under the graph of } f \qquad \textbf{(8.8)}$$

holds for all a and b with $a \leq b$.

The notation $\int_a^b f(x)dx$ is used as shorthand for the area from a to b under f; it is pronounced as the *integral from a to b under f*. From the above – rather abstract – definition of continuous pdf, several properties follow immediately.

First notice that the choice $b = a$ in (8.8) yields that $P(X = a)$ is equal to the area from a to a, which indeed is 0. The choice $a = -\infty$ and $b = \infty$ yields 1 at the left-hand side of (8.8), so apparently the total area under a pdf of a continuous random variable has to be equal to 1.

Elementary properties of the pdf *f* of a continuous *X*

$$f(x) \geq 0 \quad \text{for all real numbers } x \qquad \textbf{(8.9)}$$

$$\int_{-\infty}^{\infty} f(x)dx = \text{total area under the graph of } f = 1 \qquad \textbf{(8.10)}$$

Any function f that satisfies (8.9) and (8.10) is generally called a (continuous) probability density function, without mentioning the rv to which it belongs. A (continuous) pdf is also termed *(continuous) density*.

Let F be the distribution function (cdf) of X. The choice $a = -\infty$ in (8.8) yields $F(b)$ at the left-hand side. It follows that $F(b)$ is just the area at the left-hand side of b under f. So, F follows from f. On the other hand, f also follows from F: without mentioning the details, we just note that f is equal to the mathematical derivative F' of F.

Properties of the cdf *F* of a continuous *X*

- It is a non-decreasing continuous function that is strictly increasing on an interval where the pdf is positive.
- It is completely determined by the pdf f since for all real numbers b:

$$F(b) = \int_{-\infty}^b f(x)dx = \text{area under } f \text{ at the left side of } b \qquad \textbf{(8.11)}$$

- It completely determines the pdf f since:

$$f(x) = F'(x) \quad \text{for all real numbers } x \qquad \textbf{(8.12)}$$

A good view on the rather abstract concept 'continuous pdf' can be obtained by setting $b = a + h$ in Figure 8.7 and letting h become a very small positive number. Notice that the shaded area then becomes very narrow; its area is almost the same as the area of the rectangle with width h and height $f(a)$. That is, $P(a \leq X \leq a + h) \approx f(a)h$ and hence:

$$f(a) \approx \frac{P(a \leq X \leq a + h)}{h} = \frac{F(a + h) - F(a)}{h} \quad \text{for } h \text{ close to 0} \qquad \textbf{(8.13)}$$

On the small interval $[a, a + h]$, the (non-decreasing) distribution function F grows by the amount $F(a + h) - F(a)$. By dividing this amount by the width h of the interval, the mean growth during the interval arises; this prompts the name 'density'. So, $f(a)$ is **not** a probability, but for small positive h the product $f(a)h$ **is** a probability. If an outcome a of X has a relatively large (compared with other outcomes) pdf value $f(a)$, then this implies that the probability that X falls in a small interval $[a, a + h]$ is relatively large when compared with similar intervals of other outcomes. That is why $f(a)$ is described as the **_likelihood_** of the outcome a.

Example 8.7

Reconsider the waiting time X of Mrs Doubledutch in Example 8.6. Below, a function f will be defined that turns out to be the pdf of this continuous rv. It will be used to find the cdf of X.

Let the function f be defined by

$$f(x) = \frac{1}{60} \quad \text{for } 0 \le x < 60$$

$$f(x) = 0 \text{ otherwise}$$

The heavy lines in Figure 8.8 constitute the graph of f.

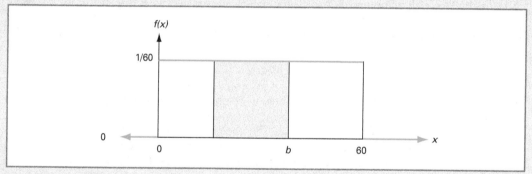

FIGURE 8.8 Graph of f

Notice that (8.9) is satisfied. But (8.10) is also valid, since the total area under f is $60 \times 1/60 = 1$. Hence, f is a continuous pdf. In Example 8.6 it was found that – because of the fact that all outcomes of X are equally likely – the probability that X falls in the interval $[a, b]$ is just the width of this interval divided by the width of the interval $[0, 60]$. That is:

$$P(a \le X \le b) = \frac{b - a}{60} \quad \text{for all } a \text{ and } b \text{ in } [0, 60] \text{ with } a \le b$$

However, the ratio $(b - a) / 60$ is also equal to the area from a to b under the graph of f, as follows from Figure 8.8 above. Hence, (8.8) is satisfied and f is the pdf of the rv $X = $ 'waiting time for the train'.

As a consequence, the probability $P(X \le x) = P(-\infty < X \le x)$ is equal to the total area from $-\infty$ to x under the graph of f, for all x in the interval $[0, 60]$. But this area is just the area from **0** to x under the graph of f, which is $x/60$. See Figure 8.9. Moreover, $F(x) = 0$ for $x < 0$, $F(x) = 1$ for $x > 60$ and:

$$F(x) = \frac{x}{60} \quad \text{for all } x \text{ with } 0 \le x \le 60$$

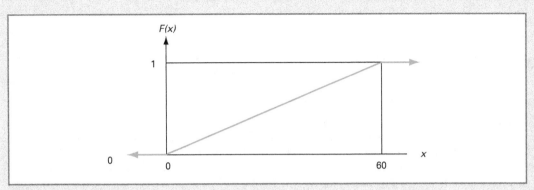

FIGURE 8.9 Graph of the cdf of *X*

Notice that between 0 and 60, *F* is a straight line.

For *x* between 0 and 60, the derivative of $F(x) = x/60$ is 1/60. Since *F* is constant over the intervals $(-\infty, 0)$ and $[60, \infty)$, its derivative is 0 over these intervals. Indeed, the derivative of *F* is equal to *f*.

It is possible to create a pdf *f* for which $f(x)$ is larger than 1 for some outcomes *x*. (Notice that this demonstrates the fact that $f(x)$ is **not** a probability.) If *X* is your waiting time (measured in **hours**, with decimals) for a bus that leaves (strictly on time) every 30 minutes, then the following pdf is a good model if you are unaware of the schedule:

$f(x) = 2$ for $0 \leq x < 0.5$

$f(x) = 0$ otherwise

Notice that this function *f* is indeed a pdf (since (8.9) and (8.10) are satisfied) and that, even for all *x* in [0, 0.5), $f(x)$ is larger than 1.

Probability density functions are often used to model a reality that is not known precisely. Ideas about this reality often determine which pdf to take, as in the following example.

Example 8.8

Reconsider the 'heavy' Sudoku puzzle that can be solved online; see Example 8.2. So far, the participants needed on average 25.5 minutes to solve it; they all needed at least 5 minutes. Let *X*, measured in minutes (with decimals), be the time that it takes a new participant to solve the puzzle. Our purpose is to create a model that can be used to determine probabilities regarding this continuous rv *X*.

Because of the threshold of 5 minutes, it seems reasonable to allow only values that are larger than or equal to 5. Apart from the mean value 25.5, no further facts are given. We need assumptions to get forward in our search for a good model. As a starting point, we assume additionally that the probability distribution is such that the mean value 25.5 is situated in the middle of the interval of possible outcomes of *X*. Hence, the interval of possible outcomes of *X* is [5, 46].

Assume, first, that all outcomes in [5, 46] are equally likely. Then the model is essentially known. The pdf *f* has to be constant over [5, 46] since all outcomes have the same likelihood. Since the total area has to be 1, it follows that:

$f(x) = \dfrac{1}{41}$ for all *x* in [5, 46]

Probabilities can be calculated easily. For instance:

$$P(15 < X < 30) = \frac{15}{41} = 0.3659$$

The cdf of X follows by similar arguments. For x between 5 and 46, the probability $F(x)$ is just the area of a rectangle with width $x - 5$ and height 1/41. Hence,

$$F(x) = \frac{1}{41}(x - 5) \quad \text{for all real numbers } x \text{ in } [5, 46]$$

Furthermore, $F(x) = 0$ if $x < 5$ and $F(x) = 1$ for $x > 46$.

Assume now that the outcomes are **not** equally likely, that outcomes close to the mean value 25.5 are much more likely than outcomes close to the boundaries of the interval [5, 46]. To be more precise: assume that when going from 5 to the centre 25.5, the likelihood of the outcomes increases linearly from 0 onwards; when going from 25.5 to the boundary 46, the likelihood goes down linearly to 0. That is, the continuous pdf f we are looking for must have a triangular graph, as shown in Figure 8.10.

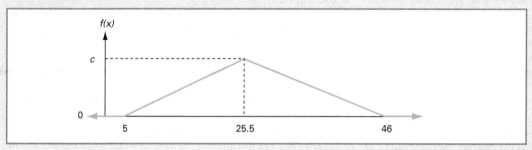

FIGURE 8.10 Triangular graph of f

The constant $c = f(25.5)$ has to be determined such that the total area under the graph of f is 1. Hence, $20.5 \times c = 1$. The conclusion is that $c = 1/20.5$. The pdf f is constituted of two straight lines; its mathematical form can be derived (can you do it?):

$$f(x) = \frac{1}{20.5^2}(x - 5) \qquad \text{if } 5 \leq x \leq 25.5$$

$$f(x) = \frac{1}{20.5^2}(-x + 46) \quad \text{if } 25.5 \leq x \leq 46$$

$$f(x) = 0 \qquad\qquad\qquad \text{otherwise}$$

Probabilities can be determined by calculating the corresponding area under the graph with the 'rectangle method'. For instance, since $f(15) = 10/(20.5)^2 = 0.0238$, the probability $P(X \leq 15)$ is just the **half** of the area of the rectangle formed by the interval [5, 15] and the height 0.0238; make a graph yourself. Since, furthermore, $f(30) = 16/(20.5)^2 = 0.0381$, it follows that:

$$P(X \leq 15) = (10 \times 0.0238)/2 = 0.119$$

$$P(X \geq 30) = (16 \times 0.0381)/2 = 0.305$$

$$P(15 < X < 30) = P(X \leq 30) - P(X \leq 15) = (1 - 0.305) - 0.119 = 0.576$$

Notice that the last answer is considerably larger than the answer for the first model.

We next calculate the cdf that belongs to f.

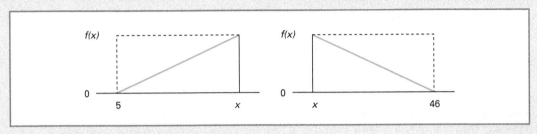

FIGURE 8.11 Rectangle method for finding the cdf of f

For x between 5 and 25.5, the rectangle method goes as follows; see Figure 8.11. The area needed for $P(X \le x)$ is just the **half** of the area of the left-hand rectangle with width $x - 5$ and height $f(x)$. Hence:

$$F(x) = \frac{1}{2 \times (20.5)^2} (x - 5)^2 \quad \text{if } 5 \le x \le 25.5$$

If $25.5 \le x \le 46$, the right-hand rectangle can be used to find $P(X > x) = 1 - F(x)$. Since $P(X > x)$ is half of the area of the rectangle with width $46 - x$ and height $f(x)$, it follows that:

$$F(x) = 1 - \frac{1}{2 \times (20.5)^2} (46 - x)^2 \quad \text{if } 25.5 \le x \le 46$$

And, of course, $F(x) = 0$ for $x < 5$ and $F(x) = 1$ for $x > 46$. Notice that indeed $F(25.5) = 0.5$, as it should because of the symmetry of the pdf.

The former two examples used pictures of rectangles to calculate probabilities. Of course, this is possible only for simple models. For more complicated models, recall that probabilities are areas and areas are integrals. The mathematical theory of integration and differentiation has to be used. The following and subsequent examples show some of the ideas.

Example 8.9

The managing director (MD) of a company believes that next year's profit X (a continuous rv measured in millions of euro) of the company will be between 0 and 2 million. In her opinion, it is likely that this profit will be somewhere around 1 million. In the **current** year, the profit of the company will be at least 0.5 million, probably even between 0.75 and 1.20 million. That is why the MD is, for **next** year, especially interested in the probabilities $P(X > 0.5)$ and $P(0.75 < X < 1.20)$.

The MD adopts the following model for the pdf f of X:

$$f(x) = \frac{3}{4} x (2 - x) = \frac{3}{2} x - \frac{3}{4} x^2 \quad \text{for } 0 \le x \le 2$$

(And $f(x) = 0$ for $x < 0$ and $x > 2$.) Note that f is a parabolic function on $[0, 2]$, that intersects the x-axis at 0 and 2; it takes its maximum value 0.75 at $x = 1$.

To find out whether f is indeed a density, we have to check whether the total area under f is 1. Since $f(x) = 0$ outside the interval $[0, 2]$ and hence no area is contributed there, it has to be checked whether the area above $[0, 2]$ under f is 1. In mathematical terms this means that the following has to be checked:

$$\int_0^2 \left(\frac{3}{2} x - \frac{3}{4} x^2 \right) dx = 1$$

The reader familiar with integration is invited to show that this is indeed correct.

To calculate the probabilities $P(X > 0.5)$ and $P(0.75 < X < 1.20)$, we need the areas under f at the right-hand side of 0.5 and between 0.75 and 1.20. That is, the integrals

$$\int_{0.5}^{2}\left(\frac{3}{2}x - \frac{3}{4}x^2\right)dx \quad \text{and} \quad \int_{0.75}^{1.2}\left(\frac{3}{2}x - \frac{3}{4}x^2\right)dx$$

have to be calculated. Those familiar with integration are invited to check that the answers are respectively 0.8438 and 0.3316. According to the model, the probability that next year's profit will be larger than 0.5 million is fairly high.

8.3 Functions of random variables

If X is a (discrete or continuous) random variable and $h(x)$ is a function, then $h(X)$ is another random variable. For instance: the function $h(x) = x^2$ yields the rv $h(X) = X^2$; the linear transformation $h(x) = ax + b$ gives the rv $h(X) = aX + b$. Since the randomness of $V = h(X)$ comes from the randomness of X, the probability distribution of V will be completely determined by the probability distribution of X. But **how** can the pdf and the cdf of V be determined from the pdf and the cdf of X?

The general theory will not be presented. Instead, two examples will be considered: one for a discrete and one for a continuous random variable.

Example 8.10

Let X be the number of eyes obtained by a single throw of the false die of the swindler. The pdf (called f_X) and the cdf (called F_X) of this discrete rv are in the table:

x	1	2	3	4	5	6
$f_X(x)$	1/12	1/6	1/6	1/6	1/6	3/12
$F_X(x)$	1/12	3/12	5/12	7/12	9/12	1

Suppose that, for reasons that will become clear later, interest is in the function $h(x) = (x - 4)^2$ and the random variable $V = h(X) = (X - 4)^2$. We will determine the pdf f_V and the cdf F_V from f_X and F_X.

Notice that V can take only the values 0, 1, 4 and 9; this follows by substituting $x = 1, 2, ..., 6$ for x in $h(x)$:

x	1	2	3	4	5	6
$v = (x - 4)^2$	9	4	1	0	1	4

The pdf of V follows easily from the pdf of X:

$f_V(0) = P((X - 4)^2 = 0) = P(X = 4) = 1/6$

$f_V(1) = P((X - 4)^2 = 1) = P(X = 5 \text{ or } X = 3) = P(X = 5) + P(X = 3) = 2/6$

$f_V(4) = P((X - 4)^2 = 4) = P(X = 6 \text{ or } X = 2) = P(X = 6) + P(X = 2) = 5/12$

$f_V(9) = P((X - 4)^2 = 9) = P(X = 1) = 1/12$

The cdf F_V follows from f_V. The table summarizes the results:

v	0	1	4	9
$f_V(v)$	2/12	4/12	5/12	1/12
$F_V(v)$	2/12	6/12	11/12	1

Things are not always as easy as in the above example. However, the linear transformations $h(x) = ax + b$ are exceptions, especially the case that $a > 0$. Such a linear transformation transforms an rv X into an rv $V = aX + b$. If X is discrete, then V is discrete; if X is continuous, then V is continuous. Here is an example where X is continuous.

Example 8.11

The manager of an electronics store is interested in the relationship between the weekly profit Y and the gross weekly sales X, both measured in units of €10000. From long experience he knows that X varies between 10 and 40 and that all outcomes between are equally likely. Furthermore, he finds out that, apart from the weekly constant costs of €50000, 25% of the weekly gross sales can be considered profit. That is why he models the weekly profit Y of the store by $Y = 0.25X - 5$, where X is assumed to have the following pdf f_X:

$$f_X(x) = \frac{1}{30} \quad \text{for } 10 \leq x \leq 40$$
$$= 0 \quad \text{for } x \text{ otherwise}$$

Below, the cdf of the weekly profit is calculated from the cdf of the gross weekly sales; the accompanying pdf is derived by taking the derivative.

The linear transformation $y = 0.25x - 5$ transforms the interval [10, 40] into the interval [−2.5, 5]. For y between −2.5 and 5 it holds that:

$$0.25X - 5 \leq y \Leftrightarrow 0.25X \leq y + 5 \Leftrightarrow X \leq 4y + 20$$

Hence, the occurrence of the event $\{0.25X - 5 \leq y\}$ implies the occurrence of the event $\{X \leq 4y + 20\}$, and vice versa. That is, the two events coincide and have the same probability:

$$P(Y \leq y) = P(0.25X - 5 \leq y) = P(X \leq 4y + 20) = F_X(4y + 20)$$

So, F_X is needed first. For $10 \leq a \leq 40$ it follows from the rectangle method that:

$$F_X(a) = P(X \leq a) = \frac{a - 10}{30}$$

Since $4y + 20$ indeed lies between 10 and 40, we obtain:

$$F_Y(y) = F_X(4y + 20) = \frac{4y + 10}{30} = \frac{4}{30}y + \frac{1}{3}$$

As a check, note that $F_Y(-2.5) = 0$ and $F_Y(5) = 1$; as it should. The following probabilities are immediate consequences:

$$P(Y < 1) = F_Y(1) = \frac{7}{15}$$

$$P(-1 < Y < 1) = F_Y(1) - F_Y(-1) = \frac{7}{15} - \frac{3}{15} = \frac{4}{15}$$

Of course, the probability that the weekly profit is positive is of special interest:

$$P(Y > 0) = 1 - F_Y(0) = 1 - \frac{1}{3} = \frac{2}{3}$$

The pdf f_Y of Y is just the derivative of F_Y. For $-2.5 < y < 5$ we obtain:

$$f_Y(y) = F'_Y(y) = \frac{4}{30} = \frac{2}{15}$$

Of course, $f_Y(y) = 0$ for y otherwise. Notice that the pdf of Y is also constant, over the interval $(-2.5, 5)$ with width 7.5.

8.4 Expectation, variance and standard deviation

The actual outcome of an rv X is not known in advance; it depends on chance, and chance is driven by the probability distribution of X. But what value of X can be **expected**? Probably, the actual outcome of X will deviate from that expected value. What are the expectations about this deviation? These problems are the topics of the present section.

Start with a random experiment and some random variable X. The probability distribution of X is characterized by its pdf f or its cdf F. **Statistics** are numbers or qualifications that describe specific features of the probability distribution of X. (See also the similar definition for (population) variables in descriptive statistics; Section 1.4.)

8.4.1 Discrete random variables

Suppose that X is discrete with probability density function f. Hence, $f(x) = P(X = x)$. Each possible outcome x of X is accompanied by $f(x)$ which – in a sense – determines the importance of x. If $f(x)$ is relatively large, then it is (when compared with other outcomes) relatively likely that x will become the realization. That is why $f(x)$ is used as the weight of x in the following definition.

Expectation or *expected value* or *mean* of a discrete X

$$E(X) = \sum_x xf(x) \qquad (8.14)$$

In this expression, the summation concerns all possible outcomes x of X, which makes sense since the set of outcomes is finite or at most countable-infinite. Notice that the weights $f(x)$ are non-negative and add up to 1; the right-hand side of (8.14) is a weighted mean of the outcomes x. (See also (3.5).) Formula (8.14) states that the possible outcomes x of X have to be multiplied by the accompanying probabilities $f(x)$ and that the results of these multiplications have to be added up.

The definition in (8.14) is rather abstract. However, $E(X)$ can be interpreted intuitively as a **long-run average**: the average of a ceaseless sequence of realizations $x_1, x_2, \ldots$ of X that result from an ongoing sequence of independent repetitions of the random experiment. The idea will be illustrated with the die of the swindler.

Example 8.12
Let X be the number of eyes that shows up with the false die of the swindler. The table below describes the probability distribution of X.

x	1	2	3	4	5	6
$f(x)$	1/12	1/6	1/6	1/6	1/6	1/4

To determine the expectation of X with (8.14), values and accompanying probabilities have to be multiplied and the results added up:

$$E(X) = 1 \times \frac{1}{12} + 2 \times \frac{1}{6} + \ldots + 5 \times \frac{1}{6} + 6 \times \frac{1}{4} = \frac{47}{12} \approx 3.917$$

(Check yourself that a fair die has expectation $42/12 = 3.5$.) But why can $47/12$ be considered as a long-run average? We will see that this is a consequence of the law of large numbers.

Suppose that the experiment is repeated n times (with n large) and that the respective numbers of eyes are $x_1, x_2, \ldots, x_n$. Let g_j be the number of times (among the n trials) that j eyes were counted, for $j = 1, 2, \ldots, 6$. So, the g_j are frequencies. To find the mean value $\bar{x}$ of the n throws, all x_i have to be summed and the result divided by n. That is, the value 1 has to be added up g_1 times, the value 2 has to be added up g_2 times, etc. At the end, the total has to be divided by n. This calculation is expressed in the first part of the equation below. The second part rewrites things:

$$\bar{x} = \frac{1}{n}(1 \times g_1 + 2 \times g_2 + \ldots + 6 \times g_6) = 1 \times \frac{g_1}{n} + 2 \times \frac{g_2}{n} + \ldots + 6 \times \frac{g_6}{n} \qquad \textbf{(8.15)}$$

Since the ratio g_j/n is just the proportion (relative frequency) of j eyes in n throws, the right-hand expression informs us that $\bar{x}$ can also be calculated by multiplying the numbers of eyes j by the corresponding proportions g_j/n and adding the results. But according to the law of large numbers as expressed in the empirical probability definition in (6.10), these proportions g_j/n will approximate the probabilities $f(j) = P(X = j)$ as n grows larger and larger. That is:

$$\frac{g_1}{n} \to f(1), \frac{g_2}{n} \to f(2), \ldots, \frac{g_6}{n} \to f(6) \quad \text{as } n \to \infty$$

Hence, $\bar{x}$ as expressed by the right-hand side of (8.15) tends to

$$1 \times f(1) + 2 \times f(2) + \ldots + 6 \times f(6) \quad \text{as } n \to \infty$$

which is just $E(X)$. That is, if we would have an **ongoing** sequence $x_1, x_2, \ldots$ of realizations of X, then the mean value of this sequence would be equal to $E(X)$. Hence, $E(X)$ can indeed be considered as a long-run average. Indeed, the swindler succeeded in constructing a die that on average gives a higher result.

Recall from Example 6.6 that the swindler did the experiment only $n = 10\,000$ times. The table below restates his results.

Number of eyes	1	2	3	4	5	6
Observed frequency	830	1603	1598	1670	1620	2679
Observed relative frequency	0.0830	0.1603	0.1598	0.1670	0.1620	0.2679

The average value $\bar{x}$ (after $10\,000$ throws) can be calculated easily with the right-hand side of (8.15):

$$\bar{x} = 1 \times 0.0830 + 2 \times 0.1603 + \ldots + 6 \times 0.2679 = 3.9684$$

Recall that $E(X) \approx 3.92$. Although $\bar{x}$ is already rather close to $E(X)$, we would need more repetitions of the experiment to get a better approximation.

$E(X)$ is also denoted by μ, or even μ_x if more random variables are involved. Recall that this notation was also used in descriptive statistics for the population mean of a variable. In Section 8.6 it will be explained why it is permissible to use it here again.

If X is a constant b and hence is degenerated at b, then the summation in (8.14) has only the term $bf(b)$, which equals b since $f(b) = 1$. Hence, the expectation of this rv X is equal to b. Or, to put it another way, $E(b) = b$. This is in accordance with intuition: the die that has the number b on all sides will always yield b as outcome.

As was already mentioned in Section 8.3, there are many occasions where the rv of interest is a function of a basic rv X. If $h(x)$ denotes the function, then interest is in $V = h(X)$. Let f_X and f_V be the respective pdfs of X and V. According to the definition in (8.14), the expectations of X and V are respectively:

$$E(X) = \sum_x x f_X(x) \quad \text{and} \quad E(V) = \sum_v v f_V(v) \tag{8.16}$$

Since V is a function of X, the pdf of V is completely determined by the pdf of X. That is why the expectation of $h(X)$ can also be obtained by multiplying the outcomes $h(x)$ by $f_X(x)$ and adding up the results of the multiplications:

Expectation of $V = h(X)$ for X discrete

$$E(V) = E(h(X)) = \sum_x h(x) f_X(x) \tag{8.17}$$

Usually, it is easier to use (8.17) for the calculation of $E(V)$ than (8.16). This is because for the right-hand formula of (8.16) the pdf of V has to be calculated first. Notice that (8.17) also gives the definition of $E(X)$ since the use of the function $h(x) = x$ immediately gives (8.14).

Example 8.13

To stimulate the sales of beer, an advertising agency advises the manager of a wine and beer supermarket to choose one of the following two advertising campaigns and to adopt it for a long period.

Campaign 1: Customers who buy a crate of beer are invited to buy at most two extra crates with 20% discount per extra crate.
Campaign 2: Each customer receives a coupon worth 50 cents, and for each crate of beer that is bought an extra coupon of €1.

We compare the mean advantage per customer for the two campaigns. Of course it is unknown how the campaigns will affect the number of crates of beer that will be sold. That is why the present situation (that is, without a campaign) is taken as the starting point. Here are the facts:

- A crate of beer costs €10.
- Each customer buys at most three crates.
- The frequency distribution of the variable 'number of crates of beer bought by a customer' is as follows:

Number of crates	0	1	2	3
Relative frequency	0.60	0.16	0.10	0.14

Let X be the number of crates of beer that is bought by an arbitrary customer. Since the effects of the campaigns are unknown, the above relative frequency distribution is taken as the model for X. Let V and W be the respective price advantages to the arbitrary customer when the campaigns 1 and 2 are launched. Below, the relations between X and V and between X and W will be investigated and $E(X)$, $E(V)$ and $E(W)$ will be calculated.

The expectation of X can be calculated from the pdf:

$E(X) = 0 \times 0.6 + 1 \times 0.16 + 2 \times 0.10 + 3 \times 0.14 = 0.78$

On average, the customers will buy 0.78 crates of beer. Let $h_1(x)$ and $h_2(x)$ respectively be the price advantages of campaigns 1 and 2 for a customer who buys x crates. Hence, $V = h_1(X)$ and $W = h_2(X)$. If $x = 0$ (that is, no crates of beer are bought), then $h_1(x) = 0$ and $h_2(x) = 0.5$. If $x = 2$ (two crates are bought), then the campaigns give the respective advantages $h_1(2) = 2$ and $h_2(2) = 2.5$; etc. The following table gives overviews of the functions $h_1(x)$, $h_2(x)$ and the pdf $f_X(x)$ of X.

x	0	1	2	3
$f_X(x)$	0.60	0.16	0.10	0.14
$h_1(x)$	0	0	2	4
$h_2(x)$	0.5	1.5	2.5	3.5

Notice that $h_2(x) = 0.5 + x$, which is a linear transformation. So, $W = 0.5 + X$. By (8.17), the expectation of V (respectively W) can be obtained by determining the products of $h_1(x)$ (respectively $h_2(x)$) and $f_X(x)$ and adding up the results:

$E(V) = 0 \times 0.6 + 0 \times 0.16 + 2 \times 0.10 + 4 \times 0.14 = 0.76$

$E(W) = 0.5 \times 0.6 + 1.5 \times 0.16 + 2.5 \times 0.10 + 3.5 \times 0.14 = 1.28$

If campaign 1 (respectively 2) is adopted, then the average price advantage for customers is €0.76 (respectively €1.28). Campaign 2 is the most favourable.

Denote the expectation of X by μ; the outcomes of X are denoted by x. There is one function $h(x)$ that has special importance and will be used frequently: the function $h(x) = (x - \mu)^2$. Since $x - \mu$ is a deviation, the square $(x - \mu)^2$ measures the squared deviation of the outcome x with respect to μ. The rv $h(X) = (X - \mu)^2$ is called the **squared random deviation**. The expectation of this rv is called the **variance** of X and is denoted by σ^2 or even σ_X^2. However, the notation $V(X)$ will also be used frequently.

Variance of a discrete *X*

$$V(X) = \sigma^2 = E((X - \mu)^2) = \sum_x (x - \mu)^2 f(x) \qquad (8.18)$$

Standard deviation of a discrete *X*

$$SD(X) = \sigma = \sqrt{\sigma^2} \qquad (8.19)$$

Notice that the right-hand side of (8.18) is a consequence of (8.17).

The summation in (8.18) is used to **calculate** σ^2. It expresses that the following has to be done:

■ For each outcome x, the squares $(x - \mu)^2$ have to be calculated and multiplied by $f(x)$.

■ All resulting products have to be added up.

The abstract expression $E((X - \mu)^2)$ in (8.18) can be used to **interpret** σ^2. It expresses that the variance of X is the **expectation** of a special function of X. Since an expectation can be considered as a long-run average, σ^2 can be interpreted as the mean value of an ongoing sequence

$(x_1 - \mu)^2, (x_2 - \mu)^2, \ldots$

that comes from a sequence $x_1, x_2, \ldots$ of realizations due to ongoing repetitions of the random experiment. From that point of view it is obvious that σ^2 (and hence σ too) is indeed a measure of variation. A small value of σ^2 means that the deviations $x_i - \mu$ in an ongoing sequence are usually close to 0 and hence the x_i close to μ; the random variable tends to be stable in the sense that its realizations do not have much variation. A large value of σ^2 means that the deviations $x_i - \mu$ can be away from 0 and the x_i away from μ; the random variable is less stable. The validity of the long-run interpretation of σ^2 is illustrated in the example below.

Example 8.14

Recall that the function f in the third row of the table below represents the pdf of the random variable X = 'number of eyes' when throwing the die of the swindler:

x	1	2	3	4	5	6
$(x - \mu)^2$	8.5069	3.6736	0.8403	0.0069	1.1736	4.3403
$f(x)$	1/12	1/6	1/6	1/6	1/6	1/4

In Example 8.12 it was found that $\mu = 47/12 \approx 3.917$. The variance of X is calculated.

For each outcome x of X, the squared deviations $(x - 47/12)^2$ have to be calculated; see the second row for the results. Next, $(x - 47/12)^2$ has to be multiplied by $f(x)$ for each x; the products have to be added up:

$$\sigma^2 = 8.5069 \times \frac{1}{12} + 3.6736 \times \frac{1}{6} + \ldots + 4.3403 \times \frac{1}{4} = 2.7431$$

$$\sigma = \sqrt{2.7431} = 1.6562$$

(For a fair die, $\mu = 3.5$, $\sigma^2 = 2.9167$ and $\sigma = 1.7078$; check it yourself.) Apparently, the swindler succeeded in constructing a die that not only has a higher expected number of eyes but also has less variation.

To illustrate the long-run interpretation of σ^2, reconsider from Example 8.12 the frequencies g_j and the relative frequencies g_j/n of throwing j eyes in n realizations $x_1, \ldots, x_n$ of X from n independent and identical repetitions of the random experiment. Notice that g_1 of the squared deviations $(x_i - \mu)^2$ are equal to $(1 - \mu)^2$, g_2 are equal to $(2 - \mu)^2$, etc. Hence, the mean of the squared deviations $(x_i - \mu)^2$ equals:

$$\frac{1}{n}\sum_{i=1}^{n}(x_i - \mu)^2 = \frac{1}{n}((1 - \mu)^2 \times g_1 + (2 - \mu)^2 \times g_2 + \ldots + (6 - \mu)^2 \times g_6)$$

$$= (1 - \mu)^2 \times \frac{g_1}{n} + (2 - \mu)^2 \times \frac{g_2}{n} + \ldots + (6 - \mu)^2 \times \frac{g_6}{n} \qquad \textbf{(8.20)}$$

Now recall that, because of the empirical definition of probability:

$$\frac{g_1}{n} \to f(1), \frac{g_2}{n} \to f(2), \ldots, \frac{g_6}{n} \to f(6) \quad \text{as } n \to \infty$$

Hence, the right-hand side of (8.20) tends to

$$(1 - \mu)^2 \times f(1) + (2 - \mu)^2 \times f(2) + \ldots + (6 - \mu)^2 \times f(6) \quad \text{as } n \to \infty$$

Consequently, the mean of the squared deviations $(x_i - \mu)^2$ for an **ongoing** sequence $x_1, x_2, \ldots$ is equal to

$$(1 - \mu)^2 f(1) + (2 - \mu)^2 f(2) + \ldots + (6 - \mu)^2 f(6)$$

which is just the right-hand expression of (8.18). The final conclusion is that the summation in (8.18) can indeed be interpreted as the long-run average of the squared deviations.

Notice that the variance of a discrete rv X can never be negative: since all terms in the summation of (8.18) are non-negative, the sum is non-negative itself. Furthermore, the variance of an rv that is degenerated at a constant b is equal to 0, since then the summation in (8.18) has only the term $(b - \mu)^2$ which is 0 because $\mu = E(X) = b$. But if there is more than one outcome possible, then μ cannot be equal to all outcomes at the same time, so at least one of the terms $(x - \mu)^2 f(x)$ in the summation of (8.18) is larger than 0 and hence the summation itself is larger than 0.

Properties of the concept 'variance' of an rv X

$V(X) \geq 0$

$V(X) = 0 \Leftrightarrow X$ is degenerated   **(8.21)**

The right-hand side of (8.21) is in accordance with intuition: if a random variable is constant, it does not vary so the variance should be 0.

CASE 8.1 EXPECTED RETURN AND RISK OF CARLSBERG BREWERIES STOCK – SOLUTION, PART 2

The discrete model can be used to calculate the accompanying expected (daily) return for a day in the future:

$E(R) = (-15) \times 0.0011 + (-9) \times 0.0006 + \ldots + (12) \times 0.0006 = 0.0802$

Based on this discrete model, it is expected that tomorrow's return will be 0.0802%. For a year with 250 business days, this comes down to about 20% per year.

Standard deviation is often used to measure the risk of a stock. The variance of the adopted model equals:

$\sigma_R^2 = (-15 - 0.0802)^2 \times 0.0011 + \ldots + (12 - 0.0802)^2 \times 0.0006 = 3.3882$

The risk of the daily return is measured as the square root, which is 1.84%.

In Section 10.3.5 another model will be considered.

8.4.2 Continuous random variables

The concepts $E(X)$ and $V(X)$ will also be defined for continuous random variables. The definitions arise from the definitions for the discrete case by replacing the discrete pdf by the continuous pdf and the summation operator by the integration operator.

In the case of a discrete rv X with (discrete) pdf f, **summations** play important roles:

1. The sum of all $f(x)$ has to be equal to 1; see (8.1).
2. $F(a) = P(X \leq a)$ is the sum of all $f(x)$ with $x \leq a$; see (8.3).
3. $E(X)$ is the sum of all products $xf(x)$ that weigh the outcomes x with the weights $f(x)$; see (8.14).
4. $E(h(X))$ is the sum of all products $h(x)f(x)$ that weigh the $h(x)$ with the weights $f(x)$; see (8.17).
5. $V(X)$ is the sum of all products $(x - \mu)^2 f(x)$ that weigh the $(x - \mu)^2$ with the weights $f(x)$; see (8.18).

In the case of a continuous rv X with (continous) pdf f, **areas** play important roles:

1. The total area under $f(x)$ has to be equal to 1; see (8.10).
2. $F(b) = P(X \leq b)$ is the area under $f(x)$ for $x \leq b$; see (8.11).

Without giving the details, it is just stated that, for a continuous rv X with (**continuous**) pdf f, the expectations $E(X)$ and $E(h(X))$ and the variance $V(X)$ follow from (3), (4) and (5) above by replacing 'sum' by 'area':

3. The **expectation** $E(X)$ of X is the total area under the function $xf(x)$ that weighs x with $f(x)$.
4. The expectation $E(h(X))$ of $h(X)$ is the total area under the function $h(x)f(x)$ that weighs $h(x)$ with $f(x)$.
5. The **variance** $V(X)$ of X is the total area under the function $(x - \mu)^2 f(x)$ that weighs $(x - \mu)^2$ with $f(x)$.

Notice that (3), (4) and (5) all use $f(x)$ to denote the importance of, respectively, x, $h(x)$ and $(x - \mu)^2$. This intuitively makes sense since $f(x)$ represents the likelihood (not probability) that x will be the realization of the rv X. Here is a summary:

Expectation or expected value or mean of a continuous X

$$E(X) = \int_{-\infty}^{\infty} xf(x)dx = \text{total area under the function } xf(x) \qquad \text{(8.22)}$$

Expectation of $V = h(X)$ when X is continuous

$$E(V) = E(h(X)) = \int_{-\infty}^{\infty} h(x)f(x)dx = \text{total area under } h(x)f(x) \qquad \text{(8.23)}$$

Variance of a continuous X

$$V(X) = \sigma^2 = E((X - \mu)^2) = \int_{-\infty}^{\infty} (x - \mu)^2 f(x)dx \qquad \text{(8.24)}$$

Standard deviation of a continuous X

$$SD(X) = \sigma = \sqrt{\sigma^2} \qquad \text{(8.25)}$$

The calculation of $E(X)$, $E(h(X))$ and $V(X)$ requires knowledge about calculations of areas with integrals. Only in isolated cases can a rectangle method be used too, as in Example 8.15 below.

Many conventions and properties for expectation and variance valid for the discrete case, remain valid in the continuous case. They are summarized below.

- $E(X)$ can be considered as the long-run average of an ongoing sequence of realizations x_1, x_2, ... of X.
- $E(X)$ is also denoted by μ or μ_x.
- $E(b) = b$ for all constants b.
- $E(h(X))$ can be considered as the long-run average of $h(x_1)$, $h(x_2)$, ... which belong to an ongoing sequence of realizations x_1, x_2, ... of X.
- $V(X) = E((X - \mu)^2)$, the expectation of the squared random deviation with respect to μ.
- $V(X)$ can be considered as the long-run average of the squared deviations with respect to μ of an ongoing sequence of realizations x_1, x_2, ... of X.
- $V(X)$ is also denoted as σ^2 or σ_X^2.
- $V(X)$ is always non-negative and usually strictly positive.
- $V(X) = 0 \Leftrightarrow X$ is degenerated.

Example 8.15

Recall the waiting time X (in minutes, with decimals) of Mrs Doubledutch for the high-speed train from Amsterdam to Paris. The model of Example 8.7 assumes that all outcomes in the range $[0, 60)$ are equally likely, which leads to the following (continuous) pdf f:

$$f(x) = \frac{1}{60} \quad \text{if } 0 \le x < 60$$

$$= 0 \quad \text{otherwise}$$

Since $E(X)$ can be considered the long-run average of an ongoing sequence of waiting times, it is intuitively obvious that $E(X)$ is just the centre of the interval. Hence, $E(X) = 30$ minutes. According to the definition in (8.22), $E(X)$ can be obtained by calculating the total area under the function $xf(x)$. Now notice that:

$$xf(x) = \frac{x}{60} \quad \text{if } 0 \le x < 60$$

$$= 0 \quad \text{otherwise}$$

The graph of this function is pictured in Figure 8.12.

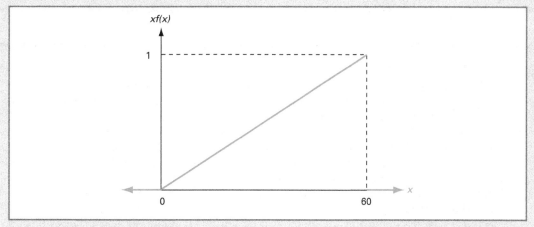

FIGURE 8.12 Graph of the function $xf(x)$.

The area under the graph of $xf(x)$ is just the half of the area of the rectangle, which yields $(60 \times 1)\,/\,2 = 30$. The conclusion is that $\mu = 30$.

To determine $V(X)$, intuition does not help us much. According to (8.24), the variance of X is the total area under the function $(x - 30)^2 f(x)$. But notice that:

$$(x - 30)^2 f(x) = \frac{(x - 30)^2}{60} \quad \text{if } 0 \le x < 60$$

$$= 0 \quad \text{otherwise}$$

In Figure 8.13, this function is sketched for x in $[0, 60]$; beyond this interval it is 0.

Since this function is 0 outside the interval $[0, 60)$, the total area under the function is equal to the area above $[0, 60)$. One really needs integration to calculate that area; readers familiar with integration are invited to check the last equality below:

$$\sigma^2 = \int_0^{60} \frac{(x - 30)^2}{60}\, dx = \int_0^{60} \left(\frac{1}{60} x^2 - x + 15 \right) dx = 300$$

So, $\sigma^2 = 300$ minutes^2 and $\sigma = 17.32$ minutes.

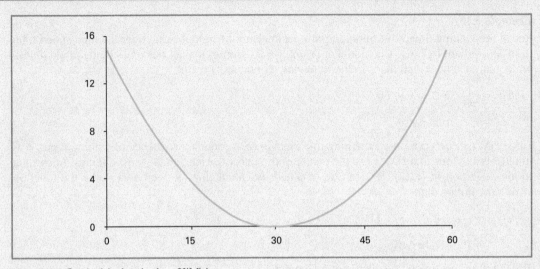

FIGURE 8.13 Graph of the function $(x - 30)^2 f(x)$

Example 8.16

Reconsider the two pdfs used in Example 8.8 to model the probability distribution of the time X (in seconds) that it takes a puzzler to solve an online 'heavy' Sudoku puzzle. Recall that both pdfs are concentrated on the interval [5, 46]; they are both 0 beyond this interval. The first model assumes that all outcomes in [5, 46] are equally likely, so the pdf has a constant height of 1/41 above the interval. The second model is based on the assumption that the likelihood of the outcomes x increases linearly when going from $x = 5$ to $x = 25.5$ and decreasing linearly when going from $x = 25.5$ to $x = 46$, which leads to the triangular pdf in Figure 8.10.

Since both pdfs are symmetric around the centre 25.5 of the interval, they have the same expected value: $\mu = 25.5$.

Notice that near the boundaries of the interval [5, 46], the constant pdf has higher likelihood than the triangular pdf (although both total areas are equal to one). The triangular pdf is more concentrated around the centre, so it has less variation. That is why it is expected that the variance of the constant pdf will be larger than the variance of the triangular pdf.

Using integration, it can be shown that the variance of the constant pdf is 140.0833. The variance of the triangular pdf turns out to be 70.0417; the calculation is very laborious. But note that the variance of the triangular model is indeed smaller than the variance of the constant model.

8.5 Rules for expectation and variance

This section is about rules to calculate for linear transformations $Y = a + bX$ the expectation, variance and standard deviation of Y when the expectation, variance and standard deviation of X have already been calculated. Furthermore, the insights obtained are used to show that an rv with expectation μ and variance σ^2 can, by way of a special linear transformation, be transformed into an rv with mean 0 and variance 1. Furthermore, a short-cut formula for the variance of an rv is derived and Chebyshev's rule is generalized to the field of probability.

Suppose, first, that X is discrete with pdf f. By the linear transformation $h(x) = a + bx$ (with a and b constants) the random variable X transforms into the random variable $Y = h(X) = a + bX$. The following results can be proved:

$$E(Y) = E(a + bX) = a + bE(X) \tag{8.26}$$

$$V(Y) = V(a + bX) = b^2V(X) \tag{8.27}$$

Derivation of transformation rules for mean and variance for discrete X

The summation rules of Appendix A2 are used in the following derivations.

$$E(Y) = \sum_x h(x)f(x) = \sum_x (a + bx)f(x) = \sum_x (af(x) + bxf(x))$$

$$= \sum_x af(x) + \sum_x bxf(x) = a\sum_x f(x) + b\sum_x xf(x)$$

Since summation of the $f(x)$ yields 1, equation (8.26) follows.

For (8.27), recall that $V(Y)$ is the expectation of the squared random deviation of Y. That is, $V(Y) = E((Y - \mu_y)^2)$. By substitution of $Y = a + bX$ and $\mu_y = a + b\mu_x$ and by getting rid of the parentheses, it follows that:

$$V(Y) = E((Y - \mu_y)^2) = E((a + bX - (a + b\mu_x))^2) = E((a + bX - a - b\mu_x)^2)$$

$$= E((bX - b\mu_x)^2) = E(b^2(X - \mu_x)^2) = b^2E((X - \mu_x)^2) = b^2V(X)$$

Note that (8.26), with $a = 0$ and b replaced by b^2, was used in the second-last equality.

It turns out that (8.26) and (8.27) are also valid for a continuous rv X; this will not be proved.

As an immediate consequence of (8.27), notice that the choices $a = 0$ and $b = -1$ yield that the rv's X and $-X$ have the same variance and standard deviation:

$$V(-X) = V((-1)X) = (-1)^2V(X) = V(X)$$

Also note that the constant a **does** influence $E(Y)$. However, it is of no importance for $V(Y)$, as is intuitively clear since constants have no variation. The choice $a = 0$ in (8.26) and (8.27) yields:

$$E(bX) = bE(X) \quad \text{and} \quad V(bX) = b^2V(X) \tag{8.28}$$

For $E(a + bX)$, the constant term a **does** contribute, while in $E(bX)$ the b can freely be placed in front of the E operator. To determine the variance of $a + bX$, the constant term a does **not** contribute, while in $V(bX)$ the b has to be squared when placed in front of the V operator.

To determine the consequences for the standard deviation $SD(Y)$, square roots have to be taken. But notice that the square root of b^2 is a non-negative number that is equal to b only for non-negative b; for negative b it is equal to $|b|$. The conclusion is that:

$$V(Y) = V(a + bX) = b^2V(X) \quad \text{and} \quad SD(Y) = |b|SD(X) \tag{8.29}$$

Results (8.27)–(8.29) are summarized below:

Properties of linear transformations of *X* (discrete and continuous)

$$Y = a + bX \quad \Rightarrow \quad \begin{cases} \mu_Y = a + b\mu_X \\ \sigma_Y^2 = b^2\sigma_X^2 \\ \sigma_Y = |b|\sigma_X \end{cases} \tag{8.30}$$

Example 8.17

Reconsider the two promotion campaigns in Example 8.13. The manager of the wine and beer supermarket wants to adopt one of the two campaigns for a long period, to promote the sales of crates of beer. Recall that the rv X = 'number of crates of beer' bought by a customer during the campaign has expectation 0.78 and that the function f in the table below was chosen as a model for the probability distribution of X. Furthermore, recall that random variables V and W measure the price advantages of the customer for the campaigns 1 and 2, that V is a function $h_1(X)$ of X with $h_1(x)$ as in the table, and that $W = 0.5 + X$.

x	0	1	2	3
$h_1(x)$	0	0	2	4
$f(x)$	0.60	0.16	0.10	0.14

Also recall that $E(V) = 0.76$. Below, the variances of V and W are calculated and compared.

Firstly, notice that W is a linear transformation of X, but V is not. By applying (8.30) with $a = 0.5$ and $b = 1$ it follows immediately that:

$$E(W) = 0.5 + E(X) = 0.5 + 0.78 = 1.28 \quad \text{and} \quad \sigma_W^2 = \sigma_X^2$$

For the calculation of $V(X)$, the squared deviations $(x - \mu_X)^2$ have to be multiplied by the probabilities $f(x)$ and the results have to be added:

x	0	1	2	3
$(x - 0.78)^2$	0.6084	0.0484	1.4884	4.9284
$f(x)$	0.60	0.16	0.10	0.14

$$V(X) = E((X - 0.78)^2) = 0.6084 \times 0.60 + \ldots + 4.9284 \times 0.14 = 1.2116$$

Hence, σ_W^2 is also equal to 1.2116 and $\sigma_W = 1.1007$. To calculate the variance of the rv V, (8.30) cannot be used. However:

$$\sigma_V^2 = E((V - \mu_V)^2) = E((h_1(X) - 0.76)^2)$$

To calculate the right-hand side, the terms $(h_1(x) - 0.76)^2$ have to be multiplied by the $f(x)$ and the results have to be added up; see (8.17) and the table below:

x	0	1	2	3
$h_1(x)$	0	0	2	4
$(h_1(x) - 0.76)^2$	0.5776	0.5776	1.5376	10.4976
$f(x)$	0.60	0.16	0.10	0.14

It follows that:

$$\sigma_V^2 = 0.5776 \times 0.60 + \ldots + 10.4976 \times 0.14 = 2.0624 \quad \text{and} \quad \sigma_V = 1.4361$$

Since $\mu_V < \mu_W$, campaign 2 is on average more favourable for the customers. Furthermore, since $\sigma_V^2 > \sigma_W^2$, the pdf of W has the smallest variance, which means that the price advantages to customers due to campaign 2 show less variation. From the point of view of the customers, campaign 2 seems to be the better.

If $h_1(x)$ and $h_2(x)$ are two general functions (not necessarily linear transformations), then $V = h_1(X)$ and $W = h_2(X)$ are two random variables. The summation $h(x) = h_1(x) + h_2(x)$ of the two functions gives another random variable $h(X) = h_1(X) + h_2(X)$.

Suppose first that X is a discrete rv. The expectation of $h(X)$ turns out to be the sum of the expectations of $h_1(X)$ and $h_2(X)$. This follows by using the summation rules of Appendix A2 again:

$$E(h(X)) = \sum_x h(x)f(x) = \sum_x (h_1(x) + h_2(x))f(x)$$

$$= \sum_x (h_1(x)f(x) + h_2(x)f(x)) = \sum_x h_1(x)f(x) + \sum_x h_2(x)f(x)$$

It follows that:

$$E(V + W) = E(V) + E(W)$$

It turns out that this important result is valid not only for discrete random variables but also for continuous ones, and not only for functions of X but also for general random variables V and W that belong to a common random experiment.

Expectation rule for the sum of two general random variables

$$E(V + W) = E(V) + E(W) \tag{8.31}$$

This rule has nice applications. Earlier, it was calculated that the rv X = 'number of eyes when throwing one fair die' and the rv Y = 'the total number of eyes when throwing a fair red and a fair blue die' have 3.5 and 7, respectively, as expectations. That is, $E(X) = 3.5$ and $E(Y) = 7$. To obtain this last expectation, the pdf of Y was needed. It will now be shown that the result that $E(Y) = 7$ can also be obtained by using (8.31).

Let X_1 and X_2 be the numbers of eyes with the red and the blue die, respectively. Then, of course, $E(X_1) = E(X_2) = 3.5$. But notice that $Y = X_1 + X_2$ and that by (8.31):

$$E(Y) = E(X_1 + X_2) = E(X_1) + E(X_2) = 3.5 + 3.5 = 7$$

With (8.31), the result that $E(Y) = 7$ is an immediate consequence of the fact that one throw with a fair die has 3.5 as expected number of eyes.

The rule in (8.31) is also valid when more than two random variables are involved. If Y is the total number of eyes when throwing ten fair dice and X_i is the number of eyes of die i, then:

$$Y = X_1 + X_2 + \ldots + X_{10} \quad \text{and} \quad E(Y) = E(X_1) + E(X_2) + \ldots + E(X_{10}) = 35$$

Example 8.18

Let us now compare the two campaigns in Example 8.17 from the manager's point of view. Let Y_1 (and Y_2) be the sales (in euro) of the manager of the wine and beer supermarket due to one customer under the campaign 1 (campaign 2) regime. Since a crate of beer costs €10, the sales due to that customer would have been $10X$ without a promotion campaign. Hence, $Y_1 = 10X - V$ and $Y_2 = 10X - W$. It follows from (8.31) that:

$$E(Y_1) = E(10X) + E(-V) = 10E(X) - E(V) = 7.8 - 0.76 = €7.04$$

$$E(Y_2) = E(10X) + E(-W) = 10E(X) - E(W) = 7.8 - 1.28 = €6.52$$

On average, the sales per customer are higher for campaign 1. This is to be expected, since the interests of the manager and the customers are complementary.

Let X be a general rv with expectation μ and variance σ^2. The above rules for expectation can be used to derive a short-cut formula (calculation formula) for the variance of X. First notice that the squared random deviation $(X - \mu)^2$ is equal to:

$$(X - \mu)^2 = X^2 - 2\mu X + \mu^2$$

Taking the expectation of both sides, applying the generalization of (8.31) for three terms and applying the rules in (8.30) yields:

$$\sigma^2 = E(X^2 + (-2\mu X) + \mu^2) = E(X^2) + E(-2\mu X) + E(\mu^2)$$

$$= E(X^2) - 2\mu E(X) + \mu^2 = E(X^2) - 2\mu^2 + \mu^2$$

It follows that the variance of X can also be obtained by first calculating the expectation of X^2 and next subtracting μ^2:

Short-cut formula for variance of X

$$\sigma^2 = E(X^2) - \mu^2 \qquad\qquad (8.32)$$

The variance of X is just the expectation of the square minus the square of the expectation. (Recall that a similar short-cut formula was found in descriptive statistics; see (4.4) and (4.5).) This formula also illustrates that $E(X^2)$ is usually **un**equal to $(E(X))^2$. Equality holds only if $\sigma^2 = 0$, so if X is degenerated.

Example 8.19

Reconsider the rv $X =$ 'total number of eyes' when throwing the red and the blue die of the swindler; see Example 8.5. The variance of X will be calculated with the short-cut formula. The pdf of X is given in the table:

x	2	3	4	5	6	7	8	9	10	11	12
$g(x)$	1/144	4/144	8/144	12/144	16/144	22/144	24/144	20/144	16/144	12/144	9/144

For one throw with the die of the swindler, we expect 47/12 eyes. Hence, by (8.31):

$$E(X) = \frac{47}{12} + \frac{47}{12} = 94/12 \approx 7.83$$

To calculate $E(X^2)$, the squared outcomes x^2 have to be multiplied by the accompanying probabilities $g(x)$ and the results have to be added up:

$$E(X^2) = 2^2 \times \frac{1}{144} + 3^2 \times \frac{4}{144} + \ldots + 12^2 \times \frac{9}{144} = 66.84722$$

The short-cut formula yields that:

$$\sigma^2 = E(X^2) - \mu^2 = 66.84722 - (7.83333)^2 = 5.4861$$

$$\sigma = 2.3422$$

If the dice had been fair, the results would have been: $E(X) = 7$, $V(X) = 5.8333$ and $SD(X) = 2.4152$.

There exists a special linear transformation that transforms X into a random variable with expectation 0 and variance 1.

Standardized version or z-score of random variable X

$$Z = \frac{X - \mu_X}{\sigma_X} = \frac{X - E(X)}{SD(X)} \tag{8.33}$$

Properties of Z

$$\mu_Z = E(Z) = 0 \quad \text{and} \quad \sigma_Z^2 = V(Z) = 1 \tag{8.34}$$

To prove (8.34), note that:

$$Z = a + bX \quad \text{with} \quad b = \frac{1}{\sigma_X} \quad \text{and} \quad a = -\frac{\mu_X}{\sigma_X}$$

Hence:

$$E(Z) = a + b\mu_X = \left(-\frac{\mu_X}{\sigma_X}\right) + \frac{1}{\sigma_X}\mu_X = 0$$

$$V(Z) = b^2\sigma_X^2 = \frac{1}{\sigma_X^2}\sigma_X^2 = 1$$

Chebyshev's inequality is an important result for variables in descriptive statistics; see Section 4.3. In the field of probability, such an inequality also exists. It is of special importance since it sheds light on the concept of standard deviation. For an rv X with expectation μ and standard deviation σ it states that the probability that X will fall between $\mu - k\sigma$ and $\mu + k\sigma$ is at least $1 - 1/k^2$.

Chebyshev's inequality for a random variable X

$$P(\mu - k\sigma < X < \mu + k\sigma) \geq 1 - \frac{1}{k^2} \quad \text{for all real } k \geq 1 \tag{8.35}$$

Application of Chebyshev's inequality for $k = 2$ yields that the probability that X will fall in the interval $(\mu - 2\sigma, \mu + 2\sigma)$ is at least 0.75.

Notice that:

$$\mu - k\sigma < X < \mu + k\sigma \Leftrightarrow -k\sigma < X - \mu < k\sigma \Leftrightarrow -k < \frac{X - \mu}{\sigma} < k$$

(The first equivalence follows by subtracting μ in all three terms; the second equivalence follows by dividing all three terms by σ.) Hence, the inequality can also be interpreted in terms of the standardized rv Z. The probability that Z falls in the interval $(-k, +k)$ is at least $1 - 1/k^2$. For instance, the probability that Z falls in the interval $(-3, +3)$ is at least $8/9 = 0.8889$.

Example 8.20

The table below gives Chebyshev's lower bound and the precise values of the probabilities that the rv X = 'total number of eyes in two throws' falls in $(\mu - 2\sigma, \mu + 2\sigma)$, both for the fair case and the false case.

	μ	σ	Interval	k = 2 Chebyshev	Precise
Fair	7	2.4152	(2.17, 11.83)	$\geqslant 0.75$	0.9444
False	7.8333	2.3422	(3.15, 12.52)	$\geqslant 0.75$	0.9722

Notice that the interval for the fair case excludes the outcomes 2 and 12, while the false case excludes the outcomes 2 and 3. Hence:

Fair case: $P(2.17 < X < 11.83) = 1 - P(X = 2 \text{ or } 12) = 34/36 = 0.9444$

False case: $P(3.15 < X < 12.52) = 1 - P(X = 2) - P(X = 3) = 139/144 = 0.9653$

The precise probabilities are considerably larger than the lower bound offered by Chebyshev.

Example 8.21
The manager of the electronics store in Example 8.11 took the pdf f with $f(x) = 1/30$ for $10 \leqslant x \leqslant 40$ (and $f(x) = 0$ for x otherwise) to model the gross weekly sales X (measured in units of €10000). He also came to the conclusion that the rv Y representing the weekly profit can be written as $0.25X - 5$. We will calculate $E(X)$, $SD(X)$, $E(Y)$ and $SD(Y)$ and determine the standardized versions of both X and Y. Moreover, Chebyshev's lower bound for $k = 1.5$ will be compared with the actual probability that X will lie less than 1.5 standard deviations from $E(X)$.

By the symmetry of f it follows immediately that $\mu_X = 25$, the centre of the interval [10, 40]. For the calculation of the variance of X, the short-cut formula can most easily be used. To calculate $E(X^2)$, the area of the function $x^2 f(x) = x^2/30$ above the interval [10, 40] is wanted, which is an integral. The integration yields:

$E(X^2) = 700$

Hence,

$$\sigma_X^2 = E(X^2) - 25^2 = 700 - 625 = 75 \quad \text{and} \quad \sigma_X = 8.6603$$

Since Y is a linear transformation of X, it follows that:

$$\mu_Y = 0.25\mu_X - 5 = 1.25, \quad \sigma_Y^2 = (0.25)^2\sigma_X^2 = 4.6875, \quad \sigma_Y = 2.1651$$

The expected weekly profit is €12500, with a standard deviation of €21651.
To obtain the standardized versions Z_1 and Z_2 of X and Y, we have to subtract the respective expectation and divide the results by the respective standard deviation:

$$Z_1 = \frac{X - 25}{8.6603} = 0.1155X - 2.8867 \quad \text{and} \quad Z_2 = \frac{Y - 1.25}{2.1651} = 0.4619Y - 0.5773$$

The bounds $\mu \pm 1.5\sigma$ can also be calculated, for X as well as for Y:

$$\mu_X - 1.5\sigma_X = 12.0096 \quad \text{and} \quad \mu_X + 1.5\sigma_X = 37.9905$$

$$\mu_Y - 1.5\sigma_Y = -1.9977 \quad \text{and} \quad \mu_Y + 1.5\sigma_Y = 4.4977$$

By Chebyshev it follows that:

$$P(12.0096 < X < 37.9905) \geq 0.5556 \quad \text{and} \quad P(-1.9977 < Y < 4.4977) \geq 0.5556$$

But since the pdf of X is given, these probabilities can be calculated precisely; of course they are equal. The rectangle method can most easily be used to obtain the area under f and above (12.0096, 37.9905):

$$P(-1.9977 < Y < 4.4977) = P(12.0096 < X < 37.9905) = \frac{37.9905 - 12.0096}{30} = 0.8660$$

Note that this answer is considerably larger than Chebyshev's lower bound.

8.6 Random observations

This section is a kind of interlude. It explains the relationship between concepts from probability theory – such as random variable and its probability distribution, its expectation and its variance – and similar concepts in descriptive statistics.

Recall that the experiment 'random drawing' bridges the gap between descriptive statistics and probability theory; see Section 7.3. The sample space of this experiment is just the population from which an element is drawn randomly; the proportion p_a of the population elements with property a is just the probability of the event $A =$ 'the randomly drawn element has property a'. That is:

$$P(A) = p_a$$

Start with a population and a discrete population variable called X with population mean μ and population variance σ^2. Suppose that the population is finite, with N population elements. As in Section 4.6, the notations $x_{(j)}$ and f_j are used to denote the different values of X and the accompanying frequencies. By (4.8) we have:

$$\mu = \frac{1}{N} \sum f_j x_{(j)} = \sum \frac{f_j}{N} x_{(j)} \tag{8.36}$$

$$\sigma^2 = \frac{1}{N} \sum f_j (x_{(j)} - \mu)^2 = \sum \frac{f_j}{N} (x_{(j)} - \mu)^2 \tag{8.37}$$

So far, there is no random experiment and things are considered in the field of descriptive statistics.

Here is the experiment. From this population an element is drawn randomly, which brings us into the field of probability. The sample space of this random experiment is just the population at the start. Next, consider the random variable X' that measures the observation of X for this randomly drawn element. This random variable X' is called **random observation** of the (population) variable X. Note that the $x_{(j)}$ are the outcomes of the rv X'. In the population, the proportion f/N of the elements has the property that the value of X is $x_{(j)}$ (property a). In the sample space, the probability of the event $A = \{X' = x_{(j)}\}$ is $P(X' = x_{(j)})$. Since $P(A) = p_a$, it follows:

$$\frac{f_j}{N} = P(X' = x_{(j)}) \tag{8.38}$$

But this holds for all different values $x_{(j)}$ of X. Hence, the pdf of the random observation of X is equal to the relative frequency distribution of X.

Substitution of (8.38) in (8.36) and (8.37) yields:

$$\mu = \sum x_{(j)} P(X' = x_{(j)})$$

$$\sigma^2 = \sum (x_{(j)} - \mu)^2 P(X' = x_{(j)})$$

It seems that μ can be obtained by multiplying each possible outcome of X' by the accompanying value of the pdf and adding up all results. But this is precisely what has to be done to calculate the expectation $E(X')$; see also (8.14). Apparently, the population mean of the variable X is equal to the expectation of the random variable X'. Similarly, σ^2 is just the variance of the random variable X', as follows from the definition in (8.18). These results are summarized in the box:

Properties of the random observation of a discrete variable X

The pdf of the random observation of X is equal to the relative frequency distribution of X. Hence:

population mean of X = expectation of random observation of X **(8.39)**
population variance of X = variance of random observation of X **(8.40)**

These results connect concepts from descriptive statistics to concepts from probability theory. At least for the discrete case, the concepts 'expectation', 'variance' and 'standard deviation' of a random variable and its probability distribution are generalizations of the concepts 'population mean', 'variance' and 'standard deviation' of a population variable and its frequency distribution. The above arguments justify the general use of the notations μ and σ^2 for **random** variables.

Example 8.22

Consider the (population) variable X = 'number of cars per household' on a population of households. The table gives the relative frequency distribution of this variable.

Value	0	1	2	3
Rel. freq.	0.230	0.559	0.190	0.021

The following table summarizes, in the first two rows, the ingredients needed to calculate the population mean μ and population variance σ^2 by way of the relative frequencies f/N:

$x_{(i)}$	0	1	2	3
f_i/N	0.230	0.559	0.190	0.021
$(x_{(i)} - \mu)^2$	1.0040	0.000004	0.9960	3.9920

By multiplying corresponding numbers, the population mean μ of X is obtained:

$$\mu = 0 \times 0.230 + \ldots + 3 \times 0.021 = 1.0020$$

Multiplying $(x_{(i)} - \mu)^2$ and f/N and adding up the results yields the population variance σ^2 and the population standard deviation of X:

$$\sigma^2 = (0 - 1.0020)^2 \times 0.230 + \ldots + (3 - 1.0020)^2 \times 0.021$$

$$= 1.0040 + \ldots + 3.9920 = 0.5040$$

$$\sigma = 0.7099$$

Now assume that one household is drawn randomly and that the rv X' measures the number of cars owned by this arbitrarily chosen household. Then this random observation X' is a discrete rv and, by the above arguments, its probability density function $f(x) = P(X' = x)$ is just the relative frequency distribution of the population variable X (given in the table above). That is to say, the pdf of the random observation of X is:

x	0	1	2	3
$f(x)$	0.230	0.559	0.190	0.021

Furthermore, the expectation and the variance of the random observation X' of X are respectively equal to the population mean and the population variance of X. That is:

$$E(X') = 1.0020 \quad \text{and} \quad V(X') = 0.5040 \quad \text{and} \quad SD(X') = 0.7099$$

Now suppose that a **continuous** (population) variable X is of interest. As before, the rv X' denotes the random observation of X. Then, the probability that X' falls in an interval I (say, event A) is equal to the relative frequency of the population elements that have their observations of X in the interval I (say, property a).

Recall from Section 2.3.3 that in the continuous case **classified** frequency distributions are used and that different classifications usually yield different frequency distributions. We need a new concept:

Population distribution of a continuous variable X

The *population distribution* of a continuous variable X is defined as the overview of **all** relative frequencies for X falling in intervals I.

But this population distribution is usually unknown and hence a model is needed. Under these circumstances, a continuous probability density function f is often proposed as a model for the random observation X' of the variable X. That is, for all intervals I, the area above I and under f is used to describe the probability that X' falls in I and hence to describe the proportion of the population elements with value of X in I (which is a relative frequency in the overview offered by the population distribution). The following example illustrates this.

Example 8.23
The first two columns of Table 8.2 present a classified relative frequency distribution of the continuous variable X = 'last year's income (in thousands of euro)' for the population of all individuals with annual incomes above €30000.

Class [a, b) (× €1000)	Rel. frequency (observed)	Rel. freq. density (observed)	Rel. frequency (model)
[30, 33)	0.2726	0.0909	0.3038
[33, 36)	0.1928	0.0643	0.1960
[36, 40)	0.1808	0.0452	0.1650
[40, 45)	0.1224	0.0245	0.1209
[45, 50)	0.0704	0.0141	0.0707
[50, 60)	0.0780	0.0078	0.0717
[60, 70)	0.0365	0.0037	0.0318
[70, 80)	0.0170	0.0017	0.0159
[80, 90)	0.0128	0.0013	0.0087
[90, 100)	0.0045	0.0005	0.0051
[100, 150)	0.0102	0.0002	0.0081
[150, 200)	0.0013	0.0000026	0.0015
[200, 500)	0.0007	0.0000023	0.0008

TABLE 8.2 Distribution of incomes above €30000 (individuals)

Notice that the classification has unequal widths, so for a histogram the relative frequency **density** (that arises by dividing the relative frequencies by the widths of the classes) has to be used; see the third column. Figure 8.14 shows (among other things) the histogram of the relative frequency density for the classes up to and including [100, 150).

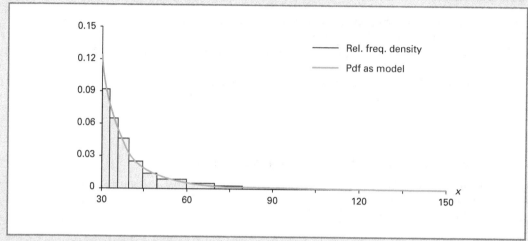

FIGURE 8.14 Histogram of relative frequency density with a model included

Below, a model is presented for the underlying population distribution of X. The pdf of the model is sketched in the figure.

Although the classified frequency distribution as presented in the table gives valuable information about the population distribution, it does not tell the whole story. For intervals that are not in the table, the relative frequencies remain unknown. The so-called **economic theory of Pareto** is used to find a model. From this theory, the following (continuous) probability density function arises:

$$f(x) = \frac{c}{x^{4.8}} \quad \text{for all } x \geq 30$$

$$f(x) = 0 \quad \text{otherwise}$$

The constant c is chosen such that the total area under the graph of f is equal to 1 (so that f is indeed a pdf) and turns out to be equal to 1 558 993.715 701 6 (to be precise). The graph of this pdf is included in Figure 8.14; it seems to fit quite well to the relative frequency density of the histogram. Calculation of mean and standard deviation of the classified frequency distribution yields respectively 41.8554 and 17.1146; see Sections 3.3.3 and 4.6 for details about their calculations. However, calculation of the total area under the function $xf(x)$ yields the expectation $\mu = 40.7143$, while the square root of the total area under the function $(x - \mu)^2 f(x)$ yields the standard deviation $\sigma = 15.5675$. (The calculation of μ and σ is omitted.)

It seems reasonable to use the pdf f as a model for the random observation X' of X, so probabilities regarding X' and hence relative frequencies regarding X become areas under the graph. The fourth column of Table 8.2 lists the areas of f above the respective intervals (and hence are probabilities); notice that the fit with the observed relative frequencies in the second column is rather good. But other relative frequencies can be calculated with the model too. For instance, the unknown relative frequency of all annual incomes between €100 000 and €125 000 can be approximated by calculating the area under f for the interval [100, 125), which yields the result 0.0059. (The calculation is left as an exercise for those who are familiar with integration.) That is, 0.59% of the incomes above €30 000 fall in the interval [100 000, 125 000), at least according to the model.

Although the population variable X and the random variable X' clearly are different concepts (for X' a random experiment is needed, for X not), they certainly are hard to distinguish. They attach the same values to all population (= sample space) elements, so they are the same from a functional point of view. That is why the distinction in notation will not be maintained in what follows. Hence, we will talk about the random variable X (if X' is meant) and the variable X (if the population variable X is meant).

8.7 Other statistics of probability distributions

In Section 8.4 the important statistics expectation, variance and standard deviation were introduced. As shown in Section 8.6, they can be considered as generalizations of similar statistics in descriptive statistics. In the present section, statistics such as median, quartiles, percentiles – also introduced in Part 1 'Descriptive Statistics' for (population) variables and frequency distributions – are generalized. For these generalizations, the distribution function turns out to play an important role.

Recall that the first, second and third quartiles of a population dataset roughly divide the dataset into four parts. These concepts are generalized first. Start with a random variable (rv) X with probability density function (pdf) f and distribution function (cdf) F. For a real number p in the interval $(0, 1)$, the p-quantile ξ_p of X and its probability distribution divides the possible outcomes of X into two groups in such a way that:

- the probability that X will be smaller than ξ_p, is at most p, and
- the probability that X will be smaller than or equal to ξ_p, is at least p

The *p-quantile* ξ_p of X and its distribution is such that:

$$P(X < \xi_p) \leq p \quad \text{and} \quad P(X \leq \xi_p) \geq p \tag{8.41}$$

Since F is a non-decreasing function, the p-quantile ξ_p is such that at ξ_p the distribution function F **reaches or passes** the level p. For continuous rvs, F is strictly increasing and continuous on the relevant interval, so F actually reaches the level p and ξ_p is unique; see Figure 8.15(a).

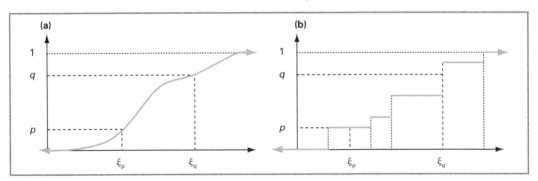

FIGURE 8.15 Distribution functions and quantiles

For discrete rvs, F is a step function; see Figure 8.15(b). Notice that the indicated level q on the vertical axis is passed (not reached) and that the indicated ξ_q is indeed the unique q-quantile since any real number smaller (larger) than ξ_q does not satisfy the right-hand (left-hand) condition in (8.41). However, there is a whole interval of values on the horizontal axis where F reaches the indicated level p. In this book, the convention is that the centre (ξ_p) of that interval is taken as p-quantile.

To find the p-quantile of the probability distribution of a continuous rv X, it suffices to solve the equation $F(x) = p$. That is, the p-quantile ξ_p is such that:

$$P(X \le \xi_p) = p$$

But probabilities regarding X are areas under the continuous pdf f. Hence, the area at the left-hand sight of ξ_p is just p. Since the total area is 1, the area at the right-hand side of ξ_p is $1 - p$. Figure 8.16(a) offers an alternative view of the concept 'p-quantile' in the continuous case.

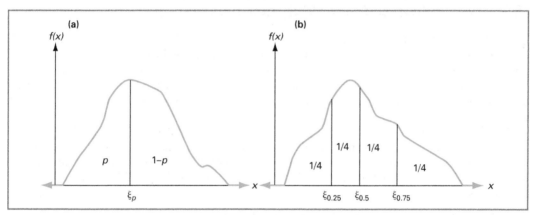

FIGURE 8.16 Continuous pdf, quantiles and quartiles

p-Quantiles for which p is a multiple of 0.10 are also called **deciles**; p-quantiles with p a multiple of 0.01 are called **percentiles**. For instance, the quantile $\xi_{0.40}$ is the fourth decile and the 40th percentile. Instead of using a proportion p to denote a quantile, one also uses the percentage $100p$. For instance, the 0.42-quantile is also called 42% quantile or even 42-quantile.

The quantiles $\xi_{0.25}, \xi_{0.5}, \xi_{0.75}$ are called **first**, **second** and **third quartile**, and are also denoted as κ_1, μ_{median} (or κ_2) and κ_3; $\xi_{0.5}$ is also called the **median**. Figure 8.16(b) illustrates their meaning for the continuous case. Furthermore, the distance between $\xi_{0.25}$ and $\xi_{0.75}$ is called the **interquartile range** of the random variable X and its probability distribution, and is denoted by δ. Notice that these definitions and notations are obvious generalizations of similar ones in descriptive statistics. By dividing δ by the absolute value of the median, the **relative interquartile range** arises. That is:

$$IQR = \xi_{0.75} - \xi_{0.25} \quad \text{and} \quad rel\ IQR = \frac{\delta}{|\mu_{median}|} \tag{8.42}$$

For a continuous rv X, the IQR covers an interval with area exactly 0.5 under the pdf; the probability that X falls in that interval is **exactly** 0.5. For a discrete random variable this probability is roughly equal to 0.5.

Let X be the number of eyes when throwing a fair die. The cdf F makes a jump of size 1/6 at each integer 1, 2, ..., 6. At $x = 2$, the cdf jumps from 1/6 to 1/3 and passes the level 0.25. Hence, $\xi_{0.25} = 2$ and even $\xi_p = 2$ for **all** p in the interval (1/6, 1/3). Since the levels 1/6 and 1/3 are reached (not passed) by F, it follows that $\xi_{1/6} = 1.5$ and $\xi_{1/3} = 2.5$. Since $p = 0.75$ falls between the levels 4/6 and 5/6, the cdf passes the level 0.75 at $x = 5$. Hence, $\xi_{0.75} = 5$ and the interquartile range δ is equal to $5 - 2 = 3$. Notice that the probabilities that X will fall in the interval [2, 5], respectively (2, 5] – both covered by the interquartile range – are 4/6 = 0.6667 and 0.5. Also notice that the 84th percentile is 6, while the 34th percentile is 3. The level 0.50 is actually **reached** by F, at $x = 3$. Hence, the median falls in the centre of the interval (3, 4), so $\mu_{median} = 3.5$. Recall that the expectation μ is also 3.5, which stresses that the probability distribution of X is symmetric.

Example 8.24

For one discrete and two continuous probability distributions that were considered before, several statistics regarding quantiles are calculated.

Let X be the number of cars owned by a randomly drawn household; see also Example 8.22. The pdf and cdf of X are repeated below:

x	0	1	2	3
$f(x)$	0.230	0.559	0.190	0.021
$F(x)$	0.230	0.789	0.979	1

It follows immediately that $\xi_{0.25} = \xi_{0.50} = \xi_{0.75} = 1$ since the graph of $F(x)$ passes all levels 0.25, 0.50 and 0.75 at $x = 1$. Hence, the interquartile range $\delta = 0$, which means that (at least) 50% of the households have one car. Recall that μ, σ^2 and σ are respectively 1.0020, 0.5040 and 0.7099. Hence, μ is slightly larger than μ_{median} which stresses that the probability distribution of X is slightly skewed to the right.

Let X be the time (in seconds) that it takes to solve the Sudoku puzzle. Reconsider the two continuous models in Example 8.8, both concentrated on the interval [5, 46]: model 1 with the constant pdf and linear cdf (now respectively called f and F), and model 2 with the triangular pdf and quadratic cdf (now called g and G). Since both models are symmetric, it follows that both medians fall exactly in the centre of the interval. Indeed, $F(25.5) = G(25.5) = 0.5$. To determine $\xi_{0.25}$ of model 1, the equation $F(x) = 0.25$ has to be solved:

$$F(x) = 0.25 \quad \Leftrightarrow \quad \frac{1}{41}(x - 5) = 0.25 \quad \Leftrightarrow \quad x - 5 = 10.25$$

So, $\xi_{0.25} = 15.25$ minutes. Similarly, $\xi_{0.75} = 35.75$ minutes. It follows that $IQR = 20.50$ minutes and that the (dimensionless) relative IQR is 0.8039. The interval [15.25, 35.75] has an area 0.5 under f; the probability that X will fall in this interval is 0.5.

To determine $\xi_{0.25}$ of the second model, the equation $G(y) = 0.25$ has to be solved:

$$G(y) = 0.25 \quad \Leftrightarrow \quad \frac{1}{2 \times (20.5)^2}(y - 5)^2 = 0.25 \quad \Leftrightarrow \quad (y - 5)^2 = 210.125$$

$$\Leftrightarrow \quad y - 5 = \pm 14.4957$$

The two solutions are equal to 19.4957 and −9.4957. However, the second solution does not belong to the interval [5, 46] and hence is not valid. The conclusion is that $\xi_{0.25} = 19.4957$ minutes. Since the pdf g is symmetric around 25.5 and hence $\xi_{0.75}$ has the same distance to 25.5 as does $\xi_{0.25}$, it follows easily that $\xi_{0.75} = 31.5043$ minutes.

Table 8.3 lists some statistics of the two continuous models; the results for σ^2 and σ are included for completeness (they have not yet been obtained). It can be seen from this overview that model 2 is more concentrated around the centre 25.5 of [5, 46] than model 1. Not only is the standard deviation of model 2 smaller but so too is the interquartile range.

	μ	σ^2	σ	κ_1	μ_{median}	κ_3	δ
Model 1	25.5	140.083	11.836	15.25	25.5	35.75	20.5
Model 2	25.5	70.042	8.369	19.496	25.5	31.504	12.009

TABLE 8.3 Some statistics of the two Sudoku models

Summary

In this chapter emphasis was on random variables and their probability distributions. By considering only events that concern a random variable, the probability measure on the sample space Ω is simplified to a probability distribution on the set of real numbers. This probability distribution can always be characterized by the cumulative distribution function.

The difference between discrete and continuous random variables is essential since the two cases have their own approach. Moreover, the differences between the following concepts are important:

- Probability measure (or model)
- Probability distribution
- Cumulative distribution function
- Discrete probability density function
- Continuous probability density function

Random observations of variables are important random variables. In a sense, they insert descriptive statistics into the field of probability. But they are also crucial for inferential statistics.

Several statistics of random variables and their probability distributions were defined. Most important are the expectation, the variance and the standard deviation. The definition of the expectation of a function of a random variable is also very useful.

List of symbols

Symbol/abbreviation	Description
X, Y	Random variables
rv	Random variable
x, y	Outcomes of X, Y
$E(X), E(Y)$	Expectations of X, Y
$V(X), V(Y)$	Variances of X, Y
$SD(X), SD(Y)$	Standard deviations of X, Y
F	cdf; (cumulative) distribution function
f	pdf; probability density function

🔑 Key terms

budget constraint **286**
Chebyshev's inequality **265**
continuous **235**
continuous probability
   density function **245**
cumulative distribution
   function **238**
decile **272**
degenerated at a
   constant **242**
degenerated random
   variable **241**
density **237**

discrete **235**
discrete probability density
   function **237**
distribution **236**
distribution function **238**
economic theory of
   Pareto **270**
expectation **252, 258**
expected value **252, 258**
first quartile **272**
integral from a to b
   under f **245**
interquartile range **272**

likelihood **246**
mean **252, 258**
median **272**
outcomes **234**
p-quantile **271**
percentile **272**
population
   distribution **269**
probability density
   function **237**
probability distribution **236**
qualitative random
   variable **234**

Exercises

Exercise 8.1

Give, for an rv *X*, the definitions and the interrelationships for the concepts 'probability distribution', 'cumulative distribution function' (cdf) and 'probability density function' (pdf) if:

  a *X* is discrete;
  b *X* is continuous.

Exercise 8.2

For an rv *X*, give the definitions of $\mu = E(X)$, $\sigma^2 = V(X)$ and $\sigma = SD(X)$ if:

  a *X* is discrete;
  b *X* is continuous.

Exercise 8.3

  a What is the difference between a random variable and a (population) variable?
  b The notations μ, σ^2 and σ were used in Chapters 1–5 to denote the population mean, variance and standard deviation. Explain why it is permissible to use these notations in Chapter 8 for the expectation, variance and standard deviation of a random variable.

Exercise 8.4

Outcome *x*	0	2	3.5	5	6
f(x)	0.20	0.05	0.30	0.25	0.20

A discrete rv *X* has the following probability density function:

  a Determine the distribution function *F* of *X*.
  b Use *f* to calculate $P(2 < X \leq 5)$.
  c Use *F* to calculate $P(2 < X \leq 5)$.
  d Calculate the expectation, the variance and the standard deviation of *X*.

Exercise 8.5

A continuous rv X has a pdf that is constant over the interval (1, 10) and 0 beyond it.

a Determine this pdf f.

b Determine the distribution function F of X.

c Use F to calculate $P(2 < X \leq 5)$.

d Calculate the expectation of X.

e (if familiar with integration) Calculate the variance and the standard deviation of X.

Exercise 8.6

a The distribution function F of an rv X is defined as follows:

$$
\begin{aligned}
F(x) &= 0 && \text{if } x \leq 0 \\
&= \frac{1}{9}x^2 && \text{if } 0 < x \leq 3 \\
&= 1 && \text{if } x > 3
\end{aligned}
$$

Is X discrete or continuous? Determine the pdf.

b The distribution function F of an rv X is defined as follows:

$$
\begin{aligned}
F(x) &= 0 && \text{if } x < 0 \\
&= 0.25 && \text{if } 0 \leq x < 1 \\
&= 0.60 && \text{if } 1 \leq x < 2 \\
&= 0.80 && \text{if } 2 \leq x < 3 \\
&= 1 && \text{if } x \geq 3
\end{aligned}
$$

Is X discrete or continuous? Determine the pdf.

Exercise 8.7

A retailer records the amounts spent by his customers. For Mondays, he obtains the following frequency distribution for the (rounded) amounts – in euros – spent by a customer.

Value	5	10	20	50	100	200	Total
Rel. frequency	0.08	0.15	0.20	0.30	0.20	0.07	1

a Define a random experiment and an rv X such that the probability density function of X is just the above relative frequency distribution.

b Calculate the expectation, the variance and the standard deviation of X.

Exercise 8.8

In each of the cases below, first describe the random experiment and give a suitable sample space; next show that the prescript indeed concerns a random variable. Is the random variable discrete or continuous?

a The number of heads when flipping three fair coins.

b The number of driving tests that someone will take before getting the driving licence.

c The age of a randomly drawn employee of a certain company.

d Next week's sales (in euros) of a clothing shop.

e Tomorrow's value of the Dow Jones Index.

Exercise 8.9

The management team of a company wants to choose between two investment projects. Let X and Y be the one-year profit per invested euro for the projects I and II, respectively. The two pdfs below are used to model the probability distributions of X and Y.

a	-0.20	-0.10	0	0.10	0.20
$f_X(a)$	0.02	0.20	0.56	0.20	0.02
$f_Y(a)$	0.01	0.04	0.10	0.50	0.35

a Determine the outcomes of X and of Y.

b Calculate the expectation, the variance and the standard deviation of both X and Y. For the calculation of the variances, use the definition of this concept.

c Calculate the variances again with the short-cut formula.

d First, use Chebyshev's rule to find lower bounds for the probabilities that X and Y will fall within three standard deviations from their means. Next, calculate these probabilities precisely.

e Find arguments for choosing project I. Also find arguments for II.

Exercise 8.10

Let X (in euro) be the payoff of the game in Exercise 7.34.

a Determine the probability density function of X.

b Extend the overview offered in part (a) with the values of the distribution function at the crucial values.

c Regarding the probabilities below, first write them down in an appropriate way. Next, calculate the probabilities that

i X will fall in (0, 4]

ii X will be at most 1

iii X will be at least 1.5

iv X will be larger than 2 but smaller than 5

v X is negative and/or X is at least 3.

d Make a graph of the cdf.

e Calculate the expectation of X. Interpret your answer as a long-run average.

Exercise 8.11

Consider a 'die' for which the 1 and the 2 are equally likely; the 3 and the 4 have both probability 1/6. However, the probabilities of throwing 6 and 5 are respectively four and three times greater than throwing 1. Let X be the number of eyes when throwing this die.

a Determine the probability density function f of X.

b Also determine the (cumulative) distribution function F of X.

c Determine the graph of the cdf F of X.

d Determine $P(4 < X \leq 6)$, first with the pdf and next with the cdf.

e Determine $P(\{X \leq 3\} \cap \{1 < X \leq 4\})$, first with the pdf and next with the cdf.

f Determine $P(\{X \leq 3\} \cup \{1 < X \leq 4\})$, first with the pdf and next with the cdf.

Exercise 8.12

a A certain rv X has the following probability density function f:

$$f(x) = \frac{a}{20} x^2 \quad \text{if } x = 0, \pm 1, \pm 2, \pm 3$$

(and $f(x) = 0$ otherwise).

   i Is X discrete or continuous? Determine the constant a.

   ii Calculate $E(X)$.

b A certain rv Y has the following probability density function g:

$$g(y) = by \quad \text{if } 0 \le y \le 4$$

(and $g(y) = 0$ otherwise).

   i Is Y discrete or continuous? Determine the constant b.

   ii (if familiar with integration) Calculate $E(Y)$.

c (if familiar with integration) A certain rv V has the following probability density function h:

$$h(v) = \frac{3}{8} v^2 \quad \text{if } 0 \le v \le 2$$

(and $h(v) = 0$ otherwise).

   i Calculate μ_V.

   ii Calculate σ_V^2 with the original definition of variance.

   iii Calculate σ_V^2 again with the short-cut formula.

Exercise 8.13

In a recent study of ten European countries, VSNU reports ten classified relative frequency distributions of the variable X = 'time taken (in months) by a masters student to find his/her first job after graduation'; see also Exercise 4.12. Unfortunately, there is no distribution for Denmark. Since it is expected that the situation in Denmark is similar to that in Norway, we take the classified frequency distribution of Norway and posit the following discrete model for X:

Outcome	0.5	2.5	5	10
Probability	0.58	0.30	0.07	0.05

a Calculate the mean and variance of this probability distribution. For the variance, use the short-cut formula.

b Determine the 0.90-quantile of this distribution.

c Also determine its interquartile range.

Exercise 8.14

For two random variables X and Y it is given that $E(X) = 4$, $E(X^2) = 17$ and $E(Y) = 8.5$. Calculate:

a $E(2X - 7Y + 8)$

b $E(Z)$ for $Z = (X - E(X) + Y - E(Y))/2$

c $E((X - 7)^2 + 3Y - 6.5)$

d σ_X^2 and the variance of $U = -4X - 3$

e the standard deviation of $-4X + 3$.

Exercise 8.15

An electronics firm sells TVs of the type Ph81-E for €500. The manager of the firm is wondering whether this price is too low when compared with the relatively high costs. For an analysis, she adopts the following pdf for the daily demand X for this type of TV:

x	0	1	2	3	4	5
$f(x)$	0.1	0.2	0.2	0.3	0.1	0.1

The daily sales Y (in euro) and the daily costs C (in euro) of this type of TV are related to X, in the following way:

$$Y = 500X \quad \text{and} \quad C = 460 + 50X$$

a Calculate the expectation, the variance and the standard deviation of X and C.
b Calculate the probability that the daily costs C are larger than the daily sales Y.
c To which level should the daily fixed costs (i.e. €460) be reduced to ensure that the probability in part (b) decreases to 0.1?
d Calculate the 0.9- and the 0.05-quantiles of X by using the cdf of X.

Exercise 8.16

Reconsider Exercise 8.15.

a Determine the pdf and the cdf of C. Use this cdf to calculate the 0.9- and the 0.05-quantiles of C.
b Calculate the first, second and third quartiles of both X and C. Also calculate the interquartile ranges of X and C.
c Explain how quantiles of C can directly be calculated from similar quantiles of X. Also explain how the interquartile range of C follows from the interquartile range of X.

Exercise 8.17

An American plans to visit Madrid on 21 June, at noon. He obtains from a European tourist agency a good probability distribution for the temperature X at noon (in °C) on 21 June, shown in the following pdf:

x	22	23	24	25	26	27	28
$f(x)$	0.05	0.1	0.1	0.3	0.2	0.2	0.05

a Calculate $P(23 \le X < 27.5)$.
b Expand the above table with a row for $F(x)$, the cdf of X. Use F to calculate $P(23 \le X < 27.5)$ again.
c Calculate $E(X)$, $V(X)$ and $SD(X)$.

The American is used to measuring temperatures in °F. Let Y denote the temperature on 21 June in Madrid (at noon) measured in °F. Hence, $Y = \frac{9}{5}X + 32$.

d Determine the pdf f_Y and the cdf F_Y of Y.
e Calculate μ_Y, σ_Y^2 and σ_Y from the pdf of Y.

f Calculate μ_Y, σ_Y^2 and σ_Y again by making use of part (c) and the linear relationship between Y and X.

Exercise 8.18

A European plans to visit Singapore on 21 June for the whole day. According to a tourist agency in Singapore, a good probability distribution for the maximum temperature X (in °F) on 21 June in Singapore is given by the following pdf:

$$f(x) = \frac{1}{16} \quad \text{if } 82 < x \le 86$$
$$= \frac{3}{16} \quad \text{if } 86 < x < 90$$
$$= 0 \quad \text{otherwise}$$

a Create the graph of f. Is X discrete or continuous? Check that f indeed is a pdf.
b Calculate the probability $P(83 < X < 88)$.
c Determine the cdf F of X. Use it to calculate the probability in part (b) again.
d Recall that $E(X)$ is just the total area under the function $xf(x)$. Determine this function, graph it and show that $E(X) = 87$ °F.

The European likes to have temperatures in °C. The rv Y that satisfies $X = \frac{9}{5}Y + 32$ measures the maximum temperature on 21 June in Singapore in °C.

e Express Y in X and calculate the expectation, the variance and the standard deviation of Y. Use that $V(X) = 4.3333$.

Exercise 8.19

A survey of the webshop Ebook.com yields the following pdf for the number of books purchased per hit:

x	0	1	2	3	4	5	6
$f(x)$	0.35	0.25	0.15	0.09	0.07	0.05	0.04

a What is the probability that a visitor to Ebook.com will buy at most three books?
b What is the probability that a visitor to Ebook.com will buy at least three books?
c What is the probability that a visitor to Ebook.com will buy more than three books?
d What is the probability that a visitor to Ebook.com will buy at most five books if it is already given that he/she buys at least two books?
e What is the probability that a visitor to Ebook.com will buy at least two books if it is already given that he/she buys at most five books?

Exercise 8.20

Consider the following distribution function:

$$F(x) = 0 \quad \text{if } x < -1$$
$$= 1/6 \quad \text{if } -1 \le x < 0$$
$$= 1/3 \quad \text{if } 0 \le x < 2$$
$$= 2/3 \quad \text{if } 2 \le x < 3$$
$$= 5/6 \quad \text{if } 3 \le x < 10$$
$$= 1 \quad \text{if } x \ge 10$$

a Create a picture of the graph of F.

b Suppose that the rv X has F as cdf. Is X discrete or continuous? Determine the (possible) outcomes of X.

c Calculate $P(-1 \leq X \leq 2)$ and $P(X > 0)$.

d Determine the pdf f of the rv X.

e Explain how f can be obtained from F.

f Explain how F can be obtained from f.

Exercise 8.21

In a certain random experiment there are only three possible outcomes, all **equally likely**. Furthermore, it is given that the outcomes are 3-tuples: $\Omega = \{(1, 0, 3), (2, 1, 1), (2, 1, 4)\}$. Consider the rv X that adds up the three positions of these outcomes.

a Explain why X is indeed a random variable.

b Determine the pdf f and the cdf F of X. Carefully explain why $f(4) = 2/3$.

c Calculate the expectation, the variance and the standard deviation of X.

Exercise 8.22

A machine produces small axles whose diameter should be 2.4 cm. However, it turns out that the actual diameter varies between 2.35 and 2.45 cm. For 10 000 recently produced axles a histogram of the relative frequencies of their precise diameter is given below.

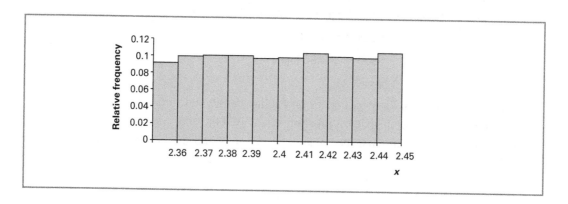

Let X denote the diameter (in cm, with decimals) of an axle produced by that machine. From the histogram it seems reasonable to conclude that all values in the interval $(2.35, 2.45)$ are equally likely to be the realization of X.

a Is X discrete or continuous?

b Use the above information to construct a probability density function f for X. Also make its graph.

c Use this pdf as a model and calculate $P(2.38 < X < 2.41)$.

d Determine the percentage of all axles larger than 2.37.

e Determine the distribution function F of X. Also draw its graph.

f Determine the answers to parts (c) and (d) again by using F.

Exercise 8.23

A skydiver tries to land on a circular island with a radius of 40 metres. From the many attempts done before she knows that the function f below is a good model for the pdf of the rv X that measures the distance (in metres) between the landing position and the centre of the island:

$$f(x) = \frac{100 - x}{5000} \quad \text{if } 0 \leq x \leq 100$$

$$= 0 \qquad \text{otherwise}$$

a Create the graph of f.

b Is X discrete or continuous? Determine all (possible) outcomes of X.

c Show that f is indeed a probability density function.

d Calculate the probability that she will land within 10 m of the centre of the island.

e Calculate the probability that the skydiver will land in the ocean.

Exercise 8.24

The weight of the contents of packs of coffee of a certain brand is 500 g, as stated on the packaging. A consumer organization weighs the contents of 200 packs and notices that on average they indeed contain 500 g coffee, but their weights vary between 495 and 505 g. By far the majority of the weights are close to 500 g, while only relatively few weights have deviations close to -5 or $+5$.

a Let X (in grams) be the weight of the contents of a pack of coffee and let a be a positive constant. Create a rough sketch of the graph of the function f below.

$$f(x) = a(x - 495) \quad \text{if } 495 < x \leq 500$$

$$= a(505 - x) \quad \text{if } 500 < x < 505$$

$$= 0 \qquad \text{otherwise}$$

Determine a such that f is a continuous probability density function.

b Determine the probability that X is precisely equal to 500.

c Calculate the probability that the deviation between X and the indicated 500 g is at most ± 2 g.

d Determine the expectation of X.

Exercise 8.25

a Use the short-cut formula to calculate the variance of the payoff X for the game in Exercise 8.10. Interpret your answer in terms of an ongoing sequence of realizations of X.

b Reconsider the number of cars X of a randomly drawn household. Recall that the pdf f of this random variable is as follows:

x	0	1	2	3
$f(x)$	0.230	0.559	0.190	0.021

Calculate the expectation, variance and standard deviation of X.

Exercise 8.26

For the incomes of the 100 000 employees of a large company, the following holds for the population variable 'gross annual income (in thousands of euro)':

$$\mu = 40 \quad \text{and} \quad \sigma = 2.5$$

a Determine an upper bound for the probability that the gross annual income of an arbitrarily chosen employee is at least €45 000. How many of the employees will have an income of at least €45 000?

b Suppose, additionally, that a histogram of the 100 000 incomes shows a nice symmetric picture. Answer the questions in part (a) again.

Exercise 8.27 (only if familiar with integration)

At a service station, the contents of the tank for unleaded petrol is 40 000 litres. To model the daily amount of petrol X (measured in units of 10 000 litres) sold during one day, the following pdf is used:

$$f(x) = \frac{3}{32} x(4 - x) \quad \text{for } 0 \le x \le 4$$

Furthermore, $f(x) = 0$ for $x < 0$ and for $x > 4$.

a Calculate μ and σ^2.

b Use Chebyshev's inequality to find a lower bound for $P(\mu - 2\sigma < X < \mu + 2\sigma)$.

c Calculate the probability $P(\mu - 2\sigma < X < \mu + 2\sigma)$ precisely.

d Let Z be the standardized version of X. Express Z in X and calculate the probability that Z is less than 3.

Exercise 8.28 (computer)

Let X be the number of eyes when rolling a fair die.

a (without computer) Determine the probability density function of X. Calculate μ, σ^2 and σ.

This experiment will be simulated several times and the probabilities $P(X = k)$ of part (a) will be approximated (for $k = 1, 2, \ldots, 6$).

b Simulate 5000 observations of X; if necessary, see Appendix A1.8.

c For this sample of 5000 observations, determine the relative frequencies of the outcome k (for $k = 1, 2, \ldots, 6$). Also calculate the mean, variance and standard deviation of the sample.

d Compare the relative frequency distribution and the sample statistics of part (c) with the pdf and the statistics of part (a). Give your comments.

Exercise 8.29

You have inherited €1000 and are wondering what to do with it. There are three options:

Option 1: Accept the money and use it.
Option 2: Put the money in an investment that has a 50% chance of doubling it but also a 50% chance of losing the whole amount.
Option 3: Put the money in an investment that has a 50% chance of making it €2500 but also a 50% chance of losing the whole amount.

You decide to use only rational arguments and firstly calculate the expected payoff and accompanying standard deviations. Which option will you choose? Why?

Exercise 8.30

You bought a new car, worth €20000. Suppose that you are free to decide whether you take out car insurance or not. To be precise, you can choose between two options:

Option 1: Take no insurance, thus hoping it will cost you nothing (probability 0.99) and accepting the fact that in the coming year you may, in the case of an accident, be confronted with a loss of €20000 that happens with probability 0.01.

Option 2: Take out insurance for which you have to pay €500 per year.

Let X denote the loss, the (random) amount of money it will cost you in the coming year. Which option do you choose? Why?

CASE 8.2 A THREE-QUESTION QUIZ ABOUT RISK (PART II)

The three questions in Case 6.3 about uncertainty and risk are reconsidered, but now only rational arguments will be used that are based on expected payoff and accompanying standard deviation. Let X denote the payoff, the amount of money that (finally) will be yours.

a For each of the three questions in Case 6.3, calculate the expectations and the standard deviations of X for the two options.

b If your arguments to answer the three questions are based only on these expectations and standard deviations, what will your answers be then?

A larger standard deviation will probably decrease your lust to invest your money, whereas a larger expectation will increase it. Suppose now that your private lust to invest money in an investment with expected payoff μ and standard deviation σ is determined by the function $u(\mu, \sigma)$ = $\mu - 0.1\sigma$. (This function is your private **utility function**.)

c For each of the three questions in Case 6.3, which option would you choose?

d Which of the six options would you choose?

CASE 8.3 INTRODUCTION TO MARKOWITZ'S PORTFOLIO THEORY (PART I)

If you want to invest, a wide variety of investment opportunities are available. You can hold your money in a bank deposit with a certain interest rate payment, you can invest in risky projects (such as the development of a new factory), or you can invest in financial assets such as stocks, bonds and options. The choice among these possibilities is not trivial and depends, among other things, upon their riskiness and upon your personal preferences. When considering the riskiness of a particular investment, it is useful to take its relationship with other available investments into account. This important idea underlies the so-called modern portfolio theory developed in the early 1950s by Nobel prize laureate Markowitz. Below, only some simple elementary principles are considered.

Individual investments are often compared on the basis of their periodic returns. The ***return*** of an investment is the proportional change of the price (value) of the investment in the period of interest when compared with the price in the former period. (Here, the 'period' is some fixed time entity such as day, month, quarter or year.) If p_0 and p_1 are the prices of an investment in two successive periods 0 and 1, then the return in period 1 (when compared with the former period 0) is:

$$\frac{p_1 - p_0}{p_0}$$

Hence, a return 0.06 (respectively −0.02) means that the price of the investment has increased 6% (decreased 2%) when compared with the price in the former period.

Investments are made to earn money in an uncertain future. Of main interest is the return R of an investment during the forthcoming period when compared with the previous period. But then R is a random variable and its expectation μ, variance σ^2 and standard deviation σ can be considered. In financial theory the standard deviation of the return of an investment is typically referred to as the **volatility** of the return. It measures the **risk** of the investment.

Of course, investors like high expected returns μ and dislike high variances σ^2 and volatility σ. The asset you would invest your money in will, therefore, depend upon **how much** you like expected return and dislike variance. To model this trade-off, it is often assumed that the preferences of investors can be characterized by a **utility function** $U(\mu, \sigma^2)$, which gives the utility for each combination of expected return μ and variance σ^2. Below, we assume that

$$U(\mu, \sigma^2) = \mu - \frac{1}{2}\alpha\sigma^2$$

gives the utility of an investor for an investment with expected return μ and variance σ^2. Here $\alpha \geqslant 0$ is the investor's personal **risk aversion parameter**. An investor with $\alpha = 0$ is called **risk indifferent** since his/her utility is determined only by the expected return and not at all by its variance. An investor with positive α is called **risk averse** since his/her utility of the investment diminishes by the positive amount $\alpha\sigma^2/2$ because of the variance σ^2 of the return of the investment. Notice that for two investments with the same μ but different σ^2, a risk-averse investor assigns the larger utility to the investment with the smaller variance. Note also that the larger α is, the more risk averse an investor is.

In modern portfolio theory it is assumed that a clever investor will try to invest his money in such a way that the utility of the investment is maximal. The questions below will give you the opportunity to discover some of the principles of investment theory.

In Simpleland there only are three possible assets, assets 1, 2 and 3. Asset 1 is riskless since the money is hold in a bank deposit and gives a yearly return of 6%; there is no risk involved. Assets 2 and 3 are both risky; their returns depend on the uncertain state of the economy in the coming year. There are three possible states – recessive (r), neutral (n) and expansive (e) – which can all occur with probability 1/3. The table below gives the returns of the assets for each of the three states of the economy. Notice that assets 2 and 3 tend to behave oppositely with respect to the states of the economy.

	Recessive (r)	Neutral (n)	Expansive (e)
Return asset 1	0.06	0.06	0.06
Return asset 2	0.11	0.02	0.05
Return asset 3	0.01	0.11	0.21

a Calculate the expectations and variances of the returns of the three individual assets. Comment on your answers.

b Investor A is risk averse, with risk aversion coefficient $\alpha = 10$. Calculate for this investor the personal utilities of each of the three assets. Give your comments.

c Investor B is more risk averse, with risk aversion coefficient $\alpha = 20$. Calculate for this investor the personal utilities of each of the three assets. Give your comments.

Of course, the investment problem considered above is rather limited. It seems natural to allow an investor to choose a combination of assets 1, 2 and 3, instead of selecting a single asset. To allow this we assume that an investor can divide his money in an arbitrary way over the three assets. Let us denote the proportion of the total investment invested in asset 1 by w_1, and define w_2 and w_3 similarly. Of course, the **budget constraint** $w_1 = 1 - w_2 - w_3$ has to be valid. The proportions w_1, w_2 and w_3 denote the relative weights of the three assets in the investor's portfolio. The question arises whether an investor can determine the weights such that the utility of the accompanying portfolio is larger than all three individual utilities. The answers to the questions below will show that this is indeed possible.

Consider the portfolio with respective weights 0.1, 0.5 and 0.4. That is, 10% of the money is invested in the riskless bank deposit, while respectively 50% and 40% of the money is invested in the risky assets 2 and 3. The return of this portfolio is $R = 0.1R_1 + 0.5R_2 + 0.4R_3$, where the random variable R_j measures the return of asset j for $j = 1$, 2 and 3. Having answered the questions below, compare your answers with those of parts (a)–(c) and give your comments.

d Calculate the expectation of the return of this portfolio. Also calculate the variance by first determining the probability density function of R.

e Determine the utility of the portfolio for investor A.

f Determine the utility of the portfolio for investor B.

The answers to these questions show an interesting and important fact. Although no investor would prefer asset 2 to asset 1 when the choice is between either of these two assets, investors do better to include asset 2 in their portfolio if they can choose a combination of assets. The reason for this is that the assets 2 and 3 behave 'conversely' in the sense that they tend to react oppositely to the states of the economy. Including both assets in a portfolio seems to eliminate a part of the risk associated with investing in asset 3.

But the answers also raise new questions. For instance, is it possible to determine the weights w_j such that the resulting portfolio yields the maximum utility? And if this is possible, are these weights for a less risk-averse investor different from the weights for a more risk-averse investor? In Case 11.3 these questions and Markowitz's portfolio theory will be considered in more detail.

CHAPTER
09

Families of discrete distributions

Chapter contents

The present chapter is interested in some special 'families' of discrete probability density functions that are used frequently in practice. Each family of pdfs yields a family of probability distributions that is suitable for a specific type of practical situation.

CASE 9.1 DEFECTIVE COMPUTER CHIPS

In spite of many precautionary measures, it can hardly be avoided that the hourly production of a special type of computer chip contains defective chips. Due mostly to microscopic particles of dust, about 1% of the hourly production is faulty, even though the process is what is termed 'under control'. At the end of each hour, 100 chips are chosen at random from the production of the last hour and each is tested. For a process that is under control, it is expected that 1 of the 100 chips is defective. If the sample has at most two defective chips, no action is taken. However, if the sample contains three or more defectives, the whole production of that hour is destroyed. But how great is the probability that the complete hourly production is destroyed although the production process is under control since the proportion of defectives in the whole hourly production is (at most) 0.01? See the solution in Section 9.2.1.

9.1 Bernoulli distributions

A random experiment for which only two outcomes are possible is called a ***Bernoulli experiment***, named after famous generations of scientists from the seventeenth, eighteenth and nineteenth centuries. The outcomes are often denoted as *S* (***success***) and *F* (***failure***), usually coded as 1 and 0, respectively. Notice that this special random experiment only has four events: ∅, {*S*},

$\{F\}$, $\Omega = \{S, F\}$. Hence, a probability model is completely determined by the probability $P(S)$: if this probability is known, then $P(F) = 1 - P(S)$ is also known. The probability $P(S)$ is called the **probability of success**.

Now consider a general random experiment, but suppose that interest is in a random variable (rv) X that attaches the value 1 to an outcome (of the experiment) qualified as 'success' and the value 0 to an outcome not qualified as such (a 'failure'). Hence, X itself only has two possible outcomes (0 and 1). Although these outcomes are indeed codes and accordingly X is qualitative, X takes an intermediate position and its expectation and variance do make sense; see also Sections 3.3.1 and 4.5. In fact, $E(X)$ and $V(X)$ are defined in the usual way, as if X were quantitative.

Below, the probability $P(X = 1)$ will often be denoted by p and will also be called a **probability of success**. In this context, the probability $1 - p$ is occasionally called the **probability of failure**. The pdf f of X is completely determined by the probability of success:

x	0	1
$f(x)$	$1 - p$	p

Such rvs are called **Bernoulli random variables**; their probability distributions are **Bernoulli distributions**. It is usual to write:

$X \sim Bern(p)$ meaning: X has the Bernoulli distribution with parameter p

Notice that each p from the interval $(0, 1)$ generates precisely one probability distribution, so indeed the Bernoulli distributions constitute a whole family of probability distributions. In this context, p is called the **parameter of the family**.

The expectation, variance and standard deviation of a Bernoulli rv can easily be calculated:

$E(X) = 0 \times (1 - p) + 1 \times p = p$

$V(X) = E(X^2) - (E(X))^2 = 0^2 \times (1 - p) + 1^2 \times p - p^2 = p - p^2 = p(1 - p)$

$SD(X) = \sqrt{p(1 - p)}$

The variance is the product of the probability of success and the probability of failure.

Expectation, variance and standard deviation of *Bern(p)*

$$\mu = p; \quad \sigma^2 = p(1 - p); \quad \sigma = \sqrt{p(1 - p)} \tag{9.1}$$

For instance, the Bernoulli distribution with parameter 1/8 has expectation 1/8, variance 7/64 = 0.1094 and standard deviation 0.3307.

Example 9.1

Consider one throw with the false die of the swindler; this die has probability ¼ of throwing 6. This random experiment has 1, 2, ..., 6 as outcomes. Now suppose that emphasis is on throwing 6 or not, so the rv (called X) that attaches a 0 to all outcomes unequal 6 and a 1 to the outcome 6, is of interest. In this context, 'success' means throwing a 6 and the probability of success is ¼. The rv X has two outcomes: 0 and 1, with probabilities ¾ and ¼. As a consequence, the expectation, variance and standard deviation of X are respectively equal to ¼, ³⁄₁₆ = 0.1875 and 0.4330.

Next, consider the rv $Y = 1 - X$. Notice that Y also has two outcomes: the outcome 0 (respectively 1) occurs if X has 1 (respectively 0) as realization. A failure for X is a success for Y; the success probability of Y is the failure probability of X. Hence, expectation, variance and standard deviation of Y are respectively ¾, ³⁄₁₆ = 0.1875 and 0.4330. In particular, X and Y have the same variance and standard deviation. Check yourself that these statistics of Y also follow from the linear transformation rules in (8.30).

Suppose that – within a population of interest – the proportion p_a of the population elements has a certain property a. That is, the (population) variable X that attaches a 1 if an element has this property and a 0 if it does not have the property, has mean p_a and variance $p_a(1 - p_a)$; see Section 4.5. Next, one population element is drawn randomly. Consider the random observation (also called X) of the variable. Then

$$P(X = 1) = p_a; \quad E(X) = p_a; \quad V(X) = p_a(1 - p_a)$$

which is in accordance with (9.1). Indeed, the results in (9.1) can be considered to be generalizations of results from descriptive statistics.

Example 9.2
In a certain country, the gross labour force participation by men and women is respectively 78.2% of the male labour force and 58.7% of the female labour force. Moreover, 55% of the total labour force is male. What is the probability that a person that is arbitrarily chosen from the total labour force actually participates in the labour market (that is, has a job)?

Let the rv X indicate whether this arbitrarily chosen person participates in the labour market (outcome 1) or not (outcome 0). Then it is the probability $P(X = 1)$ that is sought.

Denote the size of the total labour force by N. Since the sizes of the male and female labour forces are respectively $0.55N$ and $0.45N$, the total number of participants in the labour market is:

$$0.782 \times 0.55N + 0.587 \times 0.45N = 0.6943N$$

To put it another way, 69.43% of the total labour force has the property (say, a) of participating in the labour market. Hence, the probability that this randomly chosen person has property a is 0.6943; this is the answer to the question.

9.2 Binomial and hypergeometric distributions

Both families of probability distributions considered in this chapter are used to model discrete random variables that have only the non-negative integers 0, 1, ..., n as outcomes. Here n often refers to the largest number of successes that is possible.

9.2.1 Binomial distributions

A **binomial experiment** with parameter p is a series of independent and identical repetitions of a Bernoulli experiment with probability p of success. The repetitions of the experiment are called **trials**. In this context, 'independent' means that the result of a trial is not influenced by the results of the other trials; 'identical' means that all trials are done under completely the same circumstances, so that they all have the same success probability p.

Since a Bernoulli experiment has the two outcomes S and F, the outcomes of a binomial experiment with n trials are n-tuples that have S or F on its positions. The number of successes in a binomial experiment with n trials is a random variable; call it Y. Its possible outcomes are 0, 1, ..., n. The outcomes 0 and n respectively imply that all trials yield failures and all trials yield successes. For $k = 0, 1, ..., n$, the event $\{Y = k\}$ means that k trials yield S and $n - k$ trials yield F. Interest is in the probability distribution of Y.

To get an idea, suppose first that $n = 3$. The outcome (F, S, F) is an ordered 3-tuple that informs us that the first trial yields a failure, the second a success and the third another failure. But this outcome is the only outcome of the intersection of the events:

F_1: first trial of the binomial experiment yields failure
S_2: second trial yields success
F_3: third trial yields failure

That is, $\{(F, S, F)\} = F_1 \cap S_2 \cap F_3$. So, the probability that this outcome will occur can be written as a product of conditional probabilities (see (7.15)):

$$P((F, S, F)) = P(F_1)P(S_2 \mid F_1)P(F_3 \mid F_1 \cap S_2) = (1 - p)p(1 - p) = p(1 - p)^2$$

But notice that changing the order of the S and the Fs in the 3-tuple does not change the probability. Hence, the probability $P(Y = 1)$ is equal to $3p(1 - p)^2$. Here, 3 comes from the binomial coefficient '3 choose 1', the number of 3-tuples that have precisely 1 success.

More generally, for a binomial experiment with n trials, one of the outcomes of the event $\{Y = k\}$ is the n-tuple $(S, ..., S, F, ..., F)$ that has the successes on the first k positions and the failures on the last $n - k$; this outcome has probability $p^k(1 - p)^{n-k}$. But the event $\{Y = k\}$ has 'n choose k' equally likely outcomes, all with k successes and $n - k$ failures. The resulting probability $f(k) = P(Y = k)$ comes from a so-called binomial distribution:

Binomial distribution

The **binomial distribution** with parameters n and p has the following pdf:

$$f(k) = \binom{n}{k}p^k(1 - p)^{n-k} \quad \text{for all } k = 0, 1, ..., n \tag{9.2}$$

Notation for rv Y with this pdf: $Y \sim Bin(n, p)$

A random variable Y that has the above pdf is called a **binomial variable**. Each choice of n (positive integer) and p (between 0 and 1) in (9.2) yields precisely one pdf. Notice that for $n = 1$ the pdf of Y is just the pdf of the Bernoulli distribution with parameter p. Hence, the family of Bernoulli distributions is a sub-family of the family of binomial distributions. Since the probabilities $P(Y = k)$ add up to 1, it follows that:

$$\sum_{k=0}^{n} \binom{n}{k}p^k(1 - p)^{n-k} = 1$$

Since Y counts the number of successes in a binomial experiment of n trials, the rv $X = n - Y$ counts the number of failures. By temporarily considering a failure to be a success, and vice versa, it follows from the above that X also has a binomial distribution, with parameters n and $1 - p$. That is, $X \sim Bin(n, 1 - p)$.

Example 9.3

Consider five consecutive throws of the die of the swindler; let Y be the number of 6s. Obviously, the five trials are independent and identical repetitions of a Bernoulli experiment with parameter $p = \frac{1}{4}$. Success corresponds to throwing 6. Y has the binomial distribution with parameters $n = 5$ and $p = \frac{1}{4}$; notation: $Y \sim Bin(5, 0.25)$. That is:

$f(k) = P(Y = k) = \binom{5}{k}(0.25)^k(0.75)^{5-k}$ for all $k = 0, 1, ..., 5$

For instance, $f(3) = 10 \times (0.25)^3 \times (0.75)^2 = 0.0879$ since '5 choose 3' equals 10. The complete overviews of the pdf f and the cdf F of Y are presented in the table below.

k	0	1	2	3	4	5
f(k)	0.2373	0.3955	0.2637	0.0879	0.0146	0.0010
F(k)	0.2373	0.6328	0.8965	0.9844	0.9990	1

For instance, the probability that the number of 6s in five trials will be at most 2 is equal to $F(2) = 0.8965$, while the probability of at least two 6s is $1 - F(1) = 0.3672$. Since Y is discrete, F is a step function that jumps at k with jump size $f(k)$.

Let X be the number of throws for which the outcome is **un**equal to 6. Then $X = 5 - Y$ and the possible outcomes of X are also $0, 1, ..., 5$. For $k = 0, 1, ..., 5$ it follows that:

$P(X = k) = P(5 - Y = k) = P(Y = 5 - k)$

Hence, the pdf of X follows by reading the probabilities of the pdf of Y from the right to the left, which results in the table below.

k	0	1	2	3	4	5
P(X = k)	0.0010	0.0146	0.0879	0.2637	0.3955	0.2373
P(X ≤ k)	0.0010	0.0156	0.1035	0.3672	0.7627	1

Let X_i be the Bernoulli random variable that equals 1 (respectively 0) if the ith trial in a binomial experiment yields success (respectively failure). Then it holds that:

$E(X_i) = p$ and $V(X_i) = p(1 - p)$ for all $i = 1, 2, ..., n$

Since X_i counts the number of successes in the ith trial (0 or 1), it follows that the total number of successes Y in the binomial experiment is equal to the sum of $X_1, ..., X_n$. That is, $Y = X_1 + ... + X_n$. Hence, by (8.31):

$E(Y) = E(X_1 + ... + X_n) = E(X_1) + ... + E(X_n) = p + ... + p = np$

Hence Y can be written as the sum of n Bernoulli random variables $X_1, X_2, ..., X_n$ that have their own small (at least, for large n and when compared with the expectation np) and independent contributions. In Section 10.3 this fact will be used to explain why, in the case of large n, binomial distributions can be approximated by normal distributions.

Notice that the above result about $E(Y)$ is in accordance with intuition: in a sequence of five throws with the false die that has probability 0.25 of throwing 6, the expected total number of 6s thrown is just $5 \times 0.25 = 1.25$. Notice that five throws with a **fair** die has the expected total $5 \times 1/6 = 5/6 = 0.8333$.

For the variance of Y, it will later (in Section 11.4) be proved that $V(Y)$ is equal to n times the variance of the number of successes in one single trial. This result is nevertheless listed here:

Expectation, variance and standard deviation of *Bin(n, p)*

$$\mu = np, \quad \sigma^2 = np(1 - p), \quad \sigma = \sqrt{np(1 - p)} \tag{9.3}$$

Figure 9.1 contains the graphs of pdfs of binomial distributions. Notice that the pdfs in the upper row all have the same (odd) *n* and that the probability of success increases from the left-hand graph to the right-hand graph. The left-hand graph is slightly skewed to the right, the graph in the middle is symmetric around $E(Y) = 4.5$, and the right-hand graph is skewed to the left.

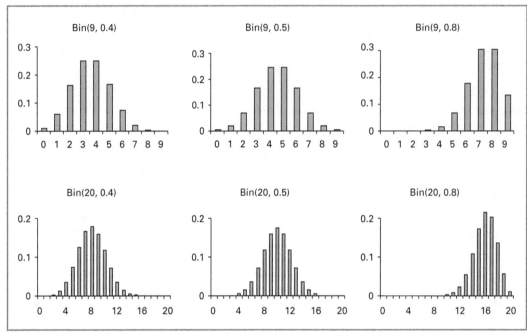

FIGURE 9.1 Pdfs of some binomial distributions

In the lower row of binomial pdfs, the left-hand graph pictures a right-skewed distribution (with $\mu = 8$), the middle graph a symmetric distribution (with $\mu = 10$), and the right-hand graph a left-skewed distribution (with $\mu = 16$).

Table 9.1 gives an overview of expectations, variances and standard deviations of the six binomial distributions in Figure 9.1. They are calculated with the formulae in (9.3).

	Bin(9, 0.4)	*Bin*(9, 0.5)	*Bin*(9, 0.8)	*Bin*(20, 0.4)	*Bin*(20, 0.5)	*Bin*(20, 0.8)
Exp.	3.6	4.5	7.2	8	10	16
Var.	2.16	2.25	1.44	4.80	5.00	3.20
St. dev.	1.4697	1.5000	1.2000	2.1909	2.2361	1.7889

TABLE 9.1 Expectations, variances and standard deviations of the binomial distributions of Figure 9.1

Many computer packages are able to calculate the probabilities $P(Y = k)$ and $P(Y \leq k)$ for values of *n*, *p* and *k* specified by the user. For Excel, see Appendix A1.9.

Example 9.4

Several probabilities and statistics of the distribution *Bin*(20, 0.4) will be calculated, partially by hand and partially with a computer.

First notice that the probabilities of seven and nine successes are respectively equal to:

$\binom{20}{7}(0.4)^7(0.6)^{13} = 0.1659$ and $\binom{20}{9}(0.4)^9(0.6)^{11} = 0.1597$

From the corresponding graph in Figure 9.1 it follows immediately that the pdf of $Bin(20, 0.4)$ takes its maximum at 8. Hence, the mode of the probability distribution is 8. From (9.3) it follows that the mean of the distribution is also 8. The table below offers the complete overview of the pdf f and the cdf F. A computer was used for the calculations. Notice that the numbers 0.0000 in the table correspond to positive numbers for which the first four decimals are all zero. The ones in the table are (apart from the last one) also rounded numbers.

x	0	1	2	3	4	5	6	7	8	9	10
$f(x)$	0.0000	0.0005	0.0031	0.0123	0.0350	0.0746	0.1244	0.1659	0.1797	0.1597	0.1171
$F(x)$	0.0000	0.0005	0.0036	0.0160	0.0510	0.1256	0.2500	0.4159	0.5956	0.7553	0.8725

x	11	12	13	14	15	16	17	18	19	20
$f(x)$	0.0710	0.0355	0.0146	0.0049	0.0013	0.0003	0.0000	0.0000	0.0000	0.0000
$F(x)$	0.9435	0.9790	0.9935	0.9984	0.9997	1.0000	1.0000	1.0000	1.0000	1.0000

To find the median of $Bin(20, 0.4)$, notice that the cdf **passes** the level 0.5 at $x = 8$. Hence, the median is also equal to 8.

The conclusion is that mean, median and mode coincide. Still the pdf is **not** symmetric since $f(7)$ and $f(9)$ are unequal. The distribution looks slightly skewed to the right, caused by the fact that only 8 outcomes are situated below its centre (at 8) while above this central position there are 12 possible outcomes.

Often, the proportion of successes in the n trials of a binomial experiment is also of interest. This proportion is just the total number of successes divided by n and is denoted by $\hat{P}$. That is, $\hat{P} = Y/n$. Notice that $\hat{P}$ is a linear transformation of Y, so $E(\hat{P})$ and $V(\hat{P})$ can easily be calculated from (9.3), with the help of (8.30):

$$E(\hat{P}) = E\left(\frac{1}{n} Y\right) = \frac{1}{n} E(Y) = \frac{1}{n} np = p;$$

$$V(\hat{P}) = V(\frac{1}{n} Y) = \frac{1}{n^2} V(Y) = \frac{1}{n^2} np(1 - p) = p(1 - p)/n$$

Expectation, variance and standard deviation of $\hat{P}$ = proportion successes in binomial experiment

$$\mu_{\hat{p}} = p, \quad \sigma_{\hat{p}}^2 = p(1 - p)/n, \quad \sigma_{\hat{p}} = \sqrt{p(1 - p)/n} \tag{9.4}$$

In Section 6.3.2 the law of large numbers was briefly discussed. Recall that this law states that the relative frequency $n(A)/n$ of the trials for which an event A occurs approaches $P(A)$ as the number of trials (n) becomes larger and larger. By taking A as the event $S = $ 'success', it follows that $n(S) = Y$ and $n(S)/n = \hat{P}$. Hence, from the law of large numbers it follows that the proportion of successes within n identical and independent trials approaches $P(S) = p$ as n grows larger and larger.

Example 9.5

The subway trains stop ten times per hour at a certain subway station. Because of this high frequency, there is no further schedule. Consider 100 arbitrary passengers that arrive independently of each other at that subway station. Determine the probability that the larger part of them has to wait more than four minutes for the train.

Assume that the time interval between two subsequent trains is exactly six minutes. So, the probability that **one** passenger will have to wait more than four minutes is 1/3. It is the rv $Y =$ 'number of passengers among these 100 that have to wait more than four minutes' and the accompanying proportion $\hat{P} = Y/100$ we are interested in. To be precise, it is the probability

$$P(\hat{P} > 0.5) = P(Y > 50)$$

that is wanted. We will show that Y is $Bin(100, 1/3)$ distributed.

Let a success mean that a passenger has to wait more than four minutes. Y measures the number of successes in 100 repetitions of the random experiment 'arriving at that subway station and measuring the waiting time for the first train'. According to the facts above, it seems reasonable to assume that these repetitions are independent and identical, so these 100 trials constitute a binomial experiment. Hence, Y is a binomial random variable with $n = 100$ and success probability 1/3; its possible outcomes are 0, 1, ..., 100.

The probability can now be calculated:

$$P(\hat{P} > 0.5) = 1 - P(Y \le 50) = 1 - 0.999801 = 0.000199 \qquad (*)$$

(The indicator (*) informs the reader that a computer package is used here.) Note that this probability is small. The reason is that the variation of $\hat{P}$ (as measured by the standard deviation) around the mean 1/3 is small when compared with the proportion 0.5 where the probability is about:

$$\sigma_{\hat{p}} = \sqrt{\frac{(1/3) \times (2/3)}{100}} = 0.04714$$

From Chebyshev's inequality it follows that at least 91.8% of the total probability (which is 1) is concentrated between $\mu_{\hat{p}} - 3.5\sigma_{\hat{p}} = 0.1683$ and $\mu_{\hat{p}} + 3.5\sigma_{\hat{p}} = 0.4983$. Indeed, the probability $P(\hat{P} \le 0.5)$ is larger than 0.918.

Example 9.6

During the flight from London to New York all aeroplane engines have the same probability p of failing. An aeroplane can land safely if no more than 50% of the engines fails; otherwise it will crash. Determine the values of p for which an aeroplane with two engines is safer than one with four engines.

Consider the rv $Y =$ 'number of engines that fail during a flight from London to New York'. Furthermore, set:

$p_2 =$ probability that an aeroplane with two engines will crash

$p_4 =$ probability that an aeroplane with four engines will crash

Assume that the engines work (and fail) independently. Then it follows that Y is $Bin(2, p)$ distributed for aeroplanes with two engines and $Bin(4, p)$ for aeroplanes with four engines. (Notice that in the present context the success probability is just the fail probability of an engine.) Hence:

$$p_2 = P(Y = 2) = p^2$$

$$p_4 = P(Y \geq 3) = P(Y = 3) + P(Y = 4) = 4p^3(1 - p) + p^4$$

$$= p^3(4(1 - p) + p) = p^3(4 - 3p)$$

An aeroplane with two engines is safer than one with four engines if:

$$p_2 < p_4 \Leftrightarrow p^2 < p^3(4 - 3p) \Leftrightarrow 1 < p(4 - 3p) \Leftrightarrow 3p^2 - 4p + 1 < 0 \Leftrightarrow (3p - 1)(p - 1) < 0$$

But note that the function $(3p - 1)(p - 1)$ is negative only for p between 1/3 and 1. Since p is automatically smaller than 1, it follows that $p_2 < p_4 \Leftrightarrow p > 1/3$. Since a fail probability of 1/3 is large and hence not very realistic, it can be concluded that aeroplanes with four engines are safer than aeroplanes with two engines.

We next consider an important application of binomial models when sampling from a population of interest.

Suppose that n elements are drawn randomly and with replacement from a population of size N that contains M successes. Since the n population elements are chosen arbitrarily and with replacement, this combined experiment can be considered as a series of n independent and identical Bernoulli experiments. Notice that it is not necessary that n should be smaller than N. Since M out of N elements are successes, the success probability p is equal to the ratio M/N. But then the rv Y = 'number of successes among the n sampled elements' is a binomial variable; the parameters are n and $p = M/N$. Notation: $Y \sim Bin(n, M/N)$. Hence:

$$P(Y = k) = \binom{n}{k}\left(\frac{M}{N}\right)^k \left(1 - \frac{M}{N}\right)^{n-k} \quad \text{for all } k = 0, 1, \ldots, n \tag{9.5}$$

$$\mu_Y = n\frac{M}{N} \quad \text{and} \quad \sigma_Y^2 = n\frac{M}{N}\left(1 - \frac{M}{N}\right) \tag{9.6}$$

Example 9.7

According to Example 9.2, in a certain country 78.2% of the male labour force and 58.7% of the female labour force actually participate in the labour market. Consider 1000 men and 1000 women from the respective labour forces. For each of the two samples, what is the probability that at least 75% participate in the labour market?

Since no further information is given, we assume that the two groups of people are randomly drawn. Since the sample sizes 1000 are small when compared with the population sizes, it is hardly a restriction to assume that the two samples are chosen with replacement. Let X (respectively Y) be the number of men (respectively women) – among the 1000 sampled persons – who are participating the labour market. By writing $\hat{P}_1 = X/1000$ and $\hat{P}_2 = Y/1000$ for the respective proportion of successes in the two samples, it becomes clear that the probabilities $P(\hat{P}_1 \geq 0.75)$ and $P(\hat{P}_2 \geq 0.75)$ are sought.

According to the arguments above, X and Y are both binomially distributed:

$$X \sim Bin(1000, 0.782) \quad \text{and} \quad Y \sim Bin(1000, 0.587).$$

By (9.4) we have:

$$E(\hat{P}_1) = 0.782; \quad SD(\hat{P}_1) = 0.0131; \quad E(\hat{P}_2) = 0.587; \quad SD(\hat{P}_2) = 0.0156$$

Because of the large sample sizes, the standard deviations are small and the two pdfs are strongly concentrated around the respective means 0.782 and 0.587. That is why it is expected that the above probability for the men will be large, but for the women it will be small.

But we can calculate these probabilities. Notice that the events $\{\hat{P}_1 \geq 0.75\}$ and $\{\hat{P}_2 \geq 0.75\}$ are equal to the events $\{X \geq 750\}$ and $\{Y \geq 750\}$, respectively. So:

$$P(\hat{P}_1 \geq 0.75) = P(X \geq 750) = 1 - P(X \leq 749) = 0.9929 \qquad (*)$$

$$P(\hat{P}_2 \geq 0.75) = P(Y \geq 750) = 1 - P(Y \leq 749) \approx 0 \qquad (*)$$

Indeed, for the men the probability is almost 1 and for the women it is almost 0.

CASE 9.1 DEFECTIVE COMPUTER CHIPS – SOLUTION

Denote the number of defective chips in the hourly sample by Y. Since the 100 chips are randomly drawn from the hourly production, the rv Y has a binomial distribution with parameters $n = 100$ and p (the proportion of defectives in the whole hourly production). If the process is under control, p is equal to 0.01 and it is indeed expected that there is one defective: $E(Y) = np = 1$. The complete hourly production is destroyed if Y is 3 or more, which would be incorrect (i.e. a false move) if the proportion p is still equal to 0.01. It is the probability $P(Y \geq 3)$ that is wanted, for the case that $p = 0.01$:

$$P(Y \geq 3) = 1 - P(Y = 0) - P(Y = 1) - P(Y = 2)$$

$$= 1 - 0.99^{100} - (100 \times 0.01 \times 0.99^{99}) - (4950 \times 0.01^2 \times 0.99^{98})$$

$$= 1 - 0.3660 - 0.3697 - 0.1849 = 0.0794$$

There is a 7.94% chance that the complete hourly production is destroyed unnecessarily.

9.2.2 Hypergeometric distributions

The present subsection reconsiders the last topic of subsection 9.2.1 but now sampling is done **without** replacement.

Consider a population that has N elements, numbered 1, 2, …, N. Suppose that M of the elements are successes (hence, $M \leq N$) and $N - M$ of them are failures. From this population, n elements are drawn randomly and **without** replacement. (Hence, n is at most N.) Notice that the outcomes of this random experiment are unordered n-tuples of different elements from the population. Recall from Section 7.2 that there are 'N choose n' outcomes in the sample space, all equally likely since the experiment chooses one n-tuple arbitrarily. Hence, the classical probability definition can be used to determine probabilities regarding the random experiment 'randomly drawing n elements without replacement'.

Interest is in the probability density function of the rv Y = 'number of successes among the n sampled elements'. Notice that the possible outcomes of Y are 0, 1, …, n. To determine the probability of the event $\{Y = k\}$, the number of enclosed n-tuples is needed. Here are the arguments:

- To choose k successes from the total number of M successes in the population, there are 'M choose k' possible k-tuples.
- To choose $n - k$ failures from the total number of $N - M$ failures in the population, there are '$N - M$ choose $n - k$' possible $(n - k)$-tuples.
- Each k-tuple of successes can be combined with each $(n - k)$-tuple of failures to obtain an n-tuple of the event $\{Y = k\}$.

Hence, the total number of n-tuples in $\{Y = k\}$ is equal to the product of 'M choose k' and '$N - M$ choose $n - k$'. The classical definition yields the probabilities; they come from a hypergeometric distribution:

Hypergeometric distribution

The **hypergeometric distribution** with parameters n, M and N has the following pdf:

$$f(k) = \frac{\binom{M}{k}\binom{N - M}{n - k}}{\binom{N}{n}} \quad \text{for all } k = 0, 1, \ldots, n \tag{9.7}$$

Notation for rv Y with this pdf: $Y \sim H(n; M, N)$

Notice that each choice of n, M and N in (9.7) under the restrictions $n \le N$ and $M \le N$ gives a pdf, so (9.7) in essence defines a whole family of pdfs.

It is **not** excluded that $n > M$ or $n > N - M$. Hence, in (9.7), k might be larger than M or $n - k$ larger than $N - M$. In those cases, 'M choose k' and '$N - M$ choose $n - k$' are both defined as 0, which is natural since it is not possible to choose more than M successes or more than $N - M$ failures.

Without showing the validity, it is just stated here that:

Expectation and variance of $H(n; M, N)$

$$\mu = n\frac{M}{N} \quad \text{and} \quad \sigma^2 = n\frac{M}{N}\frac{N - M}{N}\frac{N - n}{N - 1} \tag{9.8}$$

In these expressions, the ratios M/N and $(N - M)/N = 1 - M/N$ are the probabilities of success and failure, respectively. The ratio $(N - n)/(N - 1)$ in the above variance is called the **variance correction factor**. Notice that it is this correction factor that makes the variance in (9.8) for the without-replacement case different from the variance in (9.6) for the with-replacement case.

In practice, the calculation of the probabilities in (9.7) can be very laborious. Fortunately, many computer packages can do the work for us. For Excel, see Appendix A1.9.

Example 9.8

In the game Scrabble, all players take – before the game starts and completely arbitrarily – seven letter-blocks from a total of 102. Among these 102 letter-blocks, there are six As. Below, the pdf is determined of the random variable Y = 'number of As drawn by the player who chooses the letter-blocks first'.

Seven blocks are taken from a total of 102 blocks, without replacement. Since the seven blocks are chosen arbitrarily, it follows immediately that Y has a hypergeometric distribution. A success corresponds to drawing a letter-block with A. Hence, M is the number of As among all $N = 102$ letter-blocks and $Y \sim H(7; 6, 102)$. To demonstrate the mechanics, the calculation of $P(Y = 2)$ is shown:

$$P(Y = 2) = \frac{\binom{6}{2}\binom{96}{5}}{\binom{102}{7}} = \frac{15 \times 92 \times \ldots \times 96/120}{96 \times \ldots \times 102/5040} = \frac{15 \times 61\,124\,064}{18\,466\,953\,120} = 0.0496$$

The table below presents **all** probabilities $f(k) = P(Y = k)$. Notice that $f(7)$ is equal to 0, while $f(5)$ and $f(6)$ are only close to 0.

k	0	1	2	3	4	5	6	7
f(k)	0.6454	0.3012	0.0496	0.0036	0.0001	0.0000	0.0000	0
g(k)	0.6542	0.2862	0.0537	0.0056	0.0003	0.0000	0.0000	0.0000

Since the success and failure probabilities are respectively $6/102 = 0.0588$ and $96/102 = 0.9412$, it follows by (9.8) that

$$E(Y) = 7 \times (6/102) = 0.4118$$

$$V(Y) = 7 \times (6/102) \times (96/102) \times \frac{102 - 7}{102 - 1} = 0.3645$$

$$SD(Y) = 0.6038$$

The third row of the table gives the accompanying probabilities for the unrealistic case that sampling had been done **with** replacement. Hence, g is the pdf of the binomial distribution $Bin(7, 6/102)$ For instance:

$$P(Y = 2) = \binom{7}{2}\left(\frac{6}{102}\right)^2\left(1 - \frac{6}{102}\right)^5 = 0.0537$$

The probabilities $g(5)$, $g(6)$ and even $g(7)$ are not precisely 0, but only close to 0.

The differences between the corresponding probabilities in the two pdfs turn out to be small. Also notice that, by (9.6), mean, variance and standard deviation of g are respectively 0.4118, 0.3875 and 0.6225. The pdfs f and g have the same mean, but the variance and standard deviation of g are slightly larger than those of f. This is caused by the correction factor.

If the sample size n is at least 2, then the correction factor $(N - n)/(N - 1)$ in (9.8) is smaller than 1. Hence, the variance of the hypergeometric distribution is smaller than the variance of the corresponding binomial distribution.

If the sample size n is small compared with the population size N and sampling is done with replacement, then there is only a small probability that the n sampled elements are not all different. That is, if the ratio n/N is close to zero, then sampling with replacement and sampling without replacement are almost the same, and, as a consequence, the hypergeometric distribution is nearly the same as the binomial distribution with success probability M/N. In this context, n should be at most 10% of N. This result is formulated as follows:

Approximation of hypergeometric by binomial

If $\frac{n}{N} \le 0.1$, then $\quad H(n; M, N) \approx Bin\left(n, \frac{M}{N}\right)$ (9.9)

The symbol $\approx$ means 'is nearly the same as'. Since sample sizes are often much smaller than population sizes, the practical consequence of (9.9) is that a suitable binomial distribution can often be used to approximate a hypergeometric distribution.

Example 9.9

Consider a very large population of multi-person households. In 2002, 1.83% of these households had disposable income of less than €10 000 per year. But this was in 2002. What about the current year?

To study whether nowadays the (current) proportion p of all multi-person households with disposable income under €10 000 is different from 0.0183, a random sample of 200 multi-person households will be drawn at the end of the present year. The 200 annual disposable incomes will be recorded, and the proportion of these incomes below the level 10 000 will be calculated and used as an approximation of the – still unknown – present proportion p. But before this practical part of the research is done, we **can** ask ourselves the question whether a sample score such as 2% can be interpreted as contradictory to the population score 1.83% of 2002. That is, if $p = 0.0183$ and hence p is **the same** as the value of 2002, would a sample proportion 0.02 then be extraordinary? Below, the probability of getting a sample proportion of at least 0.02 is calculated for the case that $p = 0.0183$.

Suppose that $p = 0.0183$. With respect to the variable $Y = $ 'number of households in the sample with income below €10 000', it is the probability $P(Y \geq 4)$ that is wanted. Of course, the sample will be drawn without replacement. However, since the population is very large and hence the ratio n/N is close to zero, it hardly makes a difference to consider the sample as being sampled **with** replacement. Although the rv $Y = $ 'number of households in the sample with income below €10 000' is, strictly speaking, $H(200; M, N)$ with $M/N = 0.0183$, this distribution is practically the same as the distribution $Bin(200, 0.0183)$. Hence:

$$P(Y \geq 4) = 1 - P(Y < 4) = 1 - P(Y \leq 3) = 1 - 0.50125 = 0.49875 \qquad (*)$$

A sample score of at least 2% out of 200 has probability almost 0.5 if $p = 0.0183$. So, the score 2% for the sample is not counter to the population proportion 0.0183.

9.3 Poisson distributions

The pdfs of the last section were all concentrated on non-negative integers with an obvious upper bound (which was n). The family of pdfs in the present section is also concentrated on non-negative integers, but now such an upper bound does **not** exist.

Poisson distribution

The **Poisson distribution** with parameter μ has the following pdf:

$$f(k) = \frac{\mu^k}{k!} e^{-\mu} \quad \text{for all } k = 0, 1, \ldots \qquad (9.10)$$

Here μ is a positive real number.

Notation for rv Y with this pdf: $Y \sim Po(\mu)$

(That f is indeed a pdf is not proved; e is Euler's number, approximately 2.7183.) Each choice of $\mu > 0$ gives a pdf, so (9.10) in essence defines a whole family of pdfs. Poisson distributions are named after Siméon Denis Poisson (1781–1840).

The name of the parameter suggests that it will be the mean. This indeed is the case. But Poisson distributions have the property that mean and variance coincide:

Expectation and variance of *Po(μ)*

$$Y \sim Po(\mu) \Rightarrow E(Y) = V(Y) = \mu \tag{9.11}$$

Figure 9.2 shows graphs of the pdfs of $Po(1)$, $Po(2.5)$ and $Po(5)$:

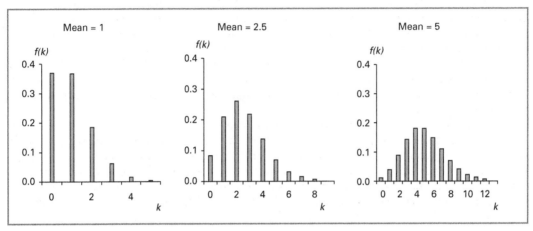

FIGURE 9.2   Some pdfs of Poisson distributions

For each pdf only a part can be graphed. For instance for $Po(1)$, the probabilities $P(Y = k)$ for $k \geq 6$ are not included. The reason is that they are all very close to 0. All distributions are skewed to the right.

For given μ and k, the probabilities $P(Y = k)$ can easily be calculated with manual calculators. For instance, for $\mu = 5$ we obtain:

$$P(Y = 8) = \frac{5^8}{8!}\, e^{-5} = 0.0653$$

Many computer packages will calculate the probabilities of the pdf and the cdf of a Poisson distribution. See also Appendix A1.9.

Poisson distributions are frequently used as models for random variables that count the number of **occurrences** of a certain type in a fixed period of time. Such occurrences might be: orders of a product, arrivals of customers, login-times of website visitors, times of deliveries, times of accidents, times of disasters, etc.

Example 9.10
A factory for agricultural machines receives on average 1.2 orders per (business) day for a cultivator. Calculate the probability of at least three orders on one day and of at least ten orders in one week (of five days).

Let Y and S respectively be the demand for cultivators on one day and in one week, discrete random variables with non-negative integer-valued outcomes. Note that $P(Y \geq 3)$ and $P(S \geq 10)$ are wanted. There is no obvious upper bound for the set of outcomes of Y, so a binomial or hypergeometric model is not the first choice. Let us adopt Poisson models for the periodized demand for cultivators. (In the context of this example, an occurrence is the arrival of an order.) Since it is given that the average demand for a period of one day is 1.2 and 'hence' for a period of five days is 6, Poisson distributions with parameters 1.2 and 6 will be used to calculate $P(Y \geq 3)$ and $P(S \geq 10)$, respectively:

$$P(Y \geq 3) = 1 - P(Y \leq 2) = 1 - P(Y = 0) - P(Y = 1) - P(Y = 2)$$

$$= 1 - e^{-1.2} - \frac{1.2}{1!} e^{-1.2} - \frac{1.2^2}{2!} e^{-1.2} = 0.1205$$

$$P(S \geq 10) = 1 - P(S \leq 9) = 1 - 0.9161 = 0.0839 \tag{$*$}$$

In this example, an incoming order is the 'occurrence' of interest. It was assumed that $Y =$ 'number of orders in one day' is $Po(1.2)$ distributed, but also that $S =$ 'number of orders in 5 days' is $Po(5 \times 1.2)$ distributed. The second is not a consequence of the first (as might be wrongly suggested). In fact, it is a consequence of an **extension of the model** in the sense that it is assumed that **all** variables $N(t) =$ 'number of orders in t days' are $Po(t \times 1.2)$ distributed. (Here, it is allowed that t is not an integer.) That is, not only is the rv $Y = N(1)$ Poisson distributed, but a whole family $\{N(t): t \geq 0\}$ of rvs is Poisson distributed too. Under such circumstances, it is common to talk about a **Poison process**; the parameter of the Poisson distribution of $N(1)$ is called the **parameter of the Poisson process**. In Example 9.10, this parameter is 1.2.

Without giving the details, we just mention the following important properties of a Poisson process:

- The numbers of occurrences in two different time intervals of the same length have the same Poisson distribution.
- The numbers of occurrences in two disjoint time intervals are independent.

If, in Example 9.10, the Poisson process with parameter 1.2 is adopted to model the process of orders, then the variables 'number of orders in the period Monday–Wednesday' and 'number of orders in the period Wednesday–Friday' both have the distribution $Po(3.6)$. Furthermore, the variables 'number of orders on Monday–Tuesday' and 'number of orders on Wednesday–Friday' are independent.

Whether a Poisson process is suitable to model the numbers of occurrences is in practice often determined by considering the **index of dispersion** $I(t)$: the function that measures the ratio of the variance and the expectation of $N(t)$. That is:

$$I(t) = \frac{V(N(t))}{E(N(t))} \text{ for all } t \geq 0$$

Since, by (9.11), for a Poisson process the index of dispersion is identically equal to 1, the decision about the suitability of the Poisson process model is often based on the comparison of the (estimated) index of dispersion with the level 1.

Another reason for the importance of Poisson distributions is that they are occasionally used to approximate binomial distributions with small success probabilities. It can be proved that for $p \leq 0.1$ the pdf of $Bin(n, p)$ is close to a Poisson distribution. Since $Bin(n, p)$ has mean np, it will be obvious that the parameter of the approximating Poisson distribution has to be np. That is:

Approximation of *Bin(n, p)* by *Po(np)*

$$p \leq 0.1 \Rightarrow Bin(n, p) \approx Po(np) \tag{9.12}$$

Example 9.11

The email virus Tax Return Hoax has infected 0.1% of all computers. How many computers have to be selected to have a 25% chance that at least one of them is infected with this virus? With this sample size, calculate the probability that in the sample more than two computers are infected.

Assume that the n computers are selected randomly (and of course with replacement). The random variable Y = 'number of infected computers' has the distribution $Bin(n, 0.001)$, where n has to satisfy the equation $P(Y \geq 1) = 0.25$. Since the success probability 0.001 is smaller than 0.1, the approximation with the distribution $Po(0.001n)$ can be used. Hence:

$$0.25 = P(Y \geq 1) = 1 - P(Y = 0) = 1 - e^{-0.001n}$$

$$1 - e^{-0.001n} = 0.25 \Leftrightarrow e^{-0.001n} = 0.75 \Leftrightarrow -0.001n = \log(0.75) \Leftrightarrow n = 287.68$$

(It was used that $\log(e^b) = b$.) It follows that $n = 288$ (rounded up!).

We use $n = 288$ to calculate $P(Y > 2)$, exactly with the binomial distribution as well as approximated by the Poisson distribution:

$$P(Y > 2) = 1 - P(Y \leq 2) = \begin{cases} 0.003186 & \text{with } Bin(288, 0.001) \\ 0.003213 & \text{with } Po(0.288) \end{cases} \qquad (*)$$

Summary

Four families of discrete distributions have been introduced. The Bernoulli distributions are used for random variables that have only two possible outcomes ('success' and 'failure'). Binomial distributions and hypergeometric distributions are often used as models for numbers of successes in samples, respectively for samples drawn with and without replacement. The family of the Poisson distributions is frequently used to model numbers of occurrences (in a period) in cases where no obvious upper bound exists for the set of outcomes. The boxes below restate some results.

Facts of three families of discrete distributions

Distribution	Notation	$f(k)$	Expectation	Variance
Hypergeometric	$H(n, M, N)$	$\binom{M}{k}\binom{N-M}{n-k} / \binom{N}{n}$	$n\dfrac{M}{N}$	$n\dfrac{M}{N}\dfrac{N-M}{N}\dfrac{N-n}{N-1}$
Binomial	$Bin(n, p)$	$\binom{n}{k}p^k(1-p)^{n-k}$	np	$np(1-p)$
Poisson	$Po(\mu)$	$\dfrac{\mu^k}{k!}e^{-\mu}$	μ	μ

Relationships among the three families of distributions

Relationships among the distributions	Relationships among the parameters	Requirement
$H(n, M, N)$		
$\approx$		$n/N \leqslant 0.1$
$Bin(n, p)$	$p = M/N$	
$\approx$		$p \leqslant 0.1$
$Po(\mu)$	$\mu = np$	

List of symbols

Symbol	Description
~	'has the distribution'
≈	'has approximately the distribution'
$Bin(n, p)$	Binomial distribution
$H(n, M, N)$	Hypergeometric distribution
$Po(\mu)$	Poisson distribution

🔑 Key terms

Bernoulli distribution **288**
Bernoulli experiment **287**
Bernoulli random
 variables **288**
binomial distribution **290**
binomial experiment **289**
binomial variable **290**
failure **287**

hypergeometric
 distribution **297**
index of dispersion **301**
parameter of the
 family **288**
parameter of the Poisson
 process **301**
Poisson process **301**

Poisson distribution **299**
probability of failure **288**
probability of success **288**
success **287**
trial **289**
variance correction
 factor **297**

? Exercises

Exercise 9.1

Consider a population of 1000 elements for which 350 have a certain quality called 'success'. X is the dummy variable that takes the value 1 for population elements that have the quality.

a One element of the population is randomly drawn and the accompanying value of X is observed. What is the type of the distribution of the random observation X? What is the parameter? Determine $E(X)$ and $V(X)$.

Next, a sample of 100 population elements is randomly drawn. Let Y be the number of successes in the sample. Determine the type of the distribution of Y, the parameters, and the expectation and variance for the case that:

b the sample is drawn with replacement;
c the sample is drawn without replacement.

Exercise 9.2

A random variable Y has the binomial distribution with parameters $n = 60$ and $p = 0.1$.

a Calculate the expectation and variance of Y.
b Calculate $P(Y = 5) + P(Y = 6)$ by hand.
c Calculate the probabilities $P(Y \le 4)$ and $P(Y \ge 7)$ with a computer.
d Check the answer to part (b) from the answer to part (c).

Exercise 9.3

A company has 200 employees, 115 women and 85 men. For a certain task, five of the employees are randomly drawn. Let Y denote the number of women in the sample.

  a What type of distribution does Y have? Determine the parameters.
  b Calculate the expectation and variance of Y.
  c Calculate $P(Y = 3)$ by hand.
  d Calculate $P(Y = 2)$ and $P(Y \leq 1)$ with a computer.
  e Which binomial distribution approximates the pdf of Y? Find approximations for the probabilities in parts (c) and (d); use a computer.

Exercise 9.4

A random variable Y has the Poisson distribution with parameter 5.8.

  a Determine the expectation and variance of Y.
  b Calculate the probability $P(Y = 5)$ by hand.
  c Use a computer to calculate the probabilities $P(Y = 6)$ and $P(Y > 6)$.

Exercise 9.5

For each experiment below, conclude whether it concerns a binomial experiment or not. Explain your conclusions. If the experiment is binomial, describe the probability of success.

  a To test the quality of a large lot of apples, 100 apples are chosen randomly and qualified as 'good' or 'bad'.
  b To test the quality of a lot of 100 television sets, 25 TVs are chosen randomly and qualified as 'OK' or 'defective'.
  c A die is thrown 20 times while recording whether a throw yields six eyes.
  d On a meteorological station, it is recorded daily (for seven consecutive days) whether it rained.
  e For 15 consecutive days it is recorded whether the price of a stock exceeds the level of €10 (the mean price of the last month).
  f For 15 consecutive days it is recorded whether the return of a stock with respect to the day before is positive.

Exercise 9.6

Product control is an important activity in many factories. Suppose that 2% of the massive production of a certain product is defective. Each hour, n products are chosen randomly from the very large production of the last hour and the rv Y = 'number of defectives in the sample' is measured.

  a Suppose that $n = 1$. Determine the probability density function and the distribution function of Y.
  b Calculate the expectation and the variance of Y.
  c Suppose from now on that $n = 100$. To what family of distributions does the probability distribution of Y belong? Also determine the parameters of the distribution.
  d Calculate the expectation and the variance of Y.
  e Calculate the probabilities $P(Y < 2)$, $P(Y \leq 3)$ and $P(2 < Y < 5)$.

Exercise 9.7

A quick poll on the website of Mercer Human Resource Consulting concludes that 34% of the senior managers believe that engaging employees is the most critical business theme for the HR organization. If this is true, what are the probabilities of the following events?

a One out of the next four senior managers responding to the poll has that same opinion.

b Two out of the next eight senior managers responding have that same opinion.

c Three out of the next twelve senior managers responding have that same opinion.

d Ten out of the next 40 senior managers responding have that same opinion.

Exercise 9.8

In June 2007, the operating system Windows Vista had a market share of 4.52%. Suppose that this is still true on the day you are doing this exercise.

a Determine the expectation and standard deviation of the number of Windows Vista operating systems among 100 computers.

b Determine the probability that at least 4 out of 100 computers have Windows Vista as operating system.

c Use a computer to calculate the probability that at least 4% of 1000 computers has Windows Vista as operating system.

Exercise 9.9

The midterm exam of a course consists of five multiple-choice questions. Each question has four possible answers, but only one is correct. A lazy student decides to do the midterm although he did not study at all and did not attend any classes. He is sure that, for each question, he will manage to cross off one of the three incorrect choices. By choosing one of the three remaining possible choices randomly, he believes that his chances of answering at least three of the five questions correctly are not too bad.

   Suppose that the above considerations of the student are correct. Let X be the number of questions (out of five) that will be answered correctly. Interest is in the pdf of X.

a To which family of distributions does this pdf belong? Determine the parameters.

b Also determine the expectation, variance and standard deviation of X.

c Calculate (by hand) all probabilities of the pdf.

d Determine the probability that the student will indeed answer at least three of the questions correctly.

Exercise 9.10

Since the academic year 2007/08 the number of female first year university students in the Netherlands has outstripped the number of male first year students. Nowadays, 51% of Dutch first year students are female.

a Use the above information to describe a random experiment and a random variable X that has the distribution $Bern(0.51)$.

b Determine its expectation and variance.

   There are 40 000 first year university students. Let Y be the number of female students among 100 randomly chosen first year students. Interest is in the probability distribution of Y.

c Explain why the distribution $Bin(100, 0.51)$ will be a good model for Y.

d Calculate the expectation, the variance and the standard deviation of Y.

e Use a computer to calculate the probabilities: $P(Y < 48)$, $P(Y > 52)$ and $P(45 < Y < 55)$.

Exercise 9.11

Among the 200 students of the Econometrics and Operations Research study stream, 40 are first year students. Suppose that 20 E&OR students are sampled randomly; they will be asked to evaluate a course. Let Y denote the number of first year students in the sample.

a To which family of distributions does the exact probability distribution of Y belong? Determine the parameters.

b Also determine the expectation, the variance and the standard deviation of Y.

c Notice that the ratio n/N equals 0.1. Which binomial distribution can be used to approximate the exact distribution of Y?

d Interest is in the probabilities below. First determine them precisely by using the exact distribution of part (a); next, approximate them by using the approximating distribution of (c). (Determine the first probability by hand; use a computer for the others.)

$$P(Y = 6); \quad P(Y \leq 6); \quad P(Y > 3); \quad P(2 < Y < 6)$$

Exercise 9.12 (computer)

In a production process of katkit cans with salmon, 5% of the cans contain less salmon than prescribed. A sample of 100 cans is drawn from the (large) production of one day.

a Let the rv X be the number of the cans in the sample that contain less salmon than prescribed. Mention the name and the parameters of the probability distribution of X.

b Make a bar chart of the pdf of this distribution.

c Determine the probability that in the sample at most four cans do not satisfy the prescribed level of salmon.

d Determine the probability that in the sample at least three cans do not satisfy the prescribed level of salmon.

e Determine the probability that in the sample at least one and at most four cans do not satisfy the prescribed level of salmon.

f Determine the expectation and the variance of X.

Let the rv $\hat{P} = X/100$ be the proportion of the cans in the sample that contain less salmon than prescribed.

g Calculate the expectation, the variance and the standard deviation of $\hat{P}$.

h Determine all possible outcomes of $\hat{P}$. Calculate the probabilities $P(\hat{P} > 0.075)$ and $P(0.03 < \hat{P} \leq 0.081)$.

Exercise 9.13

Within a group of 100 unemployed job-seekers, 40 people have been looking for work for more than a year; the other 60 persons have been unemployed for less than a year.

A project for unemployed persons offers work for 10 persons. These persons are chosen randomly from the group of 100. Let X denote the number of the sampled persons who have been unemployed for more than a year.

a Suppose that the sample is drawn without replacement. What is the name of the probability distribution of X? Also mention the parameters!

b What is the name of the probability distribution of X if the sample is drawn with replacement? Again, also mention the parameters.

c For both cases (a) and (b), determine μ and σ^2.

d For both cases, calculate by hand: $P(X = 2)$, $P(X \geq 3)$ and $P(X \geq 2)$.

Exercise 9.14

At a particular service station, cars arrive for fuel at a mean rate of 7 per hour.

a What model do you advise for the number of cars arriving for fuel over the next two hours?
b Calculate the probability that over the next hour two cars will arrive for fuel.
c (computer) Calculate the probability that in the next three hours at most 15 cars will arrive for fuel.

Exercise 9.15

Microsoft is often criticized because its showpiece Windows is too powerful in the operating systems market. According to the website http://marketshare.hitslink.com, the Linux operating system had, in June 2007, a market share of only 0.71%. Suppose that this percentage is still valid and consider the number (Y) of Linux operating systems among 10 000 computers.

a Calculate the expectation and the standard deviation of Y. Use Chebyshev's rule with $k = 3$ to find out which of the possible outcomes of Y are the most important for the probability distribution of Y.
b Use a suitable Poisson approximation for the distribution of Y to calculate the three probabilities $P(Y = k)$ for $k = 70, 71, 72$.
c (computer) Also approximate $P(Y \leq 71)$.
d (computer) Approximate the probability that Y will fall within 3 standard deviations of the mean.

Exercise 9.16

A certain website is visited an average of 60 times per four-week period. Consider the rv $X =$ 'number of visits in one week'.

a Which model do you suggest for X?
b Determine the expectation and the variance of X.
c (computer) Calculate the probability that X is larger than 14.
d Calculate $P(10 < X < 14)$.

Exercise 9.17 (computer)

The distributions $Bin(20, 0.25)$, $Po(5)$ and $H(20; 60, 240)$ have the same mean, 5. To compare them, create bar charts of these distributions in one figure. Give your comments.

CASE 9.2 THE NON-BUSINESS MOBILE PHONE MARKET

The market for mobile phones is a rapidly growing and quickly changing market. It is hard to obtain reliable information about this market: when the results of a study are published, the market has already changed and the results may be out of date. Here is some quick research from 2000.

The Dutch have a reputation for being an economical people. This is reflected in the non-business mobile phone market. Although 59% of all Dutch people over 12 years of age have a mobile phone that is not paid for by their employer, these phones are mainly pre-paid phones which are hardly ever used to call someone. The table below gives some information about the final quarter of 2000.

Type of phone	Market share (%)	Average spent per user (incl. VAT) (€)
Pre-paid	76.5	8.74
Subscriber	23.5	91.84

Non-business market for mobile phones, 2000Q4
Source: Tijdschrift voor Marketing, June 2001; GfK Panel Services Benelux

When studying these data, some questions arise:

- Why do so many people buy a mobile phone if they hardly ever use it?
- How can the phone companies increase the usage (and by that their profit)?

To find answers to these questions we first want to conduct a market survey. Our aim is to have 300 respondents that own a private mobile phone. We define the variable X associated with the sampling:

X = 'the number of people (in our random sample of 300 mobile phone users) that have a pre-paid phone'

Suppose that the situation now is unchanged when compared with 2000.

a What is the probability distribution of this variable X? Give both the type (family) of the distribution and its parameters; so we want all the information necessary to be able to calculate probabilities.

b Calculate the probability that fewer than 225 people in our sample have a pre-paid mobile phone.

c Calculate the probability that between 230 and 250 people in our sample (boundaries included) have a pre-paid mobile phone.

d What is the expected total spent on mobile phones (including VAT) in the 4th quarter of 2000 by the 300 people in our sample?

e What can be said about the percentage that the pre-payers spend when compared with the total expected amount? Give your comments.

Families of continuous distributions

Chapter contents

In the present chapter we are interested in some special families of continuous probability density functions that are frequently used in practice. Each family of pdfs yields a family of probability models suitable for a specific type of situation. We introduce uniform, exponential and normal distributions, respectively.

CASE 10.1 EU LIMIT FOR CARBON DIOXIDE EMISSIONS

The increased concentration of carbon dioxide (CO_2) in the atmosphere is a serious problem because of global warming. The European Commission has proposed a strategy to limit the average CO_2 emissions from new cars in the EU to 130 grams per kilometre by 2012.

A car manufacturer is developing a model that should be ready for the market at the beginning of 2012. This new model should comply with the new emissions limit of 130 g/km. However, for various reasons it will be impossible to ensure that all cars of this new model will emit precisely the same amount of CO_2 per kilometre. In fact, the variable X = 'CO_2 emissions (grams) per kilometre' has a standard deviation of 2 g. What should be the target mean emission to ensure that only 5% of the cars of this new type exceed the emissions limit of 130 g? See the answer in Section 10.3.5.

10.1 Uniform distributions

Uniform distributions are used as models for continuous random variables in cases where the possible outcomes fall in some bounded interval and are all equally likely. Such distributions have already been encountered in Examples 8.7 (waiting time of Mrs Doubledutch), 8.8 (time it takes to solve Sudoku puzzle), 9.5 (waiting time for subway trains) and 8.11 (weekly sales of electronics store). The major advantage of these distributions is their simplicity. Furthermore, they can easily be used for simulation purposes, as will be shown below.

Suppose that α and β are real constants with $\alpha < \beta$.

Uniform distribution

The **uniform distribution** with parameters α and β is the probability distribution with the following pdf:

$$f(y) = \frac{1}{\beta - \alpha} \quad \text{for all } y \text{ in the interval } (\alpha, \beta) \tag{10.1}$$
$$= 0 \qquad \text{for } y \text{ otherwise}$$

If an rv Y has this pdf, it is written: $Y \sim U(\alpha, \beta)$. Notice that each choice of α and β yields one pdf, so in essence (10.1) defines a whole family of probability distributions. Since α and β are constants, the pdf is constant over the interval (α, β); Y has the interval (α, β) as its set of possible outcomes and all outcomes have the same likelihood. Incidentally, notice that the total area under f is 1, so f is indeed a continuous pdf. See also Figure 10.1.

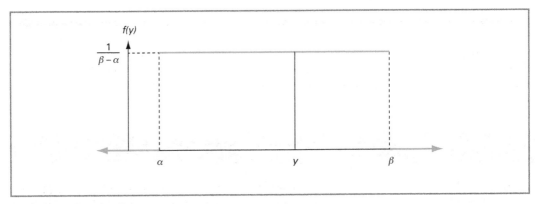

FIGURE 10.1 Pdf of $U(\alpha, \beta)$

The definition of the uniform distribution as it is given in (10.1) is based on the **open** interval (α, β), that is, α and β are not included. However, the intervals $[\alpha, \beta)$, $(\alpha, \beta]$ or even $[\alpha, \beta]$ might have been taken as well. The reason is that a uniform pdf is a **continuous** pdf, so the probability that Y will be precisely equal to the boundaries α or β of these intervals is zero; see (8.6).

Probabilities concerning Y are areas under its pdf. To find $P(Y \leq y)$ for y in (α, β), the area under f and above the interval $(\alpha, y]$ has to be calculated; see Figure 10.1. Since the corresponding rectangle has sides of lengths $y - \alpha$ and $1/(\beta - \alpha)$, this area equals $(y - \alpha)/(\beta - \alpha)$. Hence, the cdf F of $U(\alpha, \beta)$ satisfies:

$$F(y) = \frac{y - \alpha}{\beta - \alpha} \quad \text{for all } y \text{ in } (\alpha, \beta) \tag{10.2}$$

Of course, $F(y) = 0$ for $y \leq \alpha$ and $F(y) = 1$ for $y \geq \beta$. Because of the symmetry of f, it is obvious that $E(Y)$ falls precisely in the centre of the interval (α, β). To derive $V(Y)$, integration has to be used; we only give the result:

Expectation, variance and standard deviation of $U(\alpha, \beta)$

$$\mu = \frac{\alpha + \beta}{2}, \quad \sigma^2 = \frac{(\beta - \alpha)^2}{12}, \quad \sigma = \frac{\beta - \alpha}{\sqrt{12}} \qquad \text{(10.3)}$$

Example 10.1
Reconsider Examples 8.7, 8.8 and 9.5. The waiting time X for the train to Paris (in minutes) of Mrs Doubledutch has – equally likely – outcomes between 0 and 60, so $\alpha = 0$ and $\beta = 60$. It follows by (10.3) that:

$$E(X) = 30 \text{ min}; \quad V(X) = (60 - 0)^2/12 = 300 \text{ min}^2; \quad \sigma = 17.3205 \text{ min}$$

The time X (in minutes) it takes to solve the online Sudoku puzzle was modelled by a constant pdf concentrated on the interval [5, 46]; see the first model in Example 8.8. Hence, $\alpha = 5$ and $\beta = 46$. So: $\mu = 25.5$ min, $\sigma^2 = 140.0833$ and $\sigma = 11.8357$ min.
The waiting time X (in minutes) for the subway train in Example 9.5 has its outcomes in the interval [0, 6). All possible outcomes in [0, 6) are equally likely. Hence, $X \sim U(0, 6)$. We obtain: $E(X) = 3$ min, $V(X) = 3$ min² and $SD(X) = 1.7321$ min. Since $\alpha = 0$ and $\beta = 6$, it follows immediately from (10.2) that:

$$F(x) = x/6 \quad \text{for all } x \text{ in } [0, 6)$$

Of course, $F(x) = 0$ for $x < 0$ and $F(x) = 1$ for $x \geq 6$. This cdf can be used to calculate several probabilities. For instance, the probability that the waiting time is between 2.25 and 3.75 minutes:

$$\begin{aligned} P(2.25 < X < 3.75) &= P(X < 3.75) - P(X \leq 2.25) \\ &= P(X \leq 3.75) - P(X \leq 2.25) = F(3.75) - F(2.25) \\ &= 3.75/6 - 2.25/6 = 0.25 \end{aligned}$$

One of the family members has a special name: the probability distribution with $\alpha = 0$ and $\beta = 1$ is called the **standard uniform distribution**. Notice that this distribution has mean $\mu = \frac{1}{2}$, variance 0.0833, standard deviation 0.2887 and $F(x) = x$. It is this representative of the family that plays a crucial role when simulating observations from some probability distribution.
Suppose that you want to simulate a realization of an rv X with cdf F. That is, you want to generate a real number that could be a typical realization of X. This is easy if X is standard uniformly distributed and hence $F(x) = x$ for all x in (0, 1), since many computer packages have commands to simulate such a realization. We will call this command RAND; see Appendix A1.10 for Excel. But what to do if F is some general continuous cdf (Figure 10.2a) or discrete cdf (Figure 10.2b)?

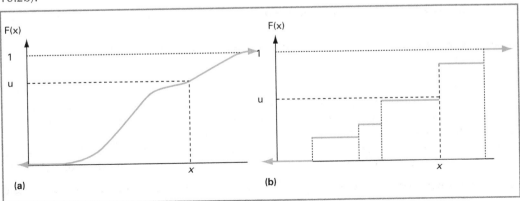

Figure 10.2 Simulation of a realization of X: (a) X continuous, (b) X discrete

In a sense, it is *F* that determines which realization of *X* will occur. Outcomes of *X* in intervals where *F* is very steep have a greater likelihood of becoming the realization than outcomes in intervals where *F* increases only slowly. That is why the following two-step simulation procedure intuitively makes sense:

1. Use RAND to simulate a random observation *u* from the interval (0, 1).
2. Use the cdf *F* to determine *x* such that $F(x) = u$.

Figure 10.2(a) illustrates the procedure if *X* is a continuous rv, Figure 10.2(b) if *X* is discrete. Regarding the discrete case, your question might be: what to do if the step function *F* actually **reaches** this randomly selected value *u*? That is, what to do if this value *u* happens to fall precisely at the level of a step? The answer is evasive but obvious: it will not happen. Since *u* is a realization of a continuous random variable *U*, the probability that it will be precisely equal to one of the values taken by the step function *F*, is zero.

Example 10.2

The yearly profit *X* (in millions of euro) of a company is modelled by way of the following pdf:

x	−1	0	1	2	3	4
f(x)	0.1	0.2	0.3	0.2	0.1	0.1
F(x)	0.1	0.3	0.6	0.8	0.9	1

The table below gives twenty simulations *u* from the standard uniform distribution *U*(0, 1), all obtained by way of the command RAND. The corresponding solutions of the equation $F(x) = u$ are also included.

u	*x*	*u*	*x*	*u*	*x*	*u*	*x*	*u*	*x*
0.6490	2	0.6153	2	0.6403	2	0.7775	2	0.4995	1
0.0891	−1	0.8645	3	0.3477	1	0.7523	2	0.8429	3
0.9224	4	0.1057	0	0.9463	4	0.6814	2	0.4552	1
0.5515	1	0.5316	1	0.6883	2	0.6215	2	0.6291	2

For instance, to solve $F(x) = 0.6490$, notice that *F* passes this level at 2; hence, *x* = 2. Although you might predict that a relatively large number of the 20 simulations will be equal to 1, it turns out that this is not the case. It would appear that 20 simulations is not enough to support such conjectures.

Example 10.3

The function *f* defined by $f(x) = 2x$ for *x* in [0, 1) and $f(x) = 0$ otherwise, is a pdf since it is non-negative and its total area equals 1 (create the graph yourself). But what do typical realizations look like of an rv *X* that has *f* as pdf? To obtain an impression, 10 000 simulations of *X* will be derived by way of the above two-step simulation procedure; see Figure 10.2(a).

The cdf *F* of *X* is determined first. For *x* between 0 and 1, *F(x)* is equal to $P(0 \le X \le x)$, the area of *f* above [0, *x*]. Since *f* takes the value 2*x* at *x*, the rectangle method yields the area $2x^2$ for the rectangle with sides x and $2x$ (create a picture yourself). Since the area of f above $[0, x]$ is just half of the area of this rectangle, it follows that $F(x) = x^2$ for all x in $[0, 1]$.

First consider the two-step simulation procedure in abstract terms. Suppose that step 1 yields the value u in the interval $(0, 1)$ on the vertical axis. The solution of the equation $F(x) = u$ can easily be obtained:

$$F(x) = u \Leftrightarrow x^2 = u \Leftrightarrow x = -\sqrt{u} \text{ or } x = \sqrt{u}$$

Since $x = -\sqrt{u}$ is negative, the only solution in $[0, 1]$ is $x = \sqrt{u}$. It follows that the value u obtained by RAND has to be transformed into $\sqrt{u}$ to obtain a simulation of X. This can be done with a computer; see also Appendix A1.10.

The file Xmp10-03.xls contains 10 000 simulations of X that were obtained by way of the above procedure. Figure 10.3 presents the classified frequency density for the classification $(0, 0.1]$, $(0.1, 0.2]$, ..., $(0.9, 1]$. Notice the similarity between the vertical levels of the pdf and the frequency density.

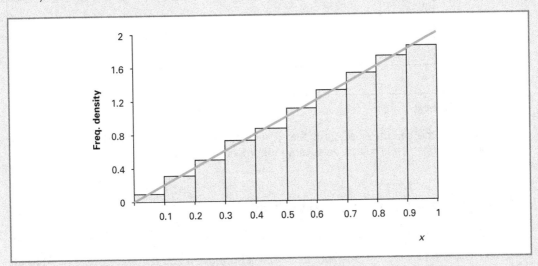

FIGURE 10.3 Histogram of frequency density of simulated data, with cdf included

10.2 Exponential distributions

If the set of outcomes of a continuous rv is the half-line $(0, \infty)$ and if the likelihood of the outcomes goes down exponentially fast when going from 0 to ∞, then the family of **exponential** pdfs may be a good model.

Suppose that μ is a **positive** real constant.

Exponential distribution

The *exponential distribution* with parameter μ is the probability distribution with the following pdf:

$$f(y) = \frac{1}{\mu} e^{-y/\mu} \quad \text{for all } y \text{ in the interval } (0, \infty) \tag{10.4}$$

$$= 0 \qquad\qquad \text{for } y \text{ otherwise}$$

If an rv Y has this pdf, it is written: $Y \sim Expo(\mu)$. Each choice of positive μ yields one pdf; so (10.4) defines a whole family of probability distributions. Figure 10.4 shows the exponential decay of the function and that $f(0) = 1/\mu$.

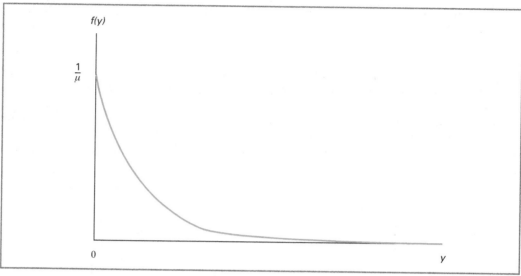

FIGURE 10.4 Pdf of $Expo(\mu)$

The name of the parameter suggests that it will be the mean of the distribution. Indeed, if $Y \sim Expo(\mu)$ then $E(Y) = \mu$. The box gives more (unproved) properties:

Some properties of $Y \sim Expo(\mu)$ with cdf F

$$E(Y) = \mu, \quad V(Y) = \mu^2, \quad F(y) = 1 - e^{-y/\mu} \quad \text{for all } y \geq 0 \tag{10.5}$$

$$P(Y > a + h \mid Y > a) = P(Y > h) \qquad \text{for all positive } a \text{ and } h \tag{10.6}$$

Because of the last property, the exponential distribution is said to be **_memoryless_**: if it is already given that Y is larger than a, then the probability that Y will even be larger than $a + h$ depends only on the increase h; the a is forgotten.

This family of models is often used for random variables that measure some kind of service time, waiting time or lifetime, for instance the waiting time until the next customer enters a shop, the waiting time for the next breakdown of a machine, the lifetime of a lamp or a tyre, etc.

Example 10.4

Suppose that the service time Y (in minutes) of a customer at the information desk of an airport is exponentially distributed with parameter 3.6. At the moment you arrive at the information desk, two customers (say, A and B) are being served at the only two counters of the desk. You will take the counter that becomes free first. What is the probability that your service time will be more than 5 minutes? What is the probability that you will leave the information desk **after** both persons A and B?

Since your service time Y is $Expo(3.6)$ distributed, it follows from (10.5) that:

$$P(Y > 5) = 1 - F(5) = 1 - (1 - e^{-5/3.6}) = e^{-5/3.6} = 0.2494$$

The answer to the second question is rather intuitive and is strongly based on the lack of memory of the exponential distribution. From the moment that a counter becomes free, the remaining service time of the customer at the other counter has again the original distribution $Expo(3.6)$. This is because of (10.6). So, the answer to the second question is 0.5.

The family of exponential distributions is especially important in situations where some type of occurrence is studied (for instance, arrival times of customers or visitors, transaction times of a stock, occurrences of disasters). If the numbers of occurrences in intervals are modelled with Poisson distributions, then the exponential distribution can often be used to model the waiting time **between** consecutive occurrences. We will not go into detail but will illustrate things with an example.

Example 10.5

In Example 9.10 the occurrences of interest were the arrivals of orders for the cultivator. If X denotes the number of occurrences in b days, then the expectation of X is $1.2b$ and the distribution of X is $Po(1.2b)$. Furthermore, let Y be the waiting time (in days) for the first order. But notice that the events $\{X = 0\}$ and $\{Y > b\}$ are the same: if there is no order in a period of b days, then the waiting time for the first order is more than b days (and vice versa). Hence, by (9.10) with $k = 0$ and $\mu = 1.2b$:

$$P(Y > b) = P(X = 0) = e^{-1.2b} \quad \text{and} \quad P(Y \le b) = 1 - e^{-1.2b}$$

This is valid for all non-negative b. By (10.5) it follows that Y is exponentially distributed with parameter $\mu = 1/1.2 = 0.8333$.

10.3 Normal distributions

A normal distribution is often used as a model for random variables for which the likelihood of the outcomes has a bell-shaped curve: the likelihood being highest in the centre and gliding down mountain-wise when going away from the centre.

10.3.1 The general normal distribution

Suppose that μ and σ are real constants; μ may be any real number but σ is strictly positive.

Normal distribution

The *normal distribution* with parameters μ and σ^2 is the probability distribution that belongs to the following pdf:

$$f(y) = \frac{1}{\sigma\sqrt{2\pi}} \exp(-(y - \mu)^2/(2\sigma^2)) \quad \text{for all real numbers } y \qquad (10.7)$$

Notation: $Y \sim N(\mu, \sigma^2)$ if Y has this density

Here, π is the constant 3.14159... and $\exp(x) = e^x$ is the exponential function with base $e = 2.71828...$. The parameters are respectively called the **mean parameter** and the **variance parameter**. Again, each choice of μ and positive σ^2 leads to one pdf, so in essence (10.7) defines a whole family of pdfs. Here are some properties; the first three follow from (10.7), the last is not proved:

Some properties of the pdf of $N(\mu, \sigma^2)$

- $f(\mu) = 1/(\sigma\sqrt{2\pi})$ is the maximum value of f.
- $f(\mu - a) = f(\mu + a)$ for all positive a; so f is **symmetric** around μ.
- If $y \to -\infty$ or $y \to \infty$, then $f(y) \to 0$.
- At $y = \mu - \sigma$ and $y = \mu + \sigma$ the graph of f has a turning point in the sense that the decline decreases when going further from μ.

Figure 10.5 presents graphs of three members of the family of normal distributions.

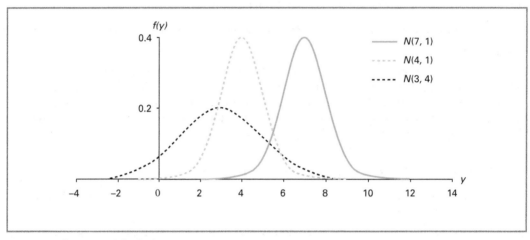

FIGURE 10.5 Some normal distributions

Notice that the shapes of $N(4, 1)$ and $N(7, 1)$ are the same; only their centres of location (respectively 4 and 7) differ. However, the pdf of $N(3, 4)$ is considerably more spread out than the other two pdfs.

The parameters of the normal distribution are just the expectation and the variance, as was already indicated by the labels μ and σ^2.

Expectation and variance of $N(\mu, \sigma^2)$

If $Y \sim N(\mu, \sigma^2)$, then $\begin{cases} E(Y) = \mu \\ V(Y) = \sigma^2 \end{cases}$ (10.8)

Unfortunately, the cdf of Y cannot be written in a closed form (as was the case for the cdf of the uniform distribution; see (10.2)). To determine $F(b) = P(Y \le b)$, we need a computer to calculate this area under the pdf of Y on the left-hand side of b. Many computer packages will calculate such cumulative probabilities; see also Appendix A1.10. (Special care has to be taken with respect to the parameter for the variation, since many packages want the standard deviation and not the variance.) Other probabilities can be derived from such cumulative probabilities by using probability rules and the special symmetry of the normal density. Table 3 in Appendix A4 can be used only if $\mu = 0$ and $\sigma^2 = 1$, and only for special positive values of b.

Things will be illustrated with an example.

Example 10.6

Suppose that the yearly returns (%) of investments I and II respectively have the distributions $N(10, 25)$ and $N(10, 100)$. Calculate the probabilities of negative returns, of returns less than 10%, of returns between 5% and 15%, and of returns more than 20%. The rv R denotes the yearly return as a percentage with respect to the year before.

With most computer packages only probabilities of the form $P(R \le b)$ can be calculated. Here, we need such probabilities for $b = 0, 5, 10, 15$ and 20. The table lists them all, for both investments (check them). See Appendix A1.10.

	$P(R \le b)$	
b	Investment I	Investment II
0	0.0228	0.1587
5	0.1587	0.3085
10	0.5000	0.5000
15	0.8413	0.6915
20	0.9772	0.8413

With this table the probabilities that are wanted can be calculated:

$P(R < 0)$ = 0.0228, respectively 0.1587

$P(R < 10)$ = 0.5, in both cases (by symmetry)

$P(5 < R < 15) = P(R \le 15) - P(R \le 5) = 0.6826$, respectively 0.3830

$P(R > 20)$ = $1 - P(R \le 20) = 0.0228$, respectively 0.1587

The two investments have the same expected return. However, investment II is more risky. The larger standard deviation has several consequences in terms of probabilities, for instance that the probability of a negative return is, for investment II, almost 7 times as large as for investment I.

But what to do if interest is the other way around, if the area at the left-hand side of a constant b is given and b itself has to be calculated? This 'inverse' problem will be considered now.

Figure 10.6 shows the pdf of $X \sim N(3, 4)$. An area 0.35 at the left-hand side of b is given and indicated. How to determine b?

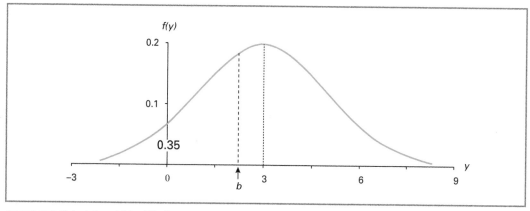

FIGURE 10.6 Calculation of b for $N(3, 4)$

It appears that we have to solve b in the equation $P(X \leq b) = 0.35$. Most computer packages have an option to determine such b; the name of the command often refers to the 'inverse' operation. For Excel, see Appendix A1.10. It turns out that $b = 2.2294$ (∗).

Example 10.7

An investor is thinking of putting money into one of the investment opportunities I and II of Example 10.6. Since he accepts a 10% chance that his return will be low, he decides to determine – for I as well as for II – the constant b in the equation $P(R < b) = 0.10$ and uses the answers to take a decision, bearing in mind that a bank account would offer a 3% guaranteed return.

Here are the results (check them):

Investment I: $b = 3.5922$,   investment II: $b = -2.8155$ **(∗)**

When investing his money in I (respectively II), he runs a 10% risk that the return will be less than 3.59% (respectively −2.82%). In view of the 3% interest rate offered by a bank account, he will choose to invest his money in I.

10.3.2 The standard normal distribution

One representative of the normal family is of special importance:

Standard normal or z-distribution

The normal distribution with $\mu = 0$ and $\sigma^2 = 1$ is called **standard normal distribution**. Its pdf follows from (10.7):

$$f(z) = \frac{1}{\sqrt{2\pi}} \exp(-z^2/2) \quad \text{for all real numbers } z$$ **(10.9)**

A standard normally distributed random variable will usually be denoted by Z.

The pdf of Z is symmetric around 0. This property is frequently used when calculating probabilities concerning Z. The area at the right-hand side of 0 is equal to the area at the left-hand side of zero, so the corresponding probabilities are both ½. In fact **all** probabilities $P(Z > a)$ and $P(Z < -a)$ are equal; see the areas at the right of a and at the left of $-a$ in Figure 10.7.

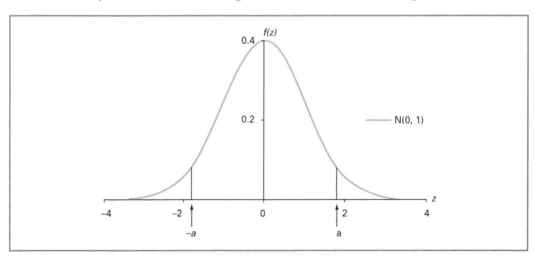

FIGURE 10.7 Pdf of $N(0, 1)$

The standard normal **distribution function** (cdf) has its own name:

$\Phi(z) = P(Z \leq z)$ for all real numbers z

(See Appendix A3 for the Greek letter Φ.) It is the cdf Φ that is tabulated in Table 3 of Appendix A4. Figure 10.8 depicts the graph of Φ:

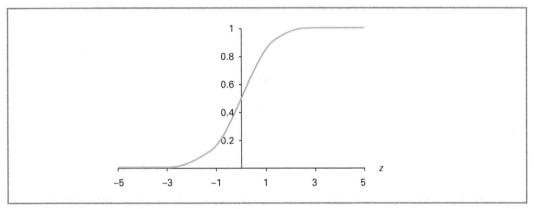

FIGURE 10.8 The cdf Φ of the standard normal distribution

The following table summarizes some of the values $\Phi(z)$. The reader is invited to check them (∗).

z	−4	−3	−2	−1	0	1	2	3	4
$\Phi(z)$	3.1686E-05	0.00135	0.02275	0.15866	0.5	0.84134	0.97725	0.99865	0.99997

The box contains some further properties of the standard normal distribution, all consequences of the symmetry of the pdf around 0; a is a real number.

Some properties of $Z \sim N(0, 1)$

(All properties can be remembered easily by drawing graphs.)

1 $P(Z < 0) = P(Z > 0) = \frac{1}{2}$.
2 $P(Z > a) = P(Z < -a)$.
3 $\Phi(a) = 1 - \Phi(-a)$.
4 $P(|Z| < a) = P(-a < Z < a) = 2\Phi(a) - 1$ for $a > 0$.
5 $P(|Z| < 1) = 0.6827$; $P(|Z| < 2) = 0.9545$; $P(|Z| < 3) = 0.9973$.   (∗)

Properties (1) and (2) have already been proved above; (3) follows immediately from (2), since:

$\Phi(a) = 1 - P(Z > a) = 1 - P(Z < -a) = 1 - \Phi(-a)$

The first equality in (4) follows from the fact that the events $\{|Z| < a\}$ and $\{-a < Z < a\}$ are the same. From Figure 10.7 it follows that the area between $-a$ and a is just 1 – the area in the two tails, which by (2) equals

$1 - 2P(Z > a) = 1 - 2 \times (1 - \Phi(a)) = 2\Phi(a) - 1$

The results in (5) are immediate consequences of (4).

Some computer packages have short-cut options for the calculation of probabilities of a **standard normal** rv and for accompanying inverse problems. As an exercise, the reader is invited to solve the equation $\Phi(b) = 0.6823$, to find the constant b that has, under the standard normal pdf, an area 0.6823 at its left-hand side.

Example 10.8
To get some practice with the standard normal distribution, the reader is invited to check the following probabilities (always draw the graph first) (∗).

a	b	$P(Z > a)$	$P(a < Z < b)$
−3	1	0.9987	0.8400
−2.5	0	0.9938	0.4938
−1	1	0.8413	0.6827
0	2.3	0.5000	0.4893
2	2	0.0228	0.0000

10.3.3 Relationships between $N(\mu, \sigma^2)$ and $N(0, 1)$

Without giving the formal details, it is just stated here that a linear transformation $Y = a + bX$ of an $N(\mu, \sigma^2)$ distributed rv X is again normally distributed. To find the parameters, recall from (8.30) that:

$$E(a + bX) = a + bE(X) = a + b\mu \quad \text{and} \quad V(a + bX) = b^2 V(X) = b^2 \sigma^2$$

Hence:

> If $X \sim N(\mu, \sigma^2)$ and $Y = a + bX$, then $Y \sim N(a + b\mu, b^2\sigma^2)$ **(10.10)**

The standardized version $Z = (X - \mu)/\sigma$ of X is a special linear transformation of X; see (8.33). It has expectation 0 and variance 1. Hence:

Standardization of $N(\mu, \sigma^2)$

The standardized version of a $N(\mu, \sigma^2)$ random variable is standard normally distributed.

Or, to say it in a more formal way:

> If $X \sim N(\mu, \sigma^2)$ and $Z = \dfrac{X - \mu}{\sigma}$, then $Z \sim N(0, 1)$ **(10.11)**

On the other hand, for all μ and all positive σ, the linear transformation $Y = \sigma Z + \mu$ of a $N(0, 1)$ rv Z is again normally distributed with parameters:

$$E(Y) = \sigma E(Z) + \mu = \mu \quad \text{and} \quad V(Y) = \sigma^2 V(Z) = \sigma^2$$

Hence:

> If $Z \sim N(0, 1)$ and $Y = \sigma Z + \mu$, then $Y \sim N(\mu, \sigma^2)$ **(10.12)**

From the above considerations it follows that each probability regarding a normal rv can be transformed into a probability regarding its standardized version. To illustrate this, assume that Y is $N(\mu, \sigma^2)$ and that $Z = (Y - \mu)/\sigma$. Since the event $\{Y \leqslant y\}$ remains unchanged when μ is subtracted from both sides of the inequality sign and the results are divided by σ, the accompanying probabilities do not change either:

$$P(Y \leq y) = P\left(\frac{Y - \mu}{\sigma} \leq \frac{y - \mu}{\sigma}\right) = P\left(Z \leq \frac{y - \mu}{\sigma}\right) = \Phi\left(\frac{y - \mu}{\sigma}\right) \tag{10.13}$$

Hence, if the parameters μ and σ^2 are known, then cumulative probabilities of Y can be determined with the cdf of the standard normal distribution.

Example 10.9

To comply with the weekly demand of a certain product, a company has a stock capacity of 500. On average, the company sells 400 items of that product per week with a standard deviation of 80. Assuming that the weekly demand D of the product is normally distributed, what is the probability that in a certain week the company will run out of stock? How much additional capacity is needed to decrease that probability to only 5%?

From the above facts it follows that $D \sim N(400, 6400)$ and that the probability $P(D > 500)$ is wanted:

$$P(D > 500) = P\left(\frac{D - 400}{80} > \frac{500 - 400}{80}\right) = P(Z > 1.25)$$

$$= 1 - P(Z \leq 1.25) = 1 - 0.8944 = 0.1056 \tag{*}$$

For the second question, the constant s is wanted such that $P(D > s) = 0.05$:

$$0.05 = P(D > s) = P\left(\frac{D - 400}{80} > \frac{s - 400}{80}\right) = P\left(Z > \frac{s - 400}{80}\right)$$

Since, on the other hand, $0.05 = P(Z > 1.6449)$, it follows that: $\qquad$ (*)

$$(s - 400)/80 = 1.6449 \text{ and } s = 531.592$$

The extra capacity is 32.

Instead of using the empirical model as was done in Chapter 8, the daily return of the Carlsberg stock can also be modelled by a **normal** model. For the parameters μ and σ^2 we take the mean 0.0696 and variance 3.2916 of the historical data; see Case 08–01.xls. For the daily return R on a day in the future, this automatically means that the expected daily return is 0.0696% and that the risk as measured by the standard deviation σ_R is 1.8143%. However, also the probability $P(R < 0)$ measures an aspect of risk:

$$P(R < 0) = P\left(\frac{R - 0.0696}{1.8143} < \frac{-0.0696}{1.8143}\right) = P(Z < -0.0384) = 0.4847 \tag{*}$$

10.3.4 Special quantiles of $N(0, 1)$

In inferential statistics, the solution b of the equation $\Phi(b) = 0.975$ will turn out to be of special importance and is generally denoted by $z_{0.025}$. Here, 0.025 is just the complementary area of 0.975. Notice that $z_{0.025}$ is a real number that is located on the horizontal line under the graph of the pdf of $N(0, 1)$, in such a way that there is an area 0.975 to its left and hence an area 0.025 to its right.

More generally, the notation z_α is used to denote a position on the horizontal axis under the pdf of $N(0, 1)$, located in such a way that an area α is cut off at the right-hand side (and hence an area $1 - \alpha$ remains at the left-hand side). (Here, α is the Greet letter a.) In inferential statistics, α is some small, positive, fixed number, often taken as 0.10, 0.05, 0.025, 0.01 or 0.001. Note especially that the notation z_α is based on the area at the **right**-hand side, see Figure 10.9.

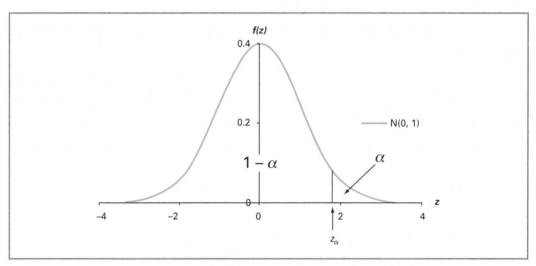

FIGURE 10.9 The location of z_α on the horizontal axis of the pdf of $N(0, 1)$

Notice that z_α is the $(1 - \alpha)$-quantile $\xi_{1-\alpha}$ of Z; see also (8.41). It splits the total area under the pdf into two parts in such a way that $1 - \alpha$ remains at the left side:

$$P(Z \leq z_\alpha) = 1 - \alpha \quad \text{and} \quad P(Z \geq z_\alpha) = \alpha \tag{10.14}$$

The table below lists z_α for some special α. For instance, the calculation of $z_{0.10}$ is an inverse operation; the equation $P(Z \leq b) = 0.90$ has to be solved. Check things yourself ($*$):

α	0.10	0.05	0.025	0.01	0.001
z_α	1.2816	1.6449	1.9600	2.3263	3.0902

Sometimes it is desirable to cut off a total area α from the two tails of the z-distribution, in such a way that one half comes from the right tail and the other half from the left. Since $z_{\alpha/2}$ denotes the position such that there is an area $\alpha/2$ at the right, the symmetry of the $N(0, 1)$ pdf guarantees that there will also be an area $\alpha/2$ at the left side of $-z_{\alpha/2}$. See Figure 10.10. Since these areas refer to probabilities, it follows that:

$$P(-z_{\alpha/2} < Z < z_{\alpha/2}) = 1 - \alpha \tag{10.15}$$

But recall that Z can be taken as the standardized version of a $N(\mu, \sigma^2)$ distributed rv Y, that is, $Z = (Y - \mu)/\sigma$ with $Y \sim N(\mu, \sigma^2)$. By substituting this into the event $\{-z_{\alpha/2} < Z < z_{\alpha/2}\}$ and reorganizing terms, it follows that:

$$\{-z_{\alpha/2} < Z < z_{\alpha/2}\} = \{-z_{\alpha/2} < \frac{Y - \mu}{\sigma} < z_{\alpha/2}\} = \{-z_{\alpha/2}\sigma < Y - \mu < z_{\alpha/2}\sigma\} \tag{10.16}$$

By adding up μ to all sides of the inequalities, the right-hand event in (10.16) can be rewritten as $\{\mu - z_{\alpha/2}\sigma < Y < \mu + z_{\alpha/2}\sigma\}$, which has Y enclosed by two boundaries. By (10.15), the first equation in the box below follows.

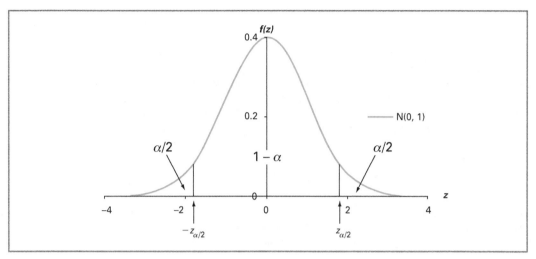

FIGURE 10.10 Cutting off a total area α in two tails

Important properties of $Y \sim N(\mu, \sigma^2)$

$$P(\mu - z_{\alpha/2}\sigma < Y < \mu + z_{\alpha/2}\sigma) \quad = 1 - \alpha \tag{10.17}$$

$$P(Y - z_{\alpha/2}\sigma < \mu < Y + z_{\alpha/2}\sigma) \quad = 1 - \alpha \tag{10.18}$$

For the second equality, rewrite the right-hand event in (10.16) such that μ is enclosed by two boundaries:

$$\{-z_{\alpha/2}\sigma < Y - \mu < z_{\alpha/2}\sigma\} = \{Y - \mu > -z_{\alpha/2}\sigma\} \cap \{Y - \mu < z_{\alpha/2}\sigma\}$$

$$= \{\mu < Y + z_{\alpha/2}\sigma\} \cap \{\mu > Y - z_{\alpha/2}\sigma\}$$

$$= \{Y - z_{\alpha/2}\sigma < \mu < Y + z_{\alpha/2}\sigma\}$$

(In the second equation, the terms are reorganized and the results are read from the opposite direction.) Equation (10.18) then follows from (10.16) and (10.15).

To get a better understanding of the importance of the above results, consider the special case that $\alpha = 0.05$, so $z_{\alpha/2} = 1.96$ and $1 - \alpha = 0.95$. Then (10.17) and (10.18) turn into:

$$P(\mu - 1.96\sigma < Y < \mu + 1.96\sigma) = 0.95$$

$$P(Y - 1.96\sigma < \mu < Y + 1.96\sigma) = 0.95 \tag{10.19}$$

The first equation informs us that an ongoing sequence of realizations of the normal rv Y has 95% of its realizations between $\mu - 1.96\sigma$ and $\mu + 1.96\sigma$. This result is important for descriptive statistics and has already been mentioned there: recall the empirical rule in Section 4.3.

The second equation states that there is a 95% chance that the mean μ is captured by the random interval $(Y - 1.96\sigma, Y + 1.96\sigma)$. This result will be crucial for inferential statistics; see especially Chapters 13–16.

Example 10.10

The manufacturer in Example 7.2 claims that only 0.5% of the packs of milk produced (1 litre, according to the label) contain less than one litre. Suppose, additionally, that the contents X (in litres) of an arbitrary pack of milk is normally distributed with mean $\mu = 1.02$ litre (case 1). If the manufacturer is right, how large is the standard deviation σ of X? Use this result to determine an interval with centre μ that contains X with probability 0.90. And if μ is **unknown** but σ is as calculated (case 2), determine a random interval such that μ is enclosed with probability 0.90.

According to the manufacturer, $P(X < 1) = 0.005$. This probability can be rewritten as the probability of the standardized version Z of X:

$$0.005 = P(X < 1) = P\left(\frac{X - \mu}{\sigma} < \frac{1 - \mu}{\sigma}\right) = P\left(Z < \frac{1 - 1.02}{\sigma}\right) = \Phi\left(\frac{-0.02}{\sigma}\right)$$

With $b = -0.02/\sigma$, the equation $\Phi(b) = 0.005$ has to be solved with a computer; an inverse operation. The answer is $b = -2.57583$ (*). Hence, $\sigma = 0.0078$ litre.

By (10.17) and (10.18) with $\alpha = 0.10$, it follows that:

$$P(\mu - 1.6449\sigma < X < \mu + 1.6449\sigma) = 0.90$$

$$P(X - 1.6449\sigma < \mu < X + 1.6449\sigma) = 0.90$$

Case 1: If $\mu = 1.02$ and $\sigma = 0.0078$, the first equation gives:

$$P(1.0072 < X < 1.0328) = 0.90$$

Case 2: If μ is unknown but $\sigma = 0.0078$, the second equation yields:

$$P(X - 0.0128 < \mu < X + 0.0128) = 0.90$$

Case 1 tell us that 90% of the packs contain between 1.0072 and 1.0328 litres, which is more than necessary. Case 2 informs us that the unknown μ will probably lie between the realizations of $X - 0.0128$ and $X + 0.0128$.

10.3.5 Normal approximation of $Bin(n, p)$

There are surprisingly many occasions where a normal probability distribution is a good model for a random variable. This holds especially for random variables that can be considered as the sum of many small and independent contributions. For instance, the height X of an adult is determined by many small, independent influences such as genetic factors, quality of nutrition when a child, physical well-being, etc. As another example, the precise contents of a pack of milk will be determined by many small, coincidental contributions during the filling process.

A **binomial** random variable X counts the number of successes in n independent trials; each trial adds 1 success or not. Hence, X is the sum of n independent contributions. In view of the above considerations, it can be expected that the probability distribution of X approaches a normal distribution as n becomes large. Recall from (9.3) that the distribution $Bin(n, p)$ has mean np and variance $np(1 - p)$, so the approximating normal distribution should be $N(np, np(1 - p))$. Indeed, the following can (but will not) be proved:

Approximation of $Bin(n, p)$ by $N(np, np(1 - p))$

If $np \geq 5$ and $n(1 - p) \geq 5$, then $Bin(n, p) \approx N(np, np(1 - p))$ **(10.20)**

The conditions in (10.20) are referred to as the **5-rule** conditions. This rule demands that the expected numbers of successes and failures are both at least 5. For each fixed p, the sample size n can always be taken so large that the two conditions are fulfilled. However, if the success probability p is small or large (for example, 0.001 or 0.999) then a large sample size is needed.

It is not yet clear what the exact meaning is of the approximation $\approx$ in (10.20). The $Bin(n, p)$ distributed rv X is discrete; it has only the integer outcomes 0, 1, ..., n. However, the $N(np, np(1 - p))$ distributed rv Y is continuous; a continuum of outcomes is possible. To approximate the probability $P(X = x)$ by the probability $P(Y = x)$ makes no sense, since the last probability is 0. But notice that events such as $\{X = 3\}$ can also be written as $\{2.5 < X < 3.5\}$ since X takes only integer values. More generally, for all $x = 0, 1, ..., n$, the event $\{X = x\}$ is the same as the event $\{x - 0.5 < X < x + 0.5\}$. That is why the probability $P(X = x)$ concerning the binomial rv X is approximated by the probability $P(x - 0.5 < Y < x + 0.5)$ concerning the normal rv Y. Adding and/or subtracting this 0.5 is called **correction for continuity**.

The use of this correction term 0.5 is extended to other events such as $\{X < 3\}$, $\{X \leq 3\}$, $\{X > 4\}$ and $\{X \geq 4\}$. Since they are equal to $\{X < 2.5\}$, $\{X \leq 3.5\}$, $\{X > 4.5\}$ and $\{X \geq 3.5\}$, their probabilities are respectively approximated by $P(Y < 2.5)$, $P(Y \leq 3.5)$, $P(Y > 4.5)$ and $P(Y \geq 3.5)$. The box below shows how the correction term is used in the normal approximation of (10.20); here $x = 0, 1, ..., n$.

Use of continuity correction for $X \sim Bin(n, p)$ and $Y \sim N(np, np(1 - p))$ ($x = 0, 1, ..., n$)

$P(X < x) \approx P(Y < x - 0.5)$ $\quad P(X > x) \approx P(Y > x + 0.5)$

$P(X \leq x) \approx P(Y \leq x + 0.5)$ $\quad P(X \geq x) \approx P(Y \geq x - 0.5)$

$P(X = x) \approx P(x - 0.5 < Y < x + 0.5)$

Example 10.11

The binomial distribution $Bin(20, 0.4)$ satisfies the 5-rule since $np = 8$ and $n(1 - p) = 12$. Hence, the probability distribution of $X \sim Bin(20, 0.4)$ can be approximated by the normal distribution of $Y \sim N(8, 4.8)$. To show how accurate this approximation is, we compare probabilities $f(x) = P(X = x)$ and $F(x) = P(X \leq x)$ with approximating probabilities of Y. The case will be demonstrated for $x = 7$:

$f(7) = P(X = 7) = P(6.5 < X < 7.5) \approx P(6.5 < Y < 7.5)$;

$F(7) = P(X \leq 7) = P(X \leq 7.5) \approx P(Y \leq 7.5)$

The right-hand normal probabilities have to be calculated with a computer. See Table 10.1 for the results, also for general k.

x	f(x)	P(x − 0.5 < Y < x + 0.5)	x	F(x)	P(Y ⩽ x + 0.5)
0	0.0000	0.0003	0	0.0000	0.0003
1	0.0005	0.0012	1	0.0005	0.0015
2	0.0031	0.0045	2	0.0036	0.0060
3	0.0123	0.0140	3	0.0160	0.0200
4	0.0350	0.0351	4	0.0510	0.0551
5	0.0746	0.0718	5	0.1256	0.1269
6	0.1244	0.1199	6	0.2500	0.2468
7	0.1659	0.1630	7	0.4159	0.4097
8	0.1797	0.1805	8	0.5956	0.5903
9	0.1597	0.1630	9	0.7553	0.7532
10	0.1171	0.1199	10	0.8725	0.8731
11	0.0710	0.0718	11	0.9435	0.9449
12	0.0355	0.0351	12	0.9790	0.9800
13	0.0146	0.0140	13	0.9935	0.9940
14	0.0049	0.0045	14	0.9984	0.9985
15	0.0013	0.0012	15	0.9997	0.9997
16	0.0003	0.0003	16	1.0000	0.9999
17	0.0000	0.0000	17	1.0000	1.0000
18	0.0000	0.0000	18	1.0000	1.0000
19	0.0000	0.0000	19	1.0000	1.0000
20	0.0000	0.0000	20	1.0000	1.0000

TABLE 10.1 Approximation of the pdf and the cdf of $Bin(20, 0.4)$

Notice that the approximations in column 3 are close to the exact probabilities in column 2; the largest absolute difference is 0.0045. The approximations in column 6 of the exact cdf $F(x)$ in column 5 are also very good: the largest absolute difference is 0.0062.

The continuity correction will only be used for calculations of binomial **probabilities**. For 'inverse' questions (when a probability is given and some quantile has to be calculated), the continuity correction will **not** be used, to avoid all kinds of problems.

Example 10.12
A large company has 10000 employees; only 440 of them are female. For a certain task, n of the employees will be sampled randomly. Interest is in the probability distribution of the rv X that counts the number of male employees in the sample. You have to do the analysis with the restriction that your computer package can only calculate **standard normal** probabilities.

How large, according to the 5-rule, should the sample be to find good approximations? Find this minimum sample size and use it to calculate the probability that more than 98% of the sampled employees are male. Also determine an approximation of the 0.90-quantile of X. That is, calculate the value (denoted $\xi_{0.90}$) such that there is a 90% chance that the number of males in the sample is below it.

The rv X counts the number of 'successes' in the sample, where a success corresponds to an employee being male. Since the sample is drawn **without** replacement, X has a hypergeometric distribution with $N = 10000$, $M = 9560$ and n is not yet chosen. As long as the ratio n/N is at most 0.1, this distribution can be approximated by the binomial distribution with parameters n and

$p = 0.956$; see (9.9). But $Bin(n, p)$ can in turn be approximated by $N(p, np(1 - p))$ **if** the 5-rule is satisfied. That is, $n(1 - 0.956)$ has to be at least 5, so n has to be at least 114. With the sample size $n = 114$, not only is the ratio n/N much smaller than 0.1 but the 5-rule holds as well. Hence:

$$Bin(114, 0.956) \approx N(108.9840, 4.7953)$$

Below, an rv Y with this normal distribution is transformed into a standard normal rv Z, to answer the last two questions above.

For the second question, notice that 98% of 114 yields 111.72, so 'more than 98%' corresponds to 'at least 112':

$$P(X \geq 112) = P(X \geq 111.5) \approx P(Y \geq 111.5) = P\left(\frac{Y - \mu}{\sigma} \geq \frac{111.5 - \mu}{\sigma}\right)$$

$$= P\left(Z \geq \frac{111.5 - 108.9840}{2.1898}\right) = P(Z \geq 1.1490)$$

$$= 1 - P(Z \leq 1.1490) = 1 - 0.8747 = 0.1253 \qquad (*)$$

For the last question, a constant b has to be determined such that the cdf of X passes or reaches the vertical level 0.90 at b. Since this cdf is approximated by the cdf of Y, the constant b has to be determined such that $P(Y \leq b) = 0.90$. This is an inverse problem; no continuity correction will be used. Notice that

$$0.90 = P(Y \leq b) = P\left(\frac{Y - \mu}{\sigma} \leq \frac{b - \mu}{\sigma}\right) = P\left(Z \leq \frac{b - 108.9840}{2.1898}\right) = \Phi(c)$$

where $c = (b - 108.9840)/2.1898$. Since $z_{0.10} = 1.2816$ and hence $P(Z \leq 1.2816) = 0.90$, it follows that $c = 1.2816$ and $b = 111.790$. The conclusion is that $\xi_{0.90} \approx 111.79$. (For comparison: the cdf of $Bin(114, 0.956)$ passes the level 0.9 at 112.)

The normal distribution is very important for inferential statistics, as will become clear from Chapter 13 onwards. In particular, the properties (10.17) and (10.18) are crucial.

There are more families of continuous probability distributions that are important for inferential statistics: the families of the Student t-distributions, the chi-square distributions and the F-distributions will respectively be introduced in Sections 16.1.1, 17.2.1 and 18.3.1.

CASE 10.1 EU LIMIT FOR CARBON DIOXIDE EMISSIONS – SOLUTION

The car manufacturer wants to determine μ, the mean CO_2 emission per kilometre for a car of the new type, such that:

$$P(X > 130) = 0.05$$

By assuming that X is normally distributed, the mean can be calculated by way of standardization:

$$0.05 = P\left(\frac{X - \mu}{2} > \frac{130 - \mu}{2}\right) = P\left(Z > \frac{130 - \mu}{2}\right)$$

However: since $z_{0.05} = 1.6449$, the standard normally distributed variable $Z = (X - \mu)/2$ also satisfies:

$$0.05 = P(Z > 1.6449)$$

Hence

$$\frac{130 - \mu}{2} = 1.6449 \text{ and } \mu = 126.7102$$

The new model of cars should have a mean CO_2 emission of 126.71 g/km.

Summary

Three types of continuous probability distributions have been considered: the uniform, the exponential and the normal distributions.

A uniform model is often used for a random variable if its possible outcomes are more or less equally likely. The standard uniform distribution is of special interest for simulation purposes.

Exponential models are frequently suitable for random variables that measure the waiting time between consecutive occurrences of a certain type, especially if the numbers of occurrences in intervals are modelled as a Poisson process. An exponential distribution is memoryless.

Normal models are frequently used for random variables that are in essence the result of many small and independent contributions. That is why a binomial variable often is approximated by a normal distribution. From Chapter 13 onwards, normal distributions will be crucial for most chapters.

Normal distributions have several nice properties:

- A normal pdf is symmetric around μ, a great help for calculating probabilities.
- A linear transformation of a normal rv is normal.
- The standardized version of a normal rv is standard normal.
- Probabilities of a normal rv can be rewritten in terms of probabilities of the accompanying standardized version.

The box below gives an overview of the distributions of this chapter:

Facts of three families of continuous distributions

Distribution	Notation	$f(y)$	Expectation	Variance
Uniform	$U(\alpha, \beta)$	$\frac{1}{\beta - \alpha}$ for $\alpha < y < \beta$	$\frac{\alpha + \beta}{2}$	$\frac{(\beta - \alpha)^2}{12}$
Exponential	$Expo(\mu)$	$\frac{1}{\mu} e^{-y/\mu}$ for $y \geq 0$	μ	μ^2
Normal	$N(\mu, \sigma^2)$	$\frac{1}{\sigma\sqrt{2\pi}} \exp(-(y - \mu)^2/(2\sigma^2))$	μ	σ^2

List of symbols

Symbol	Description
$U(\alpha, \beta)$	Uniform distribution
$Expo(\mu)$	Exponential distribution
$N(\mu, \sigma^2)$	Normal distribution
Φ	cdf of $N(0, 1)$
z_α	$(1 - \alpha)$-quantile of $N(0, 1)$

Exercises

Exercise 10.1

When rounding a real number to the nearest integer, the error that is made lies between -0.5 and 0.5. For instance: the number 9.4449 is rounded down to 9, with error 0.4449; the number 9.5 is rounded up to 10, with error -0.5.

Consider the experiment that randomly draws a real number between 0 and 100 and let X be the rounding error when rounding that number to the nearest integer.

a Find a good model for X. What is the pdf?

b Determine the expectation and the standard deviation of X.

c Calculate the probability that X lies between -0.2 and 0.4.

d Suppose that your computer package generates a random number from the interval (0, 1) by the command RAND. Explain how you can use this command to simulate a realization of X.

Exercise 10.2

An rv X is uniformly distributed over the interval $(-5, 10)$.

a Determine the pdf of X and use it to calculate $P(-2 < X < 8)$.

b Calculate the expectation and the variance of X.

c Determine the distribution function of X.

d Use it to calculate $P(-2.5 < X < 12.5)$.

Exercise 10.3

The waiting time X (in minutes) between two successive login times on a certain internet site is exponentially distributed with parameter 10.

a Determine the pdf of X.

b Determine the mean waiting time and the accompanying variance.

c Calculate the probability that this waiting time is at least 15 minutes.

d Calculate the probability that the waiting time is more than 15 minutes if it is already given that the last login took place more than 5 minutes ago.

Exercise 10.4

An rv Z is standard normally distributed. Calculate the probabilities below; always draw a graph and use a computer package.

a $P(Z > 1.4)$ and $P(Z < -2.3)$

b $P(-2.3 < Z < 1.4)$ and $P(-1.4 < Z < 1.4)$

c $P(0 < Z < 2.1)$ and $P(-1.8 < Z < 0)$

d $P(-1.8 < Z < 1.8)$

Exercise 10.5

An rv X is normally distributed with mean parameter 5 and variance parameter 9. Use graphs and a computer package to calculate the following probabilities:

a $P(X < 0)$, $P(X > 2.5)$, $P(X > 5)$ and $P(X > 11)$
b $P(2 < X < 8)$ and $P(8 < X < 11)$
c $P(X = 2.5)$ and $P(X \geq 2.5)$
d $P(2.5 < X < 7.5)$ and $P(X > 2)$

Exercise 10.6

An rv Z is standard normally distributed. Use graphs and a computer package to calculate the constants b in the following probabilities:

a $P(Z > b) = 0.25$
b $P(-b < Z < b) = 0.60$
c $P(-1.5 < Z < b) = 0.85$
d $P(0 < Z < b) = 0.30$

Exercise 10.7

The rv X is normally distributed with mean 7 and variance 16. Use graphs to solve the constants b in the following equations:

a $P(3.2 < X < b) = 0.65$
b $P(7 - b < X < 7 + b) = 0.40$

Exercise 10.8

Suppose that $X \sim Bin(120, 0.10)$.

a Which normal distribution is used to approximate $Bin(120, 0.10)$? Is the 5-rule satisfied?

Use normal approximation to calculate the following probabilities:

b $P(X = 12)$
c $P(X \leq 13)$ and $P(X > 11)$
d $P(6 < X < 15)$

Exercise 10.9

Use the standard normal table in Appendix A4 to answer the questions below:

a Suppose that $Z \sim N(0, 1)$. Calculate $P(-0.91 < Z < 1.51)$.
b Suppose that $Z \sim N(0, 1)$. Determine the constant a in $P(Z > a) = 0.66$.
c Suppose that $X \sim N(6, 25)$. Calculate $P(0.5 < X < 8.2)$.
d Suppose that $X \sim N(6, 25)$. Determine the constant b in $P(X > b) = 0.66$.

Exercise 10.10

The yearly percentage return R of an investment is normally distributed with mean 12% and variance 400.

a Calculate the probability that R will be negative.

b Putting money in a bank account yields a return of 3%. Calculate the probability that the return of the investment will be more.

c Calculate the probability that R will be more than 15 separated from the mean value.

d Find the constant b for which $P(R > b) = 0.60$.

Exercise 10.11

The owner of an appliance store uses a normal distribution with mean 10 and variance 9 to model the weekly net sales (in units of €1000).

a The owner needs €3500 a week to pay his staff. Determine the probability that the weekly net sales will not be enough to pay the whole of this amount.

b A bank that finances the appliance store measures the profitability of companies by determining the level b below which the net weekly sales fall in only 10% of the weeks. Calculate this level for the appliance store.

Exercise 10.12

For insurance companies it is very important to have an idea of the number of heavy storms per year. Suppose that such storms occur on average 2.5 times per year, and that the number of such storms in t years has a Poisson distribution with parameter $2.5t$. (Here, t can be any positive real number.) Let X be the waiting time (in years) for the next heavy storm.

a Let t be a positive number and let Y be the number of heavy storms in a period of t years. Explain why $P(X > t) = P(Y = 0)$, and calculate this probability.

b What type of distribution does X have? Why?

c Calculate the probability that the waiting time for the next heavy storm is at most 0.5 years.

d If it is given that it has been more than six months since the last heavy storm, what is the probability that it will be yet another six months before such a storm occurs?

Exercise 10.13

Consider the experiment that randomly draws a real number between 10 and 100. Let the rv X denote that number. Interest is in the probability distribution of X.

a Suppose that the command RAND of a certain computer package simulates the random drawing of a real number between 0 and 1. Transform this command into a command that can be used to simulate the random experiment of drawing a real number between 10 and 100.

b Give the name and the parameters of the probability distribution of X. Is X discrete or continuous?

c Let f be the probability density function of X. Determine $f(x)$ for all real numbers x.

d Calculate the expectation, variance and standard deviation of X.

e Calculate the following probabilities:

i $P(X = 25)$; ii $P(X \leq 25)$; iii $P(X < 25)$;

iv $P(X > 80.5)$ v $P(20.6 < X < 90.6)$ vi $P(\{X \leq 20\} \cup \{X > 35.1\})$

vii $P(\{X \leq 80\} \cap \{X > 20\})$

Exercise 10.14

Reconsider Exercise 10.13.

a Determine the 0.75-quantile of this uniform distribution. That is, find b such that $P(X \leq b) = 0.75$.

b Let F be the distribution function (cdf) of X. Determine $F(x)$ for all real numbers x. Use F to calculate $P(20.6 < X < 90.6)$ again.

c Create the graph of F.

Exercise 10.15

A service station sells between 10 000 and 35 000 litres of fuel a day; 1 litre of fuel costs €1.40. Interest is in the amount (X, in litres) of fuel and the gross sales (Y, in euro) on an arbitrary day. Below, the uniform distribution on the interval (10 000, 35 000) will be used to model X.

a Determine, according to this model, the mean daily amount of fuel sold. Also determine the pdf of X.

b Express Y in terms of X.

c Calculate the expectation, the variance and the standard deviation of both X and Y.

d Calculate the probabilities $P(X \leqslant 20 000)$ and $P(Y \leqslant 40 000)$. Also calculate $P(30 000 < Y < 40 000)$.

Exercise 10.16 (computer)

The rv Z is standard normally distributed; the rv X is $N(10, 16)$ distributed. Use a computer to answer the following questions:

a Calculate $P(Z \leq 1.23)$, $P(Z < 1.23)$, $P(Z < -0.12)$, $P(0.31 < Z \leq 1.47)$ and $P(-0.22 < Z < 2.02)$.

b Use part (a) to calculate the probabilities $P(X \leq 14.92)$, $P(X \leq 9.52)$, $P(11.24 < X < 15.88)$ and $P(9.12 < X < 18.08)$.

c Calculate $P(X > 18)$, $P(X > 22)$ and $P(X > 26)$. Use the symmetry of the normal distribution to determine the probabilities $P(X < 2)$, $P(X < -2)$ and $P(X < -6)$.

d Create the graph of the pdf $f(x)$ of X.

Exercise 10.17

Suppose that $X \sim N(10, 4)$. Determine the constants a, b and c; always make a rough picture first.

a Determine a such that $P(X > a) = 0.80$.

b Determine b such that $P(X < b) = 0.14$.

c Determine c such that $P(10 - c < X < 10 + c) = 0.40$.

Exercise 10.18

An rv X has the distribution $N(\mu, \sigma^2)$.

a Calculate $P(\mu - \sigma < X < \mu + \sigma)$.

b Calculate $P(\mu - 2\sigma < X < \mu + 2\sigma)$.

c Calculate $P(\mu - 3\sigma < X < \mu + 3\sigma)$.

d Interpret these probabilities.

e Compare these three probabilities with empirical rules from Section 4.3.

Exercise 10.19

The daily sales of a large store are between €50 000 and €100 000. Furthermore, it is known from the past that the mean of the daily sales is €75 000, that the daily sales are below this mean for 50% of days and that 42% of days have sales between €70 000 and €80 000.

In a series of three exercises, we will consider possible models for the probability distribution of the sales X (in units of €1000) on an arbitrary day.

   a Think of a suitable **uniform** model; determine its pdf f and graph it. Also calculate its mean and standard deviation.

   b Interpret, for x between 50 and 100, the value of $f(x)$.

   d Does the model fit the above empirical facts?

Exercise 10.20

Reconsider Exercise 10.19.

   a Think of a **triangular** pdf to model X; see also the graph in Example 8.8 (Figure 8.10). Determine this pdf (called g) and graph it. Also determine the mean of this probability distribution.

   b Interpret, for x between 50 and 100, the value of $g(x)$.

   c Does this model fit the empirical facts mentioned in Exercise 10.19?

Exercise 10.21

Reconsider Exercises 10.19 and 10.20.

   a Think of a suitable normal pdf to model X. To find the variance parameter, revisit the result of Exercise 10.18c.

   b Does this model fit the empirical facts mentioned in Exercise 10.19?

   c Compare the three models of Exercises 10.19, 10.20 and 10.21 in view of the empirical facts. Which model fits best?

   d Use the 'best' model to calculate the probability of high sales, that is, sales greater than €85 000.

Exercise 10.22

Let f be the pdf of a continuous random variable. Recall that $f(x)$ is **not** a probability. The meaning of a continuous pdf will be illustrated below for $X \sim N(5, \sigma^2)$ with pdf f.

   a Notice that $f(x)$ is greatest at $x = 5$. Calculate $f(5)$ for the case that $\sigma = 0.25$. How does the result illustrate that $f(5)$ is not a probability?

   b Suppose that $\sigma = 3$. Recall that $P(x \leq X \leq x + h) \approx f(x)h$ for positive h close to zero. This will be illustrated for $x = 6$. Calculate both $P(6 \leq X \leq 6 + h)$ and $f(6)h$ for $h = 1, 0.5, 0.25, 0.125, 0.0625, 0.0313, 0.0156$ and 0.0078. Put the results in a table and give your comments.

Exercise 10.23

From the very long history of a particular type of investment it is known that the mean yearly return of such an investment is 10% with a standard deviation of 5%. Let X be the yearly return (%) of an investment of this kind for the coming year. Suppose additionally that the normal distribution is used as a model, that the invested amount is €50 000, and that the transaction costs are €1000.

   a Determine the probability that the return is negative.

   b Let Y denote the net profit of the investment (in euro, after subtraction of the transaction costs). Express Y in X. Calculate the expectation and the standard deviation of Y.

   c Calculate the probability that the net profit of the investment is negative.

   d Determine the probability distribution of Y.

Exercise 10.24

How much money does a typical customer spend on a visit to a supermarket? The amount X (in euro) has a normal distribution with mean €40 and standard deviation €13.

a Calculate the probability that the customer will spend less than €45.

b Calculate the probability that the customer will spend at least €10 but at most €60.

c Interest is in the 10% of customers who spend the largest amounts of money. How much money does a customer have to spend to belong to that group?

d Within the group of part (c), determine the percentage of the customers who spend at least €70.

e Recall that $z_{0.025} = 1.96$. Use this information to determine the maximum of the amounts of money spent by 97.5% of the customers.

Exercise 10.25

Let Y be the amount of money (in euro) that is spent per visit by a typical family of four at a McDonald's restaurant; Y is normally distributed.

a Suppose that the mean amount spent by such families is μ euro, with a standard deviation of σ euro. Determine the percentage of such families that spend an amount that is less than 2.5 standard deviations separated from the mean. (Note that you do not need μ and σ to answer this question.)

b Suppose that μ is unknown and that $\sigma = €2.10$. Determine the constant a such that μ is captured between the random bounds $Y - a$ and $Y + a$ with probability 0.95.

c Suppose that the family-of-four spends €36.50. Determine, in the light of part (b), an interval that is likely to contain the unknown mean amount spent at McDonald's by families of four.

Exercise 10.26

The weights (in grams) of packs of butter have a normal distribution with mean $\mu = 252$ and standard deviation $\sigma = 2$. Below, Z is standard normally distributed.

a Calculate the probability that the weight of a pack is more than 255 g; use $P(Z > 1.5) = 0.0668$.

b Calculate the percentage of all packs produced whose weight is less than 250 g; use $P(Z > 1) = 0.1587$.

It is desirable to reduce the proportion of part (b) to 5%; this is done by adapting μ or σ.

c Determine μ if σ remains equal to 2 g; use $z_{0.05} = 1.6449$.

d Determine σ if μ remains 252 g; use $z_{0.05} = 1.6449$.

Exercise 10.27

A manufacturer of pens claims that only 0.2% of the pens produced are defective. A consumer organization wants to check this claim by testing n pens taken randomly from the total production (100 000 pens) of one day. Below, the random variable Y counts the number of defective pens in the sample.

a Suppose that $n = 500$. Calculate $P(Y = 1)$, $E(Y)$ and $V(Y)$.

b Is it wise to approximate the probability distribution of Y with a normal distribution? Why (not)?

Now suppose that 2500 pens are tested.

c Calculate $P(Y = 5)$, $E(Y)$ and $SD(Y)$.

d Use a suitable normal distribution to approximate the probability $P(Y = 5)$

e Also give approximations of the probabilities $P(Y = 6)$, $P(Y \leq 6)$, $P(Y > 7)$ and $P(4 \leq Y < 8)$.

Exercise 10.28

A manufacturer of aspirin claims that 55% of those who suffer a headache experience relief with just two aspirins. Consider a random sample of 300 headache sufferers.

a Use a normal approximation to determine the probability that less than 50% of the sampled persons experience relief with two aspirins.

b When the data are gathered, it turns out that only 145 of the 300 persons experience relief with two aspirins. What does this suggest about the claim of the manufacturer?

c Calculate $P(Y \leq 145)$, still under the assumption that the claim of the manufacturer is correct. Give your comments.

Exercise 10.29

This exercise is about the simulation from certain distributions with $U(0, 1)$.

a An rv X has the following pdf:

x	0	1	2	3
f(x)	0.375	0.375	0.188	0.062

The real numbers 0.529, 0.744 and 0.869 are random observations from the distribution $U(0, 1)$. Use them to simulate three random observations of X.

b Use the simulations 0.041, 0.290 and 0.787 of $U(0, 1)$ to obtain three simulations of the random variable Y with pdf:

$g(y) = (y + 2)/8$ for all y in $(-2, 2)$

$\quad\;\; = 0$ otherwise

Exercise 10.30

The company RentaWreck has five cars that can be hired but only for the whole day. The daily demand X (on working days) for cars has the following probability density function:

x	0	1	2	3	4	5	6	7
f(x)	0.05	0.14	0.22	0.22	0.17	0.11	0.06	0.03

It occasionally happens that the demand is greater than the supply.

a Determine the probability that on a certain day the demand is greater than the supply.

The manager of the company decides to buy another car next year if in the present year the demand is greater than the supply on at least 25 working days. Below, interest is in the probability that this car will indeed be bought. It will be assumed that this year has 250 working days and that the random experiment that checks daily (on working days, during the whole year) whether the demand is greater than the supply is a binomial experiment. The rv Y counts, for the present year, the number of the working days that the demand is greater than the supply.

b Explain what the assumption about the binomial experiment means.

c What type of probability distribution does Y have? Calculate the expectation and the standard deviation of this distribution. Also calculate the probability that the extra car will be bought.

d Also use a normal approximation to calculate the probability of part (c).

Exercise 10.31

The Intelligence Quotient (IQ) is measured by means of a psychological test. The scores for the test are scaled in such a way that the mean score for the whole population is 100 with standard deviation 16. It will be assumed that the score X of an arbitrary person has a normal distribution.

a How many of the 455 million EU25 inhabitants have an IQ of at most 80? How many have an IQ over 140?

To have the right to become a member of Mensa (an international society of people with very high IQs), you have to belong to the 2% of the population with the highest IQs.

b What is the minimum IQ demanded for potential Mensa members?

c What proportion of the (potential) Mensa members has an IQ exceeding 140?

Exercise 10.32 (computer)

Is the return of a stock really normally distributed? It is often argued that the fit of normal models to empirical returns data is not good enough near 0 and in the tails. Basically, it is argued that normal models often **underestimate** the observed relative frequencies near 0 and in the tails.

To get an idea, we consider the weekly closing prices of the stock Royal Dutch Shell during the period 7 October 2005 – 6 October 2006; see <u>Xrc10–32.xls</u>.[†]

a Create the accompanying time series of the weekly returns R (%) for the period 14 October 2005 – 6 October 2006.

b Create a histogram. Does it look like a normal distribution?

We compare the empirical distribution with the normal model with the same mean and variance, and calculate $P(-0.5 < R < 0.5)$, $P(R < -3)$ and $P(R > 5)$ for both models.

c Which normal distribution is meant?

d Calculate the three probabilities, for the empirical model and for the normal model. Do the answers support the above arguments?

CASE 10.2 THE GREEN AND RED PEOPLE

This case study is based on the paper 'Company charged with ethnic bias in hiring; test disparities need not imply racism' in *A Mathematician Reads the Newspaper* by J. A. Paulos, Part 2: Local, Business and Social Issues.

Consider a country where two groups of people are living, the Reds and the Greens. The height (cm) of the Reds is $N(174, 100)$ distributed, the height of the Greens is $N(176, 100)$ distributed. For convenience, assume that there are 1 million red and also 1 million green people in this country.

We shall consider some consequences of the difference in location of the variable 'height (cm)' for the two populations. (The class 140−<150 refers to the interval [140, 150) with 140 included and 150 not, etc.)

[†] *Source*: www.iex.nl.

a The first column of the table below contains a classification of the heights:

Height class (cm)	Number of Reds	Number of Greens	Ratio Red/Green
<140			
140 − <150			
150 − <160			
160 − <170			
170 − <180			
180 − <190			
190 − <200			
200 − <210			
≥ 210			

Fill in the other columns.

b Suppose that the members of the national basketball team are chosen purely on the basis of their height. What do you predict for the ratio Green/Red in the basketball team? Does this ratio say anything about special basketball talent of the Greens?

Moral: That the national basketball team consists mainly of Greens does not imply that the Greens are better at basketball.

Joint probability distributions

Chapter contents

Most problems in the real world deal with more than one variable. Financial advisers will generally advise clients to spread the risk and invest in **more than one** investment. A manager of a supermarket will be interested not only in the weekly sales but **also** in (their relationship with) the costs of weekly advertisements.

Usually, interest is not primarily in individual random variables but more in the relationships **between** them. In the present chapter the descriptive statistics concepts 'covariance', 'correlation' and 'linear dependence' of two population variables are generalized for the sub-field of probability.

Our starting point is a random experiment with sample space Ω and a probability model P. Furthermore, two random variables (rvs) X and Y are considered on Ω. Often, the two rvs are treated as a random pair (X, Y); possible outcomes are denoted (x, y). Special interest will be in probabilities of so-called **XY events**, events that deal with both X and Y. Examples of such events are $\{X \leq 3$ and $Y \leq 8\}$ and $\{X = 2$ and $Y = 6\}$.

Sections 11.1–11.3 are mainly about **discrete** rvs, since most of the concepts introduced are much more complicated for continuous rvs. For continuous rvs, some remarks are included at the end of Section 11.3. Section 11.4 is about rules of expectation and variance when more than one rv is concerned; for the derivation of these rules it is not necessary to restrict oneself to discrete rvs.

CASE 11.1 INSURANCE AGAINST BICYCLE THEFT

Insurance company Lowlands specializes in insurance for cars and bicycles. One type of insurance is a policy against bicycle theft covering new bicycles in the range €600–3000. In exchange for an annual premium, the insurer pays the price of the bicycle to the insured if the bicycle is stolen. In fact, the insurer takes over the **insurance risk** that is associated with the bicycle, which, from the point of view of the insurer, is a random variable X with possible outcomes 0 (if the bike is not stolen) and values within the range €600–3000.

Lowlands needs to have a good idea of the distribution of the risk X and the accompanying expected value, because these determine the level of the premium that will be charged to the insured. As a model, it is assumed that, **for the case of theft**, there is a probability of 0.04 that the insurer will have to pay $k \times 100$ euros to the insured, for all $k = 6, 7, \ldots, 30$. It is also assumed that the probability that a bike is stolen is 0.06.

Determine the premium from these homogeneous model assumptions for the risk in the case of theft, by assuming further that the company determines its premium by putting a 20% safety loading on top of the expected risk. Incidentally, what is the probability that a risk is larger than the premium that is paid for it? See the end of Section 11.3 for a solution.

11.1 Discrete joint probability density function

Suppose that X and Y are both discrete. To get some feeling for XY events, we start with an example.

Example 11.1
Reconsider the throw with the two 'obscure' dice in Example 6.1. Recall that die a has two 5s and four 2s, die b has three 4s and three 1s. The following diagram shows the four outcomes (top left of each cell) of a suitable sample space Ω jointly with their probabilities (bottom right of each cell); the outcomes are **not** equally likely.

(2, 4)	(5, 4)
12/36	6/36
(2, 1)	(5, 1)
12/36	6/36

Let X and Y be the total number of eyes and the maximum of the two dice, respectively. Notice that the possible outcomes of X are 3, 6, 9 and those of Y are 2, 4, 5. Consider the following XY events:

$$\{X \leq 6 \text{ and } Y < 5\} = \{X \leq 6\} \cap \{Y < 5\} = \{(2, 1), (5, 1), (2, 4)\} \cap \{(2, 1), (2, 4)\}$$
$$= \{(2, 1), (2, 4)\}$$

$$\{X = 6 \text{ or } Y \geq 4\} = \{X = 6\} \cup \{Y \geq 4\} = \{(2, 4), (5, 1)\} \cup \{(2, 4), (5, 1), (5, 4)\}$$
$$= \{(2, 4), (5, 1), (5, 4)\}$$

$$\{X = 3 \text{ and } Y = 2\} = \{(2, 1)\}$$

$$\{X = 3 \text{ and } Y = 4\} = \varnothing$$

$$\{X = 3 \text{ and } Y = 5\} = \varnothing$$

$$\{X = 6 \text{ and } Y = 2\} = \varnothing$$

$\{X = 6 \text{ and } Y = 4\} = \{(2, 4)\}$

$\{X = 6 \text{ and } Y = 5\} = \{(5, 1)\}$

$\{X = 9 \text{ and } Y = 2\} = \varnothing$

$\{X = 9 \text{ and } Y = 4\} = \varnothing$

$\{X = 9 \text{ and } Y = 5\} = \{(5, 4)\}$

Probabilities can now be determined:

$P(X \leq 6 \text{ and } Y < 5) = 12/36 + 12/36 = 2/3;$

$P(X = 6 \text{ or } Y \geq 6) = 12/36 + 6/36 + 6/36 = 2/3$

The probabilities $P(X = x \text{ and } Y = y)$ are in the table below. For instance, $P(X = 6 \text{ and } Y = 5) = P((5, 1)) = 6/36$.

y	2	4	5	$f(x)$
x				
3	12/36	0	0	12/36
6	0	12/36	6/36	18/36
9	0	0	6/36	6/36
$g(y)$	12/36	12/36	12/36	1

Notice that $P(X \leq 6 \text{ and } Y < 5)$ and $P(X = 6 \text{ or } Y \geq 4)$ can also be obtained from this table. In the **margins** of the table (that is, in the last column and the last row) the probability density functions f and g of X and Y are presented, where $f(x) = P(X = x)$ and $g(y) = P(Y = y)$. The pdf f arises by adding up the rows, whereas g arises by adding up the columns. For instance, the X event $\{X = 6\}$ is just the union of the XY events $\{X = 6 \text{ and } Y = 4\}$ and $\{X = 6 \text{ and } Y = 5\}$; the probabilities 12/36 and 6/36 of these XY events add up to 1/2.

In this section, interest is in the probabilities of XY events. The overview of all XY events accompanied by their probabilities is called the **joint probability distribution** of X and Y.

Joint probability density function *h* of discrete *X* and *Y*

$h(x, y) = P(X = x \text{ and } Y = y)$ for all outcomes x of X and y of Y

The joint pdf is also called the **joint (probability) density** of X and Y, for short. Notice that this joint pdf gives only a part of the overview offered by the joint probability distribution. Nevertheless, the joint probability distribution is completely determined by the joint pdf. That is why the joint pdf itself is often called the joint probability distribution. In contrast to this joint pdf, the individual pdfs f and g of, respectively, X and Y are called **marginal pdfs** or **marginal distributions** of X and Y.

By the axioms of Kolmogorov (6.11)–(6.13) that hold for the probability model P, it follows that $h(x, y)$ is non-negative and that the summation of $h(x, y)$ over all pairs of outcomes has to be 1. That is:

Discrete joint probability density

A *discrete joint probability density* is a function $h(x, y)$ that satisfies:

$$h(x, y) \geq 0 \quad \text{and} \quad \sum_{(x, y)} h(x, y) = 1 \tag{11.1}$$

As in Example 11.1, the joint pdf of X and Y can be presented in a two-dimensional table (at least, if the numbers of the possible outcomes of the two rvs are both finite). Such a table is called a **cross table**. Put the outcomes x of X in the rows and the outcomes y of Y in the columns. If x is a fixed outcome of X, then the X event $\{X = x\}$ is just the union (over all outcomes y of Y) of the XY events $\{X = x$ and $Y = y\}$. Hence, $f(x) = P(X = x)$ can be obtained from the joint pdf of X and Y by adding up all probabilities in the row that belongs to the outcome x of X. Analogously, for a fixed outcome y of Y, the Y event $\{Y = y\}$ is the union (over x) of all XY events $\{X = x$ and $Y = y\}$, and $g(y) = P(Y = y)$ arises from the joint pdf by adding up all probabilities in the column that belongs to the outcome y.

Example 11.2

An investor wants to invest €1000. She can choose between the shares of the electronic company PE (Phil Electronics) and the food supply chain AF (Albert Food). The price of the first share is €1 at the time the investment has to be made; the price of the second share is €2. How should the investor divide the €1000 between the two types of shares? In other words, how should she choose her portfolio?

Information about the joint probability density of the random prices X and Y of the shares PE and AF after one year is set out in the table.

x \ y	1.9	2	2.1
0.9	0.1	0.1	0
1	0	0.2	0.1
1.1	0	0.2	0.3

So, one year after making the investment the price X of one share PE can take the values 0.9, 1 and 1.1 euro. The price Y of one share AF can also take three values: 1.9, 2 and 2.1 euro. Below, the marginal distributions of X and Y will be considered. Their expectations and variances will be calculated and the individual performances of the two shares will be compared.

Firstly, notice that the function $h(x, y)$ that is presented in the table is indeed a joint pdf; (11.1) is satisfied since the numbers in the interior part of the table are non-negative and add up to 1. Next, the two marginal pdfs are obtained by adding up the rows and the columns of the table. They are presented below; here $f(x) = P(X = x)$ and $g(y) = P(Y = y)$.

x	0.9	1	1.1
$f(x)$	0.2	0.3	0.5

y	1.9	2	2.1
$g(y)$	0.1	0.5	0.4

Simple calculations show that $E(X) = 1.03$ and $E(Y) = 2.03$. Furthermore:

$V(X) = E(X^2) - (E(X))^2 = 1.0670 - 1.0609 = 0.0061$ and $SD(X) = 0.0781$

$V(Y) = E(Y^2) - (E(Y))^2 = 4.1250 - 4.1209 = 0.0041$ and $SD(Y) = 0.0640$

Each investment of €1 in PE yields an expected profit of 3 cents, while each investment of €1 in AF yields an expected profit of $3/2 = 1.5$ cents. Hence, at first sight, investing the money in PE shares seems to be the better choice. But on the other hand, the accompanying standard deviations of the profit per euro invested are 0.0781 for PE and $0.0640/2 = 0.0320$ for AF. Since these standard deviations represent the risks of the respective shares, the risk of investing all the money in PE shares is larger than the risk of investing all the money in AF shares.

Of course, it is also possible to invest, say, half of the €1000 in PE and the other half in AF. To determine the risk of that portfolio, the probabilistic interrelationship between X and Y will play a role too. To be continued.

Let $v(x, y)$ be some function of x and y. If x and y are replaced by the rvs X and Y, respectively, then a new rv $V = v(X, Y)$ arises. The expectation of V will depend on the probabilistic interrelationship of X and Y. That is: $E(V)$ will depend on the joint pdf $h(x, y)$ of X and Y, as follows:

$$E(V) = E(v(X, Y)) = \sum_{(x, y)} v(x, y)h(x, y) \tag{11.2}$$

Here, the summation concerns all possible pairs of outcomes (x, y). That is, for each cell in the interior part of the cross table, $v(x, y)$ has to be calculated and multiplied by the local probability $P(X = x$ and $Y = y)$; the results for all cells have to be added up.

In Example 11.2, the function $v(x, y) = xy$ yields the new rv XY. The table below gives, for all nine cells, the products xy and the accompanying joint probabilities.

xy h(x, y)	1.9	y 2	2.1
0.9	1.71	1.80	1.89
	0.1	0.1	0
x 1	1.90	2.00	2.10
	0	0.2	0.1
1.1	2.09	2.20	2.31
	0	0.2	0.3

To obtain $E(XY)$, the xy and $h(x, y)$ have to be multiplied for each cell; the results have to be summed:

$E(XY) = 1.71 \times 0.1 + 1.80 \times 0.1 + \ldots + 2.31 \times 0.3$

It follows that $E(XY) = 2.0940$.

11.2 Covariance and correlation

Recall from (5.2) that the definition of the concept 'covariance' of two **population** variables X and Y was based on the deviations $x - \mu_x$ and $y - \mu_y$ measured at the N population elements; this covariance is just the mean of all the products $(x - \mu_x)(y - \mu_y)$. The generalization of this concept for **random** variables X and Y is based on the **random** deviations $X - E(X)$ and $Y - E(Y)$. Below, this generalization will be defined as the expectation of the product $(X - \mu_x)(Y - \mu_y)$.

Consider the function $v(x, y) = (x - \mu_x)(y - \mu_y)$ with $\mu_x = E(X)$ and $\mu_y = E(Y)$. By replacing x and y by the discrete rvs X and Y, the new rv $v(X, Y) = (X - \mu_x)(Y - \mu_y)$ arises. Taking the expectation yields the covariance. By (11.2) it follows that:

Covariance of *X* and *Y* (discrete)

$$\sigma_{x,y} = E((X - \mu_x)(Y - \mu_y)) = \sum_{(x, y)} (x - \mu_x)(y - \mu_y)h(x, y) \tag{11.3}$$

Instead of $\sigma_{x,y}$, the notation cov(X, Y) will also be used frequently. For each cell, the product $(x - \mu_x)(y - \mu_y)$ has to be multiplied by the joint probability $h(x, y)$; the results have to be summed. Notice that cov(X, X) is just $V(X)$. That is, $\sigma_{x,x} = \sigma_x^2$. It would appear that the concept 'covariance' is a generalization of the concept 'variance'.

If there is a relatively large probability that the realizations of the rvs X and Y will fall on the same side of μ_x and μ_y respectively, and that the random deviations $X - \mu_x$ and $Y - \mu_y$ are either both positive or both negative, then the positive terms in the summation of (11.3) are dominant, $\sigma_{x,y}$ will be positive, and Y and X are positively linearly related. But if there is a relatively large probability that the realizations of X and Y will fall on opposite sides of μ_x and μ_y respectively, then the negative terms in (11.3) are dominant, $\sigma_{x,y}$ will be negative, and Y and X are negatively linearly related.

Example 11.3

Recall (from Example 11.2) the joint pdf of next year's prices X and Y of the shares PE and AF. Also recall that $\mu_x = 1.03$ and $\mu_y = 2.03$. The probability that X and Y fall on the same side of μ_x and μ_y, is equal to:

$P(X < \mu_x \text{ and } Y < \mu_y) + P(X > \mu_x \text{ and } Y > \mu_y) =$

$\quad = P(X < 1.03 \text{ and } Y < 2.03) + P(X > 1.03 \text{ and } Y > 2.03)$

$\quad = P(X \leq 1 \text{ and } Y \leq 2) + P(X = 1.1 \text{ and } Y = 2.1)$

$\quad = (0.1 + 0.1 + 0.2) + 0.3 = 0.7$

This probability is relatively large; Y and X are expected to be positively linearly related. The calculation of $\sigma_{x,y}$ can be done with a table similar to the last table in Section 11.1, with xy replaced by $(x - \mu_x)(y - \mu_y)$.

$(x-1.03)(y-2.03)$			y			
	$h(x, y)$	1.9		2		2.1
	0.9	0.0169		0.0039		−0.0091
			0.1		0.1		0
x	1	0.0039		0.0009		−0.0021
			0		0.2		0.1
	1.1	−0.0091		−0.0021		0.0049
			0		0.2		0.3

By multiplying the two numbers in all nine cells and adding the results, it follows that:

$\sigma_{X,Y} = 0.00169 + 0.00039 + 0.00018 - 0.00021 - 0.00042 + 0.00147 = 0.0031$

The covariance of X and Y is 0.0031 euro². Indeed, Y and X are positively linearly related.

The dimension (unity) of $\sigma_{X,Y}$ is just the product of the dimensions of X and Y. To make it dimensionless, the covariance is divided by the product of the two standard deviations:

Correlation coefficient of *X* and *Y*

$$\rho = \rho_{X,Y} = \frac{\sigma_{X,Y}}{\sigma_X \sigma_Y} \qquad (11.4)$$

It can be shown that ρ always lies between -1 and 1. If $\rho = 1$, then Y can be written as $a + bX$ for some constants a and b with b positive, so Y and X are strictly positively linearly related. If $\rho = -1$, then Y can be written as $a + bX$ for some constants a and b with b **negative**, so Y and X are strictly negatively linearly related. If $\rho = 0$ and hence $\sigma_{X,Y} = 0$, then Y and X are called **uncorrelated**; they are not linearly related. If ρ is close to 1 (or -1), then Y and X strongly tend to have their realizations at the same (or opposite) sides of their respective expectations; Y and X are strongly positively (or negatively) linearly related.

Example 11.4
The pair (X, Y) in Example 11.3 has $\sigma_X = 0.0781$, $\sigma_Y = 0.0640$ and $\sigma_{X,Y} = 0.0031$. Hence, the correlation coefficient ρ satisfies:

$$\rho = \frac{0.0031}{0.0781 \times 0.0640} = 0.6202$$

(Since the numerator and the denominator are both measured in euro squared, the above ratio is indeed dimensionless.) There is a positive linear relationship between the prices of the shares PE and AF.

Even in fairly simple examples like Example 11.3, the calculation of $\sigma_{X,Y}$ is quite a job. That is why it is natural to look (as was also done for the variance, in (8.32)) for a formula that makes life slightly easier. For that, the expectation rules of (8.30) and (8.31) are convenient:

$$\sigma_{X,Y} = E((X - \mu_X)(Y - \mu_Y)) = E(XY - X\mu_Y - \mu_X Y + \mu_X\mu_Y)$$
$$= E(XY) + E(-\mu_Y X) + E(-\mu_X Y) + E(\mu_X\mu_Y)$$

Since $\pm\mu_X$ and $\pm\mu_Y$ are constants, they can be put in front of the expectation operator:

$$\sigma_{X,Y} = E(XY) - \mu_Y E(X) - \mu_X E(Y) + \mu_X\mu_Y$$

Since $E(X) = \mu_X$ and $E(Y) = \mu_Y$, it follows that:

Short-cut formula for covariance of *X* and *Y*

$$\sigma_{X,Y} = E(XY) - \mu_X\mu_Y \qquad (11.5)$$

This formula expresses that the covariance of X and Y is just the **expectation of their product minus the product of their expectations**.

Regarding the prices X and Y of the shares PE and AF, it has already been calculated that $E(XY) = 2.0940$, $E(X) = 1.03$ and $E(Y) = 2.03$. By (11.5) it follows that $\sigma_{X,Y} = 2.0940 - (1.03 \times 2.03) = 0.0031$, which is indeed the same as the answer obtained in Example 11.3.

How do we know that the notations $\sigma_{X,Y}$ and ρ for the covariance and correlation of two random variables are justified, that they are indeed generalizations of concepts for two **population** variables that were introduced in Section 5.1? We will not go into detail, but only mention that they **are** justified and that again – similar to Section 8.6 – it is because of the special experiment 'random drawing' that these notations are allowed. Indeed, the covariance and correlation coefficient of the **random observations** of two population variables are equal to the covariance and correlation coefficient of those two population variables.

And how do the concepts 'covariance' and 'correlation coefficient' for random variables behave under linear transformations? Let $V = a + bX$ and $W = c + dY$ be linear transformations of the rvs X and Y. Because of the above considerations it may be expected that the relationships between $\sigma_{V,W}$ and $\sigma_{X,Y}$, and between $\rho_{V,W}$ and $\rho_{X,Y}$ will be identical to the relationships for population variables as described in Table 5.9. Indeed, this is the case, as can be demonstrated rather easily.

Since $\mu_V = a + b\mu_X$ and $\mu_W = c + d\mu_Y$, it follows that:

$$V - \mu_V = b(X - \mu_X) \quad \text{and} \quad W - \mu_W = d(Y - \mu_Y)$$

Hence,

$$\sigma_{V,W} = E((V - \mu_V)(W - \mu_W)) = E(bd(X - \mu_X)(Y - \mu_Y))$$
$$= bdE((X - \mu_X)(Y - \mu_Y)) = bd\sigma_{X,Y}$$

By (8.30) it follows that $\sigma_V = |b| \sigma_X$ and $\sigma_W = |d| \sigma_Y$, so:

$$\rho_{V,W} = \frac{\sigma_{V,W}}{\sigma_V \sigma_W} = \frac{bd\sigma_{X,Y}}{|bd| \sigma_X \sigma_Y}$$

Table 11.1 summarizes these results.

Covariance	$\sigma_{V,W} = bd\sigma_{X,Y}$
Correlation coefficient	If $bd > 0$: $\rho_{V,W} = \rho_{X,Y}$
	If $bd < 0$: $\rho_{V,W} = -\rho_{X,Y}$

TABLE 11.1 Covariance and correlation of $V = a + bX$ and $W = c + dY$

Example 11.5

Recall that the correlation coefficient of next year's prices X and Y of one share PE and one share AF is 0.6202. Now suppose that 500 shares PE (present price of one share is €1) and 250 shares AF (present price of one share is €2) are bought, and that for both investments €10 transaction costs have to be paid. Let V and W be the net profits after one year (after subtraction of the transaction costs) of the shares PE and AF, respectively. What can be said about the covariance and the correlation coefficient of V and W?

First, notice that V and W are linear transformations of X and Y, respectively:

$$V = 500X - 500 - 10 = 500X - 510 \quad \text{and} \quad W = 250Y - 510$$

Since both slopes are positive, it follows from Table 11.1 that $\rho_{V,W}$ is equal to $\rho_{X,Y} = 0.6202$. To determine $\sigma_{V,W}$ from $\rho_{V,W}$, the standard deviations σ_V and σ_W are calculated first. By (8.30) it follows that:

$$\sigma_V = 500\sigma_X = 500 \times 0.0781 = 39.05 \text{ and } \sigma_W = 250 \times 0.0640 = 16.00$$

Hence, $\sigma_{V,W} = \rho_{V,W} \times \sigma_V \times \sigma_W = 387.50$. According to Table 11.1, this result has to be equal to $500 \times 250 \times \sigma_{X,Y}$. Since $\sigma_{X,Y} = 0.0031$, this is indeed the case.

11.3 Conditional probabilities and independence of random variables

Again, X and Y are both discrete random variables that are considered in a random experiment with sample space Ω. Interest will be in probabilities of X events, Y events and XY events when some information on the outcome of the experiment is already available.

As in Section 7.4, 'pre-information' about the outcome of the random experiment means that it is already known that some event B has occurred. Interest is in the conditional probabilities:

$$f(x \mid B) = P(X = x \mid B) = \frac{P(\{X = x\} \cap B)}{P(B)}$$

$$g(y \mid B) = P(Y = y \mid B) = \frac{P(\{Y = y\} \cap B)}{P(B)}$$

$$h(x, y \mid B) = P(X = x \text{ and } Y = y \mid B) = \frac{P(\{X = x\} \cap \{Y = y\} \cap B)}{P(B)}$$

Example 11.6

The SPSS printout below concerns a population of 10 847 households. The variables are X = 'number of children (<18 years) per household' and Y = 'number of adults ($\geq$18 years) per household'. For a randomly drawn household, the corresponding random observations are also denoted by X and Y. Interest is in the influence of event B = {the income of the randomly drawn family is at least €20 000} on the constitution of the family.

(a) Count

		N adults 1	N adults 2	N adults 3	Total
N kids	0	605	2810	480	3895
	1	74	1116	213	1403
	2	22	1476	255	1753
	3	5	692	126	823
	4	3	184	27	214
	5	0	52	15	67
	6	0	10	0	10
	7	0	6	0	6
Total		709	6346	1116	8171

(b) Count

		N adults 1.00	N adults 2.00	N adults 3.00	Total
N kids	.00	463	1208	78	1749
	1.00	45	344	24	413
	2.00	11	380	18	409
	3.00	1	94	0	95
	4.00	0	10	0	10
Total		520	2036	120	2676

Table (a) gives the counts for the 8171 households with incomes of at least €20 000. For instance, there are 1476 of such households with 2 children and 2 adults. Table (b) gives similar counts for the 2676 households with incomes of less than €20 000; 380 of them have 2 children and 2 adults.

These data will be transformed into probabilities. For instance, the probabilities

$$h(2, 2 \mid B) = P(X = 2 \text{ and } Y = 2 \mid B) \quad \text{and} \quad h(2, 2 \mid B^c) = P(X = 2 \text{ and } Y = 2 \mid B^c)$$

can be obtained by dividing 1476 by 8171 and 380 by 2676, respectively. The cells of Table 11.2 below contain the probabilities $h(x, y \mid B)$ and $h(x, y \mid B^c)$.

$h(x, y \mid B)$ / $h(x, y \mid B^c)$	$y=1$	$y=2$	$y=3$	$f(x \mid B)$ / $f(x \mid B^c)$
$x=0$	0.0740	0.3439	0.0587	0.4766
	0.1730	0.4514	0.0291	0.6535
1	0.0091	0.1366	0.0261	0.1718
	0.0168	0.1286	0.0090	0.1544
2	0.0027	0.1806	0.0312	0.2145
	0.0041	0.1420	0.0067	0.1528
3	0.0006	0.0847	0.0154	0.1007
	0.0004	0.0351	0	0.0355
4	0.0004	0.0225	0.0033	0.0262
	0	0.0037	0	0.0037
5	0	0.0064	0.0018	0.0082
	0	0	0	0
6	0	0.0012	0	0.0012
	0	0	0	0
7	0	0.0007	0	0.0007
	0	0	0	0
$g(y \mid B)$	0.0868	0.7766	0.1365	1
$g(y \mid B^c)$	0.1943	0.7608	0.0448	1

TABLE 11.2 Probabilities $h(x, y \mid B)$ and $h(x, y \mid B^c)$

Several conclusions can be drawn; only a few are mentioned. Within the lower-income category, a relatively high number of households have no children and few (1 or 2) adults. Within the higher-income category, a relatively high number of households have children.

For a fixed event B, overviews of the probabilities $f(x \mid B)$, $g(y \mid B)$ and $h(x, y \mid B)$ are respectively called: **conditional pdf of X given B**, **conditional pdf of Y given B** and **conditional pdf of X and Y given B**.

The pre-information B is often in terms of random variables. If it is already known that $B = \{Y = y\}$ occurred, then the conditional probability $f(x \mid B)$ is:

$$P(X = x \mid Y = y) = \frac{P(X = x; Y = y)}{P(Y = y)} = \frac{h(x, y)}{g(y)} \tag{11.6}$$

For a fixed y, these probabilities are denoted by $f(x \mid Y = y)$. Notice that the joint density at (x, y) has to be divided by the marginal density of Y at y. Analogously, for a fixed x, the probabilities $P(Y = y \mid X = x)$ are denoted by $g(y \mid X = x)$:

$$g(y \mid X = x) = P(Y = y \mid X = x) = \frac{P(X = x; Y = y)}{P(X = x)} = \frac{h(x, y)}{f(x)} \tag{11.7}$$

For a fixed outcome y of Y, the conditional pdf of X given $\{Y = y\}$ is just the overview of all probabilities $f(x \mid Y = y)$. Notice that this conditional pdf is a pdf, since $f(x \mid Y = y) \geq 0$ and $\sum_x f(x \mid Y = y) = 1$; see also (8.1). Furthermore, the conditional expectation of a function $V = v(X)$ of X when it is given that $\{Y = y\}$ occurred, can be obtained from (8.17) by replacing the pdf by the conditional pdf:

Conditional expectation of $V = v(X)$ given that $\{Y = y\}$

$$E(V \mid Y = y) = E(v(X) \mid Y = y) = \sum_x v(x) f(x \mid Y = y) \tag{11.8}$$

Example 11.7

Reconsider the throw with the two obscure dice in Example 11.1. Recall that the rvs X and Y count the total number of eyes and the maximum of the two dice, respectively. The following joint and marginal pdfs were obtained:

y x	2	4	5	$f(x)$
3	12/36	0	0	12/36
6	0	12/36	6/36	18/36
9	0	0	6/36	6/36
$g(y)$	12/36	12/36	12/36	1

To determine the conditional pdf of X given $\{Y = 5\}$, notice that:

$$f(x \mid Y = 5) = \frac{h(x, 5)}{g(5)} = \frac{h(x, 5)}{12/36}$$

Hence, this conditional pdf is as follows:

x	3	6	9
$f(x \mid Y = 5)$	0	1/2	1/2

To obtain the conditional expectation of X given $\{Y = 5\}$, the outcomes x have to be multiplied by the corresponding conditional probabilities:

$E(X \mid Y = 5) = 3 \times 0 + 6 \times 1/2 + 9 \times 1/2 = 15/2$

To obtain the conditional expectation of $V = X^2$ given $\{Y = 5\}$, the squared outcomes x^2 have to be multiplied by the corresponding conditional probabilities:

$E(X^2 \mid Y = 5) = 3^2 \times 0 + 6^2 \times 1/2 + 9^2 \times 1/2 = 117/2$

Hence, the conditional variance of X given $\{Y = 5\}$ equals:

$V(X \mid Y = 5) = E(X^2 \mid Y = 5) - (E(X \mid Y = 5))^2 = 117/2 - 225/4 = 9/4$

Notice that the conditional pdf of X given $\{Y = 2\}$ is degenerated: if it is given that $\{Y = 2\}$ has occurred, then the event $\{X = 3\}$ has conditional probability 1. The table below gives three conditional pdfs of X:

x	3	6	9	Total
$f(x \mid Y = 2)$	1	0	0	1
$f(x \mid Y = 4)$	0	1	0	1
$f(x \mid Y = 5)$	0	1/2	1/2	1

It follows immediately that $E(X \mid Y = 2) = 3$ and $E(X \mid Y = 4) = 6$, while the two corresponding conditional variances are both 0 (as they should be).

For fixed x, the conditional pdf of Y given $\{X = x\}$ can be determined analogously. For instance,

$$g(y \mid X = 6) = \frac{h(6, y)}{f(6)} = \frac{h(6, y)}{18/36} \quad \text{and}$$

y	2	4	5
$g(y \mid X = 6)$	0	2/3	1/3

It follows that $E(Y \mid X = 6) = 13/3$, $E(Y^2 \mid X = 6) = 57/3$ and $V(Y \mid X = 6) = 2/9$.

Incidentally, notice that the pre-information that $\{X = 6\}$ has occurred **does** influence the probability of $\{Y = 4\}$:

$2/3 = g(4 \mid X = 6) \neq g(4) = 1/3$

Recall from Section 7.4.2 that, for fixed x and y, the **events** $\{X = x\}$ and $\{Y = y\}$ are independent if:

$P(X = x; Y = y) = P(X = x) P(Y = y)$

The rvs X and Y are called independent if the above is valid for **all** x and y.

(Stochastically) independent rvs *X* and *Y*

X and Y are *(stochastically) independent* if the joint pdf is equal to the product of the two marginal pdfs:

$h(x, y) = f(x) g(y) \quad \text{for all } x, y$ (11.9)

By (11.6) and (11.7) it follows that:

$$h(x, y) = f(x \mid Y = y) \, g(y) \quad \text{and} \quad h(x, y) = g(y \mid X = x) \, f(x)$$

Substitution in (11.9) yields that independence of X and Y can also be proved by checking one of the following two equalities:

$$f(x \mid Y = y) \overset{?}{=} f(x) \qquad \text{for all outcomes } x, y$$

$$g(y \mid X = x) \overset{?}{=} g(y) \qquad \text{for all outcomes } x, y$$

It can (but will not) be proved that independence can also be expressed in terms of cumulative probabilities:

$$X \text{ and } Y \text{ independent} \Leftrightarrow P(X \le x; Y \le y) = P(X \le x) \, P(Y \le y) \quad \text{for all } x, y$$

It is this equivalence which can be taken as the definition of independence for the case that X and Y are continuous rvs.

Example 11.8

Let U and V be the number of eyes shown by die a and die b respectively, when throwing the two obscure dice of Example 11.1. Check that the interior part of the table below indeed represents the joint pdf of U and V; the margins present the two marginal pdfs called f_U and f_V.

v / u	4	1	$f_U(u)$
5	6/36	6/36	12/36
2	12/36	12/36	24/36
$f_V(v)$	18/36	18/36	1

To check independence of U and V, the joint probabilities in each of the four cells has to be equal to the product of the two marginal probabilities that are in the same row or the same column. This is indeed the case: for instance, the probability 12/36 in the cell of the pair (2, 1) is equal to the product of 24/36 and 18/36.

Next, the covariance of U and V is calculated by way of the short-cut formula. From the marginal distributions it follows easily that $E(U) = 3$ and $E(V) = 2.5$. For cov(U, V) it follows that:

$$\sigma_{U,V} = E(UV) - E(U)E(V) = 20 \times 1/6 + 5 \times 1/6 + 8 \times 1/3 + 2 \times 1/3 - 3 \times 2.5 = 0$$

If X and Y are independent, then the functions $V = v(X)$ and $W = w(Y)$ of X and Y are independent too. This is intuitively obvious since the randomness of V and W is completely determined by X and Y, respectively. Furthermore, if X and Y are independent, then the expectation of their product is just the product of their expectation (no proof added). Since cov$(X, Y) = E(XY) - E(X)E(Y)$, it follows that independent rvs are uncorrelated.

Properties of two independent rvs X and Y

$$E(XY) = E(X)E(Y) \quad \text{and} \quad \text{cov}(X, Y) = 0 \tag{11.10}$$

Independence implies being uncorrelated; not the other way around. This **one-way** implication makes sense intuitively: independence can be considered as 'overall' independence, whereas the correlation coefficient measures only **linear** dependence and being uncorrelated means being **linearly** independent.

Example 11.9

Notice that Example 5.8 gives a counterexample for the reversed implication: the height X and the weight Y of the man in a randomly drawn household in Baker Street are uncorrelated, but Y depends quadratically on X since $Y = (X - 1)^2$.

CASE 11.1 INSURANCE AGAINST BICYCLE THEFT – SOLUTION

We have to make the model more concrete. Consider the following random variables:

X = insurance risk for the insurer due to an arbitrary policy (in units of 100 euro)
Y = a dummy variable that is 1 if the bike is stolen and 0 if not

The risk X has 0 and the 25 integers 6, 7, ..., 30 as possible outcomes. In the case of a theft, there is a probability 0.04 that the insurer will have to pay $k \times 100$ euro to the insured, for all k = 6, 7, ..., 30. Furthermore, the probability that a bike is stolen is 0.06.
These model assumptions can formally be written as follows:

$P(X = k \mid Y = 1) = 0.04$ for all k = 6, 7, ..., 30

$P(Y = 1) = 0.06$

The expected risk in the case of theft is a conditional expectation:

$E(X \mid Y = 1) = 6 \times 0.04 + 7 \times 0.04 + \dots + 30 \times 0.04 = 18$

For k = 6, 7, ..., 30, the joint (unconditional) probabilities $P(X = k; Y = 1)$ follow from the conditional probabilities by multiplying by 0.06; the probability $P(X = 0; Y = 1)$ is 0. Furthermore, the probabilities $P(X = k; Y = 0)$ are only different from 0 for k = 0, in which case it is 0.94. The cross table of the joint pdf is shown below.

x / y	0	6	7	.	.	30	$g(y)$
0	0.94	0	0	.	.	0	0.94
1	0	0.0024	0.0024	.	.	0.0024	0.06
$f(x)$	0.94	0.0024	0.0024	.	.	0.0024	1

The expected risk can now be calculated from the pdf of X; it follows that $E(X) = 1.08$. Since Lowlands takes a safety loading of 20%, the premium is $1.2 \times 108 = €129.60$. Also notice that $P(X > 1.296) = 0.06$, which answers the last question.

11.3.1 When *X* and *Y* are continuous

If *X* and *Y* are continuous it **is** possible to define the concept '**continuous** joint pdf' $h(x, y)$ of *X* and *Y*. But this topic is beyond the scope of the book. However, we can (and will) talk about independence of continuous rvs; in fact, (11.9) gives a definition. See also the remark preceding Example 11.8. Furthermore, the formal definition of covariance for the continuous case is:

$$\sigma_{X,Y} = E((X - \mu_X)(Y - \mu_Y))$$

which is the same as for the discrete case. However, strictly speaking, for the continuous case this formally written expectation is not yet defined. Also, the definition of correlation coefficient in (11.4) and the short-cut formula (11.5) remain valid, albeit only in the formal notations. The same holds for Table 11.1, since the derivation of the results was based only on the rules of the expectation operator *E* (and these rules do not bother about discrete or continuous).

11.4 Linear combinations of random variables

Remember that $v(x) = a + bx$ (with *a* and *b* constants) is a linear function **in *x***. Analogously, a **linear function in *x* and *y*** is as follows: $v(x, y) = a + bx + cy$ (with *a, b, c* constants). If *x* and *y* are replaced by random variables *X* and *Y*, a new random variable *W* (say) arises and is called a *linear combination* of *X* and *Y*:

$$W = a + bX + cY$$

For instance, if *X* = 'annual expenditure on food and clothing' of a household and *Y* = 'annual expenditure on housing and recreation' of this household, then the linear combination $V = X + Y$ measures the 'total' annual expenditure of that household – at least, it does if *X* and *Y* are both measured in euros. However, if *X* is measured in euros and *Y* in dollars, then the linear combination $W = X + 08Y$ measures the total annual expenditure in euro (assuming €1 = $1.25).

The present section is about linear combinations of random variables; these rvs can be **discrete or continuous**. Rules are derived for the expectation and the variance of such linear combinations. Furthermore, for some special cases, their pdf is studied.

11.4.1 Expectation and variance

First, consider the simple linear combination $W = U + V$ of the random variables *U* and *V*. Equation (8.31) states that:

$$E(W) = E(U) + E(V) \tag{11.11}$$

(Although this result was not proved rigorously, it **is** valid and will be taken as the starting point of our considerations.) Hence, $\mu_W = \mu_U + \mu_V$. That is, the expectation of a sum of random variables is just the sum of the individual expectations. To find out what the consequences are for the variance of *W*, recall that this variance equals $\sigma_W^2 = E((W - \mu_W)^2)$. The squared random deviation $(W - \mu_W)^2$ satisfies:

$$(W - \mu_W)^2 = (U + V - (\mu_U + \mu_V))^2 = ((U - \mu_U) + (V - \mu_V))^2$$
$$= (U - \mu_U)^2 + (V - \mu_V)^2 + 2(U - \mu_U)(V - \mu_V)$$

Its expectation is the sum of the three individual expectations:

$$E((W - \mu_W)^2) = E((U - \mu_U)^2) + E((V - \mu_V)^2) + 2E((U - \mu_U)(V - \mu_V))$$

Hence:

$$W = U + V \Rightarrow \begin{cases} \mu_W = \mu_U + \mu_V \\ \sigma_W^2 = \sigma_U^2 + \sigma_V^2 + 2\sigma_{U,V} \end{cases} \tag{11.12}$$

To find the expectation and variance of the general linear combination $W = a + bX + cY$, take $U = a + bX$ and $V = cY$ and recall that by (8.30) and Table 11.1:

$$\mu_U = a + b\mu_X; \quad \sigma_U^2 = b^2\sigma_X^2; \quad \mu_V = c\mu_Y; \quad \sigma_V^2 = c^2\sigma_Y^2; \quad \sigma_{U,V} = bc\sigma_{X,Y} \tag{11.13}$$

By substituting the results of (11.13) into (11.12), it follows that:

Expectation and variance of a linear combination of *X* and *Y*

$$W = a + bX + cY \Rightarrow \begin{cases} \mu_W = a + b\mu_X + c\mu_Y \\ \sigma_W^2 = b^2\sigma_X^2 + c^2\sigma_Y^2 + 2bc\sigma_{X,Y} \end{cases} \tag{11.14}$$

This result for the variance of $a + bX + cY$ is very important; it is used, for instance, to determine the risk of a portfolio of investments; see Example 11.10 below. Notice that the constant a does not influence the variance, which is not surprising since a constant has no variation. In general, the variance of $a + bX + cY$ is **not** equal to the sum of the variances of bX and cY. Equality holds if X and Y are uncorrelated:

$$W = a + bX + cY \text{ with } X, Y \text{ uncorrelated} \Rightarrow \sigma_w^2 = b^2\sigma_X^2 + c^2\sigma_Y^2 \tag{11.15}$$

Example 11.10

Reconsider the investor of Example 11.2. She wants to invest €1000 to create a portfolio of shares Phil Electronics (PE) and Albert Food (AF), which at present cost respectively €1 and €2 per share. The interior part of the table below contains the joint pdf $h(x, y)$ of next year's prices X and Y of one share of PE and AF, respectively. The marginal pdfs are also included.

x \ *y*	1.9	2	2.1	*f(x)*
0.9	0.1	0.1	0	0.2
1	0	0.2	0.1	0.3
1.1	0	0.2	0.3	0.5
g(y)	0.1	0.5	0.4	1

See Examples 11.2–11.4 for the expectations, variances, covariance and correlation coefficient. Notice that X and Y are dependent, as follows, for instance, from the fact that $h(1, 2)$ is unequal to the product of $f(1)$ and $g(2)$.

There are many ways to put together a portfolio of the shares PE and AF. Three possibilities are considered below:

1 The complete amount of €1000 is invested in shares PE.

2 The complete amount of €1000 is invested in shares AF.

3 50% is invested in PE and 50% is invested in AF.

Denote next year's value of portfolio i as W_i ($i = 1, 2, 3$). Since one share PE costs €1, portfolio 1 contains 1000 of them and $W_1 = 1000X$. Analogously, $W_2 = 500Y$ and $W_3 = 500X + 250Y$. Of course, the interrelationship between X and Y, as measured in terms of the covariance $\sigma_{X,Y}$, only plays a role for W_3. Hence, the expectations and variances of W_1 and W_2 can easily be calculated from (8.30). However, for the expectation and variance of W_3, (11.14) has to be used:

$W_1 = 1000X$, $\quad E(W_1) = 1000 \times E(X) = 1030; V(W_1) = 1000^2 \times V(X) = 6100$

$W_2 = 500Y$, $\quad E(W_2) = 500 \times E(Y) = 1015; V(W_2) = 500^2 \times V(Y) = 1025$

$W_3 = 500X + 250Y$, $\quad E(W_3) = 500 \times E(X) + 250 \times E(Y) = 1022.50$

$$V(W_3) = 500^2 \times V(X) + 250^2 \times V(Y) + 2 \times 500 \times 250 \, cov(X, Y)$$
$$= 2556.25$$

The **return** (%) R of a portfolio (a random percentage) is:

$$R = 100 \times \frac{\text{new value} - \text{old value}}{\text{old value}}$$

So, in the present situation, where W is next year's value of a portfolio that consists of PE and AF shares, R is equal to:

$$R = 100 \times \frac{W - 1000}{1000} = 0.1W - 100$$

Hence, $E(R) = 0.1E(W) - 100$ and $V(R) = 0.01 \times V(W)$. The table below summarizes the expected returns and the accompanying risks (**measured by the standard deviation**).

Portfolio	1	2	3
Expected return (%)	3	1.5	2.25
Risk (st. dev. return)	7.8102	3.2016	5.0559

As was to be suspected, portfolio 1 yields the largest expected profit (€30) but it also has the largest risk.

Results can easily be generalized for more than two random variables. For three random variables X_1, X_2, X_3 the following is valid:

$$E(X_1 + X_2 + X_3) = E(X_1) + E(X_2) + E(X_3)$$
$$V(X_1 + X_2 + X_3) = V(X_1) + V(X_2) + V(X_3) + 2cov(X_1, X_2) + 2cov(X_1, X_3) + 2cov(X_2, X_3)$$

For each pair (i, j), the term $cov(X_i, X_j)$ has to be included. For instance: the pair (1, 2) yields $cov(X_1, X_2)$ and the pair (2, 1) yields $cov(X_2, X_1)$, which explains the term $2cov(X_1, X_2)$ in the last equation above. That is why the generalization for n random variables $X_1, X_2, \ldots, X_n$ is as follows:

Expectation and variance of $X_1 + \ldots + X_n$

$$E\left(\sum_{i=1}^{n} X_i\right) = \sum_{i=1}^{n} E(X_i)$$

$$V\left(\sum_{i=1}^{n} X_i\right) = \sum_{i=1}^{n} V(X_i) + 2\sum_{i<j} cov(X_i, X_j) \tag{11.16}$$

(The summation for the covariance terms includes all pairs (i, j) with $i < j$.) The expectation of a sum of random variables is equal to the sum of all individual expectations. The variance of a sum of rvs is in general **un**equal to the sum of the individual variances; the sum of all covariance terms

$\text{cov}(X_i, X_j)$ has to be added. However, notice that the summation of covariance terms will disappear if the X_i are uncorrelated or even independent. Such a special case will be considered now.

From Chapter 12 onwards, a sequence of n random observations $X_1, X_2, ..., X_n$ of one underlying population variable X will be very important. It will turn out that all random observations X_i have the same expectation (μ) and variance (σ^2). Furthermore, if the random drawing is done **with** replacement, then the X_i are independent and hence $\text{cov}(X_i, X_j) = 0$ for $i \neq j$. The following result summarizes (11.16) for these circumstances. It is a basic result for inferential statistics and will be used frequently:

Expectation and variance of $X_1 + ... + X_n$ for independent X_i with the same mean and variance

$$\left.\begin{array}{l} E(X_i) = \mu \text{ for all } i \\ V(X_i) = \sigma^2 \text{ for all } i \\ \text{all } X_i \text{ independent} \end{array}\right\} \Rightarrow E\left(\sum_{i=1}^{n} X_i\right) = n\mu \quad \text{and} \quad V\left(\sum_{i=1}^{n} X_i\right) = n\sigma^2 \tag{11.17}$$

Example 11.11

Suppose that ten adult men accidentally travel together in an elevator that has a cargo capacity of 750 kg. What can be said about the probability that the total weight of the ten men exceeds this limit?

Let $X_1, X_2, ..., X_{10}$ be the individual weights (in kg) of these ten men and let Y be their total weight. Hence, $Y = \sum_{i=1}^{10} X_i$. Since no further information is given, it seems natural to assume that these ten men are randomly drawn and that the ten weights are random observations of the population variable X = 'weight (kg) of an adult man'. Hence, all expectations $E(X_i)$ and variances $V(X_i)$ are equal to the population mean μ and the population variance σ^2, respectively. According to Statistics Netherlands' *Statistical Yearbook 2005*, the mean weight of the adult men is 82.7 kg; assume that the accompanying variance is 99.94. Since the ten men come together accidentally, it can be assumed that their weights $X_1, X_2, ..., X_{10}$ are independent rvs. By (11.17) it follows that:

$$E(Y) = \sum_{i=1}^{10} E(X_i) = \sum_{i=1}^{10} 82.7 = 827 \text{ kg}$$

$$V(Y) = \sum_{i=1}^{10} V(X_i) = \sum_{i=1}^{10} 99.94 = 999.40 \text{ kg}^2 \quad \text{and} \quad SD(Y) = 31.6133 \text{ kg}$$

But what about the probability that Y is larger than 750 kg? Nothing is known yet about the type of the probability distribution of Y, so Chebyshev's inequality has to be used. This rule states that:

$$P(\mu_Y - k\sigma_Y < Y < \mu_Y + k\sigma_Y) \geq 1 - \frac{1}{k^2}$$

Solving $\mu_Y - k\sigma_Y = 750$ with $\mu_Y = 827$ and $\sigma_Y = 31.6133$ yields $k = 2.4357$. This choice of k leads to:

$$P(750 < Y < 904) \geq 0.8314$$

Since $P(Y > 750) \geq P(750 < Y < 904)$, the probability $P(Y > 750)$ is also at least 0.8314. However, see also Example 11.14.

Relation (11.17) also yields an important property for the binomial distribution that was mentioned in (9.3) but not proved: that the variance of $Bin(n, p)$ is $np(1 - p)$. Recall from Section 9.2.1 that a $Bin(n, p)$ distributed rv Y can be written as $X_1 + \ldots + X_n$ where X_i counts the number of successes (0 or 1) in the ith trial of the n independent trials of the underlying binomial experiment. In particular, $X_1, \ldots, X_n$ are independent. Also recall that $X_i \sim Bern(p)$ and that $E(X_i) = p$ and $V(X_i) = p(1 - p)$. Hence, Y is the sum of n independent rvs that all have the same expectation $\mu = p$ and variance $\sigma^2 = p(1 - p)$. Hence, by (11.17): $\sigma_Y^2 = np(1 - p)$.

11.4.2 The pdf of a linear combination

This subsection is about the probability distributions of linear functions $V = v(X, Y)$ of rvs X and Y. If X and Y are both discrete, then the rv V is also discrete and its pdf follows from the joint pdf of X and Y. We treat a few rather elementary examples.

Example 11.12[†]

If a European football match is still undecided after extra time, both teams shoot five penalties and the team that scores more wins the match. In 32 matches that had to be decided by a penalty shoot-out the variables X_1 = 'number of penalties missed by the team that starts penalty shooting' and X_2 = 'number of penalties missed by the team that did not start penalty shooting' were measured. The table gives an overview of the relative frequencies. For instance, the number in cell (1, 2) indicates that in 5 of the 32 matches the team that took the first penalty missed 1 out of 5 and the other team missed 2 out of 5.

X_2 X_1	0	1	2	3	Total
0	0	3/32	2/32	0	5/32
1	5/32	0	5/32	0	10/32
2	8/32	6/32	0	1/32	15/32
3	0	1/32	1/32	0	2/32
Total	13/32	10/32	8/32	1/32	1

The referee tosses a coin, and the winner of the toss chooses whether to take the first penalty or not. The general feeling is that the best strategy is to **take** the first penalty, since it gives a psychological advantage if the other team goes behind from the start. But is this indeed the case?

Consider a European football match that is still undecided after extra time and has to be decided by ten penalties in total. Let X_1 and X_2 be the number of penalties missed by the starting team and the other team, respectively. Take the joint frequencies in the interior part of the above table as a model for the joint pdf $h(x_1, x_2)$ of the rvs X_1 and X_2. Interest is in the pdf of $W = X_1 - X_2$. The probability $P(W < 0)$ will be calculated and compared with the 'general feeling' above.

The table below contains for each cell (x_1, x_2) the values $w = x_1 - x_2$:

X_2 X_1	0	1	2	3
0	0	-1	-2	-3
1	1	0	-1	-2
2	2	1	0	-1
3	3	2	1	0

[†] R. Van Gelder and P. Kindt, 'The art of shooting penalties', (in Dutch), *Psychology* 5 (1991).

The possible outcomes of W follow from the interior part of the above table. The following table presents the pdf of W:

w	-3	-2	-1	0	1	2	3
$P(W = w)$	0	2/32	9/32	0	12/32	9/32	0

For instance:

$P(W = 1) = h(1, 0) + h(2, 1) + h(3, 2) = 5/32 + 6/32 + 1/32 = 12/32$

The probability that the team that starts penalty shooting will eventually win, is equal to:

$P(X_1 < X_2) = P(W < 0) = 2/32 + 9/32 = 11/32 = 0.3438$

Notice that this probability is smaller than 0.5. Although the general feeling is that it is better to **start** penalty shooting, the model suggests the contrary.

It is of practical interest to know whether the probability distribution of the sum of two independent rvs whose distributions are of the same type, is again of that type.

If $X \sim Bin(n_1, p)$ and $Y \sim Bin(n_2, p)$, and X and Y are independent, then $S = X + Y$ is a binomial rv too. This is not surprising, since the n_1 and n_2 independent trials of the two underlying binomial experiments can – because of the independence of the two rvs and the common success probability – be combined to $n_1 + n_2$ independent and identical trials with probability p of success; in this large binomial experiment, S counts the total number of successes. Hence:

Property of the sum of two independent binomial rvs

$$\left.\begin{array}{l} X \sim Bin(n_1, p) \\ Y \sim Bin(n_2, p) \\ X, Y \text{ independent} \end{array}\right\} \quad \Rightarrow \quad X + Y \sim Bin(n_1 + n_2, p) \tag{11.18}$$

A similar result holds for Poisson rvs. If X counts the number of occurrences of type I (say) in a certain period of time and Y counts (independently) the number of occurrences of type II in that period, and $X \sim Po(\mu_1)$ and $Y \sim Po(\mu_2)$, then $X + Y$ counts the number of occurrences of the **combined** type in that period; the distribution $Po(\mu_1 + \mu_2)$ of $X + Y$ seems obvious (but it will not be proved):

$$\left.\begin{array}{l} X \sim Po(\mu_1) \\ Y \sim Po(\mu_2) \\ X, Y \text{ independent} \end{array}\right\} \quad \Rightarrow \quad X + Y \sim Po(\mu_1 + \mu_2)$$

Recall from Section 10.3.5 that normally distributed rvs are often the results of many small and independent contributions. If two independent and normally distributed rvs X and Y are summed, then the summation $S = X + Y$ contains even more small independent contributions and hence can be expected to be normal too. This indeed is the case. The parameters of S follow immediately from (11.14):

$$\mu_S = E(X + Y) = E(X) + E(Y)$$

$$\sigma_S^2 = V(X) + V(Y) + 2\text{cov}(X, Y) = V(X) + V(Y)$$

More generally, each linear combination $W = a + bX + cY$ of two independent, normally distributed rvs is again normally distributed. Hence:

Property for two independent normal rvs:

$$\left. \begin{array}{l} X \sim N(\mu_1, \sigma_1^2) \\ Y \sim N(\mu_2, \sigma_2^2) \\ X, Y \text{ independent} \end{array} \right\} \Rightarrow a + bX + cY \sim N(a + b\mu_1 + c\mu_2, b^2\sigma_1^2 + c^2\sigma_2^2) \qquad \textbf{(11.19)}$$

A similar result holds for more than two independent rvs: each linear combination of independent and normal rvs is again normal.

Example 11.13

Consider the population of all families with five children. For an arbitrary family, let X be the number of girls among the three first-born children of this family and Y the number of girls among the two last-born children. The experiment of the births of the first three children can be considered as a combination of three sub-experiments that all have success probability ½ of having a girl. Hence, this binomial experiment leads to the model $Bin(3, 0.5)$ for X. Analogously, the distribution $Bin(2, 0.5)$ seems to be a good model for Y. Since the experiments of the births of the first three and the last two children can be considered as independent, it follows that X and Y are independent too. Because of the common success probability, it follows by (11.18) that the total number of girls $W = X + Y$ in this randomly chosen family is $Bin(5, 0.5)$ distributed (as it should be).

Now suppose that the rvs U and W respectively count the total number of boys and the total number of girls in this randomly chosen family. Again, binomial models are suitable: U and W are both $Bin(5, 0.5)$ distributed. However, their sum $T = U + W$ is **not** binomially distributed since T is degenerated: the number of boys plus the number of girls is always equal to 5. This result does not contradict (11.18), since U and W are not independent. Since $W = 5 - U$, the rv W is even completely determined by U.

Example 11.14

Reconsider the ten adult men who accidentally travel together in the elevator; see Example 11.11. Since the weight X of an adult man is the result of many small and independent contributions, the model $N(82.7, 99.94)$ is chosen. Since the weights $X_1, ..., X_{10}$ of the ten men can be considered as independent, their total weight Y also has a normal distribution. Since the parameters were calculated in Example 11.11, we have:

$$Y \sim N(827, 999.4)$$

The probability that the total weight of these ten men exceeds the cargo capacity of 750 kg of the elevator can now be calculated more precisely:

$$P(Y > 750) = 1 - P(Y \leq 750) = 1 - 0.0074 = 0.9926 \qquad (*)$$

The probability that the total weight will exceed the cargo limit is 0.9926, considerably larger than the lower bound of Example 11.11.

In the context of adding up n random variables, there is one mistake that is often made and that should be cleared up once and for all. The mistake arises since the difference is overlooked between a sum of n independent random measurements of a variable X and a sum that adds up one random measurement n times.

Suppose that $X_1, \ldots, X_n$ are n independent random observations of a population variable X. Then all X_i have the same expectation μ and variance σ^2. Let $V = X_1 + \ldots + X_n$ be the sum of these random observations and let $W = nX_1$. Notice the difference between V and W; they are often mixed up. Note especially that:

$$\mu_V = n\mu \quad \text{and} \quad \sigma_V^2 = n\sigma^2; \quad \mu_W = n\mu \quad \text{and} \quad \sigma_W^2 = n^2\sigma^2 \tag{11.20}$$

It would appear that V and W have the same expectation but different variance. The following example illustrates this in a less formal context.

Example 11.15

Reconsider the former example. Now suppose that first, only one Dutch adult man is chosen randomly. Next, nine other men are sought that have precisely the same weight as this man. These ten men are all put in the elevator. What is the probability that the cargo capacity will be exceeded?

Let W denote the total weight of these ten men. Notice that $W = 10X_1$ if X_1 denotes the weight of the randomly chosen man. Since X_1 is $N(82.7, 99.94)$ distributed, the linear combination $W = 10X_1$ is normal too, with parameters:

$$\mu_W = 10 \times 82.7 = 827 \quad \text{and} \quad \sigma_W^2 = 10^2 \times 99.94 = 9994$$

The probability $P(W > 750)$ can be calculated:

$$P(W > 750) = 1 - P(W \leq 750) = 1 - 0.2206 = 0.7794 \tag{*}$$

This answer is considerably smaller than the answer of Example 11.14.

Summary

This chapter was about pairs of random variables. If the rvs are discrete, the discrete joint pdf describes their joint probabilistic behaviour; if they are continuous, the continuous joint pdf does so, but this concept was not considered.

From the joint pdf the two marginal pdfs can be derived. Also the conditional distribution of one of the two rvs when the realization of the other is already given, follows from this joint pdf. If the joint pdf is just the product of the two marginal pdfs, the two rvs are called **independent**. The concepts of covariance and correlation coefficient for **random** variables were defined; see (11.3) and (11.4). A short-cut formula for the covariance was derived; see (11.5). Below, an overview is given of some of these concepts and their interrelationships. Notice the directions of the arrows! Here, h is the joint pdf of X and Y, and f and g are the marginal pdfs of respectively X and Y:

Relationships between independence, pdfs and covariance

$$X, Y \text{ independent} \begin{cases} \Leftrightarrow h(x, y) = f(x)g(y) \text{ for all } x, y \\ \Leftrightarrow P(X \leq x; Y \leq y) = P(X \leq x)P(Y \leq y) \text{ for all } x, y \\ \Leftrightarrow f(x \mid Y = y) = f(x) \text{ for all } x, y \\ \Leftrightarrow g(y \mid X = x) = g(y) \text{ for all } x, y \\ \Rightarrow E(XY) = E(X)E(Y) \\ \Rightarrow \rho_{X,Y} = \sigma_{X,Y} = 0 \end{cases}$$

Several rules for linear combinations of rvs were considered. The variance rule is especially important for portfolio theory.

Rules for linear combinations of rvs

$$W = a + bX + cY \quad \Rightarrow \quad \begin{cases} \mu_W = a + b\mu_X + c\mu_Y \\ \sigma_W^2 = b^2\sigma_X^2 + c^2\sigma_Y^2 + 2bc\sigma_{X,Y} \end{cases}$$

$$\left.\begin{array}{l} X \sim N(\mu_1, \sigma_1^2) \\ Y \sim N(\mu_2, \sigma_2^2) \\ X, Y \text{ independent}\} \end{array}\right\} \quad \Rightarrow \quad a + bX + cY \sim N(a + b\mu_1 + c\mu_2, b^2\sigma_1^2 + c^2\sigma_2^2)$$

Probability distribution of $X + Y$ for independent X and Y

Distribution of X	Distribution of Y	$\Rightarrow$	Distribution of $X + Y$
$Bin(n_1, p)$	$Bin(n_2, p)$		$Bin(n_1 + n_2, p)$
$Po(\mu_1)$	$Po(\mu_2)$		$Po(\mu_1 + \mu_2)$
$N(\mu_1, \sigma_1^2)$	$N(\mu_2, \sigma_2^2)$		$N(\mu_1 + \mu_2, \sigma_1^2 + \sigma_2^2)$

List of symbols

Symbol	Description
$cov(X, Y)$	Covariance
$E(X \mid Y = y)$	Conditional expectation of X given that $Y = y$
$V(X \mid Y = y)$	Conditional variance of X given that $Y = y$
$h(x, y)$	Joint pdf
$f(x \mid Y = y)$	Conditional pdf of X given that $Y = y$
$\rho_{X,Y}$	Correlation coefficient
$\sigma_{X,Y}$	Covariance

🔑 Key terms

conditional pdfs **348**
correlation coefficient **344**
covariance **343**
cross table **341**
discrete joint probability density **341**
independent **349**

insurance risk **339**
joint (probability) density **340**
joint probability density function **340**
joint probability distribution **340**

linear combination **352**
marginal distribution **340**
marginal pdf **340**
stochastically independent **349**
uncorrelated **344**
XY event **338**

? Exercises

Exercise 11.1

The joint pdf $h(x, y)$ of the rvs X and Y is given in the following table.

y x	1	2	3
0	0.1	0.1	0.2
1	0	0.2	0
2	0.3	0.1	0

  a Explain why $h(x, y)$ indeed is a pdf.
  b Calculate the probabilities $P(X \leq 1$ and $Y = 2)$ and $P(X \leq 1$ or $Y = 2)$
  c Determine the two marginal distributions.
  d Calculate the covariance and the correlation coefficient of X and Y.

Exercise 11.2

Reconsider Exercise 11.1.

  a Determine the conditional pdf of X given that $Y = 2$.
  b Are X and Y independent? Why (not)?
  c Calculate $P(X \leq 1 \mid Y = 2)$ from part (a).
  d Calculate the conditional expectation and the conditional variance of X given that $Y = 2$.

Exercise 11.3

Let X and Y respectively be next month's prices of one share of a stock and a bond. It is given that X and Y are independent. Their marginal pdfs f and g are:

x	10	11	12	13	y	5	6	7	8
$f(x)$	0.1	0.3	0.4	0.2	$g(y)$	0.5	0.3	0.1	0.1

  a Determine the joint pdf of X and Y.
  b Determine the covariance and the correlation coefficient of X and Y.
  c Calculate $E(XY)$.
  d Calculate $E(X^2 Y^2)$.

Exercise 11.4

In a small town there are only two electronics stores. Let X and Y be the daily number of TVs sold by stores I and II, respectively. The table below gives the joint pdf of X and Y.

y x	3	4	5
2	0	0	0.1
3	0.1	0.3	0.1
4	0.4	0	0

a Calculate the covariance and the correlation coefficient of X and Y. Can you understand their signs?

b Calculate $E(XY)$.

c Determine the pdf of the daily number of TVs sold in that town. Calculate the mean and the variance of this distribution.

d Calculate the mean and the variance of part (c) again by making use of rules for linear combinations of rvs.

Exercise 11.5

For the rvs X and Y it is given that:

$$E(X) = 10; \quad E(Y) = 20; \quad V(X) = 9; \quad V(Y) = 16; \quad \rho = 0.8$$

a Calculate the covariance of X and Y.

Calculate the expectation and variance of the following linear combinations of X and Y:

b $3X$ and $-3X + 1.5$

c $X - Y + 2$

d $4X - 2Y + 5$

e $-3X - 4Y - 5$

Exercise 11.6

After having analysed several months of sales data, a manager of an appliance store posits the following joint density for the number of refrigerators (X) and deep freezers (Y) sold daily:

y / x	0	1	2
0	0.02	0.10	0.08
1	0.12	0.15	0.03
2	0.16	0.25	0.09

a Determine the marginal pdfs of X and Y.

b Calculate the expectations and the standard deviations of X and Y.

c Determine the covariance and the correlation coefficient of X and Y.

d Determine the conditional pdf of Y given that $X = 1$.

Exercise 11.7

An investor has the following information about the monthly returns (%) of two stocks:

Stock	1	2
Mean	1.25	1.50
Standard deviation	2.25	2.75

a In which stock will she invest if she wants to maximize her expected returns? And if she wants to minimize her risk?

b Calculate the expected return and the variance of the portfolio that is composed of 20% stock 1 and 80% stock 2; assume that the correlation coefficient of the returns of the two stocks is 0.4.

c Do the same as in part (b) for the portfolio composed of 70% stock 1 and 30% stock 2 (same correlation coefficient).

Exercise 11.8

A financial analyst uses the historical weekly returns (%) of three stocks to model the future weekly returns. He obtains the following means and covariances (the covariance σ_{ij} is in cell (i, j) of the right-hand matrix, for $i, j = 1, 2, 3$):

	Means	j	1	2	3
		i			
1	0.5239	1	9.4218		
2	0.6430	2	1.6632	4.2507	
3	0.6718	3	2.9718	0.3460	11.4175

a What are the three variances? Fill in the empty cells of the matrix.

b The choice between investing in stock 1 and investing in stock 2 is easy. Why?

c Which stock maximizes the expected return? Which stock minimizes the risk?

d Determine the expected return and variance of the portfolio that invests equally in the three stocks.

e Compare the expected return and the variance of the portfolio in part (d) with the expected returns and variances of the individual stocks. Give your comments.

Exercise 11.9

For two rvs X and Y it is given that $E(X) = 10$, $E(Y) = 20$, $V(X) = 25$, $V(Y) = 16$ and $\text{cov}(X, Y) = -12.8$.

a Calculate the expectation and the variance of $U = -X + 3$ and $W = 7Y - 5$.

b Calculate the expectation and the variance of $T = -2X + 3Y - 12$.

Exercise 11.10

For two rvs X and Y it is given that $\mu_X = 5$, $\mu_Y = 10$, $\sigma_X^2 = 9$, $\sigma_Y^2 = 25$ and $\rho = -0.82$.

a Calculate the expectation and the standard deviation of $T = 4X - 3Y - 6.1$.

b Calculate $E(X^2)$ and $E(XY)$. Are X and Y independent?

c Let Z_1 and Z_2 be the standardized versions of X and Y, respectively. Express Z_1 in X and Z_2 in Y.

d Calculate the covariance and the correlation coefficient of Z_1 and Z_2.

Exercise 11.11

For a population of 100 students it is investigated whether the monthly amount of money earned with a job is related to the number of DVDs bought monthly. For each of the students, the variables 'income of the past month (in euro, rounded to the nearest multiple of 50)' and 'number of DVDs bought during the past month' are both measured. Here are the resulting frequencies:

DVDs Income	0	1	2	3
150	40	8	2	0
200	0	20	5	0
250	1	0	0	9
300	0	0	5	10

For a further analysis, suppose that one student of the population will be drawn randomly. Let X be the (rounded) monthly income of this student and Y the number of DVDs bought by him/her during the past month.

a Determine the joint pdf $h(x, y)$ of the rvs X and Y. Also determine the marginal pdfs f and g of X and Y.

b Are X and Y independent? Explain.

c Calculate $P(X > 150$ and $Y \geq 1)$ and $P(X > 150$ or $Y \geq 1)$.

d Calculate the covariance of X and Y by using the original definition. Also calculate the correlation coefficient. Interpret your results.

e Use part (d) to answer the question in part (b) again.

Exercise 11.12

Suppose that the joint pdf $h(x, y)$ of Exercise 11.11 is used as a model for the joint pdf of the variables $X =$ 'monthly income of an arbitrary student' and $Y =$ 'monthly number of DVDs bought by that student'.

a Determine the conditional pdf of Y when it is already given that $X = €150$. Check that your result is indeed a pdf. Also calculate the expectation and the variance of this conditional pdf.

b Redo part (a) when the condition is respectively $X = 200$, $X = 250$ and $X = 300$.

Suppose that a DVD costs €20. Let W be the monthly disposable income of this arbitrary student after subtraction of the DVD expenditure.

c Write W as a linear combination of X and Y. Calculate the expectation and the variance of W.

Exercise 11.13

At the moment, the prices of the shares I and II are $20 and $40, respectively. An investor wants to invest $10000 in these two shares, but he is hesitating about the constitution of the portfolio. He asks a specialist for advice about the prices X (of I) and Y (of II) that these shares will have next month. Obviously, X and Y are random variables. The adviser believes that the joint pdf of X and Y is as follows:

x \ y	39	40	41	42
19	0.10	0.10	0	0
20	0.05	0.05	0.20	0
21	0	0	0.20	0.30

a Determine the marginal pdfs of both X and Y.

b Calculate $E(X)$, $E(Y)$, $V(X)$ and $V(Y)$.
c Calculate $P(X \geq 20$ and $Y \geq 40)$ and $P(X \geq 20$ or $Y \geq 40)$.
d Calculate $P(X = 20 \mid Y = 40)$ and $P(Y = 40 \mid X = 20)$.
e Are the rvs X and Y independent? Why (not)?

Exercise 11.14

Reconsider Exercise 11.13.

a Calculate the pdf of the rv $U = X + Y$. Use it to calculate the expectation and the standard deviation of U.

b Calculate the covariance and the correlation coefficient of X and Y; use the short-cut formula. Interpret your results.

c Calculate the expectation and the standard deviation of U again by using calculation rules for expectation and variance of a linear combination.

d The investor believes that there are three possibilities to compose his portfolio: he can spend the whole amount of $10 000 on shares I; he can spend the whole amount on shares II; he can use the half of the money for shares I and the other half for shares II. Denote next month's values of these portfolios by W_1, W_2 and W_3, respectively. Calculate the expectations and the standard deviations of W_1, W_2 and W_3.

e Which portfolio do you prefer and why?

Exercise 11.15

Consider two types of investment, a and b. Denote next year's value of one unit of investment a (and b) by X (and Y). Suppose that these types of investment are special, since their price movements are independent. The marginal pdfs of X and Y are respectively:

x	24	25	26		y	48	49	50	51
$P(X = x)$	0.2	0.3	0.5		$P(Y = y)$	0.1	0.2	0.4	0.3

a Determine the joint pdf of X and Y.
b Calculate $E(X)$, $E(Y)$, $V(X)$ and $V(Y)$.
c Determine the correlation coefficient of X and Y.
d Calculate $E(XY)$ and $E(X^2 Y^2)$.
e Set $U = 50X - 30Y - 15$. Calculate the expectation and the standard deviation of U.

Exercise 11.16

The house numbers of the houses in Baker Street are 1, 2, ..., 5. Suppose that two households in Baker Street are chosen at random (without replacement), to represent the street at a meeting. Let X and Y respectively denote the house numbers of the first and the second chosen household. Determine the joint pdf of X and Y. (*Hint*: Determine, for instance, $P(X = 1; Y = 2)$ by writing it in terms of a conditional probability.)

Exercise 11.17

The table below gives, for eight students, the frequency distribution of the variable $X =$ 'number of TVs owned'.

Number of TVs	0	1	2
Number of students	2	5	1

From this small population, two students are subsequently drawn without replacement. Let X_1 be the number of TVs of the first student and X_2 the number of TVs of the second student.

a Determine all probabilities $P(X_2 = y \mid X_1 = x)$.

b Determine the joint pdf of X_1 and X_2.

c Determine the marginal pdf of X_2 from the joint pdf in part (b). Compare this pdf with the pdf of X_1 and give your comments.

d Let U be the total number of TVs of the two sampled students. Determine the pdf of U. Also calculate the expectation and the variance of U by first determining the expectations, the variances and the covariance of X_1 and X_2.

Exercise 11.18

Let X and Y be two rvs. Precisely one of the following statements is correct. Which one?

a $\text{cov}(X, Y) = 0 \Rightarrow X$ and Y are independent

b $\text{cov}(X, Y) = 0 \Leftrightarrow X$ and Y are independent

c X and Y are independent $\Rightarrow \text{cov}(X, Y) = 0$

Exercise 11.19

Regarding the production process of a certain product it is known that some of the products are defective. You and a colleague have to check the quality of the products by taking random samples. It is your task to check production line 1 and your colleague's to check production line 2.

Suppose that, on a certain day, 1% of the products are defective; this holds for both production lines. You select 240 products at random from the large daily production of line 1 while your colleague selects 260 products at random from the large daily production of line 2. Let X and Y be the number of defective products in your sample and in your colleague's sample, respectively.

a What type of probability distributions do X and Y have? Determine the parameters.

b Calculate the probability that both you and your colleague have at least 1% defectives in the sample.

c Is it wise to approximate the probability distributions of X and Y by a normal distribution? Why (not)?

d You and your colleague decide to combine the two samples into one large sample of size 500. Calculate the probability that at least 1% of this combined sample are defective. Also calculate the expectation and the variance of the total number of defectives in the combined sample.

e Use a normal approximation to calculate the probability of part (d) again. Don't forget the continuity correction.

Exercise 11.20

The mean height of adult Dutchmen is 180.4 cm with a variance of 61.5911. Assume that the height of an adult man is normally distributed.

a Suppose that 10 adult Dutchmen are sampled randomly and that their heights $X_1, \ldots, X_{10}$ (in cm) are measured. Let T be their total height. What kind of distribution does T have? Determine the parameters.

b Calculate the probability that the mean height of the ten men is greater than 185 cm.

Exercise 11.21

The daily sales X (in euro) of an ice cream vendor is normally distributed with mean parameter 300 euro and variance parameter 2500 euro^2. His daily profit R arises by subtracting the daily variable costs (40% of the sales) and the daily constant costs (135 euro) from his daily sales.

a Determine the probability distribution of R.

b Determine the probability of having a negative daily profit.

c Suppose that the sales for subsequent days are independent. Calculate the probability that the weekly profit (five working days) will be negative.

d The rv $U = 5R$ measures the total profit on five days with precisely the same profit R. Determine the probability distribution of U and calculate $P(U < 0)$. Compare your answer with the answers of parts (b) and (c).

Exercise 11.22

Precisely one of the following statements is correct. Which one?

a $V(X - Y) \geq V(X) + V(Y)$

b $V(X - Y) \leq V(X) + V(Y)$

c If cov$(X, Y) > 0$, then $V(X - Y) < V(X) + V(Y)$

d If cov$(X, Y) < 0$, then $V(X - Y) < V(X) + V(Y)$

e None of the above four statements is correct.

Exercise 11.23

Consider a portfolio that consists of 400 units of a certain stock and 200 units of a (government) bond. Let X and Y denote next year's price of one unit of the stock and the bond, respectively. Assume that X and Y are negatively correlated with correlation coefficient $\rho = -0.3850$. Furthermore, assume that the expectations of X and Y are respectively €10.15 and €5.25; the accompanying standard deviations are €0.08 and €0.05.

a Determine the expectation and the standard deviation of next year's price W of the portfolio.

b Now suppose that the bond is an industrial bond, and that X and Y are positively correlated: $\rho = 0.3850$. Repeat part (a).

c Give your comments.

Exercise 11.24

The following game of chance is based on the minimum X and the maximum Y of the (two) numbers of eyes when throwing two fair dice. The player receives $2Y$ euro but pays $3X$ euro, so the profit is $W = 2Y - 3X$.

a Find the pdf of W.

b Calculate $P(W < 0)$, $P(W = 0)$ and $P(W > 0)$

c Will this game in the long run be profitable?

Exercise 11.25

The monthly gross salaries of the 100 employees of a certain company are given in the table below.

Salary (in euro)	1500	1800	2000	2500
Frequency	18	45	27	10

All employees have to pay for their industrial disability insurance: 1% of the gross monthly salary if this salary is below €1900; all other employees pay €20 per month. For an arbitrary employee, let X and Y respectively be his/her gross monthly salary and the monthly amount paid for the disability insurance. Determine the pdf of the 'net' monthly salary $V = X - Y$ of this employee.

Exercise 11.26 (computer)

The dataset Xrc11-26.xls contains the weekly closing prices of the funds ASML Holding, Postbank IT Fund and Akzo Nobel, for the period 7 October 2005 – 13 October 2006.

a Create, for the period 14 October 2005 – 13 October 2006, for each of the three funds the weekly percentage return with respect to the week before.

The historical data are used to model the future weekly returns (%) of the three funds. Suppose that the return of the risk-free rate is 0.06% per week.

b Determine the expected weekly returns and the accompanying volatilities (i.e. standard deviations) of the three funds.

c Create the so-called covariance matrix, the 3 × 3 matrix that contains the covariances $\sigma_{i,j}$ of the three funds for $i, j = 1, 2, 3$.

d Place the four assets (risk-free, ASML, Postbank and Akzo) in increasing order, first by way of expected return and second by way of volatility. Give your comments.

e Consider an investment that invests 25% in each of the four assets. Determine the expected return and the volatility of this portfolio.

f Consider an investment that invests one-third in each of the three risky assets. Determine the expected return and the volatility of this portfolio.

CASE 11.2 PORTFOLIOS OF THE STOCKS PHILIPS AND AHOLD

Historical data can be used to model the future weekly returns of the stocks in a portfolio. This will be illustrated for portfolios which, apart from a risk-free investment, may contain only the stocks Philips Electronics and Ahold. The expected returns and risks of some of these portfolios will be compared.

The datasets Case11-02.xls and Case11-02.sav both contain returns (%) of the stocks Philips and Ahold for the period 31 October 2005 – 2 October 2006, weekly observations (with respect to the week before) that are **rounded** to the nearest integer.[†] It turns out that the data of Philips are within the range −7 to 8 and the data of Ahold in the range from −8 to 9. The idea is to use the joint frequencies as an (empirical) model for the joint future weekly returns of the two stocks.

a Determine the joint frequency table and use it as a model for the joint pdf of the future weekly returns (%) of R_{ph} (for Philips) and R_{ah} (for Ahold).

b The expectations, variances, covariance and correlation coefficient of the rvs R_{ph} and R_{ah} are, of course, equal to the (population) statistics of the dataset. Calculate them.

[†] *Source*: www.belegger.nl.

c Calculate the expectations and standard deviations of the returns (%) of the following portfolios:

    i the complete amount is invested in the stock Philips;

    ii the complete amount is invested in the stock Ahold;

    iii the complete amount is invested in a risk-free investment that returns 0.06% per week;

    iv half in Philips, half in Ahold;

    v 25% risk-free, 25% in Philips and 50% in Ahold;

    vi 25% risk-free, 50% in Philips and 25% in Ahold;

    vii 33.33% risk-free, 33.33% in Philips and 33.33% in Ahold.

d Compare the expected returns and the risks of the above portfolios. Which one would you prefer? Why?

CASE 11.3 INTRODUCTION TO MARKOWITZ'S PORTFOLIO THEORY (PART II)

Reconsider Case 8.2. For a risk-averse investor with a **fixed** (positive) risk aversion coefficient α, we will determine the weights w_1, w_2 and w_3 of the portfolio with maximal utility. Let the rvs R_1, R_2 and R_3 denote the (yearly) returns of the assets 1, 2 and 3. Then

$$R_p = w_1 R_1 + w_2 R_2 + w_3 R_3 \quad \text{with the budget constraint } w_1 = 1 - w_2 - w_3$$

gives the return of the portfolio.

a Calculate the covariances $\sigma_{i,j}$ of the individual returns R_i and R_j.

b Express $\mu_p = E(R_p)$ and $\sigma_p^2 = V(R_p)$ in the weights w_2 and w_3.

c Express $U(\mu_p, \sigma_p^2)$ in w_2 and w_3.

d Find the optimal weights by equating the two partial derivatives to 0.

e Investor C has risk aversion coefficient $\alpha = 50$. Determine and interpret the weights of the optimal portfolio of investor C.

f Investors A and B respectively have risk aversion coefficient 10 and 20. Determine and interpret the optimal portfolios of investors A and B.

g Calculate the utilities of the optimal portfolios of the investors A, B and C. Compare with parts (e) and (f) of Case 8.2.

PART 3
Sampling Theory

Part Contents

In a sense, sampling theory is intermediate between probability and inferential statistics. Since samples are used in practice to draw conclusions about the whole population, it is essential to draw the samples in a careful and thought-out manner. It is important to have a good knowledge of the sampling method that is used to gather the data in order to know what can be expected and what cannot.

In the simplest set-up, the starting point is a population and a quantitative population variable X. The purpose will be to obtain a thought-out sequence $X_1, \ldots, X_n$ of n random observations of X. Based on this random sequence, new random variables (called **estimators**) will be defined that can serve as formulae to generate estimates (approximations) of unknown statistics such as the population mean, the population variance or a population proportion. As soon as the values of the random sequence are observed, the data can be plugged into these formulae to obtain such estimates. To get an idea of the precision of these estimates, the properties of the estimators (the generators of the estimates) are studied. In particular, the expectation and variance of the estimators are of interest, to learn about their estimating precision.

It will turn out that sampling without replacement and sampling with replacement are often good sampling methods. That is, the random sequences $X_1, \ldots, X_n$ that arise from these methods have desirable properties. Although sampling without replacement yields more information, sampling **with** replacement has nicer theoretical properties. Fortunately, in cases where the sample size n is small when compared with the population size N, sampling without replacement can hardly be distinguished from sampling with replacement. That is why in practice, sampling is often done without replacement, while to ascertain the precision of the resulting estimates the theory of sampling **with** replacement is used.

In Chapter 12, random samples with replacement and random samples without replacement are the main topics. The properties of the two sampling methods are considered and compared. Several sample counterparts of population statistics are defined. In inferential statistics they will be used to draw conclusions about the whole population. Chapter 13 is about the sample mean, the sample counterpart of the population mean μ. Several important properties of this estimator are considered, for instance, the famous central limit theorem. Chapter 14 is about the sample proportion (the sample counterpart of the population proportion) and other sample counterparts of population statistics. Since population proportions and sample proportions are nothing but special population **means** and sample **means** (see Section 3.3.1), Chapter 14 is to some extent an application of Chapter 13.

Random samples

12

Chapter contents

This chapter is about general sampling methods, but especially about random sampling with or without replacement and properties. Furthermore, sample counterparts of population statistics are introduced. These are formulae that – when the data are available – can be used to generate approximations of the accompanying unknown population statistics.

CASE 12.1 DEFECTS ON ELECTRONIC CIRCUIT BOARDS

An investigator wants to estimate the average number of defective components on electronic circuit boards manufactured for installation in computers. Since the testing of such components takes much time, a sample of four boards is used to check the daily production of $N = 10$ boards. However, there may be one complication: the investigator feels that the number of defects on a board will be positively related to the number of components on that board. Which sampling method should be chosen if:

  a all boards have the same number of components;

  b the boards contain varying numbers of components?

The solution is given at the end of Section 12.1.

12.1 Sampling methods

A **sampling method** is a regimen that describes precisely how the sample has to be chosen from the population of interest. The actual execution is called the **drawing** of the sample.

A sampling method can depend on chance or not. For research among the households of a large city, the sampling method that uses as a sample only the 50 households living in the same street as the researcher does not depend on chance as far as the drawing is concerned. However, if this researcher uses a die to determine whether the households living in the 25 even- or the 25 odd-numbered houses constitute the sample, then this sampling method does depend on chance.

In inferential statistics, samples are used to draw conclusions about the whole population. It will turn out that the sampling method that is used to obtain the sample determines to a large extent the conclusions that can and may be drawn. To demonstrate how careful a researcher has to be, consider the following example.

Example 12.1

A manufacturer wants to know how long visitors to the company's website stay logged in and what they think of the attractiveness of the products on offer. Arbitrary logged-in visitors are asked about the duration of their visit and to assign a score (scale 0–10) to the attractiveness of the products. In total, 522 visitors are interviewed. Their mean length of stay on the site is 10.12 minutes; the mean score that these people assign to the products is 7.5. Do these results justify the conclusion that the mean length of visit for the population of **all** visitors to the site is about 10.12 minutes and that they evaluate the attractiveness of the products on average as about 7.5?

The sampling method randomly chooses logged-in visitors. Hence, visitors who are logged in for a longer period have a larger chance of being selected than visitors who are only briefly logged in. This simply follows from the fact that someone who is logged in for 30 minutes can more easily be chosen than someone who stays only 20 seconds. That is, in the resulting sample the people who stay longer (and probably are more seriously interested) are over-represented. Hence, it is likely that the sample mean 10.12 overestimates the true mean length of stay. Furthermore, because the people who are seriously interested are probably over-represented in the sample, the score 7.5 for the attractiveness of the products probably is larger than the true mean score for the whole population of visitors. The conclusion is that the chosen sampling method puts a stamp on the conclusions that can be drawn about the whole population.

In the future, only sampling methods that depend on chance will be considered.

In practice, there exist several sampling methods. The most important ones, **random sampling with replacement** and **random sampling without replacement**, have already been considered briefly in Section 7.3. Although many more sampling methods are important in practice, only two will be looked at here. In the context of any of these more complicated sampling methods, the two sampling methods of Section 7.3 are often referred to as **simple random sampling** (with/without replacement).

In **stratified random sampling**, the population of interest is divided into sub-populations (called **strata**) and independent simple random samples are drawn from all sub-populations. For instance, to estimate the mean profitability of all EU25 electronics companies, a researcher will probably divide the population of all electronics companies in the EU25 into the 25 natural sub-populations and gather data about profitability for 25 independent random samples.

In **cluster sampling**, the population is also divided into sub-populations, called **clusters**. However, the sample now consists of all population elements of the clusters chosen in a random sample of clusters. For instance, to find an estimate of the average household income in a large city, we could divide the city into districts (the clusters), select a simple random sample of districts, and put all households in these selected districts into the sample.

The choice of a sample element costs money. That is why those costs are important items when taking the decision about the sampling method that will be used. Another reason that drives the choice of the sampling method is the reduction of variability.

Although these are important topics for practice, we will not go into detail. In this book, only the sampling methods random sampling with / without replacement will be considered and their properties will be studied.

The random experiment of drawing a sample of size n from a population that consists of N elements can be a very complex affair. To reduce complexity, this experiment is often split into the n sub-experiments of the drawings of the n successive elements. A ***random sample*** of size n is a sample that results from a sampling method for n successive drawings done in such a way that, for each of the individual drawings, all population elements have the same chance of being sampled. If, in each sub-experiment, the drawing is done from all N population elements, the random sample is drawn **with replacement**. A random sample **without replacement** arises if an element, having once been drawn, cannot be selected in other sub-experiments.

The sub-experiments of the individual drawings in a random sample **with** replacement are all identical and independent. This makes this sampling method theoretically very tractable. The whole experiment consists of n subsequent random drawings, each of them drawn from all N population elements, as described in Section 7.3. For each sub-experiment there are N possibilities, so there are N^n possible random samples of size n. Since none of the elements is favoured above other elements, each of the N^n possible random samples has the same chance N^{-n} of becoming the sample that eventually occurs. This is the reason why such samples are recorded as ordered n-tuples: unordered n-tuples are not all equally likely.

A random sample **without** replacement is also obtained from n successive random drawings, but now from a population that becomes 1 smaller after each drawing. As a result, the sub-experiments are not independent and also not identical. When sampling is done without replacement, the order of the sampled elements is usually not important. There are 'N choose n' possible (unordered) random samples of this type; because of symmetry arguments, they are all equally likely.

You may be wondering whether the without-replacement sampling method indeed returns **random** samples. Of course, the probability that population element a will be drawn in the first sub-experiment is $1/N$. But what about the probability that a will be drawn in the third (say) sub-experiment? To show that this probability is also $1/N$, let A_j be the event that element a is drawn in the jth drawing ($j = 1, 2, 3$). According to the product rule (7.15) it follows that:

$$P(A_3) = P(A_1^c \cap A_2^c \cap A_3) = P(A_1^c)P(A_2^c \mid A_1^c)P(A_3 \mid A_1^c \cap A_2^c)$$
$$= \frac{N-1}{N} \times \frac{N-2}{N-1} \times \frac{1}{N-2} = \frac{1}{N}$$

Example 12.2

Suppose that 100 households are randomly drawn from a population of 300 households. The family Doubledutch is one of these 300 households. Determine the probability p that this family will belong to the sampled households, both for the with-replacement and the without-replacement case. For the with-replacement case, also determine the probability that, among the 100 sub-experiments, the family Doubledutch is chosen at least twice.

Suppose, first, that sampling is done with replacement. Since there are in total 300^{100} possible samples (ordered 100-tuples) and 299^{100} of them do **not** contain the family Doubledutch, the probability p that the family Doubledutch is sampled is equal to:

$$p = 1 - \frac{299^{100}}{300^{100}} = 1 - \left(\frac{299}{300}\right)^{100} = 1 - 0.7161 = 0.2839$$

Now suppose that sampling is done without replacement. By (a generalization of) the product rule (7.16) it follows that:

$$p = 1 - \frac{299}{300} \times \frac{298}{299} \times \ldots \times \frac{200}{201} = 1 - \frac{299!/199!}{300!/200!} = 1 - \frac{299! \times 200!}{300! \times 199!} = 1 - \frac{200}{300} = 1/3$$

When drawing is done with replacement, the 100 sub-experiments are identical and independent; for each sub-experiment, the probability that the family Doubledutch is chosen is $1/300$. Hence, the random experiment is a binomial experiment consisting of 100 trials, where a success occurs if the family Doubledutch is drawn and the probability of success is 1/300. So, the rv $Y =$ 'number of successes among the 100 trials' is $Bin(100, 1/300)$ distributed. The probability that the family Doubledutch is chosen at least twice equals:

$$P(Y \geq 2) = 1 - P(Y = 0) - P(Y = 1) = 1 - \left(\frac{299}{300}\right)^{100} - 100 \times \frac{1}{300} \times \left(\frac{299}{300}\right)^{99}$$
$$= 1 - 0.7161 - 0.2395 = 0.0444$$

The conclusion is that the probability that a fixed population element is chosen more than once is relatively small.

In the above example, the sample size is one-third of the population size; the ratio n/N equals 1/3. It is intuitively clear that in with-replacement random sampling the probability of getting multiple occurrences of population elements will be small if the ratio n/N is small. If only a very small part of the population is to be sampled (which in practice is often the case), then random sampling without replacement is hardly distinguishable from random sampling with replacement. This important conclusion justifies the practical situation that the without-replacement random sampling approach is used to collect the sample and the with-replacement approach to analyse the properties of the sample.

Often, interest will be in one or more population variables. To start with, suppose that interest is in one variable called X. If n population elements are drawn randomly (with or without replacement), then their values of X can be measured and a sequence of random observations arises. Here, the rv X_1 measures X for the first sampled element, X_2 measures X for the second sampled element, etc. This sequence $X_1, ..., X_n$ of random observations is also called a ***random sample***, with replacement if the underlying random experiment is the with-replacement random sampling method and without replacement if the underlying random experiment is the without-replacement random sampling method.

This subsection ends with some remarks and warnings as the naming of some concepts may cause confusion. Firstly, a random sample can be a set of arbitrarily chosen population elements or a sequence of accompanying observations of a population variable. It will become clear from the context which of the two meanings is meant. Secondly, the term 'random' in a random sample of observations of X does not necessarily imply that the sequence of random variables $X_1, ..., X_n$ is meant. That is, speaking about a random sample of observations tells us only that the sample is drawn randomly (i.e. arbitrarily); whether a sequence $x_1, ..., x_n$ of realizations is meant or a sequence $X_1, ..., X_n$ of random variables is not immediately obvious but will become clear from the context.

From now on, random samples with replacement are called random samples, for short. A random sample $X_1, ..., X_n$ (so, drawn **with** replacement) is also called an **iid-sample**. The following section explains why this is an informative name.

CASE 12.1 DEFECTS ON ELECTRONIC CIRCUIT BOARDS – SOLUTION

It is intuitively obvious that a 'good' sample should give a board with many electronic components a larger chance of being selected than a board with few components.

In the case of (a), there is no problem since all boards have the same number of components. Without replacement will be used to obtain the four boards; in each of the four sub-experiments, all ten boards of the day's production have the same change of being sampled. To estimate the average number of defects per board, the numbers of defects of the four sampled boards are determined and the mean of these four numbers is the estimate.

In the case of (b), simple random sampling would give each board the same chance of being sampled. But now this is not what is wanted; now, the probability that a board is sampled should be proportional to its number of components. To illustrate a good sampling method, suppose that the ten boards contain the following numbers of components, respectively:

$$8 \quad 10 \quad 21 \quad 15 \quad 9 \quad 24 \quad 12 \quad 18 \quad 16 \quad 17$$

In total, there are 150 components on the ten boards. Components 1, 2, ..., 8 are on board 1; components 9, 10, ..., 18 on board 2; ...; components 134, 135, ..., 150 on board 10. By randomly drawing four **components** (with or without replacement, the methods are hard to distinguish) and noting to which boards they belong, board 1 gets probability 8/150 of being sampled in a sub-experiment, board 2 gets probability 10/150, etc. (Note that this sampling method allows a board to be sampled more than once.)

Suppose that the components 13, 58, 115 and 23 are sampled. Since these components belong to boards 2, 5, 8 and 3, all components on these boards are tested and the numbers of defects are determined. The average of these four results is the estimate that is wanted.

12.2 Random samples with replacement (*iid* samples)

Start with a population and a population variable X. As usual in statistics, it is the population distribution of X and its statistics such as the population mean μ and population variance σ^2 that are of interest. One way to look at this population distribution is via its overview of relative frequencies for intervals I.

Consider a random sample $X_1, ..., X_n$ of observations of that population variable X. (Since capitals are used, the sequence of random **variables** is meant.) Recall that the underlying random sampling procedure samples n population elements arbitrarily and with replacement, and that the n sub-experiments are independent and identical. Since, for each $i = 1, ..., n$, the random observation X_i depends only on the ith sub-experiment, the important consequence is that the random variables $X_1, ..., X_n$ are independent and have a 'common probabilistic behaviour'. This common probabilistic behaviour means that probabilities regarding X_1 are exactly the same as the corresponding probabilities regarding X_2, etc. In particular, $X_1, ..., X_n$ have a common distribution function:

$$P(X_1 \leq x) = P(X_2 \leq x) = ... = P(X_n \leq x) \quad \text{for all real numbers } x \tag{12.1}$$

The independence and the identical probabilistic behaviour of the sequence of random observations is expressed by saying that $X_1, ..., X_n$ are **independently and identically distributed**; or $X_1, ..., X_n$ are **iid**, for short. The phrase 'identically distributed' expresses that all probabilities of X_1 events are the same as the corresponding probabilities of X_2 events, etc.

Properties of an *iid* sample $X_1, ..., X_n$

- *i property*: $X_1, ..., X_n$ are independent.
- *id property*: $X_1, ..., X_n$ have the same probability distribution.

In this context, the random sample $X_1, ..., X_n$ is often defined as a random sample **drawn from the population (distribution)**. The *i* and *id* properties of $X_1, ..., X_n$ have several important consequences. For instance, because of the *i* property it follows that:

$$P(X_{13} \leq 2.5; X_{28} > 1.5) = P(X_{13} \leq 2.5) \times P(X_{28} > 1.5)$$

while the *id* property ensures that the right-hand side equals $P(X_1 \leq 2.5) \times P(X_1 > 1.5)$. Because of the independence, X_i and X_j are (for $i \neq j$) uncorrelated and have covariance 0. Furthermore, being identically distributed implies especially that $X_1, \ldots, X_n$ have the same marginal statistics. Their expectations are all equal to the population mean μ and their variances are all equal to the population variance σ^2. That is, for all $i, j = 1, \ldots, n$ with $i \neq j$ it holds that:

$$E(X_i) = \mu; \quad V(X_i) = \sigma^2; \quad SD(X_i) = \sigma; \quad \text{cov}(X_i, X_j) = 0 \tag{12.2}$$

Example 12.3

The mean size of the households is 2.4 with accompanying standard deviation 1.23. Furthermore, 34.4% of the households are single households (that is, only one person) and 7.1% of the households consist of at least five persons. Suppose that 100 households are to be drawn randomly and their sizes measured. What can be said in advance about probabilities concerning the sizes of the households in the sample?

The population of interest is the set of all households; the population variable of interest is X = 'size of the household'. A random sample $X_1, \ldots, X_{100}$ will be drawn. Although this will probably be done without replacement, the sample size is so small when compared with the population size that it can freely be assumed that the sample is drawn **with** replacement. It is given that the mean μ and the standard deviation σ of the population distribution of X are respectively 2.4 and 1.23. Furthermore, the few facts known about the population distribution of X can be formulated as follows: the value 1 has relative frequency 0.344 and the value 4 has a cumulative relative frequency $1 - 0.071 = 0.929$.

According to (12.2), all individual random observations X_i have expectation 2.4 and standard deviation 1.23. For instance:

$$E(X_{76}) = E(X_2) = E(X_{16}) = E(X_{97}) = \ldots = 2.4$$

$$V(X_{76}) = V(X_2) = V(X_{16}) = V(X_{97}) = \ldots = (1.23)^2 = 1.5129$$

Furthermore, the *id* property yields:

$$P(X_{76} = 1) = P(X_2 = 1) = P(X_{16} = 1) = P(X_{97} = 1) = \ldots = 0.344$$

$$P(X_{76} \leq 4) = P(X_2 \leq 4) = P(X_{16} \leq 4) = P(X_{97} \leq 4) = \ldots = 0.929$$

The *i* property and the *id* property jointly yield:

$$P(X_{53} > 1; X_{88} > 4) = P(X_{53} > 1)P(X_{88} > 4)$$
$$= P(X_1 > 1)P(X_1 > 4) = 0.656 \times 0.071 = 0.047$$

12.3 Random samples without replacement

Again, start with a population and a population variable X of interest; μ and σ^2 again are the population mean and population variance. But now a random sample $X_1, \ldots, X_n$ of observations of that population variable X is drawn **without** replacement. Notice that under these circumstances the sample size n cannot be larger than the population size N. Recall from Section 12.1 that the n sub-experiments are not independent and not identical. Since the individual random observation X_i only depends on the ith sub-experiment, it follows that the random observations $X_1, \ldots, X_n$ will not be independent either. Hence, the *i* property is no longer valid. Indeed, it can (but will not) be proved that, for $i \neq j$, the covariance of X_i and X_j is unequal to 0. To be precise, it can be shown that it equals $-\sigma^2/(N-1)$. Although the sub-experiments are not identical, the without-replacement random sample $X_1, \ldots, X_n$ still has the *id* property.

Properties of a without-replacement random sample $X_1, ..., X_n$

■ $\text{cov}(X_i, X_j) = -\dfrac{\sigma^2}{N-1}$ for $i \neq j$ (12.3)

■ *id* property.

To make this last property believable, it will now be demonstrated that in a very simple example the random observations X_i indeed have the same pdf.

Example 12.4

Reconsider the population of the five households in Baker Street; see Example 5.8. The population variable X = 'number of television sets' is considered; it has the following frequency distribution:

Value	0	1	2
Frequency	1	3	1
Rel. frequency	1/5	3/5	1/5

Suppose that a without-replacement random sample of size 2 is drawn from this population; let X_1, X_2 denote the two subsequent random observations of X. Of course, the pdf of the first random observation X_1 is equal to the relative frequency distribution presented in the table. By conditioning on the different values of X_1, it will now be shown that X_2 has the same pdf.

The table below gives the conditional probabilities $P(X_2 = j \mid X_1 = i)$.

i \ j	0	1	2
0	0	3/4	1/4
1	1/4	2/4	1/4
2	1/4	3/4	0

For instance, to determine $P(X_2 = 1 \mid X_1 = 1)$ in the cell with $i = 1$ and $j = 1$, note that the condition that the first sampled household has 1 TV implies that for the second sampled element only four households are left and two of them have 1 TV. To obtain the pdf of X_2, notice that:

$$P(X_2 = 0) = P(X_2 = 0; X_1 = 0) + P(X_2 = 0; X_1 = 1) + P(X_2 = 0; X_1 = 2)$$

$$= P(X_2 = 0 \mid X_1 = 0) \times \frac{1}{5} + P(X_2 = 0 \mid X_1 = 1) \times \frac{3}{5} + P(X_2 = 0 \mid X_1 = 2) \times \frac{1}{5}$$

$$= 0 + \frac{1}{4} \times \frac{3}{5} + \frac{1}{4} \times \frac{1}{5} = 1/5$$

$$P(X_2 = 1) = \frac{3}{4} \times \frac{1}{5} + \frac{2}{4} \times \frac{3}{5} + \frac{3}{4} \times \frac{1}{5} = 3/5$$

$$P(X_2 = 2) = \frac{1}{4} \times \frac{1}{5} + \frac{1}{4} \times \frac{3}{5} + 0 = 1/5$$

Indeed, X_2 has the same pdf as X_1.

As a consequence of the *id* property, the probabilistic behaviour of each of the random observations $X_1, ..., X_n$ is identical; all probabilities of X_1 events are the same as the corresponding probabilities of X_2 events, etc. Being identically distributed implies again that $X_1, ..., X_n$ have the same marginal statistics. Their expectations are all equal to the population mean μ and their variances are all equal to the population variance σ^2. That is, for all $i = 1, 2, ..., n$ it holds that:

$$E(X_i) = \mu, \quad V(X_i) = \sigma^2, \quad SD(X_i) = \sigma$$ (12.4)

Example 12.5

Interest is in a group of 150 pharmaceutical companies with at least 100 employees. Last year, their mean investment was €13.4 million with a standard deviation of €8.1 million. Moreover, 25 of these companies invested more than €20 million. Suppose that 50 of the companies in this group are to be drawn randomly and their investments last year recorded. What can be said in advance about the probabilistic behaviour of the investments of the individual companies in the sample?

Of interest is the population of the 150 companies with at least 100 employees and the population variable X = 'last year's investments (in millions of euro)'. A random sample $X_1, ..., X_{50}$ of size 50 will be drawn. Of course, this sample will be drawn without replacement. Since the sample is a rather large part of the population, we really have to consider the without-replacement case. It is given that the mean μ and the standard deviation σ of the population distribution of X are respectively €13.4 million and €8.1 million. Furthermore, regarding the population distribution of X, it is also given that the relative frequency of the companies with more than €20 million of investments is 1/6.

Because of the *id* property, we have (for instance):

$$E(X_{12}) = E(X_2) = E(X_{16}) = E(X_{47}) = ... = 13.4$$

$$V(X_{12}) = V(X_2) = V(X_{16}) = V(X_{47}) = ... = (8.1)^2 = 65.61$$

$$P(X_1 > 20) = P(X_{46} > 20) = P(X_{34} > 20) = ... = 1/6$$

By (12.3), the correlation coefficient of (for instance) X_{23} and X_{49} follows immediately:

$$\rho = \frac{\text{cov}(X_{23}, X_{49})}{\sigma_{X_{23}} \sigma_{X_{49}}} = \frac{-\sigma^2/149}{\sigma^2} = -\frac{1}{149} = -0.0067$$

For $i \neq j$, the random observations X_i and X_j are very weakly linearly related.

12.4 Sample statistics and estimators

In inferential statistics, samples from a population are used to draw conclusions about the whole population. Since interest is often in such statistics as the population mean, proportion, variance, covariance and correlation coefficient, it seems natural to study the constitution of the formulae of these statistics and to construct similar sample counterparts that can be used to estimate them.

Start with a (finite) population $\{1, 2, ..., N\}$ and a quantitative population variable X. Although the set of **all** population observations of X is usually unobserved, we can denote them by $x_1, x_2, ..., x_N$. Based on these notations, the population mean, variance and standard deviation were defined as (see (3.1), (4.1), (4.2)):

$$\mu = \frac{1}{N} \sum_{i=1}^{N} x_i; \quad \sigma^2 = \frac{1}{N} \sum_{i=1}^{N} (x_i - \mu)^2; \quad \sigma = \sqrt{\sigma^2} \tag{12.5}$$

They are respectively:

- the mean of the population observations x_i;
- the mean of the squares of the deviations $x_i - \mu$;
- its square root.

But these statistics are usually unknown; they have to be approximated by way of a sample from the population. Suppose that such a sample, of size n, is drawn randomly and that the values of X for the sampled elements are measured. In terms of random variables, the resulting random observations are denoted by $X_1, X_2, ..., X_n$ (with capitals); the resulting data are denoted by $x_1, x_2, ..., x_n$ (with small letters). Notice that, in this sample context, X_i denotes the random variable that represents the random observation of the ith sampled element while x_i is its realized value. Natural methods for estimating μ, σ^2 and σ are:

- the mean $\bar{X}$ of the sample observations X_i;
- the mean of the squares of the deviations $X_i - \bar{X}$;
- its square root.

These considerations lead to **estimation methods** or **estimators** $\bar{X}$, S^2 and S for μ, σ^2 and σ; see (12.6) in the box below. They are nothing but formulae that can be used to generate approximations of μ, σ^2 and σ. (Strictly speaking, S^2 is not a mean but a corrected mean; it turns out that using $n - 1$ instead of n yields a better method; see Section 14.2 for more details.) When the sample data x_1, x_2, ..., x_n are observed, they can be substituted into the estimators $\bar{X}$, S^2 and S to obtain **estimates** $\bar{x}$, s^2 and s of μ, σ^2 and σ:

Estimators of μ, σ^2 and σ

$$\bar{X} = \frac{1}{n}\sum_{i=1}^{n} X_i; \quad S^2 = \frac{1}{n-1}\sum_{i=1}^{n}(X_i - \bar{X})^2; \quad S = \sqrt{S^2} \tag{12.6}$$

Estimates of μ, σ^2 and σ

$$\bar{x} = \frac{1}{n}\sum_{i=1}^{n} x_i; \quad s^2 = \frac{1}{n-1}\sum_{i=1}^{n}(x_i - \bar{x})^2; \quad s = \sqrt{s^2} \tag{12.7}$$

Notice that the estimators $\bar{X}$, S^2 and S are random variables that do not depend on unknown parameters: as soon as the data are collected, they can be substituted into the formulae to obtain concrete estimates. These estimators are special **sample statistics**, random variables that are based only on the random observations X_1, ..., X_n and not on unknown parameters (such as μ, σ^2 and σ).

Both $\bar{X}$ and $\bar{x}$ are called **sample mean**; the first as estimator, the second as its realized value. Analogously, S^2 and s^2 are both called **sample variance**; S and s are both called **sample standard deviation**.

If the population variable is qualitative and takes only two values called success (coded as 1) and failure (coded as 0), then the population proportion p of successes – a real number between 0 and 1, usually unknown – is of interest. Recall that this qualitative variable behaves like a **quantitative** variable, X (say). Mean μ, variance σ^2 and standard deviation σ of X can be considered and are equal to p, $p(1 - p)$ and $\sqrt{p(1 - p)}$, respectively; see Sections 3.3.1 and 4.5. In particular, p is the mean of the population observations x_i that are either 1 or 0.

Drawing a random sample of size n yields $Bern(p)$ distributed random variables X_1, ..., X_n for the random observations. The accompanying sample statistic $\bar{X}$ is an estimator of p; it measures the proportion of successes in the sample and will be denoted by $\hat{P}$. When the sample data x_1, ..., x_n are observed and substituted into the formula that the estimator actually is, then the resulting value $\bar{x}$ is just the realized sample proportion denoted by $\hat{p}$ (small letter). Both $\hat{P}$ and $\hat{p}$ are called **sample proportions**, the first as estimator and the second as its realized value. In the box below, Y is the number of successes in the sample.

Estimator of p (population proportion successes)

$$\hat{P} = \frac{Y}{n} = \text{sample proportion successes}$$

Estimate of p

$$\hat{p} = \frac{y}{n} = \text{realized sample proportion successes}$$

Suppose, next, that **two** quantitative population variables X and Y are of interest, especially the degree of linear relationship between them. The (usually unobserved) population observations can be denoted by $(x_1, y_1), (x_2, y_2), \ldots, (x_N, y_N)$, where the pair (x_i, y_i) is the pair of observations of (X, Y) measured at population element i. Based on these notations, the population covariance and correlation coefficient of X and Y were defined as (see Section 5.1):

$$\sigma_{X,Y} = \frac{1}{N}\sum_{i=1}^{N}(x_i - \mu_X)(y_i - \mu_Y) \quad \text{and} \quad \rho_{X,Y} = \frac{\sigma_{X,Y}}{\sigma_X \sigma_Y} \tag{12.8}$$

Notice that the covariance is nothing but the mean of the products of the x deviations $x_i - \mu_X$ and the y deviations $y_i - \mu_Y$. Since $\sigma_{X,Y}$ is usually unknown, it has to be estimated by way of a sample from the population. Suppose that such a sample, of size n, is drawn at random and that the values of X and Y for the sample elements are measured. In terms of random variables, the resulting random observations are denoted by $(X_1, Y_1), (X_2, Y_2), \ldots, (X_n, Y_n)$; with capitals. The resulting data are denoted by $(x_1, y_1), (x_2, y_2), \ldots, (x_n, y_n)$; with small letters. Notice that, in this sample context, (X_i, Y_i) denotes the pair of random variables that represents the two random observations of the ith sampled element while (x_i, y_i) is its realized value. A natural estimation method for $\sigma_{X,Y}$ arises by constructing the 'mean' of the products of the random x deviations $X_i - \bar{X}$ and the random y deviations $Y_i - \bar{Y}$. An estimation method for $\rho_{X,Y}$ arises by replacing all factors on the right-hand side of (12.8) by the respective estimators. The following estimators $S_{X,Y}$ (**sample covariance**) and $R_{X,Y}$ (**sample correlation coefficient**) of $\sigma_{X,Y}$ and $\rho_{X,Y}$ are usually used in practice:

Estimators of covariance $\sigma_{X,Y}$ and correlation coefficient $\rho_{X,Y}$

$$S_{X,Y} = \frac{1}{n-1}\sum_{i=1}^{n}(X_i - \bar{X})(Y_i - \bar{Y}) \quad \text{and} \quad R_{X,Y} = \frac{S_{X,Y}}{S_X S_Y} \tag{12.9}$$

Estimates of covariance $\sigma_{X,Y}$ and correlation coefficient $\rho_{X,Y}$

$$s_{X,Y} = \frac{1}{n-1}\sum_{i=1}^{n}(x_i - \bar{x})(y_i - \bar{y}) \quad \text{and} \quad r_{X,Y} = \frac{s_{X,Y}}{s_X s_Y}$$

Here, S_X and S_Y are the sample standard deviation of X and Y, respectively. In particular, S_Y is the square root of S_Y^2, which in turn is the second formula in (12.6) with X replaced by Y. Again, notice that $S_{X,Y}$ is not a strict mean; see Section 14.2 for more details about the use of $n - 1$ instead of n.

Example 12.6

It goes without saying that the level of the annual income of a household determines to a large extent the level of expenditure on food, housing, clothing and recreation. To say it in more statistical terms: within populations of households, we predict a strong positive linear relationship between the variable HINC = 'annual household income' on the one hand, and each of the four variables 'annual expenditure on food / housing / clothing / recreation' (FOODEXP, HOUSEXP, CLOTEXP and RECREXP) on the other.

The strengths of these four linear relationships are measured by the population correlation coefficients. In Table 12.1, estimates of these coefficients and other statistics are given on the basis of a random sample of 300 households from a large population of households.

	FOODEXP	HOUSEXP	CLOTEXP	RECREXP	HINC	DF
Mean	8644.20	11 486.70	2608.70	2640.50	31 551.80	0.20
St. dev.	4078.56	6890.83	1435.87	1506.41	17 732.48	0.403
Correlation with HINC	0.98	0.98	0.99	0.99	1	0.05

TABLE 12.1 Estimates of mean, standard deviation and correlation coefficient

The sample mean of the 300 household incomes is €31 551.80, with a sample standard deviation of €17 732.48. According to Table 12.1, the mean expenditure on housing is by far the largest of the four mean expenditures. However, the standard deviation of this category is also the largest. It is interesting that the sample means of the expenditures on clothing and recreation are practically the same, as are the corresponding sample standard deviations.

Table 12.1 also contains sample statistics of the variable DF (dummy female). This variable equals 1 (or 0), if the reference person (head) of the household is a woman (or a man). The sample mean 0.20 informs us that 20% of the heads of the 300 households are female, which indicates that about the same will hold for the whole population.

The sample correlation coefficients suggest that there is a very strong positive linear relationship between the variable HINC and each of the four types of expenditure. However, from the slightly positive sample correlation coefficient 0.05 it can hardly be concluded that female heads of households often lead households with higher incomes.

In Chapters 13 and 14, important properties of some of the above estimators will be investigated. This is necessary to find out how close an estimate approximates the population statistic of interest.

Summary

Random sampling with and without replacement are the only two important sampling methods that are considered in this book. A random sample can refer to randomly sampled population elements, but more often to a sample of random observations of a population variable. If drawn with replacement, these random observations have the *i* property **and** the *id* property; if drawn without replacement, only the *id* property holds.

To draw conclusions about unknown population statistics, similarly constituted sample statistics are used. An estimator of such a population statistic is a formula that generates estimates when the data are substituted. The list of symbols below gives an overview of the notations for estimators and estimates.

List of symbols

Symbol	Description
$\hat{P}, \hat{p}$	Sample proportion (random, realization)
$R_{X,Y}, r_{X,Y}$	Sample correlation coefficient (random, realization)
S, s	Sample standard deviation (random, realization)
S^2, s^2	Sample variance (random, realization)
$S_{X,Y}, s_{X,Y}$	Sample covariance (random, realization)
$\bar{X}, \bar{x}$	Sample mean (random, realization)

🔑 Key terms

cluster sampling **374**
clusters **374**
drawing **373**
drawn from the
    population **377**
drawn from the population
    distribution **377**
estimate **380**
estimation method **380**
estimator **380**
i property **377**

id property **377**
iid **377**
iid sample **376**
independently and
    identically
    distributed **377**
random sample **375, 376**
sample correlation
    coefficient **382**
sample covariance **382**
sample mean **381**

sample proportion **381**
sample standard
    deviation **381**
sample statistic **381**
sample variance **382**
sampling method **373**
strata **374**
stratified random
    sampling **374**
with replacement **375**
without replacement **375**

? Exercises

Exercise 12.1

To study the opinion of consumers about a new snack, 100 people will be selected and asked to give their opinion in terms of a score between 0 and 10. Explain what it means to say that the researchers will use the sampling method 'random sampling without replacement'.

Exercise 12.2

From the 500 steelworkers in a factory, 250 are randomly sampled for a certain task. The sampling is done from a list that contains the names of these 500 people.

a The sampling is done with replacement. Explain how it will be conducted.

b The sampling is done without replacement. Explain how it will be conducted.

c Suppose that steelworker Bush is one of the 500. If the sampling is done with replacement, what is the probability that Bush will be chosen more than once?

Exercise 12.3

What are the similarities and the differences between stratified random sampling and cluster sampling?

Exercise 12.4

In the first week of April, the mean sales of shoe shops was €60 000 with standard deviation €20 000; for 18% of the shoe shops the sales were above €80 000. A tax inspector randomly draws, with replacement, ten shoe shops and measures the sales $X_1, \ldots, X_{10}$ (in euros) of the first week of April.

  a Determine $E(X_1)$ and $V(X_{10})$.

  b Calculate $P(X_1 > 80\,000; X_{10} \leq 80\,000)$.

  c Calculate $P(X_1 > 80\,000; X_5 > 80\,000; X_{10} \leq 80\,000)$.

  d Calculate the probability that two of the ten shops had sales above €80 000.

Exercise 12.5

  a Explain why a random sample $X_1, \ldots, X_n$ of random observations of a variable will not have the *i* property if it is drawn without replacement.

  b Explain why a random sample $X_1, \ldots, X_n$ of random observations of a variable will have the *i* property and the *id* property if it is drawn with replacement.

Exercise 12.6

The table below contains summarized results of three random samples of size 300 from a large population of households; for the names of the variables, see Example 12.6.

Sample 1	FOODEXP	HINC	DF
Mean	8178.20	29916.80	0.1733
St. dev.	4189.60	20511.40	0.3792
Correlation with HINC	0.9711	1	−0.0474

Sample 2	FOODEXP	HINC	DF
Mean	8644.20	31551.80	0.2033
St. dev.	4078.60	17732.50	0.4023
Correlation with HINC	0.9759	1	0.0500

Sample 3	FOODEXP	HINC	DF
Mean	8352.30	29903.70	0.1900
St. dev.	3344.50	13263.00	0.3930
Correlation with HINC	0.9716	1	0.0066

  a Write down the estimators that generated these estimates.

  b Calculate three estimates for the population covariance of FOODEXP and HINC.

  c Combine the three samples for an estimate of the mean household food expenditure that is based on a random sample of 900 households.

  d (more advanced) Explain how the three estimates of the population standard deviation of DF can be obtained from the corresponding estimates of the population mean.

Exercise 12.7

To do a certain task, 80 employees are sampled randomly from a group of 250 employees. You are one of the group of 250 employees.

a Suppose that the sampling is done with replacement. Determine the probability that you will belong to the sampled group.

b Also calculate the probability that you will be sampled at least twice.

c Now suppose that the sampling is done without replacement. Determine the probability that you will belong to the sampled group.

Exercise 12.8

The table below gives a classified relative frequency distribution of the variable X = 'disposable income (in euro) in 2002 of a one-person household'.

Disposable income	Relative frequency
Less than 10 000	0.172
10 000 —< 20 000	0.594
20 000 —< 30 000	0.174
30 000 —< 40 000	0.039
40 000 —< 50 000	0.010
At least 50 000	0.011

a Use the table to find an approximation of the mean μ and the variance σ^2 of the population distribution of X. (Use 55 000 as centre of the last class.)

Suppose that 200 one-person households will be drawn randomly and with replacement. Let the rvs $X_1, \ldots, X_{200}$ measure their disposable incomes in the present year. When answering the questions below, assume that the relative frequencies of the table are still valid for the present year, and that the present population mean and population variance of X are equal to the respective results of part (a).

b Explain why all rvs $X_1, \ldots, X_{200}$ have the same expectation and variance. Determine this common expectation and variance.

c Explain why the probabilities $P(20\,000 \leq X_i < 30\,000)$, for all $i = 1, \ldots, 200$, are all the same. Determine this common probability.

d With respect to the rvs X_1, X_{100} and X_{200}, calculate the following probabilities:

i $P(X_1 < 10\,000; X_{100} \geq 50\,000)$

ii $P(X_{100} < 10\,000; X_1 \geq 50\,000)$

iii $P(X_{100} < 30\,000; X_1 \geq 20\,000)$

iv $P(X_{100} < 30\,000 \text{ or } X_{200} \geq 20\,000)$

Exercise 12.9

Two households are drawn randomly and without replacement from a population of households. Interest is in the variable X = 'number of cars per household'. The rvs X_1 and X_2 denote the number of cars of the first and second household, respectively.

Suppose that the sample X_1, X_2 is drawn from a **small** population that contains only 10 households, and that the frequency distribution of X is as follows:

Value	0	1	2	3
Frequency	2	5	2	1

a Determine the marginal probability distributions of X_1 and X_2. Also calculate the expectations and the variances.

b Determine the respective conditional pdfs of X_2 given that $X_1 = 0$, that $X_1 = 1$, that $X_1 = 2$ and that $X_1 = 3$. Create one suitable table to demonstrate them all.

c Determine the joint pdf of X_1 and X_2.

d Calculate the covariance of X_1 and X_2, and show that it indeed equals $-\sigma^2/(N - 1)$; see (12.3).

Exercise 12.10

Reconsider Exercise 12.9 but suppose now that the population is a set of 7.1 million households. The relative frequency distribution is given in the table below.

x	0	1	2	3
Rel. freq.	0.230	0.559	0.190	0.021

a Calculate the covariance of X_1 and X_2 from (12.3). Give your comments.

The result of part (a) is one reason to treat the random sample (of size 2) as being drawn **with replacement**.

b Determine the marginal probability distributions of X_1 and X_2. Also calculate the expectations and the variances.

c Determine the joint pdf of X_1 and X_2.

Exercise 12.11

The following table comes from Statistics Netherlands' *Statistical Yearbook 2005* and is about the year 2004.

Employed labour force (15–64 years) by gender and age	%
men	58.1
women	41.9
15 – <25	10.9
25 – <35	25.5
35 – <45	28.8
45 – <55	24.1
55 – <65	10.7

Assume that the present situation regarding the variables 'gender' and 'age' of employed people is unchanged. Suppose that 500 employed people are drawn at random; let $X_1, \ldots, X_{500}$ be their gender (0 is male, 1 is female) and let $Y_1, \ldots, Y_{500}$ be their ages.

a Explain why $X_1, \ldots, X_{500}$ and $Y_1, \ldots, Y_{500}$ can be considered as two *iid* samples.

b Use the above table to determine $E(X_1)$, $V(X_{100})$, $E(Y_{200})$ and $V(Y_{500})$.

c Calculate the probabilities below; explain where you use the *i* and *id* property.

    i $P(X_{200} = 1; X_{400} = 0)$ iii $P(Y_7 \geq 25; Y_{237} < 35)$ v $P(X_1 = 0; Y_{34} < 35)$

    ii $P(Y_1 < 45; Y_2 < 25)$ iv $P(Y_{12} < 55; Y_{12} \geq 25)$

Exercise 12.12 (computer)

Interest is in a large population of households, especially in the variables X = 'annual household income ($\times$ 1000)', Y = 'size of the household' and D = 'gender (1 = female, 0 is male) of the head of the household'. The purpose is to estimate the unknown statistics μ_X, σ_X^2, σ_X, μ_Y, σ_Y^2, σ_Y, $\sigma_{X,Y}$, $\rho_{X,Y}$ and p, the population proportion of the households with a female head.

A random sample of 300 households will be drawn and for each of the sampled households the variables X, Y and D will be observed. As a result, three random samples (*iid* samples) arise: $X_1, \ldots, X_{300}$ and $Y_1, \ldots, Y_{300}$ and $D_1, \ldots, D_{300}$.

  a Formulate the respective estimators that will be used to generate estimates for the above statistics.

    The file Xrc12-12.sav contains the actual sample observations.

  b Determine estimates of μ_X, σ_X^2, σ_X, μ_Y, σ_Y^2 and p.

  c Use the estimate for p to estimate $V(D_1)$.

  d Find estimates for $\sigma_{X,Y}$ and $\rho_{X,Y}$.

  e Interpret the results of part (d).

Exercise 12.13

Consider a random sample of size n (with replacement) from a population of size N. Let A_i be the event that population element i is part of the sample ($i = 1, 2, \ldots, N$). Show that:

$$P(A_i \cap A_j) = 1 - P(A_i^c) - P(A_j^c) + P(A_i^c \cap A_j^c) \quad \text{for } i \neq j$$

Exercise 12.14

The administrative department of a large company has 400 staff: 300 women and 100 men. In view of the future policy with respect to new personnel, the management wants to know whether women can type faster than men. To investigate this qustion, 20 members of the department's staff are asked to type the same text. Several sampling methods are possible:

1 the 20 employees are chosen by random sampling without replacement;

2 the 20 employees are chosen by random sampling with replacement;

3 ten women and ten men are randomly chosen (without replacement);

4 fifteen women and five men are randomly chosen (without replacement).

  a For each of the four sampling methods, determine the total number of possible samples.

  b Does each sampling method yield a random sample?

  c Mr A and Mrs B work in the administrative department. Calculate for Mr A, for each of the four sampling methods, the probability that he will be chosen. Do the same for Mrs B. Also calculate the probability that A and B are both chosen (for method 3, see Exercise 12.13).

  d What is the name of the sampling methods 3 and 4?

Exercise 12.15

Reconsider Exercise 12.14.

  a For sampling methods 1 and 2, calculate the probability that at most one man is part of the sample.

  b Order the four sampling methods according to their suitability. Comment on your choice.

Exercise 12.16

Reconsider Exercise 12.14. The 20 administrative employees are selected with sampling method 4 and asked to type the same text. The time it takes them to do so (in minutes) and the number of typing errors are recorded.

Women															
Time	15.1	15.2	15.8	14.9	15.0	16.0	15.9	16.1	14.9	15.6	15.7	15.4	16.2	16.1	15.9
Number of errors	10	15	10	18	19	17	10	10	18	12	12	18	9	10	10

Men					
Time	16.8	16.3	17.2	15.9	17.4
Number of errors	6	7	5	14	5

Denote the sample of the times by $X_1, \ldots, X_{15}, X_{16}, \ldots, X_{20}$ and the sample of the number of errors by $Y_1, \ldots, Y_{15}, Y_{16}, \ldots, Y_{20}$. (A computer can be used for the calculations; see the data in Xrc12-16.xls.)

a The management wants to compare the means and variances of the data for the men with those of the data for the women. Which sample statistics are needed? Calculate the realizations. What are your conclusions?

b The two samples of size 20 are used to investigate the linear relationship between 'typing time' and 'number of errors'. Which sample statistic can be used? Calculate the realization and give your comments.

CASE 12.2 HOUSEHOLDS STATISTICS (PART I)

Statistics of the composition, incomes and expenditures of households are important for many studies in economics. The dataset Case12-02.sav contains data of 17 variables for a random sample of 300 heads (that is, reference persons) of households from a large population of households. Below, 'family' is used as a synonym for household. The variables are defined as follows:

DS   dummy single (= 1 for a household consisting of one adult only, the respondent. The respondent might be the head of a household with one or more minors. All other adults are coded DS = 0)

DH   dummy head of household (= 1 if the respondent is not the only adult in the household but is head of the household. By definition this household consists of two or more adults. Observe that DS × DH = 0)

DP   dummy partner (= 1 if the respondent is the partner of the head of the household. By definition we have DP × DH = 0 and DP × DS = 0. 'Partner' is defined as being married to the head of the household. Persons cohabiting are not considered partners)

DF   dummy female (= 1 if the family head is female)

WEIGHT   weight of the family head in kilograms

LENGTH   height of the family head in cm

AGE   age of the family head in years (not rounded)

EDU   level of education of the family head (1 = low, 5 = high)

WAGE   hourly wage of the family head in euro

HOURS   weekly number of hours worked by the family head

NKIDS   number of children (< 18 years) of the family

FS   family size (= number of family members, i.e. number of adults and children under 18)

FINC   net family income, in units of €1000 (including Family Allowance of €1600 per child annually)

FOODEXP   family expenditure on food, in units of €1000

HOUSEXP   family expenditure on housing, in units of €1000

CLOTEXP   family expenditure on clothing, in units of €1000

RECREXP   family expenditure on recreation, in units of €1000

Notice that the combination DS = 1 and DH = 0 refers to a single head of family, living on his/her own or with children. Furthermore, the variable DP will always be 0 since all respondents are heads of the households, so DH = 1 or DS = 1. In essence, DP is superfluous.

a For which variables is a population proportion p of interest? For which a population mean μ?

b With respect to estimation purposes, for which variables will the sample proportion $\hat{P}$ be used? For which the sample mean $\overline{X}$?

c For all variables, determine estimates of the population proportions or population means.

d For all quantitative variables, determine estimates of the population standard deviation.

e The qualitative (ordinal) variable EDU is occasionally treated as a quantitative variable. Determine estimates for its population mean and population variance.

f For the estimation of the degree of linear relationship between two quantitative variables, the estimator $R_{X,Y}$ is used. Use it (and a computer) to determine the matrix of sample correlation coefficients of all pairs of household income and expenditure variables. Give your comments.

CHAPTER

13

The sample mean

Chapter contents

The subject of the present chapter is the estimator $\bar{X}$, the random variable that can be used to generate estimates of the mean μ of a population variable X. It will turn out that only in special cases can the exact pdf of $\bar{X}$ be determined. However, for random samples (with and without replacement) the expectation and variance of $\bar{X}$ can simply be expressed in the population mean μ and population variance σ^2. Furthermore, the pdf of $\bar{X}$ is normal if the random sample is drawn from a normal distribution. If the sample is drawn from a non-normal distribution but the sample size is large, then the pdf of $\bar{X}$ is **approximately** normal. As a consequence, there are many practical situations where probabilities of $\bar{X}$ events can be calculated, or at least approximated.

CASE 13.1 THE RUIN PROBABILITY OF INSURANCE COMPANY LOWLANDS

Insurance company Lowlands, which we met in Case 11.1, also sells liability insurances with a maximum coverage of €10 000 (= 10^4). In total, the company has sold 100 000 (= 10^5) of such insurance contracts. The yearly claim amount X (discrete, measured in euro) that a contract generates has the following probability distribution:

$$
f(x) = \begin{cases} 0.90 & \text{if } x = 0 \\ 0.008 & \text{if } x = 1000, 2000, \ldots, 9000 \\ 0.028 & \text{if } x = 10\,000 \end{cases}
$$

Suppose that the yearly claim amounts $X_1, \ldots, X_{100000}$ of the 100 000 contracts are independent and identically distributed, that Lowlands has reserved an initial capital of €50 000 for the liability department to partially meet the claims of these contracts, and that each owner of a contract has to pay the yearly premium of b euros at the start of the year. If S denotes the total claim amount generated by these contracts and $U = 50\,000 + 100\,000b - S$ the capital of the liability department after one year, then the **ruin probability** $P(U < 0)$ is of interest. How large should the premium b be so that the probability that the liability department gets ruined is only 0.01? See Section 13.4 for a solution.

Hint: The ruin probability can be rewritten as $P(S > 50\,000 + 100\,000b)$, which in turn can be rewritten as a probability for the sample mean $\overline{X}$ of the *iid* sample $X_1, ..., X_{100000}$.

13.1 Expectation, variance and Chebyshev's rule

Interest is in a population variable X and its population distribution with **unknown** mean μ and variance σ^2. The intention is to find 'good' estimates of μ, on the basis of data $x_1, ..., x_n$ that come from the random sample $X_1, ..., X_n$ of random observations of X. As mentioned in Chapter 12, the mean $\bar{x}$ of the sample data $x_1, ..., x_n$ is used as an estimate of the unknown μ. Of course, it is not possible to decide whether the calculated value $\bar{x}$ is a good and precise estimate in the sense that it lies close to μ. This is because μ is unknown, so it is not obvious what 'close to' means. That is why the performance of the estimation **method** (estimator) $\overline{X}$ is used to say more about the precision when estimating μ. In particular, the expectation and variance of the estimator $\overline{X}$ are of interest, since $E(\overline{X})$ and $V(\overline{X})$ can be considered as the mean and the variance, respectively, of an ongoing sequence of measurements of $\overline{X}$.

13.1.1 Random samples (with replacement)

Suppose first that the random sample $X_1, ..., X_n$ is drawn **with** replacement. That is: $X_1, ..., X_n$ is an *iid* sample; the random observations X_i not only have the same probability distribution but are also independent. As a consequence of the *id* property, all X_i have the same expectation (μ) and variance (σ^2). Furthermore, the *i* property implies that the covariance of X_i and X_j ($i \neq j$) is 0. These properties are important in the determination of $E(\overline{X})$ and $V(\overline{X})$.

Start with $E(\overline{X})$. Below, the result (11.17) is used:

$$\mu_{\overline{X}} = E(\overline{X}) = E\left(\frac{1}{n}\sum_{i=1}^{n}X_i\right) = \frac{1}{n}E(X_1 + X_2 + ... + X_n) = \frac{1}{n} \times n\mu = \mu$$

It follows that the mean $\mu_{\overline{X}}$ of the probability distribution of $\overline{X}$ is equal to the mean μ of the population distribution of X. It would appear that the mean of all results $\bar{x}$ generated by the estimator $\overline{X}$ is just the unknown μ. That is, if the means $\bar{x}$ are calculated for all random samples $x_1, ..., x_n$ in an ongoing sequence of random samples, then the mean of the resulting ongoing sequence of measurements $\bar{x}$ is just the unknown value (μ) we are looking for.

Since $E(\overline{X}) = \mu$, it is said that $\overline{X}$ is an **unbiased** estimator of μ: the expectation of the estimator is precisely equal to the parameter it is estimating.

To find $V(\overline{X})$, the result (11.17) is used again:

$$\sigma_{\overline{X}}^2 = V(\overline{X}) = V\left(\frac{1}{n}\sum_{i=1}^{n}X_i\right) = \frac{1}{n^2}V\left(\sum_{i=1}^{n}X_i\right) = \frac{1}{n^2} \times n\sigma^2 = \frac{\sigma^2}{n}$$

(Notice the importance of the *i* property here: the covariance terms are all 0.) The variance of the probability distribution of $\overline{X}$ is equal to the variance of the population distribution divided by n. To understand the consequences, recall that $V(\overline{X}) = E((\overline{X} - \mu_{\overline{X}})^2)$, which equals $E((\overline{X} - \mu)^2)$. Apparently, $E((\overline{X} - \mu)^2) = \sigma^2/n$. That is, the expectation of the square of the random deviation $\overline{X} - \mu$ is equal to σ^2/n. If, for all samples in an ongoing sequence of random samples, the corresponding squared deviations $(\bar{x} - \mu)^2$ are calculated, then the mean of all squared deviations is equal to σ^2/n.

Now notice that σ^2/n gets close to 0 as n grows larger and larger; the expected value of the squared deviation $(\overline{X} - \mu)^2$ is close to 0 as n gets large. It would appear that the variation in the results $\bar{x}$ around the fixed (but unknown) value μ dies out as n tends to infinity.

Since $E(\overline{X})$, $V(\overline{X})$ and $SD(\overline{X})$ are also denoted by $\mu_{\overline{X}}$, $\sigma^2_{\overline{X}}$ and $\sigma_{\overline{X}}$, respectively, the above results can be summarized as follows:

Expectation, variance and standard deviation of the sample mean $\overline{X}$ of a random sample $X_1, ..., X_n$

$$\mu_{\overline{X}} = E(\overline{X}) = \mu; \quad \sigma^2_{\overline{X}} = V(\overline{X}) = \frac{\sigma^2}{n}; \quad \sigma_{\overline{X}} = SD(\overline{X}) = \frac{\sigma}{\sqrt{n}} \tag{13.1}$$

In particular, $\overline{X}$ is an **unbiased** estimator of μ.

The following example emphasizes the differences between the underlying population distribution and the probability distribution of $\overline{X}$, and between the population mean μ, variance σ^2 and standard deviation σ on the one hand, and $\mu_{\overline{X}}$, $\sigma^2_{\overline{X}}$ and $\sigma_{\overline{X}}$ on the other. It is about a large number of random samples from an underlying discrete population distribution.

Example 13.1

Interest is in a large population of households and in the variable $X =$ NKIDS = 'number of kids (under 18)'. Since the population is too large to observe X at all households, the precise population distribution of X is unknown. Furthermore, the accompanying mean μ and standard deviation σ are not known precisely. However, it is known from extensive studies that μ and σ are close to 0.9 and 1.2, respectively.

Five households are chosen randomly; the random variables $X_1, ..., X_5$ measure their respective numbers of children. According to the above theory the estimator $\overline{X}$ has the following properties:

$$E(\overline{X}) = \mu \approx 0.9; \quad V(\overline{X}) = \sigma^2/5 \approx 0.288; \quad SD(\overline{X}) = \sigma/\sqrt{5} \approx 0.537$$

For the interpretations, consider an ongoing sequence of samples of size 5 and calculate the sample mean of each of the samples. The first property above guarantees that the mean of the resulting

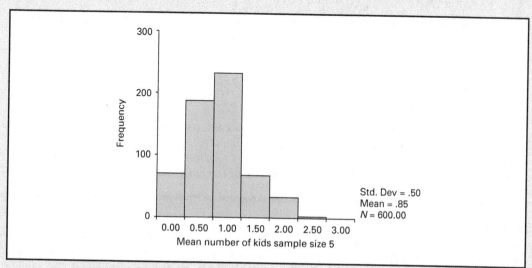

FIGURE 13.1 SPSS–Histogram of 600 sample means of size 5

ongoing sequence of sample means is equal to μ, while the second property states that the mean of the squares of the resulting ongoing sequence of deviations (with respect to μ) is equal to $\sigma^2/5$.

The dataset Xmp13-01.sav contains the means of 600 random samples of size 5. Although this sequence of sample means is finite and not ongoing, the idea is that the average of these 600 sample means must be close to μ, and that the variance and standard deviation of these 600 sample means must be close to $\sigma^2/5$ and $\sigma/\sqrt{5}$, respectively. The histogram in Figure 13.1, which was created with SPSS, presents the frequency distribution of the 600 observations of $\bar{X}$.

This histogram gives a first impression of the unknown probability distribution of $\bar{X}$. The average 0.85 of the 600 sample means must be close to the expectation $E(\bar{X}) = \mu$; the standard deviation 0.50 of the 600 observations of $\bar{X}$ must be close to the standard deviation $SD(\bar{X}) = \sigma/\sqrt{5}$. This seems to be the case since 0.9 is close to 0.85 and 0.537 is not far from the standard deviation 0.50 of these 600 observations of $\bar{X}$.

Recall from (8.35) that, for each rv Y with expectation μ_Y and standard deviation σ_Y, Chebyshev's inequality states that:

$$P(\mu_Y - k\sigma_Y < Y < \mu_Y + k\sigma_Y) \geq 1 - \frac{1}{k^2} \text{ for all real } k \geq 1$$

This rule can be applied for $\bar{X}$ instead of Y. By (13.1) it follows that:

$$P\left(\mu - k\frac{\sigma}{\sqrt{n}} < \bar{X} < \mu + k\frac{\sigma}{\sqrt{n}}\right) \geq 1 - \frac{1}{k^2} \text{ for all real } k \geq 1$$

To obtain a good interpretation of the above result, write ε for $(k\sigma)/\sqrt{n}$. So, $k\sigma = \varepsilon\sqrt{n}$, $k = (\varepsilon\sqrt{n}/\sigma)$ and hence: $1/k^2 = \sigma^2/(n\varepsilon^2)$. With these substitutions, Chebyshev's inequality turns into:

$$P(\mu - \varepsilon < \bar{X} < \mu + \varepsilon) \geq 1 - \frac{\sigma^2}{n\varepsilon^2} \text{ for all } \varepsilon > 0$$

Notice that the left-hand side is just $P(|\bar{X} - \mu| < \varepsilon)$, the probability that the estimator $\bar{X}$ is less than ε separated from μ, and that the right-hand side tends to 1 as n tends to ∞.

Restatement of Chebyshev's inequality for $\bar{X}$

$$P(|\bar{X} - \mu| < \varepsilon) \geq 1 - \frac{\sigma^2}{n\varepsilon^2} \text{ for all } \varepsilon > 0 \tag{13.2}$$

Since this lower bound tends to 1 as n gets larger and larger, the estimator $\bar{X}$ is called a **consistent** estimator of μ.

Hence, even for ε close to 0, it holds that:

$$P(|\bar{X} - \mu| < \varepsilon) \to 1 \text{ as } n \text{ becomes larger and larger} \tag{13.3}$$

The result (13.2) states that the sample mean $\bar{X}$ is a 'good' estimator of the unknown μ, in the sense that it generates estimates that probably lie as close to the unknown μ as wanted provided that n is taken large enough. Because of the validity of the law (13.2), the estimator $\bar{X}$ is called **consistent in estimating** μ; the law itself is called the (**general**) **law of large numbers**.

Example 13.2

A consumer organization is interested in the true mean contents μ of milk in litre packs of a certain brand. Regarding the population distribution of the variable X = 'contents (in litres) of a pack' it is only known that $\sigma = 0.0078$ litres; the distribution itself and its mean μ are not known. To obtain a good estimate of the unknown μ, the contents of n packs – drawn randomly from the daily production – will be measured. But how large should n be to ensure that the probability is at least 0.95 that the estimator $\bar{X}$ will yield a realization that is less than 0.001 litres separated from the true (but unknown) mean μ?

The above claim can be written as:

$$P(|\bar{X} - \mu| < 0.001) \geq 0.95$$

The sample size n has to be chosen so large that the lower bound of (13.2), with $\varepsilon = 0.001$, equals 0.95:

$$0.95 = 1 - \frac{\sigma^2}{n\varepsilon^2} = 1 - \frac{0.0078^2}{n \times 0.001^2} = 1 - \frac{60.84}{n}$$

It follows that $n = 1216.8$, so the sample size should be at least 1217.

Notice that equation (13.2) gives only a **lower** bound for the left-hand probability. In Section 13.3 it will be shown that probabilities as in the left-hand part of (13.2) are often considerably larger than Chebyshev's lower bound. As a consequence, the sample size in the last example does not really have to be that large. This follows from the famous Central Limit Theorem that will be the topic of Section 13.3 below.

13.1.2 Random samples without replacement

Now suppose that the random sample $X_1, \ldots, X_n$ is drawn **without** replacement from the population distribution of X that has mean μ and variance σ^2. In Section 12.3 it was noticed that, under these circumstances, the X_i still have the *id* property, but the *i* property is not valid any more. As a consequence, all X_i have the same expectation (μ) and variance (σ^2). Furthermore, by (12.3) the covariance of X_i and X_j ($i \neq j$) is $-\sigma^2/(N-1)$.

To determine the expectation of $\bar{X}$, the proof of the with-replacement case above can be repeated. However, the variance of $\bar{X}$ is slightly more complicated:

Expectation and variance of the sample mean $\bar{X}$ of a without replacement random sample $X_1, \ldots, X_n$

$$\mu_{\bar{X}} = E(\bar{X}) = \mu; \quad \sigma^2_{\bar{X}} = V(\bar{X}) = \frac{\sigma^2}{n}\frac{N-n}{N-1} \tag{13.4}$$

In particular, $\bar{X}$ is an **unbiased** estimator of μ.

Proof of the result for $\sigma^2_{\bar{X}}$
We have to be careful since the covariance terms are non-zero in this without-replacement case. However, by equation (11.16) and the remark thereafter we have:

$$V(\bar{X}) = \frac{1}{n^2}V\left(\sum_{i=1}^{n}X_i\right) = \frac{1}{n^2}\sum_{i=1}^{n}V(X_i) + \frac{2}{n^2}\sum_{i<j}\text{cov}(X_i, X_j)$$

$$= \frac{n\sigma^2}{n^2} - \frac{2}{n^2}\frac{n^2-n}{2}\frac{\sigma^2}{N-1} = \frac{\sigma^2}{n} - \frac{\sigma^2}{n}\frac{n-1}{N-1} = \frac{\sigma^2}{n}\left(1 - \frac{n-1}{N-1}\right) = \frac{\sigma^2}{n}\frac{N-n}{N-1}$$

Which prove the result.

Since the **variance correction factor** $(N-n)/(N-1)$ is smaller than 1, it follows that $V(\bar{X})$ is smaller than σ^2/n, the variance of $\bar{X}$ for the with-replacement case. It would appear that the use of the estimator $\bar{X}$ for estimation of μ leads to less variation if the sample is drawn without replacement.

This is intuitively clear: if sampling is done without replacement, the possibility of drawing a population element twice is excluded and the amount of information about μ that is present in the sample is larger.

However, if the sample size n is much smaller than the population size N, then the correction factor is close to 1 and the without-replacement variance is nearly the same as the with-replacement variance σ^2/n. That is why, in practice, the drawing of samples with $n/N < 0.1$ (that is, the sample is less than 10% of the population) is actually done without replacement but its result is analysed as if it were done with replacement.

Relationship between random samples with and without replacement

If $\dfrac{n}{N} < 0.1$, then random sampling 'without' $\approx$ random sampling 'with' $\hspace{2em}$ **(13.5)**

Example 13.3

In Example 13.2, the n packs of milk were drawn randomly from the large production of one day. Of course, this is done **without** replacement. However, in the analysis it was assumed that the sample size – which turned out to be at least 1217 – is at most 10% of the production.

Now suppose that indeed 1217 packs of milk are drawn randomly, but the daily production is only 2500 packs. If the sample is drawn with replacement, the lower bound 0.95 for $P(|\bar{X} - \mu| < 0.001)$ is still valid, since the derivation of Example 13.2 uses only the with-replacement assumption and not the size of the population. But what can be said about the probability $P(|\bar{X} - \mu| < 0.001)$ if the sample is drawn without replacement?

According to Chebyshev's inequality it holds that:

$$P(\mu - k\sigma_{\bar{X}} < \bar{X} < \mu + k\sigma_{\bar{X}}) \geq 1 - \frac{1}{k^2} \qquad \text{for all real } k \geq 1$$

A lower bound can be obtained by choosing k such that $\mu + k\sigma_{\bar{X}} = \mu + 0.001$. By (13.4) and the fact that σ is known to be equal to 0.0078, it follows that:

$$0.001 = k\frac{\sigma}{\sqrt{n}}\sqrt{\frac{N-n}{N-1}} = k\frac{0.0078}{\sqrt{1217}}\sqrt{\frac{1283}{2499}} = 0.00016021k$$

So $k = 6.242$. It follows that:

$$P(|\bar{X} - \mu| < 0.001) \geq 1 - \frac{1}{(6.242)^2} = 0.9743$$

which is larger than the lower bound 0.95 in Example 13.2. Drawing the sample without replacement yields more information and increases the precision of the estimator $\bar{X}$.

Unless stated otherwise, it will be assumed (without mentioning) that a random sample is drawn **with** replacement or that the ratio n/N is smaller than 0.1, so that the properties of the without-replacement sample $X_1, \ldots, X_n$ are those of a with-replacement sample.

13.2 Concerning the exact probability distribution of the sample mean

Only in special cases can the precise pdf of the sample mean $\bar{X}$ be obtained easily; see Example 13.4 for one such instance. In cases where the underlying population variable can take many different values, it is hardly possible (and sensible) to find this pdf. However, there is one important exception.

Consider a random sample (so, drawn with replacement) $X_1, \ldots, X_n$ from a population distribution with mean μ and variance σ^2. Hence, $X_1, \ldots, X_n$ is an *iid* sample, and the expectation and variance of the (probability distribution of the) sample mean $\bar{X}$ can easily be expressed in the population statistics μ and σ^2. Unfortunately, it is much more complicated (and often impossible) to determine the pdf of $\bar{X}$. Here is an example where it **is** possible.

Example 13.4

Suppose that a without-replacement random sample X_1, X_2 of size two is drawn from the population distribution of X = 'number of cars per household'. In this special case, the population distribution is known:

x	0	1	2	3
Rel. freq.	0.230	0.559	0.190	0.021

Since the population is large, the sample X_1, X_2 can freely be assumed to be drawn with replacement and to be *iid*. Hence, X_1 and X_2 are independent with a common pdf that equals the above population distribution. In particular (see also Example 8.22):

$$E(X_1) = E(X_2) = \mu = 1.0020 \quad \text{and} \quad V(X_1) = V(X_2) = \sigma^2 = 0.5040$$

By (13.1) it follows that:

$$E(\bar{X}) = 1.0020 \quad \text{and} \quad V(\bar{X}) = 0.5040/2 = 0.2520$$

To find the pdf of $\bar{X} = (X_1 + X_2)/2$, first notice that the possible outcomes are 0, 0.5, 1, 1.5, 2, 2.5, 3. Table 13.1 contains per cell the local mean $(x_1 + x_2)/2$ and the local joint probability $h(x_1, x_2)$ that, because of the i property, is just:

$$h(x_1, x_2) = P(X_1 = x_1; X_2 = x_2) = P(X_1 = x_1)\, P(X_2 = x_2)$$

$\bar{x}$					x_2				
$h(x_1, x_2)$		0		1		2		3	$P(X_1 = x_1)$
0	0		0.5		1		1.5		
		0.0529		0.1286		0.0437		0.0048	0.230
1	0.5		1		1.5		2		
		0.1286		0.3125		0.1062		0.0117	0.559
x_1 2	1		1.5		2		2.5		
		0.0437		0.1062		0.0361		0.0040	0.190
3	1.5		2		2.5		3		
		0.0048		0.0117		0.0040		0.0004	0.021
$P(X_2 = x_2)$		0.230		0.559		0.190		0.021	1

TABLE 13.1 Values of $\bar{x}$ and joint probabilities

From this table, the pdf of $\bar{X}$ can be calculated easily: for each possible outcome, add up all probabilities in the cells with that outcome. For instance:

$$P(\bar{X} = 1) = P(X_1 = 2; X_2 = 0) + P(X_1 = 1; X_2 = 1) + P(X_1 = 0; X_2 = 2)$$

$$= 0.0437 + 0.3125 + 0.0437 = 0.3999$$

The results are set out in Table 13.2. $E(\bar{X})$ and $V(\bar{X})$ can be checked with this pdf; apart from rounding errors, the same results should be found.

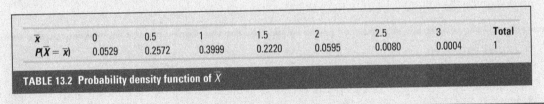

$\bar{x}$	0	0.5	1	1.5	2	2.5	3	Total
$P(\bar{X} = \bar{x})$	0.0529	0.2572	0.3999	0.2220	0.0595	0.0080	0.0004	1

TABLE 13.2 Probability density function of $\bar{X}$

Notice that $\bar{X} = X_1/n + \ldots + X_n/n$ is a linear combination of $X_1, \ldots, X_n$. By the i property, $\bar{X}$ is normal if the X_i are normal; see (11.19) and the remark thereafter. If the random sample is drawn from a normal population distribution, then the X_i are normal and hence $\bar{X}$ is also normal with parameters as in (13.1):

Property of random sample $X_1, \ldots, X_n$

Random sample from $N(\mu, \sigma^2) \Rightarrow \bar{X} \sim N(\mu, \sigma^2/n)$ **(13.6)**

The following example demonstrates the plausibility of (13.6). It is about many small random samples (size 5) from a normal distribution. If (13.6) is valid, a histogram of the means of these samples must look normal.

Example 13.5

It is well known that a normal distribution is a suitable model for the population distribution of the variable X = 'height (cm) of an adult man'. As a consequence of (13.6), the sample mean $\bar{X}$ of a random sample of size 5 from such a population distribution should also be normally distributed. We will illustrate this.

The dataset Xmp13-05.sav contains the sample means of 972 random samples of the heights (in cm) of $n = 5$ men from a certain population of adult men. We expect the histogram of these data to be more or less mound-shaped.

The mean value of the 972 sample means turns out to be 178.3725, with standard deviation 3.5479 cm; the corresponding normal pdf is included in Figure 13.2. Although the distribution $N(178.3725, 13.5879)$ does a rather good job, the mode of the histogram seems to differ slightly from the mode of the normal pdf. This may be caused by the fact that 972 is still too small a sample size to obtain an excellent fit.

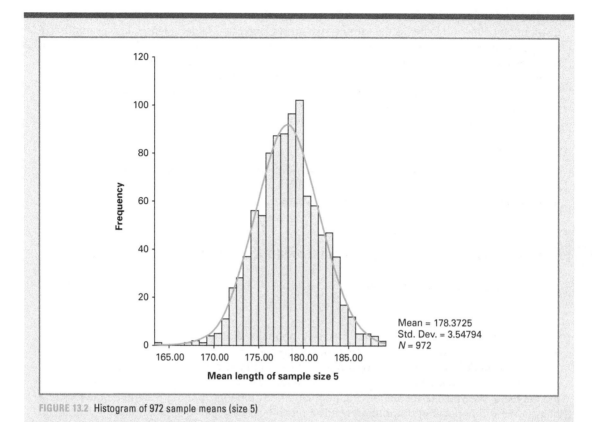

Mean = 178.3725
Std. Dev. = 3.54794
N = 972

FIGURE 13.2 Histogram of 972 sample means (size 5)

Example 13.6

Reconsider the ten adult men that accidentally travel together in the elevator; see Examples 11.11 and 11.14. Since $N(82.7, 99.94)$ is a good model for the weight X of an adult man, the weights X_1, ..., X_{10} of the ten men can be considered as an *iid* sample from this normal distribution. By (13.6), the sample mean $\overline{X}$ is also normal, with mean parameter 82.7 and variance parameter $99.94/10 = 9.994$. Hence, $\overline{X} \sim N(82.7, 9.994)$. As a consequence, probabilities concerning $\overline{X}$ events can easily be calculated.

Suppose that the elevator will stall if the total weight of the passengers exceeds 850 kg. The probability that this will happen is equal to $P(\overline{X} > 85)$:

$$P(\overline{X} > 85) = 1 - P(\overline{X} \leq 85) = 1 - 0.7666 = 0.2334 \qquad (*)$$

13.3 The central limit theorem

In general, it is not possible to determine probabilities of $\overline{X}$ events precisely, especially not if the underlying population distribution is unknown. However, for large sample sizes such probabilities can be **approximated**.

Consider a random sample X_1, ..., X_n from some population distribution with mean μ and variance σ^2. There are no further assumptions about this underlying distribution; it may be discrete or continuous, non-normal, heavily skewed to the right or left, or completely unknown. Interest is in the probability distribution of $\overline{X}$.

The famous **central limit theorem** (CLT) states that the pdf of $\overline{X}$ approaches the pdf of a normal distribution better and better as the sample size grows larger and larger. Hence, if the sample size is large, probabilities of $\overline{X}$ events can be approximated by similar probabilities of a normal distribution. The parameters of this approximating normal distribution are obvious: they follow from (13.1). It turns out that in many (but not all) practical situations a sample size 30 is enough.

To get an initial understanding of the CLT, recall that a normal distribution is often a good model for a population variable that is the result of many small and independent contributions. However, the sample mean $\overline{X}$ of a large random sample can be written as:

$$\overline{X} = \frac{1}{n} X_1 + \frac{1}{n} X_2 + \dots + \frac{1}{n} X_n$$

which **is** a summation of many (because of the large n), small (because of dividing by n), and independent (because of the i property) contributions. This makes it plausible that the probability distribution of $\overline{X}$ approaches a normal distribution as n tends to infinity. Here is the theorem:

> ### *Central limit theorem* for the sample mean $\overline{X}$ of a random sample $X_1, \dots, X_n$
>
> If n is large, then $\overline{X} \approx N\left(\mu, \frac{\sigma^2}{n}\right)$ **(13.7)**

Here, the sign $\approx$ denotes that $\overline{X}$ is not precisely normally distributed, but that probabilities of $\overline{X}$ events can be approximated by similar probabilities of a normal distribution. It is important to note that the underlying population distribution does not play a role as far the limiting normality is concerned; it can even be completely unknown. However, the rate of approximation depends on the degree of skewness of the underlying population distribution. A random sample drawn from a heavily skewed population distribution needs a much larger sample to approximate the $N(\mu, \sigma^2/n)$ distribution than does a random sample drawn from a population distribution that is itself already symmetric (such as a uniform distribution). Figure 13.3 shows graphs of pdfs of $\overline{X}$ (for $n = 1, 2$ and 3) if the underlying population distribution is $U(0, 1)$. (Recall from Section 10.1 that this distribution has $\mu = 0.5$ and $\sigma^2 = 1/12$.) The pdf of $N(0.5, 1/36)$ is also included.

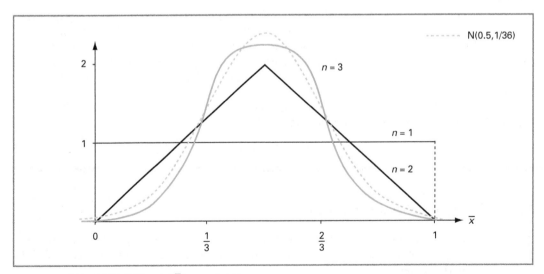

FIGURE 13.3 Pdfs of $N(0.5, 1/36)$ and of $\overline{X}$ for $n = 1, 2, 3$

In the case of a sample of size 1 from $U(0, 1)$, the pdf of $\overline{X}$ is just the pdf of one random observation, so it is $U(0, 1)$. For $n = 2$, the pdf of $\overline{X}$ turns out to be triangular (a first step towards a normal distribution). If $n = 3$, then $E(\overline{X}) = \mu = 0.5$ and $V(\overline{X}) = \sigma^2/3 = 1/36$. Already for $n = 3$, the pdf of $\overline{X}$ approximates the corresponding pdf of $N(0.5, 1/36)$.

In Example 13.7, interest is in the population variable $X = $ 'number of children per household'. The population distribution of X is slightly skewed to the right. To make the CLT plausible, 100 samples of 30 observations are considered. A histogram of the 100 sample means should look like a normal pdf since $n = 30$ should be enough.

Example 13.7

In Example 13.1, the 600 samples of size (only) 5 yielded a histogram that gives a first impression of the pdf of $\overline{X}$ when $n = 5$. From Figure 13.1 it can be seen that the accompanying frequency distribution is still skewed to the right.

To illustrate the CLT, the 3000 observations of the variable $X = $ NKIDS are divided into 100 samples of the larger size 30 and the corresponding 100 sample means $\bar{x}$ are calculated; see the file Xmp13-07.sav. Figure 13.4 graphs the frequency distribution of these 100 observations of $\overline{X}$. Although 100 is not that large, this histogram can, because of the CLT, be expected to look roughly like a normal distribution.

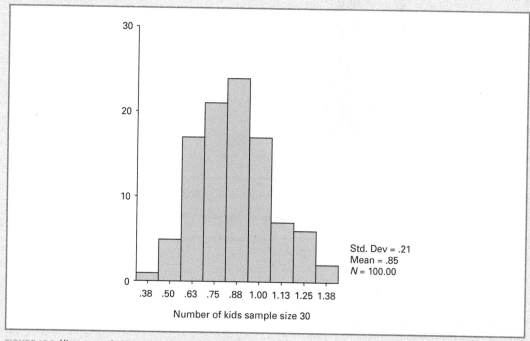

FIGURE 13.4 Histogram of 100 sample means of size 30

Indeed, the picture is not too far from being mound-shaped. It is much more symmetric than the histogram of Figure 13.1.

The approximation of the precise pdf of $\overline{X}$ by $N(\mu, \sigma^2/n)$ becomes better as n becomes larger. As a rule of thumb, $n = 20$ is usually considered enough to approximate the pdf of $\overline{X}$ in cases where the underlying population distribution is symmetric or only weakly skewed. However, for medium-skewed population distributions $n = 30$ is used, while heavily skewed population distributions may need sample sizes of 100 or more.

Example 13.8

It is well known that the population distributions of variables that measure hourly wages or incomes are often skewed to the right. This is caused by the fact that most populations of people contain 'elements' with relatively large earnings, which makes the distribution right-skewed. Figure 13.5 shows a frequency distribution of the household incomes of a population of 10 847 households with mean €30 980.10 and standard deviation €15 588.60.

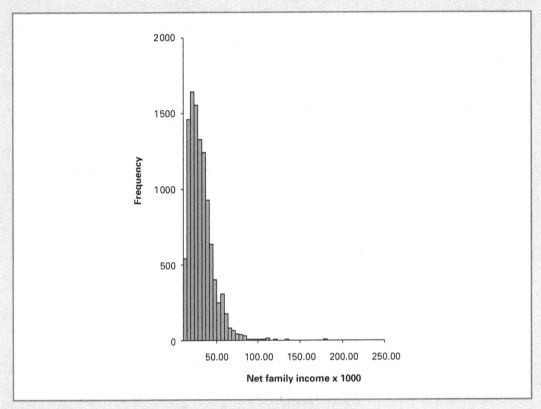

Figure 13.5 Histogram of 10 847 annual incomes (× €1000)

The underlying population distribution obviously is not normal; it is heavily skewed to the right.

A random sample of 50 households will be drawn from this population and the annual incomes (in units of €1000) will be measured. The rvs $X_1, \ldots, X_{50}$ denote the accompanying random observations. Below, the probabilities $P(\overline{X} \geq 30)$, $P(\overline{X} < 29)$ and $P(28 < \overline{X} < 34)$ will be calculated.

Since $\mu = 30.9801$ and $\sigma = 15.5886$, it follows that $E(\overline{X}) = 30.9801$ and $SD(\overline{X}) = 15.5886/\sqrt{50} = 2.2046$. The CLT yields that the probability distribution of $\overline{X}$ is approximately $N(30.9801, 4.8601)$. With this result, the three probabilities can be derived immediately; the respective answers are 0.6717, 0.1845 and 0.8264 (∗). The last answer is of particular interest: 82.6% of the total probability regarding $\overline{X}$ falls between 28 and 34.

The probabilities can also be obtained by first transforming $\overline{X}$ into a standard normal rv Z:

$$P(\overline{X} \geq 30) = P\left(Z \geq \frac{30 - 30.9801}{2.2046}\right) = P(Z \geq -0.4446) = 1 - 0.3283 = 0.6717 \qquad (*)$$

$$P(\bar{X} < 29) \qquad = P(Z < -0.8982) = 0.1845 \qquad\qquad (*)$$

$$P(28 < \bar{X} < 34) \qquad = P(-1.3518 < Z < 1.3698) = 0.9146 - 0.0882 = 0.8264 \qquad (*)$$

Example 13.9

In 2002, the mean annual personal income of male students was €6900. For female students the corresponding mean was somewhat smaller, €6700, but the corresponding standard deviations were the same: €2500. In the **current** year, the mean annual personal income of female students is €6900 but the corresponding mean (μ) for male students is not yet known.

To estimate μ, a sample of 2000 male students is chosen randomly and their annual personal incomes are recorded. It turns out that the average income is €7050, which seems to suggest that income dissimilarity between male and female students is still present. Or is €7050 just an accidental result that is valid only for this sample? To study this problem, it will be assumed that the present standard deviation of the annual personal incomes of male students is still €2500.

To analyse the problem theoretically, let the rvs $X_1, \ldots, X_{2000}$ denote a random sample from the population distribution (with unknown μ and known standard deviation $\sigma = 2500$) of the variable $X =$ 'annual personal income of a male student'. By the CLT, it follows that the sample mean $\bar{X}$ is approximately normal with variance $\sigma_{\bar{X}}^2 = 2500^2/2000 = 3125$. Recall from (10.19) that a normal random variable Y with mean μ_Y and variance σ_Y^2 satisfies:

$$P(Y - 1.96\sigma_Y < \mu_Y < Y + 1.96\sigma_Y) = 0.95$$

Application with $\bar{X}$ instead of Y, making use of the facts that $\sigma_{\bar{X}} = 55.9017$ and that $\mu_{\bar{X}} = \mu$, yields:

$$P(\bar{X} - 109.5673 < \mu < \bar{X} + 109.5673) \approx 0.95$$

This result can be interpreted as follows: for (about) 95% of all random samples of size 2000 it holds that (the unknown) μ falls between $\bar{x} - 109.5673$ and $\bar{x} + 109.5673$. As mentioned above, the outcome $\bar{x} = 7050$ was obtained for one of such samples. So it is likely that μ falls between $7050 - 109.5673 = 6940$ and $7050 + 109.5673 = 7160$. Since the lower bound 6940 is larger than 6900, it is justified to state that it is likely that the mean income of male students is larger than the mean income of female students.

13.4 Consequences of the CLT

In the foregoing sections it was concluded that the sample mean $\bar{X}$ of a random sample is, in many important cases, normally distributed or approximately normally distributed. $\bar{X}$ is exactly normal if the population distribution is normal; $\bar{X}$ is approximately normal if the sample size is large. In this section, interest is in the standardized version Z of the sample mean $\bar{X}$. What are the consequences for Z? (These consequences are fundamental for the coming chapters since they are starting points in developing confidence intervals and hypothesis tests.)

Start with a population and a population variable X; the population distribution of X has mean μ and variance σ^2. As before, $X_1, \ldots, X_n$ is a random sample (so, *iid*) from the population distribution of X. Suppose, additionally, that at least one of the two following circumstances is valid:

Situation 1 The normal distribution $N(\mu, \sigma^2)$ is a good model for (the population distribution of) X.

Situation 2 The sample size n is large.

If situation 1 holds, then $\bar{X} \sim N(\mu,\ \sigma^2/n)$; if situation 2 holds, then $\bar{X} \approx N(\mu,\ \sigma^2/n)$. In both cases, the distribution $N(\mu,\ \sigma^2/n)$ is used to calculate probabilities concerning $\bar{X}$. So, the notation $\bar{X} \approx N(\mu,\ \sigma^2/n)$ can be used for **both** cases. If situation 1 is valid, the approximation in $\approx$ can be read as a consequence of the fact that choosing a model for a variable usually already implies a simplification of reality. For situation 2, the approximation meant by $\approx$ is a consequence of the CLT.

Recall that the standardized version of a random variable is obtained by subtracting the expectation and dividing the result by the standard deviation. Since the rv $\bar{X}$ has expectation μ and standard deviation $\sigma/\sqrt{n}$, it follows that the standardized version Z of $\bar{X}$ satisfies the equality below. Since $\bar{X}$ is approximately normal, Z is approximately **standard** normal (see also (10.11)):

Normality of $\bar{X}$ and Z if situation 1 and/or 2 is valid

$$\bar{X} \approx N(\mu,\ \sigma^2/n) \quad \text{and} \quad Z = \frac{\bar{X} - \mu}{\sigma/\sqrt{n}} \approx N(0,\ 1) \tag{13.8}$$

CASE 13.1 THE RUIN PROBABILITY OF INSURANCE COMPANY LOWLANDS – SOLUTION

Since the random claim amounts $X_1 \ldots, X_{100000}$ are *iid*, they can be considered as a random sample drawn from the distribution of X. The mean and variance can easily be calculated:

$$\mu = 0 \times 0.90 + 1000 \times 0.008 + \ldots + 9000 \times 0.008 + 10000 \times 0.028 = 640$$

$$\sigma^2 = E(X^2) - 640^2$$

$$= 1000^2 \times 0.008 + \ldots + 9000^2 \times 0.008 + 10000^2 \times 0.028 - 409600$$

$$= 4670400$$

Since $n = 10^5$, it follows that:

$$E(\bar{X}) = 640,\ V(\bar{X}) = 46.7040,\quad SD(\bar{X}) = 6.83403$$

The premium b has to be determined such that the ruin probability is 0.01:

$$0.01 = P(U < 0) = P(50000 + 100000b - S < 0)$$

$$= P(S > 50000 + 100000b) = P(\bar{X} > 0.5 + b)$$

By transforming $\bar{X}$ into the standardized version Z we obtain:

$$0.01 = P\left(\frac{\bar{X} - 640}{6.83403} > \frac{0.5 + b - 640}{6.83403} \right) = P\left(Z > \frac{0.5 + b - 640}{6.83403} \right)$$

Since $P(Z > 2.3263) = 0.01$ (*), we have to solve b from the equation:

$$\frac{0.5 + b - 640}{6.83403} = 2.3263$$

It follows that the premium is €655.40, which corresponds to a 'safety loading' of 2.41% upon the expected claim amount per contract.

What are the consequences for $\bar{X}$ if, as in Section 10.3.4, an area $\alpha/2$ is cut off from each of the two tails of the density of the standardized version? Figure 13.6 pictures the situation. Recall that probabilities of Z events are areas under the pdf. Since the positions $z_{\alpha/2}$ and $-z_{\alpha/2}$ cut off areas $\alpha/2$ in, respectively, the right-hand and the left-hand tail of the density, it follows that:

$$P(Z \leq -z_{\alpha/2}) \approx \alpha/2; \quad P(-z_{\alpha/2} < Z < z_{\alpha/2}) \approx 1 - \alpha; \quad P(Z \geq z_{\alpha/2}) \approx \alpha/2 \qquad \textbf{(13.9)}$$

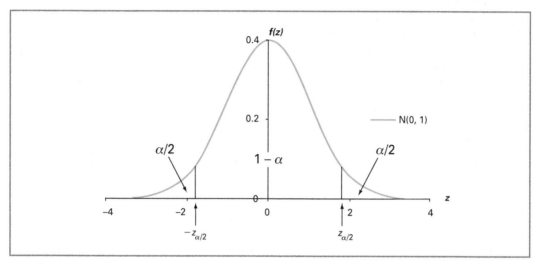

Figure 13.6 The pdf of Z and the positions of $z_{\alpha/2}$ and $-z_{\alpha/2}$

Note that in (13.9) the sign $\approx$ has a different meaning from that in (13.7). In (13.7) it means that two distributions are approximately the same, in (13.9) that two numbers are approximately the same.

As in (10.17) and (10.18) with $\overline{X}$ instead of Y, the second result in (13.9) can be rewritten in two ways.

Important results if situation 1 and/or 2 is valid

$$P\left(\mu - z_{\alpha/2} \frac{\sigma}{\sqrt{n}} < \overline{X} < \mu + z_{\alpha/2} \frac{\sigma}{\sqrt{n}}\right) \approx 1 - \alpha \qquad \textbf{(13.10)}$$

$$P\left(\overline{X} - z_{\alpha/2} \frac{\sigma}{\sqrt{n}} < \mu < \overline{X} + z_{\alpha/2} \frac{\sigma}{\sqrt{n}}\right) \approx 1 - \alpha \qquad \textbf{(13.11)}$$

These equations are fundamental for the coming chapters. Equation (13.10) states that a proportion of (about) $1 - \alpha$ of all realized random samples have the sample means $\bar{x}$ between $\mu - z_{\alpha/2}\, \sigma/\sqrt{n}$ and $\mu + z_{\alpha/2}\, \sigma/\sqrt{n}$. For instance, for $\alpha = 0.05$, it holds that the sample means $\bar{x}$ of 95% of all random samples fall between $\mu - 1.96\sigma/\sqrt{n}$ and $\mu + 1.96\sigma/\sqrt{n}$. If μ and σ are known, equation (13.10) informs us that the mean $\overline{X}$ of a random sample that still has to be drawn, will probably fall between two known boundaries.

Equation (13.11) is valuable if μ is **un**known. It states that the probability is $1 - \alpha$ that μ is captured between $\overline{X} - z_{\alpha/2} \frac{\sigma}{\sqrt{n}}$ and $\overline{X} + z_{\alpha/2} \frac{\sigma}{\sqrt{n}}$. For instance, for $\alpha = 0.05$, the unknown μ will with probability 0.95 be enclosed by the random variables $\overline{X} - 1.96 \frac{\sigma}{\sqrt{n}}$ and $\overline{X} + 1.96 \frac{\sigma}{\sqrt{n}}$. If σ is known, these (upper and lower) bounds can be calculated as soon as the data $x_1, \ldots, x_n$ are available; the confidence that μ is contained in the interval between the two bounds is $1 - \alpha$. Such an interval is often useful for estimating the unknown μ.

If σ is **un**known too, the random boundaries

$$\overline{X} - z_{\alpha/2} \frac{\sigma}{\sqrt{n}} \quad \text{and} \quad \overline{X} + z_{\alpha/2} \frac{\sigma}{\sqrt{n}}$$

are still important: replacing σ by its usual estimator S yields adapted boundaries that **can** be measured as soon as the data are available. Although the likelihood of enclosing μ may have decreased slightly, the resulting interval is valuable for estimating μ. See the forthcoming chapters for details.

Example 13.10

In 2003, the mean hourly wages of employees in the sectors 'commercial services' and 'non-commercial services' were, respectively, €16.92 and €19.12.

From past years it is known that the mean hourly wage in the non-commercial services sector has not changed at all, a consequence of the government's economic policies; last year, this mean was still equal to €19.12. To find out whether the mean hourly wage μ of the commercial services sector **has** increased, 500 employees are randomly drawn from that sector and their last year's hourly wages are recorded; the sample mean turns out to be €17.99. Below, it will be shown that this sample mean justifies the conclusion that the population mean μ for the commercial sector is larger than the 2003 mean of €16.92. In fact, it will be shown that:

- if μ were unchanged compared with 2003 then this realized sample mean 17.99 would be very exceptional;
- it is very likely that μ belongs to a certain interval that lies completely to the right of 16.92.

It will be assumed that last year's standard deviations for both sectors were €8.20. In the statistical considerations below, take $\alpha = 0.10$.

Assume that last year's mean hourly wage μ in the commercial services sector is still €16.92. We will study whether this assumption is compatible with the realization 17.99. Let the rv $\bar{X}$ denote the mean hourly wage of 500 sampled employees. Recall that $z_{\alpha/2} = z_{0.05} = 1.6449$ and that $z_{0.05}$ cuts off an area of 0.95 at the **left**-hand side of the pdf of the z-distribution. Since the sample size is large, equation (13.10) is valid. It yields:

$$P(16.3168 < \bar{X} < 17.5232) \approx 0.90 \qquad \text{(13.12)}$$

If the assumption is valid, it is likely (about 90% certainty) that the realization of $\bar{X}$ will fall between 16.31 and 17.53. But the actual realization is 17.99, which falls at the right-hand side of the interval (16.31, 17.53). So, the starting point that μ is unchanged is not compatible with this realization.

Assume now that μ is unknown. According to (13.11), again with $\alpha = 0.10$ and $\sigma = 8.20$, the following holds for the rv $\bar{X}$:

$$P(\bar{X} - 0.6032 < \mu < \bar{X} + 0.6032) \approx 0.90 \qquad \text{(13.13)}$$

In particular, it is likely that the unknown μ will be larger than the realization of $\bar{X} - 0.6032$. But since the realization 17.99 of $\bar{X}$ is already given, it can be concluded that it is likely (certainty about 90%) that μ is larger than $17.99 - 0.6032 = 17.3868$. But this lower bound is larger than the population mean 16.92 of 2003.

It follows that last year's mean hourly wage in the commercial services sector is probably larger than the mean of 2003. Although the certainty of this conclusion according to the above approach is (about) 90%, it can easily be shown that even the choice $\alpha = 0.01$ in (13.11) leads to the same conclusion; check it yourself.

In Exercise 13.16 below, similar arguments will be used to show that this sample mean 16.92 makes it very implausible that nowadays the mean hourly wage μ in the commercial services sector is equal to the mean hourly wage in the non-commercial services sector.

Summary

Properties of the (random) sample mean $\bar{X}$ were discussed, for random samples with **and** without replacement. The table below gives the expectations and variances.

Expectation and variance of $\overline{X}$

	With replacement	Without replacement
$E(\overline{X})$	μ	μ
$V(\overline{X})$	$\dfrac{\sigma^2}{n}$	$\dfrac{\sigma^2}{n}\dfrac{N-n}{N-1}$

The variance of $\overline{X}$ is smallest if the sample is drawn without replacement. This is not surprising since under these circumstances it is guaranteed that each new sample observation leads to more information about μ. However, in practical situations the sample size n is usually small compared with the population size N and then the two variances are almost equal. In the coming chapters, random samples are understood to be random samples with replacement.

The **exact** probability distribution of the sample mean $\overline{X}$ of a random sample can be determined only in some very special cases. The distribution is (approximately) normal in the following two situations: (1) the population variable X itself is (approximately) normal; (2) the sample size n is sufficiently large. The result under (2) is the central limit theorem; it is the reason why the normal distribution is so important in statistics. The table below summarizes results of $\overline{X}$ that are very important for inferential statistics.

Important properties of $\overline{X}$ if (1) and/or (2) is valid

$\overline{X} \approx N\left(\mu, \dfrac{\sigma^2}{n}\right)$	$Z = \dfrac{\overline{X} - \mu}{\sigma/\sqrt{n}} \approx N(0, 1)$

$$P\left(\overline{X} - z_{\alpha/2}\frac{\sigma}{\sqrt{n}} < \mu < \overline{X} + z_{\alpha/2}\frac{\sigma}{\sqrt{n}}\right) \approx 1 - \alpha$$

🔑 Key terms

central limit theorem **400**	(general) law of large numbers **394**	variance correction factor **395**
consistent **394**	unbiased **392**	

? Exercises

Exercise 13.1
Let the variable X represent the number of eyes when throwing a fair die. For the mean $\bar{X}$ of n throws, calculate $E(\bar{X})$, $V(\bar{X})$ and $P(\bar{X} \leq 2)$ if: $n = 1$, $n = 2$ and $n = 30$.

Exercise 13.2
A random sample (so, with replacement) of size 12 is drawn from a normal population distribution with mean 500 and standard deviation 100. Let $\bar{X}$ be the (random) sample mean. Calculate: $P(\bar{X} < 425)$, $P(\bar{X} > 550)$ and $P(|\bar{X} - 500| < 75)$.

Exercise 13.3
A random sample of size 5 is drawn from a population distribution, with $\mu = 20$ and $\sigma^2 = 9$, on a population of size 100. Let $\bar{X}$ represent the sample mean. Calculate $E(\bar{X})$ and $V(\bar{X})$ if the sample is drawn: (a) with replacement; (b) without replacement.

Exercise 13.4
A random sample of size 4 is drawn from a population distribution with mean 30 and variance 9. Let $\bar{X}$ be the (random) sample mean.

  a If it is additionally given that the population distribution is normal, calculate $P(28 < \bar{X} < 32)$.
  b Suppose now that nothing is known about the population distribution other than that $\mu = 30$ and $\sigma^2 = 9$. Use Chebyshev's inequality to find a lower bound for the probability $P(28 < \bar{X} < 32)$.

Exercise 13.5
The amount of time that business students devote to their study each week is normally distributed with a mean of 30 hours and a standard deviation of 5 hours.

  a Calculate the probability that a business student devotes more than 32 hours per week to his/her study.
  b Calculate the probability that the mean amount of time that four randomly selected business students devote each week to their study is more than 32 hours.
  c Calculate the probability that the mean amount of time that nine randomly selected business students devote each week to their study is more than 32 hours.

Exercise 13.6
Fifty arbitrary male university graduates aged 22–27 are internet-dating with fifty arbitrary female university graduates in the same age category. Let $\bar{X}$ and $\bar{Y}$ denote the mean annual income of the 50 men and the 50 women, respectively. Suppose that the mean annual income for male (female) university graduates in the age category 22–27 is €24 000 (€23 000), and that the accompanying standard deviation for both populations is €3000.

  a Calculate the expectation and standard deviation of both $\bar{X}$ and $\bar{Y}$.

Calculate the probability:

  b that the mean annual income of the 50 men is at least €24 500;
  c that the mean annual income of the 50 women is at least €24 500;
  d that the means of the annual incomes of the men and the women are both at least €24 500.

Exercise 13.7

Consider three throws with the die below; interest is in the mean number of eyes per throw.

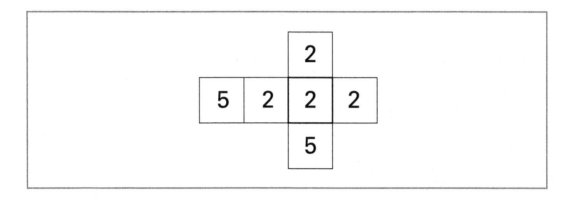

Let X_1, X_2, X_3 be the respective numbers of eyes. Since the X_i are independent and identically distributed, the random sequence X_1, X_2, X_3 is *iid* and can be considered a random sample of size 3 from the population of all throws with this die. Interest is in the sample mean $\bar{X}$.

   a Determine the underlying population distribution. Also calculate the population mean μ and the population variance σ^2.

   b Write down all outcomes of the random experiment of throwing the die three times.

   c Is $\bar{X}$ discrete or continuous? Is it normally distributed?

   d Determine the pdf of $\bar{X}$. Use it to calculate $E(\bar{X})$ and $V(\bar{X})$.

   e Calculate $E(\bar{X})$ and $V(\bar{X})$ again, but now by using (13.1).

Exercise 13.8

Consider a large population of households where the variable X = 'number of children under 18' has the following relative frequency distribution:

x	0	1	2	3	4
Rel. frequency	0.06	0.50	0.31	0.08	0.05

A random sample of two households is drawn from this population; let X_1 and X_2 be the numbers of children (<18) of these households.

   a Determine the marginal pdfs and the joint pdf of X_1 and X_2.

   b Determine the pdf of the sample mean $\bar{X}$ of this small sample. Use it to calculate $E(\bar{X})$ and $V(\bar{X})$.

   c Calculate $E(\bar{X})$ and $V(\bar{X})$ again, but now from the results in (13.1).

   d Determine $\text{cov}(X_1, X_2)$.

Exercise 13.9

A new soft drink is offered in all ten supermarkets of a town. The table below shows the frequency distribution of the variable X = 'number of bottles sold during the first day per supermarket'.

x	0	2	3	4	5
Rel. frequency	0.1	0.1	0.4	0.2	0.2

For a quick analysis, two of these supermarkets are drawn randomly and the respective numbers of bottles sold during the first day are denoted by X_1 and X_2. Suppose that the sample is drawn with replacement.

a Determine the marginal pdfs and the joint pdf of X_1 and X_2.

b Determine the pdf of the sample mean $\bar{X}$ of this small sample. Use it to calculate $E(\bar{X})$ and $V(\bar{X})$.

c Calculate $E(\bar{X})$ and $V(\bar{X})$ again, but now from the results in (13.1).

d Determine $\text{cov}(X_1, X_2)$.

Exercise 13.10

Answer the questions in Exercise 13.9 again if the sample is drawn without replacement; use (13.4) for part (c). [*Hint*: For part (a), first determine (for each $x = 0, 2, 3, 4, 5$) the conditional probabilities $g(y \mid X_1 = x) = P(X_2 = y \mid X_1 = x)$.]

Exercise 13.11

Let $X_1, \ldots, X_{10}$ be a random sample from a normal population distribution with mean 4 and variance 3.

a What type of probability distribution does the sample mean $\bar{X}$ have? Also determine the parameters.

b Calculate $P(\bar{X} < 4)$, $P(\bar{X} > 5)$ and $P(|\bar{X} - 4| < 1)$; you may use only that an $N(0, 1)$ rv Z satisfies $P(Z > 1.8257) = 0.0339$.

c Suppose now that the population distribution is **not** normal. Determine a lower bound for the probability $P(|\bar{X} - 4| < 1)$ and compare that bound with the last answer of part (b).

Exercise 13.12

The mean weight of adult women is 69.4 kg, with standard deviation 8.9 kg.

a Which normal distribution can be taken as a model for the population distribution of the variable 'weight (kg) of an adult woman'?

b Calculate the probability that the weight of one randomly chosen woman is at most 70 kg.

Let $\bar{X}$ be the mean weight (kg) of ten randomly chosen adult women.

c Determine the probability distribution of $\bar{X}$.

d Calculate the probability that this sample mean is at most 70 kg.

e Calculate $P(65 < \bar{X} < 75)$.

Exercise 13.13

The die in Exercise 13.7 is thrown 50 times. Let $\bar{X}$ be the mean number of eyes per throw. Use the CLT to answer the following questions.

a Find the normal distribution that can be used to calculate probabilities of $\bar{X}$ events.

b Calculate $P(\bar{X} > 2.5)$.

c Calculate $P(|\,\bar{X} - 3\,| < 0.25)$ and $P(|\,\bar{X} - 3.1\,| < 0.25)$.

d Calculate a such that $P(\bar{X} > a) = 0.02$.

e Calculate b such that $P(|\,\bar{X} - 3\,| < b) = 0.90$.

Exercise 13.14

In a large city, 575 traffic accidents with one or more deaths occurred during the most recent 250 working days. To learn more about the daily numbers of such serious accidents (with one or more deaths), they will be recorded on 30 randomly chosen working days. Interest is in probabilities regarding the sample mean that measures the mean daily number of such serious accidents during these 30 days.

a What is the underlying population variable in which we are interested? Do you think that a normal distribution is suitable to model the population distribution of this variable?

b To calculate probabilities concerning the sample mean, the population mean μ and population variance σ^2 are needed. Given the facts above, which value will be used for μ?

In the questions below, use $\sigma^2 = 2.5$.

c Determine a normal distribution that can be used to approximate the probability distribution of the sample mean.

d Calculate the probability that the mean daily number of such serious accidents during these 30 days is at least 2.

e Of course, the probability that the sample mean is smaller than 0, should be 0. Use a normal approximation to calculate this probability. Give your comments.

Exercise 13.15

For a certain country, interest is in the variable X = 'annual income ($\times$ 1000) per household'. In 2003 the mean annual household income was 20.1 ($\times$ 1000) with accompanying standard deviation 12.8 ($\times$ 1000). A researcher wants to know whether the mean annual household income (μ; in units of 1000) of the present year will be larger than 20.1. That is why she plans to choose 200 households randomly (at the end of the current year) and measure their annual incomes (in units of 1000). The questions below will give an analysis of the probabilistic expectations regarding the sample mean $\bar{X}$. It will always be assumed that σ for the present year is unchanged from 2003. Suppose first that μ is unchanged too.

a Determine the probability distribution of $\bar{X}$.

b Calculate $P(\bar{X} > 20)$.

c Calculate the probability $P(\mu_{\bar{X}} - 2.5\sigma_{\bar{X}} < \bar{X} < \mu_{\bar{X}} + 2.5\sigma_{\bar{X}})$. Interpret the result. Do you really need the value of μ and σ to calculate this probability?

Now suppose (more realistically) that μ is unknown.

d Determine random boundaries that capture μ with probability 0.98.

e Suppose that, at the end of this year, it will turn out that the realization of the sample mean is 22.5. Determine an interval that is likely to include μ. What can be concluded about the conjecture that μ is larger than 20.1 (the mean annual household income in 2003)?

Exercise 13.16

Reconsider Example 13.10 Use similar arguments to those in the example to show that the sample mean 16.92 makes it very implausible that nowadays the mean hourly wage μ in the commercial services sector is equal to the mean hourly wage (€19.12) in the non-commercial services sector. Use $\alpha = 0.01$.

Exercise 13.17 (partially computer)

It will be shown that typical observations of a standard normally distributed rv can be simulated by first adding up 12 simulations from the uniform distribution $U(0, 1)$ and next subtracting 6. This result is a consequence of the central limit theorem.

Let $X_1, ..., X_{12}$ be *iid* observations from $U(0, 1)$ with sum $S = X_1 + ... + X_{12}$.

a Calculate $\mu = E(X_i)$ and $\sigma = SD(X_i)$. (*Hint*: see Section 10.1.)

b Calculate $E(\bar{X})$ and $SD(\bar{X})$.

Since the pdf of $U(0, 1)$ is symmetric and constant over $(0, 1)$, the sample size 12 is enough to apply the CLT. (From Figure 13.3 it follows that even the pdf of $\bar{X}$ for $n = 3$ seems to approximate a normal distribution rather well.)

c Show that $\dfrac{\bar{X} - 0.5}{1/\sqrt{144}} = \dfrac{S - 6}{1}$ and that $S - 6 \approx N(0, 1)$.

It would appear that adding up 12 simulated $U(0, 1)$ observations and subtracting 6 is a good way to simulate from $N(0, 1)$.

d (computer) Use this result to simulate with a computer package 1000 observations from $N(0, 1)$. Also create a nice histogram of these simulated observations. (The file Xrc13-17.xls contains 1000 simulations that are created in that way.)

Exercise 13.18

A consumer organization and the manufacturer of a certain brand of margarine are in dispute. The manufacturer states that his packs of margarine contain at least 250g of margarine whereas the consumer organization claims that the content is at most 249 g. A judge has to decide the issue.

The judge decides to base her judgment on a random sample of ten packs of margarine.

■ The consumer organization will be adjudged to be right if the random sample has a mean weight of less than 249.5 g.

■ The manufacturer will be adjudged to be right if this mean weight is at least 249.5 g.

The starting point for the judge is that $N(\mu, 4)$ is a good model for the weight X of a pack of margarine of that brand. Below, $\bar{X}$ denotes the mean weight (in grams) of the ten packs of margarine.

a Determine the probability distribution of $\bar{X}$.

b Calculate the probability that the judge decides (incorrectly) that the consumer organization is right while the true μ is equal to 251.

c Suppose that the manufacturer is right, so μ is at least 250. Calculate $P(\bar{X} < 249.5)$ if the true population mean μ is precisely 250. Explain why this probability will be smaller if μ is larger than 250.

d Interpret the results of part (c).

e Suppose now that the consumer organization is right, so μ is at most 249. Calculate $P(\bar{X} \geq 249.5)$ if the true population mean μ is precisely 249. Explain why this probability will be smaller if μ is smaller than 249.

f Interpret the results of part (e).

Exercise 13.19 (computer)

From Chapter 15 onwards, the relationship between sample results and the underlying population is investigated. In this exercise we will use simulated random samples of size 100 from the so-called chi-square distribution with 4 degrees of freedom (defined in Chapter 17) to get an idea of the population mean μ and the population variance σ^2 of that distribution. The pdf of the distribution is depicted in the figure below. The file <u>Xrc13-19.xls</u> contains in its 20 columns the data of 20 (simulated) random samples of size 100 from that distribution.

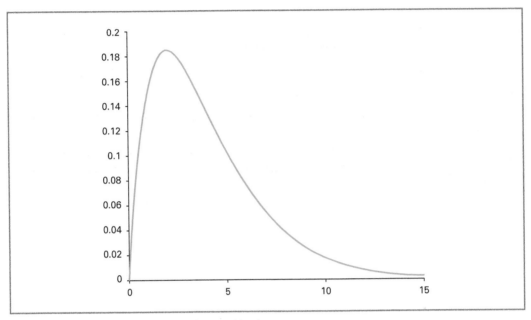

The pdf of the chi-square distribution with 4 degrees of freedom

    a Use the first column of the dataset to obtain estimates of μ and σ^2.

    b Do the same for the other 19 columns.

    c Since the expected values of the estimators $\bar{X}$ and S^2 are, respectively, μ and σ^2 (see also (14.12)), the average of the 20 realized sample means and the average of the 20 realized sample variances should also (and probably better) approximate μ and σ^2. Find such approximations. What do you think of the actual values of μ and σ^2?

    d Recall that the variance of the estimator $\bar{X}$ is $\sigma^2/100$. Since each of the 20 samples yields an estimate of this variance, the average of the resulting 20 estimates should approximate the variance of $\bar{X}$. Calculate this average. What do you think of $V(\bar{X})$?

    e In Chapter 17 it will become clear that $\mu = 4$ and $\sigma^2 = 8$. Compare the above estimation results with this new fact.

CASE 13.2 HOUSEHOLDS STATISTICS (PART II)

This case is a continuation of Case 12.2. The table below lists the population standard deviations of the 12 quantitative variables that were concluded from a more extensive study (with much larger sample size) carried out two years previously.

Variable	WEIGHT	LENGTH	AGE	WAGE	HOURS	NKIDS
σ	12.02	9.41	15.10	6.11	19.70	1.18

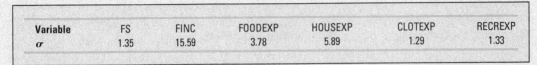

Variable	FS	FINC	FOODEXP	HOUSEXP	CLOTEXP	RECREXP
σ	1.35	15.59	3.78	5.89	1.29	1.33

It is assumed that the population standard deviations for the present year are unchanged from two years ago. Hence, the present (theoretical) probabilities

$$P(\bar{X} - 2\sigma_{\bar{X}} < \mu < \bar{X} + 2\sigma_{\bar{X}})$$

can be calculated for each of the 12 random samples $X_1, \ldots, X_{300}$ of the present year.

a Calculate these probabilities. Do you really have to repeat the calculation 12 times? Interpret your answers.

b Use your results of Case 12.2 to calculate the realizations of the 12 random intervals $(\bar{X} - 2\sigma_{\bar{X}}, \bar{X} + 2\sigma_{\bar{X}})$. What can be concluded about the 12 population means of the present year?

CHAPTER 14

Sample proportion and other sample statistics

In Chapter 12, several sample statistics were introduced. Their definitions were chosen in such a way that they are just sample counterparts of population statistics. The sample statistic called 'sample mean' was the topic of the previous chapter; its properties were considered. The present chapter is about properties of other sample statistics, especially of the sample proportion and the sample variance.

Since the sample proportion is nothing but a special sample mean, its properties follow from Chapter 13. However, since the underlying population mean μ and population variance σ^2 can now be expressed in terms of the population proportion p, the properties will be reformulated in terms of this population proportion.

Properties of other sample statistics are considered too. In particular, the sample variance and the sample standard deviation receive attention. Since estimators are random variables, it makes sense to consider their standard deviations. (For instance, recall the standard deviation $\sigma_{\bar{X}} = \sigma/\sqrt{n}$ of the estimator $\bar{X}$ of μ.) Since standard deviations of estimators play important roles in forthcoming chapters and since they usually depend on unknowns, a uniform approach to estimate such standard deviations is considered briefly. The resulting sample statistics are called standard **errors**.

CASE 14.1 APPROVAL PROBABILITIES IN QUALITY CONTROL

The sample proportion $\hat{P}$ plays an important role in the **quality control** of a batch of goods (containing one type of product) delivered by a supplier. The quality of such a batch is tested by determining the proportion of defectives in a sample of the products drawn randomly from the batch. In fact, the customer and the supplier agree upon a so-called **approval plan** (n, c). That is, they agree that a random sample of n products will be tested; if the sample proportion of defectives is at most c, then the customer approves the whole batch, otherwise the whole batch is rejected and returned to the supplier.

If the approval plan is known, the probability that the batch will be approved depends only on the proportion p of defectives in the whole batch. To express this dependence on p, the **approval probability** $P(\hat{P} \leq c)$ is often written as $P(\hat{P} \leq c \mid p)$. Hence, the approval probability can be considered as a function of p, the so-called **approval function**. Its graph depicts, for the agreed approval plan, the approval probabilities for batches of several levels of quality.

a A retailer orders a large batch of products. She agrees with the supplier upon the approval plan (50, 0.1). Explain the meaning of this approval plan.

b Calculate the approval function $P(\hat{P} \leq c \mid p)$ at $p = 0.05, 0.10, 0.15, 0.20, 0.25$ and 0.30; use a binomial distribution. Graph the function.

c Do part (b) again, but now use the normal approximation of the distribution of $\hat{P}$ (also for $p = 0.05$). Compare the answers.

For the determination of the approval plan, the interests of supplier and customer are, in a sense, opposite: the supplier wants to run only a small risk that a batch with few defectives will be rejected; the customer wants small approval probabilities for batches with many defectives.

d In the approval plan (50, 0.1), the (exact) approval probability for $p = 0.125$ is about 40% while for $p = 0.075$ the (exact) disapproval probability is 17%. The supplier believes that the last probability is too large, the retailer that the first probability is too large. They both want a new approval plan, with $n = 100$. However, the supplier wants, for $p = 0.075$, the disapproval probability 0.08, and the retailer, for $p = 0.125$, the approval probability 0.10. Determine the approval plans suggested by the supplier and the retailer. See the solution at the end of Section 14.1

14.1 Properties of the sample proportion

Start with a qualitative variable that takes only the values yes (success) and no (failure). Interest is in the proportion p of the population elements with yes. Recall that p is usually unknown and that it falls between 0 and 1. By coding a success as 1 and a failure as 0, a variable X arises that behaves as a **quantitative** variable. In Sections 3.3.1 and 4.5 it has been shown that the mean μ and the variance σ^2 of X are equal to p and $p(1 - p)$, respectively.

Now suppose that a random sample (so, with replacement) of n population elements is drawn and that this random sample has Y successes and a proportion of $\hat{P}$ successes. Note that the rv $\hat{P}$ equals Y/n and has its possible outcomes between 0 and 1. Recall from Section 12.4 that $\hat{P}$ is called sample proportion, the usual estimator of p. If, for each $i = 1, \ldots, n$, the rv X_i takes the value 1 if the ith sampled element is a success and the value 0 if the ith sampled element is a failure, then the sequence $X_1, \ldots, X_n$ of 0-1 random variables is an *iid* sequence of random observations of X. Hence,

$$E(X_i) = \mu = p; \quad V(X_i) = \sigma^2 = p(1 - p); \quad SD(X_i) = \sigma = \sqrt{p(1 - p)} \tag{14.1}$$

Moreover, summation of the X_i counts the number of ones in the sample (since the zeros do not contribute), so $Y = \sum_{i=1}^{n} X_i$ and $\hat{P} = (\sum_{i=1}^{n} X_i)/n$. Consequently, $\hat{P}$ is just the sample mean $\overline{X}$ of the random sample $X_1, \ldots, X_n$ from the Bernoulli distribution with parameter p; see also Section 9.1. That is:

$$\hat{P} = \frac{1}{n}\sum_{i=1}^{n} X_i \quad \text{with } X_1, \ldots, X_n \text{ random sample from } Bern(p) \tag{14.2}$$

Since the sample proportion $\hat{P}$ is the sample mean of a random sample from a (special) population distribution with mean $\mu = p$ and variance $\sigma^2 = p(1 - p)$, several properties of $\hat{P}$ follow immediately from Chapter 13. The results in the box below are immediate consequences of (13.1), which states that the expectation and variance of a sample mean are equal to μ and σ^2/n, respectively. In particular, $E(\hat{P}) = p$, so $\hat{P}$ is an unbiased estimator of p: the expectation of the estimator is precisely equal to the parameter it is estimating.

Mean, variance and standard deviation of the sample proportion $\hat{P}$

$$\mu_{\hat{P}} = E(\hat{P}) = p; \quad \sigma_{\hat{P}}^2 = V(\hat{P}) = \frac{p(1 - p)}{n}; \quad \sigma_{\hat{P}} = \sqrt{\frac{p(1 - p)}{n}} \tag{14.3}$$

In particular, $\hat{P}$ is an **unbiased** estimator of p.

As in (13.2) and (13.3), the (general) law of large numbers for the sample proportion is obtained from Chebyshev's inequality:

$$P(p - \varepsilon < \hat{P} < p + \varepsilon) \geq 1 - \frac{p(1 - p)}{n\varepsilon^2} \quad \text{for all } \varepsilon > 0 \tag{14.4}$$

$$P(|\hat{P} - p| < \varepsilon) \to 1 \quad \text{as } n \text{ becomes larger and larger} \tag{14.5}$$

That is:

Restatement of Chebyshev's inequality for the sample proportion $\hat{P}$

$$P(|\hat{P} - p| < \varepsilon) \geq 1 - \frac{p(1 - p)}{n\varepsilon^2} \quad \text{for all } \varepsilon > 0$$

Since this lower bound tends to 1 as n gets larger and larger, the estimator $\hat{P}$ is a **consistent** estimator of p.

By choosing a very large random sample, it can be arranged that the probability is near 1 that the estimator $\hat{P}$ will yield a realization that lies arbitrarily close to the unknown p. That is why the estimator $\hat{P}$ is **consistent** in estimating p.

Example 14.1

In a certain production process, 3% of the products are defective. Suppose that 200 products are to be drawn randomly from tomorrow's large daily production. Determine the expectation and variance of the proportion of defectives that will be found in that sample. Also determine, with (14.4), an upper bound for the probability that this sample proportion will be at least 0.02 separated from 0.03. This probability can in fact be calculated precisely; calculate it by using a suitable binomial distribution.

Regarding the sample of 200 products, interest is in the proportion $\hat{P}$ of defectives in the sample. Since the population proportion of defectives is 0.03, it follows from (14.3) that

$$E(\hat{P}) = 0.03; \quad V(\hat{P}) = \frac{0.03 \times 0.97}{200} = 0.0001455; \quad SD(\hat{P}) = 0.0121$$

By (14.4) with $p = 0.03$ and $\varepsilon = 0.02$, we have:

$$P(0.01 < \hat{P} < 0.05) = P(|\hat{P} - 0.03| < 0.02) \geq 1 - \frac{0.03 \times 0.97}{200 \times (0.02)^2} = 0.6363$$

Hence, the complementary probability $P(\hat{P} \leq 0.01 \text{ or } \hat{P} \geq 0.05)$ is at most 0.3637, which is the upper bound that is asked for.

But the probability $P(\hat{P} \leq 0.01 \text{ or } \hat{P} \geq 0.05)$ can be calculated precisely since it can be rewritten in terms of Y, the number of defectives in the sample. It is equal to:

$$P(Y \leq 2 \text{ or } Y \geq 10) = P(Y \leq 2) + P(Y \geq 10) = P(Y \leq 2) + 1 - P(Y \leq 9)$$

Since Y has the distribution $Bin(200, 0.03)$, we obtain:

$$P(\hat{P} \leq 0.01 \text{ or } \hat{P} \geq 0.05) = 0.0593 + 1 - 0.9192 = 0.1401 \tag{*}$$

The number of successes Y in a random sample of size n has $0, 1, 2, \ldots, n-1, n$ as possible outcomes and is binomially distributed with parameters n and p. As a consequence, the sample proportion $\hat{P} = Y/n$ has the possible outcomes $0, 1/n, 2/n, \ldots, (n-1)/n, 1$. Probabilities of $\hat{P}$ events follow from this binomial distribution:

$$P(\hat{P} = \tfrac{k}{n}) = P(Y = k) = \binom{n}{k}p^k(1-p)^{n-k} \quad \text{for } k = 0, 1, \ldots, n$$

If the sample size n is large, it is usually not the binomial distribution that is used to calculate probabilities of $\hat{P}$ events. Since $\hat{P}$ is just the sample mean of the random sample $X_1, \ldots, X_n$ drawn from the Bernoulli distribution with parameter p, the central limit theorem (13.7) is applicable with μ replaced by p and σ^2 by $p(1-p)$. That is:

If n is large, then $\hat{P} \approx N(p, \dfrac{p(1-p)}{n})$ $\hfill$ **(14.6)**

That is, for large samples, probabilities of $\hat{P}$ events can be approximated by way of the right-hand normal distribution. The only problem here is the sample size n that is needed to obtain good approximations. If p is close to 0, then the distribution $Bern(p)$ from which the sample is drawn, is extremely skewed to the right and $n = 30$ is not enough to obtain good approximations. Similarly, if p is close to 1, the distribution $Bern(p)$ is extremely skewed to the left and $n = 30$ is again not enough. It turns out that the **5-rule** has to be valid: n has to be so large that np and $n(1-p)$ are both at least 5. That is why (14.6) is restated as follows:

Central limit theorem for $\hat{P}$

If both np and $n(1-p)$ are at least 5, then $\hat{P} \approx N(p, \dfrac{p(1-p)}{n})$ $\hfill$ **(14.7)**

See also the 5-rule and the normal approximation in (10.20). In normal approximations of probabilities of $\hat{P}$ events, no continuity correction is used!

Example 14.2

Reconsider the previous example, where a sample of 200 products is drawn from a population with 3% defectives. Notice that the 5-rule is satisfied since $np = 6$ and $n(1-p) = 194$. Below, the probability $P(\hat{P} \leq 0.01 \text{ or } \hat{P} \geq 0.05)$ that was calculated exactly in Example 14.1, will be approximated using the CLT.

Recall that $E(\hat{P}) = 0.03$ and $V(\hat{P}) = 0.0001455$. Hence, $\hat{P} \approx N(0.03, 0.0001455)$. It follows that:

$$P(\hat{P} \le 0.01 \text{ or } \hat{P} \ge 0.05) = P(\hat{P} \le 0.01) + P(\hat{P} \ge 0.05) \approx 2 \times 0.0487 = 0.0973 \qquad (*)$$

This approximation is not far from the exact answer in Example 14.1.

Next, the consequences for the standardised version Z of $\hat{P}$ are considered; they again follow from Chapter 13 since $\hat{P}$ is a special sample mean. By (14.3) it follows that:

$$Z = \frac{\hat{P} - p}{\sqrt{p(1-p)/n}} \tag{14.8}$$

Similarly to (13.8), we have:

Normality of $\hat{P}$ and Z if the 5-rule is valid

$$\hat{P} \approx N(p, \frac{p(1-p)}{n}) \quad \text{and} \quad Z \approx N(0, 1) \tag{14.9}$$

As in (13.10), the **constant** bounds $\mu_{\hat{p}} - z_{\alpha/2}\sigma_{\hat{p}}$ and $\mu_{\hat{p}} + z_{\alpha/2}\sigma_{\hat{p}}$ capture the random variable $\hat{P}$ with approximate probability $1 - \alpha$. Furthermore, as in (13.11), the **random** bounds $\hat{P} - z_{\alpha/2}\sigma_{\hat{p}}$ and $\hat{P} + z_{\alpha/2}\sigma_{\hat{p}}$ capture the constant $\mu_{\hat{p}} = p$ with approximate probability $1 - \alpha$. That is:

Important results if the 5-rule is valid

$$P\left(p - z_{\alpha/2}\frac{\sqrt{p(1-p)}}{\sqrt{n}} < \hat{P} < p + z_{\alpha/2}\frac{\sqrt{p(1-p)}}{\sqrt{n}}\right) \approx 1 - \alpha \tag{14.10}$$

$$P\left(\hat{P} - z_{\alpha/2}\frac{\sqrt{p(1-p)}}{\sqrt{n}} < p < \hat{P} + z_{\alpha/2}\frac{\sqrt{p(1-p)}}{\sqrt{n}}\right) \approx 1 - \alpha \tag{14.11}$$

Equation (14.10) states that a proportion of (about) $1 - \alpha$ of all realized random samples has the sample proportion $\hat{p}$ between the constants $p - z_{\alpha/2}\sqrt{p(1-p)/n}$ and $p + z_{\alpha/2}\sqrt{p(1-p)/n}$.

Equation (14.11) is valuable if p is unknown. It states that the probability is about $1 - \alpha$ that p is captured between $\hat{P} - z_{\alpha/2}\sqrt{p(1-p)/n}$ and $\hat{P} + z_{\alpha/2}\sqrt{p(1-p)/n}$. However, these random bounds cannot be observed since they contain unknown parameters. By replacing the ps in these random bounds by their estimator $\hat{P}$, random boundaries arise that **can** be observed and that have a high probability of capturing the unknown p. The actual outcomes of **these** random boundaries can be calculated as soon as the data are available, and it can be concluded with a high degree of confidence that the resulting concrete boundaries will include the unknown p.

Example 14.3

In recent years the government has made much effort to increase the participation of women in the labour market. The table below gives an overview of the percentages of women in the employed labour force (15–64 years).

Year	1995	2000	2003	2004
Women (%)	37	40	41.5	42

The government's efforts seem to have been successful during 1995–2004. The question arises whether these positive developments continued in subsequent years, when the economy stagnated. To put it more formally: the question is whether the current proportion p of the female participation in the employed labour force is greater than 0.42. To answer this question, 400 employed persons (aged 15–64) will be chosen randomly at the end of the present year and the proportion $\hat{P}$ of women in that sample will be measured. Although this research is scheduled for the end of this year, the theoretical analysis will be done now. The prospects regarding $\hat{P}$ will be considered:

1 for the case that p is the same as in 2004;

2 for the case that p is unknown, with the presupposition that the sample will contain 190 women.

Both cases will be confronted with the presupposition that the sample will contain 190 women.

Assume that the present proportion p is unchanged from 2004, so $p = 0.42$. Hence, $E(\hat{P}) = 0.42$ and $SD(\hat{P}) = \sqrt{0.42 \times 0.58/400} = 0.0247$. With $\alpha = 0.05$ and $z_{0.025} = 1.96$, it follows from (14.10) that the sample proportion $\hat{P}$ will fall between

$$p - 1.96\sqrt{p(1-p)/400} = 0.3716 \quad \text{and} \quad p + 1.96\sqrt{p(1-p)/400} = 0.4684$$

with probability 0.95. That is, we can be pretty sure that the number of women in the sample will fall between 148 and 188 – at least, if p is the same as in 2004. Finding 190 women in the sample is not compatible with p being unchanged.

Now assume that the present proportion p is unknown. Suppose that the sample will contain 190 women, a proportion of 0.4750. Although this sample proportion is larger than 0.42, it does not immediately guarantee that the **population** proportion p is larger than 0.42 too. But (14.11) does ensure that the probability is about 0.95 that the unknown p falls between

$$\hat{P} - 1.96\sqrt{p(1-p)/400} \quad \text{and} \quad \hat{P} + 1.96\sqrt{p(1-p)/400}$$

so there is a high probability that p is larger than the random lower bound $\hat{P} - 1.96\sqrt{p(1-p)/400}$. Since this lower bound contains unknown constants, these unknowns are replaced by their natural estimator (which is $\hat{P}$). Although this may affect the level of probability, this will be only slightly since $\hat{P}$ is a consistent estimator and n is large. It follows that, with a high probability, p is larger than the random lower bound $\hat{P} - 1.96\sqrt{\hat{P}(1-\hat{P})/400}$. For the realization 0.4750 of $\hat{P}$, this lower bound takes the value 0.4261. Hence, if the sample contains 190 women, we can be confident that p is larger than 0.4261, and hence that p is larger than 0.42.

CASE 14.1 APPROVAL PROBABILITIES IN QUALITY CONTROL – SOLUTION

The supplier and the retailer can resolve their dilemma as follows.

a The approval plan (50, 0.1) means that supplier and retailer agree that a random sample of 50 products will be used to decide the quality of the whole batch. If the sample proportion of defectives is 0.1 or less, then the whole batch will be approved; if not, the batch will be rejected and returned to the supplier.

b The case $p = 0.20$ will be used as illustration. Let Y denote the number of defectives in the sample of size 50; so $\hat{P} = Y/50$. Then:

$$P(\hat{P} \leq 0.1 \mid p = 0.20) = P(Y \leq 5 \mid p = 0.20)$$

Since $p = 0.20$, Y has the distribution $Bin(50, 0.2)$; see Section 9.2.1. With a computer it follows easily that the answer is 0.0480.

The table below lists the probabilities; the graph depicts the approval function.

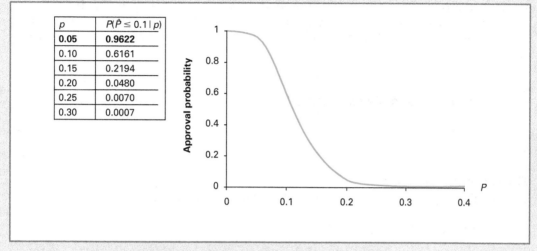

p	$P(\hat{P} \le 0.1 \mid p)$
0.05	**0.9622**
0.10	0.6161
0.15	0.2194
0.20	0.0480
0.25	0.0070
0.30	0.0007

c   For $p = 0.20$, $\hat{P}$ is approximately $N(0.20, 0.0032)$ distributed. Hence:

$$P(\hat{P} \le 0.1 \mid p = 0.20) = P(\frac{\hat{P} - 0.20}{0.05657} \le \frac{0.10 - 0.20}{0.05657} \mid p = 0.20)$$
$$= P(Z \le -1.7677) = 0.0386 \qquad (*)$$

Here Z is a standard normal rv. The following table gives the approximations:

p	0.05	0.10	0.15	0.20	0.25	0.30
Approximated probability	0.9476	0.5000	0.1611	0.0386	0.0072	0.0010

Even for $p = 0.10$ (when $50p$ satisfies the 5-rule for the first time), the difference between the exact and the aproximated probability is still rather large; for $p \ge 0.15$ the approximations become better.

d   The supplier wants c such that:

$$0.08 = P(\hat{P} > c \mid p = 0.075) = P(Z > \frac{c - 0.075}{\sqrt{0.075 \times 0.925/100}}) = P(Z > \frac{c - 0.075}{0.0263})$$

Since $P(Z > 1.4051) = 0.08$ (*), it follows that $(c - 0.075)/0.0263 = 1.4051$. The supplier wants $c = 0.1120$.

The retailer wants c such that:

$$0.10 = P(\hat{P} \le c \mid p = 0.125) = P(Z \le \frac{c - 0.125}{\sqrt{0.125 \times 0.875/100}}) = P(Z \le \frac{c - 0.125}{0.0331})$$

Since $P(Z \le -1.2816) = 0.10$ (*), it follows that $(c - 0.125)/0.0331 = -1.2816$. The retailer wants $c = 0.0826$.

14.2 Properties of other sample statistics

Apart from the sample mean and sample proportion, other sample statistics were considered in Section 12.4: the sample variance, the sample standard deviation, the sample covariance and the sample correlation coefficient. Below, some properties of these estimators are considered briefly.

Start with a random sample $X_1, \ldots, X_n$ from a population distribution with mean μ and variance σ^2. Below, it will be proved that the sample variance S^2 is an **unbiased** estimator of the population variance σ^2. That is:

Property of the sample variance S^2

$$E(S^2) = \sigma^2 \tag{14.12}$$

Proof of unbiasedness of S^2

To prove this result, the short-cut formula for S^2 will be used (see (4.4)):

$$S^2 = \frac{1}{n-1} \sum_{i=1}^{n}(X_i - \overline{X})^2 = \frac{1}{n-1}\left(\sum_{i=1}^{n}X_i^2 - n\overline{X}^2\right)$$

Moreover, we will make use of the fact that each rv Y satisfies $\sigma_Y^2 = E(Y^2) - \mu_Y^2$ and hence $E(Y^2) = \sigma_Y^2 + \mu_Y^2$.

From the expectation rules we obtain:

$$E\left(\sum_{i=1}^{n}X_i^2 - n\overline{X}^2\right) = E\left(\sum_{i=1}^{n}X_i^2\right) - E(n\overline{X}^2) = \sum_{i=1}^{n}E(X_i^2) - nE(\overline{X}^2) \tag{14.13}$$

Since $X_1, \ldots, X_n$ is a random sample, it is *iid* and, in particular, all expectations $E(X_i^2)$ are the same and equal to $\sigma^2 + \mu^2$ (as follows easily from the above by taking Y as X_i). Furthermore, recall that $\mu_{\overline{X}} = \mu$ and $V(\overline{X}) = \sigma^2/n$, so $E(\overline{X}^2) = \sigma^2/n + \mu^2$ (take Y as $\overline{X}$). By (14.13) it follows that:

$$E\left(\sum_{i=1}^{n}X_i^2 - n\overline{X}^2\right) = \sum_{i=1}^{n}(\sigma^2 + \mu^2) - n\left(\frac{\sigma^2}{n} + \mu^2\right)$$
$$= n\sigma^2 + n\mu^2 - \sigma^2 - n\mu^2 = (n-1)\sigma^2$$

By dividing both sides by $n - 1$, equation (14.12) follows immediately.

Result (14.12) finally explains why the factor $1/(n - 1)$ was used in the definition of the sample variance: if the factor $1/n$ had been used, the resulting estimator of σ^2 would not be an unbiased estimator of σ^2.

The sample variance S^2 is a random variable that is based on the random sequence $X_1, \ldots, X_n$, so it is legitimate to ask for its pdf. However, in general it is not possible to determine this pdf easily. In the following example, this pdf can be obtained since the sample is small and the underlying population distribution is simple.

Example 14.4

Consider a large population of households where the variable X = 'number of cars per household' has the following relative frequency distribution:

Value	0	1	2
Rel. frequency	0.3	0.6	0.1

It follows easily that $\mu = 0.8$ and $\sigma^2 = 0.36$. From this population, two households are drawn randomly and with replacement, and the numbers of cars are measured. This yields the random sample X_1, X_2 of size 2. Since $\bar{X} = X_1/2 + X_2/2$, it follows that:

$$(X_1 - \bar{X})^2 = (X_1 - \tfrac{1}{2}X_1 - \tfrac{1}{2}X_2)^2 = (\tfrac{1}{2}X_1 - \tfrac{1}{2}X_2)^2 = \tfrac{1}{4}(X_1 - X_2)^2$$

$$(X_2 - \bar{X})^2 = (\tfrac{1}{2}X_2 - \tfrac{1}{2}X_1)^2 = \tfrac{1}{4}(X_1 - X_2)^2$$

Hence:

$$S^2 = (X_1 - \bar{X})^2 + (X_2 - \bar{X})^2 = \tfrac{1}{2}(X_1 - X_2)^2$$

Since X_1 and X_2 take only the values 0, 1 and 2, the possible outcomes of the random variable S^2 are 0, 0.5 and 2. The table presents the pdf of S^2:

b	0	0.5	2
$P(S^2 = b)$	0.46	0.48	0.06

For instance, by the i and id properties it follows that:

$$P(S^2 = 0) = P(X_1 = X_2)$$
$$= P(X_1 = 0; X_2 = 0) + P(X_1 = 1; X_2 = 1) + P(X_1 = 2; X_2 = 2)$$
$$= P(X_1 = 0)P(X_2 = 0) + P(X_1 = 1)P(X_2 = 1) + P(X_1 = 2)P(X_2 = 2)$$
$$= (0.3)^2 + (0.6)^2 + (0.1)^2 = 0.46;$$

$$P(S^2 = 2) = P(X_1 = 0; X_2 = 2) + P(X_1 = 2; X_2 = 0) = 2 \times 0.3 \times 0.1 = 0.06$$

Since S^2 is an unbiased estimator of σ^2, we have $E(S^2) = \sigma^2 = 0.36$. But $E(S^2)$ can also be calculated from the pdf:

$$E(S^2) = 0 \times 0.46 + 0.5 \times 0.48 + 2 \times 0.06 = 0.36$$

This result indeed is in accordance with (14.12).

The sample standard deviation S is the usual estimator of the population standard deviation σ. This sample statistic is **not** unbiased. In general, the expectation of S is strictly **smaller** than σ, so the average realization of S falls below σ. It is said that S **underestimates** σ.

The proof that $E(S) \leq \sigma$ is simple when it is noticed that $V(S)$ is at least 0 and when the short-cut formula for $V(S)$ is used:

$$0 \leq V(S) = E(S^2) - (E(S))^2 = \sigma^2 - (E(S))^2$$

So, $(E(S))^2 \leq \sigma^2$ and hence $E(S) \leq \sigma$. Since $V(S)$ is equal to 0 only if S is non-random and hence degenerated, it is generally valid that $E(S) < \sigma$.

Now suppose that two variables X and Y are present and that interest is in the strength of their linear relationship. Recall that the statistics $\sigma_{X,Y}$ and ρ, the covariance and the correlation coefficient, respectively, measure this strength. However, in practice these statistics are unknown and have to be estimated. Hence, a random sample of n population elements is drawn, and X and Y are both measured for each of these sampled elements. As a consequence, a random sample of pairs of random observations arises: $(X_1, Y_1), \ldots, (X_n, Y_n)$; see also Section 12.4. The estimators – the sample covariance $S_{X,Y}$ and the sample correlation coefficient $R_{X,Y}$ – of, respectively, $\sigma_{X,Y}$ and ρ were defined in (12.9).

Concerning the properties of these estimators, it is stated here only that $S_{X,Y}$ is unbiased and that $R_{X,Y}$ is **not** unbiased:

$$E(S_{X,Y}) = \sigma_{X,Y} \tag{14.14}$$

One final remark, about consistency. It can be proved that all estimators considered so far are **consistent** in estimating the corresponding parameters. For instance, S^2 and S are consistent in estimating the respective parameters σ^2 and σ. That is, by taking the sample size n large enough, it can be arranged that the estimator lies arbitrarily close to the parameter it is estimating with a probability that is as close to 1 as wanted.

14.3 Standard errors

In general, the standard deviations of estimators contain unknown parameters. For instance, the standard deviation of the estimator $\bar{X}$ of μ is equal to $\sigma/\sqrt{n}$, where the parameter σ is usually unknown; the standard deviation of the estimator $\hat{P}$ of p is equal to $\sqrt{p(1-p)/n}$ and contains the unknown p. In the present section, interest is in estimators of such standard deviations. They will turn out to be very valuable in inferential statistics.

A **standard error** (SE) of an estimator is a sample statistic that can be used to estimate the standard deviation of that estimator. Although this definition may look rather cryptic, practice is simple. To illustrate this, standard errors of the estimators $\bar{X}$ and $\hat{P}$ will be considered. They will be denoted by $SE(\bar{X})$ and $SE(\hat{P})$, respectively.

Consider the sample mean $\bar{X}$ (the estimator of μ) and recall that $SD(\bar{X}) = \sigma/\sqrt{n}$. Suppose that σ is unknown. Recall that

$$P(\bar{X} - z_{\alpha/2}\frac{\sigma}{\sqrt{n}} < \mu < \bar{X} + z_{\alpha/2}\frac{\sigma}{\sqrt{n}}) \approx 1 - \alpha \tag{14.15}$$

so μ is captured between the random bounds $\bar{X} - z_{\alpha/2}\sigma/\sqrt{n}$ and $\bar{X} + z_{\alpha/2}\sigma/\sqrt{n}$ with probability (about) $1 - \alpha$. However, since the value of $SD(\bar{X}) = \sigma/\sqrt{n}$ cannot be calculated from the data because σ is unknown, an estimator of this standard deviation is needed. That is why σ is replaced by its (consistent) estimator S and the standard deviation $SD(\bar{X}) = \sigma/\sqrt{n}$ turns into the standard error $SE(\bar{X}) = S/\sqrt{n}$. If n is large, the probability in (14.15) hardly changes. So:

$$P(\bar{X} - z_{\alpha/2}\frac{S}{\sqrt{n}} < \mu < \bar{X} + z_{\alpha/2}\frac{S}{\sqrt{n}}) \approx 1 - \alpha \tag{14.16}$$

In isolated cases the parameter σ in $SD(\bar{X}) = \sigma/\sqrt{n}$ **is** known. If this is the case, do **not** replace σ by S, to avoid the introduction of a new source of variation. **The best estimator of a known parameter is the parameter itself**, so under such circumstances $SE(\bar{X}) = SD(\bar{X}) = \sigma/\sqrt{n}$.

Example 14.5

Recall Example 13.10, where interest was in the current mean hourly wage μ of all employees in the commercial services sector. The idea was to obtain information about this unknown μ by drawing a random sample of 500 employees, under the assumption that the population standard deviation is the same as in 2003; so $\sigma = €8.20$. Since the random boundaries $\bar{X} \pm 1.6449 \times \sigma/\sqrt{n}$ capture the unknown μ with probability (about) 0.90, they can be used to calculate a concrete interval as soon as the data are available. It is not necessary – in fact it would be bad practice – to replace the standard deviation $SD(\bar{X}) = \sigma/\sqrt{n}$ by any standard error other than $\sigma/\sqrt{n}$ itself: never estimate a known parameter.

For instance, suppose that the realization of $\bar{X}$ is 17.99. That is, suppose that the data will be such that they yield the estimate 17.99 of μ. Then the realizations of the two random boundaries can be calculated and it can be concluded with a high level of confidence that μ falls in the interval (17.3868, 18.5932).

Now suppose that σ is unknown. Then the random boundaries still capture the unknown μ with probability (about) 0.90, but $SD(\bar{X}) = \sigma/\sqrt{n}$ is unknown so the realizations of the boundaries cannot be calculated when the data are available. That is why $SD(\overline{X})$ has to be replaced by $SE(\overline{X}) = S/\sqrt{n}$, thus introducing a new source of variation. However, since the sample size $n = 500$ is large and S is a consistent estimator of σ, the probability 0.90 hardly changes; the random boundaries $\overline{X} \pm 1.6449 \times S/\sqrt{n}$ still include μ with probability about 0.90. If the data yield $\overline{x} = €17.99$ and $s = €8.55$, then the realizations of the new boundaries **can** be calculated and it can be concluded with much confidence that μ falls in the interval (17.3610, 18.6190).

Next, consider the sample proportion $\hat{P}$ (the estimator of p) and recall that $SD(\hat{P}) = \sqrt{p(1 - p)/n}$. Recall also that:

$$P\left(\hat{P} - z_{\alpha/2}\,\frac{\sqrt{p(1 - p)}}{\sqrt{n}} < p < \hat{P} + z_{\alpha/2}\,\frac{\sqrt{p(1 - p)}}{\sqrt{n}}\right) \approx 1 - \alpha \tag{14.17}$$

so p is captured between the random bounds $\hat{P} \pm z_{\alpha/2}\sqrt{p(1 - p)/n}$. However, the realizations of these bounds cannot be calculated when the data are available, since these bounds depend on the unknown p. That is why the standard deviation $SD(\hat{P})$ has to be estimated, and the best way to do that is by replacing p by its estimator $\hat{P}$. That is, $SD(\hat{P})$ is replaced by $SE(\hat{P}) = \sqrt{\hat{P}(1 - \hat{P})/n}$. If n is large, the probability in (14.17) hardly changes, so:

$$P\left(\hat{P} - z_{\alpha/2}\,\frac{\sqrt{\hat{P}(1 - \hat{P})}}{\sqrt{n}} < p < \hat{P} + z_{\alpha/2}\,\frac{\sqrt{\hat{P}(1 - \hat{P})}}{\sqrt{n}}\right) \approx 1 - \alpha \tag{14.18}$$

Example 14.6
Recall Example 14.3, where interest was in the current proportion p of women in the employed labour force. The idea was to estimate p on the basis of a random sample of 400 employed persons. This sample will be drawn at the end of the present year, but the analysis is done now. By (14.11) we have:

$$P\left(\hat{P} - 1.96\,\frac{\sqrt{p(1 - p)}}{\sqrt{400}} < p < \hat{P} + 1.96\,\frac{\sqrt{p(1 - p)}}{\sqrt{400}}\right) \approx 0.95$$

Unfortunately, the realizations of the two bounds $\hat{P} \pm 1.96\sqrt{p(1 - p)/n}$ cannot be calculated when the data are obtained since $SD(\hat{P}) = \sqrt{p(1 - p)/n}$ depends on the unknown p. That is why $SD(\hat{P})$ is replaced by an estimator and $SE(\hat{P}) = \sqrt{\hat{P}(1 - \hat{P})/n}$ is a good candidate (since $\hat{P}$ is consistent in estimating p). The adapted random bounds $\hat{P} \pm 1.96\sqrt{\hat{P}(1 - \hat{P})/400}$ still include p with probability (about) 0.95. So, if at the end of the year it turns out that 190 of the 400 sampled persons are female, then it can be concluded that there is much confidence that p falls in the interval (0.426, 0.524).

The above ideas about standard errors will be reconsidered in Chapters 15 and 16. In later chapters, new statistics will enter the scene and new estimators will be considered. Since estimators are random variables, their standard deviations are important but will usually depend on unknown parameters. Such parameters will be replaced by suitable consistent estimators, thus transforming the standard deviations into standard errors. If the sample size is large, usually no problems arise. However, for small sample sizes one has to be careful. To be continued.

Summary

Properties of some sample statistics were considered, especially of the (random) sample proportion $\hat{P}$. The tables below give some important results; the properties of the second table are important for inferential statistics.

Expectation and variance of two estimators

Estimator	Expectation	Variance
S^2	σ^2	–
$\hat{P}$	p	$p(1-p)/n$

Properties of $\hat{P}$ if n is so large that the 5-rule holds

$$\hat{P} \approx N\left(p, \frac{p(1-p)}{n}\right) \qquad Z = \frac{\hat{P} - p}{\sqrt{p(1-p)/n}} \approx N(0, 1)$$

$$P\left(\hat{P} - z_{\alpha/2}\frac{\sqrt{p(1-p)}}{\sqrt{n}} < p < \hat{P} + z_{\alpha/2}\frac{\sqrt{p(1-p)}}{\sqrt{n}}\right) \approx 1 - \alpha$$

In this chapter, the only new symbol was SE; it denotes the standard error of an estimator and arises from the standard deviation SD of that estimator by replacing unknown parameters (if any) by estimators.

🔑 Key terms

5-rule **418**	standard error **424**
central limit theorem for $\hat{P}$ **418**	underestimate **423**

❓ Exercises

Exercise 14.1

It turns out that 30% of consumers prefer to buy their groceries in supermarkets of a chain called A. Determine (a) exactly, and (b) with a normal aproximation for the sample proportion, the probability that in a random sample of 1000 consumers more than 31% prefer supermarket A.

Exercise 14.2

For the production process of a mass-produced product it is known that 0.5% of the products are defective. Consider a sample of n products drawn randomly from the (large) daily production. Interest is in the probability that more than 1% of the sampled products are defective.

a Calculate this probability if $n = 500$.

b Use normal approximation to determine this probability if $n = 1000$.

c Why is it not wise to use normal approximation in part (a)?

Exercise 14.3

A recent publication states that 42.5% of entrepreneurs have positive expectations regarding the future development of the economy. However, an (even more recent) investigation among 100 randomly chosen entrepreneurs yields a proportion of only 40%. Does this sample proportion contradict the result of the recent publication? Answer this question by investigating whether realizations of at most 0.40 are exceptional when $p = 0.425$.

Exercise 14.4

The variable X = 'weekly sales (in thousands of euro) of a computer shop' has mean μ and variance σ^2. Let $X_1, \dots, X_8$ denote the weekly sales for eight arbitrary weeks.

  a Formulate the usual estimators of μ and σ^2.

  b Write the expectation and the standard deviation of the estimator for μ in terms of μ and σ^2. Do the same for the expectation of the estimator of σ^2.

  c Construct the usual estimator of the standard deviation of the estimator of μ. What is the name of this estimator?

The table below gives the data of the random sample:

Week	1	2	3	4	5	6	7	8
Sales ($\times$ 1000)	49.2	34.8	32.7	51.5	44.8	43.2	49.7	39.8

  d Use the data to determine estimates for μ and σ^2 that follow with the two estimators in part (b).

  e Determine the realized standard error of the estimator of μ.

Exercise 14.5

The variable X = 'answer (1 = yes / 0 = no) to the question: are you satisfied with your current job?' on a population of adult women, has mean μ and variance σ^2; the population proportion 'yes' is denoted by p. Let $X_1, \dots, X_{10}$ denote the observations of X for a random sample of 10 women.

  a Express μ and σ^2 in terms of p.

  b Formulate the usual estimator of p and use it to construct an estimator of σ^2.

  c Write the expectation and the variance of the estimator of p in terms of p.

  d Construct the usual estimator of the standard deviation of the estimator of p. What is the name of this estimator?

The table below gives the data of the random sample:

Woman	1	2	3	4	5	6	7	8	9	10
Answer	1	0	1	1	0	1	1	1	0	0

  e Use the data to determine the estimates for p and σ^2 that follow with the two estimators in part (b). Also determine the realised standard error of the estimator of p.

Exercise 14.6

In a study about the hourly wages (X, in euro) paid to young people aged 14–18 with a holiday job, the following summarized results were measured for a random sample of 400 young people:

$$\sum_{i=1}^{400} x_i = 3296 \quad \text{and} \quad \sum_{i=1}^{400} x_i^2 = 27833.35$$

Use these results to find estimates for the mean and the variance of the hourly wage paid to young people with a holiday job.

Exercise 14.7

Two years ago, the political party Democratic Front for Animals (DFA) received 9.8% of the votes in an election. To study the **current** political preferences, a statistical research institute plans to organize a poll by the end of the present year. In this study, n voters will be interviewed about the political party they prefer.

Below, p denotes the proportion of the voters that would vote DFA if the elections were held now. Furthermore, $\hat{P}$ denotes the (random) sample proportion of the DFA voters.

a Suppose that $n = 500$. Determine an interval that with probability 0.90 will contain the (not yet observed) realization of $\hat{P}$ if p were the same as two years ago.

b Find random bounds (depending on p) that will include the proportion p with probability 0.95. Express the width of the accompanying interval in terms of p and n.

c The interval in part (b) will be the starting point to create, at the end of the current year, when the data are observed, an interval that will probably contain the proportion p. How large should the sample size be to obtain an interval of width about 0.02? (*Hint*: Substitute the former proportion 0.098 for p in the width of part (b).)

d At the end of the year, n randomly chosen voters are interviewed with n as calculated in part (c); the realisation of $\hat{P}$ turns out to be 0.853. Use the interval in part (b) as a starting point to create an interval that will probably contain the population proportion p. Is the width of that interval indeed about 0.02?

Exercise 14.8

A random sample of ten visitors to the restaurant Apples and Nuts were asked the following questions as they were leaving the restaurant:

Q1: Assign a score between 0 and 10 to the quality of your dinner.
Q2: Do you expect that you will soon pay another visit to this restaurant (1 = yes; 0 = no)?

a Describe the variables that are of interest. Are they quantitative or qualitative?

b With respect to the second variable, formulate the usual estimator of the proportion of all visitors who will declare that they will soon come back.

c For the mean and the variance of variable 1, formulate the usual estimators. Also give the estimator for the mean of variable 2. Which estimator do you propose for the variance of variable 2?

d Express the expectations and the variances of the estimators for the means of the two variables in terms of population means, population variances and/or population proportions.

The table below gives the data.

Visitor	1	2	3	4	5	6	7	8	9	10
Variable 1	5	8.5	7.5	6	8	9	5	7.5	8	7
Variable 2	0	1	1	1	1	1	0	1	1	0

e Determine estimates of the means and the variances of the two variables.

Exercise 14.9

Suppose that 70% of the Dutch do **not** like the 'kroket', a typical Dutch delicacy. A producer puts more meat than usual in his kroketten to stimulate sales. Denote the proportion of the Dutchmen who do not like this new kroket, by p.

To estimate p, a random sample of 200 Dutchmen is asked to taste the kroket. Let $\hat{P}$ denote the (random) sample proportion of the Dutchmen in the sample who do not like the kroket. Assume that the population proportion p is unchanged, so $p = 0.70$.

  a Determine $E(\hat{P})$, $V(\hat{P})$ and $SD(\hat{P})$.

  b Determine the exact probability that $\hat{P}$ is larger than 0.75, by using a suitable binomial distribution.

  c Also determine a normal approximation of the probability in part (b) (at least, if the 5-rule is satisfied).

  d Find an approximation for the probability that $\hat{P}$ falls between 0.68 and 0.72.

  e Calculate the exact probability that fewer than 134 of the sampled persons do not like the kroket. Also use the CLT to find an approximation.

Exercise 14.10

Reconsider Exercise 14.9, but assume now that p is unknown.

  a Explain why the random interval with lower bound $\hat{P} - 2\sqrt{p(1-p)/200}$ and upper bound $\hat{P} + 2\sqrt{p(1-p)/200}$ will probably capture the unknown p. Determine the probability that p is captured.

  b Unfortunately, the random interval cannot be observed since it depends on the unknown p. To overcome this problem, replace p by $\hat{P}$. Write down the resulting random interval.

  c When the data are known, it turns out that 125 of the 200 sampled persons do not like the kroket. Determine an estimate of p. Also determine an interval that probably contains the unknown p.

  d The producer believes that p is smaller than 0.70. Is it likely that he is right?

Exercise 14.11

It is given that 20% of the 6000 first year students of a university smoke cigarettes. Suppose that 100 of them visit this month's beer party. Let $\hat{P}$ denote the proportion of these 100 students who smoke cigarettes.

  a Calculate $E(\hat{P})$, $V(\hat{P})$ and $SD(\hat{P})$. What did you implicitly assume?

  b Calculate $P(\hat{P} < 0.25)$ by using a suitable binomial distribution. Also approximate this probability by applying the CLT.

Exercise 14.12

Reconsider Exercise 4.11, but suppose now that the proportion p of cigarette smokers within the population of first year students is unknown.

  a Determine the probability that the random interval with lower bound $\hat{P} - 1.8 \times SD(\hat{P})$ and upper bound $\hat{P} + 1.8 \times SD(\hat{P})$ will capture p.

  b Unfortunately, $SD(\hat{P})$ cannot be observed since it contains unknown parameters. Find its estimator. What is it called and how is it written down?

  c In the random interval in part (a), the unobservable $SD(\hat{P})$ is replaced by its estimator. Formulate the resulting (observable) random interval that probably captures p. Determine the probability that p is included.

d It turns out that only 13 of the 100 visitors to the party are smokers. Can it be concluded that p is smaller than 0.20?

Exercise 14.13

In the year 2000, 13.5% of the population were 65 or older. As a part of a study about the ageing process of the population and its consequences for the economy, a researcher wants to investigate whether this percentage has changed in recent years. She wants to draw a random sample of size n and determine the proportion $\hat{P}$ of the people aged 65+ in the sample, to estimate the corresponding population proportion p. The sample size must be so large that, for the case that the population proportion p is still 0.135, the probability is (only) 0.02 that the sample proportion $\hat{P}$ will deviate more than 0.01 from the population proportion p. That is, n has to be so large that

$$P(|\hat{P} - p| > 0.01) = 0.02 \qquad \text{for } p = 0.135$$

a Use the CLT to determine n.

b It is decided to use the sample size of part (a). When the study is done, it turns out that 15% of the sampled persons are 65+. Note that this sample proportion 0.15 is larger than 0.135. But does this justify the conclusion that the population proportion p nowadays is larger than 0.135? To answer that question, use part (a) to show that the realization 0.15 would be very uncommon if $p = 0.135$ were still valid.

Exercise 14.14

Most production processes generate defectives. However, the percentage of defectives will usually be kept as low as possible. To check the production process, random samples of products are usually drawn and tested.

Consider a certain mass-produced item. Suppose that the production process yields at most 2% defectives, at least when the process is what is termed 'under control'. To check the quality of the production process, 400 of the items are randomly selected each hour. If at most 9 defectives are detected, then it is concluded that the process is still under control and nothing is done. But if 10 or more defectives are found, then it is concluded that the production process is, in the terminology, 'out of control', and the process is stopped and investigated. Below, p denotes the proportion of defectives in the large hourly production and $\hat{P}$ denotes the (random) sample proportion of the defectives in an hourly sample of size 400.

a Determine the (approximate) probability distribution of $\hat{P}$.

b Calculate the probability that the production process is stopped because of the detection of too many defectives in the sample while this was unnecessary since the proportion p was only 0.015.

c The same as in part (b), but now for the cases that p was 0.0175 and 0.02. Interpret the results of parts (b) and (c).

d Suppose now that during a certain hour when the production process goes out of control, $p = 0.03$. Calculate $P(\hat{P} \le 9/400)$. Interpret it.

Exercise 14.15 (computer)

From Chapter 15 onwards, the relationship between sample results and the underlying population is investigated. In the present exercise we will use sequences of 200 (simulated) tosses of an unfair coin (1 = head and 0 = tail) to estimate the overall proportion p of heads generated with this coin. (If you prefer a less prosaic scenario, consider these sequences of two hundred 1/0 observations as realized random samples of size 200 that contain the answers (1 = yes and 0 = no) of consumers to the question: do you prefer soft drink C to soft drink P; then p is the proportion of the consumers preferring C.)

The file <u>Xrc14-15.xls</u> contains in its 20 columns the data of 20 (simulated) sequences of 200 tosses of the coin.

  a Use the first column of the dataset to obtain an estimate of p. Use this estimate to estimate the variance $\sigma^2 = p(1 - p)$.

  b Do the same for the other 19 columns.

  c Since the expected value of the estimator $\hat{P}$ is p, the average of the 20 realized sample proportions should also (and, probably, better) approximate p. Determine that approximation. What do you think of the actual value of p and σ^2?

  d Recall that the variance of the estimator $\hat{P}$ is $p(1 - p)/200$. Since each of the 20 samples yields an estimate of this variance, the average of the resulting 20 estimates should approximate the variance of $\hat{P}$. Calculate this average. What do you think of $V(\hat{P})$?

  e You can now be told that $p = 0.70$ was used to simulate the 20 random samples of size 200. Compare your estimation results with this new fact.

CASE 14.2 HOUSEHOLDS STATISTICS (PART III)

This case is a continuation of Cases 12.2 and 13.2. Similar exercises will be carried out to those in Case 13.2, but now for the population **proportions** of ones for the dummies DS, DH, DP, DF and for the population proportions of the five levels of EDU. For all nine population proportions p, the (theoretical) probabilities

$$P(\hat{P} - 2\sigma_{\hat{p}} < p < \hat{P} + 2\sigma_{\hat{p}})$$

can be calculated for each of the five random samples (for DS, DH, DP, DF, EDU) and each of the nine proportions (four for the dummies and five for EDU).

  a Calculate these probabilities. Do you really have to repeat the calculation nine times? Interpret your answers.

  b Express $\sigma_{\hat{p}}$ in terms of the corresponding proportion p. What is the difference between the standard deviations $\sigma_{\bar{x}}$ of Case 13.2 and the standard deviations $\sigma_{\hat{p}}$ of the present case?

  c To overcome the problem mentioned in part (b), the parameters p in the expression of $\sigma_{\hat{p}}$ are replaced by the corresponding estimators $\hat{P}$. Formulate the resulting expression and compare it with the probabilities of part (a). Formulate random (and observable) random variables that probably include the population proportions p.

  d Use your results from Case 12.2 to calculate the realizations of the nine random intervals of part (c). What can be concluded about the nine population proportions?

PART 4
Inferential Statistics

Studies in economics often are about large populations and unknown population statistics (parameters) concerning one or more variables. Observing the whole population is impossible, too expensive and/or too time consuming; tractable samples drawn from the population should give the researcher the information that is wanted. Several questions arise:

- How large should the sample be?
- Which methods should be used to obtain reliable conclusions about the parameters?
- What **is** the confidence level or precision of these conclusions?

Of course, the methods will depend on the statistical problems of interest. In practice, there are four types of statistical problem:

Four types of statistical problems

1 Obtaining estimates of unknown parameters.
2 Obtaining intervals that 'probably' contain the unknown parameters.
3 Concluding about conjectures concerning the parameters.
4 Predicting the value of a variable for a population element.

Inferential statistics draws conclusions about the whole population by studying samples drawn from that population. These conclusions should give answers to the above statistical problems. Consequently, it is important to have good statistical methods to generate reliable conclusions. The topics of the coming chapters aim to develop and apply such methods. It will turn out that some of the results obtained in Part 2 'Probability Theory' and Part 3 'Sampling Theory' (such as (10.18), (13.11) and (14.11)) are fundamental to the development of statistical methods for problems 1–3 above.

Chapters 15–25 will mainly be about the statistical problems 1–3. Problem 4 will be considered mainly in the context of linear regression, in Chapters 19–23.

Interval estimation and hypothesis testing: a general introduction

Chapter contents

This chapter gives an intuitive introduction to statistical problems, to statistical procedures for solving such problems and to statistical conclusions that can be drawn from the procedures. The abstract ideas about a general parameter θ are illustrated for the parameters μ (population mean) and p (population proportion). For μ, standard statistical procedures are developed for the case that the accompanying population standard deviation σ is known. Such procedures – for cases with known σ – open the way of thinking for developing statistical procedures in coming chapters, where more realistic situations will be studied. For p, some examples are already considered here; they will help to find standard statistical procedures in Chapter 16.

CASE 15.1 SHOULD INDUSTRIAL ACTIVITIES BE MOVED TO ANOTHER COUNTRY?

A German manufacturing firm with headquarters near Berlin believes that the mean gross per capita wage in Germany is too high to maintain the firm's excellent competitiveness in the coming years. Because of the lower wages paid by manufacturing firms in Italy, the company thinks of moving its industrial activities to the neighbourhood of Rome.

The mean gross per capita wage in Berlin in the manufacturing industry is €28 900 per year; the accompanying standard deviation σ is €1300. The company believes that moving to the region near Rome would be worth considering if the mean μ of the yearly gross per capita wage in manufacturing in that region is less than €21 000.

> The company asks the National Statistical Institute of Italy (NSII) for help. The NSII takes a random sample of 200 people working in the manufacturing industry in the region of Rome and finds that the average yearly gross wage equals €20 700. Indeed, this sample mean is smaller than the bound 21 000 imposed by the company. But (and this is what it is all about), is this value of the sample mean small enough to conclude that the population mean μ lies below this bound too? See the end of Section 15.3.3.

15.1 Initial approach to statistical procedures

Our starting point is a population and one or more population variables. Suppose that interest is in some general – **unknown** – statistic called θ (theta). In the present and coming chapters, θ will sometimes be a population mean μ or a population proportion p. However, in later chapters θ can also be a variance σ^2, a standard deviation σ, a correlation coefficient ρ or even regression coefficients β_0, β_1 (see Chapter 5). Since the present chapter is a general chapter, we use an all-purpose label θ. So, θ stands for some general population statistic, the parameter of interest.

Regarding this parameter of interest, three **statistical problems** will be considered:

1 Finding an approximation of θ.

2 Finding an interval that 'probably' contains θ.

3 Choosing between two opposite statements about θ.

Problem 1 means that a real number is wanted that gives a 'good' indication of the **precise** value of the unknown θ; the problem is also called a (**point**) **estimation problem**. Problem 2 is an **interval estimation problem**: an interval of real numbers is wanted that 'probably' contains θ. In the third statistical problem, the purpose is to choose between two opposite opinions about the parameter of interest. For instance, in Example 13.10 the actual purpose was to choose between two opposite statements about the mean hourly wage (μ) in the commercial services sector: 'μ is larger than €16.92' and 'μ is at most €16.92'. Problems of type 3 are called **testing problems**, where **testing** refers to methods that are used to tackle such problems (see later). The two opposite statements are called **hypotheses**.

In the theory below, methods (so-called **statistical procedures**) will be developed that are based on a random sample drawn (with replacement) from the population and offer the possibility to draw conclusions (so-called **statistical conclusions**) about the statistical problems above. However, it is important to note that such statistical conclusions can be incorrect although the practical purpose is to develop procedures leading to conclusions that can be trusted with a high level of confidence.

Example 15.1

In a certain milk factory, the daily production is 100 000 one-litre packs of milk. Interest is in the unknown number (say, γ) of packs produced daily that contain less than 1 litre. Although it is unavoidable that the filling machines cause some variation in the contents of the packs, the manufacturer is keen that γ should not exceed 7500. Every day the contents of 1000 packs, randomly chosen from that day's production, are measured and the number of packs that do not have enough content is recorded. If necessary, the machines are repaired or adjusted at the start of the production on the next day.

In this example, all three types of statistical problem occur. Not only is an approximation of γ wanted, but an interval that might contain γ is also interesting. Furthermore, the manufacturer wants to know which of the two statements '$\gamma > 7500$' (the filling process is out of control) and '$\gamma \leq 7500$' (the filling process is under control) is true.

On a certain day, 100 of the 1000 sampled packs do not have enough content, which is 10% of the sampled packs. This leads to the statistical conclusion that about 10% of the daily production does not have enough content, so γ is about 10000. However, this conclusion is rather groundless, at least at the moment. The conclusion that γ lies in the interval (7560, 12440) is more cautious, but still its level of correctness is not clear. Note that adopting this conclusion automatically leads to a statistical conclusion regarding the third problem: $\gamma > 7500$.

On the other hand, it **is** 100% true that γ lies in the interval [100, 99100]. But this conclusion does not help much.

Statistical procedures for problems 1–3 will usually be based on (suitable) estimators for θ. If interest is in some population mean $\theta = \mu$, then the procedures will often be based on the sample mean $\bar{X}$; if some population proportion p is the parameter of interest, then the sample proportion $\hat{P}$ will be an important constituent of the statistical procedures.

For statistical problem 1, the estimators themselves are the statistical procedures. The formula of the estimator tells us how to obtain the estimate when the data are available. The realization that comes out immediately yields the statistical conclusion: it states that θ is (about) equal to that realization.

For statistical problem 2, the idea is to construct **two** suitable sample statistics (so, depending only on the random sample observations and not on unknown parameters): a random lower bound L and a random upper bound U with $L \leq U$. The actual statistical procedure is the random interval (L, U). The quality of this procedure will be determined by its ability to capture the unknown θ; it does a good job if the probability $P(L < \theta < U)$ is large. When the data are available, they can be substituted into the formulae L and U and lead to a concrete interval estimate (l, u); the statistical conclusion then is that $l < \theta < u$ (so, θ lies between l and u). See Section 15.2.

Statistical procedures for problems of type 3 are called (**hypothesis**) **tests** or **test procedures**. In essence, they consist of two steps:

a The formulation of a sample statistic (called **test statistic**) that is based on an estimator of θ.

b Some partition of the set of all possible outcomes of the test statistic into two subsets (regions) I and II, with the addition that the first statement of the statistical problem will be accepted if the realization of the test statistic falls in I and the second statement if the realization falls in II.

When the data are given, the actual realization of the test statistic can be calculated and it can be determined in which of the two subsets it falls; step (b) of the procedure then automatically yields the statistical conclusion; see Section 15.3.

The following examples are meant to stimulate your intuition to construct statistical procedures for statistical problems 1–3. In forthcoming chapters these procedures will be studied more thoroughly.

Example 15.2

In Example 15.1, the parameter of interest was γ (the number of packs in the daily production that do not have enough content), but we could have taken the corresponding population proportion p as well; notice that $p = \gamma / 100000$. The statistical problems can be formulated as follows:

1 Find an estimate of p.

2 Find an interval estimate of p.

3 Find out whether $p \leq 0.075$ or $p > 0.075$.

Suitable statistical procedures for these problems will be based on the estimator $\hat{P}$, the random variable that measures the proportion of the packs in the sample that do not have enough content.

For problem 1, the estimator $\hat{P}$ itself is the statistical procedure. To obtain, for problem 2, random lower and upper bounds L and U which capture the unknown p, we could try to find a non-negative H such that the bounds $L = \hat{P} - H$ and $U = \hat{P} + H$ capture p with probability 0.99. This idea is inspired by the fact that the random bounds

$$\hat{P} - 2.5758 \sqrt{p(1-p)/1000} \quad \text{and} \quad \hat{P} + 2.5758 \sqrt{p(1-p)/1000}$$

include p with probability 0.99, as follows from (14.18) since $z_{0.005} = 2.5758$ (*). But these bounds are no **sample** statistics; they cannot be observed since they contain the unknown p. That is why, in these expressions, p is replaced by its estimator $\hat{P}$, so H is taken as $2.5758 \sqrt{\hat{P}(1-\hat{P})/1000}$. As a consequence, the random bounds L and U become

$$L = \hat{P} - 2.5758 \sqrt{\hat{P}(1-\hat{P})/1000} \quad \text{and} \quad U = \hat{P} + 2.5758 \sqrt{\hat{P}(1-\hat{P})/1000}$$

and the statistical procedure for problem 2 is just the random interval (L, U).

When the data are observed, it follows that 100 of the 1000 sampled packs do not have enough content. Having substituted these data into the formulae of the statistical procedures, the statistical conclusions for problems 1 and 2 are respectively:

'p is (about) 0.10'   and   'p lies between 0.0756 and 0.1244'

Notice that these conclusions are equivalent to:

'γ is (about) 10 000'   and   'γ lies between 7560 and 12 440'

which are the conclusions already mentioned in Example 15.1. The level of confidence concerning the conclusion for problem 2 is rather large: the conclusion comes from a statistical procedure that captures p with probability 0.99.

For problem 3, a statistical conclusion actually follows from the conclusion for problem 2. Since the location of 0.0750 is at the left-hand side of the interval (0.0756, 0.1244), the conclusion that p lies between 0.0756 and 0.1244 immediately yields the statistical conclusion '$p > 0.075$'. However, it is not yet clear how this conclusion follows from steps (a) and (b) above.

A statistical procedure based on steps (a) and (b) will be introduced now, in an intuitive way. The two hypotheses of problem 3 are $p \leq 0.075$ and $p > 0.075$. It is intuitively clear that a statistical procedure based on $\hat{P}$ should choose $p > 0.075$ if the realization of $\hat{P}$ is suitably large, but it is not yet obvious **how** large. It is also intuitively obvious that the probability that the procedure will draw the wrong conclusion in the sense that '$p > 0.075$' is concluded when in fact the true p is at most 0.075 (so the opposite statement '$p \leq 0.075$' is true), will be largest if the true p is precisely equal to 0.075. In other words, for concluding the hypothesis '$p > 0.075$' while it is not valid, the **worst case scenario** is $p = 0.075$.

The starting point for finding a suitable test statistic is the standardized version of the estimator $\hat{P}$:

$$\frac{\hat{P} - p}{\sqrt{p(1-p)/1000}}$$

Unfortunately, this random variable depends on unknown parameters and hence cannot be taken as a test statistic. It is common practice to adapt it in accordance with the worst case scenario, so p is replaced by 0.075 and the test statistic (called Z) becomes:

$$Z = \frac{\hat{P} - 0.075}{\sqrt{0.075 \times 0.925/1000}} = \frac{\hat{P} - 0.075}{0.0083}$$

This gives step (a) of the statistical procedure. Next, step (b). According to (14.9), this test statistic is standard normally distributed if the true p equals 0.075. In particular, since $z_{0.005} = 2.5758$:

if $p = 0.075$, then $P(Z \geq 2.5758) = 0.005$ **(15.1)**

The conclusion '$p > 0.075$' has to be drawn for large realizations of $\hat{P}$ and hence for large realizations of Z. But how large? The answer is: larger than or equal to 2.5758, as will be explained now.

Consider the rule that concludes '$p > 0.075$' if $Z \geq 2.5758$ and '$p \leq 0.075$' if $Z < 2.5758$. For this rule, concluding '$p > 0.075$' is equivalent to the occurrence of the event $\{Z \geq 2.5758\}$. So, according to (15.1) this rule has probability at most 0.005 of concluding '$p > 0.075$' when the truth is that $p \leq 0.075$. Furthermore, the rule gives a partition $(-\infty, 2.5758)$ and $[2.5758, \infty)$ of the set of all real numbers, so it yields step (b).

The complete statistical procedure that follows from these intuitive considerations is therefore:

a Test statistic: $Z = \dfrac{\hat{P} - 0.075}{0.0083}$.

b Choose '$p > 0.075$' if the realization of Z falls in $[2.5758, \infty)$.

   Choose '$p \leq 0.075$' if the realization of Z falls in $(-\infty, 2.5758)$.

With this statistical procedure, the statistical conclusion for problem 3 follows immediately. The realization 0.10 of $\hat{P}$ yields the realization 3.0120 of Z. Since 3.0120 falls in $[2.5758, \infty)$, the statistical conclusion is that p is larger than 0.075.

Example 15.3

Interest is in the mean hourly wage μ (in euro) of the employees in the commercial services sector. In Example 13.10 it was noted that this mean was €16.92 in 2003; the conjecture is that nowadays the mean is larger. To form a conclusion about this conjecture, a sample of 500 employees working in this sector will be drawn randomly and their hourly wages measured. As in Example 13.10, it will be assumed that σ is known; it is equal to €8.20.

For estimation of μ, the estimator $\bar{X}$ will be used. To find an interval estimate for μ, our starting point is that:

$$P(\bar{X} - z_{\alpha/2}\tfrac{\sigma}{\sqrt{n}} < \mu < \bar{X} + z_{\alpha/2}\tfrac{\sigma}{\sqrt{n}}) \approx 1 - \alpha$$

(see (13.11)). With $\alpha = 0.10$ and $z_{0.05} = 1.6449$ (*), it follows that:

$$P(\bar{X} - 1.6449\tfrac{\sigma}{\sqrt{n}} < \mu < \bar{X} + 1.6449\tfrac{\sigma}{\sqrt{n}}) \approx 0.90$$

Since σ is known, the lower and upper bounds are obvious:

$$L = \bar{X} - 1.6449 \times \frac{8.20}{\sqrt{500}} = \bar{X} - 0.6032 \quad \text{and} \quad U = \bar{X} + 0.6032$$

L and U capture the unknown μ with probability 0.90.

To choose between the statements '$\mu \leq 16.92$' and '$\mu > 16.92$', we need a test statistic (step (a)) and a decision rule connected to a partition of the set of real numbers (step (b)). The test statistic will be based on the standardized version of the estimator $\bar{X}$:

$$\frac{\bar{X} - \mu}{\sigma/\sqrt{500}}, \quad \text{which equals} \quad \frac{\bar{X} - \mu}{8.20/\sqrt{500}} \quad \text{since } \sigma \text{ is known to be 8.20}$$

If the true mean is 16.92 – the worst case scenario – then this standardized version is standard normally distributed. After adapting the standardized version for the worst case scenario, the test statistic becomes:

$$Z = \frac{\bar{X} - 16.92}{8.20/\sqrt{500}} = \frac{\bar{X} - 16.92}{0.3667}$$

If $\mu = 16.92$, then Z is (approximately) standard normally distributed and hence $P(Z \geq 1.6449) = 0.05$. The conclusion '$\mu > 16.92$' will be drawn for relatively large realizations of the estimator $\overline{X}$ and hence for relatively large realizations of Z. But notice that the rule that chooses '$\mu > 16.92$' if Z is at least 1.6449 and '$\mu \leq 16.92$' if Z is smaller than 1.6449, has probability at most 0.05 of drawing the conclusion '$\mu > 16.92$' unjustly. This rule yields a partition $(-\infty, 1.6449)$ and $[1.6449, \infty)$ of the set of all real numbers, so step (b) follows.

Here is the complete statistical procedure:

a Test statistic: $Z = \dfrac{\overline{X} - 16.92}{0.3667}$.

b Choose '$\mu \leq 16.92$' if the realization of Z falls in $(-\infty, 1.6449)$.

Choose '$\mu > 16.92$' if the realization of Z falls in $[1.6449, \infty)$.

Now suppose that the 500 hourly wages have the average value €17.99. Then the realization of $\overline{X}$ is 17.99, so it is concluded that the mean hourly wage μ is about 17.99. More precisely, it is concluded that μ lies in the interval $(l, u) = (17.3868, 18.5932)$, a conclusion that has 90% confidence since the underlying random interval captures μ with probability 0.90. The realization of the test statistic is 2.9179, which lies in $[1.6449, \infty)$. Hence, it is concluded that nowadays μ is larger than 16.92.

15.2 Point and interval estimation

Some basic theory is developed concerning statistical procedures for problems 1 and 2 of the previous section, for the general parameter θ and more specifically for a population mean μ when σ^2 is known. As always, the procedures are based on a random sample from the population.

For problem 1, things are clear. The possible statistical procedures are just the estimators of θ. In principle, many sample statistics can be used as estimator of θ. For instance, when interest is in a population mean μ and a random sample $X_1, \dots, X_{41}$ is given, many estimators are possible: the median of $X_1, \dots, X_{41}$ (the 21st of the ordered observations), the sample mean $\overline{X}$ of the whole sample, the sample mean of only the first 30 observations $X_1, \dots, X_{30}$, etc. However, it is common practice to take **consistent** estimators. In the case of μ or p, the sample mean $\overline{X}$ and the sample proportion $\hat{P}$ are usually taken as estimators. See Chapters 13 and 14 for their properties. In the context of problem 1, the statistical procedures, that is, the estimators, are sometimes called **point estimators**, to distinguish them from interval estimators (see below).

In the rest of the present section, statistical problems of type 2 will be considered. As was mentioned in the previous section, statistical procedures for such problems are random intervals (L, U) where the sample statistics L and U are such that $L \leq U$; they are called **interval estimators** for the parameter θ. They are based on the estimator (say, E) of θ. When the data are given, they can be substituted into the formulae L and U to give the realized interval (l, u), a so-called **interval estimate**. This interval estimate is called a $(1 - \alpha)$ **confidence interval** (or a $100 \times (1 - \alpha)\%$ confidence interval) if the probability that the underlying interval estimator (L, U) captures θ is **at least** $1 - \alpha$.

In this book, the reader will come across two formats of interval estimators:

Two formats of interval estimators

Format 1   Random intervals (L, U) with $L = E - H$ and $U = E + H$ where E is the estimator and $H \geq 0$ is some other sample statistic.

Format 2   Random intervals (L, U) with $L = aE$ and $U = bE$ where a and b are constants such that $a \leq 1$ and $b \geq 1$.

Notice that a format 1 interval estimator is centred round the point estimator. The rv H is called the **half-width** of the interval estimator. It measures the precision of the interval estimator (L, U) when estimating θ.

Interval estimators of format 2 are in general not centred around the point estimator although $L \leq E \leq U$. In the coming chapters, standard interval estimators of format 1 will be used for population means and population proportions. For population variances σ^2 and population standard deviations σ, interval estimators of format 2 will be considered.

Example 15.4

The interval estimator in Example 15.2 (for the proportion p of the packs of milk that do not have enough content) is the random interval (L, U) with $L = \hat{P} - H$ and $U = \hat{P} + H$, where the half-width H equals $2.5758\sqrt{\hat{P}(1 - \hat{P})/1000}$. Notice that H is a sample statistic, that the estimator $\hat{P}$ is indeed the centre of the random interval, and that this interval captures p with probability (about) 0.99. The interpretation is that 99% of all times that the interval estimator is applied as a method to generate intervals, the resulting interval indeed contains the unknown p.

Since 100 of the 1000 randomly sampled packs of milk do not have enough content, the estimate of p is $\hat{p} = 0.10$. Hence, the realization of (L, U) is $(l, u) = (0.0756, 0.1244)$, which is a 99% confidence interval (also called, 0.99 confidence interval) for p. The interpretation is that there is 99% **confidence** that p falls in this concrete interval. For this interpretation, do not use the word 'probability' instead of 'confidence', since whether the unknown number p does or does not fall in the interval $(0.0756, 0.1244)$ has probability 0 or 1; we do not know. Finally, notice that $(0.0756, 0.1244)$ is also a 96% confidence interval (or an 80% confidence interval) since it is a realization of an interval estimator that captures p also with probability **at least** 0.96 (or 0.80).

In Example 15.3, the population standard deviation σ was known and the interval estimator (L, U) for the mean hourly wage μ in the commercial services sector satisfied $L = \bar{X} - 0.6032$ and $U = \bar{X} + 0.6032$. Hence, $H = 0.6032$ is a degenerated rv. Also, this random interval is of format 1 and the estimator $\bar{X}$ indeed falls in the centre. Recall that this interval captures μ with probability 0.90. That is, 90% of the applications of (L, U) yield an interval that indeed contains the unknown μ. The data give $\bar{x} = 17.99$ as the estimate of μ. Substitution in L and U leads to the realized interval $(17.3868, 18.5932)$, which is a 90% confidence interval since it comes from an interval estimator that generates intervals that catch μ in 90% of the cases. There is 90% confidence that the interval $(17.3868, 18.5932)$ contains μ. The interval can also be considered as, say, an 85% confidence interval, since the underlying interval estimator captures μ also with probability at least 0.85.

For a worked-out example of an interval estimator of format 2, you will have to wait until Chapter 17, when the parameter of interest is σ^2. However, to give some idea, think of the random interval $(0.8 \times S^2, 1.1 \times S^2)$ that has $a = 0.8$ and $b = 1.1$. Notice that this random interval captures the population variance σ^2 with probability:

$$P(0.8 \times S^2 < \sigma^2 < 1.1 \times S^2) = P\left(\frac{1}{1.1} < \frac{S^2}{\sigma^2} < \frac{1}{0.8}\right) = P\left(0.9091 < \frac{S^2}{\sigma^2} < 1.25\right)$$

In Chapter 17, probabilities of the right-hand type will be studied in more detail. For the time being, only interval estimators of format 1 will be considered.

For most interval estimators of format 1 considered in this book, the half-width H is proportional to the standard deviation of the estimator (if this SD does not contain unknown parameters) or to the standard error of the estimator (if the SD of the estimator does contain unknown parameters). That is,

$$H = a \times SD(\text{estimator}) \quad \text{or} \quad H = b \times SE(\text{estimator}) \tag{15.2}$$

where a and b are positive constants. In Example 15.3, the standard deviation of the estimator $\bar{X}$ is equal to $\sigma/\sqrt{500} = 8.20/\sqrt{500} = 0.3667$, so it is known; the constant a equals $z_{0.05} = 1.6449$.

In Example 15.4, the standard deviation of the estimator $\hat{P}$ is equal to $SD(\hat{P}) = \sqrt{p(1-p)/1000}$, which cannot be observed since p is unknown. That is why p in $SD(\hat{P})$ is replaced by $\hat{P}$, yielding the standard error $SE(\hat{P}) = \sqrt{\hat{P}(1-\hat{P})/1000}$. Note that in this example the constant b is $z_{0.005}$ = 2.5758.

15.2.1 Standard interval estimator for μ when σ is known

In this subsection, interest is in the unknown population mean μ of a variable X. The purpose is to derive a general formula for the interval estimator of μ, for the case that the population variance σ^2 is known. Although this case is less relevant for practical situations (where σ^2 is generally **un**known), this interval estimator is still important because it is basic for finding interval estimators for more general cases.

Suppose that σ^2 is known and that a random sample $X_1, \ldots, X_n$ of random observations of X is given. As in Section 13.4, at least one of the following two situations is assumed to be valid:

Situation 1   The normal distribution $N(\mu, \sigma^2)$ is a good model for X.
Situation 2   The sample size n is large.

Under these circumstances, we have:

$$P(\overline{X} - z_{\alpha/2}\frac{\sigma}{\sqrt{n}} < \mu < \overline{X} + z_{\alpha/2}\frac{\sigma}{\sqrt{n}}) \approx 1 - \alpha \qquad (15.3)$$

(see (13.11)). Recall that the $\approx$ sign can be replaced by the $=$ sign if situation 1 holds; if situation 2 holds, the $\approx$ sign means 'is approximately equal to' (but is often written as $=$). Since σ is known, the lower and upper bounds L and U of the interval estimator follow.

Interval estimator (L, U) for μ when σ is known

$$L = \overline{X} - z_{\alpha/2}\frac{\sigma}{\sqrt{n}} \quad \text{and} \quad U = \overline{X} + z_{\alpha/2}\frac{\sigma}{\sqrt{n}} \qquad (15.4)$$

The parameter μ is captured by (L, U) with probability $1 - \alpha$. Hence, when the data are plugged into (15.4), the resulting interval is a $(1 - \alpha)$ confidence interval.

Notice that the half-width $H = z_{\alpha/2}\sigma/\sqrt{n}$ of the interval estimator is a constant (so, non-random) that gets smaller as n becomes larger. If n becomes larger but α is kept fixed, the width of the realized $(1 - \alpha)$ confidence interval becomes smaller and hence the approximation of μ by $\overline{x}$ becomes more precise. On the other hand, keeping n fixed and taking a smaller number for α means that $z_{\alpha/2}$ gets larger, the half-width increases and hence the width of the $(1 - \alpha)$ confidence interval increases.

Example 15.5

A machine produces small axles. Although the diameter of the axles can be adjusted, there is always some variation in the diameters of the final product: the standard deviation is 0.2 mm. From the daily production of axles with a fixed adjusted value, fifteen are drawn at random. Here are the diameters (in mm):

24.31   24.18   23.64   23.95   24.11   24.03   23.86   24.01

23.98   24.26   24.10   23.71   23.98   23.86   24.32

Determine a 95% confidence interval for the mean diameter μ (in mm) of the whole daily production of axles.

Since the precise diameter of an axle is determined by many small technical influences, the **normal** model with parameters μ and $\sigma^2 = 0.04$ is used, so situation (1) holds. Notice that this normality assumption is really needed, since the small sample size does not admit the application of the central limit theorem. It can easily be calculated that the (realized) sample mean and sample variance are equal to:

$$\bar{x} = 24.02 \quad \text{and} \quad s^2 = 0.0408$$

The sample variance is indeed close to the known population variance 0.04, in agreement with the adopted model $N(\mu, 0.04)$. The lower and upper bounds of the 95% confidence interval follow easily; note that $z_{0.025} = 1.96$ (*):

$$\bar{x} \pm z_{0.025} \frac{\sigma}{\sqrt{n}} = 24.02 \pm 1.96 \times \frac{0.2}{\sqrt{15}} = 24.02 \pm 0.1012$$

Hence, $l = 23.9188$ and $u = 24.1212$. The conclusion is that μ lies between these bounds, a statement that is valid with 95% confidence. The resulting interval has half-width 0.1012, so the 95% confidence interval estimates μ with a precision of ± 0.1012. Whether μ indeed falls between 23.9188 and 24.1212 is not known, but the statistical procedure (15.4) with $\alpha = 0.05$ gives high hopes since 95% of the intervals generated by it do contain μ.

Notice that the standard deviation $SD(\bar{X}) = \sigma / \sqrt{n}$ is used to calculate the bounds and not the standard error $SE(\bar{X}) = S / \sqrt{n}$. That is, σ is not replaced by s. (Never estimate something that is already known.)

15.2.2 Sample size needed for a confidence interval of prescribed width

Notice that the (format 1) interval estimator in (15.4) is of the form described in the left-hand part of (15.2). In particular, the half-width H is non-random:

$$H = z_{\alpha/2} \frac{\sigma}{\sqrt{n}} \tag{15.5}$$

This is an important observation, since it makes it possible to determine – before the actual drawing of the random sample is done – the sample size n that is needed to obtain a $(1 - \alpha)$ confidence interval of a prescribed width, as in the following example.

Example 15.6
Recall Example 15.5. Notice that the 95% confidence interval (23.9188, 24.1212) for the mean diameter μ of axles has width 0.2024. The management of the company finds the interval too wide and wants a 95% confidence interval that estimates μ with a precision of ± 0.03. Of course, a larger sample will be needed then. How large should this sample be?

The management wants a 95% confidence interval with half-width 0.03. So, n has to satisfy:

$$1.96 \times \frac{\sigma}{\sqrt{n}} = 0.03$$

Since $\sigma = 0.2$, it follows easily that $\sqrt{n} = 13.0667$ and $n = 170.7$. Hence, the sample size should be 171.

Example 15.7
In a recent survey of consumer ethics, 151 Danish business students were asked to rate (among others) the following statement from 1 to 5, where 1 means 'I strongly believe it is wrong' and 5 means 'I strongly believe it is **not** wrong':

Statement (question 9): Not saying anything when the waitress miscalculates the bill in your favour

Although this variable is actually ordinal, we consider it as continuous. The data are in the file Xmp15-07.sav. The original purpose of the research was to estimate the mean μ of the ratings of **all** Danish business students to within 0.2 by way of a 98% confidence interval. However, not all questionnaires were returned. Is the resulting sample size 151 enough to achieve this purpose? If not, how large a sample would have been needed?

Let X denote the rating assigned by a Danish student and suppose that its standard deviation is known to be 1.1. With a computer package it follows easily that:

$$\bar{x} = 3.3377 \quad \text{and} \quad s = 1.1424$$

Note that σ is known, so we will not make use of s. Since the sample size is rather large, situation 2 holds. So, we do not need any further assumptions regarding the underlying population distribution, thanks to the central limit theorem. Since $\alpha = 0.02$, it is $z_{0.01}$ that is needed for the half-width. Since $z_{0.01} = 2.3263$ (∗), it follows that:

$$\bar{x} - z_{0.01} \frac{\sigma}{\sqrt{n}} = 3.3377 - 2.3263 \times \frac{1.1}{\sqrt{151}} = 3.3377 - 0.2082 = 3.1295$$

$$\bar{x} + z_{0.01} \frac{\sigma}{\sqrt{n}} = 3.3377 + 0.2082 = 3.5459$$

There is 98% confidence that the mean rating of all Danish business students falls between 3.1295 and 3.5459. Since the half-width is 0.2082, the purpose of the research to estimate μ to within 0.2, is not achieved.

For a half-width 0.2, the following equation has to be solved:

$$0.2 = z_{0.01} \frac{\sigma}{\sqrt{n}} = 2.3263 \times \frac{1.1}{\sqrt{n}} = \frac{2.5589}{\sqrt{n}}$$

Hence, $\sqrt{n} = 12.7947$ and $n = 164$. If 164 of the distributed questionnaires had been filled in and returned, the 0.2-purpose for the half-width would have been fulfilled.

15.3 Hypothesis testing

Some basic theory is now developed concerning procedures for the statistical problem 3 of Section 15.1, for the general parameter θ and more specifically for a population mean μ when σ is known. The procedures are based on a random sample from the population.

Reconsider the general testing problem (3) concerning the parameter θ:

Choosing between two opposite statements about θ.

In practice, one of the two statements often represents a conjecture while the other is its opposite. The two statements are called H_0 (**null hypothesis**) and H_1 (**alternative hypothesis**), and it is said that H_0 is **tested against** (or **versus**) H_1. Here, the statement that represents the conjecture is taken as H_1. Recall that statistical procedures for this type of problem are called (hypothesis) tests. They describe how to choose between the two hypotheses. When the data become available, a test can be conducted; it leads to one of two possible statistical conclusions:

H_0 is rejected   or H_0 is not rejected

Notice that the conclusion 'H_0 is rejected' is the stronger of the two conclusions since it is equivalent to concluding that the alternative H_1 is true. The conclusion 'H_0 is not rejected' is a bit soft: it does **not** say that H_0 is true, only that the data do not indicate that the opposite (H_1) is true. This explains why the conjecture (the statement that you want to prove) is put in H_1: a conclusion 'the conjecture is true' sounds more convincing than the conclusion 'the data do not contradict the validity of the conjecture'.

As mentioned before, a statistical procedure consists of two steps:

a The formulation of a test statistic G (say).

b A partition of the set of all possible outcomes of the test statistic into two regions R and R^c, jointly with the rules:

- reject H_0 if the realization of G falls in R;
- don't reject H_0 if the realization of G falls in R^c.

The region R is called **rejection region** or **critical region**. The following example will give you a feel for good and bad test procedures.

Example 15.8

A margarine factory produces packs of margarine that, according to the label, have a net weight of 250 g. However, there is some variation during the production process, so not all packs weigh precisely the same. This variation is measured by the standard deviation σ, which is 4 g. To control the production process and the filling machines, a random sample of 25 packs of margarine is drawn from the daily production to test the concern that the mean net weight μ of all produced packs is less than 250 g. So, the parameter of interest is μ, while σ is known. It seems reasonable to adopt a normal model for the net weight X of a pack of margarine. Below, $\overline{X}$ denotes the mean net weight (in grams) of the packs in the sample.

To formulate the statistical problem, the concern that μ is smaller than 250 g is put in the alternative hypothesis and its opposite becomes the null hypothesis. That is, the testing problem of interest is:

i Test H_0: $\mu \geq 250$ against H_1: $\mu < 250$

The test procedure will be based on the standardized version of the estimator $\overline{X}$:

$$\frac{\overline{X} - \mu}{4/\sqrt{25}} = \frac{\overline{X} - \mu}{0.8}$$

This random variable is $N(0, 1)$ distributed, but cannot be used as a test statistic since it depends on the unknown μ. Replacing μ by the worst case scenario 250 yields:

ii Test statistic: $G = \dfrac{\overline{X} - 250}{0.8}$

Note that G is $N(0, 1)$ distributed **if $\mu = 250$**.

A first idea might be to formulate the test procedure such that it rejects H_0 precisely if the realized sample mean is smaller than 250, so if the realized value g of the test statistic is smaller than 0:

iii Reject $H_0 \Leftrightarrow \overline{x} < 250 \Leftrightarrow g$ falls in $(-\infty, 0)$

Rejecting H_0 is equivalent to $g < 0$, and the occurrence of the event $\{G < 0\}$ is equivalent to the occurrence of the statistical conclusion 'H_0 is rejected'. In particular:

$P(H_0 \text{ is rejected}) = P(G < 0)$

But note that this probability is 0.5 **if $\mu = 250$**. That is, if $\mu = 250$ and hence H_0 is true, the test procedure will reject H_0 (incorrectly) with probability 0.5. Of course, using a test procedure that has a probability 0.5 of rejecting H_0 incorrectly has to be discouraged.

A safety margin must be taken into account. To get an idea, consider the test procedure that rejects H_0 if the realized sample mean is at most 249, so if the realization g of the test statistics is at most -1.25:

iii Reject $H_0 \Leftrightarrow \bar{x} \leq 249 \Leftrightarrow g$ falls in $(-\infty, -1.25)$

Notice that the rejection region R of this test is $(-\infty, -1.25]$, so $R^c = (-1.25, \infty)$. If H_0 is valid since the true mean μ is 250, then:

$$P(H_0 \text{ is rejected}) = P(G \leq -1.25) = 0.1056 \tag{*}$$

A test procedure that has probability 0.1056 of rejecting H_0 incorrectly is more acceptable.

Let us go back to the general testing problem (3). Similar to the above example, the testing problem (one step) and the test procedure (two steps) can be formulated as follows:

i Test H_0 against H_1

ii Test statistic: G

iii Reject $H_0 \Leftrightarrow g$ falls in R

where g denotes a realization of the test statistic. Notice that these formal steps (i)–(iii) are not yet dealing with observed data; step (i) describes the testing problem in an accurate way and steps (ii)–(iii) state the test procedure that will be applied as soon as the data become available. To conduct the test, two further steps are needed:

iv Calculate the realization of G

v Draw the conclusion by comparing this realization with step (iii)

The realization of G calculated in step (iv) is denoted by **val**, the **value of the test statistic**.

There are two possibilities concerning the actual (true) value of θ: H_0 is valid or H_1 is valid. There are also two possibilities regarding the conclusion that follows from the test: H_0 is not rejected or H_0 is rejected. Of course, it is hoped that the test procedure will yield a correct conclusion:

Correct conclusions

- H_0 is rejected while the true θ is such that H_1 is true.
- H_0 is not rejected while the true θ is such that H_0 is true.

Incorrect conclusions

- H_0 is rejected while the true θ is such that H_0 is true (**type I error**)
- H_0 is not rejected while the true θ is such that H_1 is true (**type II error**)

These four combinations can also be expressed in a 2×2 table.

	Statistical conclusion	
Actual situation	**Do not reject H_0**	**Reject H_0**
H_0 is true	Correct conclusion	Incorrect; type I
H_1 is true	Incorrect; type II	Correct conclusion

TABLE 15.1 Possible situations for hypothesis tests

The quality of a test procedure is determined by its probabilities of giving incorrect conclusions. The following example is long but very illustrative.

Example 15.9

Reconsider Example 15.8 and the steps (i)–(iii) of the second test procedure:

 i Test H_0: $\mu \geq 250$ against H_1: $\mu < 250$

 ii Test statistic: $G = \dfrac{\overline{X} - 250}{0.8}$

 iii Reject $H_0 \Leftrightarrow g$ falls in $(-\infty, -1.25]$

Below, probabilities of type I and type II errors are calculated. Furthermore, the test is conducted for the realization 248.8 of the sample mean $\overline{X}$.

In Example 15.8 we calculated (without knowing) the probability of a type I error if the worst case scenario (i.e., $\mu = 250$) holds:

If $\mu = 250$, then $P(H_0$ is rejected$) = P(G \leq -1.25) = 0.1056$

But not only $\mu = 250$ can lead to a type I error: for each $\mu \geq 250$ the probability of the accompanying type I error can be calculated. For instance, if $\mu = 250.2$, then $E(\overline{X}) = 250.2$ and $\overline{X} \sim N(250.2, 0.64)$. Hence:

$$
\begin{aligned}
\text{If } \mu = 250.2, \text{ then } P(H_0 \text{ is rejected}) \ &= P(G \leq -1.25) = P(\overline{X} \leq 249) \\
&= P(\tfrac{\overline{X} - 250.2}{0.8} \leq \tfrac{249 - 250.2}{0.8}) = 0.0668 \qquad (*)
\end{aligned}
$$

This is another probability of a type I error. Notice that the last probability is smaller than the worst case probability. In fact, if H_0 is true and the true μ lies further from 250, then the accompanying type I error probability becomes smaller; the worst case scenario has the largest accompanying type I error probability:

$$
\begin{aligned}
\text{If } \mu \leq 250, \text{ then } P(H_0 \text{ is rejected}) \ &= P(G \leq -1.25) = P(\overline{X} \leq 249) \\
&= P(\tfrac{\overline{X} - \mu}{0.8} \leq \tfrac{249 - \mu}{0.8}) \leq 0.1056
\end{aligned}
$$

It is important to record the above results:

- The test statistic is $N(0, 1)$ distributed if $\mu = 250$.
- The largest probability of a type I error comes from the worst case scenario (at $\mu = 250$).
- This maximal type I probability is just the area under the $N(0, 1)$ density on the left-hand side of -1.25.

Type II errors occur when it is incorrect that H_0 is **not** rejected. For instance, if $\mu = 249.5$, the probability of the accompanying type II error can be calculated:

$$
\begin{aligned}
P(H_0 \text{ is not rejected}) \ &= P(G > -1.25) = P(\overline{X} > 249) \\
&= P(\tfrac{\overline{X} - 249.5}{0.8} > \tfrac{249 - 249.5}{0.8}) = 0.7340 \qquad (*)
\end{aligned}
$$

It would appear that this test procedure has relatively small probabilities for type I errors (maximal 0.1056), but relatively large probabilities for type II errors.

From the above it follows that the probabilities

$$P(H_0 \text{ is rejected}) \ = P(\overline{X} \leq 249)$$

depend on μ. For general μ, the sample mean $\overline{X}$ is approximately $N(\mu, 0.8)$ and:

$$
\begin{aligned}
P(H_0 \text{ is rejected}) \ &= P(G \leq -1.25) = P(\overline{X} \leq 249) \\
&= P(\tfrac{\overline{X} - \mu}{0.8} \leq \tfrac{249 - \mu}{0.8}) = P(Z \leq \tfrac{249 - \mu}{0.8})
\end{aligned}
$$

(Here, Z is $N(0, 1)$ distributed.) As a consequence, $P(H_0$ is rejected$)$ is a function of μ; it is called the ***power*** of the test procedure. The graph of the present example is shown in Figure 15.1.

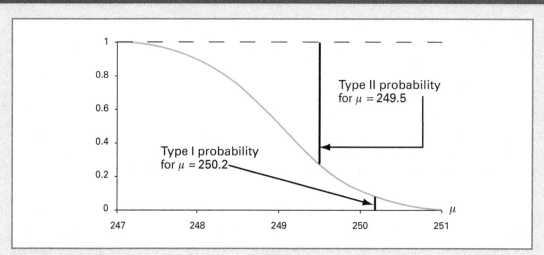

Figure 15.1 Graph of the power of the test procedure

The graph shows not only the probabilities of type I errors (the values of the power for $\mu \geq 250$), but also the probabilities of the type II errors (the complementary probabilities of the power for $\mu < 250$). The above calculated probabilities of the type I error for $\mu = 250.2$ and the type II error for $\mu = 249.5$ are indicated.

When the data are collected, it turns out that $\bar{x} = 248.8$. So, the value of the test statistic is -1.5. The test can now be conducted, which leads to the steps (iv) and (v):

iv **val** $= -1.5$

v Reject H_0 since -1.5 falls in the rejection region $R = (-\infty, -1.25]$

It is concluded that the production process is 'out of control' since the mean net weight of packs of margarine is below 250 g. This conclusion can be stated with at least 89.44% confidence that it is correct, since the probability that the underlying test procedure rejects H_0 wrongly is at most 0.1056.

Unfortunately, it is in general not possible to control both types of error, to obtain test procedures that keep all probabilities of incorrect conclusions of both types below some fixed small threshold. That is why test procedures will be considered that control all probabilities of **type I errors** in the sense that the maximum probability of the type I errors is at most equal to a prescribed fixed number called α, the **significance level** of the test. This significance level is often chosen as 0.10, 0.05, 0.025, 0.01 or 0.005. A test procedure for which the maximum probability of type I errors is at most α, is called an **α-test**.

15.3.1 Standard hypothesis tests for μ when σ is known

In this sub section, interest is in the unknown population mean μ of some variable X. The purpose is to derive standard test procedures for testing problems concerning μ, for the case that the population variance σ^2 is known. Although this case is less relevant for practical situations (where σ^2 is generally **un**known), these test procedures are still important because they are basic for finding test procedures for more general cases.

Suppose that σ^2 is known and that a random sample $X_1, \ldots, X_n$ of random observations of X is given. As in Section 13.4, at least one of the following two situations is assumed to be valid:

Situation 1 The normal distribution $N(\mu, \sigma^2)$ is a good model for X.
Situation 2 The sample size n is large.

Three types of testing problem will be considered:

a Test H_0: $\mu \leq \mu_0$ against H_1: $\mu > \mu_0$.

b Test H_0: $\mu \geq \mu_0$ against H_1: $\mu < \mu_0$.

c Test H_0: $\mu = \mu_0$ against H_1: $\mu \neq \mu_0$.

Here, μ_0 is a fixed and known constant called **hinge** or **test value** (SPSS). (For instance, in Example 15.9 above, the hinge μ_0 was 250.) Problems (a) and (b) are called **one-sided**, since the alternative hypotheses are unbounded in only one direction. Problem (a) is called **upper-tailed**, since H_1 lies to the right (upper) of H_0; problem (b) is called **lower-tailed**, since H_1 lies to the left (lower) of H_0. Testing problem (c) is called **two-sided**, since H_1 is unbounded in two directions.

Test procedures for type (a) problems will be considered first. Suppose that some significance level α is fixed in advance. The purpose is to develop test procedures for type (a) problems in such a way that the maximum probability of type I errors is α. That is, the purpose is to obtain α-tests. But recall that the maximal probability of type I errors is achieved for the worst case scenario, that is, for the case that $\mu = \mu_0$. To find a suitable test statistic, we start with the standardized version of the sample mean:

$$\frac{\bar{X} - \mu}{\sigma/\sqrt{n}} \tag{15.6}$$

Although σ is known, the above random variable is not a test statistic since it depends on the unknown μ. It is common practice to adapt it for the worst case scenario, for the case that μ is equal to the hinge. The resulting test statistic becomes:

$$Z = \frac{\bar{X} - \mu_0}{\sigma/\sqrt{n}} \tag{15.7}$$

It is standard normally distributed if $\mu = \mu_0$. To find a suitable rejection region, notice that hypothesis H_1 of the upper-tailed testing problem (a) describes relatively large possible values for the mean μ, while H_0 describes the relatively small values. It is intuitively clear that a good test procedure must reject H_0 (and conclude that H_1 is true) for relatively large realizations of $\bar{X}$ and hence for relatively large realizations of Z. Let us say that H_0 will be rejected for realizations z larger than a constant k_l (that has to be determined later). The second step of the test procedure is:

Reject $H_0 \Leftrightarrow z \geq k_l$

Now recall that the worst case scenario $\mu = \mu_0$ has the largest probability of a type I error, so this probability has to be (at most) α:

If $\mu = \mu_0$ then $P(H_0$ is rejected$) = P(Z \geq k_l) = \alpha$

However, if $\mu = \mu_0$ then Z in (15.7) is (approximately) standard normally distributed and $P(Z \geq z_\alpha) = \alpha$. Hence, it follows that k_l has to be taken as z_α. The steps (i)–(v) are now clear:

Five-step procedure for testing whether $\mu > \mu_0$ when σ is known:

i Test H_0: $\mu \leq \mu_0$ against H_1: $\mu > \mu_0$

ii Test statistic: $Z = \dfrac{\bar{X} - \mu_0}{\sigma/\sqrt{n}}$

iii Reject $H_0 \Leftrightarrow z \geq z_\alpha$

iv Calculate the realization of Z

v Draw the conclusion of the test in steps (ii) and (iii)

Notice that the rejection region of the test is $R = [z_\alpha, \infty)$. The lower bound z_α is called the **critical value** of the test. It would appear that the lower bound of the upper-tailed testing problem is such that under the $N(0, 1)$ density (the density of the test statistic if $\mu = \mu_0$) there is an area α at its right; see Figure 15.2.

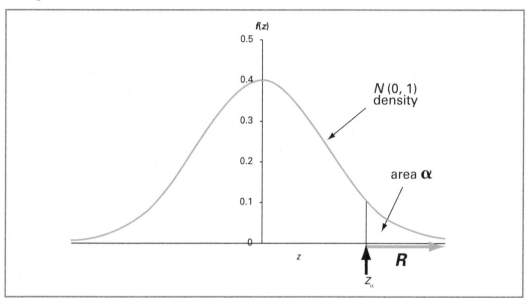

Figure 15.2 Location of the rejection region R for the upper-tailed problem

Example 15.10

It is known from long experience that the time (in minutes) it takes to assemble a caravan in a large caravan factory is well modelled by $N(89, 144)$. However, the conjecture is that on Monday mornings the mean time μ to assemble a caravan is greater than 89 minutes. Reasons might be that the workers talk too much about their weekend exploits, that they are suffering a lack of sleep because of that heavy weekend or that they just have to find a good working cadence. However, there is no reason to doubt the fact that the variance σ^2 on Monday mornings is equal to 144. The model for the time X (in minutes) that it takes to assemble a complete caravan on a Monday morning is $N(\mu, 144)$ and the testing problem of interest is:

  i $H_0: \mu \leq 89$ against $H_1: \mu > 89$

To draw a conclusion, the assembly times of 20 randomly chosen caravans are carefully measured; see the file Xmp15-10.sav. Below, $\alpha = 0.05$ is taken as significance level.

Since $z_\alpha = z_{0.05} = 1.6449$ (*), the standard test procedure is as follows:

  ii Test statistic: $Z = \dfrac{\overline{X} - 89}{12/\sqrt{20}} = \dfrac{\overline{X} - 89}{2.6833}$

  iii Reject $H_0 \Leftrightarrow z \geq 1.6449$

The rejection region is $R = [1.6449, \infty)$, so 1.6449 is the critical value of the test.

From the data it follows that $\bar{x} = 92.2000$ and $s = 10.7488$; if necessary, see Appendix A1.15. Notice that the sample standard deviation is not far from the population standard deviation 12. But s will not be used below. (Never approximate a number that is already known.) The last two steps follow immediately:

  iv $\textbf{\textit{val}} = \dfrac{92.2 - 89}{2.6833} = 1.1926$

  v H_0 is not rejected since 1.1926 does not fall in the rejection region

The conclusion is that the data do not support (at least not at significance level 0.05) the validity of the conjecture that on Monday mornings the mean assembly time is greater than at other times of the week.

To learn more about the validity of this conclusion, it is interesting to determine – afterwards – the probability that the test procedure in the steps (ii)–(iii) does not reject H_0 when it should have done so since the true μ is larger than 89. We will consider the case $\mu = 92$. For that, notice that step (iii) can also be formulated as:

iii Reject $H_0 \Leftrightarrow z \geq 1.6449 \Leftrightarrow \bar{x} \geq 93.4137$

Hence, the events 'H_0 is not rejected' can also be written as $\{\bar{X} < 93.4137\}$. If $\mu = 92$, it follows that:

$$P(H_0 \text{ is not rejected}) = P(\bar{X} < 93.4137) = P(\frac{\bar{X} - 92}{2.6833} < \frac{93.4137 - 92}{2.6833})$$
$$= P(Z < 0.5269) = 0.7009 \qquad (*)$$

This probability of a type II error is rather large.

Next, consider a **lower**-tailed testing problem of type (b):

b Test $H_0: \mu \geq \mu_0$ against $H_1: \mu < \mu_0$

Notice that the hinge is still μ_0, so the worst case scenario for the type I errors is still the case that $\mu = \mu_0$ since this case again has the largest probability for a type I error. As a consequence, the standardized version of $\bar{X}$ in (15.6) is adapted in the same way as for the upper-tailed testing problem (a), so the test statistic Z is unchanged; see (15.7). But the alternative hypothesis of testing problem (b) now describes relatively **small** possible values of the population mean, so the test procedure must reject H_0 for relatively small realizations z of the test statistic. For the time being, H_0 will be rejected for realizations $z \leq k_u$ where the constant k_u still has to be determined. That is:

Reject $H_0 \Leftrightarrow z \leq k_u$

The constant k_u has to be determined from the maximal allowable probability of type I errors, which is α. That is, k_u has to be such that:

$P(Z \leq k_u) = \alpha$ for the case that $\mu = \mu_0$

But in the case that $\mu = \mu_0$, the test statistic Z is standard normally distributed. Hence:

$P(Z \leq -z_\alpha) = \alpha$

It follows that the upper bound k_u has to be taken as $-z_\alpha$. Steps (i)–(v) can now be written down:

Five-step procedure for testing whether $\mu < \mu_0$ when σ is known

i Test $H_0: \mu \geq \mu_0$ against $H_1: \mu < \mu_0$

ii Test statistic: $Z = \dfrac{\bar{X} - \mu_0}{\sigma/\sqrt{n}}$

iii Reject $H_0 \Leftrightarrow z \leq -z_\alpha$

iv Calculate the realization of Z

v Draw the conclusion of the test in steps (ii) and (iii)

As a consequence, the rejection region R of the test procedure in steps (ii) and (iii) is $(-\infty, -z_\alpha]$; the critical value is $-z_\alpha$. This is shown in Figure 15.3.

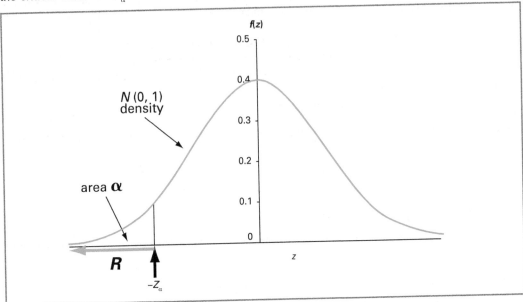

Figure 15.3 Location of the rejection region R for the lower-tailed problem

Example 15.11

The government states that the mean age of the scientific personnel at universities is growing to alarming levels and that she might feel the need to intervene if things remain unchanged during the coming years. The claim is that the mean age μ of tenured personnel is nowadays at least 46.8 years with standard deviation $\sigma = 8.6$ years. The head of an economics department at a certain university believes the claim about the standard deviation, but she has doubts about that concerning μ and wants to test her own conjecture that $\mu < 46.8$. For the test, the 48 tenured scientific personnel of the economics department are considered to be a random sample. The ages are in the file Xmp15-11.sav. Below, the significance level is 0.06.

Of course, you may have some doubts about this sample being random (why?). But let us assume that it can indeed be considered as a random sample.

The three steps below formulate the testing problem (step (i)) and the test procedure (steps (ii) and (iii)); notice that the hinge is 46.8. Since $z_{0.06} = 1.5548$ (*), we have:

   i Test H_0: $\mu \geq 46.8$ against H_1: $\mu < 46.8$; take $\alpha = 0.06$

   ii Test statistic: $Z = \dfrac{\overline{X} - 46.8}{8.6/\sqrt{48}}$

   iii Reject $H_0 \Leftrightarrow z \leq -z_\alpha = -1.5548$

Hence, the rejection region of the test is $(-\infty, -1.5548]$ and the critical value is -1.5548. The data yield $\overline{x} = 46.33$ and $s = 8.7649$, as can easily be computed. Notice that the observed sample standard deviation will not be used for the test since the population standard deviation is known. The realized value of the test statistic can be calculated and the test can be conducted:

   iv **val** $= \dfrac{46.33 - 46.8}{8.6/\sqrt{48}} = -0.3786$

   v Do not reject H_0 since **val** does not fall in the rejection region

The conclusion is that the data do not support the doubt of the head of the department.

Next, consider the two-tailed testing problem (c):

c   Test H_0: $\mu = \mu_0$ against H_1: $\mu \neq \mu_0$

Notice that there is only one possible type I error now: H_0 can be rejected wrongly only for the true mean being μ_0. The test statistic is again the sample statistic Z of (15.7), the same as in the tests for testing problems (a) and (b). To find the rejection region, notice first that H_1 describes a set of possible values for the true mean which is the union of $(-\infty, \mu_0)$ and (μ_0, ∞). That is, H_1 is about relatively small **and** relatively large values. That is why the test procedure will reject H_0 for relatively small **and** relatively large realizations of the test statistic Z. For the time being, the test procedure is formulated as follows:

Reject $H_0 \Leftrightarrow z \leq k_u$ or $z \geq k_l$

The constants k_u and k_l will again be determined from the demand that the probability of the type I error has to be equal to the significance level α. That is:

$P(Z \leq k_u) + P(Z \geq k_l) = \alpha$ for the case that $\mu = \mu_0$

But if $\mu = \mu_0$, then Z is (approximately) standard normally distributed and

$P(Z \leq -z_{\alpha/2}) + P(Z \geq z_{\alpha/2}) = \alpha/2 + \alpha/2 = \alpha$

Using the symmetry of the standard normal distribution, it follows that $k_u = -z_{\alpha/2}$ and $k_l = z_{\alpha/2}$. All steps can now be written down:

Five-step procedure for testing whether $\mu \neq \mu_0$ when σ is known

i   Test H_0: $\mu = \mu_0$ against H_1: $\mu \neq \mu_0$

ii Test statistic: $Z = \dfrac{\bar{X} - \mu_0}{\sigma / \sqrt{n}}$

iii Reject $H_0 \Leftrightarrow z \leq -z_{\alpha/2}$ or $z \geq z_{\alpha/2}$

iv Calculate the realization of Z

v   Draw the conclusion of the test in steps (ii) and (iii)

The test has two critical values: $-z_{\alpha/2}$ and $z_{\alpha/2}$. The rejection region R is the union of $(-\infty, -z_{\alpha/2}]$ and $[z_{\alpha/2}, \infty)$; see Figure 15.4.

Example 15.12

Reconsider the machine in Example 15.5 that produces small axles, where the diameter of the axles is adjustable. However, the diameters of the final product show some variation around the adjusted value; this variation is measured by the standard deviation σ that is known to be 0.2. As a model for the diameter of an axle produced by that machine, the distribution $N(\mu, 0.04)$ is used.

Suppose that the adjusted value for a large batch of axles is 24 mm. However, the manufacturer wants to test whether the mean diameter μ of the batch is unequal to 24; just to be sure. The significance level is 0.05. Fifteen axles are chosen at random from the whole batch and their diameters (in mm) are measured carefully. Suppose now that it turns out that $\bar{x} = 24.14$. Does this result support the hypothesis that μ is unequal to 24? Take $\alpha = 0.05$.

Recall that $z_{0.025} = 1.96$. The five steps of the procedure are:

i   Test H_0: $\mu = 24$ against H_1: $\mu \neq 24$; take $\alpha = 0.05$

ii Test statistic: $Z = \dfrac{\bar{X} - 24}{0.2/\sqrt{15}}$

iii Reject $H_0 \Leftrightarrow z \leq -1.96$ or $z \geq 1.96$

iv **val** $= (24.14 - 24)/0.0516 = 2.7111$

v H_0 is rejected since 2.7111 falls in the right-hand part of the rejection region

The conclusion is that the mean of the diameters of the whole batch of axles is unequal to 24. This conclusion has confidence 0.95 of being valid

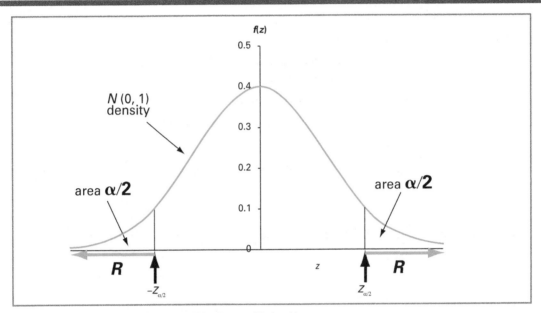

Figure 15.4 Location of the rejection region R for the two-sided problem

15.3.2 Hypothesis tests based on *p*-values when σ is known

Test procedures are often reformulated in terms of a concept called p-value (especially when data-sets and statistical packages are involved). As you will see, the trick is simple but the concept itself is quite complicated. Interest is again in testing problems concerning some population parameter μ and it is still assumed that σ is known. The statistical procedures are based on a random sample.

For a general hypothesis test, the **p-value** or **observed significance level** is the smallest significance level α that allows the conclusion of rejecting H_0.

Example 15.13

Reconsider Example 15.10, where it was tested whether the mean time μ (in minutes) it takes to assemble a caravan on Monday mornings is larger than 89. Below, the five-step procedure of the test is repeated:

i Test $H_0: \mu \leq 89$ against $H_1: \mu > 89$; take $\alpha = 0.05$

ii Test statistic: $Z = \dfrac{\bar{X} - 89}{12/\sqrt{20}}$

iii Reject $H_0 \Leftrightarrow z \geq 1.6449$

iv **val** $= (92.2 - 89)/2.6833 = 1.1926$

v H_0 is not rejected since 1.1926 does not fall in the rejection region

Recall that the rejection region R of the test procedure is $[1.6449, \infty)$ and that the critical value $k_j = 1.6449$ was chosen such that the area to its right under the standard normal density is just the significance level $\alpha = 0.05$; see Figure 15.5. Note that **val** does not belong to the rejection region. To find the p-value, we have to search for the smallest significance level for which H_0 **is** rejected, so the rejection region has to be made larger and the boundary has to be drawn to the left (as indicated in Figure 15.5).

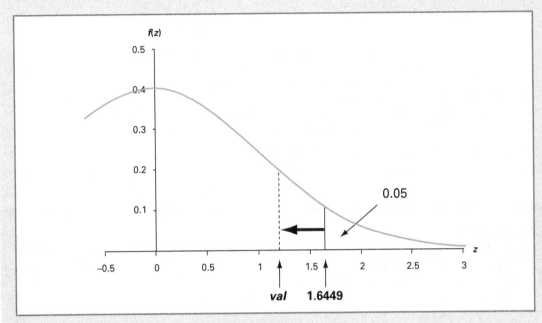

Figure 15.5 Illustration of the p value for the upper-tailed problem

However, making the rejection region larger until it just includes **val** $= 1.1926$ means that the new boundary k_j becomes equal to **val** and that the new α is just the area to the right of **val**, which is:

$$P(Z \geq \textbf{val}) = P(Z \geq 1.1926) = 0.1165 \tag{$*$}$$

So, the p-value is 0.1165; the choice $\alpha = 0.1165$ for the significance level would have meant that H_0 was just rejected.

But we could also have used the p-value to test, with significance level 0.05, whether $\mu > 89$: just by checking whether this p-value is at most $\alpha = 0.05$ (if so, reject H_0) or greater than α (if so, do not reject H_0).

If σ is known and the testing problem is an upper-tailed problem concerning μ, then – as in the above example – the p-value is the total area to the right of **val** under the standard normal density. That is, for testing problem (a):

a p-value $= P(Z \geq \textbf{val})$ for the case that $\mu = \mu_0$

This p-value can be used to conduct a test for a testing problem of type (a). Here are the five steps to follow:

Five-step procedure for testing whether $\mu > \mu_0$ with the p-value

i Test $H_0: \mu \leq \mu_0$ against $H_1: \mu > \mu_0$

ii Test statistic: $Z = \dfrac{\overline{X} - \mu_0}{\sigma/\sqrt{n}}$

iii Calculate the realization of Z

iv Calculate the p-value

v Reject $H_0 \Leftrightarrow p$-value $\leq \alpha$

To find the definition of the p-value of the lower-tailed testing problem (b), recall that the critical value $k_u = -z_\alpha$ of the rejection region $R = (-\infty, k_u]$ is chosen to ensure that the area to its **left** under the normal density is precisely α. The conclusion to reject H_0 or not depends on whether the realization **val** of the test statistic falls in R or not. To make the rejection region such that H_0 is only just rejected, its upper bound k_u has to coincide with **val**. But then the significance level of the test turns into $P(Z \leq \textbf{val})$ with Z standard normal. That is, for testing problem (b):

b p-value $= P(Z \leq \textbf{val})$ for the case that $\mu = \mu_0$

The p-value can also be used to conduct a test for a testing problem of type (b). The steps are the same as for testing problem (a), but the p-value refers to another probability.

Example 15.14

Reconsider Example 15.11. Use the five-step procedure with the p-value to conclude whether the mean age μ of the scientific personnel at universities is smaller than 46.8. Recall that $\sigma = 8.6$ and that a sample of size 48 has mean age 46.33.

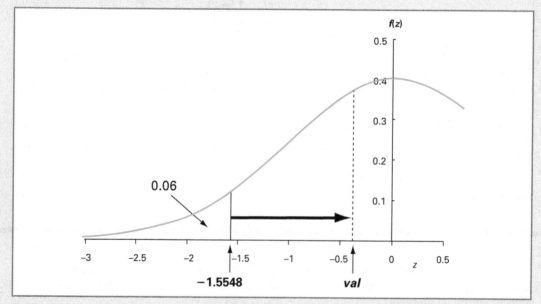

Figure 15.6 Illustration of the p-value for the lower-tailed problem

i Test $H_0: \mu \geq 46.8$ against $H_1: \mu < 46.8$; take $\alpha = 0.06$

ii Test statistic: $Z = \dfrac{\overline{X} - 46.8}{8.6/\sqrt{48}}$

iii **val** $= \dfrac{46.33 - 46.8}{8.6/\sqrt{48}} = -0.3786$

iv p-value $= P(Z \le -0.3786) = 0.3525$ for the case that $\mu = 46.8$ (∗)

v Do not reject H_0 since 0.3525 is larger than the prescribed α

Of course, the conclusion is the same as in Example 15.11

   Figure 15.6 illustrates the meaning of the p-value for this lower-tailed testing problem. It is the area that arises on the left-hand side if the upper bound of the rejection region is drawn to the location of the **val**, as indicated in Figure 15.6. It is the smallest significance level that would allow the conclusion 'reject H_0'.

To formulate the definition of the p-value for a two-sided testing problem (c), it is helpful first to consider an example.

Example 15.15

Reconsider Example 15.12, where it was tested whether the mean diameter of a batch of axles is unequal to 24 mm. Recall that, with significance level $\alpha = 0.05$, the null hypothesis H_0: $\mu = 24$ was rejected in favour of the alternative hypothesis H_1: $\mu \ne 24$ on the basis of the realization 2.7111 of the test statistic. Determine the p-value of that test. That is, determine the smallest significance level that allows the conclusion 'reject H_0'.

   Recall that the rejection region R, the union of the intervals $(-\infty, -1.96]$ and $[1.96, \infty)$, was chosen to ensure that the total area above R and under the standard normal density was $\alpha = 0.05$. To find the smallest rejection region that still rejects H_0, the lower bound k_l of the right-hand part of R has to increase from 1.96 to **val** $= 2.7111$; see Figure 15.7.

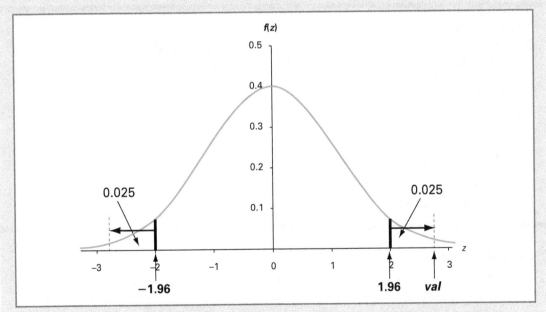

Figure 15.7 Illustration of the p-value for the two-sided problem

Because of the symmetry of the test procedure, the upper bound k_u of the left-hand part of R will accordingly decrease from -1.96 to -2.7111. The significance level that belongs to the new rejection region is just the area above that new (also two-sided) rejection region and under the normal density. That is,

p-value $= P(Z \leq -2.7111) + P(Z \geq 2.7111) = P(|Z| \geq$ **val**$)$

where $|Z|$ is the absolute value of the test statistic Z considered for the case that $\mu = 24$ (so Z is $N(0, 1)$).

The example above gives a good basis for the mathematical definition of the p-value for the two-sided testing problem. The only thing is that you can also think of examples where **val** becomes negative. That is why the p-value for two-sided testing problems (c) is based on the absolute value of the **val**, as follows:

c p-value $= P(|Z| \geq |$**val**$|)$ for the case that $\mu = \mu_0$

The following box summarizes the definitions of the p-values for the three testing problems (a), (b) and (c).

Definitions of p-values when σ is known (here $Z \sim N(0, 1)$)

- For testing whether $\mu > \mu_0$: $P(Z \geq$ **val**$)$
- For testing whether $\mu < \mu_0$: $P(Z \leq$ **val**$)$
- For testing whether $\mu \neq \mu_0$: $P(|Z| \geq |$**val**$|)$

Example 15.16

In the former example, formulate the five-step procedure with the p-value, that can be used to test whether the mean diameter of the whole batch of axles is unequal to 24 mm.
Here are the five steps:

i Test $H_0: \mu = 24$ against $H_1: \mu \neq 24$; take $\alpha = 0.05$

ii Test statistic: $Z = \dfrac{\bar{X} - 24}{0.2/\sqrt{15}}$

iii **val** $= \dfrac{24.14 - 24}{0.0516} = 2.7111$

iv p-value $= P(|Z| \geq |2.7111|) = P(|Z| \geq 2.7111) = 2 \times P(Z \geq 2.7111) = 0.0067$ (∗)

v H_0 is rejected since $0.0067 \leq \alpha$

Concluding remarks

- Notice that the equality sign $=$ is always included in the null hypothesis.
- The null hypotheses of the one-sided tests are respectively $H_0: \mu \leq \mu_0$ and $H_0: \mu \geq \mu_0$. The test procedures developed above are constructed in such a way that the type I error probabilities are controlled for the worst case (the case that the true mean is μ_0). That is why the upper-tailed and the lower-tailed test procedures can also be used for the respective testing problems $H_0: \mu = \mu_0$ against $H_1: \mu > \mu_0$ and $H_0: \mu = \mu_0$ against $H_1: \mu < \mu_0$ with a **single null hypothesis.**

■ The rejection regions of the test procedures above all **include** the critical values. However, sometimes it is preferred to exclude them from the rejection regions, that is, to write $(-\infty, k_u)$ and (k_l, ∞) instead of $(-\infty, k_u]$ and $[k_l, \infty)$. But this is not really different: the considered test statistics are continuous, so the probability of taking one particular value (such as k_u or k_l) is 0.

15.3.3 Using confidence intervals to draw conclusions about conjectures

Confidence intervals can be used to draw conclusions about conjectures concerning unknown parameters. The methods below will be based on intuition only. As a consequence, the level of confidence of the conclusions will not become completely clear.

The trick is simple. If the data yield a 95% confidence interval (l, u) for the unknown parameter θ, the interpretation is as follows: there is 95% confidence that θ lies between l and u. Now suppose that the conjecture is that $\theta < \theta_0$, where θ_0 is a fixed and known number (the hinge). If the confidence interval (l, u) is such that θ_0 lies at the right-hand side of the interval, the conclusion about the conjecture follows easily from intuitive arguments: since there is 95% confidence that θ lies between l and u, there is (at least) 95% confidence that θ is smaller than u and hence (since $u \le \theta_0$) that $\theta < \theta_0$. In this way, both one-sided and two-sided testing problems can be tackled with a **two-sided** confidence interval.

Example 15.17
Reconsider the Examples 15.2 and 15.4 about the population proportion p of the one-litre packs of milk that do not have enough content. Consider two employees of the milk factory, persons A and B. Person A believes that p is larger than 0.07 while person B fears that p is even larger than 0.08. Below, the 99% confidence interval that follows from the random sample of 1000 packs will be used to draw conclusions about both conjectures.

Recall that this confidence interval was (0.0756, 0.1244). Hence, there is 99% confidence that the proportion of packs in a day's production that does not have enough content lies between 0.0756 and 0.1244. In particular, there is (at least) 99% confidence that p is larger than 0.0756 and hence that $p > 0.07$ (since 0.07 lies on the left-hand side of the lower bound of the interval). As a consequence, the conjecture of person A has a confidence of (at least) 99%. With respect to the conjecture of person B, things are a bit more complicated since 0.08 lies **in** the confidence interval. That is why the confidence interval does not support the conclusion that the fear of person B is correct.

Example 15.18
Suppose that the machine that should produce axles with a diameter of 24 mm has some technical problems. Because of this disturbance it is feared that the mean diameter μ of the whole day's production is smaller than 24. The 95% confidence interval (23.9188, 24.1212) that results from the random sample of 15 axles from this day's production will be used to draw conclusions about the conjecture; see also Example 15.5.

There is 95% confidence that μ lies between 23.9188 and 24.1212 while the fear is that μ is smaller than 24. Since 24 lies **in** the interval, the conjecture '$\mu < 24$' is not supported by this confidence interval.

**CASE 15.1 SHOULD INDUSTRIAL ACTIVITIES BE MOVED TO ANOTHER COUNTRY? –
SOLUTION**

Below, it is assumed that the population standard deviation in the region of Rome is also €1300. Two statistical procedures will be considered, to draw conclusions about the hypothesis that $\mu < 21\,000$.

The interval estimator (L, U) with

$$L = \overline{X} - z_{0.025}\,\frac{\sigma}{\sqrt{n}} \quad \text{and} \quad U = \overline{X} + z_{0.025}\,\frac{\sigma}{\sqrt{n}}$$

captures μ with probability 0.95. Since $z_{0.025} = 1.96$, $\sigma = 1300$, $n = 200$ and $\bar{x} = 20700$, it generates the 95% confidence interval $(20\,519.83,\,20\,880.17)$. Hence, μ is probably (at least 95% confidence) smaller than 20 880.17, and, in particular, smaller than 21 000.

The five-step procedure,

  i Test H_0: $\mu \geq 21000$ against H_1: $\mu < 21000$; take $\alpha = 0.05$

 ii Test statistic: $Z = \dfrac{\overline{X} - 21000}{1300 / \sqrt{n}}$

iii Reject $H_0 \Leftrightarrow z \leq -z_{0.05} = -1.6449$

iv $\mathbf{val} = \dfrac{20700 - 21000}{91.9239} = -3.2636$

 v H_0 is rejected since $-3.2636 \leq -1.6449$

yields the same conclusion, that the data give evidence that the mean yearly gross wage in the manufacturing industry in the region of Rome is smaller than €21 000.

Summary

In this chapter some general ideas were presented about **interval estimation** and **hypothesis testing**.

For interval estimation of a parameter, the statistical procedure is called an **interval estimator**, a random interval (L, U) where the sample statistics L and U satisfy $L \leq U$. A frequently used format (format 1) is as follows:

  $L = E - H$ and $U = E + H$ with $H = \text{constant} \times SD(E)$ or $H = \text{constant} \times SE(E)$

where E is a point estimator of the parameter. When the data are substituted into the interval estimator, a confidence interval arises. The confidence level is equal to the probability that the interval estimator will capture the parameter.

For hypothesis testing, the statistical procedure is called a **test procedure** and consists of two parts: the **test statistic** and the **rejection region**. The test statistic is a sample statistic that often (but not always) has the form:

  $\dfrac{\text{estimator} - \text{hinge}}{SE(\text{estimator})}$

The rejection region is chosen such that the maximal probability of type I errors does not exceed a prescribed level α (the significance level).

The table below summarises the standard interval estimators and test procedures if the parameter of interest is a population mean μ and the accompanying population variance σ^2 is known.

Point estimator, interval estimator and test procedures

Parameter	μ (while σ^2 is known)
Point estimator	$\bar{X} = \frac{1}{n}\sum_{i=1}^{n} X_i$
Interval estimator (L, U)	$L = \bar{X} - z_{\alpha/2}\frac{\sigma}{\sqrt{n}}$ and $U = \bar{X} + z_{\alpha/2}\frac{\sigma}{\sqrt{n}}$
Alternative hypothesis H_1	$\mu > \mu_0 \qquad \mu < \mu_0 \qquad \mu \neq \mu_0$
Test statistic	$Z = \frac{\bar{X} - \mu_0}{\sigma/\sqrt{n}}$
Rejection region	$[z_\alpha, \infty) \qquad (-\infty, -z_\alpha] \qquad (-\infty, -z_{\alpha/2}]\cup[z_{\alpha/2}, \infty)$
Requirement	normal random sample $X_1, \ldots, X_n$ and/or large n

Also recall that:

- this test statistic is $N(0, 1)$ distributed if $\mu = \mu_0$;
- the critical values are chosen such that, under the $N(0, 1)$ density, the total area above the rejection region is α;
- the largest probability of a type I error is equal to α.

List of symbols

Symbol	Description
θ	Notation for a general parameter
E	Notation for a general estimator
H	Half-width of an interval estimator of format 1

🔑 Key terms

α-test **448**
alternative hypothesis **444**
confidence interval **440**
critical region **445**
critical value **449**
estimation problem **436**
half-width **441**
hinge **449**
hypothesis **436**
hypothesis test **437**
interval estimate **440**
interval estimation problem **436**
interval estimator **440**
lower-tailed **449**

null hypothesis **444**
observed significance level **454**
one-sided **449**
point estimation problem **436**
point estimator **440**
p-value **454**
power **447**
rejection region **445**
significance level **448**
single null hypothesis **458**
statistical conclusion **436**
statistical problem **436**
statistical procedure **436**

test **437**
test procedure **437**
test statistic **437**
test value **449**
tested against (or versus) **444**
testing problem **436**
two-sided **449**
type I error **446**
type II error **446**
upper-tailed **449**
value of the test statistic **446**

? Exercises

Advice:

- hypothesis testing: write down all five steps;
- interval estimation: write down the interval estimator **and** the realized interval.

Exercise 15.1

Consider a factory where a cheap type of ballpoint pen is produced, 80 000 per day. Although the machines are very modern, defectives cannot be avoided. However, the number of defective pens γ in the daily production has to be at or below the level 400; otherwise, the process is said to be out of control.

To monitor the production, the management orders that each day 300 pens have to be chosen at random and carefully inspected. If the number G of defectives in such a sample is 2 or more, it is concluded that the process is out of control and the machines have to be checked; otherwise nothing has to be done.

a Write down the testing problem (concerning γ) that is of interest.

b Take G as test statistic. Write down the test procedure that is used.

c Calculate the probability of the type I error associated with the worst case scenario.

d Calculate the probability of the type II error associated with the case that the true γ is 500.

e On a certain day, 1% of the sampled pens are defective. Conduct the test to find the conclusion that will be reported to the management.

Exercise 15.2

Since the management of a large company is worried about the proportion of employees on sick leave (p) each day in the present busy period, it is decided that, tomorrow, 200 employees will be chosen at random and the accompanying proportion $\hat{P}$ of absence due to sickness will be measured. To find an interval estimator for p, the random variables $\hat{P} - 1.96 \times SD(\hat{P})$ and $\hat{P} + 1.96 \times SD(\hat{P})$ are taken as starting points.

a Calculate the probability that p will be captured between these random bounds.

b Explain why these random bounds do not constitute an interval estimator. How can they be transformed into an interval estimator?

c When tomorrow has arrived, it turns out that 9% of the 200 employees are absent because of sickness. Use the interval estimator of part (b) to determine an interval estimate for p. What is the (approximate) confidence level of that confidence interval?

d Determine the precision of the estimate 0.09 of p.

Exercise 15.3

Write down the standard five-step procedures in the following cases:

a H_0: $\mu = 1000$ and H_1: $\mu \neq 1000$, while $\sigma = 200$, $n = 100$, $\bar{x} = 980$ and $\alpha = 0.01$.

b H_0: $\mu \leq 50$ and H_1: $\mu > 50$, while $\sigma = 5$, $n = 9$, $\bar{x} = 51$ and $\alpha = 0.03$.

c H_0: $\mu \geq 15$ and H_1: $\mu < 15$, while $\sigma = 2$, $n = 25$, $\bar{x} = 14.3$ and $\alpha = 0.10$.

Exercise 15.4

a For each of the tests in Exercise 15.3, calculate the p-value.

b Check for all parts whether you implicitly used assumptions about the population distribution from which the sample was drawn.

Exercise 15.5

For the testing problem $H_0: \mu = 90$ against $H_1: \mu \neq 90$, it is given that $\sigma = 8$ and $\bar{x} = 89$. Determine the realized value of the test statistic and the accompanying p-value that result from the standard test procedure if:

 a $n = 100$
 b $n = 50$
 c $n = 20$

What is the effect on the value of the test statistic and the p-value if n decreases?

Exercise 15.6

To test $H_0: \mu \leq 700$ against $H_1: \mu > 700$, it is given that $\bar{x} = 710$ and $n = 100$. Determine the realized value of the test statistic and the accompanying p-value that result from the standard test procedure if:

 a $\sigma = 25$
 b $\sigma = 50$
 c $\sigma = 100$

What is the effect on the value of the test statistic and the p-value if σ increases?

Exercise 15.7

To test $H_0: \mu \geq 80$ against $H_1: \mu < 80$, it is given that $n = 25$ and $\sigma = 20$. Determine the realized value of the test statistic and the accompanying p-value that result from the standard test procedure if:

 a $\bar{x} = 68$
 b $\bar{x} = 72$
 c $\bar{x} = 76$

What is the effect on the value of the test statistic and the p-value if the realized value of $\bar{X}$ increases?

Exercise 15.8

The standard interval estimator (for the case that σ is known) is used to find a confidence interval for μ. Explain what the consequences are for the interval if:

 a $\bar{x}$, α and σ remain unchanged but n increases;
 b $\bar{x}$, n and σ remain unchanged but α increases;
 c $\bar{x}$, α and n remain unchanged but σ increases;
 d α, n and σ remain unchanged but $\bar{x}$ increases.

Exercise 15.9

In each of the statements below, indicate whether it concerns a statistical procedure or a statistical conclusion. In the case of a statistical procedure, indicate the type (point estimation, interval estimation or hypothesis testing).

 a About 30% of the consumers prefer soft drink C to soft drink P, since 30 out of 100 interviewed consumers preferred C to P.
 b The production process of a certain article will be stopped if, in a sample of size 200, at least 5% of the products are defective.

Chapter 15 Interval estimation and hypothesis testing: a general introduction

c In part (b), 12 of the 200 products are defective so the production process is stopped.

d In Luxembourg, the mean hourly wage falls between €23.50 and €25.50.

e The concentration of benzene in mineral water is not allowed to be more than 10 ppm (parts per million). If a sample of 50 bottles of mineral water of a certain brand yields a concentration of at least 11 ppm, the manufacturer will be prosecuted.

Exercise 15.10 (advanced)

According to the theory of Villanova, the universe is spherical and centred on the Earth. Unfortunately, the radius θ (in 10^6 light years) is unknown. Furthermore, it is assumed that the distance X (in 10^6 light years) from an arbitrary star to the Earth is $U(0, \theta)$ distributed (see Section 10.1; it would appear that α is 0 while β is now θ).

To estimate θ, a random sample of ten stars is drawn from the population of all known stars and their distances $X_1, \ldots, X_{10}$ are measured. Two sample statistics are considered: the sample mean $\overline{X}$ and the sample statistic $\max\{X_1, \ldots, X_{10}\}$ that measures the maximum of the distances $X_1, \ldots, X_{10}$. With respect to the second sample statistic, it can (but will not) be proved that its expectation is $10\theta/11$.

a Transform these two sample statistics into unbiased estimators G_1 and G_2 of θ.

b It can (but will not) be shown that:

$$P(0.7365G_1 < \theta < 1.5573G_1) \approx 0.95$$

$$P(0.9114G_2 < \theta < 1.3147G_2) \approx 0.95$$

Use these results to obtain two interval estimators for θ that capture this parameter with probability (about) 0.95.

c The realized distances (in 10^6 light years) to the Earth are given in the table. Use the data to obtain two interval estimates for θ. Interpret them. Which interval do you prefer? Why?

6990.5	3020.1	11070.5	4980.8	4029.2
1020.7	6830.3	9918.4	8970.9	7082.7

Exercise 15.11

In a factory where light bulbs are produced, the mean lifespan of a certain type of bulb was 1000 hours with variance 10^6 hours^2. However, a new and faster production technique was introduced recently and there is some fear that this has reduced the mean lifespan. To investigate this concern, the plan is to measure the lifespan of 200 light bulbs produced with the new technique. If the mean lifespan of these bulbs is 980 hours or less it will be decided that the overall mean lifespan μ of the new production technique is smaller than 1000 and the new technique will be revised; but if the mean lifespan of the 200 bulbs is more than 980, nothing will be done. When answering the questions below, assume that the variance for the new technique is unchanged since the old technique.

a Determine the null hypothesis H_0 and the alternative hypothesis H_1.

b Determine the statistical procedure (test statistic and rejection region) that apparently is used.

c Calculate the probability that this statistical procedure will reject H_0 while this is incorrect since the true μ is still 1000 hours. What is the name of the accompanying error?

d Calculate the probability that this statistical procedure will not reject H_0 while this is incorrect since the true μ is 990 hours. What is the name of the accompanying error?

Exercise 15.12

In a study, the mean concentration (μ) of benzene in bottles of mineral water of a certain brand is investigated. The admissible boundary is 10 ppm (parts per million). There are two possible choices for the null hypothesis: $\mu \geq 10$ or $\mu \leq 10$.

a Suppose that the study is commissioned by a rather aggressive consumer organization. What will be the testing problem? Why?

b Suppose that the study is commissioned by the producer. What will be the testing problem? Why?

Exercise 15.13

A consumer organization accuses the manufacturer of a certain brand of margarine of selling short, claiming that the packs of margarine contain on average less than 250 g margarine. A judge has to decide the issue.

The judge decides to base her judgment on a random sample of ten packs of margarine. The consumer organization will be judged to be right if the random sample has a mean weight of at most 249.5 g; otherwise, the judge will acquit the manufacturer.

The starting point of the judge is that $N(\mu, 4)$ is a good model for the weight X of a pack of margarine of that brand.

a Formulate the testing problem and the test procedure that the judge will use.

b Calculate the probability of the type I error associated with $\mu = 250$.

c Calculate the probability of the type II error associated with $\mu = 249$.

Exercise 15.14

In a study about the hourly wage (X, in euro) paid to young people aged 14–18 with a holiday job, the following summarized results were measured for a random sample of 400 young people (see also Exercise 14.6):

$$\sum_{i=1}^{400} x_i = 3296 \quad \text{and} \quad \sum_{i=1}^{400} x_i^2 = 27\,833.35$$

Two years ago, the mean hourly wage paid to young people with a holiday job was €8.05, with standard deviation €1.25. The question arises whether the current hourly wage μ is greater than €8.05. In the questions below, assume that the variation in the hourly wages has not changed.

a Use a 95% confidence interval to answer the question.

b Use a hypothesis test with significance level $\alpha = 0.05$ to answer the question.

c Do you need to know whether X is normally distributed? Why (not)?

Exercise 15.15

You work at an airline company and want to find out the mean time (in minutes) taken by a DC-8 to fly from Amsterdam to Paris. This mean flight-time is important for the planning of arrival and departure times. A random sample of ten flights from Amsterdam to Paris with a DC-8 yields a realized sample mean of 40 minutes. Assume that the population distribution of the flight-time Amsterdam–Paris with a DC-8 is normal with mean μ and standard deviation $\sigma = 3$.

a Determine a 95% confidence interval for μ.

b Discuss the conjecture that the mean flight-time Amsterdam–Paris is less than 42 minutes.

Exercise 15.16

Reconsider Exercise 15.15.

a Conduct a hypothesis test to examine the conjecture in part (b); take $\alpha = 0.05$.

b Determine the p-value of the test in part (a). Use it to conduct the test once again.

c The general manager of the airline company thinks that the confidence interval of part (a) of Exercise 15.15 is too wide; he wants to take a new sample of flight-times in order to construct a 95% confidence interval with width 1 minute. Calculate the sample size that is needed.

Exercise 15.17

A certain population variable is assumed to be normally distributed with unknown mean μ and known standard deviation $\sigma = 8$. The following hypotheses will be tested:

$H_0: \mu = 100$ versus $H_1: \mu \neq 100$

A sample of size 25 from the population yields the realized sample mean 96.1.

a If the level of significance is $\alpha = 0.05$, should H_0 be rejected?

b What is your conclusion if $\alpha = 0.01$?

c Also calculate the p-value and interpret your answer in terms of significance levels.

Exercise 15.18

Consider a random sample of n observations from a population with $\sigma = 2$. Suppose that the mean of the sample equals 5.

a Determine a 95% confidence interval for the population mean μ if $n = 100$. Do the same if n doubles to 200. What are the consequences for the precision of the estimation procedure? Is it in accordance with your intuition?

b Also determine 90% confidence intervals for μ if $n = 100$ and $n = 200$, respectively. Compare the results with the results in part (a), and give your comments.

Exercise 15.19

Someone has to choose between two statements H_0 and H_1 regarding an unknown population mean μ:

$H_0: \mu = 5$ against $H_1: \mu \neq 5$

a Suppose that the corresponding population standard deviation σ equals 3. Below, take $\alpha = 0.05$ as significance level.

i Conduct the standard test if it is additionally given that the sample size n is 200 and the sample mean is 4.85.

ii Conduct the test once again if $n = 2000$ and the sample mean is unchanged. Which of the two tests do you prefer? Why?

b Calculate the p-values of the above tests.

c When the above circumstances are valid, is it necessary to assume that the population distribution is normal?

Exercise 15.20

A consumer organization wants to determine a confidence interval for μ, the mean number of drawing pins in the boxes of a certain brand which, according to the label, should contain 100 drawing pins. Assume that the population variance is known: $\sigma^2 = 36$. In nine randomly chosen boxes this organization finds the following numbers of drawing pins:

| 90 | 94 | 88 | 92 | 90 | 86 | 94 | 90 | 86 |

a Calculate the sample mean and the sample standard deviation.

b Determine a 95% confidence interval for μ. Which assumption underlies the method you used?

c Give your opinion about the statement on the label concerning the mean number of drawing pins.

d Perform a hypothesis test to examine whether, on average, there are not enough drawing pins in the boxes. Take $\alpha = 0.01$.

e Determine the smallest significance level for which the null hypothesis of the testing problem of part (d) is rejected.

Exercise 15.21

Reconsider Exercise 15.20.

a Recall that the rejection region of the five-step procedure of part (d) of Exercise 15.20 is in terms of the values of $Z = (\bar{X} - 100)/2$. It can equivalently be written in terms of values of $\bar{X}$. Do so.

b Suppose that the true mean number of drawing pins per box is precisely 100. Determine the probability that the test procedure of part (d) of Exercise 15.20 incorrectly rejects H_0: $\mu \geq 100$. What kind of error is this?

c Suppose that the true mean number of drawing pins per box is 100.8. Calculate the probability that the test procedure of part (d) of Exercise 15.20 incorrectly rejects H_0: $\mu \geq 100$. What kind of error is this?

d Suppose that the true mean number of drawing pins per box is 94.2. Determine the probability that the test procedure of part (d) of Exercise 15.20 incorrectly **does not** reject H_0: $\mu \geq 100$. What kind of error is this?

e Suppose that the true mean number of drawing pins per box is μ; that is, leave it general. Express the probability that the test procedure of part (d) of Exercise 15.20 incorrectly rejects 'H_0: $\mu \geq 100$' in terms of μ.

f This function of μ is called the **power** of the test procedure. Sketch its graph and indicate on it the probabilities of parts (b)–(d).

Exercise 15.22

The head of a department wants to investigate whether the introduction of a new scheme of coffee breaks changes the mean daily production of his department. Under the old scheme, which included a morning break of 30 minutes, daily production was normally distributed with mean $\mu = 1000$ and standard deviation $\sigma = 22.4$. The new scheme, which allows two breaks of 15 minutes in the morning, has been trialled for one week and each day's production has been measured. Consider the daily production in that week as a random sample (of size 5) from all daily production under the new regime. Here are the data:

| 990 | 1010 | 1030 | 1040 | 1030 |

It is assumed that the daily production is still normally distributed and σ is unchanged.

a Determine a 99% confidence interval for μ (under the new scheme). Does it seem likely that μ has changed by introducing the new regime?

b The management of the company finds the confidence interval in part (a) far too wide and, consequently, the interval estimate in part (a) far too inaccurate to estimate the unknown μ. A 99% confidence interval is wanted that estimates μ with a precision of 1 production unit. Determine the corresponding sample size and give your comments.

c Do you think that it is reasonable to assume that σ has not changed by changing the regime?

Exercise 15.23

A machine fills small bags with a sort of powder. The machine is adjusted so that the amount X of powder in one randomly chosen bag (in grams) has the following continuous pdf:

$f(x) = 1/6$ for all real x between $\mu - 3$ and $\mu + 3$

(and $f(x) = 0$ otherwise). Inspection of 25 bags yields the average amount of 17.2 g per bag.

a Describe the population distribution from which the sample is drawn; if necessary, see Section 10.1 (α and β are now respectively $\mu - 3$ and $\mu + 3$).

b Is this population distribution a normal distribution? What is the population mean?

c Calculate the population variance σ^2; see (10.3).

d Do you think that the sample size is large enough to use a normal distribution for the estimator $\bar{X}$ of μ? Which normal distribution?

e Use the standard hypothesis test to draw a conclusion about the conjecture that μ is smaller than 18 g. Take 0.02 as significance level.

f Suppose that the true value of μ is 18.1 g. Determine the type of the error that is made in drawing the conclusion of part (e). Calculate the probability that the test procedure of part (e) leads to that error if $\mu = 18.1$.

Exercise 15.24 (computer)

The files Xrc15-24.xls and Xrc15-24.sav contain information about 300 adults drawn randomly from a large population of adults. Three variables are measured:

X = 'weight (in kg)'

Y = 'height (in cm)'

U = 'age (in years, two decimals)'

For the whole population, it is given that the standard deviations of X, Y and U are respectively 12.0179 kg, 9.4116 cm and 15.1033 years.

a Determine 95% confidence intervals for each of the three population means. Is μ_U larger than 37?

b Use a hypothesis test to draw a conclusion about the conjecture that the mean weight of all adults in the population is larger than 70 kg. Take $\alpha = 0.05$.

c Ten years ago, the mean height of the people in the population was 175.20 cm. Can it be concluded that the mean height nowadays is different? Calculate the p-value to answer this question.

Exercise 15.25

Consider a random sample of size 16 from a normal population with unknown mean μ and known variance $\sigma^2 = 400$. The following testing problem is considered:

$H_0: \mu \le 30$ versus $H_1: \mu > 30$

The sample mean $\bar{X}$ is taken as test statistic.

a Consider the test procedure with rejection region $[35, \infty)$. Calculate the probability that this test procedure will lead to a wrong conclusion if the true population mean is 30. Is this a type I or a type II error?

b Answer the questions in part (a) again if the true population mean is 40.

c Answer the questions in part (a) again when $[37, \infty)$ is the rejection region of the test procedure. And again when the rejection region is $[39, \infty)$.

d What are the effects on the probabilities of these type I and type II errors when the critical value of the rejection region grows from 35 via 37 to 39?

Exercise 15.26 (computer)

A firm that markets goods to young people wants to identify whether young people of different countries have different views with respect to ethical questions. The firm commissions MarketView to investigate.

MarketView's researchers ask business students from eight countries to answer (among others) the four questions below, about 'actively benefiting from illegal activity'. The answers are on the scale 1, 2, 3, 4, 5 where 1 = strongly believe it is wrong, and 5 = strongly believe it is not wrong.

Q1: Changing price tags on merchandise in a retail store
Q2: Drinking a can of cola in a supermarket without paying for it
Q3: Reporting a lost item as 'stolen' to an insurance company in order to collect the insurance money
Q4: Giving misleading price information to a checkout assistant for an unpriced item

The dataset Xrc15-26.sav contains not only the answers given by 948 students, but also the ID numbers (1–948) of the students, their country (Portugal = 1, Spain = 2, Denmark = 3, Scotland = 4, Germany = 5, Italy = 6, Greece = 7, Netherlands = 8) and gender (male = 0 and female = 1).

The objective of this exercise is to find out whether young business students, from different countries or with different gender, have different views with respect to 'actively benefiting from illegal activity'. For that, the four question variables are transformed into one standardized variable called X (so, with $\mu = 0$ and $\sigma = 1$) that results from the following steps:

▪ Consider the variable that measures the average of the answers to the four questions.

▪ Centre it by subtracting its mean.

▪ Divide the centred variable by its standard deviation.

The resulting variable X is considered to be quantitative. Note that, on the population of all business students from all eight countries, its mean is 0 and standard deviation is 1. It seems reasonable to assume that this variable also has standard deviation 1 on the subpopulations of the female and male business students and on the eight subpopulations of business students from the different countries.

a Determine yourself the 948 observations of this variable X; if necessary, see Appendix A1.15.

b Test with significance level 0.05, for each of the eight countries, whether the mean opinion among business students of that country is different from the overall mean opinion of business students from the eight countries. Apart from giving the test statistics, create a table that contains the **vals** and the p-values of the (eight) tests.

c Use a suitable 95% confidence interval to conclude whether, in Europe, female business students have a lower mean score than male business students.

CASE 15.2 PERSONALITY TRAITS OF GRADUATES (PART I)

Recently the Faculty of Economics of a university for professional education has conducted a survey among its graduates. The main goal was to evaluate the effectiveness of the faculty's career coaching programme on the success of the graduates as starters on the labour market.

In the present case study we will focus on just a part of the data that is related to the Big Five personality traits, five broad dimensions of personality discovered through empirical research by L.R. Goldberg.[†] These factors and their constituent traits can be summarized as follows:

Openness	appreciation of art, emotion, adventure, unusual ideas, imagination, curiosity and variety of experience
Conscientiousness	a tendency to show self-discipline, act dutifully and aim for achievement; planned rather than spontaneous behaviour
Extraversion	energy, positive emotions, surgency, and the tendency to seek stimulation and the company of others
Agreeableness	a tendency to be compassionate and cooperative rather than suspicious and antagonistic towards others
Neuroticism	a tendency to experience unpleasant emotions easily, such as anger, anxiety, depression or vulnerability; sometimes called emotional instability[‡]

The five dimensions are abbreviated as OCEAN; jointly they form the Five Factor Model (FFM). In the survey, each of the five traits has been measured using Likert scales (see Section 2.2) consisting of 12 items with five answer options (scale 1, 2, 3, 4, 5 with 1 = low level and 5 = high level). A reliability analysis has been conducted to measure the similarity between the 12 items. The Big Five personality traits result in five variables, one for each trait. These variables are the five standardized total scores of the 12 items (so, they have $\mu = 0$ and $\sigma = 1$). They are considered to be quantitative.

With these standardized variables, different groups of graduates can be compared with respect to the five dimensions of OCEAN. For each of the graduates, the values of the five standardized variables, and the values of gender (1 = male and 2 = female) and area of specialization (study stream) are recorded; see the datasets Case15-02.xls and Case15-02.sav. There are four study streams in the datasets:

1 = MEL (management, economics and law)
2 = FM (facility management)
3 = PM (personnel management)
4 = marketing

Recall that the population standard deviations are 1 (by definition), for each of the five traits. Furthermore, it seems reasonable to assume that all standard deviations of interesting subpopulations are 1 too: the subpopulations of the male and female graduates, and the subpopulations of the four streams. Hence, z-tests can be used to compare the graduates from a specific study stream with the overall population mean, which is equal to 0. Similarly, male/female graduates can be compared with the overall population mean 0. Several questions arise. For example:

■ Are marketing graduates more extravert than average?

■ Do MEL graduates score higher on conscientiousness than average (due to their specialization in law)?

[†] See "The structure of phenotypic personality traits', *American Psychologist*, 48/1 (1993), 26–34.

[‡] http://en.wikipedia.org/wiki/Big_Five_personality_traits.

- Do PM graduates differ from average as far as openness is concerned?
- Are facility management graduates more agreeable than average?
- Are male graduates less neurotic than average?

Answer these questions.

Confidence intervals and tests for μ and p

The purpose of this chapter is to develop standard interval estimators and test procedures for population means μ and for population proportions p. The reason that emphasis on these two parameters is combined in one chapter is that the formats for their interval estimators as well as for their test statistics are similar. Furthermore, for large sample sizes the statistical procedures for both parameters lean heavily on the central limit theorem.

Our starting point is a population with a population variable of interest. If this variable is quantitative, then its mean μ is important. But if this variable is qualitative with only two possible values (success and failure), then it is the proportion of successes p that is usually of interest.

In the present chapter, statistical procedures will be considered for interval estimation and for testing problems, both for μ and for p. These procedures are, as always, based on a random sample of size n. With respect to p, only the large sample case will be considered in this book. Regarding μ, the small sample case will be considered too, albeit that the underlying population distribution has then to be normal.

When the inference concerns a population mean, then it will, unless stated otherwise, be assumed that the accompanying population variance is **unknown**. The rather unrealistic (but illustrative) case that the population variance is known was treated in the previous chapter.

CASE 16.1 IF IT SAYS MCDONALD'S, THEN IT MUST BE GOOD

This was the title of an article in the *New York Times* of 14 August 2007. It was about a small study that suggests that hamburgers, French fries, chicken nuggets, and even milk and carrots all taste better to children if they think they come from McDonald's, the consequence of advertising seen by young children. The study was originally published in the paper 'Effects of fast food branding on young children's taste preferences' which appeared in the August 2007 issue of *Archives of Pediatrics & Adolescent Medicine*.

Much money is spent by the food and drinks industry on food marketing to children, also in the EU. This fact conflicts with the widespread belief that advertising to young children is inherently unfair. The study is just a small contribution to throw some light on this matter.

The participants of the research were 63 children, aged 3–5 years, who were asked to taste two **equal** samples (one in the well-known McDonald's wrapper and one in a plain white wrapper) of each of the five food/drink items mentioned above and to express their preferences. A part of the results is presented in the table below.

Food/Drink item	Preference			Total
	Plain (coded: −1)	Taste the same or no answer (coded: 0)	McDonald's (coded: 1)	
Hamburger	22	9	29	60
Chicken nuggets	11	14	36	61
French fries	8	6	46	60
Milk	13	11	38	62
Carrots	14	14	33	61

Children's taste preferences (number of responses)

Do these results justify the conclusion that young children prefer food in McDonald's wrappers to (the same) food in plain white wrappers? See the solution at the end of Section 16.3.2.

16.1 Standardized sample mean and *t*-distribution

The standardized version $Z = \dfrac{\overline{X} - \mu}{\sigma/\sqrt{n}}$ of $\overline{X}$ (the z-score) is very important for three reasons:

■ If the sample is large, then Z is approximately standard normally distributed.

■ If the sample is drawn from a normal distribution, then Z is exactly standard normally distributed.

■ It is the starting point for the creation of interval estimators and test statistics for μ.

Unfortunately, Z depends on the parameter σ, which is unknown now. If the sample size is large, then replacing the unknown σ by its (consistent) estimator S hardly affects the probabilistic behaviour of Z. But things are different for small samples. If the sample is drawn from a normal population distribution, replacing σ by S in Z will introduce a new source of variation that really matters and it can be expected that the resulting random variable will show more variation than Z. Indeed, the probability distribution of this random variable belongs to a new family of probability distributions, the *t*-distributions. Below, a random sample drawn from a normal population distribution will sometimes be called a ***normal random sample***, for short.

The (ordinary) standardized version Z of the sample mean $\overline{X}$ is repeated here:

$$Z = \frac{\overline{X} - \mu}{\sigma/\sqrt{n}} \tag{16.1}$$

Recall that its denominator is just the standard deviation of $\overline{X}$, that is, $SD(\overline{X}) = \sigma/\sqrt{n}$. Although Z in (16.1) depends on the parameters μ and σ, its probability distribution does not **if** the sample is drawn from a normal distribution with these parameters:

Basic result (1) for inferential statistics on μ

If the random sample is drawn from $N(\mu, \sigma^2)$, then $Z = \dfrac{\overline{X} - \mu}{\sigma / \sqrt{n}} \sim N(0, 1)$ **(16.2)**

It is this rv Z that was basic for the construction of the interval estimators and the test statistics for the case that σ is known; recall Chapter 15. Since σ is unknown now, it is necessary to replace σ in (16.1) by its estimator S. Then, a new standardized random variable arises:

$$T = \frac{\overline{X} - \mu}{S / \sqrt{n}}$$ **(16.3)**

Notice that this adapted standardized version has the standard **error** of the sample mean as denominator. It was William Gosset who (at the start of the twentieth century) discovered, for the case of a normal random sample, the form of the probability distribution of T and, in particular, that it does not depend on the parameters μ and σ^2 either:

Basic result (2) for inferential statistics on μ

If the random sample is drawn from $N(\mu, \sigma^2)$, then $T = \dfrac{\overline{X} - \mu}{S / \sqrt{n}} \sim t_{n-1}$ **(16.4)**

Here, t_{n-1} denotes the so-called t-distribution with $n - 1$ degrees of freedom. So, let us briefly study this family of distributions, as an intermezzo.

16.1.1 t-Distributions

t-Distribution or *Student distribution*

The **t-distribution with parameter v** is the probability distribution that belongs to the following pdf:

$$f(y) = c_v \frac{1}{\sqrt{(1 + y^2 / v)^{(v+1)/2}}} \quad \text{for all real numbers } y$$ **(16.5)**

(Although the mathematical form of the pdf is not very useful for economists, it is included for completeness.) Here, v (Greek letter nu) is a positive integer, the **number of degrees of freedom**. The constant c_v (depending on the parameter v) is not really important, apart from the fact that it has to be such that f is a continuous pdf, so the total area under f is 1. Each choice of v yields a pdf, so (16.5) in essence defines a whole family of probability distributions: the family of the t-distributions. The graphs of pdfs of this family look, at first sight, like the standard normal pdf since they are also symmetric around 0 and also concentrated on the set of all real numbers. However, the tails are fatter. Figure 16.1 shows the graphs of the pdfs of t_1, t_5 and t_{30}. To compare things, the standard normal pdf is also included.

Figure 16.1 suggests that t_v approaches $N(0, 1)$ for larger v. This is indeed the case.

Property of the family of t-distributions

$t_v \approx N(0, 1)$ as v gets large **(16.6)**

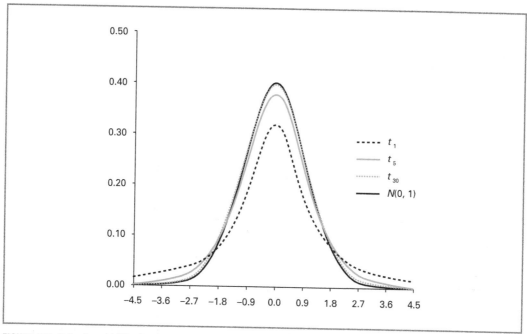

FIGURE 16.1 Pdfs of t_v and $N(0, 1)$

16.1.2 The adapted standardized version

Let us now go back to (16.4), where t_{n-1} entered the scene as the pdf of T if the sample is drawn from a normal population. But this means that probabilities regarding T are areas under the pdf of t_{n-1}. (For instance, if the sample size is 9 then t_8 has to be used.) Hence, the probability distribution of T does not depend on μ and σ^2, but only on the sample size.

The reason that $n - 1$ degrees of freedom have to be used instead of n is that the rv T depends on the sample standard deviation $S = \sqrt{S^2}$. Recall that S^2 is based on the random deviations

$$X_1 - \overline{X}, X_2 - \overline{X}, \ldots, X_{n-1} - \overline{X}, X_n - \overline{X}$$

that add up to 0. As a consequence, only the first $n - 1$ random deviations can vary freely: the last random deviation is determined by its $n - 1$ predecessors. That is why it is said that the sum of squares $\sum_{i=1}^{n}(X_i - \overline{X})^2$, and hence S^2 and the *t*-distribution of T, have $n - 1$ degrees of freedom.

For large sample sizes it holds that $t_{n-1} \approx N(0, 1)$. Already for $n = 31$, the distributions t_{30} and $N(0, 1)$ are hard to distinguish; see Figure 16.1. As a consequence, for normal random samples, the probability distributions of Z and T are approximately equal for large n.

Notice that t_{n-1} has more variation than Z. Replacing σ in (16.1) by its estimator S introduces a new source of variation.

Recall from Chapter 10 that the number z_α deals with the pdf of the *z*-distribution (the $N(0, 1)$ distribution) and is such that an area α is cut off in the right-hand tail. For the coming sections, it is important to have a similar notation for the pdfs of *t*-distributions. Figure 16.2 shows things graphically.

The $(1 - \alpha)$-quantile, defined as $t_{\alpha;n-1}$, of the distribution t_{n-1} is a real number such that under the pdf an area $1 - \alpha$ is cut off at the left-hand side and hence an area α is cut off at the right-hand side. If the random sample is drawn from a normal population, this area α is equal to the probability of the T event $\{T \geq t_{\alpha;n-1}\}$ with T as in (16.4). That is:

$$P(T > t_{\alpha;n-1}) = \alpha \tag{16.7}$$

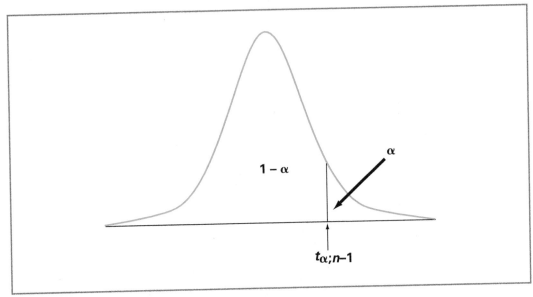

FIGURE 16.2 $(1 - \alpha)$-quantile of t_{n-1}

Probabilities concerning t-distributions and such quantiles can be calculated with computer packages; see also Appendix A1.16. (Only some probabilities and quantiles can be obtained with Table 4 in Appendix A4.) Such calculations can be practised in the following exercise.

Example 16.1
Suppose that a random sample of size 10 will be drawn from the population of all adult men and that their height (in cm) will be measured. Recall from Section 10.3 that a normal distribution is a good model for the population distribution of X = 'height (in cm) of an adult man'. Denote the ten random observations by $X_1, \ldots, X_{10}$ and let Z and T be the two standardized versions of the sample mean $\bar{X}$. Below, some probabilities concerning Z and T are calculated. (Always draw graphs first.) With respect to T, the distribution t_9 is needed.

$$P(Z > 2.1) = 1 - 0.9821 = 0.0179 \qquad (*)$$

$$P(T > 2.1) = 0.0326 \qquad (*)$$

$$P(Z \leq 0.5) = 0.6915 \qquad (*)$$

$$P(T \leq 0.5) = 1 - 0.3145 = 0.6855 \qquad (*)$$

The numbers $z_{0.02}$, $t_{0.02;9}$, $z_{0.15}$ and $t_{0.15;9}$ have to do with inverse operations: areas are given while numbers are asked for. If necessary, see Appendix A1.16.

$$z_{0.02} = 2.0537 \qquad (*)$$

$$t_{0.02;9} = 2.3984 \qquad (*)$$

$$z_{0.15} = 1.0364 \qquad (*)$$

$$t_{0.15;9} = 1.0997 \qquad (*)$$

At the right-hand side of the number $t_{\alpha/2;n-1}$ an area $\alpha/2$ is cut off. Because of the symmetry of the pdf of t_{n-1}, it follows that the area between $-t_{\alpha/2;n-1}$ and $t_{\alpha/2;n-1}$ is $1 - \alpha$. The following equations are valid for the case of normal random samples:

Properties of the adpated standardized version *T* of a normal random sample

$$P(T \le -t_{\alpha;n-1}) = \alpha; \quad P(T \ge t_{\alpha;n-1}) = \alpha; \quad P(-t_{\alpha/2;n-1} < T < t_{\alpha/2;n-1}) = 1 - \alpha \qquad \textbf{(16.8)}$$

(Compare these results with (13.9).) By substituting (16.3) into the last equation of (16.8) and rewriting things in such a way that only μ remains in the middle (as was similarly done between (10.15) and (10.18)), it follows that, for normal random samples:

Basic result for normal random samples

$$P(\overline{X} - t_{\alpha/2;n-1}\frac{S}{\sqrt{n}} < \mu < \overline{X} + t_{\alpha/2;n-1}\frac{S}{\sqrt{n}}) = 1 - \alpha \qquad \textbf{(16.9)}$$

Notice that this result is similar to (13.11). It states that μ is captured between two random bounds, with probability $1 - \alpha$. In the following section, (16.9) and (16.8) will be used to create interval estimators and test statistics.

16.2 Confidence intervals and tests for μ

Regarding the population distribution of the variable X, it is assumed that both μ and σ^2 are unknown and that μ is the parameter of interest. A random sample $X_1, \dots, X_n$ of random observations of X is given.

When making inferences from hypothesis tests, the same testing problems as in Chapter 15 will be considered but now for the realistic case that σ^2 is unknown. The problems are repeated here:

a Test $H_0: \mu \le \mu_0$ against $H_1: \mu > \mu_0$ (one-sided, upper-tailed).
b Test $H_0: \mu \ge \mu_0$ against $H_1: \mu < \mu_0$ (one-sided, lower-tailed).
c Test $H_0: \mu = \mu_0$ against $H_1: \mu \ne \mu_0$ (two-sided).

As in Section 13.4 and Chapter 15, at least one of the following two situations is assumed to be valid:

Situation 1 The normal distribution $N(\mu, \sigma^2)$ is a good model for X.
Situation 2 The sample size n is large.

Although the resulting statistical procedures will turn out to be the same, these two situations are considered separately.

16.2.1 Normal random samples

Suppose that the random sample is a normal random sample, so situation 1 is valid: the model $N(\mu, \sigma^2)$ describes the population distribution well.

Recall that the two random bounds in (16.9) capture the unknown μ with probability $1 - \alpha$. Hence:

Interval estimator (L, U) for μ when σ is unknown

$$L = \overline{X} - t_{\alpha/2;n-1}\frac{S}{\sqrt{n}} \quad \text{and} \quad U = \overline{X} + t_{\alpha/2;n-1}\frac{S}{\sqrt{n}} \qquad \textbf{(16.10)}$$

Notice that this interval estimator is centred on the point estimator $\overline{X}$ of μ and that it has the **random** half-width:

$$H = t_{\alpha/2;n-1}\frac{S}{\sqrt{n}}$$

This half-width has the form $H = b \times SE(\text{estimator})$; see the right-hand side of (15.2). When data are observed and plugged into the random interval (L, U), a $(1 - \alpha)$ confidence interval (l, u) arises. It is said that this interval contains the unknown μ with a confidence of $1 - \alpha$ (or $100 \times (1 - \alpha)\%$), which is prompted by the fact that the proportion $1 - \alpha$ of all applications of (L, U) will contain μ.

As in Section 15.3.3, a confidence interval can be used to draw a conclusion about a conjecture regarding μ. Although we will not go into the details, the procedure is intuitively obvious. The first example below illustrates things.

To obtain test procedures for the above testing problems (a)–(c), the starting point is that (16.4) yields that the random variable

$$\frac{\bar{X} - \mu}{S / \sqrt{n}}$$

has the distribution t_{n-1}. To make it a test statistic, it has to be adapted for the worst case scenario $\mu = \mu_0$. This yields the following test statistic:

$$T = \frac{\bar{X} - \mu_0}{S / \sqrt{n}} \tag{16.11}$$

If $\mu = \mu_0$, then T has the probability distribution t_{n-1}. The alternative hypotheses of the one-sided testing problems (a) and (b) describe, respectively, relatively large and relatively small values for μ, so the corresponding test procedures must reject H_0 for relatively large and relatively small realizations t of T. That is,

For (a):   reject $H_0 \Leftrightarrow t \geq k_l$

For (b):   reject $H_0 \Leftrightarrow t \leq k_u$

for suitably chosen critical values k_l and k_u. In fact, k_l and k_u have to be chosen such that the respective probabilities of the worst case type I error is α, so they have to satisfy:

If $\mu = \mu_0$, then $P(T \geq k_l) = \alpha$ and $P(T \leq k_u) = \alpha$

But if $\mu = \mu_0$, then T has the distribution t_{n-1} and the first two equations in (16.8) yield that $k_l = t_{\alpha;n-1}$ and $k_u = -t_{\alpha;n-1}$. Hence:

For (a):   reject $H_0 \Leftrightarrow t \geq t_{\alpha;n-1}$

For (b):   reject $H_0 \Leftrightarrow t \leq -t_{\alpha;n-1}$

To find the test procedure for the two-sided testing problem (c), notice first that the alternative hypothesis now describes relatively small **and** relatively large values of μ. As a consequence, the procedure has to reject H_0 for relatively small **and** relatively large realizations t of T:

For (c):   reject $H_0 \Leftrightarrow t \leq k_u$ or $t \geq k_l$

Here k_l and k_u have to be chosen such that the probability of the worst case type I error is α, so they have to satisfy:

If $\mu = \mu_0$, then $P(T \leq k_u$ or $T \geq k_l) = \alpha$ and hence $P(k_u < T < k_l) = 1 - \alpha$

But if $\mu = \mu_0$, then T has the distribution t_{n-1} and the last equation in (16.8) yields that $k_u = -t_{\alpha/2;n-1}$ and $k_l = t_{\alpha/2;n-1}$. Hence:

For (c):   reject $H_0 \Leftrightarrow t \leq -t_{\alpha/2;n-1}$ or $t \geq t_{\alpha/2;n-1}$

Hypothesis testing can equivalently be conducted with the p-value method. Again, the p-value is defined as the smallest significance level that would have yielded the conclusion 'reject H_0' once the value (**val**) of the test statistic is given. But this means that, for each of the testing problems (a) and (b), the critical value of the rejection region has to be shifted to **val** and that the resulting significance level is just the area above this new rejection region under the density of t_{n-1}. Hence:

For (a): p-value $= P(T \geq \textbf{val})$

For (b): p-value $= P(T \leq \textbf{val})$

For the p-value of a two-sided test with realization **val** of the test statistic, the two critical values of the rejection region have to be shifted symmetrically until one of them coincides with **val**. The area under t_{n-1} and above this new rejection region is just the p-value. That is:

For (c): p-value $= P(|T| \geq |\textbf{val}|) = 2 \times P(T \geq |\textbf{val}|)$

For each of the three types of problems, the p-value can again be used to draw the conclusion in step five: just check whether the p-value is $\leq \alpha$; if so, reject H_0.

Example 16.2

The general manager of a large chain of hotels wants information about the profitability of the chain. She needs this information to prove to her investors that the chain is creditworthy and that the mean profitability of the hotels in the chain is even larger than it was five years ago, when it was 10%.

To obtain her information, she randomly chooses 12 hotels, measures for each hotel last year's profitability (X) as the profit percentage of the gross annual sales, and obtains the average value 18.2% with accompanying standard deviation 8.1%. What can be concluded about the mean profitability μ of **all** hotels when compared with the score obtained five years ago? Assume that the variable that measures the profitability of hotels is normally distributed. Use a 95% confidence interval and a hypothesis test with significance level 0.05 to answer the question (twice). Also determine the p-value.

Nothing is said about σ, so we assume that it is unknown. Since the population distribution is normal, the random sample is a normal random sample and the sample mean $\bar{X}$ is normally distributed with mean parameter μ and variance parameter $\sigma^2 / 12$. The conjecture is that $\mu > 10$.

According to the above theory, the 95% confidence interval follows from the interval estimator $\bar{X} \pm t_{0.025;11} S / \sqrt{12}$. Since $t_{0.025;11} = 2.2010$ (∗), substitution of $\bar{x} = 18.2$ and $s = 8.1$ yields:

18.2 ± 5.1465 and hence $(13.0535, 23.3465)$

The conclusion is that, with 95% confidence, the mean profitability of **all** hotels in the chain is between 13.05% and 23.35%. In particular, there is (at least) 95% confidence that it is more than 13.05% and hence more than 10%. There is much confidence that the conjecture is true.

Next, a hypothesis test is used to answer the question. The five-step procedure is as follows:

i Test H_0: $\mu \leq 10$ against H_1: $\mu > 10$; use $\alpha = 0.05$

ii Test statistic: $T = \dfrac{\bar{X} - 10}{S / \sqrt{n}}$

iii Reject $H_0 \Leftrightarrow t \geq t_{0.05;11} = 1.7959$ \hfill (∗)

iv $\textbf{val} = \dfrac{18.2 - 10}{8.1 / \sqrt{12}} = 3.5069$

v Conclusion: reject H_0 since **val** falls in the rejection region $R = [1.7959, \infty)$

The data indicate that the mean profitability of all hotels in the chain is more than 10%; this conclusion has confidence 95%.

The test can also be conducted with the p-value method. Here is the five-step procedure:

i Test H_0: $\mu \leq 10$ against H_1: $\mu > 10$; use $\alpha = 0.05$

ii Test statistic: $T = \dfrac{\overline{X} - 10}{S / \sqrt{n}}$

iii $\textbf{\textit{val}} = \dfrac{18.2 - 10}{8.1 / \sqrt{12}} = 3.5069$

iv p-value $= P(T \geq \textbf{\textit{val}}) = 0.0025$ (∗)

v Conclusion: reject H_0 since p-value ≤ 0.05

Afterwards, it can be stated that we might even have used the significance level 0.0025 to obtain the same conclusion.

Example 16.3

Reconsider the machine that produces axles; see Example 15.5. Suppose that the machine is adjusted to 24 mm precisely. However, the 15 randomly sampled axles yield $\overline{x} = 24.02$ and $s^2 = 0.0408$. Can it be concluded that the mean diameter μ of all axles in the batch is different from 24? Use a hypothesis test to draw the conclusion. Also calculate the p-value and comment on the result.

Again, we assume that the diameter (in mm) of a randomly drawn axle is normally distributed. Since nothing is said about it, σ is now unknown. Take 0.10 as significance level. Check that $t_{0.05;14} = 1.7613$ (∗). The steps of the five-step procedure are:

i Test H_0: $\mu = 24$ against H_1: $\mu \neq 24$; take $\alpha = 0.10$

ii Test statistic: $T = \dfrac{\overline{X} - 24}{S / \sqrt{n}}$

iii Reject $H_0 \Leftrightarrow t \leq -1.7613$ or $t \geq 1.7613$

iv $\textbf{\textit{val}} = \dfrac{24.02 - 24}{0.0522} = 0.3835$

v H_0 is not rejected since 0.3835 does not fall in one of the two parts of the rejection region

To find the p-value, notice that:

p-value $= P(|T| \geq |\textbf{\textit{val}}|) = P(|T| \geq 0.3835) = 2P(T \geq 0.3835) = 2 \times 0.3536 = 0.7072$ (∗)

The smallest α that would have yielded the conclusion 'reject H_0', is 0.7072. This makes the alternative hypothesis very implausible.

16.2.2 Large random samples

The discussion in this section assumes that the population standard deviation σ is unknown. Since n is large, the CLT states that the sample mean $\overline{X}$ is approximately normal with mean parameter μ and variance parameter σ^2 / n. As a consequence, both standardized versions of $\overline{X}$ are approximately standard normal:

Basic result for a large random sample $X_1, \ldots, X_n$

For large n: $Z = \dfrac{\overline{X} - \mu}{\sigma / \sqrt{n}}$ and $T = \dfrac{\overline{X} - \mu}{S / \sqrt{n}}$ are both $\approx N(0, 1)$ (16.12)

That T is approximately $N(0, 1)$ follows from the fact that S is a consistent estimator of σ, so replacing σ by S in Z does not change the limiting distribution. As a consequence, the interval estimator (15.4) for the case that σ is known can be transformed into an interval estimator for the case that σ is **un**known simply by replacing σ by its estimator S, so leaving $z_{\alpha/2}$ unchanged. Now, for large n the distribution t_{n-1} is practically the same as the $N(0, 1)$ distribution. That is why the choice is made nevertheless to replace $z_{\alpha/2}$ by $t_{\alpha/2;n-1}$, so the interval estimator for the case that the sample size is large is precisely the same as the interval estimator (16.10) for the case that the random sample is normal. The only difference is that it captures the unknown μ with probability **approximately** equal to $1 - \alpha$.

Similar remarks can be made for the test procedures for the testing problems (a)–(c). The five-step procedures for the large n case are the same as the corresponding procedures for the normal random sample case. The only difference is that the probabilities that the procedures will lead to type I errors are now **approximately** equal to α. The larger n is, the better the approximations are.

Example 16.4

The dataset Xmp16-04.xls contains the heights (in cm) of 243 randomly chosen adult male inhabitants of Tilburg. The data will be used to check the conjecture that the mean height of male inhabitants of Tilburg is smaller than the national mean height of 180.4 cm. This conjecture is in line with the fact that going from the north of Europe to the south, people get smaller on average.

The variable 'height of an adult man' is modelled well by a normal distribution. Hence, the random sample of size 243 is a normal random sample. The population standard deviation is unknown. With a computer it follows that:

$$\bar{x} = 177.9144 \text{ cm} \quad \text{and} \quad s = 8.0805 \text{ cm}$$

For a 99% confidence interval we need $t_{0.005;242} = 2.5963$ (∗). (But notice that this value of the distribution t_{242} is almost the same as $z_{0.005} = 2.5758$ (∗).) Note that:

$$\bar{X} \pm t_{0.005;242} \times \frac{S}{\sqrt{243}} \quad \text{yields the interval } (176.5686, 179.2602)$$

Since 180.4 lies to the right of the upper bound of this interval, it can be concluded – with much confidence – that the mean height of adult male Tilburgers is smaller than the national mean.

The conjecture can also be tested with a hypothesis test; we take $\alpha = 0.005$. Here are the five steps:

i Test H_0: $\mu \geq 180.4$ against H_1: $\mu < 180.4$; take $\alpha = 0.005$

ii Test statistic: $T = \dfrac{\bar{X} - 180.4}{S/\sqrt{n}}$

iii Reject $H_0 \Leftrightarrow t \leq -2.5963$

iv **val** $= \dfrac{177.9144 - 180.4}{0.5184} = -4.7948$

v H_0 is rejected since -4.7948 falls in the rejection region $R = (-\infty, -2.5963]$

The same conclusion follows. To be complete,

$$p\text{-value} = P(T \leq \textbf{val}) = P(T \leq -4.7948) = 1.42 \times 10^{-6} \tag{∗}$$

This small p-value indicates that there is much confidence in the conclusion that the conjecture is true.

Example 16.5

The file Xmp16-05.sav contains last year's annual expenditure on food (in thousands of euro) for a random sample of 300 households drawn from a large population of households. With a computer it follows easily that the sample mean and standard deviation are equal to 8.3523 and 3.3445, respectively. Five years ago, it was predicted that within a few years the mean annual household food expenditure would pass €8700. However, a researcher believes that last year's mean annual expenditure on food did stay below that level. He uses the sample to check this belief.

The variable X = 'last year's annual food expenditure of a household ($\times$ €1000)' cannot – as so many variables that measure expenses or incomes – be modelled with a normal distribution. The reason is that such variables usually are skewed to the right because of relatively large observations. However, since the sample size is rather large, confidence intervals can still be calculated and hypothesis tests can be conducted.

Firstly, a 95% confidence interval is calculated for μ, the mean annual food expenditure for the whole population. Check that $t_{0.025; 299} = 1.9679$ (*). Hence:

$$\overline{X} \pm t_{0.025;299} \frac{S}{\sqrt{300}} \quad \text{yields} \quad 8.3523 \pm 0.3800 \quad \text{and} \quad (7.9723, 8.7323)$$

There is 95% confidence that μ lies between 7.9723 and 8.7323. But the predicted value 8.7 also lies in this interval, so this interval does not allow the conclusion that the belief is true. However, it can be shown (do it yourself) that a 90% confidence interval would have supported the belief.

A formal hypothesis test with significance level 0.05 has the following steps (notice that $t_{0.05;299}$ = 1.6500 (*)):

i   Test H_0: $\mu \geq 8.7$ against H_1: $\mu < 8.7$; take $\alpha = 0.05$

ii   Test statistic: $T = \dfrac{\overline{X} - 8.7}{S / \sqrt{n}}$

iii   Reject $H_0 \Leftrightarrow t \leq -1.6500$

iv   **val** $= \dfrac{8.3523 - 8.7}{0.1931} = -1.8007$

v H_0 is rejected since **val** does fall in the rejection region $R = (-\infty, -1.6500]$

The conclusion of this test with $\alpha = 0.05$ seems to contradict the conclusion that follows from the above 95% confidence interval. This is caused by the fact that using a (two-sided) 95% confidence interval to draw conclusions about a one-sided conjecture in essence raises the confidence level of the conclusion to 97.5%, which obviously was too much here. (We will not go into detail about the precise reasons.)

The p-value of the hypothesis test can also be calculated:

$$p\text{-value} = P(T \leq \textbf{val}) = P(T \leq -1.8007) = 0.0364 \tag{*}$$

Of course, since the p-value is ≤ 0.05, the conclusion of the hypothesis test follows again.

The researcher is not happy with the level of this p-value and especially not with the fact that his conjecture could not be confirmed with a 95% confidence interval. That is why he wants a larger sample and a 95% confidence interval that estimates the mean annual food expenditure to within €250. That is, he wants a 95% confidence interval with half-width $h = 0.25$. But then it should hold that

$$0.25 = h = t_{0.025;n-1} \times \frac{s}{\sqrt{n}}$$

where not only can s not be measured (since the sample is not drawn yet) but also $t_{0.025;n-1}$ is unknown since it depends on n. However, the researcher argues that $t_{0.025;n-1}$ will be almost equal to $z_{0.025} = 1.96$ (since n will be large, even larger than 300) and s can be guessed by the sample standard deviation 3.3445 that came from the former sample. That is why he solves the equation

$$0.25 = 1.96 \times \frac{3.3445}{\sqrt{n}}$$

which yields (rounded up) $n = 688$.

One last remark, which is also valid when interest is in parameters other than μ. Before actually drawing a random sample, it is often desired to know about the sample size n that is needed to obtain a certain precision for the estimate. (See, for instance, the previous example.) Here, the formulations of the precise intentions are often unclear. In the present book the following phrases will be used to describe the wish to obtain a $(1 - \alpha)$ confidence interval with **half**-width h:

Synonyms for 'create a $(1 - \alpha)$-confidence interval with half-width h'

- Create a $(1 - \alpha)$ confidence interval with width $2h$.
- Estimate the unknown parameter with a precision of h.
- Estimate the unknown parameter to within h.
- The (absolute) estimation error is at most h.

16.3 Confidence intervals and tests for p: large sample approach

Our starting point is a population with a qualitative variable that has only two possible values, denoted success and failure. Recall that this variable can (and will) be considered as quantitative, by coding a success as 1 and a failure as 0. The parameter of interest is the unknown proportion p of successes in the population.

To learn about p, a random sample of size n is taken from the population and the Bernoulli variable X (which is 1 in the case of a success) is measured for each sampled population element. The resulting random observations are denoted by $X_1, \ldots, X_n$, which is an *iid* sequence of $Bern(p)$-distributed random variables. Furthermore, let Y be the number of successes in the sample and $\hat{P}$ the proportion of successes. That is, $\hat{P} = Y / n$. $\hat{P}$ is the usual estimator of p; it has the nice properties of being unbiased and consistent in estimating p; see Section 14.1.

Below, procedures will be considered for the statistical problems 'interval estimation' and 'hypothesis testing'. The following testing problems will be considered:

a Test $H_0: p \leq p_0$ against $H_1: p > p_0$ (one-sided, upper-tailed).
b Test $H_0: p \geq p_0$ against $H_1: p < p_0$ (one-sided, lower-tailed).
c Test $H_0: p = p_0$ against $H_1: p \neq p_0$ (two-sided).

Here, p_0 is the hinge, some fixed (known) proportion between 0 and 1 that in a sense lies between the two hypotheses.

Below, only the large sample case will be considered. The required sizes of the samples will become clear from the 5-rule; see below.

In Chapter 14 some basic results were obtained for the probability distribution of $\hat{P}$; see (14.7)–(14.11). They are summarized here:

Basic results for the sample proportion $\hat{P}$ if $np \geq 5$ and $n(1 - p) \geq 5$ hold

$$Z = \frac{\hat{P} - p}{\sqrt{p(1 - p) / n}} \approx N(0, 1) \qquad (16.13)$$

$$P\left(\hat{P} - z_{\alpha/2} \frac{\sqrt{p(1-p)}}{\sqrt{n}} < p < \hat{P} + z_{\alpha/2} \frac{\sqrt{p(1-p)}}{\sqrt{n}}\right) \approx 1 - \alpha \qquad \textbf{(16.14)}$$

16.3.1 Standard interval estimators

Relation (16.14) gives the basic format to obtain an interval estimator. It states that the unknown p is captured by the random interval with lower bound $\hat{P} - z_{\alpha/2}\sqrt{p(1-p)/n}$ and upper bound $\hat{P} + z_{\alpha/2}\sqrt{p(1-p)/n}$. But these random bounds depend on the unknown p, so they cannot be observed when the data are available. That is why p is replaced by its (consistent) estimator $\hat{P}$ that **is** observable. As a result, an interval estimator (L, U) arises:

Interval estimator (L, U) for p if the 5-rule is valid

$$L = \hat{P} - z_{\alpha/2}\sqrt{\hat{P}(1-\hat{P})/n} \quad \text{and} \quad U = \hat{P} + z_{\alpha/2}\sqrt{\hat{P}(1-\hat{P})/n} \qquad \textbf{(16.15)}$$

This random interval captures p with a probability that is approximately equal to $1 - \alpha$. When the data are available, the realizations of L and U can be calculated and a $1 - \alpha$ confidence interval (l, u) arises that contains the unknown p with **confidence** $1 - \alpha$.

One problem remains: how to check the validity of the **5-rule** '$np \geq 5$ and $n(1 - p) \geq 5$'? Since p is unknown, this seems to be an impossible job. But when the data are available, the sample proportion $\hat{p}$ can be calculated and the **adapted 5-rule** '$n\hat{p} \geq 5$ and $n(1 - \hat{p}) \geq 5$' **can** be checked. It is common practice to assume the validity of the 5-rule when the validity of the adapted 5-rule has been shown. (In essence, it is again the fact that $\hat{P}$ is consistent in estimating p that makes this a correct step for large n; we will not go into the details.)

The above considerations can, in terms of realizations $\hat{p}$, be summarized as follows:

$(1 - \alpha)$-confidence interval for p if $n\hat{p} \geq 5$ and $n(1 - \hat{p}) \geq 5$ are valid

$$\hat{p} \pm z_{\alpha/2}\sqrt{\hat{p}(1-\hat{p})/n} \qquad \textbf{(16.16)}$$

Notice that the half-width of the interval estimator in (16.15) is again random. As a consequence, a prescribed half-width h does not immediately yield a sample size n. This is because the equation

$$h = z_{\alpha/2}\frac{\sqrt{\hat{p}(1-\hat{p})}}{\sqrt{n}}$$

depends on a realization $\hat{p}$ that is not yet observed. If it is still required to find a sample size that will generate a $(1 - \alpha)$ confidence interval of width approximately $2h$, then the following remarks may be helpful:

- The starting point is the half-width h of the (non-observable) random interval that follows from (16.14):

$$h = z_{\alpha/2}\frac{\sqrt{p(1-p)}}{\sqrt{n}} \qquad \textbf{(16.17)}$$

- Notice that the function $x(1 - x)$ is a mountain-shaped parabola that takes its maximal value 0.25 at $x = 0.5$. Hence, $p(1 - p)$ in (16.17) is at most 0.25. By substituting 0.25 for $p(1 - p)$ in (16.17), a value of n can be found. This approach is used when nothing is known about p.

- If something **is** known in advance about p, then this information can be used. For instance, assume that it is known that p is at most 0.3. Since the function $x(1 - x)$ with the constraint that $x \leq 0.3$ is at most $0.3 \times 0.7 = 0.21$, it follows that 0.21 can be substituted for $p(1 - p)$ in (16.17) and n can be found.

- Sometimes an estimate of p is known from former research. This estimate can be substituted for p in (16.17) and n found.

16.3.2 Standard hypothesis tests

To obtain test procedures for the above testing problems (a)–(c), property (16.13) of the estimator $\hat{P}$ is the starting point. Suppose that the significance level is α, so the largest probability of a type I error is controlled at the level α. Of course, the random variable

$$\frac{\hat{P} - p}{\sqrt{p(1 - p) / n}}$$

cannot be taken as test statistic since it contains the unknown p. It has to be adapted for the case that the true value of p is p_0 (the hinge), since it is this worst case scenario that again will yield the largest probability of a type I error. Hence, in the above standardized random variable, p has to be replaced by p_0 in both the numerator and the denominator. The following test statistic arises:

$$Z = \frac{\hat{P} - p_0}{\sqrt{p_0(1 - p_0) / n}} \tag{16.18}$$

Having determined the test statistic, it is not too complicated to find suitable rejection regions for the testing problems (a)–(c). By (16.18) it follows that, **if the true value of p is p_0**, then:

$$Z \approx N(0, 1) \text{ and: } P(Z \geq z_\alpha) \approx \alpha; \quad P(Z \leq -z_\alpha) \approx \alpha; \quad P(Z \leq -z_{\alpha/2} \text{ or } Z \geq z_{\alpha/2}) \approx \alpha$$

provided that $np_0 \geq 5$ and $n(1 - p_0) \geq 5$. (In practice, the above approximations '$\approx \alpha$' are written '$= \alpha$'.) As a consequence, the following steps determine the rejection regions for the three testing problems:

For (a): reject $H_0 \Leftrightarrow z \geq z_\alpha$

For (b): reject $H_0 \Leftrightarrow z \leq -z_\alpha$

For (c): reject $H_0 \Leftrightarrow z \leq -z_{\alpha/2} \text{ or } z \geq z_{\alpha/2}$

If the realization **val** of the test statistic is observed, then the respective p-values can be calculated:

For (a): p-value $= P(Z \geq \textbf{val})$

For (b): p-value $= P(Z \leq \textbf{val})$

For (c): p-value $= P(|Z| \geq |\textbf{val}|) = 2P(Z \geq |\textbf{val}|)$

There is one thing that always has to be checked: the proviso that $np_0 \geq 5$ and $n(1 - p_0) \geq 5$.

Example 16.6

Reconsider the milk factory of Examples 15.1 and 15.2. Recall that 1000 packs of milk were taken from the daily production of 100 000 one-litre packs of milk, to conclude about the concern that the proportion p of packs that do not have enough content is above 0.075. Also recall that 100 of the sampled packs contain less than one litre of milk. Use this result to draw a conclusion about the concern, first with a 99% confidence interval and next with a hypothesis test with significance level 0.01.

Since $\alpha = 0.01$ and $z_{0.005} = 2.5758$ (*), it follows that:

$\hat{P} \pm 2.5758\sqrt{\hat{P}(1-\hat{P})/1000}$ yields $0.1 \pm 2.5758\sqrt{0.1 \times 0.9/1000}$

Hence, the 99% confidence interval is (0.0756, 0.1244). Indeed, this interval was also obtained in Example 15.2. There is 99% confidence that p falls in this interval, so there is (at least) 99% confidence that p is larger than 0.0756 and, in particular, larger than 0.075. So, the concern seems to be justified. Notice that $n\hat{p} = 100$ and $n(1-\hat{p}) = 900$, so the 5-rule is satisfied.

For a hypothesis test, notice that the hinge of the testing problem is 0.075. Since $np_0 = 1000 \times 0.075 = 75$ and $n(1-p_0) = 925$, the 5-rule for hypothesis tests is satisfied. Here are the five steps of the test procedure:

i Test $H_0: p \leq 0.075$ against $H_1: p > 0.075$; take $\alpha = 0.01$

ii Test statistic: $Z = \dfrac{\hat{P} - 0.075}{\sqrt{0.075 \times 0.925/1000}}$

iii Reject $H_0 \Leftrightarrow z \geq z_{0.01} = 2.3263$ (∗)

iv **val** $= \dfrac{0.1 - 0.075}{0.00833} = 3.0015$

v H_0 is rejected at the 1% significance level since **val** does fall in the rejection region $R = [2.3263, \infty)$

To learn more about the conclusion of this test, the observed significance level is calculated:

p-value $= P(Z \geq 3.0015) = 1 - 0.9987 = 0.0013$ (∗)

The conclusion of the test is in favour of the alternative hypothesis. The level of the p-value guarantees that this conclusion is very convincing.

Example 16.7

A large company wants to obtain information about the sales possibilities of a newly developed product. The market research department has to study the inclination of consumers to buy the product at a given price. Below, p is the proportion of the consumers in the population of interest that is inclined to buy the product when it is marketed for that price.

The management of the company wants a 90% confidence interval that estimates p to within 0.04, so the estimation error may be at most 0.04. How large a sample is needed if the management does not have the slightest idea about the true value of p? And if it is already known (from a former study) that at most 25% of the consumers in the population of interest will be inclined to buy the product?

Recall that $\alpha = 0.1$, so $z_{\alpha/2} = z_{0.05} = 1.6449$ (∗). The management wants a 90% confidence interval with half-width 0.04. That is, n has to be such that:

$$1.6449 \times \frac{\sqrt{p(1-p)}}{\sqrt{n}} = 0.04$$

If nothing is known about p, then it is best to take $p(1-p)$ maximal, that is, replace $p(1-p)$ by 0.25. Then the above equation becomes:

$$1.6449 \times \frac{\sqrt{0.25}}{\sqrt{n}} = 0.04$$

It follows that $\sqrt{n} = 20.5613$, so $n = 423$ (rounded up).

If it is known in advance that at most 25% of the customers will buy the product, then it can be used that $p \leq 0.25$ and that $p(1-p)$ is at most $0.25 \times 0.75 = 0.1875$ (sketch the graph of the function $x(1-x)$). The equation

$1.6449 \times \dfrac{\sqrt{0.1875}}{\sqrt{n}} = 0.04$

yields $\sqrt{n} = 17.8066$ and $n = 318$ (again, rounded up). Notice that (because of the 5-rule) this sample size suffices to obtain a 90% confidence interval with (16.16) if at least 5 and at most 313 of the sampled customers are inclined to buy the new product.

Example 16.8

The Treasury defines the 'head' of a household as the adult with the highest personal annual income within the household. The dataset <u>Xmp16-08.sav</u> contains, for a random sample of 300 households, information about the gender of the heads. (The value 1 means that the head is a woman.) This sample will be used to draw a conclusion about the conjecture that the proportion *p* of all female heads is unequal to 0.20.

The sample proportion of the female heads is $\hat{p} = 0.19$; see also Appendix A1.16. Since $n\hat{p} = 57$ and $n(1 - \hat{p}) = 243$, the 5-rule is satisfied and (16.16) can be used to obtain a 95% confidence interval:

$\hat{p} \pm 1.96 \sqrt{\hat{p}(1 - \hat{p})/300}$ yields 0.19 ± 0.0444

There is 95% confidence that the proportion *p* of the female heads in the whole population is in (0.1456, 0.2344). Since 0.20 falls in this interval, it cannot be concluded that *p* is unequal to 0.20.

For a formal hypothesis test with hinge 0.20, notice that $np_0 = 300 \times 0.20 = 60$ and $n(1 - p_0) = 240$, so the 5-rule is satisfied. Here is the five-step test procedure for a two-sided test with significance level 0.05:

i Test H_0: $p = 0.20$ against H_1: $p \neq 0.20$; take $\alpha = 0.05$

ii Test statistic: $Z = \dfrac{\hat{P} - 0.20}{\sqrt{0.20 \times 0.80/300}}$

iii Reject $H_0 \Leftrightarrow z \leq -1.96$ or $z \geq 1.96$ (∗)

iv **val** $= \dfrac{0.19 - 0.20}{0.0231} = -0.4330$

v H_0 is not rejected at the 5% significance level since **val** does not fall in one of the two parts of the rejection region

To find the smallest significance level for which the null hypothesis **would** have been rejected, the *p*-value is calculated:

p-value $= P(|Z| \geq |\textbf{val}|) = P(|Z| \geq 0.4330) = 2P(Z > 0.4330) = 2 \times 0.3325 = 0.6650$ (∗)

See the Summary at the end of this chapter for an overview of all interval estimators and test statistics that were considered in Chapters 15 and 16.

CASE 16.1 IF IT SAYS MCDONALD'S, THEN IT MUST BE GOOD – SOLUTION

It was the intention of the researchers to investigate whether the McDonald's wrapper – known from television – influences the choice of a child between two presentations of some food/drink

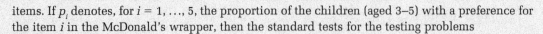

items. If p_i denotes, for $i = 1, \ldots, 5$, the proportion of the children (aged 3–5) with a preference for the item i in the McDonald's wrapper, then the standard tests for the testing problems

$$H_0: p_i \le 0.5 \text{ against } H_1: p_i > 0.5$$

yield the results shown in the table below.

Item i	1	2	3	4	5
val	−0.2583	1.4319	4.8837	1.8251	0.6423
p-value (∗)	0.6015	0.0787	0.000004	0.0364	0.2615
Conclusion (at 0.05 level)	Don't reject	Don't reject	Reject	Reject	Don't reject

The items French fries and milk obviously taste better in the red packaging with the yellow M. The category 'Taste the same or no Answer' seems to prevent drawing the same conclusion for the other three categories.

There is another approach that can be used to compare, for the combined group of food/drink items, the McDonald's wrapping and the plain white wrapping. By averaging the numbers −1, 0 or 1 assigned by a child to the five items, a variable Y arises with values between −1 and 1. Although this might look slightly disputable, this variable is considered to be quantitative. For the 63 children, the sample mean and standard deviation of the observations of Y turn out to be respectively equal to:

$$\bar{y} = 0.37 \quad \text{and} \quad s = 0.45$$

By conducting the standard test procedure for the testing problem $H_0: \mu \le 0$ against $H_1: \mu > 0$, the following results are obtained:

$$\textit{val} = \frac{0.37 - 0}{0.45/\sqrt{63}} = 6.5262 \quad \text{and} \quad p\text{-value} = 3.4 \times 10^{-11} \tag{∗}$$

The null hypothesis that there is either no preference or a preference for the plain white wrapper is convincingly rejected.

16.4 Common formats so far

Having arrived at the end of Chapter 16, it is important to establish the common formats of the interval estimators and the test statistics that have been considered so far. These formats will be used later, when parameters other than μ and p are of interest.

16.4.1 Interval estimators so far

So far, only interval estimators of format 1 have been considered; see Section 15.2. These interval estimators are always based on the (point) estimator of the parameter of interest. If the standard deviation of this estimator is observable (that is, if this standard deviation does not contain unknown parameters), then the type 1 format for interval estimators is as follows:

estimator $\pm a \times SD$(estimator)   **(16.19)**

Of course, this is the preferred format since no further estimations are needed. Format (16.19) was applied in Section 15.2.1, the interval estimator for μ when σ is known. Notice that the constant a is such that the interval estimator captures the parameter of interest with the prescribed probability.

Usually, the standard deviation of the estimator of the parameter of interest depends on unknown parameters and is not observable. In such cases these unknowns have to be replaced by suitable estimators (thus introducing one or more new sources of variation), to yield the standard **error** of the estimator of the parameter of interest; see also Section 14.3. So, here is the **general** format for the type 1 interval estimators:

$$\text{estimator} \pm b \times SE(\text{estimator}) \tag{16.20}$$

The constant b is such that the interval estimator captures the parameter of interest with the pre-scribed probability. Notice that all interval estimators that were considered in this chapter have this format.

16.4.2 Test statistics so far

The test statistics are likewise based on the unbiased and consistent estimator of the parameter of interest. To find the test statistic, the starting point is the basic format

$$\frac{\text{estimator} - \text{parameter}}{SD(\text{estimator})} \tag{16.21}$$

which can never be used immediately since it depends on the parameter of interest. The next step is to adapt this basic format for the case that the hinge of the testing problem is the true value of the parameter. It is important to notice that this step may also affect the denominator, as happened in (16.18). The following format for test statistics arises:

$$\frac{\text{estimator} - \text{hinge}}{SD(\text{estimator})} \tag{16.22}$$

Often, however, the standard deviation of the estimator, after being adapted for the hinge case, will still depend on one or more unknown parameters; so these parameters have to be replaced by suitable estimators and the standard deviation of the estimator turns into the standard **error** of the estimator. So, here is the **general format** for the type of test statistics considered so far:

$$\frac{\text{estimator} - \text{hinge}}{SE(\text{estimator})} \tag{16.23}$$

Summary

In this chapter, standard statistical procedures were considered for the parameters μ (if the variable of interest (X) is quantitative) and p (if it is qualitative with only two outcomes). The procedures are based on the respective estimators $\overline{X}$ and $\hat{P}$. The parameters μ and p (and many other parameters that will be considered later) have common basic formats for interval estimators and test statistics, respectively. These common formats are given below, where $q_{\alpha/2}$ stands for $z_{\alpha/2}$ or $t_{\alpha/2}$:

$$\text{estimator} \pm q_{\alpha/2} \times SD(\text{estimator}) \quad \text{and} \quad \frac{\text{estimator} - \text{parameter}}{SD(\text{estimator})}$$

For inference about μ when σ^2 is known, the left-hand basic format can immediately be used as interval estimator while, in the right-hand basic format, the parameter has to be replaced by the worst case scenario (hinge μ_0) of the testing problem.

For inference about μ when σ^2 is **un**known, the parameter in the right-hand format is replaced by the hinge μ_0 **and**, in both formats, the standard deviation of the estimator has to be replaced by the standard error at the cost of working with the distribution t_{n-1} instead of the z-distribution.

For inference about p, only the large sample approach has been considered. In the right-hand format the parameter p is replaced by the hinge p_0 of the testing problem, in the numerator **and** the denominator. In the left-hand format the standard deviation of the estimator is replaced by the standard error. Since n is large, the z-distribution can be maintained.

The table below summarizes the interval estimators and the test statistics that have been considered so far.

Overview of interval estimators and test statistics for μ and p

	Interval estimator	Test statistic	Requirements
For μ if σ known	$\bar{X} \pm z_{\alpha/2}\dfrac{\sigma}{\sqrt{n}}$	$Z = \dfrac{\bar{X} - \mu_0}{\sigma/\sqrt{n}}$	Normal random sample and / or large n
For μ if σ unknown	$\bar{X} \pm t_{\alpha/2;n-1}\dfrac{S}{\sqrt{n}}$	$T = \dfrac{\bar{X} - \mu_0}{S/\sqrt{n}}$	Normal random sample and / or large n
For p	$\hat{P} \pm z_{\alpha/2}\dfrac{\sqrt{\hat{P}(1-\hat{P})}}{\sqrt{n}}$ (*)	$Z = \dfrac{\hat{P} - p_0}{\sqrt{p_0(1-p_0)/n}}$ (**)	n has to be so large that: (*) $n\hat{p} \geq 5;\ n(1-\hat{p}) \geq 5$ (**) $np_0 \geq 5;\ n(1-p_0) \geq 5$

In practice, it is often desired to find a sample size n such that a parameter can be estimated to within a prescribed precision h. That is, n is wanted that satisfies:

$$z_{\alpha/2} \times SD(\text{estimator}) = h$$

Since $SD(\bar{X}) = \sigma/\sqrt{n}$ and $SD(\hat{P}) = \sqrt{p(1-p)}/\sqrt{n}$, a sample size n can be obtained by replacing the unknown parameters in these standard deviations by suitable approximations, for instance from former studies.

List of symbols

Symbol	Description
t_ν	t-distribution with parameter ν
$t_{\alpha;n-1}$	Quantile of the t_{n-1}-distribution

🔑 Key terms

adapted 5-rule **484**   number of degrees of   Student distribution **474**
normal random sample **473**   freedom **474** t-distribution **474**

Exercises

Exercise 16.1

Write down the formal test procedure (so, steps (ii) and (iii)) to test with significance level α, on the basis of a random sample of size n, the conjecture that the population mean μ is:

a larger than a fixed value μ_0;
b smaller than a fixed value μ_0;
c unequal to a fixed value μ_0.

What are the precise underlying assumptions about the random sample?

Exercise 16.2

Write down the formal test procedure (so, steps (ii) and (iii)) to test with significance level α, on the basis of a random sample of size n, the conjecture that a population proportion p is:

a larger than a fixed value p_0;
b smaller than a fixed value p_0;
c unequal to a fixed value p_0.

What are the precise underlying assumptions about the random sample?

Exercise 16.3

Suppose that $n = 9$, $\bar{x} = 85$ and $s = 2$. Use the standard five-step test procedure to test with significance level 0.05:

a H_0: $\mu = 83$ against H_1: $\mu \neq 83$
b H_0: $\mu \leq 83$ against H_1: $\mu > 83$
c H_0: $\mu = 83$ against H_1: $\mu > 83$.

Did you implicitly assume something about the underlying population?

Exercise 16.4

Suppose that $n = 100$, $\bar{x} = 119.5$ and $s = 2$.

a Use the standard five-step test procedure to test (with significance level 0.05) H_0: $\mu \geq 120$ against H_1: $\mu < 120$.
b Also use a 95% confidence interval to draw a conclusion about the conjecture that $\mu < 120$.
c Did you implicitly assume something about the underlying population?
d Also determine the p-value of the test in part (a). What does it add to the conclusion of the test?

Exercise 16.5

In a random sample of 500 observations, 290 successes and 210 failures were detected.

a Estimate with 95% confidence the population proportion of successes.
b Use a hypothesis test with significance level 0.05 to test whether the population proportion of successes is less than 0.6.
c Does the conclusion of part (b) also follow from part (a)?

Exercise 16.6

Reconsider Exercise 16.5. How large a sample is needed for a 95% confidence interval that estimates the population proportion to within 0.02? To answer this question, you can make use of the estimate of Exercise 16.5.

Exercise 16.7

Suppose you want to estimate a population proportion p with a 98% confidence interval of width 0.03. How large should your sample be?

Exercise 16.8

Reconsider Exercise 15.15, but assume now that σ is unknown. Suppose that a random sample of 10 flights from Amsterdam to Paris yielded the following flight-times (in minutes):

41   43   44   35   40   40   39   42   38   38

a Determine a 95% confidence interval for μ. Interpret it.

b Use it to discuss the conjecture that flying from Amsterdam to Paris takes on average less than 42 minutes.

c Also test this conjecture with a formal hypothesis test; take $\alpha = 0.05$. Give the complete five-step test procedure.

d Determine the p-value of the test in part (c) and use it to test again; give all five steps.

Exercise 16.9 (computer)

The files Xrc16-09.xls and Xrc16-09.sav list, for 63 companies, the rates of return (%) of their stocks in 1995. Let μ be the mean of **all** 1995 returns of the stocks listed on the stock exchange.

a Create a histogram of the data. Do you think it is reasonable to assume that the population variable 'rate of return' on the population of all companies listed on the stock exchange in 1995 is normally distributed?

b Produce a table with the most important sample statistics.

c Test whether $\mu > 30\%$, where 30 is the average rate of return in **1993** of **all** companies listed on the stock exchange. Mention all five steps. Use, as much as possible, the sample statistics calculated in part (b). Is it necessary in this particular case to assume that the population distribution is normal?

d Calculate the p-value of this test. Interpret the answer.

Exercise 16.10

Before a new medicine against a certain disease can be put on the market, the following experiment is done. The medicine is given to 280 randomly selected patients who suffer from this disease. After the treatment, 196 persons are cured. Let p_1 be the population proportion of the patients in the whole population that will be cured after the treatment. The former, traditional medicine had a cure rate of 60%.

a Test whether the new medicine is better than the traditional one. Take $\alpha = 0.05$ as level of significance. Do not forget to check the requirements.

As a control group, 140 new patients who suffer from the same disease are selected at random, but the patients in that group are given only (without knowing) vitamin C. After this 'treatment' 70 people are cured. Let p_2 be the population proportion that is cured after receiving the vitamin.

b Test whether the vitamin is worse than the traditional medicine. Take as level of significance $\alpha = 0.05$.

c Determine a 95% confidence interval for p_1 and also for p_2. Use these intervals to give intuitive arguments for a conclusion that $p_1 > p_2$.

Exercise 16.11

(Compare with Exercise 15.20.) A consumer organization wants to obtain information about μ, the mean number of drawing pins in the boxes of a certain brand which, according to the label, should contain 100 pins. In nine randomly chosen boxes this organization finds the following numbers of drawing pins:

90   94   88   92   90   86   94   90   86

a Test whether $\mu < 100$. Take as level of significance $\alpha = 0.05$. Which assumption do you need?

b If the true value of μ is 101, what type of error did you make in part (a)?

c Determine a 90% confidence interval for μ. Is it likely that $\mu = 101$?

Exercise 16.12

The shopkeepers in a shopping centre ask the local government for permission to extend the number of parking places. The actual capacity is based on, among other things, the average time of 45 minutes that a car is parked in one place. The shopkeepers believe that this average parking time has now increased. The responsible authority decides to take a sample of 16 parked cars to see whether the claim of the shopkeepers is justified. It turns out that the average parking time of these cars is 48 minutes with standard deviation 5 minutes. Below, you may assume that the parking times are normally distributed.

a Because 48 minutes is more than 45 minutes the authority granted the request of the shopkeepers to enlarge the number of parking places. Is this a good decision if we approach this problem statistically? (First with $\alpha = 0.10$ and next with $\alpha = 0.01$.)

b What would be your conclusion if the sample size were only 2?

c What is your conclusion in part (a) in the case that the sample size is 50?

d What can you say about the method of sampling that is used? Suggest how these problems might be solved.

Exercise 16.13

A publisher of pulp novels has performed an observational study among the buyers of his books. This study yielded the following data (in euro) on the amount of money spent per week on these booklets:

	Age $\leq$ 30 yrs	Age $>$ 30 years
n	200	400
$\bar{x}$	6.40	7.00
s	5.47	4.61

Let μ_1 denote the mean amount spent per week among buyers aged 30 or less, and accordingly, μ_2 the mean for the older buyers.

a Determine 99% confidence intervals for μ_1 and μ_2.

b Do you think that it is reasonable to conclude that $\mu_1 = \mu_2$?

The editor is not very much interested in a division according to age and prefers an interval estimate for μ, the mean amount spent per week, irrespective of age. Therefore, all observations are combined into a large sample of size $n = 600$. Let $\bar{x}$ and s denote the sample mean and the sample standard deviation of this combined sample.

c Show that $\bar{x} = 6.8$ and $s = 4.9170$.

d Determine a 99% confidence interval for μ.

e Do you think that it is unrestrictedly allowed to combine both samples into one large sample? Under which additional condition does it seem to be reasonable?

Exercise 16.14

For the production process of ballpoint pens of a certain type, it is known that 5% of the pens are defective. The manager believes that this percentage is too high and hence changes the production process. To test the merits of this new process, a random sample of 350 pens is taken; only 9 of them are defective.

a Determine a 90% confidence interval for p, the population proportion of defective pens produced with the new process. Interpret it. Do you think that the new process is better?

b Conduct a two-sided test to conclude whether revising the process has indeed **changed** the population proportion of defective pens; take $\alpha = 0.04$.

c The manager thinks that the interval in part (a) is too wide; she wants a new 90% confidence interval with width at most 0.02. Determine the sample size that is necessary if you use the conclusion of part (a) that p is smaller than 0.05.

d For part (c), also determine the sample size if the manager wants to estimate p with a precision of 0.005 without any prior information about p.

Exercise 16.15

According to a dietician, adult north European men are on average more than 10 kg overweight. To test this statement, twenty north European men are chosen at random. Their weights (kg) are measured and their respective ideal weights subtracted. The results are:

8	11.5	9	20.5	11	9	11.5	9	11	7.5
9	17.5	8	7.5	8.5	9.5	11.5	7.5	8	13

It follows that $\bar{x} = 10.4$ kg and $s = 3.3896$ kg. Investigate the statement of the dietician:

a With a 95% confidence interval; write down the formula and the realized interval estimate.

b With a one-sided test; take significance level 0.025. Also calculate the p-value of this test. Explain what this p-value adds to the conclusion of the test.

Exercise 16.16 (computer)

When evaluating the main lecture of the course Statistics 1, students answer the question 'the lecturer explains the topics very clearly' by assigning 1, 2, 3, 4 or 5. Here 1 means 'strongly disagree' and 5 means 'strongly agree'. The lecturer of this course has a lot of experience; usually, his scores with respect to the above question are rather good: 70% of the students score his performance as 3, 4, or 5.

With respect to his most recent course Statistics 1, a sample of 60 students also answered the above question. The results are in the file Xrc16-16.xls.

a Construct a 94% confidence interval for p, the population proportion of the students who (regarding the most recent course in Statistics 1) rated the lecturer's ability to explain topics as 3, 4 or 5. Also, mention the interval estimator that you use. Do not forget to check the 5-rule.

b The lecturer thinks that his mean score for the above question was higher than usual. Use the above confidence interval to discuss the lecturer's conjecture.

Exercise 16.17 (computer)

The director of a firm that produces a brand of coffee wants to know whether it would be worthwhile to advertise his product in local newspapers. From former research it is known that **without advertisements in local newspapers**, the mean market share of the brand is 16.75%.

The director selects a random sample of 33 cities of more or less the same size, places advertisements for the coffee in local newspapers, and (after some time) measures the market shares in these cities. See the file Xrc16-17.xls.

  a Use a hypothesis test (with $\alpha = 0.04$) to investigate whether the mean market share μ **with** advertising in local newspapers is larger than without.

  b Did you implicitly make an assumption about the population distribution of the market shares? If yes, mention it. If no, why not?

Exercise 16.18

A market researcher wants a 95% confidence interval for the population proportion p of all households that want to take up the offer to receive for one week a certain newspaper free of charge. He wants to use that interval to estimate p to within 0.02. Determine the required size of the random sample of households if:

  a the researcher does not want to assume anything about the true value of p;

  b it is practically out of the question that p is larger than 0.3;

  c it is practically out of the question that p will be smaller than 0.05 or larger than 0.60.

  d Do not forget to compare your answers to parts (a)–(c) with the 5-rule.

Exercise 16.19

A civil servant of the Ministry of Health wants to examine the number of hours per week that children between two and five years of age watch television. A preliminary investigation among 100 children of this age category yielded an average of 12 hours per week with standard deviation 5 hours. However, the researcher wants a 99% confidence interval that estimates the mean number of hours television-watching per week to within 30 minutes.

  a Find an approximation of the required sample size.

  b When the sample is drawn (with the size that follows from part (a)), it turns out that the sample mean is 12.6 hours and the sample standard deviation 4.8. Determine a 99% confidence interval for the mean number of hours of television-watching per week within this age category. Is its half-width indeed (about) 30 minutes?

Exercise 16.20

A marketing team wishes to evaluate the opinion of consumers about a new product. A random survey of 500 consumers indicates that 287 like the new product, 123 dislike the product, and the remaining 90 have no opinion.

  a Is there evidence that more than 50% of the consumers like the product?

  b Determine a 95% confidence interval for the proportion of the consumers that like the new product.

  c Suppose that the marketing team had wished to achieve a margin of error less than or equal to 0.02 with 95% confidence. How large a sample would be needed?

Exercise 16.21 (computer)

The Greenway wood factory produces plywood for the furniture industry. One of the products is oak-veneered panels with a thickness of 0.75 inches. It is of course important that these panels meet the quality demands of the customers; for instance the mean thickness has to be 0.75 inches. That is why, every hour, five panels are chosen and tested. After a production time of 20 hours, the thickness measurements of 100 panels are available. See the file Xrc16-21.sav.

a Formulate the testing problem that belongs to this quality check.

b Test with the p-value approach whether the mean thickness of the panels deviates from the corresponding quality demand; use the suitable t-test option of your statistical package to determine the value of the test statistic and the p-value directly.

Exercise 16.22 (computer)

The Cell Tone Company sells mobile phones and phone credits in a part of the USA. In a recent meeting, a market researcher indicated that the mean age of Cell Tone customers is less than 40 years. This statement was made in view of a proposed advertising campaign aimed at young people. But before giving the go-ahead to the campaign, Cell Tone decided to do a quick study among its customers. The file Xrc16-22.xls contains data of 50 respondents:

Phone Survey Data
Column A: Customer gender
Column B: City
Column C: Age
Column D: Phone type
Column E: Price
Column F: Location of use
Column G: Monthly plan price
Column H: Total dollars spent

Row 1 of the file contains the names of the above variables. The rows 2 – 51 contain the data.

a Open this file in your statistical package. Take care that the headings appear above the columns.

b Conduct a test to check the statement of the market researcher; use the suitable t-test option of your statistical package to determine the value of the test statistic and the p-value (watch out!) directly.

c Translate the conclusion of the test into advice to Cell Tone with respect to starting the advertising campaign or not.

d Remember the type I and type II errors that can be made when test procedures are conducted. Regarding the conclusion of the present test, which error might have been made?

Exercise 16.23 (computer)

The EU has decided that, by 2012, the carbon dioxide (CO_2) emissions of all new cars should be at most 130 g/km. In 1997 the mean CO_2 emission of new cars was 184.4. In 2005 – being halfway – it might be expected that the mean emission should be smaller than 157.2. Is this indeed the case? To answer that question statistically, the CO_2 emissions were measured for 20 new cars (from 2005); the data are in the file Xrc16-23.xls.[†] Do these data give evidence that the automobile branch follows the right course?

Exercise 16.24 (computer)

Reconsider Exercise 15.26, the study about young people's views with respect to 'actively benefiting from illegal activity'. The researchers are wondering whether the assumption that all variances for the eight country subpopulations are 1, is satisfied. That is why they decide to redo the study **without** taking this assumption as the starting point. Instead, nothing is assumed about the population variances for the eight subpopulations. See Xrc15-26.sav for the data.

[†] *Source:* Data adapted from RL Polk Marketing Systems GmbH.

Redo part (b) of Exercise 15.26 without the assumption. Compare the conclusions with those in Exercise 15.26(b).

CASE 16.2 THE EFFECTS OF AN INCREASE IN THE MINIMUM WAGE (PART I)

According to conventional economic theory, perfectly competitive employers will always cut their workforce in response to any rise in the minimum wage. In practice, however, employer reactions are not so clear-cut. While some studies confirm the predictions of the theory, other – often more recent – studies conclude that employment is unaffected by increases in the minimum wage.

The present investigation intends to clarify the issue by studying the consequences for employment and price levels of the increase (from $4.25 to $5.05) of the New Jersey minimum wage of 1 April 1992.[†]

Because (a) the fast-food industry employs predominantly low-wage workers, (b) the absence of tips simplifies the measurement of wages and (c) fast-food restaurants are relatively easy to sample, the researchers chose to assess the effects of the minimum wage increase on a random sample of Burger King, Wendy's, KFC and Roy Rogers restaurants in New Jersey and eastern Pennsylvania. The dataset <u>Case16-02.sav</u> contains, for 410 restaurants, observations of the following variables, eight months after the wage increase came into effect:

EMPFT   number of full-time employees
EMPPT   number of part-time employees
NMGRS   number of managers/assistant managers
PSODA   price of medium soda, including tax
PFRY   price of small fries, including tax
PENTREE price of entrée, including tax

For the situation of one month **before** 1 April 1992, the means and standard deviations of these variables for fast-food restaurants were as follows:

Variable	EMPFT	EMPPT	NMGRS	PSODA	PFRY	PENTREE
Mean	8.20	18.83	3.42	1.04	0.92	1.32
Standard deviation	8.62	10.08	1.02	0.09	0.11	0.64

Can it be concluded that, for one of the three categories of employees, the mean number of employees has changed? Did the mean of the total number of employees per restaurant decrease? And can any effect of the minimum wage increase be detected in the prices? Give your comments on the results.

[†] *Source*: Original study by D. Card and A.B. Krueger, 'Minimum wages and employment: a case study of the fast-food industry in New Jersey and Pennsylvania', *The American Economic Review*, 84/4 (1994), 772–93.

Statistical inference about σ^2

This chapter considers inferences about the population variance σ^2, the variance of the variable X that is of interest. Again, the purpose is twofold. On the basis of a random sample $X_1 \ldots, X_n$ of observations of X we want to:

■ construct an interval estimator (L, U) that encloses σ^2 with probability $1 - \alpha$;

■ construct a test statistic that can be used to draw conclusions about the validity of hypotheses concerning σ^2.

After a recapitulation that is meant to 'line up' the things we did before when dealing with the parameters μ and p, a special property of the estimator S^2 of σ^2 will be considered. It is this property that is fundamental for the statistical procedures of the present chapter. There is one drawback: the property is valid under the requirement that the random sample is a **normal** random sample. That is, the model $N(\mu, \sigma^2)$ has to fit the variable X well.

CASE 17.1 STANDARD DEVIATION AS A MEASURE OF DISUNITY

Many evaluation forms contain questions (or statements) that have to be answered with (for instance) the following codes:

1 = strongly disagree; 2 = disagree; 3 = neutral; 4 = agree; 5 = strongly agree

Although, strictly speaking, a variable that underlies such a question is qualitative (ordinal), the dataset of answers to the question is usually summarized by way of the mean and standard deviation. Apparently, the underlying variable is silently replaced by a similar but quantitative variable with values in the range {1, 2, 3, 4, 5} or even the continuous interval [1, 5], where the outcomes 1 and 5 refer, respectively, to the extreme opinions 'strongly disagree' and 'strongly agree'. The mean and standard deviation then are – whether you like it or not – considered as measures for, respectively, the mean opinion and the disunity with respect to the question. An application follows.

Instructor B is a young, enthusiastic, but inexperienced teacher of a statistics course for BBA students. His experienced and popular colleague, instructor A, is used to receiving the average score 4.4 and standard deviation 0.8 for the variable X = 'answer to the question: instructor explains things very clearly' with possible answers 1 to 5 as above. Will the mean opinion of the BBA students about the performance of instructor B be less positive? And will the degree of disunity be larger? Use the dataset Case17-01.xls to answer these questions statistically. See the solution at the end of Section 17.4.

17.1 Recap and introduction

Before setting up a theory to find an interval estimator and a test statistic for σ^2, we firstly line up the approaches followed in earlier chapters for the parameters μ and p. This recapitulation is also important for later chapters.

17.1.1 Recap regarding μ

We briefly repeat the approach for μ that was successful in earlier chapters and try to find the aspects that may also be successful in the present chapter. The theoretical steps A–D below were important for statistical inference about the parameter μ:

A The statistical procedures regarding the parameter are based on a 'good' estimator of the parameter.

For μ, this estimator is the sample mean $\overline{X}$. It is unbiased and consistent; its expectation and standard deviation are respectively μ and $\sigma/\sqrt{n}$.

B The estimator is transformed into a basic variable (called **pivot**) that may depend on unknown parameters but its (approximate) probability distribution does not.

Pivot

A **pivot** is a random variable that may depend on unknown parameters but its (approximate) distribution does not.

For μ, the basic variable that is used is the standardized version of $\overline{X}$ (also called z-score):

$$Z = \frac{\overline{X} - \mu}{\sigma/\sqrt{n}}$$

Indeed, Z is a pivot if the random sample is normal and/or the sample size is large. It depends on the unknown parameters μ and σ, but its (approximate) probability distribution is the standard normal distribution, which does not depend on these or other parameters.

C The pivot and its probability distribution are used to construct an interval estimator for the parameter.

For μ, it was used that $P(-z_{\alpha/2} < Z < z_{\alpha/2}) = 1 - \alpha$. From this result it was derived that the rvs

$$\overline{X} - z_{\alpha/2}\frac{\sigma}{\sqrt{n}} \quad \text{and} \quad \overline{X} + z_{\alpha/2}\frac{\sigma}{\sqrt{n}}$$

enclose μ with probability (approximately) $1 - \alpha$. By replacing σ by its estimator S and the z-distribution by the t_{n-1}-distribution, the lower and upper bounds L and U of the interval estimator arise:

$$L = \bar{X} - t_{\alpha/2;n-1}\frac{S}{\sqrt{n}} \quad \text{and} \quad U = \bar{X} + t_{\alpha/2;n-1}\frac{S}{\sqrt{n}}$$

When the sample data are observed, the realization of (L, U), the $(1 - \alpha)$ confidence interval, can be calculated.

   D The pivot can be transformed into a test statistic for a testing problem about the parameter, by the following two adaptations:

      a replace the parameter in the pivot by the hinge;

      b if other unknown parameters are present, replace them by consistent estimators.

When a testing problem for μ is considered and this testing problem has hinge μ_0, the pivot Z turns into

$$T = \frac{\bar{X} - \mu_0}{S/\sqrt{n}}$$

which has (approximately) the t_{n-1}-distribution if $\mu = \mu_0$ and the random sample is normal and/or n is large.

17.1.2 Recap regarding p

Similar steps A–D were followed for the parameter p (population proportion):

   A The sample proportion $\hat{P}$ was taken as estimator.

This estimator is unbiased and consistent; its expectation and standard deviation are respectively p and $\sqrt{p(1 - p)/n}$.

   B For large n, the pivot variable $Z = \dfrac{\hat{P} - p}{\sqrt{p(1 - p)}/\sqrt{n}}$ was taken.

Recall that Z is approximately $N(0, 1)$ distributed if n is large enough.

   C We used Z and its (approximate) distribution to construct an interval estimator for p.

Since $P(-z_{\alpha/2} < Z < z_{\alpha/2}) \approx 1 - \alpha$ for n large enough, it followed that the rvs

$$\hat{P} - z_{\alpha/2}\frac{\sqrt{p(1 - p)}}{\sqrt{n}} \quad \text{and} \quad \hat{P} + z_{\alpha/2}\frac{\sqrt{p(1 - p)}}{\sqrt{n}}$$

include p with probability (approximately) $1 - \alpha$. By replacing p in these rvs by its estimator $\hat{P}$, it follows that

$$L = \hat{P} - z_{\alpha/2}\frac{\sqrt{\hat{P}(1 - \hat{P})}}{\sqrt{n}} \quad \text{and} \quad U = \hat{P} + z_{\alpha/2}\frac{\sqrt{\hat{P}(1 - \hat{P})}}{\sqrt{n}}$$

are the (random) lower and upper bounds of an interval estimator for p.

   D Z was transformed into a test statistic for a testing problem with hinge p_0.

Since p is present in both the numerator and the denominator, the basic variable turns into:

$$Z = \frac{\hat{P} - p_0}{\sqrt{p_0(1 - p_0)}/\sqrt{n}}$$

(Notice that the second adaptation of D is no longer necessary since there are no further unknown parameters in this rv.) Recall that this rv is, for large n, approximately $N(0, 1)$ distributed if $p = p_0$, so it is also denoted by Z.

   Notice that the interval estimators for μ and p are both of format 1 (see Section 15.2), since $L = E - H$ and $U = E + H$ where E is the estimator of the parameter and H is the half-width. Both the interval estimators and the test statistics arise from a common format of the pivot variable, which is the standardized version of the respective estimators:

$$Z = \frac{\text{estimator} - \text{parameter}}{SD(\text{estimator})}$$

17.1.3 Introduction regarding σ^2

A 'good' estimator of σ^2 based on the random sample $X_1, \ldots, X_n$ was defined in (12.6): the sample variance

$$S^2 = \frac{1}{n-1}\sum_{i=1}^{n}(X_i - \bar{X})^2$$

Notice that this is step A. This estimator is unbiased and consistent; its expectation is σ^2 and (without proof) its standard deviation approaches 0 if n is large.

The above steps A–D will be part of the theoretical considerations regarding the parameter σ^2. This is a very important message! However, it will also turn out that the interval estimator is **not** of format 1 but of format 2 (see Section 15.2). The lower bound L and the upper bound U are respectively of the form aS^2 and bS^2 where a and b are constants such that $a \leq 1$ and $b \geq 1$. Furthermore, both the interval estimator and the test statistic will be derived from a pivot that is **not** the standardized version of S^2 but is proportional to the ratio S^2 / σ^2 of estimator and parameter.

It is this ratio S^2 / σ^2, or more specifically $W = (n-1)S^2 / \sigma^2$, that will be studied in detail in the following section. Notice that the expectation of W is $n-1$, since:

$$E(W) = E(\frac{n-1}{\sigma^2}S^2) = \frac{n-1}{\sigma^2}E(S^2) = \frac{n-1}{\sigma^2}\sigma^2 = n - 1$$

Although the rv W does depend on the parameter σ^2, it will turn out that its probability distribution does **not** (which makes it a pivot, a desirable property according to (B)). There is one proviso: the random sample $X_1, \ldots, X_n$ has to be a **normal** random sample.

Example 17.1

A large population of adult men has mean 178.3712 cm and variance 60.9691 for the variable $X =$ 'height (cm)'. As noted before, this variable can be modelled as a normal variable. To learn more about the probability distribution of the rv $W = (n-1)S^2 / \sigma^2$, we consider its values for a large number of samples.

The 480 rows of the dataset <u>Xmp17-01.sav</u> contain in the first five columns the results of measurements of X for 480 random samples of size $n = 5$. Each row has a random sample on the first five positions. The means and sample variance of all samples are in columns 6 and 7, while the last column contains the results of 4 times the ratio of sample variance and the population variance 60.9691. That is, the last column (called w) contains 480 observations of the ratio $4S^2 / \sigma^2$. A histogram is presented in Figure 17.1.

The mean of the 480 observations is 3.896, which is close to 4 (as expected), while the standard deviation is about 2.85. The best-fitting normal curve is also included. However, the fit is bad. Apart from the fact that the histogram is concentrated on only non-negative numbers and the normal curve is not, the fit is also not good in the right-hand tail.

Since a normal curve does not fit well to this histogram of 480 observations of $4S^2 / \sigma^2$, it is suspected that we need another type of distribution for this rv.

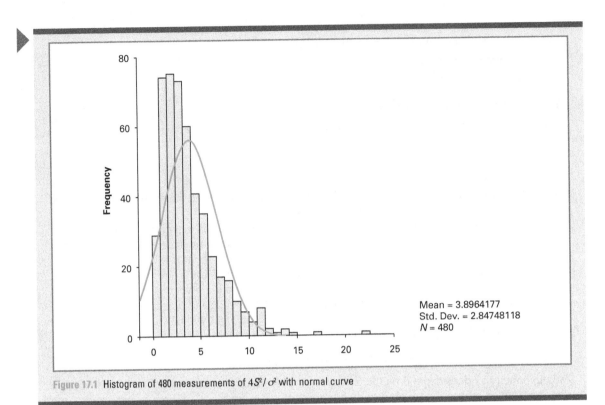

Figure 17.1 Histogram of 480 measurements of $4S^2/\sigma^2$ with normal curve

17.2 A property of the estimator

We start with a new family of continuous probability densities, the family of the χ^2 distributions (chi-square distributions). Next, we mention a result that states that the sum of the squared random variables in a standard-normal random sample $Z_1, ..., Z_n$ has a χ^2 distribution. This result will be used to make it plausible that the variable $(n-1)S^2/\sigma^2$ also has a χ^2 distribution.

17.2.1 χ^2 Distributions

Here is the definition of this family of distributions.

χ_ν^2 Distribution

The **χ^2 distribution with parameter ν** – denoted by χ_ν^2 – is the continuous probability distribution with probability density function f defined by:

$$f(y) = c_\nu y^{(\nu/2)-1} e^{-y/2} \quad \text{for all positive real numbers } y \tag{17.1}$$

(and $f(y) = 0$ for non-positive y).

Property: If an rv Y is χ_ν^2-distributed, then $\mu_Y = \nu$ and $\sigma_Y^2 = 2\nu$ $\quad$ **(17.2)**

(The mathematical form of the pdf is less important.) Here, the parameter ν (Greek letter nu) is a positive integer, the **number of degrees of freedom** of the distribution. It is also denoted as df. The constant c_ν (depending on the parameter ν) is not really important, apart from the fact that it has to be such that f is a continuous pdf, so the total area under f is 1. Each choice of ν yields a pdf, so (17.1) in essence defines a whole family of probability distributions: the family of the χ^2 distributions. If an rv Y has the distribution χ_ν^2, we write $Y \sim \chi_\nu^2$ and say that Y is χ_ν^2-distributed. It is important to notice that the densities are concentrated on the **positive** half-line. Figure 17.2 shows the graphs of some pdfs of the family.

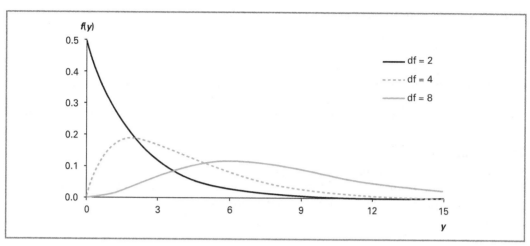

Figure 17.2 Graphs of the densities of χ_2^2, χ_4^2 and χ_8^2

Note that these distributions are not symmetric. Furthermore, it turns out that $f(0) = 0$ for 3 or more degrees of freedom; for 1 or 2 degrees of freedom this is not the case. By (17.2), a χ_4^2-distributed rv Y has $E(Y) = 4$ and $V(Y) = 8$; a χ_8^2-distributed rv has $E(Y) = 8$ and $V(Y) = 16$.

The family of χ^2 distributions was historically detected when studying the probability distribution of a sum of squared standard normal rvs. The elementary theorems below describe this origin of the family.

Elementary results for χ^2 distributions

- If $Z \sim N(0, 1)$, then $Z^2 \sim \chi_1^2$.
- If $Z_1, \ldots, Z_n$ is a **standard** normal random sample, then $Y = \sum_{i=1}^{n} Z_i^2 \sim \chi_n^2$.

- If $X_1, \ldots, X_n$ is a **normal** random sample, then $Y = \sum_{i=1}^{n} \left(\frac{X_i - \mu}{\sigma}\right)^2 \sim \chi_n^2$

It would appear that the square of a standard normal rv is χ_1^2-distributed and the sum of the squares of n iid and standard normal rvs is χ_n^2-distributed. (Recall that a random sample (with replacement) is iid and vice versa; see Section 12.2.) The third result is an immediate consequence of the second, as will be shown now.

If $X_1, \ldots, X_n$ is a normal random sample with $E(X_i) = \mu$ and $SD(X_i) = \sigma$, then the X_i are independent and are all $N(\mu, \sigma^2)$-distributed. Hence, the standardized versions

$$Z_1 = \frac{X_1 - \mu}{\sigma}, \ Z_2 = \frac{X_2 - \mu}{\sigma}, \ \ldots, \ Z_n = \frac{X_n - \mu}{\sigma}$$

of the X_i are also independent and are all **standard** normally distributed. That is, they constitute an iid sample from a standard normal distribution and hence a standard normal random sample. The third result then follows from the second by substituting the above standardized versions for the Z_i.

Notice that the randomness of Y in the third result comes from the random terms $(X_1 - \mu)/\sigma$, $(X_2 - \mu)/\sigma, \ldots, (X_n - \mu)/\sigma$. It would appear that each of these terms independently contributes – because of the first result – one degree of freedom, which leads to n degrees of freedom for Y. If μ in the expression of Y in the third result is replaced by its estimator $\bar{X}$, then the rv W arises:

$$W = \sum_{i=1}^{n} \left(\frac{X_i - \bar{X}}{\sigma}\right)^2$$

Notice that W is the sum of the squares of the random terms

$$\frac{X_1 - \overline{X}}{\sigma}, \ \frac{X_2 - \overline{X}}{\sigma}, \ ..., \ \frac{X_n - \overline{X}}{\sigma}$$

that are **not** independent since their sum is 0 (check it!). It would appear that the first $n - 1$ of these terms can vary freely and each of them contributes one degree of freedom, but not the last term: since the summation of the terms is 0, the last term is completely determined by the first $n - 1$ terms. Indeed, it can be proved that W is also χ^2-distributed but with $n - 1$ degrees of freedom.

17.2.2 Property of S^2

The variable of interest is called X, is quantitative and has mean μ and variance σ^2. It is assumed that the model $N(\mu, \sigma^2)$ fits the variable well. Since emphasis is on the population variance σ^2, a random sample $X_1, ..., X_n$ is drawn and the sample variance

$$S^2 = \frac{1}{n-1}\sum_{i=1}^{n}(X_i - \overline{X})^2$$

is used as its estimator. The *rv*

$$\frac{(n-1)S^2}{\sigma^2} = \frac{1}{\sigma^2}\sum_{i=1}^{n}(X_i - \overline{X})^2 = \sum_{i=1}^{n}\left(\frac{X_i - \overline{X}}{\sigma}\right)^2$$

has a χ_{n-1}^2 distribution (see the considerations following the third elementary result above). Here is the result:

Property of a random sample from $N(\mu, \sigma^2)$

$W = \dfrac{(n-1)S^2}{\sigma^2}$ has the χ_{n-1}^2 distribution   **(17.3)**

This is an important result! W is the pivot that will be used in step B. Although it depends on the unknown parameter σ^2, probabilities regarding W **can** be calculated since they are areas under the pdf of the χ_{n-1}^2 distribution that does **not** depend on σ^2.

Statistical packages can be used to calculate probabilities regarding W; see also Appendix A1.17. If you prefer to use a table, see Appendix A4.

Example 17.2

Reconsider Example 17.1, where the variable X = 'height (cm) of an adult man' was considered on a large population. The model $N(178.3712, 60.9691)$ describes the population distribution of X well.

Suppose that a random sample $X_1, ..., X_5$ of size 5 is drawn. By (17.3) and (17.2) it follows that:

$$W = \frac{4S^2}{60.9691} \sim \chi_4^2; \quad E(W) = 4; \quad V(W) = 8; \quad SD(W) = 2.83$$

The second row in Table 17.1 gives an overview of the probabilities $P(W \le k)$, for $k = 1, 2, 3, 4, 5, 6, 8, 10$; see Appendix A1.17 for their calculation. The third row contains the proportions (called g_k) of the 480 observations w in the file Xmp17-01.sav that are $\le k$. Because of (17.3), the results of the third row should be close to the corresponding results in the second row.

---

k	1	2	3	4	5	6	8	10
$P(W \leq k)$ (*)	0.0902	0.2642	0.4422	0.5940	0.7127	0.8009	0.9084	0.9596
g_k	0.0813	0.2896	0.4688	0.6208	0.7333	0.8125	0.9104	0.9583

TABLE 17.1 Some probabilities regarding χ_4^2

The third row follows most easily by ordering the 480 w data. For instance, 390 of these 480 data are ≤ 6, which yields $g_6 = 390/480 = 0.8125$. Indeed, the observed proportions in the third row are close to the probabilities in the second. This result supports the validity of (17.3).

Recall from Chapter 10 that the number z_α is dealing with the pdf of the z-distribution (the $N(0, 1)$ distribution) and is such that an area α is cut off in the right-hand tail. It is useful to have a similar notation for the pdfs of χ^2 distributions. Figure 17.3 shows things graphically.

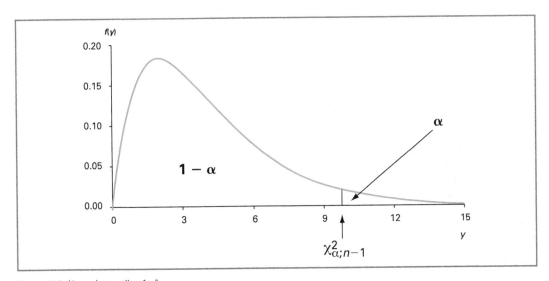

Figure 17.3 $(1 - \alpha)$-quantile of χ_{n-1}^2

For a given α, the $(1 - \alpha)$-quantile (defined as $\chi_{\alpha;n-1}^2$) of the χ_{n-1}^2 distribution is a real number on the horizontal axis such that an area $1 - \alpha$ is cut off at its left-hand side and hence an area α at its right-hand side; see Figure 17.3. The notation appears to 'look to the **right**'. If the random sample is drawn from a normal population, this area α is, by (17.3), equal to the probability of the W event $\{W \geq \chi_{\alpha;n-1}^2\}$. That is:

$$P(W \geq \chi_{\alpha;n-1}^2) = \alpha \tag{17.4}$$

When α is given, the calculation of $\chi_{\alpha;n-1}^2$ is an inverse operation: a probability (area) α is already known and a real number on the horizontal axis has to be determined such that at its right-hand side there is an area α. See Appendix A1.17 for the calculation of $\chi_{\alpha;n-1}^2$.

In later sections we occasionally want to cut off a total area α in such a way that one half is cut off in the right-hand tail and the other half in the left-hand tail. That is, two real numbers are wanted on the horizontal axis under the pdf of χ_{n-1}^2: the first real number has an area $\alpha/2$ on its right, the second has an area $\alpha/2$ on its **left** and hence an area $1 - (\alpha/2)$ on its right; see Figure 17.4. But this means that these real numbers are respectively denoted as $\chi^2_{\alpha/2;n-1}$ and $\chi^2_{1-\alpha/2;n-1}$. Between these numbers there is an area $1 - \alpha$:

$$P(\chi^2_{1-\alpha/2;n-1} < W < \chi^2_{\alpha/2;n-1}) = 1 - \alpha \qquad \qquad \text{(17.5)}$$

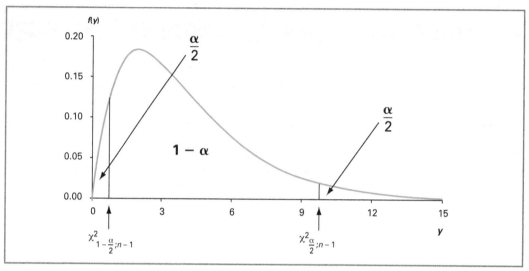

Figure 17.4 The quantiles $\chi^2_{1-\alpha/2,n-1}$ and $\chi^2_{\alpha/2,n-1}$ of χ^2_{n-1}

Example 17.3

Reconsider Example 17.1. When sorting the data in column 8 of the dataset Xmp17-01.sav, it follows easily that the smallest of the 5% largest observations of W is 9.37577 and the largest of the smallest 5% is 0.71918. So, we predict that $\chi^2_{0.05;4}$ and $\chi^2_{0.95;4}$ will be close to these values.

With a computer it follows that:

$$\chi^2_{0.05;4} = 9.4877 \quad \text{and} \quad \chi^2_{0.95;4} = 0.7107 \qquad \qquad (*)$$

Indeed, these quantiles are close to the above empirical values. For $W = 4S^2/\sigma^2$ we have:

$$P(W \le 0.7107) = 0.05; \quad P(0.7107 < W < 9.4877) = 0.90; \quad P(W \ge 9.4877) = 0.05$$

Notice that, since the population variance $\sigma^2 = 60.9691$ is a known number, these probabilities of W events can be rewritten as probabilities of S^2 events. For instance:

$$\{W \le 0.7107\} = \{\frac{4S^2}{60.9691} \le 0.7107\} = \{S^2 \le \frac{0.7107 \times 60.9691}{4}\} = \{S^2 \le 10.8327\}$$

It follows that:

$$P(S^2 \le 10.833) = 0.05; \quad P(10.833 < S^2 < 144.614) = 0.90; \quad P(S^2 \ge 144.614) = 0.05$$

17.3 Confidence intervals

The equation in (17.5) will be used to determine an interval estimator that captures the unknown σ^2 with probability $1 - \alpha$. When the data are given, this interval estimator generates a $1 - \alpha$ confidence interval for σ^2.

Consider a random sample $X_1, \ldots, X_n$ from the distribution $N(\mu, \sigma^2)$. Assume that σ^2 is unknown (as is usually the case). Under these requirements the estimator S^2 has the property that $(n - 1)S^2 / \sigma^2$ is χ^2_{n-1}-distributed.

Recall that we are searching for sample statistics L and U (with $L \leq U$) such that they capture σ^2 with probability $1 - \alpha$:

$$P(L < \sigma^2 < U) = 1 - \alpha$$

It seems natural to start with the W event in (17.5), substitute $(n - 1)S^2 / \sigma^2$ for W, and rewrite things such that σ^2 remains in the middle:

$$\{\chi^2_{1-\alpha/2;n-1} < W < \chi^2_{\alpha/2;n-1}\} \quad = \{\chi^2_{1-\alpha/2;n-1} < \frac{(n-1)S^2}{\sigma^2} < \chi^2_{\alpha/2;n-1}\}$$

$$= \{\sigma^2\chi^2_{1-\alpha/2;n-1} < (n-1)S^2 < \sigma^2\chi^2_{\alpha/2;n-1}\}$$

$$= \{\sigma^2 < \frac{(n-1)S^2}{\chi^2_{1-\alpha/2;n-1}} \text{ and } \sigma^2 > \frac{(n-1)S^2}{\chi^2_{\alpha/2;n-1}}\}$$

$$= \{\frac{(n-1)S^2}{\chi^2_{\alpha/2;n-1}} < \sigma^2 < \frac{(n-1)S^2}{\chi^2_{1-\alpha/2;n-1}}\}$$

(The second equality follows by multiplying the three terms by σ^2, the third by rewriting the two inequalities, and the fourth by reorganizing terms.) But the left-hand W event has probability $1 - \alpha$, so the right-hand event has too. Hence, we just finished step C. It follows that:

Basic result for normal random samples

$$P\left(\frac{(n-1)S^2}{\chi^2_{\alpha/2;n-1}} < \sigma^2 < \frac{(n-1)S^2}{\chi^2_{1-\alpha/2;n-1}} \right) = 1 - \alpha \qquad \textbf{(17.6)}$$

Interval estimator (L, U) for σ^2:

$$L = \frac{n-1}{\chi^2_{\alpha/2;n-1}} S^2 \quad \text{and} \quad U = \frac{n-1}{\chi^2_{1-\alpha/2;n-1}} S^2 \qquad \textbf{(17.7)}$$

Notice that this interval estimator is of format 2 (see Section 15.2), since $L = aS^2$ and $U = bS^2$ with $a = (n-1)/\chi^2_{\alpha/2;n-1}$ and $b = (n-1)/\chi^2_{1-\alpha/2;n-1}$.

When the data are observed, they can be substituted into the formulae of L and U to obtain a $(1 - \alpha)$ **confidence interval**, a realized interval (l, u) that contains σ^2 with confidence $1 - \alpha$.

Example 17.4

A machine that produces small axles is temporarily adjusted for axles of 24 mm. There always is some variation in the diameters of the axles produced, but this is not a problem as long as they are between 23.6 and 24.4 mm. Since the precise diameter X (in mm) is determined by many small technical influences, the **normal** model with parameters μ and variance σ^2 fits well. Because of the quality bounds 23.6 and 24.4, the boundary 0.13 is taken as the largest allowable standard deviation. There can be two causes to conclude that the production process is 'out of control': $\mu \neq 24$ or $\sigma > 0.13$. Interest is now in the second cause.

To check the quality of the daily production, the diameters of 15 randomly drawn axles are measured. Here are the results:

24.31 24.18 23.64 23.95 24.11 24.03 23.86 24.01 23.98 24.26 24.10
23.71 23.98 23.86 24.32

A 95% confidence interval for the population variance σ^2 will be used to check whether the production process is out of control as far as the variation is concerned.

The formulae for the boundaries l and u of a 95% confidence interval follow from (17.7); we determine all the constituents. Notice that $\alpha = 0.05$, so $\alpha/2 = 0.025$ and $1 - \alpha/2 = 0.975$. With a computer it can easily be calculated that:

$$\chi^2_{0.975;14} = 5.6287 \quad \text{and} \quad \chi^2_{0.025;14} = 26.1189 \tag{*}$$

Since the observed sample variance s^2 equals 0.0408, we obtain:

$$l = \frac{14}{\chi^2_{0.025;14}} s^2 = \frac{14}{26.1189} \times 0.0408 = 0.0219$$

$$u = \frac{14}{\chi^2_{0.975;14}} s^2 = \frac{14}{5.6287} \times 0.0408 = 0.1015$$

Hence, the 95% confidence interval for σ^2 is (0.0219, 0.1015). There is 95% confidence that σ^2 falls in it, so there is (at least) 95% confidence that σ^2 is larger than 0.0219.

The production process is deemed to be out of control if $\sigma > 0.13$. But $\sigma > 0.13$ is equivalent to $\sigma^2 > 0.13^2$ and hence to $\sigma^2 > 0.0169$. There is 95% confidence that σ^2 is larger than 0.0219. But 0.0219 is larger than 0.0169, so there is (at least) 95% confidence that $\sigma^2 > 0.0169$. The conclusion is that the production process is out of control since the degree of variation is unacceptably large.

Example 17.5

Investment advisers frequently claim that the mean return on investments in stocks is often larger than the mean return of investments in bonds. On the other hand, they also claim that investments in stocks are more risky than investments in bonds. Recall that the risk of an investment is usually measured by the standard deviation of the returns.

To get an idea of the risk of stocks **funds** when compared with bonds **funds**, we will consider the daily returns (%) of the SNS Euro Stocks Fund and compare the standard deviation of its returns for the period 23 November 2004–17 November 2006 with the risk of the SNS Euro Bonds Fund. From historical data it is known that the population standard deviation of the daily returns of this bonds fund is 0.005%.

The file Xmp17-05.sav contains, for the relevant period, the daily returns of the SNS Euro Stocks Fund as quoted on Euronext. Is this fund more risky than the SNS Euro Bonds Fund? Use a suitable 99% confidence interval to answer this question.

The variable of interest is the daily return R of the stocks fund, measured as a percentage with respect to the business day before. The file contains observations of R for 513 successive business days. To apply the above theory, we need that these observations come from a normal random sample $R_1, \ldots, R_{513}$. That is, it has to be reasonable to assume that the R_i are independent and that they all have the same distribution $N(\mu, \sigma^2)$. The independence is not a big problem; returns often behave rather independently. To check normality, a histogram is constructed with the best-fitting normal curve included; see Figure 17.5.

It seems that the histogram is not too far from being normal. (Whether all R_i have the same expectation μ and the same variance σ^2 is a complicated problem that we will evade. In a first analysis, we assume that the 513 observations come from a normal and *iid* sample $R_1, \ldots, R_{513}$.)

The data yield that the sample standard deviation s equals 0.008968. Since $\alpha = 0.01$, $\chi^2_{0.995;512}$ = 433.3317 (*) and $\chi^2_{0.005;512} = 598.1784$ (*), it follows easily that:

$$l = \frac{512}{598.1784} \times (0.008968)^2 = 0.000069$$

$$u = \frac{512}{433.3317} \times (0.008968)^2 = 0.000095$$

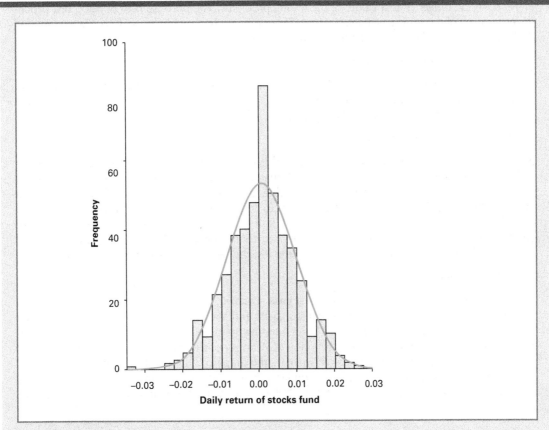

Figure 17.5 Histogram of daily returns, with best-fitting normal curve

Hence, there is 99% confidence that the population variance σ^2 of the stocks fund falls in the interval (0.000069, 0.000095). Notice that the population variance of the bonds fund is 0.000025, which does not belong to the interval but lies at its left-hand side. That is why we can conclude that σ^2 is larger than 0.000025 and hence that σ is larger than 0.005%. The risk of the stocks fund is larger than the risk of the bonds fund.

17.4 Tests

Hypothesis tests are considered for three testing problems. The test procedures are based on the pivot $W = (n - 1)S^2 / \sigma^2$ introduced in Section 17.2, by adapting it according to the worst case scenario. The underlying requirement is that the random sample is drawn from a normal population.

The starting point is again a variable X that can be modelled well by the distribution $N(\mu, \sigma^2)$. Furthermore, a dataset of n observations of X is given, that are the realizations of a random sample $X_1, \ldots, X_n$ from this distribution.

In this context, test procedures will be developed for the following three elementary testing problems:

a Test H_0: $\sigma^2 \leq \sigma_0^2$ against H_1: $\sigma^2 > \sigma_0^2$ (one-sided, upper-tailed).

b Test H_0: $\sigma^2 \geq \sigma_0^2$ against H_1: $\sigma^2 < \sigma_0^2$ (one-sided, lower-tailed).

c Test H_0: $\sigma^2 = \sigma_0^2$ against H_1: $\sigma^2 \neq \sigma_0^2$ (two-sided).

Here, σ_0^2 is the **hinge** (or **test value**) of the testing problems. The purpose is to develop **test procedures** for each of the testing problems (a)–(c). These procedures have to be such that, when applied to the data, they lead to one of the following two (statistical) conclusions:

H_0 is rejected   or H_0 is not rejected

Just as we did in Chapters 15 and 16, when the parameters of interest were μ and p, we will develop the test procedures such that incorrectly rejecting H_0 is avoided by controlling the probability that this happens at a maximal level α (the significance level):

$P(H_0$ is incorrectly rejected $) \leq \alpha$

In other words, the probability is at most α that the test procedure will reject H_0 while H_0 **is** valid since the true value of σ^2 belongs to H_0. Recall that incorrectly rejecting H_0 corresponds to **type I errors** and that there is a type I error for each value of σ^2 in H_0. It is intuitively obvious that it will be hardest to control the type I error that belongs to σ_0^2, since the hinge separates H_0 from H_1. The probability that the procedure will reject H_0 while the true value of σ^2 is σ_0^2, is the largest probability of a type I error. That is why it is sufficient to control the probability that the procedure will reject H_0 while the true value of σ^2 is σ_0^2. So, the test procedures for (a)–(c) have to satisfy:

$P(H_0$ is rejected while $\sigma^2 = \sigma_0^2) \leq \alpha.$

Again, the hinge σ_0^2 corresponds to the **worst case scenario.**

Test procedures consist of two steps: a test statistic and a rejection region. For the test statistic, it is natural to take the **pivot** $W = (n-1)S^2/\sigma^2$ as starting point; recall that although W itself depends on the parameter σ^2, the probability distribution of W does not since it is the χ_{n-1}^2 distribution. Being loyal to D in Section 17.1, we replace σ^2 in W by the hinge σ_0^2: the pivot is adapted according to the worst case scenario. (Notice that no second step is needed, since there are no further unknown parameters left.) This leads to the test statistic below; we again use the notation W to denominate it:

Test statistic: $W = \dfrac{n-1}{\sigma_0^2}\, S^2$ ..... **(17.8)**

Property of the test statistic: if $\sigma^2 = \sigma_0^2$, then $W \sim \chi_{n-1}^2$ ..... **(17.9)**

Next, rejection regions are needed for the testing problems (a)–(c). Because of (17.8), the rejection regions of the test procedures will be in terms of values w of the test statistic W. We start with (a).

Since the alternative hypothesis H_1 of (a) describes relatively large values of σ^2, H_0 will be rejected – and hence H_1 accepted – for relatively **large** values of the test statistic; to be precise, for values w that are $\geq \chi_{\alpha;n-1}^2$. The reason for the last step is clear: if the true value of σ^2 is σ_0^2 (the worst case scenario), then W has the distribution χ_{n-1}^2 and the probability that the realization of W will be $\geq \chi_{\alpha;n-1}^2$, is precisely equal to α (see also (17.4)). That is:

if $\sigma^2 = \sigma_0^2$, then $P(H_0$ is rejected$) = P(W \geq \chi_{\alpha;n-1}^2) = \alpha$

Similar arguments can be used for testing problem (b): H_0 will be rejected for realizations w of the test statistic that are $\leq \chi_{1-\alpha;n-1}^2$. The reason is:

if $\sigma^2 = \sigma_0^2$, then $P(H_0$ is rejected$) = P(W \leq \chi_{1-\alpha;n-1}^2) = \alpha$

For testing problem (c), equation (17.5) is used. H_0 will be rejected for relatively small **and** relatively large realizations w; to be precise, for values w that are $\leq \chi_{1-\alpha/2;n-1}^2$ and for values w that are $\geq \chi_{\alpha/2;n-1}^2$. The reason is:

if $\sigma^2 = \sigma_0^2$, then $P(H_0$ is rejected$) = P(W \leq \chi_{1-\alpha/2;n-1}^2$ or $W \geq \chi_{\alpha/2}^2)$

$$= 1 - P(\chi_{1-\alpha/2;n-1}^2 < W < \chi_{\alpha/2;n-1}^2) = 1 - (1 - \alpha) = \alpha$$

For each of the testing problems, steps (i)–(v) of the hypothesis tests are now clear:

Five-step procedures for testing H_0 against H_1

i Testing problems: **a** test H_0: $\sigma^2 \leq \sigma_0^2$ against H_1: $\sigma^2 > \sigma_0^2$

      b test H_0: $\sigma^2 \geq \sigma_0^2$ against H_1: $\sigma^2 < \sigma_0^2$

      c test H_0: $\sigma^2 = \sigma_0^2$ against H_1: $\sigma^2 \neq \sigma_0^2$

ii Test statistic: $W = \dfrac{n-1}{\sigma_0^2} S^2$

iii Rejection: **a** reject $H_0 \Leftrightarrow w \geq \chi^2_{\alpha;n-1}$

            b reject $H_0 \Leftrightarrow w \leq \chi^2_{1-\alpha;n-1}$

            c reject $H_0 \Leftrightarrow w \leq \chi^2_{1-\alpha/2;n-1}$ or $w \geq \chi^2_{\alpha/2;n-1}$

iv Calculate **val**, the realization of W

v Draw the conclusion by comparing **val** with step (iii)

Requirement: *normal* random sample

The respective rejection regions of the tests for (a)–(c) are:

Rejection regions: $[\chi^2_{\alpha;n-1}, \infty)$, $[0,\chi^2_{1-\alpha;n-1}]$ and $[0,\chi^2_{1-\alpha/2;n-1}]\cup[\chi^2_{\alpha/2;n-1}, \infty)$

(The lower bounds of the last two rejection regions are taken to be 0, since negative outcomes w of W cannot occur.) Notice that the bounds of the (closed) rejection regions, the **critical values**, are included. However, they might be excluded as well (making the rejection regions open) since the probability that the realization of W will precisely fall at the boundaries of the rejection regions is 0 because W is a continuous random variable.

As soon as the **val** of a test is obtained, it is interesting to learn about the minimal significance level that will lead to the conclusion to reject H_0. That is, the **p-value** is of interest. Things will be illustrated by examples.

Example 17.6

(See also Example 17.4.) The fear that the production process of the axles with diameter 24 mm is 'out of control' since $\sigma > 0.13$, can also be checked with a hypothesis test. As always, the fear or conjecture is put in the alternative. Recall that a random sample of 15 axles yielded $s^2 = 0.0408$ and that the variable X = 'diameter (mm) of an axle' can be modelled well with a normal distribution. For the test, take 0.05 as significance level.

Since the requirement that the random sample is normal is satisfied, the above test procedure for problem (a) can be used. Here are the five steps:

i Test H_0: $\sigma^2 \leq 0.0169$ against H_1: $\sigma^2 > 0.0169$; $\alpha = 0.05$

ii Test statistic: $W = \dfrac{n-1}{0.0169} S^2$

iii Reject $H_0 \Leftrightarrow w \geq \chi^2_{0.05;n-1} = \chi^2_{0.05;14} = 23.6848$ (*)

iv **val** $= \dfrac{14}{0.0169} \times 0.0408 = 33.7988$

v Reject H_0 since **val** is ≥ 23.6848

The rejection region is $[23.6848, \infty)$, so the critical value is 23.6848. The conclusion is that the fear is justified; indeed, σ is larger than 0.13.

Once the *val* is observed, we can ask for the *p*-value: the **minimal** significance level that would have led to the conclusion to reject H_0. But then the critical value has to coincide with *val* = 33.7988, and we want to calculate the area under the χ^2_{14} density at the right side of 33.7988:

$$p\text{-value} = P(W \geq 33.7988) = 0.0022 \tag{*}$$

(Of course, a computer is used for the last equality; see also Appendix A1.17.) After all, we even might have taken $\alpha = 0.0022$ to come to the same conclusion as above.

Notice that the test in Example 17.6 is upper-tailed. In the following examples, tests of types (b) and (c) will be considered.

Example 17.7

Suppose that the population variance of the daily return (%) on the SNS Euro Stocks Fund is 0.000081 and that the population variance σ^2 of the SNS **Bonds** Fund is unknown. The file Xmp17-07.sav contains the daily returns (%) of this bonds fund for the period 23 November 2004–17 November 2006 as quoted on Euronext. The conjecture is that σ^2 is smaller than 0.000081; check it with a hypothesis test with $\alpha = 0.01$.

As in Example 17.5, we assume that the 513 data points come from a **normal** *iid* sample. Since $\alpha = 0.01$, it follows that $1 - \alpha = 0.99$ and the critical value of the lower-tailed test is $\chi^2_{0.99;512}$ (which has to be calculated with a computer). The dataset has standard deviation $s = 0.005059$; check it. Here are the five steps:

   i   Test H_0: $\sigma^2 \geq 0.000081$ against H_1: $\sigma^2 < 0.000081$; $\alpha = 0.01$

   ii Test statistic: $W = \dfrac{n-1}{0.000081} S^2$

  iii Reject $H_0 \Leftrightarrow w \leq \chi^2_{0.99;n-1} = \chi^2_{0.99;512} = 440.5099$ (*)

  iv *val* $= \dfrac{512}{0.000081} \times (0.005059)^2 = 161.7761$

   v   Reject H_0 since *val* is ≤ 440.5099

The conclusion is that the variance of the daily returns of SNS Bonds Fund is indeed smaller than the variance 0.000081 of the SNS Stocks Fund, so investing in the bonds fund is less risky.

Having done the test, we can ask for the minimal significance level that would have given the conclusion to reject H_0. For that, the critical value of the test has to coincide with the *val* and it is the area at the left-hand side of the *val* that is asked for:

$$p\text{-value} = P(W \leq 161.7761)$$

It turns out that the *p*-value is very close to 0.   (*)

Example 17.8

Consider a large population of adults with a job, working at least 32 hours per week and having a middle income. Interest is in the variable X = 'gross hourly wage (euro)'. Five years ago, when this population was studied extensively, it was found that the mean gross hourly wage was €20.50 with standard deviation €10.02. The present study is aimed at the situation nowadays.

The file Xmp17-08.sav contains the present gross hourly wages of a random sample of 150 adults from this population of people with middle incomes. The conjecture is that nowadays the **mean** hourly wage of these people is larger than its level €20.50 five years ago. But before using the sample to check the validity of this conjecture, it is tested, with $\alpha = 0.05$, whether the population **variance** is different from its level of five years ago. The conclusion of the test will be used to test the conjecture.

Since the population consists of people with middle incomes, it seems reasonable to use the model $N(\mu, \sigma^2)$ for X. We want to test whether $\sigma \neq 10.02$; so a two-sided test is needed. With a computer it follows (check it) that the sample mean is €21.78, that the sample standard deviation is €10.282, and that the critical values are $\chi^2_{0.975;n-1} = 117.0980$ (*) and $\chi^2_{0.025;n-1} = 184.6870$ (*). Here are the five steps of the test:

  i Test H_0: $\sigma^2 = 100.4004$ against H_1: $\sigma^2 \neq 100.4004$; $\alpha = 0.05$

  ii Test statistic: $W = \dfrac{n-1}{100.4004}S^2$

  iii Reject $H_0 \Leftrightarrow w \leq 117.0980$ or $w \geq 184.6870$

  iv **val** $= \dfrac{149}{100.4004} \times (10.2820)^2 = 156.8939$

  v Do **not** reject H_0 since **val** lies **between** the two critical values

Notice that the rejection region R of the test is: $(-\infty, 117.0980] \cup [184.6870, \infty)$, but, since negative values w of W are not possible, it is better to take:

$R = [0, 117.0980] \cup [184.6870, \infty)$

The determination of the p-value of this test is slightly complicated since the test is two-sided. Given the realized value 156.8939 of the test statistic, we have to shift the right-hand critical value (which is the closest) to 156.8939. Since $P(W \geq 156.8939) = 0.3128$(*), it follows that the minimal significance level for which the two-sided test rejects H_0 is:

p-value $= 2 \times 0.3128 = 0.6256$

Next, we test the conjecture. Since H_0 is not rejected, there is no good reason to say that nowadays σ^2 is unknown. That is why we will use $\sigma^2 = (10.02)^2 = 100.4004$. But this means that we can test the conjecture with a test about μ while assuming that σ^2 is known; see Section 15.3.1. Since the underlying random sample is normal, the sample mean $\overline{X}$ and the test statistic Z are normal too. Taking 0.01 as significance level and writing H_0' and H_1' for the hypotheses, the five-step procedure goes as follows:

  i Test H_0': $\mu \leq 20.50$ against H_1': $\mu > 20.50$

  ii Test statistic: $Z = \dfrac{\overline{X} - 20.50}{10.02/\sqrt{n}}$

  iii Reject $H_0 \Leftrightarrow z \geq z_{0.01} = 2.3263$ (*)

  iv **val** $= \dfrac{21.78 - 20.50}{10.02/\sqrt{150}} = 1.5645$

  v Do not reject H_0'

The conclusion of the test does not support the conjecture. However, the p-value of this test is $P(Z \geq 1.5645) = 0.0589$ (*). Hence, $\alpha = 0.06$ would have given the conclusion that the conjecture is valid.

CASE 17.1 STANDARD DEVIATION AS A MEASURE FOR DISUNITY – SOLUTION

The two questions can be restated as follows in terms of the mean opinion μ and the accompanying standard deviation σ:

Is μ smaller than 4.4? Is σ larger than 0.8?

To answer the first question, the alternative hypothesis H_1: $\mu < 4.4$ will be tested. From the dataset it follows that $n = 69$, $\bar{x} = 3.1739$ and $s = 1.11086$. The **val** of the test statistic T and the accompanying p-value follow immediately from these results:

$$val = \frac{3.1739 - 4.4}{1.11086 / \sqrt{69}} = -9.1683$$

$$p\text{-value} = P(T \leq -9.1683) = 8.4 \times 10^{-14} \tag{*}$$

There is sufficient evidence to conclude that the mean opinion about the performance of instructor B is smaller than 4.4.

For the second question, we test the alternative hypothesis H_1: $\sigma^2 > 0.64$ with a hypothesis test with significance level 0.02. The **val** of the test statistic W is:

$$val = \frac{68 \times (1.11086)^2}{0.64} = 131.1136$$

Since the critical value (lower bound) of the rejection region is $\chi^2_{0.02;68} = 94.0370(*)$, it is concluded that the degree of disunity among the BBA students with respect to instructor B's performance is larger than that for instructor A.

Summary

In this chapter, statistical inference concerned the population variance σ^2. If the random sample is drawn from a normal population distribution, then the sample variance S^2 has the property that $(n - 1)S^2 / \sigma^2$ has a χ^2 distribution with $n - 1$ degrees of freedom, a distribution that does **not** depend on the unknown parameter σ^2. This last argument enabled us to construct interval estimators and hypothesis tests. The following tables summarize things.

Point estimator, interval estimator and test procedures

Parameter	σ^2
Point estimator	$S^2 = \dfrac{1}{n-1} \sum_{i=1}^{n}(X_i - \bar{X})^2$
Interval estimator (L, U)	$L = \dfrac{n-1}{\chi^2_{\alpha/2;n-1}} S^2$ and $U = \dfrac{n-1}{\chi^2_{1-\alpha/2;n-1}} S^2$
Alternative hypothesis H_1	$\sigma^2 > \sigma^2_0$ $\quad\quad$ $\sigma^2 < \sigma^2_0$ $\quad\quad$ $\sigma^2 \neq \sigma^2_0$
Test statistic	$W = \dfrac{n-1}{\sigma^2_0} S^2$
Rejection region	$[\chi^2_{\alpha;n-1}, \infty)$ $\quad$ $[0, \chi^2_{1-\alpha;n-1}]$ $\quad$ $[0, \chi^2_{1-\alpha/2;n-1}] \cup [\chi^2_{\alpha/2;n-1}, \infty)$
Requirement	*normal* random sample $X_1, ..., X_n$

List of symbols

Symbol	Description
χ^2_v	χ^2 distribution with parameter v
$\chi^2_{\alpha,n-1}$	quantile of the χ^2_{n-1} distribution

🔑 Key terms

χ^2 distribution **502**

number of degrees of freedom **502**

pivot **499**

❓ Exercises

Exercise 17.1

Consider an rv W that is χ^2_{40}-distributed. To calculate the probabilities below, you will need a computer (see also Appendix A1.17).

a Calculate $P(W > 40)$ and $P(W > 70)$.

b Determine the expectation and the standard deviation of W.

c Calculate real numbers a and b such that at their right-hand sides the density of the χ^2_{40} distribution allows areas of 0.10 and 0.05, respectively. What are the standard notations for a and b?

Exercise 17.2

Let S^2 be the sample variance of a random sample $X_1, \ldots, X_{25}$ drawn from a normal population with mean μ and variance σ^2.

a Determine an rv U that is based on S^2 and has a χ^2 distribution. How many degrees of freedom does this distribution have?

b Explain why U is a pivot.

c Suppose now that $\sigma^2 = 36$. Calculate $P(S^2 > 37.5)$ by rewriting the event $\{S^2 > 37.5\}$ as a U event.

d Again for $\sigma^2 = 36$, calculate the constant b such that $P(S^2 > b) = 0.05$.

Exercise 17.3

For a certain population of people with middle incomes it seems reasonable to assume that the variable $X = $ 'gross hourly wage' is normally distributed. A random sample of 300 people yields the following results about their gross hourly wages:

$\bar{x} = 18.24$ and $s = 6.36$

a Determine a 95% confidence interval for the variance of X. Interpret it.

b Five years ago, a study about income dissimilarities concluded that the standard deviation of the gross hourly wages in this population of middle incomes was 5.8455. Can it be concluded that nowadays the spread of the middle incomes has increased? Use part (a) to answer this question.

Exercise 17.4 (computer, but not necessarily)

To evaluate the opinion of his students about the course he has taught over the past 10 weeks, a lecturer asks a sample of 30 students to assign a mark between 0 and 10 to their individual overall satisfaction with respect to the course. The lecturer expects that the mean satisfaction of all his students will be at least 7 (which was the mean score last year). Moreover, he believes that the consensus of the students about the course will be large; he expects that the population standard deviation will be less than 1.5.

The results are in the dataset Xrc17-04.sav. (If you do not have a computer: it follows that $\bar{x} = 7.6$ and $s = 1.2825$.)

a Use a hypothesis test with significance level 0.02 to test the conjecture of the lecturer about the mean satisfaction.

b Use a hypothesis test with significance level 0.10 to test the conjecture of the lecturer about the consensus.

c Calculate the p-value of the test in part (b). Interpret the answer.

Exercise 17.5 (computer, but not necessarily)

In July/August 2007 the US banking sector was confronted with liquidity problems as a result of subprime lending and increasing mortgage rates. As a consequence, many stock exchanges lost a considerable part of their value. This also was the case at the Amsterdam Stock Exchange (AEX).

Some people argue that from 24 July 2007 onwards a 'restless' period started, with market returns showing higher volatility than usual. We will study this for the AEX index.

The dataset Xrc17-05.xls contains the daily returns (%) of the AEX for the period 25 July 2007–1 October 2007. We will assume that these data are typical for the restless period and that they come from a normal random sample. By studying the historical returns of the AEX index over a long period ending just before 24 July 2007, it was found that the volatility of the daily returns, as measured by the (population) standard deviation, was smaller than 0.95%.

a Use a hypothesis test with significance level 0.01 to investigate whether the volatility of the AEX returns in the period following 15 July 2007 was larger than 0.95%. (If you do not have a computer at your disposal: $n = 49$ and $s = 1.452172$.)

b Also investigate it with a 99% confidence interval.

Exercise 17.6

Consider an rv W that is χ^2_9-distributed.

a Determine the expectation μ and the standard deviation σ of W.

b Will the peak of the density be to the left or to the right of μ? Why?

c Calculate $P(W < k)$ for $k = 2, 5, 8, 11, 14, 17$. Also calculate $P(2 < W < 12)$.

d For $\alpha = 0.10$, calculate the quantiles $\chi^2_{\alpha;9}$, $\chi^2_{1-\alpha;9}$, $\chi^2_{\alpha/2;9}$ and $\chi^2_{1-\alpha/2;9}$. Make rough pictures to indicate the meaning of these quantiles.

Exercise 17.7

As mentioned in earlier chapters, the weight X (in kg) of an adult man can be modelled very well by a normal distribution. Consider a large population of adult men with mean weight $\mu = 82.7$ kg and standard deviation $\sigma = 9.9970$ kg.

Twenty men will be chosen from this population at random and their weights (in kg) $X_1, \ldots, X_{20}$ will be measured. Interest is in probabilities with respect to the sample variance S^2 and the rv $W = (n - 1)S^2 / \sigma^2$.

a Determine the probability distribution of W. Also determine $E(W)$ and $SD(W)$.

b Calculate the probabilities $P(18 < W < m)$ for $m = 19, 22, 25, 28, 31$. What are the implications for S^2?

c Calculate the real number c such that $P(W < c) = 0.75$.

d Determine constants a and b with $a < b$ such that the density of W has an area 0.005 at the right-hand side of b and also at the left-hand side of a. What are the formal notations for a and b?

e Calculate $P(90 < S^2 < 100)$. Also determine $E(S^2)$.

Exercise 17.8

To guarantee the wages of their staff it is important for shopkeepers to receive a regular stream of income from sales. Too much variation in weekly sales is often unwanted.

The manager of a large fashion shop wants to check whether the standard deviation of the weekly sales (in thousands of euro) is larger than 10 (which is the preferred level). To find out, she treats the weekly sales of the most recent five weeks as being the realization of a normal random sample. Here are the data:

75   95   110   80   115

a Use a 95% confidence interval to find out whether the population standard deviation of the weekly sales is larger than 10.

b Also conduct a hypothesis test with $\alpha = 0.025$ to check it.

c Calculate the p-value of the test in part (b). Interpret your answer.

d Do you believe that the sales of the most recent five weeks indeed come from a normal random sample?

Exercise 17.9 (computer, but not necessarily)

In the inventory control of a product, the variation of the daily demand is important, especially when the firm claims to have the product in stock.

A firm advertises that it has the product that it sells always in stock. On average, 200 items of the product are sold per day. On every business day, in the evening, the stock is replenished to 300. The firm's management says that this storage amount is precisely right since the standard deviation of the demand is 33. However, an employee believes that this claim about the standard deviation is incorrect. To check his belief, he gathers the daily demands of the most recent 300 business days; the data are stored in the file Xrc17-09.sav. When answering the questions below, assume that the daily demand is normally distributed. (If you do not have a computer at your disposal: the sample mean and standard deviation are respectively 204.50 and 42.238.)

a Can you explain why it is 'precisely right' if the standard deviation is 33?

b Check the belief of the employee with a hypothesis test with significance level 0.01.

c Calculate the p-value of the test in part (b).

d Also check (with a hypothesis test with $\alpha = 0.05$) whether the mean daily demand is unequal to 200.

Exercise 17.10 (computer, but not necessarily)

Mrs Doubledutch is very economical, but likes new things. When buying a new Peugeot 307 SW, the car salesman has told Mrs Doubledutch that this type of car:

■ does on average more than 12 kilometres per litre petrol;

■ is very constant in using petrol; the standard deviation is less than 0.7 km/litre.

To check these claims, Mrs Doubledutch writes down the number of kilometres driven **and** the number of litres used during the first 21 fill-ups. The resulting measurements of the variable $X =$ 'number of kilometres per litre petrol' are in the file Xrc17-10.sav. (If you do not have a computer at your disposal: the sample mean and standard deviation are, respectively, 12.193105 and 0.639236.)

a Use a 95% confidence interval to check the salesman's first claim.

b Use a hypothesis test with significance level 0.05 to check the salesman's second claim.

c Also calculate the p-value of the test in part (b) and interpret it.

Exercise 17.11 (computer)

The aim of the present study is to compare the daily returns (%) at the Amsterdam Stock Exchange (AEX) with the daily returns (%) at the NASDAQ. On each business day during the period 1 May 2002–30 April 2004 the daily returns (%) at both exchanges (when compared to the day before) were measured; see the file Xrc17-11.sav. It seems reasonable to assume that daily returns at one exchange are independent. Of course, the two returns (on one day) at the two exchanges are **not** independent.

a Create a new column that contains the differences between the daily returns of the AEX index and the corresponding daily returns of the NASDAQ. Call this column D.

b Create a histogram for the data of D and compare it with the best-fitting normal curve. Report your findings.

Assume that the difference D between the daily returns (%) of the AEX index and the NASDAQ index is normally distributed.

c Test whether the mean μ_D of **all** differences is negative; use $\alpha = 0.05$. Conduct a formal hypothesis test by using the computer in two ways:

i Calculate only the mean and standard deviation of the data in column D. Use these results to calculate the value of the test statistic.

ii Use a procedure that immediately returns the value of the test statistic. What does your conclusion imply for the means μ_A and μ_N of the returns at AEX and NASDAQ?

d Use the second approach of part (c) to determine the p-value of the test.

Suppose that the population standard deviation of the daily returns of the NASDAQ index is known: it is equal to 2.24%.

e Use a hypothesis test with significance level 0.05 to test whether the population standard deviation of the returns of the AEX index is smaller than 2.24.

f Give your final comments.

Exercise 17.12 (computer)

Reconsider Exercises 15.26 and 16.24, about young people's views with respect to 'actively benefiting from illegal activity'. In Exercise 15.26 it was assumed that all population variances of the eight country subpopulations are 1, but in Exercise 16.24 nothing was assumed. We will investigate for Germany which of the two approaches is best. See Xrc15-26.sav.

Use a hypothesis test with $\alpha = 0.05$ to check whether the population variance of the variable X for the subpopulation of German business students is different from 1.

CASE 17.2 FTSE 100 AND THE COLLAPSE OF THE US HOUSING MARKET

In July/August 2007, the US housing market collapsed because of the liquidity problems of US banks and increasing mortgage rates. As a consequence, the value of many (US and European) stock exchanges dropped dramatically.

After that event, an uncertain and restless period started; optimistic days were followed by pessimistic days. Newspaper articles claimed that, during that period, the volatility of many share prices was much higher than usual, which reflects this uncertainty.

You are invited to show that, statistically, the FTSE 100 at London Stock Exchange experienced larger volatility than usual in the period after 25 July 2007. Show that:

a The standard deviation of the daily returns (%) in the period after 25 July 2007 was larger than 0.90, an upper bound for the standard deviation over a long period before that date.

b The mean of the daily range (= max − min) of the price of the FTSE 100 in the period after 25 July 2007 was larger than 50, an upper bound for the mean of the daily ranges over a long period before that date.

Base your study on the dataset Case17-02.sav (from www.Euronext.com) that contains the daily returns and daily ranges for the period 25 July–2 October 2007.

CHAPTER

18

Confidence intervals and tests to compare two parameters

Chapter contents

The topic of the present chapter is the difference between two unknown parameters of the same type: two population means μ_1 and μ_2, two population proportions p_1 and p_2, or two population variances σ_1^2 and σ_2^2. For statistical inference, two samples are available. Hypothesis testing and interval estimation will be considered for the difference parameters $\mu_1 - \mu_2$ and $p_1 - p_2$. To study the difference between σ_1^2 and σ_2^2, the ratio parameter σ_1^2 / σ_2^2 will be compared with 1. Our starting point is that we now have **two** populations and one variable.

CASE 18.1 DID THE CHANGES TO STATISTICS 2 INCREASE THE GRADES?

The instructor of the course Statistics 2 for Business and Economics was very dissatisfied about the grades (scale 0–10) and the study attitude of his students. To improve things, he took some actions for the following year's version of the course:

■ making use of another – more challenging – book;

■ maintaining the Team Assignment (weight 0.15) and the Final Exam (weight 0.45), but offering two Midterms (weights 0.2) instead of one;

■ a threshold of 5, in the following sense: if the final score (calculated with these weights) is at least 5.5 but the score for the Final Exam is **less than** 5, then the final grade becomes 5.

As soon as the final grades of the altered course are available, the instructor compares them with the final grades of the year before; see the file Case18-01.sav. Did the changes increase the mean grade? For the solution, see the end of Section 18.3.2.

18.1 Some problems with two parameters

The following problems all belong to the present chapter.

1 Within the first-year education program of Economics, is the mean grade for the course Mathematics 1 more than 0.5 larger than the mean grade for the course Statistics 1?

2 For the course Statistics 1, is the mean grade of students in International Business different from the mean grade of students in Economics?

3 Is the mean profitability of the companies in the construction sector larger than the mean profitability of the companies in the textiles sector?

4 For a certain group of working people, is the mean hourly wage of the men greater than the mean hourly wage of the women?

5 For all husband–wife couples where both partners have a job, is the mean hourly wage of the husbands greater than the mean hourly wage of the wives?

6 Is the lifespan (measured in number of kilometres until worn out) of a newly developed tyre greater than the lifespan of the older type of tyres?

7 Is the unemployment rate in Germany different from the unemployment rate in Denmark?

8 Do consumers prefer cola P or cola C?

9 Has the percentage of people that approves a certain government measure changed after the discussion on television?

10 Is the risk of portfolio I larger than the risk of portfolio II?

11 Is the new machine more productive than the old one?

12 Is the new machine more precise in creating axles of length 5 cm than the old one?

Questions 1–7 and 11 are in essence all about a difference $\mu_1 - \mu_2$ of two population means. Questions 8 and 9 are about a difference $p_1 - p_2$ of two proportions; questions 10 and 12 can be reformulated in terms of the comparison of two population variances.

Our starting point is that we now have **two** populations – population 1 and population 2 – and **one** variable X that is considered on both. If X is quantitative, we are interested in comparing the means μ_1 and μ_2 of X or the variances σ_1^2 and σ_2^2 of X on the two populations. If X is a qualitative variable that takes only two values (success and failure), then the success-proportions p_1 and p_2 on the two populations are compared.

The comparisons will be done on the basis of **two** random samples, one from each population. Interval estimators and test statistics will be constructed for the parameters of interest: $\mu_1 - \mu_2$, $p_1 - p_2$ and σ_1^2/σ_2^2. Again, the approaches for each of these parameters will be based on the steps A–D of Section 17.1:

- Find a good point estimator (step A).
- Transform this estimator into a suitable pivot (step B).
- Construct an interval estimator from that pivot and its distribution (step C).
- Construct a test statistic by first replacing the parameter of interest by the hinge of the testing problem and next replacing any further unknown parameters by point estimators (step D).

18.2 The difference between two population means

To compare μ_1 and μ_2, their difference $\mu_1 - \mu_2$ is taken as parameter. For statistical inference about it, two random samples are drawn: one from population 1 and one from population 2. For both random samples the variable X is observed, which yields two random samples of X observations.

The **precise** description of the way the sampling has to be done, is called the ***experimental design***. In this section two experimental designs will be considered:

a The two random samples are drawn independently of each other.

b The two random samples are paired, for instance by way of another variable.

In problem 1 of Section 18.1, both designs are technically possible. Two samples can be drawn independently of each other (which is design (a)): one from the population of the Mathematics 1 participants and one from the population of the Statistics 1 participants. However, it is also possible to use design (b): choose students at random from the list of the participants of **both** exams and record (as pairs) their grades for Mathematics 1 and Statistics 1.

In problem 6, the design can be as follows: 50 cars are chosen at random; 25 get tyres of the new type and 25 get the tyres of the old type. Finally, when the tyres are worn out, this experimental design will give the researchers two sets of data (one for the new type and one for the old type) by recording for each car the average lifespan of the four tyres. The important point of this design is that the two samples are independent of each other since the cars of the first sample are chosen independently of the cars of the second sample. So, it is design (a). But problem 6 can also be tackled with experimental design (b): randomly select 25 cars; for each car, put two tyres of the new type on the wheels left-front and left-rear, and two tyres of the old type on the wheels right-front and right-rear. This design will also finally yield two samples (columns) of data, but now the samples are dependent because they are paired.

It will be intuitively clear that, for problem 6, the experimental design (b) is preferable. Since the two types of tyres are tested on the same cars, the variation due to other factors (such as different driving habits of the drivers) is diminished.

Both designs yield two sets of sample data. In design (a) the two sample sizes can be unequal; in design (b) the two sample sizes have to be equal because of the pairing.

The choice of the design depends on the problem of interest. Sometimes only one of the two types is technically possible; sometimes both designs are possible although often one of the two is preferred.

Design (a) is often called the ***independent samples design***. Design (b) is often referred to as the ***matched-pairs design*** or the ***paired observations design***.

The sample data can be stored in two ways:

- the data of sample 1 in column 1 of a dataset, those of sample 2 in column 2;
- the data of both samples in column 1 while column 2 indicates whether the observations come from sample 1 or sample 2.

In the case of design (b), the data are usually stored as in the first approach: the two columns give the two random samples while the pairs can be read horizontally. In the case of design (a), both approaches are possible. The second approach is preferred if the computer package is ***row-oriented*** (like SPSS). That is, if the default statistical procedures of the package assume that the observations in a row of the spreadsheet all come from one population element. The first approach is preferred if the package is ***cell-oriented*** (like Excel), that is, if the **cells** represent population elements. See Appendix A1.18 for more information about this topic.

18.2.1 Two independent samples

In this subsection we analyse the independent samples design. This random experiment draws two random samples – one from each population – in such a way that the two samples are themselves independent. This mutual independence turns out to be crucial for obtaining good interval estimators and hypothesis tests.

For statistical inference about $\mu_1 - \mu_2$, two random samples are drawn: sample 1 is of size n_1 and is taken from population 1; sample 2 is of size n_2 and is taken from population 2. There is one important addendum: the two samples are drawn **independently of each other**. For each sampled population element, the variable X is observed. This leads to two independent random samples of observations of X; see below. The first of the two subscripts refers to the number (1 or 2) of the sample:

$X_{11}, X_{12}, ..., X_{1n_1}$ denotes random sample 1

$X_{21}, X_{22}, ..., X_{2n_2}$ denotes random sample 2

To estimate $\mu_1 - \mu_2$, we start with the two sample means:

$$\bar{X}_1 = \frac{1}{n_1}\sum_{i=1}^{n_1} X_{1i} \quad \text{and} \quad \bar{X}_2 = \frac{1}{n_2}\sum_{i=1}^{n_2} X_{2i}$$

The usual estimator of $\mu_1 - \mu_2$ is just the difference $\bar{X}_1 - \bar{X}_2$. We derive its properties.

Since both samples are random samples, the estimators $\bar{X}_1$ and $\bar{X}_2$ are unbiased and consistent estimators of, respectively, μ_1 and μ_2; see (13.1) and (13.2). From the rules of the expectation operator (see (8.31)) it then follows that $\bar{X}_1 - \bar{X}_2$ is also unbiased:

$$E(\bar{X}_1 - \bar{X}_2) = E(\bar{X}_1) - E(\bar{X}_2) = \mu_1 - \mu_2 \tag{18.1}$$

Because of consistency, the estimators $\bar{X}_1$ and $\bar{X}_2$ approximate the parameters μ_1 and μ_2 arbitrarily closely if n_1 and n_2 are large enough. Hence, the estimator $\bar{X}_1 - \bar{X}_2$ will approximate the parameter $\mu_1 - \mu_2$ very precisely if **both** sample sizes are large enough. That is, $\bar{X}_1 - \bar{X}_2$ is a consistent estimator of $\mu_1 - \mu_2$.

To obtain the variance of this estimator, first notice that – again thanks to the fact that the two samples are random samples – the variances of the two sample means $\bar{X}_1$ and $\bar{X}_2$ can be expressed in the two population variances; see (13.1):

$$V(\bar{X}_1) = \frac{\sigma_1^2}{n_1} \quad \text{and} \quad V(\bar{X}_2) = \frac{\sigma_2^2}{n_2}$$

Now recall from (11.12) that, for two general **independent** rvs X and Y, the variance of their difference is equal to the **sum** of the two individual variances:

$$V(X - Y) = V(X) + V(Y) \tag{18.2}$$

Since the two random samples are independent, the two sample means are independent too. Hence, (18.2) can be applied:

$$V(\bar{X}_1 - \bar{X}_2) = V(\bar{X}_1) + V(\bar{X}_2)$$

Former results can be substituted. The following box summarizes things:

Basic properties of the estimator $\bar{X}_1 - \bar{X}_2$ of $\mu_1 - \mu_2$

The estimator is unbiased and consistent. (18.3)

$$V(\bar{X}_1 - \bar{X}_2) = \frac{\sigma_1^2}{n_1} + \frac{\sigma_2^2}{n_2} \tag{18.4}$$

$$SD(\bar{X}_1 - \bar{X}_2) = \sqrt{\frac{\sigma_1^2}{n_1} + \frac{\sigma_2^2}{n_2}} \tag{18.5}$$

Notice that (18.5) reflects the consistence of the estimator $\bar{X}_1 - \bar{X}_2$: as the two sample sizes grow large, the standard deviation becomes small.

To create interval estimators and test statistics, we again follow the steps A–D. We have just finished step A, the search for a good estimator. The next step is to find a suitable pivot. For that, it will – similar to Section 13.4 – always be assumed that at least one of the following two situations is valid:

Situation 1   The distributions $N(\mu_1, \sigma_1^2)$ and $N(\mu_2, \sigma_2^2)$ are good models for the variable X on, respectively, population 1 and population 2.

Situation 2   Both sample sizes n_1 and n_2 are large.

By (13.8) it then follows that:

$$\overline{X}_1 \approx N(\mu_1, \frac{\sigma_1^2}{n_1}) \quad \text{and} \quad \frac{\overline{X}_1 - \mu_1}{\sigma_1/\sqrt{n_1}} \approx N(0, 1)$$

$$\overline{X}_2 \approx N(\mu_2, \frac{\sigma_2^2}{n_2}) \quad \text{and} \quad \frac{\overline{X}_2 - \mu_2}{\sigma_2/\sqrt{n_2}} \approx N(0, 1)$$

Since $\overline{X}_1$ and $\overline{X}_2$ are independent, it follows from (11.19) that $\overline{X}_1 - \overline{X}_2$ is also (approximately) normal. Here is the precise result:

Results that are essential for statistical inference about $\mu_1 - \mu_2$

$$\overline{X}_1 - \overline{X}_2 \approx N\left(\mu_1 - \mu_2, \frac{\sigma_1^2}{n_1} + \frac{\sigma_2^2}{n_2}\right)$$

$$Z = \frac{\overline{X}_1 - \overline{X}_2 - (\mu_1 - \mu_2)}{\sqrt{\frac{\sigma_1^2}{n_1} + \frac{\sigma_2^2}{n_2}}} \approx N(0, 1) \tag{18.6}$$

Note that Z is a pivot. We have just finished step B.

To obtain an interval estimator for $\mu_1 - \mu_2$, the unknown population variances in the denominator of the pivot have to be replaced by good estimators. An obvious choice is to use the two sample variances:

$$S_1^2 = \frac{1}{n_1 - 1} \sum_{i=1}^{n_1}(X_{1i} - \overline{X}_1)^2 \quad \text{and} \quad S_2^2 = \frac{1}{n_2 - 1} \sum_{i=1}^{n_2}(X_{2i} - \overline{X}_2)^2 \tag{18.7}$$

Indeed, this sometimes will be the approach that is followed. But be aware of the fact that replacing the population variances by the respective sample variances introduces **two** new sources of variation. That is why it is natural to give special attention to the case that the two population variances are equal and hence only one parameter has to be replaced by an estimator.

The case that the two population variances are equal

Suppose that the two population variances σ_1^2 and σ_2^2 are equal; denote them both by σ^2, that is, $\sigma_1^2 = \sigma_2^2 = \sigma^2$. Then the denominator of Z in (18.6) can be rewritten as:

$$SD(\overline{X}_1 - \overline{X}_2) = \sqrt{\frac{\sigma^2}{n_1} + \frac{\sigma^2}{n_2}} = \sqrt{\sigma^2\left(\frac{1}{n_1} + \frac{1}{n_2}\right)} \tag{18.8}$$

Only one unknown parameter (σ^2) has to be replaced by a good estimator. Since the two population variances are equal, both samples yield information about the common population variance. So, it is wise to use **both** samples to obtain a good estimator of σ^2. The mean of all squared random deviations seems to be a good candidate. The $n_1 + n_2$ random deviations are respectively equal to:

$$X_{11} - \overline{X}_1, X_{12} - \overline{X}_1, \ldots, X_{1n_1} - \overline{X}_1 \quad \text{and} \quad X_{21} - \overline{X}_2, X_{22} - \overline{X}_2, \ldots, X_{2n_2} - \overline{X}_2$$

Notice that the first n_1 of these deviations add up to 0; the same holds for the last n_2. In total, this costs 2 degrees of freedom. That is why the 'mean' of the squared deviations is taken as follows:

$$S_p^2 = \frac{\sum_{i=1}^{n_1}(X_{1i} - \bar{X}_1)^2 + \sum_{i=1}^{n_2}(X_{2i} - \bar{X}_2)^2}{n_1 + n_2 - 2} \tag{18.9}$$

Note that S_p^2 has $n_1 + n_2 - 2$ degrees of freedom. Because of (18.7), the two summations in the numerator are respectively equal to $(n_1 - 1)S_1^2$ and $(n_2 - 1)S_2^2$. Hence, S_p^2 can also be written as:

$$S_p^2 = \frac{n_1 - 1}{n_1 + n_2 - 2}S_1^2 + \frac{n_2 - 1}{n_1 + n_2 - 2}S_2^2 \tag{18.10}$$

This sample variance is called the **pooled sample variance**. Since S_1^2 and S_2^2 are both unbiased estimators of σ^2, the pooled variance is also an unbiased estimator of σ^2. Since S_p^2 is based on both samples, it is intuitively clear that the pooled sample variance is a better estimator than each of the estimators S_1^2 and S_2^2.

Equation (18.10) guarantees that the pooled variance falls between S_1^2 and S_2^2. If the two sample sizes are equal (that is, if the design is a **balanced design**), the pooled variance is precisely the average of the variances S_1^2 and S_2^2, and is equally distanced from them. However, if the first sample is larger than the second sample, the pooled variance will lie closer to S_1^2 than to S_2^2.

By replacing the common population variance σ^2 in (18.8) by the estimator S_p^2, the SD transforms into SE (standard error). The rv Z in (18.6) transforms into T:

$$T = \frac{\bar{X}_1 - \bar{X}_2 - (\mu_1 - \mu_2)}{\sqrt{S_p^2(\frac{1}{n_1} + \frac{1}{n_2})}} \tag{18.11}$$

This variable has – if situation 1 and/or situation 2 is valid – the (approximate) distribution $t_{n_1+n_2-2}$ and can be used to construct an interval estimator that captures $\mu_1 - \mu_2$ with probability $1 - \alpha$. Here is the result:

Interval estimator (L, U) for $\mu_1 - \mu_2$ in the case of equal variances:

$$L = \bar{X}_1 - \bar{X}_2 - t_{\alpha/2;n_1+n_2-2}\sqrt{S_p^2(\frac{1}{n_1} + \frac{1}{n_2})} \tag{18.12}$$

$$U = \bar{X}_1 - \bar{X}_2 + t_{\alpha/2;n_1+n_2-2}\sqrt{S_p^2(\frac{1}{n_1} + \frac{1}{n_2})}$$

We have just finished step C. Notice that this interval estimator is of format 1; see Section 15.2.

When the data are observed, they can be substituted into (18.12). The result (l, u) is a $(1-\alpha)$ confidence interval for $\mu_1 - \mu_2$.

Next, we consider testing problems concerning $\mu_1 - \mu_2$, with hinge h. A test statistic follows from the pivot in (18.6) and its rewritten denominator of (18.8). According to step D, we first replace the parameter by the hinge and next replace the remaining unknown parameter σ^2 by the estimator. The resulting test statistic is $t_{n_1+n_2-2}$-distributed if the hinge h is the true value of the parameter $\mu_1 - \mu_2$. Here is the resulting five-step test procedure:

Five-step test procedures for $\mu_1 - \mu_2$: *equal variance test*

i Testing problems: (a) test $H_0: \mu_1 - \mu_2 \leq h$ against $H_1: \mu_1 - \mu_2 > h$
(b) test $H_0: \mu_1 - \mu_2 \geq h$ against $H_1: \mu_1 - \mu_2 < h$
(c) test $H_0: \mu_1 - \mu_2 = h$ against $H_1: \mu_1 - \mu_2 \neq h$

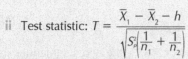

ii Test statistic: $T = \dfrac{\bar{X}_1 - \bar{X}_2 - h}{\sqrt{S_p^2\left(\dfrac{1}{n_1} + \dfrac{1}{n_2}\right)}}$

iii Rejection: (a) reject $H_0 \Leftrightarrow t \geq t_{\alpha;n_1+n_2-2}$
                  (b) reject $H_0 \Leftrightarrow t \leq -t_{\alpha;n_1+n_2-2}$
                  (c) reject $H_0 \Leftrightarrow t \leq -t_{\alpha/2;n_1+n_2-2}$ or $t \geq t_{\alpha/2;n_1+n_2-2}$

iv Calculate **val**

v Draw the conclusion by comparing **val** with step (iii)

Statistical procedures that are based on the assumption that the two population variances are equal are said to follow the **equal variance approach**.

Example 18.1

Employees in the non-commercial services sector often claim that their annual income is too low by comparison with the annual incomes of employees in the commercial services sector, although the hourly wage in the first sector is **larger** than the hourly wage in the second. The reason is that the non-commercial services sector does not offer enough opportunities for paid overtime.

A researcher has been commissioned by a civil service union to investigate whether this complaint is justified. She starts by comparing the mean gross hourly wages of the two sectors. For that, she randomly selects 150 employees from the non-commercial sector and 100 from the commercial sector; for each employee she records the gross hourly wage X.

Population 1: employees of the non-commercial services sector
Population 2: employees of the commercial services sector

Here are the summarized results:

$$\bar{x}_1 = 19.20 \quad \text{and} \quad s_1^2 = 67.2400; \quad \bar{x}_2 = 17.10 \quad \text{and} \quad s_2^2 = 73.1025$$

Is the mean hourly wage μ_1 of the non-commercial services sector indeed larger than the mean hourly wage μ_2 of the commercial services sector? Answer this question in two ways: with a 90% confidence interval and with a hypothesis test with significance level 0.05.

The question can be restated as: is $\mu_1 - \mu_2 > 0$? So, this statement is taken as the alternative hypothesis. Since the two sample variances are not far apart, the researcher decides to take the equality of the corresponding population variances as her starting point. (Later, in Section 18.3, a test will be considered that will justify that decision.) To calculate the confidence interval with (18.12), the pooled variance has to be calculated first:

$$s_p^2 = \frac{n_1 - 1}{n_1 + n_2 - 2}s_1^2 + \frac{n_2 - 1}{n_1 + n_2 - 2}s_2^2 = \frac{149}{248} \times 67.24 + \frac{99}{248} \times 73.1025 = 69.5803$$

Hence, the realized value of the standard error of $\bar{X}_1 - \bar{X}_2$ is equal to:

$$\sqrt{s_p^2\left(\frac{1}{n_1} + \frac{1}{n_2}\right)} = \sqrt{69.5803 \times 0.0167} = 1.0769$$

Since $t_{0.05;248} = 1.6510$ (*) and $\bar{x}_1 - \bar{x}_2 = 2.10$, we obtain:

$l = 2.10 - 1.6510 \times 1.0769 = 0.322$

$u = 2.10 + 1.6510 \times 1.0769 = 3.878$

The 90% confidence interval is (0.322, 3.878); there is 90% confidence that $\mu_1 - \mu_2$ lies in this interval. In particular, there is (at least) 90% confidence that $\mu_1 - \mu_2$ is larger than 0.322 and hence that it is larger than 0. Indeed, we can be rather certain that the mean hourly wage in the non-commercial services sector is larger than the mean hourly wage in the commercial services sector.

The question can also be answered on the basis of a formal hypothesis test. Here is the five-step test procedure:

i Test $H_0: \mu_1 - \mu_2 \leq 0$ against $H_1: \mu_1 - \mu_2 > 0$; $\alpha = 0.05$

ii Test statistic: $T = \dfrac{\bar{X}_1 - \bar{X}_2 - 0}{\sqrt{S_p^2(\frac{1}{n_1} + \frac{1}{n_2})}}$

iii Reject $H_0 \Leftrightarrow t \geq t_{0.05;248} = 1.6510$

iv $val = \dfrac{2.10}{1.0769} = 1.95$

v Reject H_0 since val is larger than the critical value 1.6510

The hypothesis test yields the same conclusion.

Example 18.2

Some courses in statistics have – apart from the Final Exam – one or more midterms, to encourage students to work regularly. In general, the level of these midterms is lower than the level of the Final Exam.

The course Statistics 2 for International Business has two midterms, and the instructors attempt to make these exams slightly easier than the Final Exam. The question then arises: is the mean score for a midterm higher than the mean score for the Final Exam?

The file Xmp18-02.sav contains grades (scale 0–10) of a random sample of 60 International Business students. Note that the data are stored according to the second storage approach (described at the start of Section 18.2). For the first 30 students the grades for Midterm 2 are recorded; for the last 30 students the grades for the Final Exam. The first column lists the 60 grades; the second column gives the type of the exam (mid or final). Below, it will be investigated whether the conjecture is true that the (overall) mean grade for Midterm 2 is higher than the (overall) mean grade for the Final Exam.

Since the 60 students are sampled randomly, it follows immediately that the first and the second group of 30 students constitute **independent** random samples. The two samples are both of size 30, which in the present situation is enough to allow us to use the central limit theorem and take situation 2 above as our starting point. With the notations μ_{fin} and μ_{mid} for the overall means of the grades for the Final Exam and Midterm 2, the conjecture is restated as $\mu_{fin} - \mu_{mid} < 0$. This will be tested with $\alpha = 0.025$.

The table below sets out descriptive sample statistics obtained with SPSS.

Grade

TypeTest	Mean	N	Std. Deviation
final	5.3400	30	2.52007
mid	6.4333	30	2.52058
Total	5.8867	60	2.55896

Notice that the two sample standard deviations are rather close, so it seems reasonable to assume that the two population variances are equal. Hence, the equal variance test procedure will be used.

▶

Note that $n_1 + n_2 - 2 = 58$ and that $t_{0.025;58} = 2.0017$ (∗). The first three steps of the test procedure are as follows:

  i   Test $H_0: \mu_{fin} - \mu_{mid} \geq 0$ against $H_1: \mu_{fin} - \mu_{mid} < 0$

  ii   Test statistic: $T = \dfrac{\overline{X}_{fin} - \overline{X}_{mid} - 0}{\sqrt{S_p^2\left(\dfrac{1}{n_1} + \dfrac{1}{n_2}\right)}}$

  iii   Reject $H_0 \Leftrightarrow t \leq -2.0017$

The pooled variance is equal to:

$$s_p^2 = \tfrac{1}{2}(2.52007^2 + 2.52058^2) = 6.3520$$

Hence:

  iv $\mathbf{val} = \dfrac{5.3400 - 6.4333}{\sqrt{6.3520 \times (1/30 + 1/30)}} = \dfrac{-1.0933}{0.6507} = -1.680$

  v   Do not reject H_0 since **val** does not belong to the rejection region

We cannot conclude that the mean grade for Midterm 2 is higher than the mean grade for the Final Exam, at least not with significance level 0.025.

With many computer packages it is possible to run a default procedure that yields **val** immediately. The printout of a standard SPSS procedure that does this is presented below (see Appendix A1.18 for the explanation of its generation).

Independent Samples Test

		Levene's Test for Equality of Variances		t-test for Equality of Means						
		F	Sig.	t	df	Sig. (2-tailed)	Mean Difference	Std. Error Difference	95% Confidence Interval of the Difference	
									Lower	Upper
Grade	Equal variances assumed	.000	.998	−1.680	58	.098	−1.09333	.65075	−2.39594	.20928
	Equal variances not assumed			−1.680	58.000	.098	−1.09333	.65075	−2.39594	.20928

Indeed, the **val** (−1.680) appears opposite 'equal variances assumed' in the column headed by 't'. Notice that the printout also gives the 95% confidence interval for $\mu_{fin} - \mu_{mid}$. Since 0 is included, the conjecture that $\mu_{fin} < \mu_{mid}$ is not supported.

The case that the two population variances are unequal

For this case, as for the last, we conduct steps C and D. If the two population variances are unequal (or when it is unknown whether they are equal), we have to replace the two unknown parameters σ_1^2 and σ_2^2 of the pivot Z in (18.6) by their natural estimators S_1^2 and S_2^2, thus introducing **two** new sources of variation. Unfortunately, the probability distribution of the resulting random variable (say G) still depends on σ_1^2 and σ_2^2, and hence is not a t-distribution. One of the most cautious 'solutions' turns out to be to use a special t-distribution to **approximate** the distribution of G: the

t-distribution with m degrees of freedom where m is obtained by subtracting 1 from the smallest of the two sample sizes, that is: $m = \min(n_1, n_2) - 1$.

Below, the resulting interval estimator and the five-step procedures are formulated. Notice, however, that these statistical procedures are approximations (that become better when the two sample sizes are large).

Interval estimator (L, U) for $\mu_1 - \mu_2$ in the case of unequal population variances

$$L = \bar{X}_1 - \bar{X}_2 - t_{\alpha/2;m}\sqrt{\frac{S_1^2}{n_1} + \frac{S_2^2}{n_2}}$$

$$U = \bar{X}_1 - \bar{X}_2 + t_{\alpha/2;m}\sqrt{\frac{S_1^2}{n_1} + \frac{S_2^2}{n_2}} \quad \text{where } m = \min(n_1, n_2) - 1$$

(18.13)

Five-step test procedures for $\mu_1 - \mu_2$: *unequal variance test*

i Testing problems: (a) test $H_0: \mu_1 - \mu_2 \leq h$ against $H_1: \mu_1 - \mu_2 > h$
   (b) test $H_0: \mu_1 - \mu_2 \geq h$ against $H_1: \mu_1 - \mu_2 < h$
   (c) test $H_0: \mu_1 - \mu_2 = h$ against $H_1: \mu_1 - \mu_2 \neq h$

ii Test statistic: $G = \dfrac{\bar{X}_1 - \bar{X}_2 - h}{\sqrt{\frac{S_1^2}{n_1} + \frac{S_2^2}{n_2}}}$

iii Rejection: (a) reject $H_0 \Leftrightarrow g \geq t_{\alpha;m}$
   (b) reject $H_0 \Leftrightarrow g \leq -t_{\alpha;m}$
   (c) reject $H_0 \Leftrightarrow g \leq -t_{\alpha/2;m}$ or $g \geq t_{\alpha/2;m}$

iv Calculate *val*

v Draw the conclusion by comparing *val* with step (iii)

In a balanced design, with $n_1 = n_2 = n$, the denominator of G equals:

$$SE(G) = \sqrt{(S_1^2 + S_2^2)\frac{1}{n}} = \sqrt{\frac{(S_1^2 + S_2^2)}{2}\frac{2}{n}} = \sqrt{S_p^2(\frac{1}{n} + \frac{1}{n})}$$

Hence, the test statistic G of the unequal variance test for a balanced design coincides with the test statistic T of the equal variance test. Still, since $m = n - 1$, it is the distribution t_{n-1} that is used for the unequal variance test and not t_{2n-2}. This illustrates that the unequal variance test is indeed cautious: null hypotheses are less quickly rejected.

Statistical procedures that are based on the assumption that the two population variances are unequal are said to follow the **unequal variance approach**.

Example 18.3
The increased concentration of carbon dioxide (CO_2) in the atmosphere is a serious problem contributing to global warming. The European Commission has proposed a strategy to force car manufacturers to reduce the emissions from their new cars. As of 2012, CO_2 emissions must be no more than 130 g/km. By comparison, in 1997 **new** cars emitted between 158 and 223 g/km, in 2005 between 139 and 195 g/km.

Suppose that in 1997 for a random sample of 100 cars, a mean CO_2 emission of 215 g/km was measured, with a standard deviation of 21 g. In 2005, a random sample of 150 cars yielded a mean CO_2 emission of 190 g/km with standard deviation 31 g. Can it be concluded that the CO_2 emission has decreased by more than 20 g?

Denote the sets of all cars in 1997 and 2005 by population 1 and population 2, respectively. The variable of interest is $X = $ 'CO_2 emission (gram) per kilometre'. The question is whether the population mean μ_2 of X in 2005 is more than 20 smaller than the population mean μ_1 of X in 1997, that is, $\mu_2 < \mu_1 - 20$ or, to put it another way, $\mu_1 - \mu_2 > 20$.

About the population variance, the expectancy is that it will have increased due to the fact that nowadays there are new, 'clean' cars **and** old, polluting cars whereas in 1997 most cars were polluting. This belief is supported by the increase of the two sample variances from $21^2 = 441$ to $31^2 = 961$. That is why the unequal variance approach will be used to answer the question by way of a 95% confidence interval and a hypothesis test with significance level 0.025.

First notice that the minimum of 150 and 100 is 100, so $m = 99$. Furthermore, $t_{0.025;99} = 1.9842$ (*). The use of the interval estimator of $\mu_1 - \mu_2$ in (18.13) for the summarized results

$$\bar{x}_1 = 215; \quad s_1 = 21; \quad \bar{x}_2 = 190; \quad s_2 = 31$$

yields

$$\bar{x}_1 - \bar{x}_2 \pm t_{0.025;99} \sqrt{\frac{s_1^2}{100} + \frac{s_2^2}{150}} \Leftrightarrow 25 \pm 1.9842 \sqrt{\frac{21^2}{100} + \frac{31^2}{150}} \Leftrightarrow 25 \pm 6.5258$$

Hence, there is 95% confidence that $\mu_1 - \mu_2$ lies in the interval $(18.4742, 31.5258)$. But the number 20 falls in this interval, so the belief $\mu_1 - \mu_2 > 20$ is not supported by the data.

Here is the five-step procedure of the hypothesis test:

i Test $H_0: \mu_1 - \mu_2 \leq 20$ against $H_1: \mu_1 - \mu_2 > 20$; $\alpha = 0.025$

ii Test statistic: $T = \dfrac{\bar{X}_1 - \bar{X}_2 - 20}{\sqrt{\dfrac{S_1^2}{n_1} + \dfrac{S_2^2}{n_2}}}$

iii Reject $H_0 \Leftrightarrow t \geq t_{0.025;99} = 1.9842$

iv $val = \dfrac{5}{3.2889} = 1.5203$

v Do not reject H_0 since val is smaller than the critical value

The p-value of the test is 0.0658 (*). A significance level of 0.07 would have yielded the **rejection** of H_0.

18.2.2 Two paired samples

Again, one variable X is considered on two populations called 1 and 2. But now experimental design (b) (the matched-pairs design) is used: two random samples of equal size are drawn, one from each population, in such a way that the two samples are paired. In particular, the two random samples are **not** independent. For statistical inference about $\mu_1 - \mu_2$, the trick turns out to be to use the corresponding **differences** between the two samples and to forget about the two original samples. This will transform the statistical problems concerning $\mu_1 - \mu_2$ into statistical problems that were already considered in Chapter 16.

Suppose that the two random samples have equal size n and that they are paired. The two random samples can again be denoted as $X_{11}, X_{12}, \ldots, X_{1n}$ and $X_{21}, X_{22}, \ldots, X_{2n}$, but since they are paired, they can also be organized in pairs, as follows:

$$(X_{11}, X_{21}), (X_{12}, X_{22}), \ldots, (X_{1n}, X_{2n})$$

For convenience, we denote X on the populations 1 and 2 by V and W, respectively. Then the above sample of pairs of random variables can be considered as a random sample of n observations of (V, W). From this point of view, the parameter of interest $\mu_1 - \mu_2$ can also be written as $\mu_V - \mu_W$.

Now consider the variable $D = V - W$, the difference of V and W. Then the differences between the first and the second coordinates of the above sample pairs, that is

$$D_1 = X_{11} - X_{21}, \ D_2 = X_{12} - X_{22}, \ ..., \ D_n = X_{1n} - X_{2n}$$

can be considered as a random sample of observations of D. Denote the population mean and the population variance respectively by μ_D and σ_D^2. The mean of D is equal to the difference of the means of V and W, that is, $\mu_D = \mu_V - \mu_W$, which equals $\mu_1 - \mu_2$. Hence, the parameter of interest is μ_D and $D_1, D_2, ..., D_n$ is a random sample from the distribution of D, which has variance σ_D^2.

This puts us back in the set-up of Chapter 16: statistical interest is in **one** unknown population mean while the population variance is also unknown; statistical inference about this population mean is made on the basis of **one** random sample of size n. The only difference from the situation in Chapter 16 is that the population mean is now denoted by μ_D instead of μ, that the population variance is denoted by σ_D^2 instead of σ^2, and that the random sample is denoted by $D_1, ..., D_n$ instead of $X_1, ..., X_n$.

Because of Section 12.4, we take

$$\overline{D} = \frac{1}{n}\sum_{i=1}^{n} D_i \quad \text{and} \quad S_D^2 = \frac{1}{n-1}\sum_{i=1}^{n}(D_i - \overline{D})^2$$

as estimators of μ_D and σ_D^2. As in (13.1) it holds that:

$$E(\overline{D}) = \mu_D; \quad V(\overline{D}) = \frac{\sigma_D^2}{n}, \quad SD(\overline{D}) = \frac{\sigma_D}{\sqrt{n}}$$

It is assumed (as in Sections 16.2 and 16.3) that at least one of the following two situations is valid:

Situation 1 The normal distribution $N(\mu_D, \sigma_D^2)$ is a good model for D.
Situation 2 The sample size n is large.

Then we have

$$\overline{D} \approx N(\mu_D, \frac{\sigma_D^2}{n}) \quad \text{and} \quad Z = \frac{\overline{D} - \mu_D}{\sigma_D/\sqrt{n}} \approx N(0, 1)$$

These results are basic for the construction of interval estimators and hypothesis tests for $\mu_D = \mu_1 - \mu_2$. Again, the standard deviation $\sigma_D/\sqrt{n}$ has to be replaced by the standard error

$$SE(D) = \frac{S_D}{\sqrt{n}}$$

But then the (standard normal) z-distribution has to be replaced by the t_{n-1}-distribution. Here are the results:

Interval estimator (L, U) for $\mu_D = \mu_1 - \mu_2$ in case of paired observations

$$L = \overline{D} - t_{\alpha/2;n-1}\frac{S_D}{\sqrt{n}} \quad \text{and} \quad U = \overline{D} + t_{\alpha/2;n-1}\frac{S_D}{\sqrt{n}} \tag{18.14}$$

Five-step test procedure for $\mu_D = \mu_1 - \mu_2$ in the case of paired observations

i Testing problems: (a) test $H_0: \mu_D \leq h$ against $H_1: \mu_D > h$
(b) test $H_0: \mu_D \geq h$ against $H_1: \mu_D < h$
(c) test $H_0: \mu_D = h$ against $H_1: \mu_D \neq h$

ii Test statistic: $T = \dfrac{\overline{D} - h}{S_D / \sqrt{n}}$

iii Rejection: (a) reject $H_0 \Leftrightarrow t \geq t_{\alpha;n-1}$
(b) reject $H_0 \Leftrightarrow t \leq -t_{\alpha;n-1}$
(c) reject $H_0 \Leftrightarrow t \leq -t_{\alpha/2;n-1}$ or $t \geq t_{\alpha/2;n-1}$

iv Calculate **val**

v Draw the conclusion by comparing **val** with step (iii)

Example 18.4

Although the mean grade of a sample of 30 students for Midterm 2 is almost 1.1 higher than the mean grade of another sample of 30 students for the Final Exam, this difference was not enough to conclude that the overall mean score for Midterm 2 is higher than the overall mean score for the Final Exam; see Example 18.2. The reason for not rejecting $H_0: \mu_{fin} - \mu_{mid} \geq 0$ is that the numerator of the test statistic, the standard error of the difference between the two sample means, was relatively large.

To decrease variation, we now take the following experimental design. First, randomly draw 30 students who participated in both Midterm 2 and the Final Exam. Next, record for each of these students the pair of grades. The resulting dataset in Xmp18-04.sav consists of two columns of observations: the first contains the grades for Midterm 2 of these 30 students; the second contains the accompanying grades for the Final Exam. Notice that the two columns are paired: the two numbers in each row belong to one student.

The data are used for a paired-samples t-test. First, the column with the differences $d_1, \ldots, d_{30}$ has to be created; take the population of Final Exam grades as population 1, so subtract the Midterm 2 grades from the grades of the Final Exam. Only this new column is used for statistical inference. The table below contains its relevant statistics obtained with SPSS.

Descriptive Statistics

	N	Minimum	Maximum	Mean	Std. Deviation
D	30	−4.40	2.70	−1.0167	1.69910
Valid N (listwise)	30				

It follows that $\overline{d} = -1.0167$ and $s_D = 1.69910$. The testing problem is the same as in Example 18.2. Note that $\mu_D = \mu_{fin} - \mu_{mid}$. The critical value of the test procedure is now $t_{0.025;29} = 2.0452$ (*). Here is the five-step procedure:

i Test $H_0: \mu_D \geq 0$ against $H_1: \mu_D < 0$; $\alpha = 0.025$

ii Test statistic: $T = \dfrac{\overline{D} - 0}{S_D / \sqrt{n}}$

iii Reject $H_0 \Leftrightarrow t \leq -2.0452$

iv $val = \dfrac{-1.0167}{1.69910/\sqrt{30}} = \dfrac{-1.0167}{0.3102} = -3.278$

v Reject H_0 since **val** belongs to the rejection region

Since the standard error in the denominator of the test statistic is now considerably smaller, the **val** is small enough to reject H_0 at significance level 0.025.

Again, most computer packages have default procedures that generate **val** immediately. The printout generated by SPSS is shown below. See also Appendix A1.18. SPSS took the Midterm 2 grades as population 1.

Paired Samples Test

		Paired Differences					t	df	Sig. (2-tailed)
		Mean	Std. Deviation	Std. Error Mean	95% Confidence Interval of the Difference				
					Lower	Upper			
Pair 1	Mid2 – FinExam	1.01667	1.69910	.31021	.38221	1.65112	3.277	29	.003

Example 18.5

For a large population of people working at least 32 hours per week, interest is in wage differentials between men and women. Of course, it is possible to draw a random sample of men and, independently of that sample, a sample of women as well. However, since it is suspected that wage differentials (if present) might partially be caused by variation in age, the variable 'age' is used to create paired observations.

For each age between 18 and 60, one woman and one man are randomly drawn from the population and asked to tell their hourly wages. Unfortunately, not all 43 pairs did respond. The dataset Xmp18-05.sav contains 27 pairs of data; the ages are in column 1. Columns 2 and 3 contain the wages of the women and the men, paired according to their common ages in column 1. It is obvious that, because of this pairing, the data in the columns 2 and 3 are not independent. Hence, a paired t-test has to be conducted to test whether the mean hourly wage μ_f of women is smaller than the mean hourly wage μ_m of men who are of the same age. It will be tested whether $\mu_f - \mu_m$ is negative, with significance level 0.01.

Since $n = 27$, the test uses 26 degrees of freedom. Since the parameter is $\mu_f - \mu_m$, the variable D is the difference between the hourly wage of a woman and a man of the same age. The printout created with SPSS is shown below.

Paired Samples Test

		Paired Differences					t	df	Sig. (2-tailed)
		Mean	Std. Deviation	Std. Error Mean	95% Confidence Interval of the Difference				
					Lower	Upper			
Pair 1	Wage_Fem - Wage_Male	−4.53444	6.75423	1.29985	−7.20633	−1.86256	−3.488	26	.002

The formal five-step procedure below uses $t_{0.01;26} = 2.4786$ (∗). The **val** follows immediately from the printout.

  i Test H_0: $\mu_f - \mu_m \geq 0$ against H_1: $\mu_f - \mu_m < 0$; $\alpha = 0.01$

  ii Test statistic: $T = \dfrac{\overline{D} - 0}{S_D / \sqrt{n}}$

  iii Reject $H_0 \Leftrightarrow t \leq -2.4786$

iv **val** = −3.488

v Reject H_0 since **val** belongs to the rejection region

The mean of the hourly wages of the women is less than the mean of the hourly wages of the men who are of the same age.

The printout also gives a 95% confidence interval for $\mu_f - \mu_m$. There is 95% confidence that $\mu_f - \mu_m$ lies in the interval (−7.2063, −1.8626). In particular, there is (at least) 95% confidence that μ_f is more than €1.80 smaller than μ_m.

18.3 The ratio of two population variances

As before, one variable (called X) is considered on two populations called 1 and 2. In the present section the respective population variances σ_1^2 and σ_2^2 are compared. Instead of the difference parameter, the ratio parameter σ_1^2/σ_2^2 is the parameter of interest. Theory will be developed only for the case that the two **independent** random samples are **normal** random samples. Again, steps A–D will be followed.

Statistical procedures for comparing two population variances are frequently used in practice, for instance, to detect income dissimilarities between regions or countries, or to find out whether a new type of machine is more accurate than the old type, or to compare the risks of two financial portfolios. However, statistical procedures for comparing two population variances can also be used to choose between the equal variance procedure and the unequal variance procedure when an inference is to be drawn on the difference between two population means; see also Section 18.2.1.

Suppose that $N(\mu_1, \sigma_1^2)$ and $N(\mu_2, \sigma_2^2)$ are good models for the distribution of the variable X on population 1 and population 2, respectively. We consider only the experimental design of independent random samples.

Two random samples of size n_1 and n_2 are drawn from the populations 1 and 2, respectively. This is done in such a way that the two random samples are independent. For both samples, the variable X is observed. As in Section 18.2.1 we denote the two random samples of observations of X by $X_{11}, X_{12}, \ldots, X_{1n_1}$ and $X_{21}, X_{22}, \ldots, X_{2n_2}$, respectively. Since the estimators S_1^2 and S_2^2 (see (18.7)) of σ_1^2 and σ_2^2 are based only on sample 1 and sample 2, respectively, an important consequence is that these estimators are **independent** random variables.

Of course, it is the sample statistic S_1^2/S_2^2 that is used as point estimator of σ_1^2/σ_2^2. Since S_1^2 and S_2^2 are both consistent and hence can approximate σ_1^2 and σ_2^2 as closely as wanted (just by taking the sample sizes large enough), their ratio S_1^2/S_2^2 can approximate σ_1^2/σ_2^2 as closely as wanted to and hence is consistent too. This makes S_1^2/S_2^2 a good estimator of σ_1^2/σ_2^2. We have now finished step A.

To find a pivot in step B, the fact that S_1^2 and S_2^2 are independent is important, for the following essential result.

Important result for two independent random samples

If sample 1 is drawn from $N(\mu_1, \sigma_1^2)$ and sample 2 from $N(\mu_2, \sigma_2^2)$, then:

$$F = \frac{S_1^2/\sigma_1^2}{S_2^2/\sigma_2^2} \sim F_{n_1-1;\, n_2-1} \tag{18.15}$$

Here, $F_{n_1-1;\, n_2-1}$ denotes the so-called **F-distribution** with n_1-1 and n_2-1 degrees of freedom. Let us briefly study this new family of distributions, as an intermezzo.

18.3.1 Intermezzo: *F*-distributions

The definition of this family is as follows:

F_{ν_1, ν_2}-distribution

The **F-distribution with parameters ν_1 and ν_2** – denoted F_{ν_1, ν_2} – is the continuous probability distribution with probability density function f defined by:

$$f(y) = c_{\nu_1, \nu_2} \frac{y^{(\nu_1/2)-1}}{(1 + \nu_1 y/\nu_2)^{(\nu_1+\nu_2)/2}} \quad \text{for all positive real numbers } y \tag{18.16}$$

(and $f(y) = 0$ for non-positive y).

(The mathematical form of the pdf is less important.) The parameters ν_1 and ν_2 are positive integers; they are called the **numbers of degrees of freedom** of the distribution. The constant c_{ν_1, ν_2} (which depends on the two parameters) is not really important, but it is such that the total area under the graph of f is 1. Each choice of ν_1 and ν_2 yields a pdf, so (18.16) in essence defines a whole family of probability distributions: the family of the *F*-distributions. If an rv Y has the distribution F_{ν_1, ν_2}, we write $Y \sim F_{\nu_1, \nu_2}$ and say that Y is F_{ν_1, ν_2}-distributed. Notice that all densities in this family are concentrated on the **positive** half-line. Figure 18.1 shows the graphs of two pdfs of the family. As the χ^2-distributions, these densities are also **not** symmetric and they are also concentrated on $(0, \infty)$.

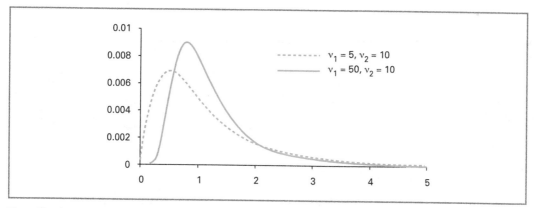

Figure 18.1 Graphs of the densities of $F_{5, 10}$ and $F_{50, 10}$

In the course of the present chapter we will need quantiles of *F*-distributions. Recall that the $(1 - \alpha)$-quantile of a continuous probability distribution is a real number located on the horizontal axis of the graph of the accompanying pdf in such a way that there is an area $1 - \alpha$ to its left and hence an area α to its right. So, the $(1 - \alpha)$-quantile of the distribution F_{ν_1, ν_2} is located on the horizontal axis of the accompanying density such that there is an area α to its right; the notation for this quantile is $F_{\alpha; \nu_1, \nu_2}$. Notice that the quantile $F_{1-\alpha; \nu_1, \nu_2}$ has an area $1 - \alpha$ to its **right**, so it is the α-quantile. Quantiles can easily be calculated with a statistical package; see also Appendix A1.18. If you prefer to use a table, see Appendix A4.

The distribution F_{ν_1, ν_2} has some properties that are worth mentioning. The first is that the mean of this distribution depends only on ν_2: it equals $\nu_2/(\nu_2 - 2)$ if $\nu_2 > 2$. It follows that the means of $F_{5, 10}$ and $F_{50, 10}$ in Figure 18.1 are both equal to 1.25. For large ν_2, the mean of the F_{ν_1, ν_2}-distribution is close to 1.

Another property of F_{ν_1, ν_2} is used frequently. It says that the inverse $U = 1/Y$ of an F_{ν_1, ν_2}-distributed rv Y also has an *F*-distribution, with the two degrees of freedom in opposite order. This property has consequences for quantiles. Since there is an area α at the right-hand side of $F_{\alpha; \nu_1, \nu_2}$, it follows that:

$$\alpha = P(Y > F_{\alpha; \nu_1, \nu_2}) = P\left(\frac{1}{Y} < \frac{1}{F_{\alpha; \nu_1, \nu_2}}\right) = 1 - P\left(\frac{1}{Y} \geq \frac{1}{F_{\alpha; \nu_1, \nu_2}}\right)$$

(In the second equality the inverses are taken, so 'smaller than' becomes 'larger than'.) Hence:

$$P\left(U \geq \frac{1}{F_{\alpha; \nu_1, \nu_2}}\right) = 1 - \alpha$$

On the other hand, the rv U is F_{ν_2, ν_1}-distributed and there is an area $1-\alpha$ at the right-hand side of $F_{1-\alpha; \nu_2, \nu_1}$, so:

$$P\left(U \geq F_{1-\alpha; \nu_2, \nu_1}\right) = 1 - \alpha$$

From the last two equations it follows that $F_{1-\alpha; \nu_2, \nu_1} = 1/F_{\alpha; \nu_1, \nu_2}$.

The above properties are summarized in the following box.

Properties of F_{ν_1, ν_2}

If an rv Y is F_{ν_1, ν_2}-distributed with $\nu_2 > 2$, then $E(Y) = \dfrac{\nu_2}{\nu_2 - 2}$ **(18.17)**

If Y is F_{ν_1, ν_2}-distributed, then $U = \frac{1}{Y}$ is F_{ν_2, ν_1}-distributed   **(18.18)**

$F_{1-\alpha; \nu_2, \nu_1} = \dfrac{1}{F_{\alpha; \nu_1, \nu_2}}$ **(18.19)**

Example 18.6

Suppose that an rv Y has the probability distribution $F_{40,40}$. It is required to cut off areas 0.025 in each of the two tails of the accompanying density, that is, the quantiles $F_{0.025; 40,40}$ and $F_{0.975; 40,40}$ are required.

With a computer it follows that $F_{0.025; 40,40} = 1.8752$ (*); see Appendix A1.18. Then the other quantile can easily be calculated, by (18.19):

$$F_{0.975; 40,40} = \frac{1}{F_{0.025; 40,40}} = \frac{1}{1.8752} = 0.5333$$

It follows that $P(0.5333 < Y < 1.8752) = 0.95$, so 95% of all the area under the pdf of $F_{40,40}$ lies between 0.5333 and 1.8752.

Remark. Historically, the family of the F-distributions was detected when scientists were searching for the probability distribution of random variables that were ratios of two sums of squared deviations. It was proved that, under some additional requirements, such a ratio can be transformed into an F-distributed rv after having divided the two sums of squared deviations by their respective numbers of degrees of freedom.

18.3.2 Interval estimator and tests for σ_1^2/σ_2^2

Let us now go back to (18.15), where F_{n_1-1, n_2-1} was introduced as the pdf of

$$F = \frac{S_1^2/\sigma_1^2}{S_2^2/\sigma_2^2} = \frac{S_1^2}{S_2^2} \div \frac{\sigma_1^2}{\sigma_2^2}$$

According to (18.7), the estimator S_1^2/S_2^2 is based on the ratio of two sums of squared deviations; it arises by dividing the numerator and the denominator of that ratio by their respective number of degrees of freedom $n_1 - 1$ and $n_2 - 1$. Because of the remark at the end of the previous subsection, it then becomes less surprising that S_1^2/S_2^2 can be transformed into an F-distributed rv; it would appear that it has to be divided by σ_1^2/σ_2^2.

Notice that both the numerator and the denominator of F in (18.15) are ratios of sample variance and population variance, the numerator for 1 and the denominator for 2. It would appear that the ratio of the two variance ratios has an F-distribution and hence is a pivot; thanks to the fact that the two normal random samples are independent. The two types of degrees of freedom $n_1 - 1$ and $n_2 - 1$ come, respectively, from the numerator and the denominator. According to (18.18), exchanging numerator and denominator of F yields a random variable (say) G that is F_{n_2-1,n_1-1} -distributed. F and G are both pivots (step B).

The rvs $F \sim F_{n_1-1,n_2-1}$ and $G \sim F_{n_2-1,n_1-1}$ can be rewritten in the following ways:

$$F = \frac{S_1^2/\sigma_1^2}{S_2^2/\sigma_2^2} = \frac{S_1^2}{S_2^2} \div \frac{\sigma_1^2}{\sigma_2^2} \quad \text{and} \quad G = \frac{S_2^2/\sigma_2^2}{S_1^2/\sigma_1^2} = \frac{S_2^2}{S_1^2} \times \frac{\sigma_1^2}{\sigma_2^2} \tag{18.20}$$

The right-hand expressions of F and G will be used to construct, respectively, a test statistic and an interval estimator for σ_1^2/σ_2^2.

First the interval estimator. Note that $F_{1-\alpha/2;n_2-1,n_1-1}$ and $F_{\alpha/2;n_2-1,n_1-1}$ are positive real numbers such that at their right-hand sides the respective areas under F_{n_2-1,n_1-1} are $1 - \alpha/2$ and $\alpha/2$. Hence, between these numbers an area $1 - \alpha$ remains (make a picture). That is:

$$1 - \alpha = P\left(F_{1-\alpha/2;n_2-1,n_1-1} < \frac{S_2^2}{S_1^2} \times \frac{\sigma_1^2}{\sigma_2^2} < F_{\alpha/2;n_2-1,n_1-1}\right)$$

Since we only want σ_1^2/σ_2^2 to remain in the middle, we multiply by S_1^2/S_2^2. This leads to rvs L and U that capture σ_1^2/σ_2^2 with probability $1 - \alpha$. Here is the result:

Interval estimator (L, U) for σ_1^2/σ_2^2

$$L = F_{1-\alpha/2;n_2-1,n_1-1}\frac{S_1^2}{S_2^2} \quad \text{and} \quad U = F_{\alpha/2;n_2-1,n_1-1}\frac{S_1^2}{S_2^2} \tag{18.21}$$

Requirement: two independent, normal random samples

We have just finished step C. By (18.19) it follows that:

$$F_{1-\alpha/2;n_2-1,n_1-1} = \frac{1}{F_{\alpha/2;n_1-1,n_2-1}} \tag{18.22}$$

That is, $F_{1-\alpha/2}$ is just the inverse of $F_{\alpha/2}$ **after exchanging** the degrees of freedom. Also notice that the above interval estimator is of format 2; see Section 15.2.

When the data are available, they can be substituted to obtain a $(1 - \alpha)$ confidence interval (l, u) for σ_1^2/σ_2^2. This interval can be used to draw conclusions about hypotheses concerning σ_1^2/σ_2^2. Since the hypothesis $H: \sigma_1^2 \neq \sigma_2^2$ can equivalently be written as $H: \sigma_1^2/\sigma_2^2 \neq 1$, the inequality of the two population variances can be checked by considering the position of 1 (**not** 0) when compared with the location of the confidence interval (l, u). See also the examples below.

We next construct hypothesis tests for the following testing problems concerning σ_1^2/σ_2^2; again, the hinge is denoted by h.

ia Test $H_0: \frac{\sigma_1^2}{\sigma_2^2} \leq h$ against $H_1: \frac{\sigma_1^2}{\sigma_2^2} > h$

ib Test $H_0: \frac{\sigma_1^2}{\sigma_2^2} \geq h$ against $H_1: \frac{\sigma_1^2}{\sigma_2^2} < h$

ic Test $H_0: \dfrac{\sigma_1^2}{\sigma_2^2} = h$ against $H_1: \dfrac{\sigma_1^2}{\sigma_2^2} \neq h$

To find the test statistic, we follow step D: the right-hand expression of the pivot F in (18.20) is adapted in accordance to the hinge (the worst case scenario). The result does not depend on any further parameters, so it is taken as test statistic and also denoted by F:

ii Test statistic: $F = \dfrac{S_1^2/S_2^2}{h}$

If σ_1^2/σ_2^2 is equal to the hinge, then the test statistic is F_{n_1-1,n_2-1}-distributed. The rejection regions for testing problems (ia), (ib) and (ic) follow immediately from the probabilities below, which are valid if σ_1^2/σ_2^2 is equal to the hinge:

$$P(F \geq F_{\alpha;n_1-1,n_2-1}) = \alpha; \quad P(F \leq F_{1-\alpha;n_1-1,n_2-1}) = \alpha; \quad P(F_{1-\alpha/2;n_1-1,n_2-1} \leq F \leq F_{\alpha/2;n_1-1,n_2-1}) = 1-\alpha$$

The following box gives the five-step procedures:

Five-step procedures for testing problems concerning σ_1^2/σ_2^2

  i Testing problem (ia), (ib), (ic)

 ii Test statistic: $F = \dfrac{S_1^2/S_2^2}{h}$

iii Rejection: (a) reject $H_0 \Leftrightarrow f \geq F_{\alpha;n_1-1,n_2-1}$
      (b) reject $H_0 \Leftrightarrow f \leq F_{1-\alpha;n_1-1,n_2-1}$
      (c) reject $H_0 \Leftrightarrow f \leq F_{1-\alpha/2;n_1-1,n_2-1}$ or $f \geq F_{\alpha/2;n_1-1,n_2-1}$

 iv Calculate **val**

  v Draw the conclusion by comparing **val** with step (iii)

Requirement: two independent, normal random samples

The hypothesis $H: \sigma_1^2 \neq \sigma_2^2$ can be tested with the test for the two-sided testing problem (ic) by taking $h = 1$.

Example 18.7

In Examples 18.1 and 18.2 the equal variance approach was used, in Example 18.3 the unequal variance approach. In the present example it will be checked (afterwards) whether we have chosen the correct approaches.

In Example 18.1, the following statistics were given:

$$n_1 = 150 \quad \text{and} \quad s_1^2 = 67.2400; \quad n_2 = 100 \quad \text{and} \quad s_2^2 = 73.1025$$

To test whether $\sigma_1^2 \neq \sigma_2^2$ (or, equivalently, whether $\sigma_1^2/\sigma_2^2 \neq 1$) with significance level 0.05, we need the critical values $F_{0.975;149,99}$ and $F_{0.025;149,99}$, which are equal to 0.7015 and 1.4446 (*), respectively. Here is the five-step test procedure:

  i Test $H_0: \dfrac{\sigma_1^2}{\sigma_2^2} = 1$ against $H_1: \dfrac{\sigma_1^2}{\sigma_2^2} \neq 1$; $\alpha = 0.05$

 ii Test statistic: $F = \dfrac{S_1^2}{S_2^2}$

iii Reject $H_0 \Leftrightarrow f \leq 0.7015$ or $f \geq 1.4446$

 iv **val** $= \dfrac{67.2400}{73.1025} = 0.9198$

  v Do not reject H_0 since **val** does not belong to the rejection region

The data do not give evidence that the two population variances are unequal, which is in accordance with the equal variance approach that was chosen in Example 18.1.

Let us use a suitable 95% confidence interval to find out whether we have also chosen the correct approach in Example 18.2. Recall that:

$$n_1 = 30 \text{ and } s_{fin} = 2.52007; \quad n_2 = 30 \text{ and } s_{mid} = 2.52058$$

According to (18.21), we need the quantile $F_{0.025;29,29} = 2.1010$ (∗). The other quantile can now be calculated:

$$F_{0.975;29,29} = \frac{1}{F_{0.025;29,29}} = 0.4760$$

The bounds of the 95% confidence interval follow immediately:

$$l = 0.4760 \times \frac{2.52007^2}{2.52058^2} = 0.4758 \quad \text{and} \quad u = 2.1010 \times \frac{2.52007^2}{2.52058^2} = 2.1001$$

Hence, we are 95% confident that σ_1^2/σ_2^2 lies in (0.4758, 2.1001). Since the number 1 falls in this interval, it cannot be concluded that $\sigma_1^2/\sigma_2^2 \neq 1$. Hence, the data do not indicate that the wrong approach was chosen.

In Example 18.3 the following statistics were used:

Sample 1 (1997): $n_1 = 100$ and $s_1 = 21$
Sample 2 (2005): $n_2 = 150$ and $s_2 = 31$

For a 95% confidence interval, we need $F_{0.975;149,99}$ and $F_{0.025;149,99}$, which are, respectively, 0.7015 and 1.4446 (calculated above). Since $s_1^2/s_2^2 = 0.4589$, we obtain:

$$l = 0.7015 \times 0.4589 = 0.322 \quad \text{and} \quad u = 1.4446 \times 0.4589 = 0.663$$

There is 95% confidence that σ_1^2/σ_2^2 lies in (0.322, 0.663). In particular, it is rather certain that σ_1^2/σ_2^2 is smaller than 1 and hence $\neq 1$. It seems that the unequal variance approach was the correct approach.

Example 18.8

Investors always are eager to learn about mean returns and risks of portfolios. The dataset Xmp18-08.sav contains weekly returns (%) of two portfolios for a one-year period.

An investor has studied the developments of two portfolios for a while. She believes that the portfolios move quite independently and that the mean return of portfolio 2 is larger than the mean return of portfolio 1. However, she also believes that portfolio 2 is much riskier: that the standard deviation of the return of portfolio 2 is in fact more than 40% greater than the standard deviation of the return of portfolio 1. Do the data support these conjectures? Use significance level 0.05.

Using a statistical package, the correlation coefficient between the observations of the two portfolios can easily be calculated: $r = 0.060$. This low value is in accordance with the belief of the investor that the two portfolios behave independently. Below, it will be assumed that the two samples in the dataset come from two independent random samples.

Columns 1 and 2 of the dataset contain the returns of, respectively, portfolio 1 and portfolio 2. In particular, the data in the file are not organized in the way that row-oriented packages (such as SPSS) would like them to be. Hence, with such packages we cannot use the standard procedures that immediately yield the value of the test statistics. But descriptive statistics **can** be obtained, so the values of the test statistics can be calculated from them. Below, the parameters μ_1, μ_2 and σ_1^2,

σ_2^2 denote, respectively, the (population) means and the variances of the returns of portfolios 1 and 2.

The printout below contains the relevant statistics obtained with SPSS.

Descriptive Statistics

	N	Minimum	Maximum	Mean	Std. Deviation
portfol_1	52	-.05	.59	.2188	.16141
portfol_2	52	-.32	.83	.3319	.29580
Valid N (listwise)	52				

Let us first test the conjecture about the risks. It states that $\sigma_2 > 1.4\sigma_1$ or (equivalently) that $\sigma_2^2/\sigma_1^2 > 1.4^2 = 1.96$. Hence, the alternative hypothesis of the F-test is:

$$H_1: \frac{\sigma_2^2}{\sigma_1^2} > 1.96$$

Since the ratio parameter of interest is now σ_2^2/σ_1^2, the estimator is S_2^2/S_1^2. The test statistic is:

$$F = \frac{S_2^2/S_1^2}{1.96}$$

If we take 0.05 as significance level, the critical value and the rejection region are:

$$F_{0.05;51,51} = 1.5920 \; (*) \quad \text{and} \quad [1.5920, \infty)$$

The **val** of the test can easily be calculated:

$$\textbf{val} = \frac{0.29580^2/0.16141^2}{1.96} = 1.7135$$

Since this value is larger than 1.5920, it seems that the standard deviation of portfolio 2 is indeed more than 40% larger than the standard deviation of portfolio 1.

From this test it follows that the two population variances are unequal; in fact, the second variance is more than 1.96 times the first. So, for the conjecture about the means of the returns, we need the unequal variance approach.

The testing problem of the first conjecture has $H_1: \mu_2 - \mu_1 > 0$ as alternative hypothesis. The test statistic and rejection region are, respectively:

$$T = \frac{\bar{X}_2 - \bar{X}_1}{\sqrt{\frac{S_2^2}{n_2} + \frac{S_1^2}{n_1}}} \quad \text{and} \quad [t_{0.05;51}, \infty) = [1.6753, \infty) \tag{*}$$

The calculation of the **val** is straightforward:

$$\textbf{val} = \frac{0.3319 - 0.2188}{\sqrt{0.29580^2/52 + 0.16141^2/52}} = 2.4203$$

Indeed, the data support the conjecture that the mean return of portfolio 2 is larger than the mean return of portfolio 1.

There is one thing more that has to be done: we have to check whether it is reasonable to assume that the two samples are **normal**. This can be done visually by creating histograms; see Figure 18.2.

The histograms do not fit well to the normal distributions. As a consequence, we must have some doubt about the confidence level of the hypothesis test of σ_2^2/σ_1^2.

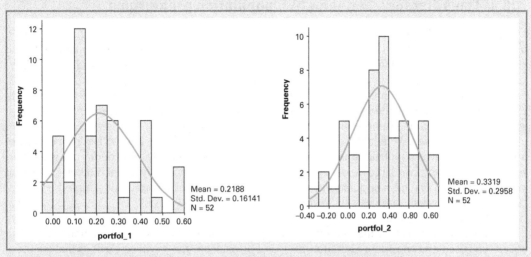

Figure 18.2 Histograms with best-fitting normal curves for the returns of the portfolios

CASE 18.1 DID THE CHANGES TO STATISTICS 2 INCREASE THE GRADES? – SOLUTION

It seems reasonable to assume that the before and after grades in the dataset are the realizations of two independent random samples from all participants of the course Statistics 2 respectively before and after the alterations. So, the experimental design is of the type 'independent random samples'. Let 1 and 2 refer to the before and the after measurements, respectively.

To find out whether the unequal variances approach should be used, we first test whether the two population variances are different. This is done by way of a 95% confidence interval:

Interval estimator: (L, U) with $L = F_{1-\alpha/2;n_2-1,n_1-1} \dfrac{S_1^2}{S_2^2}$ and $U = F_{\alpha/2;n_2-1,n_1-1} \dfrac{S_1^2}{S_2^2}$

Note that $\alpha = 0.05$, and that $n_1 = 277$ and $n_2 = 252$. From the data it follows that

$$\bar{x}_1 = 5.157 \quad \text{and} \quad s_1 = 2.1183; \quad \bar{x}_2 = 5.881 \quad \text{and} \quad s_2 = 1.9869$$

Since $F_{0.975;251,276} = 0.784282$ (∗) and $F_{0.025;251,276} = 1.273255$ (∗), we obtain:

$$l = 0.784282 \times \frac{2.1183^2}{1.9869^2} = 0.8914 \quad \text{and} \quad u = 1.4472$$

Since 1 belongs to the interval (0.8914, 1.4472), there is not enough evidence to conclude that the corresponding population variances are different. That is why we will use the **equal** variances approach.

Here is the five-step procedure (with $\alpha = 0.01$) for testing whether the mean grade after the changes is higher than the mean grade before:

  i Test $H_0: \mu_1 - \mu_2 \geq 0$ against $H_1: \mu_1 - \mu_2 < 0$; $\alpha = 0.01$

  ii Test statistic: $T = \dfrac{\bar{X}_1 - \bar{X}_2}{\sqrt{S_p^2 \left(\frac{1}{n_1} + \frac{1}{n_2} \right)}}$

  iii Reject $H_0 \Leftrightarrow t \leq -2.3334$ (∗)

  iv **val** = −4.043 (obtained directly from printout; check it)

  v H_0 is rejected

The mean grade after the changes is higher than the mean grade before.

18.4 The difference between two population proportions

Again, one variable (called X) is considered on two populations called 1 and 2. However, in the present section X is a 0-1 variable where 1 corresponds to success and 0 to failure. Under these circumstances the difference $\mu_1 - \mu_2$ between the two population means turns into the difference $p_1 - p_2$ between the proportions of successes in population 1 and population 2. Although it is possible to consider both experimental designs – independent samples and paired samples – in the present context, only the independent samples design will be considered here. Additionally, it will be assumed that the samples from both populations are large.

Suppose that the 0-1 variable X is considered on the populations 1 and 2. The proportions of successes in the two populations are denoted by p_1 and p_2, respectively. In order to draw an inference about $p_1 - p_2$, two large random samples are drawn independently of each other – one from population 1 and one from population 2 – and the variable X is observed at all sampled elements. Again the two random samples can be denoted as $X_{11}, X_{12}, \ldots, X_{1n_1}$ and $X_{21}, X_{22}, \ldots, X_{2n_2}$. Note that all rvs X_{ij} take only the values 0 and 1.

Recall from Section 14.1 that the two sample proportions of successes, now denoted as $\hat{P}_1$ and $\hat{P}_2$, are unbiased and consistent estimators of p_1 and p_2, respectively. It will be obvious that the usual estimator of $p_1 - p_2$ is $\hat{P}_1 - \hat{P}_2$, the difference between the two corresponding sample proportions. This estimator is unbiased and consistent. This follows immediately from (18.3) since $\hat{P}_1$ and $\hat{P}_2$ are respectively equal to the sample means $\bar{X}_1$ and $\bar{X}_2$ of the random samples above. Since, in the present circumstances, $\sigma_1^2 = p_1(1 - p_1)$ and $\sigma_2^2 = p_2(1 - p_2)$, equations (18.4) and (18.5) can be applied here as well. The box summarizes things.

Basic properties of the estimator $\hat{P}_1 - \hat{P}_2$ of $p_1 - p_2$

The estimator is unbiased and consistent **(18.23)**

$$V(\hat{P}_1 - \hat{P}_2) = \frac{p_1(1 - p_1)}{n_1} + \frac{p_2(1 - p_2)}{n_2} \quad \textbf{(18.24)}$$

$$SD(\hat{P}_1 - \hat{P}_2) = \sqrt{\frac{p_1(1 - p_1)}{n_1} + \frac{p_2(1 - p_2)}{n_2}} \quad \textbf{(18.25)}$$

As in Section 18.2.1, the expression for the variance of $\hat{P}_1 - \hat{P}_2$ illustrates the consistence property of the estimator. We have finished step A now.

Next, step B: find a pivot. For large n_1 and n_2, the estimator $\hat{P}_1 - \hat{P}_2$ is approximately normally distributed. This follows immediately from (18.6). The only additional requirement is that the sample sizes have to be so large that the expected numbers of successes and failures in both random samples have to be at least 5.

Results that are essential for statistical inference about $p_1 - p_2$

$$\hat{P}_1 - \hat{P}_2 \approx N\left(p_1 - p_2, \frac{p_1(1 - p_1)}{n_1} + \frac{p_2(1 - p_2)}{n_2}\right)$$

$$Z = \frac{\hat{P}_1 - \hat{P}_2 - (p_1 - p_2)}{\sqrt{\frac{p_1(1 - p_1)}{n_1} + \frac{p_2(1 - p_2)}{n_2}}} \approx N(0, 1) \quad \textbf{(18.26)}$$

Requirements: $n_1 p_1 \geq 5$, $n_1(1 - p_1) \geq 5$, $n_2 p_2 \geq 5$ and $n_2(1 - p_2) \geq 5$ (5-rule)

The pivot in (18.26) is the starting point for developing interval estimators and hypothesis tests for $p_1 - p_2$.

To create an interval estimator that captures $p_1 - p_2$ with probability $1 - \alpha$, the denominator in (18.26) – which is the standard deviation of the estimator – is an obstacle since it contains the unknown parameters p_1 and p_2. That is why these parameters are replaced by their estimators $\hat{P}_1$ and $\hat{P}_2$, thus transforming the standard deviation in the denominator into a standard error:

$$SE(\hat{P}_1 - \hat{P}_2) = \sqrt{\frac{\hat{P}_1(1 - \hat{P}_1)}{n_1} + \frac{\hat{P}_2(1 - \hat{P}_2)}{n_2}}$$

After these transformations, you might expect that the rv Z turns into an rv called T for which a t-distribution is used. However, it is usual to keep on using the standard normal z-distribution under these circumstances.

The box below gives the random bounds of the interval estimator.

Interval estimator (L, U) for $p_1 - p_2$

$$L = \hat{P}_1 - \hat{P}_2 - z_{\alpha/2} \sqrt{\frac{\hat{P}_1(1 - \hat{P}_1)}{n_1} + \frac{\hat{P}_2(1 - \hat{P}_2)}{n_2}}$$

$$U = \hat{P}_1 - \hat{P}_2 + z_{\alpha/2} \sqrt{\frac{\hat{P}_1(1 - \hat{P}_1)}{n_1} + \frac{\hat{P}_2(1 - \hat{P}_2)}{n_2}}$$

(18.27)

When the data are available, they can be substituted into the interval estimator to obtain a realized interval (l, u) that contains the difference parameter $p_1 - p_2$ with a confidence of $1 - \alpha$. At least, if n_1 and n_2 are large enough. However, checking the 5-rule is impossible since p_1 and p_2 are unknown. That is why, in practice, it is checked whether

$$n_1\hat{p}_1 \geq 5, \quad n_1(1 - \hat{p}_1) \geq 5, \quad n_2\hat{p}_2 \geq 5 \quad \text{and} \quad n_2(1 - \hat{p}_2) \geq 5$$

where $\hat{p}_1$ and $\hat{p}_2$ are the realized sample proportions of samples 1 and 2, respectively.

Just as in Section 18.2.1, three types of testing problem will be considered. Again, the hinge is denoted by h. Here is the upper-tailed type:

ia   Test H_0: $p_1 - p_2 \leq h$ against H_1: $p_1 - p_2 > h$

Remember from the recap in Section 17.1 (step D) that the test statistic should arise from the pivot Z in (18.26) by the following two steps:

a   Replace the parameter in the pivot by the hinge.

b   If unknown parameters remain, replace them by consistent estimators.

The first step yields that the term $p_1 - p_2$ in Z is replaced by h. However, the second step depends on the value of the hinge. If $h = 0$, the two parameters p_1 and p_2 are equal in the worst case scenario and estimation of this common parameter in the denominator of Z can best be done by combining the two samples. The cases $h = 0$ and $h \neq 0$ will be considered separately.

The case that $h = 0$

Adapting Z according to the worst case scenario implies that the adaptation should be done in accordance with p_1 and p_2 being equal. But if $p_1 = p_2$, then the denominator in Z has only one unknown parameter left. Since both samples contain information about this common population proportion, we should use both samples to estimate it. That is why the number of successes (say, Y_1 and Y_2) in the two samples are now added up and the result is divided by the size $n_1 + n_2$ of the combined sample. The resulting random variable is denoted by $\hat{P}$ and is called the **pooled sample proportion**.

$$\hat{P} = \frac{Y_1 + Y_2}{n_1 + n_2}$$

Since this estimator does a better job than the two original estimators $\hat{P}_1$ and $\hat{P}_2$ when estimating the common population proportion, it is $\hat{P}$ that is used to replace this proportion in the denominator of Z in (18.26). Here is the resulting test statistic:

ii $$Z = \frac{\hat{P}_1 - \hat{P}_2}{\sqrt{\dfrac{\hat{P}(1-\hat{P})}{n_1} + \dfrac{\hat{P}(1-\hat{P})}{n_2}}} = \frac{\hat{P}_1 - \hat{P}_2}{\sqrt{\hat{P}(1-\hat{P})\left(\dfrac{1}{n_1} + \dfrac{1}{n_2}\right)}}$$

The case that h ≠ 0

If $h \neq 0$, the adaptation of the pivot Z in (18.26) in step D is straightforward: firstly, the term $p_1 - p_2$ is replaced by the hinge h; next, the two remaining unknown parameters p_1 and p_2 in the denominator are replaced by their respective estimators $\hat{P}_1$ and $\hat{P}_2$. Here is the resulting test statistic:

ii $$Z = \frac{\hat{P}_1 - \hat{P}_2 - h}{\sqrt{\dfrac{\hat{P}_1(1-\hat{P}_1)}{n_1} + \dfrac{\hat{P}_2(1-\hat{P}_2)}{n_2}}}$$

Overview

The box below summarizes things:

Five-step test procedures for $p_1 - p_2$

i Testing problems: (a) test $H_0: p_1 - p_2 \leq h$ against $H_1: p_1 - p_2 > h$
(b) test $H_0: p_1 - p_2 \geq h$ against $H_1: p_1 - p_2 < h$
(c) test $H_0: p_1 - p_2 = h$ against $H_1: p_1 - p_2 \neq h$

ii Test statistic: if $h = 0$: $Z = \dfrac{\hat{P}_1 - \hat{P}_2}{\sqrt{\hat{P}(1-\hat{P})\left(\dfrac{1}{n_1} + \dfrac{1}{n_2}\right)}}$ with $\hat{P} = \dfrac{Y_1 + Y_2}{n_1 + n_2}$

if $h \neq 0$: $Z = \dfrac{\hat{P}_1 - \hat{P}_2 - h}{\sqrt{\dfrac{\hat{P}_1(1-\hat{P}_1)}{n_1} + \dfrac{\hat{P}_2(1-\hat{P}_2)}{n_2}}}$

iii Rejection: (a) reject $H_0 \Leftrightarrow z \geq z_\alpha$
(b) reject $H_0 \Leftrightarrow z \leq -z_\alpha$
(c) reject $H_0 \Leftrightarrow z \leq -z_{\alpha/2}$ or $z \geq z_{\alpha/2}$

iv Calculate **val**

v Draw the conclusion by comparing **val** with step (iii)

How large should n_1 and n_2 be? In practice, the requirement is the same as for confidence intervals:

$$n_1\hat{p}_1 \geq 5, \quad n_1(1-\hat{p}_1) \geq 5, \quad n_2\hat{p}_2 \geq 5 \quad \text{and} \quad n_2(1-\hat{p}_2) \geq 5$$

Example 18.9

In the report series *Flash Eurobarometer*, the European Commission publishes results of opinion polls conducted throughout the EU. For instance, in the report *Introduction of the Euro in the New Member States* (June 2006) random samples of citizens from the ten new member states were asked to give their opinion on the following question:

Q17: Do you think the euro will help to maintain price stability or, on the contrary, increase inflation in your country?

The possible answers were:

will help maintain price stability
will increase inflation
no impact
don't know or don't want to answer

Hence, the nominal variable Q17 has four possible values. In the present example, this variable will be transformed into the 0-1 variable (called X) that assigns the values 1 and 0 to the responses as follows:

1 = 'will help maintain price stability'
0 = 'will increase inflation (including: no impact, don't know or don't want to answer)'

Table 18.1 lists the results for four of the new member states.

Country	Sample size	Will help maintain price stability (%)	Otherwise (%)
Estonia	1015	17.4	82.6
Lithuania	1026	18.7	81.3
Hungary	1016	44.7	55.3
Poland	1011	28.6	71.4

TABLE 18.1 Results for four new member states

Let p_E, p_L, p_H and p_P denote for these four countries the respective proportions of the citizens that believe that the euro 'will help maintain price stability'. The data will be used to answer the following questions: Are the proportions of Estonia and Lithuania different? Is the proportion in Hungary more than 0.10 larger than in Poland? Is the proportion of Estonia more than 0.20 less than the proportion of Hungary?

The first question can be reformulated as: is $p_E - p_L \neq 0$? Below, a hypothesis test with significance level 0.01 will be used to answer this question. Note that the hinge of the testing problem is 0. Since the testing problem is two-sided, we need the critical value $z_{0.005} = 2.5758$ (*). Here are the five steps:

i Test H_0: $p_E - p_L = 0$ against H_1: $p_E - p_L \neq 0$; $\alpha = 0.01$

ii Test statistic: $Z = \dfrac{\hat{P}_E - \hat{P}_L}{\sqrt{\hat{P}(1 - \hat{P})\left(\dfrac{1}{n_E} + \dfrac{1}{n_L}\right)}}$

iii Reject $H_0 \Leftrightarrow z \leq -2.5758$ or $z \geq 2.5758$

iv To find the combined sample proportion $\hat{p}$, notice that Estonia and Lithuania respectively yielded $y_E = 0.174 \times 1015 \approx 177$ and $y_L = 192$ successes. Hence:

$$\hat{p} = (177 + 192) / (1015 + 1026) = 0.1808$$

$$\textbf{val} = \frac{0.174 - 0.187}{\sqrt{0.1808 \times 0.8192 \times (1/1015 + 1/1026)}} = -0.76$$

v It cannot be concluded that the two proportions are different

The second question can be reformulated as: is $p_H - p_P > 0.10$? Hence, the hinge of the testing problem is 0.10. In the five-step procedure we again take $\alpha = 0.01$.

▶

i Test $H_0: p_H - p_P \leq 0.1$ against $H_1: p_H - p_P > 0.1$; $\alpha = 0.01$

ii $Z = \dfrac{\hat{P}_H - \hat{P}_P - 0.1}{\sqrt{\dfrac{\hat{P}_H(1 - \hat{P}_H)}{n_H} + \dfrac{\hat{P}_P(1 - \hat{P}_P)}{n_P}}}$

iii Reject $H_0 \Leftrightarrow z \geq z_\alpha = 2.3263$ $\hspace{3cm}$ (*)

iv **val** $= \dfrac{0.447 - 0.286 - 0.1}{\sqrt{\dfrac{0.447 \times 0.553}{1016} + \dfrac{0.286 \times 0.714}{1011}}} = 2.8908$

v H_0 is rejected since **val** is larger than the critical value

Indeed, the proportion of Hungary is more than 0.1 larger than the proportion of Poland.

The third question can be reformulated as: is $p_E - p_H < -0.20$? A 98% confidence interval for $p_E - p_H$ will be created to answer it. According to (18.27), the realized standard error is equal to:

$$\sqrt{\dfrac{0.174 \times 0.826}{1015} + \dfrac{0.447 \times 0.553}{1016}} = 0.0196$$

Since $\hat{p}_E - \hat{p}_H = -0.2730$ and $z_{0.01} = 2.3263$, it follows that:

$$-0.2730 \pm 2.3263 \times 0.0196 \quad \text{yields} \quad (-0.3186, -0.2274)$$

There is 98% confidence that $p_E - p_H$ is smaller than -0.2274, so there is (at least) 98% confidence that $p_E - p_H < -0.20$. The answer to the last question is yes.

Summary

This chapter was about statistical procedures for differences between two population means, two population variances and two population proportions. Interval estimators and test statistics were developed by using suitable pivots as the starting point. For the difference $\mu_1 - \mu_2$ between two population means, the design of the experiment determines the statistical procedures that have to be used. In the case of an independent samples design, the equal variance case and the unequal variance case have their own standard interval estimators and hypothesis tests.

The table below summarizes the statistical procedures covered in this chapter. The following chapter is the first in a series of chapters about linear regression.

Overview of interval estimators and test statistics in Chapter 18

Parameter	Experimental design	Interval estimator	Test statistic; hinge h
$\mu_1 - \mu_2$	Independent random samples; equal variance approach	$\bar{X}_1 - \bar{X}_2 \pm t_{\alpha/2; n_1 + n_2 - 2} \sqrt{S_p^2(\frac{1}{n_1} + \frac{1}{n_2})}$	$\dfrac{\bar{X}_1 - \bar{X}_2 - h}{\sqrt{S_p^2(\frac{1}{n_1} + \frac{1}{n_2})}}$
	Independent random samples; unequal variance approach	$\bar{X}_1 - \bar{X}_2 \pm t_{\alpha/2; m} \sqrt{\dfrac{S_1^2}{n_1} + \dfrac{S_2^2}{n_2}}$ with $m = \min(n_1, n_2) - 1$	$\dfrac{\bar{X}_1 - \bar{X}_2 - h}{\sqrt{\dfrac{S_1^2}{n_1} + \dfrac{S_2^2}{n_2}}}$
	Requirement: normal random samples and/or large sample sizes		
	Paired random samples	$\bar{D} \pm t_{\alpha/2; n-1} \dfrac{S_D}{\sqrt{n}}$	$\dfrac{\bar{D} - h}{S_D / \sqrt{n}}$
	Requirement: differences $D_1, \dots, D_n$ normal random sample and/or large n		

$p_1 - p_2$	Independent random samples	$\hat{P}_1 - \hat{P}_2 \pm z_{\alpha/2;} \sqrt{\dfrac{\hat{P}_1(1 - \hat{P}_1)}{n_1} + \dfrac{\hat{P}_2(1 - \hat{P}_2)}{n_2}}$	If $h = 0$: $$\dfrac{\hat{P}_1 - \hat{P}_2}{\sqrt{\hat{P}(1 - \hat{P})\left(\dfrac{1}{n_1} + \dfrac{1}{n_2}\right)}}$$ If $h \neq 0$: $$\dfrac{\hat{P}_1 - \hat{P}_2 - h}{\sqrt{\dfrac{\hat{P}_1(1 - \hat{P}_1)}{n_1} + \dfrac{\hat{P}_2(1 - \hat{P}_2)}{n_2}}}$$
	Requirement: sample sizes so large that the 5-rule is valid		
$\dfrac{\sigma_1^2}{\sigma_2^2}$	Independent random samples	$L = F_{1-\alpha/2;n_2-1,n_1-1} \dfrac{S_1^2}{S_2^2}$ $U = F_{\alpha/2;n_2-1,n_1-1} \dfrac{S_1^2}{S_2^2}$	$\dfrac{S_1^2/S_2^2}{h}$
	Requirement: normal random samples		

List of symbols

Symbol	Description
F_{ν_1,ν_2}	F-distribution with parameters ν_1 and ν_2
F_{α,ν_1,ν_2}	Quantile of the F_{ν_1,ν_2}-distribution

🔐 Key terms

balanced design **525**
cell-oriented **522**
equal variance
   approach **526**
equal variance test **525**
experimental design **522**
F_{ν_1,ν_2}-distribution **535**

independent samples
   design **522**
matched-pairs design **522**
paired observations
   design **522**
pooled sample
   proportion **543**

pooled sample
   variance **525**
row-oriented **522**
unequal variance
   approach **529**
unequal variance
   test **529**

❓ Exercises

Exercise 18.1

A variable X is measured for two random samples, one drawn from population 1 and the other from population 2. Sample sizes, sample means and sample variances are listed in the table:

	Size	Mean	Variance
Sample 1	16	20.1	9
Sample 2	25	18.3	8.9

The question is whether these sample results allow the conclusion that the two corresponding population means are different. Use a hypothesis test with $\alpha = 0.10$.

   a Describe the experimental design in more detail than done in the above text.

   b Will you use the equal variance test or the unequal variance test? Why?

   c To answer the question, write down the five-step procedure.

   d What implicit assumption did you make?

Exercise 18.2

To find out whether, for adult man–woman couples, the means of the IQs of the men and the women are different, 200 of such couples were randomly selected and the IQs of the 400 people measured. The means and the variances of the IQs for the 200 men and the 200 women are given in the table, jointly with the variance for the 200 differences between the IQs of the men and the accompanying women.

	Mean	Variance	Variance of differences
Sample men	102.9	99.2	120.3
Sample women	104.1	101.3	

   a What is the name of this experimental design?

   b Use a hypothesis test with significance level 0.05 to test whether the means of the IQs of the men and the women in man–woman couples are different.

   c Determine the minimal significance level that would have been necessary to conclude that the two population means are different.

   d In parts (b) and (c), did you use an extra, implicit assumption about normality?

Exercise 18.3

For Exercise 18.1, use a suitable 95% confidence interval to justify the choice made in part (b).

Exercise 18.4

In a recent enquiry, 500 men and 500 women were asked whether they approve government measures to make the teaching profession more attractive by raising teachers' salaries. It turned out that 150 of the men and 180 of the women agreed with this measure. Can it be concluded that the population proportion of the women approving the measure is larger than the population proportion of the men?

   a Use a hypothesis test with significance level 0.05 to answer the question.

   b Answer the question again with a 95% confidence interval.

Exercise 18.5

To compare the hourly wages for manual workers in an urban and a rural region, two random samples of 30 workers aged 40–50 are drawn (one from each region) and their hourly wages are recorded. Let 1 refer to the urban region and 2 to the rural region. Suppose that the two population standard deviations of the hourly wages in the two regions are respectively €2.80 and €3.05.

   a Transform the rv $W = S_1^2/S_2^2$ into an rv with an F-distribution. Which F-distribution? What did you implicitly assume?

b For this *F*-distribution, calculate the 0.95-quantile and use it to calculate the 0.05-quantile.

c Calculate the probability $P(W \leq 0.9)$.

Exercise 18.6

Reconsider Exercise 18.5, but now suppose that the two population standard deviations are unknown. The table summarizes the hourly wages of the two samples of manual workers.

	Size	Mean	Standard deviation
Urban sample	30	19.20	2.50
Rural sample	30	18.90	3.10

a Conduct a suitable *F*-test to find out whether the hourly wages in the rural region are more spread out than in the urban region. Use $\alpha = 0.05$.

b Use a hypothesis test with significance level 0.05 to check whether the mean hourly wage in the urban region is larger than the mean hourly wage in the rural region. Did you use the equal or unequal variance test? Why?

Exercise 18.7

In Section 18.2.1 the following was noted for the equal variance case: 'Since S_1^2 and S_2^2 are both unbiased estimators of σ^2, the pooled variance is also an unbiased estimator of σ^2.' Give the formal proof of this result.

Exercise 18.8

Consider the twelve examples in Section 18.1. For each example, think of two populations and a variable of interest. Also express the questions themselves in terms of parameters $\mu_1 - \mu_2, p_1 - p_2$ or σ_1^2/σ_2^2.

Exercise 18.9

A variable X is considered on two populations. Let μ_1 and μ_2 denote the mean of X on population 1 and population 2, respectively. To find out whether μ_1 and μ_2 are different, two random samples of sizes 40 and 60 are, independently of each other, drawn from, respectively, 1 and 2, and for each sampled population element the value of X is observed. The sample means and standard deviations turn out to be equal to:

$\bar{x}_1 = 5.80;\quad s_1 = 1.30;\quad \bar{x}_2 = 6.60;\quad s_2 = 1.25$

a Use a suitable hypothesis test with significance level 0.05 to conclude whether $\mu_1 \neq \mu_2$.

b Determine the *p*-value of this test.

c Also use a 95% confidence interval to draw a conclusion about the statement that the two population means are different.

Exercise 18.10

In a matched-pairs experiment, the following data are observed:

Sample 1	75	65	80	70	80
Sample 2	80	90	75	85	95

Interest is in the corresponding population means μ_1 and μ_2.

a Use a suitable hypothesis test with significance level 0.05 to conclude whether $\mu_1 < \mu_2$.

b Determine the p-value of this test.

c Also use a 90% confidence interval to draw a conclusion about the statement that the two population means are different.

Exercise 18.11

Use a suitable hypothesis test with $\alpha = 0.10$ to test whether $\mu_2 - \mu_1 > 3$, on the basis of the following observations:

Sample 1	17	23	24	17	24
Sample 2	21	29	28	26	31

a if sample 1 (respectively 2) contains the heights (cm) of plants that daily receive 1 g (respectively 3 g) of fertilizer;

b if each column contains the heights (cm) of a plant after, respectively, two and three weeks.

c For both (a) and (b), mention all your assumptions.

Exercise 18.12

To find out whether, nowadays, young adult Italian men are taller than their fathers, a random sample of size 5 is drawn from the population of all Italian fathers with adult sons. Here are the data:

Height father (cm)	167	170	172	168	173
Height son (cm)	169	173	174	172	172

a Carefully describe the experimental design.

b Is the mean height of Italian fathers smaller than the mean height of their (adult) sons? Use a formal hypothesis test with significance level 0.05.

c Carefully formulate the requirements you implicitly assumed.

Exercise 18.13

The fuel consumption of two makes of cars (A and B) are compared. For nine cars of type A and six cars of type B, the numbers of kilometres per litre of fuel are recorded:

Make A	16	18	15	19	17	14	19	16	16
Make B	13	15	11	17	12	13			

a Describe the two populations and the variable of interest.

b Carefully formulate the assumptions.

c Use a suitable 95% confidence interval to draw a conclusion about the statement that type A cars use less fuel than type B cars.

Exercise 18.14 (computer, but not necessarily)

Sonja Butcher claims that her reducing weight programme reduces weight by (on average) more than 6.5 kg in two months. The dataset Xrc18-14.sav contains the weights (kg) of 20 women just before the programme was started and two months later. The mean and standard deviation of the differences between the before and after measurements turn out to be, respectively, 7.95000 and 2.98196.

  a Conduct a hypothesis test (with $\alpha = 0.05$) to check the claim.

  b Calculate the p-value of the test and interpret it.

The printout of the standard SPSS paired samples test with hinge 0 is shown below.

		Paired Differences					t	df	Sig. (2-tailed)
		Mean	Std. Deviation	Std. Error Mean	95% Confidence Interval of the Difference				
					Lower	Upper			
Pair 1	Before - After	7.95000	2.98196	.66679	6.55440	9.34560	11.923	19	.000

  c Formulate the testing problem that can be tackled with this printout. Give the meaning of all numbers in this printout.

Exercise 18.15

Suppose that the rv Y is $F_{3,34}$-distributed.

  a Determine $E(Y)$. What are the possible outcomes of Y?

  b Calculate $P(Y > m)$ for $m = 1, 2, ..., 7$.

  c Set $U = \frac{1}{Y}$. What is the distribution of U? Calculate $P(U > m)$ for $m = 1, 2, ..., 7$.

Exercise 18.16

Reconsider Exercise 18.15.

  a Calculate $F_{\alpha;3,34}$ for $\alpha = 0.20, 0.10, 0.05, 0.025$ and 0.01. What do these numbers indicate?

  b Calculate $F_{\alpha;34,3}$ for $\alpha = 0.20, 0.10, 0.05, 0.025$ and 0.01.

  c Use part (b) to calculate $F_{1-\alpha;3,34}$ for $\alpha = 0.20, 0.10, 0.05, 0.025$ and 0.01. What do these numbers indicate?

Exercise 18.17

A factory has installed a new machine for the production of axles of 5 cm length. Although this new machine is said to reduce variation by more than 60%, the axles produced still show variation with respect to the adjusted length.

    To test the precision of the new machine when compared with the old one, 100 axles are drawn from the daily production of both the new machine (machine 1) and the old machine (machine 2). The lengths of all 200 axles are measured carefully and recorded. Here are the summarized results:

$$\bar{x}_1 = 5.0145; \quad s_1 = 0.0103; \quad \bar{x}_2 = 5.0010; \quad s_2 = 0.0201$$

Denote the population means and variances for the two machines by μ_1, σ_1^2 and μ_2, σ_2^2, respectively.

a Use a formal hypothesis test with significance level 0.05 to find out whether σ_1^2 is smaller than $0.4\sigma_2^2$.

b Which implicit assumption did you use?

c Use the conclusion of the test in part (a) to find out whether $\mu_1 \neq \mu_2$; use $\alpha = 0.05$.

Exercise 18.18

In Exercises 18.9, 18.11a and 18.13 you probably used the equal variance approach. Use significance level 0.05 to check, for all cases, whether this choice was statistically correct.

Exercise 18.19

A variable X is considered on two populations. Suppose that the distributions $N(4, 9)$ and $N(6, 9)$ are good models for the distribution of X on, respectively, population 1 and population 2. From each of the populations a random sample is drawn and X is observed for each sampled element. This is done in such a way that the two samples are independent. The sample sizes are, respectively, $n_1 = 10$ and $n_2 = 20$.

For this independent samples design, let $\bar{X}_1$, S_1^2 and $\bar{X}_2$, S_2^2 respectively denote the corresponding sample means and sample variances.

a Determine the probability distribution of $\bar{X}_1 - \bar{X}_2$.

b Calculate $P(\bar{X}_1 - \bar{X}_2 > 0)$.

c Calculate $P(\bar{X}_1 \leq 3 \text{ and } \bar{X}_2 \leq 5)$.

Exercise 18.20

Reconsider Exercise 18.19. Interest is also in the proportional difference between S_2^2 and S_1^2 when compared with S_1^2, that is, in $V = (S_2^2 - S_1^2)/S_1^2$.

a Calculate $P(S_2^2/S_1^2 > 1)$.

b Calculate $P(V > 1)$.

Exercise 18.21

Has a debate on television changed the proportion of the voters who intend to vote for the political party SCVDP in the upcoming elections? Below, p_1 and p_2 denote the proportions of SCVDP-voters before and after the debate, respectively.

Before the debate, 80 of 400 randomly chosen voters said they would vote SCVDP. After the debate, 168 of 600 randomly selected voters declared that they would vote SCVDP.

a Use this information to answer the question posed at the start. Use a hypothesis test with significance level 0.05.

b Use a suitable 95% confidence interval to find out whether $p_1 - p_2 < -0.02$.

Exercise 18.22

For the transportation of a certain (voluminous) product, two types of packaging material are available. To find out which of the two types is best, 100 products are transported in material I and 200 products in material II. After transportation, it turns out that 12 of the products packed in material I and 6 of the products packed in material II have some damage. Is the proportion of damaged products for material I more than 0.04 larger than the proportion of damaged products for material II?

a Use a hypothesis test with significance level 0.05 to answer this question.

b Determine the p-value of that test. Interpret the answer.

Exercise 18.23

The International Air Transport Association (IATA) investigates the quality of airports as experienced by businesspeople. The grades range from 10 (highest) to 1 (lowest).

Column 1 of the dataset Xrc18-23.sav contains the grades for the overall quality of two airports (called a and b) as assigned by 100 randomly and independently sampled businesspeople; the second column indicates the airport to which the grade applies. To answer the following questions you can use the SPSS printout below.

a Before considering the problem of differentials between mean quality for the two airports, first decide whether the equal or the unequal variance approach is the better in this instance. Notice that the value of the F-test statistic differs from the value of 'Levene's test' in the printout. Use $\alpha = 0.10$.

b Test whether the mean scores of all businesspeople for the overall quality of airports a and b, are different. Use significance level 0.05.

Group Statistics

	airport	N	Mean	Std. Deviation	Std. Error Mean
score	a	50	6.34	2.163	.306
	b	50	6.72	2.374	.336

Independent Samples Test

		Levene's Test for Equality of Variances		t-test for Equality of Means					95% Confidence Interval of the Difference	
		F	Sig.	t	df	Sig. (2-tailed)	Mean Difference	Std. Error Difference	Lower	Upper
score	Equal variances assumed	.078	.781	−.837	98	.405	−.380	.454	−1.281	.521
	Equal variances not assumed			−.837	97.164	.405	−.380	.454	−1.281	.521

Exercise 18.24 (computer)

This exercise investigates whether adult men are on average more than 10 cm taller than adult women. The data are in the file Xrc18-24.sav; DFem is a dummy that equals 1 if the person is female.

a Before starting this investigation, firstly study the distributions of the height of a man (woman) by considering histograms; see Appendix A1.18.

b Do the investigation, but first check the statement that the mean height of adult men is larger than the mean height of adult women. Use hypothesis tests with significance level 0.05. Did you use the equal variance or the unequal variance approach? Why?

Exercise 18.25

A company wants to market a new medicine to combat high blood pressure. However, the drug has to be tested first.

Twenty-two arbitrarily chosen patients who suffer from high blood pressure are chosen randomly; thirteen of them receive the new drug while nine receive a placebo. The blood pressure of all 22 patients is measured twice: before the drug / placebo was administered and after. The data are in two files: the file Xrc18-25a.sav contains the data of the patients who received the drug; the file Xrc18-25b.sav contains the data of the patients who received the placebo.

a Test whether the drug, on average, decreases blood pressure; use significance level 0.05. In the printout below, the variable D_drug refers to the difference between the after and the before measurements of a patient who received the drug.

Descriptive Statistics

	N	Minimum	Maximum	Mean	Std. Deviation
Drug_before	13	137.00	163.00	147.9231	8.12877
Drug_after	13	120.00	155.00	138.9231	12.52638
D_drug	13	−21.00	6.00	−9.0000	8.51469
Valid N (listwise)	13				

b Use a 90% confidence interval to conclude whether the placebo decreases blood pressure. See the printout below; D_placebo refers to the difference between the after and the before measurements of a patient who received the placebo.

Descriptive Statistics

	N	Minimum	Maximum	Mean	Std. Deviation
Plac_before	9	136.00	161.00	145.6667	9.34077
Plac_after	9	127.00	160.00	142.6667	11.33578
D_placebo	9	−11.00	8.00	−3.0000	5.97913
Valid N (listwise)	9				

It is a well-known phenomenon that patients tend to react positively to drugs, even if this drug is only a placebo. To exclude this so-called placebo effect, the differences between the after and before measurements of the drug are compared with the differences between the after and before measurements of the placebo. The transformed data are in the file Xrc18-25c.sav; notice that column 1 contains the drug-differences and column 2 the placebo-differences, so this dataset is not 'row-organized'.

c Test (with $\alpha = 0.05$) whether the mean change in blood pressure is larger for the drug than for the placebo.

Descriptive Statistics

	N	Minimum	Maximum	Mean	Std. Deviation
D_drug	13	−21.00	6.00	−9.0000	8.51469
D_placebo	9	−11.00	8.00	−3.0000	5.97913
Valid N (listwise)	9				

Exercise 18.26

A firm that markets goods targeting youth in Denmark, Germany, Italy and The Netherlands asked the market research company MarketView to investigate whether young consumers in different countries nowadays have a common view with respect to ethical questions. (Prior to the EU the firm had different policies for each country, as it had found that, in some countries, its consumers were less ethical in their dealings with the firm than in others.)

In MarketView's research business students at universities in the above countries were used as a proxy for consumers in general. One of the questions that MarketView put to randomly selected business students was:

X = **quest2** = 'opinion about the ethical question "drinking a can of cola in a supermarket without paying for it", with values on the scale 1 to 5 where 1 corresponds to *strongly believe it is wrong* and 5 to *strongly believe it is **not** wrong*'.

country = 'country the student lives'; 3 = Denmark, 5 = Germany, 6 = Italy and 8 = The Netherlands

See the dataset Xrc18-26.sav. SPSS produced the printout shown below.

country * quest2 Crosstabulation

Count

		quest2					Total
		1	**2**	**3**	**4**	**5**	
country	Denmark	114	29	6	1	1	151
	Germany	54	19	10	4	4	91
	Italy	73	24	7	7	8	119
	The Netherlands	78	11	2	0	0	91
Total		319	83	25	12	13	452

a Can it be concluded that the proportion of business students in Denmark that answered 1 to quest2 is larger than the proportion of business students in Italy that answered 1? Use a hypothesis test with $\alpha = 0.02$.

b Can it be concluded that the proportion of business students in Germany that answered 1 to quest2 is different from the proportion of business students in Italy that answered 1? Use a 95% confidence interval to answer this question.

Exercise 18.27 (computer)

In the article 'Kicks from the penalty mark in soccer' (*Journal of Sports Statistics*, 15 January 2007), the outcomes of penalty kicks in the international tournaments World Cup (WC), European Championships (EC) and Copa America (CA) were compared. One of the purposes was to study the role of stress. The dataset Xrc18-27.sav contains the results of 409 penalty kicks taken during 1976–2004 in these tournaments when, after 120 minutes, there was no winner. Apart from the outcome of the kick (1 for goal and 0 for miss), the type of tournament (WC, EC, CA) is recorded.

a If stress plays a role, it can be expected that the proportion of hits decreases with the importance of the tournament. Is the proportion of hits for the European Championships greater than the proportion of hits for the World Cup tournaments? Use a hypothesis test with significance level 0.05 to answer this question.

b Is the proportion of hits of the Copa America tournament greater than the proportion of hits of the World Cup tournaments? Use a 95% confidence interval to answer this question.

Exercise 18.28 (computer)

It is well known that the returns of stocks can be larger than the return of bonds but that, on the other hand, the risk of investing in stocks is often greater than the risk of investing in bonds. But what about **funds** of stocks and bonds?

In the present exercise the mean daily return of SNS Euro Bonds Fund and SNS Euro Stocks Fund are compared. The file Xrc18-28.sav contains the returns (measured as proportions with respect to the business day before, not as percentages) of these funds for the period 23 November 2004–17 November 2006. From financial theory it is known that dependences between returns on an investment measured on different days are often hard to prove. However, returns for different investments measured on the same day are often dependent.

a Which experimental design best suits the present study, the independent samples design or the paired-samples design?

b The conjecture is that the mean daily return of the stocks fund is larger than the mean daily return of the bonds fund. Use a hypothesis test with significance level 0.05 to draw a conclusion about this conjecture.

c Determine the minimal significance level that supports the conjecture.

d Let σ_1^2 and σ_2^2 be the respective population variances of the returns of the bonds fund and the stocks fund. Is it wise to use the test procedure (F-test) of Section 18.2 to check whether $\sigma_1^2 < \sigma_2^2$? Why (not)?

Exercise 18.29 (computer)

Many investments are directly based on the returns of exchange markets. For such investments it is important to learn more about differences (if any) between the means of returns and the risks (volatilities) at different exchanges.

The dataset Xrc18-29.sav contains, for the business days in the period 1 May 2002–30 April 2004, the daily returns (%) of the AEX and the NASDAQ. Below, the conjecture will be investigated that the means of the returns at AEX and NASDAQ are different.

a It seems reasonable to assume that daily returns at one exchange are independent. Of course, the two returns (on one day) at the two exchanges are not independent. Which experimental design best fits these reflections?

b Use a suitable 95% confidence interval to draw a conclusion about the conjecture.

Exercise 18.30 (computer)

For businesses it is becoming more and more interesting to move activities to countries with low labour costs. China is of special interest since this country is developing very fast and its market is growing rapidly.

The World Bank Group published a study on labour costs in ten industrial cities in the province of Sichuan (87 million inhabitants) in south-west China. In the publication *Snapshot Sichuan: Benchmarking FDI Competitiveness in China's Sichuan Province* (March 2006), the researchers report the average gross annual salaries (in US dollars) that companies pay to their technical workers. The dataset Xrc18-30.sav contains, for those ten cities, data for the sectors 'Electronics' and 'Machinery and Machine Building'. Is the mean average annual salary in Sichuan for technical workers in the Machinery and Machine Building sector higher than for technical workers in the Electronics sector?

a Use a hypothesis test with significance level 0.05 to answer this question.

b Determine the p-value of the test in part (a). Formulate precisely what this number means.

Exercise 18.31 (computer)

Is the mean gross hourly wage of men higher than the mean gross hourly wage of women? If so, to what extent is this caused by the 'fact' (is it?) that the mean education level of men is higher than the mean education level of women and/or by differences between the age distributions of working men and working women?

These questions will be considered in the coming chapters, for a large population of people working at least 32 hours per week. As a preparation, we will in the present exercise compare working men and working women on the basis of their mean (gross) hourly wage, mean education level and mean age. See the file Xrc18-31.sav, with data of 75 men and 75 women. In the tests below, use significance level 0.05.

a Use a statistical package to generate for the variables W = 'gross hourly wage (in euro)', EDL = 'education level' (1 = lowest level and 5 = highest (university) level) and AGE = 'age (in years)', estimates of the means and the standard deviations for the subpopulations of the men (FEM = 0) and the women (FEM = 1).

b Can it be concluded that the mean hourly wage of men is more than €3 greater than the mean hourly wage of women?

c Are the means of the education levels of men and women different? Answer this question with the standard test, although actually EDL is not a quantitative variable.

d Is the mean age of the working men greater than the mean age of the working women?

CASE 18.2 FUND ABN AMRO AEX FD9 AND BEATING THE MARKET

Many financial institutions have created stocks funds, often with the underlying intention to offer an investment opportunity that 'does better than the market'. ABN AMRO created the investment product ABN AMRO AEX FD9. But does it do better than the AEX market?

a First investigate whether the expected return of the fund is greater than the expected return of the AEX index. Use the file Case18-02.xls, which contains the daily returns (%, closing prices) of the fund and the AEX index for the period 18 October 2005–17 October 2007.

b Next, investigate whether the risk of the fund is smaller than the risk of the AEX index. Measure the risk by way of the variance of the daily return. For this second investigation you will need two independent samples. To approximate that situation use the dataset Case18-02a.xls which contains the returns of the fund over the period 18 October 2005–17 October 2006 and the returns of the AEX index over the period 18 October 2006–17 October 2007 (which is disjoint with the former period).

CASE 18.3 AMBITIOUSNESS OF STUDENTS

In WO-monitor 2004/2005 of VSNU (2007), a total of 12 855 recently graduated European masters students from ten countries were asked the following question about their study:

Did you (yes/no) try to get the highest grades possible?

The data are summarized in the table below.

Country	Percentage Yes	Sample size
Austria	54	990
Finland	44	890
France	62	937
Germany	68	901
Italy	63	1558
Netherlands	34	930
Norway	65	667
Spain	71	1969
Switzerland	58	2642
UK	64	1371

For two countries of your choice, conduct a hypothesis test to check whether the two corresponding proportions 'Yes' are different. Use the *p*-value approach.

CASE 18.4 THE EFFECTS OF AN INCREASE IN THE MINIMUM WAGE (PART II)

This case is a continuation of Case 16.2. In the present case, the effects of the increase in the the minimum wage on 1 April 1992 will be studied by comparing, for the random sample of 410 restaurants, the measurements of the six variables **before and** after that date. (Notice that in Case 16.2 the population means of the situation before 1 April 1992 were assumed to be known.) In the dataset Case18-04.sav, the extension 1 to the six variables refers to the original variables but measured **before** the increase in the minimum wage.

For each of the six variables, investigate whether the increase of the minimum wage affected the mean of the variable.

CHAPTER 19

Simple linear regression

Chapter contents

This chapter is the first in a series of chapters about relationships between variables. The purpose is to find out whether and how one variable – often called Y, the ***research variable*** or ***dependent variable*** – is related to one or more other variables called ***independent variables*** or ***predictors*** or ***explanatory variables***.

To have an example in mind, think of a large company that wants to know whether and how the dependent variable Y = 'weekly sales of a certain product' is related to the independent variables X_1 = 'weekly amount spent on advertising that product', X_2 = 'type of advertising' (possibilities: on television, radio or in newspapers) and X_3 = 'weekly price of that product'. Knowledge of a mathematical function that expresses Y in terms of X_1, X_2, X_3 and that describes reality well, is important for the company. It helps the company to choose the best advertising medium and a price that maximises the profit. It also helps to get an idea of next week's sales; having taken the decision about next week's values of X_1, X_2 and X_3, the mathematical function can be used to **predict** next week's sales.

Chapter 19 considers the case that only one independent variable is involved. Chapter 20 is about the general case, but only elementary, quantitative independent variables are included in the model. Chapter 21 extends the model by including more advanced quantitative and/or qualitative independent variables. Chapter 22 explains what can be done if some of the model requirements are not satisfied. Also Chapter 23 (time series) makes use of linear regression techniques.

Apart from Section 22.6, the dependent variable is always **quantitative**.

CASE 19.1   WAGE DIFFERENTIALS BETWEEN MEN AND WOMEN (PART I)

This case (the 'red-wire' application) will be considered many times throughout Chapters 20–22. It is about a large population of people working at least 32 hours per week. The intention is to investigate **whether** and **to what extent** human capital variables such as 'age', 'education level' and 'gender' influence the (dependent) variable 'gross hourly wage'. The study will be based on the dataset <u>Case19–01.sav</u>, which contains for 75 men and 75 women the observations for the following variables:

  W gross hourly wage of a person (in euro)
  FEM   gender of the person; 1 if female and 0 if male
  AGE   age of the person (in years)
  EDL   level of the person's highest – finished – education
        1 = primary school or less
        2 = secondary school; lower professional education
        3 = secondary school; otherwise
        4 = higher professional education
        5 = university education

One of the aims of the study is to find answers to the following questions:

1   Is the mean gross hourly wage of men greater than the mean gross hourly wage of women? If so, to what extent is this caused by the 'fact' (is it?) that the mean education level of men is higher than the mean education level of women and/or by differences between the age distributions of working men and working women?

2   Compare the means of the hourly wages of university-educated persons and level 1 educated persons having the same age and gender.

3   Are the (mathematical) relationships between 'gross hourly wage' on the one hand and 'age' and 'education level' on the other, different for men and women?

According to the economic theory, there are three factors that determine a person's wage:

■   qualities of the person (so, the human capital variables; for instance, the three variables AGE, EDL, Gender, but also 'experience (in years)');

■   qualities of the job (description of the job: is it demanding, dirty or unpleasant work, what is the added value of the job);

■   qualities of the matching process of the person and her job (to what extent do the qualities of the person fit to the qualities of her job).

In the present study, only the human capital variables AGE, EDL and Gender are used to explain the variation of the dependent variable 'gross hourly wage'. The data have been collected by interviewing the persons at home, not at work.

Note that question 1 was briefly considered in Exercise 18.31. Within the sample, the difference between the means of the hourly wages of men and women is 24.90 − 18.67 = €6.23. For the whole population, the statistical conclusion was that the mean hourly wage of the men is more than €3.00 greater than the mean hourly wage of the women.

At the end of Section 19.6, the relationships between W and respectively AGE and EDL will be studied, for the whole population and for the subpopulations of the men and the women; this has to do with question 3.

19.1 Relating a variable to other variables

The search for variables that influence a dependent variable Y and for mathematical relationships between Y and such variables is generally called **regression analysis**. There can be several reasons why clarity about the relationships between the dependent variable and relevant independent variables is wanted. For instance:

1 for small populations, the wish *to* **position** the population elements with respect to each other; the regression equation (see Section 5.2) can be used to detect 'abnormal' population elements;

2 for time series, the wish *to* **forecast** future observations of Y;

3 for large populations, the wish *to* **predict** the (still unobserved) value of Y for a population element for which the values of the independent variable are observed.

For situation 1, hardly more than the descriptive statistics techniques of Chapter 5 is needed; as an application, see Example 5.6. For situation 2, special concepts such as trend and seasonal component are often needed (see, for instance, Example 2.9); forecasting will be considered in Chapter 23. The regression analyses of Chapters 19–22 usually concern large populations (situation 3).

The common feature of most regression studies is the wish to understand where the variation of the dependent variable Y comes from. To explain that variation, models are formulated that assume a linear relationship between Y and some variables. The unknown parameters of such a model will be estimated on the basis of sample observations, which leads to an estimated model that, in a sense, fits best to the sample data. The estimated model is used to make predictions. Several questions arise:

- How precise are the estimates? What do they tell about the whole population?
- Which part of the variation in the y data comes from the variation of the independent variables in the model?
- How precise are the predictions?

As an introductory illustration of the above and as a repetition of Section 5.2, consider the following example.

Example 19.1

For companies in the clothing sector it is important to know which variables influence household expenditure on clothing and to what extent. Such research yields information about the consumers who buy clothing products. In the present study, interest is especially in a population of households for which the reference person (called 'head of the household') has a job, with a **positive** gross hourly wage.

It is obvious that the observations of $Y = $ 'annual expenditure on clothing per household ($\times$ €1000)' vary: not all households spend the same amount on clothing. But what are the sources of this variation, which variables (factors) cause it? Not all heads of the household have the same hourly wage, so this may be a factor. To put it another way, it is expected that the variation of the variable $X = $ 'gross hourly wage (in euro) of the head of the household' will partially cause the variation of Y. But more variables may be important; some are listed here (you can probably think of others):

X_2 = 'number of hours worked weekly by the head of the household'
X_3 = 'total annual income of all household members'
X_4 = 'size of the household'
X_5 = 'education level of the head of the household'
X_6 = 'age of the head of the household'

Let us start with only one independent variable. We will consider the relationship between Y (in thousands of euro) and X (in euro), on a large population of households with positive gross hourly wage for the heads. We expect that, on average, a fixed percentage of each increase in x will be used for an increase in y, so a linear relationship $y = \beta_0 + \beta_1 x$ with positive β_1 is expected to describe the average relationship between observations of Y and X. Of course, this relationship will not hold precisely for an arbitrary household. So, we write:

$$Y = \beta_0 + \beta_1 X + \varepsilon \qquad (19.1)$$

Here, the ***error variable*** $\varepsilon = Y - (\beta_0 + \beta_1 X)$ measures the **deviation** between Y and $\beta_0 + \beta_1 X$.

The file Xmp19-01.sav contains the observations of these two variables for a random sample of 236 households from the population of interest. The scatter plot in Figure 19.1 gives a first impression of the relationship between the x data and the y data. The best-fitting line, the **regression** or **least squares line**, is included; see Chapter 5 for details.

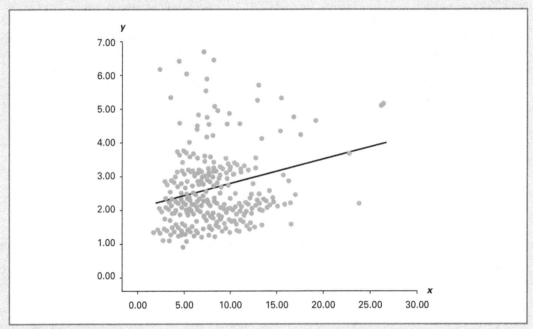

Figure 19.1 Scatter plot of y on x
Source: Personal archives (2005)

With a computer it can easily be determined that the equation of the regression line and the correlation coefficient of the y data and the x data respectively are:

$$\hat{y} = 2.107 + 0.070x \quad \text{and} \quad r = 0.2396$$

(Recall from Chapter 5 that $\hat{y}$ is used (instead of y) to distinguish the sample regression line from the population regression line $y = \beta_0 + \beta_1 x$.) The sample cloud is heavily spread out and the y data are only weakly (positively) linearly dependent on the x data. The variation around the regression line still is considerable, although, when compared to the variation of the y data around their average $\bar{y}$, it has reduced slightly.

It would appear that the respective least squares estimates of β_0 and β_1 are 2.107 and 0.070; an increase of €1 in hourly wage leads in the sample on average to an increase of 0.070×1000

= €70 in annual clothing expenditure. But **how close** are these estimates 2.107 and 0.070 to the population regression coefficients β_0 and β_1?

- How precise are these estimates?

Furthermore, the sample regression line $\hat{y} = 2.107 + 0.070x$ can be considered as an estimate of the population regression line $y = \beta_0 + \beta_1 x$; the sample regression line can be used to **predict** the actual situation. If the head of a household has gross hourly wage €13.20, then it is predicted that the annual clothing expenditure of this household is

$$2.107 + 0.070 \times 13.20 = 3.031$$

which corresponds to €3031. But how close is that prediction – obtained with the sample regression line – to the actual annual expenditure of that household?

- How precise are such predictions?

To measure the estimation precision of the least squares method and the prediction precision of the regression line method, standard deviations will play an important role.

This chapter will lean heavily on Chapter 5, especially Section 5.2. The reader is invited to read this chapter again if necessary.

To say more about the precision of estimation and prediction methods, several model assumptions are needed.

19.2 The simple linear regression model

Interest is in a quantitative variable Y, especially in whether and how it depends on another variable X. In the present chapter, X is also quantitative. Both variables are considered on one (usually large) population. It is the linear relationship between Y and X that will be studied.

Of course, we would prefer to measure the pairs of observations (x, y) for **all** population elements, create a scatter plot of them and find the mathematical equation of the straight line that fits best through the cloud (the population regression line). The resulting equation would be the relationship between Y and X that we are looking for. This **population** cloud – with the best-fitting straight line included – would tell the whole story.

Unfortunately, things are not that easy. Usually it is impossible, too time consuming and/or too expensive to measure X and Y for all population elements. As a consequence, the population cloud with the best-fitting straight line remains unobserved and, at most, we have a vague idea of the cloud and its line. Instead, n population elements – usually only a very small part of the population – are sampled (often randomly) and their values of X and Y are measured. Next, the resulting n pairs of observation (x, y) are depicted in a scatter plot and the equation of the best-fitting straight line (the sample regression line) is determined. This sample equation is used as an estimate of the equation of the unknown straight line in the population plot. It is the task of the statistician to determine the precision of this method of finding estimates.

To illustrate things, reconsider the population of households for which the heads have positive gross hourly wage. Consider the variables:

X = 'gross hourly wage (in euros) of the head of the household'
Y = 'annual expenditure ($\times$ €1000) on clothing of the household'

The scatter plot in Figure 19.2(a) refers to the unobserved population cloud with the unknown population regression line. The vagueness of the picture is meant to illustrate that cloud and line are not known; at best, we have only a vague idea.

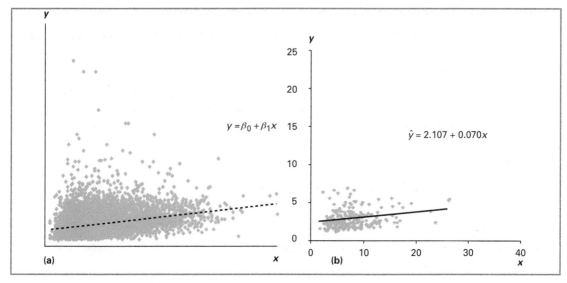

Figure 19.2 Unobserved population cloud and observed sample cloud

The file Xmp19-01.xls contains the data of a random sample of 236 households from that population. The scatter plot of y on x and the sample regression line are included in Figure 19.2(b). Here, cloud and line **are** known; they are used to get an idea of the unobserved population cloud and to estimate the unknown population line.

Before discussing the model assumptions it is important to notice a few things right from the start:

■ It has to be clear whether notations refer to the population or to the sample.

■ It also has to be clear whether notations refer to random variables or to realizations.

■ Model requirements deal with the population. They stylize the (unknown) reality of the population cloud.

■ Estimators and estimates deal with the sample.

To be able to discuss the precision of statistical procedures later, several model requirements (model assumptions) are formulated **now**. Some of them will be illustrated in a population cloud (although such a cloud cannot be observed).

Our starting point is that $\varepsilon = Y - (\beta_0 + \beta_1 X)$. That is, for each pair of observations (x, y) in the population cloud the accompanying observation of ε is just the difference between y and the vertical level $\beta_0 + \beta_1 x$ of the (unknown) population regression line; see Figure 19.3. Notice that ε takes positive **and** negative values: if (x, y) lies below the unknown population regression line, then ε takes a negative value.

Things can also be considered in terms of random observations. That is, for a randomly chosen population element, the random variables X and Y are the random observations (for that element) of the independent and the dependent variable; the random variable $\varepsilon = Y - (\beta_0 + \beta_1 X)$ is the random observation (for that element) of the error variable.

The formulation of the model will be done in terms of the n random pairs $(X_1, Y_1), \ldots, (X_n, Y_n)$ that belong to the n population elements in the sample. These pairs yield the n error random variables $\varepsilon_1, \ldots, \varepsilon_n$ with $\varepsilon_i = Y_i - (\beta_0 + \beta_1 X_i)$. Occasionally, the index i will be omitted, for convenience. All model requirements are about the error terms $\varepsilon_1, \ldots, \varepsilon_n$. They will be considered now, jointly with the consequences for $Y_1, \ldots, Y_n$.

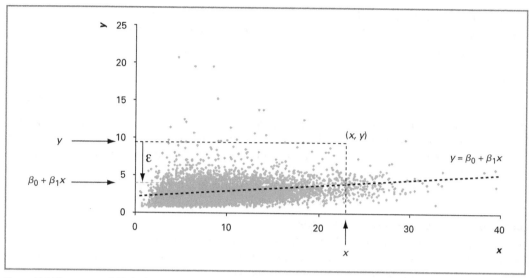

Figure 19.3 The error variable in the population cloud

19.2.1 Requirements of the model

Two assumptions about independence

We first discuss two technical requirements about independence:

- The random variables $\varepsilon_1, \ldots, \varepsilon_n$ have to be independent.
- Each ε_i has to be independent of all $X_1, \ldots, X_n$

Not only is ε_i independent of X_i but ε_i is also independent of X_j for all $j \neq i$.

These two technical assumptions have consequences for the variable Y, more specifically for the random variables $Y_1, Y_2, \ldots, Y_n$. These consequences are formulated as **conditional** statements, as follows. If it is already given that:

$$C = \{X_1 = x_1, X_2 = x_2, \ldots, X_n = x_n\}$$

has occurred, then it follows that:

$$Y_1 = \beta_0 + \beta_1 x_1 + \varepsilon_1, Y_2 = \beta_0 + \beta_1 x_2 + \varepsilon_2, \ldots, Y_n = \beta_0 + \beta_1 x_n + \varepsilon_n$$

so the randomness of $Y_1, Y_2, \ldots, Y_n$ only comes from $\varepsilon_1, \ldots, \varepsilon_n$. Hence, the random variables $Y_1, Y_2, \ldots, Y_n$ are conditionally independent, given that $X_1 = x_1, \ldots, X_n = x_n$.

The first of these technical requirements is always satisfied if the n random pairs form a **random** (hence, *iid*) sample. This is because the *iid* property yields that the n pairs (X_i, Y_i) are independent and identically distributed, and hence – since ε_i is a function of X_i and Y_i – the n random variables ε_i are also independent and identically distributed. Problems can arise in the case of time series; see Section 22.4 for details.

Zero error-expectation along all vertical lines

The third – more intuitive – requirement states that, **if it is already given that $X = x$**, the expectation of ε is 0 (so, irrespective of the value x). In terms of the population: along each vertical straight line in the population cloud, the mean of all observations of ε has to be 0. Here is the complete requirement: for all values x of X it holds that

$$Y = \beta_0 + \beta_1 x + \varepsilon \text{ conditionally on } X = x \quad \text{and} \quad E(\varepsilon \mid x) = 0 \tag{19.2}$$

The right-hand part $E(\varepsilon \mid x) = 0$ has to be read as follows: given that X is fixed at x, the expectation of ε is 0.

Assumption (19.2) has consequences for Y. If it is already given that $X = x$, then $\beta_0 + \beta_1 X$ is equal to $\beta_0 + \beta_1 x$ and hence the expectation of Y under that condition satisfies:

$$E(Y \mid x) = E(\beta_0 + \beta_1 x + \varepsilon \mid x) = \beta_0 + \beta_1 x + E(\varepsilon \mid x) = \beta_0 + \beta_1 x$$

Hence:

$$E(Y \mid x) = \beta_0 + \beta_1 x \qquad \qquad \textbf{(19.3)}$$

This result implies that, along all vertical lines in the population cloud, the mean of all observations of Y is precisely the corresponding vertical level of the unknown population regression line. To put it another way, the mean of Y for the subpopulation with $X = x$ falls precisely at the population regression line.

It is (19.3) that explains why the population regression line is often called the **line of means**: the means of Y along all vertical lines fall precisely on it.

Common variance along all vertical lines

The fourth requirement states that, **if it is already given that $X = x$**, the variance of ε is a common number called σ_ε^2 and hence does not depend on x. For the population cloud this means that the variance of ε measured along all vertical lines is the same. That is:

$$V(\varepsilon \mid x) = \sigma_\varepsilon^2 \quad \text{for all values } x \text{ of } X \qquad \qquad \textbf{(19.4)}$$

($V(\varepsilon \mid x)$ should be read as the variance of ε if it is already given that $X = x$.) If this assumption is satisfied, we talk about ***homoskedasticity***. Since $V(\varepsilon \mid x)$ in a sense measures the fatness of the population cloud along the vertical line at x, requirement (19.4) roughly states that the population cloud is equally fat along all vertical lines.

In addition, (19.4) has consequences for Y. If $X = x$, then $Y = \beta_0 + \beta_1 x + \varepsilon$, so the variance of Y is the same as the variance of the sum of the constant $\beta_0 + \beta_1 x$ and the error variable ε. But constants do not vary, so $V(Y \mid x) = V(\varepsilon \mid x)$. Hence:

$$V(Y \mid x) = \sigma_\varepsilon^2 \quad \text{for all values } x \text{ of } X \qquad \qquad \textbf{(19.5)}$$

Normality along all vertical lines

The model requirements (19.2) and (19.4) are not strong enough for our purposes, at least not for smaller sample sizes. We also need an assumption that guarantees that in the population plot, for each fixed level x of X, a histogram of the observations of Y along the vertical line at x fits well to a normal curve. This is illustrated in Figure 19.4.

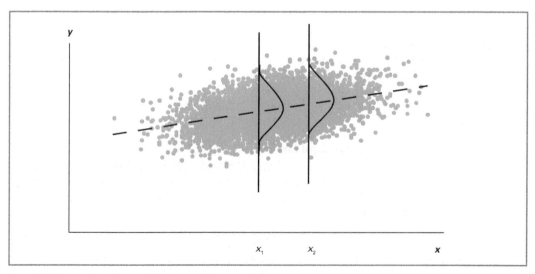

Figure 19.4 Illustration of the conditional normality assumption

This normality along vertical lines follows from the fifth model requirement, which, jointly with the requirements (19.2) and (19.4), is formulated as follows:

$$\varepsilon \sim N(0, \sigma_\varepsilon^2) \quad \text{for all values } x \text{ of } X \tag{19.6}$$

It states that, along all vertical lines, the histograms of the errors have a good fit to the (common) distribution $N(0, \sigma_\varepsilon^2)$.

If it is given that $X = x$, then $Y = \beta_0 + \beta_1 x + \varepsilon$, which is normal since ε is normal. Combining this result with (19.3) and (19.5) yields:

$$Y \sim N(\beta_0 + \beta_1 x, \sigma_\varepsilon^2) \text{ for all values } x \text{ of } X \tag{19.7}$$

So, the normality requirement for ε indeed has the consequence for Y that is illustrated in Figure 19.4. When intersecting the (unobserved) population cloud from the left to the right with vertical lines and creating histograms of the observations of Y along all lines, there is one type of normal curves that fits well to all these histograms. These normal curves have their central positions on the population regression line and they differ only because of their means.

19.2.2 The simple linear regression model of *Y* on *X*

The box below gives the definition of the simple linear regression model in terms of random observations (X, Y). Here, 'simple' refers to the fact that only one independent variable is involved.

Simple linear regression model (for *Y* on *X*)

The combination of the requirements:

- $Y = \beta_0 + \beta_1 X + \varepsilon$ and $E(\varepsilon \mid x) = 0$ for all values x of X,
- $\varepsilon \sim N(0, \sigma_\varepsilon^2)$ conditionally on $X = x$ (for all x),
- random observations of ε are independent,
- random observations of ε are independent of all random observations of X,

is called a ***simple linear regression model*** of Y on X. The assumption $E(\varepsilon \mid x) = 0$ jointly with $Y = \beta_0 + \beta_1 X + \varepsilon$ is called the ***basic assumption*** of the model. The standard deviation σ_ε is called ***standard deviation of the model***.

In the literature, these model requirements are called the **strong** conditions. This is because, for large n, they can often be relaxed to weak conditions by omitting the normality assumption but maintaining (19.2) and (19.4); see also Section 22.3.2.

Notice that the basic assumption can equivalently be formulated as:

$$E(Y \mid x) = \beta_0 + \beta_1 x \tag{19.8}$$

The model has consequences for the whole sample $(X_1, Y_1), \ldots, (X_n, Y_n)$. Suppose that it is already given that $C = \{X_1 = x_1, X_2 = x_2, \ldots, X_n = x_n\}$ has occurred. What are the consequences for the distributions of ε_i and Y_i? Since ε_i is independent of **all** X_j, the pre-information in C about $X_1, \ldots, X_{i-1}, X_{i+1}, \ldots, X_n$ has no influence on the distributions of ε_i and $Y_i = \beta_0 + \beta_1 x_i + \varepsilon_i$, so this part of the pre-information can be omitted. But then the same holds for the conditional expectations and variances of ε_i and Y_i. Table 19.1 summarizes these results.

Model requirements for $\epsilon_1, \dots, \epsilon_n$	Consequences for $Y_1, \dots, Y_n$
$\varepsilon_1, \dots, \varepsilon_n$ are independent	
	Conditionally on C it holds that: $Y_i = \beta_0 + \beta_1 x_i + \varepsilon_i$
Each ε_i is independent of all $X_1, \dots, X_n$	Conditionally on C it holds that: $Y_1, \dots, Y_n$ are independent
$E(\varepsilon_i \mid C) = E(\varepsilon_i \mid x_i) = 0$	$E(Y_i \mid C) = \beta_0 + \beta_1 x_i$
$V(\varepsilon_i \mid C) = V(\varepsilon_i \mid x_i) = \sigma_\varepsilon^2$	$V(Y_i \mid C) = \sigma_\varepsilon^2$
Conditionally on C it holds that: $\varepsilon_i \sim N(0, \sigma_\varepsilon^2)$	Conditionally on C it holds that: $Y_i \sim N(\beta_0 + \beta_1 x_i, \sigma_\varepsilon^2)$

TABLE 19.1 Conditional results for $(X_1, Y_1), \dots, (X_n, Y_n)$, given that $C = \{X_1 = x_1, \dots, X_n = x_n\}$ occurred

The results about $E(\varepsilon_i \mid C)$ and $V(\varepsilon_i \mid C)$ can in fact be sharpened. Since ε_i and X_i are independent, even the pre-information in C that is about X_i has no influence on the expectation and the variance of ε_i. Hence:

$$0 = E(\varepsilon_i \mid C) = E(\varepsilon_i \mid x_i) = E(\varepsilon) \quad \text{and} \quad \sigma_\varepsilon^2 = V(\varepsilon_i \mid C) = V(\varepsilon_i \mid x_i) = V(\varepsilon)$$

It is important to notice that (19.7), in particular, stylizes reality. Practice is that this requirement will never be completely valid. Reality is often different from the model assumptions since most populations contain 'extreme' elements that cause extreme observations (or even outliers) in the dataset. Violation of assumption (19.7) will almost always occur along the right- and left-hand borders of the population cloud and cannot be avoided or solved. However, the central – 'crowded' – part of the cloud should more or less satisfy the assumption. More will be said about model departures in Chapter 22.

If the (virtual) population cloud of Figure 19.2(a) **does** describe reality, then assumption (19.7) does not seem to be satisfied exactly. When intersecting the central part of the cloud by vertical lines, the accompanying histograms of the observations of the error variable will be (slightly) skewed to the right because of extreme observations of the variable 'annual expenditure on clothing'. This is a problem that is often encountered if the dependent variable has to do with expenditure or income. Fortunately, this problem is less important for larger samples.

The following example illustrates the above conditional normality assumption in the rather unrealistic situation that the population **is** observed.

Example 19.2

Interest is in the variation of the variable $Y = $ 'weight (kg)' on a population of 5330 adult men. Since $X = $ 'height (cm)' may partially cause this variation, we will regress Y on X. In this case, the population data **are** observed; see Xmp19-02a.xls. With a computer the equation of the population regression line follows easily:

$$y = \beta_0 + \beta_1 x \quad \text{with } \beta_0 = -16.606 \text{ and } \beta_1 = 0.5287$$

An increase of 1 cm in height causes, on average, an increase of 0.5287 kg in weight in the population. Furthermore, it turns out that $\sigma_\varepsilon = 9.65$ kg (details are omitted). Figure 19.5 shows the population cloud with the population regression line.

The first impression is that, apart from the left-hand and right-hand borders of the cloud, assumptions (19.2) and (19.4) are satisfied. The line of means seems to cut the cloud more or less

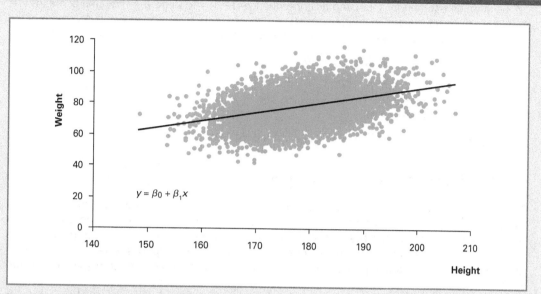

Figure 19.5 Population scatter plot of 'weight' on 'height' with line of means

into two symmetric parts. The cloud is more or less 'equally fat' along vertical lines, at least if the edges are omitted.

We will check things for $x = 175$ and for $x = 185$. By (19.3) the following should be valid:

$$E(Y \mid 175) = -16.606 + 0.5287 \times 175 = 75.9 \quad \text{and} \quad V(Y \mid 175) = (9.65)^2$$

$$E(Y \mid 185) = -16.606 + 0.5287 \times 185 = 81.2 \quad \text{and} \quad V(Y \mid 185) = (9.65)^2$$

The datasets Xmp19-02b.xls and Xmp19-02c.xls contain the weights of the adult men in the subpopulations with (rounded) heights 175 cm and 185 cm, respectively. These datasets are the y observations encountered in the population cloud along the vertical lines with $x = 175$ and $x = 185$, respectively. With a computer it follows easily that these datasets have respective means 77.1 kg and 81.2 kg, and respective standard deviations 9.8 kg and 9.4 kg. Notice that these means are not far from the two means 75.9 and 81.2 that follow from the model. Also, the two standard deviations are close to the common standard deviation 9.65 of the model. According to the conditional normality assumption, histograms of the two datasets should fit well to normal curves that belong to $N(75.9, 9.65^2)$ and $N(81.2, 9.65^2)$. Figure 19.6 shows such histograms; they do indeed suggest a normal distribution.

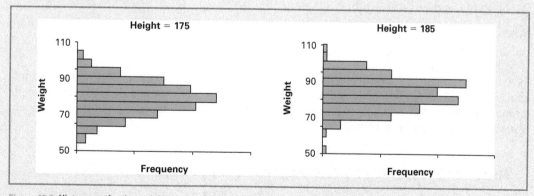

Figure 19.6 Histograms for the weights at $x = 175$ and $x = 185$

Because of assumption (19.2) and its consequence (19.3), the slope β_1 of the line of means $y = \beta_0 + \beta_1 x$ can be interpreted in an elegant way. Of course, β_1 is the (mathematical) derivative of the function $\beta_0 + \beta_1 x$, but thanks to (19.3) we additionally have that, for each fixed value x of X:

$$E(Y \mid x + 1) \qquad = \beta_0 + \beta_1 x + \beta_1$$
$$\underline{E(Y \mid x) \qquad\qquad = \beta_0 + \beta_1 x}$$
$$E(Y \mid x + 1) - E(Y \mid x) = \beta_1$$

That is, β_1 is the change (positive or negative) of the expectation of Y if x increases by 1.

19.3 Point estimators of β_0, β_1 and σ_ε^2

The simple linear regression model has three unknown parameters: the population regression coefficients β_0, β_1 and the common variance σ_ε^2 of the error variables ε. These parameters have to be estimated on the basis of a sample of n population elements. The **definition** of the estimators (in the present section) does not require the model assumptions to be valid. However, the model assumptions **are** important for the validity of the properties of the estimators; see Section 19.4.

Again, interest is in regression of Y on X. Suppose that a sample of n population elements yields the following sample of **random** pairs:

$$(X_1, Y_1), (X_2, Y_2), \ldots, (X_n, Y_n)$$

This sample will be the input for the estimators.

19.3.1 Estimators of β_0 and β_1

Recall that β_0 and β_1 are the intercept and the slope of the population regression line. They can be considered as the results of the application of the least-squares method to the **population** cloud; see Section 5.2. Although this cloud is unobserved and hence β_0 and β_1 are unknown, their formulae could be derived; see (5.7):

$$\beta_1 = \frac{\sigma_{X,Y}}{\sigma_X^2} \quad \text{and} \quad \beta_0 = \mu_Y - \beta_1 \mu_X$$

Notice especially that β_1 has the same numerator as the population correlation coefficient ρ but that the denominators of β_1 and ρ are different; see also Section 5.1.

Estimates of the unknown parameters β_1 and β_0 can be obtained by applying the least-squares method to a **sample** cloud. For a **realized** sample $(x_1, y_1), \ldots, (x_n, y_n)$, the least-squares estimates b_0 and b_1 of β_0 and β_1 were derived in Section 5.2 (by using the least-squares method):

$$b_1 = \frac{s_{X,Y}}{s_X^2} \quad \text{and} \quad b_0 = \bar{y} - b_1 \bar{x}$$

where

$$s_{X,Y} = \frac{1}{n-1} \sum_{i=1}^n (x_i - \bar{x})(y_i - \bar{y}) \quad \text{and} \quad s_X^2 = \frac{1}{n-1} \sum_{i=1}^n (x_i - \bar{x})^2$$

(Recall that $s_{X,Y}$ and s_X^2 respectively are the sample covariance of the (x, y) data and the sample variance of the x data.) To obtain estimat**ors** – that is, sample statistics that depend on the sample of the **random** pairs $(X_1, Y_1), \ldots, (X_n, Y_n)$ – we just replace the realizations x_i and y_i by the random variables X_i and Y_i that they come from.

> ### Least-squares (point) estimators B_0 and B_1 of β_0 and β_1
>
> $$B_1 = \frac{S_{X,Y}}{S_X^2} \quad \text{and} \quad B_0 = \bar{Y} - B_1 \bar{X} \tag{19.9}$$

Here,

$$S_{X,Y} = \frac{1}{n-1}\sum_{i=1}^n (X_i - \bar{X})(Y_i - \bar{Y}) \quad \text{and} \quad S_X^2 = \frac{1}{n-1}\sum_{i=1}^n (X_i - \bar{X})^2 \qquad \textbf{(19.10)}$$

The estimators B_0 and B_1 are also called **sample regression coefficients**. $S_{X,Y}$ and S_X^2 are, respectively, the sample covariance and the sample variance of $X_1, ..., X_n$ (written as random variables), the usual estimators of the population covariance $\sigma_{X,Y}$ and the population variance σ_X^2; see (12.9) and (12.6). Hence, the estimator of β_1 is just the ratio of the estimators of the covariance of (X, Y) and the variance of X.

Notice that B_0 and B_1 are random variables and that their formulae do not depend on parameters. Hence, B_0 and B_1 are indeed sample statistics. As soon as the data are available, they can be substituted into the estimators B_0 and B_1 to obtain the estima**tes** b_0 and b_1. The realized sample regression line $\hat{y} = b_0 + b_1 x$ can then be considered as an approximation (estimate) of the population regression line $y = \beta_0 + \beta_1 x$.

Recall from Section 5.2 that the sample regression line $\hat{y} = b_0 + b_1 x$ can be used to obtain predictions $\hat{y}_p$ of y for population elements for which only the x-coordinate is known. By substituting x_i for x, the result $\hat{y}_i = b_0 + b_1 x_i$ is also called a **prediction of** y_i. The predictions $\hat{y}_1, ..., \hat{y}_n$ of $y_1, ..., y_n$ in the sample can be used to obtain information about the predicting performance of the sample regression line. Also recall that the deviations $y_i - \hat{y}_i$ between the y_i and their predictions $\hat{y}_i$ are called **residuals**; $y_i - \hat{y}_i$ is denoted by e_i.

The following example demonstrates these matters and repeats techniques that were used in Section 5.2.

Example 19.3

Advertising is important for supermarkets, to stimulate sales and to stress a product's distinctive features. Table 19.2 shows, for the most recent seven weeks, the sales and the amounts spent on advertising by a supermarket. Of course, 'sales' is the dependent variable and 'advertising costs' the independent variable. To be precise, $Y =$ 'weekly sales ($\times$ €1000)' and $X =$ 'weekly advertising costs ($\times$ €1000)'.

Week i	y_i	x_i	$x_i - \bar{x}$	$y_i - \bar{y}$	$(x_i - \bar{x})^2$	$(x_i - \bar{x})(y_i - \bar{y})$	$\hat{y}_i$	e_i
1	80.2	2.0	−0.243	−4.400	0.059	1.069	80.180	0.020
2	85.6	2.2	−0.043	1.000	0.002	−0.043	83.814	1.786
3	74.3	2.0	−0.243	−10.300	0.059	2.502	80.180	−5.880
4	93.9	2.5	0.257	9.300	0.066	2.391	89.265	4.635
5	77.8	1.8	−0.443	−6.800	0.196	3.012	76.546	1.254
6	87.0	2.3	0.057	2.400	0.003	0.137	85.631	1.369
7	93.4	2.9	0.657	8.800	0.432	5.782	96.533	−3.133
Total	592.2	15.7	0.000	0.000	0.817	14.850		0.051 ≈ 0

$\bar{x} = 2.2429;\quad \bar{y} = 84.6;\quad s_X^2 = 0.1362;\quad s_{X,Y} = 2.4750$

$b_1 = \frac{2.4750}{0.1362} = 18.17;\quad b_0 = 84.6 - 18.172 \times 2.2429 = 43.84$

TABLE 19.2 Calculation scheme for the sample regression line of y on x

The totals of columns 2 and 3 yield the sample means $\bar{y}$ and $\bar{x}$. Columns 4 and 5 arise by subtracting the corresponding mean from columns 3 and 2, respectively. Column 6 contains the squares of column 4; its total yields the sample variance s_x^2 after division by $n - 1 = 6$. Column 7 contains the products of the columns 4 and 5; its total gives the sample covariance $s_{x,y}$ after division by 6.

The slope of the sample regression line is obtained by dividing the sample covariance by the sample variance of the x data, the intercept also follows easily; see the calculations in the last rows of the table. The equation of the sample regression line becomes:

$$\hat{y} = 43.84 + 18.17x$$

This regression line estimates the (unknown) population regression line $y = \beta_0 + \beta_1 x$ of the unobserved population cloud of pairs (x, y) for **all** weeks (not just the seven in the sample). The estimates of β_0 and β_1 are, respectively, 43.84 and 18.17. The interpretation of the slope of the sample regression line is as follows: spending an extra €1000 each week on advertising leads on average to an extra €18 170 each week in sales. It is not permissible to interpret the intercept as the average weekly sales for weeks without advertising costs; this is because 0 does not belong to the range of the x data.

The sample regression line can be used to predict weekly sales for coming weeks. If the advertising costs for the coming week are €2100, then:

$$\hat{y} = 43.84 + 18.17 \times 2.1 = 81.997$$

It is predicted that in the coming week the sales will be €81 997.

To get an idea of the predicting performance of the sample regression line, we next calculate the residuals $e_i = y_i - \hat{y}_i$ where $\hat{y}_i$ can be interpreted as the predicted sales in weeks with the same level of advertising costs as in week i of the sample. For instance:

$$\hat{y}_4 = 43.84 + 18.17 \times 2.5 = 89.265, \quad \text{so } e_4 = 93.9 - 89.265 = 4.635$$

See the second last and the last columns of the table for the predictions $\hat{y}_i$ and the residuals e_i.

Recall from (5.8) that the total of all residuals has to be 0; the total 0.051 of the last column is unequal to 0, but this is due to rounding. Also recall that the sum of squared errors SSE measures the variation of the dots in the sample cloud around the sample regression line and hence measures the predicting performance of this line. By straightforward calculations it follows that:

$$\text{SSE} = \sum_{i=1}^{7}(y_i - \hat{y}_i)^2 = \sum_{i=1}^{7}e_i^2 = (0.020)^2 + \ldots + (-3.133)^2 = 72.510$$

19.3.2 Estimator of σ_ε^2

The variance σ_ε^2 of the error terms measures the overall variation around the population regression line in the population cloud. Hence, it is natural to estimate σ_ε^2 by means of a measure for the overall variation around the **sample** regression line in the **sample** cloud.

From the sample data $(x_1, y_1), \ldots, (x_n, y_n)$ the sample regression line $\hat{y} = b_0 + b_1 x$ can be determined. For the pair (x_i, y_i), the (vertical) deviation $e_i = y_i - \hat{y}_i$ between y_i and $\hat{y}_i = b_0 + b_1 x_i$ gives information about the variation locally at (x_i, y_i); see Figure 19.7.

To measure the **overall** variation, it makes no sense to take the sum $\sum_{i=1}^{n} e_i$ of the n residuals as measure since this summation always is 0. The sum of the **squared** deviations does a better job in measuring this variation. Recall from Section 5.2 that this last summation is called SSE, the sum of the squared errors:

$$\text{SSE} = \sum_{i=1}^{n}e_i^2 = \sum_{i=1}^{n}(y_i - \hat{y}_i)^2 = \sum_{i=1}^{n}(y_i - b_0 - b_1 x_i)^2 \tag{19.11}$$

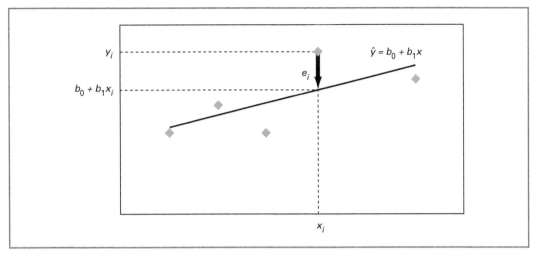

Figure 19.7 Sample scatter plot with sample regression line and residuals

Although SSE can be used to measure the variation around the sample regression line, it is not used to estimate the **variance** σ_ε^2. Instead, a 'mean' of the squared deviations, called s_ε^2, is used:

$$s_\varepsilon^2 = \frac{1}{n-2} \sum_{i=1}^{n} (y_i - \hat{y}_i)^2 = \frac{\text{SSE}}{n-2}$$

To transform this expression into an estimat**or** of σ_ε^2, it has to be based on the random pairs (X_1, Y_1), ..., (X_n, Y_n). Hence, not only do the realized y_i have to be replaced by the random variables Y_i but also the realized $\hat{y}_i$ by the random variables $\hat{Y}_i = B_0 + B_1 X_i$.

Estimator of σ_ε^2

$$S_\varepsilon^2 = \frac{1}{n-2} \sum_{i=1}^{n} (Y_i - \hat{Y}_i)^2 \text{ with } \hat{Y}_i = B_0 + B_1 X_i \qquad \textbf{(19.12)}$$

Estimator of σ_ε

$$S_\varepsilon = \sqrt{S_\varepsilon^2} = \sqrt{\frac{1}{n-2} \sum_{i=1}^{n} (Y_i - \hat{Y}_i)^2} \qquad \textbf{(19.13)}$$

The estimators S_ε and S_ε^2 are called, respectively, ***standard error of the estimated model*** and the ***sample variance of the estimated model***.

There are two reasons why the summation in (19.12) is divided by $n - 2$ instead of n or $n - 1$. The first is that the n random variables $Y_1 - \hat{Y}_1, ..., Y_n - \hat{Y}_n$ on which the summation in S_ε^2 is based, cannot vary freely since they all depend on the estimators B_0 and B_1. In essence, the estimation of the unknown parameters β_0 and β_1 has put **two** conditions on the sample of the n random pairs. That is, the summation has only $n - 2$ degrees of freedom instead of n. Another (related) reason is that the estimator with the factor $1/(n - 2)$ on average does better when estimating σ_ε^2 than an estimator with the factor $1/n$ or $1/(n - 1)$. See also Section 19.4.

In the last example we obtained SSE = 72.510. Hence,

$$S_\varepsilon^2 = \frac{72.510}{7-2} = 14.502 \quad \text{and} \quad s_\varepsilon = \sqrt{14.502} = 3.808$$

The standard error of the estimated model is 3.808. That is, the estimate of the standard deviation σ_ε of the model is 3.808 units of a thousand euro.

The estimator S_ε^2 measures the variation **per degree of freedom** in sample (x, y) plots around the accompanying sample regression lines. Being an estimator, it is a **method** to obtain estimates for the variance of the error variable.

Example 19.4

The estimators B_0, B_1 and S_ε^2 are applied in two examples.

File Xmp19-04a.sav contains the annual expenditure on clothing of a sample of 236 households, jointly with the gross hourly wages of their reference persons (called heads); see also Example 19.1. The sample is a random sample from a large population of households with employed heads. This sample will be used to explain the variation of $Y =$ 'annual clothing expenditure ($\times$ €1000) of a household' from $X =$ 'gross hourly wage (in euro) of its head'. The partial SPSS printout below is obtained.

Model Summary

Model	R	R Square	Adjusted R Square	Std. Error of the Estimate
1	.240(a)	.057	.053	1.14606

a Predictors: (Constant), x

Coefficients(a)

Model		Unstandardized Coefficients		Standardized Coefficients	t	Sig.
		B	Std. Error	Beta		
1	(Constant)	2.107	.165		12.768	.000
	x	.070	.019	.240	3.775	.000

a Dependent Variable: y

For the time being only the fat numbers are of interest. The estimates b_0 and b_1 of the population regression coefficients β_0 and β_1 can be found in the Coefficients part of the printout, under the heading B. The standard error s_ε of the estimated model can be found in the Model Summary. It follows that:

$$\hat{y} = 2.107 + 0.070x \quad \text{and} \quad s_\varepsilon = 1.14606, \quad \text{so } S_\varepsilon^2 = 1.3135$$

If the gross hourly wage increases by €1, then the average annual expenditure on clothing is estimated to increase by €70; the intercept 2.107 cannot be interpreted since 0 is not in the range of the x data. The variance of the error variable for the whole population is estimated by 1.3135.

File Xmp19-04b.sav contains the observations of $Y =$ 'weight (kg)' and $X =$ 'height (cm)' for a random sample of 242 persons from a population of 5330 adult men; see also Example 19.2.

Model Summary

Model	R	R Square	Adjusted R Square	Std. Error of the Estimate
1	.434(a)	.188	.185	9.48609

a Predictors: (Constant), Height

Coefficients(a)

Model		Unstandardized Coefficients		Standardized Coefficients	t	Sig.
		B	**Std. Error**	**Beta**		
1	(Constant)	−26.707	13.920		−1.919	.056
	Height	.584	.078	.434	7.465	.000

a Dependent Variable: Weight

The printout yields:

$$\hat{y} = -26.707 + 0.584x \qquad \text{and} \qquad s_\varepsilon = 9.48609, \quad \text{so } s_\varepsilon^2 = 89.9859$$

Hence, the intercept β_0 and the slope β_1 of the population regression line in Figure 19.5 of Example 19.2 are estimated to be equal to −26.707 and 0.584, respectively. Being 1 cm taller will increase the average weight by 0.584 kg. The intercept −26.707 cannot be interpreted since 0 is not part of the range of the x data. The standard deviation σ_ε of the model is estimated by 9.48609 kg.

Recall that this population of 5330 adult men is special in the sense that X and Y were observed over **all** population elements, so β_0 and β_1 are already known; they are equal to −16.606 and 0.5287, respectively. The estimate of β_1 is slightly too large; the deviation is about 10.5% of the actual value of β_1.

For the calculation of b_1, the sample covariance of the (x, y) data and the sample variance of the x data have to be calculated. For calculations with a simple calculator, the short-cut formulae are convenient; see also (5.4). For later purposes we also repeat the short-cut formula for the sample variance of the y data:

Short-cut formulae for $s_{X,Y}$, and for s_X^2 and s_Y^2

$$s_{X,Y} = \frac{1}{n-1}\left(\sum_{i=1}^{n} x_i y_i - n\overline{x}\overline{y}\right)$$

$$s_X^2 = \frac{1}{n-1}\left(\sum_{i=1}^{n} x_i^2 - n\overline{x}^2\right) \quad \text{and} \quad s_Y^2 = \frac{1}{n-1}\left(\sum_{i=1}^{n} y_i^2 - n\overline{y}^2\right)$$

Notice that the formulae for the sample variances actually follow from the formula for the sample covariance: replace y by x and x by y, respectively.

19.4 Properties of the estimators

If the simple linear regression model is valid, then the estimators B_0, B_1 and S_ε^2 of β_0, β_1 and σ_ε^2 turn out to have nice properties that will be used in Section 19.5 to derive interval estimators and hypothesis tests. In the present section, emphasis is on these properties.

Suppose that the n pairs of random variables $(X_1, Y_1), \ldots, (X_n, Y_n)$ satisfy the simple linear regression model. For the accompanying error variables ε_i it holds that $\varepsilon_i = Y_i - (\beta_0 + \beta_1 X_i)$. Recall that some of the results in Table 19.1 are conditional ones, under the condition $C = \{X_1 = x_1; \ldots; X_n = x_n\}$.

The estimators B_0, B_1 and S_ε^2 turn out to be **unbiased**; see the proof below. Furthermore, they are all **consistent** (no proof given): the deviation between the estimator and the parameter becomes arbitrarily small if the sample size n is taken larger and larger.

B_0, B_1 and S_ε^2 are unbiased and consistent estimators of, respectively, β_0, β_1 and σ_ε^2. **(19.14)**

Proofs of the unbiasedness of B_1 and B_0

In essence, the proofs are immediate consequences of Table 19.1 and the rules for expectations and summations. Since conditional expectations **are** expectations, the rules in (11.14) remain valid.

Both proofs lean heavily on the result that $E(\bar{Y} \mid C) = \beta_0 + \beta_1\bar{x}$, which will be proved first:

$$E(\bar{Y} \mid C) = E(\frac{1}{n}\sum_{i=1}^{n}Y_i \mid C) = \frac{1}{n}\sum_{i=1}^{n}E(Y_i \mid C) = \frac{1}{n}\sum_{i=1}^{n}(\beta_0 + \beta_1 x_i)$$

$$= \frac{1}{n}\sum_{i=1}^{n}\beta_0 + \beta_1\frac{1}{n}\sum_{i=1}^{n}x_i = \beta_0 + \beta_1\bar{x}$$

To prove that $E(B_1) = \beta_1$, notice that – under the condition that C has occurred – B_1 can be written in terms of the **realized** observations x_i and the **random** observations Y_i; in particular, s_X^2 is then not random:

$$B_1 = \frac{1}{(n-1)s_X^2}\sum_{i=1}^{n}(x_i - \bar{x})(Y_i - \bar{Y}) \quad \text{conditionally on } C$$

Hence:

$$E(B_1 \mid C) = \frac{1}{(n-1)s_X^2}E\left(\sum_{i=1}^{n}(x_i - \bar{x})(Y_i - \bar{Y}) \mid C\right) = \frac{1}{(n-1)s_X^2}\sum_{i=1}^{n}((x_i - \bar{x})E(Y_i - \bar{Y} \mid C))$$

Since:

$$E(Y_i - \bar{Y} \mid C) = E(Y_i \mid C) - E(\bar{Y} \mid C) = \beta_0 + \beta_1 x_i - (\beta_0 + \beta_1\bar{x}) = \beta_1(x_i - \bar{x})$$

it follows that:

$$E(B_1 \mid C) = \frac{1}{(n-1)s_X^2}\sum_{i=1}^{n}(x_i - \bar{x})\beta_1(x_i - \bar{x}) = \beta_1\frac{1}{(n-1)s_X^2}\sum_{i=1}^{n}(x_i - \bar{x})^2 = \beta_1$$

It would appear that $E(B_1 \mid C) = \beta_1$ whatever the pre-information about the x data in C is. But then the result also holds unconditionally.

The proof of $E(B_0) = \beta_0$ follows from $E(B_1 \mid C) = \beta_1$:

$$E(B_0 \mid C) = E(\bar{Y} - B_1\bar{x} \mid C) = E(\bar{Y} \mid C) - \bar{x}E(B_1 \mid C) = \beta_0 + \beta_1\bar{x} - \beta_1\bar{x} = \beta_0$$

Since this result is valid irrespective of the pre-information in C, it also holds unconditionally.

The above proofs of the unbiasedness of the estimators B_0 and B_1 did not make use of the model requirements about the normality and independence of the error variables. Recall that, because of the normality assumption, the Y_i are conditionally normally distributed. If it is given that C has occurred, then it can be shown (but we won't) that both B_1 and B_0 are linear combinations of $Y_1, Y_2, \ldots, Y_n$, so they are linear combinations of conditionally independent and normally distributed random variables. Hence, B_1 and B_0 are, by (11.19), conditionally normal themselves.

The box below summarizes the important properties of the three estimators. Notice especially the expressions for the conditional variances $V(B_1 \mid C)$ and $V(B_0 \mid C)$.

Properties of B_1 and B_0 (in the simple linear regression model)

$$B_1 \sim N\left(\beta_1, \frac{\sigma_\varepsilon^2}{(n-1)s_X^2}\right) \qquad \text{conditionally on } C \qquad \textbf{(19.15)}$$

$$B_0 \sim N\left(\beta_0, \sigma_\varepsilon^2\left(\frac{1}{n} + \frac{\bar{x}^2}{(n-1)s_X^2}\right)\right) \qquad \text{conditionally on } C \qquad \textbf{(19.16)}$$

Property of S_ε^2

$$E(S_\varepsilon^2) = \sigma_\varepsilon^2 \qquad \textbf{(19.17)}$$

It would appear that the conditional variance of B_1 **does** depend on the x sample, by way of n and the sample variance s_X^2; the conditional variance of B_0 depends on the x sample by way of n, $\bar{x}$ and s_X^2. For instance (19.15) has to be interpreted as follows: B_1 has the prescribed normal distribution under the condition that $x_1, \ldots, x_n$ are the actually observed realizations of the random variables $X_1, \ldots, X_n$.

In this book, (19.16) will not be used. It is included only for completeness.

Since the conditional normality of B_1 is not proved here, the normality of B_1 will be made plausible by comparing a histogram of observations of B_1 with the best-fitting normal density.

Example 19.5

The file Xmp19-05.xls contains the observations of $Y =$ 'weight (kg)' and $X =$ 'height (cm)' for 473 random samples of size 5 from a large population of adult men; each row of the dataset contains one sample of pairs. Recall from Example 19.2 that the simple linear regression model fits well to these variables.

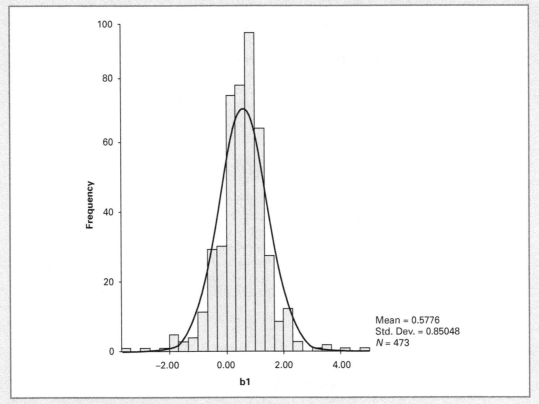

Figure 19.8 Histogram of the b_1 data with best-fitting normal curve

For each sample (row) the accompanying value of B_1 is calculated. The resulting 473 observations of B_1 are recorded in the file Xmp19-05a.sav. According to the above theory, a histogram of these b_1 data should approximate a **normal** distribution. Figure 19.8 contains the histogram that SPSS creates automatically.

Although the fit is not perfect, the histogram indeed resembles a normal curve. For a better fit, more samples of size 5 are needed.

From now on, we will often (not always) follow the general convention to omit the condition in our notations. For instance, instead of writing $V(B_1 \mid C)$ we will often write:

$$V(B_1) = \frac{\sigma_\varepsilon^2}{(n-1)s_X^2}$$

Note that this is the variance of B_1 if it is already given that $C = \{X_1 = x_1; \ldots; X_n = x_n\}$ has occurred. That the above equality indeed concerns a **conditional** variance then follows from the fact that the right-hand expression depends on the constant term $(n-1)s_X^2 = \sum_{i=1}^n (x_i - \bar{x})^2$ (and **not** on the random term $(n-1)S_X^2 = \sum_{i=1}^n (X_i - \bar{X})^2$).

It is important to notice that the results so far and the statistical conclusions in the coming sections have to be read in a conditional sense: under the condition that it is already known that C has occurred. We call this the **conditional method**. This conditional method implies that it is already known that the random variable X_1 has the realized value x_1, that X_2 has the realized value x_2, etc. As a consequence, the sample mean and the sample variance are no longer random, so we can write $\bar{x}$ and s_X^2. Moreover, the randomness of the sample covariance only comes from the random variables $Y_1, \ldots, Y_n$ since the realizations of the X_i in the formula are already known to be equal to x_i. Hence, the sample covariance is now equal to

$$S_{X,Y} = \frac{1}{n-2} \sum_{i=1}^n (x_i - \bar{x})(Y_i - \bar{Y})$$

which contains the non-random x_i and $\bar{x}$, and the random Y_i and $\bar{Y}$.

As a consequence of the conditional method, the statistical conclusions of the coming sections are, strictly speaking, only conditionally valid, under the condition that the random variables $X_1, \ldots, X_n$ take the actual realisations $x_1, \ldots, x_n$. However, since these realizations $x_1, \ldots, x_n$ are usually the results of chance and the standard statistical procedures will be built for general $x_1, \ldots, x_n$, this conditional validity is hardly a restriction. That is why the condition is often omitted when the statistical conclusions are interpreted.

19.5 Inference about the parameter β_1

This section is about confidence intervals and hypothesis tests with respect to the parameter β_1. Statistical procedures will be developed that can be used to create such intervals and tests. Again, our approach will follow the theoretical steps A–D of Section 17.1.

Our starting point is B_1, an unbiased and consistent estimator of β_1 (which is step A). The next step (step B) is to find a suitable pivot. Since the estimator B_1 is conditionally normally distributed, its standardized version is conditionally **standard** normally distributed and the standard normal distribution does not depend on any unknown parameters. Since the standard deviation of B_1 equals

$$SD(B_1) = \sqrt{\frac{\sigma_\varepsilon^2}{(n-1)s_X^2}} = \frac{\sigma_\varepsilon}{\sqrt{(n-1)s_X^2}}$$

it follows by (19.15) that:

$$Z = \frac{B_1 - \beta_1}{\sigma_\varepsilon / \sqrt{(n-1)s_X^2}} \text{ is standard normally distributed.}$$

Although Z depends on the unknown parameters β_1 and σ_ε, its probability distribution does not. Indeed, Z is a pivot. We have completed step B.

Unfortunately, Z cannot be used immediately to construct interval estimators and test statistics. This is because it contains the unknown parameter σ_ε. However, thanks to the strong model requirements it is permissible to replace σ_ε by its **consistent** estimator S_ε at the cost of getting a t-distributed random variable. Since S_ε has $n - 2$ degrees of freedom, we get a t_{n-2}-distribution. Here are the summarized results with some consequences:

Basic results for inferential statistics about β_1

If the simple linear regression model requirements hold, then:

$$Z = \frac{B_1 - \beta_1}{\sigma_\varepsilon / \sqrt{(n-1)s_X^2}} \sim N(0, 1) \quad \text{and} \quad T = \frac{B_1 - \beta_1}{S_\varepsilon / \sqrt{(n-1)s_X^2}} \sim t_{n-2} \tag{19.18}$$

Consequences

$$P(T \le -t_{\alpha;n-2}) = \alpha; \quad P(T \ge t_{\alpha;n-2}) = \alpha; \quad P(-t_{\alpha/2;n-2} < T < t_{\alpha/2;n-2}) = 1 - \alpha \tag{19.19}$$

Notice that the denominator of T is the **standard error** of B_1. It arises from the standard **deviation** of B_1 by replacing the unknown parameter σ_ε by the estimator S_ε. Hence:

$$SE(B_1) = S_{B_1} = \frac{S_\varepsilon}{\sqrt{(n-1)s_X^2}} \tag{19.20}$$

Note that the numerator of $SE(B_1)$ is random; the denominator is not. The random standard error $SE(B_1)$ is also denoted by S_{B_1}; its realizations are written s_{B_1}.

The above arguments can be used to construct interval estimators (step C) and test statistics (step D).

19.5.1 Interval estimator for β_1

By substituting the right-hand side of (19.18) into the right-hand equation in (19.19) we obtain:

$$1 - \alpha = P\left(-t_{\alpha/2;n-2} < \frac{B_1 - \beta_1}{S_\varepsilon / \sqrt{(n-1)s_X^2}} < t_{\alpha/2;n-2}\right)$$

$$= P\left(-t_{\alpha/2;n-2} \frac{S_\varepsilon}{\sqrt{(n-1)s_X^2}} < B_1 - \beta_1 < t_{\alpha/2;n-2} \frac{S_\varepsilon}{\sqrt{(n-1)s_X^2}}\right)$$

Rewriting things such that only β_1 remains in the middle yields:

$$P\left(B_1 - t_{\alpha/2;n-2} \frac{S_\varepsilon}{\sqrt{(n-1)s_X^2}} < \beta_1 < B_1 + t_{\alpha/2;n-2} \frac{S_\varepsilon}{\sqrt{(n-1)s_X^2}}\right) = 1 - \alpha$$

With probability $1 - \alpha$ it holds that β_1 is captured between a random lower bound L and a random upper bound U. But notice that these bounds are sample statistics; they do not depend on unknown parameters, so their values can be calculated as soon as the data are available.

Interval estimator (L, U) for β_1

$$L = B_1 - t_{\alpha/2;n-2} \frac{S_\varepsilon}{\sqrt{(n-1)s_X^2}} \quad \text{and} \quad U = B_1 + t_{\alpha/2;n-2} \frac{S_\varepsilon}{\sqrt{(n-1)s_X^2}} \tag{19.21}$$

This random interval captures β_1 with probability $1 - \alpha$. When the data are available and substituted into the formulae of L and U, a $(1 - \alpha)$ confidence interval (l, u) arises. It encloses β_1 with **confidence** $1 - \alpha$. We have just finished step C.

Notice that the interval estimator is of format 1; see Section 15.2. The half-width H is random because of S_ε. The half-width is:

$$H = t_{\alpha/2;n-2} \frac{S_\varepsilon}{\sqrt{(n-1)s_X^2}}$$

Example 19.6

To learn the mechanics of the formulae of L and U, reconsider Example 19.3 where the variation of Y = 'weekly sales ($\times$ €1000)' is (partially) explained from the variation of X = 'weekly advertising costs ($\times$ €1000)'. On the basis of the (x, y) data for the most recent seven weeks it was calculated that:

$$b_1 = 18.17; \quad s_X^2 = 0.1362; \quad s_\varepsilon = 3.808$$

To obtain a 95% confidence interval (l, u) for the slope of the population regression line, check yourself that $t_{0.025;n-2} = t_{0.025;5} = 2.5706$ (*). The calculation of l and u follows below:

$$b_1 \pm t_{0.025;n-2} \frac{s_\varepsilon}{\sqrt{(n-1)s_X^2}} = 18.17 \pm 2.5706 \frac{3.808}{\sqrt{6 \times 0.1362}} = 18.17 \pm 2.5706 \times 4.2124$$

Hence, the realized standard error of B_1 is equal to 4.2124, the lower bound l is $18.17 - 10.8285 = 7.342$, and the upper bound u is $18.17 + 10.8285 = 28.999$. There is 95% confidence that β_1 lies in the interval (7.342, 28.999). To put it another way, there is 95% confidence that spending an extra €1000 on advertising will, on average, increase the weekly sales by an amount between €7342 and €28999. The interval is very wide due to the fact that the sample size is only 7.

Since the 95% confidence interval is situated completely to the right of 0, we are (at least) 95% certain that β_1 is positive: it is likely that advertising makes sense, that – on average – the weekly sales will be increased by an increase in advertising.

Example 19.7

We want to find out whether the variation of the variable 'annual clothing expenditure ($\times$ €1000) per household' is, at least partially, explained by the variable 'gross hourly wage (in euro) of the head of the household'. Denote the first variable by Y and the second by X, and adopt the simple linear regression model to explain Y from X; see also Example 19.1. The file Xmp19-07.sav contains the observations of X and Y for a sample of 236 households. This dataset will be used to construct a 98% confidence interval for the slope of the population regression line. The interval will be used to conclude whether X and Y are linearly related.

The SPSS printout below contains ingredients for the formulae of the lower and upper bound. It is important to learn how such printouts can be created; see also Appendix A1.19.

Notice that $\alpha = 0.02$ and $t_{\alpha/2;n-2} = t_{0.01;234} = 2.3424$ (*). From the columns headed B and Std. Error it follows that $b_1 = 0.070$ and the accompanying standard error is 0.019. Hence:

$$b_1 \pm t_{\alpha/2;n-2}s_{B_1} = 0.070 \pm 2.3424 \times 0.019 = 0.070 \pm 0.0445$$

So, $l = 0.0255$ and $u = 0.1145$; we are 98% confident that the slope β_1 of the population regression line lies in the interval (0.0255, 0.1145). If the hourly wage of the head of household increases by

Coefficients(a)

Model		Unstandardized Coefficients		Standardized Coefficients	t	Sig.
		B	Std. Error	Beta		
1	(Constant)	2.107	.165		12.768	.000
	x	.070	.019	.240	3.775	.000

a Dependent Variable: y

€1, then it is likely that the expected annual clothing expenditure of the household will increase by an amount between €25.50 and €114.50.

In particular, there is (at least) 98% confidence that β_1 is larger than the lower bound 0.0255 of the interval. And since this lower bound is positive, there is (at least) 98% confidence that β_1 is positive. So we can conclude, with (at least) 98% confidence, that Y and X are linearly related, even **positively** linearly related.

19.5.2 Hypothesis tests for β_1

The pivot variable Z will be transformed into a test statistic, following the approach of step D of Section 17.1. Test procedures for β_1 will be developed for the following three testing problems:

a   Test $H_0: \beta_1 \leq b$ against $H_1: \beta_1 > b$ (one-sided, upper-tailed).

b   Test $H_0: \beta_1 \geq b$ against $H_1: \beta_1 < b$ (one-sided, lower-tailed).

c   Test $H_0: \beta_1 = b$ against $H_1: \beta_1 \neq b$ (two-sided).

Here, b is the hinge of the testing problems. It is a fixed and known real number that bounds the null hypotheses H_0.

As in the testing problems for other parameters, the test procedures for β_1 will be such that

$$P(H_0 \text{ is rejected but incorrectly }) \leq \alpha$$

for a prescribed significance level α. That is, the probabilities for all type I errors have to be controlled at the level α. Again, the hinge has the largest type I error, so it is sufficient to control this worst case scenario:

$$P(H_0 \text{ is rejected while } \beta_1 = b) \leq \alpha.$$

That is, the probability is at most α that the test procedure rejects H_0 when it should not do so since the true β_1 equals b. To find the test statistic, the pivot

$$Z = \frac{B_1 - \beta_1}{\sigma_\varepsilon / \sqrt{(n-1)s_X^2}}$$

has (by step D) to be adapted in accordance with the hinge, and remaining unknown parameter(s) have to be replaced by consistent estimator(s). That is, β_1 has to be replaced by the hinge b and σ_ε by the estimator S_ε. These steps lead to the following test statistic:

$$T = \frac{B_1 - b}{S_\varepsilon / \sqrt{(n-1)s_X^2}}$$

By the right-hand side of (19.18) this sample statistic is t_{n-2}-distributed in the case that β_1 is equal to b, so for the worst case scenario.

Since the alternative hypothesis of testing problem (a) is about relatively large β_1, the null hypothesis will be rejected for relatively large realizations of B_1 and hence for relatively large realizations of T. In fact, H_0 is rejected for realizations t that are $\geq$ the real number $t_{\alpha;n-2}$ since the t_{n-2}

density has an area α at the right-hand side of this number. Similarly, the null hypothesis of testing problem (b) is rejected for realizations of T that are $\leq -t_{\alpha;n-2}$; the null hypothesis of (c) is rejected for realizations of T that are $\leq -t_{\alpha/2;n-2}$ and for realizations that are $\geq t_{\alpha/2;n-2}$.

Five-step procedures for testing H_0 against H_1

i Testing problems: (a) test $H_0: \beta_1 \leq b$ against $H_1: \beta_1 > b$
(b) test $H_0: \beta_1 \geq b$ against $H_1: \beta_1 < b$
(c) test $H_0: \beta_1 = b$ against $H_1: \beta_1 \neq b$

ii Test statistic: $T = \dfrac{B_1 - b}{S_e / \sqrt{(n-1)s_X^2}}$

iii Rejection: (a) reject $H_0 \Leftrightarrow t \geq t_{\alpha;n-2}$
(b) reject $H_0 \Leftrightarrow t \leq -t_{\alpha;n-2}$
(c) reject $H_0 \Leftrightarrow t \leq -t_{\alpha/2;n-2}$ or $t \geq t_{\alpha/2;n-2}$

iv *Calculate* **val**, the realization of T

v Draw the conclusion by comparing **val** with step (iii)

Requirement: simple linear regression model (with strong requirements)

The respective rejection regions of the tests for (a)–(c) are:

Rejection regions: $[t_{\alpha;n-2}, \infty), (-\infty, -t_{\alpha;n-2}]$ and $(-\infty, -t_{\alpha/2;n-2}] \cup [t_{\alpha/2;n-2}, \infty)$

The critical values of the three tests are, respectively, $t_{\alpha;n-2}, -t_{\alpha;n-2}$ and the pair $-t_{\alpha/2;n-2}, t_{\alpha/2;n-2}$.

Having calculated **val**, it is interesting to find the most extreme significance level (the **p-value**) for which H_0 is still rejected. For problem (a) this means that we are interested in the α that arises when letting the critical value $t_{\alpha;n-2}$ coincide with **val**. For (b), we are looking for the α such that $-t_{\alpha;n-2}$ coincides with **val**. For (c), we want the α that arises if we let **val** coincide with the closest of the two critical values $-t_{\alpha/2;n-2}$ and $t_{\alpha/2;n-2}$.

---

p-values for testing problems (a)–(c)

For (a): $P(T \geq \textbf{val})$

For (b): $P(T \leq \textbf{val})$

For (c): $P(|T| \geq |\textbf{val}|) = 2P(T \geq |\textbf{val}|)$

19.5.3 A test for the usefulness of the simple linear regression model

The two-sided testing problem $H_0: \beta_1 = 0$ against $H_1: \beta_1 \neq 0$ with hinge 0 deserves special attention because of its interpretation of H_0 and H_1. To understand its importance, consider the basic assumption $E(Y \mid x) = \beta_0 + \beta_1 x$ of the simple linear regression model. If H_0 is true, then $E(Y \mid x) = \beta_0 + 0x = \beta_0$ for all values x, so $E(Y \mid x)$ is not influenced at all by realizations x of X; the model is **useless**. But if H_1 is true and hence β_1 is unequal to 0, then different values of x give different values of $E(Y \mid x)$ and the model makes sense. It is then said that the model is **useful** (although it would be better to say 'it has some usefulness') and that the independent variable X is **significant**. So, the test of the above testing problem tests whether the model is useful and whether the variable X is **significant** in explaining the variation of Y.

Having proposed a simple linear regression model for a study of interest, it is the test for the above testing problem that is first conducted. If H_0 is rejected, then the researcher goes on with detecting the qualities of the model. But if H_0 is not rejected, the researcher will usually stop working on that model and will try instead to find an independent variable and a linear regression model that **is** useful.

Since β_1 and the population correlation coefficient ρ have the same numerator, it follows that, if $\beta_1 = 0$ then $\rho = 0$, and vice versa. Hence, the testing problems

$$H_0: \beta_1 = 0 \text{ vs } H_1: \beta_1 \neq 0 \quad \text{and} \quad H_0: \rho = 0 \text{ vs } H_1: \rho \neq 0$$

are equivalent. Hence, the test procedure for the first testing problem is also a test procedure for the second.

Example 19.8

To explain, for a supermarket, the variation of Y = 'weekly sales ($\times$ €1000)', a simple linear regression model with independent variable X = 'weekly advertising costs ($\times$ €1000)' was proposed. But is this model useful? If so, does an increase of €1000 on advertising lead, on average, to an increase of more than €15 000 in weekly sales? To answer the questions, use the small dataset ($n = 7$) of Example 19.3 that yielded the results $b_1 = 18.17$, $s_X^2 = 0.1362$ and $s_e = 3.808$.

The first question is about the testing problem:

  i Test $H_0: \beta_1 = 0$ against $H_1: \beta_1 \neq 0$

We will use $\alpha = 0.05$. Here are the steps (ii)–(v) of the test; check yourself that $t_{\alpha/2;n-2} = t_{0.025;5} = 2.5706$ (*).

  ii Test statistic: $T = \dfrac{B_1}{S_e/\sqrt{(n-1)s_X^2}}$

  iii Reject $H_0 \Leftrightarrow t \leq -2.5706$ or $t \geq 2.5706$

  iv **val** $= \dfrac{18.17}{3.808/\sqrt{6 \times 0.1362}} = 4.3134$

  v Reject H_0 since 4.3134 is larger than 2.5706

It is concluded that the model is useful. To put it another way, it is concluded that the population correlation coefficient ρ is unequal to 0. Since $\beta_1 \neq 0$, we can go on with studying the model.

The second question is about the hypothesis $H: \beta_1 > 15$; take it as alternative hypothesis. Here are the five steps (take $\alpha = 0.05$, check yourself that $t_{0.05;5} = 2.0150$ (*)):

  i Test $H_0: \beta_1 \leq 15$ against $H_1: \beta_1 > 15$; $\alpha = 0.05$

  ii Test statistic: $T = \dfrac{B_1 - 15}{S_e/\sqrt{(n-1)s_X^2}}$

  iii Reject $H_0 \Leftrightarrow t \geq 2.0150$

  iv **val** $= \dfrac{18.17 - 15}{3.808/\sqrt{6 \times 0.1362}} = 0.7525$

  v Do not reject H_0 since 0.7525 is **smaller** than 2.0150

The data do not give sufficient evidence that the statement in the second question is true.

Notice that the p-values of the two tests are respectively:

$$P(|T| \geq 4.3134) = 2P(T \geq 4.3134) = 2 \times 0.0038 = 0.0076; \qquad (*)$$

$$P(T > 0.7525) = 0.2428 \qquad (*)$$

(Check these results yourself.) The first p-value makes even more credible the conclusion that the model is useful: even the significance level 0.0076 would have given the same conclusion. The second p-value informs us that significance levels of at least 0.2428 would have been needed to accept the statement of the second question.

Example 19.9

Reconsider the regression of Y = 'weight (kg)' on X = 'height (cm)' for a population of adult men and adopt the simple linear regression model; see also Example 19.4 (second part). Is the model useful? The conjecture is that if the height is 1 cm more, then the weight will on average be less than 0.75 kg more. Check it.

The file Xmp19-04b.sav contains the observations of X and Y for a sample of 242 men. This dataset will be used to conduct two hypothesis tests, both with α = 0.03, to answer the two problems that were raised above. The Coefficients part of the printout (below) is needed for the ingredients of the two test statistics.

Coefficients(a)

Model		Unstandardized Coefficients		Standardized Coefficients	t	Sig.
		B	Std. Error	Beta		
1	(Constant)	−26.707	13.920		−1.919	.056
	Height	.584	.078	.434	7.465	.000

a Dependent Variable: Weight

Here are the five steps to check the usefulness of the model; it is used that $t_{0.015;240}$ = 2.1831 (∗).

i Test H_0: β_1 = 0 against H_1: $\beta_1 \neq 0$; α = 0.03

ii Test statistic: $T = \dfrac{B_1 - 0}{S_\varepsilon / \sqrt{(n-1)s_X^2}}$

iii Reject $H_0 \Leftrightarrow t \leq -2.1831$ or $t \geq 2.1831$

iv $val = \dfrac{0.584}{0.078}$ = 7.4872

v Reject H_0 since 7.4872 is larger than the critical value 2.1831

The conclusion is that the model is useful. Since the p-value equals 1.3×10^{-12} (∗), this conclusion is very convincing.

In fact, the val is already present in the printout. Under the heading 't' SPSS printouts give the values of the test statistics for the t-tests (one-sided or two-sided) **with hinge 0**. In the present example val is equal to 7.465, which is more precise than the value calculated before. The printout also gives, under the heading Sig., the p-value of the **two-sided** test with hinge 0, but only the first three decimals are mentioned.

Since the model is useful, we can go on with studying it. The above conjecture states that increasing the height x by 1 cm will on average lead to a weight increase of less than 0.75 kg. That is, the conjecture is that β_1 < 0.75; take it as H_1. Here are the five steps of a test; check that $t_{\alpha;n-2}$ = 1.8897 (∗).

i Test H_0: $\beta_1 \geq 0.75$ against H_1: $\beta_1 < 0.75$; α = 0.03

ii Test statistic: $T = \dfrac{B_1 - 0.75}{S_\varepsilon / \sqrt{(n-1)s_X^2}}$

iii Reject $H_0 \Leftrightarrow t \leq -1.8897$

iv $val = \dfrac{0.584 - 0.75}{0.078}$ = −2.1282

v Reject H_0 since −2.1282 is smaller than −1.8897

It is likely that the conjecture is valid. Since the maximal probability of a type I error is $\alpha = 0.03$, the test procedure in steps (ii) and (iii) guarantees that the probability that the procedure rejects H_0 incorrectly is at most 0.03. Hence, there is only 3% chance that the conclusion to reject H_0 is incorrect.

19.6 ANOVA table and degree of usefulness

Having concluded that the regression model is useful (that is, has some statistical usefulness), the immediate next question is: **how** useful is the model? In the present section the intention is to measure the **degree** of usefulness of the estimated model.

19.6.1 SST, SSE and SSR

Recall that interest is in the variation of Y. Starting with the observations $y_1, \ldots, y_n$ of Y we want to compare the situation before regression with the situation after regression. **Before** regression, the data $y_1, \ldots, y_n$ are the only available data; x data cannot be used yet. The (unexplained) variation within the y data is measured by the sample variance (around the sample mean $\bar{y}$):

$$s_Y^2 = \frac{1}{n-1}\sum_{i=1}^{n}(y_i - \bar{y})^2$$

After regression on X, the variable X is also measured at the n population elements in the sample, so the pairs of observation $(x_1, y_1), \ldots, (x_n, y_n)$ are available. Furthermore, the sample regression equation $\hat{y} = b_0 + b_1 x$ is determined, and the residuals $e_i = y_i - \hat{y}_i$ and the sample variance s_ε^2 of the esti-mated model are calculated. After regression it is the variation around the sample regression line that is still unexplained; it is measured by:

$$s_\varepsilon^2 = \frac{1}{n-2}\sum_{i=1}^{n}e_i^2 = \frac{1}{n-2}\sum_{i=1}^{n}(y_i - \hat{y}_i)^2$$

To determine the degree of usefulness of the estimated linear regression model, we have to compare s_Y^2 with s_ε^2. However, since the two measures for variation have different denominators, they are less suitable for comparison. That is why it is common to use

$$SST = \sum_{i=1}^{n}(y_i - \bar{y})^2 \quad \text{and} \quad SSE = \sum_{i=1}^{n}(y_i - \hat{y}_i)^2$$

as measures of variation, respectively before and after regression. SST stands for **sum of squares total**; recall that SSE stands for sum of the squared errors. So:

  SST measures the variation before regression;
  SSE measures the variation that is left after regression of Y on X.

It turns out that the difference $SSR = SST - SSE$ is always non-negative, that a regression model always reduces the variation. SSR is the amount of the variation of the y data that is explained by regressing Y on X. Before regression, SST is the amount of variation that has to be explained. After regression we know that $SST - SSE$ is the amount that is explained by the regression model; SSE measures the amount of variation that is still unexplained after regression. The following results are valid.

Important results for the sums of squares

  $$SST = SSR + SSE$$

  $$SSR = \sum_{i=1}^{n}(\hat{y}_i - \bar{y})^2 = b_1^2(n-1)s_X^2 \tag{19.22}$$

Proof of the decomposition SST = SSR + SSE

Recall from Section 5.2 and (5.5) that the least-squares estimates b_0 and b_1 are such that the function

$$v(a, b) = \sum_{i=1}^{n}(y_i - (a + bx_i))^2$$

reaches its minimum at (b_0, b_1). But notice that:

$$v(\bar{y}, 0) = \sum_{i=1}^{n}(y_i - (\bar{y} + 0))^2 = \text{SST}$$

$$v(b_0, b_1) = \sum_{i=1}^{n}(y_i - (b_0 + b_1x_i))^2 = \sum_{i=1}^{n}(y_i - \hat{y}_i)^2 = \text{SSE}$$

Since $v(b_0, b_1)$ is the minimum of the function $v(a, b)$, it follows especially that:

$$v(b_0, b_1) \le v(\bar{y}, 0), \text{ so SSE} \le \text{SST}$$

Indeed, SSR (the *sum of squares due to regression*) is always non-negative.

To learn more about these 'squares' of SSR, notice that the deviation $y_i - \bar{y}$ of y_i with respect to $\bar{y}$ decomposes into two other types of deviations:

$$y_i - \bar{y} = y_i - \hat{y}_i + \hat{y}_i - \bar{y} = (y_i - \hat{y}_i) + (\hat{y}_i - \bar{y}) = e_i + (\hat{y}_i - \bar{y}) \tag{19.23}$$

Figure 19.9 illustrates this decomposition.

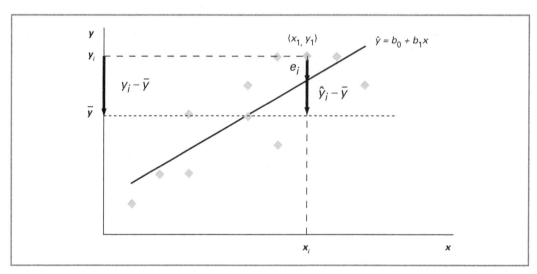

Figure 19.9 Decomposition of the deviations $y_i - \bar{y}$

Note also that:

$$e_i = y_i - \hat{y}_i = y_i - (b_0 + b_1x_i) = y_i - (\bar{y} - b_1\bar{x} + b_1x_i)$$

So:

$$e_i = (y_i - \bar{y}) - b_1(x_i - \bar{x}) \tag{19.24}$$

But the decomposition (19.23) yields that e_i is also equal to $(y_i - \bar{y}) - (\hat{y}_i - \bar{y})$. So, $\hat{y}_i - \bar{y} = b_1(x_i - \bar{x})$ and

$$\sum_{i=1}^{n}(\hat{y}_i - \bar{y})^2 = b_1^2 \sum_{i=1}^{n}(x_i - \bar{x})^2 = b_1^2(n - 1)s_x^2 \tag{19.25}$$

which proves the second equality in (19.22).

Taking squares in (19.24) yields:

$$e_i^2 = (y_i - \bar{y})^2 + b_1^2(x_i - \bar{x})^2 - 2b_1(x_i - \bar{x})(y_i - \bar{y}) \tag{19.26}$$

Notice that $b_1 = s_{X,Y}/s_X^2$, so $s_{X,Y} = b_1 s_X^2$. Since

$$\sum_{i=1}^{n}(x_i - \bar{x})(y_i - \bar{y}) = (n-1)s_{X,Y} = (n-1)b_1 s_X^2$$

it follows, by taking summations in (19.26), that:

$$\begin{aligned}
\text{SSE} &= \text{SST} + b_1^2(n-1)s_X^2 - 2b_1(n-1)s_{X,Y} \\
&= \text{SST} + b_1^2(n-1)s_X^2 - 2b_1^2(n-1)s_X^2 = \text{SST} - b_1^2(n-1)s_X^2
\end{aligned}$$

Since SSR = SST – SSE, we obtain SSR = $b_1^2(n-1)s_X^2$.

19.6.2 Coefficient of determination

The use of the regression model reduces the unexplained variation from SST to SSE. To put it another way, the benefit of using regression is SSR = SST – SSE. However, it is common to measure the benefit as a **proportion** of the variation before regression. Here is the precise definition.

Coefficient of determination

The *coefficient of determination* (notation r^2) of the estimated model is defined as:

$$r^2 = \frac{SSR}{SST} = \frac{SST - SSE}{SST} = 1 - \frac{SSE}{SST}$$

Notice that the above arguments about the sums of squares were all in terms of realizations. So, the above definition is also a realization of a random variable. Most computer printouts write R^2 or R Square instead of r^2.

The coefficient of determination is the ratio of SSR and SST; it measures the proportion of the reduction of the variation of the y data due to the regression model when compared with the variation of the y data before the regression. It has the following properties:

- $0 \le r^2 \le 1$.
- If $r^2 \approx 0$, then the reduction in variation is only a small proportion.
- If $r^2 \approx 1$, then the reduction in variation is a large proportion.
- If (e.g.) $r^2 = 0.85$, then 85% of the variation in the y data is explained by the regression model and 15% remains unexplained.

In Chapter 5 the notation $r = r_{X,Y}$ was introduced to denote the sample correlation coefficient. Indeed, the coefficient of determination of a **simple** linear regression model is equal to the square of the sample correlation coefficient. This follows easily from (19.22) and the definition of b_1:

$$r^2 = \frac{SSR}{SST} = \frac{b_1^2(n-1)s_X^2}{(n-1)s_Y^2} = \frac{b_1^2 s_X^2}{s_Y^2} = \frac{s_{X,Y}^2}{s_X^4} \frac{s_X^2}{s_Y^2} = \frac{s_{X,Y}^2}{s_X^2 s_Y^2} = (r_{X,Y})^2$$

19.6.3 ANOVA table

The sums of squares that were discussed before are important not only because they are ingredients of the formula for the coefficient of determination. They are important in their own right too, for instance because they occur in computer printouts.

A main part of the regression printouts of statistical packages is the *ANOVA table*. ANOVA stands for analysis of variation. The table tells us about the decomposition SST = SSR + SSE. It especially tells us about SSR, the reduction of the unexplained variation due to the use of the regression model. But it also contains the degrees of freedom of the three sums of squares and even the value of s_ε^2 (the sample variance of the estimated model). The printouts of most statistical packages include the information listed in Table 19.3.

	Sum of squares	Degrees of freedom	Mean square
Regression	$SSR = \sum_{i=1}^{n}(\hat{y}_i - \bar{y})^2$	1	$\dfrac{SSR}{1} = SSR$
Residual	$SSE = \sum_{i=1}^{n}(y_i - \hat{y}_i)^2$	$n - 2$	$\dfrac{SSE}{n-2} = s_\varepsilon^2$
Total	$SST = \sum_{i=1}^{n}(y_i - \bar{y})^2$	$n - 1$	$\dfrac{SST}{n-1} = s_Y^2$

TABLE 19.3 ANOVA table for simple linear regression

'Residual' is usually used instead of 'error'. Recall that SST has $n - 1$ degrees of freedom and SSE has $n - 2$, so one degree of freedom is left for SSR. Notice that the last positions in columns 2 and 3 are the totals of the columns, but this does not hold for column 4.

Example 19.10

Reconsider the seven pairs of observations of $X =$ 'weekly advertising costs ($\times$ €1000)' and $Y =$ 'weekly sales ($\times$ €1000)' for a supermarket. In Example 19.3 it was calculated that SSE = 72.510. Adding up the squares of the numbers in column 5 of Table 19.2 in Example 19.3 yields SST = 342.38. But then SSR follows immediately: SSR = 342.38 – 72.51 = 269.87.

This last result also follows from (19.22) and the fact that (see Example 19.3) $s_X^2 = 0.1362$ and $b_1 = 18.17$:

$$SSR = b_1^2(n - 1)s_X^2 = (18.17)^2 \times 6 \times 0.1362 = 269.80$$

(The answer is slightly different because of rounding errors.)

	Sum of squares	Degrees of freedom	Mean square
Regression	269.87	1	269.87
Residual	72.51	5	14.50
Total	342.38	6	57.06

TABLE 19.4 ANOVA table

The determination coefficient can easily be calculated:

$$r^2 = \frac{SSR}{SST} = \frac{269.87}{342.38} = 0.788$$

About 78.8% of the variation in the y data is explained by regressing Y on X.

Example 19.11

In former examples the relationship between Y = 'annual clothing expenditure ($\times$ €1000) of a household' and 'gross hourly wage (in euro) of the head of that household' was studied. For a large population of households, a weak linear relationship was detected. However, the variable X = 'annual income ($\times$ €1000) of that household' might have more influence since the total income of a household will often be a better indicator of the expenses of the household than just the salary of one of its members (albeit the head).

To find out, we will regress Y on X on the basis of a sample of 300 households from a large population of households. See the file Xmp19-11.sav. The printout is shown below.

Model Summary

Model	R	R Square	Adjusted R Square	Std. Error of the Estimate
1	.996(a)	.992	.992	.10651

a Predictors: (Constant), FINC

ANOVA(b)

Model		Sum of Squares	df	Mean Square	F	Sig.
1	Regression	396.770	1	396.770	34971.811	.000(a)
	Residual	3.381	298	.011		
	Total	400.150	299			

a Predictors: (Constant), FINC
b Dependent Variable: CLOTEXP

Coefficients(a)

Model		Unstandardized Coefficients		Standardized Coefficients	t	Sig.
		B	Std. Error	Beta		
1	(Constant)	.017	.015		1.133	.258
	FINC	.083	.000	.996	187.008	.000

a Dependent Variable: CLOTEXP

It follows immediately that the model with basic assumption $E(Y \mid x) = \beta_0 + \beta_1 x$ is useful; the hypothesis $H_0: \beta_1 = 0$ is rejected in favour of $H_1: \beta_1 \neq 0$ at a very small significance level, as follows from the p-value under Sig. So, it is statistically concluded that there is a linear relationship between Y and X; the (only) independent variable in the model is significant. The equation of the sample regression line is $\hat{y} = 0.017 + 0.083x$. In the dataset, an increase of €1000 in the annual household income leads on average to an increase of €83 in the annual household clothing expenditure.

To estimate the **degree** of usefulness, the coefficient of determination will be calculated next. The ANOVA table yields that the total variation in the y data is SST = 400.150; before regression 400.150 was unexplained. Having done the regression, the amount 3.381 remains unexplained. Put another way, this linear regression model explains 396.770 of the amount of variation 400.150 measured before regression, which corresponds to the proportion:

$$\frac{396.770}{400.150} = 0.9916$$

Hence, $r^2 = 0.992$ and the correlation coefficient r equals 0.996. These numbers can also be found in the first part of the printout, where they are called R Square and R, respectively. The interpretation is that 99.2% of the variation in the y data is explained by the regression model. There is a very strong linear relationship between Y and X.

The column (in the ANOVA table) with the heading df gives the respective degrees of freedom. Starting with $n = 300$, SST costs 1 df (so 299 are left) while the estimation of the population regression equation costs 2 (so 298 are left). Hence, compared with the 'before regression situation', the simple regression is done at the cost of 1 degree of freedom.

Each position in the column with the heading Mean Square contains the average of n squared deviations per degree of freedom; the type of deviation differs per row. (Notice that, for some unclear reason, the table does not include SST/$(n - 1)$.) In particular, the second position in this column yields the sample variance of the estimated model: $s_\varepsilon^2 = 0.011$. As a matter of fact, the **standard error** of the estimated model is also part of the printout: $s_\varepsilon = 0.10651$.

Does an increase of €1000 in the annual household income of the **whole population** on average lead to an increase of more than €80 in clothing expenditure? You might be tempted to say yes immediately since this answer is suggested by the slope 0.083 of the **sample** regression line. However, a thought-out answer has to be based on a hypothesis test or a confidence interval for the **population** slope β_1. Below, the last approach is followed; a 98% confidence interval is constructed.

Since $\alpha = 0.02$, we need $t_{0.01;298} = 2.3389$ (*). Notice that, according to the printout, the standard error of B_1 equals 0.000, which does **not** mean that it is 0 but only that the first three decimals are 0. By double-clicking a few times on this number in the original printout, more decimals appear: $s_{B_1} = 0.000442$. The bounds of the interval can now be calculated:

$$b_1 \pm t_{\alpha/2;n-2}s_{B_1} = 0.083 \pm 2.3389 \times 0.000442 = 0.083 \pm 0.00103$$

We have 98% confidence that the slope β_1 lies in the interval (0.082, 0.084). Since this interval lies completely to the right of 0.080, there is (at least) 98% confidence that β_1 is larger than 0.080. Hence, the answer to the question is: yes.

19.6.4 Application: the market model

The **market model** is the simple linear regression model that relates the daily percentage return R of a stock to the daily percentage return R_m of the 'market' (usually a stock exchange). Here is the basic assumption of the model, written down in a rather 'popular' way:

$$E(R) = \beta_0 + \beta_1 R_m$$

In financial texts, the slope β_1 is often called the **beta** or **beta coefficient** of the stock. It describes the sensitivity of the stock's rate of return to changes in the overall level of the market. If, for example, the beta of a certain stock is 0.8, then a 1% increase of the market index will on average lead to a 0.8% increase in the price of the stock. Stocks with a beta greater than 1 are even more sensitive than the market (which represents the 'average' stock); it is more **volatile** than the market. When the model is estimated, the resulting sample coefficient b_1 is – not very wisely – also called beta. It gives information about the **market-related risk** of the stock, since it measures the stock's volatility with respect to the market.

The coefficient of determination is also an important statistic in financial analysis. It measures the proportion of the risk due to the market.

Example 19.12

The Heineken stock is part of the Amsterdam Exchange (AEX) index. To learn about the relationship of the daily returns (also described as rate of returns) of the Heineken stock and the AEX index, the market model is considered.

The dataset Xmp19-12.sav contains the daily returns (%) of the Heineken stock and the accompanying returns (%) of the AEX index for the period 7 January 2005 – 3 January 2007. The returns are measured as the percentage change in the daily closing price with respect to the day before. (Notice that the AEX column contains two missing data points.) The printout is presented below.

Model Summary

Model	R	R Square	Adjusted R Square	Std. Error of the Estimate
1	.410(a)	.168	.166	.88900

a Predictors: (Constant), Ret_AEX

ANOVA(b)

Model		Sum of Squares	df	Mean Square	F	Sig.
1	Regression	80.601	1	80.601	101.985	.000(a)
	Residual	399.901	506	.790		
	Total	480.502	507			

a Predictors: (Constant), Ret_AEX

b Dependent Variable: Ret_Heineken

Coefficients(a)

Model		Unstandardized Coefficients		Standardized Coefficients	t	Sig.
		B	Std. Error	Beta		
1	(Constant)	.045	.040		1.135	.257
	Ret_AEX	.527	.052	.410	10.099	.000

a Dependent Variable: Ret_Heineken

First notice that the model is useful; test it yourself. Note that Heineken's (sample) beta is equal to 0.527. If the AEX index increases 1%, then the Heineken stock on average increases 0.527%. The Heineken stock is less sensitive than the market.

Also notice that $r^2 = 0.168$. That is, 16.8% of the risk is market-related; the remaining 83.2% of the risk is due to factors that are specific to Heineken rather than the market (the **_firm-specific risk_**).

CASE 19.1 WAGE DIFFERENTIALS BETWEEN MEN AND WOMEN (PART I) – SOLUTION

Models with basic assumption $E(W \mid x) = \beta_0 + \beta_1 x$ are estimated, respectively for values x of AGE and EDL, for the whole population and for the subpopulations of the men and the women. Some estimation results are presented in the table below.

	Independent variable	r^2	Regression line
Whole population	AGE	0.363	$\hat{w} = 3.068 + 0.565x_1$
	EDL	0.244	$\hat{w} = 9.567 + 4.416x_2$
Female population	AGE	0.469	$\hat{w} = 3.900 + 0.482x_1$
	EDL	0.253	$\hat{w} = 9.154 + 3.532x_2$
Male population	AGE	0.277	$\hat{w} = 4.555 + 0.571x_1$
	EDL	0.255	$\hat{w} = 11.259 + 4.804x_2$

Summarized results for simple regressions of W on AGE and EDL

It turns out that all models are useful, even for small significance levels α. With regard to the slopes of the sample regression lines, the following results are notable:

- An increase of 1 year in AGE will on average increase the hourly wages of men and women by €0.57 and €0.48, respectively.

- A step of one level in EDL will on average increase the hourly wages of men and women by €4.80 and €3.53, respectively.

Both results are to the disadvantage of the women. Whether these sample results imply similar results for the two populations will be studied later.

19.7 Conclusions about Y and $E(Y)$

A sample regression line can be used to predict Y **and** to estimate $E(Y)$. The present section will work this out carefully. Before starting the discussion, it is important to obtain a good understanding of the difference between estimation and prediction.

19.7.1 Estimation and prediction

An estimator is a sample statistic that can be used to approximate the parameter, that is, to approximate a fixed but unknown **real number**. A *point predictor* is a sample statistic that can be used to approximate the not yet observed outcome of a **random variable**. A $(1 - \alpha)$ confidence interval is an interval that encloses with confidence $1 - \alpha$ the unknown (but fixed) real number; a $(1 - \alpha)$ *prediction interval* encloses the actual outcome of the random variable with confidence $1 - \alpha$.

For instance, the manager of a furniture store might be interested in the mean weekly sales μ (an unknown but fixed real number) and/or in the not yet observed outcome of next week's sales Y. In both cases, it seems reasonable to take the random variable $\bar{Y}$ that measures the mean sales of the most recent n weeks, as a method to approximate. In essence, the sample mean $\bar{Y}$ then has two duties:

- it is the estimator of μ;
- it is also the point predictor of Y.

When the data of the most recent n weeks are available, the observation of $\bar{Y}$ can be calculated, and is used to **estimate** the unknown number μ and to **predict** the not yet observed outcome of Y. It is intuitively clear that the estimation duty of $\bar{Y}$ will show less variation than the prediction duty. This is because the first duty has the random deviation $\bar{Y} - \mu$ with only $\bar{Y}$ as source of variation whereas the second duty has the random deviation $\bar{Y} - Y$ with the two variation sources $\bar{Y}$ **and** Y.

Things will now be considered in the field of linear regression. Recall that the sample regression line is determined on the basis of the (x, y) observations of a sample of n population elements.

Now suppose that there is another population element for which the observation x_p of X is already measured. However, its observation of Y is not yet measured, so Y is still random. When denoting this random observation of Y by Y_p, the pair (x_p, Y_p) arises; the first coordinate is a realization and the second is a random variable.

Assume that this extra pair of observations also satisfies the model requirements. In particular, the basic assumption yields:

$$Y_p = \beta_0 + \beta_1 x_p + \varepsilon_p \quad \text{and} \quad E(Y_p) = \beta_0 + \beta_1 x_p \tag{19.27}$$

We have two purposes:

- To determine an interval that is very likely to contain the (unknown) expected value $E(Y_p)$; it is a confidence interval that is wanted.

- To determine an interval that is very likely to contain the (still unknown) actual outcome of Y_p itself; it is a prediction interval that is wanted.

Here is a regression example to illustrate the difference between the two purposes; see also Example 19.3. To explain Y = 'weekly sales' from X = 'weekly advertising costs', a sample of 7 past weeks has yielded the sample regression line. Suppose now that the manager of the supermarket is interested in weeks with the level 1.9 ($\times$ €1000) for X. The reason might be that:

- She wants to estimate the mean value of Y over **all** weeks with $x = 1.9$, so she wants (as in the first purpose) an interval for $E(Y)$ when $x = 1.9$.

- She wants to spend €1900 on advertisements during the coming week, and hence is interested in the outcome of Y for the coming week and, more specifically, in an interval that is likely to contain this outcome. In this case the purpose is the second one.

Let us go back to the general theory. The substitution of x_p for x in the **sample** regression line yields $\hat{y}_p = b_0 + b_1 x_p$. Notice that $b_0 + b_1 x_p$ is the natural estimate of $\beta_0 + \beta_1 x_p$. It follows from the right-hand side in (19.27) that $\hat{y}_p$ estimates the unknown but fixed number $E(Y_p)$. But from the left-hand side it follows that $\hat{y}_p$ also predicts the not yet observed outcome of the random variable Y_p.

Hence, the sample statistic $\hat{Y}_p = B_0 + B_1 x_p$ has two duties:

i estimator of $E(Y_p)$;

ii point predictor of Y_p itself.

Notice that each duty has its own random deviation: duty (i) has $\hat{Y}_p - E(Y_p)$ and duty (ii) has $\hat{Y}_p - Y_p$. Since B_0 and B_1 are unbiased estimators of β_0 and β_1, the above random deviations both have expectation 0, as follows from the following arguments:

$$E(\hat{Y}_p - E(Y_p)) = E(\hat{Y}_p) - E(Y_p) = E(B_0 + B_1 x_p) - (\beta_0 + \beta_1 x_p)$$

$$= E(B_0) + E(B_1)x_p - (\beta_0 + \beta_1 x_p) = \beta_0 + \beta_1 x_p - (\beta_0 + \beta_1 x_p) = 0$$

and

$$E(\hat{Y}_p - Y_p) = E(\hat{Y}_p) - E(Y_p) = 0$$

$\hat{Y}_p$ is not only an unbiased **estimator** of $E(Y_p)$ but also an unbiased **point predictor** of Y_p. However, it is suspected that $\hat{Y}_p - Y_p$ will have a larger variance than $\hat{Y}_p - E(Y_p)$, since $\hat{Y}_p - Y_p$ has not only $\hat{Y}_p$ as variation source but also Y_p. This is indeed the case; we will not prove it rigorously. Notice that the variance of $\hat{Y}_p - E(Y_p)$ is equal to the variance of $\hat{Y}_p$ since $E(Y_p)$ is a constant and hence has no variation.

Results for the standard deviation of $\hat{Y}_p$ and $\hat{Y}_p - Y_p$

$$SD(\hat{Y}_p) = \sigma_\varepsilon \sqrt{\frac{1}{n} + \frac{(x_p - \bar{x})^2}{(n-1)s_X^2}} \tag{19.28}$$

$$SD(\hat{Y}_p - Y_p) = \sigma_\varepsilon \sqrt{1 + \frac{1}{n} + \frac{(x_p - \bar{x})^2}{(n-1)s_X^2}} \tag{19.29}$$

Note that the distinction between the two formulae is the 1 in the square root of the second standard deviation, which makes that standard deviation larger than the first. (It is the extra variation due to Y_p that causes this 1.)

19.7.2 Interval estimator and interval predictor

Two random intervals are constructed that capture, both with probability $1 - \alpha$, the fixed number $E(Y_p)$ and the random variable Y_p, respectively.

$\hat{Y}_p = B_0 + B_1 x_p$ is a linear combination of the random variables B_0 and B_1 which, according to (19.15) and (19.16), are both normally distributed. Hence, $\hat{Y}_p$ is normally distributed itself, with $E(\hat{Y}_p) = \beta_0 + \beta_1 x_p$ as mean parameter and the square of (19.28) as variance parameter. Since also Y_p is normal, it follows that $\hat{Y}_p - Y_p$ is normal too, with expectation 0 and the square of (19.29) as variance.

But for our purposes we are more interested in **standardized** versions of $\hat{Y}_p$ and $\hat{Y}_p - Y_p$. So, from $\hat{Y}_p$ we subtract $E(Y_p)$ and divide the result by the standard deviation in (19.28). Similarly, we subtract $E(\hat{Y}_p - Y_p) = 0$ from $\hat{Y}_p - Y_p$ and divide the result by the standard deviation of (19.29). Notice that both resulting random ratios are standard normal and that both denominators contain the unknown parameter σ_ε. This last fact is unwanted. That is why σ_ε is replaced by its estimator S_ε, at the cost of having left t-distributed random variables. Since S_ε^2 has $n - 2$ degrees of freedom, the two resulting random variables are t_{n-2}-distributed.

Properties of $\hat{Y}_p$ and $\hat{Y}_p - Y_p$

$$\hat{Y}_p \sim N\left(\beta_0 + \beta_1 x_p, \sigma_\varepsilon^2\left[\frac{1}{n} + \frac{(x_p - \bar{x})^2}{(n-1)s_X^2}\right]\right)$$

$$\hat{Y}_p - Y_p \sim N\left(0, \sigma_\varepsilon^2\left[1 + \frac{1}{n} + \frac{(x_p - \bar{x})^2}{(n-1)s_X^2}\right]\right)$$

$$\frac{\hat{Y}_p - (\beta_0 + \beta_1 x_p)}{S_\varepsilon\sqrt{\frac{1}{n} + \frac{(x_p - \bar{x})^2}{(n-1)s_X^2}}} \quad \text{and} \quad \frac{\hat{Y}_p - Y_p}{S_\varepsilon\sqrt{1 + \frac{1}{n} + \frac{(x_p - \bar{x})^2}{(n-1)s_X^2}}} \text{ are both } t_{n-2} \tag{19.30}$$

Notice that the denominators of the two t_{n-2}-distributed random variables are, respectively, $SE(\hat{Y}_p)$ and $SE(\hat{Y}_p - Y_p)$, the standard **errors** of $\hat{Y}_p$ and $\hat{Y}_p - Y_p$.

The last results in the box are suitable to derive random intervals (L, U) and (L_p, U_p) that capture, respectively, the unknown number $E(Y_p) = \beta_0 + \beta_1 x_p$ and the still unobserved random variable Y_p with probability $1 - \alpha$. Since

$$P(-t_{\alpha/2;n-2} < T < t_{\alpha/2;n-2}) = 1 - \alpha \quad \text{for } T \sim t_{n-2}$$

we can – similarly to the approach preceding (19.21) – successively take for T the left-hand and the right-hand random variable in (19.30), and rewrite things in such a way that $E(Y_p) = \beta_0 + \beta_1 x_p$ and Y_p, respectively, remain in the middle. Here are the results; the random interval (L_p, U_p) is called an **interval predictor**:

Interval estimator (L, U) for $E(Y_p)$

$$L = \hat{Y}_p - t_{\alpha/2;n-2}S_\varepsilon \sqrt{\frac{1}{n} + \frac{(x_p - \bar{x})^2}{(n-1)s_x^2}} \; ; \qquad U = \hat{Y}_p + t_{\alpha/2;n-2}S_\varepsilon \sqrt{\frac{1}{n} + \frac{(x_p - \bar{x})^2}{(n-1)s_x^2}}$$

Interval predictor (L_p, U_p) for Y_p

$$L_p = \hat{Y}_p - t_{\alpha/2;n-2}S_\varepsilon \sqrt{1 + \frac{1}{n} + \frac{(x_p - \bar{x})^2}{(n-1)s_x^2}} \; ; \qquad U_p = \hat{Y}_p + t_{\alpha/2;n-2}S_\varepsilon \sqrt{1 + \frac{1}{n} + \frac{(x_p - \bar{x})^2}{(n-1)s_x^2}}$$

Notice that both random intervals are of format 1; see also Section 15.2. The half-width H is equal, respectively, to the product of $t_{\alpha/2;n-2}$ and the **standard error** of $\hat{Y}_p$, and the product of $t_{\alpha/2;n-2}$ and the **standard error** of $\hat{Y}_p - Y_p$. Also notice that the intervals are most narrow if x_p coincides with the mean $\bar{x}$ of the x data.

When the data and x_p are available, they can be substituted into the formulae of the two random intervals. The resulting concrete intervals (l, u) and (l_p, u_p) are, respectively, a $(1 - \alpha)$ confidence interval for $E(Y_p)$ and a $(1 - \alpha)$ **prediction interval** for Y_p. The second interval will be wider than the first because of the extra 1 in the underlying formula.

Example 19.13

Financial market analysts claim that tomorrow the AEX index will decrease by 2%. What will be the consequences for tomorrow's closing price of the Heineken stock? Will it decrease too?

Notice that the questions have to do with the value -2 for R_m, the return of the market (see Example 19.12). From the printout in Example 19.12 it follows immediately that the predicted value of tomorrow's return on the Heineken stock will be:

$$0.045 + 0.527 \times (-2) = -1.009$$

So, the (point) prediction is that the stock will lose slightly more than 1%. However, an interval that very likely contains tomorrow's value of R is more interesting. We construct a 98% prediction interval.

Although the calculation of such an interval can easily be done with a statistical package, we will – just for once – illustrate the use of the formula for (L_p, U_p) and calculate its **ingredients** with a computer. See the file Xmp19-13.xls for the data. It is already known from the printout in Example 19.12, that:

$$\hat{y}_p = -1.0090; \quad x_p = -2; \quad s_\varepsilon = 0.88900; \quad n = 508$$

From the x data (the AEX returns) in the dataset it follows that (check yourself):

$$\bar{x} = 0.07046; \; (x_p - \bar{x})^2 = 4.28682; \; s_x^2 = 0.57280$$

Since $t_{0.01;506} = 2.33374$ (*), the half-width h of the interval can be calculated:

$$h = t_{0.01;506}s_\varepsilon \sqrt{1 + \frac{1}{508} + \frac{(x_p - \bar{x})^2}{507 \times s_x^2}} = 2.33374 \times 0.88900 \times 1.00833 = 2.0920$$

Hence:

$$l_p = -1.0090 - 2.0920 = -3.1010 \quad \text{and} \quad u_p = -1.0090 + 2.0920 = 1.0830$$

The statistical conclusion is that there is 98% confidence that tomorrow's return on the Heineken stock will be between -3.1010 and 1.0830. Since 0 falls in this interval, it cannot be concluded that tomorrow's return on Heineken's stock will be negative. Check yourself that the smallest α with upper bound u_p equal to 0, is $\alpha = 0.261$.

Also check the bounds of the 98% prediction interval with a statistical package; see also Appendix A1.19.

Example 19.14

Reconsider Example 19.11 where linear regression of Y = 'annual clothing expenditure ($\times$ €1000)' on X = 'annual income ($\times$ €1000)' was studied for a large population of households. The model was estimated on the basis of a random sample of 300 households. See also Xmp19-11.sav for the data.

In the present example, a statistical package is used to calculate 92% confidence intervals for $E(Y_p)$ and 92% prediction intervals for Y_p itself, for several choices of x_p. Since $\bar{x} = 30.4265$ (check it), the x_p are chosen symmetrically around this value. Notice that the choices $\bar{x} - a$ and $\bar{x} + a$ for x_p yield the same half-widths (for all values of a); this follows immediately from the formulae for the interval estimator and interval predictor. Check yourself (some of) the bounds in Table 19.5.

x_p	l	u	l_p	u_p
$\bar{x} - 18$	1.0259	1.0612	0.8556	1.2315
$\bar{x} - 16$	1.1923	1.2252	1.0210	1.3966
$\bar{x} - 14$	1.3587	1.3894	1.1863	1.5618
$\bar{x} - 12$	1.5250	1.5536	1.3516	1.7270
$\bar{x} - 10$	1.6913	1.7179	1.5170	1.8921
$\bar{x} - 8$	1.8574	1.8823	1.6823	2.0573
$\bar{x} - 6$	2.0233	2.0468	1.8476	2.2226
$\bar{x} - 4$	2.1891	2.2116	2.0129	2.3878
$\bar{x} - 2$	2.3547	2.3765	2.1782	2.5530
$\bar{x} = 30.4265$	2.5200	2.5416	2.3434	2.7183
$\bar{x} + 2$	2.6852	2.7070	2.5087	2.8835
$\bar{x} + 4$	2.8501	2.8726	2.6739	3.0488
$\bar{x} + 6$	3.0148	3.0384	2.8391	3.2141
$\bar{x} + 8$	3.1794	3.2043	3.0043	3.3794
$\bar{x} + 10$	3.3438	3.3704	3.1695	3.5447
$\bar{x} + 12$	3.5081	3.5366	3.3347	3.7100
$\bar{x} + 14$	3.6723	3.7030	3.4999	3.8754
$\bar{x} + 16$	3.8364	3.8693	3.6651	4.0407
$\bar{x} + 18$	4.0005	4.0358	3.8302	4.2061

TABLE 19.5 92% confidence intervals and 92% predication intervals

For instance:

- The **mean** annual clothing expenditure of all households with an annual income of €4000 above average will probably fall between €2850 and €2873.
- The annual clothing expenditure of one (individual) household with an annual income of €4000 above average will probably fall between €2674 and €3049.

For both intervals the level of confidence is 92%; the second interval is the widest.

The intervals in the last two columns are much wider than the corresponding intervals in columns 2 and 3. Furthermore, if x_p goes further away from $\bar{x}$ then the confidence intervals becomes wider. The same holds for the prediction intervals.

The interval estimator (L_p, U_p) for Y_p gives rise to an interpretation of s_ε, the estimate of σ_ε on the basis of the sample. For large sample sizes n it holds that the sample variance s_X^2 is close to the population variance σ_X^2 and that $\frac{(x_i - \bar{x})^2}{n-1}$ is close to 0. Hence, for suitable α (for instance 0.05) and large sample sizes n, the constant

$$t_{\alpha/2;n-2}\sqrt{1 + \frac{1}{n} + \frac{(x_i - \bar{x})^2}{(n-1)s_X^2}}$$

is at most 3. Hence, it follows from the normality property of $\hat{Y}_i - Y_i$ that it is very likely that a large majority of the y data $y_1, \ldots, y_n$ fall within a distance of $3s_\varepsilon$ from the sample regression line. Put another way, a channel of width $6s_\varepsilon$ with the sample regression line in its centre encloses the large majority of the n dots in the sample scatter plot.

Unfortunately, the term **predictor** is used in literature for two different concepts, both in the field of regression. It is used to denote both an independent variable and a point predictor as has been defined in this section. Usually, however, it is clear from the context which of the two is meant.

19.8 Residual analysis

The residuals $e_i = y_i - \hat{y}_i$ will be considered as realizations of the error term ε. Since the model requirements are about the error term, the residuals are used to check these requirements. This analysis is called *residual analysis*.

The simple linear regression model (with the strong assumptions) comes down to the following five requirements, which will be looked at in turn below:

1 Y is linearly related to X;
2 homoskedasticity;
3 independence of $\varepsilon_1, \ldots, \varepsilon_n$;
4 independence of ε and X;
5 normality.

In essence, all the requirements are about the error variables ε_i. Since $\varepsilon = Y - (\beta_0 + \beta_1 X)$, the realizations of ε are $y_i - (\beta_0 + \beta_1 x_i)$ and hence they are unobservable because of the unknown parameters. However, when the data are available, β_0 and β_1 can be approximated by b_0 and b_1, which yields:

$$y_i - (b_0 + b_1 x_i) = y_i - \hat{y}_i = e_i$$

That is why the residuals $e_1, \ldots, e_n$ are generally used as realizations of the variable ε.

19.8.1 The five requirements of the model

Linearity

The basic assumption of the model states that, for each fixed value x of X, the random variable $Y = \beta_0 + \beta_1 x + \varepsilon$ satisfies

$$E(\varepsilon \mid x) = 0 \quad \text{so} \quad E(Y \mid x) = \beta_0 + \beta_1 x$$

That is, in the subpopulation of the population elements for which X takes the fixed value x, the mean of the observations of Y has to be $\beta_0 + \beta_1 x$ and hence has to reach the level of the population regression line. Notice that this requirement $E(\varepsilon \mid x) = 0$ cannot be checked by way of $\bar{e}$. This is because the $e_1, \ldots, e_n$ do not belong to a fixed value of x. As a matter of fact, it is not at all useful to consider the mean of the residuals since this mean is **always** 0.

If the assumption $E(\varepsilon \mid x) = 0$ is valid, the residuals should fluctuate whimsically around the value 0. That is why this assumption can be checked graphically by way of a scatter plot of the pairs $(x_1, e_1), \ldots, (x_n, e_n)$. This scatter plot has to look fickle; it must not have an obvious pattern.

Figure 19.10 shows scatter plots of residuals on x data for two different regressions. In plot (a) an obvious pattern can be recognized: the rough impression of the picture seems to show a curvature. For small and large values of x, the residuals are often negative; for median values of x, the residuals are positive. These facts demonstrate that the relationship between Y and X is more complicated than just linear; a parabolic relationship seems to do better.

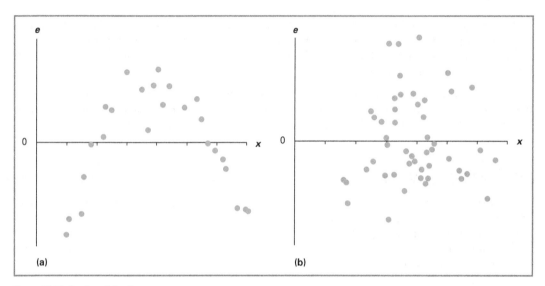

Figure 19.10 Scatter plots of e on x

The plot in Figure 19.10(b) is rather wild. It is hard to detect an obvious pattern in it. This plot seems to be in accordance with the linearity assumption.

If the relationship between Y and X is other than stated in the model requirement, one calls this **misspecification**. In the case of Figure 19.10(a), the simple linear regression model does not specify the relationship between Y and X correctly.

As a matter of fact, misspecification can often be detected by studying the scatter plot of the y data on the x data. In coming chapters more advanced models will be considered, to overcome problems like those in Figure 19.10(a).

Homoskedasticity

Recall that the homoskedasticity assumption requires that $V(\varepsilon \mid x) = \sigma_\varepsilon^2$, whatever the value of x. The variation around the population regression line has to be the same along all vertical lines. In a sense, it means that the population cloud of all (x, y) observations has to be equally 'fat' for all vertical intersections.

In practice, aberrations of this model requirement are often such that the variation around the regression line increases with the vertical level y. That explains why homoskedasticity can be checked with a scatter plot of the residuals e_i on the predictions $\hat{y}_i$. The scatter plot in Figure 19.11 obviously comes from a regression where the homoskedasticity requirement is **not** satisfied. If the homoskedasticity assumption does not hold, one talks about **heteroskedasticity**.

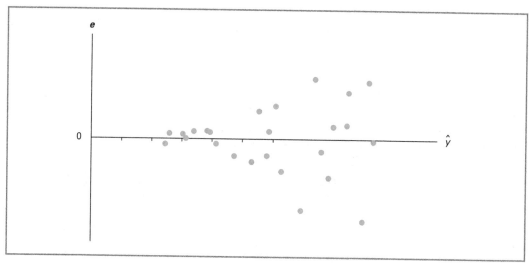

Figure 19.11 Scatter plot of *e* on *ŷ*

For small values of $\hat{y}$ the residuals *e* tend to lie closer to 0 than they do for larger values of $\hat{y}$. Thus the variation around 0 is larger for large $\hat{y}$ than for small $\hat{y}$.

Independence of $\varepsilon_1, ..., \varepsilon_n$

The error random variables $\varepsilon_1, ..., \varepsilon_n$ have to be independent. This independence assumption is automatically satisfied if the regression is based on a **randomly** chosen sample of population elements. Under these circumstances the random pairs $(X_1, Y_1), ..., (X_n, Y_n)$ are independent. Since, for each *i*, the error variable $\varepsilon_i = Y_i - (\beta_0 + \beta_1 X_i)$ depends only on the pair (X_i, Y_i), it follows that $\varepsilon_1, ..., \varepsilon_n$ are independent too.

Problems with the independence requirement often arise if the observations come from time series. The point is that then the population elements are periods (time units such as years, months, days). As a consequence, the pair (X_{t+1}, Y_{t+1}) that belongs to period $t + 1$ will often be dependent of the pair (X_t, Y_t) that belongs to period t, since t and $t + 1$ are usually successive, not randomly chosen, periods. Hence, the error terms are often dependent too.

Example 19.15

The dataset Xmp19-15.sav contains, for business days, the prices and the daily returns (%) of the Heineken stock and the AEX index (closing prices) for the period 6 December 2005 – 3 January 2007, so the elements are business days and the data come from time series. (Note that in the dataset some data are missing.) Two regressions are performed: the prices of the Heineken stock are regressed on the prices of the AEX and the returns of the Heineken stock are regressed on the returns of the AEX. For both regressions the residuals are also calculated (check them, see also Appendix A1.19); see the last two columns of the file.

Both scatter plots in Figure 19.12 plot residuals (vertically) against time (business day, horizontally). Plot (a) is about the residuals of the regression of the Heineken prices on the AEX prices and plot (b) is about the residuals of the Heineken returns on the AEX returns.

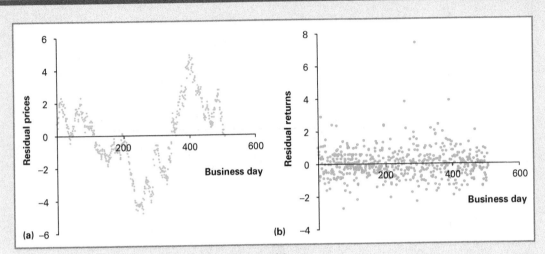

Figure 19.12 Scatter plots for the prices (a) and for the returns (b)
Source: Euronext (2007)

The differences are striking. Plot (a) shows an obvious pattern, apparently caused by the fact that any two successive residuals e_t and e_{t+1} tend to stay at the same side of the null-line. The **successive** residuals e_t and e_{t+1} are obviously positively **correlated**; the residuals are *first-order autocorrelated*. Here, 'first-order' refers to the fact that **successive** residuals are considered and 'auto' to the fact that it is about correlation within the sequence $e_1, ..., e_n$ itself. Things are completely different in plot (b). The residuals (of the regression of the returns) seem to behave independently of their predecessors, which makes the plot very wild and fickle.

The behaviour of the two types of residuals can also be demonstrated by creating all pairs (e_t, e_{t+1}) and placing them in respective scatter plots. For instance, to create the pairs for the residuals of the regression of the prices, the **first** $n - 1$ residuals are placed in one column and the **last** $n - 1$ in another column; the pairs arise in the rows. See Figure 19.13 for the two scatter plots. In plot (a), a large majority of the dots fall in quadrants I and III (see Section 5.1), which will yield a positive covariance and hence a positive correlation coefficient. In plot (b) the dots in quadrants II and IV seems to cancel out the dots in quadrants I and III; the correlation coefficient will be close to 0.

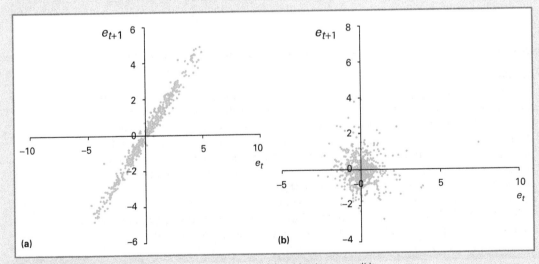

Figure 19.13 Scatter plots of the pairs (e_t, e_{t+1}) for the prices (a) and for the returns (b)

Of course, the correlation coefficients can be calculated. They are, respectively, 0.99 and 0.04. (Check all the above results yourself.)

Apparently, when regressing the price of the Heineken stock on the price of the AEX index, the independence requirement is not satisfied. However, when regressing the daily returns of the Heineken stock on the daily returns of the AEX index, there seems to be no problem with the independence assumption.

Independence of ε and X
The discussion of this requirement is postponed until Section 22.5.

Normality
The validity of the normality assumption for the error variables ε_i can be checked graphically by way of a histogram of the residuals $e_1, \ldots, e_n$. The histogram has to be more or less mountain-shaped, like a normal density. Heavy skewness or the presence of (too many) outliers is an indication that the normality assumption is not valid.

Histograms for the two types of residuals of Example 19.15 are presented in Figure 19.14; the best-fitting normal curve is included. The histogram in part (a) is not well described by the normal curve whereas that in part (b) fits much better to the normal curve that is included.

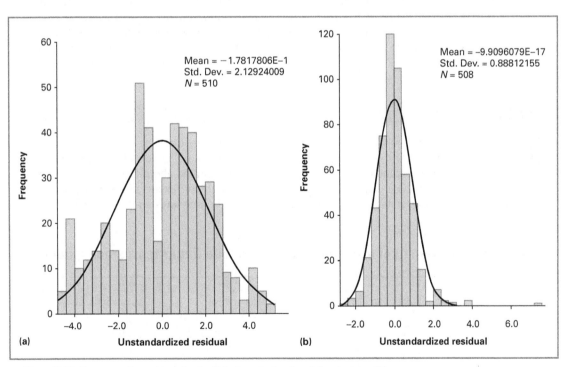

Figure 19.14 Histograms for residuals for the Heineken stock; prices (a) and returns (b)

To detect outliers within the sequence $e_1, \ldots, e_n$ of residuals (which might conflict with the normality assumption) it is wise to study the histogram of the **standardized** residuals. The *standardized residuals* are the z-scores of the ordinary residuals: they arise by subtracting the mean of the residuals (which is 0) and dividing the results by some accompanying standard error (no further details are given). Most computer packages have options to calculate the standardized residuals directly; see also Example A1.19.

The advantage of changing over to the standardized residuals is that their histogram should look like a **standard** normal curve, at least if the normality requirement of the simple linear regression model is satisfied. Since the standard normal density has 99.7% of the total area between -3 and $+3$, the search for outliers within the sequence $e_1, \ldots, e_n$ simplifies to searching standardized residuals that lie outside the interval $(-3, 3)$.

19.8.2 Consequences of model violations

Of course, the reliability of statistical conclusions will be affected by model violations. To what degree such conclusions are affected is a problem on its own that can be studied theoretically and practically. We will consider it only very briefly.

As long as the linearity requirement is valid, the point estimators B_0 and B_1 remain unbiased. However, many violations of the model requirements cause S_ε^2 to become biased. In the case of positive autocorrelation for $\varepsilon_1, \ldots, \varepsilon_n$, it can be shown that s_ε^2 is typically smaller than σ_ε^2. As a consequence, replacing σ_ε by s_ε leads to confidence intervals that are too narrow and to values of test statistics that are too large. Hence, the confidence level of confidence intervals which, according to the theory, should be 0.95 is actually less than 0.95; the t-test for checking the significance of the independent variable concludes too easily that the model is useful.

For larger sample sizes, violations of the normality requirement hardly affect confidence intervals and hypothesis tests. This is because the central limit theorem can then be used. However, for interval **predictions**, the normality of $\hat{Y}_p - Y_p$ also needs the approximate normality of the individual random variable Y_p, which is **not** guaranteed by the CLT. As a consequence, the confidence level of prediction intervals has to be distrusted in case of non-normality, even for large n.

In the case of serious violation of the model requirements it is wise to adapt the model. There are many ways to do so, for instance by replacing the variables X and Y by transformations; more about this in Chapter 22.

Summary

The simple linear regression model is meant to describe the linear relationship between a dependent variable Y and one independent variable X. It is a collection of requirements constructed such that several standard statistical procedures are justified about the slope of the linear relation, and about the observation $Y = Y_p$ and its accompanying expectation for a population element with $X = x_p$. Important ingredients of these statistical procedures are in the table.

Overview of statistical procedures for the simple linear regression model

	Estimator / point predictor	Random interval	Test statistic; hinge b
Parameter β_1	B_1	$B_1 \pm t_{\alpha/2,n-2} \dfrac{S_\varepsilon}{\sqrt{(n-1)s_X^2}}$	$\dfrac{B_1 - b}{S_\varepsilon/\sqrt{(n-1)s_X^2}}$
Parameter σ_ε^2	$S_\varepsilon^2 = \dfrac{1}{n-2}\sum_{i=1}^{n}(Y_i - \hat{Y}_i)^2$		
Parameter $E(Y_p)$	$\hat{Y}_p = B_0 + B_1 x_p$	$\hat{Y}_p \pm t_{\alpha/2,n-2}\, S_\varepsilon \sqrt{\dfrac{1}{n} + \dfrac{(x_p - \bar{x})^2}{(n-1)s_X^2}}$	
Random variable Y_p	$\hat{Y}_p = B_0 + B_1 x_p$	$\hat{Y}_p \pm t_{\alpha/2,n-2}\, S_\varepsilon \sqrt{1 + \dfrac{1}{n} + \dfrac{(x_p - \bar{x})^2}{(n-1)s_X^2}}$	

If the independent variables are random (which generally is the case), then the conditional method has to be applied for the interpretation of the results. As a consequence, statistical conclusions have to be interpreted conditionally, under the – usually not very impressive – condition (called C) that the realizations of the random variables $X_1, \ldots, X_n$ are $x_1, \ldots, x_n$.

The ANOVA table gives sample information about the performance of the model to explain the variation of Y. It decomposes the measured variation (SST) of the y data into a part that is explained by the model (SSR) and a part that remains unexplained (SSE). It contains all the ingredients to calculate the coefficient of determination which, in a sense, measures the degree of usefulness of the model.

If a residual analysis indicates that not all model assumptions are satisfied, then the statistical conclusions have to be interpreted cautiously. Sometimes it may be better to change the model; more about that in Chapter 22.

List of symbols

Symbol	Description
β_0	Intercept of the population regression line
β_1	Slope of the population regression line
B_0	Least-squares estimator of β_0
B_1	Least-squares estimator of β_1
b_0	Least-squares estimate of β_0
b_1	Least-squares estimate of β_1
ε_i	Error variable
e_i	Residual $y_i - \hat{y}_i$
r^2	Coefficient of determination
σ_ε	Standard deviation of ε
S_ε	Standard error of the estimated model
SSE	Sum of squares of the error terms
SSR	Sum of squares due to regression
SST	Sum of squares total
$\hat{Y}_p$	Point predictor of Y_p
$\hat{y}_p$	Point prediction of Y_p

🔑 Key terms

ANOVA table **587**
basic assumption **567**
beta (coefficient) **590**
coefficient of
    determination **587**
conditional method **578**
dependent variable **559**
error variable **562**
explanatory variable **559**
firm-specific risk **591**
first-order
    autocorrelated **600**
heteroskedasticity **598**
homoskedasticity **566**
independent variable **559**

interval predictor **595**
least-squares (point)
    estimators **570**
market model **590**
market-related
    risk **590**
misspecification **598**
point predictor **592**
prediction interval **592**
predictor **559**
regression analysis **561**
research variable **559**
residual analysis **597**
sample regression
    coefficients **571**

sample variance of the
    estimated model **573**
significant **582**
simple linear regression
    model **567**
standard deviation of the
    model **567**
standard error of the
    estimated model **573**
standardized
    residuals **601**
sum of squares due to
    regression **586**
sum of squares total **585**
volatile **590**

? Exercises

Exercise 19.1

(Repetition of the techniques of Chapter 5.) For a quick impression of the relationship between variables X and Y, both are observed for a random sample of five population elements:

i	1	2	3	4	5
x_i	8	9	7	10	12
y_i	80	85	79	90	90

a Calculate the covariance and the correlation coefficient. Interpret the results.

b Determine the equation of the sample regression line of y on x. Interpret the regression coefficients.

c Calculate $\hat{y}_1, \ldots, \hat{y}_5$.

d Calculate the residuals $e_1, \ldots, e_5$.

Exercise 19.2

Reconsider Exercise 19.1.

a Calculate SSE, SST and SSR.

b Calculate the standard error of the estimated model. Also calculate the realized standard error of B_1.

c Calculate the coefficient of determination. Interpret the answer.

Exercise 19.3

To explain the variation of $Y = $ 'annual food expenditure ($\times$ €1000)' from $X = $ 'annual income ($\times$ €1000)' on a large population of households, a simple linear regression model is used. Data are available for a random sample of 300 households; see the file Xrc19-03.sav.

a With respect to the population cloud of the (x, y) pairs, only a vague idea is present; see the scatter plot. Give interpretations of all model requirements in terms of this population plot.

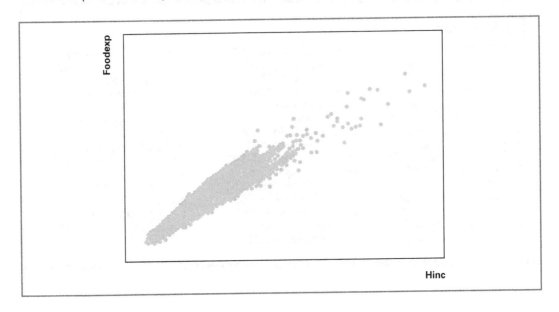

b Running the regression yields the following printout; find estimates for β_0, β_1 and σ_ε^2. Also find the realized standard error of B_1.

Model Summary

Model	R	R Square	Adjusted R Square	Std. Error of the Estimate
1	.965(a)	.932	.932	.90926

a Predictors: (Constant), HINC

ANOVA(b)

Model		Sum of Squares	df	Mean Square	F	Sig.
1	Regression	3378.591	1	3378.591	4086.593	.000(a)
	Residual	246.371	298	.827		
	Total	3624.962	299			

a Predictors: (Constant), FINC
b Dependent Variable: FOODEXP

Coefficients(a)

Model		Unstandardized Coefficients		Standardized Coefficients		
		B	Std. Error	Beta	t	Sig.
1	(Constant)	1.151	.126		9.122	.000
	HINC	.241	.004	.965	63.926	.000

a Dependent Variable: FOODEXP

c Use the printout to determine SST, SSE, SSR and the coefficient of determination.
d Is the model useful? Test it with significance level 0.05.
e Give interpretations of the answers to parts (b) and (c).

Exercise 19.4

You want to explain the variation in the variable Y = 'daily number of hours that you study' from the variation in the variable X = 'daily number of hours that you watch television' by doing a simple linear regression. For a random sample of 7 days you measure both X and Y. Here are the results:

x	3	5	6	4	3	7	6
y	4	2	0	2	2	0	1

a Calculate:
  i the sample means $\bar{x}$ and $\bar{y}$;
  ii the sample variances s_x^2 and s_y^2;
  iii the sample covariance and the correlation coefficient.
b Determine the slope and the intercept of the sample regression line; interpret the slope.
c Calculate the residual e_3 (that belongs to the third pair of observations).

d Use the results so far to calculate SSE.

e Calculate the standard error of the estimated model (s_e). Also calculate the realized standard error of B_1, that is, calculate s_{B_1}.

Exercise 19.5

The manager of a supermarket wants to regress Y = 'sales per week (in units of 10^4 euro)' on X = 'costs of advertisements in that week (in units of 10^3 euro)'. A quick study, for six arbitrary weeks, gives the following results for X and Y:

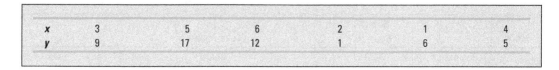

x	3	5	6	2	1	4
y	9	17	12	1	6	5

Below, you can use (without proving) that $\bar{x}$ = 3.5, s_X^2 = 3.5, $\bar{y}$ = 8.3333, s_Y^2 = 31.8667 and SSE = 81.1049.

a Calculate the slope of the (sample) regression line. Interpret your answer.

b Calculate the coefficient of determination from SST and SSE. Interpret it.

c Calculate the standard error of the estimated model.

Exercise 19.6

Reconsider Exercise 19.5. In a larger study, observations of X and Y for 60 arbitrary weeks are used. For these data, the equation of the regression line and other summarized results are listed below:

$\hat{y}$ = 0.85 + 1.95x; $\bar{x}$ = 3.9; s_X^2 = 4.2; s_e = 5.1

a One of these 60 weeks has x = 3 and y = 9. Calculate the accompanying residual.

b Next week, the advertising costs will be €4000. Determine a point prediction for next week's sales Y. Also determine a 95% prediction interval for next week's sales Y when the advertising costs will be €4000.

The manager now wants to know whether X and Y are linearly related. To decide this she uses the above summarized results to create a suitable 95% confidence interval.

c Write down: the interval estimator; the resulting 95% confidence interval; the conclusion about the question: are X and Y linearly related?

Exercise 19.7

Reconsider Exercises 19.5 and 19.6. The manager claims that an increase of €1000 in the weekly advertising costs will on average lead to more than €15 000 extra sales. To test (with α = 0.05) whether this claim is valid, use the results below that come from an extensive study of the results of 100 weeks:

$\hat{y}$ = 0.8439 + 1.9238x;   standard error of B_1: 0.2883

a Write down the complete five-step procedure.

b Calculate the accompanying p-value. Interpret the answer.

Exercise 19.8

In all studies below, indicate which variable will be the **dependent** variable.

a The Bureau for Economic Policy Analysis of Sweden wants to improve the model that relates the yearly gross domestic product and the yearly amount of new investment (both in billions of euro).

b A company is interested in the relationship between its yearly net profit and its yearly sales, both in millions of euro.

c A company wants to analyse the relationship between its yearly net profit and the dividend per unit of shares (in euro) paid to its shareholders.

d A consumer keeps a cashbook to compare the monthly household expenditure on food to the monthly household income, both measured in euro.

e The business manager of the football club Olympique Lyon is analysing the relationship between the gross yearly sales (in millions of euro) and the position of the team in the French ranking at the end of the season.

f A financial analyst at Heineken considers the relationship between the daily return (%) of the AEX index and the daily return (%) of the stock Heineken.

g A financial analyst of a bank studies the relationship between the interest rate at the end of a quarter and the interest rate at the end of the previous quarter.

h An insurance agent investigates, for car insurance, the relationship between the type of insurance held and the claim amount (in euro).

Exercise 19.9

(Repetition of the techniques of Chapter 5.) Many studies in economics are about the relationship between sales and price. In such studies, 'sales' often is the dependent variable and 'price' the independent variable. Even if a product has a linear price-sales curve, the relationship between sales and price will – due to several confounding factors – not be precisely linear.

Let the variable X be the weekly price (in euro) of a certain product and let Y be the number of items sold in that week. The table below gives, for seven arbitrary weeks, the prices and the sales. We will adopt the simple linear regression model with basic assumption $E(Y \mid x) = \beta_0 + \beta_1 x$ to analyse the data.

Week i	Price x_i	Sales y_i
1	16	23
2	18	21
3	19	20
4	14	21
5	17	20
6	19	18
7	16	24

a Create (by hand) the scatter plot of y on x. Give your first impression of the relationship between Y and X.

b Determine the equation of the sample regression line by using a table similar to that in Example 19.3 that also includes the squares of the y deviations with respect to $\bar{y}$. Interpret the estimates of β_1 and β_0.

c Determine the predictions $\hat{y}_i$ and the residuals e_i.

Exercise 19.10

Reconsider Exercise 19.9.

a Determine the ANOVA table.

b Calculate the standard error of the estimated model. Also calculate the (realized) standard error of B_1.

c Is the model useful? Use $\alpha = 0.05$ to test it. Give your comments.

Exercise 19.11

On the website www.m-a.org.uk of the Mathematical Association, a study is presented on the relationship between the variables 'age (in years)' and 'mileage (in miles)' of cars. 'Mileage' is taken as dependent variable. The file Xrc19-11.sav contains data for a sample of 22 cars.

The SPSS printout below shows the estimation results of a simple linear regression of the mileage data on the age data.

ANOVA(b)

Model		Sum of Squares	df	Mean Square	F	Sig.
1	Regression	4392448134.328	1	4392448134.328	118.968	.000(a)
	Residual	738426865.672	20	36921343.284		
	Total	5130875000.000	21			

a Predictors: (Constant), Age

b Dependent Variable: Mileage

Coefficients(a)

Model		Unstandardized Coefficients		Standardized Coefficients	t	Sig.
		B	Std. Error	Beta		
1	(Constant)	3222.388	4195.714		.768	.451
	Age	10241.791	938.991	.925	10.907	.000

a Dependent Variable: Mileage

a Determine the equation of the sample regression line.

b Is the model useful? Test it with $\alpha = 0.01$.

c Determine the coefficient of determination and the coefficient of correlation. Interpret the answers.

d Interpret the slope of the regression line in part (a).

Exercise 19.12

Reconsider Exercise 19.11.

a Someone claims that in a period of one year the expected mileage of a car increases by more than 8000 miles. Use a hypothesis test with $\alpha = 0.01$ to check that claim.

b Determine the p-value of the test in part (a). Interpret the answer.

c The regression is done again but now the dependent variable is measured in kilometres instead of miles. What are the consequences for the sample regression line? Answer this question without actually transforming the miles data into kilometres data. (1 mile = 1609 metres.)

Exercise 19.13

Nowadays, oil is becoming a scarce product. The time-plots in the figure below show the developments over time of oil imports and oil demand (both in millions of tonnes) in the EU15 during the period 1971–2003. The figure shows that over the past 30 years the demand and import curves for oil have separated. Still, the two curves seem to decrease and increase together, thus suggesting a positive linear relationship.

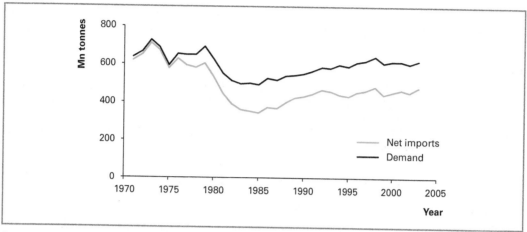

Source: World Trade Organization (2005)

At first sight, a regression of oil imports on oil demand looks promising. However, since the data come from time series, it can hardly be expected that, for the simple linear regression model of imports on demand, the requirement about the independence of the error variables will be satisfied.

That is why the original data are transformed into growth data (%) with respect to the year before, yielding growth data for the period 1972–2003. The file Xrc19-13.sav contains the data. The next step is to regress Y = 'yearly change (%) in oil imports' in the EU15 on X = 'yearly change (%) in oil demand'. Parts of the printout are given below.

ANOVA(b)

Model		Sum of Squares	df	Mean Square	F	Sig.
1	Regression	1304.091	1	1304.091	190.790	.000(a)
	Residual	205.056	30	6.835		
	Total	1509.147	31			

a Predictors: (Constant), Demand_Change

b Dependent Variable: Imports_Change

Coefficients(a)

Model		Unstandardized Coefficients		Standardized Coefficients	t	Sig.
		B	Std. Error	Beta		
1	(Constant)	−.646	.462		−1.398	.172
	Demand_Change	1.179	.085	.930	13.813	.000

a Dependent Variable: Imports_Change

a Write down the basic assumption of the simple linear regression model that regresses Y on X.

b Write down the equation of the sample least-squares line.

c Are the two variables positively linearly related? Use a 98% confidence interval to answer this question. Also use it to draw a conclusion about the conjecture that an increase in the demand for oil of 1% will on average lead to an increase in oil imports of more than 0.9%.

d In 2003 and 2004 the demand was, respectively, 622 million tonnes and 630 million tonnes; imports in 2003 were 480 million tonnes. Use this information to predict the imports in 2004.

Exercise 19.14

A shareholder of Royal Philips Electronics investigates the variation in the yearly observations of the variable Y = 'yearly dividend (in millions of euro) paid out by Philips'. Since the yearly profit of the company seems to be an important factor, he decides to adopt the simple linear regression model of Y on X = 'yearly net profit (in billions of euro) of Philips'. For the most recent eight years he gets the following results:

$$\bar{x} = 3.67; \quad \bar{y} = 1.70; \quad \sum_{i=1}^{8}(x_i - \bar{x})(y_i - \bar{y}) = 1.101; \quad \sum_{i=1}^{8}(x_i - \bar{x})^2 = 9.821; \quad \sum_{i=1}^{8}(y_i - \bar{y})^2 = 0.18$$

a Write down the basic assumption of the model.

b For each of the following symbols, indicate whether it denotes a parameter, a random variable or a realization of a random variable:

$$Y \quad \bar{y} \quad \beta_0 \quad b_0 \quad \sigma_\varepsilon^2$$

c Estimate the population regression coefficients β_0 and β_1.

d Determine SST, SSR and SSE.

e Measure the degree of usefulness of the estimated model; interpret your answer. Also find an estimate of σ_ε^2, the variance of the error variable in the model.

f Conduct a hypothesis test with significance level 0.01 to find out whether Y and X are positively linearly related.

Exercise 19.15

A real estate agent is interested in the variation of the selling prices of the houses sold by her agency. Since the volume of a house will influence the selling price, she takes the variable X = 'volume (in m³) of the house' as independent variable in a simple linear regression model for Y = 'selling price (in thousands of euro) of the house'.

For a sample of 30 houses, the agent obtains the following results:

$$\bar{x} = 478.5000; \quad \bar{y} = 426.3333; \quad \sum_{i=1}^{30}x_i y_i = 6\,630\,850; \quad \sum_{i=1}^{30}x_i^2 = 7\,539\,475; \quad \sum_{i=1}^{30}y_i^2 = 5\,897\,200$$

a Use these results to determine the equation of the sample regression line. Interpret the sample regression coefficients.

b Determine the ANOVA table. Use it to calculate the standard error of the estimated model.

c The agent is especially interested in the mean selling price for houses of 500 m³. Determine an interval that contains this mean with confidence 95%.

d The agent has to sell a house with a volume of 480 m³. Determine an interval that contains the selling price of that house with a confidence of 95%.

Exercise 19.16 (computer)

The dataset Xrc19-16.sav contains the volumes and selling prices for the sample of 30 houses of Exercise 19.15. Check the results of that exercise with the help of a statistical package.

Exercise 19.17

Reconsider Exercise 19.11, where mileage of a car is explained from the age of the car. You are asked to use the printout of Exercise 19.11 to calculate some intervals. It is additionally given that the mean mileage of the 22 cars in the sample is 46 750 miles.

a Determine the sample mean and the sample variance of the x data.

b Determine an interval that, with a confidence of 98%, contains the mileage of a car that is 5 years old. Interpret the interval.

c Determine an interval that, with a confidence of 98%, contains the mean mileage of all 5-year-old cars. Interpret the interval.

d Determine an interval that, with a confidence of 98%, contains the expected mileage of a 2-year-old car. Interpret the interval.

Exercise 19.18 (computer)

The owner of a large computer shop believes that there are two important factors that influence the weekly sales of Inspiration notebooks: the price (which is modified each week) and the weekly number of advertisements. For 20 arbitrary weeks, he measures the following variables:

Y = 'number of Inspiration notebooks sold in a week'
X_1 = 'price (in euro) of the notebook in that week'
X_2 = 'number of advertisements (in local newspapers) in that week'

The dataset Xrc19-18.sav contains these measurements.

a Use the data to estimate two simple linear regression models: the model that explains Y from X_1 and the model that explains Y from X_2.

b Use p-values of the printouts to decide whether the models are useful.

c Which of the two models do you prefer? Why?

Exercise 19.19 (computer)

A firm markets goods targeting youths in Spain. Because the activities involve a high degree of trust on the part of both parties, the company needs information about the opinion of young people on ethical questions. MarketView is asked to do a research.

MarketView interviews 84 Spanish university students about several ethical problems. One of the questions (question 20) asks the student's opinion, on a five-point scale, about:

Recording a compact disc/album instead of buying it.

On the five-point scale, 1 means 'strongly believe it is wrong' and 5 'strongly believe it is **not** wrong'. The dataset Xrc19-19.sav contains not only the answers of the 84 students but also their ages.

Is the answer to this question statistically related to the age of the student? To study this, the variable Q20 = 'answer to question 20' is considered as quantitative and regressed on X = 'age of the student'.

a To get a first impression of the relationship between the two variables, create a scatter plot with sample regression line.

b Estimate the linear regression model that explains Q20 from X. Is the model useful? Use $\alpha = 0.05$ to decide about it.

The results in the scatter plot and the printout seem to be heavily determined by one relatively old student. For the following questions, omit this student from the dataset and run the regression again.

c Is the model still useful? If so, interpret the slope of the sample regression line.

d Interpret the three sums of squares in the ANOVA table. Also interpret the coefficient of determination and the standard error of the estimated model.

Exercise 19.20 (computer)

Reconsider Exercise 19.19; use the dataset **without** the 'extreme' observation.

a Test (with $\alpha = 0.01$) whether, on average, 'older' students have a stronger belief that it is wrong than 'younger' students. Also determine the p-value of this hypothesis test, in eight decimals; use the printout to determine it. Interpret your answer.

b Calculate an interval that is likely to contain the slope of the population regression line; use a confidence level of 96%. Interpret your result.

c Use a statistical package to obtain an interval that is likely to include the opinion of Jorge Gonzales, a 20-year-old student. Take 95% as confidence level.

d Use a statistical package to obtain an interval that is likely to include the expected opinion of an arbitrary 20-year-old student. Take 95% as confidence level.

e Use a statistical package to obtain an interval that is likely to include the mean opinion of 21-year-old Spanish university students. Take 98% as confidence level.

Exercise 19.21 (computer)

Compare with Exercise 19.13, which is about the EU15. Interest is now in the relationship between the imports of, and demand for, oil in the **USA**. The dataset Xrc19-21.sav gives, for the period 1971–2004, the annual import figures and the annual demand (both in millions of tonnes).

a Create, in one figure, the time-plots (that is, put year horizontally) of the imports and the demand data. Use these plots to explain why it does not seem to be a good idea to regress the imports on the demand.

b The **growth data** (in percentages with respect to the previous year, for the period 1972–2004) are in columns 4 and 5 of the dataset. Regress the growth data for the imports on the growth data for the demand.

c Is the model useful? Also interpret the coefficient of determination.

d Conjecture: if the growth (%) of the demand is 1 extra, then the growth (%) of the imports is on average more than 3 extra. Test it with $\alpha = 0.05$.

e For 2005, the change in demand was (compared with 2004) 2.5% but the change in imports was not known. Use this information to determine a 96% prediction interval for the growth of the oil imports in 2005.

Exercise 19.22 (computer)

Reconsider the regression model of Exercise 19.21.

a Create the sequence of the 33 residuals e_t.

b Check the model requirements graphically, in the way indicated in Section 19.8. Use the **standardized** residuals to check the normality assumption. To check independence, also create two columns that contain, respectively, the last 32 and the first 32 of the 33 residuals; use these new columns for the scatter plot of e_t on e_{t-1}. Interpret all plots.

c Regress e_t on e_{t-1}. Use the printout to test whether e_t is linearly related to e_{t-1}. What is the consequence for the independence requirement of the model we started with?

Exercise 19.23 (computer)

To study the sensitivity of the stock Numico for changes at the market (AEX), the daily returns (%, against closing prices) of this stock are considered for the period 01 February 2005 – 29 January 2007; see the file Xrc19-23.sav. The market model (see Section 19.6.4) will be used to estimate the beta coefficient of Numico and to discuss the implications. In the second part of this exercise, a residual analysis will be conducted to check the model requirements.

a Write down the basic assumption of the model. Determine the sample regression equation.

b Determine the (estimate of the) beta coefficient of Numico. Interpret it. Give your (statistical) comments on the conjecture that the stock Numico is less volatile than the market.

c Check the model requirements graphically, as indicated in Section 19.8.

Exercise 19.24 (computer)

Check the model requirements in Exercise 19.3 graphically.

Exercise 19.25 (computer)

The dataset Xrc19-25.sav contains the prices of ladies' diamond rings and the carat size of their stones. The source of the data is a full-page advertisement placed in the *Straits Times* of 29 February 1992, by a Singapore-based retailer of diamond jewellery. The rings are made with 20 carat gold and are mounted with a single diamond. For each ring, the following variables are measured:

CARAT: Size of diamond in carats (1 carat = 0.2 gram)
PRICE: Price of ring in Singapore dollars

a Create a scatter plot of the PRICE data on the CARAT data. What kind of relationship between the variables PRICE and CARAT do you expect?

b Use a simple linear regression model to explain the variation of PRICE from the variation of CARAT. Estimate the model and report your statistical findings.

c Check the model assumptions.

Exercise 19.26 (computer)

Having been commissioned by a firm that markets goods to young people in several European countries, market research company MarketView interviewed 948 business students from eight countries about their opinion on the following questions (see also Exercise 15.26):

Q1: Changing price tags on merchandise in a retail store
Q2: Drinking a can of cola in a supermarket without paying for it
Q3: Reporting a lost item as 'stolen' to an insurance company in order to collect the insurance money
Q4: Giving misleading price information to a checkout assistant for an unpriced item

Possible responses were on a scale of 1 to 5 where 1 = strongly believe it is wrong and 5 = strongly believe it is not wrong. Also the ages of the respondents have been recorded. The file Xrc19-26.sav lists the ages and the observations of the standardized version Y of the variable that measures for a student the average of the answers to these four questions. The variable Y is considered as quantitative; it measures the response of a student to 'actively benefiting from illegal activity'.

The relationship between Y and the variable AGE seems to be interesting. You are invited to regress Y on AGE. Conduct a statistical analysis and give your comments.

CASE 19.2 PROFITS OF TOP CORPORATIONS IN THE USA (PART I)

It will be investigated whether and to what degree the profits of a large company are influenced by: its revenues, the state in which the company is situated and the gender of the CEO. In part I of the study only the influence of 'revenues' will be considered.

For the data, we use the Fortune 1000 listing in the USA; see the file Case19-02.sav.[†] The following variables are observed:

---

† *Original source*: http://money.cnn.com.

Y: profits of a company (in millions of dollars)

X: revenues (in millions of dollars) of that company

The data in the dataset are ranked on the basis of the observations of X.

Since the data are ordered Top 1000 data, we cannot consider them to come from an ordinary random sample of 1000 corporations. Instead, a more abstract approach is chosen. It is assumed that the 1000 observations are typical 'states of nature' of a general top corporation in the USA. That is, we forget about the accompanying names of the corporations in the dataset and consider the data to give the ordered states of nature (the revenues and the profits) of 1000 random observations from the distribution of (X, Y). Thus the simple linear regression model of Y on X can be used for statistical inference about the development of the profits of a top corporation.

a Find the linear relationship between Y and X, and give an analysis of the regression.

b Interpret the sample regression coefficients.

c A top corporation is wondering whether $1 million of extra revenues will yield more than $80 000 extra profits. Use the simple linear regression model to test it.

CASE 19.3 INCOME AND EDUCATION LEVEL OF IDENTICAL TWINS (PART I)

This study is one of the many studies in economics that want to find an answer to a question that, at first sight, seems to be rather simple:

By how much will another year of schooling raise one's income?

Previous estimates of the value of a year's education may have been imprecise for two reasons. The first is the difficulty of isolating education's effect on income from the effect that other variables related to education have on income. That is, a worker's natural ability, his family background and his innate intelligence are all possibly confounding factors that must be controlled for in order to estimate accurately the effect of education on income. The second difficulty has to do with the false reporting of education levels. Since people are more likely to report a higher education level than they have actually attained, especially in face-to-face interviews, the data will contain a number of people with lower education levels in the higher education categories. As a consequence, since education usually increases income, estimates for the precise amount of this increase will be too low.

Thus, the present study interviewed twins, collecting information about education, income and background. Because monozygotic twins (twins from a single egg) are genetically identical and have similar family backgrounds, they provide an excellent control for confounding variables. Furthermore, to correct for the bias in years of education, the researchers interviewed the twins separately and recorded two entries for each individual's education level: the self-reported education level and the education level reported by his/her twin.[‡]

The data in the file Case19-03.sav were collected by a team of five interviewers at the 16th Annual Twins Day Festival in Twinsburg, Ohio, in August 1991. In total, 495 individuals over the age of 18 were interviewed.

Each row contains information on a pair of twins. Note that each individual in a pair was **randomly** assigned a number: twin 1 or twin 2. The extensions L and H of some variables refer, respectively, to twin 1 and twin 2.

[‡] The published study is available on the website of JSTOR at www.jstor.org. *Original source*: O. Ashenfelter and A. Krueger, 'Estimates of the economic return to schooling from a new sample of twins', *The American Economic Review*, 84/5 (1994), 1157–73.

HRWAGEH	hourly wage of twin 2 (in dollars)
EDUCH	self-reported education (in years) of twin 2
HRWAGEL	hourly wage of twin 1 (in dollars)
EDUCL	self-reported education (in years) of twin 1
DEDUCxx	the difference (twin 1 minus twin 2) in cross-reported education; for example, twin 1's cross-reported education is the number of years of schooling completed by twin 1 as reported by twin 2

a Find out whether one extra year of education on average increases hourly wage by more than $0.50. Run two regressions, one with the data of twins 1 and one with the data of twins 2, to answer this question. Give your comments.

b For the regressions of part (a), check the model assumptions.

c Run another regression to find out whether one extra year for the difference between the numbers of years of education of twin 1 and twin 2 will on average cause a difference in hourly wage of more than $0.70. Answer this question with the difference between the self-reported EDUC as well as with the difference between the cross-reported EDUC.

d Ashenfelter and Krueger, the authors of the original study, want to interpret the result of part (c) in terms of (general) people with similar family backgrounds. Formulate that interpretation.

Multiple linear regression: introduction

Chapter contents

In economics studies, a research variable Y can sometimes be explained very well by only one independent variable X. This is the case if X and Y are strongly linearly related. However, it often happens that the independent variable explains only a small part of the variation of the dependent variable Y and there is then a need to extend the model with more independent variables.

In the present chapter, regression models will be considered with k independent variables called $X_1, \ldots, X_k$ where k is at least 1. The approach is similar to the approach of the previous chapter. As a 'red-wire' application, the case about the wage differentials between men and women, introduced in Case 19.1 – will be considered.

A few things have to be said before starting:

- The independent variables are usually not stochastically independent in the sense of (11.9). Within the setting of regression models, the phrase 'independent variables' is historically used as a counterpart of 'dependent variable'.

- In the present chapter, the notations $X_1, \ldots, X_k$ are used to denote the k independent variables of the model. That is, they are **not** a sample of random observations of one variable X, so (and importantly) they are not a random sample.

- Be aware of the fact that many formulae of Chapter 19 are no longer valid for regression models with two or more independent variables.

CASE 20.1 POLLUTION DUE TO TRAFFIC (PART I)

Pollution is a serious problem in many European cities. In particular, the 'fine particulate matter' produced by cars is an enormous threat to health. In the present study we will investigate whether, and to what degree, the emission of this fine particulate matter is determined by the population size of the city, its diameter and area, the wind and the country where the city is situated. Here is the list of variables:

Y: Emission due to road transport (tonnes/year) in a city

X_1:   Number of citizens (in units of 1000)

X_2:   Diameter of the city (in km)

X_3:   Area of the city (in km²)

X_4:   Annual mean wind speed (in m/s)

X_5:   Low wind speed days (number of days per year)

Country: Country where the city is situated

We will use the dataset <u>Case20-01.sav</u> that comes from the IIASA Interim Report IR–07–01 entitled *Estimating Concentrations of Fine Particulate Matter in Urban Background Air of European Cities* (2007) by Amann et al. It contains measurements of the above variables for the 473 cities in the EU25 with at least 100 000 inhabitants. In the first part of this study, a possible country effect will not yet be considered.

After having estimated (and compared) the five obvious simple linear regression models for Y, some models will be estimated with more than one independent variable. See the end of Section 20.5.3 for the solution.

20.1 The multiple linear regression model

Recall that SST measures the variation before regression within the data of the dependent variable. By using the linear regression model with one independent variable, we managed to reduce this unexplained variation to SSE, so the amount SSR = SST − SSE is explained **after** simple linear regression. It is the challenge of the researcher to reduce the unexplained variation further by introducing one or more new independent variables into the model. Such **multiple** linear regression models are the topic of the present section.

Let us start with a situation where **two** independent variables are included, so $k = 2$. To have an example in mind, think of explaining the hourly wage Y of a person from his/her age X_1 and education level X_2 (possible values 1, ..., 5 with 1 the lowest level and 5 the highest). If we were able to observe the three variables on the whole population, a triple (x_1, x_2, y) would be measured for each person in the population. The scatter plot of all triples would yield a dense population cloud of dots in a three-dimensional system of axes, with (x_1, x_2) forming the horizontal base plane and y the vertical direction. Instead of finding the best-fitting straight **line** (the least-squares line) with equation $y = \beta_0 + \beta_1 x_1$, the objective now is to find the best-fitting two-dimensional **plane** with equation $y = \beta_0 + \beta_1 x_1 + \beta_2 x_2$. Here, the best fitting plane is the least-squares plane that arises by generalizing the LS method of (5.5). Of course, an arbitrary dot with coordinates (X_1, X_2, Y) in this three-dimensional population cloud will usually not fall precisely in this best-fitting plane; in the direction of the y-axis an error ε will be present. That is why our starting point is that

$$Y = \beta_0 + \beta_1 X_1 + \beta_2 X_2 + \varepsilon$$

where the coefficients $\beta_0, \beta_1, \beta_2$ are the least-squares coefficients.

However, the population is too large to observe all (x_1, x_2, y), so the population cloud remains largely unobserved and the three coefficients β_0, β_1, β_2 are unknown real numbers. What **can** be done is observing the three variables (X_1, X_2, Y), for a **sample** of population elements and finding the equation $y = b_0 + b_1x_1 + b_2x_2$ of the two-dimensional plane that fits best through the three-dimensional **sample** cloud. Here, the coefficients b_0, b_1, b_2 are the least-squares coefficients. These coefficients can be considered as estimates of the unknown coefficients β_0, β_1, β_2; the sample plane with equation $y = b_0 + b_1x_1 + b_2x_2$ approximates the population plane with equation $y = \beta_0 + \beta_1x_1 + \beta_2x_2$.

The objective of the present section is to formulate things for the general situation, with k independent variables $X_1, X_2, ..., X_k$. In this general set-up, the population cloud of the $(k + 1)$-tuples $(x_1, ..., x_k, y)$ lies in a $(k + 1)$-dimensional space and interest is in the best-fitting k-dimensional 'plane'. This so-called **hyperplane** has equation $y = \beta_0 + \beta_1x_1 + ... + \beta_kx_k$ and β_0, β_1, ..., β_k are the least-squares coefficients that result from the LS method. The coordinates $(X_1, ..., X_k, Y)$ of an arbitrary population element will in general not fall precisely in this hyperplane; an error ε in the y direction will be present. Hence:

$$Y = \beta_0 + \beta_1X_1 + ... + \beta_kX_k + \varepsilon \qquad \textbf{(20.1)}$$

Since the coefficients β_0, β_1, ..., β_k are usually unknown, they are estimated on the basis of a sample of $(k + 1)$-tuples $(x_1, ..., x_k, y)$ that form a sample cloud. The least-squares equation $y = b_0 + b_1x_1 + ... + b_kx_k$ is the equation of the hyperplane that fits best through this sample cloud. The coefficients b_0, b_1, ..., b_k are estimates of the unknown coefficients β_0, β_1, ..., β_k of the population cloud.

The estimates b_0, b_1, ..., b_k of the unknown β_0, β_1, ..., β_k are the results of the application of the LS method to the sample of the $(k + 1)$-tuples $(x_1, ..., x_k, y)$. Apparently, the estima**tors** B_0, B_1, ..., B_k are formulae that depend on the sample of n random $(k + 1)$-tuples $(X_1, ..., X_k, Y)$. Although it is possible to write down those formulae, we will not do so. Just remember that they arise from the LS method. As before, the coefficients β_0, β_1, ..., β_k are called the ***population regression coefficients***; the LS estimates b_0, b_1, ..., b_k are called the ***sample regression coefficients***.

Least-squares (point) estimators of β_0, β_1, ..., β_k

The ***least-squares (point) estimators*** B_0, B_1, ..., B_k of the population regression coefficients β_0, β_1, ..., β_k are the unique functions of the n random $(k + 1)$-tuples $X_1, ..., X_k$, Y that arise from the LS method.

Example 20.1 (red-wire, part II)

See also Case 19.1. For a large population of people who work at least 32 hours per week, interest is in the variation of the variable $Y = W$ = 'gross hourly wage'. It is suspected that the variable X_1 = 'age of the person' will partially cause that variation. That is why the equation $y = \beta_0 + \beta_1x_1$ of the accompanying population regression line is estimated by way of the LS method, on the basis of a sample of 150 persons from the population; see <u>Xmp20–01.sav</u> for the data. The printout below gives estimation results.

ANOVA(b)

Model		Sum of Squares	df	Mean Square	F	Sig.
1	Regression	5722.780	1	5722.780	84.448	.000(a)
	Residual	10029.517	148	67.767		
	Total	15752.297	149			

a Predictors: (Constant), Age
b Dependent Variable: W

Coefficients(a)

Model		Unstandardized Coefficients		Standardized Coefficients	t	Sig.
		B	Std. Error	Beta		
1	(Constant)	3.068	2.145		1.430	.155
	Age	.565	.062	.603	9.190	.000

a Dependent Variable: W

The sample regression line has the equation $\hat{y} = 3.068 + 0.565x$. The variation before regression is measured by SST = 15752.297. Having done this simple linear regression, the amount 10029.517 remains unexplained. It follows that the proportion

$$\frac{\text{SST} - \text{SSE}}{\text{SST}} = \frac{5722.780}{15752.297} = 0.363$$

of the variation in the hourly wage data is explained by taking 'age' as independent variable.

To explain a larger part of the variation of the y-data, the variable X_2 = 'education level of the person' is included as a second independent variable. Hence, it is the population equation $y = \beta_0 + \beta_1 x_1 + \beta_2 x_2$ that has to be estimated with the LS method. Although the coefficients β_0 and β_1 of this equation are not the same as the coefficients of the above population regression **line**, we still use the same notation for them. Observations of X_2 for the 150 sampled persons are also given in the dataset. The SPSS printout is presented below.

ANOVA(b)

Model		Sum of Squares	df	Mean Square	F	Sig.
2	Regression	7672.184	2	3836.092	69.789	.000(a)
	Residual	8080.113	147	54.967		
	Total	15752.297	149			

a Predictors: (Constant), Edl, Age
b Dependent Variable: W

Coefficients(a)

Model		Unstandardized Coefficients		Standardized Coefficients	t	Sig.
		B	Std. Error	Beta		
2	(Constant)	-3.060	2.189		-1.398	.164
	Age	.478	.057	.510	8.352	.000
	Edl	3.254	.546	.364	5.955	.000

a Dependent Variable: W

The sample regression equation turns out to be: $\hat{y} = -3.060 + 0.478x_1 + 3.254x_2$. Notice especially that the new coefficients b_0 and b_1 are different from the former ones. For instance, b_1 decreases by the amount 0.087 from 0.565 to 0.478. Of course, the value of SST in the last printout is the same as the value of SST in the first printout. The value 8080.113 for SSE in the last printout is **smaller than** the value of SSE in the first printout. It would appear that including the new variable X_2 has decreased the unexplained variation from 10029.517 to 8080.113. Compared with the situation before regression, the proportion

$$\frac{15\,752.297 - 8080.113}{15\,752.297} = 0.487$$

of the variation in the hourly wage data is explained by taking 'age' **and** 'education level' as independent variables. The inclusion of the extra independent variable has **de**creased SSE and **in**creased the coefficient of determination.

So far there are no model requirements. However, to learn more about the precision of the LS estimators B_0, B_1, ..., B_k we do need assumptions similar to the assumptions in the simple linear regression case. Our starting point is that the $(k + 1)$-tuple $(X_1, ..., X_k, Y)$ is observed n times; the accompanying Y observations and error variables are again denoted by Y_1, ..., Y_n and ε_1, ..., ε_n, respectively.

As in Chapter 19, a part of the randomness is taken away by making the random variables $(X_1, ..., X_k)$ non-random. This is done by using the conditional method again, which in this more advanced situation means that the pre-information $D = \{X_1 = x_1; X_2 = x_2; ...; X_k = x_k\}$ is already known: it is already known that D has occurred, that the random variables X_1, ..., X_k have realisations x_1, ..., x_k. Under this condition, (20.1) turns into:

$$Y = \beta_0 + \beta_1 x_1 + ... + \beta_k x_k + \varepsilon$$

Hence, Y is random only because of the random variable ε at the right-hand side. Again, it will be assumed that the conditional expectation of ε (given that D has occurred) is equal to 0, whatever the realizations x_1, ..., x_k are. This means that the mean of all error terms for the subpopulation with $X_1 = x_1$, ..., $X_k = x_k$ is equal to 0, for all realizations x_1, ..., x_k of the independent variables. To put it another way, along all intersections of the population cloud with straight lines parallel to the y-axis, the dots have mean error 0 with respect to the least-squares hyperplane. Here is the mathematical way of describing this:

$$E(\varepsilon \mid D) = E(\varepsilon \mid x_1, ..., x_k) = 0 \quad \text{for all realizations } x_1, ..., x_k \qquad \textbf{(20.2)}$$

But more assumptions are needed about such intersections with straight lines. Along all intersections of the population cloud with straight lines parallel to the y-axis, the variance of the errors of the dots with respect to this least-squares hyperplane is always the same real number: σ_ε^2. That is:

$$V(\varepsilon \mid D) = V(\varepsilon \mid x_1, ..., x_k) = \sigma_\varepsilon^2 \quad \text{for all realizations } x_1, ..., x_k \qquad \textbf{(20.3)}$$

Furthermore, all histograms of the errors ε along such intersecting lines parallel to the y-axis have to look like a normal distribution. Jointly with (20.2) and (20.3), this can be written down as follows:

$$\varepsilon \sim N(0, \sigma_\varepsilon^2) \quad \text{conditionally on } X_1 = x_1, ..., X_k = x_k \qquad \textbf{(20.4)}$$

Also, some technical independence requirements (similar to the ones in Chapter 19) are needed. Here is the overview of all model requirements; jointly they form the multiple linear regression model.

Multiple linear regression model (for Y on X_1, ..., X_k)

The combination of the requirements

- $Y = \beta_0 + \beta_1 X_1 + ... + \beta_k X_k + \varepsilon$ and $E(\varepsilon \mid x_1, ..., x_k) = 0$ for all realizations x_1, ..., x_k of X_1, ..., X_k;
- $\varepsilon \sim N(0, \sigma_\varepsilon^2)$ conditionally on $X_1 = x_1$, ..., $X_k = x_k$ (for all x_1, ..., x_k);

■ random observations of ε are independent; and
■ random observations of ε are independent of all random observations of $X_1, ..., X_k$,

is called a ***multiple linear regression model*** (with the strong requirements) of Y on $X_1, ...,$ X_k.

The assumption $E(\varepsilon \mid x_1, ..., x_k) = 0$ jointly with $Y = \beta_0 + \beta_1 X_1 + ... + \beta_k X_k + \varepsilon$ is called the **basic assumption** of the model. The standard deviation σ_ε is called **standard deviation of the model**.

If it is already given that D has occurred, then the first model requirement yields that

$$Y = \beta_0 + \beta_1 x_1 + ... + \beta_k x_k + \varepsilon \quad \text{and} \quad E(\varepsilon \mid x_1, ..., x_k) = 0$$

which is the basic assumption of the model. That is, conditionally on D the random variable Y is the sum of the real number $\beta_0 + \beta_1 x_1 + ... + \beta_k x_k$ and the random variable ε. By the conditional normality assumption of ε it follows that Y is also conditionally normal with mean parameter and variance parameter equal, respectively, to:

$$E(Y \mid x_1, ..., x_k) = \beta_0 + \beta_1 x_1 + ... + \beta_k x_k \quad \text{and} \quad V(Y \mid x_1, ..., x_k) = \sigma_\varepsilon^2$$

Along each intersection of the population cloud with a straight line parallel to the y-axis, the mean of the y values of the dots falls precisely in the LS hyperplane of the population cloud; the accompanying variance of these y values is just the variance of ε. Table 20.1 gives an overview of all model requirements and their consequences.

Model requirements	Consequences for Y
$\varepsilon_1, ..., \varepsilon_n$ are independent	
Each ε_i is independent of all n random observations of $(X_1, ..., X_k)$	Conditionally on all sample observations of $(X_1, ..., X_k)$ it holds that $Y_1, ..., Y_n$ are independent
It holds for all $x_1, ..., x_k$ that conditionally on $X_1 = x_1, ..., X_k = x_k$ we have $\varepsilon \sim N(0, \sigma_\varepsilon^2)$	It holds for all $x_1, ..., x_k$ that conditionally on $X_1 = x_1, ..., X_k = x_k$ we have $Y \sim N(\beta_0 + \beta_1 x_1 + ... + \beta_k x_k, \sigma_\varepsilon^2)$

TABLE 20.1 Model assumptions and their consequences for Y

In particular, the basic assumption of the model can equivalently be written as:

$$E(Y \mid x_1, ..., x_k) = \beta_0 + \beta_1 x_1 + ... + \beta_k x_k \tag{20.5}$$

It is this assumption that yields the interpretations of the population regression coefficients $\beta_1, ..., \beta_k$. Below, only the interpretation of β_1 is discussed.

Consider two population elements – say, A and B – that have the same values $x_2, ..., x_k$ for the independent variables $X_2, ..., X_k$ but B's value of X_1 is x_1 and A's value of X_1 is 1 more. (It is said that A and B **have the same characteristics** as far as $X_2, ..., X_k$ are concerned.) We compare their expected values for Y, called, respectively, μ_A and μ_B:

for A: $\mu_A = E(Y \mid x_1 + 1, x_2, ..., x_k) = \beta_0 + \beta_1 + \beta_1 x_1 + ... + \beta_k x_k$

for B: $\mu_B = E(Y \mid x_1, x_2, ..., x_k) \quad = \beta_0 \qquad + \beta_1 x_1 + ... + \beta_k x_k$

$$\overline{\mu_A - \mu_B \qquad\qquad\qquad = \beta_1}$$

It follows that β_1 is the difference between the expectations of Y for two population elements A and B that have the same characteristics for $X_2, ..., X_k$ but the value of X_1 for A is 1 more than the value of X_1 for B. To put it another way: if x_1 increases by 1, then, ***ceteris paribus*** (i.e. leaving the other variables unchanged), the expectation of Y changes by β_1. Notice the importance of this ceteris paribus condition for the above subtraction: because of this condition, the term $\beta_2 x_2 + \ldots + \beta_k x_k$ drops out.

As in the simple regression case, we will not always use the neat (but complicated) conditional expectations to write things down. We will often (not always) follow the common habit of writing something unconditionally although it is meant conditionally. For instance, the conditional result in (20.5) is often written as:

$$E(Y) = \beta_0 + \beta_1 x_1 + \ldots + \beta_k x_k$$

Example 20.2

To explain the height Y (in cm) of an adult man, several factors will be of interest. These factors may have to do with height of father, height of mother, heights of grandmothers and grandfathers, the extent of playing sports, food habits, etc. Here are some variables that might be important to explain the variation in the heights of adult men:

X_1 : height (cm) of father
X_2 : height (cm) of mother
X_3 : average height (cm) of the two grandfathers
X_4 : average height (cm) of the two grandmothers
X_5 : mean number of hours sports per week during the ages 12–18

An interesting simple linear regression model that can be a starting point for more advanced research is the model with the following basic assumption:

$$Y = \beta_0 + \beta_1 x_1 + \varepsilon \quad \text{with} \quad E(\varepsilon) = 0$$

This model has only one independent variable; the variation due to other factors is included in the error variable ε. The interpretation of the population regression coefficient β_1 is as follows: if the height of the father becomes 1 cm more, then the height of the son will on average change by β_1. The intercept β_0 cannot be interpreted.

Another interesting model is the linear regression model with independent variables X_1 and X_2, with the following basic assumption:

$$E(Y) = \beta_0 + \beta_1 x_1 + \beta_2 x_2$$

(Notice that the choice is made here to write the basic assumption in expectation form; since $E(\varepsilon) = 0$, the error variable has disappeared.)

The general model based on **all** variables proposed above has the following basic assumption:

$$E(Y) = \beta_0 + \beta_1 x_1 + \beta_2 x_2 + \beta_3 x_3 + \beta_4 x_4 + \beta_5 x_5$$

This model has $k = 5$ independent variables and 6 regression coefficients. Although the notation β_i for some of the population regression coefficients in the last model was also used in previous models, their (unknown) values will probably be different. The interpretation of β_1 in the last model is as follows:

β_1 is the average change in the height of the son if the height of the father is 1 cm more while the mean number of hours sports, the height of the mother, the average heights of the grandfathers and the average heights of the grandmothers remain unchanged.

With the definition of the multiple linear regression model, another parameter entered the stage: the variance σ_ε^2 of the error variable ε. This parameter also has to be estimated on the basis of the n sample observations of $(X_1, \ldots, X_k, Y)$. This will be discussed now.

Recall from (20.3) that σ_ε^2 measures the variation in the population cloud along all straight lines parallel to the y-axis. That is why it will be estimated by a measure for the variation of the **sample** cloud in the direction of the y-axis. By substituting the observations $x_1, \ldots, x_k$ of the ith sampled element into the sample regression equation $\hat{y} = b_0 + b_1 x_1 + \ldots + b_k x_k$, the prediction $\hat{y}_i$ and the accompanying residual $e_i = y_i - \hat{y}_i$ arises. So, the ith dot in the sample cloud yields the deviation e_i in the direction of the y-axis. But then

$$SSE = \sum_{i=1}^{n} e_i^2 = \sum_{i=1}^{n} (y_i - \hat{y}_i)^2$$

measures the overall variation in the sample cloud. Again, it is not SSE that is used to estimate σ_ε^2 but a 'mean' of the squared deviations. The box below defines the estimators of σ_ε^2 and σ_ε. Being estimators, they are based on the **random** variables Y_i and $\hat{Y}_i$, written as **capitals**.

Estimator of σ_ε^2

$$S_\varepsilon^2 = \frac{1}{n - (k+1)} \sum_{i=1}^{n} (Y_i - \hat{Y}_i)^2 = \frac{SSE}{n - (k+1)} \qquad (20.6)$$

Estimator of σ_ε

$$S_\varepsilon = \sqrt{S_\varepsilon^2} = \sqrt{\frac{1}{n - (k+1)} \sum_{i=1}^{n} (Y_i - \hat{Y}_i)^2} \qquad (20.7)$$

In the present **multiple** linear regression setting, the $k+1$ parameters $\beta_0, \beta_1, \ldots, \beta_k$ had to be estimated first to obtain the n building blocks $Y_1 - \hat{Y}_1, \ldots, Y_n - \hat{Y}_n$ of SSE, leading to a loss of $k+1$ degrees of freedom. That is, only $n - (k+1)$ degrees of freedom are left, the difference between n and the number of β parameters in the model that had to be estimated. By dividing SSE by $n - (k+1)$, the resulting random variable becomes unbiased; see also the following section.

Again, S_ε and S_ε^2 are called the **standard error** and the **sample variance of the estimated model**, respectively.

Example 20.3 (red-wire, part III)
Reconsider Example 20.1. We adopt the multiple linear regression model with basic assumption:

$$E(Y) = \beta_0 + \beta_1 x_1 + \beta_2 x_2$$

This model has $k = 2$ independent variables, and $k + 1 = 3$ regression coefficients have to be estimated. That is why the sample variance of the estimated model is equal to:

$$S_\varepsilon^2 = \frac{1}{n-3} \sum_{i=1}^{n} (Y_i - \hat{Y}_i)^2$$

The estimation is done on the basis of a sample of $n = 150$ persons from the population. A part of the printout was considered in the second half of Example 20.1. Below, another part is shown.

Model Summary

Model	R	R Square	Adjusted R Square	Std. Error of the Estimate
2	.698(a)	.487	.480	7.41396

a   Predictors: (Constant), Edl, Age

It follows that s_ε, the standard error of the estimated model, is equal to 7.41396 and hence $s_\varepsilon^2 = 54.9668$.

20.2 Properties of the point estimators

This section gives an overview of the properties of the estimators $B_0, B_1, ..., B_k, S_\varepsilon^2$ and S_ε. These properties will turn out to be in line with simple linear regression.

Thanks to the strong assumptions of the multiple linear regression model, the least-squares estimators $B_0, B_1, ..., B_k$ are – just as in the previous chapter – unbiased, consistent and conditionally normally distributed (given the x data, the sample observations of all independent variables). Hence, the mean parameter of the normal distribution of B_i is β_i. The variance parameter, however, is less tractable than previously. That is why we will not give its formula here; we just state that it again depends on σ_ε^2 and on the x data. The variance parameter will shortly (and formally) be written as $V(B_i)$ or as $\sigma_{B_i}^2$; its square root will be denoted by $SD(B)$ or σ_{B_i}.

Because of the normality property, the standardized version of B_i is conditionally **standard** normally distributed and hence it is a **pivot**. That is why this z-score will be used to create interval estimators and test statistics; see Section 20.5.

Also the estimator S_ε^2 of σ_ε^2 is unbiased and consistent. The estimator S_ε of σ_ε is not unbiased, but it **is** consistent.

The box below summarizes the above properties.

Properties of the consistent estimators $B_0, B_1, ..., B_k, S_\varepsilon^2$ and S_ε

Conditionally given the x data, it holds that:

$$E(B_i) = \beta_i \quad \text{and} \quad B_i \text{ is a normal random variable for all } i = 0, 1, ..., k \tag{20.8}$$

$$\frac{B_i - \beta_i}{SD(B_i)} \sim N(0, 1) \quad \text{for all } i = 0, 1, \cdots, k \tag{20.9}$$

$$E(S_\varepsilon^2) = \sigma_\varepsilon^2 \quad \text{and} \quad E(S_\varepsilon) < \sigma_\varepsilon \tag{20.10}$$

The last property states that S_ε on average underestimates σ_ε.

[The proof is elegant and short but tricky, and is given here only for devotees. First note that the variance of S_ε is positive (see also (8.21)). Next, use the short-cut formula (8.32) for the variance of S_ε and the fact that S_ε^2 is unbiased:

$$0 < V(S_\varepsilon) = E(S_\varepsilon^2) - (E(S_\varepsilon))^2 = \sigma_\varepsilon^2 - (E(S_\varepsilon))^2$$

It follows that $(E(S_\varepsilon))^2 < \sigma_\varepsilon^2$ and hence (by taking square roots) that $E(S_\varepsilon) < \sigma_\varepsilon$.]

Since $SD(B_i)$ is unknown because of the fact that it depends on σ_ε, it cannot be measured. That is why we will have to replace σ_ε by the estimator S_ε. But then the standard deviation $SD(B_i)$ changes and turns into the standard error $SE(B_i)$; see also Section 14.3. When the data are observed and the regression is run with the computer, then the realization of $SE(B_i)$ can be read from the printout.

Example 20.4

The dataset Xmp20-04.sav contains, for a sample of 11 adult sons, the observations of the following variables; see also Example 20.2:

Y : height (cm) of the son
X_1: height (cm) of his father
X_2: height (cm) of his mother
X_5: mean number of hours sports per week during the ages 12–18

To explain Y from X_1, X_2 and X_5, the linear regression model with basic assumption

$$E(Y) = \beta_0 + \beta_1 x_1 + \beta_2 x_2 + \beta_3 x_5$$

is estimated on the basis of the sample. Part of the printout is presented below.

Coefficients(a)

Model		Unstandardized Coefficients		Standardized Coefficients	t	Sig.
		B	**Std. Error**	**Beta**		
1	(Constant)	**5.864**	35.063		.167	.872
	X1	**.772**	.163	1.092	4.739	.002
	X2	**.223**	.089	.394	2.504	.041
	X5	**-.275**	.205	-.308	-1.346	.220

a   Dependent Variable: Y

The regression equation is $\hat{y} = 5.864 + 0.772x_1 + 0.223x_2 - 0.275x_5$. For instance, the coefficient 0.223 is an estimate of β_2. It informs us that an increase of 1 cm in the height of the mother while leaving the other independent variables unchanged, will on average lead to an increase of 0.223 cm in the height of the son.

At first sight, the negative coefficient of X_5 seems to be strange: spending more time on sports would, at least on average, not stimulate the growth of the son. But in Section 20.5 we will see that the variable X_5 is not even significant within the model, so the interpretation of its sample coefficient is not valid.

The column with the heading Std. Error gives the realized standard errors of the least-squares estimators. For instance, the realized standard error of B_1 is 0.163, which is rather small when compared with the realization of B_1 itself.

20.3 ANOVA table

The objective of regression is to explain the variation of Y. It can be shown that the introduction of an extra independent variable into the linear regression model will always reduce the unexplained variation. As in the simple linear regression case, progress is again recorded in an ANOVA table.

Before regression, the (unexplained) variation of the y data is measured by SST. Having done a (possibly multiple) linear regression, the unexplained variation has decreased from SST to SSE, by the amount SSR = SST − SSE. Compared with the simple linear regression model, the definitions and some of the results for these sums of squares are formally unchanged. They are repeated here:

$$\text{SST} = \sum_{i=1}^{n} (y_i - \bar{y})^2; \quad \text{SSE} = \sum_{i=1}^{n} (y_i - \hat{y}_i)^2; \quad \text{SSR} = \sum_{i=1}^{n} (\hat{y}_i - \bar{y})^2$$

Note, however, that the predictions $\hat{y}_i$ in the multiple regression case arise by substituting for the ith sampled element the data of the independent variables into the **multiple** linear regression equation.

Also for the multiple linear regression case the ANOVA table is important; see Table 20.2.

	Sum of squares	Degrees of freedom	Mean square	F-ratio
Regression	$SSR = \sum_{i=1}^{n} (\hat{y}_i - \bar{y})^2$	k	$\dfrac{SSR}{k}$	$\dfrac{SSR/k}{SSE/(n-(k+1))}$
Residual	$SSE = \sum_{i=1}^{n} (y_i - \hat{y}_i)^2$	$n-(k+1)$	$\dfrac{SSE}{n-(k+1)} = s_\varepsilon^2$	
Total	$SST = \sum_{i=1}^{n} (y_i - \bar{y})^2$	$n-1$	$\dfrac{SST}{n-1} = s_Y^2$	

TABLE 20.2 ANOVA table for multiple linear regression ($k \geq 1$)

As before, the first two positions in columns 2 and 3 add up to give the third position. Starting with the original n degrees of freedom, the estimation of the population coefficients $\beta_0, \beta_1, \dots, \beta_k$ costs $k + 1$ degrees of freedom, so $n - (k + 1)$ are left. (Notice that in the before-regression situation, only 1 degree of freedom was lost because of the estimation of the mean of Y by the sample mean $\bar{y}$.)

The general ANOVA table ($k \geq 1$) usually includes a column that has not yet been discussed. This column contains the **F-ratio**, whose meaning and importance are the topic of Section 20.4.

Example 20.5

The sales of a product will depend not only on the price of the product itself but also on the prices of products that can be used as substitutes. For instance, tea is traditionally a substitute for coffee; if coffee becomes too expensive, consumers tend to drink more tea. If Y is the quarterly sales of coffee, and X_1 and X_2 are the prices of coffee and tea, respectively, in that quarter, then interest is in the following basic assumption:

$$E(Y) = \beta_0 + \beta_1 x_1 + \beta_2 x_2$$

The table below contains results for seven arbitrary quarters; see also Xmp20-05.sav.

Quarter i	1	2	3	4	5	6	7
Sales y_i	23	21	20	21	20	18	24
Price coffee x_{1i}	16	18	19	14	17	19	16
Price tea x_{2i}	15	12	12	13	14	10	15

Table 20.3 Quarterly data for coffee sales and the prices of coffee and tea

Running the regression with SPSS gives the (partial) printout shown below.

ANOVA(b)

Model		Sum of Squares	df	Mean Square	F	Sig.
1	Regression	18.087	2	9.043	6.117	.061(a)
	Residual	5.913	4	1.478		
	Total	24.000	6			

a Predictors: (Constant), x2, x1
b Dependent Variable: y

Coefficients(a)

Model		Unstandardized Coefficients		Standardized Coefficients	t	Sig.
		B	Std. Error	Beta		
1	(Constant)	10.082	9.778		1.031	.361
	x1	−.056	.358	−.051	−.157	.883
	x2	.913	.358	.834	2.553	.063

a   Dependent Variable: y

The sample regression equation is $\hat{y} = 10.082 - 0.056x_1 + 0.913x_2$. The ANOVA table in the printout can now be filled in; see Table 20.4.

	Sum of squares	Degrees of freedom	Mean square	F-ratio
Regression	SSR = 18.087	2	9.043	6.117
Residual	SSE = 5.913	4	1.478	
Total	SST = 24.000	6	4.000	

Table 20.4 ANOVA table for Example 20.5

Before regression, the unexplained variation in the y data was 24.000; after regression 5.913 is still unexplained. Hence, this regression reduced the unexplained variation by 18.087. It also follows that the sample variance of the estimated model is 1.478, so the standard error of the estimated model is $\sqrt{1.478} = 1.216$.

20.4 Usefulness of the model

The first step after having chosen the independent variables of the linear regression model is to find out whether the model makes sense statistically. To decide that, an F-test will be introduced. If the test concludes that the model has some usefulness, the **degree** of usefulness is also wanted.

Suppose that we have decided to use the variables $X_1, ..., X_k$ as independent variables for a linear regression meant to explain the variation of the research variable Y. Then the basic assumption

$$E(Y) = \beta_0 + \beta_1 x_1 + ... + \beta_k x_k \tag{20.11}$$

represents the model. The question arises how the data can be used to decide the usefulness of the model.

If all independent variables are useless in this model, then changing any of the $x_1, ..., x_k$ should not have any influence on the expectation $E(Y)$. But this can only happen if all the coefficients of $x_1, ..., x_k$ in the basic assumption are 0, so if $\beta_1 = \beta_2 = ... = \beta_k = 0$. It would appear that uselessness of the model is equivalent to all $\beta_1, ..., \beta_k$ being 0. On the other hand, if at least one of the independent variables is useful within the model (say X_j is), then it follows that changing x_j in the basic assumption will lead to a change in $E(Y)$, and hence the coefficient β_j has to be unequal to 0. But this means that usefulness of the model is equivalent to at least one of the population coefficients $\beta_1, ..., \beta_k$ being **unequal** to 0. The hypothesis 'the model is useless' is equivalent to '$\beta_1 = ... = \beta_k = 0$'; the hypothesis 'the model is useful' is equivalent to 'at least one of the $\beta_1, ..., \beta_k$ is unequal to 0'. Notice that both hypotheses are about $\beta_1, ..., \beta_k$, not about the intercept β_0.

20.4.1 Testing the usefulness of the model

The above arguments explain why a test is wanted for the following testing problem:

   i Test $H_0: \beta_1 = \ldots = \beta_k = 0$ against H_1: at least one of the $\beta_1, \ldots, \beta_k$ is $\neq 0$

The null hypothesis states that the model is useless. If it is valid, it is expected that running the regression will hardly change the unexplained variation; the variation that is unexplained after this regression (measured by SSE) will be almost the same as the variation before regression (measured by SST). The alternative hypothesis states that the model is **useful**. If this hypothesis is valid, it is expected that running the regression really will reduce the variation in the y data; SSE really will be smaller than SST.

The above considerations indicate that the ratio

$$\frac{SST - SSE}{SSE} = \frac{SSR}{SSE} \tag{20.12}$$

can help to solve this useless/useful problem. Both its numerator and its denominator are sums of squared deviations and hence are both non-negative. If the model is useless, SSE will hardly be different from SST and the above ratio is close to 0. But if the model is useful, then SSE will be smaller than SST and the ratio is positive and really away from 0. It would appear that relatively large values of the ratio correspond with the validity of H_1. That is why the test statistic for testing problem (i) will be based on (20.12).

Firstly, (20.12) is rewritten as the random variable

$$\frac{\sum_{i=1}^{n}(\hat{Y}_i - \bar{Y})^2}{\sum_{i=1}^{n}(Y_i - \hat{Y}_i)^2}, \tag{20.13}$$

a ratio of sums of squared deviations. It is a theoretical fact that the probability distributions of such ratios are often related to the family of F-distributions; see Section 18.3.1. According to the remark at the end of that section, both the numerator and the denominator have to be divided by the respective numbers of degrees of freedom, respectively k and $n - (k + 1)$. The following random variable (called F) arises:

$$F = \frac{\frac{1}{k}\sum_{i=1}^{n}(\hat{Y}_i - \bar{Y})^2}{\frac{1}{n-(k+1)}\sum_{i=1}^{n}(Y_i - \hat{Y}_i)^2}$$

Unfortunately, the probability distribution of F turns out to depend on the unknown parameters $\beta_0, \beta_1, \ldots, \beta_k$ and σ_ε^2. However, it can be proved that F is $F_{k,n-(k+1)}$-distributed **if all $\beta_1, \ldots, \beta_k$ are 0**. Hence, if H_0 is valid, then the rv F has an F-distribution with, respectively, k degrees of freedom because of the numerator and $n - (k + 1)$ degrees of freedom because of the denominator:

     If H_0 is valid, then $F \sim F_{k,n-(k+1)}$ (20.14)

That is why F is taken as test statistic. Usually, F is not written in the above, rather formal way. Instead, the following 'looser' expression is generally used for the test statistic of the testing problem (i). It is often denoted as the **f-ratio**:

   ii Test statistic: $F = \dfrac{(SST - SSE)/k}{SSE/(n - (k + 1))}$

It follows from above arguments that H_0 has to be rejected – and hence H_1 accepted – if the realization of F is relatively large, so SST $-$ SSE is seriously away from 0. The only question is: what is meant by 'relatively large'? Since F is $F_{k,n-(k+1)}$-distributed if H_0 is true, it follows that:

If H_0 is true, then $P(F \geq F_{\alpha;k,n-(k+1)}) = \alpha$

So, if the test procedure is such that H_0 is rejected for realizations $\geq F_{\alpha;k,n-(k+1)}$, then the probability that the procedure rejects H_0 incorrectly (concludes that the model is useful when it is not) is α. This determines the rejection region as $[F_{\alpha;k,n-(k+1)}, \infty)$:

iii Reject $H_0 \Leftrightarrow f \geq F_{\alpha;k,n-(k+1)}$

(Here, f denotes a possible value of the test statistic F.) But this completes the test procedure. When the data are known, the **val** can be calculated (step (iv)) and it can be checked whether it falls in the rejection region or not (step (v)). Here is the complete procedure of this **model test**:

Five-step procedure of the *model test for usefulness*:

i Test $H_0: \beta_1 = \ldots = \beta_k = 0$ against H_1: at least one of the $\beta_1, \ldots, \beta_k$ is $\neq 0$

ii Test statistic: $F = \dfrac{(\text{SST} - \text{SSE})/k}{\text{SSE}/(n - (k + 1))}$

iii Reject $H_0 \Leftrightarrow f \geq F_{\alpha;k,n-(k+1)}$

iv Calculate **val**, the realization of F

v Draw the conclusion by confronting **val** to step (iii)

Note that this test is one-sided although the testing problem looks two-sided. The test allows rejection of the null hypothesis only for large values of F. Indeed, for relatively **small** values of F (so, positive values close to 0) the null hypothesis must **not** be rejected since such small values imply that SSE and SST are almost equal.

The p-value provides some additional information for this test:

$$p\text{-value} = P(F \geq \textbf{\textit{val}}) \tag{20.15}$$

Having observed the **val** of the test, the p-value gives the smallest significance level α that would have given the statistical conclusion that the model is useful.

Example 20.6 (red-wire, part IV)

To explain the hourly wage Y from $X_1 = $ 'age' and $X_2 = $ 'education level', we adopted the linear regression model with basic assumption:

$$E(Y) = \beta_0 + \beta_1 x_1 + \beta_2 x_2$$

The model is estimated on the basis of a sample of 150 persons. According to the printout in Example 20.1, the following holds:

$$\hat{y} = -3.060 + 0.478x_1 + 3.254x_2; \quad \text{SST} = 15752.297; \quad \text{SSE} = 8080.113$$

Before interpreting the regression coefficients and the sums of squared deviations, we first check whether the model is useful. As significance level we take 0.05. The five-step procedure is as follows:

i Test $H_0: \beta_1 = \beta_2 = 0$ against H_1: at least one of the β_1, β_2 is $\neq 0$

ii Test statistic: $F = \dfrac{(\text{SST} - \text{SSE})/2}{\text{SSE}/(n - 3)}$

iii Reject $H_0 \Leftrightarrow f \geq F_{\alpha;2,n-3} = F_{0.05;2,147} = 3.0576$ (*)

iv $\textbf{\textit{val}} = \dfrac{(15752.297 - 8080.113)/2}{8080.113/147} = \dfrac{3836.0920}{54.9668} = 69.7893$

v Since 69.7893 is larger than the critical value 3.0576, H_0 is rejected

We conclude that the model is useful; at least one of the two independent variables helps to explain the variation of Y. But the *val* of this test is already available (under the heading F) in the ANOVA table of Example 20.1, which is repeated below. Notice that the numerator and denominator of the ratio in step (iv) are also present, under the heading Mean Square.

ANOVA(b)

Model		Sum of Squares	df	Mean Square	F	Sig.
2	Regression	7672.184	2	3836.092	69.789	.000(a)
	Residual	8080.113	147	54.967		
	Total	15752.297	149			

a   Predictors: (Constant), Edl, Age
b   Dependent Variable: W

Also the (first three decimals of the) *p*-value of the test is included, under the heading Sig. Next, we interpret the sample regression coefficients b_1, b_2 and b_0:

- If the age is 1 more but the education level is unchanged, then the hourly wage will on average be 47.8 cents more;
- One step up in education level with age unchanged causes on average an increase of €3.254 in hourly wage.
- The value −3.060 of b_0 cannot be interpreted since the pair of values (0, 0) for (x_1, x_2) is not included in the dataset.

By running this regression, the unexplained variation of the hourly wage data decreases by the amount SSR = 7672.184, from SST = 15752.297 to SSE = 8080.113. This corresponds to the proportional decrease:

$$\frac{SST - SSE}{SST} = \frac{SSR}{SST} = \frac{7672.184}{15752.297} = 0.4871$$

It follows that 48.71% of the variation of the hourly wage data is explained by the model with independent variables 'age' and 'educational level'.

20.4.2 Equivalence of the *F*-test and the (hinge 0) two-sided *t*-test if $k = 1$

Suppose now that $k = 1$. In that case (the case of **simple** linear regression) the above *F*-test for checking the usefulness can still be used. However, in Section 19.5.3 we discussed another test to decide the usefulness of a simple linear regression model. Does this mean that there are two tests that may lead to different conclusions for a common significance level?

The answer is no, since the two tests are equivalent. The test procedure of Section 19.5.3 is a two-sided *t*-test (based on a *t*-distribution), with the following steps (ii) and (iii):

ii $T = \dfrac{B_1}{S_\varepsilon / \sqrt{(n-1)s_X^2}}$ and   iii   Reject $H_0 \Leftrightarrow t \le -t_{\alpha/2;n-2}$ or $t \ge t_{\alpha/2;n-2}$

The $k = 1$ version of the test procedure of Section 20.4.1 is an *F*-test, with the steps:

ii $F = \dfrac{SSR/1}{SSE/(n-2)}$ and   iii   Reject $H_0 \Leftrightarrow f \ge F_{\alpha;1,n-2}$

However, F is equal to T^2. This follows from (19.22):

$$F = \frac{SSR}{SSE/(n-2)} = \frac{SSR}{S_\varepsilon^2} = \frac{B_1^2(n-1)s_X^2}{S_\varepsilon^2} = T^2$$

As a consequence,

$$\alpha = P(T \le -t_{\alpha/2;n-2} \text{ or } T \ge t_{\alpha/2;n-2}) = P(|T| \ge t_{\alpha/2;n-2}) = P(T^2 \ge (t_{\alpha/2;n-2})^2)$$

and

$$\alpha = P(F \ge F_{\alpha;1,n-2})$$

Since $F = T^2$, it follows that $F_{\alpha;1,n-2} = (t_{\alpha/2;n-2})^2$. But this means that the t-test rejects H_0 if and only if:

$$t \le -t_{\alpha/2;n-2} \text{ or } t \ge t_{\alpha/2;n-2} \Leftrightarrow |t| \ge t_{\alpha/2;n-2} \Leftrightarrow t^2 \ge F_{\alpha;1,n-2} \Leftrightarrow f \ge F_{\alpha;1,n-2}$$

The realization of the test statistic T falls in the rejection region of the t-test if and only if the squared realization falls in the rejection region of the F-test, which in turn happens if and only if the realization of the test statistic F falls in the rejection region of the F-test. Although the two test procedures have different test statistics, they always will yield the same statistical conclusions.

20.4.3 Coefficients of determination

Go back to the general case that $k \ge 1$. Having concluded that the model is useful, the next step is to measure the **degree** of usefulness. As illustrated in Example 20.6, the degree of usefulness is the proportional reduction of the unexplained variation realized by the model. As in Section 19.6, it is called the coefficient of determination.

Coefficient of determination

The *coefficient of determination* (notation r^2) of the estimated model is defined as:

$$r^2 = \frac{SSR}{SST} = \frac{SST - SSE}{SST} = 1 - \frac{SSE}{SST}$$

Adjusted coefficient of determination

$$r_{adj}^2 = 1 - \frac{SSE/(n - (k + 1))}{SST/(n - 1)} = 1 - \frac{s_\varepsilon^2}{s_Y^2} = 1 - \frac{SSE}{SST} \times \frac{n - 1}{n - (k + 1)}$$

The **adjusted** r^2 arises from (the ordinary) r^2 by dividing the terms in the ratio SSE/SST by the respective numbers of degrees of freedom. The reason for doing that is that r^2 may be unrealistically high in cases where k is large compared with the sample size n. By the adjustment the coefficient becomes smaller, to avoid a wrong impression. If n is large relative to k, then r_{adj}^2 and r^2 are almost the same and they give more or less the same information about the proportion of the variation of Y explained by the model. But if k is large relative to n, then r_{adj}^2 really is smaller than r^2 and serves as a warning for not being too optimistic.

The adjusted r^2 is of some importance for the process of building a model. By including an extra independent variable in the model, the ordinary r^2 will always increase, even if the new variable does not really make the model better. The **adjusted** r^2 does not always increase if an extra variable is added. A decrease might be a reason to have doubts about the usefulness of this new variable.

If $k > 1$, then r^2 is – in contrast with the simple linear regression case – **not** equal to the square of the correlation coefficient of 'x data' and y data. This is because there are now at least two independent variables.

Example 20.7 (red-wire, part V)

Reconsider the previous example. The part of the SPSS printout that is given below is about the degree of usefulness of the model.

Model Summary

Model	R	R Square	Adjusted R Square	Std. Error of the Estimate
2	.698(a)	.487	.480	7.41396

a   Predictors: (Constant), Edl, Age

The ordinary r^2 and the adjusted r^2 are close to each other: about 48% of the variation in the y data is explained by the model.

20.4.4 Alternative expression for the test statistic

When looking at the expressions of the test statistic F and the coefficient of determination r^2, you will get the feeling that they are somehow mathematically related. Indeed, F can be expressed in terms of the random variable R^2 that belongs to realizations r^2. To see that, notice that the random part of F is:

$$\frac{SST - SSE}{SSE} = \frac{SST - SSE}{SST} \times \frac{SST}{SSE} = \frac{SST - SSE}{SST} \bigg/ \frac{SSE}{SST} = R^2/(1 - R^2) = \frac{R^2}{1 - R^2}$$

It follows that:

$$F = \frac{R^2/k}{(1 - R^2)/(n - (k + 1))} \tag{20.16}$$

20.5 Inference about the individual regression coefficients

When it is concluded that the model is useful and hence that at least one of the independent variables in the model makes sense, then it may be required to **find** those useful independent variables. In this section, interval estimators and t-tests will be considered for an individual regression coefficient β_i (for $i = 1, \ldots, k$). Again, we will follow steps A–D of Section 17.1.

Recall from Section 20.2 that the LS estimator B_i is a good estimator of β_i (step A) since it is unbiased and consistent, and that it can (by (20.9)) easily be transformed into a pivot (step B):

$$Z = \frac{B_i - \beta_i}{SD(B_i)} \sim N(0, 1)$$

The pivot Z and its distribution will be used to construct an interval estimator and a test statistic.

20.5.1 Interval estimator for β_i

$SD(B_i)$, the standard deviation of B_i, depends on the unknown parameter σ_ε^2. For statistical inference, the factor σ_ε^2 in $SD(B_i)$ has to be replaced by its estimator S_ε^2. This converts $SD(B_i)$ into the standard error $SE(B_i)$, Z into T, and $N(0, 1)$ into $t_{n-(k+1)}$. This $t_{n-(k+1)}$-distributed random variable T leads immediately to an interval estimator; see also (19.18) and (19.21).

Basic results for inferential statistics about β_i

If the multiple linear regression model-requirements hold, then:

$$Z = \frac{B_i - \beta_i}{SD(B_i)} \sim N(0, 1) \quad \text{and} \quad T = \frac{B_i - \beta_i}{SE(B_i)} \sim t_{n-(k+1)} \tag{20.17}$$

$(1 - \alpha)$ interval estimator (L, U) for β_i

$$L = B_i - t_{\alpha/2;n-(k+1)}SE(B_i) \quad \text{and} \quad U = B_i + t_{\alpha/2;n-(k+1)}SE(B_i) \qquad (20.18)$$

$SD(B_i)$ and $SE(B_i)$ are also written as σ_{B_i} and S_{B_i}. This interval estimator (step C) captures the unknown β_i with probability $1 - \alpha$; it is of format 1 (see Section 15.2). When the data are given, they can be plugged into the formulae of L and U to yield a $(1 - \alpha)$ **confidence interval** for β_i.

20.5.2 Hypothesis tests for β_i

The pivot Z can also be used to construct a test statistic for each of the following three testing problems with hinge b:

a Test $H_0: \beta_i \leq b$ against $H_1: \beta_i > b$ (one-sided, upper-tailed).
b Test $H_0: \beta_i \geq b$ against $H_1: \beta_i < b$ (one-sided, lower-tailed).
c Test $H_0: \beta_i = b$ against $H_1: \beta_i \neq b$ (two-sided).

According to the guidelines for step D in Section 17.1, the pivot Z first has to be adapted according to the hinge and next the unknown σ_ε^2 has to be replaced by its estimator S_ε^2. Here is the result:

Test statistic: $T = \dfrac{B_i - b}{SE(B_i)}$

If b is the true β_i, then T has a $t_{n-(k+1)}$-distribution. This result is basic for the five-step procedures.

Five-step procedures for testing H_0 against H_1

i Testing problems: (a) test $H_0: \beta_i \leq b$ against $H_1: \beta_i > b$
  (b) test $H_0: \beta_i \geq b$ against $H_1: \beta_i < b$
  (c) test $H_0: \beta_i = b$ against $H_1: \beta_i \neq b$

ii Test statistic: $T = \dfrac{B_i - b}{SE(B_i)}$

iii Rejection: (a) reject $H_0 \Leftrightarrow t \geq t_{\alpha;n-(k+1)}$
  (b) reject $H_0 \Leftrightarrow t \leq -t_{\alpha;n-(k+1)}$
  (c) reject $H_0 \Leftrightarrow t \leq -t_{\alpha/2;n-(k+1)}$ or $t \geq t_{\alpha/2;n-(k+1)}$

iv Calculate **val**, the realisation of T

v Draw the conclusion by confronting **val** to step (iii)

Requirement: multiple linear regression model (with strong requirements)

Remark. The procedures are based on the $t_{n-(k+1)}$-distribution. So, for $k > 1$ they are **not** based on the t_{n-2}-distribution! This is because $SE(B_i)$ depends on S_ε^2 which has $n - (k + 1)$ degrees of freedom.

20.5.3 A test for the individual significance of an independent variable

The two-sided testing problem with hinge $b = 0$, deserves special attention:

Test $H_0: \beta_i = 0$ against $H_1: \beta_i \neq 0$

To put emphasis on the ith independent variable, we write the basic assumption of the model as follows:

$$E(Y) = \beta_0 + \beta_1 x_1 + \ldots + \beta_i x_i + \ldots + \beta_k x_k$$

If H_0 is true, then the value x_i does not influence $E(Y)$ since $0 x_i = 0$. Hence, if H_0 is true, the variable X_i is useless within the model. But if H_1 is true, then the term $\beta_i x_i$ in the basic assumption changes when x_i changes. Hence, if H_1 is true, the variable X_i is use**ful** within the model. So, H_1 states that the independent variable X_i is useful (**significant**) within the model and H_0 states that X_i is useless (**insignificant**). It would appear that t-tests can be used to conclude for each $i = 1, \ldots, k$ whether X_i is significant for the model. But be aware that this conclusion will depend on the significance level α that is fixed in advance. To decide the significance of the individual independent variables, the p-values of the t-tests are valuable.

To emphasize the fact that the above t-tests for the hinge 0 case draw conclusions about the **individual** significance of one independent variable within the model, we will occasionally talk about the **individual significance** or **individual usefulness** of that variable. In this context the conclusion (from the F-test) that the model is useful is often expressed as: the independent variables are **jointly useful**.

Example 20.8 (red-wire, part VI)

In Example 20.6 it was concluded that the linear regression model with the independent variables X_1 and X_2 is statistically useful, so at least one of these two variables should be significant within the model. Which one? Or both? How significant?

The Coefficients part of the printout is repeated below.

Coefficients(a)

Model		Unstandardized Coefficients		Standardized Coefficients	t	Sig.
		B	Std. Error	Beta		
2	(Constant)	−3.060	2.189		−1.398	.164
	Age	.478	.057	.510	8.352	.000
	Edl	3.254	.546	.364	5.955	.000

a   Dependent Variable: W

To decide about the significance of the variable 'education level', conduct the following hypothesis test – with $\alpha = 0.05$ – about the accompanying regression coefficient β_2. Note that $t_{0.025;147} = 1.9762$ (*) and that **val** can be found in the column with heading 't'.

　i Test H_0: $\beta_2 = 0$ against H_1: $\beta_2 \neq 0$; $\alpha = 0.05$

　ii Test statistic: $T = \dfrac{B_2}{SE(B_2)}$

　iii Reject $H_0 \Leftrightarrow t \leq -1.9762$ or $t \geq 1.9762$

　iv **val** = 5.955 (from the printout)

　v Since 5.955 is larger than the critical value 1.9762, H_0 is rejected

The statistical conclusion is that 'education level' is significant at significance level 0.05. The p-value of this **two-sided** test is

$$2P(T \geq 5.955) = 1.835 \times 10^{-8} \tag{*}$$

which makes the conclusion very convincing. The variable 'education level' is significant within the linear regression model that also includes the variable 'age'.

In a similar way it can be concluded that the variable 'age' is significant in the regression model which also includes 'education level' as independent variable. The first three decimals of the p-value of the underlying two-sided test (with hinge 0) are all 0, so this conclusion is also very convincing. Both independent variables are individually significant.

A further conjecture is that one extra year of age will (ceteris paribus, so leaving the education level unchanged) on average lead to less than 60 cents change in the gross hourly wage. We will use a 96% confidence interval to draw a conclusion about that.

First notice that the conjecture states that $\beta_1 < 0.60$. Since $\alpha = 0.04$, we need $t_{0.02;147}$ = 2.0721(*). The interval estimator

$$B_1 \pm t_{0.02;147}SE(B_1)$$

gives the realization $0.478 \pm 2.0721 \times 0.057$, from which the lower bound $l = 0.360$ and the upper bound $u = 0.596$ follow. We are 96% confident that β_1 is smaller than 0.596, so we are (at least) 96% confident that $\beta_1 < 0.60$.

CASE 20.1: POLLUTION DUE TO TRAFFIC (PART I) – SOLUTION

Firstly, the five simple models are estimated. The following table lists the equations of the regression lines and the coefficients of determination.

Independent variable	Regression line	r^2
X_1	$\hat{y} = -28.549 + 0.764x$	0.798
X_2	$\hat{y} = -326.230 + 74.064x$	0.404
X_3	$\hat{y} = -32.853 + 2.418x$	0.482
X_4	$\hat{y} = 376.592 - 18.200x$	0.0005
X_5	$\hat{y} = 295.514 + 1.001x$	0.002

With t-tests it can be shown that the variables X_1, X_2 and X_3 are all positively linearly related to Y. However, for the variables X_4 and X_5 the data do not give enough evidence that the variables are linearly related to Y; it seems that wind is hardly a factor when explaining the variation of Y. When comparing the five simple linear regression models, it becomes clear that X_1 explains by far the largest part of the variation of Y.

Next, more independent variables are included in the model. Of course, X_1 will be included. Since X_2 and X_3 in a sense measure the same thing, only one of them is included; we take X_3. Because of the above analysis of the simple models, we have doubts about the inclusion of the variables X_4 and/or X_5. That is why four models are estimated. The table below gives an overview of some statistical conclusions.

Independent variables	Significant variables ($\alpha = 0.05$)	r^2
X_1, X_3	X_1, X_3	0.8018
X_1, X_3, X_4	X_1, X_3	0.8020
X_1, X_3, X_5	X_1, X_3, X_5	0.8037
X_1, X_3, X_4, X_5	X_1, X_3, X_5	0.8043

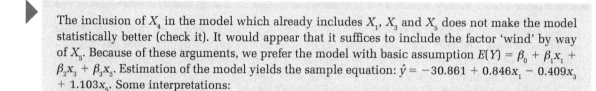

The inclusion of X_4 in the model which already includes X_1, X_3 and X_5 does not make the model statistically better (check it). It would appear that it suffices to include the factor 'wind' by way of X_5. Because of these arguments, we prefer the model with basic assumption $E(Y) = \beta_0 + \beta_1 x_1 + \beta_2 x_3 + \beta_3 x_5$. Estimation of the model yields the sample equation: $\hat{y} = -30.861 + 0.846x_1 - 0.409x_3 + 1.103x_5$. Some interpretations:

- If city A has 1000 inhabitants more than a city B which is comparable with A as far as the area and the number of low wind days per year are concerned, then it is expected that the yearly urban emission of fine particulate matter due to transportation is about 0.846 tonnes more in city A than in city B.

- If A and B have the same number of inhabitants and the same number of low wind days per year but the area of A is 1 km² more, then the expected yearly urban emission of fine particulate matter due to transportation is about 0.409 tonnes **less** in city A than in city B.

20.6 Conclusions about Y and E(Y)

The estimated linear regression model can be used to predict the value of Y and to estimate $E(Y)$ for a (new) population element for which the independent variables are observed (but the dependent variable is not). Two types of interval will be constructed: one that probably includes the unknown real number $E(Y)$ and one that probably includes the (unobserved) value of the random variable Y for this population element, both with a prescribed level of confidence.

In the theory below, the index p will be attached in the notations, to denote our prediction purposes. Let $x_{p1}, \ldots, x_{pk}$ denote the observations (the **x characteristics**) of the independent variables for this new population element and let Y_p be the accompanying (still random) observation of Y. The random variable

$$\hat{Y}_p = B_0 + B_1 x_{p1} + \ldots + B_k x_{pk} \tag{20.19}$$

is based on the least-squares estimators and has two duties:

 i estimator of $E(Y_p)$;
 ii point predictor of Y_p itself.

See also Section 19.7.1. Having used a sample of size n to determine estimates $b_0, \ldots, b_k$ of the population regression coefficients, the observation $\hat{y}_p = b_0 + b_1 x_{p1} + \ldots + b_k x_{pk}$ of $\hat{Y}_p$ can be considered as a **point estimate** of the unknown number $E(Y_p)$, but also as a **point prediction** of the not yet observed random variable Y_p.

Both duties have accompanying random deviations. The random deviation of duty (i) is $\hat{Y}_p - E(Y_p)$ and the random deviation of duty (ii) is $\hat{Y}_p - Y_p$. As in Section 19.7.1, the second duty brings the largest variation.

Formulae for the interval estimator of $E(Y_p)$ and the interval predictor of Y_p follow the formats:

$$\hat{Y}_p \pm t_{\alpha/2;n-(k+1)} SE(\hat{Y}_p) \quad \text{and} \quad \hat{Y}_p \pm t_{\alpha/2;n-(k+1)} SE(\hat{Y}_p - Y_p) \tag{20.20}$$

Substitution of the estimation results (due to the sample) and the x characteristics of the new population element into (20.19) and into the standard errors $SE(\hat{Y}_p)$ and $SE(\hat{Y}_p - Y_p)$, yields realizations of the random intervals in (20.20): respectively, a **confidence interval** for $E(Y_p)$ and a **prediction interval** for Y_p. Of course, such concrete intervals are calculated with a statistical package; see also Appendix A1.20. Incidentally, the formulae in Section 19.7.2 for the interval estimator and the interval predictor are valid only for the case $k = 1$. For the general linear regression model, corresponding formulae are not given here; only the formats are given.

Example 20.9 (red-wire, part VII)

Reconsider the model that explains Y = 'gross hourly wage' from X_1 = 'age' and X_2 = 'education level'. Recall that the sample regression equation is:

$$\hat{y} = -3.060 + 0.478x_1 + 3.254x_2$$

Now consider the subpopulation of the persons who are 30 years old and have completed a university education; that is, the x characteristics are $x_1 = 30$ and $x_2 = 5$. Substitution into the regression equation yields:

$$\hat{y} = -3.060 + 0.478 \times 30 + 3.254 \times 5 = 27.55 \text{ euro}$$

This number can be considered as:

- an estimate of the mean hourly wage in that subpopulation (duty (i)); and
- a prediction of the hourly wage of one individual in this subpopulation (duty (ii)).

To obtain two intervals that, with 95% confidence, include, respectively, $E(Y)$ (the mean hourly wage within the subpopulation) and Y (the hourly wage of one individual in that subpopulation), a computer is needed. Here are the results:

95% confidence interval for $E(Y)$:   (24.77, 30.36)

95% prediction interval for Y: (12.65, 42.48)

(Of course, the above point prediction $\hat{y} = 27.55$ falls precisely in the centre of both intervals.) There is 95% confidence that the mean hourly wage of the subpopulation will lie between €24.77 and €30.36. There also is 95% confidence that the hourly wage of Jensen, a 30-year-old person with education level 5, will lie between €12.65 and €42.48. Notice that the last interval is much wider than the first, which reflects the fact that the deviations of duty (ii) have more variation.

20.7 Residual analysis

As in Section 19.8, the validity of the model requirements can be checked with the residuals $e_i = y_i - \hat{y}_i$, for $i = 1, 2, \ldots, n$.

The model requirements can be briefly formulated as follows: linearity in each of the independent variables $X_1, \ldots, X_k$, homoskedasticity, independence of $\varepsilon_1, \ldots, \varepsilon_n$, independence of ε and $X_1, \ldots, X_k$ and normality.

Linearity

The linearity in the independent variable X_i can be checked by way of a scatter plot of e on x_i. In total, k scatter plots can be studied. If the model specifies the relationship between Y and X_i correctly, the accompanying scatter plot should be wild and fickle. If an obvious pattern is detected, the model is not specified correctly and has to be reconsidered. See also Section 22.3.1.

Homoskedasticity

This requirement can be checked by considering the scatter plot of e on $\hat{y}$. See also Sections 19.8 and 22.2.

Independence of $\varepsilon_1, \ldots, \varepsilon_n$

Usually this is a problem only in the case of time series; see Sections 19.8.1 and 22.4.

Independence of ε and $X_1, \ldots, X_k$

More will be said about this requirement in Section 22.5.

Normality

Normality can be studied by way of a histogram of the residuals. To detect outliers, it is better to make a histogram of the **standardized** residuals. See also Section 22.3.2.

Example 20.10 (red-wire, part VIII)

We will check the model requirements for the wages example. First, we study the linearity in the two independent variables. The two scatter plots of e on AGE and e on EDL are shown in Figure 20.1.

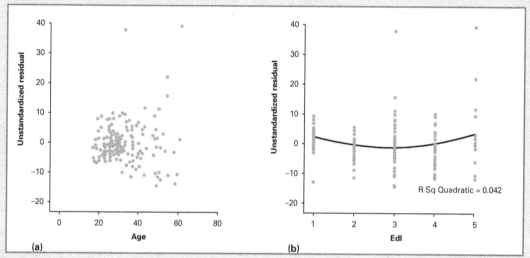

Figure 20.1 Scatter plots of e on AGE and e on EDL

At first sight, plot (a) does not seem to show an obvious pattern. However, after having omitted the four dots with rather extreme positive residuals, it looks like relatively many of the dots with ages above 40 lie below the level 0. That is why it might be interesting to find out whether the inclusion of the variable 'AGE^2' (the square of AGE) will make the model better. In plot (b) the dots are concentrated in five vertical lines, which is a consequence of the fact that EDL has only five possible values. It looks as if the plot shows some weak evidence of a parabolic pattern (valley-shaped), which is provocatively emphasized by fitting a parabola to the data. Compared with the vertical level 0, relatively many dots fall above that level for the education levels 1 and 5. It might suggest that the square of the variable EDL also has to be included in the model.

Figure 20.2(a) plots the standardized residuals on the predictions $\hat{y}$. The variation along vertical lines seems to increase slightly from 25 onwards, which suggests a weak heteroskedasticity problem. Also notice that there are two or three outliers with standardized residual at least 3.

To study normality, a histogram is shown of the standardized residuals. Notice that this histogram also shows the outliers. But apart from these outliers, the normality assumption does not seem to be a problem.

It is not necessary to study independence since the data come from persons that were randomly chosen within the population. In particular, there is no time involved; the data were observed at one moment in time.

The above residual analysis shows that there are no very serious conflicts with the model requirements. However, there are some outliers and the plots give weak indications of misspecification and heteroskedasticity. With respect to the misspecification: it might be better to take a model that is **quadratic** in AGE and EDL instead of linear. The outlier and heteroskedasticity problems might be solved by taking log(W) instead of W as independent variable; see also Sections 21.4 and 22.2.

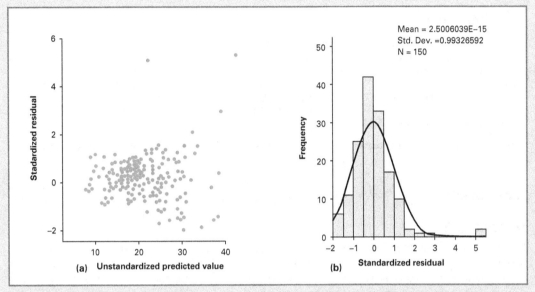

Figure 20.2 (a) Scatter plot of the standardized residuals on $\hat{y}$; (b) histogram of the standardized residuals

Summary

Multiple linear regression is an extension of simple linear regression; instead of one independent variable, **at least one** independent variable is included. The multiple linear regression model is a set of requirements which, if satisfied, guarantees that the least squares estimators B_i are unbiased, consistent and conditionally normal. It is this normality property that yielded an F-test to decide the overall usefulness of the model. Furthermore, this property yielded interval estimators and hypothesis tests to draw statistical conclusions about an individual independent variable. The statistical procedures are summarized below; recall that S_{B_i} is another notation for $SE(B_i)$.

Overview of statistical procedures for the multiple linear regression model

		Statistical procedure	
	Statistical problem	**Random interval**	**Test statistic**
$\beta_1, \ldots, \beta_k$	Usefulness of the model		$\dfrac{(SST - SSE)/k}{SSE/(n - (k + 1))}$
Parameter β_i	• Usefulness of X_i within the model • Conjecture about β_i	$B_i \pm t_{\alpha/2; n-(k+1)} S_{B_i}$	$\dfrac{B_i - b}{S_{B_i}}$ if hinge is b
Parameter $E(Y_p)$	Interval estimation	$\hat{Y}_p \pm t_{\alpha/2; n-(k+1)} SE(\hat{Y}_p)$	
Random variable Y_p	Interval prediction	$\hat{Y}_p \pm t_{\alpha/2; n-(k+1)} SE(\hat{Y}_p - Y_p)$	

The inclusion of more independent variables in the linear regression model implies that SSE decreases. Again, the ANOVA table illustrates the reduction of the unexplained variation from SST (before regression) to SSE (after the regression with k independent variables).

🔑 Key terms

adjusted coefficient of determination **631**	individual usefulness **634**	model test (for usefulness) **629**
ceteris paribus **621**	insignificant **634**	multiple linear regression model **621**
coefficient of determination **631**	jointly useful **634**	significant **634**
F-ratio **628**	least-squares (point) estimators **618**	x characteristics **636**
individual significance **634**		

❓ Exercises

Exercise 20.1

A real estate agent in the USA explains the selling price Y (in units of $1000) of a house from the area X_1 (in square feet) of the lot and the condition X_2 of the house (1 = in a very bad state, 10 = in an excellent state) on the basis of a sample of 10 houses. Here are some summarized results with respect to the regression equation and the sums of squares:

$$\hat{y} = 9.7823 + 1.8709x_1 + 1.2781x_2; \quad SST = 827.50100; \quad SSE = 8.17305$$

a Write down the basic assumption of the model.

b Is the model useful? Conduct the model test for usefulness; take $\alpha = 0.05$.

c Determine the ordinary and the adjusted coefficient of determination; interpret them.

d Interpret the sample regression coefficients.

Exercise 20.2

Reconsider Exercise 20.1. It is additionally given that the realized standard errors of B_1 and B_2 are, respectively, 0.076174 and 0.144400.

a Is the variable X_1 useful within the model? Use a 95% confidence interval to answer this question.

b Also investigate the individual significance of the variable X_2 by conducting a hypothesis test with $\alpha = 0.05$.

c Calculate the p-value of the test of part (b). Interpret your answer

Exercise 20.3

Two variables X_2 and X_3 are added to a simple linear regression model of Y on X_1. The new model will be estimated on the basis of a sample of size $n = 20$.

a Explain what the consequences of the model extension are for SST, SSE and SSR.

b What are the consequences for r^2 and r^2_{adj}?

c Write down the testing problem of the model test for usefulness. What do the hypotheses mean? Also write down the test statistic in terms of the coefficient of determination.

d How large must r^2 be to conclude that the new model is useful at significance level 0.05?

Exercise 20.4

Answer part (d) of Exercise 20.3 again, with $n = 30, 50$ and 100. Give your comments.

Exercise 20.5

For agricultural co-operatives the number of **members** may be important. To investigate this, the variation of the variable Y = 'net turnover (in millions of euro)' is investigated by means of a linear regression model with the following independent variables:

X_1: number of employees

X_2: number of members

The model is estimated from data for 29 co-operatives; see Xrc20-05.sav. The table below contains some results.

	Estimated regression coefficient	Realized standard error	Sums of squares
constant	260.158	109.328	SST = 31611752
X_1	0.303	0.030	SSE = 5865375
X_2	−0.010	0.016	

a Are the two variables X_1 and X_2 jointly significant? Conduct one test with significance level 0.02.

b Use p-values to investigate the (individual) significance of the two independent variables.

c Test whether, on average and ceteris paribus, one extra employee will increase the net turnover by more than €200 000. Use a hypothesis test with $\alpha = 0.05$.

Exercise 20.6

The final grade for the course Statistics 2 is determined by four examinations:

Midterm 1 (counts 20%)
Team Assignment (15%, created in teams of three)
Midterm 2 (20%)
Final Exam (45%)

It is suspected that the grade for the Final Exam is, at least partially, determined by the grades for the two midterms and the Team Assignment. To find out to what extent this is the case, the grade Y for the Final Exam is regressed on the grade X_1 for Midterm 1, the grade X_2 for the Team Assignment and the grade X_3 for Midterm 2. The dataset Xrc20-06.sav contains the data for a sample of 132 International Business students. All grades are on the scale 0–10.

a Write down all model requirements and interpret them.

b Interpret the population regression coefficient of X_1.

The model is estimated on the basis of the results for the 132 students; see the printout.

Model Summary

Model	R	R Square	Adjusted R Square	Std. Error of the Estimate
1	.804(a)	.646	.637	1.55099

a Predictors: (Constant), X3, X2, X1

ANOVA(b)

Model		Sum of Squares	df	Mean Square	F	Sig.
1	Regression	561.213	3	187.071	77.765	.000(a)
	Residual	307.914	128	2.406		
	Total	869.128	131			

a   Predictors: (Constant), X3, X2, X1
b   Dependent Variable: Y

Coefficients(a)

Model		Unstandardized Coefficients		Standardized Coefficients	t	Sig.
		B	Std. Error	Beta		
1	(Constant)	−1.402	1.102		−1.272	.206
	X1	.313	.084	.273	3.724	.000
	X2	.125	.138	.049	.907	.366
	X3	.620	.078	.578	7.989	.000

a   Dependent Variable: Y

   c Determine the regression equation.
   d Is the model useful? Test it with $\alpha = 0.02$.
   e Determine the coefficient of determination and interpret it.

Exercise 20.7

Reconsider Exercise 20.6, especially the printout.

   a Which variables are individually significant at the 0.05 level?
   b Interpret the sample regression coefficient of X_1.
   c Two students (A and B) have precisely the same grade for Midterm 1 and the Team Assignment. However, B scored 1 point higher than A for Midterm 2. The conjecture is that for the Final Exam the expected grade of B will be more than 0.4 higher than the expected grade of A. Use a hypothesis test with $\alpha = 0.02$ to draw a conclusion about this conjecture.

Exercise 20.8

The coefficient of determination r^2 will increase if more independent variables are added to the model. That is why it seems to be wise to include as many independent variables as possible. However, for fixed n this will **de**crease the number of degrees of freedom $n - (k + 1)$ of the estimator of σ_ε^2, which in turn influences the confidence level of the statistical procedures.

Suppose that we want to use 18 economic indicators to predict next year's GDP by way of a linear regression model. So, the basic assumption of the proposed model is:

$$E(Y) = \beta_0 + \beta_1 x_1 + \ldots + \beta_{18} x_{18}$$

However, we have data for only the most recent 20 years. These data yield that $r^2 = 0.95$. Test (with $\alpha = 0.05$) whether this – large – value of r^2 is enough to conclude that the model is useful.

Exercise 20.9

Suppose that as many as 25 independent variables are used in a linear regression model to explain the variation of the quarterly return (%) of a certain stock. The model is estimated on the basis of the quarterly returns for the most recent ten years. The determination coefficient turns out to be 0.80. Is this rather large value enough to conclude at significance level 0.05 that the model is useful?

Exercise 20.10

To investigate whether the asking price Y (in thousands of euro) of houses in a middle-sized city is related to the size X_1 (in m²) of the house and the area X_2 (in m², excluding the house itself) of the plot, a sample of 51 houses, with asking prices in the range €400 000–600 000, is studied on the website of a large conglomeration of estate agents. See the file Xrc20-10.sav. We first regress Y on X_1 only.

  a Running the regression of Y on X_1 yields the following partial printout. Use an F-test to find out whether the accompanying simple linear regression model is useful; take 0.05 as significance level. Also determine and interpret the accompanying p-value.

ANOVA(b)

Model		Sum of Squares	df	Mean Square	F	Sig.
1	Regression	16423.313	1	16423.313	4.549	.038(a)
	Residual	176886.097	49	3609.920		
	Total	193309.410	50			

a Predictors: (Constant), SizeHouse
b Dependent Variable: AskPrice

  b Do the test of part (a) again, but now with a t-test.
  c Calculate the coefficient of determination and the adjusted coefficient of determination. Interpret your answers.

Exercise 20.11

Reconsider Exercise 20.10. The model is extended by including X_2; see the partial printout below.

ANOVA(b)

Model		Sum of Squares	df	Mean Square	F	Sig.
2	Regression	16988.206	2	8494.103	2.312	.110(a)
	Residual	176321.204	48	3673.358		
	Total	193309.410	50			

a Predictors: (Constant), SizeLot, SizeHouse
b Dependent Variable: AskPrice

  a Write down the basic assumption of the model. Is the model useful? Again, take 0.05 as significance level.
  b Calculate r^2 and r^2_{adj}.
  c Give your comments on the results of this and the previous exercise.

Exercise 20.12

This exercise is meant to check some presuppositions about weight:

  1 Taller people are heavier
  2 Growing older makes you fatter
  3 People with lower education level are fatter

4 Working harder is a good remedy to lose weight

5 Earning more makes you fatter

Their validity will be checked by a linear regression study for a large, western population of adult people on the basis of a random sample of size 300. The following variables are measured:

WEIGHT (kg), HEIGHT (cm), AGE (years), EDU (education level, values 1–5), HOURS (number of hours worked weekly), WAGE (gross hourly wage in euro)

See Xrc20-12.sav for the data. Running the regression with WEIGHT as dependent variable and the other variables as independent variables, gives the following printout:

ANOVA(b)

Model		Sum of Squares	df	Mean Square	F	Sig.
1	Regression	15165.465	5	3033.093	32.273	.000(a)
	Residual	27630.629	294	93.982		
	Total	42796.094	299			

a Predictors: (Constant), HOURS, HEIGHT, EDU, AGE, WAGE
b Dependent Variable: WEIGHT

Coefficients(a)

Model		Unstandardized Coefficients		Standardized Coefficients	t	Sig.
		B	Std. Error	Beta		
1	(Constant)	−55.256	11.431		−4.834	.000
	HEIGHT	.714	.063	.547		.000
	AGE	.128	.044	.152	2.913	.004
	EDU	.611	.550	.057	1.110	.268
	WAGE	−.348	.157	−.140	−2.212	.028
	HOURS	.012	.040	.019	.301	.764

a Dependent Variable: WEIGHT

a Determine the sample regression equation. Are the signs of the sample regression coefficients in accordance with the above presuppositions?

b It can be seen at a glance that the model is useful at significance level 0.01. How? Determine the ordinary and adjusted coefficients of determination. Interpret them.

c Use hypothesis tests with $\alpha = 0.05$ to check presuppositions 1–3.

d Use suitable 90% confidence intervals to check presuppositions 4 and 5.

Exercise 20.13 (computer)

Reconsider Exercise 20.12. Check the model requirements.

Exercise 20.14 (computer)

It seems to be obvious that the annual household expenditure on food (Y, in thousands of euro) will depend on the annual income of the household (X_1, in thousands of euros). Other important predictors might be the number of children (X_2) under 18 and the size (X_3) of the household. This research is meant to detect whether and to what extent Y is determined by X_1, X_2 and X_3.

We consider the linear regression model that explains Y from X_1, X_2 and X_3. The model is estimated on the basis of a random sample of size 300; see Xrc20-14.sav.

a Run the regression and write down the sample regression equation.

b Test whether the model is useful; use significance level 0.01.

c Quantify the degree of usefulness. Interpret your answer.

d Which variables are individually significant at the 0.01 level?

Exercise 20.15 (computer)

Reconsider Exercise 20.14.

a The conjecture is that an increase of €1000 in the annual income of a household, while household size and number of children remain unchanged, will on average lead to an increase of more than €220 in annual food expenditure. Use $\alpha = 0.01$ to test the conjecture.

b Interpret the sample regression coefficient of X_2; don't forget the ceteris paribus restriction. Is such a restriction possible in practice?

c Determine the correlation coefficient of the data of X_2 and X_3. (This large sample correlation coefficient makes the interpretation of b_2 (and b_3) barely possible. In Chapter 21 this unwanted phenomenon will be called collinearity.)

d Check the model requirements.

e Consider a family of four people; two of them are children. The annual income of this household is €30 000, but the annual expenditure Y on food is unknown. Determine intervals that enclose $E(Y)$ and Y with a confidence of 99%.

Exercise 20.16 (computer)

Reconsider Exercise 19.18 and Xrc19-18.sav, about a large computer shop. Y = 'number of Inspiration notebooks sold in a week' was regressed on, respectively:

X_1 = 'price (in euro) of the notebook in that week'

X_2 = 'number of advertisements (in local newspapers) in that week'

Although both models did rather well, the simple linear regression model with X_2 yielded the largest coefficient of determination. We will now study the model that encloses both variables X_1 and X_2.

a Use the data to determine a 90% confidence interval for the mean number of notebooks sold in weeks when the price is €800 and five advertisements are placed.

b Next week the price of the notebook will be €800 and five advertisements will be placed. Determine an interval that, with a confidence of 90%, will contain the number of notebooks sold next week.

The transition from the model with only X_2 to the model with X_1 **and** X_2 reveals several interesting things.

c Report your findings with respect to SST, SSE, SSR and r^2 when the simple model with only X_2 is extended by including X_1.

d Is X_1 significant within the model, at significance level 0.10?

e Which of the two models do you prefer? Why?

Exercise 20.17 (computer)

The following question has been asked for several years: *will an increase in the hourly wage have a positive or a negative effect on production?* Note that this question not only has economic importance, but also has social, and hence political, aspects.

In the second half of the twentieth century many studies addressed this question. Some used regression methods to find the relationship between the variable 'hours worked' (which represents production) on the one hand, and 'hourly wage' and other variables, on the other hand. Always, interest is in the regression coefficient of 'hourly wage': will it be (significantly) positive or negative, and what are the circumstances?

The data in the file Xrc20-17.sav come from a rather famous study: *Income Guarantees and the Working Poor*, by D. H. Greenberg and M. Kosters (Rand Corporation Report R-579-OEO, December 1970; see also http://lib.stat.cmu.edu/DASL/, case 'Wages and Hours'). The dataset is based on a national sample of 6000 households in the USA with a male head and earnings of less than $15 000 annually in 1966. Thirty-nine demographic subgroups were formed for analysis of the relation between average hours worked during the year and average hourly wages and other variables. So, the dataset has $n = 39$ (and missing values!). The variables are:

1 Y = HRS: Average hours worked during the year
2 X_1 = WAGE: Average hourly wage ($)
3 X_2 = ERSP: Average yearly earnings of spouse ($)
4 X_3 = ERNO: Average yearly earnings of other family members ($)
5 X_4 = NEIN: Average yearly non-earned income ($)
6 X_5 = ASSET: Average family asset holdings (Bank account, etc.) ($)
7 X_6 = AGE: Average age of respondent
8 X_7 = DEP: Average number of dependents
9 X_8 = RACE: Percentage of white respondents
10 X_9 = SCHOOL: Average highest grade of school completed

a Regress Y on X_1 only. Conduct a statistical analysis and give your comments. In particular, show that the slope of the population regression line is significantly positive at significance level 0.01.

b Regress Y on $X_1, \ldots, X_9$. Conduct a statistical analysis and give your comments.

c Omit X_5, X_6 and X_9 from the model of part (b). Run the regression again and give your comments.

Exercise 20.18 (computer)

The question of Exercise 20.17 is also central to this exercise: *will an increase in the hourly wage have a positive or a negative effect on production?* The file Xrc20-18.sav contains data of a random sample of size 300 from a large population of households, about the following variables:

HOURS weekly number of hours worked by the head of the household
AGE age of the head of the household, in years (not rounded off)
EDU level of education of the head of the household (1 = low, 5 = high)
WAGE hourly wage of the head of the household, in euros
NKIDS number of children (< 18 years) of the household
HINC net household income, in units of €1000 (including Family Allowance of €1600 per child annually)

Since HOURS is an important economic indicator for production, we take HOURS as dependent variable.

a Regress HOURS on WAGE. Conduct a statistical analysis and give your comments. In particular, show that the slope of the population regression line is significantly positive at significance level 0.01.

b Regress HOURS on the five other variables. Conduct a statistical analysis and give your comments.

c Omit HINC and EDU from the model of part (b). Run the regression again and give your comments.

CASE 20.2 INCOME AND EDUCATION LEVEL OF IDENTICAL TWINS (PART II)

This case is a continuation of Case 19.3, where differences in hourly wages were explained from differences in number of years of education. One advantage of making use of twins is that a pair of twins has similar family background; another advantage is that the self-reported education can be checked by asking the other twin.

The dataset Case20-02.sav contains, apart from the observations of the variables in Case 19.3, also the observations of the following variables:

AGE   Age in years of twin 1
DTENU   Difference (twin 1 minus twin 2) in tenure, or number of years at current job
DUNCOVE   Difference (twin 1 minus twin 2) in union coverage, where 1 signifies 'covered'
          and 0 'not covered'.

a Explain 'hourly wage' from 'age' and 'number of years of education'. Estimate the value (that is, extra hourly wage) of one extra year of education if age is kept unchanged. Do it twice, with the data of twin 1 and with the data of twin 2. Is this value significantly larger than $0.60?

b Find out whether one extra year for the **difference** between the numbers of years of education of twin 1 and twin 2 will on average increase the corresponding difference in hourly wage by more than $0.60 if age, difference in tenure and difference in union coverage remain unchanged. Answer this question with the difference between the self-reported education as well as with the difference between the cross-reported education.

c Compare in part (b) the estimates of the regression coefficients of the two variables that measure the difference in education for twin 1 and twin 2.

d The authors of the original paper, Ashenfelter and Krueger, want to interpret the result of part (b) in terms of people with similar family and working backgrounds. Formulate that interpretation.

Multiple linear regression: extension

The topics of the present chapter are aimed at improving the model. We will discuss the introduction into the regression model of more complicated independent variables.

- Qualitative independent variables can be included by way of **dummy variables**.
- If the relationship between Y and X_1 is expected to be different for different levels of X_2, an **interaction term** $X_1 \times X_2$ can be included.
- Instead of X, a function $h(X)$ can be used. In particular, second- and third-order functions, and logarithmic functions are important.
- Instead of Y, the **natural logarithm** $\log(Y)$ can be used.

A special test can decide the usefulness of a set of (new) independent variables. But including new independent variables may cause complications; we will discuss them.

- A **partial *F*-test** will be constructed, to find out whether a set of (new) independent variables really improves the model.
- It is often unwanted if one of the independent variables is highly correlated with a linear combination of the others. This **collinearity** problem will be discussed.

Model building is the collective noun for all activities leading to the model that is eventually used for statistical inference and for prediction purposes. The complete process of the building of the model is summarized in Section 21.6.

CASE 21.1 PRICING DIAMOND STONES

This case is about diamonds and about factors that determine the price of a diamond. The model was originally meant to help jewellers to price a diamond. However, the approach that is followed to find the 'final' model can also be useful for other businesses since pricing has to be done in many fields.

The study comes from the paper 'Pricing the C's of diamond stones' by Singfat Chu (*Journal of Statistics Education*, 9/2 2001) and is based on a dataset Case21-01.sav about 308 diamond stones. The data appeared in an advertisement in Singapore's *Business Times* of 18 February 2000.

The price Y of a diamond stone (in Singapore $) is determined by the four Cs: carat, clarity, colour and cut.

- **Carat** refers to the weight of a diamond stone; 1 carat is equivalent to 0.2 grams.
- **Clarity** has to do with the natural 'birthmarks' of diamonds. If no inclusions are observed under a loupe with a 10 × magnification, the diamond is labelled IF (internally flawless). Lesser diamonds are – in descending order – labelled as 'very very slightly imperfect' (VVS1 or VVS2) and 'very slightly imperfect' (VS1 or VS2).
- **Colour** purity is in descending order labelled as: D, E, F, G, ….
- The **cut** of a raw diamond is determined by the craftsmanship of the diamond cutter, but the dataset does not contain information on this.

However, the dataset does contain information about the **certification body** (Gemmological Institute of America (GIA), International Gemmological Institute from Antwerp (IGI) or the Hoge Raad voor Diamant (HRD)) that has assayed the diamond stone independently, to help purchasers to choose when confronted with so much brilliance.

In our models, we will have to find solutions to the following problems:

- A scatter plot of Y on 'Carat' shows some heteroskedasticity. What to do?
- 'Clarity', 'colour' and 'certification body' are qualitative variables. How can such variables be included in the model?
- It will turn out that the obvious model underestimates prices at both ends of the price range. What can be done?

See the end of Section 21.6 for a solution.

21.1 Usefulness of portions of a model

In the coming sections, new types of independent variables are introduced. But whether such variables indeed improve the model is not clear in advance. In the present section an *F*-test is considered that can be used to check the usefulness of a subset of independent variables. The test will be applied frequently in coming sections.

Consider k variables that might possibly be used as independent variables in a linear regression model. At the moment we do not doubt the usefulness of g of them (with $g < k$), but we want to discuss the usefulness of $k - g$ of these variables. To makes things more tractable, let us split the sequence of variables into two parts:

$$X_1, \ldots, X_g, X_{g+1}, \ldots, X_k$$

The sequence $X_1, \ldots, X_g$ concerns variables that (at least at the moment) are not the subject of discussion. However, we are wondering whether the set of $k - g$ variables $\{X_{g+1}, \ldots, X_k\}$ improves the model statistically. In essence, we are wondering whether the **complete model** with basic assumption

$$E(Y) = \beta_0 + \beta_1 x_1 + \ldots + \beta_g x_g + \beta_{g+1} x_{g+1} + \ldots + \beta_k x_k$$

is '**better**' than the **reduced model** with basic assumption

$$E(Y) = \beta_0 + \beta_1 x_1 + \ldots + \beta_g x_g$$

in the sense that the complete model explains a significantly larger part of the variation of Y than the reduced model. That is, the question is whether the variables $X_{g+1}, \ldots, X_k$ jointly contribute to the explanation of the variation of Y.

If the variables $X_{g+1}, \ldots, X_k$ do not play a role in the complete model, the accompanying coefficients $\beta_{g+1}, \ldots, \beta_k$ are all 0 since 'playing no role' means that changing $x_{g+1}, \ldots, x_k$ in the basic assumption does not change E(Y). It would appear that the joint use**lessness** of $X_{g+1}, \ldots, X_k$ within the complete model is equivalent to the validity of the hypothesis:

$$H_0: \beta_{g+1} = \ldots = \beta_k = 0$$

On the other hand, validity of the hypothesis

$$H_1: \text{at least one of the } \beta_{g+1}, \ldots, \beta_k \text{ is} \neq 0$$

means that E(Y) will change under changes of at least one of the independent variables $X_{g+1}, \ldots, X_k$. We are interested in testing H_0 against H_1, since H_0 states that $X_{g+1}, \ldots, X_k$ are useless within the model whereas H_1 states that at least one of these variables is useful.

Both the reduced model and the complete model have to be estimated, so two regressions have to be run. Let SSE_r and SSE_c be the SSE of the reduced and the complete model, respectively. It is obvious that $SSE_r - SSE_c \geq 0$, since the SSE will always decrease if new independent variables are added to the reduced model. If H_0 is valid, the model does not improve by including the variables $X_{g+1}, \ldots, X_k$, the difference $SSE_r - SSE_c$ will be close to 0 and hence the ratio $(SSE_r - SSE_c) / SSE_c$ too. But if H_1 is valid, the model becomes better by including the variables $X_{g+1}, \ldots, X_k$; the difference $SSE_r - SSE_c$ and the ratio $(SSE_r - SSE_c) / SSE_c$ will be away from 0. That is why the test statistic F is based on that ratio. Its precise definition arises by dividing the numerator and the denominator by the respective numbers of degrees of freedom.

Notice that (since $g + 1$ regression coefficients have to be estimated in the reduced model) SSE_r has $n - (g + 1)$ degrees of freedom and SSE_c has $n - (k + 1)$, so $SSE_r - SSE_c$ has $n - (g + 1) - (n - (k + 1)) = k - g$ degrees of freedom. Here is the resulting test statistic:

ii $\quad F = \dfrac{(SSE_r - SSE_c)/(k - g)}{SSE_c/(n - (k + 1))}$

It turns out that F has the distribution $F_{k-g, n-(k+1)}$ if H_0 is true. The number of degrees of freedom in the numerator is just the number of zero restrictions in the null hypothesis.

H_0 will be rejected (and hence H_1 accepted) if the realization of F is relatively large. In fact, H_0 will be rejected if the realization is $\geq F_\alpha$:

iii $\quad$ Reject $H_0 \Leftrightarrow f \geq F_{\alpha;k-g, n-(k+1)}$

The reason for choosing $F_{\alpha;k-g, n-(k+1)}$ as critical value is that this boundary guarantees that the probability that the test procedure (ii)–(iii) will incorrectly reject H_0, is equal to α:

If H_0 is true, then $P(H_0 \text{ is rejected}) = P(F \geq F_{\alpha;k-g, n-(k+1)}) = \alpha$

Here is the complete five-step procedure for testing whether the variables $X_{g+1}, \ldots, X_k$ jointly contribute to the explanation of the variation of Y. The test is called **partial F-test**.

Five-step procedure for testing the usefulness of the subset $\{X_{g+1}, \ldots, X_k\}$:

i Test $H_0: \beta_{g+1} = \ldots = \beta_k = 0$ against H_1: at least one of $\beta_{g+1}, \ldots, \beta_k$ is $\neq 0$

ii Test statistic: $F = \dfrac{(SSE_r - SSE_c)/(k - g)}{SSE_c/(n - (k + 1))}$

iii Reject $H_0 \Leftrightarrow f \geq F_{\alpha;k-g,n-(k+1)}$

iv Calculate **val**, the realization of F

v Draw the conclusion by comparing **val** with step (iii)

Example 21.1

Many courses have midterms, intermediate exams that have to stimulate the students to work regularly and not only during the last two weeks before the final exam takes place. But do these midterms really contribute to the grade of the final exam?

To check things for a course called Statistics 1 for Economics, we use a linear regression model to regress Y = 'grade for the Final Exam' on the variables:

X_1 = 'grade for Midterm 1' X_2 = 'grade for the Team Assignment'
X_3 = 'grade for the Mid-Exam' X_4 = 'grade for Midterm 2'

(All grades use the scale 0–10; the respective weights for the determination of the final grade are 0.1, 0.1, 0.3, 0.1 and 0.4.) The question is whether the set $\{X_1, X_4\}$ of independent variables really contributes to explain the variation of Y. See Xmp21–01.sav for the data of 201 students who participated in all of the four parts.

Two models have to be compared, the 'large' model with all four independent variables and the 'small' model with only X_2 and X_3 included:

Complete model: $E(Y) = \beta_0 + \beta_1 x_1 + \beta_2 x_2 + \beta_3 x_3 + \beta_4 x_4$
Reduced model: $E(Y) = \beta_0 + \beta_2 x_2 + \beta_3 x_3$

(Note that the two models make use of the same notations for population regression coefficients, but they may have different values.) The question is to find out whether the term $\beta_1 x_1 + \beta_4 x_4$ in the complete model really contributes. So, the testing problem is:

i Test $H_0: \beta_1 = \beta_4 = 0$ against H_1: at least one of β_1, β_4 is $\neq 0$

We take $\alpha = 0.05$. The reduced model contains the $g = 2$ independent variables X_2 and X_3, while the complete model contains all $k = 4$ independent variables. Since $k - g = 2$, $n - (k + 1) = 196$ and $F_{0.05;2,196} = 3.0420$ (*), the test procedure in the steps (ii)–(iii) is:

ii Test statistic: $F = \dfrac{(SSE_r - SSE_c)/2}{SSE_c/(n - 5)}$

iii Reject $H_0 \Leftrightarrow f \geq 3.0420$

We have to estimate the two models. The two ANOVA parts are shown below. (Note that model 1 is the reduced model.)

ANOVA(b)

Model		Sum of Squares	df	Mean Square	F	Sig.
1	Regression	906.527	2	453.264	123.638	.000(a)
	Residual	725.877	198	3.666		
	Total	1632.404	200			

a   Predictors: (Constant), Mid_Exam, TA
b   Dependent Variable: Final_Exam

		Sum of Squares	df	Mean Square	F	Sig.
Model						

ANOVA(b)

Model		Sum of Squares	df	Mean Square	F	Sig.
2	Regression	1060.645	4	265.161	90.898	.000(a)
	Residual	571.759	196	2.917		
	Total	1632.404	200			

a Predictors: (Constant), Mid2, TA, Mid1, Mid_Exam
b Dependent Variable: Final_Exam

It follows that: $SSE_r = 725.877$ and $SSE_c = 571.759$. It would appear that the inclusion of X_1 and X_4 in the model that already includes X_2 and X_3 has further reduced the unexplained variation by the amount:

$725.877 - 571.759 = 154.118$ (which **cannot** be found in these printouts)

We have to conduct the test in (ii)–(iii) to find out whether this reduction is enough to conclude that the extension is useful. We next calculate **val**:

iv $val = \dfrac{154.118/2}{571.759/196} = 26.4160$

v Reject H_0 since **val** is larger than the critical value 3.0420

The extension is useful; inclusion of X_1 and X_4 improves the reduced model. Put another way, the data give evidence for the joint significance of the variables X_1 and X_4 in the complete model.

If $g = k - 1$ and hence the significance discussion is about only one independent variable, then the testing problem is about only one population regression coefficient (one β) and it is also possible to use a t-test. Using a similar procedure to that used in Section 20.4.2, it can be shown that the partial F-test and this t-test are equivalent.

Now suppose that the significance discussion is about at least two independent variables, so $k - g$ is at least 2. It is tempting to replace the partial F-test of this section by $k - g$ t-tests for H_0: $\beta = 0$ against H_1: $\beta \neq 0$, one for each of the regression coefficients $\beta_{g+1}, \ldots, \beta_k$. However, conducting one F-test is not equivalent to conducting the $k - g$ t-tests. Firstly, each individual t-test has probability α of rejecting H_0 incorrectly (type I error) and hence the group of t-tests has a probability (much) larger than α of rejecting at least one of the $k - g$ null hypotheses incorrectly. So, conducting the group of t-tests is more risky in the sense of making type I errors. But secondly, the possible conclusion after conducting the group of t-tests that at least one of the independent variables is significant is not precisely the same as the conclusion of the F-test that the group of independent variables are **jointly** significant. The first conclusion refers to the significance of one variable in a model that also contains all other variables of the group of variables; the second conclusion refers to the significance of the group of variables as a whole.

Although the group of t-tests is closely related to the (single) F-test, there are differences. If a researcher has the idea (for instance because of individual t-tests) that a group of independent variables can be omitted from the model without making the model worse, this can be checked statistically by conducting one test: the above partial F-test.

It is possible to express the test statistic of the partial F-test in terms of the coefficients of determination of the two models that are compared. If R_r^2 and R_c^2 indicate the (random) coefficients of determination of the reduced and the complete model, respectively, then it is not difficult to rewrite the test statistic as:

$$F = \frac{(R_c^2 - R_r^2)/(k - g)}{(1 - R_c^2)/(n - (k + 1))}$$

(21.1)

21.2 Collinearity

A regression can be seriously disturbed if an independent variable is strongly linearly related to a linear combination of other independent variables. This problem is called **collinearity** and is studied below.

Suppose that you are interested in the variation of the variable Y = 'annual clothing expenditure of a household'. Since the variables

X_1 = 'annual income of the household'
X_2 = 'number of children (< 18) of the household'
X_3 = 'size of the household'
X_4 = 'number of adults (≥ 18) of the household'
X_5 = 'number of parents of the households'

seem to be relevant for explaining the variation of Y, you gather data of Y and all these five variables. But then you did too much since X_3 is precisely equal to the sum of X_2 and X_4. It is said that the model has **strict collinearity**: one independent variable is a linear combination of other independent variables; here $X_3 = X_2 + X_4$. The point is not only that statistical packages usually cannot handle such a dataset but also that the underlying theory cannot be applied: although it was not mentioned up to now, there is a regularity condition that states that strict collinearity among the independent variables is not allowed.

Taking the above into account, you decide to regress Y on X_1, X_2, X_3, X_5 thus excluding X_4. Although the new model does not have a **strict** collinearity problem, it remains that for many households the observation of X_3 will be the same as the observation of $X_2 + X_5$ which may cause X_3 to be strongly linearly related to $X_2 + X_5$. Although this not always is a 'problem', a strong linear relationship among the independent variables is called a **problem of collinearity**.

Collinearity or **multicollinearity** means that one independent variable is strongly related to a linear combination of the other independent variables. It can be detected by regressing individual independent variables on the set of all other independent variables and by studying the resulting coefficients of determination. In practice, the **correlation matrix** – the matrix of the correlation coefficients of all pairs of independent variables – is also used to obtain information about collinearity. If a pair of independent variables has a large positive (close to 1) or negative (close to -1) correlation coefficient, then problems with collinearity may arise. But what are these problems?

Suppose that the independent variable X_j is strongly linearly related to the other independent variables. This, for instance, may have been concluded from the fact that a regression of X_j on the other independent variables yielded a determination coefficient r_j^2 (say) that is rather close to 1. This collinearity turns out to have the following consequences:

- interpretation of β_j and its estimate b_j is complicated;
- the standard deviation of the estimator B_j – and hence the standard error of B_j – is larger than it would have been in case of no (or hardly any) collinearity.

The first problem is obvious. The interpretation of β_j presupposes that 'x_j can be increased by 1 while leaving the values of the other independent variables unchanged'. But this ceteris paribus presupposition is now hard to realize because of the fact that r_j^2 is close to 1 and hence changing x_j will probably lead to changes in the values of other independent variables.

Owing to the second problem, a confidence interval for β_j will become wider and the value of the test statistic for a testing problem about β_j will become smaller. Statistical inference about β_j is seriously sabotaged by this collinearity. For instance, in a t-test about the significance of X_j, since the denominator of the test statistic is now larger, the **val** is pushed in the direction of 0 and it is harder to prove the individual significance of X_j.

Whether collinearity is experienced as a serious problem depends on the example you are considering. It sometimes happens that an independent variable is strongly linearly related to other independent variables, but its significance can still be proved. It would appear that even though the collinearity pushes the **val** of the significance test in the direction of 0, the **val** is still far enough away from 0 to conclude that the independent variable is significant. However, it also happens that a researcher may be pretty sure that a certain independent variable should be significant, but it turns out to be insignificant due to collinearity. If this is the case, a solution to the problem might be to omit an independent variable that causes the collinearity.

Example 21.2

We want to explain, for a large population of households, the variation of Y = 'annual clothing expenditure of a household' from the variables

X_1 = HINC: 'annual income of the household'
X_2 = NKIDS: 'number of children (< 18) of the household'
X_3 = HS: 'size of the household'

on the basis of a random sample of 300 households. The data are in the file Xmp21–02.sav. The basic assumption of the model is:

$$E(Y) = \beta_0 + \beta_1 x_1 + \beta_2 x_2 + \beta_3 x_3$$

We will show with a partial F-test not only that the variables NKIDS and HS are jointly significant within this model but also that they are both individually insignificant. This 'discrepancy' turns out to be caused by collinearity.

Running the regression yields that the linear regression model with these three independent variable is useful (the *val* of the model test is 12 458.179) and that $r^2 = 0.992142$ (check it). See the first part of the printout below.

ANOVA(b)

Model		Sum of Squares	df	Mean Square	F	Sig.
1	Regression	397.006	3	132.335	12458.179	.000(a)
	Residual	3.144	296	.011		
	Total	400.150	299			

a   Predictors: (Constant), HS, HINC, NKIDS
b   Dependent Variable: CLOTEXP

Coefficients(a)

Model		Unstandardized Coefficients		Standardized Coefficients	t	Sig.
		B	Std. Error	Beta		
1	(Constant)	−.013	.023		−.541	.589
	HINC	.082	.000	.991	185.861	.000
	NKIDS	.011	.014	.011	.787	.432
	HS	.012	.012	.014	1.028	.305

a   Dependent Variable: CLOTEXP

However, the second part of the printout shows that neither of the two variables NKIDS and HS is individually significant within the model, which is unexpected. To find out whether these two variables jointly do contribute, we test:

  $H_0: \beta_2 = \beta_3 = 0$ against H_1: at least one of β_2, β_3 is $\neq 0$; $\alpha = 0.01$

For the partial F-test, we also have to estimate the model that includes only HINC as independent variable. The ANOVA part of the printout of that model (model 2), is shown below.

ANOVA(b)

Model		Sum of Squares	df	Mean Square	F	Sig.
2	Regression	396.770	1	396.770	34971.811	.000(a)
	Residual	3.381	298	.011		
	Total	400.150	299			

a   Predictors: (Constant), HINC
b   Dependent Variable: CLOTEXP

This reduced model has 0.991551 as coefficient of determination, which is only slightly less than for the complete model. Still, the difference is large enough to conclude that NKIDS and HS jointly do contribute, as follows from steps (ii)–(v) of the partial F-test:

ii   Test statistic: $F = \dfrac{(SSE_r - SSE_c)/2}{SSE_c/(n-4)}$

iii   Reject $H_0 \Leftrightarrow f \geq F_{0.01;2,296} = 4.6776$ (*)

iv $val = \dfrac{(3.381 - 3.144)/2}{3.144/296} = 11.1565$

v   Reject H_0, since *val* is larger than 4.6776

Since the individual insignificance of both NKIDS and HS is very surprising, we compute the sample correlation coefficient of NKIDS and HS, and find that it is 0.927. So, NKIDS and HS are heavily linearly related. As a consequence, the *val*s of the two *t*-tests are both forced in the direction of 0; so much so that it cannot be concluded that they are individually significant. One solution to this problem might be to omit one of the two variables, for instance NKIDS.

21.3 Higher-order terms and interaction terms

Up to now it was implicitly assumed that each of the independent variables brings its own dimension. However, this is not necessarily the case since X_2 (for instance) is also allowed to be a function of X_1.

The first part of the present section is devoted to the consequences of including squared variables (or even higher-order terms) in the model. The second part is about so-called **interaction terms**: product variables such as X_1X_2 that are included if it is suspected that the relationship between Y and X_1 is different for different levels of X_2.

A simple linear regression model with independent variable X can be extended to a model that also includes X^2 or even to a model that also includes X^2 and X^3. A multiple linear regression model with the predictors X_1 and X_2 can be extended by also including X_1^2 and X_2^2. Such extended models have their own names. An overview is given in Table 21.1, where, instead of the shortcut $E(Y)$, the correct notations $E(Y \mid x)$ and $E(Y \mid x_1, x_2)$ are used again.

The first-order model with one predictor describes the relationship between $E(Y \mid x)$ and x as a straight line in x. The **second**-order model with one predictor describes the relationship as a parabola; if $\beta_2 < 0$ the parabola is mountain-shaped, but if $\beta_2 > 0$ it is valley-shaped. The third-order model with one predictor describes the relationship between $E(Y \mid x)$ and x as a third-order function in x. The curve of such functions can change twice, as shown in Figure 21.1.

Basic assumption of the model	Name of the model
$E(Y \mid x) = \beta_0 + \beta_1 x$	First-order model with one predictor
$E(Y \mid x) = \beta_0 + \beta_1 x + \beta_2 x^2$	Second-order model with one predictor
$E(Y \mid x) = \beta_0 + \beta_1 x + \beta_2 x^2 + \beta_3 x^3$	Third-order model with one predictor
$E(Y \mid x_1, x_2) = \beta_0 + \beta_1 x_1 + \beta_2 x_2$	First-order model with two predictors
$E(Y \mid x_1, x_2) = \beta_0 + \beta_1 x_1 + \beta_2 x_2 + \beta_3 x_1 x_2$	First-order model with interaction
$E(Y \mid x_1, x_2) = \beta_0 + \beta_1 x_1 + \beta_2 x_2 + \beta_3 x_1^2 + \beta_4 x_2^2$	Second-order model with two predictors
$E(Y \mid x_1, x_2) = \beta_0 + \beta_1 x_1 + \beta_2 x_2 + \beta_3 x_1^2 + \beta_4 x_2^2 + \beta_5 x_1 x_2$	Second-order model with interaction

Table 21.1 Some models with higher-order terms

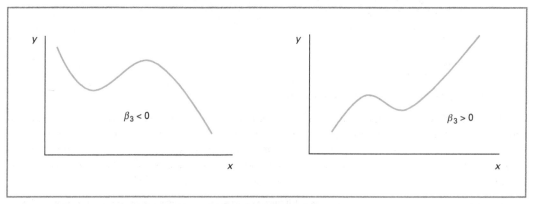

FIGURE 21.1 Curves associated with the third-order model with one predictor

The model $E(Y \mid x_1, x_2) = \beta_0 + \beta_1 x_1 + \beta_2 x_2$, the first-order model with two predictors, is used if the researcher believes that the relationship between $E(Y \mid x_1, x_2)$ and x_1 as well as the relationship between $E(Y \mid x_1, x_2)$ and x_2 are basically linear. Notice that this model implicitly assumes that – apart from the intercept – the linear relationship between $E(Y \mid x_1, x_2)$ and x_1 is the same for each fixed level of x_2; that the two predictors **do not interact**. This follows by fixing x_2 at different levels, for instance at 1, 2, 3 respectively:

$$E(Y \mid x_1; x_2 = 1) = (\beta_0 + \beta_2) + \beta_1 x_1$$

$$E(Y \mid x_1; x_2 = 2) = (\beta_0 + 2\beta_2) + \beta_1 x_1$$

$$E(Y \mid x_1; x_2 = 3) = (\beta_0 + 3\beta_2) + \beta_1 x_1$$

The three equations describe parallel straight lines, since only the intercepts may differ.

If it is expected that the relationship between Y and X_1 is different for different levels of X_2, it is better to take a first-order model with interaction term $X_1 X_2$. If the coefficient β_3 of this term is **unequal** to 0, then different levels of x_2 lead to linear relationships with different slopes. For instance, $x_2 = 2$ yields $\beta_0 + \beta_1 x_1 + 2\beta_2 + 2\beta_3 x_1$, which is equal to $(\beta_0 + 2\beta_2) + (\beta_1 + 2\beta_3)x_1$. Here are the results for $x_2 = 1, 2, 3$:

$$E(Y \mid x_1; x_2 = 1) = (\beta_0 + \beta_2) + (\beta_1 + \beta_3)x_1$$

$$E(Y \mid x_1; x_2 = 2) = (\beta_0 + 2\beta_2) + (\beta_1 + 2\beta_3)x_1$$

$$E(Y \mid x_1; x_2 = 3) = (\beta_0 + 3\beta_2) + (\beta_1 + 3\beta_3)x_1$$

Indeed, the slopes of these three lines are different if and only if $\beta_3 \neq 0$, so if and only if the interaction term X_1X_2 is significant within this model.

In the second-order model with two predictors, the relationship between $E(Y \mid x_1, x_2)$ and x_1 as well as the relationship between $E(Y \mid x_1, x_2)$ and x_2 are basically parabolic. This model assumes implicitly that – apart from the intercept – the parabolic relationship between $E(Y \mid x_1, x_2)$ and x_1 is the same for each fixed level of x_2; it assumes that the two predictors **do not interact**:

$$E(Y \mid x_1; x_2 = 1) = (\beta_0 + \beta_2 + \beta_4) + \beta_1 x_1 + \beta_3 x_1^2$$

$$E(Y \mid x_1; x_2 = 2) = (\beta_0 + 2\beta_2 + 4\beta_4) + \beta_1 x_1 + \beta_3 x_1^2$$

$$E(Y \mid x_1; x_2 = 3) = (\beta_0 + 3\beta_2 + 9\beta_4) + \beta_1 x_1 + \beta_3 x_1^2$$

But if it is expected that the quadratic (parabolic) relationship between Y and X_1 is different for different levels of X_2, it is better to include the interaction term X_1X_2, thus making the model a second-order model with interaction term. That such a model indeed leads to different (non-parallel) parabolic curves for different levels of X_2 follows easily by filling in such levels. For instance, $x_2 = 2$ yields;

$$\beta_0 + \beta_1 x_1 + 2\beta_2 + \beta_3 x_1^2 + 4\beta_4 + 2\beta_5 x_1 = (\beta_0 + 2\beta_2 + 4\beta_4) + (\beta_1 + 2\beta_5)x_1 + \beta_3 x_1^2$$

The different levels 1, 2 and 3 of X_2 lead to:

$$E(Y \mid x_1; x_2 = 1) = (\beta_0 + \beta_2 + \beta_4) + (\beta_1 + \beta_5)x_1 + \beta_3 x_1^2$$

$$E(Y \mid x_1; x_2 = 2) = (\beta_0 + 2\beta_2 + 4\beta_4) + (\beta_1 + 2\beta_5)x_1 + \beta_3 x_1^2$$

$$E(Y \mid x_1; x_2 = 3) = (\beta_0 + 3\beta_2 + 9\beta_4) + (\beta_1 + 3\beta_5)x_1 + \beta_3 x_1^2$$

They describe non-parallel parabolic functions if $\beta_5 \neq 0$, so if the interaction variable is significant in the model.

Example 21.3 (red-wire, part IX)

To explain the variation of Y = 'hourly wage', we regressed Y on X_1 = 'age' and X_2 = 'education level' by using a dataset of size $n = 150$; see Examples 20.1, 20.3, 20.6 and 20.7. The model was a first-order model with two predictors: $E(Y) = \beta_0 + \beta_1 x_1 + \beta_2 x_2$. But this model implicitly assumes that different levels for x_2 lead – apart from the intercept – to equal linear relationships between $E(Y)$ and x_1. To get an idea of whether this is indeed the case, we create sample regression lines for the five sub-datasets that belong to each of the education levels $x_2 = 1, 2, 3, 4, 5$. See Xmp20-01. sav for the data and Figure 21.2 for the lines.

For education level 5 especially, the slope of the sample regression line for the accompanying sub-dataset is very different from the slopes for the other sub-datasets. Note, however, that this level 5 regression line is based on only 13 sample elements.

We try to improve the model by including the interaction term X_1X_2. Hence, we will estimate the model with basic assumption:

$$E(Y) = \beta_0 + \beta_1 x_1 + \beta_2 x_2 + \beta_3 x_1 x_2$$

(See Appendix A1.21 for the creation of the new column with observations of the interaction variable.) The printout is shown below.

Model Summary

Model	R	R Square	Adjusted R Square	Std. Error of the Estimate
3	.720(a)	.518	.508	7.21339

a Predictors: (Constant), X1X2, X1, X2

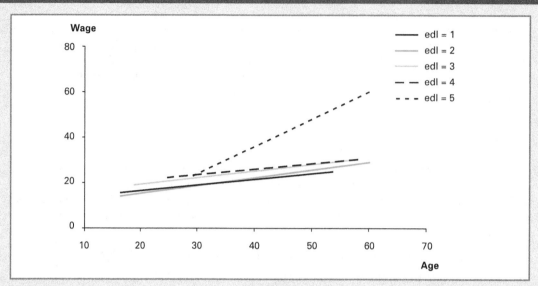

FIGURE 21.2 Combined scatter plots of Y on X_1 for the fixed education levels 1, 2, 3, 4, 5

ANOVA(b)

Model		Sum of Squares	df	Mean Square	F	Sig.
3	Regression	8155.480	3	2718.493	52.246	.000(a)
	Residual	7596.818	146	52.033		
	Total	15752.297	149			

a   Predictors: (Constant), X1X2, X1, X2
b   Dependent Variable: Y

Coefficients(a)

Model		Unstandardized Coefficients		Standardized Coefficients	t	Sig.
		B	Std. Error	Beta		
3	(Constant)	9.069	4.514		2.009	.046
	X1	.127	.128	.136	.993	.322
	X2	−1.530	1.657	−.171	−.923	.357
	X1X2	.134	.044	.748	3.048	.003

a   Dependent Variable: Y

The model is (very) useful; r^2 and r^2_{adj} increased from 0.487 (see Example 20.3) to 0.518 and from 0.480 to 0.508, respectively. The equation of the sample regression line is:

$$\hat{y} = 9.069 + 0.127x_1 - 1.530x_2 + 0.134x_1x_2$$

For $x_2 = 1$ and 5, the equation respectively turns into

$$\hat{y} = 7.539 + 0.261x_1 \quad \text{and} \quad \hat{y} = 1.419 + 0.797x_1$$

which follows by substituting 1 and 5 for x_2.

However, when studying the individual significance of the three independent variables it follows that only the interaction variable is significant at the 0.05 level. It would appear that X_1X_2 takes over the roles of X_1 and X_2. This is unwanted and probably due to collinearity since the correlation coefficient of X_2 and X_1X_2 is rather large (0.843; check it). In Section 22.1 a method (other than excluding X_1X_2) will be considered that occasionally works to solve such collinearity problems.

Example 21.4

Reconsider Example 21.2 and the dataset <u>Xmp21-02.sav</u>. It will be investigated whether the relationship between Y = CLOTEXP and, respectively, X_1 = HINC and X_3 = HS can better be described by a quadratic model. In both cases we expect a mountain-shaped parabola ('enough is enough'). The following three models will be estimated; in the last one the interaction term is also included:

$$E(Y) = \beta_0 + \beta_1 x_1 + \beta_2 x_1^2 \text{ and } E(Y) = \beta_0 + \beta_1 x_3 + \beta_2 x_3^2$$
$$E(Y) = \beta_0 + \beta_1 x_1 + \beta_2 x_3 + \beta_3 x_1^2 + \beta_4 x_3^2 + \beta_5 x_1 x_3$$

(If necessary, see Appendix A1.21 for the creation of the X_1^2, X_3^2 and X_1X_3 columns.) The coefficients of determination turn out to be 0.991763, 0.069049 and 0.992419, respectively The printouts below show the coefficient parts of the three models.

With respect to model 1, it seems that the relationship between CLOTEXP and HINC is indeed better described by this quadratic model. Indeed, β_2 is significantly negative (even at level 0.01) since the estimate of β_2 is negative and the p-value of the corresponding t-test is $0.006 / 2 = 0.003$. Hence, the quadratic relationship of model 1 is mountain-shaped. Incidentally, the realized standard error is **not** 0; only the first three decimals are 0. (In fact, it turns out that $s_B = 0.00002177$.)

For model 2, the variable X_3^2 is not significant, not even at significance level 0.10. The model does not become worse if X_3^2 is omitted.

Things are more complicated for model 3. Firstly, notice that the interaction term is not significant at the 0.05 level. Also, within this model the relationship between CLOTEXP and HINC seems to be best described by a quadratic function: the coefficient of X_1^2 is still significantly negative. The relationship between CLOTEXP and HS is no longer clear. It seems that the influence of HS on CLOTEXP is now divided among the variables X_3, X_3^2 and X_1X_3. Probably, the collinearity between these variables will be the problem. We had better omit the variables X_3^2 and X_1X_3, thus maintaining the model that includes only X_1, X_1^2 and X_3. Investigate that model yourself.

Coefficients(a)

Model		Unstandardized Coefficients		Standardized Coefficients	t	Sig.
		B	Std. Error	Beta		
1	(Constant)	−.061	.032		−1.919	.056
	HINC	.087	.002	1.053	49.231	.000
	HINC2	−6.02E-005	.000	−.059	−2.764	.006

a   Dependent Variable: CLOTEXP

Coefficients(a)

Model		Unstandardized Coefficients		Standardized Coefficients	t	Sig.
		B	Std. Error	Beta		
2	(Constant)	1.628	.278		5.852	.000
	HS	.467	.192	.562	2.432	.016
	HS2	−.040	.029	−.320	−1.386	.167

a   Dependent Variable: CLOTEXP

Coefficients(a)

Model		Unstandardized Coefficients		Standardized Coefficients	t	Sig.
		B	Std. Error	Beta		
3	(Constant)	−.029	.040		−.731	.465
	HINC	.086	.002	1.031	47.442	.000
	HS	−.020	.019	−.024	−1.041	.299
	HINC2	−6.36E-005	.000	−.063	−2.895	.004
	HS2	.003	.003	.026	1.211	.227
	HINCxHS	.001	.000	.034	1.796	.074

a   Dependent Variable: CLOTEXP

21.4 Logarithmic transformations

Regression models in economics and business sometimes use $\log(Y)$ and/or $\log(X)$ instead of Y and/or X. (Here, log denotes the **natural** logarithm.) A reason is that the model with $\log(Y)$ often better satisfies the model requirements. Another reason is that such log-models allow interpretations **in terms of percentages**. This will be illustrated below, but only for simple linear regression models. Things are similar for multiple linear regression models.

For a simple linear regression model with basic assumption $E(Y) = \beta_0 + \beta_1 x$, the consequences of replacing Y and/or X by $\log(Y)$ and/or $\log(X)$ will be demonstrated. In this context, the starting point $E(Y) = \beta_0 + \beta_1 x$ is termed the level-level model. The model with basic assumption $E(\log(Y)) = \beta_0 + \beta_1 x$ is termed the log-level model, etc. Table 21.2 gives an overview of the results that will be proved below; the interpretations will be illustrated in Example 21.5. In the last column of the table, $\%\Delta x$ and $\%\Delta y$ should be read as small percentage changes at, respectively, x and y.

Model	Dependent variable	Independent variable	Interpretation of β_1
1 Level–level	Y	X	$\Delta y \approx \beta_1 \Delta x$ (slope)
2 Level–log	Y	$\log(X)$	$\Delta y \approx \left(\dfrac{\beta_1}{100}\right) \times (\%\Delta x)$
3 Log–level	$\log(Y)$	X	$\%\Delta y \approx (100\beta_1)\Delta x$ (semi-elasticity)
4 Log–log	$\log(Y)$	$\log(X)$	$\%\Delta y \approx \beta_1 \times (\%\Delta x)$ (elasticity)

Table 21.2 Log/level models for simple linear regression

Example 21.5

For each of the four types of model, an example will be considered and the estimate b_1 of β_1 will be interpreted.

Level–level $Y =$ 'hourly wage (in euro)', $X =$ 'number of years of education'
Model: $E(Y) = \beta_0 + \beta_1 x$
Sample regression line: $\hat{y} = -0.905 + 0.541x$
Interpretation b_1: One more year of education will, on average, increase hourly wage by €0.541

Level–log $Y =$ 'number of hours worked per week', $X =$ 'hourly wage'
Model: $E(Y) = \beta_0 + \beta_1 \log(x)$
Sample regression line: $\hat{y} = 0.33 + 45.1 \log(x)$
Interpretation b_1: An increase in hourly wage of 1% will, on average, lead to an increase of $\frac{45.1}{100} = 0.451$ in hours worked per week

Log–level $Y =$ 'hourly wage (in euro)', $X =$ 'number of years of education'
Model: $E(\log(Y)) = \beta_0 + \beta_1 x$
Sample regression line: $\log(y) = 0.584 + 0.083x$
Interpretation b_1: One more year of education will, on average, increase the hourly wage by $100 \times 0.083 = 8.3\%$

Log–log $Y =$ 'demand for a product', $X =$ 'price of the product'
Model: $E(\log(Y)) = \beta_0 + \beta_1 \log(x)$
Sample regression line: $\log(y) = 4.7 - 1.25 \log(x)$
Interpretation b_1: An increase in price of 1% will, on average, lead to a fall of 1.25% in demand

Proofs of the interpretations in Table 21.2

1 Level–level

This interpretation is clear.

2 Level–log

Let $y(x) = \beta_0 + \beta_1 \log(x)$, then:

$$\Delta y = y(x + \Delta x) - y(x) = \beta_1 [\log(x + \Delta x) - \log(x)]$$

Since the derivative of $\log(x)$ is $\frac{1}{x}$, it holds for small Δx that:

$$\log(x + \Delta x) - \log(x) \approx \frac{\Delta x}{x} = \frac{1}{100} \times (100 \frac{\Delta x}{x}) = \frac{1}{100} \times (\% \Delta x)$$

Hence, $\Delta y \approx \dfrac{\beta_1}{100} \times (\% \Delta x)$

3 Log–level

Let $\log(y(x)) = \beta_0 + \beta_1 x$. Then

$$\log(y(x + \Delta x)) - \log(y(x)) = \beta_1 \Delta x$$

Since, because of the chain rule, the derivative of $\log(y(x))$ equals $\frac{y'(x)}{y(x)}$, it holds for small Δx that:

$$\log(y(x + \Delta x)) - \log(y(x)) \approx \frac{y'(x) \times \Delta x}{y(x)} \approx \frac{y(x + \Delta x) - y(x)}{y(x)} = \frac{1}{100} \times (\% \Delta y)$$

Hence, $\% \Delta y \approx (100 \beta_1) \Delta x$.

4 Log–log
Let $\log(y(x)) = \beta_0 + \beta_1 \log(x)$. Taking the derivative on both sides, yields

$$\frac{y'(x)}{y(x)} = \beta_1 \frac{1}{x} \quad \text{and} \quad \beta_1 = \frac{xy'(x)}{y(x)}$$

which is indeed the elasticity of y with respect to x. It follows that:

$$\beta_1 \approx \frac{x}{y(x)} \times \frac{y(x + \Delta x) - y(x)}{\Delta x} = \frac{x}{\Delta x} \times \frac{y(x + \Delta x) - y(x)}{y(x)}$$

So:

$$\beta_1 \times \frac{100\Delta x}{x} \approx \frac{100(y(x + \Delta x) - y(x))}{y(x)} \quad \text{and} \quad \%\Delta y \approx \beta_1 \times (\%\Delta x)$$

21.5 Analysis of variance by way of dummy variables

Originally, *analysis of variance* (*ANOVA*) was a separate statistical application that studied the means of one research variable Y under different circumstances (**treatments**). A classical example is the study of the yield Y of similar pieces of land that were treated with different fertilizers. In such studies the objective is to compare the means of Y under these different treatments in order to conclude whether these means are different and to find out which treatment has the largest (or smallest) mean.

Although the classical theory of analysis of variance has its own merits, we will not study that theory. Instead, dummy variables will be introduced into linear regression models, thus enabling us to answer questions that originally belonged to the field of analysis of variance.

It is possible to take 0–1 (dummy) variables as independent variables. Below, it will turn out that the population regression coefficients of dummy variables in linear regression models can be interpreted in terms of means or expectations. More general qualitative variables can also be included: for a qualitative variable with m possible values, $m - 1$ dummy variables will be needed.

21.5.1 Simple linear regression model with a dummy as independent variable
Suppose that the (only) independent variable X is a dummy variable, so only the values 0 and 1 are possible. Then the interpretation of the slope β_1 turns out to be special, as follows from the following arguments.

Recall from Section 19.2.2 that the slope β_1 in the model with basic assumption $E(Y) = \beta_0 + \beta_1 x$ can be interpreted as the change (positive or negative) of the expectation of Y **if x increases by 1**. However, X is a dummy, so it can only increase 1 unit by growing from 0 to 1. But this means that, for simple linear regression models with a dummy as independent variable, the slope β_1 is the difference between the expectations of Y when comparing the level 1 of the dummy with the level 0. The following more formal arguments illustrate this interpretation. Once again it is necessary to use the formal way to describe the basic assumption, that is, $E(Y \mid x) = \beta_0 + \beta_1 x$.

If μ_1 and μ_0 denote the mean of the observations of Y in the subpopulations with $x = 1$ and $x = 0$, respectively, then it follows that the slope β_1 is just the difference of these two means:

$$\mu_1 = E(Y \mid x = 1) = \beta_0 + \beta_1$$
$$\mu_0 = E(Y \mid x = 0) = \beta_0$$
$$\overline{} \quad -$$
$$\mu_1 - \mu_0 \qquad = \beta_1$$

Hence, statements and hypotheses concerning the difference $\mu_1 - \mu_0$ can be transformed into statements and hypotheses about β_1.

Similar remarks can be made for the **sample** regression line. It can be proved that substitution of $x = 1$ and $x = 0$ into the sample regression equation $\hat{y} = b_0 + b_1 x$ yields, respectively, the sample mean $\overline{y}_1$ of the y data with $x = 1$ and the sample mean $\overline{y}_0$ of the y data with $x = 0$. Furthermore, it can be proved that the difference between these two sample means is just the slope of the sample regression line: $b_1 = \bar{y}_1 - \bar{y}_0$.

Example 21.6 (red-wire, part X)

In former examples we detected that the variation of Y = 'gross hourly wage' is partially explained by the variation of the variables 'age' and 'education level'. To find out whether the variable 'gender' is also a factor, we start by explaining Y only from the dummy variable X = *fem* that takes the value 1 if the person is female. The data are in the file Xmp21-06.sav. Compare the results below with your solution to Exercise 18.31, which is based on the same dataset.

Since the level 1 of the dummy corresponds to the females and 0 to the males, the slope β_1 is now the difference between the means of the hourly wages for the females and the males. The regression can be run in the usual way. The printout is shown below.

Model Summary

Model	R	R Square	Adjusted R Square	Std. Error of the Estimate
4	.304(a)	.093	.086	9.82783

a   Predictors: (Constant), fem

ANOVA(b)

Model		Sum of Squares	df	Mean Square	F	Sig.
4	Regression	1457.540	1	1457.540	15.091	.000(a)
	Residual	14294.757	148	96.586		
	Total	15752.297	149			

a   Predictors: (Constant), fem
b   Dependent Variable: w

Coefficients(a)

Model		Unstandardized Coefficients		Standardized Coefficients	t	Sig.
		B	Std. Error	Beta		
4	(Constant)	24.901	1.135		21.943	.000
	fem	−6.234	1.605	−.304	−3.885	.000

a   Dependent Variable: w

It turns out that the model is useful at a very low significance level. The coefficient of determination is equal to 0.093, so 9.3% of the variation of the wage data is explained by gender. The sample regression equation is $\hat{y} = 24.901 - 6.234x$.

By substituting $x = 1$ into this equation we get the mean hourly wage $24.901 - 6.234 = €18.667$ of the 75 females in the sample; by substituting $x = 0$ we obtain the mean hourly wage €24.901 of the 75 men in the sample. (Check yourself that these two substitutions do indeed give these results.) The slope -6.234 of the sample regression equation is the difference between the sample means of the hourly wages of the females and the males.

The conjecture is that in this population the mean hourly wage of the men is more than €3 larger than the mean hourly wage of the women. With the notations μ_0 and μ_1 for the mean hourly wages of the men and the women in the population, the conjecture states that $\mu_0 > \mu_1 + 3$ or (equivalently) that $\mu_1 - \mu_0 < -3$. But recall that $\mu_1 - \mu_0 = \beta_1$, so the conjecture can be written as

$\beta_1 < -3$. In Chapter 19 we learned how to draw conclusions about such conjectures by conducting a hypothesis test. We take 0.05 as significance level. Notice that we need a one-sided test and that $t_{0.05;148} = 1.6552$ (*). Here is the five-step procedure:

   i Test $H_0: \beta_1 \geq -3$ against $H_1: \beta_1 < -3$; $\alpha = 0.05$

   ii Test statistic: $T = \dfrac{B_1 - (-3)}{S_{B_1}}$.

   iii Reject $H_0 \Leftrightarrow t \leq -1.6552$

   iv ***val*** $= \dfrac{-6.234 + 3}{1.605} = -2.015$

   v Reject H_0 since ***val*** is smaller than -1.6552

The statistical conclusion is that the conjecture is valid. The p-value of the test is equal to:

   p-value $= P(T \leq -2.015) = 0.0229$ (*)

Even $\alpha = 0.0229$ would have yielded the same conclusion, that the mean hourly wage of the men in the population is more than €3 larger than the mean hourly wage of the women in the population.

21.5.2 Multiple linear regression with a dummy

It is also possible to include a dummy in a model that already contains 'ordinary' independent variables. Again, the population regression coefficient of the dummy can be interpreted as a difference of two population means. However, things are formulated under a *ceteris paribus* restraint.

Example 21.7 (red-wire, part XI)

Suppose we want to explain Y = 'gross hourly wage' not only from the quantitative variables X_1 = 'age' and X_2 = 'education level', but also from the gender dummy X_3 = *fem*. Then, the basic assumption of the model is:

   $E(Y \mid x_1, x_2, x_3) = \beta_0 + \beta_1 x_1 + \beta_2 x_2 + \beta_3 x_3$

Let us compare women and men of the same age and the same level of education, and denote the respective means of the hourly wages by μ_f and μ_m. Substitution of $x_3 = 1$ and $x_3 = 0$ yields respectively:

   $\mu_f = E(Y \mid x_1, x_2; x_3 = 1) = \beta_0 + \beta_1 x_1 + \beta_2 x_2 + \beta_3$

   $\mu_m = E(Y \mid x_1, x_2; x_3 = 0) = \beta_0 + \beta_1 x_1 + \beta_2 x_2$

   -- $-$

   $\mu_f - \mu_m$ $= \beta_3$

Hence, β_3 is just the difference between the means of the hourly wages of the women and the men under the *ceteris paribus* restraint that women and men are of the same age and level of education. Indeed, this *ceteris paribus* restriction **is** important for the above subtraction since otherwise the term $\beta_1 x_1 + \beta_2 x_2$ would not have cancelled.

   Of course, the model can be estimated on the basis of the dataset of the 150 people; see Xmp21-07.sav. The SPSS-printout is shown below.

Model Summary

Model	R	R Square	Adjusted R Square	Std. Error of the Estimate
5	.718(a)	.516	.506	7.23007

a Predictors: (Constant), fem, edl, age

ANOVA(b)

Model		Sum of Squares	df	Mean Square	F	Sig.
5	Regression	8120.316	3	2706.772	51.781	.000(a)
	Residual	7631.982	146	52.274		
	Total	15752.297	149			

a Predictors: (Constant), fem, edl, age
b Dependent Variable: w

Coefficients(a)

Model		Unstandardized Coefficients		Standardized Coefficients	t	Sig.
		B	Std. Error	Beta		
5	(Constant)	−.036	2.371		−.015	.988
	age	.441	.057	.471	7.709	.000
	edl	3.245	.533	.363	6.089	.000
	fem	−3.551	1.213	−.173	−2.928	.004

a Dependent Variable: w

By conducting the model F-test it follows that the model is useful. From the coefficient of determination it follows that 51.6% of the variation in the wage data is explained by the three independent variables. Furthermore, all three independent variables are individually significant within the model, even at the significance level 0.01. In particular, it is concluded that $\beta_3 \neq 0$.

Notice that the statistical conclusion that $\beta_3 = \mu_f - \mu_m$ is unequal to 0 has a rather deep meaning. It tells the researcher that, within the population of interest, mean wage differentials between women and men are still present even after having taken into account that these differentials might be partially caused by differences in age and education level.

Notice that, in the last example, $b_2 = 3.245$. That is, a step of 1 in education level will, *ceteris paribus* and on average, lead to an increase of €3.245 in hourly wage. However, this interpretation is about a **general** jump in education level, not specifically from (for instance) level 4 to level 5. Since it also is interesting to know the mean change in hourly wage when going from level 4 (higher professional education) to level 5 (university education), we will in coming paragraphs adapt the above model in order to be able to consider such problems. In fact, we will replace the variable 'education level' by four dummy variables.

21.5.3 One-factor ANOVA

A basically qualitative variable with m levels (called **treatments**) can be represented in a model by $m - 1$ dummies. For instance, if, at an agricultural research centre, 100 small pieces of land are sprinkled with one out of ten possible fertilizers to compare their yields Y, then the following **nine** dummy variables can be used:

$$D_1 = \begin{cases} 1 & \text{if fertilizer 1 is used} \\ 0 & \text{otherwise} \end{cases}$$

$$D_2 = \begin{cases} 1 & \text{if fertilizer 2 is used} \\ 0 & \text{otherwise} \end{cases}$$

$$\vdots$$

$$D_9 = \begin{cases} 1 & \text{if fertilizer 9 is used} \\ 0 & \text{otherwise} \end{cases}$$

Notice that fertilizer 10 corresponds to the situation that all nine dummies are 0; this level is called the **base level**. The resulting dataset has 100 rows and ten columns; a row is about a piece of land. The first column contains the yields of the pieces of land. The second column contains zeros (for the pieces of land that were not treated with fertilizer 1) and ones (for the pieces of land treated with fertilizer 1); etc. The rows that contain only zeros in columns 2 to 10 are about pieces of land that were treated with fertilizer 10.

The (multiple) linear regression model that belongs to that example is as follows:

$$E(Y) = \beta_0 + \beta_1 D_1 + \ldots + \beta_9 D_9$$

For μ_0, the mean yield of the pieces of land treated with fertilizer 10, it follows that:

$$\mu_0 = \beta_0 + 0 + \ldots + 0 = \beta_0$$

For μ_j, the mean yield of the pieces of land treated with fertilizer j (for $j = 1, 2, \ldots, 9$), it follows that:

$$\mu_j = \beta_0 + \beta_j$$

Hence, $\beta_j = \mu_j - \mu_0$. The population regression coefficient β_j is, for $j \neq 0$, equal to the difference between the mean yield of pieces of land treated with fertilizer j and the mean yield of pieces of land treated with the base-level fertilizer (which is taken as 10). The population regression coefficient β_0 is equal to the mean yield of pieces of land treated with the base-level fertilizer.

There is another important consequence for models that have only dummy variables derived from one qualitative variable. In the above fertilizer example, the usefulness of the model can be checked by conducting the model test (the F-test of Section 20.4.1) for the testing problem:

$H_0: \beta_1 = \ldots = \beta_9 = 0$ against $H_1:$ at least one of $\beta_1, \ldots, \beta_9$ is $\neq 0$

However, since $\beta_j = \mu_j - \mu_0$ this testing problem can be written in terms of the $\mu_1, \ldots, \mu_9, \mu_0$:

$H_0: \mu_1 = \ldots = \mu_9 = \mu_0$ against $H_1:$ at least one of the $\mu_1, \ldots, \mu_9$ is $\neq \mu_0$

If H_0 is true, then the treatments cannot be distinguished as far their mean effects on Y are concerned. If H_1 is true, the treatments do **not** all have the same effect.

Example 21.8

The management of a bicycle factory is thinking of introducing a bonus system to stimulate productivity. In that system, welders would receive extra salary for each bicycle frame that they weld over and above the standard of 1000 frames per week for each welder. Possible bonuses are 15, 25 or 35 cents for each additional frame. However, the management has doubts whether such a system of bonuses is effective. To find out, 36 welders are arbitrarily divided into four groups of nine. Group 4 does not receive any bonus. The welders in groups 1, 2 and 3 are told that they will receive a bonus of 15, 25 and 35 cents, respectively, for each additional frame over 1000. After one week, the number of frames is counted for each welder. The results are in the file Xmp21-08.sav.

In this example, the treatments are the four types of bonus system. Three dummies are needed to incorporate the systems into the model; we will take the 0 bonus system as base level. The dummies

D_1, D_2 and D_3 get the value 1 if the welder is paid according to the 15 cents, 25 cents or the 35 cents system, respectively. The researcher has to create the dummies herself; see also Appendix A1.21. Note that the model is $E(Y) = \beta_0 + \beta_1 D_1 + \beta_2 D_2 + \beta_3 D_3$.

When the three columns for the dummies are created, the regression can be done; see the printouts below.

ANOVA(b)

Model		Sum of Squares	df	Mean Square	F	Sig.
1	Regression	224926.556	3	74975.519	15.304	.000(a)
	Residual	156772.000	32	4899.125		
	Total	381698.556	35			

a   Predictors: (Constant), D3, D2, D1
b   Dependent Variable: Nr_Frames

Coefficients(a)

Model		Unstandardized Coefficients		Standardized Coefficients	t	Sig.
		B	Std. Error	Beta		
1	(Constant)	1254.556	23.331		53.771	.000
	D1	107.333	32.995	.451	3.253	.003
	D2	139.556	32.995	.587	4.230	.000
	D3	220.444	32.995	.927	6.681	.000

a   Dependent Variable: Nr_Frames

According to the above theory, the test of whether there is a treatment effect is nothing but the F-test of Section 20.4.1. In the five-step procedure below, the parameters μ_1, μ_2 and μ_3 are the mean weekly number of frames for welders who respectively get paid according to the 15, 25 and 35 cents system. The mean number of frames for welders who get no bonus at all is denoted by μ_0. Notice that there are $k = 3$ independent variables in the model, that $n - (k + 1) = 36 - 4 = 32$ and that $F_{0.05;3;32} = 2.9011$ (*).

i   Test H_0: $\mu_1 = \mu_2 = \mu_3 = \mu_0$ against H_1: at least one of μ_1, μ_2, μ_3 is $\neq \mu_0$; $\alpha = 0.05$

ii   Test statistic: $F = \dfrac{SSR/3}{SSE/(n-4)}$

iii   Reject $H_0 \Leftrightarrow f \geq 2.9011$

iv   **val** = 15.304

v   Reject H_0 since **val** is larger than 2.9011

The conclusion is that there is a treatment effect; the means μ_1, μ_2, μ_3, μ_0 are not all equal.

To find out whether a bonus of 15 cents is enough to drive up productivity, we have to test whether $\mu_1 > \mu_0$ and hence that $\beta_1 > 0$. A t-test as in Chapter 20 can be used. Notice that $t_{0.05;32} = 1.6939$. (*) Here is the five-step procedure:

i   Test H_0: $\beta_1 \leq 0$ against H_1: $\beta_1 > 0$; $\alpha = 0.05$

ii   Test statistic: $T = \dfrac{B_1}{S_{B_1}}$

iii   Reject $H_0 \Leftrightarrow t \geq 1.6939$

iv   **val** = 3.253 (see the above printout)

v   Reject H_0 since **val** is larger than 1.6939

668

668 Chapter 21 Multiple linear regression: extension

Already the 15 cents system raises productivity. To compare the mean productivity of the 15 cents and the 25 cents system with the present model, a more advanced statistical method is needed that is beyond the scope of this book. However, this comparison can also be done by using a model similar to the present one but with the 15 cents system as base level. With that model these two mean productivities **can** be compared.

Example 21.9 (red-wire, part XII)

The variable 'education level' has five possible levels; we need four dummies to represent it in a linear regression for Y = 'gross hourly wage'. Taking level 1 as base level, we need the dummies:

$$D_2 = \begin{cases} 1 & \text{if educational level is 2} \\ 0 & \text{otherwise} \end{cases} \quad \dots \quad D_5 = \begin{cases} 1 & \text{if education level is 5} \\ 0 & \text{otherwise} \end{cases}$$

We will consider the model with the basic assumption $E(Y) = \beta_0 + \beta_1 D_2 + \dots + \beta_4 D_5$. The file Xmp21-09.sav contains only the 150 observations of the variables Y and 'education level'. The dummy columns again have to be created by you.

ANOVA(b)

Model		Sum of Squares	df	Mean Square	F	Sig.
6	Regression	4866.356	4	1216.589	16.205	.000(a)
	Residual	10885.942	145	75.075		
	Total	15752.297	149			

a Predictors: (Constant), D5, D4, D2, D3
b Dependent Variable: w

Coefficients(a)

Model		Unstandardized Coefficients		Standardized Coefficients	t	Sig.
		B	Std. Error	Beta		
6	(Constant)	17.162	1.769		9.704	.000
	D2	−.359	2.283	−.015	−.157	.875
	D3	4.984	2.132	.232	2.338	.021
	D4	7.582	2.476	.276	3.062	.003
	D5	21.039	3.063	.557	6.868	.000

a Dependent Variable: w

The model test tells us not only that the model is useful but also that the means of the hourly wages for the five subpopulations with education levels 1, 2, 3, 4 and 5 are not all the same. The standard t-tests yield that the dummies D_3, D_4 and D_5 are individually significant at the 0.05 significance level, but the dummy D_2 is not. These results can also be transformed into conclusions about means of hourly wages. The means of the hourly wages for the subpopulations with education levels 3, 4 and 5 are all unequal to the mean hourly wage of the subpopulation with education level 1. The data do not support the statement that the means of the hourly wages of the subpopulations with levels 1 and 2 are different.

From the column headed Sig. even stronger conclusions can be drawn. For instance, the mean of the hourly wages in the level 3 subpopulation is **larger** than the mean of the hourly wages in the level 1 subpopulation, as follows from a one-sided t-test for the testing problem with $H_1: \beta_2 > 0$. The p-value of this test is $0.021 / 2 = 0.0105$.

It is interesting to compare the coefficient of determination of this model (0.309) with the coefficient of determination of the simple linear regression model that explains Y from the variable 'education level' only. It turns out that the latter model has $r^2 = 0.244$. It would appear that replacing an ordinal variable by dummies can be fruitful.

21.5.4 Two-factor ANOVA (with or without interaction)

It is also possible to consider regression models with **two** qualitative independent variables (say, A and B) by including dummy variables. If A and B have, respectively, a and b levels, then $a - 1$ dummies are needed for A and $b - 1$ for B. This leads to a model with $a + b - 2$ dummy variables. If it is also required to take account of possible interactions, the $(a - 1)(b - 1)$ interaction terms have to be included too and a total of $ab - 1$ independent variables (all dummies) results. It will be clear that the number of independent variables can become very large.

Example 21.10

In the bicycle factory of Example 21.8, three different types of welding torch are used. Since one type of welding torch might be faster than another type, it is wise to take account of that. In essence, this was (without mentioning) already taken care of in the experimental design of Example 21.8: within all four groups of welders, welding torches of each of the three types were assigned arbitrarily to three welders.

The dataset Xmp21-10.sav records in column 1 the number of frames after one week, in column 2 the number of the payment group (1, 2, 3 or base level 4) and in column 3 the type of welding torch that was used (I, II or base level III).

The linear regression model contains the dummies D_1, D_2 and D_3 that belong to factor $A =$ 'bonus system of the welder' and the dummies E_1 and E_2 of the factor $B =$ 'type welding torch used by the welder'. To take account of interaction, another 6 interaction terms D_iE_j can be included. In total, there will be then 11 independent variables. Notice that the interaction terms are also dummies. For instance, D_2E_1 is 1 if both D_2 and E_1 are 1, so if the welder gets the 25 cents bonus and uses a type I welding torch; D_2E_1 is 0 if the welder does not get paid according to the 25 cents system and/or does not use a type I welding torch.

Firstly, consider the model **without** interaction terms:

$$E(Y) = \beta_0 + \beta_1D_1 + \beta_2D_2 + \beta_3D_3 + \beta_4E_1 + \beta_5E_2$$

The printout is shown below. By conducting the F-test of Section 20.4.1, it follows immediately that the model is useful and hence that the $4 \times 3 = 12$ different treatments do not all yield the same mean number of welded frames. It can also be tested whether there is a **welding torch effect** by conducting the partial F-test for the following testing problem:

  i Test H_0: $\beta_4 = \beta_5 = 0$ against H_1: at least one of β_4, β_5 is $\neq 0$; $\alpha = 0.05$

For the partial F-test, we also need the ANOVA printout of Example 21.8, for the model without the dummies E_1, E_2. Steps (ii)–(v) of the test can be written down. It is used that $F_{0.05;2,30} = 3.3158$ (*).

  ii Test statistic: $F = \dfrac{(SSE_r - SSE_c)/2}{SSE_c/(n - 6)}$

  iii Reject $H_0 \Leftrightarrow f \geq 3.3158$

  iv **val** $= \dfrac{(156772.000 - 91314.278)/2}{91314.278/30} = 10.7526$

  v Reject H_0 since **val** is larger than 3.3158

There is a significant welding torch effect. Moreover, when adding the dummies E_1, E_2, the coefficient of determination grows from 0.589 to 0.761.

ANOVA(b)

Model		Sum of Squares	df	Mean Square	F	Sig.
2	Regression	290384.278	5	58076.856	19.080	.000(a)
	Residual	91314.278	30	3043.809		
	Total	381698.556	35			

a   Predictors: (Constant), E2, D3, D2, E1, D1
b   Dependent Variable: Nr_Frames

Coefficients(a)

Model		Unstandardized Coefficients		Standardized Coefficients	t	Sig.
		B	Std. Error	Beta		
2	(Constant)	1302.250	22.523		57.818	.000
	D1	107.333	26.008	.451	4.127	.000
	D2	139.556	26.008	.587	5.366	.000
	D3	220.444	26.008	.927	8.476	.000
	E1	−103.500	22.523	−.474	−4.595	.000
	E2	−39.583	22.523	−.181	−1.757	.089

a   Dependent Variable: Nr_Frames

Next, consider the model **with** the six interaction terms. The printout is shown below. It turns out that the introduction of the six interaction terms forces the coefficient of determination to increase from 0.761 to 0.801. However, the adjusted coefficient of determination **de**creases from 0.721 to 0.709. This last fact, jointly with a quick glance at the printout, is the reason that it is barely expected that the inclusion of the interaction terms made the model better. But this also can be tested with a partial F-test. We mention only the steps (iv) and (v):

iv $val = \dfrac{(91314.278 - 76088.000)/6}{76088.000/24} = 0.8005$

v The data do not support the hypothesis that the interaction terms improve the model since 0.8005 is **not** larger than the critical value $F_{0.05;6,24} = 2.5082$ (*)

ANOVA(b)

Model		Sum of Squares	df	Mean Square	F	Sig.
3	Regression	305610.556	11	27782.778	8.763	.000(a)
	Residual	76088.000	24	3170.333		
	Total	381698.556	35			

a   Predictors: (Constant), D3E2, D3E1, D2E2, D2E1, D1E2, D1E1, D3, D2, D1, E2, E1
b   Dependent Variable: Nr_Frames

Coefficients(a)

Model		Unstandardized Coefficients		Standardized Coefficients	t	Sig.
		B	**Std. Error**	**Beta**		
3	(Constant)	1280.000	32.508		39.375	.000
	D1	137.000	45.973	.576	2.980	.007
	D2	166.667	45.973	.701	3.625	.001
	D3	252.667	45.973	1.063	5.496	.000
	E1	−73.667	45.973	−.337	−1.602	.122
	E2	−2.667	45.973	−.012	−.058	.954
	D1E1	−64.000	65.016	−.172	−.984	.335
	D1E2	−25.000	65.016	−.067	−.385	.704
	D2E1	4.632E-13	65.016	.000	.000	1.000
	D2E2	−81.333	65.016	−.218	−1.251	.223
	D3E1	−55.333	65.016	−.149	−.851	.403
	D3E2	−41.333	65.016	−.111	−.636	.531

a   Dependent Variable: Nr_Frames

21.5.5 General linear regression model

The general model is about qualitative **and** quantitative variables. Such generalizations are straight-forward. We consider an example.

Example 21.11 (red-wire, part XIII)

Suppose you believe that, ceteris paribus, university-educated men have, on average, higher hourly wages than university-educated women of the same age. Which model would you need to test it? Dataset: Xmp21-11.sav.

Let the dummies $D_2, \ldots, D_5$ be as in Example 21.9. Your belief concerns the question whether the relationship between $E(Y)$ and D_5 is different for males and females. Hence, it is about the inter-action between the dummies *fem* and D_5. In any case, a model is needed that also includes the interaction terms $femD_i$ of *fem* and the dummies D_i of 'education level'. We will also include the quantitative variable $X_1 = $ 'age'.

The first idea is to consider the following model:

$$E(Y) = \beta_0 + \beta_1 x_1 + \beta_2 fem + \beta_3 D_2 + \ldots + \beta_6 D_5 + \beta_7 femD_2 + \ldots + \beta_{10} femD_5$$

Your belief is about a comparison – for a fixed level x_1 of age – of the mean hourly wage of the sub-population with *fem* = 1 and education level = 5 and the mean hourly wage of the subpopulation with *fem* = 0 and education level = 5. That is, the following two means have to be compared:

$$E(Y \mid x_1; edl = 5; fem = 1) = \beta_0 + \beta_1 x_1 + \beta_2 + \beta_6 + \beta_{10}$$

$$E(Y \mid x_1; edl = 5; fem = 0) = \beta_0 + \beta_1 x_1 + 0 + \beta_6 + 0$$

Your belief states that the first mean is smaller than the second, so $\beta_2 + \beta_{10} < 0$. Unfortunately, we do not have the expertise to test the testing problem with $H_1: \beta_2 + \beta_{10} < 0$ as alternative hypothesis.

However, your belief **can** be tested if we take level 5 as base level and include the dummies $D_1, ..., D_4$ and the accompanying interaction terms in the model. (The dummy D_1 is 1 if the person has education level 1.) Here is the result:

$$E(Y) = \beta_0 + \beta_1 x_1 + \beta_2 fem + \beta_3 D_1 + ... + \beta_6 D_4 + \beta_7 femD_1 + ... + \beta_{10} femD_4$$

Notice that, for this model,

$$E(Y \mid x_1; edl = 5; fem = 1) = \beta_0 + \beta_1 x_1 + \beta_2$$

and

$$E(Y \mid x_1; edl = 5; fem = 0) = \beta_0 + \beta_1 x_1$$

so your belief states that the β_2 **of this model** is negative. And we do know how to test that: the printout is shown below.

ANOVA(b)

Model		Sum of Squares	df	Mean Square	F	Sig.
7	Regression	8691.806	10	869.181	17.112	.000(a)
	Residual	7060.492	139	50.795		
	Total	15752.297	149			

a   Predictors: (Constant), femD4, age, D1, femD2, femD3, femD1, D3, D4, D2, fem
b   Dependent Variable: w

Coefficients(a)

Model		Unstandardized Coefficients		Standardized Coefficients	t	Sig.
		B	Std. Error	Beta		
7	(Constant)	22.129	3.437		6.438	.000
	age	.430	.057	.458	7.477	.000
	fem	−8.049	4.759	−.393	−1.691	.093
	D1	−17.235	3.126	−.617	−5.514	.000
	D2	−17.493	3.073	−.729	−5.692	.000
	D3	−12.785	2.779	−.596	−4.601	.000
	D4	−9.842	3.236	−.358	−3.042	.003
	femD1	5.751	5.582	.146	1.030	.305
	femD2	6.500	5.328	.220	1.220	.225
	femD3	5.536	5.140	.204	1.077	.283
	femD4	2.914	5.560	.083	.524	.601

a   Dependent Variable: w

Notice that $t_{0.05;139} = 1.6559$ (*). Here is the five-step procedure:

i Test $H_0: \beta_2 \geq 0$ against $H_1: \beta_2 < 0$; $\alpha = 0.05$

ii Test statistic: $T = \dfrac{B_2}{S_{B_2}}$

iii Reject $H_0 \Leftrightarrow t \leq -1.6559$

iv **val** = −1.691

v Reject H_0 since **val** is smaller than −1.6559 (albeit narrowly)

The *p*-value of this test is 0.093 / 2 = 0.0465. The conclusion is in favour of your belief.

21.6 Model building

Model building is the name given to the collection of all activities that will finally lead to a model that can be used for statistical inference and prediction purposes. Below, the complete process of the building of such a model is summarized, but first, a word about stepwise regression.

21.6.1 Stepwise regression

Most statistical packages offer computerized approaches to determine suitable independent variables from a list of potential predictors. The result of such selection methods is that only variables that are individually significant within the chosen model are selected. See also Appendix A1.21.

Suppose there are k potential predictors. One type of *stepwise regression* more or less goes as follows. In the first step, the stepwise regression procedure runs all k **simple** linear regressions and selects the independent variable with the largest t-value (value of the t-test with hinge 0); call it X_1. If this variable is significant at significance level 0.05, the second step starts. All $k − 1$ linear regression models with **two** independent variables – X_1 and one of the 'others' – are estimated and the one (call it X_2) with the largest 'other' t-value is selected. Only if both variables are individually significant at level 0.05, step 3 starts. All models with X_1, X_2 and one of the $k − 2$ 'other' predictors are estimated. And so on. The procedure stops as soon as a step does not yield a new model with only individually significant variables; in that case, the model that resulted from the former step is taken as final model.

Stepwise regression procedures have **disadvantages** that must be mentioned.

It is tempting to treat the predictors that are not chosen by the procedure as unimportant. However, the model that finally results is the consequence of many t-tests. In each test, there is a positive probability that the conclusion is incorrect (errors of both types are possible). In particular, when the initial number of the predictors is rather large, there is a very high probability that somewhere in the process of steps an unimportant predictor is included (type I error) and an important predictor is eliminated (type II error).

Another disadvantage is that the procedure selects only predictors that are individually significant. But, occasionally, insignificant variables really have to be present in the model. For instance, if you started a study to find out whether wage differentials between men and women still exist nowadays, the gender dummy in your model should not be omitted as soon as it turns out to be insignificant. You would erase your proofs.

Warning!

It is not wise to let only the stepwise regression procedure decide which predictors are important for explaining the variation of a research variable. Prior theoretical reasons should drive the model choice, not the computer. Use the procedure only as a comparison afterwards, to compare your list of independent variables created on the basis of your knowledge (of economics) and ideas with the list coughed up by a computer. And if you decide to apply a stepwise regression procedure, use only basic predictors (no squares, dummies or interaction terms).

21.6.2 Steps to build a model

1 *Identify the dependent variable precisely.* For instance, if you are interested in 'weekly sales of a product', possible choices are 'number of units sold', 'gross sales (in euro)' or 'net profit (in euro)'.

2 *Determine the list of potential predictors.* Create this list of possible quantitative and qualitative predictors on the basis of your knowledge of the dependent variable, economic theory and other studies.

3 *Gather the data.* Take care that you get enough observations. Rule of thumb (although premature): take at least five observations for each independent variable.

4 *Make scatter plots for first impressions.* Create the scatter plots of the dependent variable on each of the possible predictors. Ask yourself the following questions. Do I see a linear relationship? Is there a more advanced relationship (second-order, third order, or even more complicated)?

5 *Formulate and estimate some basic linear regression models.* Use your findings in the previous step to formulate and estimate some basic models (first-order, second-order, only with dummies of one qualitative variable, etc.). Make use of your general knowledge of the dependent variable. A stepwise regression could be part of this step, but use it only for basic predictors (not for dummies or quadratic terms or interaction terms).

6 *Formulate your model.* This is the most difficult part of the process. The basic models and the findings of step 5 are combined into a larger model. However, things that seemed obvious in step 5 may become less obvious within this combined model. Collinearity may enter the stage; see also Section 22.1. Here is some further advice:

 a In the case of collinearity, intervene only if it causes problems. For instance, if X and X^2 are highly correlated but are (still) both individually significant, leave them both in. However, if the introduction of X^2 makes both X and X^2 individually insignificant, you might decide to leave X^2 out (or to solve the problem otherwise).

 b Insignificant independent variables should not automatically be removed from the model. In particular, when a variable turns out to be insignificant but this is unexpected, it is wise to leave it in.

 c If only one dummy of a group of dummies of one qualitative variable is significant, leave the whole group in.

 d If X^2 is significant but X is not, do not remove X from the model.

 e Try to find a balance between keeping the model simple and finding a model that fits 'best'.

7 *Check the model requirements.* Conduct a residual analysis to check the model assumptions for the model of step 6. If necessary (and possible), transform your variables and adapt the model. It may turn out that you have to go back to step 5 or even to choose another dependent variable. See also Chapter 22.

8 *Interpret the printout and formulate the consequences.* Having tested the usefulness of the 'final' model, the estimated model can now be used: to interpret the sample regression coefficients and to draw statistical conclusions about the population regression coefficients. The model can also be used to make (interval) predictions.

Example 21.12

Companies in the food sector are very keen on controlling the number of calories in their products. Burger King (www.BK.com) started a study of the number of calories and the nutritional content of its products; see the file Xmp21-12.sav. For a sample of 116 of its products the variable Calories and the following 11 other variables are observed:

Type Product
  1 = Whopper Sandwich
  2 = Fire-Grilled Burger
  3 = Chicken, Fish and Veggie
  4 = Side Orders
  5 = Salads (without dressing or garlic parmesan croutons)
  6 = Salad Dressings and Toppings
  7 = Desserts
  8 = Breakfast
  9 = Shakes

Serving Size (g)   Total fat (g)   Saturated fat (g)   Trans fat (g)   Cholesterol (mg)
Sodium (mg)   Total Carbs. (g) Dietary Fibre (g)   Total Sugars (g)   Protein (g)

The purpose is to create a model to explain the dependent variable Calories.

As a first attempt (but not being familiar with such food problems), the first-order model with the following independent variables is considered: Serving Size, Total fat, Saturated fat, Trans fat, Cholesterol, Total Carbs and Total Sugars. The printout is shown below.

ANOVA(b)

Model		Sum of Squares	df	Mean Square	F	Sig.
1	Regression	9560620.804	7	1365802.972	3964.328	.000(a)
	Residual	37208.506	108	344.523		
	Total	9597829.310	115			

a Predictors: (Constant), Total_Sugars_g, Total_fat_g, Trans_fat_g, Cholesterol_mg, Serving_Size_g, Total_Carbs_g, Saturated_fat_g
b Dependent Variable: Calories

Coefficients(a)

Model		Unstandardized Coefficients		Standardized Coefficients	t	Sig.
		B	Std. Error	Beta		
1	(Constant)	−2.668	3.566		−.748	.456
	Serving_Size_g	.319	.038	.187	8.433	.000
	Total_fat_g	8.554	.406	.490	21.093	.000
	Saturated_fat_g	3.824	.976	.098	3.917	.000
	Trans_fat_g	−7.311	1.409	−.049	−5.189	.000
	Cholesterol_mg	.205	.046	.054	4.451	.000
	Total_Carbs_g	4.475	.193	.564	23.208	.000
	Total_Sugars_g	−1.865	.160	−.248	−11.666	.000

a Dependent Variable: Calories

The model is useful, $r^2 = 0.996$ and all variables are individually significant.

To find out whether Type Product is a useful factor too, we next include eight dummies and take Salads as base level. In the printout below, D_j refers to the dummy belonging to level j of Type Product.

Coefficients(a)

Model		Unstandardized Coefficients		Standardized Coefficients	t	Sig.
		B	Std. Error	Beta		
2	(Constant)	−29.463	10.519		−2.801	.006
	Serving_Size_g	.357	.048	.210	7.503	.000
	Total_fat_g	8.148	.392	.467	20.787	.000
	Saturated_fat_g	4.389	1.098	.112	3.996	.000
	Trans_fat_g	−3.590	1.512	−.024	−2.374	.019
	Cholesterol_mg	.202	.051	.054	3.948	.000
	Total_Carbs_g	4.009	.210	.505	19.130	.000
	Total_Sugars_g	−1.478	.187	−.196	−7.893	.000
	D1	31.241	11.557	.037	2.703	.008
	D2	40.053	13.021	.039	3.076	.003
	D3	48.559	10.909	.071	4.451	.000
	D4	19.852	12.505	.023	1.588	.116
	D6	21.279	12.651	.016	1.682	.096
	D7	21.053	17.446	.010	1.207	.230
	D8	22.512	11.990	.030	1.878	.063
	D9	15.508	13.868	.018	1.118	.266

a   Dependent Variable: Calories

Furthermore, this model has SSE = 23271.502 and r^2 = 0.9998. The partial F-test to find out whether the eight dummies jointly are useful, has **val** = 7.4850 (check it) while $F_{0.05;8,100}$ = 2.0323 (*). So, the dummies significantly improve the model.

Next, stepwise regression is used to find a set of predictors from the list at the start; see also Appendix A1.21. A part of the (extensive) printout is shown below.

Coefficients(a)

Model		Unstandardized Coefficients		Standardized Coefficients	t	Sig.
		B	Std. Error	Beta		
1	(Constant)	158.954	17.771		8.944	.000
	Saturated_fat_g	35.327	1.584	.902	22.304	.000
2	(Constant)	92.395	13.558		6.815	.000
	Saturated_fat_g	27.858	1.273	.711	21.883	.000
	Total_Carbs_g	2.921	.258	.368	11.322	.000
3	(Constant)	4.376	6.801		.643	.521
	Saturated_fat_g	3.981	1.157	.102	3.441	.001
	Total_Carbs_g	3.776	.114	.476	33.244	.000
	Total_fat_g	10.870	.468	.623	23.224	.000
4	(Constant)	−1.850	.820		−2.257	.026

Model		Unstandardized Coefficients		Standardized Coefficients	t	Sig.
		B	Std. Error	Beta		
	Saturated_fat_g	.820	.144	.021	5.714	.000
	Total_Carbs_g	4.097	.014	.516	290.108	.000
	Total_fat_g	8.788	.061	.503	144.022	.000
	Protein_g	3.769	.043	.218	87.513	.000
5	(Constant)	−1.537	.758		−2.027	.045
	Saturated_fat_g	.027	.219	.001	.123	.903
	Total_Carbs_g	3.976	.030	.501	134.345	.000
	Total_fat_g	9.074	.084	.520	107.673	.000
	Protein_g	3.873	.046	.224	84.745	.000
	Total_Sugars_g	.166	.036	.022	4.556	.000
6	(Constant)	−1.551	.746		−2.077	.040
	Total_Carbs_g	3.974	.024	.501	164.218	.000
	Total_fat_g	9.083	.041	.520	222.205	.000
	Protein_g	3.876	.039	.225	100.241	.000
	Total_Sugars_g	.169	.022	.022	7.729	.000

a   Dependent Variable: Calories

In the first step, the variable Saturated fat is included. In step 2, the variable Total Carbs is added; etc. In step 5, the variable Total Sugars is added, but the variable Saturated fat becomes insignificant so it is excluded in Step 6. Step 6 describes the four variables that are the results of stepwise regression. Notice that this set of independent variables differs from the set considered before.

This last model yields $r^2 = 0.998$ and SSE = 1732.625. Adding the eight dummies does not improve the model: $SSE_c = 1499.801$ and **val** of partial F-test is 1.9987 (while $F_{0.05;8,103} = 2.0295$ (*)).

CASE 21.1 PRICING DIAMOND STONES – SOLUTION

The scatter plot of Y = 'price' on 'carat' uncovers the first problem: the homoskedasticity requirement of models based on Y will not be satisfied since heavier stones have considerably more price variation; see plot (a) below.

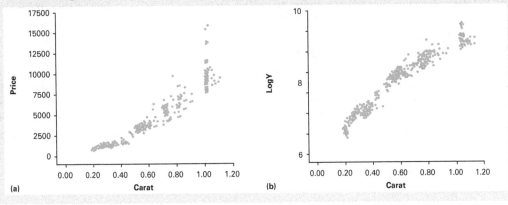

(a)

(b)

That is why, instead of Y, the natural logarithm logY is taken as dependent variable. Indeed, the scatter plot of logY on 'carat' (plot (b)) is more promising with respect to homoskedasticity.

To introduce the factors 'clarity', 'colour' and 'certification body' into the model, we include three groups of dummies: IF, VVS1, VVS2 and VS1; D, E, F, G and H; GIA and IGI, respectively. Hence, we take VS2 as base level for 'clarity', I for 'colour' and HRD for 'certification body'. Inclusion of all these dummies yields a first-order model for logY that contains 'carat' and 11 dummies. The sample regression equation is:

$$logY = 6.077 + 2.855Carat + 0.299IF + 0.298VVS1 + 0.202VVS2 + 0.097VS1 +$$
$$+ 0.417D + 0.387E + 0.310F + 0.210G + 0.129H + 0.009GIA - 0.174IGI$$

The underlying model seems to do well. Furthermore, the normality requirement does not seem to be a problem (check it). However, the scatter plot of the standardized residuals on 'carat' contains a very obvious mountain-shaped pattern (plot (a) below), which tells us that the model is 'mis-specified' (see also Section 19.8). Apparently, the model overestimates light and heavy stones and underestimates medium stones. There are several approaches to overcome this problem; all have to do with the relationship between logY and 'carat'. We will mention two of them.

Because of the mountain-shaped pattern of plot (a) below, it seems most natural to include the square of 'carat' in the model. If this is done, the regression equation becomes:

$$logY = 5.306 + 5.671Carat + 0.320IF + 0.226VVS1 + 0.143VVS2 + 0.076VS1 + 0.443D +$$
$$+ 0.363E + 0.287F + 0.198G + 0.104H + 0.006GIA - 0.019IGI - 2.103Caratsq$$

This model does very well, with $r^2 = 0.995$. The scatter plot of the standardized residuals on 'carat' looks fine (plot (b) below).

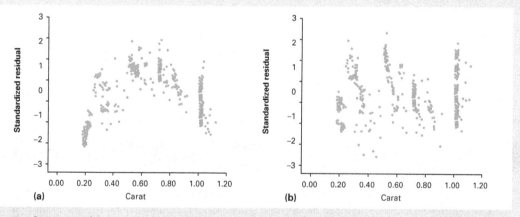

Furthermore, the scatter plot of the standardized residuals on the predictions does not show any problem regarding heteroskedasticity; check it yourself.

Another possible model follows from the scatter plot of Y on 'carat', which clearly shows a division of the stones into three clusters: less than 0.5 carats ('small'), 0.5 to less than 1 carat ('medium') and at least 1 carat ('large'). Inclusion of the dummies medium and large (taking small as base level) might improve things if it is taken into account that the relationship between logY and 'carat' might be different for the three clusters. That is why the interaction terms medium*carat and large*carat have also to be included. The reader is invited to estimate that model too and to compare it with the previous 'squared' model. Although the two models turn out to be more or less equally good, the squared model has some advantages because it is less complicated to interpret.

Summary

Several ways have been considered to generalize and extend the basic multiple regression model of Chapter 20. If a scatter plot of y on x gives cause for it, the variable X in the model can be replaced by a function $h(X)$ of X. Examples are:

$$\beta_0 + \beta_1 X + \beta_2 X^2 \quad \text{or} \quad \beta_0 + \beta_1 X + \beta_2 X^2 + \beta_3 X^3 \quad \text{or} \quad \log(X)$$

Also, the dependent variable Y can be transformed, for instance into $\log(Y)$. Such log transformations of X and/or Y are often used in the case of income or expenditure variables since the distributions of such variables tend to be right-skewed and may deviate significantly from normality assumptions. Such log transformations mean that regression coefficients can be interpreted in terms of percentages; see Section 21.4.

If it is suspected that the relationship between Y and an independent variable X_1 is different for different levels of X_2, the interaction term $X_1 X_2$ may be useful. For instance, the relationship between Y = 'gross hourly wage' and X_1 = 'age' might be different for the levels 1 and 5 of X_2 = 'education level', so it makes sense to investigate whether the interaction term is significant.

But be aware of collinearity! The inclusion of higher-order terms and interaction terms, especially, may cause collinearity problems.

Qualitative variables can be included into the model by way of dummy variables. It is important to establish that the regression coefficients of dummy variables can be interpreted in terms of differences between means or expectations under *ceteris paribus* restrictions.

The usefulness of a portion of the model can be tested with the partial F-test. For that test, two linear regression models – the complete model and the reduced model – have to estimated, since their SSEs are needed for the test statistic. The crucial point here seems to be **to find** the reduced model. It may help to remember that the reduced model corresponds to the null hypothesis ('that portion of the model is useless within the model') and the complete model to the alternative hypothesis ('that portion of the model is useful within the model').

🔒 Key terms

analysis of variance **662**	correlation matrix **653**	reduced model **650**
ANOVA **662**	interaction term **655**	stepwise regression **673**
base level **666**	model building **673**	strict collinearity **653**
collinearity **653**	multicollinearity **653**	treatments **665**
complete model **650**	partial F-test **650**	

❓ Exercises

Exercise 21.1

To explain the variation of a certain variable Y, observations of five potential predictors X_1, X_2, ..., X_5 are available. Suppose that $n = 100$ and that SST = 100000. For each of three first-order linear regression models, the SSE is given in the following table:

First-order model with predictors	SSE
X_1, X_2	60000
X_3, X_4, X_5	70000
X_1, X_2, X_3, X_4, X_5	55000

Test with $\alpha = 0.05$ whether the variables X_3, X_4 and X_5 are jointly useful within the model that includes X_1, X_2, X_3, X_4, X_5.

Exercise 21.2

A researcher wants to explain the variation of a research variable Y by way of a second-order linear regression model (with interaction) with predictors X_1 and X_2. She bases her research on a dataset of 150 observations of Y, X_1 and X_2. With respect to the y data, it is given that the sample variance is 3078.0261.

a She first estimates the first-order model with X_1 and X_2 only. It turns out that this model explains the amount SSR = 245716.2346 of the total variation. Test whether this model is useful; use $\alpha = 0.01$.

b She adds the squares of the variables X_1 and X_2 to the model of part (a). As a consequence, SSR increases (compared to part (a)) by the amount 89172.1826. Do these two squares jointly improve the model of part (a)? Use $\alpha = 0.01$.

c Finally, she adds the interaction term to the model of part (b) and ends up with the unexplained variation 85666.1832. Did the inclusion of the interaction term improve the model of part (b)? Use $\alpha = 0.01$.

Exercise 21.3

The owner of a fashion shop wants to find the relationship between the daily sales Y (in euro) and the daily number of people (X) who visit the shop. Since she believes that too many visitors will slow down the sales, she makes use of a second-order linear regression model.

a Formulate the basic assumption of the model. What will be the sign of the population regression coefficient of X^2 if the belief of the owner is true? Why?

The model is estimated on the basis of a sample of 250 days. Here are some of the results:

$\hat{y} = -23.758 + 98.742x - 0.199x^2$; $s_{B_1} = 10.1555$; $s_{B_2} = 0.0734$

b Use a hypothesis test with significance level 0.02 to draw a conclusion about the belief of the owner.

Exercise 21.4

The managing director of a large furniture factory wants to know whether, on average, the male and female workers need different times to assemble a special type of chair. She randomly chooses 20 female and 30 male workers and records their assembly times (in minutes); see the dataset Xrc21-04.sav. Here, DF = 1 if the worker is female.

a Write down the basic assumption of a linear regression model that is suitable for this study. Explain the meaning of this assumption in the present context. Do the same for the homoskedasticity requirement.

The data are used to estimate the model. A part of the printout is given below.

Coefficients(a)

Model		Unstandardized Coefficients		Standardized Coefficients	t	Sig.
		B	Std. Error	Beta		
1	(Constant)	24.800	.486		51.068	.000
	DF	−.200	.768	−.038	−.260	.796

a   Dependent Variable: AssTime

b Use these results to test whether the data allow the conclusion that the means of the assembly times for the male and female workers are different.

c Instead of using a linear regression model, it is also possible to use the independent samples approach of Section 18.2.1. From the data it follows that:

$\bar{x}_1 = 24.60;\ s_1 = 3.033;\ \bar{x}_2 = 24.80;\ s_2 = 2.384$

(Here, 1 refers to the female workers.) **Conduct the equal variances *t*-test** to find out whether the means of the assembling times for the male and female workers are different.

d But are the population variances **un**equal? Test it with $\alpha = 0.05$.

Exercise 21.5

A real estate agent selling holiday homes in the cities W and C in Costa Rica wants to explain the selling price Y (in units of $1000) of a holiday home from the following variables:

X_1: land area of the lot (in units of 1000 square feet)
X_2: 1 if the holiday home is in C (and 0 if it lies in W)
X_3: number of rooms in the holiday home

The agent has data for 90 homes (50 in C and 40 in W) and uses them to estimate the first-order linear regression model with the predictors X_1, X_2 and X_3. Here are some of the results:

SST $= 56080.65000$ and   SSR $= 38414.30838$

$b_0 = 10.7237;\quad b_1 = 32.3748;\quad b_2 = -11.5826;\quad b_3 = 0.9210;\quad s_{B_2} = 3.1594$

a Write down the basic assumption of the model. Interpret the coefficient of X_2 in terms of means of selling prices.

b The conjecture is that the expected selling price of a holiday home in W is more than $5000 higher than the expected selling price of a holiday home in C with the same land area and the same number of rooms. Test it with $\alpha = 0.05$.

Exercise 21.6

In this investigation we study the effects of several factors on the management performance of executive managers within a certain large company. The variables below were measured for 100 managers from several departments:

Y: grade for overall performance on scale 1–5 (1 = very bad; 5 = excellent)
X_1:   1 if the manager is male and 0 if female
X_2:   number of years experience in that occupation
X_3:   grade (scale 1–5) for contacts with subordinates (1 = very bad; 5 = excellent)
X_4:   number of hours spent weekly on the job
X_5:   1 if middle or senior manager and 0 if not
X_6:   grade for efforts to motivate subordinates

a Firstly, the model $E(Y) = \beta_0 + \beta_1 x_1 + \beta_2 x_2 + \beta_3 x_3 + \beta_4 x_4$ is estimated. It turns out that SSE $= 352$ and $r^2 = 0.11$. Is this model useful to explain the variation of Y? Take $\alpha = 0.05$.

b Next, the variables X_5 and X_6 are added to the model in part (a). For the resulting model, it turns out that SSE $= 341$ and $r^2 = 0.14$. Do X_5 and X_6 jointly contribute to explain the variation of Y? Take $\alpha = 0.05$.

c Finally, the interaction term $X_5 X_6$ is included in the model of part (b); it turns out that SSE $= 321$ and $r^2 = 0.19$. Is the interaction term significant within this model? Take $\alpha = 0.05$.

d Could you have answered the questions in parts (b) and (c) if the facts about SSE had not been given? Explain.

Exercise 21.7

Prove the validity of (21.1).

Exercise 21.8

Reconsider the research variable Y = 'annual household expenditure on food (in thousands of euro)'; see also Exercises 20.14 and 20.15. Interest is in the relationship between Y and the variables:

X_1 = 'annual household income (in thousands of euro)'
X_2 = 'number of children in the household'
X_3 = 'size of the household'

The file Xrc21-08.sav contains data for a random sample of 300 households.

a The scatter plots below give first impressions of the relationships between Y and each of the three above variables. The best fitting (least-squares) quadratic curves are included. By studying these plots, what kinds of relationship do you expect between Y and respectively X_1, X_2, X_3?

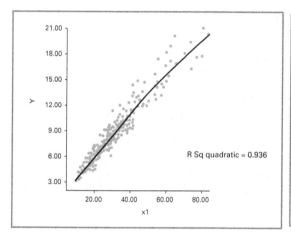

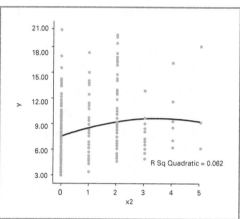

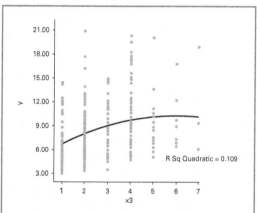

b The three printouts below are the coefficients parts of printouts for the three models with basic assumption $E(Y) = \beta_0 + \beta_1 x_i + \beta_2 x_i^2$. Do these printouts confirm your findings in part (a)? Use the p-values of suitable t-tests (with $\alpha = 0.05$) to answer this question.

Model		Unstandardized Coefficients		Standardized Coefficients	t	Sig.
		B	Std. Error	Beta		
1	(Constant)	.126	.264		.478	.633
	X1	.304	.015	1.218	20.448	.000
	X1X1	−.001	.000	−.260	−4.373	.000

Model		Unstandardized Coefficients		Standardized Coefficients	t	Sig.
		B	Std. Error	Beta		
2	(Constant)	7.757	.257		30.174	.000
	X2	1.279	.428	.429	2.987	.003
	X2X2	−.182	.124	−.210	−1.465	.144

Model		Unstandardized Coefficients		Standardized Coefficients	t	Sig.
		B	Std. Error	Beta		
3	(Constant)	5.105	.819		6.234	.000
	X3	1.737	.565	.694	3.072	.002
	X3X3	−.146	.085	−.389	−1.722	.086

c Consider the second-order model (without interaction) in X_1, X_3. From the printouts below, determine the sample regression equation and test (with $\alpha = 0.05$) whether the quadratic terms improve the **first**-order model in X_1, X_3.

ANOVA(b)

Model		Sum of Squares	df	Mean Square	F	Sig.
a	Regression	3413.749	2	1706.875	2400.147	.000(a)
	Residual	211.213	297	.711		
	Total	3624.962	299			

a Predictors: (Constant), X3, X1

ANOVA(b)

Model		Sum of Squares	df	Mean Square	F	Sig.
b	Regression	3424.752	4	856.188	1261.554	.000(a)
	Residual	200.210	295	.679		
	Total	3624.962	299			

a Predictors: (Constant), X3X3, X1X1, X1, X3

Coefficients(a)

Model		Unstandardized Coefficients		Standardized Coefficients	t	Sig.
		B	Std. Error	Beta		
b	(Constant)	−.318	.295		−1.079	.281
	X1	.288	.014	1.154	20.426	.000
	X3	.337	.143	.135	2.354	.019
	X1X1	−.001	.000	−.218	−3.900	.000
	X3X3	−.015	.021	−.040	−.711	.477

Exercise 21.9

Reconsider Exercise 21.8. We are wondering whether the relationship between Y and X_1 is different for different levels of X_3. That is why we extend the model of part (c) by including the interaction term X_1X_3.

a Formulate the basic assumption of this new model. Do X_1 and X_3 interact? Use $\alpha = 0.05$ and the following printout.

Coefficients(a)

Model		Unstandardized Coefficients		Standardized Coefficients	t	Sig.
		B	Std. Error	Beta		
c	(Constant)	.225	.313		.720	.472
	X1	.272	.014	1.090	19.148	.000
	X3	.115	.148	.046	.775	.439
	X1X1	−.001	.000	−.284	−5.030	.000
	X3X3	−.033	.021	−.088	−1.560	.120
	X1X3	.011	.003	.214	4.312	.000

b It seems that 'X_1X_3 takes over the role of X_3'. Explain what is meant by this statement. What could be the reason?

c Which of the models in this exercise and Exercise 21.8 do you prefer? Why?

Exercise 21.10

In Exercise 19.11 the relationship was studied between the mileage Y (in miles) of a car and its age X_1 (in years). However, it turns out that the first half of the 22 measurements in the dataset come from petrol cars and the second half from diesel cars. This offers the opportunity to deepen the study of the variation of Y, by extending the model with the dummy variable X_2 which takes the value 1 for petrol cars. The file Xrc21-10.sav contains the extended dataset.

a Write down the basic assumption of the model. Interpret the coefficient of X_2 in terms of means.

After having run the regression, it turns out that:

$\hat{y} = 7114.799 + 10120.783x_1 - 6756.257x_2; \quad s_{B_1} = 784.113; \quad s_{B_2} = 2163.588;$
$SST = 5130875000.0; \quad SSR = 4642893498.4$

b Determine the coefficient of determination and interpret it. Also interpret the coefficient of x_2 in the sample regression equation.

c Use a t-test to decide whether the extension improves the model; use 0.01 as significance level.

d Use a suitable 99% confidence interval to conclude whether the mean of the mileages of diesel cars of a certain age is more than 1000 miles greater than the mean of the mileages of petrol cars of the same age.

Exercise 21.11

In Exercise 20.12 a first-order linear regression model was considered to explain, for a large population of adult people, the variation of the variable WEIGHT (kg). The following independent variables were used:

HEIGHT (cm), AGE (years), EDU (education level, values 1–5), HOURS (number of hours worked weekly), WAGE (gross hourly wage in euro)

However, the variable 'gender' will also explain a part of the variation of the variable WEIGHT, so we should include that variable too. The file Xrc21-11.sav also contains the observations of the dummy variable DF (which takes the value 1 for females) for a random sample of 300 persons.

a Formulate the first-order linear regression model.

b The printout below gives estimation results for the model. Use it to test whether the mean weight of male adults with certain fixed characteristics for the variables HEIGHT, AGE, EDU, HOURS and WAGE is more than 3 kg greater than the mean weight of female adults with the same characteristics. Use 0.05 as significance level.

Model		Unstandardized Coefficients		Standardized Coefficients	t	Sig.
		B	Std. Error	Beta		
1	(Constant)	−31.518	12.995		−2.425	.016
	HEIGHT	.590	.071	.452	8.315	.000
	AGE	.132	.043	.157	3.074	.002
	EDU	.459	.541	.043	.848	.397
	WAGE	−.116	.167	−.047	−.697	.486
	HOURS	−.047	.042	−.074	−1.111	.267
	DF	−6.598	1.830	−.218	−3.606	.000

c Test whether the three variables EDU, WAGE and HOURS jointly contribute to the model of part (b); use significance level 0.05. You will need the following printouts:

ANOVA(b)

Model		Sum of Squares	df	Mean Square	F	Sig.
1	Regression	16339.829	6	2723.305	30.160	.000(a)
	Residual	26456.265	293	90.294		
	Total	42796.094	299			

a   Predictors: (Constant), DF, EDU, AGE, HOURS, HEIGHT, WAGE

ANOVA(b)

Model		Sum of Squares	df	Mean Square	F	Sig.
2	Regression	15953.802	3	5317.934	58.643	.000(a)
	Residual	26842.292	296	90.683		
	Total	42796.094	299			

a   Predictors: (Constant), DF, AGE, HEIGHT

Exercise 21.12

Reconsider Exercise 21.11. Because of the results so far, we exclude the variables HOURS and WAGE from the model of part (b). However, instead of EDU we include four education dummies taking education level 1 as base level. Use the printout below to test whether these education dummies improve the first-order linear regression model that includes only the independent variables HEIGHT, AGE and DF; use significance level 0.05. (You may also need a printout from Exercise 21.11.)

ANOVA(b)

Model		Sum of Squares	df	Mean Square	F	Sig.
3	Regression	16092.097	7	2298.871	25.137	.000(a)
	Residual	26703.997	292	91.452		
	Total	42796.094	299			

a   Predictors: (Constant), Dlev5, AGE, HEIGHT, Dlev2, Dlev4, DF, Dlev3

Exercise 21.13

An economist investigates the profitability of companies with at most 200 employees. Data are available for 100 such companies. For each of them, the following variables are observed:

Y: last year's profit of the company (in 10^4 euro)
NEmpl: number of employees at 31 December last year
DServ: dummy with value 1 if the company operates in the service sector (and 0 if in the non-service sector)

The economist decides to use a linear regression model with LNEmpl = log(NEmpl) and DServ as independent variables. (Here, log refers to the natural logarithm.)

a Write down the basic assumption of the model. Interpret the coefficients of LNEmpl and DServ.

The data are used to estimate the model. Here are some results:

SST = 195228.77390; SSE = 156994.41893; SSR = 38234.35496

$\hat{y}$ = 48.9948 + 17.7194LNEmpl + 38.3421DServ

s_{B_1} = 5.77333; s_{B_2} = 10.25552

b Test whether the mean profit of companies in the service sector is larger than the mean profit of companies in the non-service sector with the same number of employees. Use $\alpha = 0.05$.

c Test whether $\beta_1 > 7$; use $\alpha = 0.05$. Interpret your conclusion.

The economist decides to include the interaction term.

d Explain why this interaction term might indeed improve the model.

e The SSE of the (estimated) extended model turns out to be 155310.00842. Use this information to conclude whether the interaction term is significant within the model; use $\alpha = 0.05$.

Exercise 21.14 (red-wire, part XIV)

Consider the wages case that explains W = 'gross hourly wage' from age, fem and the education dummies D_2, D_3, D_4 and D_5; see Xrc21-14.sav.

To check homoskedasticity, we plot the standardized residuals on the predictions $\hat{w}$ and get the left-hand picture below.

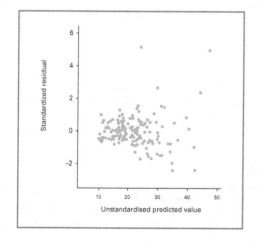 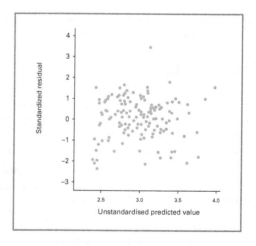

Notice that the variation along vertical lines slightly increases with $\hat{w}$, which is unwanted since it is not in accordance with the homoskedasticity requirement. As an attempt to solve that problem, the dependent variable W is replaced by $LOGW = \log(W)$. The resulting model is estimated and the scatter plot of the standardized residuals on the predictions is created; see the right-hand plot. This plot seems to be in accordance with the homoskedasticity requirement.

But several questions arise. Is it a 'good' model? How should the sample regression coefficients be interpreted? Is this model better than the original model?

a Write down the basic assumption of the log model. Interpret the coefficient of age.

The printout below gives estimation results for this model.

ANOVA(b)

Model		Sum of Squares	df	Mean Square	F	Sig.
	Regression	16.675	6	2.779	34.576	.000(a)
	Residual	11.494	143	.080		
	Total	28.169	149			

a   Predictors: (Constant), D5, D4, fem, D2, age, D3
b   Dependent Variable: LOGW

Coefficients(a)

Model		Unstandardized Coefficients		Standardized Coefficients	t	Sig.
		B	Std. Error	Beta		
	(Constant)	2.217	.097		22.884	.000
	age	.020	.002	.498	8.725	.000
	fem	−.156	.048	−.180	−3.255	.001
	D2	.015	.075	.014	.195	.846
	D3	.258	.070	.284	3.687	.000
	D4	.342	.082	.294	4.176	.000
	D5	.549	.103	.344	5.318	.000

a   Dependent Variable: LOGW

b Use the p-value of the F-test to check whether the model is useful.

c The coefficient of determination for this estimated model is 0.592 while the original model has $r^2 = 0.545$. Explain why it is problematic to compare the two determination coefficients.

d Give the interpretation of the coefficient of fem in the sample regression equation.

e Also interpret the coefficient of D_5 in the sample regression equation.

Exercise 21.15 (red-wire, part XV)

Reconsider Exercise 21.14. Mr Jansen is 30 years old and has a higher professional education (level 4).

a Determine a point prediction for his hourly wage. (*Hint*: recall that $e^{LOGW} = W$.)

b A 98% prediction interval for Mr Jansen's log-wage can easily be calculated with a computer: (2.46696, 3.83578). Determine a 98% prediction interval for Mr Jansen's hourly wage.

Exercise 21.16

The management of an amusement park wants to know each day how many additional personnel are needed. Important factors are: whether it concerns a day of the weekend or not, whether the weather forecast for that day is sunny or cloudy, and the maximum temperature forecast for that day. The management makes use of the linear regression model with the basic assumption

$$E(Y) = \beta_0 + \beta_1 x_1 + \beta_2 x_2 + \beta_3 x_3$$

where: $Y =$ number of visitors on a day

$X_1 = \begin{cases} 1 \text{ if the day is in the weekend} \\ 0 \text{ if not in the weekend} \end{cases}$

$X_2 = \begin{cases} 1 \text{ if the weather is forecast to be sunny on that day} \\ 0 \text{ if the weather is forecast to be cloudy} \end{cases}$

$X_3 =$ forecast temperature (°C) for that day

The model is estimated on the basis of a random sample of 25 days. Here are some results:

$$\hat{y} = 1000 + 100x_1 + 500x_2 + 10x_3; \quad r^2 = 0.70; \quad s_{B_1} = 40; \quad s_{B_2} = 100; \quad s_{B_3} = 5$$

a Test whether the model is useful; use $\alpha = 0.05$.

b Interpret β_1.

c Interpret $b_1 = 100$.

d Test whether the expected number of visitors on a sunny day is larger than the expected number of visitors on a cloudy day with the same characteristics for X_1 and X_3. Use $\alpha = 0.05$.

Exercise 21.17

A certain production process has always been performed at the temperature of 240°C and pressure 1.6 atm. However, recent publications suggest that other levels of temperature and/or pressure may yield higher production. The following research is meant to investigate the role of temperature.

For 12 days the production process is performed at 240°C, for 12 days at 260°C and for 12 days at 280°C. On each day, the size of the production is recorded in the file Xrc21-17.sav.

a Write down the basic assumption of the first-order regression model that explains the daily production Y directly from the temperature T.

b After having estimated this model, it turns out that $r^2 = 0.046$ (check it yourself). Is the model useful? Use significance level 0.05.

Next, consider the model that explains Y from the dummies D_2 and D_3 that correspond to the temperatures 260 and 280, respectively.

c Write down the basic assumption of the model and interpret the coefficient of D_3.

After having estimated this model, it turns out that $r^2 = 0.598$, that $b_1 = 14.000$ and that $s_{B_1} = 2.060$ (check it yourself).

d Is this model useful? Use significance level 0.01.

e Do the three temperatures on average yield different daily production sizes? Use significance level 0.01.

f Check whether performing the production process at 260°C yields a mean daily production size that is more than 8 larger than the mean daily production size at 240°C. Use a suitable 98% confidence interval to draw your conclusion.

Exercise 21.18

In Exercise 21.17, the role of pressure was not yet incorporated. However, during the first four days of each group of 12 days the pressure was set at 1.6 atm, the next four days at 1.7 atm and the last four days at 1.8 atm. The daily pressures are included in the dataset Xrc21-18.sav, along with daily production and the temperatures. A model will be constructed that includes the temperature **and** the pressure.

As in the previous exercise, temperature is included by way of the two dummies D_2 and D_3 that take the value 1 if the temperature is, respectively, 260 or 280. So, 240 is taken as base level for temperature. Similarly, pressure is included by way of the two dummies E_2 and E_3 that take the value 1 if the pressure is, respectively, 1.7 or 1.8. That is, the base level for the pressure is 1.6.

a Write down the basic assumption of the model with the two types of dummies.

The ANOVA parts of the printouts of several models are shown below.

ANOVA(b)

Model		Sum of Squares	df	Mean Square	F	Sig.
2	Regression	1560.000	4	390.000	22.898	.000(a)
	Residual	528.000	31	17.032		
	Total	2088.000	35			

a Predictors: (Constant), E3, D3, D2, E2
b Dependent Variable: Prod

ANOVA(b)

Model		Sum of Squares	df	Mean Square	F	Sig.
1	Regression	1248.000	2	624.000	24.514	.000(a)
	Residual	840.000	33	25.455		
	Total	2088.000	35			

a Predictors: (Constant), D3, D2
b Dependent Variable: Prod

ANOVA(b)

Model		Sum of Squares	df	Mean Square	F	Sig.
3	Regression	312.000	2	156.000	2.899	.069(a)
	Residual	1776.000	33	53.818		
	Total	2088.000	35			

a Predictors: (Constant), E3, E2
b Dependent Variable: Prod

b Does there exist a temperature effect on the mean daily production? Test it with $\alpha = 0.05$.

c Does there exist a pressure effect on the mean daily production? Test it with $\alpha = 0.05$.

Exercise 21.19

Reconsider Exercise 21.18. Also include the four interaction variables $D_i E_j$ (for $i = 2, 3$ and $j = 2, 3$) in the model of Exercise 21.18. The printout is shown below.

ANOVA(b)

Model		Sum of Squares	df	Mean Square	F	Sig.
4	Regression	1728.000	8	216.000	16.200	.000(a)
	Residual	360.000	27	13.333		
	Total	2088.000	35			

a Predictors: (Constant), D3E3, D3E2, D2E3, D2E2, D2, E3, D3, E2
b Dependent Variable: Prod

Coefficients(a)

Model		Unstandardized Coefficients		Standardized Coefficients	t	Sig.
		B	Std. Error	Beta		
4	(Constant)	136.000	1.826		74.490	.000
	D2	18.000	2.582	1.114	6.971	.000
	D3	3.000	2.582	.186	1.162	.255
	E2	6.000	2.582	.371	2.324	.028
	E3	−6.33E-015	2.582	.000	.000	1.000
	D2E2	−3.000	3.651	−.124	−.822	.419
	D2E3	−9.000	3.651	−.371	−2.465	.020
	D3E2	1.750E−15	3.651	.000	.000	1.000
	D3E3	3.000	3.651	.124	.822	.419

a   Dependent Variable: Prod

a Write down the basic assumption of the model. Are the variables D_iE_j dummies? Why (not)?

b Test whether there exists interaction between temperature and pressure; use $\alpha = 0.05$. You will also need a printout from Exercise 21.18.

c Find a temperature / pressure combination that seems to do better than the combination 240 / 1.6. Only use estimation arguments, no test.

Exercise 21.20 (computer)

A firm that markets goods targeting youth in several countries has commissioned the market research company MarketView to investigate whether there are differences between countries with respect to ethical questions.

MarketView undertook market research among university students across eight EU countries. Approximately 900 students were interviewed in total; the samples per country were taken from students within one university.

Several questions were asked about ethics. The respondents had to answer the questions by assigning a mark on the scale 1–5, where 1 = strongly believe it is wrong and 5 = strongly believe it is **not** wrong. Here is one of the questions:

Q27: Visiting a store to obtain information about a product knowing that you will be buying the product from a cheaper store that does not have informed sales personnel

The dataset Xrc21-20.sav gives the results of the interviews as far as question Q27 is concerned. (Note that the dataset has one missing value; case 551.) In total, four variables were measured:

1 *Country*: Portugal = 1; Spain = 2; Denmark = 3; Scotland = 4; Germany = 5; Italy = 6; Greece = 7; The Netherlands = 8

2 *Gender*: male = 0; female = 1

3 *Age of respondent*, in years

4 *Q27*

To explain the variation of Q27, consider the first-order linear regression model that explains Q27 from Country, Gender and Age.

a Is this model useful? Use $\alpha = 0.05$. Also determine the coefficient of determination.

b Give your comments about the individual significance of the three independent variables; take $\alpha = 0.05$.

c Does this model offer the opportunity to compare the different countries as far as their opinion about this ethical question is concerned?

To improve the model, replace the variable Country by dummies; take the Netherlands as base level.

d Compare the coefficient of determination of this model with the coefficient of the former model.

e Are the country dummies jointly significant? Take $\alpha = 0.05$.

f Is Age still significant?

g For each country, compare the mean score for Q27 with the mean score for the Netherlands. Give your comments.

Exercise 21.21 (computer)

The dataset Xrc21-21.sav contains intermediate grades of 131 students for a course called Statistics 1 for International Business. It contains observations of the variables

$X_1 = $ 'grade Midterm 1'; $X_2 = $ 'grade Mid-Exam'; $X_3 = $ 'grade Midterm 2';

$Y = $ 'grade Final Exam'

and also of the dummy variable DRep that takes the value 1 if the student is a repeater.

Midterms 1 and 2 are multiple choice tests; their scores each count 10% when determining the final grade. The grades for the Mid-Exam and the Final Exam have respective weights 0.3 and 0.4. (There also is a Team Assignment that counts 10%, but these grades are not included in the dataset.)

Below, it will be investigated whether the relationship between Y and X_1, X_2, X_3 is different for repeaters and regular students.

a Firstly, estimate the first-order linear regression model with predictors X_1, X_2, X_3 and DRep. Give a statistical analysis of the results.

Since it is suspected that the linear relationship between Y and each of the three variables X_1, X_2, and X_3 will be different for repeaters and regular students, we replace in the model of part (a) the dummy DRep by the three interaction terms $X_i \times$ DRep.

b Write down the basic assumption of the new model.

c Write $E(Y \mid x_1, x_2, x_3; DRep = 1)$ and $E(Y \mid x_1, x_2, x_3; DRep = 0)$ as a linear function of x_1, x_2, and x_3.

d Test whether the three interaction terms jointly contribute to the explanation of the variation of Y. Take $\alpha = 0.10$.

e Estimate the model and give a statistical analysis of the results; compare your results with part (c).

Exercise 21.22 (computer)

Do households with a female head spend more on clothing than households with a male head? This biased question will be investigated for a large population of households.

The dataset Xrc21-22.sav contains, for a random sample of 300 households, the observations of the variables

Y: CLOTEXP = 'annual clothing expenditure of a household (in €1000)'

X_1: HINC = 'annual income of the household (in €1000)'

X_2: HS = 'size of the household'

and a dummy variable $X_3 = $ DF that indicates whether the head of the household is female (value 1) or male (value 0).

a Firstly, consider the second-order model based on the (only) predictor X_1. Estimate the model and analyse it statistically. Find out whether the second-order relationship between Y and X_1 is mountain-shaped. Use $\alpha = 0.01$.

b Add the variable X_2 and X_3 to the model of part (a). Does this extension improve the model? Use $\alpha = 0.01$.

c Use the extended model to answer the question whether households with female heads spend, on average, more on clothing than households with male heads having the same household sizes and household income.

Exercise 21.23 (computer)

In Exercises 20.17 and 20.18 we studied the classical problem whether *an increase in the hourly wage will have a positive or a negative effect on production*. Exercise 20.18 was about a large population of households; the variable HOURS = 'weekly number of hours worked by the head of the household' was used to represent production. On the basis of a sample of size 300 it was obtained that the first-order linear regression model with the independent variables

AGE: age of the head of the household, in years (not rounded off)

WAGE: hourly wages of the head of the household, in euro

NKIDS: number of children (< 18 years) of the household

explains about 44% of the variation of HOURS and that an increase in WAGE has a significant **positive** effect on HOURS.

It may be interesting to introduce Gender into the model, since households with a female head may respond differently to increasing wages when compared with households with a male head. That is why we will include the dummy DF (which is 1 if the head of the household is female). See Xrc21-23.sav.

a Is DF significant within this new model?

b Give your comments about the following claim. If the hourly wage of the head increases by €1 while the other independent variables remain unchanged, then the head will, on average, work more than 1.75 hours more per week.

c Is the slope of the linear relationship between HOURS and WAGE for households with a female head smaller than for households with a male head? Investigate it by including a suitable interaction term in the model and by conditioning on, respectively, DF = 1 and DF = 0.

d With the model of part (c), find an estimate of the slope for the linear relationship between HOURS and WAGE for households with a female head. Is this estimated slope still positive?

Exercise 21.24 (computer)

MarketView interviewed 948 business students from eight countries to study possible different views with respect to ethical questions about 'actively benefiting from illegal activity'. The dataset Xrc21-24.sav contains information about age, gender and country, and about the standardized version (call it Y) of the variable that measures the average opinion of a student on four questions. See Exercise 15.26 for the four questions, and for the coding of the eight countries and male / female; see also Exercises 16.24 and 19.26.

In this exercise we will study by way of a multiple linear regression model the relationship between Y, on the one hand, and AGE, seven country dummies (take Netherlands as base level) and the gender dummy on the other hand.

a Formulate the basic assumption of that model.

b Estimate the model and give a statistical analysis. Is AGE significant?

c Do the data give evidence of a country effect? And a gender effect?

CASE 21.2 PROFITS OF TOP CORPORATIONS IN THE USA (PART II)

In Case 19.2 it was noticed that the profits of top corporations in the USA are linearly related to their revenues. However, we suspect that the relationship between 'profits' and 'revenues' can be described better by a higher-order function: second-order or maybe even third-order. In the present case such higher-order models will be considered. See Case 19.2 for the definitions of the variables; see the file Case19-02.sav for the data.

a Construct the scatter plot of 'profits' on 'revenues'.

It is hard to tell from this scatter plot whether a second- or third-order relationship between Y and X will do better. That is why the variables X^2 and X^3 are just added to the simple linear regression model; it can be determined afterwards whether they are individually significant.

b Formulate the basic assumption of the third-order linear regression model with 'revenues' as the only predictor.

c Estimate the model and give a statistical analysis.

d The model is used by a corporation with annual revenues of $9024 million. Determine a 95% confidence interval for the expected profits of this corporation.

CASE 21.3 PERSONALITY TRAITS OF GRADUATES (PART II)

This case is a continuation of Case 15.2; see the datasets Case15-02.xls and Case15-02.sav. For each of the traits, it will be investigated whether a **study stream effect** exists. That is, it will be studied whether graduates from different streams tend to respond differently with respect to the traits. Also, the presence of a **gender effect** will be investigated. Do male and female graduates tend to score differently with respect to the traits?

The research is done by applying two-factor ANOVA, with 'gender' and 'study stream' as factors. For estimation and prediction purposes, a two-factor ANOVA model with interaction is also conducted. Give your comments on the results.

Consider four graduates: a male and a female graduate both of facility management, and a male and a female graduate both of marketing. For all of them, find point estimates and 95% interval estimates for their expected score on neuroticism. Give your comments.

CASE 21.4 POLLUTION DUE TO TRAFFIC (PART II)

This case is a continuation of Case 20.1. Here, it will be investigated whether a country effect exists with respect to the level of emission of fine particulate matter due to transportation. Since, for some EU25 countries, only a few data are available, these countries are omitted. Only the data of the eight countries France, Germany, Italy, the Netherlands, Poland, Romania, Spain and the UK will be used. Thus, the data of 379 cities are left; see Case21-04.sav.

To study the presence of a country effect on the level of emission – superimposed on the effects due to variables X_1, X_3 and X_5 – the factor 'country' will be included by way of country dummies. You are invited to do so; choose your own base level.

Give a statistical analysis of the results. Does a country effect really exist? For which countries is the expected emission in a city larger (respectively smaller) than the expected emission in a city in your base level country with similar levels for x_1, x_3 and x_5? Also give your comments with respect to the individual significance of X_1, X_3 and X_5 within this model, especially when compared to the model of Case 20.1.

CHAPTER 22

Multiple linear regression: model violations

Chapter contents

Is it possible to test for the validity of the requirements? What are the consequences if one of the requirements is not satisfied? What can be done to repair things? These are the questions of the present chapter. Unfortunately, a **general** method to repair model violations does not exist. But, as we shall see, occasionally the problems can be relaxed, or even solved, by transforming the dependent and/or independent variables.

Apart from violations of the model requirements about

- homoskedasticity,
- normality,
- independence of $\varepsilon_1, \ldots, \varepsilon_n$,
- linearity,
- independence of ε_i and X_j (even for $i = j$),

we shall also consider collinearity problems and the case that the dependent variable is binary (that is, has only two possible outcomes).

Earlier chapters have already discussed how distortions of model requirements can be detected visually. In the present chapter we will search for tests, and discuss the consequences of such distortions and possible ways to tackle the problems.

CASE 22.1 INCOME AND EDUCATION LEVEL OF IDENTICAL TWINS (PART III)

This case is a continuation of Cases 19.3 and 20.2, where the relationship between hourly wage and number of years of education was studied. The special feature of the research is that the analysis is based on data for **twins**. The advantages are that twins have similar family backgrounds and that a person's overvaluation of the number of years of education can now be checked by asking his/her twin.

In the present case, the dataset <u>Case22–01.sav</u> will be used. The following problems will be considered:

 a In part (b) of Case 19.3 it was found that histograms of residuals uncover a serious problem with respect to the normality requirement. How can it be repaired? Illustrate things in the simple linear regression models that explain difference in hourly wage from difference in education. What are the consequences for the interpretation of the estimated value of a year's education?

 b In models for wage that include education X as a predictor, the model requirement that error term ε and predictor X have to be independent, or at least uncorrelated, is often violated because of measurement errors of X and omitted variables (such as 'ability') that are highly correlated with X; see Section 22.5. But the model in part (a) really is about wage and education. What can be done?

For the solutions, see the end of Sections 22.4 (for part (a)) and 22.5 (for part (b)).

22.1 Collinearity

Collinearity was defined in Section 21.2 (refer back if necessary). The presence of collinearity does not influence the level of SSE and hence does not have consequences for the value of the F-test for usefulness of the model. However, it influences the standard deviations of (some of) the least-squares estimators and hence it does have consequences for the individual significance of the independent variables. Collinearity problems can often be detected by noticing a certain degree of equivocality in the printout: a relatively large value of the F-ratio (which implies a very useful model) is not in accordance with relatively small absolute values of the *val*s in (one or more) individual t-tests. Collinearity can be the reason why a variable that was at first expected to be an important factor eventually turns out to be insignificant or only weakly significant.

What are the consequences?
We discussed in Section 21.2 some immediate consequences of collinearity:

 ■ The usual interpretation of at least one regression coefficient is problematic.
 ■ The absolute value of the *val* of a t-test is forced closer to 0.

To a certain extent, collinearity is almost always present in linear regression models since the independent (!) variables are usually not (stochastically) independent. The correlation matrix usually shows positive or negative correlations. There are examples where, at a fixed significance level, a large degree of collinearity hardly influences the individual significance of the independent variables. Then it is customary to leave things as they are (although the interpretation of regression coefficients is still a problem). But there also are examples where collinearity really is problematic since variables that should be significant are stamped as insignificant by the usual t-tests.

What can be done?
Stepwise regression can be used to select only significant variables from a list of potential predictors. But this procedure also has disadvantages, as was noted in Section 21.6.1. And do you really want your model to include only significant variables?

If two potential predictors X_1 and X_2 are highly correlated and their common presence in a regression model leads to collinearity problems, then the approach is often to **exclude** one of the two, usually the one with the highest individual p-value. The arguments for this step are often the general aspiration to keep the model simple and the fact that a variable that is highly correlated with other independent variables in the model will hardly bring new information about the dependent variable Y.

If one predictor is a function of one or more other predictors (for instance, $X_2 = X_1^2$), then the following trick occasionally works to reduce collinearity. Replace all basic variables in the model by the corresponding **centred variables** that arise by subtracting the corresponding sample means. Things are illustrated in two examples.

Example 22.1

To explain the variation of Y = 'annual household expenditure on food', the following variables seem to be important:

X_1: annual income of the household (in thousands of euro)
X_2: number of children under 18 in the household
X_3: size of the household

Running the regression on the basis of a sample of 300 households (see Xmp22–01.sav; see also Exercise 20.14) yields the printout shown below.

ANOVA(b)

Model		Sum of Squares	df	Mean Square	F	Sig.
1	Regression	3421.921	3	1140.640	1662.866	.000(a)
	Residual	203.041	296	.686		
	Total	3624.962	299			

a Predictors: (Constant), HSIZE, HINC, NKIDS
b Dependent Variable: FOODEXP

Coefficients(a)

Model		Unstandardized Coefficients		Standardized Coefficients	t	Sig.
		B	Std. Error	Beta		
1	(Constant)	.189	.187		1.007	.315
	HINC	.234	.004	.936	65.808	.000
	NKIDS	−.381	.110	−.128	−3.452	.001
	HSIZE	.552	.094	.221	5.899	.000

a Dependent Variable: FOODEXP

Although the model is useful and all variables are individually significant, the negative sign of the coefficient of NKIDS is strange and unexpected. The expected food expenditure of a household should **in**crease if NKIDS increases, not **de**crease. This anomaly is probably caused by collinearity that comes from the large positive correlation (0.927) between X_2 and X_3; see the correlation matrix below.

Correlations

		NKIDS	HSIZE	HINC	FOODEXP
NKIDS	Pearson Correlation	1	.927(**)	.169(**)	.235(**)
	Sig. (2-tailed)		.000	.003	.000
	N	300	300	300	300
HSIZE	Pearson Correlation	.927(**)	1	.229(**)	.317(**)
	Sig. (2-tailed)	.000		.000	.000
	N	300	300	300	300
HINC	Pearson Correlation	.169(**)	.229(**)	1	.965(**)
	Sig. (2-tailed)	.003	.000		.000
	N	300	300	300	300
FOODEXP	Pearson Correlation	.235(**)	.317(**)	.965(**)	1
	Sig. (2-tailed)	.000	.000	.000	
	N	300	300	300	300

** Correlation is significant at the 0.01 level (2-tailed).

It would appear that the following happened. The standard deviation of the estimator B_2 became larger because of the high correlation, making it possible for the estimate of β_2 to become negative. Because of this collinearity, we shall **exclude** NKIDS from the model.

Example 22.2 (red-wire, part XVI)

The dataset Xmp22-02.sav contains the observations of the variables Y = 'gross hourly wage', X_1 = 'age' and X_2 = 'education level'. Scatter plots of Y on X_1 and X_2 show sample clouds for which the best-fitting curves have some tendency for downward curving for the larger values of X_1 and X_2; check it yourself. These curvatures are more or less natural since hourly wages tend to grow less fast for higher levels of age and education. That is why a **second-order** linear regression model will be considered. To study the influence of the level of education on the relationship between hourly wage and age, the interaction term X_1X_2 is also included. The following model arises:

$$E(Y) = \beta_0 + \beta_1 x_1 + \beta_2 x_2 + \beta_3 x_1^2 + \beta_4 x_2^2 + \beta_5 x_1 x_2$$

Estimation of this model on the basis of the dataset yields that the conclusion that the model is useful can be called **convincing** since the value of the F-ratio (32.660) is quite large and $r^2 = 0.531$. But studying the individual significance of the independent variables leads to less convincing conclusions, as follows from the coefficients part of the printout, shown below.

Coefficients(a)

Model		Unstandardized Coefficients		Standardized Coefficients	t	Sig.
		B	Std. Error	Beta		
1	(Constant)	6.655	6.433		1.034	.303
	X1	.588	.364	.627	1.615	.109
	X2	−5.614	2.604	−.628	−2.156	.033
	X1X1	−.006	.005	−.514	−1.199	.233
	X2X2	.674	.417	.441	1.617	.108
	X1X2	.137	.052	.762	2.627	.010

a Dependent Variable: Y

The printout reveals that neither of the variables X_1 and X_1^2 is significant, not even at level 0.10. The same holds for X_2^2, while the observed significance level (0.033) of X_2 is slightly disappointing. The interaction term X_1X_2 is significant; this more complicated term seems to take over a part of the roles of the basic variables X_1 and X_2. The individual significance of the independent variables is not convincing and obviously distorted. Computation of the sample correlation coefficients of the pairs

$$X_1, X_1^2 \quad X_2, X_2^2 \quad X_1, X_1X_2 \quad X_2, X_1X_2$$

yields, respectively, 0.988, 0.974, 0.683 and 0.843. It would appear that we have collinearity problems.

Of course, we can decide to exclude the squared variables and the interaction term from the model, thus keeping a simple model with X_1 and X_2 only. And many researchers will indeed do this. But there is also a 'trick' that makes it possible to retain squared variables and an interaction term in the model while collinearity is less problematic: replace the basic variables Y, X_1 and X_2 by the respective centred variables. Since the sample means of Y, X_1 and X_2 are equal to 21.7841, 33.1133 and 2.77, respectively, the centred variables are:

$$Y' = Y - 21.7841; \quad X_1' = X_1 - 33.1133; \quad X_2' = X_2 - 2.77$$

The following second-order model with interaction term arises:

$$E(Y') = \beta_0 + \beta_1 x_1' + \beta_2 x_2' + \beta_3 (x_1')^2 + \beta_4 (x_2')^2 + \beta_5 x_1' x_2'$$

After having created (do it yourself) the six columns with observations of the variables that appear in this model, the model can be estimated. In the printout, the extension 'c' of the variables refers to the centred versions. (So, for instance X1c is just X_1'.) It turns out that the F-ratio and r^2 are the same as for the former model, a general fact that can (but will not) be proved mathematically. The printout of the coefficients part is shown below.

Coefficients(a)

Model		Unstandardized Coefficients		Standardized Coefficients	t	Sig.
		B	Std. Error	Beta		
2	(Constant)	−.544	.975		−.557	.578
	X1c	.541	.075	.577	7.174	.000
	X2c	2.656	.578	.297	4.592	.000
	X1cX1c	−.006	.005	−.096	−1.199	.233
	X2cX2c	.674	.417	.100	1.617	.108
	X1cX2c	.137	.052	.183	2.627	.010

a Dependent Variable: Yc

The variables X_1' and X_2' are both (very) significant. There also exists a significant interaction. However, the two quadratic terms are insignificant at significance level 0.10. Also, jointly the squared variables turn out to be insignificant (partial F-test), which leads to the conclusion that the data do support interaction between the centred variable but do not support the importance of the quadratic terms. Computation of the correlation coefficients of the centred pairs of variables corresponding to the above four pairs yields, respectively, 0.553, 0.121, −0.071 and 0.177. Obviously, collinearity is less a problem.

Some people argue that collinearity is not a problem at all. Collinearity does not bias the results; it only leads to big standard errors. But if your independent variables are highly correlated, **they should be big.** Intuitively, big standard errors mean that the effects of different variables are highly uncertain, and if your independent variables are highly correlated, highly uncertain is what you should be. That is why some people conclude that collinearity is simply a matter of not having enough information (not enough data).

22.2 Heteroskedasticity

In previous chapters it was noticed that heteroskedasticity can be detected visually from the scatter plot of the residuals e_i on the predictions $\hat{y}_i$. For instance, if the plot shows a cloud of dots where the variation along vertical lines increases with $\hat{y}$, at least from a certain level onwards, then there is a heteroskedasticity problem.

Hypothesis test for heteroskedasticity

Our starting point is a proposed linear regression model with k independent variables $X_1, \ldots, X_k$ that is estimated on the basis of a sample of n observations. In terms of the random sample observations, the basic assumption of the model and the homoskedasticity requirement can be formulated as follows:

$$Y_i = \beta_0 + \beta_1 x_{i1} + \ldots + \beta_k x_{ik} + \varepsilon_i \quad \text{with } E(\varepsilon_i) = 0 \text{ and } V(\varepsilon_i) = \sigma_\varepsilon^2 \text{ for all } i = 1, 2, \ldots, n$$

Here, $x_{i1}, \ldots, x_{ik}$ are the observed values of the independent variables (the x characteristics) for the ith sampled object. The model requirement about homoskedasticity can, in terms of the variances $\sigma_1^2, \ldots, \sigma_n^2$ of the error variables $\varepsilon_1, \ldots, \varepsilon_n$ of the sample, also be formulated as $\sigma_1^2 = \ldots = \sigma_n^2 = \sigma_\varepsilon^2$, where σ_ε^2 denotes the common variance. A common form of **hetero**skedasticity is that the variances σ_i^2 linearly depend on the observed values of the independent variables. That is:

$$\sigma_i^2 = \gamma_0 + \gamma_1 x_{i1} + \ldots \gamma_k x_{ik} \quad \text{for all } i = 1, 2, \ldots, n$$

Here, $\gamma_0, \gamma_1, \ldots, \gamma_k$ are unknown parameters. Since $E(\varepsilon_i) = 0$ and hence

$$\sigma_i^2 = E(\varepsilon_i^2) - (E(\varepsilon_i))^2 = E(\varepsilon_i^2)$$

things can be reformulated as:

$$E(\varepsilon_i^2) = \gamma_0 + \gamma_1 x_{i1} + \ldots + \gamma_k x_{ik} \quad \text{for all } i = 1, 2, \ldots, n$$

Note that this is the basic assumption of a linear regression model for the dependent variable ε^2, with regression coefficients $\gamma_0, \ldots, \gamma_k$. To express it otherwise, in terms of new error terms v_i (different from ε_i):

$$\varepsilon_i^2 = \gamma_0 + \gamma_1 x_{i1} + \ldots + \gamma_k x_{ik} + v_i \quad \text{with } E(v_i) = 0, \text{ for all } i = 1, 2, \ldots, n$$

This suggests another model, an **auxiliary** regression model. Although the dependent variable ε^2 is unobservable, the residuals of the original model (that is, the model we started with) can be used as observations of ε and their squares as observations of ε^2. Hence, this auxiliary model can be estimated.

If the hypothesis $H_0: \gamma_1 = \ldots = \gamma_k = 0$ is valid, then things are in line with homoskedasticity. If H_0 is not valid, then heteroskedasticity is an issue. But this means that a model F-test as in Section 20.4.1 can throw some light on the problem of homoskedasticity versus heteroskedasticity.

Example 22.3 (red-wire, part XVII)

Regression of Y = 'hourly wage' on X = AGE yields the scatter plot shown in Figure 22.1. This plot shows obvious heteroskedasticity. The auxiliary model

$$E(\varepsilon_i^2) = \gamma_0 + \gamma_1 x_i \quad \text{for all } i = 1, 2, \ldots, n$$

(with n = 150) will be estimated to test for heteroskedasticity; see Xmp22-03.sav.

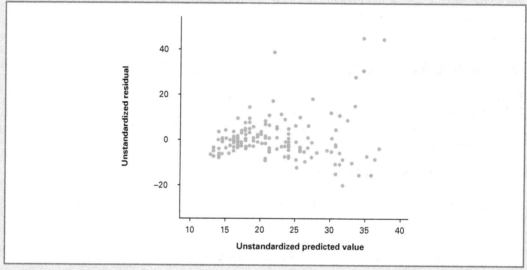

Figure 22.1 Scatter plot of Y on X

After having created the 150 residuals of the simple linear regression model of Y on AGE, the residuals are squared and used as the dependent observations for the new model. A part of the printout of that auxiliary model is shown below (Esq refers to the squared residuals).

Coefficients(a)

Model		Unstandardized Coefficients		Standardized Coefficients	t	Sig.
		B	Std. Error	Beta		
1	(Constant)	−141.698	52.677		−2.690	.008
	AGE	6.298	1.511	.324	4.169	.000

a Dependent Variable: Esq

To test homoskedasticity against heteroskedasticity, the testing problem

$H_0: \gamma_1 = 0$ against $H_1: \gamma_1 \neq 0$

is tested. Of course, the model F-test can be used but the two-sided t-test is equivalent; see also

Section 20.4.2. This *t*-test has *p*-value 0.000, so it clearly rejects H_0. It is concluded that there is sufficient evidence for the existence of heteroskedasticity.

From the printout it also follows that the slope of the linear relationship between ε^2 and AGE is positive. Hence, a larger value of AGE leads on average to a larger value of the variance of ε, which is in line with the scatter plot.

What are the consequences?

In the case of heteroskedasticity there is apparently no common σ_ε. It can (but will not) be proved that the ordinary least-squares estimators B_i of Section 20.1 are still unbiased but there are better unbiased estimators, with smaller standard deviations. As a consequence, the ordinary LS estimators B_i are less precise in their duty of estimating the β_i. Furthermore, the usual accompanying standard **errors** of B_i (so, the estimators of $SD(B_i)$) usually **under**estimate $SD(B_i)$.

A second consequence of there being no common σ_ε in the case of heteroskedasticity is that we cannot use this common standard deviation of the model in our formulae for the standard deviations of the estimators B_i of the regression coefficients β_i. If we still do that, and hence follow the 'old' statistical procedures for interval estimation and hypothesis testing, we have to be aware that the resulting confidence intervals, prediction intervals and conclusions of tests cannot be trusted. The claimed confidence levels of the statistical procedures may not be valid.

What can be done?

A general remedy does not exist. Sometimes, it works to transform the dependent variable *Y*. Here is a list of transformations of *Y* that occasionally reduce heteroskedasticity:

$\log(Y)$ if in the plot of *e* on $\hat{y}$ the vertical variation increases with $\hat{y}$

$Y^2, \sqrt{Y}$ if this vertical variation is proportional to the local mean

$\frac{1}{Y}$ if this vertical variation increases from a certain point onwards

The transformation $\log(Y)$ was already applied in Exercise 21.14 to reduce heteroskedasticity in the hourly wages example. The advantage of the transformation $\log(Y)$ is that it remains possible to interpret the regression coefficients in a natural way, in terms of percentage changes (see Section 21.4).

In the case of heteroskedasticity, the more precise *y* observations (those with less variability) should get larger weights in determining the sample regression coefficients. There exists a rather general estimation method of the population regression coefficients (the **weighted least-squares regression** method) that uses the weights $1/\sigma_i^2$ – with σ_i^2 being the variance of ε_i – in the least-squares method, leading to better **unbiased** estimators $\tilde{B}_i$ of the β_i. Unfortunately, the standard deviations of these estimators usually depend on these unknown parameters σ_i^2, which all have to be estimated. This considerably decreases the number of available degrees of freedom. However, in some cases this additional number of unknown parameters can be reduced by making assumptions about the way the variation of the errors along vertical lines within the population cloud depends on *y*.

A special application of the weighted least-squares method arises in cases where an originally homoskedastic model becomes heteroskedastic because of aggregation. **Aggregation** occurs if the available observations do not belong to the population elements of interest but instead are the averages (or totals) for **groups** of population elements. For instance, if the intention is to explain the sales of stores in a certain line of business from the numbers of employees, but the available observations are the averages per store for **chains** of stores, then the available data are aggregated data. A random observation Y_7 (say) that measures the average for a group (chain) with $m_7 = 2$ elements (stores) has a variance $\sigma_\varepsilon^2/2$, which is five times larger than the variance $\sigma_\varepsilon^2/10$ of a random observation Y_1 that measures the average for a group with $m_1 = 10$ elements. That is, the variances of the observed aggregated averages $Y_1, Y_2, \ldots, Y_n$ have the ratios

$$\frac{1}{m_1} : \frac{1}{m_2} : \dots : \frac{1}{m_n}$$

where m_j is the size of group j. It would appear that the least-squares method has to be weighted with these weights. Example 22.4 illustrates things; see also Appendix A1.22.

Example 22.4

Interest is in the relationship between the dependent variable 'last year's sales of a store (in millions of euro)' and the independent variable 'number of employees of the store at the end of last year'. It is assumed that a simple linear regression model describes this situation well.

The file Xmp22-04.sav contains sales data (Y) and data about numbers of employees (X). However, the observations in the dataset are not the observations of individual stores (as they should be) but are the observed **averages** for the stores in **chains**; see the table below.

Number of the chain	1	2	3	4	5	6	7
Number of stores per chain (m_j)	10	20	2	2	5	5	2
Average sales of chain per store	21	21	36	9	15	21	33
Average number of employees per store	6	4	11	2	9	9	10

Of course, we can deny this complication and run the ordinary least-squares regression. Some estimation results are listed below:

$\hat{y} = 7.004 + 2.097x$; $s_{B_1} = 0.845$; SST = 537.429; SSE = 240.788; $r^2 = 0.552$;
F-ratio = 6.160; p-value for X is 0.056

However, we can also use the weighted least-squares method (see Appendix A1.22 for computation) with the m_j as weights. This leads to the partial printout shown below.

Coefficients(a,b)

Model		Unstandardized Coefficients		Standardized Coefficients	t	Sig.
		B	Std. Error	Beta		
2	(Constant)	15.676	5.402		2.902	.034
	X	.887	.832	.431	1.067	.335

a Dependent Variable: y
b Weighted Least Squares Regression – Weighted by m_i

Here are estimation results for the new model:

$\hat{y} = 15.676 + 0.887x$; $s_B = 0.832$; SST = 1206.000; SSE = 982.394; $r^2 = 0.185$;
F-ratio = 1.138; p-value for X is 0.335

These results are very different from the results obtained before. For instance, the slope of the sample regression line is much smaller, and the same holds for the coefficient of determination; as a matter of fact, the model is not even useful at significance level 0.10. Notice that SST also differs: the weights m_j are used here too.

Heteroskedasticity can sometimes be ascribed to one of the independent variables. If the standard deviation along vertical lines in the scatter plot of Y on X increases proportionally to x, then X causes heteroskedasticity. A solution of the problem can be to transform the original model such that the ratio Y/X becomes the new dependent variable.

For instance, consider the simple linear model with basic assumption $Y = \beta_0 + \beta_1 x + \varepsilon$ with $E(\varepsilon \mid x) = 0$. Suppose that the standard deviation of ε is proportional to the value x of X. That is,

$V(\varepsilon \mid x) = x^2\sigma^2$ and $SD(\varepsilon \mid x) = x\sigma$, where σ is the constant of proportionality. Since the variance changes with x, this model is not homoskedastic. Division by x transforms the basic assumption into:

$$\frac{Y}{x} = \frac{\beta_0}{x} + \beta_1 + \frac{\varepsilon}{x}$$

With the transformations $Y^* = Y/x$, $x^* = 1/x$ and $\varepsilon^* = \varepsilon/x$, the following model arises:

$$Y^* = \beta_1 + \beta_0 x^* + \varepsilon^*$$

Notice especially the role exchange of β_0 and β_1. Since

$$V(\varepsilon^* \mid x) = V\left(\frac{1}{x}\varepsilon \mid x\right) = \frac{1}{x^2}V(\varepsilon \mid x) = \frac{1}{x^2}x^2\sigma^2 = \sigma^2$$

the transformed model **is** homoskedastic.

Example 22.5

Consider a large company where only three levels of monthly salary occur: €2000, €3000 and €4000. To determine the relationship between salary X and monthly amount Y (in euro) spent on travel by the employees, a random sample of 20 employees is used. See the table below and the file Xmp22-05.sav.

Salary x	Monthly travel costs y						
2000	160	170	180	200	210	230	250
3000	200	220	240	260	310	320	320
4000	280	300	310	350	400	460	

Ordinary regression of Y on X yields the printout shown below.

Coefficients(a)

Model		Unstandardized Coefficients		Standardized Coefficients	t	Sig.
		B	Std. Error	Beta		
1	(Constant)	47.876	42.591		1.124	.276
	X	.075	.014	.785	5.369	.000

a Dependent Variable: Y

The model is (very) useful; the sample regression equation is $\hat{y} = 47.876 + 0.075x$. However, the scatter plot of y on x shows obvious heteroskedasticity: the vertical variation increases with x; see Figure 22.2. As a consequence, the conclusions from the above printout cannot be trusted.

Column 2 of the table below gives the three vertical variances (of the three groups of y data) after division by x^2.

x	Variance of the y data	Variance / x^2
2000	1067	0.000267
3000	2490	0.000277
4000	4720	0.000295

But the numbers in column 3 are almost the same. That is why we follow the above theory and regress $V = Y/X$ on $U = 1/X$; see the printout below.

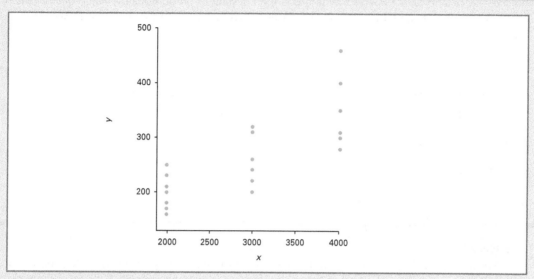

Figure 22.2 Scatter plot of y on x

Coefficients(a)

Model		Unstandardized Coefficients		Standardized Coefficients	t	Sig.
		B	Std. Error	Beta		
2	(Constant)	.073	.013		5.449	.000
	one_byX	52.857	35.180	.334	1.502	.150

a Dependent Variable: YbyX

The regression equation $\hat{v} = 0.073 + 52.857u$ gives the best-fitting linear relation between $v = y/x$ and $u = 1/x$. Things can be transformed back to the best-fitting relation between y and x by exchanging b_0 and b_1 in the equation. This yields:

$$\hat{y} = 52.857 + 0.073x$$

Of course, the accompanying standard errors also have to be exchanged. Notice that the new standard error 0.013 is slightly smaller than the original standard error 0.014.

22.3 Non-linearity and non-normality

The linearity assumption can be checked visually by way of the scatter plots of the residuals on each of the independent variables. The normality requirement of the error term can be checked by creating a histogram of the residuals or the standardized residuals. Many statistical packages offer the option to include the best-fitting normal curve, to facilitate the comparison of histogram and normality.

Our starting point is a proposed linear regression model (the **original** model)

$$Y = \beta_0 + \beta_1 x_1 + \ldots + \beta_k x_k + \varepsilon \quad \text{with } E(\varepsilon) = 0$$

that is estimated on the basis of a sample of n observations.

22.3.1 Non-linearity

Hypothesis test for non-linearity

The model requirement about linearity states that $E(Y)$ depends linearly on $x_1, \ldots, x_k$. The question arises whether this linearity assumption stylizes things too much.

Consider the following **auxiliary** model:

$$Y_i = \beta_0 + \beta_1 x_{i1} + \ldots + \beta_k x_{ik} + \gamma \hat{y}_i^2 + v_i \quad \text{with } E(v_i) = 0, \quad \text{for all } i = 1, 2, \ldots, n$$

Here, $\hat{y}_i = b_0 + b_1 x_{i1} + \ldots + b_k x_{ik}$ is the prediction of y_i that follows from the original model and γ is an additional unknown parameter. As in Section 22.2, the v_i are new error terms. If $\gamma = 0$, then $E(Y)$ depends linearly on $x_1, \ldots, x_k$. If $\gamma \neq 0$, then $E(Y)$ has a quadratic relationship to $x_1, \ldots, x_k$.

This new model can be estimated by using – apart from the data of the k original independent variables – the column with the squares of the predictions $\hat{y}_1, \ldots, \hat{y}_n$ as extra input. A two-sided t-test can be used to test $H_0: \gamma = 0$ against $H_1: \gamma \neq 0$.

Example 22.6

In Example 21.1, the grade Y for the Final Exam of Statistics 1 was explained from:

$X_1 = $ 'grade for Midterm 1' $X_2 = $ 'grade for the Team Assignment'
$X_3 = $ 'grade for the Mid-Exam' $X_4 = $ 'grade for Midterm 2'

See Xmp21-01.sav. Since we are wondering whether a second-order model might improve things, we test for non-linearity by using the squares of the predictions $\hat{y}_1, \ldots, \hat{y}_n$ of the original model as extra input for the auxiliary model with basic assumption:

$$Y_i = \beta_0 + \beta_1 x_{i1} + \ldots + \beta_k x_{i4} + \gamma \hat{y}_i^2 + v_i \quad \text{with } E(v_i) = 0, \quad \text{for all } i = 1, 2, \ldots, n$$

The estimation of this new model yields the printout shown below.

Coefficients(a)

Model		Unstandardized Coefficients		Standardized Coefficients	t	Sig.
		B	Std. Error	Beta		
1	(Constant)	−4.638	1.190		−3.897	.000
	Mid1	.269	.100	.151	2.699	.008
	TA	.156	.090	.077	1.729	.085
	Mid_Exam	.766	.140	.635	5.462	.000
	Mid2	.504	.098	.407	5.145	.000
	Yhat2	−.028	.019	−.241	−1.483	.140

a Dependent Variable: Final_Exam

It follows that the auxiliary variable $\hat{Y}^2$ is **not** significant in this new model, not even at significance level 0.10. It seems that the original model (which is linear in $x_1, \ldots, x_4$) is not misspecified in the sense that inclusion of second-order terms would improve it.

22.3.2 Non-normality

The normality requirement can be checked visually by comparing a histogram of the residuals with the best-fitting normal curve. A more convincing statistical method is based on a comparison of the discrete distribution function (see (2.1)) of the residuals after standardization and the standard normal distribution function.

Hypothesis test for non-normality

The *Lilliefors test* or **Kolmogorov–Smirnov test with Lilliefors correction** can be used to test, for a general random variable X (not necessarily in a regression context), the null hypothesis that X is normal against the alternative hypothesis that X is not normal, on the basis of a random (so, *iid*) sample $X_1, \ldots, X_n$. Below, we first present the test procedure for general X with *iid* sample $X_1, \ldots, X_n$. Next, we apply it to the error term ε of a linear regression model with *iid* sample error terms $\varepsilon_1, \ldots, \varepsilon_n$.

The statistical procedure of the **Lilliefors test** for testing

i H_0: X is normal against H_1: X is not normal

on the basis of the *iid* observations $X_1, \ldots, X_n$ can be described in four steps, with p-value approach. A statistical package has to be used for the calculation of the p-value; see also Appendix A1.22. We describe the test in four steps.

1   Determine the standardized versions of the realized observations $x_1, \ldots, x_n$ and put them in ascending order. That is:

   a   determine the mean $\bar{x}$ and standard deviation s of the sample observations;

   b   for each observation x_i, subtract $\bar{x}$ and divide the result by s;

   c   place the resulting n standardized observations in ascending order and denote them by $z_1, \ldots, z_n$ with $z_1 < z_2 < \ldots < z_n$.

2   Determine the maximum absolute discrepancy between the discrete cumulative distribution function of the dataset $\{z_1, \ldots, z_n\}$ and the cumulative distribution function Φ of the $N(0, 1)$ distribution. This maximum is the *val* of the procedure.

3   Use a computer to calculate the accompanying p-value, the probability that the maximum absolute discrepancy is larger than or equal to *val*.

4   Reject H_0 if and only if this p-value is not larger than α.

See (2.1) for the definition of the discrete cdf; see Figure 10.8 for Φ. Note that the underlying test statistic (that produces the *val*) is actually the maximal discrepancy of the procedure described in step 2. Of course, H_0 has to be rejected for large values of that test statistic. Although the distribution of the test statistic has been tabulated, we will use the p-value approach and blindly follow the computer.

For instance, consider the x dataset 1.5, 2.75, 4.9, 7.5, 9. It transforms in step 1 into the ordered z dataset −1.1553, −0.7575, −0.0732, 0.7543, 1.2317. Figure 22.3 shows the discrete cdf of the z dataset (step function, jumping $1/5 = 0.2$ at each z_i) and the continuous function Φ.

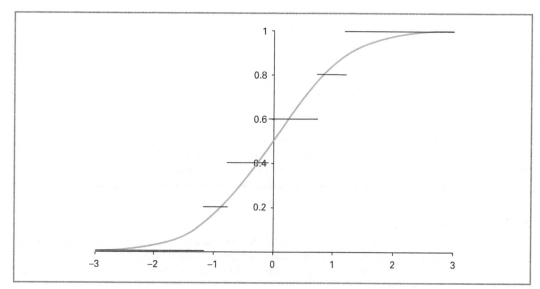

Figure 22.3 Discrete cdf of the z dataset and the continuous cdf Φ

Note that the maximal discrepancy between Φ and the step function occurs at the second data point, at $z_2 = -0.7575$. Since $\Phi(z_2) = 0.2244$ (∗) and the step function jumps at z_2 from 0.2 to 0.4, it follows that $val = 0.4 - 0.2244 = 0.1756$. With a computer package it follows more precisely that the p-value is 0.200; there is, at significance level 0.05, not enough evidence that the x dataset comes from a non-normal distribution.

Now suppose that it is the normality of the error term ε of a linear regression model we are interested in. Since the *iid* sample of the error terms $\varepsilon_1, \ldots, \varepsilon_n$ is unobserved, we use the regression residuals $\varepsilon_1, \ldots, \varepsilon_n$ to test

i H_0: ε is normal against H_1: ε is non-normal

with the Lilliefors test procedure. The four steps are similar to above.

Example 22.7 (red-wire, part XVIII)

Does evidence exist of non-normality for the error terms of the simple linear regression model that explains $Y =$ 'hourly wage' from $X =$ AGE? We will use the Lilliefors test and the residuals of that regression to find the answer to that question. See Xmp22-07.sav for the data.

By using a statistical package it follows, after having standardized the residuals as in step 1, that the val of the Lilliefors test is 0.128 and that the accompanying p-value is close to 0; see the printout below.

Tests of Normality

	Kolmogorov-Smirnov(a)		
	Statistic	df	Sig.
Zres	.128	150	.000

a Lilliefors Significance Correction

It follows that there is evidence of non-normality for the sequence of n residuals and hence it is concluded that the error term ε in the simple linear regression model is non-normal.

What are the consequences?

If the error variable is not normal, then the least-squares estimators $B_0, B_1, \ldots, B_k$ of the population regression coefficients $\beta_0, \beta_1, \ldots, \beta_k$ are not normally distributed, and hence the resulting interval estimates and conclusions of hypothesis tests cannot be trusted. Fortunately, the problems are less serious for **large sample sizes**. If n is large, then the probability distributions of the estimators are, thanks to the central limit theorem, **approximately** normal (under additional regularity conditions that are rather weak) and the formulae for interval estimates and test procedures remain approximately valid.

What can be done?

If n is small, non-normality can occasionally be repaired or partially repaired by a linear transformation of the dependent variable. Here are two transformations of Y that are used occasionally to reduce non-normality:

$\log(Y)$ if the histogram of the (standardized) residuals is skewed to the right
Y^2 if this histogram is skewed to the left

If nothing works, it is advisable to look for another dependent variable that measures more or less the same feature of interest.

22.4 Dependence of the error terms

The independence requirement for the error term may give problems if the data are time series data. In time series, the observations y_t and x_t have to be read as the observations of Y and X at the **period** t. The sequences $y_1, y_2, \ldots, y_n$ and $x_1, x_2, \ldots, x_n$ are just the observations of Y, respectively X, at **successive** periods called $1, 2, \ldots, n$. To draw a conclusion about the dependence / independence of the error sequence $\varepsilon_1, \varepsilon_2, \ldots, \varepsilon_n$, the residual sequence $e_1, e_2, \ldots, e_n$ is studied. In essence, it is the **correlation within** the sequence itself that is considered, the so-called **autocorrelation** of the sequence of residuals. If the first $n-1$ residuals $e_1, \ldots, e_{n-1}$ are (positively or negatively) correlated to the last $n-1$ residuals $e_2, \ldots, e_n$, then the sequence of residuals is said to show (positive or negative) **first-order autocorrelation**. This is because that correlation involves pairs of **successive** residuals:

$$(e_1, e_2), (e_2, e_3), \ldots, (e_{n-1}, e_n)$$

Usually, it is this first-order autocorrelation that is studied in order to ascertain dependence.

How can it be detected visually?

In Section 19.8 it was mentioned that the independence requirement can be checked by studying the scatter plot of residuals on time, so the residuals $e_1, e_2, \ldots, e_n$ on the successive periods $1, 2, \ldots, n$. Figure 22.4(a) shows an example of positive first-order autocorrelation. Successive residuals tend to be on the same side of the horizontal level 0; runs of positive residuals are followed by runs of negative residuals and vice versa.

Figure 22.4(b) shows a typical example of negative first-order autocorrelation. The residuals tend to oscillate around the level 0; positive residuals are often followed by negative ones and vice versa.

First-order autocorrelation can also be checked by studying the scatter plot of the last $n-1$ residuals $e_2, \ldots, e_n$ on the first $n-1$ residuals $e_1, \ldots, e_{n-1}$, in other words, the scatter plot of e_t on e_{t-1}. (See Figure 19.13 for typical plots of positive first-order autocorrelation and no first-order autocorrelation.) This plot visualizes the correlation coefficient $r_{e_t, e_{t-1}}$: relatively many dots in quadrants 1 and 3 (respectively, 2 and 4) correspond to positive (respectively, negative) first-order autocorrelation; the degree of concentration around an increasing (respectively, decreasing) straight line gives visual information about the strength of the linear relationship between e_t and e_{t-1}.

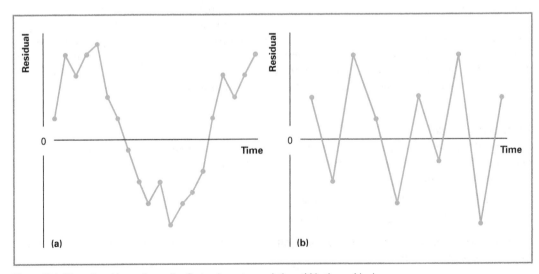

Figure 22.4 Plots of positive and negative first-order autocorrelation within the residuals

Hypothesis test for dependence of the error terms

Of course, visual perception may be deceptive. That is why a test has been invented to enable conclusions to be drawn about the following three testing problems on yes/no (positive or negative) first-order autocorrelation within the sequence of error terms:

ia H_0: there is no positive first-order autocorrelation;

 H_1: there **is** positive first-order autocorrelation.

ib H_0: there is no negative first-order autocorrelation;

 H_1: there **is** negative first-order autocorrelation.

ic H_0: there is no first-order autocorrelation;

 H_1: there **is** first-order autocorrelation.

Notice that testing problem (ic) is in a sense two-sided. The common tests for these testing problems are the **Durbin–Watson tests**. They are unusual in the sense that they allow three (instead of two) possible statistical conclusions:

 H_1 is true; H_0 is true;   the test is inconclusive

The three testing problems have the same test statistic, called D. We will not discuss the form of D (see below), but it can be proved that, for large n, it is equal to $2(1 - R_1)$ where R_1 (also denoted as $r_{e_t, e_{t-1}}$) is the correlation coefficient of two **successive** residuals:

ii $$D = \frac{\sum_{t=2}^{n}(e_t - e_{t-1})^2}{\sum_{t=1}^{n}e_t^2} \approx 2(1 - R_1)$$

The following facts are obvious for large n, but they are also valid for small n:

- D can only take values in the range $[0, 4]$.
- Realizations close to 0 indicate **positive** first-order autocorrelation.
- Realizations close to 4 indicate **negative** first-order autocorrelation.
- The closer to 0 (respectively, 4), the stronger the positive (respectively, negative) first-order autocorrelation.
- Realizations in the neighbourhood of 2 (which is the centre of the range $[0, 4]$) support the validity of 'no first-order autocorrelation'.

For tests of the three testing problems, the range $[0, 4]$ has to be divided into three regions: the region that supports the alternative hypothesis, the region that supports the null hypothesis, and the region that does not allow a choice between H_0 and H_1 (the conclusion then is 'test is inconclusive'). As always in hypothesis testing, the significance level α watches over drawing incorrect conclusions, so the boundaries (the critical values of the test) of the regions will depend on the choice of α. These critical values follow from the **Durbin–Watson (DW) bounds** d_L and d_U (with $d_L \leq d_U < 2$) that can be found in the **Durbin–Watson table** (see Table 7 in Appendix A4). The DW bounds depend on the sample size n and the number of independent variables (k) of the regression model. It can (but will not) be proved that, for fixed k, the DW bounds d_L and d_U are both increasing with n.

Testing problem (ia)

The (lower and upper) critical values of the test are the DW-bounds, now denoted by $d_{\alpha, L}$ and $d_{\alpha, U}$. The test goes as follows (d is a realization of D):

 $d \leq d_{\alpha, L} \Rightarrow$ conclude H_1

$d_{\alpha,L} < d < d_{\alpha,U} \Rightarrow$ conclude 'inconclusive'

$d \geq d_{\alpha,U} \Rightarrow$ conclude H_0

Figure 22.5 illustrates (Inc stands for 'inconclusive').

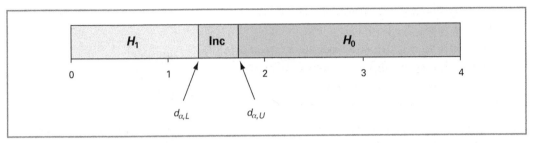

Figure 22.5 Durbin–Watson regions for testing problem (ia)

Testing problem (ib)

The critical values of the test are $4 - d_{\alpha,U}$ and $4 - d_{\alpha,L}$. The test goes as follows:

$d \geq 4 - d_{\alpha,L} \Rightarrow$ conclude H_1

$4 - d_{\alpha,U} < d < 4 - d_{\alpha,L} \Rightarrow$ conclude 'inconclusive'

$d \leq 4 - d_{\alpha,U} \Rightarrow$ conclude H_0

Figure 22.6 illustrates.

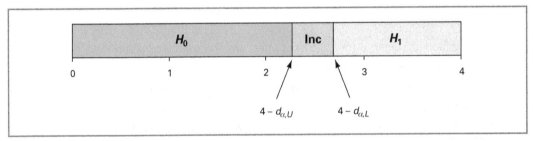

Figure 22.6 Durbin–Watson regions for testing problem (ib)

Testing problem (ic)

Since the test is 'two-sided' (positive and negative), the critical values of the test are based on $\alpha/2$; they are $d_{\alpha/2,L}$, $d_{\alpha/2,U}$, $4 - d_{\alpha/2,U}$ and $4 - d_{\alpha/2,L}$. Here is the test:

$d \leq d_{\alpha/2,L}$ or $d \geq 4 - d_{\alpha/2,L} \Rightarrow$ conclude H_1

$d_{\alpha/2,L} < d < d_{\alpha/2,U}$ or $4 - d_{\alpha/2,U} < d < 4 - d_{\alpha/2,L} \Rightarrow$ conclude 'inconclusive'

$d_{\alpha/2,U} \leq d \leq 4 - d_{\alpha/2,U} \Rightarrow$ conclude H_0

Figure 22.7 illustrates.

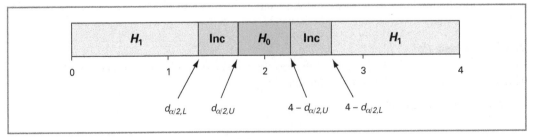

Figure 22.7 Durbin–Watson regions for testing problem (ic)

Only the steps (iv) and (v) of the test procedures are left. The **val** of the test can be obtained with the help of a statistical package (see Appendix A1.22); comparing the **val** with the critical values of step (iii) yields the conclusion of the test.

Five-step procedure of Durbin–Watson test

i Formulate the testing problem ((ia), (ib) or (ic)).

ii Test statistic: D.

iii Divide the interval $[0, 4]$ into the three regions by way of the DW bounds and indicate the accompanying conclusions.

iv Use your statistical package to calculate **val**.

v Compare **val** with step (iii) and draw the statistical conclusion.

Example 22.8

Reconsider Example 19.15, where the daily closing price of the Heineken stock was regressed on the closing price of the AEX index. Figure 19.12(a) in that example suggests that the error terms are dependent because of **positive** first-order autocorrelation. To check it statistically, we will conduct the Durbin–Watson test. See Xmp19-15.sav for the stock and AEX data.

Notice that $n = 510$ and $k = 1$. We take $\alpha = 0.01$. Since the testing problem of interest is one-sided, we do not have to split α; we need $d_{0.01,L}$ and $d_{0.01,U}$. However, the DW table only gives results for n at most 100:

$n = 100$ and $k = 1$ and $\alpha = 0.01$: $d_L = 1.52$ and $d_U = 1.56$

Hence, the DW bounds that we need will be larger: $d_{0.01,L} \geq 1.52$ and $d_{0.01,U} \geq 1.56$. With a statistical package, **val** can easily be obtained: **val** $= 0.020$. The five-step procedure is as follows:

i Test H_0: there is no positive first-order autocorrelation, against

   H_1: there **is** positive first-order autocorrelation; take $\alpha = 0.01$

ii Test statistic: D

iii $d \leq d_{0.01,L} \Rightarrow$ conclude H_1

   $d_{0.01,L} < d < d_{0.01,U} \Rightarrow$ conclude 'inconclusive'

   $d \geq d_{0.01,U} \Rightarrow$ conclude H_0

iv **val** $= 0.020$

v Since 0.020 is smaller than 1.52 and hence smaller than $d_{0.01,L}$, we conclude that there is positive first-order autocorrelation

Durbin–Watson tests will frequently be conducted in Chapter 23. In that chapter, another (more intuitive) test will also be considered to check first-order autocorrelation, a test that has the advantage of **not** having an 'inconclusive region'. For large n, this new test approximates the DW test.

What are the consequences?

It can (but will not) be proved that, in the case of autocorrelated error terms, the ordinary least-squares estimators B_i of Section 20.1 are still unbiased but their standard deviations are not minimal since there exist better unbiased estimators, with smaller standard deviations. Furthermore, the usual accompanying standard **errors** of B_i (so, the estimators of $SD(B_i)$) usually **under**estimate $SD(B_i)$.

Since the derivation of the properties of the least-squares estimators is heavily based on **in**dependence, the conclusions of statistical procedures (interval estimation, hypothesis testing, predictions) cannot be trusted in the case of autocorrelation.

What can be done?

Again, a uniform approach for solving problems concerning independence of the error terms does not exist. Sometimes it works to replace the original variables by growth variables. For instance, in the Heineken case (Example 19.15) the positive first-order autocorrelation more or less disappears if the model (based on prices of the stock and the AEX index) is replaced by the model based on the respective daily returns.

Occasionally it works to include time itself in the model. This can (for instance) be done by creating a new column called T with the values $1, 2, \ldots, n$. Sometimes, even T^2 is needed to get rid of all time effects.

If it is expected that the correlation between two successive error terms is very strong, then it occasionally works to replace the dependent and independent variables by the corresponding successive differences. In the case of simple linear regression of Y on X, this means that the variables Y_t and X_t in the model $E(Y_t) = \beta_0 + \beta_1 x_t$ are replaced by $Y_t - Y_{t-1}$ and $X_t - X_{t-1}$, respectively.

There exists a rather general approach that occasionally works if the error terms $\varepsilon_1, \ldots, \varepsilon_n$ of the original regression model are not independent but instead form an **autoregressive system**. That is, from $t = 2$ onwards, each error term ε_t is equal to a fixed proportion ρ of the previous error term ε_{t-1} plus a random component; for ε_1 a special form is chosen. We first define such systems more precisely and next consider the general approach.

Autoregressive system

The error terms $\varepsilon_1, \ldots, \varepsilon_n$ are said to form an *autoregressive system* if a constant ρ ($0 \leq \rho < 1$) and *iid*, $N(0, \sigma_v^2)$ distributed random variables $v_1, \ldots, v_n$ exist such that:

$$\varepsilon_1 = v_1/\sqrt{1 - \rho^2} \quad \text{and} \quad \varepsilon_t = \rho\varepsilon_{t-1} + v_t \quad \text{for } t = 2, \ldots, n$$

The equations $\varepsilon_t = \rho\varepsilon_{t-1} + v_t$ for $t = 2, \ldots, n$ can be considered as the basic assumption of a new linear regression model where, beforehand, the intercept is taken as 0. The new error terms v_t satisfy the assumptions of the simple linear regression model.

The autoregressive system of error terms $\varepsilon_1, \ldots, \varepsilon_n$ turns out to have some desirable properties (which will not be proved):

■ $\varepsilon_1, \ldots, \varepsilon_n$ are identically, normally distributed.
■ ρ is just the correlation coefficient of ε_t and ε_{t-1}.

If the error terms $\varepsilon_1, \ldots, \varepsilon_n$ are an autoregressive system, then there exist better estimators than the least-squares estimators of Section 20.1. Things will be shown only for the case of simple linear regression.

From the basic assumption

$$Y_t = \beta_0 + \beta_1 x_t + \varepsilon_t \quad \text{for } t = 1, 2, \ldots, n$$

it also follows that

$$\rho Y_{t-1} = \beta_0\rho + \beta_1\rho x_{t-1} + \rho\varepsilon_{t-1} \quad \text{for } t = 2, \ldots, n.$$

By subtracting the second equation from the first, we obtain (since $v_t = \varepsilon_t - \rho\varepsilon_{t-1}$):

$$Y_t - \rho Y_{t-1} = \beta_0(1 - \rho) + \beta_1(x_t - \rho x_{t-1}) + v_t \quad \text{for } t = 2, \ldots, n$$

Setting $Y_t^* = Y_t - \rho Y_{t-1}$, $X_t^* = X_t - \rho X_{t-1}$ and $\beta_0^* = \beta_0(1 - \rho)$ yields:

$$Y_t^* = \beta_0^* + \beta_1 x_t^* + v_t \quad \text{for } t = 2, \ldots, n$$

If ρ were known, then the random variables Y_t^* and X_t^* would be observable and, since the error terms $v_1, \ldots, v_n$ are *iid* and $N(0, \sigma_v^2)$, a new linear regression model would arise that satisfies the assumptions. Estimation of this new model would yield the (ordinary) least-squares estimators of β_0^* and β_1, and an estimate of the original intercept β_0 could also be calculated.

But the correlation coefficient ρ is **unknown**. That is why, in the above considerations, ρ is replaced by its estimator R_1 (also denoted by $r_{e_t, e_{t-1}}$). Then, the estimation procedure (the **Cochrane–Orcutt procedure**) goes as follows:

1 Run the original regression in the usual way; use the residuals of that model to calculate the value of R_1.

2 Calculate the observations of the transformed variables Y_t^* and X_t^* after having replaced ρ by its estimate.

3 Estimate the model $Y_t^* = \beta_0^* + \beta_1 x_t^* + v_t$ in the usual way.

Example 22.9

Labour productivity is defined as the output (measured by GDP) per unit of labour input; unit labour cost is the labour cost per unit of output. These variables are key indicators of the labour market. For the eurozone we would expect a negative relationship between the quarterly labour productivity Y and the quarterly unit labour cost X, both measured in euro. The dataset Xmp22-09.sav contains figures of X and Y published by the European Central Bank for the period 1996Q1–2007Q3. Straightforward regression of Y on X is expected to give problems with respect to the independence requirement of the error terms. We will use the Cochrane–Orcutt procedure to try to resolve these problems.

Indeed, the ordinary linear regression model of Y on X turns out not to be valid since the value of the DW statistic is 0.808 (check it yourself), which indicates a positive linear relationship. That is why the accompanying printout (see below) cannot be trusted. The regression coefficient r_1 of two successive residuals turns out to be 0.58455 (check it yourself).

Coefficients(a)

Model		Unstandardized Coefficients		Standardized Coefficients	t	Sig.
		B	Std. Error	Beta		
1	(Constant)	1.704	.098		17.425	.000
	X	−.631	.065	−.823	−9.725	.000

a Dependent Variable: Y

Regression of $Y_t^* = Y_t - 0.58455 Y_{t-1}$ on $X_t^* = X_t - 0.58455 X_{t-1}$ yields the following printout.

Coefficients(a)

Model		Unstandardized Coefficients		Standardized Coefficients	t	Sig.
		B	Std. Error	Beta		
2	(Constant)	.665	.057		11.615	.000
	Xstar	−.569	.076	−.749	−7.492	.000

a Dependent Variable: Ystar

The value of the DW statistic is now 1.876, giving evidence that there is no first-order autocorrelation problem. The new model is useful, with $r^2 = 0.561$.

Note that the realized standard error of the estimator of β_1 with the new model is larger than with the old model. This is in accordance with the fact that when there is autocorrelation within the error terms the ordinary least-squares method often leads to underestimation of the standard deviation of the estimator of β_1.

The accompanying regression equation is $\hat{y}^* = 0.66483 - 0.56856x^*$. Since $\beta_0^* = \beta_0(1 - \rho)$ is estimated as $0.66483 = b_0(1 - 0.58455)$, it follows that the linear relationship between Y and X is estimated to be $\hat{y} = 1.6003 - 0.56856x$.

CASE 22.1 INCOME AND EDUCATION LEVEL OF IDENTICAL TWINS (PART III) – SOLUTION TO PART A

From part (b) of Case 19.3 it follows that the distribution of the residuals is heavily skewed to the right. That is why from now on the **natural logarithm** of hourly wage will be used instead of hourly wage itself.

With respect to part (c) of Case 19.3, it is now the difference between the **log** hourly wages of twin 1 and twin 2 that is explained from the difference between the numbers of years of education of twin 1 and twin 2. Estimation of the simple linear regression model that has DEDUC = EDUCL − EDUCH as (the only) independent variable, yields $r^2 = 0.092$ and estimated slope 0.092 with standard error 0.024. When DEDUC is replaced by DEDUCxx (the cross-reported difference in education), then $r^2 = 0.099$ and estimated slope and standard error are practically unchanged.

The estimated slope allows the following interpretation. If the difference between the numbers of years of education of twin 1 and twin 2 increases by one year, then the **ratio** of the hourly wages of twin 1 and twin 2 is estimated to increase by 9.2%. (See also Table 21.2 and recall that $\log(a) - \log(b) = \log(a/b)$.)

Incidentally, histograms of the residuals show that the non-normality problem is solved now.

22.5 Instrumental variables

This section is about the requirement that, within a linear regression model for Y, the error variable ε and the variable(s) X have to be independent. This assumption is essential for obtaining 'good' estimators for the population regression coefficients in the sense that they are unbiased and (in any case) consistent. With inconsistent estimators, the statistical procedures cannot be trusted at all.

Although it is (under some additional regularity conditions) possible to relax the independence requirement for ε and X to the requirement that ε and X are uncorrelated, there are still many examples where even this weaker requirement is not satisfied. Two cases are considered.

Suppose we want to explain 'hourly wage' from schooling and ability. The variable X = 'number of years of education' measures schooling and seems to be a good predictor. However, which variable can be used to measure ability? Of course, we could use 'IQ', but IQ and ability are not the same. Since a variable measuring ability will probably be highly correlated with X, a solution could be to leave such a variable out since 'a part' of its information is already included in X. But in essence this means that, when regressing 'hourly wage' on X only, the variable measuring ability is an *omitted variable* that is **hidden** in the error term ε. It is very likely that ε and X are **not** uncorrelated (since 'ability' is correlated with X).

There are many economic situations where the research variable and an explanatory variable X influence **each other**. Under such circumstances, it turns out that ε and X are usually not uncorrelated. For example, if we want to find the relationship between D = 'national income' and C =

'consumption expenditure', it is clear that C is determined by D. However, D is also determined by C since C is a part (determinant) of D. The following **simultaneous equation model** describes the situation:

$$\begin{cases} C = \beta_0 + \beta_1 D + \varepsilon \\ D = C + E \end{cases}$$

Here, E denotes the non-consumption expenditure; it is assumed that E is determined outside the model. It turns out that, under these circumstances, $\text{cov}(D, \varepsilon)$ is **un**equal to 0 and that the ordinary least-squares estimators B_0 and B_1 are **biased** and **in**consistent.

What can be done?

Sometimes the problem that $\text{cov}(\varepsilon, X) \neq 0$ in a model with $Y = \beta_0 + \beta_1 X + \varepsilon$, can be solved. The trick is to try to find an *instrumental variable* (IV), a variable V that has the following relationships with ε and X:

$$\text{cov}(V, \varepsilon) = 0 \quad \text{and} \quad \text{cov}(V, X) \neq 0 \tag{22.1}$$

So, an instrumental variable is linearly related to X but not to ε. If such a variable V can be found, then the *two-stage least-squares regression* method turns out to yield estimators that **are** unbiased and consistent. Below, the observations x_i, y_i and v_i refer to observations of X, Y and V for the sampled elements $i = 1, \ldots, n$. The two stages of this regression method are as follows:

1 Regress X on V on the basis of the pairs $(v_1, x_1), \ldots, (v_n, x_n)$ and calculate the resulting predictions $\hat{x}_1, \ldots, \hat{x}_n$.
2 Regress Y on $\hat{X}$ on the basis of the pairs $(\hat{x}_1, y_1), \ldots, (\hat{x}_n, y_n)$.

It can (but will not) be shown that this method yields sample regression coefficients after stage 2 that are unbiased and consistent estimators of β_0 and β_1.

It would appear that the two-stage method estimates the relation $Y = \beta_0 + \beta_1 X + \varepsilon$ not on the basis of $(x_1, y_1), \ldots, (x_n, y_n)$ but on the basis of $(\hat{x}_1, y_1), \ldots, (\hat{x}_n, y_n)$ where the $\hat{x}_i$ are the predictions of the x_i that follow from the substitution of the v_i in the sample regression equation of stage 1. The $\hat{x}_i$ can be considered as approximations of the x_i since X and V **are** linearly related. Since the computed values $\hat{x}_i$ are based on variables that are uncorrelated with the errors, the results of the two-stage approach are optimal.

Note that the two-stage least-squares regression method is in a sense a generalization of the ordinary regression method: if X already satisfies the condition that $\text{cov}(X, \varepsilon) = 0$, then X itself can be taken as instrumental variable for X and the two-stage least-squares method is the same as the ordinary least-squares method since stage 1 yields $\hat{x}_i = x_i$.

This two-stage method can easily be generalized for models with more than one independent variables. Find instrumental variables for all independent variables X that give complications with the requirement $\text{cov}(X, \varepsilon) = 0$; for all other independent variables, let them be their own instrumental variables.

Most statistical packages will generate printouts with the two-stage least-squares regression method. See Appendix A1.22 for computational advice.

How can it be proved that a variable V is an instrumental variable? Checking whether $\text{cov}(V, X) \neq 0$ is not a problem: use the x and v data to check whether V is significant in the first-order linear regression model of X on V. However, the error term ε (of the regression of Y on X) is unobservable, so in general we cannot hope to have a test for the assumption $\text{cov}(V, \varepsilon) = 0$. In most cases one must defend the validity of $\text{cov}(V, \varepsilon) = 0$ by making use of economic arguments and introspection.

In the above example about regression of 'hourly wage' on $X =$ 'number of years of education', the variable $V =$ 'number of years education of the father' is a possible instrumental variable for X. It is well known from several studies that X and V are correlated. Whether this variable is uncorrelated with 'ability' and hence with the error term, is a problem that would need further reflection.

In the other example, it is the variable E (non-consumption expenditure) that can be used as instrumental variable for national income D. Since E is determined **outside** the model, it can be hoped that E is uncorrelated with ε. Since $D = C + E$, it follows that E is automatically correlated with D.

In the case of time series, IVs can sometimes be found by going one period back in time (lag 1). A classical example concerns the relationship between the quarterly demand (Y) for a certain commodity and the variables X_1 = 'price of the commodity in that quarter' and X_2 = 'average income of the consumers in that quarter'. Since demand and price have a reciprocal effect **on each other**, we cannot use ordinary least-squares. Instead, a two-stage least-squares regression model for Y might do well if we take X_1 and X_2 as independent (explanatory) variables and $\tilde{X}_1$ = 'price of the commodity in the **previous** quarter' and X_2 as respective instrumental variables for X_1 and X_2.

Example 22.10

According to economic theory, a high euro/dollar exchange rate is, at least in the short run, advantageous for inflation control in the euro area. The idea is that a weak dollar makes imports from the USA cheaper and hence reduces the inflation rate. The present study will try to detect this effect on the basis of data published by the European Central Bank and Eurostat.

The dataset Xmp22-10.sav contains, for the period November 2003 – October 2007, observations of the monthly inflation rate Y in the euro area and the monthly exchange rate X of the euro against the US dollar. Y is measured in percentage changes with respect to one year before. According to the above theory, it is expected that the slope in the relationship $Y = \beta_0 + \beta_1 X + \varepsilon$ is significantly negative. This theory will first be checked under the linear regression assumptions. But since the inflation and exchange rates influence **each other** and hence it cannot be expected that $\text{cov}(\varepsilon, X)$ is 0, we also apply two-stage linear regression with V = 'exchange rate in the previous month' as instrumental variable of X.

Ordinary linear regression yields the following results:

$$\hat{y} = 3.825 - 1.344x; \quad r^2 = 0.093;$$

p-value of two-sided t-test for significance of X: 0.035

It follows that the model is useful at significance level 0.05 and that an increase of 0.1 in x will on average reduce inflation by 0.1344 percentage points. This reduction is in accordance with the above theory. But the validity of $\text{cov}(\varepsilon, X) = 0$, has to be doubted. That is why we next run a two-stage regression with V as instrumental variable of X. See Appendix A1.22, also for the creation of the v data from the x data.

Two-stage linear regression yields the following results:

$$\hat{y} = 4.695 - 2.029x; \quad r^2 = 0.152;$$

p-value of two-sided t-test for significance of X: 0.007

The model is useful, a conclusion that is more convincing than the corresponding conclusion of the former model. An increase of 0.1 in x will on average reduce inflation by 0.2029 percentage points, a larger decrease than was promised by the former model. This model more convincingly supports the above theory.

Note, however, that there is another model requirement that will not be satisfied: we expect a positive linear relationship for the error terms that belong to successive months.

CASE 22.1 INCOME AND EDUCATION LEVEL OF IDENTICAL TWINS (PART III) – SOLUTION TO PART B

When explaining difference between log-hourly wages from difference in education, problems arise with respect to the model requirement that the error term ε and predictor X have to be independent or at least uncorrelated. There are two reasons for that. The first is that interviewed people tend to overestimate their personal level of education, thus confronting the researcher with (unobservable) measurement errors for X that accordingly become part of ε and cause correlation between ε and X. A second reason is that the model does not include a variable that measures 'difference in ability'. Such a variable would be correlated with the 'difference in education' predictor. But since 'difference in ability' is not represented in the model, it is part of the error term ε, which gives another reason why ε and X will not be uncorrelated.

To correct for measurement error in the education data, it has been suggested to use DEDUCxx, the difference between cross-reported education. The search for an instrumental variable for DEDUC = EDUC1 − EDUC2 might also come up with DEDUCxx, with the advice to use it in a **two-stage** linear regression. When regressing the difference between the log-hourly wages of twin 1 and twin 2 on the difference between the numbers of years of education of twin 1 and twin 2 by taking the cross-reported difference DEDUCxx as instrumental variable for DEDUC, the estimated slope turns out to be 0.167 with standard error 0.043. (Check it; see Appendix A1.22)

Notice that the estimate 0.167 is almost twice as large as the least-squares estimate 0.092 that was found in part (a). It seems that conventional methods produce serious underestimates of the economic value of one year of schooling.

22.6 Introduction to binary choice models and the logit model

So far, the regression models were applied only for continuous, quantitative dependent variables Y. The present section studies the consequences when Y is a 0–1 (dummy) variable. **Binary choice models** are regression models for 0–1 research variables Y. Only the logit model is considered in more detail.

22.6.1 Binary choice models

There are many examples where the research variable Y is a dummy variable. For instance:

- Female labour force participation: $y = 1$ if the female participates in the labour market, $y = 0$ if she does not. Potential predictors: the presence of a partner in life, number of children, their ages, income of partner, etc.
- To buy or not to buy a good (car / house / holiday, etc.): $y = 1$ if person or household buys the good, $y = 0$ if not. Potential predictors: total expenditure, composition of the household, etc.
- Satisfaction with income / safety / society / life in general: $y = 1$ if person or household is satisfied, $y = 0$ if not. Potential predictors: household income, household composition, degree of urbanization of region, etc.

First, it will be investigated what happens if we stick to the standard **linear** regression model. Problems already arise with the basic assumption. To understand that, recall from probability theory that the expectation of a 0–1 random variable Y that takes the value 1 with probability p, is equal to p. This follows immediately from the (discrete) probability density function of Y:

y	0	1
$P(Y = y)$	$1 - p$	p

Recall that:

$$E(Y) = 0 \times (1 - p) + 1 \times p = p$$

In other words, $E(Y)$ is equal to $P(Y = 1)$. Now go back to the basic assumption of the linear regression model explaining Y from $X_1, \ldots, X_k$. It states that:

$$E(Y \mid x_1, \ldots, x_k) = \beta_0 + \beta_1 x_1 + \ldots + \beta_k x_k \quad \text{for all } x_1, \ldots, x_k \tag{22.2}$$

Here, the left-hand side of the equality has to be read as the expectation of Y if it is already given that $x_1, \ldots, x_k$ are the values of the independent variables. But Y is now a 0–1 variable, so this expectation of Y is a probability of getting 1. To be precise, the left-hand expectation in (22.2) is equal to the probability that Y takes the value 1 if it is already given that $x_1, \ldots, x_k$ are the values of the independent variables:

$$P(Y = 1 \mid x_1, \ldots, x_k) = \beta_0 + \beta_1 x_1 + \ldots + \beta_k x_k \quad \text{for all } x_1, \ldots, x_k \tag{22.3}$$

The left-hand part is a probability, a number between 0 and 1. However, the right-hand side can, in principle, take any value. This means that if we were to estimate the β_i by using the least-squares method, there is no guarantee that the estimate $b_0 + b_1 x_1 + \ldots + b_k x_k$ of the right-hand side of (22.3) would lie in the interval [0, 1]. It would appear that we cannot stick to the linear regression model if Y is a 0–1 variable.

The idea is now to 'repair' the right-hand side of (22.3) by forcing it into the [0, 1] straitjacket by way of a continuous distribution function (cdf) F; see also Section 8.2.1. In fact, we want a function F that has the following properties ($\Re$ is the set of real numbers):

$F: \Re \rightarrow [0, 1]$;

F is strictly increasing and continuous;

$F(x)$ tends to 0 as x tends to $-\infty$; $F(x)$ tends to 1 as x tends to ∞.

And a continuous cdf has those properties!

Since $F(\beta_0 + \beta_1 x_1 + \ldots + \beta_k x_k)$ always belongs to the interval [0, 1], we assume that (22.4) holds:

Basic assumption of binary choice model

$$P(Y = 1 \mid x_1, \ldots, x_k) = F(\beta_0 + \beta_1 x_1 + \ldots + \beta_k x_k) \quad \text{for all } x_1, \ldots, x_k \tag{22.4}$$

Corollary of the assumption

$$P(Y = 0 \mid x_1, \ldots, x_k) = 1 - F(\beta_0 + \beta_1 x_1 + \ldots + \beta_k x_k) \quad \text{for all } x_1, \ldots, x_k \tag{22.5}$$

In this binary choice model (22.4) the probability $P(Y = 1 \mid x_1, \ldots, x_k)$ is **not** linearly related to x_i. Hence, we **cannot** use the interpretation of β_i in terms of an increase of x_i by 1 unit.

There are many choices possible for the distribution function F. Because of its special properties, a good choice is the cdf of the standard normal distribution; recall from Section 10.3 that this cdf is generally denoted by Φ. The model that is based on this choice for F is called **probit model**. We will not go into the details of that model.

22.6.2 Logit model

Instead, we will consider the model that is based on the cdf of the so-called **logistic distribution**. This cdf is denoted by Λ and its mathematical prescription is as follows:

$$\Lambda(t) = \frac{e^t}{1 + e^t} \quad \text{for all real numbers } t$$

The model based on $F = \Lambda$ is called **logit model**. The box below shows how its basic assumption depends on the given observations $x_1, \ldots, x_k$ of the independent variables.

Basic assumption of logit model

$$P(Y = 1 \mid x_1, \ldots, x_k) = \Lambda(w) = \frac{e^w}{1 + e^w} \quad \text{with} \quad w = \beta_0 + \beta_1 x_1 + \ldots + \beta_k x_k \tag{22.6}$$

Corollary of the assumption

$$P(Y = 0 \mid x_1, \ldots, x_k) = 1 - \Lambda(w) = \frac{1}{1 + e^w} \tag{22.7}$$

For the logit model, the ratio $P(Y = 1)/P(Y = 0)$ is of special importance. With respect to 0–1 variables Y, this ratio is generally called the **odds ratio**. For example, if 80% of households own a car and Y measures whether a household (yes / no) has a car, then the odds ratio is $0.80 / 0.20 = 4$. The chances are 4 to 1 that the household owns a car.

The equations (22.6) and (22.7) yield immediately that:

$$\frac{P(Y = 1)}{P(Y = 0)} = \frac{e^w}{1 + e^w} \bigg/ \frac{1}{1 + e^w} = \frac{e^w}{1 + e^w} \times \frac{1 + e^w}{1} = e^w$$

By substituting the expression of w, it follows that:

$$\frac{P(Y = 1)}{P(Y = 0)} = e^{\beta_0 + \beta_1 x_1 + \ldots + \beta_k x_k}$$

Taking the (natural) logarithm yields:

$$\log\left(\frac{P(Y = 1)}{P(Y = 0)}\right) = \beta_0 + \beta_1 x_1 + \ldots + \beta_k x_k \tag{22.8}$$

It would appear that the log of the odds ratio brings us back 'to earth': we obtain a linear function! Hence, for instance, β_1 can be interpreted by letting x_1 increase by 1: if x_1 increases by 1 while the values of the other independent variables remain unchanged, then the log of the odds ratio changes by β_1. But changes of a log can be interpreted as percentages; see Section 21.4. This yields the following more concrete interpretation of β_1:

> If x_1 increases by 1 and $x_2, \ldots, x_k$ remain unchanged, then the odds ratio changes by $100\beta_1\%$; the relative chance of the value 1 changes by $100\beta_1\%$.

Similar interpretations can be given for $\beta_2, \ldots, \beta_k$.

22.6.3 Statistical inference using the logit model

Of course, the unknown parameters $\beta_0, \ldots, \beta_k$ have to be estimated on the basis of a sample. Suppose that we are in possession of a **large random sample** of size n and that for each element the values of the dependent and independent variables are measured. Then we can, with the help of a statistical package, start an estimation procedure to estimate $\beta_0, \ldots, \beta_k$. Unfortunately, the ordinary least-squares (OLS) method does not work properly for 0–1 observations y. Instead, a method called **maximum likelihood estimation** (MLE) is used, but we will not define this method or go into details.

Having done the estimation, the (ML) estimates $b_0, \ldots, b_k$ of $\beta_0, \ldots, \beta_k$ can be read from the printout. Replacing in the right-hand expression of (22.6) the unknown parameters $\beta_0, \ldots, \beta_k$ by their estimates $b_0, \ldots, b_k$, leads to the **logit regression equation**:

$$\hat{y} = \Lambda(b_0 + b_1 x_1 + \ldots + b_k x_k) \tag{22.9}$$

If, for a new population element, the values $x_1, \ldots, x_k$ are given but the value of Y is not, then these values $x_1, \ldots, x_k$ can be substituted into the logit regression equation. The resulting value $\hat{y}$ falls (thanks to the fact that Λ is a cdf) in the interval $[0, 1]$ and can be used in two ways:

- as an estimate of $P(Y = 1)$ if the characteristics are $x_1, \ldots, x_k$;
- after rounding, as a prediction of Y if the characteristics are $x_1, \ldots, x_k$.

The first duty is clear because of (22.6). For the second duty, notice that the resulting $\hat{y}$ is not necessarily precisely equal to 0 or 1. Hence, to make it a prediction of the 0–1 random variable Y, the value $\hat{y}$ has to be rounded to the nearest integer (which is 0 or 1).

Example 22.11

In order to handle communication with tax offices in an easy way, each household has a reference person called 'head of the household'. But which factors determine whether the head of the household is female? In the present investigation, the role of the variables 'age' and 'hourly wage' will be studied for a large population of household heads.

The dependent variable Y is taken as the dummy DF, which is 1 for households with a female head. The independent variables are $X_1 = $ 'age of the head' and $X_2 = $ 'hourly wage of the head'. The dataset Xmp22-11.sav contains the observations of these variables for a sample of 300 heads. Here is the basic assumption of the model and its log-odds transformation:

$$P(Y = 1) = \Lambda(\beta_0 + \beta_1 x_1 + \beta_2 x_2) \quad \text{and} \quad \log\left(\frac{P(Y = 1)}{P(Y = 0)}\right) = \beta_0 + \beta_1 x_1 + \beta_2 x_2$$

Estimation of the model leads to the partial printout shown below; see Appendix A1.22 for a way to create it.

Variables in the Equation

		B	S.E.	Wald	df	Sig.	Exp(B)
Step 1(a)	AGE	.037	.011	10.535	1	.001	1.038
	WAGE	.133	.033	16.541	1	.000	1.142
	Constant	−3.795	.624	37.044	1	.000	.022

a Variable(s) entered on step 1: AGE, WAGE.

Hence, the logit regression equation is $\hat{y} = \Lambda(-3.795 + 0.037x_1 + 0.133x_2)$. The positive signs of the coefficients of x_1 and x_2 inform us that older heads and heads with higher wages are more likely to be female. However, for the more precise interpretations of the coefficients, the log-odds transformation has to be used.

The coefficient $b_1 = 0.037$ of x_1 has the following interpretation. If age increases by 1 year while the wage is unchanged, the odds ratio is estimated to increase by 3.7%. To put it another way, the relative probability that the head is female will on average increase by 3.7%. The coefficient $b_2 = 0.133$ of x_2 leads to the interpretation that an increase of €1 in hourly wage (while age is unchanged) will on average lead to an increase of 13.3% of the odds ratio; the relative probability that the head is female is estimated to increase by 13.3%.

The logit regression equation can be used to estimate the probability that a head is female and to predict the gender of a head from his/her age and hourly wage. Suppose that a tax office knows that the head of a certain household is 35 years old and that his/her hourly wage is €22.50. That is, this household has the characteristics $x_1 = 35$ and $x_2 = 22.50$. Substitution into the logit regression equation yields:

$$\hat{y} = \Lambda(-3.795 + 0.037 \times 35 + 0.133 \times 22.50) = \Lambda(0.4925)$$

From the definition of the logistic distribution function it follows that $\hat{y} = 0.6207$. This value can be considered as an estimate of the probability that the head of this household is female. Its rounded value 1 can be considered as a prediction of the value of Y for this head. So, the statistical conclusion is that this head is female. Note, however, that statistical conclusions can be incorrect.

Statistical inference using the logit model will be considered briefly.

The **usefulness** of the model can be statistically checked by a χ^2-test. The approach depends on the statistical package that is used. For instance, SPSS gives the **val** of the test statistic for the testing problem

i Test $H_0 : \beta_1 = \ldots = \beta_k = 0$ against H_1: at least one of the $\beta_1, \ldots, \beta_k$ is $\neq 0$

under the heading '-2 Log likelihood'. The null hypothesis is rejected (and the conclusion that the model is useful accepted) if **val** $\geq \chi^2_{\alpha;k}$.

To determine the estimated **degree of usefulness**, the coefficient of determination cannot be used any more (again, because of the fact that Y is a 0–1 variable). Of the many alternatives available we will consider only the easiest and most appealing one.

By substituting the x characteristics of each of the sample elements into the logit regression equation (22.9), the n numbers $\hat{y}_1, \ldots, \hat{y}_n$ are obtained, all falling between 0 and 1. By rounding them to the nearest integer, the predictions $\hat{p}_1, \ldots, \hat{p}_n$ (say) arise. But notice that $\hat{p}_1, \ldots, \hat{p}_n$ are the predictions of $y_1, \ldots, y_n$, so they can be used to measure the predicting performance of the model. In fact, the proportion of correct predictions measures the quality of the model. A prediction $\hat{p}_i$ is a correct prediction if the difference $y_i - \hat{p}_i = 0$; a prediction is incorrect if the absolute value of the difference equals 1, so if $| y_i - \hat{p}_i | = 1$. Since

$$\frac{1}{n} \sum_{i=1}^{n} | y_i - \hat{p}_i |$$

is the proportion of the **in**correct predictions, we arrive at the following definition of the PCP, the **proportion correctly predicted**:

PCP of the logit model

$$PCP = 1 - \frac{1}{n} \sum_{i=1}^{n} | y_i - \hat{p}_i | \tag{22.10}$$

For large sample sizes, the **individual significance** of each of the independent variables can be tested with a t-test. Things are illustrated in the continuation of Example 22.11.

Example 22.12

Although the logit regression coefficients have already been interpreted and the estimated model used for prediction purposes, we have not yet checked the usefulness of the model. The printout below belongs to the partial printout that was given in the former example.

Model Summary

Step	-2 Log likelihood	Cox & Snell R Square	Nagelkerke R Square
1	272.457(a)	.071	.114

a Estimation terminated at iteration number 5 because parameter estimates changed by less than .001

Classification Table(a)

Observed			Predicted		
			DF		Percentage Correct
			0	1	
Step 1	DF	0	242	0	100.0
		1	54	4	6.9
	Overall Percentage				82.0

a   The cut value is .500

For the testing problem

  i Test $H_0 : \beta_1 = \beta_2 = 0$ against H_1: at least one of the β_1, β_2 is $\neq 0$

with $\alpha = 0.05$, the **val** of the test statistic is 272.457; see Model Summary. (The other numbers in this summary are not interpreted.) Since $k = 2$ and $\chi^2_{0.05;2} = 5.9915$ (*), we reject H_0 for realizations ≥ 5.9915. Since **val** lies far in the rejection region, we do reject H_0. The model is useful.

    From the classification table, the percentage correctly predicted can be read: PCP = 0.82. The estimated model predicts 82% of the observations $y_1, \ldots, y_{300}$ correctly. Note, however, that for the females only 6.9% are right. On the other hand, **all** male heads in the sample are correctly predicted.

    The individual significance of the two independent variables follows from the printout of Example 22.11. For a quick view, the column with heading Sig. can be used: both p-values of the t-tests are smaller than significance level 0.05. Below, we give the formal five-step procedure of the t-test for the significance of the variable 'age'. Notice that **val** cannot be read directly from the printout. (In fact, it is the square of **val** that is given, under the heading Wald.) Here is the five-step procedure:

  i Test $H_0: \beta_1 = 0$ against $H_1: \beta_1 \neq 0$; $\alpha = 0.05$

  ii $T = \dfrac{B_1 - 0}{S_{B_1}}$ where B_1 is the ML estimator

  iii Reject $H_0 \Leftrightarrow t \leq -1.9680$ or $t \geq 1.9680$ (*)

  iv **val** $= \dfrac{0.037}{0.011} = 3.364$

  v Reject H_0 since **val** lies in the rejection region

The variable 'age' is significant.

Summary

We discussed tests for model violations and the consequences, and actions that can be taken if model requirements are not satisfied. The main conclusion was that statistical procedures cannot be trusted if the residual analysis shows that one or more of the requirements is seriously violated. Non-normality and heteroskedasticity can sometimes be repaired by using log(Y) as dependent variable instead of Y. The method of weighted least-squares may offer a solution if heteroskedasticity is caused by aggregation.

The requirement of independence of the error terms is often violated in the case of time series. The Durbin–Watson test offers a statistical tool to check things and to detect deviations from the requirement. As a remedy, it sometimes works to include time T as a new independent variable. In other cases, replacing Y by the variable that measures the growth (or the relative growth) of Y when compared with the previous period, offers a solution to such independence violations.

Strong collinearity among the independent variables may seriously disturb the statistical properties of a model. This is because collinearity increases the standard deviations of the least-squares estimators, which can sabotage the drawing of statistical conclusions about the population regression coefficients. The problem is often evaded by omitting one or more independent variables from the model.

Although the independence requirement for ε and predictor(s) X can usually be relaxed to the requirement that $cov(\varepsilon, X) = 0$, even this weakened version is sometimes not satisfied for economic variables. A rather abstract way that occasionally works is to use instrumental variables and two-stage least-squares regression. However, finding good IVs is usually not easy.

Although the ordinary linear regression model is sometimes used for situations where the dependent variable can take only two values, it has to be discouraged. Instead, the logit model often offers a good alternative.

🔑 Key terms

aggregation **701**	Durbin–Watson test **709**	odds ratio **719**
autocorrelation **708**	first-order	omitted variable **714**
autoregressive	autocorrelation **708**	probit model **718**
system **712**	instrumental variable **715**	proportion correctly
binary choice model **717**	Lilliefors test **706**	predicted of the logit
Cochrane–Orcutt	logistic distribution **718**	model **721**
procedure **713**	logit model **719**	two-stage least-squares
Durbin–Watson	logit regression	regression **715**
bounds **713**	equation **719**	weighted least-squares
Durbin–Watson	maximum likelihood	regression **701**
table **709**	estimation **719**	

❓ Exercises

Exercise 22.1

The file Xrc22-01.sav contains recent data for 71 successive business days about prices and returns of the Heineken stock and the AEX index. Two regressions have been run:

- regression of price of the stock on price of the AEX index;
- regression of return (%) of the stock on return (%) of the AEX index.

For the first regression, the value of the DW statistic turns out to be 0.191, for the second regression it is 1.789. The dataset has some missing values, so that $n = 70$ for the first regression and $n = 69$ for the second regression.

a Formulate the basic assumptions of the two models.

b Test with $\alpha = 0.05$ whether the error variables of the first model are positively first-order autocorrelated.

c Test with $\alpha = 0.02$ whether the error variables of the second model are first-order autocorrelated.

Exercise 22.2

The dataset Xrc22-02.sav contains data for Germany on the yearly growth (%) of the active population and the yearly growth (%) of the number of persons engaged in economic activity, for the period 1958–2004. The dataset has missing data for two years due to the German reunification, so $n = 45$.

To explain the yearly growth Y of the number of persons engaged in economic activity from the yearly growth X of the active population, we regress Y on the two variables X and the dummy $Dtime$ that is 1 for the years before the reunification in 1990. We suspect that successive errors will be positively correlated. The printout of the first-order linear regression model with independent variables X and $Dtime$ is given below. For statistical inference, use significance level 0.05.

Model Summary(b)

Model	R	R Square	Adjusted R Square	Std. Error of the Estimate	Durbin-Watson
1	.778(a)	.605	.586	.75927	1.098

a Predictors: (Constant), Dtime, GROWTH_pop
b Dependent Variable: GROWTH_engaged

ANOVA(b)

Model		Sum of Squares	df	Mean Square	F	Sig.
1	Regression	37.112	2	18.556	32.188	.000(a)
	Residual	24.212	42	.576		
	Total	61.324	44			

a Predictors: (Constant), Dtime, GROWTH_pop
b Dependent Variable: GROWTH_engaged

Coefficients(a)

Model		Unstandardized Coefficients		Standardized Coefficients	t	Sig.
		B	Std. Error	Beta		
1	(Constant)	−.473	.211		−2.239	.031
	GROWTH_pop	1.110	.146	.754	7.612	.000
	Dtime	.242	.255	.094	.947	.349

a Dependent Variable: GROWTH_engaged

a Is the model useful? How useful?

b Are both variables individually significant?

c Is there a significant difference between the mean of the growth figures of employment before and after reunification?

d Check whether there is positive first-order autocorrelation in the sequence of errors. What are the consequences for parts (a)–(c)?

Exercise 22.3

Is it possible to predict gender from weight and height? To answer this question, we regress the dummy variable $Y = DF$ (which takes the value 1 if the person is female) on the variables X_1 = 'weight (kg)' and X_2 = 'height (cm)' on the basis of a sample of 300 adults; see Xrc22-03.sav. We will use the logit model.

a Write down the basic assumption of the logit model. Also write down the accompanying log-odds equation (22.8).

Use the SPSS printout below to answer questions (b)–(f).

Model Summary

Step	-2 Log likelihood	Cox & Snell R Square	Nagelkerke R Square
1	187.346(a)	.301	.481

a Estimation terminated at iteration number 6 because parameter estimates changed by less than .001.

Classification Table(a)

Observed			Predicted		
			DF		Percentage Correct
			0	1	
Step 1	DF	0	233	9	96.3
		1	27	31	53.4
	Overall Percentage				88.0

a The cut value is .500

Variables in the Equation

		B	S.E.	Wald	df	Sig.	Exp(B)
Step 1(a)	WEIGHT	−.054	.019	8.006	1	.005	.947
	HEIGHT	−.174	.030	32.949	1	.000	.840
	Constant	32.235	4.863	43.944	1	.000	99898664302728.100

a Variable(s) entered on step 1: WEIGHT, HEIGHT.

b Is the model useful? Use 0.01 as significance level.
c Interpret the PCP.
d Determine the logit regression equation. Interpret the coefficients in terms of the odds ratio.
e Are the two independent variables individually significant? Use α = 0.01.
f Use your own weight and height to find out whether the model predicts your gender correctly.

Exercise 22.4 (computer)

A manufacturer of a certain commodity believes that the yearly sales Y (in units of €10000) can partially be explained by the yearly expenditure X (in units of €1000) on advertising. To estimate the relationship between Y and X, he adopts the simple linear regression model and estimates it on the basis of the data of the most recent 16 years; see the file Xrc22-04.sav.

a Estimate the model. Show that, at significance level 0.05, the error terms have positive first-order autocorrelation.
b In a first attempt to solve this violation of the requirement for independent errors, the manufacturer includes time T in the model. In the dataset, time is represented by the data 1, 2, ... 16. Write down the basic assumption of this new model, estimate it and check, at significance level 0.10, whether there is evidence that the requirement for independent errors does not hold for this new model.

c The manufacturer also tries to explain the **growth** of Y from the **growth** of X, both (as differences) when compared with the previous year. Answer the same questions as in part (b).

Exercise 22.5

From an extensive study it is known that the simple linear regression model is very suitable to explain, for the population of adult working persons with middle incomes, the variable 'monthly expenditure on recreation (in euro)' from the variable 'monthly salary (in euro)'. For this population, the validity of the model requirements is not a problem.

A researcher wants to use this model for prediction purposes, but in his dataset the observations do not come from individuals but are aggregated measurements about the employed adults in 12 **households**. The file Xrc22-05.sav lists these data as **averages** (per employed adult household member). The number of employed adults in household i is denoted by m_i.

a Set Y = 'average (per employed household member) of the monthly recreation expenditure for a household' and X = 'average (per employed member) of the monthly salary for that household'. Explain why the homoskedasticity requirement of the ordinary linear regression model is violated in the present situation.

b (computer) Use weighted least-squares to estimate $E(Y) = \beta_0 + \beta_1 x$. Is the model useful? Give the regression equation and interpret the sample regression coefficient of x.

Exercise 22.6 (computer)

The market for mobile phones is dominated by the brands A, B, C and D. The table below lists, for each brand, the number of employees and the production (units) of five random days; see also Xrc22-06.sav. Interest is in the relationship between Y = 'daily production' and X = 'number of employees' on the mobile phone market.

Brand	Number of employees	Production (units)				
A	5000	1800	2000	2000	2000	2000
B	10000	3000	3200	3500	3500	3600
C	15000	4200	4200	4500	4800	5000
D	20000	4800	5000	5700	6000	6200

a Create the scatter plot of the 20 pairs (x, y).

b A researcher wants to estimate β_0 and β_1 in the equation $Y = \beta_0 + \beta_1 x + \varepsilon$ by adopting the linear regression model. But which assumption will cause problems?

To overcome the problem of part (b), the researcher decides to use the transformations

$$Y^* = Y/x, \quad x^* = 1/x, \quad \varepsilon^* = \varepsilon/x,$$

and to change over to estimation of the equation $Y^* = \gamma_0 + \gamma_1 x^* + \varepsilon^*$.

c What is the relationship between γ_0, γ_1 and β_0, β_1? Which additional requirement about the original error term ε has to be imposed to ensure that this new model satisfies the homoskedasticity assumption?

d Create the data that are needed. Does the scatter plot of y^* on x^* support the homoskedasticity assumption? Use the data to estimate the new model by way of simple linear regression. Transform your results into estimates of β_0, β_1.

e Determine a suitable 95% confidence interval for β_1. Is $\beta_1 > 0$?

Exercise 22.7

The study in this exercise is based on a sample of 4371 recently graduated Dutch persons (interviewed one and a half years after graduation) and was carried out by J. Allen et al. (2007) for the VSNU. It is about the relationship between the binary variable Y = 'yes/no found paid work' and the factors 'highest preparatory education before going to the university', 'experience during the education', 'age', 'gender', 'sector of education' and 'level of competence (personal belief)'. The conclusions of the study were originally presented in the VSNU's *WO-monitor 2004/2005*.

The table below contains computed results (the ML estimates and the accompanying standard errors) for a logistic regression of the variable Y = 'yes/no found paid work' on 26 dummy variables. In the table, HBO is higher professional education, VWO is secondary education originally meant to prepare for university, HAVO is secondary education originally meant to prepare for HBO. 'Ref' refers to a base level used for a group of dummy variables.

a Formulate the basic assumption of the logit model that is used here.

b Is the model useful at significance level 0.01? It is given that the *val* of the test you will have to conduct is 1589.573.

c Which of the variables are significant at significance level 0.05?

d Interpret the coefficient of x_{11} in the logit regression equation.

e Below, some conclusions of the report are listed. Check them yourself; level 0.05.

- University graduates with HBO as preparatory education have a higher probability of having work after one and a half years than university graduates with VWO as preparatory education.

- Getting administrative experience while in education increases the probability of having work.

- It is more difficult to find work if the graduate is older.

- Graduates in Agriculture and Languages & Culture have a smaller probability of having work than graduates in Economics.

- Getting academic competences while in education has a negative effect on the probability of having work (?).

Also interpret the results of the other competences.

	ML estimate	Standard error
Highest preparatory education (ref: VWO)		
X_1 HAVO	−0.230	0.556
X_2 HBO	0.586	0.204
X_3 Other	0.958	1.037
X_4 : Also did other university education	0.227	0.344
Experience during education		
X_5 Less than 1 year relevant experience	−0.149	0.188
X_6 At least 1 year relevant experience	0.008	0.197
X_7 Got job training	0.178	0.165
X_8 Administrative experience	0.567	0.169
X_9 Experience abroad	0.061	0.162
Age 1.5 years after graduation (ref: 26-30)		
X_{10} 21–25	0.342	0.170
X_{11} 31–35	−1.013	0.251
X_{12} 36 years and older	−1.442	0.362
X_{13} : Gender: woman	−0.008	0.162

▶

Sector of graduation (ref: economics)			
X_{14}	Medical care	0.656	0.385
X_{15}	Agriculture	−0.918	0.326
X_{16}	Nature	−0.522	0.312
X_{17}	Law	0.133	0.305
X_{18}	Languages and culture	−0.731	0.265
X_{19}	Technology	0.790	0.328
X_{20}	Behavioural and social	−0.433	0.238
Competences			
X_{21}	Academic competences	−0.559	0.222
X_{22}	Specialized knowledge	0.432	0.128
X_{23}	Attitudes	−0.082	0.155
X_{24}	Management competences	0.985	0.175
X_{25}	ICT	0.089	0.091
X_{26}	Specialised knowledge other areas	0.199	0.106
(constant)		−0.819	0.670

Exercise 22.8 (computer)

WO-monitor 2004/2005 also reports a study among 12 855 persons (graduated five years previously, in 1999/2000) from ten European countries about being abroad during university education. Here, the research variable *Y* is 'did (yes/no) spend some time on a foreign university'. The results are in the file Xrc22-08.sav. The variable 'country' has the following values:

1 = Austria	2 = France	3 = Finland	4 = Germany
5 = Italy	6 = Netherlands	7 = Norway	8 = Spain
9 = Switzerland	10 = UK		

This variable will be included in a logit model for *Y* by way of nine dummies that are already created in the dataset. Note that the Netherlands is taken as base level.

a Write down the basic assumption of the model.

b Is the model useful? Use significance level 0.01. Measure the degree of usefulness by way of the PCP.

c Which variables are individually significant at significance level 0.05? Give the precise interpretation of the coefficient of the dummy for Germany, in terms of the odds ratio.

d Herr Frankel graduated five years ago from a university in Germany. Find an estimate for the probability that during his study he spent some time at a foreign university. Also **predict** Herr Frankel's answer to *Y*.

Exercise 22.9 (computer)

We study, for the euro area, the relationship between economic quarterly growth *Y* and quarterly unit labour costs *X*. Both variables are measured as a percentage with respect to one year before. It is suspected that *Y* tends to decrease as *X* increases.

To check that conjecture, *Y* is regressed on *X* on the basis of quarterly data of European Central Bank (Statistical Data Warehouse) for the period 1996Q1–2007Q2; see Xrc22-09.sav. Since the two variables *X* and *Y* influence **each other**, we may not expect that cov(ε, *X*) will be 0 when using the linear regression model for the equation $Y = \beta_0 + \beta_1 X + \varepsilon$. Below, we will compare the results of an ordinary regression and a two-stage regression with instrumental variable *V* = 'unit labour costs in the **previou**s quarter' for *X*.

a Use ordinary linear regression to find estimates for β_0 and β_1 in the above equation. Write down the equation of the regression line and the coefficient of determination. Is β_1 significantly negative?

b Answer the same questions in the case of two-stage regression with V as instrumental variable for X.

c What do you think about the requirement in part (b) that $cov(\varepsilon, V) = 0$? Do you think that first-order autocorrelation of the error terms may be a problem?

Exercise 22.10 (computer)

As studied in Example 22.10, a strong euro/dollar exchange rate can be advantageous for inflation control in the euro area, at least in the short run. However, we might also argue that, for the long run, a strong euro will have a negative effect on exports.

This last statement will be investigated now. The long-run effects will be incorporated into the model by regressing Y = 'exports (in billions of euro) in a quarter' on X = 'exchange rate of the last quarter before that quarter'. Since there are also many other factors that influence Y, we first consider the relationship between Y and the variable T = 'time' that is measured by the number of the quarter since the introduction of the euro in 2002. The quarterly data (period 2002 Q1–2007 Q2) are in the file Xrc22-10.sav.

a Create the scatter plot of the y data on 'time'. Give your comments.

Because of the strongly linear relationship between Y and T (as observed in part (a)), it is suspected that the relationship between Y and X has to be studied in the presence of T. Thus, T will incorporate all other factors that influence Y. That is, we will study the equation: $Y = \beta_0 + \beta_1 T + \beta_2 X + \varepsilon$.

b Use ordinary linear regression to estimate the population regression coefficients. Are T and X individually significant? Interpret the estimates of the coefficients.

c Also use two-stage linear regression to estimate the population regression coefficients; take T and V = 'exchange rate of the second-last quarter before that quarter' as instrumental variables of T and X, respectively. Are T and X individually significant? Interpret the estimates of the coefficients that follow with this model.

Exercise 22.11 (computer)

We want to estimate the **value** (the increase of hourly wage) of one extra year of education. For that, the dataset Xrc22-11.sav is used; a dataset with observations of identical twins. There are two reasons for using data of twins. The first is that (identical) twins have similar personal characteristics and similar family backgrounds, which facilitates the isolation of education's effect on income. The second reason is that the usual overvaluation of the self-recorded education can be checked by asking the person's twin.

The dataset contains, for a random sample of 183 twins, observations of the variables:

HRWAGEH   hourly wage of twin 2
EDUCH   self-reported education (in years) of twin 2
HRWAGEL   hourly wage of twin 1
EDUCL   self-reported education (in years) of twin 1
DEDUCxx   difference (twin 1 minus twin 2) in cross-reported education; for example, twin 1's cross-reported education is the number of years of schooling completed by twin 1 as reported by twin 2
DTENU   difference (twin 1 minus twin 2) in tenure, or number of years at current job.

(The ordering 1 and 2 is randomly assigned; note that the dataset has many missing values.) To use the advantage of having paired observations as much as possible, we are going to explain the **difference** between the log-hourly wages of twin 1 and twin 2 from the difference between the numbers of years of education of twin 1 and twin 2. Since the factor 'work' may also influence the variation of the difference between the log-hourly wages, the variable DTENU will also be included in the linear regression model. See Cases 19.3, 20.2 and 22.1 for references and if you want more information.

a Create the observations of the variables DEDUC = EDUCL − EDUCH and DLHRWAG = log(HRWAGEL) − log(HRWAGEH). Here, log refers to the natural logarithm (also called LN).

b Regress DLHRWAG on DTENU and DEDUC. Give a statistical analysis of the results.

c Regress DLHRWAG on DTENU and DEDUCxx. Give a statistical analysis of the results.

d The model in part (b) will probably violate the model requirement that error term and predictor DEDUC have to be independent or at least uncorrelated. That is why DEDUCxx is used as instrumental variable for DEDUC in a two-stage regression. Do the regression and report your findings.

e Compare the estimated regression coefficients of DEDUC and DEDUCxx in parts (b), (c) and (d) in terms of the value of one extra year of difference in education. What do you conclude?

Exercise 22.12

Reconsider Exercise 22.1. Consider the model that explains the return (%) of the stock Heineken from the return (%) of the AEX index. Use the printouts below to test:

a for heteroskedastictity;

b for non-normality of the error terms.

c In part (a) carefully describe the auxiliary model that is used.

Coefficients(a)

Model		Unstandardized Coefficients		Standardized Coefficients	t	Sig.
		B	Std. Error	Beta		
	(Constant)	.389	.082		4.731	.000
	Ret_AEX	−.037	.136	−.033	−.268	.790

a Dependent Variable: Res2 = squared residual of original model.

Tests of Normality

	Kolmogorov-Smirnov(a)		
	Statistic	df	Sig.
ResStand	.077	69	.200(*)

* This is a lower bound of the true significance.
a Lilliefors Significance Correction

Exercise 22.13 (computer)

Reconsider Exercise 20.6, where the grade Y for the Final Exam of the course Statistics 2 was regressed on the grade X_1 for Midterm 1, the grade X_2 for the Team Assignment and the grade X_3 for Midterm 2. See Xrc20-06.sav for the data. In the tests below, take $\alpha = 0.05$.

a Test for heteroskedasticity. What is the auxiliary model?

b Test for non-linearity with the help of another suitable auxiliary model.

c Test for non-normality of the error terms.

Exercise 22.14 (computer)

Reconsider Exercise 22.4 and its dataset. Also, the Cochrane–Orcutt procedure is used to explain Y from X. Compare the basic assumptions of the old and the new model. Use the DW test with significance level 0.10 to test for first-order autocorrelation. Use the regression equation of the new model to estimate the old model. Compare these estimates with the regression equation of the old model.

Compare the growth model of Exercise 22.4 and the Cochrane–Orcutt model on the basis of the determination coefficients. Which model do you prefer?

CASE 22.2 PROFITS OF TOP CORPORATIONS IN THE USA (PART III)

This case is a continuation of Cases 19.2 and 21.2. The dataset <u>Case22-02.sav</u> contains not only the profits and the revenues of the Fortune 1000 corporations but also information about the gender of their CEO and about the state in which they are located. Interesting questions are: Are the profits of corporations with female CEOs different from the profits of comparable corporations with male CEOs? Are the profits of corporations in the popular states California, New York and Texas in general larger than the profits of comparable corporations in the other states?

To answer these questions, create a model that includes the first three powers of the variable 'revenues', a dummy for the variable CEO, and three dummies for the states California, New York and Texas. Formulate the basic assumption of the model and estimate it. Give a statistical analysis of the results.

Checking the model requirements and improving the model, is not easy. Below, some attempts are reported.

A serious problem is that relatively few corporations have very extreme levels of revenues. These corporations (with revenues of $40 000 million upwards) also cause a relatively large vertical variation in the scatter plot of 'profits' on 'revenues'. That is why it may be better to include LnRev = log(revenues) in the model, instead of 'revenues'. Apart from LnRev and the four gender and state dummies, the square (called LnRev2) and the third power of LnRev3 are also included. However, since a collinearity problem arises, the third power term is omitted again. But the vertical variation in the scatter plot of 'profits' on LnRev is still present. Note that replacing the dependent variable 'profits' by log(profits) is not possible since the variable 'profits' has negative observations. Of course, a dependent variable like log(profits + 20 000) could be chosen. But it turns out that this choice hardly makes things better.

It would appear that the relationship between profits and revenues changes from the revenues level 40 000 upwards. One way to incorporate that observation is to create dummies DLow and DHigh that take the value 1 for LnRev below 10.6 (which is approximately equal to log(40 000)) and LnRev above 10.6, respectively. Apart from the four gender and state dummies, the new model includes the four interaction terms of these new dummies with LnRev and LnRev2. Unfortunately, the coefficient of determination hardly increases.

CASE 22.3 A 'FINAL' MODEL FOR THE WAGE DIFFERENTIALS CASE

The case about wage differentials between men and women has been considered many times in the course of Chapters 19–22. It was used to illustrate several statistical procedures. In the present case we will try to find the 'final' model, although it will turn out that some variables have to be transformed.

See <u>Case22-03.sav</u>. It is already clear that AGE, EDL and FEM are useful predictors to explain the variation of W = 'gross hourly wage'. The first-order model with AGE, EDL and FEM can be made more flexible by including AGE2 (the square of AGE) and by replacing EDL by the four dummy variables DEDL2, DEDL3, DEDL4 and DEDL5:

DEDL2 = 1   if the person has education level 2
DEDL3 = 1   if the person has education level 3
DEDL4 = 1   if the person has education level 4
DEDL5 = 1   if the person has education level 5

The inclusion of AGE2 is proposed since it is expected that wages increase with increasing ages but that this increase of wages will slow down for larger values of age. Replacing EDL by the four dummies makes the model more flexible in the sense that it allows comparisons of mean wages for different education levels. To find out whether the relationship between WAGE and AGE is different for different levels of education, we also include the four interaction variables AGE*DEDL.

a Formulate the basic assumption of the resulting model. Estimate it and analyse it statistically.

b Is this your 'final' model for W?

c Check the model requirements of the model in part (a).

Although similar heteroskedasticity problems were ignored when applying this case in the course of Chapters 19–22, we will not ignore them now. One way to solve the problem is to use $\log(W)$ instead of W and $\log(\text{AGE})$ instead of AGE. (Although it is also possible to leave all independent variables unchanged, the advantage of using log transformation of independent variables – if possible – is that the coefficients can be interpreted as **elasticities**; see also Section 21.4.) Consider the model with dependent variable LW $= \log(W)$ and independent variables:

LAGE $= \log(\text{AGE})$; LAGE2 $=$ LAGE^2; DEDL2; DEDL3; DEDL4; DEDL5; FEM

d Formulate the basic assumption of the resulting model. Estimate it and give a statistical analysis.

e Check the homoskedasticity requirement for this model.

f Add the interaction term LAGE*FEM. Does there exist a significant interaction between LAGE and FEM?

g Add the four interaction terms LAGE*DEDL. Do they jointly improve the model of part (d)?

Because of the results so far, we adopt the model of part (d) as 'final' model.

h Use this model and your findings in parts (d)–(g) to answer the first two questions raised at the start of Case 19.1.

Time series and forecasting

In Section 2.4, **time series data** were defined as 'measurements of a single variable at successive periods or moments in time'; the sequence of those successive data was called a **time series**. One variable Y is measured over time, usually at equally distanced instants (days / months / quarters / years / ...). The data are often called $y_1, y_2, ..., y_n$ where 1, 2, ..., n denote the successive periods and n the most recent period. In the field of time series, *forecast* and *forecasting* are often used as alternative names for prediction and predicting.

In this chapter, emphasis will be on forecasting future observations of a time series. Again, linear regression plays an important role, in two ways.

The variation in Y sometimes is – at least partially – caused by developments over time and by seasonal behaviour. This is the case for the quarterly sales figures of many commodities, for instance ice cream. Under these circumstances, a linear regression model is often constructed that explains Y from a variable T representing time and some dummies representing the seasonal component. The model is estimated on the basis of historical data, and the resulting regression equation is used for forecasting – often despite the fact that the independence requirement of the error terms is violated and hence the accuracy of the forecasts is unclear.

For many time series, successive observations Y_t and Y_{t-1} are strongly correlated. If this is the case, the variation of Y_t is partially caused by the variation of Y_{t-1} and a regression model with Y_t as dependent variable and Y_{t-1} as independent variable is obvious. Such so-called **autoregressive models** will also be studied.

The dataset Case23-01.sav contains 201 weekly closing prices (Y) of Microsoft stock as listed on NASDAQ, for the period 5 January 2004 – 5 November 2007. It will be investigated whether and how these historical data can be used to forecast future prices.

For a first impression, some descriptive statistics about the data $y_1, y_2, \ldots, y_{201}$ are calculated:

$$\bar{y} = 27.1893; \qquad s_Y = 2.2384; \qquad r_1 = 0.9344$$

Here, r_1 measures the correlation coefficient between the first 200 and the last 200 observations of the dataset of size 201, so between $y_1, y_2, \ldots, y_{200}$ and $y_2, y_3, \ldots, y_{201}$. Furthermore, the scatter plot of the time series over time may be instructive. In the figure below, time is represented by the variable 'week', which refers to the number of the week to which the observation belongs (values 1 up to 201).

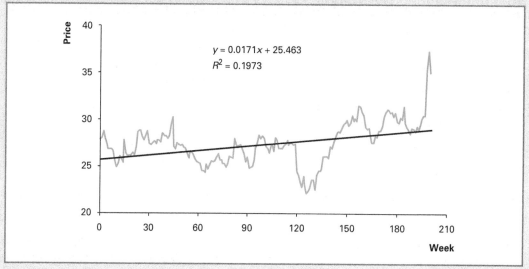

Scatter plot of the weekly price of Microsoft, 5 January 2004 – 5 November 2007

The plot also contains the least-squares line. Several questions arise:

■ Can this line be used to forecast the prices of future weeks? What about the independence requirement of the error terms?

■ The correlation 0.9344 implies that the price Y_t in week t is strongly linearly related to the price Y_{t-1} in week $t - 1$ (the week before). How can the price of the current week be used to forecast the price of the coming week?

See the end of Section 23.6 for a solution.

23.1 Introduction

The study of a time series, without involving other variables, is called **time series analysis**. Usually, the purpose is to learn about the behaviour of the variable Y over time by studying historical data and to use the knowledge gained to forecast the future behaviour of Y. This is illustrated in Example 23.1.

Example 23.1

The dataset Xmp23-01.xls contains the quarterly price indexes for sales of one-family houses in Denmark for the period 2000Q1 – 2006Q3.

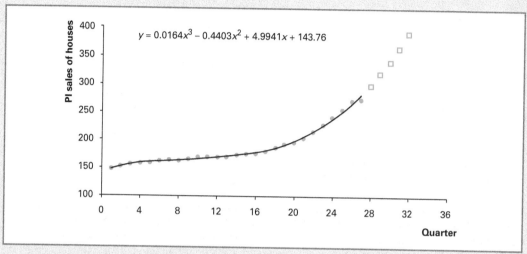

Figure 23.1 Quarterly price index (PI) for sales of one-family houses in Denmark (1995 = 100)
Source: Statistics Denmark (2007)

The variable of interest is Y = 'quarterly price index for sales of one-family houses', and y_1, y_2, ..., y_{27} are the measurements in the first quarter of 2000, the second quarter of 2000, up to the third quarter of 2006. The data $y_1, ..., y_{27}$ are used to fit a third-order equation, as shown in Figure 23.1; note that x refers to the quarter. By substituting the values 28, 29, 30, 31 and 32 into this equation, forecasts (predictions) arise for the last quarter of 2006 and all quarters of 2007; these forecasts are also included in the figure. (For instance, substitution of $x = 28$ yields the forecast 298.41.) Hence, these forecasts follow from the historical data and the choice of a model to incorporate them. Here, the model is the third-order relationship between Y and quarter.

One important remark has to be made. Although the third-order equation seems to fit very well to the historical data, the forecasts for the five coming quarters predict considerable increases in the quarterly price index that might be unrealistic. In particular, since observation y_{27} might announce a future curvature downwards, the forecasts will probably turn out to be much too optimistic.

Time series are not always as smooth as in the example. Often, a time series is wild because of unpredictable random behaviour. If that is the case, it is important to have methods to smooth out the data and to reveal the underlying pattern that can be used to make forecasts.

23.2 Components of time series

A time series can be the result of four different components: a trend component, a cyclical component, a seasonal component and a random component.

A **trend** is a relatively smooth pattern that presents the mean behaviour of Y over time; it ignores occasional and periodic influences. It often results from a natural growth process or movements of the economic climate.

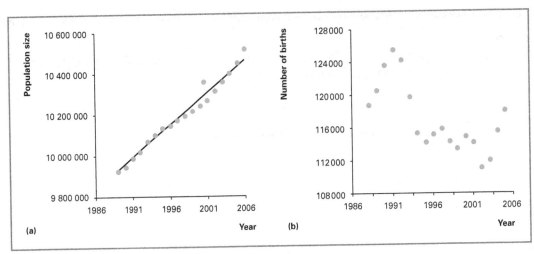

Figure 23.2 Population sizes (1989–2006) and numbers of births (1988–2005) in Belgium
Source: ECODATA (2007), Federale Overheidsdienst Economie, KMO, Middenstand & Energie

The data in Figure 23.2(a) suggest a linearly increasing trend; they tend to curl cyclically around the straight line. The graph in Figure 23.2(b) seems to represent a third-order behaviour, but for a trend a decreasing straight line is more realistic. The births data tend to spiral rather wildly.

The **cyclical component** of a time series refers to a cyclical behaviour that can sometimes be detected within long time series. The cycle time is not precisely fixed but (by definition) is at least one year. The wave-like pattern around the linear trend in Figure 23.2(a) might be caused by a cyclical component with a cycle time of about 6 or 7 years. General examples are the successive business cycles of periods of recessions and booms, product demand cycles and financial cycles. Cyclical variation is often hard to describe. In this book, variation caused by a cyclical component will be ignored, or will be taken as part of a trend.

The **seasonal component** also is a wave-like pattern, but now the cycles concern short repetitive periods (quarters, months, weeks). By definition the cycle time is not larger than one year. The demand for several products shows seasonal variation, for instance the demand for beer or ice cream.

Figure 23.3 graphs, for Sweden, the **monthly** consumer price indexes (CPI, 100 = mid-August 1980) for the sector 'clothing and footwear'.

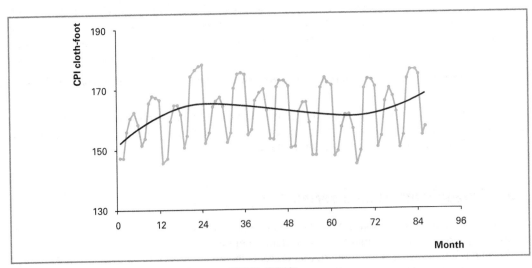

Figure 23.3 CPI (Sweden) for clothing and footwear, 2000M1–2007M2
Source: Statistics Sweden (2007)

There obviously is a seasonal component present. This component causes much variation around a curved trend: low prices in January–February are followed by increasing prices in March–May, decreasing prices in June–July, increasing prices in August–October and again decreasing prices in November–December.

The ***random component*** is 'the rest'; it causes the ***random variation***, the irregular and unpredictable changes that do not originate from any other components. Because of random variation, it is often difficult to detect an obvious trend and/or seasonal component. This is why ***smoothing techniques*** will be considered, to smooth the time series data hoping that the smoothed dataset gives – at least visually – a better view of trend and/or seasonal component.

23.3 Smoothing techniques: moving averages, exponential smoothing

To create a good model that is suitable for forecasting the future developments of a time series, it is important to know whether a trend and/or seasonal component is present. Unfortunately, the presence and form of such components is often not clear because random variation makes time plots 'turbid'. There are several ways to smooth time series data. Below, moving average smoothing and exponential smoothing are considered.

Start with a time series $y_1, y_2, \ldots, y_n$, measurements of a variable Y at the successive periods 1, 2, ..., n. The three-period ***moving average*** s_t at period t is defined as the (arithmetic) mean of the observations y_{t-1}, y_t and y_{t+1}:

$$s_t = \frac{y_{t-1} + y_t + y_{t+1}}{3}$$

Note that s_t only makes sense for the periods $t = 2, 3, \ldots, n - 1$ since at $t = 1$ there is no y_{t-1} and at $t = n$ there is no y_{t+1}. So, the three-period moving average series 'misses' the first and last observations.

Similarly, the five-period moving average at period t is the mean of the observations $y_{t-2}, y_{t-1}, y_t, y_{t+1}$ and y_{t+2}. It only makes sense at the periods $t = 3, \ldots, n - 2$, so the five-period moving average series 'misses' the first two and last two observations. Define for yourself the seven-period moving average, etc. The following example illustrates things.

Example 23.2

The file Xmp23-02.xls contains (for Norway) quarterly sales figures for electricity, for the years 1998–2000; see Table 23.1.

Year	Quarter t	Sales at quarter t	Three-quarter MA at quarter t	Five-quarter MA at quarter t
1998	1	15.1	–	–
	2	13.8	12.43	–
	3	8.4	12.03	13.10
	4	13.9	12.20	12.38
1999	5	14.3	13.23	11.52
	6	11.5	11.77	12.42
	7	9.5	11.30	12.02
	8	12.9	11.43	10.84
2000	9	11.9	11.07	10.32
	10	8.4	9.73	11.04
	11	8.9	10.13	–
	12	13.1	–	–

MA = moving average

TABLE 23.1. Quarterly electricity sales in Norway: moving average smoothing
Source: Statistics Norway (2007)

In this time series, the period is 'quarter'; the dataset contains sales data for 12 quarters. The first measurement of the three-quarter moving average sequence can be calculated for $t = 2$, the last for $t = 11$:

$$s_2 = \frac{15.1 + 13.8 + 8.4}{3} = 12.43; \quad s_3 = \frac{13.8 + 8.4 + 13.9}{3} = 12.03; \quad \ldots;$$

$$s_{11} = \frac{8.4 + 8.9 + 13.1}{3} = 10.13$$

Figure 23.4 shows that the three-quarter moving average (MA) series is much smoother than the original series of sales data.

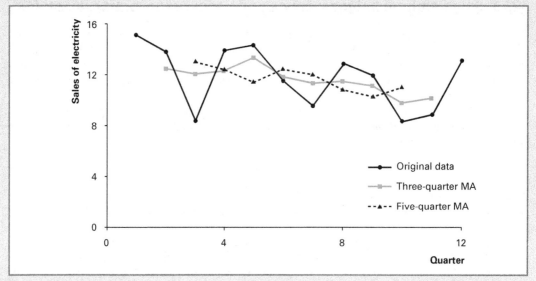

Figure 23.4 Time plots of the sales data, and the three- and five-quarter moving average series

The five-quarter moving average series (say, $\tilde{s}_t$) can only be calculated for the quarters $t = 3, 4,$ ..., 10:

$$\tilde{s}_3 = \frac{15.1 + 13.8 + 8.4 + 13.9 + 14.3}{5} = 13.10; \ldots; \tilde{s}_{10} = \frac{12.9 + 11.9 + 8.4 + 8.9 + 13.1}{5} = 11.04$$

The series $\tilde{s}_3, \ldots, \tilde{s}_{10}$ is even smoother than the three-quarter moving average series.

The three-quarter moving average series uncovers a seasonal (quarterly) variation along a decreasing trend; the cycles s_2, s_3, s_4, s_5 and s_6, s_7, s_8, s_9 look very similar. These features are less visible for the five-quarter moving average series: it is harder to detect a common behaviour between the cycles $\tilde{s}_3, \ldots, \tilde{s}_6$ and $\tilde{s}_7, \ldots, \tilde{s}_{10}$.

In general, it holds that a time series is smoothed further if the smoothing period – over which the averaging is done – is larger. If this smoothing period is too large, it may become impossible to detect seasonal variation and/or a trend component within the smoothed series.

It is also possible to take an **even** integer as smoothing period. The following three steps give the first measurement s_3 of a four-period moving average series:

1 Calculate the mean of the data $y_1, \ldots, y_4$; this mean is located at $t = 2.5$
2 Calculate the mean of the data $y_2, \ldots, y_5$; this mean is located at $t = 3.5$
3 Take the average of these two means.

Note that the resulting average is indeed located at $t = 3$. To find s_4, these steps have to be repeated for $y_2, \ldots, y_5$ and $y_3, \ldots, y_6$; etc. The last smoothed measurement is s_{n-2}.

A disadvantage of the moving average smoothing technique is that some periods, at the start and at the end, are missing moving average measurements. Furthermore, the moving average s_t depends on only a few observations of the time series, at periods around period t.

A smoothing technique exists that does not have these drawbacks. In exponential smoothing, the smoothed measurement at period t depends on **all** observations of the time series for the periods $1, \ldots, t - 1$; however, the importance of an observation decreases as the accompanying time period moves further from t. The formal definition is based on a **smoothing constant** w, with $0 < w < 1$, that is chosen in advance.

Exponentially smoothed series $s_1, \ldots, s_n$ of a time series $y_1, \ldots, y_n$

$$s_1 = y_1$$
$$s_t = wy_t + (1 - w)s_{t-1} \quad \text{for all } t = 2, 3, \ldots, n \tag{23.1}$$

It would appear that the smoothed series is initialized by setting $s_1 = y_1$. Next:

$$s_2 = wy_2 + (1 - w)s_1 = wy_2 + (1 - w)y_1$$
$$s_3 = wy_3 + (1 - w)s_2 = wy_3 + (1 - w)(wy_2 + (1 - w)y_1)$$
$$= wy_3 + w(1 - w)y_2 + (1 - w)^2 y_1$$

For general t it follows that:

$$s_t = wy_t + w(1 - w)y_{t-1} + w(1 - w)^2 y_{t-2} + \ldots + w(1 - w)^{t-2} y_2 + (1 - w)^{t-1} y_1 \tag{23.2}$$

From this last characterisation, several things can be deduced:

1 The exponentially smoothed measurement s_t at the – current – period t depends on **all** – historical – time series observations at the periods $t, t - 1, \ldots, 1$.

2 The dependence of s_t on y_{t-k} decreases as k increases, and dies out for large k.

3 A smaller smoothing constant w causes more smoothing than a larger smoothing constant.

Example 23.3

Reconsider the quarterly sales data of Example 23.2. Below, we will smooth the time series by way of exponential smoothing.

The exponentially smoothed series in column 4 of Table 23.2 makes use of $w = 0.2$. With $s_1 = y_1 = 15.1$, it follows that:

$$s_2 = 0.2 \times y_2 + 0.8 \times s_1 = 0.2 \times 13.8 + 0.8 \times 15.1 = 14.84;$$
$$s_3 = 0.2 \times y_3 + 0.8 \times s_2 = 0.2 \times 8.4 + 0.8 \times 14.84 = 13.55;$$

etc.

See Appendix A1.23 for the computation of exponentially smoothed series. The exponentially smoothed series with smoothing constant $w = 0.7$ can be created similarly; see the last column of Table 23.2. Time plots are created to compare the two smoothed series; see Figure 23.5.

Year	Quarter t	Sales at quarter t	Exponentially smoothed with $w = 0.2$	Exponentially smoothed with $w = 0.7$
1998	1	15.1	15.10	15.10
	2	13.8	14.84	14.19
	3	8.4	13.55	10.14
	4	13.9	13.62	12.77
1999	5	14.3	13.76	13.84
	6	11.5	13.31	12.20
	7	9.5	12.54	10.31
	8	12.9	12.62	12.12
2000	9	11.9	12.47	11.97
	10	8.4	11.66	9.47
	11	8.9	11.11	9.07
	12	13.1	11.51	11.89

TABLE 23.2 Quarterly electricity sales in Norway: exponential smoothing

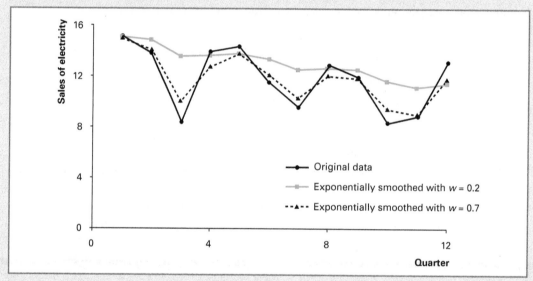

Figure 23.5 Time plots of the sales data and accompanying exponentially smoothed series

Indeed, the plot based on the smaller smoothing constant smoothes things most; the plot based on $w = 0.7$ tends to follow the plot of the original data. Note, however, that smoothing with $w = 0.2$ has smoothed the plot so much that it is hard to detect the seasonal variation.

So far, we have considered only methods that can be used to detect a trend component and/or a seasonal component. These descriptive techniques often help the researcher to find a good model that can be used to forecast future observations of the time series. In the coming sections some frequently used models will be formulated.

23.4 Exponential smoothing and forecasting

An exponentially smoothed series $s_1, \ldots, s_n$ shows the times series $y_1, \ldots, y_n$ after having removed the random variation. A simple – but also simplistic – model arises by reversing this and assuming that the time series $y_1, \ldots, y_n$ arises from the exponentially smoothed series $s_1, \ldots, s_n$ by **adding** random variation. From that point of view, the smoothed series can be used to forecast (predict) future observations of the time series.

Future observations of the time series are random variables; for the periods $n + 1, n + 2$, etc. they will be denoted by Y_{n+1}, Y_{n+2}, etc. The observation of the time period $n + k$ is denoted by Y_{n+k}, for $k = 1, 2$, etc. The forecasting is simple: since s_n is the most recent smoothed measurement, we use it as forecast for **all** future observations Y_{n+k} of the time series. We write $\hat{y}_{n+k} = s_n$ and predict that all future observations of the time series are equal to s_n.

Forecast value of Y_{n+k} with exponential smoothing

$$\hat{y}_{n+k} = s_n \text{ for all } k = 1, 2, \text{ etc.} \tag{23.3}$$

It follows immediately that this forecasting method is useful only if the time series has no trend or seasonal component.

Example 23.4

The file Xmp23-04.xls contains the euro/US dollar exchange rate for all business days between 1 April 2005 and 29 March 2007. We will use exponential smoothing with smoothing constant $w = 0.1$ to forecast the exchange rates for the (five) business days in the one-week period 30 March – 5 April 2007.

Table 23.3 contains the exchange rates for the most recent business days, in the one-week period 23–29 March 2007, jointly with the exponentially smoothed measurements. Since the smoothed value of the last business day is 1.3265, this is the forecast for all business days in the 'coming' one-week period. This is also indicated in the table.

Business day	Date	Exchange rate	Exp. smoothed with $w = 0.1$
421	23/03/07	$1.33	$1.3232
422	26/03/07	$1.33	$1.3239
423	27/03/07	$1.33	$1.3245
424	28/03/07	$1.33	$1.3250
425	29/03/07	$1.34	$1.3265
426	30/03/07	–	$1.3265
427	2/04/07	–	$1.3265
428	3/04/07	–	$1.3265
429	4/04/07	–	$1.3265
430	5/04/07	–	$1.3265

TABLE 23.3 Exchange rates, and forecasts with exponential smoothing
Source: Euronext (2007)

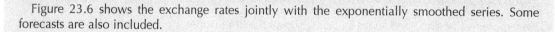

Figure 23.6 shows the exchange rates jointly with the exponentially smoothed series. Some forecasts are also included.

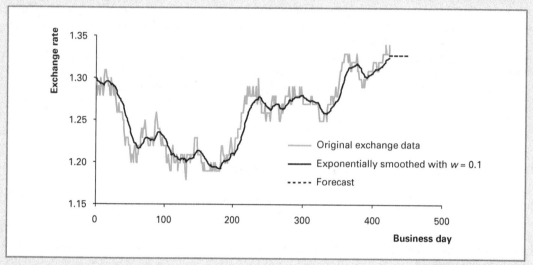

Figure 23.6 Exchange rates and exponentially smoothed series, with forecasts

These forecasts can be valuable for scenarios that deny the existence of long-term trends in exchange rate movements.

23.5 Linear regression and forecasting

The technique of linear regression can be used to obtain forecasts for future observations of a time series. The idea is to take the period (say T, representing time, with successive 'observations' 1, 2, ..., n) as predictor to represent the time evolution in the trend and to take dummy variables to represent the seasonal component. If time plots suggest a second- or even third-order trend, we might also include T^2 and even T^3. When the model is estimated, the sample regression equation can be used to forecast (predict) the values of Y for the future periods $t = n + 1, n + 2, \ldots$. But be aware of autocorrelation within the sequence of error terms!

Technically, it also is possible to create confidence intervals and conduct hypothesis tests with respect to the population regression coefficients β_i. And in practice such statistical procedures are indeed performed. Also prediction (forecast) intervals for future observations of the time series are frequently determined. However, recall that regression models for time series are often in conflict with the independence assumption for the error terms. Furthermore, when using the sample regression equation to forecast, it automatically assumes that **extrapolation** is permitted: substitution of $t = n + 1, n + 2, \ldots$ into the sample regression equation comes down to going beyond the data range (from 1 to n) of the independent variable T.

These conflicts with the model requirements are reasons why the conclusions that result from statistical procedures for a regression model of a time series should be interpreted cautiously. For instance, if the independence assumption is not satisfied, the actual confidence level of a prediction interval with a supposed confidence level of 99% might be much smaller than the 99% that is promised.

Example 23.5

The time plot of the quarterly price indexes for sales of one-family houses in Denmark (in Example 23.1) suggests a third-order relationship between Y = 'quarterly price index of one-family houses in Denmark' and T = 'time in quarters'; the 'observations' of T are t = 1, 2, ..., n. That is, the linear regression model with the following basic assumption is suggested:

$$E(Y_t) = \beta_0 + \beta_1 t + \beta_2 t^2 + \beta_3 t^3 \quad \text{for all } t = 1, 2, ..., n$$

Recall that this basic assumption can also be written as:

$$Y_t = \beta_0 + \beta_1 t + \beta_2 t^2 + \beta_3 t^3 + \varepsilon_t \text{ with } E(\varepsilon_t) = 0, \quad \text{for all } t = 1, 2, ..., n$$

In this model, the trend is represented by $\beta_0 + \beta_1 t + \beta_2 t^2 + \beta_3 t^3$. We will use the data to estimate it.

The n = 27 quarterly data points are in the file Xmp23-05.sav; they belong to the quarters 2000Q1 up to and including 2006Q3. To estimate the model, the columns for T, T^2 and T^3 have to be created by the researcher him/herself. Notice that this model has k = 3 independent variables; hence, $n - 4$ = 23 degrees of freedom have to be used in statistical procedures.

Model Summary(b)

Model	R	R Square	Adjusted R Square	Std. Error of the Estimate	Durbin-Watson
1	.997(a)	.994	.994	2.85041	1.265

a Predictors: (Constant), TTT, T, TT
b Dependent Variable: PI_houses

ANOVA(b)

Model		Sum of Squares	df	Mean Square	F	Sig.
1	Regression	33415.076	3	11138.359	1370.906	.000(a)
	Residual	186.871	23	8.125		
	Total	33601.947	26			

a Predictors: (Constant), TTT, T, TT
b Dependent Variable: PI_houses

Coefficients(a)

Model		Unstandardized Coefficients		Standardized Coefficients	t	Sig.
		B	Std. Error	Beta		
1	(Constant)	143.758	2.538		56.640	.000
	T	4.994	.771	1.103	6.479	.000
	TT	−.440	.063	−2.805	−6.950	.000
	TTT	.016	.001	2.743	11.018	.000

a Dependent Variable: PI_houses

According to the printout, the sample regression equation is

$$\hat{y} = 143.758 + 4.994t - 0.440t^2 + 0.016t^3$$

which is the same equation as in Example 23.1 (see Figure 23.1). It is tempting to use the printout for statistical inference: the model looks very useful, with coefficient of determination 0.994; the trend component seems to be of third order, since the p-value for the test of H_0: $\beta_3 = 0$ against H_1: $\beta_3 \neq 0$ is close to 0. Furthermore, substitution of t = 28, 29 and 30 into the sample regression equation yields 289.86, 308.77 and 329.58, respectively. It is tempting to use them as forecasts for the sales of one-family houses in the last quarter of 2006 and the first two quarters of 2007.

But ... unfortunately, the model is invalid! The error terms are not independent, not even uncorrelated. To see that, we conduct the Durbin–Watson test on the basis of the residuals. In the five step procedure below, we used $n = 27$, $k = 3$ and $\alpha/2 = 0.05$. The Durbin–Watson bounds from the table are $d_{0.05;L} = 1.16$ and $d_{0.05;U} = 1.65$.

i H_0: there is no first-order autocorrelation, against

  H_1: there **is** first-order autocorrelation; take $\alpha = 0.10$

ii Test statistic: D

iii $d \leq 1.16 \Rightarrow$ conclude H_1

  $1.16 < d < 1.65 \Rightarrow$ conclude 'inconclusive'

  $1.65 \leq d \leq 2.35 \Rightarrow$ conclude H_0

  $2.35 < d < 2.84 \Rightarrow$ conclude 'inconclusive'

  $d \geq 2.84 \Rightarrow$ conclude H_1

iv **val** = 1.265 (see the first part of the printout)

v Since 1.265 is between 1.16 and 1.65, the test gives the conclusion 'inconclusive'

In any case, the data do not support the conclusion that H_0 is valid. Hence, statistical conclusions that are drawn on the basis of the above printout cannot be trusted. Also, the precision of the above forecasts is unclear.

In Section 22.4 some suggestions were mentioned to repair things. Unfortunately, none of them works in the present example.

It is also possible to represent a seasonal component in a linear regression model. This can be done by introducing dummy variables into the model. If the period is a quarter, then three dummies are needed to represent the four quarters in a year; if the period is a month, then eleven dummies are needed. Things are illustrated in two examples.

Example 23.6

Reconsider the quarterly sales data for electricity in Norway. Figure 23.4 in Example 23.2 seems to suggest that a slowly decreasing linear trend and a seasonal component are present. The following model is proposed:

$$E(Y_t) = \beta_0 + \beta_1 t + \beta_2 DQ_{1t} + \beta_3 DQ_{2t} + \beta_4 DQ_{3t}$$

Here, the variable T = 'time' gives the number of the quarter that is considered, so $t = 1, ..., 12$. The variables DQ_1, DQ_2 and DQ_3 are dummy variables and, for instance, DQ_{2t} has to be read as the observation of DQ_2 in quarter t. DQ_1 takes the value 1 if the observation is a quarter 1 observation and the value 0 otherwise. Similarly, DQ_2 and DQ_3 take the value 1 if the observation belongs to quarter 2 or quarter 3, respectively. Note that the choice has been made to take quarter 4 as base level.

All observations of the variables T, DQ_1, DQ_2 and DQ_3 have to be keyed into the dataset by the researcher. This is already done in the file Xmp23-06.sav. After running the regression, the printout below is obtained.

Model Summary(b)

Model	R	R Square	Adjusted R Square	Std. Error of the Estimate	Durbin-Watson
1	.908(a)	.824	.724	1.28167	2.047

a Predictors: (Constant), DQ3, Time, DQ2, DQ1
b Dependent Variable: SalesElec

ANOVA(b)

Model		Sum of Squares	df	Mean Square	F	Sig.
1	Regression	53.870	4	13.468	8.199	.009(a)
	Residual	11.499	7	1.643		
	Total	65.369	11			

a Predictors: (Constant), DQ3, Time, DQ2, DQ1
b Dependent Variable: SalesElec

Coefficients(a)

Model		Unstandardized Coefficients		Standardized Coefficients	t	Sig.
		B	Std. Error	Beta		
1	(Constant)	15.525	1.170		13.269	.000
	Time	−.278	.113	−.411	−2.455	.044
	DQ1	−.368	1.100	−.068	−.334	.748
	DQ2	−2.623	1.071	−.487	−2.450	.044
	DQ3	−4.645	1.053	−.862	−4.413	.003

a Dependent Variable: SalesElec

We first study the independence of the error terms. From the printout it follows that the *val* of the Durbin–Watson test is 2.047, which is close to the ideal value 2. Notice that $n = 12$ and $k = 4$, while the DW table gives only DW bounds from $n = 15$ onwards. When using the 0.01 table for $n = 15$, we obtain $d_{0.01;L} = 0.49$ and $d_{0.01;U} = 1.70$. Hence, at significance level 0.02 it is concluded that there is no first-order autocorrelation since *val* lies between 1.70 and $4 - 1.70 = 2.30$. The model seems to be valid as far as the independence requirement for the error terms is concerned. We will use the model for statistical inference.

The ANOVA part yields that the model is useful; the coefficient of determination is 0.824. At significance level 0.05, the variables $T = $ 'Time', DQ_2 and DQ_3 are all individually significant (as follows immediately from the p-values of the two-sided t-tests). The sample regression equation is:

$$\hat{y}_t = 15.525 - 0.278t - 0.368DQ_{1t} - 2.623DQ_{2t} - 4.645DQ_{3t}$$

The interpretation of the coefficient −0.278 of the variable 'time' is that, apart from seasonal influences, sales of electricity decrease each quarter by 0.278 on average.

Does the model really have a seasonal component? That is, is at least one of β_2, β_3, β_4 unequal to 0? This can be tested with a partial F-test. To do so, we also need the ANOVA part of the printout for the (reduced) model that arises by omitting the three seasonal dummies; see below.

ANOVA(b)

Model		Sum of Squares	df	Mean Square	F	Sig.
2	Regression	11.841	1	11.841	2.212	.168(a)
	Residual	53.528	10	5.353		
	Total	65.369	11			

a Predictors: (Constant), Time
b Dependent Variable: SalesElec

Here are the five steps of the test:

i Test $H_0: \beta_2 = \beta_3 = \beta_4 = 0$ against H_1: at least one of $\beta_2, \beta_3, \beta_4$ is $\neq 0$; $\alpha = 0.05$

ii Test statistic: $F = \dfrac{(SSE_r - SSE_c)/3}{SSE_c/(n-5)}$

iii Reject $H_0 \Leftrightarrow f \geq F_{0.05;3;7} = 4.3468$ $\hspace{2cm}$ (∗)

iv $\mathbf{val} = \dfrac{(53.528 - 11.499)/3}{11.499/7} = 8.5284$

v H_0 is rejected

Indeed, there is a significant seasonal component. The coefficient -4.645 of DQ_3 in the sample regression equation means that (leaving aside the trend component), due to seasonal variation the mean sales of electricity is estimated to be 4.645 less in the third quarter than in the fourth (base) quarter.

Recall that the observations in the dataset come from the 12 quarters in the period 1998–2000. With the help of a computer it is possible to create forecasts and 95% prediction intervals for Norwegian electricity sales in each of the quarters of 2001. Table 23.4 contains the results. It also contains the electricity sales as they actually occurred in these quarters.

Quarter	Forecast	95% prediction interval	Actual observation
2001Q1	11.54	(7.44, 15.65)	24.3
2001Q2	9.01	(4.90, 13.11)	17.0
2001Q3	6.71	(2.60, 10.81)	15.8
2001Q4	11.08	(6.97, 15.18)	16.1

TABLE 23.4 Forecasts and actual observations of electricity sales in Norway
Source: Statistics Norway (2007)

Notice that none of the four actual observations falls in the respective 95% prediction interval. It would appear that the extrapolation step necessary to create these prediction intervals conflicts with reality. (This illustrates the danger of extrapolation.) Indeed, the time plot in Figure 23.7 shows that the time series of electricity sales in Norway (for some reason) started to behave differently from the first quarter in 2001 onwards. See also the dataset Xmp23-06b.sav.

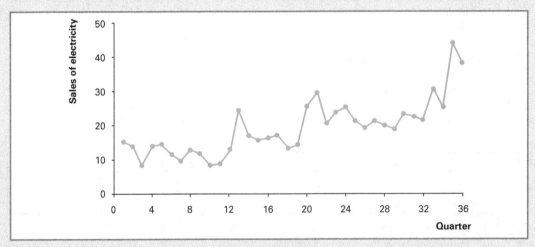

Figure 23.7 Quarterly electricity sales in Norway, 1998Q1–2006Q4
Source: Statistics Norway (2007)

Example 23.7

Reconsider the monthly data for the period 2000M1–2007M2 of the Swedish CPI for clothing and footwear; see Figure 23.3. Apart from a trend, the time plot seems to suggest the presence of seasonal variation due to month effects. The following model is chosen:

$$E(Y_t) = \beta_0 + \beta_1 t + \beta_2 t^2 + \beta_3 t^3 + \beta_4 DM_{2t} + \beta_5 DM_{3t} + \dots + \beta_{14} DM_{12t}$$

Here, Y = 'monthly CPI of clothing and footwear'; t = month 1, 2, ..., 86; DM_2 is the dummy that takes the value 1 if the observation is made in the second month (February) of any year and takes the value 0 otherwise; DM_3 is the dummy that takes the value 1 if the observation is made in March and 0 if not; etc. Notice that January is taken as base level.

Again, the columns with the observations of T = 'Number of the month' and DM_i for i = 2, 3, ..., 12 have to be created by the researcher. In the dataset Xmp23-07.sav, this is already done; see Appendix A1.23 for the creation of such columns. Running the regression yields the partial printout shown below.

Model Summary(b)

Model	R	R Square	Adjusted R Square	Std. Error of the Estimate	Durbin-Watson
1	.973(a)	.946	.935	2.37835	.575

a Predictors: (Constant), DM12, TT, DM8, DM7, DM9, DM6, DM10, DM5, DM11, DM4, DM3, DM2, T, TTT
b Dependent Variable: CPI_Cloth_Foot

According to the high level of the coefficient of determination, the model seems to be very useful. But there is an autocorrelation problem! The value 0.575 of the DW statistic is far from the ideal value 2. In fact, this low value 0.575 indicates the presence of **positive** first-order autocorrelation within the sequence of residuals.

To repair things, we try 'growth (%) of CPI' as dependent variable, instead of CPI itself. That is, we take

$$G_t = \frac{100(Y_t - Y_{t-1})}{Y_{t-1}} \quad \text{for 2, ..., 86}$$

instead of Y_t. Hence, the basic assumption turns into:

$$E(G_t) = \beta_0 + \beta_1 t + \beta_2 t^2 + \beta_3 t^3 + \beta_4 DM_{2t} + \beta_5 DM_{3t} + \dots + \beta_{14} DM_{12t}$$

After having created the column with the growth observations and having run the regression, the printout below is obtained (GGt refers to G_t).

Model Summary(b)

Model	R	R Square	Adjusted R Square	Std. Error of the Estimate	Durbin-Watson
2	.984(a)	.968	.961	1.15130	2.145

a Predictors: (Constant), DM12, TT, DM8, DM7, DM9, DM6, DM10, DM5, DM11, DM4, DM3, DM2, T, TTT
b Dependent Variable: GGt

ANOVA(b)

Model		Sum of Squares	df	Mean Square	F	Sig.
2	Regression	2789.306	14	199.236	150.311	.000(a)
	Residual	92.785	70	1.325		
	Total	2882.091	84			

a Predictors: (Constant), DM12, TT, DM8, DM7, DM9, DM6, DM10, DM5, DM11, DM4, DM3, DM2, T, TTT
b Dependent Variable: GGt

Coefficients(a)

Model		Unstandardized Coefficients		Standardized Coefficients	t	Sig.
		B	Std. Error	Beta		
2	(Constant)	−12.593	.744		−16.923	.000
	T	.026	.056	.109	.461	.646
	TT	−.001	.001	−.350	−.626	.534
	TTT	8.19E−006	.000	.259	.749	.456
	DM2	13.599	.599	.682	22.716	.000
	DM3	19.102	.622	.902	30.711	.000
	DM4	14.706	.621	.694	23.692	.000
	DM5	12.723	.620	.601	20.535	.000
	DM6	9.505	.619	.449	15.365	.000
	DM7	5.933	.618	.280	9.603	.000
	DM8	14.044	.617	.663	22.759	.000
	DM9	24.164	.616	1.141	39.196	.000
	DM10	14.115	.616	.666	22.913	.000
	DM11	12.314	.616	.581	20.000	.000
	DM12	11.805	.615	.557	19.181	.000

a Dependent Variable: GGt

The value of the DW statistics is not far from the ideal value 2, so it is believed that the model is valid as far as independence of the error terms is concerned.

The usual F-test yields that the model is useful, in fact very useful, as follows from the coefficient of determination 0.968. All month dummies are individually significant within the model. Note that they are all positively significant, which indicates that the mean growth in month 1 is significantly smaller than the mean growth in each of the other months. For instance, the estimated regression coefficient 13.599 of DM_2 informs us that, when leaving the trend out of consideration, the growth of the CPI for clothing and footwear in the second month is on average 13.6 percentage points more than in the first month (base level).

However, the trend variables T, T^2 and T^3 are **not** individually significant. This raises the question whether the data do give evidence of a trend component. To put it another way, the question is whether at least one of the parameters β_1, β_2, β_3 is unequal to 0. This yields the following testing problem:

$$H_0: \beta_1 = \beta_2 = \beta_3 = 0 \text{ against } H_1: \text{ at least one of } \beta_1, \beta_2, \beta_3 \text{ is} \neq 0$$

We use a partial F-test with significance level 0.01. This test is based on the distribution $F_{3,70}$; check yourself (∗) that the rejection region is [4.0744, ∞). For **val**, we also need the ANOVA part of the model **without** the three trend variables:

ANOVA(b)

Model		Sum of Squares	df	Mean Square	F	Sig.
3	Regression	2787.748	11	253.432	196.098	.000(a)
	Residual	94.343	73	1.292		
	Total	2882.091	84			

a Predictors: (Constant), DM12, DM11, DM10, DM9, DM8, DM7, DM6, DM5, DM4, DM3, DM2
b Dependent Variable: GGt

Straightforward calculation shows that **val** = 0.3918; there is no evidence of a trend component.

Although this last conclusion might lead us to decide to omit the trend variables from the model, we will not do that, and will use the model with the trend component to forecast the growth of the clothing and footwear CPI for March–December 2007. Table 23.5 contains the results.

Month	Forecast	Lower and upper bounds of 95% prediction intervals	
2007M3	7.22113	4.52935	9.91291
2007M4	2.87899	0.15088	5.60710
2007M5	0.95288	−1.81660	3.72236
2007M6	−2.20685	−5.02315	0.60945
2007M7	−5.71736	−8.58629	−2.84843
2007M8	2.45822	−0.46951	5.38595
2007M9	12.64481	9.65179	15.63783
2007M10	2.66546	−0.39965	5.73057
2007M11	0.93652	−2.20774	4.08078
2007M12	0.50345	−2.72725	3.73415

TABLE 23.5 Forecasts and 95% prediction intervals for the growth of clothing and footwear CPI

Note that the growth is significantly positive in March, April and September; it is significantly negative in July.

Forecasting by exponential smoothing can be effective if trend and seasonal components are not (or hardly) present. Forecasting with linear regression can be used if a significant trend and/or significant seasonal component is present. However, conflicts with the independence requirement of the error terms may sabotage the statistical conclusions. In the following section a model is considered that is not afraid of strong correlation within a time series but instead makes use of it.

23.6 Autoregressive model and forecasting

The model in the present section is especially useful if there is no obvious trend or seasonal component, but successive observations of the time series are expected to be strongly (positively or negatively) correlated.

The (first order) autoregressive model AR(1)

The basic assumption of this model is:

$$Y_t = \beta_0 + \beta_1 Y_{t-1} + \varepsilon_t \quad \text{with } E(\varepsilon_t) = 0, \quad \text{for all } t = 2, 3, \ldots, n \tag{23.4}$$

One of the model requirements is that the error terms $\varepsilon_1, \ldots, \varepsilon_n$ are independent and identically $N(0, \sigma_\varepsilon^2)$ distributed.

The model states that, apart from an error term ε_t, the observation at time t follows linearly from the most recent observation (so, at time $t-1$). It is the correlation between Y_t and Y_{t-1} that is used here as a justification for the model. The model is called AR(1) to emphasize the fact that it describes a **lag** 1 (delay 1) situation. The following example illustrates its use to forecast future observations of the time series.

Remark. In the case of a model with a lagged variable, do **not** use the DW test to check the independence requirements of the error terms.

Example 23.8

The file Xmp23-08a.sav contains the monthly inflation rates (%, compared with one year before) of the Netherlands for the period 2000M1–2007M3. The AR(1) model will be used to forecast the monthly inflation rates for the last nine months of 2007.

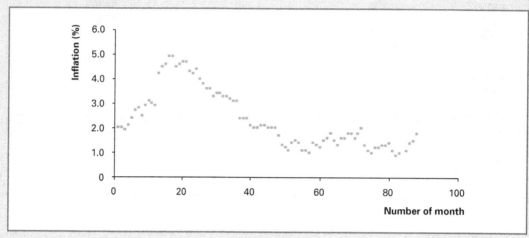

Figure 23.8 Inflation rates in the Netherlands; months 2000M1–2007M3
Source: Statistics Netherlands (2007)

The time plot in Figure 23.8 does not show an obvious trend or seasonal variation. Of course, variation is present, but there is no obvious reason to assign it to a long-term trend. Notice that the dataset contains 87 observations. To run the autoregression, the **last** 86 observations have to be regressed on the **first** 86. The dataset Xmp23-08b.sav is created from the former dataset; the first column contains the observations of the dependent variable and the second column contains the (lag 1) observations of the independent variable. Running the regression gives the printout shown below.

Model Summary(b)

Model	R	R Square	Adjusted R Square	Std. Error of the Estimate
1	.975(a)	.950	.949	.2639

a Predictors: (Constant), Y_lag1
b Dependent Variable: Y

ANOVA(b)

Model		Sum of Squares	df	Mean Square	F	Sig.
1	Regression	110.823	1	110.823	1591.605	.000(a)
	Residual	5.849	84	.070		
	Total	116.672	85			

a Predictors: (Constant), Y_lag1
b Dependent Variable: Y

Coefficients(a)

Model		Unstandardized Coefficients		Standardized Coefficients	t	Sig.
		B	Std. Error	Beta		
1	(Constant)	.05574	.064		.867	.388
	Y_lag1	.97538	.024	.975	39.895	.000

a Dependent Variable: Y

The model is very useful; $r^2 = 0.950$.

From the dataset it follows that the last observation (March 2007) is 1.8. By substituting this value into the sample regression equation

$$\hat{y}_t = 0.05574 + 0.97538y_{t-1}$$

the forecast 1.8114 for April 2007 follows. This last number can in turn be substituted to obtain a forecast 1.8226 for the inflation rate of May 2007, etc. Table 23.6 summarizes the results:

Month	4	5	6	7	8	9	10	11	12
Forecast inflation	1.81	1.82	1.83	1.84	1.85	1.86	1.87	1.88	1.89

TABLE 23.6 Forecast inflation rates for April–December 2007

Again, extrapolation is used. In fact, it is assumed that the model (and its requirements) is valid not only for $t = 2, \ldots, 87$ but also for $t = 88, 89, \ldots, 96$.

23.6.1 Using an auxiliary AR(1) model to draw conclusions about first-order autocorrelation

In Section 22.4 the Durbin–Watson test was presented to test for the presence of first-order autocorrelation within the series of residuals in a linear regression model and hence for the dependence requirement of the error terms. It is also possible to use an auxiliary AR(1) model for the error terms in order to draw conclusions about the presence of first-order autocorrelation.

The model requirements for $\varepsilon_1, \ldots, \varepsilon_n$ in a linear regression model imply that these error terms constitute an autoregressive system with $\rho = 0$; see Section 22.4. But then it seems a good idea to check the independence assumption of the linear regression model by testing whether the parameter ρ in the equations $\varepsilon_t = \rho\varepsilon_{t-1} + v_t$ (for $t = 2, \ldots, n$) is 0 or not. Since ε is unobservable, we will use the residuals $e_1, \ldots, e_n$ of the original regression to regress e_t on e_{t-1} in an auxiliary regression **without** constant.

Assume that the error terms $\varepsilon_1, \ldots, \varepsilon_n$ of the original basic assumption $Y = \beta_0 + \beta_1 X + \varepsilon$ constitute an autoregressive system. So:

$$\varepsilon_t = \rho\varepsilon_{t-1} + v_t \quad \text{for } t = 2, \ldots, n$$

But then ρ is just the correlation coefficient of ε_t and ε_{t-1}, and – possibly apart from the independence assumptions – all other linear regression model requirements about $\varepsilon_1, \ldots, \varepsilon_n$ are satisfied. The following testing problem is of interest:

 i Test $H_0: \rho = 0$ against $H_1: \rho \neq 0$

This can be tested on the basis of the residuals of the original model, with linear regression without constant. (Note that the slope is now called ρ.) When running the regression, it has to be indicated that no constant β_0 is involved.

In a sense, this testing procedure is preferable to the DW procedure: the AR(1) procedure does not make use of 'inconclusive regions'. Moreover, this AR(1) model also allows tests for positive (negative) first-order autocorrelation: just test whether $\rho > 0$ (or $\rho < 0$).

Example 23.9

In Example 23.7 the Durbin–Watson test was used to come to the conclusion that the series of the residuals of the linear regression model explaining the Swedish CPI for clothing/footwear from a third-order trend and a monthly seasonal component, shows first-order autocorrelation. In the present example it will be investigated whether the auxiliary AR(1) model for the series $\varepsilon_1, \ldots, \varepsilon_n$ leads to the same conclusion.

At first, the series of the 86 residuals have to be determined by running the regression of Example 23.7 again. Next, the **last** 85 residuals are selected and used as observations of the dependent variable in an auxiliary linear regression **without** constant; the **first** 85 residuals are used as observations of the independent variable. Notice that this simple linear regression has $n = 85$ and $k = 1$. Since only the slope has to be estimated, only 1 degree of freedom is lost and $n - 1 = 84$ degrees of freedom are used in the statistical procedures. The relevant printout of that regression is shown below.

Coefficients(a,b)

Model		Unstandardized Coefficients		Standardized Coefficients	t	Sig.
		B	Std. Error	Beta		
1	Res_lag1	.689	.075	.706	9.131	.000

a Dependent Variable: Res
b Linear Regression through the Origin

As before, let ρ be the correlation coefficient of ε_t and ε_{t-1}. Interest is in the following five-step procedure:

i Test $H_0: \rho = 0$ against $H_1: \rho \neq 0$; $\alpha = 0.01$

ii Test statistic: $T = \dfrac{B_1}{S_{B_1}}$

iii Reject $H_0 \Leftrightarrow t \leq -t_{0.005;84} = -2.6356$ or $t \geq 2.6356$ (*)

iv **val** = 9.131

v H_0 is rejected

Indeed, there is first-order autocorrelation within the series of the residuals. Since the p-value of this test is very small, the conclusion of the test is convincing.

We also can test whether $\rho > 0$, so whether there is **positive** first-order autocorrelation within the series of residuals. In that case, **val** is unchanged but the (only) critical value of the test is $t_{0.01;84} = 2.3716$ (*). The conclusion is that there is indeed positive first-order autocorrelation.

CASE 23.1 FORECASTING THE PRICE OF MICROSOFT STOCK–SOLUTION

The scatter plot of the historical prices shows that the price of Microsoft stock tends to increase over time and that finding the trend line might be helpful. From these observations, the linear regression model with basic assumption

$$E(Y_t) = \beta_0 + \beta_1 t \quad \text{for } t = 1, 2, \ldots, 201$$

arises more or less naturally. Running the regression yields:

$$\hat{y}_t = 25.463 + 0.017t \quad \text{and} \quad r^2 = 0.197$$

Blindly conducting the standard statistical procedures yields that the model is useful and that, on average, the price of the stock increases weekly by 0.017. Technically, we can use the regression line to produce forecasts for coming weeks; just substitute $t = 202, 203$, etc. into the regression equation. **However**, the scatter plot of the residuals against time shows that there is positive first-order autocorrelation within the sequence of residuals and hence that the independence requirement (about the error terms) of the model does not hold. This establishment is confirmed by the low value 0.160 of the Durbin–Watson statistic. The conclusion has to be that the model is inadequate and that its statistical conclusions cannot be trusted.

To find another model, notice that the correlation coefficient 0.9344 of $y_2, y_3, \ldots, y_{201}$ and $y_1, y_2, \ldots, y_{200}$ estimates the strength of the linear relationship between Y_t and Y_{t-1}. It would appear that the linear relationship between the price of Microsoft stock in a certain week and its price in the previous week, is very strong. This prompts the AR(1) model:

$$E(Y_t) = \beta_0 + \beta_1 y_{t-1} \quad \text{for } t = 2, 3, \ldots, 201$$

(The choice is made to use the same notations for the population regression coefficients as in the former model, but of course the coefficients are different.) This model can be estimated by regressing $y_2, y_3, \ldots, y_{201}$ on $y_1, y_2, \ldots, y_{200}$. Here are some results:

$$\hat{y}_t = 1.061 + 0.962 y_{t-1}; \quad r^2 = 0.873$$

(Regression of the accompanying error terms ε_t on ε_{t-1} (without constant) shows that the above AR(1) model does not yield problems with respect to independence of the error terms.) The model is (very) useful. A histogram of the standardized residuals shows that a normal distribution fits rather well, apart from the outliers on the dates 15 November 2004, 24 April 2006 and 22 October 2007. It follows that this AR(1) model is adequate unless the stock is confronted with drastic shocks. Unfortunately, the AR(1) model has no mechanism to predict the occurrence of such shocks.

The model can be used to find forecasts for future weeks. For 12 November 2007 it is forecast that the price of Microsoft stock will be:

$$\hat{y}_{202} = 1.061 + 0.962 \times 34.74 = 34.49$$

Furthermore, a 95% prediction interval can be computed; we obtain (32.86, 36.12). By consulting http://finance.yahoo.com we can find out that the actual value was 34.09. The model seems to be quite satisfactory in the short run. Application of the model for weeks that are further in the future will, of course, increase inaccuracy since predicted values have to be substituted into the regression equation.

By conducting a two-sided t-test with hinge 1, it follows that the slope β_1 of the AR(1) model is not significantly different from 1 (at level 0.05). Since (again at level 0.05) the intercept β_0 is not significantly different from 0, it essentially follows that the movement of the stock price can also be described satisfactorily by using the value of the previous week (shocks excluded). That is, our findings support a theory in finance that states that past values are not very useful in predicting future values; only the most recent value is relevant. See also Exercise 23.15.

Summary

In time series analysis, the developments (in the past) of a time series are studied, often with the objective of using the knowledge obtained for forecasting **future** observations. To be able to detect the underlying pattern in a sequence of historical observations, and hence to formulate a suitable model, the data sometimes have to be smoothed. We discussed two smoothing procedures: moving averages and the exponential smoothing. If there is no obvious trend or seasonal component, then the latter method can also be used to forecast. If the historical data do contain a trend and/or seasonal component, then, for forecasting purposes, a linear regression model can be used that is based on T = 'time' to describe the trend and/or special dummies to represent the seasonal component. If successive observations of the time series are strongly correlated, then the AR(1) model may be useful for forecasting future observations.

An AR(1) model without constant can be used to find out whether the residuals of a linear regression are first-order autocorrelated. The t-test for the significance of that AR(1) model offers an alternative to the DW test.

One important remark has to be made with respect to the use of regression models for time series analysis. Although it is technically possible to conduct the standard statistical procedures (of Chapters 19–22) about interval estimation, hypothesis testing and interval prediction, the **actual** confidence level of such procedures is not always clear. This is because the requirements about independent (or at least uncorrelated) error terms and about independent (or at least uncorrelated) ε and explanatory variable(s) X, are often violated. Furthermore, for forecasting purposes, extrapolation is necessary and it is not always clear whether this is allowed.

🔑 Key terms

autoregressive model **749**
cyclical component **736**
exponentially
  smoothed **739**
first-order autoregressive
  model **749**
forecast **733**

forecasting **733**
mean absolute deviation
  (MAD) **765**
mean reversion **766**
moving average **737**
persistence **766**
random component **737**

random variation **737**
seasonal component **736**
smoothing constant **739**
smoothing
  techniques **737**
time series analysis **734**
trend **735**

❓ Exercises

Exercise 23.1

The file Xrc23-01.xls contains the daily exchange rate of the Japanese yen against the euro for 2 January – 4 April 2007. Figure 1 below presents the time plot.

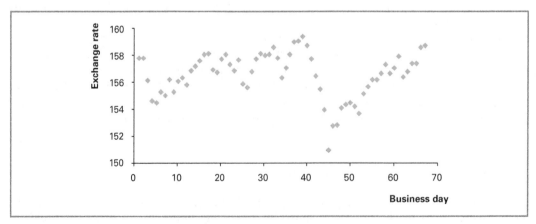

Figure 1 Daily exchange rates of the Japanese yen, 2 January–4 April 2007
Sources: European Central Bank (2007) and Netherlands Central Bank (2007)

To remove the random variation from this time series, we will create the 3-, 4- and 5-day moving average series, and the exponentially smoothed series with smoothing constants $w = 0.3$ and 0.8.

a The table below contains the exchange rates for the first seven business days of the sample period. Fill it in with the three moving average series and the two exponentially smoothed series.

		Moving average			Exponentially smoothed	
Business day	Exchange rate	3-day	5-day	4-day	$w = 0.3$	$w = 0.8$
1	157.76					
2	157.76					
3	156.11					
4	154.55					
5	154.42					
6	155.26					
7	154.99					
8	–	–	–	–	–	–

b Figure 2 presents the time plots of the time series itself, and of the 3- and 5-day moving average (MA) series. Give your comments on the degree of smoothing.

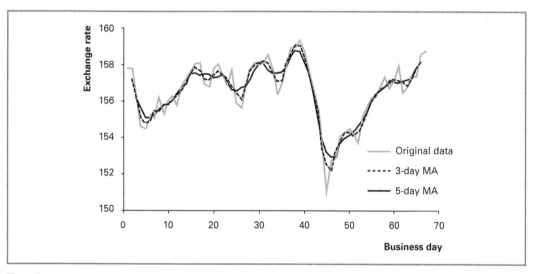

Figure 2

c Figure 3 contains the time plots of the original time series, and the exponentially smoothed series with smoothing constants 0.3 and 0.8. Give your comments on the degree of smoothing.

d Do the smoothed time plots of parts (b) and (c) suggest a trend component?

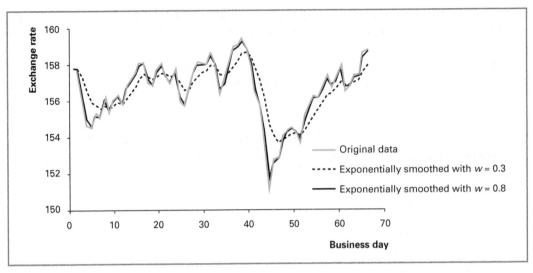

Figure 3

Exercise 23.2

The file Xrc23-02.xls contains the daily exchange rate of the Japanese yen against the euro for 4
January 1999 – 4 April 2007. The exponentially smoothed series with smoothing constant $w = 0.8$
is also included. The original and the smoothed observations of the five most recent business days
are in the table below.

Date	Business day	Exchange rate	Smoothed observation
29/3/2007	2112	156.75	156.73
30/3/2007	2113	157.32	157.20
2/4/2007	2114	157.35	157.32
3/4/2007	2115	158.53	158.29
4/4/2007	2116	158.70	158.62

a Give forecasts for the following five business days by using the exponentially smoothed
series.

This forecasting can also be done with a first-order autoregressive model; see also Xrc23-02.sav.

b Formulate the basic assumption of the model. Explain how the data in the file can be used to
estimate the model.

The resulting printout is shown below.

Model Summary(b)

Model	R	R Square	Adjusted R Square	Std. Error of the Estimate	Durbin-Watson
2	.999(a)	.997	.997	.84068	1.980

a Predictors: (Constant), Y_lag1
b Dependent Variable: Y

ANOVA(b)

Model		Sum of Squares	df	Mean Square	F	Sig.
2	Regression	541488.501	1	541488.501	766174.299	.000(a)
	Residual	1493.348	2113	.707		
	Total	542981.849	2114			

a Predictors: (Constant), Y_lag1
b Dependent Variable: Y

Coefficients(a)

Model		Unstandardized Coefficients		Standardized Coefficients		
		B	Std. Error	Beta	t	Sig.
2	(Constant)	0.06503	.144		.450	.653
	Y_lag1	0.99958	.001	.999	875.314	.000

a Dependent Variable: Y

   c Give forecasts for the following five days by using the autoregressive model.

   d Compare the forecasts of parts (a) and (c). Which forecasting model do you prefer? Why?

Exercise 23.3

It is well known that the number of job vacancies in a country shows some seasonal variation. To learn more about this, we investigate the data in the file Xrc23-03.sav which contains the monthly numbers of job vacancies in Germany during the period 1990M9–2007M3 (so, $n = 199$). To get an idea of the situation, the time plot is presented below.

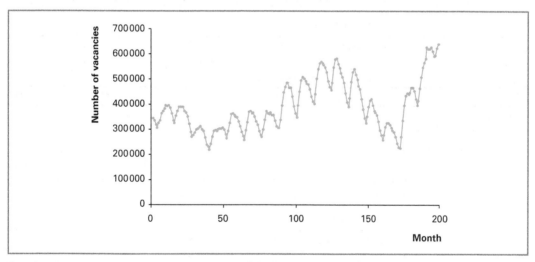

Time plot of the monthly job vacancies in Germany, 1990M9–2007M3
Source: Deutsche Bundesbank (2007)

This plot obviously shows seasonal variation on a complicated trend component.

   To forecast the numbers of job vacancies in coming months, a model is wanted that in any case does not give first-order autocorrelation problems with respect to the residuals. In this exercise, some models will be studied.

   a **Model 1.** Formulate the basic assumption of the model that explains Y_t = 'number of job vacancies in month t' from a linear trend and 11 monthly dummies; take December as base level. Carefully define your variables. Use the DW test with $\alpha = 0.02$ to draw a conclusion about the presence of first-order autocorrelation within the sequence of residuals; see the printout below. (Since the DW table does not give enough information, use $n = 100$ and $k = 5$ in that table.)

Model Summary(b)

Model	R	R Square	Adjusted R Square	Std. Error of the Estimate	Durbin-Watson
1	.604(a)	.365	.324	82432.015	.024

a Predictors: (Constant), DM11, T, DM6, DM7, DM5, DM8, DM4, DM1, DM10, DM2, DM9, DM3
b Dependent Variable: NrVacancies

b Model 2. Set $G_t = 100(Y_t - Y_{t-1})/Y_{t-1}$. (That is, G_t measures the growth (%) in month t with respect to the previous month.) Formulate the basic assumption of the model that explains G_t from a linear trend and 11 monthly dummies; again, take December as base level. Does this model satisfy the independence requirement? See the printout (GGt refers to G_t).

Model Summary(b)

Model	R	R Square	Adjusted R Square	Std. Error of the Estimate	Durbin-Watson
2	.882(a)	.779	.764	3.28109	.822

a Predictors: (Constant), DM11, T, DM7, DM6, DM8, DM5, DM9, DM4, DM1, DM2, DM10, DM3
b Dependent Variable: GGt

c Model 3. Formulate the basic assumption of the AR(1) model that explains G_t from G_{t-1} only. Use an auxiliary AR(1) model (without constant) for the error terms of the original AR(1) model, to test for first-order autocorrelation. See the printout; if Et denotes the Unstandardized Residual (of the original AR(1) model) at time t, then Et_1 refers to the Unstandardized Residual at time $t - 1$.

Coefficients(a,b)

Aux Model		Unstandardized Coefficients B	Std. Error	Standardized Coefficients Beta	t	Sig.
A	Et_1	.280	.069	.280	4.078	.000

a Dependent Variable: Unstandardized Residual
b Linear Regression through the Origin

d Model 4. Formulate the basic assumption of a 'mixed' model that explains G_t from G_{t-1} and 11 monthly dummies; take December as base level. Again, use an auxiliary AR(1) model (without constant) for the error terms of the mixed model, to test for first-order autocorrelation. See the printout (RESt_1 refers to the lag1 residuals of the mixed model).

Coefficients(a,b)

Aux Model		Unstandardized Coefficients B	Std. Error	Standardized Coefficients Beta	t	Sig.
B	RESt_1	-.047	.072	-.047	-.651	.516

a Dependent Variable: Unstandardized Residual
b Linear Regression through the Origin

Exercise 23.4

Reconsider Exercise 23.3, especially the 'mixed' model of part (d). This model did not give problems with respect to first-order autocorrelation of the residuals. We will use it to predict the number of job vacancies in Germany in April 2007.

ANOVA(b)

Model		Sum of Squares	df	Mean Square	F	Sig.
4	Regression	7690.792	12	640.899	90.477	.000(a)
	Residual	1303.372	184	7.084		
	Total	8994.163	196			

a Predictors: (Constant), DM11, DM10, DM9, DM8, DM7, DM6, DM4, DM5, DM1, DM3, DM2, GGt_1
b Dependent Variable: GGt

Coefficients(a)

Model		Unstandardized Coefficients		Standardized Coefficients	t	Sig.
		B	Std. Error	Beta		
4	(Constant)	−.702	.738		−.951	.343
	GGt_1	.592	.060	.592	9.884	.000
	DM1	12.285	.919	.511	13.371	.000
	DM2	6.850	1.283	.285	5.339	.000
	DM3	2.387	1.390	.099	1.717	.088
	DM4	−2.124	1.285	−.086	−1.653	.100
	DM5	−1.246	1.055	−.050	−1.181	.239
	DM6	.326	.983	.013	.331	.741
	DM7	−.486	.980	−.020	−.496	.620
	DM8	.036	.964	.001	.037	.970
	DM9	−2.786	.963	−.113	−2.892	.004
	DM10	−3.551	.932	−.144	−3.811	.000
	DM11	−1.362	.914	−.057	−1.490	.138

a Dependent Variable: GGt

 a Test with significance level 0.05 whether model 4 indeed contains a seasonal (monthly) component.

ANOVA(b)

Model		Sum of Squares	df	Mean Square	F	Sig.
3	Regression	3994.717	1	3994.717	155.811	.000(a
	Residual	4999.446	195	25.638		
	Total	8994.163	196			

a Predictors: (Constant), GGt_1
b Dependent Variable: GGt

 b Which months have, apart from the autoregressive effect, a significantly larger growth than December? Which months have a significantly smaller growth? Take $\alpha = 0.10$.

 c The numbers of job vacancies in Germany in February and March 2007 were 623 960 and 639 696, respectively. Use this additional information to forecast the number of job vacancies in April 2007. (For comparison, blindly using Model 1 would have given a forecast of 516 191 job vacancies.)

Exercise 23.5

Interest is in forecasting investment in the manufacturing industry of Sweden. To get an idea of the **historical** quarterly investments, the time plot for 1990Q1–2006Q3 is presented below; see also Xrc23-05.sav.

A purely visual study of this plot suggests the presence of a trend and a seasonal component with a repetition period of one year.

 a **Model 1**. Formulate the basic assumption of a linear regression model with a first-order linear trend and a seasonal component; take the fourth quarter as base level. However, this model gives autocorrelation problems. Use the DW test with significance level 0.10 to show that.

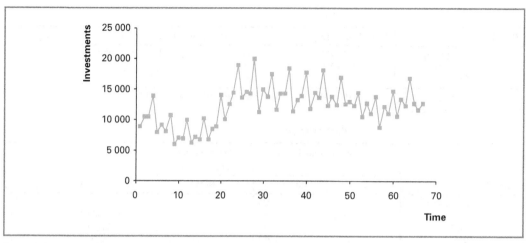

Quarterly investments in the manufacturing industry of Sweden (SEK millions)
Source: Statistics Sweden (2007)

Model Summary(b)

Model	R	R Square	Adjusted R Square	Std. Error of the Estimate	Durbin-Watson
1	.713(a)	.508	.477	2355.527	.235

a Predictors: (Constant), DQ3, T, DQ2, DQ1
b Dependent Variable: Investment

   b **Model 2**. Instead of Y_t = 'investments in quarter t', use $G_t = 100(Y_t - Y_{t-1})/Y_{t-1}$ in a model with a linear trend and quarter dummies. Again, use quarter 4 as base level. Formulate the basic assumption of that model.

   c Use the DW test with significance level 0.10 to draw a conclusion about first-order autocorrelation within the sequence of the residuals. In the printout below, GGt refers to G_t.

Model Summary(b)

Model	R	R Square	Adjusted R Square	Std. Error of the Estimate	Durbin-Watson
2	.944(a)	.891	.884	9.19723	2.249

a Predictors: (Constant), DQ3, T, DQ1, DQ2
b Dependent Variable: GGt

   d Also use a suitable auxiliary AR(1) model to draw a conclusion about first-order autocorrelation within the sequence of the residuals.

Coefficients(a,b)

Aux Model		Unstandardized Coefficients		Standardized Coefficients	t	Sig.
		B	Std. Error	Beta		
A	RES_lag1	−.146	.126	−.143	−1.157	.251

a Dependent Variable: Unstandardized Residual
b Linear Regression through the Origin

Exercise 23.6

Reconsider Exercise 23.5, especially model 2. The following additional printouts are needed.

ANOVA(b)

Model		Sum of Squares	df	Mean Square	F	Sig.
2	Regression	42055.879	4	10513.970	124.295	.000(a)
	Residual	5159.930	61	84.589		
	Total	47215.809	65			

a Predictors: (Constant), DQ3, T, DQ1, DQ2
b Dependent Variable: GGt

Coefficients(a)

Model		Unstandardized Coefficients		Standardized Coefficients	t	Sig.
		B	Std. Error	Beta		
2	(Constant)	35.958	3.061		11.746	.000
	T	−.033	.059	−.024	−.557	.580
	DQ1	−68.353	3.252	−1.095	−21.017	.000
	DQ2	−16.574	3.204	−.271	−5.174	.000
	DQ3	−37.678	3.204	−.616	−11.759	.000

a Dependent Variable: GGt

ANOVA(b)

Model		Sum of Squares	df	Mean Square	F	Sig.
3	Regression	96.348	1	96.348	.131	.719(a)
	Residual	47119.461	64	736.242		
	Total	47215.809	65			

a Predictors: (Constant), T
b Dependent Variable: GGt

a Is the model useful? Does it have a significant linear trend? Does it have a significant seasonal (quarterly) component? For all these questions, use $\alpha = 0.05$.

b In the third quarter of 2006, SEK 12 721 was invested in Sweden's manufacturing industry. Use that information to forecast the level of investment in the last quarter of 2006.

Exercise 23.7

If you manage to find a good forecasting model for the returns of a stock index, you can become rich. In this exercise and the following one, the complications of finding good forecasts will be studied and experienced. (We will only look at the returns; will **not** take transaction costs etc. into account.)

At the moment, it is 4 April 2007, 18.00 hr. You decide to invest €100 000 in the Euronext 100 Fund, a fund that precisely follows the levels of the index Euronext 100. On 4 April, the index closed at 1010.38.

To get an idea of your chances, you study the developments of the index over the past two years. The files Xrc23-07.xls and Xrc23-07.sav contain the daily returns (%, at closing prices, with respect to the previous business day) for 12 April 2005 – 4 April 2007. For a first impression, the time plot is presented below. The exponentially smoothed series with smoothing constant 0.3 is also included.

To forecast the returns for the business days in the period 5 April – 11 April 2007, two forecasting methods will be considered: forecasting by exponential smoothing and by a linear regression model with a linear trend.

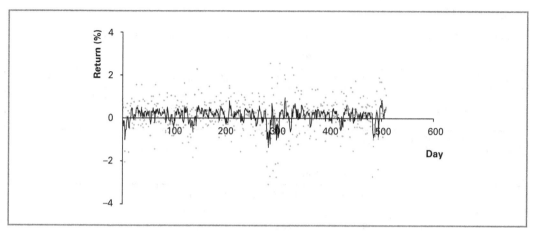

Time plot of the returns of Euronext 100 with exponentially smoothed series, 12 April 2005 – 4 April 2007

a Calculate the most recent exponentially smoothed data by filling in the table below.

Date	Day number	Index	Return index	Exp. smoothed return
29/03/07	504	992.6400	1.3477	0.3832
30/03/07	505	992.3200	−0.0322	
2/04/2007	506	995.3300	0.3033	
3/04/2007	507	1005.7400	1.0459	
4/04/2007	508	1010.3800	0.4614	

b In the third column of the following table, fill in the forecasts that result from the exponential smoothing model (only the business days are included).

Date	Day number	Exp forecast	Trend regression forecast
05/4/2007	509		
10/4/2007	510		
11/4/2007	511		

c The returns data are also used to estimate the linear regression model with the basic assumption $E(Y_t) = \beta_0 + \beta_1 t$. Here, t and Y_t are the number of the (business) day and the return (%) on that day. The printout is given below.

Model Summary(b)

Model	R	R Square	Adjusted R Square	Std. Error of the Estimate	Durbin-Watson
2	.008(a)	.000	−.002	.7891266	2.077

a Predictors: (Constant), NrDay
b Dependent Variable: Return

Coefficients(a)

Model		Unstandardized Coefficients		Standardized Coefficients	t	Sig.
		B	**Std. Error**	**Beta**		
2	(Constant)	.063	.070		.892	.373
	NrDay	4.419E-05	.000	.008	.185	.853

Answer the following questions:

    i Is the model useful?

    ii How useful?

    iii Do you expect problems with the independence requirement?

    iv Irrespective of the former answers, fill in the regression forecasts in the table of part (b).

Exercise 23.8

Reconsider Exercise 23.7. Also the autoregressive model AR(1) will be used.

    a Consider the AR(1) model that explains the return at business day t from the return of the previous business day. The data are used to estimate this model. The printout is given below.

Model Summary(b)

Model	R	R Square	Adjusted R Square	Std. Error of the Estimate	Durbin-Watson
3	.039(a)	.002	.000	.7889999	1.997

a Predictors: (Constant), Return_lag1
b Dependent Variable: Return

Coefficients(a)

Model		Unstandardized Coefficients		Standardized Coefficients	t	Sig.
		B	**Std. Error**	**Beta**		
3	(Constant)	.078	.035		2.207	.028
	Return_lag1	−.039	.044	−.039	−.879	.380

a Dependent Variable: Return

Answer the following questions:

    i Formulate the model.

    ii Is the model useful?

    iii How useful?

    iv Irrespective of the former answers, fill in the regression forecasts in an extension of the table of part (b) of Exercise 23.7. (Create a new column with heading 'AR(1) forecast'.)

    b Which of the three forecasting models do you prefer? Why?

    c Calculate your predicted profit for each of the three forecasting methods if you were to sell your shares on 11 April 2007 at 18.00 hr.

    d On 11 April 2007 at 18.00 hr – after only **three** business days because of Easter – you decide to sell your shares. The Fund closed at the level 1013.64. Calculate your profit. Which forecasting method has given the best prediction?

Exercise 23.9 (computer)

The files Xrc23-09.xls and Xrc23-09.sav contain the yearly unemployment rates (%) for Belgium for 1960–2006. The intention is to forecast the unemployment rates for 2007 and 2008.

   a Create a time plot of the data, placing 'year' along the horizontal axis.
   b Use exponential smoothing with smoothing constant 0.8 to find forecasts; see also Appendix A1.23.
   c Use a linear regression model with a linear trend to find forecasts; take 'year' as independent variable.
   d Use the autoregressive model AR(1) to find forecasts.
   e Which model do you prefer? Why?
   f According to the Belgian National Institute for Statistics (2007) the best forecasts are 8.5% and 8.4%, respectively. Which of the three forecasting methods yielded the closest results?

Exercise 23.10 (computer)

The files Xrc23-10.xls and Xrc23.10.sav contain the monthly CPIs for clothing in Denmark, from 2000M1 to 2007M3.[†] The intention is to find a trend and/or seasonal component. For tests, always use $\alpha = 0.05$.

   a Create a time plot of the data, putting the number of the month ($t = 1, 2, \ldots, 87$) along the horizontal axis. Give your comments.
   b Write down the basic assumption of the linear regression model with a second-order trend and 11 month-dummies; take January as base level.
   c i   Is the model useful?
      ii What is the degree of usefulness?
      iii Does the model have a significant trend component?
      iv Is the trend really second-order?
      v   Does the model have a significant seasonal component?
      vi Which variables are individually significant?
      vii But, what about first-order autocorrelation within the sequence of residuals? What are the consequences?

Exercise 23.11 (computer)

Reconsider Exercise 23.10. The intention is to use the model to forecast the clothing CPI for the months April–December 2007.

   a Interpret the sample regression coefficient of the dummy of April.
   b Find forecasts for the clothing CPIs in the months April–December of 2007.
   c Also determine a 95% prediction interval for the clothing CPI of April 2007.
   d What about the presence of first-order autocorrelation within the residuals? Use $\alpha = 0.10$. What are the consequences for the conclusions of the statistical procedures in this exercise (the prediction interval) and the previous exercise (the tests)?

Exercise 23.12 (computer)

With respect to the exports (goods and services) of the euro area, an increasing trend over time is expected. But will there be a seasonal component? This will be studied now.

[†] *Source*: Statistics Denmark (2007).

The datasets Xrc23–12.xls and Xrc23–12.sav contain the quarterly exports of goods and services (in millions of ECU/euro) of the euro area for the period 1995Q1–2007Q2.[‡]

a Create a time plot of the data by placing the number of the quarter ($t = 1, 2, \ldots, 50$) along the horizontal axis. Give your comments.

b Which linear regression model would you suggest?

c Estimate that model and give a statistical analysis.

d Give your statistical comments about the presence of a trend and a seasonal component.

Exercise 23.13 (computer)

On the website www.statbank.dk of Statistics Denmark, historical datasets can be found for the sales of many commodities or groups of commodities. The file Xrc23–13.sav contains the quarterly turnovers of beverages (in units of DKK 1000) in Denmark for the period 1993Q1–2007Q2. The objective is to use these data to obtain, for each quarter in the period 2007Q3–2008Q4, an interval that is likely to contain the quarterly turnover of beverages.

a Create a time plot of the data by placing the number of the quarter ($t = 1, 2, \ldots, 58$) along the horizontal axis. Give your comments.

b As a first attempt, which linear regression model do you suggest?

c Use the model and the data to obtain the standard 95% prediction intervals for the six quarters in the period 2007Q3–2008Q4.

d Do you trust the confidence level (0.95)? Give your comments.

e Also consider the model that explains the quarterly **growth** of the exports from the lag 1 growth and three quarter dummies. Does this model do better?

Exercise 23.14 (computer)

The dataset Xrc23–14.sav contains end-of-week figures for the exchange indexes Dow Jones, Nikkei, Eurotop 100 and AEX for the period 2003W1–2007W24.[§] (Note that December 1997 = 100.)

In the present study we will use the data of 2003W1–2007W24 to estimate, for each of the four exchanges, the regression model with a linear trend and the AR(1) model. Next, the data of 2007W25–2007W44 and a measure called **mean absolute deviation** (MAD) will be used to compare the forecasting performances of the two models. See the file Xrc23–14_extra.sav for these extra data. The figure below shows the time plots of the four indexes for the combined period 2003W1–2007W44.

a Write down the basic assumptions of these two models.

b For each of the four exchanges, estimate the two models on the basis of the data from the period 2003W1–2007W24; write down the regression equations and the coefficients of determination.

c As a measure of forecasting performance, we use: $\text{MAD} = \frac{1}{m}\sum_{t=1}^{m}|y_t - \hat{y}_t|$.

Here, y_t and $\hat{y}_t$ respectively denote the actual value of the time series at time t and the value as forecast by the model; m denotes the number of data points involved. For each of the four exchanges, calculate the MADs of the two models on the basis of the data in the period 2007W25–2007W44. Use your computer in a clever way.

d Which model did best for which exchanges?

[‡] *Sources*: European Central Bank (2007); Eurostat (2007).

[§] *Source*: De Nederlandsche Bank (2007).

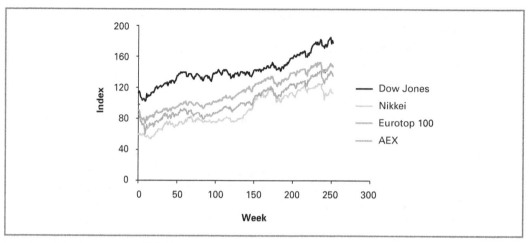

Time plots of the four indexes, 2003W1–2007W44

Exercise 23.15 (computer)

The file Xrc23–15.sav contains weekly historical data of the stock Microsoft, for the period 13 March 1986–5 November 2007. This large dataset will be used to test a general theory in finance that states that returns of many stocks cannot be predicted. (See also Case 23.1).

  a To get a first impression, determine a time plot of the data. Give your comments.

  b Compute the accompanying returns (%) and present them in a time plot.

This general theory in finance in essence states that returns of many stocks are **unpredictable** in the sense that the return in period t is independent of the returns in previous periods. Below, it will be investigated whether the return R_t of the share Microsoft in week t is **linearly** related to the return R_{t-1} in the previous week $t - 1$.

  c Formulate the basic assumption of the AR(1) model for R_t. Explain why this theory proclaims to test whether the slope of this AR(1) model is unequal to 0.

  d Conduct that test; use $\alpha = 0.10$. Give your comments, especially about the predictability of the returns of Microsoft.

CASE 23.2 PERSISTENCE OF THE CAPITAL MARKET RATE (WORKED OUT)

A natural assumption for many economic variables is that the best point predictor of the variable for forecasts in the far future is a constant, independent of the current level of the variable. This is called **mean reversion**. However, if the variable is currently above that long-term level it is likely that it remains above that level for some time; if it is currently less than that level, it is likely that **that** situation will be maintained for some time. This is called **persistence**. The capital market rate also seems to show this persistence property in the short run and the mean reversion property in the long run. This case illustrates the persistence property.

The capital market rate is the interest rate for government bonds with a remaining time to maturity of 5 to 8 years. The time plot below shows the historical development during the period 1982Q1–2003Q4. See Case23–02.xls for the data.

It follows that the interest rates have fluctuated considerably over the sample period, from 11% in 1982 to about 3.5% in 2003. These fluctuations had an enormous impact on, for example,

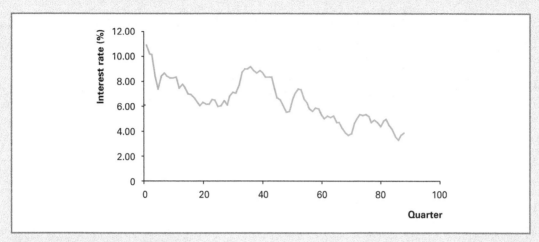

Capital market interest rate, 1982Q1–2003Q4
Source: Dutch Central Bank (2007)

mortgage costs to house buyers since mortgage rates are typically set just above capital market rates.

Let R_t be the capital market rate at quarter t. A simple but very popular model is the first-order autoregressive model:

$$R_t = \beta_0 + \beta_1 R_{t-1} + \varepsilon_t \quad \text{with } E(\varepsilon_t) = 0, \quad \text{for all } t = 2, \dots, 88$$

Recall that the error terms $\varepsilon_1, \dots, \varepsilon_{88}$ are independent and identically $N(0, \sigma_\varepsilon^2)$ distributed. By assuming that the model is also valid for the quarters 89 and 90 (the first and second quarters of 2004), forecasts for R_{89} and R_{90} will be calculated as soon as the model is estimated, just by using the last observation (3.92%, see the dataset) of the sample period. Furthermore, the persistence of the series of interest rates is demonstrated by showing that the probability that after three months the interest rate has changed by more than one percentage point, is still rather small.

At first, the AR(1) model is estimated on the basis of the 87 pairs of successive interest rates. It follows easily that $b_0 = 0.31294$, $b_1 = 0.93965$ and $s_\varepsilon^2 = (0.43820)^2$. Without giving the details, it can also be shown that the long-run mean is estimated to be 5.1854%. (To avoid misunderstandings, this long-run mean is **not** the mean of the 88 interest rates.) It is this value that is the subject of the mean reversion assumption. The economic theory assumes that – in the long run – the interest rate will reverse to this level.

As a consequence of the above estimates, the forecast interest rate for the first and the second quarter of 2004 is:

$$0.31294 + 0.93965 \times 3.92 = 3.9964$$

$$0.31294 + 0.93965 \times 3.9964 = 4.0682$$

Hence, it is predicted that the interest rates will increase slightly compared with the end-of-sample value 3.92. However, these forecasts are still below the long-run mean of 5.1854%.

Since it is given that the rate for quarter 88 is 3.92%, it follows for R_{89} that:

$$R_{89} = \beta_0 + \beta_1 \times 3.92 + \varepsilon_{89} \quad \text{with } \varepsilon_{89} \sim N(0, \sigma_\varepsilon^2)$$

Hence:

$$R_{89} \sim N(\beta_0 + 3.92\beta_1, \sigma_\varepsilon^2) \approx N(3.9964, (0.43820)^2)$$

(The right-hand approximating normal distribution follows by replacing the parameters β_0, β_1, σ_ε^2 by their estimates 0.31294, 0.93965, $(0.43820)^2$.) Consequently:

$$P(R_{89} > 4.92) \approx P(Z > \frac{4.92 - 3.9964}{0.4382}) = P(Z > 2.1077) = 0.0175 \qquad (*)$$

$$P(R_{89} < 2.92) \approx P(Z < \frac{2.92 - 3.9964}{0.4382}) = P(Z < -2.4564) = 0.0070 \qquad (*)$$

Hence, the probability that the interest rate after one quarter will have changed by more than 1 percentage point is equal to $0.0175 + 0.0070 = 0.0245$, which is rather small. This result demonstrates the persistence property referred to before.

Chi-square tests

24

Chapter contents

In all preceding chapters on inferential statistics, interest was in unknown statistics of the underlying population distribution. In the course of Chapters 15–23 the following parameters were the parameters of interest: μ, p, σ^2, μ_1, μ_2, p_1, p_2, σ_1^2, σ_2^2, β_0, β_1, ..., β_k, σ_ε^2. In the present chapter, primary interest is not in parameters but in more general aspects of population distributions.

CASE 24.1 KICKS FROM THE PENALTY MARK IN SOCCER

In the paper 'Kicks from the penalty mark in soccer' by Jordet et al. (*Journal of Sports Statistics*, 15 January 2007), the outcomes of penalty kicks in the international tournaments World Cup (WC), European Championships (EC) and Copa America (CA) were compared. One of the purposes was to study the role of stress. The dataset Case24–01.sav contains the results of 409 penalty kicks taken during the period 1976–2004 in these tournaments when, after extra time, there was no winner. Apart from the outcome of the kick (1 for goal and 0 for miss), the type of the tournament (WC, EC, CA) is recorded. Some questions arise:

■ Each tournament has its own proportion of hits. Are the three proportions different? That is, are the distributions of $Y =$ 'yes/no goal' different for kicks in the three types of tournament?

■ If stress plays a role, it can be expected that the proportion of hits decreases with the importance of the tournament. Is the proportion of hits for the European Championships larger than the proportion of hits for the World Cup tournaments?

See the end of Section 24.3 for answers.

24.1 Introduction

In the present chapter, one or two variables (say, X and Y) are considered on one or more populations. These variables can be qualitative or quantitative.

Interest is in certain general null hypotheses H_0 regarding the population distribution of X or the joint population distribution of X and Y. Four categories of null hypothesis will be considered:

1 H_0: X has a certain completely specified distribution.

2 H_0: X has a certain distribution that is **in**completely specified since it has unknown parameters.

3 H_0: X and Y are independent.

4 H_0: Y has a common distribution on all populations of interest.

In each of the four categories, the alternative hypothesis H_1 is the logical opposite of the statement in H_0. Tests for testing problems where H_0 is of category 1 or 2 are called **goodness of fit tests** and where H_0 is of category 3 or 4 are called **tests for association**.

Examples of null hypotheses of categories 1 and 2 are, respectively:

■ X is normally distributed with mean parameter 25 and variance parameter 100;

■ X is normally distributed.

In the first example the distribution is indeed completely specified. The second example specifies only the type; the parameters μ and σ^2 are unknown.

Important applications of category 1 can be found in process control, where the quality X of the produced item has to be controlled. As an example, consider the production process of axles of 5 cm length. The production is said to be 'under control' if the length X of the axles produced has a normal distribution with mean 5 cm and standard deviation 0.1 cm. The process is 'out of control' if the mean length is different from 5 cm, if the standard deviation is different from 0.1 cm, and also if normality no longer holds. That is, the null hypothesis H_0: length of axle is $N(5, 0.01)$ corresponds to 'under control' and the opposite statement to 'out of control'.

Tests for category 1 and 2 problems also occur in market research, for instance to find out whether advertising campaigns have influenced market share. See also Example 24.2.

Testing problems of category 2 are often encountered when primary interest is in studies that need a normality assumption. For instance;

■ The error term ε in a linear regression study has to be normal, at least for small n. To check this normality, the regression residuals $e_1, ..., e_n$ can be used.

■ When primary interest is in the mean μ of a variable X but the sample size n is small, then the standard interval estimators and hypothesis tests assume normality for X; see Section 16.2. It is often desired to test this normality requirement.

■ When primary interest is in the variance σ^2 of some variable X or in a comparison of the variances σ_1^2 and σ_2^2 of X on two populations, then the standard statistical procedures assume normality for X; see Sections 17.3, 17.4 and 18.3.2.

Testing problems of categories 3 and 4 occur when comparing the behaviours/opinions/preferences of people.

■ Is the buying behaviour of people different for different income groups? Or, to put it another way, are the variables 'consumer behaviour' and 'income level' associated?

■ Is 'political view' related to 'income level'?

■ Within the Faculty of Economics and Business, is the choice of master programme influenced by the choice of bachelor programme?

It will turn out that the test statistics for the four testing problems are basically the same, all based on a division of the possible values of X (or X and Y) into (say) k classes. On the basis of a random sample of n observations of X (or X and Y), the frequencies $N_1, \ldots, N_k$ of the classes $1, \ldots, k$ are determined and compared with the frequencies $e_1, \ldots, e_k$ that are expected **if the null hypothesis is valid**. If the null hypothesis is not valid, it is expected that (at least some of) the deviations

$$N_1 - e_1, N_2 - e_2, \ldots, N_k - e_k$$

will be away from 0, so the sample statistic

$$G = \sum_{i=1}^{k} \frac{(N_i - e_i)^2}{e_i} = \sum_{i=1}^{k} \left(\frac{N_i - e_i}{\sqrt{e_i}} \right)^2 \tag{24.1}$$

will take a relatively large value. It turns out that, under the null hypothesis, G is approximately χ^2-distributed for large n, so G is used as test statistic and H_0 will be rejected for values $\geq \chi^2_\alpha$. The only problem is the number of degrees of freedom, which will depend on which of the four categories is considered. This will be dealt with in what follows.

Many statistical packages have standard procedures for the χ^2-tests that are presented in this chapter. See also Appendix A1.24.

24.2 Goodness of fit tests

In this section, only one variable (called X) is of interest. In particular, interest is in whether the variable X has a certain completely specified population distribution (category 1) or whether the population distribution of X is specified but apart from one or more unknown parameters (category 2). Suppose that a random sample $X_1, \ldots, X_n$ is available.

24.2.1 Category 1: complete specification

Suppose that we want to know whether X has a certain completely specified population distribution, so this distribution contains no unknown parameters. Then, the category 1 testing problem is:

H_0: X does have this completely specified population distribution, against

H_1: X does **not** have this completely specified population distribution

This is step (i) of a five-step test procedure. For step (ii), the set of possible values of X is divided into a partition of k classes called $1, \ldots, k$. Next, the observations of the random sample $X_1, \ldots, X_n$ are compared with the k classes, the accompanying observed frequencies $N_1, \ldots, N_k$ are determined and compared with the frequencies $e_1, \ldots, e_k$ that are expected **if H_0 is true**. Large values of the test statistic

$$\text{ii} \quad G = \sum_{i=1}^{k} \frac{(N_i - e_i)^2}{e_i}$$

denote that observed frequencies N_i strongly deviate from the expected frequencies e_i, so for large values H_0 has to be rejected. But how large should these values be? The following (unproven) result will give the answer to this question:

$$H_0 \text{ is true and } e_i \geq 5 \text{ for all } i = 1, \ldots, k \Rightarrow G \approx \chi^2_{k-1} \tag{24.2}$$

To understand the $k - 1$ degrees of freedom, notice that according to (24.1), the test statistic G is determined by the terms $(N_i - e_i)/\sqrt{e_i}$. The 'freedom' of these k terms within G is reduced because of the restriction that the sum of $N_1 - e_1, \ldots, N_k - e_k$ is 0:

$$\sum_{i=1}^{k} (N_i - e_i) = \sum_{i=1}^{k} N_i - \sum_{i=1}^{k} e_i = n - n = 0$$

It would appear that only $k - 1$ of the k terms $N_i - e_i$ can vary freely, so the same holds for the k terms $(N_i - e_i)/\sqrt{e_i}$ in G. Hence, G itself has $k - 1$ degrees of freedom.

Steps (iii)–(v) of the five-step procedure follow immediately. The box below summarizes all steps.

Five-step procedure for goodness of fit test for category 1 testing problems

   i Test H_0: X does have this completely specified population distribution, against

       H_1: X does **not** have this completely specified population distribution

   ii Test statistic: $G = \sum\limits_{i=1}^{k} \dfrac{(N_i - e_i)^2}{e_i}$

   iii Reject $H_0 \Leftrightarrow g \geq \chi^2_{\alpha;\, k-1}$

   iv Calculate **val**

   v Conclude by comparing **val** with step (iii)

Requirement: all $e_i \geq 5$

Tests of this type are frequently conducted in cases of a completely specified **normal** population distribution, as in the following example.

Example 24.1

The Greenway wood factory produces oak tabletops of length 2 m. The production lines are automated; each hour about 500 tabletops are produced. From the many years of experience it is known that, if the process is under control, the length X (in cm) of a tabletop follows a normal distribution with mean parameter and variance parameter equal to 200 cm and 0.0009 cm², respectively.

    To check whether the process is still under control, each hour the lengths (in cm) of 26 tabletops, chosen randomly from the hourly production, are measured carefully and a χ^2 test with significance level 0.05 is conducted to check for violation of the model $N(200, 0.0009)$. In the test, a partition of four classes is considered in such a way that, after standardization, the total area under the density of $N(0, 1)$ is divided into four areas of 0.25 each. See Figure 24.1.

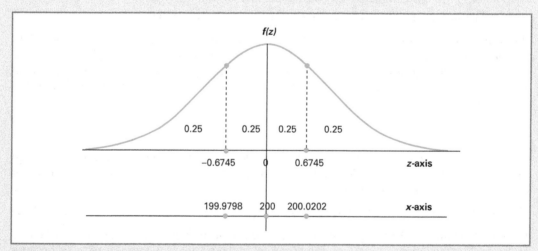

Figure 24.1 Classifications of the z- and the x-axes

    Note that the numbers -0.6745 and 0.6745 on the z-axis under the $N(0, 1)$ density divide the total area 1 into four parts with equal areas. This yields the partition (classification)

$$(-\infty, -0.6745], \quad (-0.6745, 0], \quad (0, 0.6745], \quad (0.6745, \infty)$$

of the z-axis. Since the transformation $z = (x - 200)/0.03$ transforms X into the $N(0, 1)$ distributed random variable Z, the above classification can easily be transformed back to a classification of the x-axis. For instance, the boundary of the first class is obtained by solving the equation $(x - 200)/0.03 = -0.6745$. The resulting classification of the x-axis is:

$(-\infty, 199.9798]$, $(199.9798, 200]$, $(200, 200.0202]$, $(200.0202, \infty)$

The χ^2 test is based on these classes (called, respectively, 1, 2, 3 and 4).

Below, the test will be conducted on the basis of hourly data with the following frequencies for the 26 measurements of the length X:

Class i	1	2	3	4
Frequency n_i	11	3	3	9

The frequencies n_1, n_2, n_3, n_4 have to be compared with the frequencies e_1, e_2, e_3, e_4 that are expected if the model $N(200, 0.0009)$ is valid. But if this model is valid, the probability that X falls in class 1 is 0.25 and it is expected that $26 \times 0.25 = 6.5$ measurements will fall in class 1. The same holds for the other classes, so all e_1, e_2, e_3, e_4 are equal to 6.5. The table below gives a calculation scheme for the calculation of **val** in step (iv) of the five-step procedure.

Class i	1	2	3	4
n_i	11	3	3	9
e_i	6.5	6.5	6.5	6.5
$\frac{(n_i - e_i)^2}{e_i}$	3.1154	1.8846	1.8846	0.9615

Hence:

val $= 3.1154 + 2 \times 1.8846 + 0.9615 = 7.8461$

Notice that the 4 classes lead to 3 degrees of freedom for the test statistic. So, $\chi^2_{0.05;3} = 7.8147$ (*) is the critical value of the five-step procedure:

i Test H_0: X has the distribution $N(200, 0.0009)$, against
   H_1: X does not have the distribution $N(200, 0.0009)$; $\alpha = 0.05$

ii Test statistic: $G = \sum_{i=1}^{4} \frac{(N_i - e_i)^2}{e_i}$

iii Reject $H_0 \Leftrightarrow g \geq 7.8147$

iv **val** $= 7.8461$

v Reject H_0 since **val** belongs to the rejection region

Note that the p-value of the test is only slightly smaller than 0.05:

p-value $= P(G \geq 7.8461) = 0.0493$ (*)

At significance level 0.05, the conclusion is that $N(200, 0.0009)$ is **not** the distribution of X. Whether this violation of the model $N(200, 0.0009)$ is caused by a different mean parameter and/or variance parameter or by non-normality, is not clear.

We next consider another group of examples where this type of goodness of fit test (with complete specification) is applied. In these examples, X is a **qualitative** variable that has k possible outcomes coded as $1, 2, \ldots, k$, and interest is in whether the respective probabilities $p_1, p_2, \ldots, p_k$ of these outcomes are equal to some prescribed numbers $a_1, a_2, \ldots, a_k$ (with $a_1 + a_2 + \ldots + a_k = 1$) or not. A **random** sample (so, *iid* sample) of n observations of X is used to draw a conclusion about that.

The random experiment of drawing that sample is a special example of a multinomial experiment.

Multinomial experiment

A *multinomial experiment* with parameters $p_1, p_2, \ldots, p_k$ is a series of n independent and identical repetitions of an experiment for which the outcome of each repetition is classified into one of the k categories $1, 2, \ldots, k$. Here, p_j is the probability that the outcome of a repetition falls in category j. The repetitions of the experiment are called *trials*, the categories are called *cells*.

'Independent' means that the result of a trial is not influenced by the results of the other trials; 'identical' means that all trials are done under completely the same circumstances, so that for each trial the probability of cell j is equal to p_j. Notice that $p_1 + p_2 + \ldots + p_k = 1$. A binomial experiment (see Section 9.2.1) is a special case of a multinomial experiment: the case that $k = 2$.

In the present situation of a random sample of size n, the categories are the outcomes $1, 2, \ldots, k$ of X and the parameters $p_1, p_2, \ldots, p_k$ are the corresponding probabilities. Interest is in the null hypothesis:

$$H_0: p_1 = a_1, p_2 = a_2, \ldots, p_k = a_k$$

This hypothesis indeed completely specifies a distribution of X: the probabilities $P(X = j)$ are equal to the prescribed numbers a_j. So, a test for H_0 indeed belongs to the class of tests considered in the present subsection. Typical examples are in the field of market research, for instance to study whether market shares have changed when compared with the situation some time ago.

Example 24.2

The grocery market is dominated by the grocery chains A, B, C and D. These chains recently started an aggressive campaign, with many advertisements and with sudden price reductions, to increase their market shares. Before the campaign began 40% of consumers said they preferred chain A; 25%, 15% and 10% said they preferred the chains B, C and D, respectively. The rest (10%) gave preference to other supermarkets.

To find out whether the price war has changed the market shares, 5000 randomly chosen consumers are asked about their preference: A, B, C, D or category E (the 'other' supermarkets). The results are in the second row of the table below.

Chain i	A	B	C	D	E	Total
n_i	2100	1300	700	480	420	5000
e_i	2000	1250	750	500	500	5000
$\dfrac{(n_i - e_i)^2}{e_i}$	5.0000	2.0000	3.3333	0.8000	12.8000	23.9333

Can it be concluded that the market shares have changed?
Interest is in the following testing problem:

i Test $H_0: p_A = 0.40, p_B = 0.25, p_C = 0.15, p_D = 0.10, p_E = 0.10$, against

H_1: at least one of p_A, p_B, p_C, p_D, p_E has changed; $\alpha = 0.01$

Indeed, H_0 gives a complete specification of a distribution of the variable $X =$ 'preference of a consumer' (which has possible outcomes A, B, C, D, E). The test statistic is:

ii $\quad G = \sum_{i=1}^{5} \dfrac{(N_i - e_i)^2}{e_i}$

Since there are five possible outcomes, G has four degrees of freedom. Here is step (iii):

iii $\quad$ Reject $H_0 \Leftrightarrow g \geq \chi^2_{0.01;4} = 13.2767$ $\qquad\qquad$ (∗)

To calculate **val**, the frequencies n_i have to be compared with the expected frequencies for the case that H_0 is still valid. If H_0 is still valid after the price war, it is expected that $0.40 \times 5000 = 2000$ consumers will prefer chain A and hence $e_A = 2000$. Similarly, $e_B = 0.25 \times 5000 = 1250$, etc. The third row of the table contains all these expected frequencies. In the fourth row, the calculation of **val** is started by calculating the contribution of the five individual cells; **val** is just the total. Steps (iv) and (v) follow immediately:

iv $\quad$ **val** = 23.9333

v $\quad H_0$ is rejected

The conclusion is that the market shares have indeed changed. By studying the last row in the table it follows that **val** is determined mainly by the contribution of cell E. The 'other' supermarkets 'pay the bill' for the price war.

24.2.2 Category 2: incomplete specification

Suppose, now, that the population distribution in the null hypothesis H_0 is not completely specified since it contains m unknown parameters. As a consequence, the expected frequencies $e_1, \ldots, e_k$ depend on these parameters and cannot be calculated. The best thing that can be done is to replace the m unknown parameters by 'good' estimates that follow from the sample observations and to test whether the resulting **best-fitting** distribution has to be rejected.

Note, however, that the expected frequencies for this best-fitting distribution will depend on the estimates and hence on the sample observations. This means that, in the test statistic, these expected frequencies should appear as random variables, so we write $E_1, \ldots, E_k$ instead of $e_1, \ldots, e_k$. The test statistic becomes:

$$G = \sum_{i=1}^{k} \frac{(N_i - E_i)^2}{E_i}$$

Again, a χ^2 distribution is used for G if the sample size is large. However, apart from the degree of freedom that is lost because of the restriction $\Sigma_i(N_i - E_i) = 0$, there are m more degrees of freedom lost due to the fact that the estimation of the m parameters causes m new restrictions. That is why for a χ^2 goodness of fit test with m unknown parameters in H_0, the following result is used (where $e_1, \ldots, e_k$ are the realisations of $E_1, \ldots, E_k$):

H_0 is true and $e_i \geq 5$ for all $i = 1, \ldots, k \Rightarrow G \approx \chi^2_{k-m-1}$ $\qquad$ **(24.3)**

In the five-step procedure below, step (iv) especially will take some time.

Five-step procedure for goodness of fit test for category 2 testing problems

i $\quad$ Test H_0: X does have the **in**completely specified population distribution, against

$\quad H_1$: X does not have this **in**completely specified population distribution.

▶

ii Test statistic: $G = \sum_{i=1}^{k} \dfrac{(N_i - E_i)^2}{E_i}$

iii Reject $H_0 \Leftrightarrow g \geq \chi^2_{\alpha;k-m-1}$

iv **a** Use the data to estimate the m unknown parameters under H_0

   b Use the resulting estimates to determine the realized expected frequencies $e_1, ..., e_k$ under H_0

   c Calculate **val**

v Conclude by comparing **val** with step (iii)

Requirement: all $e_i \geq 5$

Two examples are considered.

Example 24.3

There are many hypothesis tests that are based on the assumption that the random sample is drawn from a **normal** population distribution. For instance, the small-sample t-tests of Section 16.2, the χ^2 tests of Chapter 17 and the F-tests of Chapter 18 have this requirement. By using a goodness of fit test with incomplete specification, more can be said about the validity of that normality requirement.

In Example 17.5 it was investigated whether the (unknown) standard deviation σ of the daily return (%) of the SNS Euro Stocks Fund was larger than the (known) standard deviation 0.005% of the SNS Euro Bonds Fund; recall that standard deviation is a measure of risk. But is the implicit assumption about the normality of the daily returns of the SNS Euro Stocks Fund valid? The daily returns for the period 23 November 2004 – 17 November 2006 are in the file Xmp24–03.xls.

Let X be the variable 'daily return (%) of the SNS Euro Stocks Fund'. The testing problem is as follows:

i Test H_0: X is normally distributed, against H_1: X is not normally distributed; $\alpha = 0.10$

Note that the distribution in H_0 is not completely specified; there are $m = 2$ unknown parameters: the mean parameter μ and the variance parameter σ^2. Using a computer, estimates can easily be obtained:

$\bar{x} = 0.00064714$ and $s^2 = (0.00896846)^2$

(Since daily returns are often close to 0, eight decimals are used.) Below, a χ^2 test procedure will be used that is based on a division of the set of possible outcomes of X into six classes. As in Example 24.1, a partition of the horizontal axis under the z-density is transformed into a partition under the density of X. For that, the standardized version of X is used:

$Z = \dfrac{X - 0.00064714}{0.00896846}$

The following classification of the z-axis is chosen:

$(-\infty, -2], \quad (-2, -1], \quad (-1, 0], \quad (0, 1], \quad (1, 2], \quad (2, \infty)$

This is shown in Figure 24.2. The classification of the z-axis transforms into the following classification of the x-axis:

$(-\infty, -0.0172898], \qquad (-0.0172898, -0.0083213], \quad (-0.0083213, 0.0006471],$

$(0.0006471, 0.0096156], \quad (0.0096156, 0.0185841], \qquad (0.0185841, \infty)$

Using this last partition, the frequencies $n_1, n_2, ..., n_6$ can be obtained; the results are in the second row of the table below.

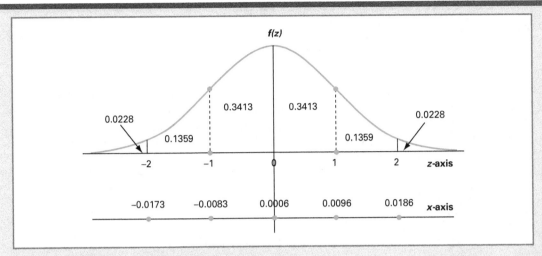

Figure 24.2 Classifications of the z- and the x-axes

Class i	1	2	3	4	5	6	Total
n_i	12	68	173	183	63	14	513
$P(X$ in class $i)$	0.0228	0.1359	0.3413	0.3413	0.1359	0.0228	1
e_i	11.6964	69.7167	175.0869	175.0869	69.7167	11.6964	513
$\dfrac{(n_i - e_i)^2}{e_i}$	0.0079	0.0423	0.0249	0.3576	0.6471	0.4537	1.5335

Next, the realizations of $E_1, \ldots, E_6$ have to be calculated. If H_0 is true and μ and σ^2 are estimated by the above estimates, X gets the distribution:

$$N(0.00064714, (0.00896846)^2)$$

Under these circumstances, the probability that X will fall in class 1 is equal to the probability that Z will fall in $(-\infty, -2]$:

$$P(Z \leq -2) = 0.0228$$

Hence, $e_1 = 513 \times 0.0228 = 11.6964$. The expectation e_2 follows similarly: it is equal to 513 times the probability that X will fall in class 2, so 513 times $P(-1 < Z \leq -2)$:

$$e_2 = 513 \times P(-1 < Z \leq -2) = 513 \times 0.1359 = 69.7167$$

And so on. The subsequent results are in the third and fourth rows of the table; note especially that all expectations are at least 5. The **val** can now be calculated. Since $k = 6$ and $m = 2$, there are $k - m - 1 = 3$ degrees of freedom left. With significance level $\alpha = 0.10$, the critical value of the test is $\chi^2_{0.10;3} = 6.2514$ (∗). Here are steps (ii)–(v):

ii $G = \displaystyle\sum_{i=1}^{k} \frac{(N_i - E_i)^2}{E_i}$

iii Reject $H_0 \Leftrightarrow g \geq 6.2514$

iv **val** $= 1.5335$

v Do not reject H_0

The data do not give evidence that the distribution of X is non-normal.

Goodness of fit tests with incomplete specification are also used in multinomial experiments.

Example 24.4

In the context of Example 24.2, it is also possible to test the null hypothesis:

$$H_0: p_D = p_E \text{ and } p_A = p_B + p_C$$

This hypothesis expresses the statement that chain D and the supermarkets that are allied in category E have the same market share, while the market share of A is equal to the sum of the market shares of B and C. Will this hypothesis also be valid after the price war?

Again, $k = 5$ since the problem has 5 'cells'. The hypothesis H_0 is not completely specified; there are two unknown parameters left. They can be taken as p_A and p_B, since (under H_0) $p_C = p_A - p_B$ and $p_A + \dots + p_E = 1$. The $m = 2$ unknown parameters are estimated respectively by:

$$\hat{p}_A = \frac{2100}{5000} = 0.42 \quad \text{and} \quad \hat{p}_B = \frac{1300}{5000} = 0.26$$

If H_0 is valid, this automatically leads to the following estimates for p_C, p_D and p_E:

$$\hat{p}_C = \hat{p}_A - \hat{p}_B = 0.16 \quad \text{and} \quad \hat{p}_D = \hat{p}_E = (1 - 0.84)/2 = 0.08$$

Hence, if H_0 is true, the following frequencies are expected:

$$e_A = \hat{p}_A \times 5000 = 2100; \quad e_B = \hat{p}_B \times 5000 = 1300;$$

$$e_C = 0.16 \times 5000 = 800; \quad e_D = e_E = 0.08 \times 5000 = 400$$

(Note especially that all expectations are at least 5.) Things are summarized in the table below.

Chain i	A	B	C	D	E	Total
n_i	2100	1300	700	480	420	5000
e_i	2100	1300	800	400	400	5000
$\frac{(n_i - e_i)^2}{e_i}$	0	0	12.5	16.0	1.0	29.5

The contributions of each of the five cells to the value of the test statistic G can easily be calculated; see the last row of the table. Since there are $k - m - 1 = 2$ degrees of freedom left, the critical value at significance level 0.05 is $\chi^2_{0.05;2} = 5.9915$ (*). Here are the five steps:

i Test $H_0: p_D = p_E$ and $p_A = p_B + p_C$ against $H_1: p_D \neq p_E$ and/or $p_A \neq p_B + p_C$

ii Test statistic: $G = \sum_{i=1}^{5} \frac{(N_i - E_i)^2}{E_i}$

iii Reject $H_0 \Leftrightarrow g \geq 5.9915$

iv **val** = 29.5

v Reject H_0

There is sufficient evidence to conclude that the statement of H_0 is not true.

The Lilliefors test can also be used to test the null hypothesis that a variable is normal; see Section 22.3.2. With a computer it follows that, for Example 24.3, the **val** of the Lilliefors test is 0.038, with accompanying p-value 0.070. At significance level 0.10, it is concluded that the variable 'daily return (%) of the SNS Euro Stocks Fund' is **not** normal, which differs from the conclusion of Example 24.3.

24.3 Tests for independence and homogeneity

Chi-square tests for independence and for homogeneity are combined in one section. The reason is that, technically, they have to be conducted in the same way, as will become clear below.

24.3.1 Category 3: tests for independence

Interest is in two variables, X and Y, and especially in whether random observations of X and Y are independent. It is assumed that X has r (possibly coded) outcomes denoted by $i = 1, 2, \ldots, r$ and Y has c (possibly coded) outcomes denoted by $j = 1, 2, \ldots, c$. In many applications X and Y are qualitative variables, but the theory below is valid for discrete quantitative variables as well.

Suppose that a random sample $(X_1, Y_1), \ldots, (X_n, Y_n)$ is available. The following testing problem will be considered:

 i Test H_0: X and Y are independent against H_1: X and Y are dependent.

Because of the definition of independence in (11.9), the null hypothesis H_0 can be rewritten as:

H_0: $P(X = i$ and $Y = j) = P(X = i) \times P(Y = j)$ for all i, j

By setting

$h(i, j) = P(X = i$ and $Y = j)$; $f(i) = P(X = i)$; $g(j) = P(Y = j)$

the null hypothesis is also equivalent to:

H_0: $h(i, j) = f(i) \times g(j)$ for all i, j

Table 24.1 characterizes the situation **if** H_0 is valid.

$h(i, j)$		j				$f(i)$
		1	2	...	c	
i	1	$f(1)g(1)$	$f(1)g(2)$	...	$f(1)g(c)$	$f(1)$
	2	$f(2)g(1)$	$f(2)g(2)$	...	$f(2)g(c)$	$f(2)$
	.	.	.	.	.	.
	.	.	.	.	.	.
	r	$f(r)g(1)$	$f(r)g(2)$	...	$f(r)g(c)$	$f(r)$
$g(j)$		$g(1)$	$g(2)$	...	$g(c)$	1

TABLE 24.1 Independence of X and Y

Each probability $h(i, j)$ in the interior part (the matrix with r rows and c columns) arises by multiplying the corresponding probabilities of the margins.

The construction of a test procedure for problem (i) will be based on the random sample of n random observations of X and Y. But notice that the random experiment that randomly draws n population elements and observes X and Y is essentially a multinomial experiment. The cells are the $k = rc$ combinations of the outcomes of X and Y; the parameters $p_1, p_2, \ldots, p_k$ are the $k = rc$ probabilities $h(i, j)$. Indeed, these probabilities add up to 1. Hence, a goodness of fit χ^2 test can be used for testing problem (i).

If H_0 is valid, this multinomial experiment has only $(r - 1) + (c - 1)$ parameters that are unknown: the probabilities $f(i)$ for $i = 1, \ldots, r - 1$ and $g(j)$ for $j = 1, \ldots, c - 1$. It would appear that H_0 is **in**completely specified since it contains $m = r + c - 2$ unknown parameters. So, a goodness of fit χ^2 test with incomplete specification can be conducted. The sample observations are used to estimate the unknown parameters. However, this will be done at the cost of losing $m = r + c - 2$ degrees of freedom from the $k - 1 = rc - 1$ that were originally present. The following number of degrees of freedom is left:

$$rc - 1 - (r + c - 2) = rc - 1 - r - c + 2 = rc - r - c + 1 = (r - 1)(c - 1)$$

Hence, the χ^2 test will make use of $(r - 1)(c - 1)$ degrees of freedom.

Step (iv) of this χ^2 test with incomplete specification needs special attention:

Step (iv)(a). The estimation under H_0 of the $m = r + c - 2$ unknown parameters

$$f(1), f(2), \ldots, f(r - 1) \quad \text{and} \quad g(1), g(2), \ldots, g(c - 1)$$

will of course be based on the n pairs of sample observations. Notice that the possible outcomes of a trial are the pairs (i, j) with $i = 1, \ldots, r$ and $j = 1, \ldots, c$. In the table below, n_{ij} denotes the frequency of the pair (i, j), while n_{+j} and n_{i+} arise by adding up the frequencies in the ith row and the jth column, respectively.

n_{ij}		j				Freq. X
		1	**2**	...	**c**	
	1	n_{11}	n_{12}	...	n_{1c}	n_{1+}
	2	n_{21}	n_{22}	...	n_{2c}	n_{2+}
i	.	.	.	...	.	.
	.	.	.	...	.	.
	r	n_{r1}	n_{r2}	...	n_{rc}	n_{r+}
Freq. Y		n_{+1}	n_{+2}	...	n_{+c}	n

As respective estimates of $f(i)$ and $g(j)$, the relative frequencies n_{i+}/n and n_{+j}/n are used.

Step (iv)(b). If H_0 is true, the probability $h(i, j) = f(i)g(j)$ is estimated by the product of the above estimates:

$$h(i, j) \approx \frac{n_{i+}}{n} \times \frac{n_{+j}}{n}$$

To determine for the $k = rc$ cells of the multinomial experiment the realized expected frequencies $e_1, \ldots, e_k$, these products are multiplied by n:

$$\text{Expected frequency for cell } (i, j): \quad n \times \frac{n_{i+}}{n} \times \frac{n_{+j}}{n} = \frac{n_{i+} \times n_{+j}}{n}$$

Hence, the realized expected frequencies $e_1, \ldots, e_k$ arise by dividing for each cell (i, j) the product of the accompanying margin frequencies by n.

Step (iv)(c). The **val** can be calculated by adding up all cell contributions for G.

Here is the complete five-step procedure.

χ^2 test for independence

   i Test H_0: X and Y are independent against H_1: X and Y are dependent.

   ii Test statistic: $G = \sum_{i=1}^{k} \frac{(N_i - E_i)^2}{E_i}$

(Here, *l* refers to a cell in the matrix with *r* rows and *c* columns; *l* runs row-wise through these *rc* cells.)

iii Reject $H_0 \Leftrightarrow g \geq \chi^2_{\alpha;(r-1)(c-1)}$

iv **a, b** Determine the $e_1, \ldots, e_k$ by dividing for each cell (i, j) the product of the margin-frequencies n_{i+} and n_{+j} by n

    c Calculate **val** by adding up all cell contributions to *G*

v Draw the conclusion by comparing **val** with step (iii)

Requirement: all $e_i \geq 5$

Remark. Notice that the frequencies n_{i+} and n_{+j} in step (iv) are realizations of random variables.

Example 24.5

Since the end of the 1990s, the European Commission has been looking closely, attentively and worriedly at the level of roaming charges in the EU. In October 2006, a Special Eurobarometer appeared entitled *Roaming*. This document contains many interesting facts about mobile communications within the EU25.

One of the studies in that document is about the variable Y = 'mobile communication service used when abroad' with the four possible outcomes:

    only voice calls;   only text messages;   both;   other (advanced mobile services)

The topic of the study was to find out whether there are differences between countries with respect to the use of these communication services. From the population of all citizens aged 15 and over from seven EU countries (Belgium, Denmark, Germany, Luxembourg, Netherlands, Finland, Sweden) who own a mobile phone and use it when abroad, 4549 people were sampled and interviewed. The observations of Y and X = 'country of the respondent' were recorded. The data were used to find out whether X and Y are dependent.

Below, it is assumed that these people were chosen randomly from the combined population of the citizens of these seven countries. The resulting frequencies are in Table 24.2.

		Only voice calls	Only text messages	Both	Other	Row Freq.
			Y			
X	BE	250	109	249	32	640
	DK	126	222	234	18	600
	DE	321	117	270	22	730
	LU	140	47	200	14	401
	NL	300	169	246	54	769
	FI	185	247	240	35	707
	SE	232	119	323	28	702
	Column freq.	1554	1030	1762	203	4549

TABLE 24.2 Frequency distributions of *Y* for seven countries

Is there evidence that the use of mobile communication service depends on the country a person comes from?

A χ^2 test for independence of 'country' and 'use of communication service' will be conducted. Since $r = 7$ and $c = 4$, the χ^2 test makes use of $6 \times 3 = 18$ degrees of freedom. The most complicated part of the test procedure is step (iv)(a) and (b): the calculation of the realised expected frequencies $e_1, \ldots, e_{28}$. Below, things are considered for two cells:

Cell 1:

Estimated **proportion** in cell 1 under H_0 : $\dfrac{640}{n} \times \dfrac{1554}{n}$

Estimated expected frequency e_1 : $n \times \dfrac{640}{n} \times \dfrac{1554}{n} = \dfrac{640 \times 1554}{n} = 218.6327$

Cell 14:

Estimated expected frequency e_{14} : $\dfrac{401 \times 1030}{n} = 90.7958$

The realized expected frequencies $e_1, \ldots, e_{28}$ are calculated by dividing the product of the two corresponding marginal **frequencies** by $n = 4549$. The left-hand matrix in Table 24.3 contains all expected frequencies.

Expectations e_i under H_0				Cell-contributions to *val*			
218.6327	144.9110	247.8962	28.5601	4.5003	8.8992	0.0049	0.4143
204.9681	135.8540	232.4027	26.7751	30.4241	54.6257	0.0110	2.8759
249.3779	165.2891	282.7566	32.5764	20.5701	14.1076	0.5755	3.4338
136.9870	90.7958	155.3225	17.8947	0.0663	21.1251	12.8512	0.8477
262.7008	174.1196	297.8628	34.3168	5.2959	0.1505	9.0302	11.2898
241.5208	160.0813	273.8479	31.5500	13.2270	47.1938	4.1836	0.3773
239.8127	158.9492	271.9112	31.3269	0.2545	10.0406	9.5990	0.3533

TABLE 24.3 Expectations and contributions per cell

Notice that all $e_1, \ldots, e_{28}$ are at least 5, so there are no problems with the 5-rule. Since the 28 frequencies n_i and the 28 expected frequencies e_i are now known, the contribution $(n_i - e_i)^2/e_i$ of cell i to the *val* can be calculated. For instance, for cell 14, this contribution is:

$$\dfrac{(47 - 90.7958)^2}{90.7958} = 21.1251$$

The right-hand matrix of Table 24.3 contains all cell contributions; adding them up yields the *val* of the χ^2 test. (It will be clear that a computer package (like Excel) is useful here.)

The complete five-step procedure can now be written down. For $\alpha = 0.01$, the critical value of the rejection region is $\chi^2_{0.01;18} = 34.8053$ (*).

i Test H_0: X and Y are independent against H_1: X and Y are dependent

ii Test statistic: $G = \displaystyle\sum_{i=1}^{k} \dfrac{(N_i - E_i)^2}{E_i}$

(Here, i refers to a cell in the matrix with 7 rows and 4 columns; i runs row-wise through these 28 cells.)

iii Reject $H_0 \Leftrightarrow g \geq 34.8053$

iv *val* = 286.3282

v H_0 is rejected convincingly

The use of mobile communication services when abroad depends on the country. From the right-hand matrix of Table 24.3 it follows that the values 1 and 2 of Y contribute especially to *val*.

24.3.2 Category 4: tests for homogeneity

Interest is in **one** variable Y that is considered on **two or more** (say r) populations called 1, 2, ..., r. Often, Y is a qualitative variable (with coded outcomes 1, 2, ..., c), but the theory below remains valid if it is a discrete quantitative variable. From each population, a random sample of observations of Y is available; the random samples are independent with fixed (but possibly different) sample sizes.

The problem now is whether (yes/no) the distributions of Y on these populations are different. The testing problem is:

i test H_0: the distribution of Y is the same for all r populations, against

H_1: the distribution of Y is **not** the same for all r populations.

Let X be the variable that measures the population to which a sampled element belongs; notice that the possible outcomes of X are 1, 2, ..., r. It will be intuitively clear that the statement in H_0 that Y has the same distribution for all populations can also be formulated as: the distribution of Y does not depend on this variable X. But that means that H_0 can be restated as: X and Y are independent. In essence, the current testing problem is the testing problem of Section 24.3.1 for this special X. However, the sampling procedure is different. In particular, the frequency of the sample observations with X equal to i is now the sample size of the sample that is drawn from population i, and it does not come from a random variable. (In Section 24.3.1 the frequency n_{i+} is the realization of the **random** variable that counts whether X is equal to i.)

Although the sampling procedures of the present and the former subsections are clearly different, it can be shown that the steps (ii)–(v) of the test of Section 24.3.1 also work in the present situation. The only thing is that the meaning of $E_1, ..., E_k$ in the test statistic is slightly different: in their realizations $e_1, ..., e_k$ the frequency n_{i+} is now the sample size drawn from population i.

χ^2 test for homogeneity

i Test H_0: the distribution of Y is the same for all r populations, against

H_1: the distribution of Y is **not** the same for all r populations.

ii Test statistic: $G = \sum_{l=1}^{k} \frac{(N_l - E_l)^2}{E_l}$

(Here, l refers to a cell in the matrix with r rows and c columns; l runs row-wise through these rc cells.)

iii Reject $H_0 \Leftrightarrow g \geq \chi^2_{\alpha;(r-1)(c-1)}$

iv **a,b** determine the $e_1, ..., e_k$ by dividing for each cell (i, j) the product of the margin frequencies n_{i+} and n_{+j} by n

c Calculate **val** by adding up all contributions to G of the cells

v Draw the conclusion by comparing **val** with step (iii)

Requirement: all $e_l \geq 5$

Example 24.6

The problem stated at the start of Example 24.5 can also be formulated as a study of whether there are differences in distribution between the seven countries in their use of communication services when abroad. That is, a homogeneity test for the variable Y = 'mobile communication service used when abroad' is asked for. It is even more realistic that the sample of size 4549 is just the union of seven random samples, one from each country. However, the test procedure is precisely the same. Hence, the conclusion will also be the same: there are indeed differences in distribution.

One last remark about the requirement that all $e_l \geq 5$. It reflects the fact that n has to be large, so large that all (realized) expected frequencies under H_0 are at least 5. If, for one or more classes, the (realized) expected frequencies are smaller than 5, then classes have to be combined to answer the requirement.

CASE 24.1 KICKS FROM THE PENALTY MARK IN SOCCER – SOLUTION

Let p_{wc}, p_{ca} and p_{ec} denote the proportions of hits in the tournaments of the World Cup, the Copa America and the European Championships, respectively. To find out whether these proportions are different, the following testing problem is considered:

H_0: $p_{wc} = p_{ca} = p_{ec}$ against H_1: p_{wc}, p_{ca} and p_{ec} are not all equal

But these hypotheses can also be written in terms of the dummy variable Y that indicates whether a penalty yields a hit (value 1) or a miss (value 0). Since the distribution of Y on a population of matches is completely determined by the proportion of hits, the null hypothesis can also be formulated as: the distributions of Y on the three populations of all WC, CA and EC tournaments are equal. That is why a homogeneity test (with $\alpha = 0.05$) will be conducted for the following testing problem:

H_0: the three distributions are equal against H_1: they are not all equal

There are $r = 3$ populations involved and Y can take $c = 2$ values. Hence, the distribution χ^2_2 has to be used. The table below contains in the six cells the observed frequencies of hits and misses, the realized expected numbers of hits and misses for the case that H_0 is true, and all cell contributions to the value of the test statistic of the χ^2 test:

Freq. Exp. freq. Cell contribution	Y		
Population	0	1	n_{j+}
WC	44 32.1711 4.3493	109 120.8289 1.1580	153
CA	23 27.9658 0.8818	110 105.0342 0.2348	133
EC	19 25.8631 1.8212	104 97.1369 0.4849	123
n_{+j}	86	323	409

By adding up all cell contributions, it follows that **val** = 8.9300. Since the critical value of the test is $\chi^2_{0.05;2} = 5.9915$ (*), H_0 is rejected; the three proportions of hits are **not** all the same. Notice that the magnitude of **val** is mainly determined by the relatively large number of misses in the WC tournaments.

If stress plays a role, it can be expected that $p_{wc} < p_{ec}$ and that $p_{wc} < p_{ca}$. With respect to p_{ec} and p_{ca} we do not expect a difference. These presuppositions can be tested with z-tests as in Section 18.4. The respective **val**s are -2.6188, -2.2833 and -0.3085. Indeed, these expectations seem to be valid.

Summary

Chi-square distributions are important not only when statistical interest is in a population variance σ^2 (see Chapter 17). In the present chapter such distributions were also used in:

- goodness of fit tests (with or without complete specification);
- tests for independence of two variables;
- tests for homogeneity of a distribution on two or more populations.

All these new tests are based on a common form for the test statistic:

$$G = \sum_{l=1}^{k} \frac{(N_l - E_l)^2}{E_l}$$

For each cell $l = 1, \ldots, k$, the observed frequency N_l of cell l is compared with the frequency E_l that is expected if the null hypothesis is valid.

 Key terms

cells **774**	multinomial	test for association **770**
goodness of fit test **770**	experiment **774**	trials **774**

? Exercises

Exercise 24.1

It is claimed that the weekly sales X (in units of 10^5 euro) of a company is $N(5, 2.25)$ distributed. To check the claim, X is measured for a sample of 100 weeks. After standardization (subtracting 5 and dividing the result by 1.5), the resulting data are classified and the accompanying frequencies are determined; see the table below.

Class	$(-\infty, -1.5)$	$[-1.5, -0.5)$	$[-0.5, 0.5)$	$[0.5, 1.5)$	$[1.5, \infty)$
Frequency	9	19	48	17	7

Test the claim by putting it in the null hypothesis. Take $\alpha = 0.03$.

Exercise 24.2

It is claimed that the weekly sales X (in units of 10^5 euro) of a company is normally distributed. To check the claim, X is measured for a sample of 100 weeks; the **sample** mean and **sample** standard deviation turn out to be 5.1 and 1.6. After standardization (subtracting 5.1 and dividing the result by 1.6), the resulting data are classified and the accompanying frequencies are determined; see the table below.

Class	$(-\infty, -1.5)$	$[-1.5, -0.5)$	$[-0.5, 0.5)$	$[0.5, 1.5)$	$[1.5, \infty)$
Frequency	14	24	38	17	7

Test the claim by putting it in the null hypothesis. Take $\alpha = 0.03$.

Exercise 24.3

Consider a multinomial experiment involving $n = 300$ trials and $k = 4$ cells. The table below contains the observed frequencies per cell.

Cell	1	2	3	4
Frequency	62	86	72	80

Test the null hypothesis H_0: $p_1 = 0.2$, $p_2 = 0.3$, $p_3 = 0.2$, $p_4 = 0.3$ at significance level 0.01.

Exercise 24.4

Reconsider Exercise 24.3. Also test the null hypothesis H_0: $p_1 = p_3$ and $p_2 = p_4$; again with $\alpha = 0.01$.

Exercise 24.5

The annual income (X, in units of €1000) is measured for a random sample of 500 people drawn from a large population. The question is whether X can be modelled by the distribution $N(\mu, \sigma^2)$ with $\mu = 30$ and $\sigma = 5$; the data will be used to draw a conclusion about that. In the table below, the data are summarized by counting the frequencies in six classes bounded by μ, $\mu \pm \sigma$, $\mu \pm 2\sigma$ and $\pm\infty$.

Class	$(-\infty, 20)$	$[20, 25)$	$[25, 30)$	$[30, 35)$	$[35, 40)$	$[40, \infty)$
Frequency	15	60	150	175	80	20

a Calculate, for each of the classes, the probability that X will fall in it if indeed X has that normal distribution.

b Also calculate the expected frequencies per class if X has that normal distribution.

c Test the null hypothesis that states that X has that distribution; use significance level 0.05.

Exercise 24.6

Reconsider Exercise 24.5, but suppose now that μ and σ are unknown. However, the classes and the observed frequencies are the same as in Exercise 24.5. The underlying data are used to find estimates for these parameters; it turns out that sample mean and sample standard deviation are, respectively, 30.8 and 5.6. Test the null hypothesis H_0: X is normally distributed, again with $\alpha = 0.05$.

Exercise 24.7

Does the preference of a consumer depend on gender? This is an important question in marketing.

The dependence of gender X and preference Y with respect to three types of soft drink (a, b and c) will be tested. Two hundred consumers are randomly chosen; gender (m or f) and preference is recorded. The summarized results are shown below.

Frequency per cell		Y		
		a	b	c
X	m	60	38	14
	f	45	25	18

Test whether, with respect to soft drinks, the preference of a consumer depends on gender; use $\alpha = 0.10$.

Exercise 24.8

Do men and women differ as far as their preference for brand of beer is concerned? To answer that question, 100 male and 100 female beer drinkers are randomly chosen and asked to give their preference about the brands a, b and c. The results are in the table; X = 'gender' and Y = 'preference for brand of beer'.

Frequency per cell		Y		
		a	b	c
X	m	60	38	2
	f	46	27	27

a Test whether the distributions of Y for male and female beer drinkers are different; use $\alpha = 0.10$.

b When comparing the experimental designs of this exercise and Exercise 24.7, what is the striking difference?

Exercise 24.9

Goodness of fit tests can be used to check whether objects (human beings, machines, tools, etc.) behave 'as they should'. Do the ten workers in a factory have the same output? Are there differences between the quality levels of new products? Such questions are answered by conducting goodness of fit tests similar to the simple test in the present exercise.

Is a certain die unfair? To check it, the die is rolled 4500 times; here are the results:

Number of eyes	1	2	3	4	5	6	Total
Frequency	795	751	779	717	660	798	4500

Let $p_1, p_2, \ldots, p_6$ denote the probabilities that X = 'number of eyes in one throw' is equal to 1, …, 6.

a Explain why the experiment of throwing the die 4500 times is a multinomial experiment. What are the trials? What are the cells?

b Formulate the testing problem that is of interest. Is the distribution of X under H_0 completely specified?

c Conduct a χ^2 goodness of fit test to draw a conclusion about the question; use $\alpha = 0.05$.

d Which cell has the largest contribution to the **val**?

Exercise 24.10

Reconsider Exercise 24.9. Indeed, the die seems to be unfair as far as the side with 5 eyes is concerned. But is the die also unfair because of other sides?

a Formulate a suitable testing problem. Is the distribution of X under H_0 completely specified?

b Conduct a χ^2 goodness of fit test to draw a conclusion; use $\alpha = 0.05$.

Exercise 24.11

For a certain population of adult people, the weight X (in kg) is said to be $N(72, 64)$ distributed. Is this indeed the case? A random sample of 100 adults yielded the following classified results shown in the table.

Class	< 60	60 – < 70	70 – < 80	≥ 80
Frequency	11	28	37	24

Use this classified frequency distribution to draw a conclusion about the question; take $\alpha = 0.05$. Is the requirement of the test satisfied?

Exercise 24.12

Reconsider Exercise 24.11. Although 80% of the people in the population are male, the 20% females might cause non-normality of X. From the underlying 100 data points it is known that $\bar{x} = 70$ and $s = 10$.

   a Use the classified frequency distribution of Exercise 24.11 to check whether the distribution of X is non-normal; take $\alpha = 0.05$.

   b What about the requirement of the test?

Exercise 24.13

Are X = 'level of income' and Y = 'political preference' related? To get an idea of the answer, 390 people are asked about their level of income (1 = low, 4 = high) and the political party (A, B or C) of their preference. The results are shown in the table.

Frequency per cell		Political party		
		A	B	C
Level of income	1	23	11	1
	2	40	75	31
	3	16	107	60
	4	2	14	10

   a Suppose that the sample of 390 people is a random sample from the underlying population. Test whether X and Y are dependent; use $\alpha = 0.01$.

   b Suppose now that the 390 observations are the realizations of four independent random samples from each of the populations of the people with income level 1, 2, 3 and 4. The sample sizes are, respectively, 35, 146, 183 and 26. That is, the income classes are considered as populations and Y = 'political preference' is the only variable. Test the null hypothesis that states that the four distributions of Y are the same. Again, use $\alpha = 0.01$. How does this differ from the test of part (a)?

Exercise 24.14

The table below gives some (fictitious) information about smoking behaviour and alcohol consumption.

Frequency	Does not smoke	Does smoke
Does not drink alcohol	179	45
Does drink alcohol	242	185

   a Test whether the variables X = 'drinks alcohol (yes/no)' and Y = 'smoker (yes/no)' are dependent; use $\alpha = 0.02$.

   b Which cell(s) contribute most? What does it tell?

Exercise 24.15

Reconsider Exercise 24.14. Suppose that the underlying experimental design was as follows: two independent random samples were drawn, one (size 427) from the population of alcohol drinkers and one (size 224) from the population of the people who do not drink alcohol. Recall that under these circumstances steps (ii)–(v) of the test in Exercise 24.14 can also be used as a test for a homogeneity testing problem.

Let p_{na} and p_a be the proportion that smokes for the non-alcohol (*na*) drinkers and the alcohol (*a*) drinkers, respectively.

a Explain why the homogeneity testing problem can equivalently be formulated as $H_0: p_{na} = p_a$ against $H_1: p_{na} \neq p_a$.

b Use a test from Chapter 18 for this testing problem.

(*Remark*: It can be shown that the test statistic of part (a) of Exercise 24.14 is the square of the present test statistic.)

Exercise 24.16 (computer)

A midterm exam on statistics was taken by students from three courses: business administration (BA), international business (IB) and business studies (BS). This midterm consisted of ten multiple-choice questions; the grades (scale 1–10) are in the file Xrc24–16.sav.

The aim of the present study is to find out whether the variable Y = 'Does / does not pass the midterm' has a common distribution over the three populations of students. (Here, 'does pass' means that the grade is at least 6.)

a Investigate this problem by assuming that the grades in the dataset come from three random samples, one from each of the three populations. Take 0.01 as significance level. If possible, use a statistical package.

b Let p_{BA}, p_{BS} and p_{IB} denote the population proportions of the BA, BS and IB students who pass the midterm. Rewrite the testing problem in terms of these proportions.

c Which population especially contributes to the **val** of part (a)?

Exercise 24.17

In Example 21.11, a linear regression model was used to explain, for a large population of people who work at least 32 hours per week, the gross hourly wage W from (among others) *Age*, the dummy *FEM*, and the education dummies D_2, D_3, D_4 and D_5. Recall that the residuals of the linear regression are considered as realizations of the error term. To get a **visual** impression of whether the error term is indeed normally distributed (as required), histograms of the residuals and the *z*-residuals (special standardized residuals) can be created. The first histogram should look like a normal curve, the second like a standard normal curve.

In the present exercise, two χ^2 tests will be used to get a **statistical** impression about the validity of:

■ the standard normality of the *z*-residuals;

■ the normality of the (unstandardized) residuals.

To get the residuals (res) and the *z*-residuals (*z*-res), we have to run the regression; the dataset Xrc24–17.sav already contains them. Both χ^2 tests are originally based on the partition (classification) $(-\infty, -1]$, $(-1, 0]$, $(0, 1]$, $(1, \infty)$ of the axis under the $N(0, 1)$ density. Recall that the mean of the residuals is (always) 0. From the dataset it follows that the sample standard deviation of the 150 residuals is 6.93694. (To understand things better, you should know that a *z*-residual is not precisely the same as the standardization of the accompanying residual; we will not go into detail.)

a Determine the $N(0, 1)$ probabilities that belong to the above partition.

b Explain how this classification transforms into the following classification for the data of the residuals: $(-\infty, -6.93694]$, $(-6.93694, 0]$, $(0, 6.93694]$, $(6.93694, \infty)$.

These two classifications yield the following classified frequency distributions of the *z*-residuals and the ordinary residuals (check it yourself).

Classif. for *z*-res	$(-\infty, -1]$	$(-1, 0]$	$(0, 1]$	$(1, \infty)$
Classif. frequencies	13	66	59	12
Classif. for res	$(-\infty, -6.93694]$	$(-6.93694, 0]$	$(0, 6.93694]$	$(6.93694, \infty)$
Classif. frequencies	13	66	58	13

c Test, with significance level 0.05, whether the distribution of the z-residuals is standard normal on the given classification.

d Test, with significance level 0.05, whether the distribution of the (ordinary) residuals is normal on the given classification.

Exercise 24.18 (computer)

In many publications, stock returns are considered to be normally distributed. In the present exercise, the stock of Peugeot is considered. The file Xrc24–18.sav contains the daily returns (%) of this stock, as listed on Euronext, for the period 22 April 2005 – 19 April 2007.

a Create a histogram of the returns and give your comments with respect to visual non-normality.

b Check normality of the returns on the classification that originates from the classification $(-\infty, -1], (-1, -0.5], (-0.5, 0], (0, 0.5], (0.5, 1], (1, \infty)$ under the standard normal density. Use $\alpha = 0.05$.

Exercise 24.19 (computer)

Many computer packages have a random generator that generates simulations from some completely specified distribution. For instance, the Excel function rand() generates observations from the uniform distribution $U(0, 1)$.

Let the random variable X denote the result of an application of the Excel function rand(). Create 10000 simulations with rand() and use them to test, with significance level 0.01, the claim that X is $U(0, 1)$ distributed. Make use of the classification that divides the interval (0, 1) into ten sub-intervals of equal width.

Exercise 24.20 (computer)

Reconsider Exercise 23.15 and the dataset Xrc23–15.sav. In essence, the 'general theory' mentioned in that exercise states not only that the returns of many stocks are unpredictable in the sense that the return R_t in period t is independent of the returns in former periods, but also that the returns of stocks constitute a so-called normal **random walk**, an *iid* sequence of observations from a normal distribution (so, with a constant expectation and variance).

In Exercise 23.15 we concluded that, with respect to the weekly returns of the stock Microsoft, there is no evidence that R_t follows linearly from R_{t-1}, which supports the theory that R_{t-1} and R_t are independent (the *i* property). In the present exercise it is assumed that weekly returns of Microsoft stock can indeed be considered as an *iid* sequence of observations, but it will be **tested** whether this common distribution is non-normal. We will use the classification that is determined by the endpoints $-2.5, -2, -1.5, -1, -0.5, 0, 0.5, 1, 1.5, 2, 2.5$ and ∞ under the $N(0, 1)$ density.

a Create a histogram of the 1129 returns; include the best-fitting normal curve. Give your comments about normality.

b Calculate the mean and the standard deviation of the returns in the dataset.

c Calculate the 12 probabilities that belong to the classification. Do you expect problems with the requirement that all e_i have to be at least 5?

d Test the null hypothesis that states that the weekly return of Microsoft is normally distributed; use $\alpha = 0.01$. Which classes contribute most to the **val**? Compare your answers with your findings in part (a).

e Give your final conclusion about the validity of the normal random walk theory for the stock Microsoft.

CASE 24.2 DIFFERENT VIEWS IN THE EU ABOUT ILLEGAL ACTIVITY

In commission of a firm that markets goods for young people in several European countries, researchers of MarketView interviewed 948 business students from eight countries about their opinion on several questions. Here is one of the questions:

Q3: Reporting a lost item as 'stolen' to an insurance company in order to collect the insurance money

Possible responses are 1, 2, 3, 4, 5 where 1 = strongly believe it is wrong and 5 = strongly believe it is not wrong. The dataset Case24–02.sav contains not only the responses to this question but also the codes of the countries where the students live (Portugal = 1, Spain = 2, Denmark = 3, Scotland = 4, Germany = 5, Italy = 6, Greece = 7, Netherlands = 8).

Interest is in the distributions of the variable Q3 on the subpopulations of the business students from the eight countries. Use a suitable test to find out whether differences exist between the eight distributions. If necessary, combine values of Q3.

CHAPTER

Non-parametric statistics

Chapter contents

In Section 18.2, test procedures were introduced to determine whether the population means μ_1 and μ_2 of one variable on two populations are different. For the independent samples case this was done in Section 18.2.1 and for the paired samples case in Section 18.2.2. For the independent samples case, things were partially generalized to more than two populations in Section 21.5 (for instance, in Section 21.5.3 about one-factor ANOVA). Recall that it was always required that the variable of interest be **quantitative**. Furthermore, it was assumed that the samples were drawn from **normal** populations or that the sample sizes were relatively large. The question then arises whether similar tests can be developed if the variable is ordinal or quantitative but (extremely) non-normal. This is the topic of the present chapter.

However, for a **qualitative** variable the parameter μ (i.e., the population mean) has no meaning since observations of the variable cannot be added up. Although it would be possible to choose another location measure as parameter (such as the median if the variable is ordinal), we will follow an approach without any specified parameter: a so-called ***non-parametric*** approach. If the (only) variable X is considered on k populations, the hypotheses will be formulated in terms of the **locations** of the distributions of X (see also Chapter 3), thus not specifying a parameter to measure the location.

In the present chapter, the null hypothesis will be as follows:

H_0: the k population locations are the same **(25.1)**

It will be required that the variable X is ordinal or quantitative but non-normal. Furthermore, it will be assumed that the two (or k) population distributions are identical so far as shape and spread are concerned; they can differ only because of difference in location.

CASE 25.1 BUSINESS START-UPS AND LACK OF CAPITAL

Starting your own business is a complicated, laborious and expensive matter. In the report *Entrepreneurship Survey of the EU (25 Member States), United States, Iceland and Norway* (European Commission, 2007), people from 28 countries were asked several questions about obstacles to business start-ups. The results of the report are of special importance for entrepreneurs since they give information about the climate of enterprise in the 28 countries. One of the questions was:

It is difficult to start one's own business due to a lack of available financial support

with possible values 1 = strongly agree, 2 = agree, 3 = disagree, 4 = strongly disagree and 9 = DK/NA (don't know, no answer). The results are in the file <u>Case25–01.sav</u>. In the questions below, take only the answers 1, 2, 3 and 4 into account. See Section 25.4.

a Are there differences between the countries with respect to the distribution of answers to the question?

b If so, can it be shown that there exist differences in location?

c Compare the individual countries on the basis of the answers to the question. In which of the countries does the problem of lack of available financial support seem to be largest?

d In which of the EU25 countries does the problem of lack of available financial support seem to be smallest? Compare the results with Iceland, Norway and the USA.

25.1 Introduction

Figure 25.1 shows three pairs of population distributions. For each pair, the two distributions have the same shape and spread. For the first pair, the locations are also the same. However, the second and the third pairs have different locations. For the second pair the location of population distribution 2 is smaller than the location of population distribution 1; for the third pair things are the other way round. Notice that phrases such as 'smaller than' and 'larger than' make sense since the variable of interest is ordinal or quantitative.

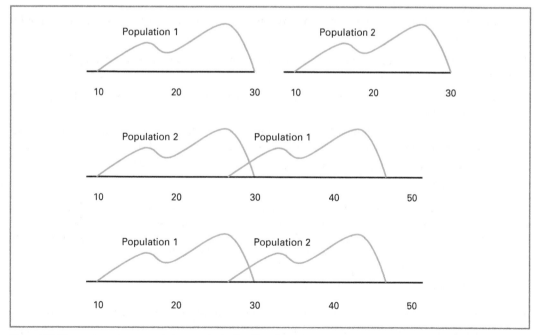

Figure 25.1 Comparison of two populations on the basis of location

If two populations are involved, the null hypothesis is:

H_0: the two population locations are the same.

Three types of alternative hypothesis will be considered:

a H_1: the location of population 1 is different from the location of population 2.

b H_1: the location of population 1 is larger than the location of population 2.

c H_1: the location of population 1 is smaller than the location of population 2.

Notice that (b) and (c) are depicted in the second and the third pictures, respectively, of Figure 25.1. Alternative (a) is just the opposite of the first pair of pictures in the figure.

The testing problem for H_0 against alternative (a) will be tested if we want to know whether there is sufficient evidence to infer that the two population locations are different. If we want to know whether in general the random variable X is larger on population 1 than on population 2, H_0 will be tested against alternative (b). But if we want to know whether in general the random variable X is larger on population 2 than on population 1, H_0 will be tested against alternative (c).

It will turn out that the test procedures of Sections 25.2, 25.3 and 25.4 are all based on ranking procedures: the n data points are ordered and their **ranks** (ranking positions, integers between 1 and n) are determined. Since nothing further is assumed about the distribution of X (in particular, normality is not assumed), the test procedures of the coming sections are – apart from the naming **non-parametric** tests – also called **distribution-free** tests.

Section 25.2 deals with the case of two independent samples. A test will be developed that can be used if the variable is ordinal or quantitative but non-normal. The paired samples case is the topic of Section 25.3. Two test procedures are considered: one for ordinal data and one for non-normal quantitative data. In Section 25.4 a test is considered for the general case of k independent samples.

25.2 Two independent samples

Starting points: The ordinal or quantitative but non-normal variable X is considered on two populations called 1 and 2. The two distributions are assumed to be identical in shape and spread, but may differ in location.

Objective: The comparison of the two locations, on the basis of two independent random samples of respective sizes n_1 and n_2.

To test H_0 in (25.1) against one of the alternatives (a), (b) and (c), a ranking procedure is used. It is based on the following steps:

1 All $n = n_1 + n_2$ sample observations of X are placed in ascending order, which is possible since the variable is at least ordinal.

2 The ranks 1, 2, …, n are assigned to the ordered observations; if a **tie** (coinciding observations) occurs, then the mean of the corresponding ranks is assigned to all observations in the tie.

3 The sums T_1 and T_2 of the ranks in sample 1 and sample 2, respectively, are calculated.

4 T_1 is the test statistic.

Check yourself that $T_1 + T_2$ is just the sum of all ranks 1, 2, …, n, which is $n(n + 1)/2$. As a consequence, the value of T_2 is known as soon as the value of T_1 is calculated. If H_0 is valid and hence the two locations are equal, it will be intuitively clear that $E(T_1)$ can be calculated by multiplying the total $n(n + 1)/2$ by n_1/n and $E(T_2)$ by multiplying the total by n_2/n. So:

$$H_0 \text{ is valid} \Rightarrow E(T_1) = \frac{n_1(n + 1)}{2} \text{ and } E(T_2) = \frac{n_2(n + 1)}{2}$$

That is why the test procedure for alternative (b) will reject H_0 for values of T_1 that are relatively **large** compared with $n_1(n + 1)/2$. Similarly, for the testing of alternative (c), H_0 will be rejected for values that are relatively **small**; for the testing of alternative (a), the null hypothesis will be rejected for relatively small **and** for relatively large values of T_1. But how small and how large? Similarly to the tests in former chapters, this will be determined by the significance level. Things will be worked out in the following simple example.

Example 25.1

The assembly department of a company that manufactures chairs makes use of two methods to put the chairs together. To reduce assembly times and costs, the management wants to check the conjecture that method 1 in general has shorter assembly times than method 2. Six experienced workers are randomly chosen and asked to assemble a chair; three of them use method 1 and the others use method 2. The assembly times (in minutes) are recorded:

Sample 1	20	21	18
Sample 2	16	25	24

The variable of interest is X = 'assembly time (in minutes) of a chair', a quantitative variable. The testing problem and the test statistic are:

i Test H_0: the locations of the two distributions are equal, against
   H_1: the location of distribution 1 is smaller than the location of distribution 2; take α = 0.05

ii Test statistic: $T = T_1$

We next calculate **val**. Note that 16 is the smallest observation, so it gets rank 1. The next observation in order is 18, so it gets rank 2; etc. The largest observation is 25 and the rank 6 is assigned to it. The table below gives the ranks of all observations.

Sample 1	Rank	Sample 2	Rank
20	3	16	1
21	4	25	6
18	2	24	5

It follows that **val**, the sum of the ranks of the sample 1 observations, is equal to 9. Since the sum of the ranks of sample 2 is 12, **val** is indeed smaller than the value of T_2. But whether **val** is small enough to reject H_0 in favour of H_1 is not yet clear. To decide about that, we have to consider the probability distribution of T_1 for the case that H_0 is true. Recall that the two distributions are identical in shape and spread, and that they can differ only because of different locations. Hence, if H_0 is valid, then the two distributions are completely the same and each of the 20 possible triples of ranks for the sample of method 1 has probability 1/20 of occurring. The left-hand part of Table 25.1 gives the overview of all possible triples of ranks and the accompanying value of T_1.

If H_0 is true, the probability distribution of T follows easily from the overview of triples and accompanying sums of ranks; see the right-hand part of Table 25.1. For instance, since there are 3 triples with sum of ranks equal to 11 it follows that the probability that T_1 takes the value 11 is 3/20.

Now it is possible to determine the rejection region. Of course, H_0 has to be rejected for values of T_1 that are relatively small, say for values $\leq t_l$. Because of the significance level 0.05, this constant t_l has to be such that, under H_0, the following holds:

$P(T_1 \leq t_l) \leq 0.05$

This is because the type I error is controlled at the level 0.05. From the probability distribution in Table 25.1 it follows that t_l can be taken as 6 (although any other real number in the interval [6, 7) would also be suitable). See step (iii) below.

Ranks sample 1	Sum of ranks		Value of T_1	Probability
1, 2, 3	6		6	1/20
1, 2, 4	7		7	1/20
1, 2, 5	8		8	2/20
1, 2, 6	9		9	3/20
1, 3, 4	8		10	3/20
1, 3, 5	9		11	3/20
1, 3, 6	10		12	3/20
1, 4, 5	10		13	2/20
1, 4, 6	11		14	1/20
1, 5, 6	12		15	1/20
2, 3, 4	9			
2, 3, 5	10			
2, 3, 6	11			
2, 4, 5	11			
2, 4, 6	12			
2, 5, 6	13			
3, 4, 5	12			
3, 4, 6	13			
3, 5, 6	14			
4, 5, 6	15			

TABLE 25.1 List of rank triples and distribution of T_1, under H_0.

iii   Reject $H_0 \Leftrightarrow t \leq 6$

Since the test procedure is known, the last two steps of the test are straightforward:

iv   **val** = 9

v H_0 is not rejected since 9 is larger than the critical value

The data do not give enough evidence that the use of method 1 will in general lead to shorter assembly times than the use of method 2.

A test as in the example is called a **Wilcoxon rank sum test**. Unfortunately, such a test procedure will become very laborious if the sample sizes n_1 and n_2 get larger. For instance, for $n_1 = n_2 = 8$, there are $\binom{16}{8} = 12\,870$ possible realizations for the 8-tuples of ranks of sample 1 (which corresponds to column 1 of Table 25.1). That is why, for the case that both n_1 and n_2 are at most 10, tables have been created to find the critical value(s); see Table 8 in Appendix A4. We next explain how, for a few significance levels, these tables can be used to find the critical value(s) of the test procedure for fixed sample sizes n_1 and n_2.

If H_0 is tested against the two-sided alternative (a) with significance level 0.05 (respectively, 0.10), then the upper bound t_l of the lower part of the rejection region and the lower bound t_u of the upper part can be found in part A (respectively, B) of the table. For instance, for $\alpha = 0.05$, $n_1 = 4$ and $n_2 = 7$, the bounds t_l and t_u are equal to 13 and 35, respectively; H_0 will be rejected if the value of the test statistic is at most 13 or at least 35. For comparison, notice that the expectation of T under H_0 is 24.

If H_0 is tested against the one-sided alternative (b), then H_0 will be rejected for relatively large values of $T = T_1$. For significance level 0.025 (respectively 0.05), the lower bound t_u of the rejection region can be found in part A (respectively, B) of the table. For instance, for $\alpha = 0.05$, $n_1 = 8$ and $n_2 = 3$, the bound t_u equals 57. For comparison, note that the expectation of T under H_0 is 48.

If H_0 is tested against the one-sided alternative (c), then H_0 will be rejected for relatively small values of $T = T_1$. For significance level 0.025 (respectively, 0.05), the upper bound t_l of the rejection region can be found in part A (respectively, B) of the table. For instance, for $\alpha = 0.025$, $n_1 = 7$ and $n_2 = 9$, the bound t_l equals 41. For comparison, note that the expectation of T under H_0 is 59.5.

What to do if both sample sizes are larger than 10? In essence, the approach of Example 25.1 is still possible although it becomes more and more tedious. Fortunately, statisticians have proved that, under H_0, the distribution of T_1 is approximately normal if both n_1 and n_2 are more than 10. Since

$$E(T_1) = \frac{n_1(n + 1)}{2} \text{ and } \sigma_{T_1} = \sqrt{\frac{n_1 n_2 (n + 1)}{12}} \tag{25.2}$$

it follows that

$$Z = \frac{T_1 - E(T_1)}{\sigma_{T_1}} \tag{25.3}$$

is approximately standard normally distributed. The box below summarizes the results.

The Wilcoxon rank sum test and its large-samples version

Below, situation (I) means that $n_1 \leq 10$ and $n_2 \leq 10$; situation (II) that $n_1 > 10$ and $n_2 > 10$.

i Test H_0: the locations of the two distributions are equal, against one of the alternatives (a), (b), (c)

ii Test statistic for (I): $T_1 =$ sum of the ranks of sample 1

Test statistic for (II): $Z = \dfrac{T_1 - E(T_1)}{\sigma_{T_1}}$

iii For (I): (a) reject $H_0 \Leftrightarrow t \leq t_l$ or $t \geq t_u$ (use Table 8 of Appendix A4 with $\alpha/2$)
(b) reject $H_0 \Leftrightarrow t \geq t_u$ (use Table 8 of Appendix A4 with α)
(c) reject $H_0 \Leftrightarrow t \leq t_l$ (use Table 8 of Appendix A4 with α)

For (II): (a) reject $H_0 \Leftrightarrow z \leq -z_{\alpha/2}$ or $z \geq z_{\alpha/2}$
(b) reject $H_0 \Leftrightarrow z \geq z_\alpha$
(c) reject $H_0 \Leftrightarrow z \leq -z_\alpha$

iv Calculate **val**.

v Draw the conclusion.

Example 25.2
The general opinion is that, in the course of time, both partners in more and more couples have decided to work. But is this notion actually true?

In 2006, Statistics Netherlands investigated this for husband–wife couples by comparing the situation of 1995 with the situation of 2001. The data in Table 25.2 were obtained from two independent samples of couples of size (say) 100, one from 1995 and the other from 2001.

Interest is in the ordinal variable $X = $ 'working status of the couple', with the three possible values of the first column. Denote the populations of all husband–wife couples in 1995 and 2001 by 1 and 2, respectively. Figure 25.2 depicts the two frequency distributions of the samples.

Value	Code	1995	2001
Neither works	1	7	4
One works	2	56	42
Both work	3	37	54
Total	–	100	100

TABLE 25.2 Comparison of 1995 and 2001 for husband–wife couples
Source: Statistics Netherlands (2006)

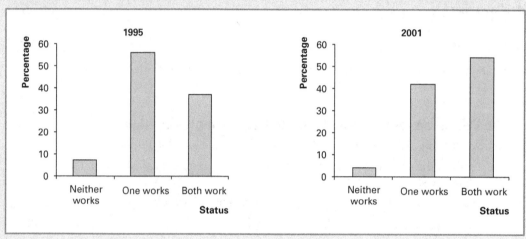

Figure 25.2 Bar charts for 1995 and 2001

These frequency distributions seem to support the conjecture that, in general, X is smaller on the 1995 population than on the 2001 population. But this visual impression is only based on samples. Can it be concluded that the conjecture is structurally valid? This will be tested below, with $\alpha = 0.05$. Since both sample sizes are larger than 10, the large-samples version of Wilcoxon's rank sum test will be used. Here are steps (i)–(iii):

i Test H_0: the locations of the two population distributions are equal, against

H_1: the location of 1 is smaller than the location of 2

ii Test statistic: $Z = \dfrac{T_1 - E(T_1)}{\sigma_{T_1}}$

iii Reject $H_0 \Leftrightarrow z \le -z_{0.05} = -1.6449$

To obtain the *val*, all 200 observations have to be ranked first and the assigned ranks have to be determined; see the table below.

Ordered observation	1	. .	1	2	. .	2	3	. .	3	Total
Size of tie		11			98			91		200
Position	1	. .	11	12	. .	109	110	. .	200	20100
Assigned rank	6	. .	6	60.5	. .	60.5	155	. .	155	20100

For instance, the tie at the coded value 2 has size 98 and is situated on the positions 12, ..., 109. Hence, the assigned rank is the average of these numbers 12, ..., 109; this average is 60.5. The sum of the **sample 1** ranks is equal to:

$7 \times 6 + 56 \times 60.5 + 37 \times 155 = 9165$

Since $E(T_1) = 100 \times 201/2 = 10050$ and $\sigma_{T_1} = \sqrt{100 \times 100 \times 201/12} = 409.2676$, we have:

iv $\quad \textbf{\textit{val}} = \dfrac{9165 - 10050}{409.2676} = -2.1624$

v $\quad$ Reject H_0 since **val** is smaller than -1.6449

Indeed, there is sufficient evidence to conclude at significance level 0.05 that the location of the 1995 population is smaller than the location of the 2001 population.

25.3 Two matched samples

$\mathbf{I}$n the present section the variable X is again considered on two populations and, once more, the objective is to compare the two population locations on the basis of two random samples, one from population 1 and one from population 2. However, now the two samples are **not** independent but instead paired and hence have the same size. Recall that a similar situation was encountered in Section 18.2.2. Interest then was in the difference $\mu_1 - \mu_2$ between the two population means, which was estimated on the basis of the differences between the corresponding paired observations of sample 1 and sample 2. However, in the present section the variable can be ordinal, in which case such differences make no sense. That is why this section has two subsections: one for the case that X is ordinal and one for the case that X is quantitative but non-normal.

25.3.1 Sign test

Starting points: The variable X is **ordinal** and is considered on two populations. The two distributions are assumed to be identical in shape and spread, but may differ in location.
Objective: The comparison of the two locations, on the basis of two paired random samples.

Since the data are ordinal, the differences between the first and the second observations in the pairs are meaningless. However, it **is** possible to decide whether the first observation in a pair is larger than, equal to or smaller than the second. For each pair, a plus ($+$) is recorded when the first observation in the pair is larger than the second and a minus ($-$) when the first observation is smaller than the second. If H_0 is true, it is expected that the number of pluses and the number of minuses are equal. But if alternative (b) (respectively, (c)) is valid, the number of pluses will in general be larger (respectively, smaller) than the number of minuses. That is why the test statistic below will be based on $Y = $ 'the number of pluses'. Furthermore, it is common practice to eliminate the pairs with equal first and second observations, and to denote the number of remaining pairs by n. Things will be organized as such.

Since the two paired samples are random samples, it follows immediately that Y is binomially distributed. The parameters are n and the probability p that within a pair the first observation will be larger than the second. In particular, if H_0 is true, Y is binomially distributed with parameters n and 0.5, and the expectation and standard deviation of Y are $0.5n$ and $\sqrt{0.25n}$, respectively; see (9.3). That is why the standardized variable

$$Z = \frac{Y - 0.5n}{\sqrt{0.25n}}$$

will be used as test statistic. Under H_0 the variable Z is approximately standard normal if the 5-rule is valid, so if $0.5n$ is at least 5 (see (10.20)). That is, under H_0 the variable Z is approximately $N(0, 1)$ distributed if n is at least 10.

From the above arguments the five steps of the test easily follow. The resulting test is called (unsurprisingly) a ***sign test***.

Sign tests for the case that n is at least 10

i Test H_0: the locations of the two distributions are equal, against one of the alternatives (a), (b), (c)

ii Test statistic: $Z = \dfrac{Y - 0.5n}{\sqrt{0.25n}}$

iii For (a): reject $H_0 \Leftrightarrow z \leq -z_{\alpha/2}$ or $z \geq z_{\alpha/2}$
For (b): reject $H_0 \Leftrightarrow z \geq z_{\alpha}$
For (c): reject $H_0 \Leftrightarrow z \leq -z_{\alpha}$

iv Calculate *val*

v Draw the conclusion

Remark: n is the number of the sampled pairs with unequal first and second observations.

Example 25.3

Brewery C has commissioned a study to find out how consumers rate its beer, beer C, against that of its rival Brewery H. Twenty randomly chosen consumers are asked to taste both beer brands (without being told which they are tasting) and to assign one of the ratings 1, 2, 3, 4 or 5 to them. These ratings have the following meaning: 1 = don't like it at all, 2 = don't like it, 3 = indifferent, 4 = like it, 5 = like it very much.

The data are in Table 25.3. The problem objective is to use these data to investigate whether consumers in general give higher ratings to beer C than to beer H.

Respondent	Beer C	Beer H	Respondent	Beer C	Beer H
1	4	3	11	4	3
2	4	5	12	4	2
3	2	2	13	3	3
4	3	1	14	5	4
5	5	4	15	5	2
6	4	4	16	1	2
7	3	3	17	3	1
8	3	4	18	4	2
9	5	4	19	4	3
10	5	5	20	3	1

TABLE 25.3 Ratings for beer C and beer H

Notice that these data are indeed paired. There are 5 pairs with equal ratings, so $n = 15$. There are 12 pairs where the rating for C is larger than the rating for H, so the value of Y is 12. Here are the five steps of the sign test:

i Test H_0: the locations of the two distributions are identical, against

H_1: the location for population C is larger than the location for population H;
$\alpha = 0.02$

ii Test statistic: $Z = \dfrac{Y - 0.5n}{\sqrt{0.25n}}$

iii Reject $H_0 \Leftrightarrow z \geq z_{0.02} = 2.0537$ (∗)

iv $\textbf{val} = \dfrac{12 - 7.5}{1.9365} = 2.3238$

v H_0 is rejected since $\textbf{val}$ is larger than the critical value 2.0537

It can be concluded at significance level 0.02 that, in general, consumers like beer C more than beer H. This conclusion is even more convincing since the p-value equals

$P(Z > 2.3238) = 0.0101$ (∗)

so even a significance level of slightly more than 0.01 would have been enough to obtain the same conclusion.

25.3.2 Wilcoxon's signed rank sum test

Starting points: The variable X is quantitative but (extremely) non-normal, and is considered on two populations. The two distributions are assumed to be identical in shape and spread, but may differ in location.
Objective: The comparison of the two locations, on the basis of two paired random samples.

Since the variable is now quantitative, the differences between the first and the second observations in the pairs **do** make sense. Ignoring this fact and counting only the positive differences (as in the sign test) would mean that not all information is taken into account; this is undesirable and it can be avoided.

Although the t-tests of Section 18.2.2 were also based on the paired differences, interest is now in different situations: the population distribution is **non**-normal and the sample size is not large enough to apply the central limit theorem for the sample mean $\overline{D}$ of the paired differences. This often occurs if the population distribution of the differences has more than one top or is extremely skewed, for instance if the variable of interest has to do with incomes or expenditures. Below, two tests will be constructed, one for the small samples case and one for the large samples case.

Both test statistics will be based on T^+, which results from the following steps:

1 Eliminate all pairs for which the paired differences are 0.
2 Compute all remaining (n) paired differences.
3 Calculate the absolute values of the paired differences, place them in ascending order and assign ranks to them (in the case of a tie, assign the average of the corresponding ranks).
4 Calculate the sum T^+ of the ranks of the positive paired differences.

For testing H_0 against alternative (b), the null will be rejected for relatively large values of T^+; for testing H_0 against alternative (c), the null will be rejected for relatively small values. Again, the question is: how large, how small?

For small samples (with n at most 30), T^+ itself is the test statistic. The critical values of the tests have to be read from a table: Table 9 in Appendix A4. For instance, if 18 of the paired differences are unequal to 0 (so $n = 18$) and the objective is to test H_0 against alternative (c) at significance level 0.05, part B of Table 9 has to be used and H_0 will be rejected for values of T^+ that are at most $t_l = 47$. But if you want to test H_0 against the two-sided alternative (a) at significance level 0.05 on the basis of 18 pairs with paired differences unequal 0, part A of Table 9 yields the critical values 40 and 131.

For large samples (with $n > 30$), statisticians have proved that T^+ is approximately normal with expectation and standard deviation – under H_0 – equal, respectively, to:

$$E(T^+) = \frac{n(n + 1)}{4} \text{ and } \sigma_{T^+} = \sqrt{\frac{n(n + 1)(2n + 1)}{24}}$$ **(25.4)**

(Check for yourself the result for $E(T^+)$.) The following test statistic is used:

$$Z = \frac{T^+ - E(T^+)}{\sigma_{T^+}}$$ **(25.5)**

The box below summarizes the procedures.

Wilcoxon's signed rank sum tests and its large-sample versions

Below, situation (I) means that $n \leq 30$, situation (II) that $n > 30$; here n is the number of the sampled pairs with unequal first and second observations.

  i Test H_0: the locations of the two distributions are equal, against one of the alternatives (a), (b), (c)

  ii Test statistic for (I): T^+ = rank sum of positive differences

    Test statistic for (II): $Z = \frac{T^+ - E(T^+)}{\sigma_{T^+}}$

  iii For (I):   reject $H_0 \Leftrightarrow t \leq t_l$ or $t \geq t_u$ (use Table 9 of Appendix A4 with $\alpha/2$)

               reject $H_0 \Leftrightarrow t \geq t_u$ (use Table 9 of Appendix A4 with α)

               reject $H_0 \Leftrightarrow t \leq t_l$ (use Table 9 of Appendix A4 with α)

    For (II): reject $H_0 \Leftrightarrow z \leq -z_{\alpha/2}$ or $z \geq z_{\alpha/2}$

              reject $H_0 \Leftrightarrow z \geq z_\alpha$

              reject $H_0 \Leftrightarrow z \leq -z_\alpha$

  iv Calculate *val*

  v Draw the conclusion

Example 25.4

In Example 18.4, a paired samples *t*-test was used to find out whether, for the course Statistics 2 of the International Business programme, the second Midterm gives on average higher grades than the Final Exam. The test was conducted on the basis of 30 paired observations; the result was affirmative.

Since the histogram of the 30 differences seems to suggest a bimodal distribution (check it yourself; use Xmp18–04.sav), there is some doubt whether the normality requirement is satisfied or whether 30 is large enough. Hence, Wilcoxon's signed rank sum test will be conducted too. Let the populations 1 and 2 denote the scores for Final Exam and Midterm 2, respectively. In the datasets, D refers to the difference between the score for the Final Exam and the score for Midterm 2.

The file Xmp25–04.xls gives the data after having included the column with the absolute differences under the heading $|D|$ and having ordered the data along the values in that column. Two of the paired differences are 0, so $n = 28$. Table 25.4 includes a column with the (assigned) ranks and a column that indicates whether the paired differences are (yes/no) positive.

Check yourself the assigned ranks, especially in case of ties. Notice that only six of the 28 paired differences are positive, and that the total of the ranks of these positive differences is 74. That is, the realized value of T^+ is 74. Here are the five steps of Wilcoxon's signed rank sum test:

  i Test H_0: the locations of the distributions for the scores of Midterm 2 and the Final Exam are equal, against

H_1:   the location of the distribution of the scores for the Final Exam is smaller than the location of the distribution of the scores for Midterm 2; take $\alpha = 0.05$.

ii Test statistic: T^+

iii Reject $H_0 \Leftrightarrow t \leq t_l = 130$

iv **val** $= 74$

v Reject H_0 in favour of H_1

D	|D|	Position	Rank	Plus? y/n	D	|D|	Position	Rank	Plus? y/n
0.2	0.2	1	1.5	y	−1.3	1.3	15	15	n
−0.2	0.2	2	1.5	n	−1.8	1.8	16	16	n
−0.3	0.3	3	3	n	−1.9	1.9	17	17	n
0.4	0.4	4	4	y	−2.2	2.2	18	18.5	n
−0.5	0.5	5	5.5	n	−2.2	2.2	19	18.5	n
−0.5	0.5	6	5.5	n	2.5	2.5	20	20	y
−0.8	0.8	7	8	n	−2.6	2.6	21	21	n
−0.8	0.8	8	8	n	2.7	2.7	22	22.5	y
−0.8	0.8	9	8	n	−2.7	2.7	23	22.5	n
−0.9	0.9	10	10	n	−2.8	2.8	24	24.5	n
−1	1	11	11	n	−2.8	2.8	25	24.5	n
−1.2	1.2	12	13	n	−3	3	26	26	n
1.2	1.2	13	13	y	−4	4	27	27	n
1.2	1.2	14	13	y	−4.4	4.4	28	28	n

TABLE 25.4 Differences and ranks

The conclusion means that there is evidence (at the 0.05 significance level) that in general the score for Midterm 2 is higher than the score for the Final Exam.

Although n is not larger than 30, we also conduct the large-sample version of the test, to demonstrate the technique. Note that then the test statistic is Z in (25.5) and the rejection region is $(-\infty, -1.6449]$. Since, under H_0, the expectation and the standard deviation of T^+ are, respectively, 203 and 43.9147, the realized value of T^+ follows easily:

$$val = \frac{74 - 203}{43.9147} = -2.9375$$

The conclusion is the same as in the former test.

25.4 Two or more independent samples

Starting points: The variable of interest is ordinal or quantitative non-normal; it is considered on two or more (say k) populations 1, 2, ..., k. The k population distributions are assumed to be identical in shape and spread, but may differ in location.

Objective: The comparison of the k locations of the populations on the basis of k independent random samples with sizes $n_1, ..., n_k$.

In Section 21.5, linear regression with dummy variables has been used to compare the means of a variable of interest on two or more populations, for instance in Section 21.5.3 on one-factor ANOVA. However, the method that was presented is valid only if the variable is quantitative and, preferably, close to normal.

In the current section, the variable of interest is ordinal or quantitative and non-normal. Even if the variable is quantitative and non-normal, it will be treated as an ordinal variable. Instead of considering the means, a non-parametric approach is chosen. Here is the testing problem:

i Test H_0: the k population locations are the same, against

H_1: at least two population locations differ.

(Notice that the case $k = 2$ yields the two-sided testing problem of Section 25.2.)

The test statistic contains terms that are constructed in a way that resembles the construction in Section 25.2. It is based on the following steps:

1 All $n = n_1 + \ldots + n_k$ sample observations of the variable are placed in ascending order (note that this is possible because the variable is at least ordinal).

2 The ranks 1, 2, ..., n are assigned to the ordered observations; if a tie occurs, then assign the mean of the corresponding ranks to all observations in the tie.

3 Calculate the sums $T_1, \ldots, T_k$ of the ranks for the samples 1, ..., k.

4 Use these rank sums $T_1, \ldots, T_k$ as ingredients of the test statistic.

Without giving the details, the test statistic (called W) is just stated here:

ii Test statistic: $\quad W = \left[\dfrac{12}{n(n+1)} \sum_{j=1}^{k} \dfrac{T_j^2}{n_j} \right] - 3(n+1)$ (25.6)

It can be shown that W can only take **non-negative** values. If H_0 is true, the rank sums are similar and – we will not prove it – the value of W will be small. Relatively large values of W support the alternative hypothesis H_1. But what is relatively large? To answer this question, the distribution of W under H_0 is needed. Statisticians have proved that, if all sample sizes n_j are at least 5, this non-negative random variable W is approximately χ^2-distributed with $k - 1$ degrees of freedom. The box gives the five steps of the resulting Kruskal–Wallis test.

Kruskal–Wallis test when all sample sizes $n_1, \ldots, n_k$ are at least 5

i Test H_0: the k population locations are the same, against

H_1: at least two population locations differ

ii Test statistic: $W = \left[\dfrac{12}{n(n+1)} \sum_{j=1}^{k} \dfrac{T_j^2}{n_j} \right] - 3(n+1)$

iii Reject $H_0 \Leftrightarrow w \geq \chi^2_{\alpha;k-1}$

iv Calculate **val**

v Draw the conclusion

It can (but will not) be shown that, for $k = 2$, the large-samples version of Wilcoxon's rank sum test gives, for two-sided testing problems, the same conclusion as this Kruskal–Wallis test. For one-sided testing problems only Wilcoxon's test of Section 25.2 can be used.

Similarly to Section 25.2, it **is** possible to formulate the Kruskal–Wallis test for smaller sample sizes. But we will not do so.

Example 25.5

On the website http://uk.shopping.com, consumer ratings (scale 1, 2, 3, 4, 5) can be found for several products. For instance, automobile GPS receivers are rated. The (ordered) ratings of three TomTom receivers are recorded here:

1 TomTom GO910 GPS receiver: 1, 1, 1, 2, 3, 3, 3, 4, 4, 5, 5, 5, 5, 5, 5
2 TomTom GO510 GPS Receiver: 1, 1, 2, 2, 3, 3, 4, 4, 4, 5, 5
3 TomTom GO300 GPS Receiver: 1, 1, 2, 3, 3, 4, 4, 4, 4, 5, 5, 5, 5, 5

Do these data provide, at the 5% significance level, evidence that consumers tend to rate these GPS receivers differently?

The variable of interest is X = 'consumer rating for the GPS receiver'. Although there is no evidence of it, we do assume that the three samples are random samples and that they are independent. Notice that $n_1 = 15$, $n_2 = 11$ and $n_3 = 14$, which are large enough. Note that $n = 40$. For the Kruskal–Wallis test, we need the values of the three rank sums T_1, T_2 and T_3; see Table 25.5.

Value	Freq.	Positions	Average position	Ranks sample 1	Ranks sample 2	Ranks sample 3
1	7	1, ..., 7	4	3×4	2×4	2×4
2	4	8, ..., 11	9.5	1×9.5	2×9.5	1×9.5
3	7	12, ..., 18	15	3×15	2×15	2×15
4	9	19, ..., 27	23	2×23	3×23	4×23
5	13	28, ..., 40	34	6×34	2×34	5×34
Total	**40**	820	–	$T_1 = 316.5$	$T_2 = 194$	$T_3 = 309.5$

TABLE 25.5 Positions and assigned ranks

For instance, in total, there are seven ratings 3. They are on positions 12, ..., 18, so their average position is 15. The three ratings 3 in sample 1 contribute $3 \times 15 = 45$ to T_1 and the two ratings 3 in both samples 2 and 3 contribute $2 \times 15 = 30$ to both T_2 and T_3.

The values of the rank sums T_1, T_2 and T_3 are just the totals of the last three columns in Table 25.5. From these ingredients the **val** of the test can be calculated:

$$val = \frac{12}{40 \times 41} \times \left(\frac{316.5^2}{15} + \frac{194^2}{11} + \frac{309.5^2}{14} \right) - 3 \times 41 = 0.9641.$$

The complete test goes as follows:

i Test H_0: the three population locations are the same, against

  H_1: at least two population locations differ.

ii Test statistic: $W = \left[\frac{12}{n(n + 1)} \sum_{j=1}^{k} \frac{T_j^2}{n_j} \right] - 3(n + 1)$

iii Reject $H_0 \Leftrightarrow w \geq \chi^2_{0.05;2} = 5.9915$ (*)

iv **val** = 0.9641

v Do not reject H_0

The data do not provide evidence that a difference exists between the opinions of consumers about the three TomTom GPS receivers.

CASE 25.1 BUSINESS START-UPS AND LACK OF CAPITAL – SOLUTION

a It is a homogeneity χ^2 test that is asked for. There are $r = 28$ populations and $c = 4$ possible values of the variable. The test uses 81 degrees of freedom. After clever use of the computer package it follows that **val** = 2508.826 and that, for all cells, the expected frequencies are at least 5. Since the critical value of the test is $\chi^2_{0.05;81} = 103.0095$, it is concluded that not all 28 distributions are the same.

b Five-step procedure:

  i Test H_0: the 28 population locations are the same, against

  H_1: at least two population locations differ

  ii Test statistic: $W = \left[\dfrac{12}{n(n+1)} \sum\limits_{j=1}^{k} \dfrac{T_j^2}{n_j}\right] - 3(n+1)$

  iii Reject $H_0 \Leftrightarrow w \geq \chi^2_{0.05;27} = 40.1133$ (*)

  iv **val** = 1616.4

  v The locations are not all the same

c, d The table summarizes the two sample proportions for 1, 2 and 3, 4.

Country	Proportion 1, 2	Proportion 3, 4
Belgium	0.7715	0.2285
Czech Republic	0.6054	0.3946
Denmark	0.6651	0.3349
Germany	0.7724	0.2276
Estonia	0.6721	0.3279
Greece	0.9205	0.0795
Spain	0.8217	0.1783
France	0.8895	0.1105
Ireland	0.7090	0.2910
Italy	0.8836	0.1164
Cyprus	0.8665	0.1335
Latvia	0.9163	0.0837
Lithuania	0.8337	0.1663
Luxembourg	0.7959	0.2041
Hungary	0.8820	0.1180
Malta	0.8074	0.1926
Netherlands	0.6007	0.3993
Austria	0.6782	0.3218
Poland	0.8573	0.1427
Portugal	0.8962	0.1038
Slovenia	0.8660	0.1340
Slovakia	0.8781	0.1219
Finland	0.5991	0.4009
Sweden	0.7740	0.2260
UK	0.7305	0.2695
Norway	0.6709	0.3291
Iceland	0.5439	0.4561
USA	0.7181	0.2819

Largest: Greece; smallest: Finland.

Summary

The **non-parametric tests** of this chapter are applied to problems where the (only) variable of interest is ordinal or quantitative but non-normal and where the data come from two or more populations. The objective is to compare the locations of the corresponding population distributions. When the data come from two independent samples, **Wilcoxon's rank sum test** can be used to compare the two locations. When the data of the two samples are paired, the **sign test** is used if the data are ordinal and the **Wilcoxon signed rank sum test** if the data are quantitative but non-normal. If two or more independent samples are involved, the **Kruskal–Wallis test** can be used to find differences between the locations.

List of test statistics, distributions and requirements

Name of test	Test statistic	E and SD	Distribution under H_0	Requirements sample sizes
Wilcoxon rank sum test	T_1	$\dfrac{n_1(n+1)}{2}$	Use Table 8 in Appendix A4	$n_1 \le 10, n_2 \le 10$
	$Z = \dfrac{T_1 - E(T_1)}{\sigma_{T_1}}$	$\sqrt{\dfrac{n_1 n_2 (n+1)}{12}}$	$N(0, 1)$	$n_1 > 10, n_2 > 10$
Sign test	$Z = \dfrac{Y - 0.5n}{\sqrt{0.25n}}$		$N(0, 1)$	$n \ge 10^{\dagger}$
Wilcoxon signed rank sum test	T^+	$\dfrac{n(n+1)}{4}$	Use Table 9 in Appendix A4	$n \le 30^{\dagger}$
	$Z = \dfrac{T^+ - E(T^+)}{\sigma_{T^+}}$	$\sqrt{\dfrac{n(n+1)(2n+1)}{24}}$	$N(0, 1)$	$n > 30^{\dagger}$
Kruskal–Wallis test	W of (25.6)		χ^2_{k-1}	All $n_j \ge 5$

† n = number of paired differences unequal to 0

List of symbols

Symbol	Description
T_j	Rank sum of sample j for $j = 1, ..., k$
T^+	Rank sum of positive differences

🔑 Key terms

distribution-free **794**
Kruskal–Wallis test **804**
non-parametric **792**

sign test **799**
Wilcoxon rank sum test **796**

Wilcoxon's signed rank sum test **802**

❓ Exercises

Exercise 25.1

Use Wilcoxon's rank sum test for the data below, to determine whether the location of population 1 is smaller than the location of population 2; use significance level 0.05.

Sample 1:   65   50   63   56   71
Sample 2:   80   62   93   72   68

Exercise 25.2

Use Wilcoxon's rank sum test on the data below to determine whether the two population locations differ; use significance level 0.10.

Sample 1:   10   2   17   15   27   13   21   12
Sample 2:   3   23   13   12   20

Exercise 25.3

Use Wilcoxon's rank sum test on the data below to determine whether the two population locations differ; use significance level 0.05.

Sample 1:   10   2   17   15   27   13   21   12   18   25   30   18
Sample 2:   3   23   13   12   20   15   11   12   5   13   18

Exercise 25.4

In a paired samples experiment with an ordinal variable, 15 minuses, 5 zeros and 30 pluses are found when comparing the first observation in the pairs with the second observation. Perform the sign test (with $\alpha = 0.05$) to determine whether the two population locations are different.

Exercise 25.5

Use the sign test with $\alpha = 0.05$ on the data below (that come from an ordinal variable with coded values 1, 2, 3, 4, 5) to find out whether the location of population 1 is larger than the location of population 2.

Pair: 1   2   3   4   5   6   7   8   9   10   11
Sample 1:   5   3   5   2   3   4   4   5   4   3   5
Sample 2:   3   2   4   3   3   1   3   4   2   5   2

Exercise 25.6

Use Wilcoxon's signed rank sum test and the respective values 890 and 60 of the statistics T^+ and n to determine, at the 5% significance level, whether the two population locations are different.

Exercise 25.7

Conduct Wilcoxon's signed rank sum test on the data below to find out whether the location of population 1 is larger than the location of population 2; use 0.05 as significance level.

Pair: 1   2   3   4   5   6   7   8   9   10
Sample 1:   10   12   15   12   15   11   11   18   12   17
Sample 2:   11   10   14   10   10   12   8   15   10   17

Exercise 25.8

Use the Kruskal–Wallis test with $\alpha = 0.05$ and the statistics in the table below to determine whether the four population locations are not all the same.

i	Sample size n_i	Rank sum T_i
1	15	320
2	20	530
3	5	120
4	10	305

Exercise 25.9

Use the Kruskal–Wallis test (with $\alpha = 0.10$) and the data below to determine whether there is enough evidence to conclude that at least two of the three population locations are different.

Sample 1:   20   15   18   25   18   22   20
Sample 2:   18   10   15   15   19   20   15
Sample 3:   15   20   19   15   15   10

Exercise 25.10

The human resource manager of a large company believes that among the university-educated personnel of the company, the business graduates tend to quit their jobs sooner than the non-business graduates. To check her belief, she randomly selects thirteen university-educated employees hired precisely four years ago (seven business and six non-business graduates) and records the numbers of months they have worked for the company. The table contains the data.

| Business | 16 | 14 | 48 | 20 | 35 | 12 | 20 |
| Non-business | 35 | 48 | 28 | 48 | 34 | 45 | |

a Explain why the data are non-normal.

b Check the belief of the human resource manager at the 5% significance level.

Exercise 25.11

A manufacturer of detergents advertises that the recently renewed formula of the detergent Colour Line has really improved the product. To find out whether the consumers agree with that claim, 300 consumers are asked to give their opinion about the quality of Colour Line. The results are compared with the results of a similar study (a few years ago) among 300 consumers about the old formula of the product. For both studies, the consumers were asked the question: Do you agree that Colour Line has great washing power and is kind to colours? Possible answers:

1 I strongly agree
2 I agree
3 indifferent
4 I do not agree
5 I do not agree at all

a Is the variable ordinal or quantitative?

b Which type of experimental design is used: independent samples or matched samples?

The table below contains summarized results of the two samples.

Value	1	2	3	4	5	Total
% old formula	25	35	20	10	10	100
% new formula	30	38	20	7	5	100

c Can it be concluded that, at the 2% significance level, the new formula has improved Colour Line?

d Find for part (c) the smallest significance level for which the test procedure allows the conclusion that the new formula has improved the product.

Exercise 25.12

Reconsider Exercise 25.11. Suppose now that 300 consumers are asked to wash with **both** formulas of Colour Line and to compare them by assigning one of the codes 1 to 5 (as in Exercise 25.11) to each of the formulas. After having summarized the data it turns out that 40% prefer the new formula, 30% prefer the old formula and 30% are indifferent. Answer the same questions as in Exercise 25.11.

Exercise 25.13

On a course evaluation form, students answer the following question (among others) by assigning a grade 1, 2, 3, 4 or 5: Is the content of the course intellectually challenging? The table contains the results for 2006 and 2007 of the course Statistics 1 for International Business.

Grade	1	2	3	4	5
Year 2006	11	14	51	45	5
Year 2007	2	4	11	26	15

a Calculate the expectation and the standard deviation of Wilcoxon's rank sum test if the two locations of the variable 'grade for challenge' are the same.

b Test whether, as far as the objective of challenging the students is concerned, the 2007 version of Statistics 1 is evaluated more highly than the 2006 version. Use $\alpha = 0.05$.

Exercise 25.14 (computer)

Do households in general spend more on housing than on food? You are asked to answer that question for a large population of households. For your analysis you make use of the dataset Xrc25–14.sav that contains, for a random sample of 25 households, the annual expenditures ($\times$ €1000) on housing and food. Since the sample size is not very large and since variables that measure expenses are generally non-normal, you are invited to use a distribution-free test, with $\alpha = 0.05$.

Exercise 25.15 (computer)

Answer the question in Exercise 25.14 again, but now on the basis of a random sample of 40 households; see Xrc25–15.sav. Again, use a non-parametric test with $\alpha = 0.05$.

Exercise 25.16 (computer)

For the same population of households as in Exercise 25.14, the annual expenditures ($\times$ €1000) on clothing and recreation per household have been on the same level for years. However, for last year the conjecture is that this situation has changed. Check this conjecture statistically on the basis of a random sample of 100 households; see Xrc25–16.sav. Use a non-parametric test with $\alpha = 0.05$.

Exercise 25.17

In the Eurobarometer report *Undeclared Work in the European Union* (European Commission, 2007), the role of the 'black economy' is studied in all EU countries. One of the questions put (to people willing to answer this precarious question) was: Have you, in the past 12 months, acquired any goods (supposedly) stemming from undeclared work? The frequencies for Belgium and Denmark are in the table.

	No	Yes	Total
Belgium	926	83	1009
Denmark	846	141	987

In this exercise, the problem of interest is whether the Belgian and Danish people react differently to the above question.

a Describe the populations and the variable.

b Use a distribution-free test (with $\alpha = 0.05$) to draw a conclusion about the problem.

Exercise 25.18

Reconsider Exercise 25.17. The problem of interest can also be tackled with two other tests. Again, use significance level 0.05

a Use a χ^2 test to draw a conclusion about the problem.

b Use a test that compares the proportions responding Yes in the two countries.

Exercise 25.19 (computer)

Is it true that the education level of older people in general is lower than the education level of younger people? To answer that question for a large population of adults, the (ordinal) variable $X =$ 'education level' (with values 1 (low), ..., 5 (high)) is observed for three random samples of people, one from each of the age categories 30–<40, 40–<55 and 55+. See the dataset Xrc25–19.sav. Use a suitable hypothesis test with significance level 0.05 to find out whether the locations of the three populations are different. For the calculations, use your computer in a clever way.

Exercise 25.20 (computer)

Interest is in the three populations of people with education levels 1, 2 and 3, within a large population of adults. Can structural differences be detected with respect to the variable $X =$ 'number of hours worked weekly'? To find at least a partial answer to that question, the accompanying three locations of the (non-normal) variable X will be compared on the basis of three random samples, one from each population; see the dataset Xrc25–20.sav. Use a hypothesis test with significance level 0.02 to find out whether the locations of the three populations differ. For the calculations, use your computer in a clever way.

Exercise 25.21

To what extent are the educational systems in the EU25 countries preparing their citizens to become entrepreneurs? In a recent study of the European Commission about entrepreneurship, people from 28 countries were asked to rate the statement

My school education made me interested to become an entrepreneur

with possible answers 1, 2, 3 or 4. Here, 1 = strongly agree, 2 = agree, 3 = disagree and 4 = strongly disagree. The table contains the results for six countries.

Country	Rate				Total
	1	**2**	**3**	**4**	**Total**
Belgium	113	237	268	337	955
Denmark	18	82	262	128	490
Germany	58	160	391	375	984
Netherlands	46	159	501	263	969
Norway	31	111	183	152	477
Sweden	26	99	160	208	493

a   Use a χ^2 test with significance level 0.05 to find out whether the distributions of the answers to this question are not the same for all six countries.

b   Are there differences among the six locations?

c   Which country is most negative about the educational role with respect to entrepreneurship?

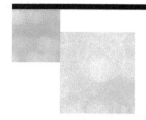

Appendix A1
Excel and SPSS (Internet)

The website for the book provides you with guidance and advice on how to use Excel for your statistics module, and also how to use the statistical software package SPSS.

To access this free resource, go to:

www.mheducation.co.uk/textbooks/nieuwenhuis

The Excel and SPSS instructions can be found on the website in the Student Centre of the Online Learning Centre.

Appendix A2
Summation operator Σ

The summation symbol Σ (pronounced sigma; it is the Greek capital S) is used to shorten complex or long notations involving summations. For instance, suppose that the variable X measures the gross annual income of an adult (in units of €1000). Consider a sample of 5 adults (numbered as 1, 2, ..., 5) and let x_1, x_2, x_3, x_4, x_5 be their (gross annual) incomes. That is, x_1 measures this income for person 1 in the sample, x_2 for person 2, and so on. This is briefly expressed as follows: x_i measures the income of person i in the sample, where i goes from 1 to 5. The total income of the five persons, i.e. $x_1 + x_2 + x_3 + x_4 + x_5$, is shortly denoted as:

$$\sum_{i=1}^{5} x_i$$

Here Σ has to be read as the statement 'this is the sum of' or 'here we add'; add the x_i, starting with $i = 1$ and ending with $i = 5$. For $x_1 = 44$, $x_2 = 47$, $x_3 = 43$, $x_4 = 71$, $x_5 = 61$ we obtain:

$$\sum_{i=1}^{5} x_i = 44 + 47 + 43 + 71 + 61 = 266$$

Hence, the total gross annual income of the persons in the sample is €266 000. But then the mean gross annual income of the persons in the sample is equal to €53 200:

$$\frac{1}{5}\sum_{i=1}^{5} x_i = \frac{266}{5} = 53.2$$

As you will find out in this book, sums of squares are important in statistical theory. Let us have a look at the total of the **squared** incomes of the five persons in the sample. So, we **first** have to take the squares and **next** add the results. That is:

$$\sum_{i=1}^{5} x_i^2 = x_1^2 + x_2^2 + x_3^2 + x_4^2 + x_5^2 = 44^2 + 47^2 + 43^2 + 71^2 + 61^2 = 14756$$

Notice that the order of the operations 'take squares' of the x_i and 'add' the results cannot be changed. If we first add the x_i and next take the square, the answer is different. You might remember from secondary school that $3^2 + 4^2 \neq (3 + 4)^2$. More general:

$$\sum_{i=1}^{5} x_i^2 \neq (\sum_{i=1}^{5} x_i)^2$$

The summation sign can also be used in cases where no measured variable is involved, like in the following situations:

$$\sum_{i=1}^{7} i = 1 + 2 + 3 + 4 + 5 + 6 + 7 = 28$$

Here, the numbers i are summed, starting with $i = 1$ and ending with $i = 7$. In the summation

$$\sum_{i=1}^{7} i^2 = 1^2 + 2^2 + \ldots + 7^2 = 140$$

the squares of the numbers i are summed, starting with $i = 1$ and ending with $i = 7$.

In a more statistical context, X is some variable and interest is in a sample of n population elements (numbered as 1, 2, ..., n). Then x_1 denotes the value of X measured at element 1, x_2 denotes the value measured at element 2, and so on. We write

$$\sum_{i=1}^{n} x_i \quad \text{and} \quad \frac{1}{n}\sum_{i=1}^{n} x_i^2$$

for the sum $x_1 + x_2 + \ldots + x_n$ of the measurements and the sum of squares $x_1^2 + x_2^2 + \ldots + x_n^2$, respectively.

In the above expressions, i was used to describe the elements at which the variable X was measured. Of course, we might equally well have used j (or t, or k) to denote the objects. (What's in a name?) That is:

$$\sum_{i=1}^{n} x_i = \sum_{j=1}^{n} x_j = \sum_{t=1}^{n} x_t$$

Sometimes, the summed x_i all have equal values. For instance, if $X =$ 'number of children per family' and the measurements x_1, x_2, x_3, x_4 are all equal to 2, then

$$\sum_{i=1}^{4} x_i = \sum_{i=1}^{4} 2 = 2 + 2 + 2 + 2 = 4 \times 2 = 8$$

In a more general setting with a being a constant (that is, it does not depend on the summation index i), we have:

$$\sum_{i=1}^{n} a = na \quad \text{and} \quad \sum_{i=1}^{n} a^2 = na^2 \tag{A2.1}$$

That is, when summing n numbers that are all equal to a, the result is na; the mean of n numbers that are all equal to a, is a.

Example A2.1 Let X denote the weekly sales (in units of $100\,000) of a supermarket in the USA that belongs to a European chain of supermarkets. During the most recent eight weeks, the observations $x_1, x_2, \ldots, x_8$ of X were:

12.0   13.2   11.1   14.6   13.9   12.3   14.8   11.9

Regarding the total of the sales in this period and the mean sales per week, it follows that

$$\sum_{i=1}^{8} x_i = 12.0 + 13.2 + \ldots + 11.9 = 103.8 \quad \text{and} \quad \frac{1}{8}\sum_{i=1}^{8} x_i = 12.9750 \approx 13.0,$$

respectively. Both quantities are measured in units of $100\,000.

However, the management of this European chain wants to have the total sales during these eight weeks in units of $100\,000$ **euro**. Assume that the exchange rate of €1 is $1.25.

Note, first, that the equivalent of $1 is $1 : 1.25 =$ €0.8. The easiest way to obtain the total sales of the supermarket in the currency the management wants it, is to multiply the above total by 0.8. This leads to

$$0.8 \times \sum_{i=1}^{8} x_i = 0.8 \times 103.8 = 83.04$$

for the total sales in units of €$100\,000$. Notice that we **first** summed the x_i and **next** multiplied by 0.8.

However, there is also another method that gives the same answer. This method **first** multiplies all individual observations x_i by 0.8 and **next** adds the results. Multiplying the values by 0.8 leads to

9.60   10.56   8.88   11.68   11.12   9.84   11.84   9.52

Summing these numbers yields

$$\sum_{i=1}^{8} 0.8x_i = 83.04$$

Consequently,

$$\sum_{i=1}^{8} 0.8x_i = 0.8 \times \sum_{i=1}^{8} x_i.$$

It follows that the factor 0.8 can be placed in front of the summation sign Σ. Of course, this result remains valid if 0.8 is replaced by another constant a; think of the exchange rate being a instead of 0.8. Apparently, summing the values ax_i gives the same result as first summing the x_i and next multiplying the result by the constant a.

Often, one is interested in more than one variable. If X measures annual household expenses on food and clothing, and Y measures annual household expenses on housing and recreation, then one might be interested in Z, the total of these annual expenditures per household. That is, $Z = X + Y$. Suppose that the households 1, 2, ..., n are sampled and that the values x_i, y_i, and z_i of X, Y, and Z for each household i are observed. Then of course the total of the annual household expenses of these n households (that is, $\sum_{i=1}^{n} z_i$) is equal to the sum of their total household expenses on food/clothing (that is, $\sum_{i=1}^{n} x_i$) and their total household expenses on housing/recreation (that is, $\sum_{i=1}^{n} y_i$).

Example A2.2 In Baker Street, life is simple. Its five households are confronted with only two types of expenditures: expenditures on food/clothing and expenditures on housing/recreation. The households are numbered $i = 1, ..., 5$. Their respective annual household expenditures $x_1, ..., x_5$ on food/clothing and their respective expenditures $y_1, ..., y_5$ on housing/recreation are given below. The respective total annual household expenditures are denoted by $z_1, ..., z_5$.

	1	2	3	4	5
x_i	5620	21370	22180	9770	8810
y_i	6730	23960	25920	12430	10070

TABLE. Household expenditures in Baker Street

Interest is in the total annual household expenditures of the five households in Baker Street. That is, interest is in $\sum_{i=1}^{5} z_i$. Below, two ways are considered to find the answer, which of course is the total of all numbers in the inner part of the above table.

One way is first to determine the $z_i = x_i + y_i$, the totals of the expenses for the five households: 12350, 45330, 48100, 22200, 18880. Next, add these five results, to find 146860. Notice that this method first sums the five columns in the inner part of the table and next sums the results. Another way is to determine first the totals of each of the two rows and then to add the two totals. That is, first determine $\sum_{i=1}^{5} x_i$ and $\sum_{i=1}^{5} y_i$ (yielding 67750 and 79110, respectively), and add them. The answer is of course the same as above. Consequently, with n equal to 5,

$$\sum_{i=1}^{n} (x_i + y_i) = \sum_{i=1}^{n} x_i + \sum_{i=1}^{n} y_i \tag{A2.2}$$

The table below illustrates the two calculations:

	Households					
	1	2	3	4	5	Total
Exp. food/clothing	x_1	x_2	x_3	x_4	x_5	Σx_i
Exp. housing/recr.	y_1	y_2	y_3	y_4	y_5	Σy_i
Total	$x_1 + y_1$	$x_2 + y_2$	$x_3 + y_3$	$x_4 + y_4$	$x_5 + y_5$	$\Sigma(x_i + y_i) = \Sigma x_i + \Sigma y_i$

TABLE. Two ways to determine the total expenditures

That is, the summation of the sample observations $x_i + y_i$ is the same as the sum of the summation of the x_i and the summation of the y_i.

In view of the above results, the following property of the summation operator Σ is intuitively obvious. Here x_i and y_i are observations of variables X and Y respectively, and a, b and c are constants (that is, do not depend on the summation index).

Property of the summation operator Σ

$$\sum_{i=1}^{n}(ax_i + by_i + c) = a\sum_{i=1}^{n}x_i + b\sum_{i=1}^{n}y_i + nc \tag{A2.3}$$

It is not difficult to prove the overall validity of this result. It follows immediately by writing the summations out, reorganizing the results and placing constants outside the brackets:

$$
\begin{aligned}
\sum_{i=1}^{n}(ax_i + by_i + c) &= (ax_1 + by_1 + c) + (ax_2 + by_2 + c) + \ldots + (ax_n + by_n + c)\\
&= (ax_1 + ax_2 + \ldots + ax_n) + (by_1 + by_2 + \ldots + by_n) + nc\\
&= a(x_1 + x_2 + \ldots + x_n) + b(y_1 + y_2 + \ldots + y_n) + nc\\
&= a\sum_{i=1}^{n}x_i + b\sum_{i=1}^{n}y_i + nc
\end{aligned}
$$

from which (A2.3) follows. Notice that the choice $a = 1$, $b = 1$ and $c = 0$ returns (A2.2); the choice $a = 0$, $b = 0$ and c replaced by a, returns (A2.1). The choice $a = 1$, $b = 0$ and $c = 0$ yields:

$$\sum_{i=1}^{n}ax_i = a\sum_{i=1}^{n}x_i \tag{A2.4}$$

That is, the constant a can be put in front of Σ. This rule was already mentioned at the end of Example A2.1.

Example A2.3 For the production of a certain good, two types of raw material (I and II) are needed. In all months the produced amounts of the good are the same. However, the prices of the raw materials vary on a monthly basis. For the (fixed) monthly production of the good, 120 units of raw material I and 90 units of raw material II are needed, while the monthly fixed costs are €26 500. If it is additionally given that, over the past 12 months, the means of the monthly prices (in euro) of one unit of the two raw materials were 1900 and 3150, respectively, then determine the total production costs of the good during the last year.

Denote the most recent 12 months by $t = 1, \ldots, 12$. Let the prices (in euro) of one unit of the raw materials I and II in month t be denoted by x_t and y_t, respectively, and let z_t be the cost of the total production of the good in that month. Note that we are interested in $\sum_{t=1}^{12}z_t$. In month t the costs regarding the raw materials I and II are $120x_t$ and $90y_t$, respectively. Hence, z_t is equal to $120x_t + 90y_t + 26\,500$ and

$$\sum_{t=1}^{12}z_t = \sum_{t=1}^{12}(120x_t + 90y_t + 26\,500) = 120\sum_{t=1}^{12}x_t + 90\sum_{t=1}^{12}y_t + 12 \times 26\,500$$

(The last equality follows from the above summation rules.) Since the mean of the monthly prices of raw materials I and II over the past 12 months were 1900 and 3150, it follows that:

$$\sum_{t=1}^{12}x_t = 12 \times 1900 = 22\,800 \quad \text{and} \quad \sum_{t=1}^{12}y_t = 12 \times 3150 = 37\,800$$

Consequently,

$$\sum_{t=1}^{12}z_t = 120 \times 22\,800 + 90 \times 37\,800 + 12 \times 26\,500 = €6\,456\,000$$

The total of the production costs of the good during the past 12 months was €6.456 million.

At the end of this subsection, another frequently used operator is studied: the **absolute value operator** $|x|$. It transforms negative numbers into their positive counterparts and leaves non-negative numbers unchanged. For instance, the absolute values of -3, 4.2 and 0 are equal to 3, 4.2 and 0, respectively. Notation:

$$|-3| = 3; \quad |4.2| = 4.2; \quad |0| = 0$$

In statistics, the absolute value operator is often used to determine distances. This is because the distance between two numbers a and b is simply equal to the absolute value of $a - b$, that is, it is equal to $| a - b |$. The following example applies this.

Example A2.4 Recall Example A2.1, in which, for a supermarket in the USA, the variable $X =$ 'weekly sales (in units of \$100 000)' was measured during each of the most recent eight weeks. Among others, it was calculated that the mean value of X during these eight weeks was equal to 12.975, which corresponds to \$1.2975 million. The question now arises how far the eight weekly observations are separated from this mean value. That is: what are the distances between these eight weekly sales and the mean 12.9750?

The first observation x_1 is equal to 12.0, while the distance between x_1 and the mean value is:

$$| x_1 - 12.9750 | = | -0.9750 | = 0.9750$$

So, x_1 is 0.9750 units separated from the mean value. Similarly:

$$| x_2 - 12.9750 | = | 13.2 - 12.9750 | = | 0.2250 | = 0.2250$$

And so on. All eight distances are reported below:

  0.9750   0.2250   1.8750   1.6250   0.9250   0.6750   1.8250   1.0750

This answers the question above and immediately raises a new one: what is the mean distance?

Note that this question is about the mean of the distances between the eight observations and their average. That is, the mean of the distances

$$| x_1 - 12.9750 |, \quad | x_2 - 12.9750 |, \ldots, \quad | x_8 - 12.9750 |$$

is asked for. But these eight distances have already been calculated above and their mean is easily found to be 1.1500. To say it in a more formal way:

$$\frac{1}{8} \sum_{i=1}^{8} |x_i - 12.9750| = 1.1500$$

This answer corresponds to \$115 000.

The summation operator and the absolute value operator are also present as functions in Excel, under **Paste Function**, category **Math & Trig**. These functions, **SUM** and **ABS**, can be used in the following exercise.

Exercise A2.1
The file XrcA2–01.xls contains the weekly gross sales x_i (in units of €1000) of the 300 shops of a large clothing chain.

  a Calculate the total sales of the chain.
  b Use the answer in part (a) to determine the mean gross sales per shop.
  c For each shop, the weekly sales are compared to this mean by calculating (in column B) of the datafile the absolute values of the deviations $x_i -$ mean.
  d Determine the sum of all absolute deviations.
  e Also calculate the sum of all squares x_i^2.

Appendix A3
Greek letters

The following Greek letters are used in this book:

α	alpha
β	beta
γ	gamma
δ	delta
ε	epsilon
θ	theta
κ	kappa
λ, Λ	lambda
μ	mu
ν	nu
ξ	xi
π	pi
ρ	rho
σ, Σ	sigma
τ	tau
Φ	capital phi
χ	chi
ω, Ω	omega

Appendix A4

p		0.05	0.10	0.15	0.20	0.25	0.30	0.35	0.40	0.45	0.50
n	y										
2	0	0.9025	0.8100	0.7225	0.6400	0.5625	0.4900	0.4225	0.3600	0.3025	0.2500
	1	0.9975	0.9900	0.9775	0.9600	0.9375	0.9100	0.8775	0.8400	0.7975	0.7500
	2	1.0000	1.0000	1.0000	1.0000	1.0000	1.0000	1.0000	1.0000	1.0000	1.0000
3	0	0.8574	0.7290	0.6141	0.5120	0.4219	0.3430	0.2746	0.2160	0.1664	0.1250
	1	0.9928	0.9720	0.9393	0.8960	0.8438	0.7840	0.7183	0.6480	0.5748	0.5000
	2	0.9999	0.9990	0.9966	0.9920	0.9844	0.9730	0.9571	0.9360	0.9089	0.8750
	3	1.0000	1.0000	1.0000	1.0000	1.0000	1.0000	1.0000	1.0000	1.0000	1.0000
4	0	0.8145	0.6561	0.5220	0.4096	0.3164	0.2401	0.1785	0.1296	0.0915	0.0625
	1	0.9860	0.9477	0.8905	0.8192	0.7383	0.6517	0.5630	0.4752	0.3910	0.3125
	2	0.9995	0.9963	0.9880	0.9728	0.9492	0.9163	0.8735	0.8208	0.7585	0.6875
	3	1.0000	0.9999	0.9995	0.9984	0.9961	0.9919	0.9850	0.9744	0.9590	0.9375
	4		1.0000	1.0000	1.0000	1.0000	1.0000	1.0000	1.0000	1.0000	1.0000
5	0	0.7738	0.5905	0.4437	0.3277	0.2373	0.1681	0.1160	0.0778	0.0503	0.0313
	1	0.9774	0.9185	0.8352	0.7373	0.6328	0.5282	0.4284	0.3370	0.2562	0.1875
	2	0.9988	0.9914	0.9734	0.9421	0.8965	0.8369	0.7648	0.6826	0.5931	0.5000
	3	1.0000	0.9995	0.9978	0.9933	0.9844	0.9692	0.9460	0.9130	0.8688	0.8125
	4		1.0000	0.9999	0.9997	0.9990	0.9976	0.9947	0.9898	0.9815	0.9688
	5			1.0000	1.0000	1.0000	1.0000	1.0000	1.0000	1.0000	1.0000
6	0	0.7351	0.5314	0.3771	0.2621	0.1780	0.1176	0.0754	0.0467	0.0277	0.0156
	1	0.9672	0.8857	0.7765	0.6554	0.5339	0.4202	0.3191	0.2333	0.1636	0.1094
	2	0.9978	0.9842	0.9527	0.9011	0.8306	0.7443	0.6471	0.5443	0.4415	0.3438
	3	0.9999	0.9987	0.9941	0.9830	0.9624	0.9295	0.8826	0.8208	0.7447	0.6563
	4	1.0000	0.9999	0.9996	0.9984	0.9954	0.9891	0.9777	0.9590	0.9308	0.8906
	5		1.0000	1.0000	0.9999	0.9998	0.9993	0.9982	0.9959	0.9917	0.9844
	6				1.0000	1.0000	1.0000	1.0000	1.0000	1.0000	1.0000
7	0	0.6983	0.4783	0.3206	0.2097	0.1335	0.0824	0.0490	0.0280	0.0152	0.0078
	1	0.9556	0.8503	0.7166	0.5767	0.4449	0.3294	0.2338	0.1586	0.1024	0.0625
	2	0.9962	0.9743	0.9262	0.8520	0.7564	0.6471	0.5323	0.4199	0.3164	0.2266
	3	0.9998	0.9973	0.9879	0.9667	0.9294	0.8740	0.8002	0.7102	0.6083	0.5000
	4	1.0000	0.9998	0.9988	0.9953	0.9871	0.9712	0.9444	0.9037	0.8471	0.7734
	5		1.0000	0.9999	0.9996	0.9987	0.9962	0.9910	0.9812	0.9643	0.9375
	6			1.0000	1.0000	0.9999	0.9998	0.9994	0.9984	0.9963	0.9922
	7					1.0000	1.0000	1.0000	1.0000	1.0000	1.0000
8	0	0.6634	0.4305	0.2725	0.1678	0.1001	0.0576	0.0319	0.0168	0.0084	0.0039
	1	0.9428	0.8131	0.6572	0.5033	0.3671	0.2553	0.1691	0.1064	0.0632	0.0352
	2	0.9942	0.9619	0.8948	0.7969	0.6785	0.5518	0.4278	0.3154	0.2201	0.1445
	3	0.9996	0.9950	0.9786	0.9437	0.8862	0.8059	0.7064	0.5941	0.4770	0.3633
	4	1.0000	0.9996	0.9971	0.9896	0.9727	0.9420	0.8939	0.8263	0.7396	0.6367
	5		1.0000	0.9998	0.9988	0.9958	0.9887	0.9747	0.9502	0.9115	0.8555
	6			1.0000	0.9999	0.9996	0.9987	0.9964	0.9915	0.9819	0.9648
	7				1.0000	1.0000	0.9999	0.9998	0.9993	0.9983	0.9961
	8						1.0000	1.0000	1.0000	1.0000	1.0000

TABLE A4.1 Distribution functions (cdfs) of binomial distributions: $F(y)$ for $Y \sim Bin(n, p)$

	p	0.05	0.10	0.15	0.20	0.25	0.30	0.35	0.40	0.45	0.50
n	y										
9	0	0.6302	0.3874	0.2316	0.1342	0.0751	0.0404	0.0207	0.0101	0.0046	0.0020
	1	0.9288	0.7748	0.5995	0.4362	0.3003	0.1960	0.1211	0.0705	0.0385	0.0195
	2	0.9916	0.9470	0.8591	0.7382	0.6007	0.4628	0.3373	0.2318	0.1495	0.0898
	3	0.9994	0.9917	0.9661	0.9144	0.8343	0.7297	0.6089	0.4826	0.3614	0.2539
	4	1.0000	0.9991	0.9944	0.9804	0.9511	0.9012	0.8283	0.7334	0.6214	0.5000
	5		0.9999	0.9994	0.9969	0.9900	0.9747	0.9464	0.9006	0.8342	0.7461
	6		1.0000	1.0000	0.9997	0.9987	0.9957	0.9888	0.9750	0.9502	0.9102
	7				1.0000	0.9999	0.9996	0.9986	0.9962	0.9909	0.9805
	8					1.0000	1.0000	0.9999	0.9997	0.9992	0.9980
	9							1.0000	1.0000	1.0000	1.0000
10	0	0.5987	0.3487	0.1969	0.1074	0.0563	0.0282	0.0135	0.0060	0.0025	0.0010
	1	0.9139	0.7361	0.5443	0.3758	0.2440	0.1493	0.0860	0.0464	0.0233	0.0107
	2	0.9885	0.9298	0.8202	0.6778	0.5256	0.3828	0.2616	0.1673	0.0996	0.0547
	3	0.9990	0.9872	0.9500	0.8791	0.7759	0.6496	0.5138	0.3823	0.2660	0.1719
	4	0.9999	0.9984	0.9901	0.9672	0.9219	0.8497	0.7515	0.6331	0.5044	0.3770
	5	1.0000	0.9999	0.9986	0.9936	0.9803	0.9527	0.9051	0.8338	0.7384	0.6230
	6		1.0000	0.9999	0.9991	0.9965	0.9894	0.9740	0.9452	0.8980	0.8281
	7			1.0000	0.9999	0.9996	0.9984	0.9952	0.9877	0.9726	0.9453
	8				1.0000	1.0000	0.9999	0.9995	0.9983	0.9955	0.9893
	9						1.0000	1.0000	0.9999	0.9997	0.9990
	10								1.0000	1.0000	1.0000
11	0	0.5688	0.3138	0.1673	0.0859	0.0422	0.0198	0.0088	0.0036	0.0014	0.0005
	1	0.8981	0.6974	0.4922	0.3221	0.1971	0.1130	0.0606	0.0302	·0.0139	0.0059
	2	0.9848	0.9104	0.7788	0.6174	0.4552	0.3127	0.2001	0.1189	0.0652	0.0327
	3	0.9984	0.9815	0.9306	0.8389	0.7133	0.5696	0.4256	0.2963	0.1911	0.1133
	4	0.9999	0.9972	0.9841	0.9496	0.8854	0.7897	0.6683	0.5328	0.3971	0.2744
	5	1.0000	0.9997	0.9973	0.9883	0.9657	0.9218	0.8513	0.7535	0.6331	0.5000
	6		1.0000	0.9997	0.9980	0.9924	0.9784	0.9499	0.9006	0.8262	0.7256
	7			1.0000	0.9998	0.9988	0.9957	0.9878	0.9707	0.9390	0.8867
	8				1.0000	0.9999	0.9994	0.9980	0.9941	0.9852	0.9673
	9					1.0000	1.0000	0.9998	0.9993	0.9978	0.9941
	10							1.0000	1.0000	0.9998	0.9995
	11									1.0000	1.0000
12	0	0.5404	0.2824	0.1422	0.0687	0.0317	0.0138	0.0057	0.0022	0.0008	0.0002
	1	0.8816	0.6590	0.4435	0.2749	0.1584	0.0850	0.0424	0.0196	0.0083	0.0032
	2	0.9804	0.8891	0.7358	0.5583	0.3907	0.2528	0.1513	0.0834	0.0421	0.0193
	3	0.9978	0.9744	0.9078	0.7946	0.6488	0.4925	0.3467	0.2253	0.1345	0.0730
	4	0.9998	0.9957	0.9761	0.9274	0.8424	0.7237	0.5833	0.4382	0.3044	0.1938
	5	1.0000	0.9995	0.9954	0.9806	0.9456	0.8822	0.7873	0.6652	0.5269	0.3872
	6		0.9999	0.9993	0.9961	0.9857	0.9614	0.9154	0.8418	0.7393	0.6128
	7		1.0000	0.9999	0.9994	0.9972	0.9905	0.9745	0.9427	0.8883	0.8062
	8			1.0000	0.9999	0.9996	0.9983	0.9944	0.9847	0.9644	0.9270
	9				1.0000	1.0000	0.9998	0.9992	0.9972	0.9921	0.9807
	10						1.0000	0.9999	0.9997	0.9989	0.9968
	11							1.0000	1.0000	0.9999	0.9998
	12									1.0000	1.0000

TABLE A4.1 (Continued) Distribution functions (cdfs) of binomial distributions: $F(y)$ for $Y \sim Bin(n, p)$

	p	0.05	0.10	0.15	0.20	0.25	0.30	0.35	0.40	0.45	0.50
n	y										
13	0	0.5133	0.2542	0.1209	0.0550	0.0238	0.0097	0.0037	0.0013	0.0004	0.0001
	1	0.8646	0.6213	0.3983	0.2336	0.1267	0.0637	0.0296	0.0126	0.0049	0.0017
	2	0.9755	0.8661	0.6920	0.5017	0.3326	0.2025	0.1132	0.0579	0.0269	0.0112
	3	0.9969	0.9658	0.8820	0.7473	0.5843	0.4206	0.2783	0.1686	0.0929	0.0461
	4	0.9997	0.9935	0.9658	0.9009	0.7940	0.6543	0.5005	0.3530	0.2279	0.1334
	5	1.0000	0.9991	0.9925	0.9700	0.9198	0.8346	0.7159	0.5744	0.4268	0.2905
	6		0.9999	0.9987	0.9930	0.9757	0.9376	0.8705	0.7712	0.6437	0.5000
	7		1.0000	0.9998	0.9988	0.9944	0.9818	0.9538	0.9023	0.8212	0.7095
	8			1.0000	0.9998	0.9990	0.9960	0.9874	0.9679	0.9302	0.8666
	9				1.0000	0.9999	0.9993	0.9975	0.9922	0.9797	0.9539
	10					1.0000	0.9999	0.9997	0.9987	0.9959	0.9888
	11						1.0000	1.0000	0.9999	0.9995	0.9983
	12								1.0000	1.0000	0.9999
	13										1.0000
14	0	0.4877	0.2288	0.1028	0.0440	0.0178	0.0068	0.0024	0.0008	0.0002	0.0001
	1	0.8470	0.5846	0.3567	0.1979	0.1010	0.0475	0.0205	0.0081	0.0029	0.0009
	2	0.9699	0.8416	0.6479	0.4481	0.2811	0.1608	0.0839	0.0398	0.0170	0.0065
	3	0.9958	0.9559	0.8535	0.6982	0.5213	0.3552	0.2205	0.1243	0.0632	0.0287
	4	0.9996	0.9908	0.9533	0.8702	0.7415	0.5842	0.4227	0.2793	0.1672	0.0898
	5	1.0000	0.9985	0.9885	0.9561	0.8883	0.7805	0.6405	0.4859	0.3373	0.2120
	6		0.9998	0.9978	0.9884	0.9617	0.9067	0.8164	0.6925	0.5461	0.3953
	7		1.0000	0.9997	0.9976	0.9897	0.9685	0.9247	0.8499	0.7414	0.6047
	8			1.0000	0.9996	0.9978	0.9917	0.9757	0.9417	0.8811	0.7880
	9				1.0000	0.9997	0.9983	0.9940	0.9825	0.9574	0.9102
	10					1.0000	0.9998	0.9989	0.9961	0.9886	0.9713
	11						1.0000	0.9999	0.9994	0.9978	0.9935
	12							1.0000	0.9999	0.9997	0.9991
	13								1.0000	1.0000	0.9999
	14										1.0000
15	0	0.4633	0.2059	0.0874	0.0352	0.0134	0.0047	0.0016	0.0005	0.0001	0.0000
	1	0.8290	0.5490	0.3186	0.1671	0.0802	0.0353	0.0142	0.0052	0.0017	0.0005
	2	0.9638	0.8159	0.6042	0.3980	0.2361	0.1268	0.0617	0.0271	0.0107	0.0037
	3	0.9945	0.9444	0.8227	0.6482	0.4613	0.2969	0.1727	0.0905	0.0424	0.0176
	4	0.9994	0.9873	0.9383	0.8358	0.6865	0.5155	0.3519	0.2173	0.1204	0.0592
	5	0.9999	0.9978	0.9832	0.9389	0.8516	0.7216	0.5643	0.4032	0.2608	0.1509
	6	1.0000	0.9997	0.9964	0.9819	0.9434	0.8689	0.7548	0.6098	0.4522	0.3036
	7		1.0000	0.9994	0.9958	0.9827	0.9500	0.8868	0.7869	0.6535	0.5000
	8			0.9999	0.9992	0.9958	0.9848	0.9578	0.9050	0.8182	0.6964
	9			1.0000	0.9999	0.9992	0.9963	0.9876	0.9662	0.9231	0.8491
	10				1.0000	0.9999	0.9993	0.9972	0.9907	0.9745	0.9408
	11					1.0000	0.9999	0.9995	0.9981	0.9937	0.9824
	12						1.0000	0.9999	0.9997	0.9989	0.9963
	13							1.0000	1.0000	0.9999	0.9995
	14									1.0000	1.0000

TABLE A4.1 (Continued) Distribution functions (cdfs) of binomial distributions: $F(y)$ for $Y \sim Bin(n, p)$

n	y	0.05	0.10	0.15	0.20	0.25	0.30	0.35	0.40	0.45	0.50
16	0	0.4401	0.1853	0.0743	0.0281	0.0100	0.0033	0.0010	0.0003	0.0001	0.0000
	1	0.8108	0.5147	0.2839	0.1407	0.0635	0.0261	0.0098	0.0033	0.0010	0.0003
	2	0.9571	0.7892	0.5614	0.3518	0.1971	0.0994	0.0451	0.0183	0.0066	0.0021
	3	0.9930	0.9316	0.7899	0.5981	0.4050	0.2459	0.1339	0.0651	0.0281	0.0106
	4	0.9991	0.9830	0.9209	0.7982	0.6302	0.4499	0.2892	0.1666	0.0853	0.0384
	5	0.9999	0.9967	0.9765	0.9183	0.8103	0.6598	0.4900	0.3288	0.1976	0.1051
	6	1.0000	0.9995	0.9944	0.9733	0.9204	0.8247	0.6881	0.5272	0.3660	0.2272
	7		0.9999	0.9989	0.9930	0.9729	0.9256	0.8406	0.7161	0.5629	0.4018
	8		1.0000	0.9998	0.9985	0.9925	0.9743	0.9329	0.8577	0.7441	0.5982
	9			1.0000	0.9998	0.9984	0.9929	0.9771	0.9417	0.8759	0.7728
	10				1.0000	0.9997	0.9984	0.9938	0.9809	0.9514	0.8949
	11					1.0000	0.9997	0.9987	0.9951	0.9851	0.9616
	12						1.0000	0.9998	0.9991	0.9965	0.9894
	13							1.0000	0.9999	0.9994	0.9979
	14								1.0000	0.9999	0.9997
	15									1.0000	1.0000
17	0	0.4181	0.1668	0.0631	0.0225	0.0075	0.0023	0.0007	0.0002	0.0000	0.0000
	1	0.7922	0.4818	0.2525	0.1182	0.0501	0.0193	0.0067	0.0021	0.0006	0.0001
	2	0.9497	0.7618	0.5198	0.3096	0.1637	0.0774	0.0327	0.0123	0.0041	0.0012
	3	0.9912	0.9174	0.7556	0.5489	0.3530	0.2019	0.1028	0.0464	0.0184	0.0064
	4	0.9988	0.9779	0.9013	0.7582	0.5739	0.3887	0.2348	0.1260	0.0596	0.0245
	5	0.9999	0.9953	0.9681	0.8943	0.7653	0.5968	0.4197	0.2639	0.1471	0.0717
	6	1.0000	0.9992	0.9917	0.9623	0.8929	0.7752	0.6188	0.4478	0.2902	0.1662
	7		0.9999	0.9983	0.9891	0.9598	0.8954	0.7872	0.6405	0.4743	0.3145
	8		1.0000	0.9997	0.9974	0.9876	0.9597	0.9006	0.8011	0.6626	0.5000
	9			1.0000	0.9995	0.9969	0.9873	0.9617	0.9081	0.8166	0.6855
	10				0.9999	0.9994	0.9968	0.9880	0.9652	0.9174	0.8338
	11				1.0000	0.9999	0.9993	0.9970	0.9894	0.9699	0.9283
	12					1.0000	0.9999	0.9994	0.9975	0.9914	0.9755
	13						1.0000	0.9999	0.9995	0.9981	0.9936
	14							1.0000	0.9999	0.9997	0.9988
	15								1.0000	1.0000	0.9999
	16										1.0000
18	0	0.3972	0.1501	0.0536	0.0180	0.0056	0.0016	0.0004	0.0001	0.0000	0.0000
	1	0.7735	0.4503	0.2241	0.0991	0.0395	0.0142	0.0046	0.0013	0.0003	0.0001
	2	0.9419	0.7338	0.4797	0.2713	0.1353	0.0600	0.0236	0.0082	0.0025	0.0007
	3	0.9891	0.9018	0.7202	0.5010	0.3057	0.1646	0.0783	0.0328	0.0120	0.0038
	4	0.9985	0.9718	0.8794	0.7164	0.5187	0.3327	0.1886	0.0942	0.0411	0.0154
	5	0.9998	0.9936	0.9581	0.8671	0.7175	0.5344	0.3550	0.2088	0.1077	0.0481
	6	1.0000	0.9988	0.9882	0.9487	0.8610	0.7217	0.5491	0.3743	0.2258	0.1189
	7		0.9998	0.9973	0.9837	0.9431	0.8593	0.7283	0.5634	0.3915	0.2403
	8		1.0000	0.9995	0.9957	0.9807	0.9404	0.8609	0.7368	0.5778	0.4073
	9			0.9999	0.9991	0.9946	0.9790	0.9403	0.8653	0.7473	0.5927
	10			1.0000	0.9998	0.9988	0.9939	0.9788	0.9424	0.8720	0.7597
	11				1.0000	0.9998	0.9986	0.9938	0.9797	0.9463	0.8811
	12					1.0000	0.9997	0.9986	0.9942	0.9817	0.9519
	13						1.0000	0.9997	0.9987	0.9951	0.9846
	14							1.0000	0.9998	0.9990	0.9962
	15								1.0000	0.9999	0.9993
	16									1.0000	0.9999
	17										1.0000

TABLE A4.1 (Continued) Distribution functions (cdfs) of binomial distributions: $F(y)$ for $Y \sim Bin(n, p)$

	p	0.05	0.10	0.15	0.20	0.25	0.30	0.35	0.40	0.45	0.50
n	y										
19	0	0.3774	0.1351	0.0456	0.0144	0.0042	0.0011	0.0003	0.0001	0.0000	0.0000
	1	0.7547	0.4203	0.1985	0.0829	0.0310	0.0104	0.0031	0.0008	0.0002	0.0000
	2	0.9335	0.7054	0.4413	0.2369	0.1113	0.0462	0.0170	0.0055	0.0015	0.0004
	3	0.9868	0.8850	0.6841	0.4551	0.2631	0.1332	0.0591	0.0230	0.0077	0.0022
	4	0.9980	0.9648	0.8556	0.6733	0.4654	0.2822	0.1500	0.0696	0.0280	0.0096
	5	0.9998	0.9914	0.9463	0.8369	0.6678	0.4739	0.2968	0.1629	0.0777	0.0318
	6	1.0000	0.9983	0.9837	0.9324	0.8251	0.6655	0.4812	0.3081	0.1727	0.0835
	7		0.9997	0.9959	0.9767	0.9225	0.8180	0.6656	0.4878	0.3169	0.1796
	8		1.0000	0.9992	0.9933	0.9713	0.9161	0.8145	0.6675	0.4940	0.3238
	9			0.9999	0.9984	0.9911	0.9674	0.9125	0.8139	0.6710	0.5000
	10			1.0000	0.9997	0.9977	0.9895	0.9653	0.9115	0.8159	0.6762
	11				1.0000	0.9995	0.9972	0.9886	0.9648	0.9129	0.8204
	12					0.9999	0.9994	0.9969	0.9884	0.9658	0.9165
	13					1.0000	0.9999	0.9993	0.9969	0.9891	0.9682
	14						1.0000	0.9999	0.9994	0.9972	0.9904
	15							1.0000	0.9999	0.9995	0.9978
	16								1.0000	0.9999	0.9996
	17									1.0000	1.0000
20	0	0.3585	0.1216	0.0388	0.0115	0.0032	0.0008	0.0002	0.0000	0.0000	0.0000
	1	0.7358	0.3917	0.1756	0.0692	0.0243	0.0076	0.0021	0.0005	0.0001	0.0000
	2	0.9245	0.6769	0.4049	0.2061	0.0913	0.0355	0.0121	0.0036	0.0009	0.0002
	3	0.9841	0.8670	0.6477	0.4114	0.2252	0.1071	0.0444	0.0160	0.0049	0.0013
	4	0.9974	0.9568	0.8298	0.6296	0.4148	0.2375	0.1182	0.0510	0.0189	0.0059
	5	0.9997	0.9887	0.9327	0.8042	0.6172	0.4164	0.2454	0.1256	0.0553	0.0207
	6	1.0000	0.9976	0.9781	0.9133	0.7858	0.6080	0.4166	0.2500	0.1299	0.0577
	7		0.9996	0.9941	0.9679	0.8982	0.7723	0.6010	0.4159	0.2520	0.1316
	8		0.9999	0.9987	0.9900	0.9591	0.8867	0.7624	0.5956	0.4143	0.2517
	9		1.0000	0.9998	0.9974	0.9861	0.9520	0.8782	0.7553	0.5914	0.4119
	10			1.0000	0.9994	0.9961	0.9829	0.9468	0.8725	0.7507	0.5881
	11				0.9999	0.9991	0.9949	0.9804	0.9435	0.8692	0.7483
	12				1.0000	0.9998	0.9987	0.9940	0.9790	0.9420	0.8684
	13					1.0000	0.9997	0.9985	0.9935	0.9786	0.9423
	14						1.0000	0.9997	0.9984	0.9936	0.9793
	15							1.0000	0.9997	0.9985	0.9941
	16								1.0000	0.9997	0.9987
	17									1.0000	0.9998
	18										1.0000

TABLE A4.1 (Continued) Distribution functions (cdfs) of binomial distributions: $F(y)$ for $Y \sim Bin(n, p)$

p	0.05	0.10	0.15	0.20	0.25	0.30	0.35	0.40	0.45	0.50
n	y										
50	0	0.0769	0.0052	0.0003	0.0000	0.0000	0.0000	0.0000	0.0000	0.0000	0.0000
	1	0.2794	0.0338	0.0029	0.0002	0.0000	0.0000	0.0000	0.0000	0.0000	0.0000
	2	0.5405	0.1117	0.0142	0.0013	0.0001	0.0000	0.0000	0.0000	0.0000	0.0000
	3	0.7604	0.2503	0.0460	0.0057	0.0005	0.0000	0.0000	0.0000	0.0000	0.0000
	4	0.8964	0.4312	0.1121	0.0185	0.0021	0.0002	0.0000	0.0000	0.0000	0.0000
	5	0.9622	0.6161	0.2194	0.0480	0.0070	0.0007	0.0001	0.0000	0.0000	0.0000
	6	0.9882	0.7702	0.3613	0.1034	0.0194	0.0025	0.0002	0.0000	0.0000	0.0000
	7	0.9968	0.8779	0.5188	0.1904	0.0453	0.0073	0.0008	0.0001	0.0000	0.0000
	8	0.9992	0.9421	0.6681	0.3073	0.0916	0.0183	0.0025	0.0002	0.0000	0.0000
	9	0.9998	0.9755	0.7911	0.4437	0.1637	0.0402	0.0067	0.0008	0.0001	0.0000
	10	1.0000	0.9906	0.8801	0.5836	0.2622	0.0789	0.0160	0.0022	0.0002	0.0000
	11		0.9968	0.9372	0.7107	0.3816	0.1390	0.0342	0.0057	0.0006	0.0000
	12		0.9990	0.9699	0.8139	0.5110	0.2229	0.0661	0.0133	0.0018	0.0002
	13		0.9997	0.9868	0.8894	0.6370	0.3279	0.1163	0.0280	0.0045	0.0005
	14		0.9999	0.9947	0.9393	0.7481	0.4468	0.1878	0.0540	0.0104	0.0013
	15		1.0000	0.9981	0.9692	0.8369	0.5692	0.2801	0.0955	0.0220	0.0033
	16			0.9993	0.9856	0.9017	0.6839	0.3889	0.1561	0.0427	0.0077
	17			0.9998	0.9937	0.9449	0.7822	0.5060	0.2369	0.0765	0.0164
	18			0.9999	0.9975	0.9713	0.8594	0.6216	0.3356	0.1273	0.0325
	19			1.0000	0.9991	0.9861	0.9152	0.7264	0.4465	0.1974	0.0595
	20				0.9997	0.9937	0.9522	0.8139	0.5610	0.2862	0.1013
	21				0.9999	0.9974	0.9749	0.8813	0.6701	0.3900	0.1611
	22				1.0000	0.9990	0.9877	0.9290	0.7660	0.5019	0.2399
	23					0.9996	0.9944	0.9604	0.8438	0.6134	0.3359
	24					0.9999	0.9976	0.9793	0.9022	0.7160	0.4439
	25					1.0000	0.9991	0.9900	0.9427	0.8034	0.5561
	26						0.9997	0.9955	0.9686	0.8721	0.6641
	27						0.9999	0.9981	0.9840	0.9220	0.7601
	28						1.0000	0.9993	0.9924	0.9556	0.8389
	29							0.9997	0.9966	0.9765	0.8987
	30							0.9999	0.9986	0.9884	0.9405
	31							1.0000	0.9995	0.9947	0.9675
	32								0.9998	0.9978	0.9836
	33								0.9999	0.9991	0.9923
	34								1.0000	0.9997	0.9967
	35									0.9999	0.9987
	36									1.0000	0.9995
	37										0.9998
	38										1.0000

TABLE A4.1 (Continued) Distribution functions (cdfs) of binomial distributions: $F(y)$ for $Y \sim Bin(n, p)$

	p	0.05	0.10	0.15	0.20	0.25	0.30		y	0.35	0.40	0.45	0.50
n	y												
100	0	0.0059	0.0000	0.0000	0.0000	0.0000	0.0000		16	0.0000	0.0000	0.0000	0.0000
	1	0.0371	0.0003	0.0000	0.0000	0.0000	0.0000		17	0.0001	0.0000	0.0000	0.0000
	2	0.1183	0.0019	0.0000	0.0000	0.0000	0.0000		18	0.0001	0.0000	0.0000	0.0000
	3	0.2578	0.0078	0.0001	0.0000	0.0000	0.0000		19	0.0003	0.0000	0.0000	0.0000
	4	0.4360	0.0237	0.0004	0.0000	0.0000	0.0000		20	0.0008	0.0000	0.0000	0.0000
	5	0.6160	0.0576	0.0016	0.0000	0.0000	0.0000		21	0.0017	0.0000	0.0000	0.0000
	6	0.7660	0.1172	0.0047	0.0001	0.0000	0.0000		22	0.0034	0.0001	0.0000	0.0000
	7	0.8720	0.2061	0.0122	0.0003	0.0000	0.0000		23	0.0066	0.0003	0.0000	0.0000
	8	0.9369	0.3209	0.0275	0.0009	0.0000	0.0000		24	0.0121	0.0006	0.0000	0.0000
	9	0.9718	0.4513	0.0551	0.0023	0.0000	0.0000		25	0.0211	0.0012	0.0000	0.0000
	10	0.9885	0.5832	0.0994	0.0057	0.0001	0.0000		26	0.0351	0.0024	0.0001	0.0000
	11	0.9957	0.7030	0.1635	0.0126	0.0004	0.0000		27	0.0558	0.0046	0.0002	0.0000
	12	0.9985	0.8018	0.2473	0.0253	0.0010	0.0000		28	0.0848	0.0084	0.0004	0.0000
	13	0.9995	0.8761	0.3474	0.0469	0.0025	0.0001		29	0.1236	0.0148	0.0008	0.0000
	14	0.9999	0.9274	0.4572	0.0804	0.0054	0.0002		30	0.1730	0.0248	0.0015	0.0000
	15	1.0000	0.9601	0.5683	0.1285	0.0111	0.0004		31	0.2331	0.0398	0.0030	0.0001
	16		0.9794	0.6725	0.1923	0.0211	0.0010		32	0.3029	0.0615	0.0055	0.0002
	17		0.9900	0.7633	0.2712	0.0376	0.0022		33	0.3803	0.0913	0.0098	0.0004
	18		0.9954	0.8372	0.3621	0.0630	0.0045		34	0.4624	0.1303	0.0166	0.0009
	19		0.9980	0.8935	0.4602	0.0995	0.0089		35	0.5458	0.1795	0.0272	0.0018
	20		0.9992	0.9337	0.5595	0.1488	0.0165		36	0.6269	0.2386	0.0429	0.0033
	21		0.9997	0.9607	0.6540	0.2114	0.0288		37	0.7024	0.3068	0.0651	0.0060
	22		0.9999	0.9779	0.7389	0.2864	0.0479		38	0.7699	0.3822	0.0951	0.0105
	23		1.0000	0.9881	0.8109	0.3711	0.0755		39	0.8276	0.4621	0.1343	0.0176
	24			0.9939	0.8686	0.4617	0.1136		40	0.8750	0.5433	0.1831	0.0284
	25			0.9970	0.9125	0.5535	0.1631		41	0.9123	0.6225	0.2415	0.0443
	26			0.9986	0.9442	0.6417	0.2244		42	0.9406	0.6967	0.3087	0.0666
	27			0.9994	0.9658	0.7224	0.2964		43	0.9611	0.7635	0.3828	0.0967
	28			0.9997	0.9800	0.7925	0.3768		44	0.9754	0.8211	0.4613	0.1356
	29			0.9999	0.9888	0.8505	0.4623		45	0.9850	0.8689	0.5413	0.1841
	30			1.0000	0.9939	0.8962	0.5491		46	0.9912	0.9070	0.6196	0.2421
	31				0.9969	0.9307	0.6331		47	0.9950	0.9362	0.6931	0.3086
	32				0.9984	0.9554	0.7107		48	0.9973	0.9577	0.7596	0.3822
	33				0.9993	0.9724	0.7793		49	0.9985	0.9729	0.8173	0.4602
	34				0.9997	0.9836	0.8371		50	0.9993	0.9832	0.8654	0.5398
	35				0.9999	0.9906	0.8839		51	0.9996	0.9900	0.9040	0.6178
	36				0.9999	0.9948	0.9201		52	0.9998	0.9942	0.9338	0.6914
	37				1.0000	0.9973	0.9470		53	0.9999	0.9968	0.9559	0.7579
	38					0.9986	0.9660		54	1.0000	0.9983	0.9716	0.8159
	39					0.9993	0.9790		55		0.9991	0.9824	0.8644
	40					0.9997	0.9875		56		0.9996	0.9894	0.9033
	41					0.9999	0.9928		57		0.9998	0.9939	0.9334
	42					0.9999	0.9960		58		0.9999	0.9966	0.9557
	43					1.0000	0.9979		59		1.0000	0.9982	0.9716
	44						0.9989		60			0.9991	0.9824
	45						0.9995		61			0.9995	0.9895
	46						0.9997		62			0.9998	0.9940
	47						0.9999		63			0.9999	0.9967
	48						0.9999		64			1.0000	0.9982
	49						1.0000		65				0.9991
	50								66				0.9996
	51								67				0.9998
	52								68				0.9999
	53								69				1.0000

TABLE A4.1 (Continued) Distribution functions (cdfs) of binomial distributions: $F(y)$ for $Y \sim Bin(n, p)$

μ	0.01	0.05	0.10	0.15	0.20	0.25	0.30	0.35	0.40	0.45	0.50
y											
0	0.9900	0.9512	0.9048	0.8607	0.8187	0.7788	0.7408	0.7047	0.6703	0.6376	0.6065
1	1.0000	0.9988	0.9953	0.9898	0.9825	0.9735	0.9631	0.9513	0.9384	0.9246	0.9098
2		1.0000	0.9998	0.9995	0.9989	0.9978	0.9964	0.9945	0.9921	0.9891	0.9856
3			1.0000	1.0000	0.9999	0.9999	0.9997	0.9995	0.9992	0.9988	0.9982
4					1.0000	1.0000	1.0000	1.0000	0.9999	0.9999	0.9998
5									1.0000	1.0000	1.0000

μ	0.55	0.60	0.65	0.70	0.75	0.80	0.85	0.90	0.95	1.00	1.05
y											
0	0.5769	0.5488	0.5220	0.4966	0.4724	0.4493	0.4274	0.4066	0.3867	0.3679	0.3499
1	0.8943	0.8781	0.8614	0.8442	0.8266	0.8088	0.7907	0.7725	0.7541	0.7358	0.7174
2	0.9815	0.9769	0.9717	0.9659	0.9595	0.9526	0.9451	0.9371	0.9287	0.9197	0.9103
3	0.9975	0.9966	0.9956	0.9942	0.9927	0.9909	0.9889	0.9865	0.9839	0.9810	0.9778
4	0.9997	0.9996	0.9994	0.9992	0.9989	0.9986	0.9982	0.9977	0.9971	0.9963	0.9955
5	1.0000	1.0000	0.9999	0.9999	0.9999	0.9998	0.9997	0.9997	0.9995	0.9994	0.9992
6			1.0000	1.0000	1.0000	1.0000	1.0000	1.0000	0.9999	0.9999	0.9999
7									1.0000	1.0000	1.0000

μ	1.10	1.20	1.30	1.40	1.50	1.60	1.70	1.80	1.90	2.00	2.10
y											
0	0.3329	0.3012	0.2725	0.2466	0.2231	0.2019	0.1827	0.1653	0.1496	0.1353	0.1225
1	0.6990	0.6626	0.6268	0.5918	0.5578	0.5249	0.4932	0.4628	0.4337	0.4060	0.3796
2	0.9004	0.8795	0.8571	0.8335	0.8088	0.7834	0.7572	0.7306	0.7037	0.6767	0.6496
3	0.9743	0.9662	0.9569	0.9463	0.9344	0.9212	0.9068	0.8913	0.8747	0.8571	0.8386
4	0.9946	0.9923	0.9893	0.9857	0.9814	0.9763	0.9704	0.9636	0.9559	0.9473	0.9379
5	0.9990	0.9985	0.9978	0.9968	0.9955	0.9940	0.9920	0.9896	0.9868	0.9834	0.9796
6	0.9999	0.9997	0.9996	0.9994	0.9991	0.9987	0.9981	0.9974	0.9966	0.9955	0.9941
7	1.0000	1.0000	0.9999	0.9999	0.9998	0.9997	0.9996	0.9994	0.9992	0.9989	0.9985
8			1.0000	1.0000	1.0000	1.0000	0.9999	0.9999	0.9998	0.9998	0.9997
9							1.0000	1.0000	1.0000	1.0000	0.9999
10											1.0000

μ	2.20	2.30	2.40	2.50	2.60	2.70	2.80	2.90	3.00	3.10	3.20
y											
0	0.1108	0.1003	0.0907	0.0821	0.0743	0.0672	0.0608	0.0550	0.0498	0.0450	0.0408
1	0.3546	0.3309	0.3084	0.2873	0.2674	0.2487	0.2311	0.2146	0.1991	0.1847	0.1712
2	0.6227	0.5960	0.5697	0.5438	0.5184	0.4936	0.4695	0.4460	0.4232	0.4012	0.3799
3	0.8194	0.7993	0.7787	0.7576	0.7360	0.7141	0.6919	0.6696	0.6472	0.6248	0.6025
4	0.9275	0.9162	0.9041	0.8912	0.8774	0.8629	0.8477	0.8318	0.8153	0.7982	0.7806
5	0.9751	0.9700	0.9643	0.9580	0.9510	0.9433	0.9349	0.9258	0.9161	0.9057	0.8946
6	0.9925	0.9906	0.9884	0.9858	0.9828	0.9794	0.9756	0.9713	0.9665	0.9612	0.9554
7	0.9980	0.9974	0.9967	0.9958	0.9947	0.9934	0.9919	0.9901	0.9881	0.9858	0.9832
8	0.9995	0.9994	0.9991	0.9989	0.9985	0.9981	0.9976	0.9969	0.9962	0.9953	0.9943
9	0.9999	0.9999	0.9998	0.9997	0.9996	0.9995	0.9993	0.9991	0.9989	0.9986	0.9982
10	1.0000	1.0000	1.0000	0.9999	0.9999	0.9999	0.9998	0.9998	0.9997	0.9996	0.9995
11				1.0000	1.0000	1.0000	1.0000	0.9999	0.9999	0.9999	0.9999
12								1.0000	1.0000	1.0000	1.0000

TABLE A4.2 Distribution functions (cdfs) of Poisson distributions: $F(y)$ for $Y \sim Po(\mu)$

μ	3.40	3.60	3.80	4.00	4.25	4.50	4.75	5.00	5.25	5.50	5.75
y											
0	0.0334	0.0273	0.0224	0.0183	0.0143	0.0111	0.0087	0.0067	0.0052	0.0041	0.0032
1	0.1468	0.1257	0.1074	0.0916	0.0749	0.0611	0.0497	0.0404	0.0328	0.0266	0.0215
2	0.3397	0.3027	0.2689	0.2381	0.2037	0.1736	0.1473	0.1247	0.1051	0.0884	0.0741
3	0.5584	0.5152	0.4735	0.4335	0.3862	0.3423	0.3019	0.2650	0.2317	0.2017	0.1749
4	0.7442	0.7064	0.6678	0.6288	0.5801	0.5321	0.4854	0.4405	0.3978	0.3575	0.3199
5	0.8705	0.8441	0.8156	0.7851	0.7449	0.7029	0.6597	0.6160	0.5722	0.5289	0.4866
6	0.9421	0.9267	0.9091	0.8893	0.8617	0.8311	0.7978	0.7622	0.7248	0.6860	0.6464
7	0.9769	0.9692	0.9599	0.9489	0.9326	0.9134	0.8914	0.8666	0.8392	0.8095	0.7776
8	0.9917	0.9883	0.9840	0.9786	0.9702	0.9597	0.9470	0.9319	0.9144	0.8944	0.8719
9	0.9973	0.9960	0.9942	0.9919	0.9880	0.9829	0.9764	0.9682	0.9582	0.9462	0.9322
10	0.9992	0.9987	0.9981	0.9972	0.9956	0.9933	0.9903	0.9863	0.9812	0.9747	0.9669
11	0.9998	0.9996	0.9994	0.9991	0.9985	0.9976	0.9963	0.9945	0.9922	0.9890	0.9850
12	0.9999	0.9999	0.9998	0.9997	0.9995	0.9992	0.9987	0.9980	0.9970	0.9955	0.9937
13	1.0000	1.0000	1.0000	0.9999	0.9999	0.9997	0.9996	0.9993	0.9989	0.9983	0.9975
14				1.0000	1.0000	0.9999	0.9999	0.9998	0.9996	0.9994	0.9991
15						1.0000	1.0000	0.9999	0.9999	0.9998	0.9997
16								1.0000	1.0000	0.9999	0.9999
17										1.0000	1.0000

μ	6.00	6.50	7.00	7.50	8.00	8.50	9.00	9.50	10.00	10.50	11.00
y											
0	0.0025	0.0015	0.0009	0.0006	0.0003	0.0002	0.0001	0.0001	0.0000	0.0000	0.0000
1	0.0174	0.0113	0.0073	0.0047	0.0030	0.0019	0.0012	0.0008	0.0005	0.0003	0.0002
2	0.0620	0.0430	0.0296	0.0203	0.0138	0.0093	0.0062	0.0042	0.0028	0.0018	0.0012
3	0.1512	0.1118	0.0818	0.0591	0.0424	0.0301	0.0212	0.0149	0.0103	0.0071	0.0049
4	0.2851	0.2237	0.1730	0.1321	0.0996	0.0744	0.0550	0.0403	0.0293	0.0211	0.0151
5	0.4457	0.3690	0.3007	0.2414	0.1912	0.1496	0.1157	0.0885	0.0671	0.0504	0.0375
6	0.6063	0.5265	0.4497	0.3782	0.3134	0.2562	0.2068	0.1649	0.1301	0.1016	0.0786
7	0.7440	0.6728	0.5987	0.5246	0.4530	0.3856	0.3239	0.2687	0.2202	0.1785	0.1432
8	0.8472	0.7916	0.7291	0.6620	0.5925	0.5231	0.4557	0.3918	0.3328	0.2794	0.2320
9	0.9161	0.8774	0.8305	0.7764	0.7166	0.6530	0.5874	0.5218	0.4579	0.3971	0.3405
10	0.9574	0.9332	0.9015	0.8622	0.8159	0.7634	0.7060	0.6453	0.5830	0.5207	0.4599
11	0.9799	0.9661	0.9467	0.9208	0.8881	0.8487	0.8030	0.7520	0.6968	0.6387	0.5793
12	0.9912	0.9840	0.9730	0.9573	0.9362	0.9091	0.8758	0.8364	0.7916	0.7420	0.6887
13	0.9964	0.9929	0.9872	0.9784	0.9658	0.9486	0.9261	0.8981	0.8645	0.8253	0.7813
14	0.9986	0.9970	0.9943	0.9897	0.9827	0.9726	0.9585	0.9400	0.9165	0.8879	0.8540
15	0.9995	0.9988	0.9976	0.9954	0.9918	0.9862	0.9780	0.9665	0.9513	0.9317	0.9074
16	0.9998	0.9996	0.9990	0.9980	0.9963	0.9934	0.9889	0.9823	0.9730	0.9604	0.9441
17	0.9999	0.9998	0.9996	0.9992	0.9984	0.9970	0.9947	0.9911	0.9857	0.9781	0.9678
18	1.0000	0.9999	0.9999	0.9997	0.9993	0.9987	0.9976	0.9957	0.9928	0.9885	0.9823
19		1.0000	1.0000	0.9999	0.9997	0.9995	0.9989	0.9980	0.9965	0.9942	0.9907
20				1.0000	0.9999	0.9998	0.9996	0.9991	0.9984	0.9972	0.9953
21					1.0000	0.9999	0.9998	0.9996	0.9993	0.9987	0.9977
22						1.0000	0.9999	0.9999	0.9997	0.9994	0.9990
23							1.0000	0.9999	0.9999	0.9998	0.9995
24								1.0000	1.0000	0.9999	0.9998
25										1.0000	0.9999
26											1.0000

TABLE A4.2 (Continued) Distribution functions (cdfs) of Poisson distributions: $F(y)$ for $Y \sim Po(\mu)$

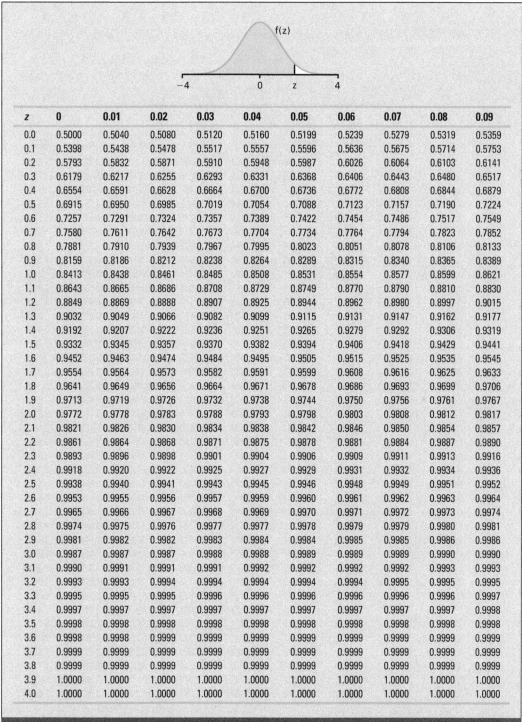

z	0	0.01	0.02	0.03	0.04	0.05	0.06	0.07	0.08	0.09
0.0	0.5000	0.5040	0.5080	0.5120	0.5160	0.5199	0.5239	0.5279	0.5319	0.5359
0.1	0.5398	0.5438	0.5478	0.5517	0.5557	0.5596	0.5636	0.5675	0.5714	0.5753
0.2	0.5793	0.5832	0.5871	0.5910	0.5948	0.5987	0.6026	0.6064	0.6103	0.6141
0.3	0.6179	0.6217	0.6255	0.6293	0.6331	0.6368	0.6406	0.6443	0.6480	0.6517
0.4	0.6554	0.6591	0.6628	0.6664	0.6700	0.6736	0.6772	0.6808	0.6844	0.6879
0.5	0.6915	0.6950	0.6985	0.7019	0.7054	0.7088	0.7123	0.7157	0.7190	0.7224
0.6	0.7257	0.7291	0.7324	0.7357	0.7389	0.7422	0.7454	0.7486	0.7517	0.7549
0.7	0.7580	0.7611	0.7642	0.7673	0.7704	0.7734	0.7764	0.7794	0.7823	0.7852
0.8	0.7881	0.7910	0.7939	0.7967	0.7995	0.8023	0.8051	0.8078	0.8106	0.8133
0.9	0.8159	0.8186	0.8212	0.8238	0.8264	0.8289	0.8315	0.8340	0.8365	0.8389
1.0	0.8413	0.8438	0.8461	0.8485	0.8508	0.8531	0.8554	0.8577	0.8599	0.8621
1.1	0.8643	0.8665	0.8686	0.8708	0.8729	0.8749	0.8770	0.8790	0.8810	0.8830
1.2	0.8849	0.8869	0.8888	0.8907	0.8925	0.8944	0.8962	0.8980	0.8997	0.9015
1.3	0.9032	0.9049	0.9066	0.9082	0.9099	0.9115	0.9131	0.9147	0.9162	0.9177
1.4	0.9192	0.9207	0.9222	0.9236	0.9251	0.9265	0.9279	0.9292	0.9306	0.9319
1.5	0.9332	0.9345	0.9357	0.9370	0.9382	0.9394	0.9406	0.9418	0.9429	0.9441
1.6	0.9452	0.9463	0.9474	0.9484	0.9495	0.9505	0.9515	0.9525	0.9535	0.9545
1.7	0.9554	0.9564	0.9573	0.9582	0.9591	0.9599	0.9608	0.9616	0.9625	0.9633
1.8	0.9641	0.9649	0.9656	0.9664	0.9671	0.9678	0.9686	0.9693	0.9699	0.9706
1.9	0.9713	0.9719	0.9726	0.9732	0.9738	0.9744	0.9750	0.9756	0.9761	0.9767
2.0	0.9772	0.9778	0.9783	0.9788	0.9793	0.9798	0.9803	0.9808	0.9812	0.9817
2.1	0.9821	0.9826	0.9830	0.9834	0.9838	0.9842	0.9846	0.9850	0.9854	0.9857
2.2	0.9861	0.9864	0.9868	0.9871	0.9875	0.9878	0.9881	0.9884	0.9887	0.9890
2.3	0.9893	0.9896	0.9898	0.9901	0.9904	0.9906	0.9909	0.9911	0.9913	0.9916
2.4	0.9918	0.9920	0.9922	0.9925	0.9927	0.9929	0.9931	0.9932	0.9934	0.9936
2.5	0.9938	0.9940	0.9941	0.9943	0.9945	0.9946	0.9948	0.9949	0.9951	0.9952
2.6	0.9953	0.9955	0.9956	0.9957	0.9959	0.9960	0.9961	0.9962	0.9963	0.9964
2.7	0.9965	0.9966	0.9967	0.9968	0.9969	0.9970	0.9971	0.9972	0.9973	0.9974
2.8	0.9974	0.9975	0.9976	0.9977	0.9977	0.9978	0.9979	0.9979	0.9980	0.9981
2.9	0.9981	0.9982	0.9982	0.9983	0.9984	0.9984	0.9985	0.9985	0.9986	0.9986
3.0	0.9987	0.9987	0.9987	0.9988	0.9988	0.9989	0.9989	0.9989	0.9990	0.9990
3.1	0.9990	0.9991	0.9991	0.9991	0.9992	0.9992	0.9992	0.9992	0.9993	0.9993
3.2	0.9993	0.9993	0.9994	0.9994	0.9994	0.9994	0.9994	0.9995	0.9995	0.9995
3.3	0.9995	0.9995	0.9995	0.9996	0.9996	0.9996	0.9996	0.9996	0.9996	0.9997
3.4	0.9997	0.9997	0.9997	0.9997	0.9997	0.9997	0.9997	0.9997	0.9997	0.9998
3.5	0.9998	0.9998	0.9998	0.9998	0.9998	0.9998	0.9998	0.9998	0.9998	0.9998
3.6	0.9998	0.9998	0.9999	0.9999	0.9999	0.9999	0.9999	0.9999	0.9999	0.9999
3.7	0.9999	0.9999	0.9999	0.9999	0.9999	0.9999	0.9999	0.9999	0.9999	0.9999
3.8	0.9999	0.9999	0.9999	0.9999	0.9999	0.9999	0.9999	0.9999	0.9999	0.9999
3.9	1.0000	1.0000	1.0000	1.0000	1.0000	1.0000	1.0000	1.0000	1.0000	1.0000
4.0	1.0000	1.0000	1.0000	1.0000	1.0000	1.0000	1.0000	1.0000	1.0000	1.0000

TABLE A4.3 Distribution function of the standard normal distribution: $\Phi(z)$ for $Z \sim N(0, 1)$

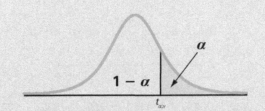

α	0.40	0.25	0.10	0.05	0.025	0.01	0.005	0.0025	0.001
1	0.3249	1.0000	3.0777	6.3138	12.7062	31.8205	63.6567	127.3213	318.3088
2	0.2887	0.8165	1.8856	2.9200	4.3027	6.9646	9.9248	14.0890	22.3271
3	0.2767	0.7649	1.6377	2.3534	3.1824	4.5407	5.8409	7.4533	10.2145
4	0.2707	0.7407	1.5332	2.1318	2.7764	3.7469	4.6041	5.5976	7.1732
5	0.2672	0.7267	1.4759	2.0150	2.5706	3.3649	4.0321	4.7733	5.8934
6	0.2648	0.7176	1.4398	1.9432	2.4469	3.1427	3.7074	4.3168	5.2076
7	0.2632	0.7111	1.4149	1.8946	2.3646	2.9980	3.4995	4.0293	4.7853
8	0.2619	0.7064	1.3968	1.8595	2.3060	2.8965	3.3554	3.8325	4.5008
9	0.2610	0.7027	1.3830	1.8331	2.2622	2.8214	3.2498	3.6897	4.2968
10	0.2602	0.6998	1.3722	1.8125	2.2281	2.7638	3.1693	3.5814	4.1437
11	0.2596	0.6974	1.3634	1.7959	2.2010	2.7181	3.1058	3.4966	4.0247
12	0.2590	0.6955	1.3562	1.7823	2.1788	2.6810	3.0545	3.4284	3.9296
13	0.2586	0.6938	1.3502	1.7709	2.1604	2.6503	3.0123	3.3725	3.8520
14	0.2582	0.6924	1.3450	1.7613	2.1448	2.6245	2.9768	3.3257	3.7874
15	0.2579	0.6912	1.3406	1.7531	2.1314	2.6025	2.9467	3.2860	3.7328
16	0.2576	0.6901	1.3368	1.7459	2.1199	2.5835	2.9208	3.2520	3.6862
17	0.2573	0.6892	1.3334	1.7396	2.1098	2.5669	2.8982	3.2224	3.6458
18	0.2571	0.6884	1.3304	1.7341	2.1009	2.5524	2.8784	3.1966	3.6105
19	0.2569	0.6876	1.3277	1.7291	2.0930	2.5395	2.8609	3.1737	3.5794
20	0.2567	0.6870	1.3253	1.7247	2.0860	2.5280	2.8453	3.1534	3.5518
21	0.2566	0.6864	1.3232	1.7207	2.0796	2.5176	2.8314	3.1352	3.5272
22	0.2564	0.6858	1.3212	1.7171	2.0739	2.5083	2.8188	3.1188	3.5050
23	0.2563	0.6853	1.3195	1.7139	2.0687	2.4999	2.8073	3.1040	3.4850
24	0.2562	0.6848	1.3178	1.7109	2.0639	2.4922	2.7969	3.0905	3.4668
25	0.2561	0.6844	1.3163	1.7081	2.0595	2.4851	2.7874	3.0782	3.4502
26	0.2560	0.6840	1.3150	1.7056	2.0555	2.4786	2.7787	3.0669	3.4350
27	0.2559	0.6837	1.3137	1.7033	2.0518	2.4727	2.7707	3.0565	3.4210
28	0.2558	0.6834	1.3125	1.7011	2.0484	2.4671	2.7633	3.0469	3.4082
29	0.2557	0.6830	1.3114	1.6991	2.0452	2.4620	2.7564	3.0380	3.3962
30	0.2556	0.6828	1.3104	1.6973	2.0423	2.4573	2.7500	3.0298	3.3852
40	0.2550	0.6807	1.3031	1.6839	2.0211	2.4233	2.7045	2.9712	3.3069
50	0.2547	0.6794	1.2987	1.6759	2.0086	2.4033	2.6778	2.9370	3.2614
60	0.2545	0.6786	1.2958	1.6706	2.0003	2.3901	2.6603	2.9146	3.2317
70	0.2543	0.6780	1.2938	1.6669	1.9944	2.3808	2.6479	2.8987	3.2108
80	0.2542	0.6776	1.2922	1.6641	1.9901	2.3739	2.6387	2.8870	3.1953
90	0.2541	0.6772	1.2910	1.6620	1.9867	2.3685	2.6316	2.8779	3.1833
100	0.2540	0.6770	1.2901	1.6602	1.9840	2.3642	2.6259	2.8707	3.1737
120	0.2539	0.6765	1.2886	1.6577	1.9799	2.3578	2.6174	2.8599	3.1595
∞	0.2533	0.6745	1.2816	1.6449	1.9600	2.3263	2.5758	2.8070	3.0902

TABLE A4.4 Quantiles of t-distributions: $t_{\alpha;v}$ such that $P(T > t_{\alpha;v}) = \alpha$ for $T \sim t_v$

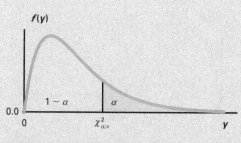

α	0.995	0.990	0.975	0.950	0.900	0.100	0.050	0.025	0.010	0.005
v										
1	0.0000	0.0002	0.0010	0.0039	0.0158	2.7055	3.8415	5.0239	6.6349	7.8794
2	0.0100	0.0201	0.0506	0.1026	0.2107	4.6052	5.9915	7.3778	9.2103	10.5966
3	0.0717	0.1148	0.2158	0.3518	0.5844	6.2514	7.8147	9.3484	11.3449	12.8382
4	0.2070	0.2971	0.4844	0.7107	1.0636	7.7794	9.4877	11.1433	13.2767	14.8603
5	0.4117	0.5543	0.8312	1.1455	1.6103	9.2364	11.0705	12.8325	15.0863	16.7496
6	0.6757	0.8721	1.2373	1.6354	2.2041	10.6446	12.5916	14.4494	16.8119	18.5476
7	0.9893	1.2390	1.6899	2.1673	2.8331	12.0170	14.0671	16.0128	18.4753	20.2777
8	1.3444	1.6465	2.1797	2.7326	3.4895	13.3616	15.5073	17.5345	20.0902	21.9550
9	1.7349	2.0879	2.7004	3.3251	4.1682	14.6837	16.9190	19.0228	21.6660	23.5894
10	2.1559	2.5582	3.2470	3.9403	4.8652	15.9872	18.3070	20.4832	23.2093	25.1882
11	2.6032	3.0535	3.8157	4.5748	5.5778	17.2750	19.6751	21.9200	24.7250	26.7568
12	3.0738	3.5706	4.4038	5.2260	6.3038	18.5493	21.0261	23.3367	26.2170	28.2995
13	3.5650	4.1069	5.0088	5.8919	7.0415	19.8119	22.3620	24.7356	27.6882	29.8195
14	4.0747	4.6604	5.6287	6.5706	7.7895	21.0641	23.6848	26.1189	29.1412	31.3193
15	4.6009	5.2293	6.2621	7.2609	8.5468	22.3071	24.9958	27.4884	30.5779	32.8013
16	5.1422	5.8122	6.9077	7.9616	9.3122	23.5418	26.2962	28.8454	31.9999	34.2672
17	5.6972	6.4078	7.5642	8.6718	10.0852	24.7690	27.5871	30.1910	33.4087	35.7185
18	6.2648	7.0149	8.2307	9.3905	10.8649	25.9894	28.8693	31.5264	34.8053	37.1565
19	6.8440	7.6327	8.9065	10.1170	11.6509	27.2036	30.1435	32.8523	36.1909	38.5823
20	7.4338	8.2604	9.5908	10.8508	12.4426	28.4120	31.4104	34.1696	37.5662	39.9968
21	8.0337	8.8972	10.2829	11.5913	13.2396	29.6151	32.6706	35.4789	38.9322	41.4011
22	8.6427	9.5425	10.9823	12.3380	14.0415	30.8133	33.9244	36.7807	40.2894	42.7957
23	9.2604	10.1957	11.6886	13.0905	14.8480	32.0069	35.1725	38.0756	41.6384	44.1813
24	9.8862	10.8564	12.4012	13.8484	15.6587	33.1962	36.4150	39.3641	42.9798	45.5585
25	10.5197	11.5240	13.1197	14.6114	16.4734	34.3816	37.6525	40.6465	44.3141	46.9279
26	11.1602	12.1981	13.8439	15.3792	17.2919	35.5632	38.8851	41.9232	45.6417	48.2899
27	11.8076	12.8785	14.5734	16.1514	18.1139	36.7412	40.1133	43.1945	46.9629	49.6449
28	12.4613	13.5647	15.3079	16.9279	18.9392	37.9159	41.3371	44.4608	48.2782	50.9934
29	13.1211	14.2565	16.0471	17.7084	19.7677	39.0875	42.5570	45.7223	49.5879	52.3356
30	13.7867	14.9535	16.7908	18.4927	20.5992	40.2560	43.7730	46.9792	50.8922	53.6720
40	20.7065	22.1643	24.4330	26.5093	29.0505	51.8051	55.7585	59.3417	63.6907	66.7660
50	27.9907	29.7067	32.3574	34.7643	37.6886	63.1671	67.5048	71.4202	76.1539	79.4900
60	35.5345	37.4849	40.4817	43.1880	46.4589	74.3970	79.0819	83.2977	88.3794	91.9517
70	43.2752	45.4417	48.7576	51.7393	55.3289	85.5270	90.5312	95.0232	100.4252	104.2149
80	51.1719	53.5401	57.1532	60.3915	64.2778	96.5782	101.8795	106.6286	112.3288	116.3211
90	59.1963	61.7541	65.6466	69.1260	73.2911	107.5650	113.1453	118.1359	124.1163	128.2989
100	67.3276	70.0649	74.2219	77.9295	82.3581	118.4980	124.3421	129.5612	135.8067	140.1695

TABLE A4.5 Quantiles of χ^2-distributions: $\chi^2_{\alpha;v}$ such that $P(W > \chi^2_{\alpha;v}) = \alpha$ for $W \sim \chi^2_v$

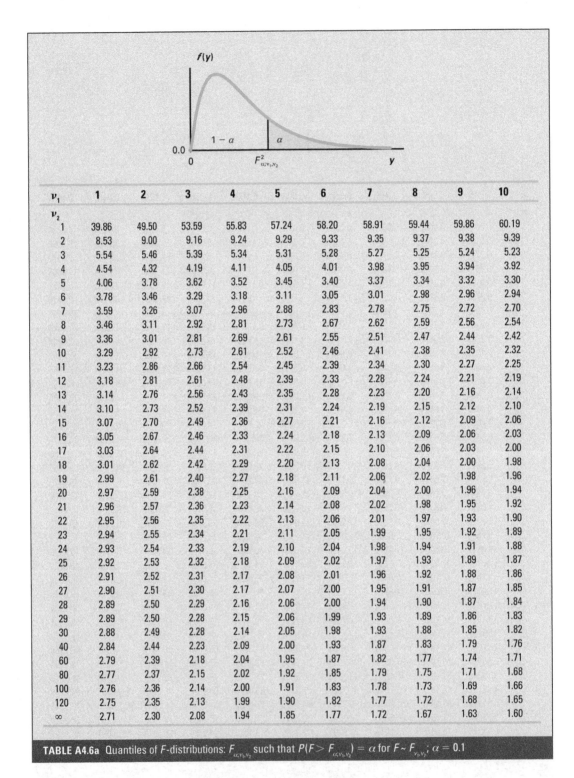

ν_1	1	2	3	4	5	6	7	8	9	10
ν_2										
1	39.86	49.50	53.59	55.83	57.24	58.20	58.91	59.44	59.86	60.19
2	8.53	9.00	9.16	9.24	9.29	9.33	9.35	9.37	9.38	9.39
3	5.54	5.46	5.39	5.34	5.31	5.28	5.27	5.25	5.24	5.23
4	4.54	4.32	4.19	4.11	4.05	4.01	3.98	3.95	3.94	3.92
5	4.06	3.78	3.62	3.52	3.45	3.40	3.37	3.34	3.32	3.30
6	3.78	3.46	3.29	3.18	3.11	3.05	3.01	2.98	2.96	2.94
7	3.59	3.26	3.07	2.96	2.88	2.83	2.78	2.75	2.72	2.70
8	3.46	3.11	2.92	2.81	2.73	2.67	2.62	2.59	2.56	2.54
9	3.36	3.01	2.81	2.69	2.61	2.55	2.51	2.47	2.44	2.42
10	3.29	2.92	2.73	2.61	2.52	2.46	2.41	2.38	2.35	2.32
11	3.23	2.86	2.66	2.54	2.45	2.39	2.34	2.30	2.27	2.25
12	3.18	2.81	2.61	2.48	2.39	2.33	2.28	2.24	2.21	2.19
13	3.14	2.76	2.56	2.43	2.35	2.28	2.23	2.20	2.16	2.14
14	3.10	2.73	2.52	2.39	2.31	2.24	2.19	2.15	2.12	2.10
15	3.07	2.70	2.49	2.36	2.27	2.21	2.16	2.12	2.09	2.06
16	3.05	2.67	2.46	2.33	2.24	2.18	2.13	2.09	2.06	2.03
17	3.03	2.64	2.44	2.31	2.22	2.15	2.10	2.06	2.03	2.00
18	3.01	2.62	2.42	2.29	2.20	2.13	2.08	2.04	2.00	1.98
19	2.99	2.61	2.40	2.27	2.18	2.11	2.06	2.02	1.98	1.96
20	2.97	2.59	2.38	2.25	2.16	2.09	2.04	2.00	1.96	1.94
21	2.96	2.57	2.36	2.23	2.14	2.08	2.02	1.98	1.95	1.92
22	2.95	2.56	2.35	2.22	2.13	2.06	2.01	1.97	1.93	1.90
23	2.94	2.55	2.34	2.21	2.11	2.05	1.99	1.95	1.92	1.89
24	2.93	2.54	2.33	2.19	2.10	2.04	1.98	1.94	1.91	1.88
25	2.92	2.53	2.32	2.18	2.09	2.02	1.97	1.93	1.89	1.87
26	2.91	2.52	2.31	2.17	2.08	2.01	1.96	1.92	1.88	1.86
27	2.90	2.51	2.30	2.17	2.07	2.00	1.95	1.91	1.87	1.85
28	2.89	2.50	2.29	2.16	2.06	2.00	1.94	1.90	1.87	1.84
29	2.89	2.50	2.28	2.15	2.06	1.99	1.93	1.89	1.86	1.83
30	2.88	2.49	2.28	2.14	2.05	1.98	1.93	1.88	1.85	1.82
40	2.84	2.44	2.23	2.09	2.00	1.93	1.87	1.83	1.79	1.76
60	2.79	2.39	2.18	2.04	1.95	1.87	1.82	1.77	1.74	1.71
80	2.77	2.37	2.15	2.02	1.92	1.85	1.79	1.75	1.71	1.68
100	2.76	2.36	2.14	2.00	1.91	1.83	1.78	1.73	1.69	1.66
120	2.75	2.35	2.13	1.99	1.90	1.82	1.77	1.72	1.68	1.65
∞	2.71	2.30	2.08	1.94	1.85	1.77	1.72	1.67	1.63	1.60

TABLE A4.6a Quantiles of F-distributions: $F_{\alpha;\nu_1,\nu_2}$ such that $P(F > F_{\alpha;\nu_1,\nu_2}) = \alpha$ for $F \sim F_{\nu_1,\nu_2}$; $\alpha = 0.1$

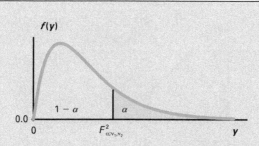

12	15	20	24	30	40	60	120	∞	ν_1
									ν_2
60.71	61.22	61.74	62.00	62.26	62.53	62.79	63.06	63.33	1
9.41	9.42	9.44	9.45	9.46	9.47	9.47	9.48	9.49	2
5.22	5.20	5.18	5.18	5.17	5.16	5.15	5.14	5.13	3
3.90	3.87	3.84	3.83	3.82	3.80	3.79	3.78	3.76	4
3.27	3.24	3.21	3.19	3.17	3.16	3.14	3.12	3.11	5
2.90	2.87	2.84	2.82	2.80	2.78	2.76	2.74	2.72	6
2.67	2.63	2.59	2.58	2.56	2.54	2.51	2.49	2.47	7
2.50	2.46	2.42	2.40	2.38	2.36	2.34	2.32	2.29	8
2.38	2.34	2.30	2.28	2.25	2.23	2.21	2.18	2.16	9
2.28	2.24	2.20	2.18	2.16	2.13	2.11	2.08	2.06	10
2.21	2.17	2.12	2.10	2.08	2.05	2.03	2.00	1.97	11
2.15	2.10	2.06	2.04	2.01	1.99	1.96	1.93	1.90	12
2.10	2.05	2.01	1.98	1.96	1.93	1.90	1.88	1.85	13
2.05	2.01	1.96	1.94	1.91	1.89	1.86	1.83	1.80	14
2.02	1.97	1.92	1.90	1.87	1.85	1.82	1.79	1.76	15
1.99	1.94	1.89	1.87	1.84	1.81	1.78	1.75	1.72	16
1.96	1.91	1.86	1.84	1.81	1.78	1.75	1.72	1.69	17
1.93	1.89	1.84	1.81	1.78	1.75	1.72	1.69	1.66	18
1.91	1.86	1.81	1.79	1.76	1.73	1.70	1.67	1.63	19
1.89	1.84	1.79	1.77	1.74	1.71	1.68	1.64	1.61	20
1.87	1.83	1.78	1.75	1.72	1.69	1.66	1.62	1.59	21
1.86	1.81	1.76	1.73	1.70	1.67	1.64	1.60	1.57	22
1.84	1.80	1.74	1.72	1.69	1.66	1.62	1.59	1.55	23
1.83	1.78	1.73	1.70	1.67	1.64	1.61	1.57	1.53	24
1.82	1.77	1.72	1.69	1.66	1.63	1.59	1.56	1.52	25
1.81	1.76	1.71	1.68	1.65	1.61	1.58	1.54	1.50	26
1.80	1.75	1.70	1.67	1.64	1.60	1.57	1.53	1.49	27
1.79	1.74	1.69	1.66	1.63	1.59	1.56	1.52	1.48	28
1.78	1.73	1.68	1.65	1.62	1.58	1.55	1.51	1.47	29
1.77	1.72	1.67	1.64	1.61	1.57	1.54	1.50	1.46	30
1.71	1.66	1.61	1.57	1.54	1.51	1.47	1.42	1.38	40
1.66	1.60	1.54	1.51	1.48	1.44	1.40	1.35	1.29	60
1.63	1.57	1.51	1.48	1.44	1.40	1.36	1.31	1.24	80
1.61	1.56	1.49	1.46	1.42	1.38	1.34	1.28	1.21	100
1.60	1.55	1.48	1.45	1.41	1.37	1.32	1.26	1.19	120
1.55	1.49	1.42	1.38	1.34	1.30	1.24	1.17	1.00	∞

TABLE A4.6a (Continued) Quantiles of F-distributions: $F_{\alpha;\nu_1,\nu_2}$ such that $P(F > F_{\alpha;\nu_1,\nu_2}) = \alpha$ for $F \sim F_{\nu_1,\nu_2}$; $\alpha = 0.1$

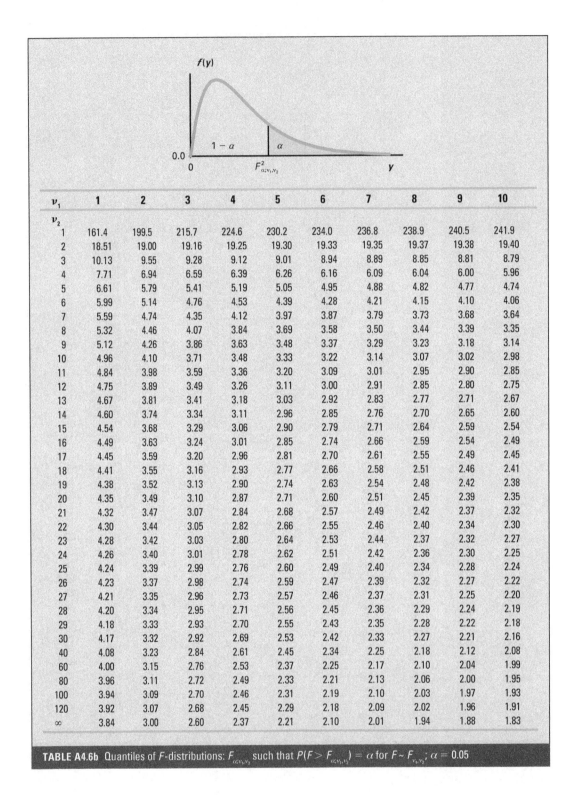

ν_1	1	2	3	4	5	6	7	8	9	10
ν_2										
1	161.4	199.5	215.7	224.6	230.2	234.0	236.8	238.9	240.5	241.9
2	18.51	19.00	19.16	19.25	19.30	19.33	19.35	19.37	19.38	19.40
3	10.13	9.55	9.28	9.12	9.01	8.94	8.89	8.85	8.81	8.79
4	7.71	6.94	6.59	6.39	6.26	6.16	6.09	6.04	6.00	5.96
5	6.61	5.79	5.41	5.19	5.05	4.95	4.88	4.82	4.77	4.74
6	5.99	5.14	4.76	4.53	4.39	4.28	4.21	4.15	4.10	4.06
7	5.59	4.74	4.35	4.12	3.97	3.87	3.79	3.73	3.68	3.64
8	5.32	4.46	4.07	3.84	3.69	3.58	3.50	3.44	3.39	3.35
9	5.12	4.26	3.86	3.63	3.48	3.37	3.29	3.23	3.18	3.14
10	4.96	4.10	3.71	3.48	3.33	3.22	3.14	3.07	3.02	2.98
11	4.84	3.98	3.59	3.36	3.20	3.09	3.01	2.95	2.90	2.85
12	4.75	3.89	3.49	3.26	3.11	3.00	2.91	2.85	2.80	2.75
13	4.67	3.81	3.41	3.18	3.03	2.92	2.83	2.77	2.71	2.67
14	4.60	3.74	3.34	3.11	2.96	2.85	2.76	2.70	2.65	2.60
15	4.54	3.68	3.29	3.06	2.90	2.79	2.71	2.64	2.59	2.54
16	4.49	3.63	3.24	3.01	2.85	2.74	2.66	2.59	2.54	2.49
17	4.45	3.59	3.20	2.96	2.81	2.70	2.61	2.55	2.49	2.45
18	4.41	3.55	3.16	2.93	2.77	2.66	2.58	2.51	2.46	2.41
19	4.38	3.52	3.13	2.90	2.74	2.63	2.54	2.48	2.42	2.38
20	4.35	3.49	3.10	2.87	2.71	2.60	2.51	2.45	2.39	2.35
21	4.32	3.47	3.07	2.84	2.68	2.57	2.49	2.42	2.37	2.32
22	4.30	3.44	3.05	2.82	2.66	2.55	2.46	2.40	2.34	2.30
23	4.28	3.42	3.03	2.80	2.64	2.53	2.44	2.37	2.32	2.27
24	4.26	3.40	3.01	2.78	2.62	2.51	2.42	2.36	2.30	2.25
25	4.24	3.39	2.99	2.76	2.60	2.49	2.40	2.34	2.28	2.24
26	4.23	3.37	2.98	2.74	2.59	2.47	2.39	2.32	2.27	2.22
27	4.21	3.35	2.96	2.73	2.57	2.46	2.37	2.31	2.25	2.20
28	4.20	3.34	2.95	2.71	2.56	2.45	2.36	2.29	2.24	2.19
29	4.18	3.33	2.93	2.70	2.55	2.43	2.35	2.28	2.22	2.18
30	4.17	3.32	2.92	2.69	2.53	2.42	2.33	2.27	2.21	2.16
40	4.08	3.23	2.84	2.61	2.45	2.34	2.25	2.18	2.12	2.08
60	4.00	3.15	2.76	2.53	2.37	2.25	2.17	2.10	2.04	1.99
80	3.96	3.11	2.72	2.49	2.33	2.21	2.13	2.06	2.00	1.95
100	3.94	3.09	2.70	2.46	2.31	2.19	2.10	2.03	1.97	1.93
120	3.92	3.07	2.68	2.45	2.29	2.18	2.09	2.02	1.96	1.91
∞	3.84	3.00	2.60	2.37	2.21	2.10	2.01	1.94	1.88	1.83

TABLE A4.6b Quantiles of F-distributions: $F_{\alpha;\nu_1,\nu_2}$ such that $P(F > F_{\alpha;\nu_1,\nu_2}) = \alpha$ for $F \sim F_{\nu_1,\nu_2}$; $\alpha = 0.05$

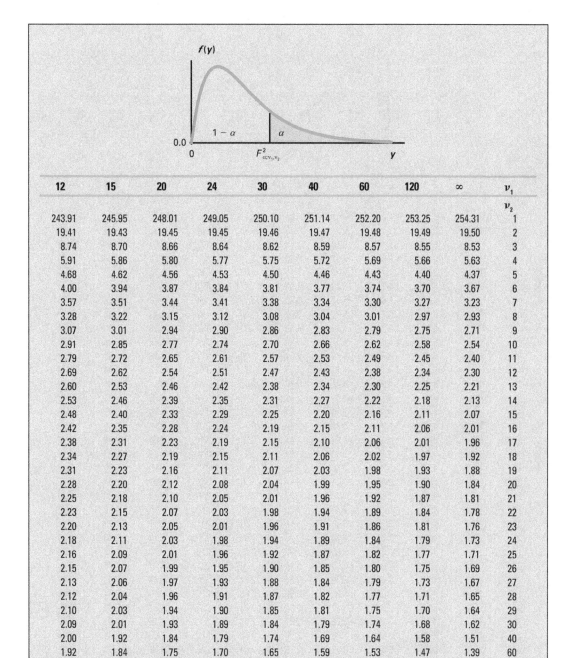

12	15	20	24	30	40	60	120	∞	ν_1
									ν_2
243.91	245.95	248.01	249.05	250.10	251.14	252.20	253.25	254.31	1
19.41	19.43	19.45	19.45	19.46	19.47	19.48	19.49	19.50	2
8.74	8.70	8.66	8.64	8.62	8.59	8.57	8.55	8.53	3
5.91	5.86	5.80	5.77	5.75	5.72	5.69	5.66	5.63	4
4.68	4.62	4.56	4.53	4.50	4.46	4.43	4.40	4.37	5
4.00	3.94	3.87	3.84	3.81	3.77	3.74	3.70	3.67	6
3.57	3.51	3.44	3.41	3.38	3.34	3.30	3.27	3.23	7
3.28	3.22	3.15	3.12	3.08	3.04	3.01	2.97	2.93	8
3.07	3.01	2.94	2.90	2.86	2.83	2.79	2.75	2.71	9
2.91	2.85	2.77	2.74	2.70	2.66	2.62	2.58	2.54	10
2.79	2.72	2.65	2.61	2.57	2.53	2.49	2.45	2.40	11
2.69	2.62	2.54	2.51	2.47	2.43	2.38	2.34	2.30	12
2.60	2.53	2.46	2.42	2.38	2.34	2.30	2.25	2.21	13
2.53	2.46	2.39	2.35	2.31	2.27	2.22	2.18	2.13	14
2.48	2.40	2.33	2.29	2.25	2.20	2.16	2.11	2.07	15
2.42	2.35	2.28	2.24	2.19	2.15	2.11	2.06	2.01	16
2.38	2.31	2.23	2.19	2.15	2.10	2.06	2.01	1.96	17
2.34	2.27	2.19	2.15	2.11	2.06	2.02	1.97	1.92	18
2.31	2.23	2.16	2.11	2.07	2.03	1.98	1.93	1.88	19
2.28	2.20	2.12	2.08	2.04	1.99	1.95	1.90	1.84	20
2.25	2.18	2.10	2.05	2.01	1.96	1.92	1.87	1.81	21
2.23	2.15	2.07	2.03	1.98	1.94	1.89	1.84	1.78	22
2.20	2.13	2.05	2.01	1.96	1.91	1.86	1.81	1.76	23
2.18	2.11	2.03	1.98	1.94	1.89	1.84	1.79	1.73	24
2.16	2.09	2.01	1.96	1.92	1.87	1.82	1.77	1.71	25
2.15	2.07	1.99	1.95	1.90	1.85	1.80	1.75	1.69	26
2.13	2.06	1.97	1.93	1.88	1.84	1.79	1.73	1.67	27
2.12	2.04	1.96	1.91	1.87	1.82	1.77	1.71	1.65	28
2.10	2.03	1.94	1.90	1.85	1.81	1.75	1.70	1.64	29
2.09	2.01	1.93	1.89	1.84	1.79	1.74	1.68	1.62	30
2.00	1.92	1.84	1.79	1.74	1.69	1.64	1.58	1.51	40
1.92	1.84	1.75	1.70	1.65	1.59	1.53	1.47	1.39	60
1.88	1.79	1.70	1.65	1.60	1.54	1.48	1.41	1.32	80
1.85	1.77	1.68	1.63	1.57	1.52	1.45	1.38	1.28	100
1.83	1.75	1.66	1.61	1.55	1.50	1.43	1.35	1.25	120
1.75	1.67	1.57	1.52	1.46	1.39	1.32	1.22	1.00	∞

TABLE A4.6b (Continued) Quantiles of F-distributions: $F_{\alpha;\nu_1,\nu_2}$ such that $P(F > F_{\alpha;\nu_1,\nu_2}) = \alpha$ for $F \sim F_{\nu_1,\nu_2}$; $\alpha = 0.05$

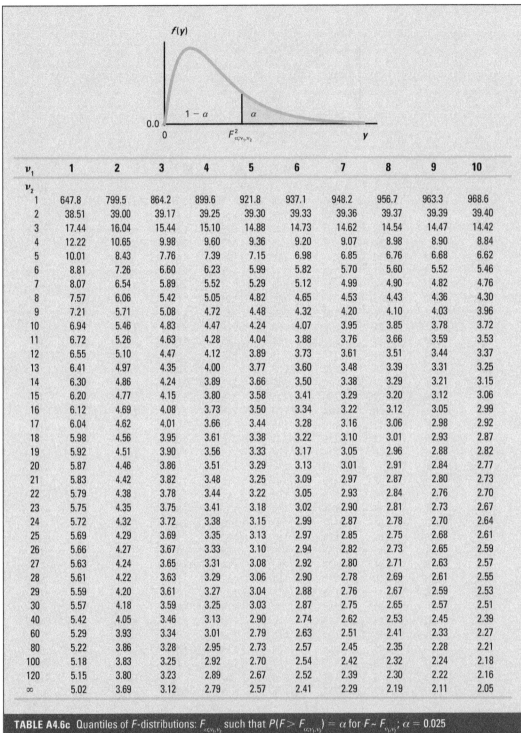

ν_1	1	2	3	4	5	6	7	8	9	10
ν_2										
1	647.8	799.5	864.2	899.6	921.8	937.1	948.2	956.7	963.3	968.6
2	38.51	39.00	39.17	39.25	39.30	39.33	39.36	39.37	39.39	39.40
3	17.44	16.04	15.44	15.10	14.88	14.73	14.62	14.54	14.47	14.42
4	12.22	10.65	9.98	9.60	9.36	9.20	9.07	8.98	8.90	8.84
5	10.01	8.43	7.76	7.39	7.15	6.98	6.85	6.76	6.68	6.62
6	8.81	7.26	6.60	6.23	5.99	5.82	5.70	5.60	5.52	5.46
7	8.07	6.54	5.89	5.52	5.29	5.12	4.99	4.90	4.82	4.76
8	7.57	6.06	5.42	5.05	4.82	4.65	4.53	4.43	4.36	4.30
9	7.21	5.71	5.08	4.72	4.48	4.32	4.20	4.10	4.03	3.96
10	6.94	5.46	4.83	4.47	4.24	4.07	3.95	3.85	3.78	3.72
11	6.72	5.26	4.63	4.28	4.04	3.88	3.76	3.66	3.59	3.53
12	6.55	5.10	4.47	4.12	3.89	3.73	3.61	3.51	3.44	3.37
13	6.41	4.97	4.35	4.00	3.77	3.60	3.48	3.39	3.31	3.25
14	6.30	4.86	4.24	3.89	3.66	3.50	3.38	3.29	3.21	3.15
15	6.20	4.77	4.15	3.80	3.58	3.41	3.29	3.20	3.12	3.06
16	6.12	4.69	4.08	3.73	3.50	3.34	3.22	3.12	3.05	2.99
17	6.04	4.62	4.01	3.66	3.44	3.28	3.16	3.06	2.98	2.92
18	5.98	4.56	3.95	3.61	3.38	3.22	3.10	3.01	2.93	2.87
19	5.92	4.51	3.90	3.56	3.33	3.17	3.05	2.96	2.88	2.82
20	5.87	4.46	3.86	3.51	3.29	3.13	3.01	2.91	2.84	2.77
21	5.83	4.42	3.82	3.48	3.25	3.09	2.97	2.87	2.80	2.73
22	5.79	4.38	3.78	3.44	3.22	3.05	2.93	2.84	2.76	2.70
23	5.75	4.35	3.75	3.41	3.18	3.02	2.90	2.81	2.73	2.67
24	5.72	4.32	3.72	3.38	3.15	2.99	2.87	2.78	2.70	2.64
25	5.69	4.29	3.69	3.35	3.13	2.97	2.85	2.75	2.68	2.61
26	5.66	4.27	3.67	3.33	3.10	2.94	2.82	2.73	2.65	2.59
27	5.63	4.24	3.65	3.31	3.08	2.92	2.80	2.71	2.63	2.57
28	5.61	4.22	3.63	3.29	3.06	2.90	2.78	2.69	2.61	2.55
29	5.59	4.20	3.61	3.27	3.04	2.88	2.76	2.67	2.59	2.53
30	5.57	4.18	3.59	3.25	3.03	2.87	2.75	2.65	2.57	2.51
40	5.42	4.05	3.46	3.13	2.90	2.74	2.62	2.53	2.45	2.39
60	5.29	3.93	3.34	3.01	2.79	2.63	2.51	2.41	2.33	2.27
80	5.22	3.86	3.28	2.95	2.73	2.57	2.45	2.35	2.28	2.21
100	5.18	3.83	3.25	2.92	2.70	2.54	2.42	2.32	2.24	2.18
120	5.15	3.80	3.23	2.89	2.67	2.52	2.39	2.30	2.22	2.16
∞	5.02	3.69	3.12	2.79	2.57	2.41	2.29	2.19	2.11	2.05

TABLE A4.6c Quantiles of F-distributions: $F_{(\alpha;\nu_1,\nu_2)}$ such that $P(F > F_{(\alpha;\nu_1,\nu_2)}) = \alpha$ for $F \sim F_{\nu_1,\nu_2}$; $\alpha = 0.025$

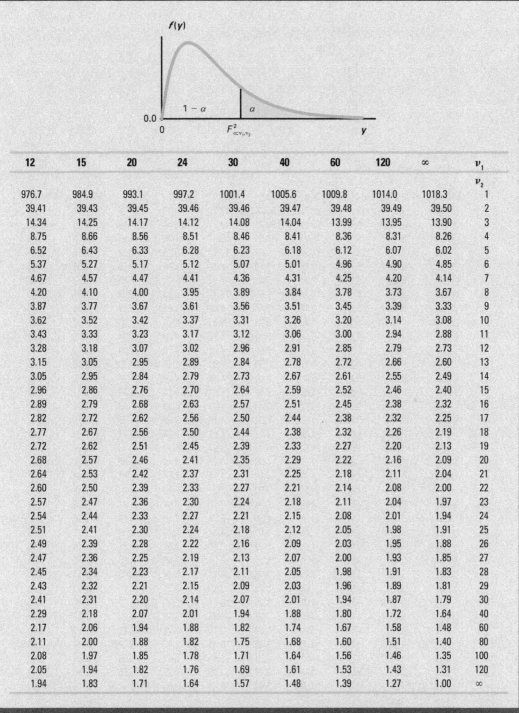

12	15	20	24	30	40	60	120	∞	v_1
									v_2
976.7	984.9	993.1	997.2	1001.4	1005.6	1009.8	1014.0	1018.3	1
39.41	39.43	39.45	39.46	39.46	39.47	39.48	39.49	39.50	2
14.34	14.25	14.17	14.12	14.08	14.04	13.99	13.95	13.90	3
8.75	8.66	8.56	8.51	8.46	8.41	8.36	8.31	8.26	4
6.52	6.43	6.33	6.28	6.23	6.18	6.12	6.07	6.02	5
5.37	5.27	5.17	5.12	5.07	5.01	4.96	4.90	4.85	6
4.67	4.57	4.47	4.41	4.36	4.31	4.25	4.20	4.14	7
4.20	4.10	4.00	3.95	3.89	3.84	3.78	3.73	3.67	8
3.87	3.77	3.67	3.61	3.56	3.51	3.45	3.39	3.33	9
3.62	3.52	3.42	3.37	3.31	3.26	3.20	3.14	3.08	10
3.43	3.33	3.23	3.17	3.12	3.06	3.00	2.94	2.88	11
3.28	3.18	3.07	3.02	2.96	2.91	2.85	2.79	2.73	12
3.15	3.05	2.95	2.89	2.84	2.78	2.72	2.66	2.60	13
3.05	2.95	2.84	2.79	2.73	2.67	2.61	2.55	2.49	14
2.96	2.86	2.76	2.70	2.64	2.59	2.52	2.46	2.40	15
2.89	2.79	2.68	2.63	2.57	2.51	2.45	2.38	2.32	16
2.82	2.72	2.62	2.56	2.50	2.44	2.38	2.32	2.25	17
2.77	2.67	2.56	2.50	2.44	2.38	2.32	2.26	2.19	18
2.72	2.62	2.51	2.45	2.39	2.33	2.27	2.20	2.13	19
2.68	2.57	2.46	2.41	2.35	2.29	2.22	2.16	2.09	20
2.64	2.53	2.42	2.37	2.31	2.25	2.18	2.11	2.04	21
2.60	2.50	2.39	2.33	2.27	2.21	2.14	2.08	2.00	22
2.57	2.47	2.36	2.30	2.24	2.18	2.11	2.04	1.97	23
2.54	2.44	2.33	2.27	2.21	2.15	2.08	2.01	1.94	24
2.51	2.41	2.30	2.24	2.18	2.12	2.05	1.98	1.91	25
2.49	2.39	2.28	2.22	2.16	2.09	2.03	1.95	1.88	26
2.47	2.36	2.25	2.19	2.13	2.07	2.00	1.93	1.85	27
2.45	2.34	2.23	2.17	2.11	2.05	1.98	1.91	1.83	28
2.43	2.32	2.21	2.15	2.09	2.03	1.96	1.89	1.81	29
2.41	2.31	2.20	2.14	2.07	2.01	1.94	1.87	1.79	30
2.29	2.18	2.07	2.01	1.94	1.88	1.80	1.72	1.64	40
2.17	2.06	1.94	1.88	1.82	1.74	1.67	1.58	1.48	60
2.11	2.00	1.88	1.82	1.75	1.68	1.60	1.51	1.40	80
2.08	1.97	1.85	1.78	1.71	1.64	1.56	1.46	1.35	100
2.05	1.94	1.82	1.76	1.69	1.61	1.53	1.43	1.31	120
1.94	1.83	1.71	1.64	1.57	1.48	1.39	1.27	1.00	∞

TABLE A4.6c (Continued) Quantiles of F-distributions: $F_{\alpha;v_1,v_2}$ such that $P(F > F_{\alpha;v_1,v_2}) = \alpha$ for $F \sim F_{v_1,v_2}$; $\alpha = 0.025$

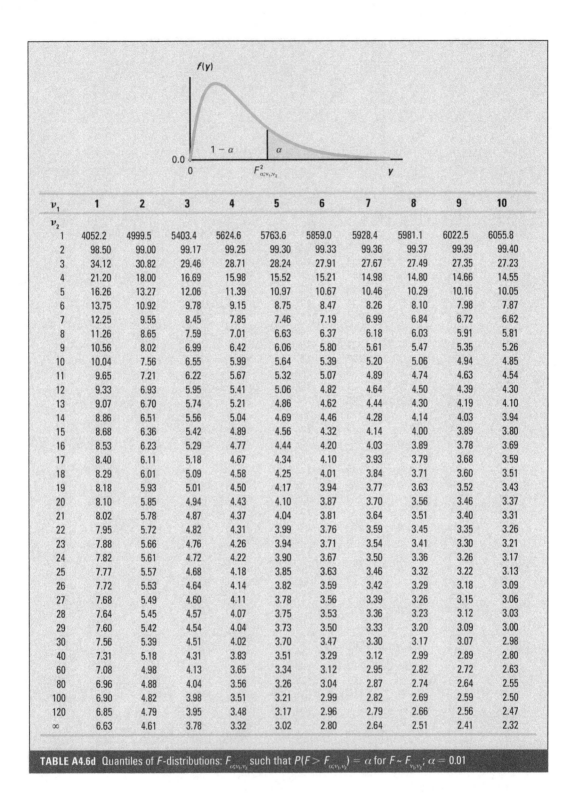

ν_1	1	2	3	4	5	6	7	8	9	10
ν_2										
1	4052.2	4999.5	5403.4	5624.6	5763.6	5859.0	5928.4	5981.1	6022.5	6055.8
2	98.50	99.00	99.17	99.25	99.30	99.33	99.36	99.37	99.39	99.40
3	34.12	30.82	29.46	28.71	28.24	27.91	27.67	27.49	27.35	27.23
4	21.20	18.00	16.69	15.98	15.52	15.21	14.98	14.80	14.66	14.55
5	16.26	13.27	12.06	11.39	10.97	10.67	10.46	10.29	10.16	10.05
6	13.75	10.92	9.78	9.15	8.75	8.47	8.26	8.10	7.98	7.87
7	12.25	9.55	8.45	7.85	7.46	7.19	6.99	6.84	6.72	6.62
8	11.26	8.65	7.59	7.01	6.63	6.37	6.18	6.03	5.91	5.81
9	10.56	8.02	6.99	6.42	6.06	5.80	5.61	5.47	5.35	5.26
10	10.04	7.56	6.55	5.99	5.64	5.39	5.20	5.06	4.94	4.85
11	9.65	7.21	6.22	5.67	5.32	5.07	4.89	4.74	4.63	4.54
12	9.33	6.93	5.95	5.41	5.06	4.82	4.64	4.50	4.39	4.30
13	9.07	6.70	5.74	5.21	4.86	4.62	4.44	4.30	4.19	4.10
14	8.86	6.51	5.56	5.04	4.69	4.46	4.28	4.14	4.03	3.94
15	8.68	6.36	5.42	4.89	4.56	4.32	4.14	4.00	3.89	3.80
16	8.53	6.23	5.29	4.77	4.44	4.20	4.03	3.89	3.78	3.69
17	8.40	6.11	5.18	4.67	4.34	4.10	3.93	3.79	3.68	3.59
18	8.29	6.01	5.09	4.58	4.25	4.01	3.84	3.71	3.60	3.51
19	8.18	5.93	5.01	4.50	4.17	3.94	3.77	3.63	3.52	3.43
20	8.10	5.85	4.94	4.43	4.10	3.87	3.70	3.56	3.46	3.37
21	8.02	5.78	4.87	4.37	4.04	3.81	3.64	3.51	3.40	3.31
22	7.95	5.72	4.82	4.31	3.99	3.76	3.59	3.45	3.35	3.26
23	7.88	5.66	4.76	4.26	3.94	3.71	3.54	3.41	3.30	3.21
24	7.82	5.61	4.72	4.22	3.90	3.67	3.50	3.36	3.26	3.17
25	7.77	5.57	4.68	4.18	3.85	3.63	3.46	3.32	3.22	3.13
26	7.72	5.53	4.64	4.14	3.82	3.59	3.42	3.29	3.18	3.09
27	7.68	5.49	4.60	4.11	3.78	3.56	3.39	3.26	3.15	3.06
28	7.64	5.45	4.57	4.07	3.75	3.53	3.36	3.23	3.12	3.03
29	7.60	5.42	4.54	4.04	3.73	3.50	3.33	3.20	3.09	3.00
30	7.56	5.39	4.51	4.02	3.70	3.47	3.30	3.17	3.07	2.98
40	7.31	5.18	4.31	3.83	3.51	3.29	3.12	2.99	2.89	2.80
60	7.08	4.98	4.13	3.65	3.34	3.12	2.95	2.82	2.72	2.63
80	6.96	4.88	4.04	3.56	3.26	3.04	2.87	2.74	2.64	2.55
100	6.90	4.82	3.98	3.51	3.21	2.99	2.82	2.69	2.59	2.50
120	6.85	4.79	3.95	3.48	3.17	2.96	2.79	2.66	2.56	2.47
∞	6.63	4.61	3.78	3.32	3.02	2.80	2.64	2.51	2.41	2.32

TABLE A4.6d Quantiles of F-distributions: $F_{\alpha;\nu_1,\nu_2}$ such that $P(F > F_{\alpha;\nu_1,\nu_2}) = \alpha$ for $F \sim F_{\nu_1,\nu_2}$; $\alpha = 0.01$

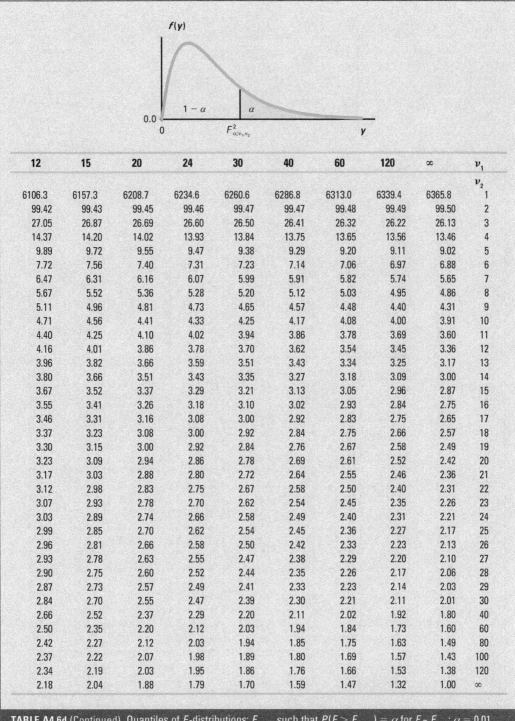

12	15	20	24	30	40	60	120	∞	ν_1
									ν_2
6106.3	6157.3	6208.7	6234.6	6260.6	6286.8	6313.0	6339.4	6365.8	1
99.42	99.43	99.45	99.46	99.47	99.47	99.48	99.49	99.50	2
27.05	26.87	26.69	26.60	26.50	26.41	26.32	26.22	26.13	3
14.37	14.20	14.02	13.93	13.84	13.75	13.65	13.56	13.46	4
9.89	9.72	9.55	9.47	9.38	9.29	9.20	9.11	9.02	5
7.72	7.56	7.40	7.31	7.23	7.14	7.06	6.97	6.88	6
6.47	6.31	6.16	6.07	5.99	5.91	5.82	5.74	5.65	7
5.67	5.52	5.36	5.28	5.20	5.12	5.03	4.95	4.86	8
5.11	4.96	4.81	4.73	4.65	4.57	4.48	4.40	4.31	9
4.71	4.56	4.41	4.33	4.25	4.17	4.08	4.00	3.91	10
4.40	4.25	4.10	4.02	3.94	3.86	3.78	3.69	3.60	11
4.16	4.01	3.86	3.78	3.70	3.62	3.54	3.45	3.36	12
3.96	3.82	3.66	3.59	3.51	3.43	3.34	3.25	3.17	13
3.80	3.66	3.51	3.43	3.35	3.27	3.18	3.09	3.00	14
3.67	3.52	3.37	3.29	3.21	3.13	3.05	2.96	2.87	15
3.55	3.41	3.26	3.18	3.10	3.02	2.93	2.84	2.75	16
3.46	3.31	3.16	3.08	3.00	2.92	2.83	2.75	2.65	17
3.37	3.23	3.08	3.00	2.92	2.84	2.75	2.66	2.57	18
3.30	3.15	3.00	2.92	2.84	2.76	2.67	2.58	2.49	19
3.23	3.09	2.94	2.86	2.78	2.69	2.61	2.52	2.42	20
3.17	3.03	2.88	2.80	2.72	2.64	2.55	2.46	2.36	21
3.12	2.98	2.83	2.75	2.67	2.58	2.50	2.40	2.31	22
3.07	2.93	2.78	2.70	2.62	2.54	2.45	2.35	2.26	23
3.03	2.89	2.74	2.66	2.58	2.49	2.40	2.31	2.21	24
2.99	2.85	2.70	2.62	2.54	2.45	2.36	2.27	2.17	25
2.96	2.81	2.66	2.58	2.50	2.42	2.33	2.23	2.13	26
2.93	2.78	2.63	2.55	2.47	2.38	2.29	2.20	2.10	27
2.90	2.75	2.60	2.52	2.44	2.35	2.26	2.17	2.06	28
2.87	2.73	2.57	2.49	2.41	2.33	2.23	2.14	2.03	29
2.84	2.70	2.55	2.47	2.39	2.30	2.21	2.11	2.01	30
2.66	2.52	2.37	2.29	2.20	2.11	2.02	1.92	1.80	40
2.50	2.35	2.20	2.12	2.03	1.94	1.84	1.73	1.60	60
2.42	2.27	2.12	2.03	1.94	1.85	1.75	1.63	1.49	80
2.37	2.22	2.07	1.98	1.89	1.80	1.69	1.57	1.43	100
2.34	2.19	2.03	1.95	1.86	1.76	1.66	1.53	1.38	120
2.18	2.04	1.88	1.79	1.70	1.59	1.47	1.32	1.00	∞

TABLE A4.6d (Continued) Quantiles of F-distributions: $F_{(\alpha;\nu_1,\nu_2)}$ such that $P(F > F_{(\alpha;\nu_1,\nu_2)}) = \alpha$ for $F \sim F_{\nu_1,\nu_2}$; $\alpha = 0.01$

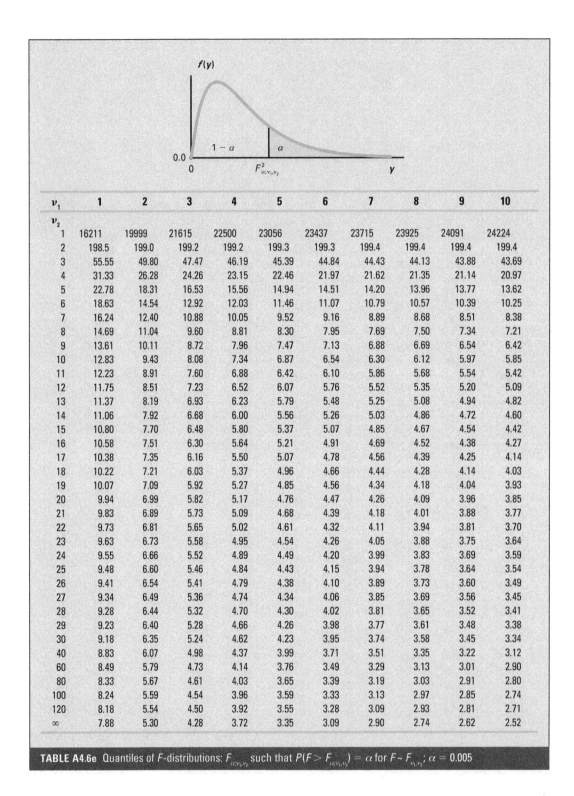

v_1	1	2	3	4	5	6	7	8	9	10
v_2										
1	16211	19999	21615	22500	23056	23437	23715	23925	24091	24224
2	198.5	199.0	199.2	199.2	199.3	199.3	199.4	199.4	199.4	199.4
3	55.55	49.80	47.47	46.19	45.39	44.84	44.43	44.13	43.88	43.69
4	31.33	26.28	24.26	23.15	22.46	21.97	21.62	21.35	21.14	20.97
5	22.78	18.31	16.53	15.56	14.94	14.51	14.20	13.96	13.77	13.62
6	18.63	14.54	12.92	12.03	11.46	11.07	10.79	10.57	10.39	10.25
7	16.24	12.40	10.88	10.05	9.52	9.16	8.89	8.68	8.51	8.38
8	14.69	11.04	9.60	8.81	8.30	7.95	7.69	7.50	7.34	7.21
9	13.61	10.11	8.72	7.96	7.47	7.13	6.88	6.69	6.54	6.42
10	12.83	9.43	8.08	7.34	6.87	6.54	6.30	6.12	5.97	5.85
11	12.23	8.91	7.60	6.88	6.42	6.10	5.86	5.68	5.54	5.42
12	11.75	8.51	7.23	6.52	6.07	5.76	5.52	5.35	5.20	5.09
13	11.37	8.19	6.93	6.23	5.79	5.48	5.25	5.08	4.94	4.82
14	11.06	7.92	6.68	6.00	5.56	5.26	5.03	4.86	4.72	4.60
15	10.80	7.70	6.48	5.80	5.37	5.07	4.85	4.67	4.54	4.42
16	10.58	7.51	6.30	5.64	5.21	4.91	4.69	4.52	4.38	4.27
17	10.38	7.35	6.16	5.50	5.07	4.78	4.56	4.39	4.25	4.14
18	10.22	7.21	6.03	5.37	4.96	4.66	4.44	4.28	4.14	4.03
19	10.07	7.09	5.92	5.27	4.85	4.56	4.34	4.18	4.04	3.93
20	9.94	6.99	5.82	5.17	4.76	4.47	4.26	4.09	3.96	3.85
21	9.83	6.89	5.73	5.09	4.68	4.39	4.18	4.01	3.88	3.77
22	9.73	6.81	5.65	5.02	4.61	4.32	4.11	3.94	3.81	3.70
23	9.63	6.73	5.58	4.95	4.54	4.26	4.05	3.88	3.75	3.64
24	9.55	6.66	5.52	4.89	4.49	4.20	3.99	3.83	3.69	3.59
25	9.48	6.60	5.46	4.84	4.43	4.15	3.94	3.78	3.64	3.54
26	9.41	6.54	5.41	4.79	4.38	4.10	3.89	3.73	3.60	3.49
27	9.34	6.49	5.36	4.74	4.34	4.06	3.85	3.69	3.56	3.45
28	9.28	6.44	5.32	4.70	4.30	4.02	3.81	3.65	3.52	3.41
29	9.23	6.40	5.28	4.66	4.26	3.98	3.77	3.61	3.48	3.38
30	9.18	6.35	5.24	4.62	4.23	3.95	3.74	3.58	3.45	3.34
40	8.83	6.07	4.98	4.37	3.99	3.71	3.51	3.35	3.22	3.12
60	8.49	5.79	4.73	4.14	3.76	3.49	3.29	3.13	3.01	2.90
80	8.33	5.67	4.61	4.03	3.65	3.39	3.19	3.03	2.91	2.80
100	8.24	5.59	4.54	3.96	3.59	3.33	3.13	2.97	2.85	2.74
120	8.18	5.54	4.50	3.92	3.55	3.28	3.09	2.93	2.81	2.71
∞	7.88	5.30	4.28	3.72	3.35	3.09	2.90	2.74	2.62	2.52

TABLE A4.6e Quantiles of F-distributions: $F_{\alpha;v_1,v_2}$ such that $P(F > F_{\alpha;v_1,v_2}) = \alpha$ for $F \sim F_{v_1,v_2}$; $\alpha = 0.005$

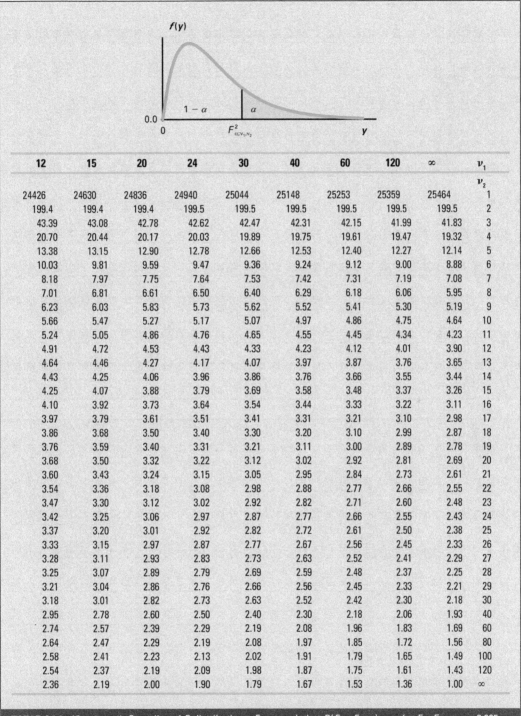

12	15	20	24	30	40	60	120	∞	ν_1
									ν_2
24426	24630	24836	24940	25044	25148	25253	25359	25464	1
199.4	199.4	199.4	199.5	199.5	199.5	199.5	199.5	199.5	2
43.39	43.08	42.78	42.62	42.47	42.31	42.15	41.99	41.83	3
20.70	20.44	20.17	20.03	19.89	19.75	19.61	19.47	19.32	4
13.38	13.15	12.90	12.78	12.66	12.53	12.40	12.27	12.14	5
10.03	9.81	9.59	9.47	9.36	9.24	9.12	9.00	8.88	6
8.18	7.97	7.75	7.64	7.53	7.42	7.31	7.19	7.08	7
7.01	6.81	6.61	6.50	6.40	6.29	6.18	6.06	5.95	8
6.23	6.03	5.83	5.73	5.62	5.52	5.41	5.30	5.19	9
5.66	5.47	5.27	5.17	5.07	4.97	4.86	4.75	4.64	10
5.24	5.05	4.86	4.76	4.65	4.55	4.45	4.34	4.23	11
4.91	4.72	4.53	4.43	4.33	4.23	4.12	4.01	3.90	12
4.64	4.46	4.27	4.17	4.07	3.97	3.87	3.76	3.65	13
4.43	4.25	4.06	3.96	3.86	3.76	3.66	3.55	3.44	14
4.25	4.07	3.88	3.79	3.69	3.58	3.48	3.37	3.26	15
4.10	3.92	3.73	3.64	3.54	3.44	3.33	3.22	3.11	16
3.97	3.79	3.61	3.51	3.41	3.31	3.21	3.10	2.98	17
3.86	3.68	3.50	3.40	3.30	3.20	3.10	2.99	2.87	18
3.76	3.59	3.40	3.31	3.21	3.11	3.00	2.89	2.78	19
3.68	3.50	3.32	3.22	3.12	3.02	2.92	2.81	2.69	20
3.60	3.43	3.24	3.15	3.05	2.95	2.84	2.73	2.61	21
3.54	3.36	3.18	3.08	2.98	2.88	2.77	2.66	2.55	22
3.47	3.30	3.12	3.02	2.92	2.82	2.71	2.60	2.48	23
3.42	3.25	3.06	2.97	2.87	2.77	2.66	2.55	2.43	24
3.37	3.20	3.01	2.92	2.82	2.72	2.61	2.50	2.38	25
3.33	3.15	2.97	2.87	2.77	2.67	2.56	2.45	2.33	26
3.28	3.11	2.93	2.83	2.73	2.63	2.52	2.41	2.29	27
3.25	3.07	2.89	2.79	2.69	2.59	2.48	2.37	2.25	28
3.21	3.04	2.86	2.76	2.66	2.56	2.45	2.33	2.21	29
3.18	3.01	2.82	2.73	2.63	2.52	2.42	2.30	2.18	30
2.95	2.78	2.60	2.50	2.40	2.30	2.18	2.06	1.93	40
2.74	2.57	2.39	2.29	2.19	2.08	1.96	1.83	1.69	60
2.64	2.47	2.29	2.19	2.08	1.97	1.85	1.72	1.56	80
2.58	2.41	2.23	2.13	2.02	1.91	1.79	1.65	1.49	100
2.54	2.37	2.19	2.09	1.98	1.87	1.75	1.61	1.43	120
2.36	2.19	2.00	1.90	1.79	1.67	1.53	1.36	1.00	∞

TABLE A4.6e (Continued) Quantiles of F-distributions: $F_{(\alpha;\nu_1,\nu_2)}$ such that $P(F > F_{(\alpha;\nu_1,\nu_2)}) = \alpha$ for $F \sim F_{\nu_1,\nu_2}$; $\alpha = 0.005$

TABLE A4.7b Critical Values for the Durbin–Watson d Statistic ($\alpha = .01$)

n	$k=1$ $d_{L,.01}$	$d_{U,.01}$	$k=2$ $d_{L,.01}$	$d_{U,.01}$	$k=3$ $d_{L,.01}$	$d_{U,.01}$	$k=4$ $d_{L,.01}$	$d_{U,.01}$	$k=5$ $d_{L,.01}$	$d_{U,.01}$
15	0.81	1.07	0.70	1.25	0.59	1.46	0.49	1.70	0.39	1.96
16	0.84	1.09	0.74	1.25	0.63	1.44	0.53	1.66	0.44	1.90
17	0.87	1.10	0.77	1.25	0.67	1.43	0.57	1.63	0.48	1.85
18	0.90	1.12	0.80	1.26	0.71	1.42	0.61	1.60	0.52	1.80
19	0.93	1.13	0.83	1.26	0.74	1.41	0.65	1.58	0.56	1.77
20	0.95	1.15	0.86	1.27	0.77	1.41	0.68	1.57	0.60	1.74
21	0.97	1.16	0.89	1.27	0.80	1.41	0.72	1.55	0.63	1.71
22	1.00	1.17	0.91	1.28	0.83	1.40	0.75	1.54	0.66	1.69
23	1.02	1.19	0.94	1.29	0.86	1.40	0.77	1.53	0.70	1.67
24	1.04	1.20	0.96	1.30	0.88	1.41	0.80	1.53	0.72	1.66
25	1.05	1.21	0.98	1.30	0.90	1.41	0.83	1.52	0.75	1.65
26	1.07	1.22	1.00	1.31	0.93	1.41	0.85	1.52	0.78	1.64
27	1.09	1.23	1.02	1.32	0.95	1.41	0.88	1.51	0.81	1.63
28	1.10	1.24	1.04	1.32	0.97	1.41	0.90	1.51	0.83	1.62
29	1.12	1.25	1.05	1.33	0.99	1.42	0.92	1.51	0.85	1.61
30	1.13	1.26	1.07	1.34	1.01	1.42	0.94	1.51	0.88	1.61
31	1.15	1.27	1.08	1.34	1.02	1.42	0.96	1.51	0.90	1.60
32	1.16	1.28	1.10	1.35	1.04	1.43	0.98	1.51	0.92	1.60
33	1.17	1.29	1.11	1.36	1.05	1.43	1.00	1.51	0.94	1.59
34	1.18	1.30	1.13	1.36	1.07	1.43	1.01	1.51	0.95	1.59
35	1.19	1.31	1.14	1.37	1.08	1.44	1.03	1.51	0.97	1.59
36	1.21	1.32	1.15	1.38	1.10	1.44	1.04	1.51	0.99	1.59
37	1.22	1.32	1.16	1.38	1.11	1.45	1.06	1.51	1.00	1.59
38	1.23	1.33	1.18	1.39	1.12	1.45	1.07	1.52	1.02	1.58
39	1.24	1.34	1.19	1.39	1.14	1.45	1.09	1.52	1.03	1.58
40	1.25	1.34	1.20	1.40	1.15	1.46	1.10	1.52	1.05	1.58
45	1.29	1.38	1.24	1.42	1.20	1.48	1.16	1.53	1.11	1.58
50	1.32	1.40	1.28	1.45	1.24	1.49	1.20	1.54	1.16	1.59
55	1.36	1.43	1.32	1.47	1.28	1.51	1.25	1.55	1.21	1.59
60	1.38	1.45	1.35	1.48	1.32	1.52	1.28	1.56	1.25	1.60
65	1.41	1.47	1.38	1.50	1.35	1.53	1.31	1.57	1.28	1.61
70	1.43	1.49	1.40	1.52	1.37	1.55	1.34	1.58	1.31	1.61
75	1.45	1.50	1.42	1.53	1.39	1.56	1.37	1.59	1.34	1.62
80	1.47	1.52	1.44	1.54	1.42	1.57	1.39	1.60	1.36	1.62
85	1.48	1.53	1.46	1.55	1.43	1.58	1.41	1.60	1.39	1.63
90	1.50	1.54	1.47	1.56	1.45	1.59	1.43	1.61	1.41	1.64
95	1.51	1.55	1.49	1.57	1.47	1.60	1.45	1.62	1.42	1.64
100	1.52	1.56	1.50	1.58	1.48	1.60	1.46	1.63	1.44	1.65

Source: J. Durbin and G. S. Watson, 'Testing for Serial Correlation in Least Squares Regression, II,' *Biometrika* 30 (1951), pp. 159–78. Reproduced by permission of the Biometrika Trustees.

TABLE A4.7a Critical Values for the Durbin–Watson d Statistic ($\alpha = .05$)

n	$k=1$ $d_{L,.05}$	$d_{U,.05}$	$k=2$ $d_{L,.05}$	$d_{U,.05}$	$k=3$ $d_{L,.05}$	$d_{U,.05}$	$k=4$ $d_{L,.05}$	$d_{U,.05}$	$k=5$ $d_{L,.05}$	$d_{U,.05}$
15	1.08	1.36	0.95	1.54	0.82	1.75	0.69	1.97	0.56	2.21
16	1.10	1.37	0.98	1.54	0.86	1.73	0.74	1.93	0.62	2.15
17	1.13	1.38	1.02	1.54	0.90	1.71	0.78	1.90	0.67	2.10
18	1.16	1.39	1.05	1.53	0.93	1.69	0.82	1.87	0.71	2.06
19	1.18	1.40	1.08	1.53	0.97	1.68	0.86	1.85	0.75	2.02
20	1.20	1.41	1.10	1.54	1.00	1.68	0.90	1.83	0.79	1.99
21	1.22	1.42	1.13	1.54	1.03	1.67	0.93	1.81	0.83	1.96
22	1.24	1.43	1.15	1.54	1.05	1.66	0.96	1.80	0.86	1.94
23	1.26	1.44	1.17	1.54	1.08	1.66	0.99	1.79	0.90	1.92
24	1.27	1.45	1.19	1.55	1.10	1.66	1.01	1.78	0.93	1.90
25	1.29	1.45	1.21	1.55	1.12	1.66	1.04	1.77	0.95	1.89
26	1.30	1.46	1.22	1.56	1.14	1.65	1.06	1.76	0.98	1.88
27	1.32	1.47	1.24	1.56	1.16	1.65	1.08	1.76	1.01	1.86
28	1.33	1.48	1.26	1.56	1.18	1.65	1.10	1.75	1.03	1.85
29	1.34	1.48	1.27	1.56	1.20	1.65	1.12	1.74	1.05	1.84
30	1.35	1.49	1.28	1.57	1.21	1.65	1.14	1.74	1.07	1.83
31	1.36	1.50	1.30	1.57	1.23	1.65	1.16	1.74	1.09	1.83
32	1.37	1.50	1.31	1.57	1.24	1.65	1.18	1.73	1.11	1.82
33	1.38	1.51	1.32	1.58	1.26	1.65	1.19	1.73	1.13	1.81
34	1.39	1.51	1.33	1.58	1.27	1.65	1.21	1.73	1.15	1.81
35	1.40	1.52	1.34	1.58	1.28	1.65	1.22	1.73	1.16	1.80
36	1.41	1.52	1.35	1.59	1.29	1.65	1.24	1.73	1.18	1.80
37	1.42	1.53	1.36	1.59	1.31	1.66	1.25	1.72	1.19	1.80
38	1.43	1.54	1.37	1.59	1.32	1.66	1.26	1.72	1.21	1.79
39	1.43	1.54	1.38	1.60	1.33	1.66	1.27	1.72	1.22	1.79
40	1.44	1.54	1.39	1.60	1.34	1.66	1.29	1.72	1.23	1.79
45	1.48	1.57	1.43	1.62	1.38	1.67	1.34	1.72	1.29	1.78
50	1.50	1.59	1.46	1.63	1.42	1.67	1.38	1.72	1.34	1.77
55	1.53	1.60	1.49	1.64	1.45	1.68	1.41	1.72	1.38	1.77
60	1.55	1.62	1.51	1.65	1.48	1.69	1.44	1.73	1.41	1.77
65	1.57	1.63	1.54	1.66	1.50	1.70	1.47	1.73	1.44	1.77
70	1.58	1.64	1.55	1.67	1.52	1.70	1.49	1.74	1.46	1.77
75	1.60	1.65	1.57	1.68	1.54	1.71	1.51	1.74	1.49	1.77
80	1.61	1.66	1.59	1.69	1.56	1.72	1.53	1.74	1.51	1.77
85	1.62	1.67	1.60	1.70	1.57	1.72	1.55	1.75	1.52	1.77
90	1.63	1.68	1.61	1.70	1.59	1.73	1.57	1.75	1.54	1.78
95	1.64	1.69	1.62	1.71	1.60	1.73	1.58	1.75	1.56	1.78
100	1.65	1.69	1.63	1.72	1.61	1.74	1.59	1.76	1.57	1.78

Source: J. Durbin and G. S. Watson, 'Testing for Serial Correlation in Least Squares Regression, II,' *Biometrika* 30 (1951), pp. 159–78. Reproduced by permission of the Biometrika Trustees.

(A) $\alpha = 0.025$ one-tail; $\alpha = 0.05$ two-tail

n_1	3		4		5		6		7		8		9		10	
n_2	t_l	t_u	t_l	t_u	t_l	t_u	t_l	t_u	t_l	t_u	t_l	t_u	t_l	t_u	t_l	t_u
4	6	18	11	25	17	33	23	43	31	53	40	64	50	76	61	89
5	6	21	12	28	18	37	25	47	33	58	42	70	52	83	64	96
6	7	23	12	32	19	41	26	52	35	63	44	76	55	89	66	104
7	7	26	13	35	20	45	28	56	37	68	47	81	58	95	70	110
8	8	28	14	38	21	49	29	61	39	73	49	87	60	102	73	117
9	8	31	15	41	22	53	31	65	41	78	51	93	63	108	76	124
10	9	33	16	44	24	56	32	70	43	83	54	98	66	114	79	131

(B) $\alpha = 0.05$ one-tail; $\alpha = 0.10$ two-tail

n_1	3		4		5		6		7		8		9		10	
n_2	t_l	t_u	t_l	t_u	t_l	t_u	t_l	t_u	t_l	t_u	t_l	t_u	t_l	t_u	t_l	t_u
3	6	15	11	21	16	29	23	37	31	46	39	57	49	68	60	80
4	7	17	12	24	18	32	25	41	33	51	42	62	52	74	63	87
5	7	20	13	27	19	36	26	46	35	56	45	67	55	80	66	94
6	8	22	14	30	20	40	28	50	37	61	47	73	57	87	69	101
7	9	24	15	33	22	43	30	54	39	66	49	79	60	93	73	107
8	9	27	16	36	24	46	32	58	41	71	52	84	63	99	76	114
9	10	29	17	39	25	50	33	63	43	76	54	90	66	105	79	121
10	11	31	18	42	26	54	35	67	46	80	57	95	69	111	83	127

TABLE A4.8 Critical values for the Wilcoxon rank sum test

Source: From F. Wilcoxon and R.A. Wilcox. 'Some Rapid Approximate Statistical Procedures' (1964). p. 28.

(A) $\alpha = 0.025$ one-tail; $\alpha = 0.05$ two-tail			(B) $\alpha = 0.05$ one-tail; $\alpha = 0.10$ two-tail	
n	t_l	t_u	t_l	t_u
6	1	20	2	19
7	2	26	4	24
8	4	32	6	30
9	6	39	8	37
10	8	47	11	44
11	11	55	14	52
12	14	64	17	61
13	17	74	21	70
14	21	84	26	79
15	25	95	30	90
16	30	106	36	100
17	35	118	41	112
18	40	131	47	124
19	46	144	54	136
20	52	158	60	150
21	59	172	68	163
22	66	187	75	178
23	73	203	83	193
24	81	219	92	208
25	90	235	101	224
26	98	253	110	241
27	107	271	120	258
28	117	289	130	276
29	127	308	141	294
30	137	328	152	313

TABLE A4.9 Critical values for the Wilcoxon signed rank sum test

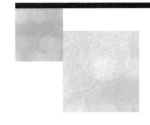

Appendix A5
Some numeric answers to exercises

Chapter 1
1.10e: 6/52; **1.10f**: 11/52

Chapter 2
2.4b: 0, 0.40, 0.40, 0.90, 1
2.5b: 0.15, 0.5667, 0.7333, 0.9125, 0.9750
2.6a: 2, 5, 8.3, 12 and 13; **2.6c**: 1/12, 31/60, 1/10, 17/90, 1/9 and 2, 5, 8.3, 12, 13; **2.6e**: 7.2911;
2.6f: 60
2.7d: 75
2.10d: 0.8825, 0.99957
2.12a: 6.465 million
2.15d: 0.516, 0.422
2.16d: 0.106
2.17d: 0.5700, 0.5300; **2.17e**: 3
2.20f: 0.7620

Chapter 3
3.1a: 22, 4.4, 4; **3.1b**: 56; **3.1c**: 56; **3.1d**: 3, 0, −2
3.2a: 108, 484; **3.2b**: 0, 11.2; **3.2c**: 1.4, 0.6, 1.4, 2.6, 0.4 and 6.4
3.3: 18
3.4a: 30.4, 3.04, 3.15; **3.4b**: 15.2, 15.2; **3.4c**: 244.8, 924.16; **3.4d**: 0.1204, 0.0329; **3.4e**: 112, 112;
3.4f: 5, 0, 0 and 5, 0, −4; **3.4g**: 152.384, 34.4
3.5a: 2.4, 3, 2.5; **3.5c**: 2.5
3.6a: 4.85; **3.6c**: 5.6 and (0, 1]
3.7: 2227.27
3.8a: 0.2, −0.125, −0.0952; **3.8b**: −0.0169; **3.8d**: mean = −0.0067
3.10b: 'no'
3.11b: 'important', 'important'
3.12a: 7.8; **3.12b**: 8
3.13a: 7.7778; **3.13b**: 8; **3.13c**: 8
3.14a: values 0, 1, 2, > 2; **3.14d**: 1; **3.14e**: 1.0020
3.16a: 881; **3.16c**: Saturday, Thursday; **3.16d**: mean = 0.4665
3.17a: 5.3721, 5.6154, (6, 8]; **3.17b**: 0.4684
3.18b: 1.501, 1.395; **3.18d**: overall mean = 1.338795
3.19b: 25.2940, 1–9; **3.19c**: 6.3847; **3.19d**: 0.831
3.20b: 3649071, 9771984; **3.20c**: 3630000, 9846800 and 1971, 1980; **3.20e**: 37.3484; **3.20f**:
2004, 1984
3.21a: 188.917, 178.9; **3.21c**: 194.95; **3.21d**: weighted mean = 133.2
3.22a: 370.624, 405.7 and 539.58 and 341.7725; **3.22b**: 120.948, 40.61 and 708.07 and 76.1244;
3.22d: 60.94
3.23a: 2.45; **3.23e**: modes: 0, 1, 0; medians: 0, 1, 0; means: 1/3, 2/3, 2/5
3.24b: 21.067
3.25a: 0.156105, 0.158468; **3.25b**: 183083×10^8 yuan

Chapter 4
4.1: $s^2 = 3.4667$, $s = 1.8619$
4.2: $\sigma^2 = 14.1094$, $\sigma = 3.7562$
4.3: 106.25
4.6a: at least 84%; **4.6b**: at most 16%; **4.6c**: about 99%, about 0.5%
4.7a: 4.6, 1.9898, 0.4326; **4.7b**: 2, 0.5
4.8a: 3.6, 1.9698, 0.5472; **4.8b**: 3.0982, 0.8995
4.9a: 4; **4.9b**: 0.25, 0.4341
4.10a: 1.1576, 0.5704 and 0.4015, 0.2144; **4.10b**: 1.3025, 0.9019 and 0.3407, 0.2856
4.11a: Spain, Netherlands; **4.11b**: 62.50, 9.75 (Excel); **4.11c**: 58.3, 10.87; **4.11d**: Netherlands and no outliers
4.12a: 1.875, 4.6719, 2.1615; **4.12b**: 0.8621, 0.4310, 2.7, 2.2690
4.13a: 51.3, 60.3; **4.13b**: 5.2, 2.1144, 3.5, 6.0, 6.5, 3 and 5.9, 1.9830, 4.5, 6.0, 7.5, 3
4.14c: 4.2274, 0.0264
4.17a: 2.8, 1.6733; **4.17b**: 25.2, 5.02; **4.17c**: (Excel) 11.23, 3.3466
4.18a: 4405.331, 66.3727; **4.18c**: 125.25, 178.90, 243.87, 118.62, 0.6631; **4.18d**: 9, 9;
4.18e: Poland and no outliers
4.19a: 0.1974, 0.4443; **4.19b**: (Excel) 0.5575, 0.295, 0.405, 0.730; **4.19c**: quarter 2
4.20a: 1, 1, 0, 0; **4.20b**: 1, 1, 2, 2; **4.20c**: 0.4231, 0.6505 and 0.4888, 0.6991 and 0.5070, 0.7120 and 0.5030, 0.7092; **4.20d**: 1.4072; **4.20e**: 14.3%, at most 44.44%
4.21a: 370.624, 25561.7781, 159.88 and 25302453.33, 670716017030489, 25898185.59;
4.21b: 13.3%, ≈ 13%, at most 44.4%; **4.21c**: 0.5333, 0.2489, 0.4989; **4.21d**: 120.9481, 27955.2022, 167.1981 and 50775970.3104, 39665462637195400, 199161900.5663 and 12.2%, ≈ 13%, at most 44.4% and 0.2927, 0.2070, 0.4350;
4.21e:

	EU10	EU15	World
μ	188.917	370.624	120.948
σ	66.373	159.880	167.198
$\frac{\sigma}{\mu}$	0.351	0.43	1.391

4.22c: 0.2990%, 18.2586 (%)^2, 4.2730%; **4.22d**: 4.185%, ≈ 4.5%
4.24a: 8.3333%, 12.00 hrs and 12.00, 12 hrs; **4.24b**: 13.45, 37.1295, 6.0933; **4.24c**: (16, 18], 14.2707, 8.9107
4.25a, b:

	1995	2000	2003	2004
Mean X	0.5253	0.5799	0.4798	0.4854
Mean Y	34.5780	36.2840	35.6270	36.1260
Stand deviation X	0.4994	0.4936	0.4996	0.4998
Stand deviation Y	10.9262	11.9921	12.2605	12.5496

4.25c: 53.7608, 58.6869
4.26a: means: 2.3140, 2.2722, 2.2577, 2.2505; stdevs: 1.2540, 1.2399, 1.2391, 1.2377;
4.26b: 2, 2; **4.26c**: 56
4.27b: 8, 8, 9, 1, 1/8; **4.27c**: 7.7778, 2.2618, 1.5039
4.28a: 1.9664; **4.28c**: at most 44 countries

Chapter 5
5.2a: –31.85, –0.8647; **5.2c**: $\hat{y}$ = 132.6075 – 2.9766x
5.3b: 3.9048, 4.8095, –3.8810; **5.3c**: –0.8956
5.4a: –20.4, 170.2, 12.8, 97.2; **5.4b**: –0.722, –25.4669
5.5a: 5100, 3000, 1800, 1250; **5.5b**: 22500000, 0.75; **5.5c**: 20000; **5.5d**: men: 14900,
17640000; women: 12000, 14062500
5.6a: 4.8571, 1.5714, 2.4762, 1.9524, –1.9048, –0.8663; **5.6b**: $\hat{y}$ = 5.3077 – 0.7692x;
5.6c: 4.1539 hrs; **5.6d**: –0.6925; **5.6e**: 2.9234
5.7a: 7.4001; **5.7b**: 2.114; **5.7c**: 2.6922, 6.4100, 2.0708, 18.8545, 4.3784, 0.7007; **5.7d**: 2.114
5.8b: 1.83333, 1.48667 and 1.98333, 0.17767; **5.8c**: 0.3787, 0.7368;
5.8d: $\hat{y}$ = 1.516 + 0.2547x
5.9a: 10, –9, 9, 8; **5.9b**: 5/6, –60, –5/6; **5.9c**: 10/9, 43/9 and –20/27, –43/27
5.10a: 0.9091; **5.10b**: $\hat{y}$ = 4.8857 + 1.2857x; **5.10c**: 9.4, –9.4, 75.6, 67.2, –64.8, –0.9091, $\hat{w}$=
–1.3433 – 0.8571v
5.11a: –460052426856, 4894.8079, 1.29808×10^{14}, –0.5772;
5.11b: $\hat{y}$ = –0.000003544x + 215.249728192; **5.11c**: 179.1009, –1.6609
5.14b: $\hat{y}$ = 4.5242 + 0.3489x; **5.14c**: 2.047; **5.14d**: 4.94, 5.64, 4.94, 4.84, 5.33, 5.47 and
–2.44, –3.94, –3.04, –2.94, –3.73, –4.67;
5.14e: 0.2185; **5.14f**: 4219.735847
5.15a: degrees Celsius, degrees Celsius, squared degrees Celsius and degrees Celsius;
5.15b: 58.8047, 58.8380, 0.1811, 0.4255; **5.15c**: 0.0159, 0.0072
5.17b: 2.05, 1.440013 and 1.825, 2.017255; **5.17c**: –1.52955, –0.52654; **5.17d**: $\hat{y}$ = 3.3371 –
0.7376x; **5.17e**: –1.825, 2.017255, 1.52955, 0.52654, $\hat{u}$ = –3.3371 + 0.7376x
5.18a: 20422.62, 0.925247; **5.18e**: 0.9412, 0.9765
5.19c: 182.0272, 0.5455; **5.19d**: $\hat{y}$ = –0.6672 + 0.5455x
5.20b: 1938.227
5.21e: 1.5150693, 0.3317449; **5.21f**: $\hat{w}$= 1.6565342 + 0.3227847v, 209.5927
5.22a: 48.75, 1208.3333 and 82.0856, 2049.0012; **5.22b**: 166760.8865, 0.9915;
5.22c: $\hat{y}$ = 1.81117 + 24.7492x
5.23b: $\hat{x}_t$ = 4.32437 + 0.98754x_{t-1}, 1812.896252; **5.23c**: 0.9876; **5.23d**: 353.13
5.24b: $\hat{r}_t$ = 0.000024421 – 0.0444225r_{t-1}, 0.015920444; **5.24c**: 0.0444; **5.24e**: 353.30

Chapter 6
6.3b: 0.5051; **6.3c**: 0.3333
6.5b: 0.4231
6.6a: 1/15; **6.6b**: 2/5; **6.6c**: 2/3
6.7b: 0.07
6.11b: 0.0278; **6.11c**: 0.5556; **6.11d**: 0.0463
6.12b: 0.90; **6.12c**: 0.10
6.14b: 0.72, 0.24, 0.04
6.15b: 0.250, 0.098, 0.392, 0.203, 0.057; **6.15c**: 0.275, 0.092, 0.121, 0.400, 0.112
6.18a: 5/7; **6.18c**: 0.5
6.21a: 8294000; **6.21g**: 0.2227, 0.7773

Chapter 7
7.1: 0.4, 0.7, 0.1
7.2a: 0.28; **7.2b**: 0.65; **7.2c**: 0.25; **7.2d**: 0.8929; **7.2e**: 0.3846
7.3b: 0.72, 0.35, 0.75, 0.1071, 0.6154
7.4a: 0.3, 0.5; **7.4b**: 0.7, 0.9; **7.4d**: 0.2, 0.3333; **7.4f**: 0.8, 0.66667
7.5a: 0.75; **7.5b**: 0.68; **7.5c**: 0.5
7.6a: 5040; **7.6b**: 1140, 3838380, 3838380; **7.6c**: 360; **7.6d**: 870912000
7.7a: 10^{10}; **7.7b**: 9034502400; **7.7c**: 75287520
7.8: 0.735

7.9: 0.9385

7.10a: 0.62; **7.10b**: 0.0645, 0.1579

7.11a: 0.4, 0.6, 0, 0.2

7.12: 0.1, 0.4, 0.3

7.13a: 0.20, 0.26, 0.54; 0.30, 0.26, 0.44; **7.13b**: 0.24, 0.26, 0.30; 0.20, 0.20, 0.34; 0.16, 0.06, 0.24; **7.13d**: 0.44, 0.46, 0.50; 0.46, 0.46, 0.60; 0.70, 0.60, 0.78

7.14a: 2.347×10^{12}; **7.14b**: 646646; **7.14c**: 0.0096, 0.0096; **7.14d**: 7.04×10^{12}, 1939938; **7.14e**: 0.0032, 0.0032

7.15: 0.3721

7.16: 0.4, 0.6667, 0.6, 0.3333

7.18a: 0.5; **7.18b**: 3/8, ¾

7.19a: 0.1; **7.19b**: 0.3625

7.20a: 17%; **7.20b**: 0.2353

7.21: 0.6

7.22a: 48%, 77.78%, 100%; **7.22c**: 0.4717, 0.3396, 0.1887; **7.22d**: 0.679; **7.22e**: 66.7%

7.23a: 0.5, 0.2809, 0.2191; 0.1750, 0.1077, 0.3604; **7.23b**: 0.1967; **7.23c**: 0.1538

7.24d: 0.1186, 0.3814, 0.3814, 0.8814

7.26a: 0.52; **7.26b**: 0.48; **7.26d**: 0.1176; **7.26e**: 0.8824; **7.26f**: 0.5; **7.26h**: 0.5, 0.5, 1/8, 7/8, 0.5

7.27a: 0.4; **7.27b**: 0.2; **7.27c**: 0.4; **7.27d**: 2/7

7.28a: $p / (0.2 + 0.8p)$; **7.28c**: 5/6

7.29a: 0.6966; **7.29b**: 0.0325

7.30a: 0.4; **7.30b**: 0.3; **7.30c**: 0.2; **7.30d**: 0.5; **7.30e**: 0.7; **7.30f**: 0.5; **7.30g**: 0.6; **7.30h**: 0.2

7.31a: 0.025; **7.31b**: 0.36

7.32a: 0.0639; **7.32b**: 0.0637; **7.32c**: 0.484; **7.32d**: 0.2472

7.34d: ¼, ½, ½; **7.34e**: ¼, 0, ½, ¾

7.35a: all 1/36, 1/36; **7.35b**: two paths yield 1/72 and 3/72, the rest 1/36; 1.5/36

7.37: 0.0894

Chapter 8

8.4d: 3.6, 4.365; 2.0893

8.5c: 1/3; **8.5d**: 5.5; **8.5e**: 6.75, 2.5981

8.7b: 54.90, 2632.99, 51.3127

8.9a: both –0.20, –0.10, 0, 0.10, 0.20; **8.9b**: 0, 0.0056, 0.0748; 0.1140, 0.0068, 0.0825; **8.9d**: both 8/9; 1, 0.99

8.10b: 6/12, 7/12, 8/12, 9/12, 10/12, 11/12, 1; **8.10c**: 3/12, 7/12, 5/12, 2/12, 10/12; **8.10e**: 0.1667

8.11d: 28/54; **8.11e**: 13/54; **8.11f**: 26/54

8.12a(i): 5/7; **8.12a(ii)**: 0; **8.12b(i)**: 1/8; **8.12b(ii)**: 8/3; **8.12c(i)**: 1.5; **8.12c(ii)**: 0.15

8.13a: 1.89, 5.1979; **8.13b**: 5; **8.13c**: 2

8.14a: –43.5; **8.14b**: 0; **8.14c**: 29; **8.14d**: 16; **8.14e**: 4

8.15a: 2.4, 2.04, 1.4283; 580, 5100, 71.4143; **8.15b**: 0.3; **8.15c**: 450; **8.15d**: 4.5, 0

8.16a: 685, 460; **8.16b**: 1, 2.5, 3; 510, 585, 610; 2, 100

8.17a: 0.9; **8.17b**: 0.05, 0.15, 0.25, 0.55, 0.75, 0.95, 1; **8.17c**: 25.3; 2.31, 1.5199; **8.17e**: 77.54, 7.4844, 2.7358

8.18b: 9/16; **8.18e**: 30.5556, 1.3374; 1.1565

8.19a: 0.84; **8.19b**: 0.25; **8.19c**: 0.16; **8.19d**: 0.90; **8.19e**: 0.375

8.20b: –1, 0, 2, 3, 10; **8.20c**: 2/3, 2/3

8.21c: 5, 2, $\sqrt{2}$

8.22c: 0.3; **8.22d**: 80%; **8.22f**: 0.3, 80%

8.23b: [0, 100]; **8.23d**: 0.19; **8.23e**: 0.36

8.24a: 1/25; **8.24b**: 0; **8.24c**: 16/25; **8.24d**: 0

8.25a: 11.9722; **8.25b**: 1.0020, 0.5040, 0.7099

8.26a: 0.25, at most 25000; **8.26b**: 0.125, at most 12500

8.27a: 2, 0.8; **8.27b**: 0.75; **8.27c**: 0.9839; **8.27d**: 1
8.28a: 3.5, 2.9167, 1.7078
8.29: 1000, 1000, 1250; 0, 1000, 1250

Chapter 9
9.1a: 0.35, 0.35, 0.2275; **9.1b**: 100, 0.35, 35, 22.75; **9.1c**: 100, 350, 1000, 35, 20.4955
9.2a: 6, 5.4; **9.2b**: 0.3355; **9.2c**: 0.2710, 0.3950
9.3a: 5, 115, 200; **9.3b**: 2.8750, 1.1973; **9.3c**: 0.3476; **9.3d**: 0.2553, 0.1047; **9.3e**: 0.3434, 0.2538, 0.1077
9.4a: 5.8, 5.8; **9.4b**: 0.1656; **9.4c**: 0.1601, 0.3616
9.6b: 0.02, 0.0196; **9.6c**: 100, 0.02; **9.6d**: 2, 1.96; **9.6e**: 0.4033, 0.8590, 0.2725
9.7a: 0.4025; **9.7b**: 0.2675; **9.7c**: 0.2055; **9.7d**: 0.0675
9.8a: 4.52, 2.0774; **9.8b**: 0.5260; **9.8c**: 0.2407
9.9a: 5, 1/3; **9.9b**: 5/3, 10/9, 1.0541; **9.9c**: 0.1317, 0.3292, 0.3292, 0.1646, 0.0412, 0.0041; **9.9d**: 0.2099
9.10b: 0.51, 0.2499; **9.10d**: 51, 24.99, 4.9990; **9.10e**: 0.2419, 0.3824, 0.6223
9.11a: 20, 40, 200; **9.11b**: 4, 2.8945, 1.7013; **9.11d**: 0.1095, 0.9243, 0.5978, 0.6232; 0.1091, 0.9133, 0.5886, 0.5981
9.12a: 100, 0.05; **9.12c**: 0.4360; **9.12d**: 0.8817; **9.12e**: 0.4301; **9.12f**: 5, 4.75; **9.12g**: 0.05; 0.000475, 0.0218; **9.12h**: 0, 0.01, 0.02, ... , 0.99, 1; 0.1280, 0.6791
9.13a: 10, 40, 100; **9.13b**: 10, 0.4; **9.13c**: 4, 2.1818; 4, 2.4; **9.13d**: 0.1153, 0.8461, 0.9614; 0.1209, 0.8328, 0.9537
9.14b: 0.0223; **9.14c**: 0.1111
9.15a: 71, 8.3962; **9.15b**: 0.0473, 0.0473, 0.0466; **9.15c**: 0.5315; **9.15d**: 0.9980
9.16b: 15, 15; **9.16c**: 0.3306; **9.16d**: 0.2448

Chapter 10
10.1b: 0, 0.2887; **10.1c**: 0.6
10.2a: 2/3; **10.2b**: 2.5, 18.75; **10.2d**: 0.8333
10.3b: 10, 100; **10.3c**: 0.2231; **10.3d**: 0.3679
10.4a: 0.0808, 0.0107; **10.4b**: 0.9085, 0.8384; **10.4c**: 0.4821, 0.4641; **10.4d**: 0.9282
10.5a: 0.0478, 0.7977, 0.5, 0.0228; **10.5b**: 0.6826, 0.1359; **10.5c**: 0, 0.0478; **10.5d**: 0.5954, 0.8413
10.6a: 0.6745; **10.6b**: 0.8416; **10.6c**: 1.3839; **10.6d**: 0.8416
10.7a: 10.6776; **10.7b**: 2.0976
10.8b: 0.1209; **10.8c**: 0.6759, 0.5605; **10.8d**: 0.7295
10.9a: 0.7531; **10.9b**: −0.4125; **10.9c**: 0.5343; **10.9d**: 3.9375
10.10a: 0.2743; **10.10b**: 0.6736; **10.10c**: 0.4532; **10.10d**: 6.9340
10.11a: 0.0151; **10.11b**: 6155
10.12a: $e^{-2.5t}$; **10.12c**: 0.7135; **10.12d**: 0.7135
10.13d: 55, 675, 25.9808; **10.13e**: 0, 0.1667, 0.1667, 0.2167, 0.7778, 0.8322, 0.6667
10.14a: 77.5; **10.14b**: 0.7778
10.15a: 22500; **10.15c**: 22500, 52083333.3333, 7216.8784; 31500, 102083333.3333, 10103.6297; **10.15d**: 0.40, 0.7429, 0.2857
10.16a: 0.8907, 0.8907, 0.4522, 0.3075, 0.5654; **10.16b**: 0.8907, 0.4522, 0.3075, 0.5654; **10.16c**: 0.0228, 0.0013, 0.000032; 0.0228, 0.0013, 0.000032
10.17a: 8.3168; **10.17b**: 7.8394; **10.17c**: 1.0488
10.18a: 0.6827; **10.18b**: 0.9545; **10.18c**: 0.9973
10.19a: 75, 14.4338; **10.19b**: 0.02
10.20a: 75
10.21d: 0.1151
10.22a: 1.5958
10.23a: 0.0228; **10.23b**: 4000, 2500; **10.23c**: 0.0548

10.24a: 0.6497; **10.24b**: 0.9275; **10.24c**: 56.66; **10.24d**: 10.5%; **10.24e**: 65.48
10.25a: 98.76%; **10.25b**: 4.12; **10.25c**: (32.38, 40.62)
10.26a: 0.0668; **10.26b**: 15.87%; **10.26c**: 253.2898; **10.26d**: 1.2159
10.27a: 0.3682, 1, 0.998; **10.27c**: 0.1756, 5, 2.2338; **10.27d**: 0.1771; **10.27e**: 0.1605, 0.7490, 0.1315, 0.6175
10.28a: 0.0360; **10.28c**: 0.0118
10.29a: 1, 1, 2; **10.29b**: −1.1901, 0.1541, 1.5485
10.30a: 0.09; **10.30c**: 22.5, 4.5249, 0.3206; **10.30d**: 0.3292
10.31a: 48.05 million, 2.82 million; **10.31b**: 132.86; **10.31c**: 0.31
10.32d: 0.2353, 0.1176, 0.0588; 0.1687, 0.1008, 0.0166

Chapter 11
11.1b: 0.3, 0.7; **11.1d**: −0.40, −0.5976
11.2c: 0.75; **11.2d**: 1, 0.5
11.3b: 0, 0; **11.3c**: 67.86; **11.3d**: 4764.42
11.4a: −0.41, −0.8199; **11.4b**: 11.8; **11.4c**: 7, 0.2
11.5a: 9.6; **11.5b**: −28.5, 81; **11.5c**: −8, 5.8; **11.5d**: 5, 54.4; **11.5e**: −115, 567.4
11.6b: 0.9, 0.7; **11.6c**: −0.10, −0.1829
11.7b: 1.45, 5.8345; **11.7c**: 1.325, 4.2008
11.8d: 0.6129, 3.8947
11.9a: 135, 784; **11.9b**: 28, 397.6
11.10a: −16.1, 25.7721; **11.10b**: 34, 37.70; **11.10d**: −0.82, −0.82
11.11c: 0.49, 0.60; **11.11d**: 52.9500, 0.8574
11.12a: 0.24, 0.2624; **11.12b**: 1.20, 0.1600; 270, 0.8100; 8/3, 2/9; **11.12c**: 173.20, 1369.76
11.13b: 20.30, 40.85, 0.61, 1.0275; **11.13c**: 0.75, 0.90; **11.13d**: 1/3, 1/6
11.14a: 61.15, 1.7110; **11.14b**: 0.6450, 0.8147; **11.14d**: 10150, 390.5125; 10212.50, 253.4142; 10181.25, 307.3958
11.15b: 25.30, 49.90, 0.61, 0.89; **11.15c**: 0; **11.15d**: 1262.47, 1595919.63;
11.15e: −247, 48.2286
11.17d: 7/4, 0.6162
11.19b: 0.2078; **11.19d**: 0.5604, 5, 4.95; **11.19e**: 0.5889
11.20b: 0.0319
11.21b: 0.0668; **11.21c**: 0.0004; **11.21d**: 0.0668
11.23a: 5110, 29.6243; **11.23b**: 1364, 37.0189
11.24b: 12/36, 4/36, 20/36
11.26b: 0.7239, 4.1038; 0.1487, 2.5947; 0.5543, 2.7949; **11.26e**: 0.3717, 1.9575;
11.26f: 0.4756, 2.6100

Chapter 12
12.2c: 0.0901
12.4a: 60000, 4×10^8; **12.4b**: 0.1476; **12.4c**: 0.0266; **12.4d**: 0.2980
12.6b: 83451052.6144, 70580771.5346, 43098333.3606; **12.6c**: 8391.57
12.7a: 0.2743; **12.7b**: 0.0412; **12.7c**: 0.32
12.8a: 16540, 74428400; **12.8c**: 0.174; **12.8d**: 0.0019, 0.0019, 0.2200, 0.9540
12.9a: 1.2, 1.2; 0.76, 0.76; **12.9d**: −0.0844
12.10a: -7.0986×10^{-8}; **12.10b**: 1.002, 1.002; 0.5040, 0.5040
12.11b: 0.419, 0.2434, 39.82, 135.9676; **12.11c**: 0.2434, 0.0711, 0.3243, 0.7840, 0.2115
12.12b: 31.5518, 314.4408, 17.73248; 2.76, 1.9016, 0.20; **12.12c**: 0.16; **12.12d**: 6.190, 0.253
12.14a: 400choose20, 400^{20}, 100choose10 times 300choose10, 100choose5 times 300choose15;
12.14c: 0.05, 0.0488, 0.1, 0.05; 0.05, 0.0488, 0.1, 0.05; 0.0024, 0.0023, 0.00333, 0.0025
12.15a: 0.022, 0.0243
12.16b: −0.1829

Chapter 13
13.1: 3.5, 2.9167, 1/3; 3.5, 1.4583, 1/6; 3.5, 0.0972, 0.000000752
13.2: 0.0047, 0.0416, 0.9906
13.3: 20, 1.8; 20, 1.7273
13.4a: 0.8176; **13.4b**: 0.4375
13.5a: 0.3446; **13.5b**: 0.2119; **13.5c**: 0.1151
13.6a: 24000, 424.2641; 23000, 424.2641; **13.6b**: 0.1193; **13.6c**: 0.0002; **13.6d**: 0.00024
13.7a: 3, 2; **13.7d**: 3, 2/3
13.8b: 1.56, 0.4132; **13.8d**: 0
13.9b: 3.2, 0.98; **13.9d**: 0
13.10b: 3.2, 0.868; **13.10d**: −0.2178
13.11b: 0.5, 0.0339, 0.9322; **13.11c**: 0.7000
13.12b: 0.5269; **13.12d**: 0.5844; **13.12e**: 0.9177
13.13b: 0.9938; **13.13c**: 0.7887, 0.7333; **13.13d**: 3.4107; **13.13e**: 0.3290
13.14b: 2.3; **13.14d**: 0.8507; **13.14e**: 8.1×10^{-16}
13.15b: 0.5440; **13.15c**: 0.9876; **13.15e**: (20.3944, 24.6056)
13.17a: 0.5, $1/\sqrt{12}$; **13.17b**: 0.5, 1/12
13.18b: 0.0089; **13.18c**: 0.2146; **13.18e**: 0.2146
13.19a: 3.78981, 6.35162; **13.19c**: 3.96, 7.53; **13.19d**: 0.0753

Chapter 14
14.1a: 0.2337; **14.1b**: 0.2451
14.2a: 0.0416; **14.2b**: 0.0125
14.4d: 43.2125, 48.8670; **14.4e**: 2.4715
14.5e: 0.6, 0.24, 0.1549
14.6: 8.24, 1.69
14.7a: (0.0761, 0.1199); **14.7c**: 3396; **14.7d**: (0.8411, 0.8649)
14.8e: 7.15, 1.9472; 0.7, 0.21
14.9a: 0.7, 0.00105, 0.0324; **14.9b**: 0.0506; **14.9c**: 0.0614; **14.9d**: 0.4629; **14.9e**: 0.1579
14.10c: 0.6250, (0.5565, 0.6935)
14.11a: 0.20, 0.0016, 0.04; **14.11b**: 0.8686, 0.8944
14.12a: 0.9281; **14.12c**: 0.9281
14.13a: 6320
14.14b: 0.0500; **14.14c**: 0.1263, 0.2375; **14.14d**: 0.1896
14.15a: 0.7750, 0.1744; **14.15c**: 0.2139; **14.15d**: 0.001059

Chapter 15
15.1c: 0.4426; **15.1d**: 0.4401; **15.1e**: *val* = 4
15.2a: 0.95; **15.2c**: (0.0503, 0.1297), 0.95; **15.2d**: 0.0397
15.3a: *val* = −1; **15.3b**: *val* = 0.6; **15.3c**: *val* = −1.75
15.4a: 0.3173, 0.2743, 0.0401
15.5a: −1.25, 0.2113; **15.5b**: −0.8839, 0.3768; **15.5c**: −0.5590, 0.5762
15.6a: 4, 0.000032; **15.6b**: 2, 0.0228; **15.6c**: 1, 0.1587
15.7a: −3, 0.0013; **15.7b**: −2, 0.0228; **15.7c**: −1, 0.1587
15.10c: (9414.5469, 19906.6856), (11098.6191, 16009.8250)
15.11c: 0.3886; **15.11d**: 0.5562
15.13b: 0.2146; **15.13c**: 0.2146
15.14a: (8.1175, 8.3625); **15.14b**: *val* = 3.04
15.15a: (38.14, 41.86)
15.16a: *val* = −2.1082; **15.16b**: 0.0175; **15.16c**: 139
15.17a: *val* = −2.4375; **15.17c**: 0.0148
15.18a: (4.6080, 5.3920), (4.7228, 5.2772); **15.18b**: (4.6710, 5.3290), (4.7674, 5.2326)
15.19a: *val* = −0.7071, *val* = −2.2361; **15.19b**: 0.4795, 0.0253

15.20a: 90, 3; **15.20b**: (86.08, 93.92); **15.15d**: *val* = −5; **15.15e**: 0.000000287
15.21b: 0.01; **15.21c**: 0.0032; **15.21d**: 0.2831
15.22a: (994.1964, 1045.8036); **15.22b**: 3330
15.23c: 3; **15.23e**: *val* = −2.3095; **15.23f**: 0.0096
15.24a: (73.3600, 76.0800), (174.7900, 176.9200), (37.7732, 41.1914); **15.24b**: *val* = 6.8026;
15.24c: 0.2280
15.25a: 0.1587; **15.25b**: 0.1587; **15.25c**: 0.0808, 0.0359; **15.25d**: 0.2743, 0.4207
15.26c: (−0.1952, −0.0098)

Chapter 16
16.3a: *val* = 3; **16.3b**: *val* = 3; **16.3c**: *val* = 3
16.4a: *val* = −2.5; **16.4b**: (119.1032, 119.8968); **16.4d**: 0.0070
16.5a: (0.5367, 0.6233); **16.5b**: *val* = −0.9129
16.6: 2340
16.7: 4269
16.8a: (38.0923, 41.9077); **16.8c**: *val* = −2.3717; **16.8d**: 0.0209
16.9c: *val* = 1.1368; **16.9d**: 0.1300
16.10a: *val* = 3.416; **16.10b**: *val* = −2.4152; **16.10c**: (0.6463, 0.7537), (0.4172, 0.5828)
16.11a: *val* = −10; **16.11c**: (88.1405, 91.8595)
16.12a: *val* = 2.4; **16.12b**: *val* = 0.8485; **16.12c**: *val* = 4.2426
16.13a: (5.3940, 7.4060), (6.4034, 7.5966); **6.13d**: (6.2813, 7.3187)
16.14a: (0.0118, 0.0396); **16.14b**: *val* = −2.0859; **16.14c**: 1286; **16.14d**: 27057
16.15a: (8.8136, 11.9864); **16.15b**: *val* = 0.5277; 0.3019
16.16a: (0.7227, 0.9106)
16.17a: *val* = 2.7779
16.18a: 2401; **16.18b**: 2017; **16.18c**: 2401
16.19a: 664; **16.19b**: (12.1202, 13.0798)
16.20a: *val* = 3.3093; **16.20b**: (0.5307, 0.6173); **16.20c**: 2401
16.21b: *val* = 0.962
16.22b: *val* = −0.805

Chapter 17
17.1a: 0.4703, 0.0023; **17.1b**: 40, 8.9443; **17.1c**: 51.8051, 55.7585
17.2a: 24; **17.2c**: 0.4058; **17.2d**: 54.6225
17.3a: (34.6750, 47.8055)
17.4a: *val* = 2.5624; **17.4b**: *val* = 21.1997; **17.4c**: 0.1482
17.5a: *val* = 112.1580; **17.5b**: (1.3151, 3.8182)
17.6a: 9, $\sqrt{18}$; **17.6c**: 0.0085, 0.1657, 0.4659, 0.7243, 0.8777, 0.9513, 0.7782;
17.6d: 14.6837, 4.1682, 16.9190, 3.3251
17.7a: 19, $\sqrt{38}$; **17.7b**: 0.0656, 0.2381, 0.3619, 0.4390, 0.4820; **17.7c**: 22.7178;
17.7d: 6.8440, 38.5823; **17.7e**: 0.1263, 99.94
17.8a: (112.175, 2580.512); **17.8b**: *val* = 12.5; **17.8c**: 0.0140
17.9b: *val* = 489.8352; **17.9c**: 4.06×10⁻¹¹; **17.9d**: *val* = 1.8453
17.10a: (11.9021, 12.4841); **17.10b**: *val* = 16.6785; **17.10c**: 0.3263
17.11c: *val* = −0.939; **17.11d**: 0.1742; **17.11e**: *val* = 495.2225
17.12: *val* = 174.9138

Chapter 18
18.1c: *val* = 1.8805
18.2b: *val* = −1.5473; **18.2c**: 0.1234
18.3: (0.4149, 2.7309)
18.4a: *val* = −2.0176; **18.4b**: (−0.1182, −0.0018)
18.5b: 1.8608, 0.5374; **18.5c**: 0.5696

18.6a: *val* = 0.6504; **18.6b**: *val* = 0.4126
18.9a: *val* = –3.0856; **18.9b**: 0.0026; **18.9c**: (–1.3145, –0.2855)
18.10a: *val* = –2.1573; **18.10b**: 0.0486; **18.10c**: (–21.87, –0.13)
18.11a: *val* = 1.2677; **18.11b**: *val* = 3.1623
18.12b: *val* = –2.3905
18.13c: (0.9902, 5.3432)
18.14a: *val* = 2.1746; **18.14b**: 0.0212
18.15a: 1.0625, (0, ∞); **18.15b**: 0.4047, 0.1325, 0.0440, 0.0153, 0.0056, 0.0021, 0.0009;
18.15c: 0.5953, 0.3152, 0.1987, 0.1392, 0.1043, 0.0819, 0.0665
18.16a: 1.6327, 2.2524, 2.8826, 3.5293, 4.4156; **18.16b**: 2.9839, 5.1642, 8.6062, 14.0599,
26.4605; **18.16c**: 0.3351, 0.1936, 0.1162, 0.0711, 0.0378
18.17a: *val* = 0.6565; **18.17c**: *val* = 5.9773
18.18: *val* = 1.0816, 0.9310, 0.6384
18.19b: 0.0426; **18.19c**: 0.0099
18.20a: 0.5274; **18.20b**: 0.1436
18.21a: *val* = –2.8699; **18.21b**: (–0.1332, –0.0268)
18.22a: *val* = 1.4425; **18.22b**: 0.0746
18.23a: *val* = 0.8301; **18.23b**: *val* = –0.837
18.24b: *val* = –3.2275
18.25a: *val* = –3.8111; **18.25b**: (–6.7061, 0.7061); **18.25c**: *val* = –1.8200
18.26a: *val* = 2.50; **18.26b**: (–0.1536, 0.1136)
18.27a: *val* = 2.6187; **18.27b**: (0.0183, 0.2110)
18.28b: *val* = –1.613; **18.28c**: 0.0535
18.29b: (–0.2870, 0.1014)
18.30a: *val* = 1.8634; **18.30b**: 0.0476
18.31b: *val* = 2.0126; **18.31c**: *val* = 0.7981; **18.31d**: *val* = 2.8589

Chapter 19
19.1a: 9.3, 0.9186; **19.1b**: $\hat{y}$ = 61.67568 + 2.51351x; **19.1c**: 81.7838, 84.2973, 79.2703,
86.8108, 91.8378; **19.1d**: –1.7838, 0.7027, –0.2703, 3.1892, –1.8378
19.2a: 17.2973, 110.8, 93.5027; **19.2b**: 2.4012, 0.6242; **19.2c**: 0.8439
19.3b: 1.151, 0.241, 0.827; 0.004; **19.3c**: 3624.962, 246.371, 3378.591; 0.932; **19.3d**: *val* =
63.926
19.4a: 4.8571, 1.5714; 2.4762, 1.9524; –1.9048, –0.8663; **19.4b**: –0.7692, 5.3077; **19.4c**:
–0.6925; **19.4d**: 2.923; **19.4e**: 0.765, 0.1984
19.5a: 2.1143; **19.5b**: 0.491; **19.5c**: 4.5029
19.6a: 2.3; **19.6b**: 8.65, (–1.6436, 18.9436); **19.6c**: (1.3015, 2.5985)
19.7a: *val* = 1.4700; **19.7b**: 0.0724
19.9b: $\hat{y}$ = 32.05 – 0.65x; **19.9c**: 21.65, 20.35, 19.70, 22.95, 21.00, 19.70, 21.65; 1.35, 0.65,
0.30, –1.95, –1.00, –1.70, 2.35
19.10b: 1.7635, 0.3943; **19.10c**: *val* = –1.648
19.11a: $\hat{y}$ = 3222.388 + 10241.791x; **19.11b**: *val* = 10.907; **19.11c**: 0.8561, 0.925
19.12a: *val* = 2.3874; **19.12b**: 0.0135
19.13b: $\hat{y}_t$ = 0.646 + 1.179x_t; **19.13c**: (0.970, 1.388); **19.13d**: 484.2 tons
19.14c: 1.289, 0.1121; **19.14d**: 0.18, 0.1234, 0.0566; **19.14e**: 0.6856, 0.0094; **19.14f**: *val* = 3.62
19.15a: $\hat{y}$ = 61.812 + 0.7618x; **19.15b**: 44.4079; **19.15c**: (425.933, 459.491); **19.15d**: (335.007,
519.945)
19.17a: 4.25, 1.9940; **19.17b**: (38624.659, 70238.027); **19.17c**: (50703.753, 58158.933);
19.17d: (17440.825, 29971.115)
19.18a: $\hat{y}$ = 943.885 – 1.049x_1, $\hat{y}$ = 68.270 + 8.362x_2; **19.18b**: both 0.000
19.19b: *val* = –4.475
19.20a: *val* = –2.964, 0.00199357; **19.20b**: (–0.4132, –0.0708); **19.20c**: (2.2, 5]; **19.20d**: (3.89,
4.46); **19.20e**: (3.45, 4.43)

19.21c: *val* = 9.365; **19.21d**: *val* = 0.8551; **19.21e**: (–5.0709, 26.4934)
19.23a: $\hat{y}$ = 0.021 + 0736x; **19.23b**: 0.736, *val* = –3.7183

Chapter 20
20.1b: *val* = 350.8663; **20.1c**: 0.9901, 0.9873
20.2a: (1.6908, 2.0510); **20.2b**: *val* = 8.8511; **20.2c**: 0.0000475
20.3d: at least 0.3778
20.4: at least 0.2556, at least 0.1547, at least 0.0778
20.5a: *val* = 57.0642; **20.5b**: 1.7×10^{-18}, 0.5374; **20.5c**: *val* = 3.4333
20.6d: *val* = 77.765; **20.6e**: 0.646
20.7c: *val* = 2.8205
20.8: *val* = 1.0556
20.10a: *val* of *F*–test is 4.549; **20.10b**: *val* = 2.133; **20.10c**: 0.0850, 0.0663
20.11a: *val* = 2.312; **20.11b**: 0.0879, 0.0499
20.12b: 0.3544, 0.3434; **20.12c**: *val* = 11.333, *val* = 2.913, *val* = 1.110; **20.12d**: (–0.6071, –0.0889), (–0.0540, 0.0780)
20.14b: *val* = 1662.866; **20.14c**: 0.944
20.15a: *val* = 3.5; **20.15c**: 0.927; **20.15e**: (8.47862, 8.82729), (6.49869, 10.80721)
20.16a: (104.4486, 112.4483); **20.16b**: (93.0097, 123.8882)

Chapter 21
21.1: *val* = 2.8485
21.2a: *val* = 84.8254; **21.2b**: *val* = 52.2476; **21.2c**: *val* = 63.9957
21.3b: *val* = –2.7112
21.4b: *val* = –0.260, *p*–value = 0.796; **21.4c**: see b.; **21.4d**: *val* = 1.6186
21.5b: *val* = –2.0835
21.6a: *val* = 2.9354; **21.6b**: *val* = 1.5000; **21.6c**: *val* = 5.7321
21.8c: *val* = 8.1062
21.9a: *val* = 4.312
21.10b: 0.905; **21.10c**: *val* = –3.1227; **21.10d**: (–12946.07, –566.45)
21.11b: *val* = –1.9661; **21.11c**: *val* = 1.4251
21.12: *val* = 0.3781
21.13b: *val* = 3.7387; **21.13b**: *val* = 1.8567; **21.13c**: 1.0412
21.14b: *p*–value= 0.000××××
21.15a: 23.55; **21.15b**: (11.7866, 46.3296)
21.16a: *val* = 16.3333; **21.16d**: *val* = 5
21.17b: *val* = 1.6394; **21.17d**: *val* = 24.5448; **21.17e**: *val* = 24.5448; **21.17f**: (8.9637, 19.0363)
21.18b: *val* = 36.6364; **21.18c**: *val* = 9.1591
21.19b: *val* = 3.15
21.20a: *val* = 9.874; **21.20e**: *val* = 11.258
21.21d: *val* = 2.234
21.22a: *val* = –2.764; **21.22b**: *val* = 9.807; **21.22c**: *val* = 0.331
21.23b: *val* = 2.9419; **21.23c**: *val* = –2.904; **21.23d**: 1.819
21.24c: *val* = 5.5987

Chapter 22
22.1b: *val* = 0.191; **22.1c**: *val* = 1.789
22.2d: *val* = 1.098
22.3b: *val* = 187.346; **22.3e**: *val* = –2.8421
22.4a: *val* = 0.989; **22.4b**: *val* = 1.141; **22.4c**: *val* = 2.263
22.5b: *p*–value = 0.007
22.6e: (0.228, 0.274)
22.8b: *val* = 15417.606, 68.9%; **22.8d**: 0.3550, no

22.9a: 3.271, −0.877, 0.499; **22.9b**: 3.537, −1.093, 0.453
22.10b: 852009.1, 18067.039, −277461
22.12a: *val* = −0.268; **22.12b**: *val* = 0.077
22.13a: *val* = 1.644; **22.13b**: *val* = 1.284; **22.13c**: *val* = 0.058

Chapter 23
23.2a: all 158.62; **23.2c**: 158.6984, 158.6968, 158.6951, 158.6935, 158.6919
23.3a: *val* = 0.024; **23.3b**: *val* = 0.822; **23.3c**: *val* = 4.078; **23.3d**: *val* = −0.651
23.4a: *val* = 47.4348; **23.4c**: 631169
23.5a: *val* = 0.235; **23.5c**: *val* = 2.249; **23.5d**: *val* = −1.157
23.6a: *val* = 124.295, *val* = 165.3466; **23.6b**: 17010×10^6
23.7a: 0.2586, 0.2720, 0.5042, 0.4913; **23.7b**: all 0.4913; 0.0855, 0.0855, 0.0856;
23.7c: *val* = 0.185
23.8a: *val* = −0.879, 0.002; **23.8c**: 1481, 257, 211; **23.8d**: 323
23.9b: both 8.56%; **23.9c**: 9.87%, 10.05%; **23.9d**: 8.65%, 8.69%
23.10c: *val* = 88.689, 0.940, *val* = 43.8534, *val* = −8.630, *val* = 93.7438
23.11b: 99.81, 99.68, 98.15, 85.93, 85.67, 97.44, 98.64, 98.24, 97.52; **23.11c**: (95.96966,
103.65857); **23.11d**: *val* = 1.228
23.12d: *val* of the *F*–test is 4.2525
23.15d: *val* = −0.225

Chapter 24
24.1: *val* = 6.5169
24.2: *val* = 10.1650
24.3: *val* = 3.7556
24.4: *val* = 2.3857
24.5a: 0.0228, 0.1359, 0.3413, 0.3413, 0.1359, 0.0228; **24.5b**: 11.4, 67.95, 170.65, 170.65,
67.95, 11.4; **24.5c**: *val* = 13.3012
24.6: *val* = 2.6987
24.7: *val* = 2.4812
24.8a: *val* = 26.0546
24.9c: *val* = 19.1466
24.10b: *val* = 6.0417
24.11: *val* = 8.9603
24.12a: *val* = 7.0016
24.13a: *val* = 69.1
24.14a: *val* = 34.73
24.15b: *val* = −5.8924
24.16a: *val* = 51.6610
24.17a: 0.1587, 0.3413, 0.3413, 0.1587; **24.17c**: *val* = 16.2297; **24.17d**: *val* = 14.9945
24.18b: *val* = 10.6161
24.19: *val* = 6.956
24.20b: 0.256997, 6.384483; **24.20d**: *val* = 192.8

Chapter 25
25.1: *val* = 19
25.2: *val* = 56
25.3: *val* = 1.3232
25.4: *val* = 2.2361
25.5: *val* = 1.8974
25.6: *val* = −0.1840
25.7: *val* = 41
25.8: *val* = 2.5490

25.9: *val* = 4.4240
25.10b: *val* = 35.5
25.11c: *val* = 2.2891; **25.11d**: 0.0110
25.12: *val* = 2.0702, 0.0192
25.13a: 5365, 335.6561; **25.13b**: *val* = 4.1650
25.14: *val* = 320.50
25.15: *val* = 5.2555
25.16: *val* = −1.5698
25.17b: *val* = −2.3459
25.18a: *val* = 18.3894; **25.18b**: *val* = −4.2887
25.19: *val* = 65.8630
25.20: *val* = 20.6399
25.21a: *val* = 215.7767; **25.21b**: *val* = 32.05

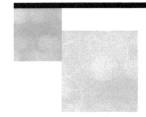

Index